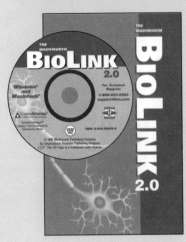

BioLink 2.0
ISBN 0-534-54227-1
With this CD-ROM presentation tool, you can easily assemble art and database files with your notes to create fluid lectures that help stimulate even the least-engaged students. *BioLink 2.0* includes illustrations from Wadsworth texts, the best animations and films from the *Interactive Concepts in Biology CD-ROM*, and material from Wadsworth laserdiscs. It also features a Kudo® (Image Browser™) with an easy drag-and-drop feature that allows file export into presentation tools such as *PowerPoint*. Upon its creation, a file or lecture with *BioLink 2.0* can be posted to the web, where students can access it for reference or study needs. Free to adopters.

to with CNN, these ... n short clips of high-interest news stories about environmental topics that will spark lively classroom discussions and generate a deeper understanding of the importance of the environment in our day-to-day lives. An accompanying workbook contains a worksheet for each video clip. Adopters can receive one video each year for three years.

Introduction to Environmental Science Presentation CD-ROMs
ISBN 0-534-23906-4
This set of cross-platform CD-ROMs contains more than 3,500 photographs and diagrams, some from Miller's text, to help you create riveting lecture presentations. (One set of five CDs per department with adoptions of 200 or more copies of Miller's *Environmental Science*, 7th Edition. Available for U.S. market only.)

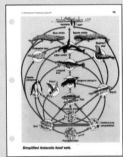

Transparency Acetates and Masters
ISBN 0-534-50625-9
Includes 75 color acetates of line art from Miller texts and nearly 600 black-and-white master sheets of key diagrams for making overhead transparencies. Free to adopters.

Video: In the Shadow of the Shuttle: Protecting Endangered Species
ISBN ISBN 0-534-53039-7

Video: Costa Rica: Science in the Rainforest
ISBN 0-534-53040-0
Contact your Wadsworth/ITP representative for availability by class size.

Videodisk: Liquid Assets: The Ecology of Water
ISBN 0-538-63066-3
Two interactive, double-sided videodisks compare the ecologies of the Everglades and San Francisco Bay. Includes a study guide. (U.S. adopters: Contact your Wadsworth/ITP representative for availability by adoption size. Non U.S. adopters: Contact your ITP representative for information on availability outside the U.S.)

Instructor's Manual with Test Items
ISBN 0-534-53543-7
Focuses on the major goal of environmental education by emphasizing the integration of concepts, values, participation, and skills! Section I contains numerous types of lab, interdisciplinary, and problem-solving activities; discussion and review questions; chapter objectives and goals; key terms; suggested term papers; references to useful films, videos, and computer programs; and references to concepts maps as they relate to each chapter of the student text. Section II contains multiple choice test questions, organized by text chapter. Section III includes three appendixes containing concept maps for each chapter, information on where to obtain audio/visual supplements, and information on computer software. Free to adopters.

Computerized Test Items
Macintosh 0-534-53544-5
Windows 0-534-53545-3
These floppy disks contain all test items from the *Instructor's Manual* in an easily customizable format. Free to adopters.

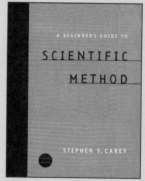

THE MOST WIDELY EMBRACED APPROACH TO ENVIRONMENTAL SCIENCE IN PRINT TODAY!

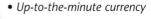

- *Up-to-the-minute currency*
- *Fair and balanced coverage*
- *Accuracy and authority backed by more than 10,000 sources*
- *Greatly expanded material on basic ecology*
- *Integration of Internet tools throughout the book*
- *State-of-the-art presentation tools for riveting lectures*
- *Student materials that make them sit up and take notice*

You are looking at the culmination of more than 30 years of research and teaching expertise in environmental sciences. Incorporating current new material, the latest in classroom technology, and many other enhancements, this Seventh Edition of Miller's *Environmental Science* is the most effective teaching and learning tool ever offered in this field.

At 566 concise pages, this briefer alternative to Miller's *Living in the Environment* uses basic scientific and ecological concepts, laws, and principles to help students understand and form their own conclusions about environmental and resource issues. Miller covers all the same topics found in the longer book, but in less detail.

Instructors throughout the continent appreciate this author's dedication and talent: Miller's texts are favored (and used) at an overwhelming ratio of **five to one** over other environmental science books! Authority, reliability, unbeatable currency, and classroom support that doesn't stop . . .

See for yourself ➤➤→

TODAY'S DEVELOPMENTS . . .
TOMORROW'S IMPLICATIONS

Whether it's the latest information on how hormone disrupters threaten wildlife and humans or the startling facts about China's coal production, this edition of *Environmental Science* supports your lectures with the most talked-about and significant developments occurring today.

tors; they fit together somewhat like a molecular key fitting into a specially shaped keyhole on a receptor molecule (Figure 8-4, left). Once bonded together, the hormone and its receptor molecule move onto the cell's nucleus to execute the chemical message carried by the hormone.

Case Study: Do Hormone Disrupters Threaten the Health of Wildlife and Humans? Over the last few years experts from a number of disciplines have been piecing together field studies on wildlife, studies on laboratory animals, and epidemiological studies of human populations indicating that a variety of human-made chemicals can act as *hormone disrupters*.

Some, called *hormone mimics*, are estrogenlike chemicals that disrupt the endocrine system by attaching to estrogen receptor molecules (Figure 8-4, center). Others, called *hormone blockers*, disrupt the endocrine system by preventing natural hormones such as

incinerated, Section 13-7), (2) ce chlorinated hydrocarbons wid environment and capable of bei fied in food chains), (3) variou plastics, (4) some pesticides, and

Numerous wildlife and lab various possible effects of estr mone blockers. Here are a few ranch minks fed Lake Michigan endocrine disrupters such as D reproduce, (2) alligators in a Flo nearby hazardous-waste site small penises and low testost couldn't reproduce, (3) rats ex womb and infancy tend to be h trogenlike p-nonylphenol (add inhibit the growth of testicles in

Most natural hormones excreted. However, many of t

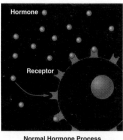
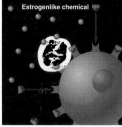

| Normal Hormone Process | Hormone Mimic | Hor |

Figure 8-4 Hormones are molecules that act as messengers in the endocrine system to regulate various bodily processes, including reproduction, growth, and development. Each type of hormone has a unique molecular shape that allows it to attach to specially shaped receptors on the surface of or inside cells and transmit its chemical message (left). Molecules of certain pesticides and other molecules have shapes similar to those of natural hormones. Some of these hormone impostors, called *hormone mimics*, disrupt the endocrine system by attaching to estrogen receptor molecules (center). Others, called *hormone blockers*, prevent natural hormones such as androgens (male sex hormones) from attaching to their receptors (right). Some pollutants called *thyroid disrupters* may disrupt hormones released by thyroid glands and cause growth and weight disorders and brain and behavioral disorders.

From page 228, *Environmental Science, 7th Edition.*

Miller's texts are praised for consistently incorporating the most current material on the global impact of environmental issues. On reading this startling discussion of coal burning in China, your students will be surprised to discover that mostly because of coal burning, China is now responsible for 10 percent of the world's total CO_2 emissions—more than the U.S., Canada, and Japan combined! This interesting information relates directly to the global warming issue and creates opportunity for lively class discussion about global consequences.

Coal Burning and Destructive Synergies in Cl

CONNECTIONS

China has huge quantities of coal—11% of the world's reserves—and is the world's largest producer of coal, accounting for 25% of global output. Currently, the country burns coal to provide 76% of its commercial energy.

Mostly because of coal burning, China now produces 10% of the world's emissions of CO_2 each year. By 2025, China is projected to emit more CO_2 than the current combined total of the United States, Canada, and Japan.

Several factors interact synergistically to cause China to burn more coal than it needs to. One is energy inefficiency. About 66% of the energy used in China is unnecessarily wasted—compared to about 43% in the United States (which is in turn only half as energy-efficient as Japan and most western European countries).

Another factor is that drinking water must usually be boiled because it is often too contaminated to drink. This uses more energy, much of it produced by the inefficient burning of coal in homes. Destruction of forests for more cropland, fuelwood, and construction material for the country's growing population leads to more soil erosion and increased use of fertilizers, further polluting the water and requiring more coal burning to sterilize water.

For China's economic future, and perhaps for the world's climate, many analysts believe that it is crucial to defuse these destructive synergistic interactions, which lead to a cycle of harmful positive

From page 547, *Environmental Science, 7th Edition*

STRONG MULTI-DISCIPLINARY ATTENTION TO CURRENCY

This book's excellent multi-disciplinary coverage extends to its presentation of current material. Miller includes new biology and ecology research, in addition to the latest environmental implications from economic and social fronts.

This ambitious 1997 appraisal attempts to place an economic value on the worth of an ecosystem.

How Much Is a Forest Worth?

SPOTLIGHT

In 1997 a group of 13 ecologists, economists, and geographers from 12 prestigious laboratories and universities attempted to estimate how much nature's ecological services (Figure 1-2) are worth. According to this ambitious and crude appraisal led by ecological economist Robert Costanza of the University of Maryland, the economic value of Mother Nature's bounty ranges from $16 trillion to $54 trillion. In other words, each year nature provides free goods and services worth somewhere between $2,800 and $9,000 per person.

Their medium estimate for nature's life-support services for humans is $33 trillion a year. This is nearly twice the $18-trillion-per-year economic output of the earth's 194 nations and almost five times the $6.9-trillion U.S. GDP in 1996.

To make these estimates, the researchers divided the earth's surface into 16 biomes (Figure 5-3) and aquatic life zones such as various types of forests, grasslands, cropland, open oceans, estuaries, wetlands, and lakes and rivers (but omitted deserts and tundra because of a lack of data). Then they agreed on a list

of 17 goods and services provided by nature (Figure 1-2) and sifted through more than 100 studies that attempted to put a dollar value on such services in the 16 different types of ecosystems.

Some analysts believe that such estimates are misleading and dangerous because they amount to putting a dollar value on ecosystem services with an infinite value because they are irreplaceable. In the 1970s economist E. F. Schumacher warned, "To undertake to measure the immeasurable is absurd" and is a "pretense that everything has a price." Biologist and conservationist David Ehrenfeld's response to this evaluation was, "I am afraid that I don't see much hope for a civilization so stupid that it demands a quantitative estimate of the value of its own umbilical cord." Those who cringe at the thought of putting a dollar value on nature's ecosystem services believe that they should be protected on *moral* grounds instead of being reduced to easily manipulated cost–benefit analyses (Spotlight, p. 202).

The researchers admit that their estimates are full of assumptions and omissions and could easily be too low by a factor of 10, 1 million, or more. For example, their calcula-

From page 442, *Environmental Science, 7th Edition.*

Students discover how one city in Brazil solves many social and environmental problems by becoming a virtual laboratory for sustainable living.

Curitiba, Brazil

SOLUTIONS

One of the world's showcase ecocities is Curitiba, Brazil, with a population of 2.1 million. It is a prosperous, booming, clean, and mostly economically self-sufficient city.

Trees are everywhere in Curitiba because city officials have given neighborhoods over 1.5 million trees to plant and care for. No tree in the city can be cut down without a permit; for every tree cut, two must be planted.

After persistent flooding in the 1950s and 1960s, city officials put certain flood-prone areas off-limits for building, passed laws protecting natural drainage systems, built artificial lakes to contain floodwaters, and converted many riverbanks and floodplains into parks. This design-with-nature strategy has made costly flooding a thing of the past and has greatly increased the amount of open or green space per capita (which increased 100-fold

The key concept was to channel the city's development along five major transportation and high-density corridors that extend outward from the central city like spokes on a bicycle wheel. Each corridor has exclusive lanes for express buses.

Curitiba probably has the world's best bus system. Each day a network of clean and efficient buses carries over 1.5 million passengers—75% of the city's commuters and shoppers—at a low cost (20–40¢ per ride with unlimited transfers) on express bus lanes. Only high-rise apartment buildings are allowed near major bus routes, and each building must devote the bottom two floors to stores, which reduces the need for residents to travel. As a result, Curitiba has one of Brazil's lowest outdoor air pollution rates.

The city recycles roughly 70% of its paper (equivalent to saving nearly 1,200 trees per day) and 60% of its metal, glass, and plastic, which is sorted by households for

From page 187, *Environmental Science, 7th Edition.*

And so many other new and expanded topics:

- Pro/con feedback on global warming
- Island biogeography
- How ecologists learn about nature
- Sea otters as keystone species
- Ecological importance of mountains
- Mechanisms of succession
- Dangers from *pfiesteria* bacteria
- Survival ability of cockroaches
- Environmental refugees
- The global TB epidemic
- The ecoindustrial revolution
- Alternatives to cars
- Ecologically sustainable development
- Health effects of airborne particles
- Above-ground storage of toxic wastes
- Overfishing and aquatic biodiversity
- Effects of wildlife habitat fragmentation
- Future options for nuclear power
- *And much more*

BECAUSE THERE ARE AT LEAST TWO SIDES TO EVERY ISSUE . . .

No other author achieves the balance—the fairness to different environmental viewpoints—achieved in this book. Describing the intensive effort to make *Environmental Science* as balanced as humanly possible, Miller writes: "With the help of reviewers and a talented copy editor and bias hunter, I have scoured every sentence in this book for even a hint of bias."

In addition, Miller consistently presents two or more viewpoints so students can learn to think critically about and decide these issues for themselves. Here, he juxtaposes opposing views on whether the Endangered Species Act should be strengthened or weakened giving students a platform of information for making up their own minds.

---

consuming and controversial, thus delaying or preventing protection. To avoid this problem the panel of scientists recommended that when a species is listed as endangered, a core amount of *survival habitat* should be set aside immediately as an emergency measure to help ensure short-term survival of the species for 25–50 years. If the critical habitat is later identified, the survival habitat area would automatically expire.

Should the Endangered Species Act Be Weakened? Since 1995 there have been intense efforts to seriously weaken the Endangered Species Act by (1) making the protection of endangered species on private land voluntary, (2) having the government pay landowners if it forces them to stop using part of their land to protect endangered species (the issue of regulatory takings, p. 216), (3) making it harder and more expensive to list newly endangered species by requiring government wildlife officials to navigate through a series of hearings and peer-review panels, and (4) giving the secretary of interior the power to permit a listed species to become extinct without trying to save it, (5) allowing the secretary of the interior to give any state, county, or landowner permanent exceptoin from the law, with no requirement for public notification or comment, (6) allowing landowners to lock in long-term endangered species management plans—known as *habitat conservation* plans (HCPs)—that exempt the owners from further obligations for 100 years or more, and (7) prohibiting the public from commenting on or bringing lawsuits on any changes in habitat conservation plans for endangered species.

Those who favor weakening or eliminating the Endangered Species Act argue that it has been a failure

because only seven species have been the list and only 20 species have recove be reclassified from endangered to thr of these critics have spread horror stor environmentalists have used endan (many of them small and unfamiliar) to ment and resource extraction, violate p rights, and waste tax dollars trying t creatures that are on the verge of extinct

The late California Representativ joked that the best way to deal wit species is to "give them all a designated blow it up." Washington Senator Slad gests that all endangered species be rem wild and bred in zoos as "a way to pr without blocking economic developme

Should the Endangered Species Ac Strengthened? Most conservation wildlife scientists contend that the En cies Act has not been a failure (Spotligh also refute the charge that the act has ca nomic losses.

The truth is that the Endangered S had virtually no impact in the nation

Has the Endang

SPOTLIGHT

Critics of the Endangered Species Act call it an expensive failure because only a few species have been removed from the endangered list. Most of these critics are ranchers, developers, and officials of timber and mining companies who want more access to resources on public lands (Spotlight, p. 217).

Most biologists strongly disagree that the act has been a failure, for several reasons. *First*, species are listed only when they are already in serious danger of extinction. This is like setting up a poorly funded hospital emergency room that takes only the most desperate cases, often with little hope for recovery, and then saying it should be shut down because it has not saved enough patients.

Second, it takes decades for most species to become endangered or

From pages 496-497, *Environmental Science, 7th Edition.*

. . . GOOD NEWS AND BAD NEWS GET EQUAL BILLING

New to the 7th Edition, this extensive, two-page table (only a portion is reproduced here) shows students in explicit, side-by-side detail, the pluses and minuses of many different environmental issues.

Table 1-1 Good and Bad News About Environmental Problems	
Some Good News	**Some Bad but Challenging News**
Annual world population growth slowed from 2.2% to 1.47% between 1960 and 1997.	The world's population is still growing rapidly and is projected to increase from 5.84 billion to 8 billion between 1997 and 2025.
Global life expectancy rose from 33 to 66 years between 1900 and 1997.	Life expectancy in developing countries is 10 years less than in developed countries.
Annual global infant mortality dropped by 65% between 1900 and 1997.	Infant mortality in developing countries is 7.1 times higher than in developed countries and could be improved in many developed countries.
Global grain production has outpaced population growth since 1978.	Since 1980 population growth has exceeded grain production in 69 developing countries and the global per capita fish catch has declined by 7.5% between 1988 and 1995. Future food production may be limited by its harmful environmental impacts.
The percentage of the world's population that are hungry has been reduced since 1960.	Because of population growth there are now more hungry people (about 1 million) than ever.
Biodiversity loss has been reduced by higher crop yields on less land.	The challenge is to increase yields while sharply reducing the growing environmental impacts of such intensive production.
Biodiversity loss has been reduced by increased urbanization, with more people living on less land	This effect is misleading because the higher resource use per person in cities must be supplied by croplands, forests, grasslands, fisheries, and mines elsewhere. This often diminishes biodiversity and increases environmental degradation in these areas that sustain otherwise unsustainable cities.
At least 5% of the world's land area has been set aside to protect wildlife and wildlife habitats.	This is half of the estimated minimum needed. Many wildlife sanctuaries exist on paper only and receive little protection and many important types of biological communities are not pro-

From page 22, *Environmental Science, 7th Edition.*

In this Pro/Con discussion of ozone depletion, Miller uses an effective "Charge" and "Response" format to ensure that students can follow the point-by-point arguments of opposing views.

Is Ozone Depletion a Hoax?

PRO/CON

Charge: There is no ozone hole, and the whole idea is merely a scientific theory.

Response: Technically, it is not correct to call this phenomenon an ozone *hole*. Instead, it is a thinning or partial depletion of normal ozone levels in the stratosphere. A scientific theory represents a widely accepted idea or principle that has a very high degree of certainty because it has been supported by a great deal of evidence. Nothing in science can be proven absolutely, but there is overwhelming scientific evidence of ozone thinning in the stratosphere.

Charge: There is no scientific proof that an ozone hole exists.

Response: Ground-based and

stratosphere than have human-caused CFC emissions.

Response: Most of the water-soluble HCl injected into the troposphere from occasional large-scale volcanic eruptions is washed out by rain before it reaches the stratosphere. A serious calculation error by one of the skeptics (Dixy Lee Ray) greatly overestimated the amount of HCl injected into the stratosphere by recent volcanic eruptions. Measurements indicate that no more than 20% of the chlorine from biomass burning (in the form of methyl chloride, CH_3Cl) reaches the stratosphere; this amount of chlorine is about five times less than the current contribution from CFCs. Even Fred Singer (who is highly skeptical about some aspects of ozone deple-

1958 ther
for a sign
(such as t
over Anta

Charg
ozone pro
estimate
should be
observe.

Respo
that their
proved. F
that their
by chlorine
ing chem
activities
of the obs

Charg
in UV-B r
spheric o
tected in
States and

From page 290, *Environmental Science, 7th Edition.*

NEW DIAGRAMS AND TEXTUAL MATERIAL CLARIFY COMPLEX CONCEPTS

An early proponent for giving students a strong grounding in basic ecology, Miller continues to stress this material in every edition. In this 7th Edition, you will find *significantly* increased coverage of basic ecological concepts. For example, students will more readily understand reproductive strategies with this new diagram and the accompanying text that effectively delineates the distinctions between r-strategists and K-strategists.

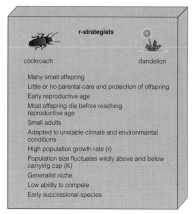
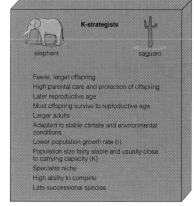

r-strategists

cockroach   dandelion

Many small offspring
Little or no parental care and protection of offspring
Early reproductive age
Most offspring die before reaching reproductive age
Small adults
Adapted to unstable climate and environmental conditions
High population growth rate (r)
Population size fluctuates wildly above and below carrying cap (K)
Generalist niche
Low ability to compete
Early successional species

K-strategists

elephant   saguaro

Fewer, larger offspring
High parental care and protection of offspring
Later reproductive age
Most offspring survive to reproductive age
Larger adults
Adapted to stable climate and environmental conditions
Lower population growth rate (r)
Population size fairly stable and usually close to carrying capacity (K)
Specialist niche
High ability to compete
Late successional species

Figure 5-31 Some generalized characteristics of r-strategists and K-strategists. Many species have characteristics between these two extremes.

population size, but a higher proportion would die if the population density were high.

What Reproductive Strategies Do Species Use to Survive? Each species has a characteristic mode of reproduction. At one extreme are species that reproduce early. They have many (usually small) offspring each time they reproduce, reach reproductive age rapidly, and have short generation times. They give their offspring little or no parental care or protection to help them survive. Species with this reproductive strategy overcome the massive loss of their offspring

They are fairly large and mature slowly. They are cared for and protected by one or both parents until they reach reproductive age. This strategy results in a few big and strong individuals that can compete for resources and reproduce a few young to begin the cycle again (Figure 5-31, right).

Such species are called K-strategists because they tend to maintain their population size near their habitat's carrying capacity (K). Their populations typically follow an S-shaped growth curve (Figure 5-28). Examples are most large mammals (such as elephants, whales, and humans), birds of prey, and large and

From page 140, *Environmental Science, 7th Edition.*

New and expanded topics:

- The nature of science
- Scientific methods
- Objectivity of scientists
- Insights into how ecologists conduct field, laboratory & systems research
- Models, systems, and feedback
- Ecological importance of parasites
- Survival strategies of plants
- Linear and exponential growth
- Species succession and the role of disturbance
- Past mass extinctions and adaptive radiations
- The importance of fossils in understanding evolution
- Population dynamics in nature
- *And much more*

EXCITING, NEW FULL-COLOR ILLUSTRATIONS CAPTURE THE DIVERSITY OF ECOSYSTEMS

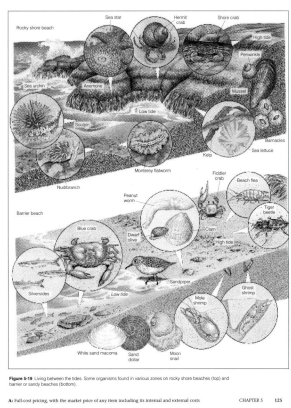

Figure 5-19 Living between the tides. Some organisms found in various zones on rocky shore beaches (top) and barrier or sandy beaches (bottom).

A: Full-cost pricing, with the market price of any item including its internal and external costs   CHAPTER 5   125

This instructive, new illustration visually and clearly differentiates organisms found in various zones on rocky shore beaches and barrier or sandy beach bottoms.

EXPANDED BASIC ECOLOGY

From page 125, *Environmental Science, 7th Edition.*

Some components and interactions in a tropical rainforest ecosystem.

Figure 5-11 Some components and interactions in a tropical rain forest ecosystem. When these organisms die, their organic matter is broken down by decomposers into minerals used by plants. Transfers of matter and energy among producers, primary consumers (herbivores), and secondary (or higher-level) consumers (carnivores) are indicated by colored arrows.

Figure 5-11, page 117, *Environmental Science, 7th Edition*

THE WADSWORTH BIOLOGY RESOURCE CENTER:

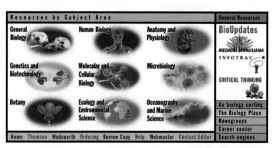

http://www.wadsworth.com/biology

This extensive website for environmental science and biology is one of the premier sites for science education on the Web with more than 100,000 visitors every month. The resource center is available FREE to your students when you adopt Miller's *Environmental Science*. The resource center offers hypercontents relevant to each of Wadsworth's texts, online study aids, web-based critical thinking questions, and career tips for science students. Reviews of current articles on engaging environmental, biological, and genetics topics plus original articles are also available to help both students and instructors keep up with the latest science news.

Chapter hyperlinks

Miller's *Environmental Science*, together with the full-service Web resource center, offers you a truly interactive learning system. Each chapter in the text is now linked to websites that expand information on specific chapter topics. "Chapter hyperlinks" are called out in the text with this icon:

In-text Internet sources for every chapter

A detailed section at the back of Miller's *Environmental Science* titled "Internet Sources and Further Readings" offers students plenty of web addresses for further topical exploration. They are conveniently organized by text chapter. In addition, Appendix One now contains web addresses for publications, environmental organizations, and federal and international agencies—updated for accuracy right up to press time.

> Woodwell, George M., and Fred T. Mackenzie, eds. 1995. *Biotic Feedbacks in the Global Climatic System. Will the Warming Feed the Warming?* New York: Oxford University Press.
>
> **11 / WATER RESOURCES AND WATER POLLUTION**
>
> Internet
>
> Clean Water Act: http://www.law.cornell.edu/uscode/33/ch26.html

Introduction to the Internet for Miller's Environmental Science Texts, 2nd Edition
ISBN 0-534-51918-0

Available exclusively when bundled with Miller's texts, this 114-page booklet is packed with essential information to help students get the most out of their explorations on the Web. It includes Internet sites and exercises for every chapter in Miller's books—plus clear and concise information on what the Internet is and how it works. Topics include: logging on, Internet services, using e-mail and other forms of electronic communication, research strategies and techniques, and search engines.

And now, you can create your own web site!

With *Thomson World Class Syllabus*, it's a quick and easy process. This software enables you to post your own course information, office hours, related Internet links, downloaded materials, lesson information, assignments, sample tests or quizzes, and its available FREE to adopters of Wadsworth texts. More information is available at:

http://www.worldclasslearning.com

INFOTRAC COLLEGE EDITION—
A FULLY SEARCHABLE ONLINE DATABASE WITH ACCESS
TO COMPLETE ARTICLES FROM OVER 600 PERIODICALS

Available FREE with each copy of Miller's *Environmental Science, 7th Edition,* this online library gives students access to full articles—not simply abstracts—from more than 600 popular and scholarly periodicals dating back as far as four years. Developed by Information Access Company, InfoTrac College Edition is available exclusively from Wadsworth. (If you are a non-U.S adopter, please check with your ITP representative for InfoTrac policy and availability.)

Free, unlimited use

When you adopt Miller's *Environmental Science,* both you and each of your students receive personal account identification numbers that allow free, unlimited use of InfoTrac College Edition for the duration of one academic term.

A virtual library

Opening the door to the full text and images of countless articles, this virtual library is expertly indexed and ready to use. Access through a simple interface quickly and seamlessly gives you and your students the answers you need.

Student Guide to InfoTrac College Edition

Located on The Wadsworth Biology Resource Center site, this online guide correlates each chapter in the book to articles from InfoTrac! Includes an introduction to InfoTrac, a set of electronic readings for each chapter, and critical thinking questions that are linked to InfoTrac articles to invite deeper examination of the issues.

Just a few of the 600+ periodicals available with InfoTrac College Edition:

AIDS Weekly	Evolution
American Forest	Harvard Medical School Health Letter
Annual Review of Genetics	Health
Annual Review of Microbiology	Health News Review
Archives of Environmental Health	Human Biology
Archives of Sexual Behavior	Human Ecology Forum
Audubon	Issues in Science and Technology
Biological Bulletin	Journal of Soil and Water Conservation
Bioscience	National Wildlife
Botanical Review	Oceania
Cancer Biotechnology Today	Oceanus
Cancer News	Popular Science
Cancer Weekly	Population Reports
Conservationist	Science
Discover	Science World
Ecologist	Sierra
Environment	World Health Organization Bulletin
Environmental Nutrition	

BioLink 2.0—An easy-to-use tool that brings your lectures to life

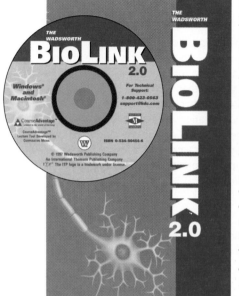

FREE to adopters, this CD-ROM presentation tool helps you to easily assemble art and database files with your lecture notes. The result? A fluid lecture that will engage even the least involved of your students. *BioLink 2.0* is such an impressive tool that it is quickly becoming the way that teachers teach. Once you've created a file or lecture with *BioLink,* you can post it to the Web for your students to use.

With BioLink, you create custom lectures using art form the CD and the Web

BioLink enables you to assemble, edit, publish, and present custom lectures built from more than one source. Use its extensive multimedia database of over 2,000 images, animations, and Quick-time movies; download art from the Internet; or use your own artwork. *BioLink's* CD-ROM contains art from all of Wadsworth's environmental science and biology textbooks—including Miller's *Environmental Science.*

 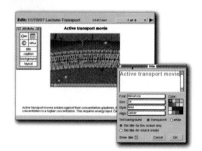 

BioLink's three modes: Sequence, Edit, and Present

The straightforward design of *BioLink* ensures great lecture presentations. The three main modes—Sequence, Edit, and Present—help you compile, manage, and adapt your lecture presentation. You can then save your *BioLink*-created lecture on disk for future use, or adapt it for a new lecture, or convert it to HTML and post it on the Web for student reference. It's that easy!
ISBN 0-534-54227-1

Introduction to Environmental Science Presentation CD-ROMs

This set of cross-platform CD-ROMs contains more than 3,500 photographs and diagrams, some from Miller's text, to help you create riveting lecture presentations. (One set of five CDs per department with adoptions of 200 or more copies of Miller's *Environmental Science,* 7th Edition. Available for U.S. market only.)
ISBN 0-534-23906-4

VIDEOS AND VIDEODISKS TO SPARK YOUR LECTURES . . .

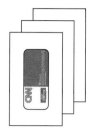

CNN Environmental Science Today
ISBN 0-534-55341-9 (Vol.1)

Updated annually and available FREE to adopters thanks to Wadsworth's exclusive agreement with CNN, these exciting videos contain short clips of high-interest news stories about environmental topics that will spark lively class discussion and generate a deeper understanding of the importance of environmental science in our day-to-day lives. Adopters can receive one CNN video each year for three years.

Also Available!

In the Shadow of the Shuttle: Protecting Endangered Species
ISBN 0-534-53039-7

Costa Rica: Science in the Rainforest
ISBN 0-534-53040-0

Contact your Wadsworth/ITP representative for availability by adoption size.

Videodisk: Liquid Assets: The Ecology of Water
ISBN 0-534-63066-3

Two interactive, double-sided videodisks compare the ecologies of the Everglades and San Francisco Bay. Includes a study guide. (U.S. adopters: Contact your Wadsworth/ITP representative for availability by adoption size. Non-U.S. adopters: Contact your ITP representative for information on this product's availability outside the U.S.)

. . . PLUS A PRESTIGIOUS TELECOURSE KEYED TO MILLER'S TEXT

Race to Save the Planet Telecourse

Designed by Annenberg/CPB specifically for the introductory environmental science course, this telecourse examines major environmental questions such as population growth, soil erosion, destruction of forests, and human-induced climate changes. Includes ten one-hour television programs that are keyed to Miller's *Environmental Science* (and *Living in the Environment*), supplemental readings, and the Telecourse Study Guide. For further information about licensing the course, call 1-800-ALS-ALS8. For general information about the course, call 1-800-LEARNER.

Study Guide for Race to Save the Planet, 1999 Edition
ISBN 0-534-53704-9
by Edward C. Wolf

Helps students organize and comprehend telecourse material! Features learning objectives, exercises, activities, glossaries, and references for each of the 10 telecourse programs—plus three print-only units to complete the course. Available from Wadsworth Publishing.

Telecourse Faculty Guide

For each telecourse lesson, the guide includes specific learning objectives, principal themes, questions for discussion, supplementary readings, and a test bank. Available from Annenberg/CPB. Call 1-800-ALS-ALS8 for details.

TOOLS TO HELP THEM THINK CRITICALLY ABOUT ENVIRONMENTAL ISSUES

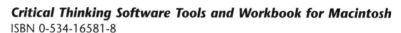

All student materials featured here are available for sale and for packaging with Miller's text in cost-saving bundles. Ask your rep or see the back cover of this book for our "best bet" bundling suggestions

Critical Thinking and the Environment: A Beginner's Guide for Environmental Science
ISBN 0-534-50619-4

by Jane Heinze-Fry & G. Tyler Miller

Students strengthen their ability to understand and apply their critical thinking skills with this 96-page booklet. Part 1 defines critical thinking as it relates to environmental issues. Part 2 offers helpful strategies for enhancing critical thinking skills. Part 3 presents questions, exercises, and scenarios that ask students to think critically about specific environmental problems, solutions, and values.

Critical Thinking Software Tools and Workbook for Macintosh
ISBN 0-534-16581-8

by Frank Draper & Steve Peterson

Using this software and its accompanying workbook, students hone their critical thinking skills as they interact with stimulating models of ecosystem dynamics, general evolution, and other important concepts and issues. Students control the learning process as they build a model, incorporate assumptions into their model, and then, through simulation and animation, discover the dynamic implications of their assumptions. The workbook contains exercises to guide students through the software. Includes an introductory user's guide, an applications guide for using the software, and complete technical documentation. Available for Macintosh only.

A Beginner's Guide to Scientific Method, 2nd Edition
ISBN 0-534-52843-0

by Stephen S. Carey

A brief, non-technical introduction to the basic methods underlying all good scientific research. Ideal as a supplementary text for any course where students need a basic understanding of how science is done.

The Game of Science, 5th Edition
ISBN 0-534-09072-9

by Garvin McCain & Erwin Segal

Exposes students to a realistic view of the scientific community—what science is, who scientists are, and what they do by approaching science as a game with rules, players, and goals. 225 pages.

HANDS-ON LABS AND ACTIVITIES TO GET THEM INVOLVED

Laboratory Manual for Environmental Science
ISBN 0-534-17809-X

by C. Lee Rockett & Kenneth J. Van Dellen

Features a variety of engaging activities, including laboratory exercises, workbook-type exercises, and projects. Designed for use with a minimum of sophisticated equipment, the exercises involve various areas of science, such as biology, chemistry, geology, physics, and meteorology to help students understand the interactions among the different environmental elements. 116 pages.

Green Lives, Green Campuses Activities Workbook
ISBN 0-534-19953-4

by Jane Heinze-Fry

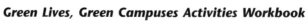

Helps students enhance their understanding of environmental concepts and strengthen their critical thinking skills with a variety of self-assessment, brainstorming, and evaluation exercises. Makes abstract concepts relevant to students' lives by offering activities that require on-campus investigation into environmental issues. 96 pages.

INTERESTING SUPPLEMENTAL READING TO DRAW THEM INTO CLASS DISCUSSION

Customizable Environmental Science Articles

assembled by Jane Heinze-Fry

Indexed by topic and geographical location, this collection offers more than 200 articles on subjects such as air pollution, the atmosphere, ecology, economics, energy, food resources, geological processes, global warming, hazardous waste, historical environmentalism, populations and dynamics, land resources, and more. Instructors may use the indices to choose any combination of articles and have them bound as customized supplements for their courses. Pricing is based on the length of the articles. Call your Wadsworth/ITP representative or the Custom Solutions Center (1-800-245-6724) for details.

Watersheds 2: Ten Cases in Environmental Ethics
ISBN 0-534-51181-3

by Lisa H. Newton & Catherine K. Dillingham

Classic cases of contemporary importance plus the detail and crucial scientific background to help students experience these serious and complex issues! Treatment is balanced, impartial, and presented engagingly to encourage critical thinking and classroom discussion. Topics include: toxic waste form nuclear weapons facilities; worldwide population growth and its consequences; pesticides, birds, and the legacy of Rachel Carson; over-exploitation of fisheries; property rights and takings; Bhopal and responsible care; tropical rainforests; the Exxon Valdez wreck; global climate change; and the north coast and the spotted owl. Each case is prefaced by questions to provoke critical thinking. 240 pages.

Environmental Ethics: An Introduction to Environmental Philosophy, 2nd Edition
ISBN 0-534-50508-2

by Joseph Desjardins

Grounded in practical environmental issues and concerns, this timely book introduces students to a wide range of ethical and philosophical issues that stem from the interaction of human beings and their natural environment. Desjardins introduces ethical theory as it applies to environmental issues and includes cases dealing with contemporary problems. 272 pages.

Radical Environmentalism: Philosophy and Tactics
ISBN 0-534-17790-5

by Peter List

Key readings covering some of the most important radical ecophilosophies being discussed today and representative examples of ecotactics that radical groups espouse. Readings are balanced between "radical" selections and critiques of radical tactics. 288 pages.

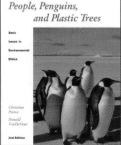

People, Penguins, and Plastic Trees: Basic Issues in Environmental Ethics, 2nd Edition
ISBN 0-534-17922-3

by Christine Pierce & Donald VanDeVeer

With the help of these essays, students learn to make their own decisions about how to lead lives that are both personally satisfying and also ecologically sound and responsible. This book's classic collection includes essays that examine disputes surrounding animals, ecosystems, the land, and the proper place for humans in the ongoing network of lives on this planet. 496 pages.

THERE'S A MILLER TEXT RIGHT FOR YOUR COURSE!

Environmental Science: Working with the Earth, 7th Edition

A briefer version of *Living in the Environment* that offers condensed coverage of important environmental concepts, problems, and solutions. 566 pages. ISBN 0-534-80545-0

Living in the Environment: Principles, Connections, and Solutions, 10th Edition

Broad and detailed discussion of environmental and resource issues. Authoritative, up-to-date. 761 pages. ISBN 0-534-51919-9

Sustaining the Earth: An Integrated Approach, 3rd Edition

The most concise of Miller's texts. A core text with a global approach that emphasizes how environmental and resource problems are interrelated. 338 pages. ISBN 0-534-52890-2

AND NOW . . . FULL-COLOR CUSTOMIZATION!

With Miller's texts, you have many options to create your own customized text. Select the chapters you prefer and let ITP Custom Solutions bind the full-color text that is right for you. Of course, you can still select any of these texts in their standard, complete forms. The choice is yours. Contact your ITP sales representative or call Custom Solutions at 1-800-245-6724.

Environmental Science
Working with the Earth

SEVENTH EDITION

G. TYLER MILLER, JR.

President, Earth Education and Research

Adjunct Professor of Human Ecology
St. Andrews Presbyterian College

Wadsworth Publishing Company
I⟨T⟩P® **An International Thomson Publishing Company**

Belmont, CA • Albany, NY • Bonn • Boston • Cincinnati • Detroit • Johannesburg • London
Madrid • Melbourne • Mexico City • New York • Paris • Singapore • Tokyo • Toronto • Washington

Two trees have been planted in a tropical rain forest for every tree used to make this book, courtesy of G. Tyler Miller, Jr., and Wadsworth Publishing Company. The author also sees that 50 trees are planted to compensate for the paper he uses and that several hectares of tropical rain forest are protected.

 This book is printed on acid-free recycled paper with the highest available content of post-consumer waste.

Biology Publisher: Jack Carey
Assistant Editor: Kristin Milotich
Editorial Assistant: Michael Burgreen
Project Management: Electronic Publishing Services Inc., NYC
Managing Designer: Carolyn Deacy
Print Buyer: Karen Hunt
Art Editor: Electronic Publishing Services Inc., NYC
Permissions Editor: Peggy Meehan
Technical Illustrators: Electronic Publishing Services Inc., NYC; Precision Graphics; Sarah Woodward; Darwin and Vally Hennings; Tasa Graphic Arts, Inc.; Alexander Teshin Associates; John and Judith Waller; Raychel Ciemma; and Victor Royer
Page Layout: Electronic Publishing Services Inc., NYC
Color Separator: H&S Graphics
Printer: Transcontinental Printing, Inc./Metropole Group

Cover Design: Carole Lawson
Cover Photograph: Galen Rowell/Mountain Light Photography. Bryce Canyon National Park, Utah
Part Opening Photographs:
Part I: Composite satellite view of the earth. © Tom Van Sant/ The GeoSphere Project
Part II: Endangered green sea turtle. © David B. Fleetham
Part III: Forest in Czechoslovakia killed by acid deposition and other air pollutants. Silvestris Fotoservice/NHPA
Part IV: Highly endangered Florida panther. John Cancalosi/ Peter Arnold, Inc.
Part V: Wind farm in California. U.S. Windpower.

Library of Congress Cataloging-in-Publication Data
Miller, G. Tyler (George Tyler)
 Environmental science : working with the earth / G. Tyler Miller, Jr. — 7th ed.
  p. cm. — (Wadsworth biology series)
 Includes bibliographical references and index.
 ISBN 0-534-53541-0 (alk. paper)
  1. Environmental science. 2. Human ecology.
 3. Environmental protection. I. Title. II. Series.
GE105.M544 1998
363.7—dc21   97-14111

For more information, contact Wadsworth Publishing Company:

Wadsworth Publishing Company
10 Davis Drive, Belmont, California 94002, USA

International Thomson Publishing Europe
Berkshire House 168-173, High Holborn
London, WC1V 7AA, England

Thomas Nelson Australia
102 Dodds Street
South Melbourne 3205, Victoria, Australia

Nelson Canada
1120 Birchmont Road
Scarborough, Ontario, Canada M1K 5G4

International Thomson Editorés
Campos Eliseos 385, Piso 7
Col. Polanco, 11560 México D.F. México

International Thomson Publishing GmbH
Königswinterer Strasse 418, 53227 Bonn, Germany

International Thomson Publishing Asia
221 Henderson Road, #05-10 Henderson Building
Singapore 0315

International Thomson Publishing Japan
Hirakawacho Kyowa Building, 3F
2-2-1 Hirakawacho, Chiyoda-ku
Tokyo 102, Japan

International Thomson Publishing South Africa
Building 18, Constantia Park, 240 Old Pretoria Road
Halfway House, 1685 South Africa

For Instructors and Students

How Did I Become Involved with Environmental Problems?   In 1966 I heard a scientist give a lecture on the problems of overpopulation and environmental abuse. Afterward I went to him and said, "If even a fraction of what you have said is true, I will feel ethically obligated to give up my research on the corrosion of metals and devote the rest of my life to research and education on environmental problems and solutions. Frankly, I don't want to believe a word you have said, and I'm going into the literature to try to prove that your statements are either untrue or grossly distorted."

After 6 months of study I was convinced of the seriousness of these problems. Since then, I have been studying, teaching, and writing about them. This book summarizes what I have learned in almost three decades of trying to understand environmental principles, problems, connections, and solutions.

What Is My Philosophy of Education?   In our lifelong pursuit of knowledge, I believe we should do three things. The first is to question everything and everybody, as any good scientist does.

Second, each of us should develop a list of principles, concepts, and rules to serve as guidelines in making decisions, and we should continually evaluate and modify this list on the basis of experience. This is based on my belief that the purpose of our lifelong pursuit of education is to learn how to sift through mountains of facts and ideas to find the few that are most useful and worth knowing. We need to be *wisdom seekers*, not information vessels. This requires a firm commitment to learning how to think logically. This book is full of facts and numbers, but they are useful only to the extent that they lead to an understanding of key ideas, scientific laws, concepts, principles, and connections.

Third, I believe in interacting with what I read as a way to use and continually sharpen my critical thinking skills. I do this by marking key sentences and paragraphs with a highlighter or pen. I put an asterisk in the margin next to something I think is important and double asterisks next to something that I think is especially important. I write comments in the margins, such as *Beautiful, Confusing, Misleading,* or *Wrong*. I fold down the top corner of pages with highlighted passages and the top and bottom corners of especially important pages. This way, I can flip through a book and quickly review the key passages. I urge you to interact in such ways with this book.

Why Is There a Need for More Scientific Content? This book is designed for introductory courses on environmental science. It treats environmental science as an *interdisciplinary* study, combining ideas and information from natural sciences (such as biology, chemistry, and geology) and social sciences (such as economics, politics, and ethics) to present a general idea of how nature works and how things are interconnected. It is a study of *connections in nature*.

Since its first edition this book has led the way in using scientific laws, principles, models, and concepts to help us understand environmental and resource problems and possible solutions, and how these concepts, problems, and solutions are connected. I have introduced only the concepts and principles necessary for understanding the material in the book, and I have tried to present them simply but accurately. The key principles and concepts used in this textbook are summarized in the Epilogue (pp. 565–566).

A growing number of critics have criticized environmental science teachers and textbooks for **(1)** not being grounded in basic science, **(2)** superficial treatment of environmental issues, **(3)** giving an unbalanced, biased view of such issues, and **(4)** focusing mostly on bad environmental news without giving the good environmental news.

I generally agree with these criticisms and believe that *there is an urgent need to expand the fundamental scientific content of environmental science courses.* In each new edition of this book I have done this. This latest edition further expands the basic scientific content (see changes in this edition on p. v).

Since the first edition of this book I have also emphasized in-depth treatment of key environmental issues. Many instructors have written me thanking me for being the only author in this field to give students an in-depth, balanced, and up-to-date view of major environmental problems such as global warming and ozone depletion (Chapter 10) and loss of biodiversity (a major integrating theme of this book and now widely considered the world's key environmental problem) without doing this in a complex and hard-to-understand manner.

To help ensure that the material is accurate and up to date, I have consulted more than 10,000 research sources in the professional literature. In writing this book, I have also benefited from the more than 200 experts and teachers (see list on pp. ix–xii) who have provided detailed reviews of the editions of this and my other three books in this field.

How Do I Deal with the Bias Dilemma? Anyone writing a textbook on environmental science has certain opinions and biases based on many years of study, research, analysis, and critical thinking. Thus, all environmental science books are biased to some extent.

However, there are at least two sides to all problems and controversies. The challenge for authors of environmental science textbooks and teachers of environmental science courses is to give a fair and balanced view of opposing viewpoints without injecting personal bias. The reason for doing this is to allow students to make up their own minds and to end up with their own opinions and biases. Studying a subject as important as environmental science and ending up with no conclusions, opinions, and beliefs means that both the teacher and student have failed. But such conclusions should be based on using critical thinking to evaluate opposing ideas.

A major goal of this new edition is to reduce bias as much as possible. With the help of reviewers and a talented editor and bias hunter (George Dyke), I have scoured every sentence in this book to help achieve this goal. Examples of my efforts to give a balanced presentation of opposing viewpoints are found in the pros and cons of reducing birth rates (pp. 169–170), the Pro/Con box on genetic engineering (p. 240), Section 10-3 on global warming, Section 10-6 on ozone depletion, discussion of acid deposition (pp. 250–256), the asbestos problem (p. 258), problems with the Superfund law (pp. 390–391), food aid (p. 416), and the pros and cons of pesticides (Sections 15-2 and 15-3). Even more important are the thousands of sentences and phrases that have been omitted or reworded to eliminate any hint of bias.

Bias is subtle, however, and I invite instructors and students to write me and point out any remaining bias.

How Do I Deal with the Bad News/Good News Dilemma? Critics have charged environmentalists with wallowing in doom and gloom and not reporting or rejoicing in the many improvements in environmental quality that have taken place over the past several decades. I generally agree with this criticism and in this edition I have emphasized the good environmental news along with the bad environmental news in Chapter 1 (Table 1-1, p. 22) and throughout the book.

However, some critics of environmentalists have gone too far in the other direction and emphasized the good news without giving a balanced presentation of the bad news. Environmental good news should inspire us and make us feel good about what has been accomplished by the hard and dedicated work of millions of people. At the same time, environmental bad news should challenge us to maintain and strengthen past environmental improvements and deal with the real and more complex environmental problems we face.

Rosy optimism and gloom-and-doom pessimism are traps; both usually lead to denial, indifference, and inaction. I have tried to avoid these two extremes and give a realistic yet hopeful view of the future. This book is filled with technological advances that have led to environmental improvements (for example, see the Individuals Matter boxes on pp. 294, 365, 461, and 533) and stories of people who have acted to help sustain the earth's life-support systems for us and for all life, and whose actions inspire us to do better (for example, see the Individuals Matter boxes on pp. 46, 215, 319, 320, 379, 385, 424, 436, and 464, and the Guest Essay on p. 418). It's an exciting and challenging time to be alive as we struggle to enter into a new, more cooperative relationship with the planet that is our only home.

What Are the Key Features of This Book? This book is divided into five major parts (see Brief Contents, p. xii). After Parts I and II have been covered, the rest of the book can be used in almost any order. In addition, most chapters and many sections within these chapters can be moved around or omitted to accommodate courses with different lengths and emphases.

After major scientific concepts and environmental problems are discussed, various solutions proposed by a variety of scientists, environmental activists, and analysts are given. A range of possible solutions, some of them highly controversial, are provided to encourage students to think critically and make up their your own minds.

Each chapter begins with a brief *Earth Story*, a case study designed to capture interest and set the stage for the material that follows. In addition to these 19 case studies, 55 other case studies are found throughout the book (some in special boxes and others within the text); they provide a more detailed look at specific environmental problems and their possible solutions. These 74 case studies give students an in-depth analysis of key environmental problems and solutions. Sixteen *Guest Essays* (four of them new) present an individual researcher's or activist's point of view, which is then evaluated through critical thinking questions.

Other special boxes found in the text include *Pro/Con boxes* that present both sides of controversial environmental issues; *Connections boxes* that show connections in nature and among environmental concepts, problems, and solutions; *Solutions boxes* that summarize a variety of solutions (some of them controversial) to environmental problems proposed by various analysts; *Spotlight boxes* that highlight and give insights into key environmental problems and concepts; and *Individuals Matter boxes* that describe what people have done to help solve environmental problems. To encourage critical thinking and integrate it throughout the book, all boxes (except Individuals Matter) end with critical thinking questions.

This book is an integrated study of environmental problems, connections, and solutions. The twelve integrating themes in this book are *biodiversity and earth capital; science and technology; input, throughput, output, and feedback systems analysis; pollution prevention and waste reduction; population and exponential growth; energy and energy efficiency; connections; solutions and sustainability; uncertainty and controversy; economics and environment; environmental politics and law; and individual action and earth citizens.*

I hope you will start by looking at the brief table of contents (p. xii) to get an overview of this book. Then I suggest that you look at the concepts and connections diagram inside the back cover, which shows the major components and relationships found in environmental science. In effect, it is a map of the book. Then, to get a summary of the book's key principles, I urge you to read the two-page list of key principles found in the Epilogue (pp. 565–566).

The book's 412 illustrations are designed to present complex ideas in understandable ways and to relate learning to the real world. They include 301 full-color diagrams (50 of them maps) and 111 carefully selected color photographs.

I have not cited specific sources of information; this is rarely done for an introductory-level text in any field, and it would interrupt the flow of the material. Instead, the Internet sites and readings listed at the end of the book for each chapter provide backup for almost all the information in this book and serve as springboards to further information and ideas. This edition also has a new interactive World Wide Web site that can be used as a source of information and ideas (see description on p. ix), a new *Ecolink* CD-ROM (p. ix), and a new *InfoTrac* online university library of articles (p. vii).

Instructors wanting books covering this material with a different emphasis, length, and organization can use one of my three other books written for various types of environmental science courses: *Living in the Environment*, 10th edition (761 pages, Wadsworth, 1998, the longest and most comprehensive book); *Sustaining the Earth: An Integrated Approach*, 3rd edition (338 pages, Wadsworth, 1998, a shorter book with a different integrated approach); and *Environment: Problems and Solutions* (150 pages, Wadsworth, 1994, a very short introduction that assumes a background in scientific concepts).

What Are Major Changes in the Seventh Edition?
Major changes in this edition include:

- Updated and revised material throughout the book.

- Phrasing subsection titles as questions. This provides students with a clear overview of what each subsection covers and provides an internal set of learning objectives.

- A major effort to reduce bias (see p. iv).

- Significantly increased coverage of basic scientific and ecological concepts, with the addition or expansion of material on linear and exponential growth; emphasis on input, output, and throughput of resources through systems; pollution prevention; nature of science; scientific methods; objectivity of scientists; models, systems, and feedback (new Section 3-2); ionic and covalent compounds; balancing chemical equations and the law of conservation of matter; difference between heat and temperature; resource partitioning; asexual and sexual reproduction; anaerobic respiration; how ecologists learn about nature through field research, laboratory research, and systems research (new Section 4-5); fundamental and realized niches; sea otters as keystone species; ecological importance of parasites; structure and function of ecosystems (new diagrams of tropical rain forest, temperate deciduous forest, and coral reef ecosystems); harmful synergistic interactions; feedback and delay in controlling carbon dioxide levels; population dynamics in nature; r-strategists and K-strategists; role of continental drift in speciation and extinction; past mass extinctions and adaptive radiations; changes during ecological succession; survival strategies of plants; types of plants; human impacts on deserts, grasslands, forests, and mountain systems; ecological importance of mountains; organisms found on rocky shores and sandy shores; time delays and feedback in ecosystems; importance of fossils in understanding evolution; natural selection; reproductive isolation and speciation; early succession, midsuccession, and late succession species; mechanisms of succession (facilitation, inhibition, tolerance); role of disturbance in succession; predictability of succession; and summary of basic ecological processes.

- Addition or expansion of many topics, including summaries of environmental accomplishments (good news), more details on various environmental worldviews, deep ecology, concern about future generations, growth and migration of environmental refugees, pros and cons of urban areas, resource and environmental problems of urban areas, pros and cons of cars, bicycles as alternatives to cars, pros and cons of mass transit, feasibility of reducing auto use, genuine progress indicator as substitute for GNP, ecologically sustainable development, improving environmental laws and regulations, comparative risks, bioaccumulation and biomagnification, hormone disrupters, global TB epidemic, health effects of airborne particulates, nature and limitations of climate models, amplifying (positive) and corrective (negative) feedbacks on global warming, validity (pros and cons) of ozone depletion, using cloud seeding and towing icebergs to increase water supplies, beneficial effects of floods, cleanup of Thames River in Great Britain, dangers from *Pfiesteria* bacteria, recovery of Seattle's Lake Washington from severe eutrophication, faster composting, green design, ecoindustrial revolution, aboveground storage of toxic and hazardous wastes, survival abilities of cockroaches, placing monetary values on ecosystem services (how much is a forest worth?), biosphere reserves, kudzu invasions, endangered turtle species, biophilia, Nature Conservancy, effects of overfishing on aquatic biodiversity, effects of wildlife habitat fragmentation, fire ants, pros and cons of weakening or strengthening the Endangered Species Act, new water-based fuel, ecocars, coal use in China, government energy research and development funding, advantages of nuclear power, nuclear waste contamination in the former Soviet Union, and future options for nuclear power.

■ Greater emphasis on *critical thinking* by addition of a new Guest Essay on the essentials of critical thinking (p. 52), critical thinking questions in all boxes (except Individuals Matter), and several new critical thinking questions and projects at the end of most chapters.

■ Addition of World Wide Web hyperlinks for material in the book marked with the icon ▨ (see description on p. vii).

■ Addition of suggested Internet and World Wide Web sites for each chapter. They are listed along with the Further Readings in the back of the book.

■ Addition of a number of new boxes.

■ Four new Guest Essays.

■ *Ecolink*, an instructor presentation tool that allows for the quick, easy assembly of media files into a multimedia presentation. Through an easy-to-use interface, users are able to assemble, edit, publish, and present custom lectures built from an extensive multimedia database. The database is made up of components that will be supplied on CD-ROM or those of the instructor residing on local hard drives or the Internet. The content consists of all the illustrations in the Starr/Taggart text, the best material from the student *Interactive Concepts in Biology* CD-ROM, and material from our laserdiscs.

■ Addition of 5 new color photos and 28 new diagrams, and improvement of 14 existing diagrams.

■ *InfoTrac*, a fully searchable online university library that gives students access to full-length articles from over 600 scholarly and popular journals, updated daily, and dating back as much as 4 years. An accompanying booklet provides guidance and suggests activities and projects using *InfoTrac*.

■ Revision and updating of the *Instructor's Manual and Test Items Booklet*.

Welcome to Uncertainty, Controversy, and Challenge   There are no easy solutions to the environmental problems and challenges we face. We will never have scientific certainty or agreement about what we should do because science provides us with probabilities, not certainties; science advances through continuous controversy. What is important is not what the experts disagree on (the frontiers of knowledge that are still being developed, tested, and argued about), but what they generally agree on—the *scientific consensus*—on concepts, problems, and possible solutions.

Despite considerable research, we still know little about how nature works at a time when we are altering nature at an accelerating pace. This uncertainty, as well as the complexity and importance of these issues to current and future generations of humans and other species, makes many of these issues highly controversial. Intense controversy also arises because environmental science is a dynamic blend of natural and social sciences that sometimes questions the ways we view and act in the world around us. This interdisciplinary attempt to mirror reality challenges us to evaluate our environmental worldviews, values, and lifestyles, as well as our economic and political systems. This can often be a threatening process.

Study Aids   Each chapter begins with a few general questions to reveal how it is organized and what students will be learning. When a new term is introduced and defined, it is printed in **boldface type**. A glossary of all key terms is located at the end of the book.

Factual recall questions (with answers) are listed at the bottom of most pages. You might cover the answer (on the right-hand page) with a piece of paper and then try to answer the question on the left-hand page. (These questions are not necessarily related to the chapter in which they are found.)

Each chapter and box ends with a question or questions to encourage students to think critically and apply what they have learned to their lives. Some ask students to take sides on controversial issues and back up their conclusions and beliefs. The end-of-chapter questions are followed by several projects that individuals or groups can carry out. Many additional projects are given in the *Instructor's Manual* and in the *Green Lives, Green Campuses* and *Critical Thinking and the Environment* supplements available with this book.

Readers who become especially interested in a particular topic can consult the list of Internet sites and further readings for each chapter, given in the back of the book. Appendix 1 contains a list of important publications and a list of some key environmental organizations and government and international agencies.

Students can also access World Wide Web for material in the book marked with the icon 🌐. Access to this system is at http://www.wadsworth.com/biology. With this new feature, students can click on a chapter in the Hypercontents listed and find resources that couldn't be listed in the book. If you can't find a topic you are looking for, try our search page. These resources are updated constantly. Because the World Wide Web is like an image gallery, library, and information booth, each link reference is accompanied by a description to help guide you in your selections. In the Web site for this text, you will also find Cool Events, Critical Thinking Questions, Tips on Surfing, Interactive Quizzes for each chapter, and much more. Happy surfing.

Help Me Improve This Book   Let me know how you think this book can be improved; if you find any errors, bias, or confusing explanations please send them to Jack Carey, Biology Publisher, Wadsworth Publishing Company, 10 Davis Drive, Belmont, CA 94002. He will forward them to me. Most errors can be corrected in subsequent printings of this edition, rather than waiting for a new edition.

Supplements   The following supplements are available (those marked with an asterisk are *new* for this edition):

- *Ecolink*, an instructor presentation tool that allows for the quick, easy assembly of media files into a multimedia presentation. Through an easy-to-use interface, users are able to assemble, edit, publish, and present custom lectures built from an extensive multimedia database. The database is made up of components that will be supplied on CD-ROM or those of the instructor residing on local hard drives or the Internet. The content consists of all the illustrations in the Starr/Taggart text, the best material from the student *Interactive Concepts in Biology* CD-ROM, and material from our laserdiscs.

- *InfoTrac*, a fully searchable online university library that gives students access to full-length articles from over 600 scholarly and popular journals, updated daily, and dating back as much as 4 years. An accompanying booklet provides guidance and suggests activities and projects using *InfoTrac*.

- *Instructor's Manual and Test Items*, written and updated by Jane Heinze-Fry (Ph.D. in science and environmental education). For each chapter, it has goals and objectives; one or more concept maps; key terms; teaching suggestions; multiple-choice test questions with answers; projects, field trips, and experiments; term-paper and report topics; and a list of audiovisual materials and computer software.

- *Interactive World Wide Web site* for this textbook (as described previously).

- *Internet Booklet*, by Daniel J. Kurland and Jane Heinze-Fry. An introduction to the Internet and World Wide Web, plus selected sites to visit and learning exercises.

- *Critical Thinking and the Environment: A Beginner's Guide*, by Jane Heinze-Fry and G. Tyler Miller, Jr. An introduction to different critical thinking approaches, with questions by chapter using these approaches.

- *Critical Thinking Software Tools and Workbook*. This special version of STELLA II® software, together with an accompanying workbook, demonstrates critical thinking exercises using models related to the text.

- *Green Lives, Green Campuses*, written by Jane Heinze-Fry. This hands-on workbook contains projects to help students evaluate the environmental impact of their own lives and to guide them in making an environmental audit of their campus.

- *Environmental Articles*, assembled by Jane Heinze-Fry. This is a collection of more than 200 articles indexed by topic and geographical location. Instructors may use the indexes to choose any combination of articles and have them bound as customized supplements for their courses.

- A set of 75 color acetates and more than 600 black-and-white transparency masters for making overhead transparencies or slides of line art (including concept maps for each chapter), available to adopters.

- *Laboratory Manual*, by C. Lee Rockett (Bowling Green State University) and Kenneth J. Van Dellen (Macomb Community College).

- *Watersheds: Classic Cases in Environmental Ethics*, by Lisa H. Newton and Catherine K. Dillingham (Wadsworth, 1994). Nine balanced case studies that amplify material in this book.

- *Environmental Ethics*, by Joseph R. Des Jardins (Wadsworth, 1993). A very useful survey of environmental ethics. Brief case studies and many specific examples are included.

- *Radical Environmentalism*, by Peter C. List (Wadsworth, 1993). A series of readings on environmental politics and philosophy.

- *A Beginner's Guide to Scientific Method*, by Stephen S. Carey (Wadsworth, 1994). A concise, hands-on introduction that helps students develop critical thinking skills essential to understanding the scientific process.

■ *The Game of Science*, 5th edition, by Garvin McCain and Erwin M. Segal (Brooks/Cole, 1988). An accurate, lively, and up-to-date view of what science is, who scientists are, and how they approach science.

Annenberg/CPB Television Course   This textbook is being offered as part of the Annenberg/CPB Project television series *Race to Save the Planet*, a 10-part public broadcasting series and a college-level telecourse examining the major environmental questions facing the world today. The series takes into account the wide spectrum of opinion about what constitutes an environmental problem and discusses the controversies about appropriate remedial measures. It analyzes problems and emphasizes the successful search for solutions. The course develops a number of key themes that cut across a broad range of environmental issues, including sustainability, the interconnection of the economy and the ecosystem, short-term versus long-term gains, and the trade-offs involved in balancing problems and solutions. A study guide and a faculty guide, both available from Wadsworth Publishing Company, integrate the telecourse and this text.

For further information about available television course licenses and duplication licenses, contact PBS Adult Learning Service, 1320 Braddock Place, Alexandria, VA 22314-1698 (1-800-ALS-AL5-8).

For information about purchasing videocassettes and print material, contact the Annenberg/CPB Collection, P.O. Box 2284, South Burlington, VT 05407-2284 (1-800-LEARNER).

Acknowledgments   I wish to thank the many students and teachers who responded so favorably to the six previous editions of *Environmental Science,* the ten editions of *Living in the Environment,* the three editions of *Sustaining the Earth: An Integrated Approach,* and the first edition of *Environment: Problems and Solutions*—and who corrected errors and offered many helpful suggestions for improvement. I am also deeply indebted to the reviewers, who pointed out errors and suggested many important improvements in this book. Any errors and deficiencies left are mine.

The members of the talented production team, listed on the copyright page, have made vital contributions as well. I especially appreciate the competence and cheerfulness of production editor Rob Anglin and the helpful inputs from copyeditor Carol Anne Peschke. My thanks also go to Wadsworth's hard-working sales staff and to Kristin Milotich, Kerri Abdinoor, and Michael Burgreen for their help and efficiency.

Thanks to C. Lee Rockett and Kenneth J. Van Dellen for developing the *Laboratory Manual* to accompany this book. I also wish to thank the people who have translated this book into five different languages for use throughout much of the world.

My special thanks go Jane Heinze-Fry for her insightful analysis, for being such a delight to work with, and for her outstanding work on the *Instructor's Manual,* concept mapping, *Green Lives, Green Campuses, Environmental Articles,* and the new *Critical Thinking and the Environment: A Beginner's Guide* and *Internet Booklet.*

My deepest thanks go to Jack Carey, biology publisher at Wadsworth, for his encouragement, help, 30 years of friendship, and superb reviewing system. It helps immensely to work with the best and most experienced editor in college textbook publishing. I dedicate this book to the earth that sustains us all.

G. Tyler Miller, Jr.

Guest Essayists and Reviewers

Guest Essayists The following are the authors of Guest Essays: **Lester R. Brown**, president, Worldwatch Institute; **Alberto Ruz Buenfil**, environmental activist, writer, and performer; **Robert D. Bullard**, professor of sociology, University of California, Riverside; **Lois Marie Gibbs**, director, Citizens' Clearinghouse for Hazardous Wastes; **Garrett Hardin**, professor emeritus of human ecology, University of California, Santa Barbara; **Paul Hawken**, environmental author and business leader; **Jane Heinze-Fry**, author and consultant in environmental education; **Amory B. Lovins**, energy policy consultant and director of research, Rocky Mountain Institute; **Bobbi S. Low**, professor of resource ecology, University of Michigan; **Lester W. Milbrath**, director of the Research Program in Environment and Society, State University of New York, Buffalo; **Peter Montague**, senior research analyst, Greenpeace, and director, Environmental Research Foundation; **Norman Myers**, consultant in environment and development; **David W. Orr**, professor of environmental studies, Oberlin College; **David Pimentel**, professor of entomology, Cornell University; **Andrew (Andy) C. Revkin**, environmental journalist, *The New York Times*, and environmental author; and **Vandana Shiva**, director, Research Foundation for Science, Technology, and Natural Resource Policy, Dehra Dun, India.

Cumulative Reviewers Barbara J. Abraham, Hampton College; Donald D. Adams, State University of New York at Plattsburgh; Larry G. Allen, California State University, Northridge; James R. Anderson, U.S. Geological Survey; Kenneth B. Armitage, University of Kansas; Gary J. Atchison, Iowa State University; Marvin W. Baker, Jr., University of Oklahoma; Virgil R. Baker, Arizona State University; Ian G. Barbour, Carleton College; Albert J. Beck, California State University, Chico; W. Behan, Northern Arizona University; Keith L. Bildstein, Winthrop College; Jeff Bland, University of Puget Sound; Roger G. Bland, Central Michigan University; Georg Borgstrom, Michigan State University; Arthur C. Borror, University of New Hampshire; John H. Bounds, Sam Houston State University; Leon F. Bouvier, Population Reference Bureau; Daniel J. Bovin, Universite´ Laval; Michael F. Brewer, Resources for the Future, Inc.; Mark M. Brinson, East Carolina University; Patrick E. Brunelle, Contra Costa College; Terrence J. Burgess, Saddleback College North; David Byman, Pennsylvania State University, Worthington Scranton; Lynton K. Caldwell, Indiana University; Faith Thompson Campbell, Natural Resources Defense Council, Inc.; Ray Canterbery, Florida State University; Ted J. Case, University of San Diego; Ann Causey, Auburn University; Richard A. Cellarius, Evergreen State University; William U. Chandler, Worldwatch Institute; F. Christman, University of North Carolina, Chapel Hill; Preston Cloud, University of California, Santa Barbara; Bernard C. Cohen, University of Pittsburgh; Richard A. Cooley, University of California, Santa Cruz; Dennis J. Corrigan; George Cox, San Diego State University; John D. Cunningham, Keene State College; Herman E. Daly, The World Bank; Raymond F. Dasmann, University of California, Santa Cruz; Kingsley Davis, Hoover Institution; Edward E. DeMartini, University of California, Santa Barbara; Charles E. DePoe, Northeast Louisiana University; Thomas R. Detwyler, University of Wisconsin; Peter H. Diage, University of California, Riverside; Lon D. Drake, University of Iowa; T. Edmonson, University of Washington; Thomas Eisner, Cornell University; Michael Esler, Southern Illinois University; David E. Fairbrothers, Rutgers University; Paul P. Feeny, Cornell University; Nancy Field, Bellevue Community College; Allan Fitzsimmons, University of Kentucky; Andrew J. Friedland, Dartmouth College; Kenneth O. Fulgham, Humboldt State University; Lowell L. Getz, University of Illinois at Urbana-Champaign; Frederick F. Gilbert, Washington State University; Jay Glassman, Los Angeles Valley College; Harold Goetz, North Dakota State University; Jeffery J. Gordon, Bowling Green State University; Eville Gorham, University of Minnesota; Michael Gough, Resources for the Future; Ernest M. Gould, Jr., Harvard University; Katharine B. Gregg, West Virginia Wesleyan College; Peter Green, Golden West College; Paul K. Grogger, University of Colorado at Colorado Springs; L. Guernsey, Indiana State University; Ralph Guzman, University of California, Santa Cruz; Raymond Hames, University of Nebraska, Lincoln; Raymond E. Hampton, Central Michigan University; Ted L. Hanes, California State University, Fullerton; William S. Hardenbergh, Southern Illinois University at Carbondale; John P. Harley, Eastern Kentucky University; Neil A. Harriman, University of Wisconsin, Oshkosh; Grant A. Harris, Washington State University; Harry S. Hass, San Jose City College; Arthur N. Haupt, Population Reference Bureau; Denis A. Hayes, environmental consultant;

A. van Hylckama, Texas Tech University; Robert R. Van Kirk, Humboldt State University; Donald E. Van Meter, Ball State University; Gary Varner, Texas A & M University; John D. Vitek, Oklahoma State University; Lee B. Waian, Saddleback College; Warren C. Walker, Stephen F. Austin State University; Thomas D. Warner, South Dakota State University; Kenneth E. F. Watt, University of California, Davis; Alvin M. Weinberg, Institute of Energy Analysis, Oak Ridge Associated Universities; Brian Weiss; Anthony Weston, SUNY at Stony Brook; Raymond White, San Francisco City College; Douglas Wickum, University of Wisconsin, Stout; Charles G. Wilber, Colorado State University; Nancy Lee Wilkinson, San Francisco State University; John C. Williams, College of San Mateo; Ray Williams, Rio Hondo College; Roberta Williams, University of Nevada Las Vegas; Samuel J. Williamson, New York University; Ted L. Willrich, Oregon State University; James Winsor, Pennsylvania State University; Fred Witzig, University of Minnesota at Duluth; George M. Woodwell, Woods Hole Research Center; Robert Yoerg, Belmont Hills Hospital; Hideo Yonenaka, San Francisco State University; Malcolm J. Zwolinski, University of Arizona.

Brief Contents

Detailed Contents

Francois Gohier/Ardea London

Saguaro cacti, Arizona

Paul W. Johnson/Biological Photo Service

Temperate deciduous forest, fall, Rhode Island

Temperate deciduous forest, winter, Rhode Island

Tree farm, North Carolina

Mangrove swamp, Colombia

National Archives/EPA Documerica

Monoculture cropland, Blythe, California

Jack Carey

Crater Lake, Oregon

Michael Grecco/Picture Group

Sulfur dioxide emissions from coal-burning power plant

Water hyacinth, Florida

Endangered cotton-top tamarin, tropical forest, South America

Invertebrate sea star

Warning coloration of poison-dart frog, South America

The environmental crisis is an outward manifesta-
tion of a crisis of mind and spirit. There could be no
greater misconception of its meaning than to believe
it is concerned only with endangered wildlife, human-

1 ENVIRONMENTAL PROBLEMS AND THEIR CAUSES

Living in an Exponential Age

Once there were two kings from Babylon who enjoyed playing chess, with the winner claiming a prize from the loser. After one match, the winning king asked the loser to pay him by placing one grain of wheat on the first square of the chessboard, two on the second, four on the third, and so on. The number of grains was to double each time until all 64 squares were filled.

The losing king, thinking he was getting off easy, agreed with delight. It was the biggest mistake he ever made. He bankrupted his kingdom and still could not produce the 2^{63} grains of wheat he had promised. In fact, it's probably more than all the wheat that has ever been harvested!

This is an example of **exponential growth**, in which a quantity increases by a fixed percentage of the whole in a given time. As the losing king learned, exponential growth is deceptive. It starts off slowly, but after only a few doublings it grows to enormous numbers because each doubling is more than the total of all earlier growth.

Here is another example. Fold a piece of paper in half to double its thickness. If you could do this 42 times, the stack would reach from the earth to the moon, 386,400 kilometers (240,000 miles) away. If you could double it 50 times, the folded paper would almost reach the sun, 149 million kilometers (93 million miles) away!

The environmental problems we face—population growth, wasteful use of resources, destruction and degradation of wildlife habitats, extinction of plants and animals, poverty, and pollution—are interconnected and are growing exponentially. For example, world population has more than doubled in only 47 years, from 2.5 billion in 1950 to 5.84 billion in 1997. Unless death rates rise sharply, it may reach 8 billion by 2025, 10–11 billion by 2050, and 14 billion by 2100 (Figure 1-1). Global economic output, much of it environmentally damaging, has increased almost sixfold since 1950.

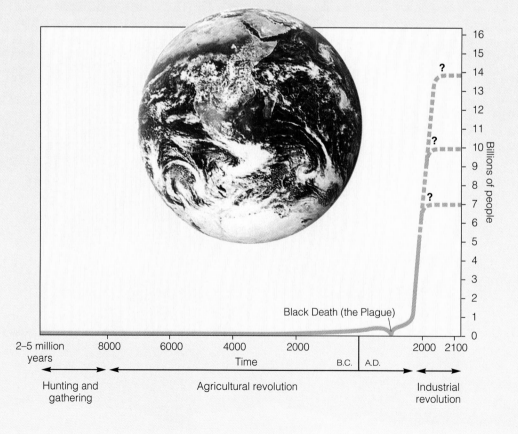

Figure 1-1 The J-shaped curve of past exponential world population growth, with projections beyond 2100. Notice that exponential growth starts off slowly, but as time passes the curve becomes increasingly steep. World population has more than doubled in only 47 years, from 2.5 billion in 1950 to 5.84 billion in 1997. Unless death rates rise sharply, it may reach 8 billion by 2025, 10–11 billion by 2050, and 14 billion by 2100. (This figure is not to scale.) (Data from World Bank and United Nations; photo courtesy of NASA)

Alone in space, alone in its life-supporting systems, powered by inconceivable energies, mediating them to us through the most delicate adjustments, wayward, unlikely, unpredictable, but nourishing, enlivening, and enriching in the largest degree— is this not a precious home for all of us? Is it not worth our love?

BARBARA WARD AND RENÉ DUBOS

This chapter is an overview of environmental problems, their root causes, and the controversy over their seriousness. It discusses these questions:

- What is earth capital? What is a sustainable society?

- How fast is the human population increasing?

- What are the earth's main types of resources? How can they be depleted or degraded?

- What are the principal types of pollution? How can pollution be reduced and prevented?

- What are the root causes of the environmental problems we face?

- How serious are environmental problems, and is our current course sustainable?

Figure 1-2 Solar and earth capital consist of the life-support resources and processes provided by the sun and the planet for use by us and other species. These two forms of capital support and sustain all life and all economies on the earth.

1-1 LIVING SUSTAINABLY

What Are Solar Capital, Earth Capital, and Sustainability? Our existence, lifestyles, and economies depend completely on the sun and the earth, a blue and white island in the black void of space. We can think of energy from the sun as **solar capital**, and we can think of the planet's air, water, soil, wildlife, minerals, and natural purification, recycling, and pest control processes as **earth capital** (Figure 1-2). The term *environment* is often used to describe these life-support systems; in effect it's another term for describing solar capital and earth capital.

Environmentalists and many leading scientists believe that we are depleting and degrading the earth's natural capital at an accelerating rate as our population

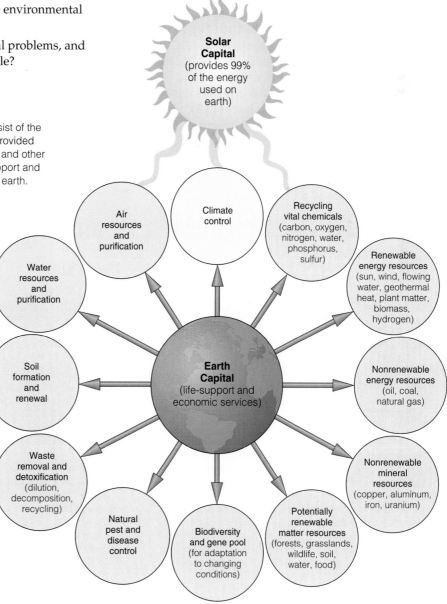

(Figure 1-1) and demands on the earth's resources and natural processes increase exponentially (Figure 1-3). Others, mostly economists, disagree. They contend that there are no limits to human population growth and economic growth that can't be overcome by human ingenuity and technology.

A **sustainable society** manages its economy and population size without exceeding all or part of the planet's ability to absorb environmental insults, replenish its resources, and sustain human and other forms of life over a specified period, usually hundreds to thousands of years. During this period, it satisfies the needs of its people without depleting earth capital and thereby jeopardizing the prospects of current and future generations of humans and other species.

Living sustainably means living off of income and not depleting the capital that supplies the income. Imagine that you inherit $1 million. If you invest this capital at 10% interest, you will have a sustainable annual income of $100,000; that is, you can spend up to $100,000 a year without touching your capital.

Suppose you develop a taste for diamonds or a yacht, or all of your relatives move in with you. If you spend $200,000 a year, your $1 million will be gone early in the seventh year; even if you spend just $110,000 a year, you will be bankrupt early in the eighteenth year. The lesson here is a very old one: *Don't eat the goose that lays the golden egg.* Deplete your capital, and you move from a sustainable to an unsustainable lifestyle.

The same lesson applies to earth capital (Figure 1-2). With the help of solar energy, natural processes developed over billions of years can indefinitely renew the topsoil, water, air, forests, grasslands, and wildlife on which we and other forms of life depend as long as we don't use these potentially renewable resources faster than they are replenished.

Some of earth's natural processes also provide flood prevention, build and renew soil, slow soil erosion, and keep the populations of at least 95% of the species we consider pests under control. Living sustainably involves not disrupting or diminishing these and other natural processes and services provided by nature (Figure 1-2).

1-2 GROWTH AND THE WEALTH GAP

What Is the Difference Between Linear Growth and Exponential Growth? Suppose you hop on a train that accelerates by 1 kilometer (0.6 mile) an hour every second. After 5 seconds, you would be traveling at 5 kilometers (3 miles) per hour; after 30 seconds, your speed would be 30 kilometers (19 miles) per hour. This is an example of **linear growth**, in which a quantity increases by a constant amount per unit of time, as 1, 2, 3, 4, 5—or 1, 3, 5, 7, 9—and so on. If plotted on a graph, such growth in speed or growth of money in a savings account yields a straight line sloping upward (Figure 1-4).

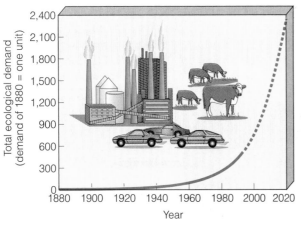

Figure 1-3 J-shaped curve of exponential growth in the total ecological demand on the earth's resources from agriculture, mining, and industry between 1880 and 1996. Projections to 2020 assume that resource use will continue to increase at the current rate of 5.5% per year. At that rate, our total ecological demand on the earth's resources doubles every 13 years. If global economic output grew by only 3% a year, resource consumption would still double every 23 years. (Data from United Nations, World Resources Institute, and Carrol Wilson, *Man's Impact on the Global Environment*, Cambridge, Mass.: MIT Press, 1970)

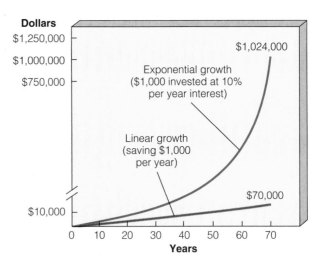

Figure 1-4 Linear and exponential growth. If you save $1,000 a year for a lifetime of 70 years, the resulting linear growth will allow you to save $70,000 (lower curve). If you invest $1,000 each year at 10% interest for 70 years and reinvest the interest, your money will grow exponentially to $1,024,000 (upper curve). If resource use, economic growth, or money in a savings account grows exponentially for 70 years (a typical human lifetime) at a rate of 10% a year, it will experience 1,024-fold increase.

Q: How many people are there in the world?

But suppose the train had a magic motor strong enough to double its speed every second. After 6 seconds, you would be moving at 64 kilometers (40 miles) per hour; after only 30 seconds, you would be traveling a billion kilometers (620 million miles) per hour!

This is an example of the astounding power of **exponential growth**, in which a quantity increases by a fixed percentage of the whole in a given time as each increase is applied to the base for further growth. Any quantity growing by a fixed percentage, even as small as 0.001% or 0.1%, is undergoing exponential growth and will experience extraordinary growth as its base of growth doubles again and again. If plotted on a graph, continuing exponential growth eventually yields a graph shaped somewhat like the letter *J* (Figure 1-4).

How long does it take to double resource use, population size, or money in a savings account that is growing exponentially? A quick way to calculate this **doubling time** in years is to use the **rule of 70**: 70/percentage growth rate = doubling time in years (a formula derived from the basic mathematics of exponential growth). For example, in 1997 the world's population grew by 1.47%. If that rate continues, the earth's population will double in about 48 years (70/1.47 = 48 years)—more growth than has occurred in *all* of human history. In the example of exponential growth of savings by 10% a year in Figure 1-4, your money would double roughly every 7 years (70/10 = 7).

Even though supplies of many resources seem large, exponential growth in their use can deplete them in a short time. For example, suppose we have an exhaustible resource so plentiful that its estimated supplies would last 1 billion years at its current rate of consumption. If the rate of use of this resource increases exponentially by 5% per year, however, the billion-year supply would last only 500 years.

How Rapidly Is the Human Population Growing?
The increasing size of the human population is one example of exponential growth. If such exponential growth continues (even at a low percentage of annual growth), eventually the population growth curve rounds a bend and heads almost straight up, creating a *J*-shaped curve (Figure 1-1 and Spotlight, right).

It took 2 million years to reach a billion people, 130 years to add the second billion, 30 years for the third, 15 years for the fourth, and only 12 years for the fifth billion. At current growth rates, the sixth billion will be added between 1986 and 1999, the seventh by 2012, and the eighth billion by 2025.

Recent studies by researchers at Conservation International suggest that roughly 48% of the earth's total land area has been partially or totally modified by human activities (Figure 1-5). If uninhabitable areas of rock and ice are excluded, *73% of the habitable area of*

SPOTLIGHT

Current Exponential Growth of the Human Population

The current growth of the world's population is the result of roughly 2.6 times as many births (about 140 million) as deaths (some 53 million) per year. The relentless ticking of this population clock means that in 1997 the world's population of 5.84 billion grew by 86.7 million people (5.84 billion × 0.0147 = 86.7 million)—an average increase of 236,000 people a day, or 9,830 an hour.

At this 1.47% annual rate of exponential growth, it takes only about

- 5 days to add the number of Americans killed in all U.S. wars
- 2 months to add as many people as live in the Los Angeles basin
- 1.9 years to add the 167 million people killed in all wars fought in the past 200 years
- 3 years to add 268 million people (the population of the United States in 1997)
- 14 years to add 1.24 billion people (the population of China, the world's most populous country, in 1997)

Critical Thinking

Some argue that current population growth is good because it provides more workers, consumers, and problem solvers to keep the global economy growing. Others argue that current population growth threatens economies and the earth's life-support systems through increased pollution and environmental degradation. What is your position? Why?

the planet has been altered by human activities. What will happen to the earth's remaining wildlife habitats, wildlife species, and biodiversity if the human population increases from 5.84 billion to 8 billion between 1997 and 2025 and perhaps to 11 billion by 2050?

What Is Economic Growth? Virtually all countries seek **economic growth**: an increase in their capacity to provide goods and services for people's final use. Such growth is normally achieved by increasing the flow or **throughput** of matter and energy resources used to produce goods and services through an economy. This is accomplished by means of population growth (more consumers and producers), or more consumption per person, or both.

Economic growth is usually measured by an increase in a country's **gross national product (GNP)**:

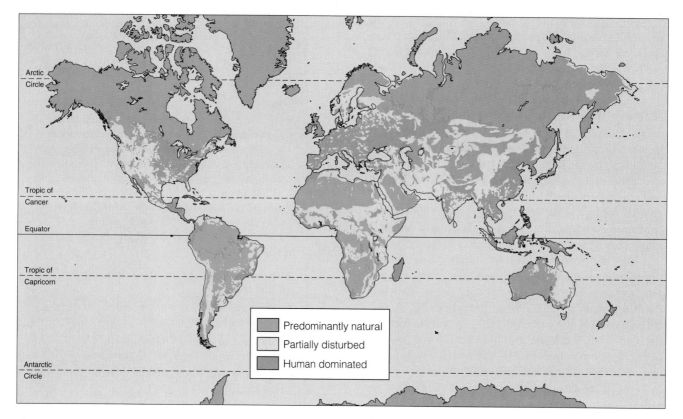

Figure 1-5 The degree of human disturbance of the earth's land area. Green indicates undisturbed areas. Yellow indicates partially disturbed areas, including secondary forest growth after cutting and shifting cultivation. Red indicates seriously disturbed natural ecosystems, including deforestation, farmland, overgrazed grasslands, and urban areas. Excluding uninhabitable areas of rock, ice, desert, and steep mountain terrain, *only about 27% of the planet's land area remains undisturbed by human activities.* (Data from Conservation International)

the market value in current dollars of all goods and services produced within *and* outside a country by the country's businesses for final use during a year. **Gross domestic product (GDP)** is the market value in current dollars of all goods and services produced *within* a country for final use during a year. To show one person's slice of the economic pie, economists often calculate the **per capita GNP**: the GNP divided by the total population.

The United Nations broadly classifies the world's countries as economically developed or developing. The **developed countries** are highly industrialized. Most (except the countries of the former Soviet Union) have high per capita GNPs (above $4,000). These countries, with 1.18 billion people (20% of the world's population in 1997), command about 85% of the world's wealth and income and use about 88% of its natural resources. They generate about 75% of its pollution and wastes (including about 90% of the world's estimated hazardous waste). Three developed countries—the United States, Japan, and Germany—together account for more than half of the world's economic output.

All other nations are classified as **developing countries**, with low to moderate industrialization

and per capita GNPs. Most are in Africa, Asia, and Latin America. Their 4.67 billion people (80% of the world's population in 1997) have only about 15% of the wealth and income and use only about 12% of the world's natural resources. In this context, **development** is the change from a society that is largely rural, agricultural, illiterate, and poor, with a rapidly growing population, to one that is mostly urban, industrial, educated, and wealthy, with a slow-growing or stable population.

Figure 1-6 compares various characteristics of developed countries and developing countries. More than 95% of the projected increase in world population is expected to take place in developing countries (Figure 1-7), *where 1 million people are added every four days.* By 2010, the combined population of Asia and Africa is projected to be 5.2 billion—almost as many as now live on the entire planet. The primary reason for such rapid population growth in developing countries (1.8% compared to 0.1% in developed countries) is the *large percentage of people who are under age 15* (35% compared to 20% in developed countries in 1997) and will be moving into their prime reproductive years over the next several decades.

Q: How many people are added to the world's population each day?

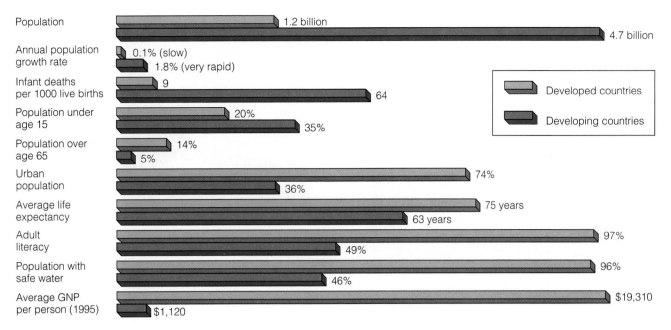

Figure 1-6 Some characteristics of developed countries and developing countries in 1997. (Data from United Nations and Population Reference Bureau)

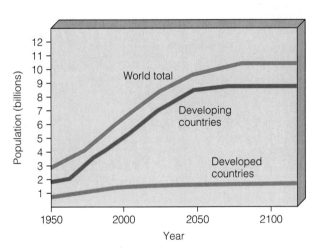

Figure 1-7 Past and projected population size for developed countries, developing countries, and the world, 1950–2120. Over 95% of the projected addition of 3.6 billion people between 1990 and 2030 is projected to occur in developing countries. (Data from United Nations)

What Is the Wealth Gap? Since 1960, and especially since 1980, the gap between the per capita GNP of the rich, middle income, and poor has widened (Figure 1-8). Today, one person in five lives in luxury, the next three get by, and the fifth struggles to survive on less than $1 a day. One person in six is hungry or malnourished (Figure 1-9) or severely undernourished and lacks clean drinking water, decent housing (Figure 1-10), and adequate health care. One of every three people lacks enough fuel to keep warm and to cook food and more than half of humanity lacks sanitary toilets.

Daily life for the almost 1 billion desperately poor people in developing countries is a harsh struggle for survival. Parents—some with nine or more children—are lucky to have an annual income of $365 ($1 a day). Having many children makes good sense to most poor parents because their children are a form of economic security, helping them grow food, gather fuel (mostly wood and dung), haul drinking water, tend livestock, work, or beg in the streets. The desperately poor tend to have many offspring because many of their children die at an early age. The two or three who live to adulthood will help their parents survive in old age (their 50s or 60s).

However, when many poor families have several children, the result is often far more people than local resources can support. To survive now, even though they know this may lead to disaster in the long run, they may deplete and degrade local forests, soil, grasslands, wildlife, and water supplies.

The poor often have little choice but to live in areas with the highest levels of air and water pollution and with the greatest risk of natural disasters such as floods, earthquakes, hurricanes, and volcanic eruptions. They also must take jobs (if they can find them) that often subject them to unhealthy and unsafe working conditions at very low pay.

Each year, at least 10 million of the desperately poor, or an average of 27,400 people per day (half of them children under age 5), die from malnutrition

A: An average of about 36,000

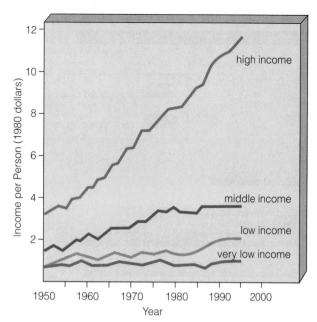

Figure 1-8 The wealth gap: changes in the distribution of global per capita GNP in high-income, middle-income, low-income, and very low-income countries, 1950–1996. Instead of trickling down, most of the income from economic growth has flowed up, with the situation worsening since 1980. In 1960, 2.3% of the global income went to the poorest 20% of the people; in 1980, this group received only 1.7%, and by 1990 their share was only 1.3%. More than 1 billion people survive on less than a dollar a day. (Data from United Nations)

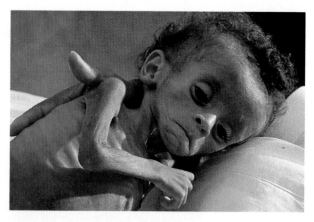

Figure 1-9 One in every three children under age five, such as this Brazilian child, suffers from malnutrition. Each day, at least 13,700 children die prematurely from hunger or hunger-related diseases—an average of 10 preventable deaths each minute. Some analysts put the estimated death toll at twice this number. (John Bryson/Photo Researchers)

Figure 1-10 One-fifth of the people in the world have inadequate housing, and at least 100 million people have no housing at all. These homeless people in Calcutta, India, must sleep on the street. (United Nations)

(lack of protein and other nutrients needed for good health) or related diseases and from contaminated drinking water. *This premature dying of human beings is equivalent to 69 jumbo jet planes, each carrying 400 passengers, crashing every day with no survivors.* Some put this annual death toll from poverty at about 20 million a year.

1-3 RESOURCES

What Is a Resource? In human terms, a **resource** is anything we get from the environment (the earth's life-support systems) to meet our needs and desires. However, all forms of life need resources such as food, water, and shelter for survival and good health. On our short human time scale, we classify material resources as renewable, potentially renewable, or nonrenewable (Figure 1-11; also see the orange boxes in the Concepts and Connections diagram inside the back cover).

Some resources, such as solar energy, fresh air, wind, fresh surface water, fertile soil, and wild edible plants, are directly available for use by us and other organisms. Other resources, such as petroleum (oil), iron, groundwater (water found underground), and

modern crops, aren't directly available. They become useful to us only with some effort and technological ingenuity. Petroleum, for example, was a mysterious fluid until we learned how to find it, extract it, and refine it into gasoline, heating oil, and other products that could be sold at affordable prices.

What Are Renewable Resources? Solar energy is called a **renewable** or **perpetual resource** because on a human time scale this solar capital (Figure 1-2) it is essentially inexhaustible. It is expected to last at least 6 billion years as the sun completes its life cycle.

Q: What is the projected population of the world in 2050?

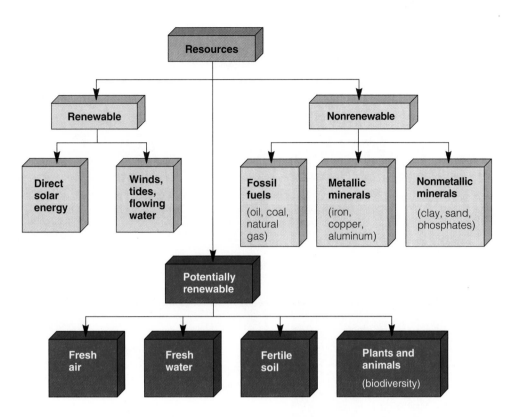

Figure 1-11 Major types of material resources. This scheme isn't fixed; potentially renewable resources can become nonrenewable resources if used for a prolonged period at a faster rate than they are renewed by natural processes.

A **potentially renewable resource**[1] can be replenished fairly rapidly (hours to several decades) through natural processes. Examples of such resources are forest trees, grassland grasses, wild animals, fresh lake and stream water, groundwater, fresh air, and fertile soil.

One important potentially renewable resource for us and other species is **biological diversity**, or **biodiversity**, which consists of the different life forms (species) that can best survive the variety of conditions currently found on the earth. Kinds of biodiversity include **(1) genetic diversity** (variety in the genetic makeup among individuals within a single species), **(2) species diversity** (variety among the species or distinct types of living organisms found in different habitats of the planet, Figure 1-12), and **(3) ecological diversity** (variety of forests, deserts, grasslands, streams, lakes, oceans, wetlands, and other biological communities).

This rich variety of genes, species, and biological communities gives us food, wood, fibers, energy, raw materials, industrial chemicals, and medicines—all of which pour hundreds of billions of dollars into the world economy each year. The earth's vast inventory of life-forms and biological communities also provides free recycling, purification, and natural pest control services (Figure 1-2).

However, potentially renewable resources can be depleted. The highest rate at which a potentially renewable resource can be used *indefinitely* without reducing its available supply is called its **sustainable yield**. If a resource's natural replacement rate is exceeded, the available supply begins to shrink, a process known as **environmental degradation**. Several types of environmental degradation can change potentially renewable resources into nonrenewable or unusable resources (Figures 1-13 and 1-14).

Connections: Renewable Resources and the Tragedy of the Commons One cause of environmental degradation is the overuse of **common-property resources**, which are owned by no one (or jointly by everyone in a country or area) but are available to all users free of charge. Most are potentially renewable. Examples include clean air, the open ocean and its fish, migratory birds, publicly owned lands (such as national forests, national parks, and wildlife refuges), gases of the lower atmosphere, and space.

In 1968, biologist Garrett Hardin (Guest Essay, pp. 172–173) called the degradation of common-property resources the **tragedy of the commons**. It

---

[1] Most sources use the term *renewable resource*. I have added the word *potentially* to emphasize that these resources can be depleted if we use them faster than natural processes renew them.

Figure 1-12 Two species found in tropical forests are part of the earth's precious biodiversity. On the left is the world's largest flower, the flesh flower (*Rafflesia arnoldi*), growing in a tropical rain forest in Sumatra. The flower of this leafless plant can be as large as 1 meter (3.3 feet) in diameter and weigh 7 kilograms (15 pounds). The plant gives off a smell like rotting meat, presumably to attract flies and beetles that pollinate its flower. After blossoming for a few weeks, the flower dissolves into a slimy black mass. On the right is a cotton top tamarin. (Left, Mitschuhiko Imanori/Nature Production; right, Gary Milburn/Tom Stack & Associates)

happens because each user reasons, "If I don't use this resource, someone else will. The little bit I use or pollute is not enough to matter." With only a few users, this logic works. However, the cumulative effect of many people trying to exploit a common-property resource eventually exhausts or ruins it. Then no one can benefit from it, and therein lies the tragedy.

One solution is to use common-property resources at rates below their sustainable yields or overload limits by reducing population, regulating access, or both. Unfortunately, it is difficult to determine the sustainable yield of a forest, grassland, or an animal population, partly because yields vary with weather, climate, and unpredictable biological factors, and because tracking such data is expensive.

These uncertainties mean that *it is best to use a potentially renewable resource at a rate well below its estimated sustainable yield*. This is a prevention or precau-

tionary approach designed to reduce the risk of environmental degradation. This approach is rarely used because it requires hard-to-enforce regulations that restrict resource use and thus conflict with the drive for short-term profit or pleasure.

Another approach is to convert common-property resources to private ownership. The reasoning behind this is that owners of land or some other resource have a strong incentive to see that their investment is protected. However, this approach is not practical for global common resources, such as the atmosphere, the open ocean, and migratory birds that cannot be divided up and converted to private property. Experience has also shown that private ownership can lead to short-term exploitation and environmental degradation instead of long-term sustainability.

Some believe that privatization is a better way to protect nonrenewable and potentially renewable

Q: Where does everything that supports your life come from?

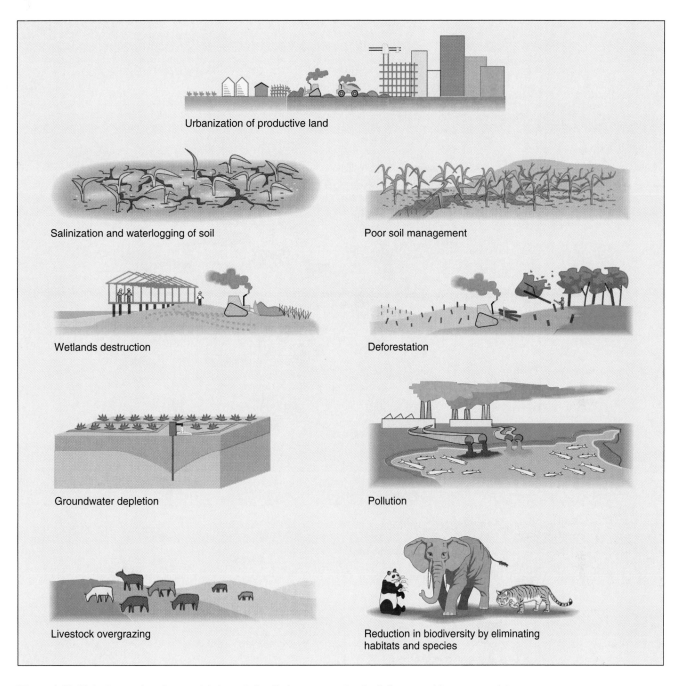

Figure 1-13 Major types of environmental degradation that can convert potentially renewable resources into nonrenewable resources.

resources found on publicly owned lands than command-and-control government regulations and bureaucracies. Most environmentalists disagree. They point out that widespread pollution and environmental degradation have resulted from the removal of nonrenewable mineral resources and unsustainable use of potentially renewable resources on privately owned lands.

Many environmentalists agree that the command-and-control approach to use of resources on public

lands (such as national parks, wildlife refuges, and wilderness areas) has some serious problems. They and some free-market economists are seeking *users-pay* solutions to replace the current *taxpayers-pay* approach to use of such publicly owned resources. This would involve a mix of marketplace incentives coupled with regulations that require users to pay a fair price for all resources extracted from public lands and to be responsible for preventing or cleaning up any environmental damage caused by resource extraction or use.

A: The sun and the earth (earth capital)

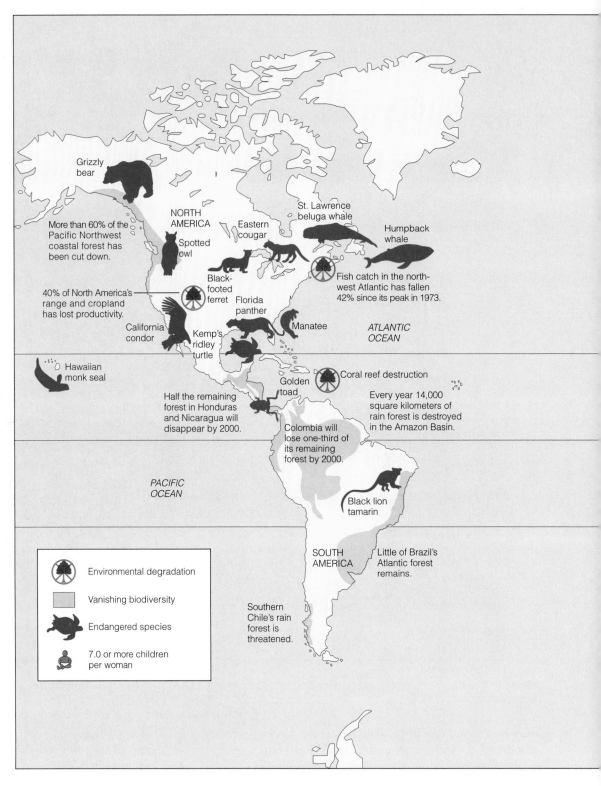

Figure 1-14 Examples of how some of the earth's natural systems that support all life and all economies are being assaulted at an accelerating rate as a result of the exponential growth of population and resource use and the resulting environmental degradation and loss of biodiversity. (Data from The World Conservation Union, World Wildlife Fund, Conservation International, United Nations, Population Reference Bureau, U.S. Fish and Wildlife Service, and Daniel Boivin)

Q: How many years does it take to add 1 billion people at current growth rates?

ASIA

Many parts of
former Soviet Union
are polluted with
industrial and radio-
active waste.

Poland is
world's most
polluted country.

Central Asia from the
Middle East to China
has lost 72% of range
and cropland.

Giant
panda

EUROPE

Imperial eagle

Japanese timber imports
are responsible for much
of the world's tropical
deforestation.

Area of
Aral Sea has
shrunk 46%.

Mediterranean

Snow leopard

640,000 square kilometers
south of the Sahara have
turned to desert since 1940.

Iraq

Deforestation in the Himalaya
causes flooding in Bangladesh.

Asian
elephant

Mali

Yemen

Ethiopia

Burkina
Faso

AFRICA

Niger

India and
Sri Lanka
have almost
no rain
forest left.

90% of the coral reefs
are threatened in the
Philippines. All virgin
forest will be gone
by 2010.

Côte
d'Ivoire

Golden lion
tamarin

Benin

Kouprey

Togo

Uganda

Somalia

68% of the
Congo's
rain forest
is slated
for clearing.

Rwanda
Burundi

Tanzania

Comoros

In peninsular Malaysia
almost all forests will
be gone by 2000.

Queen Alexandra's
birdwing butterfly

Zambia

Malawi

INDIAN OCEAN

Indonesia's
coral reefs are
threatened.

Nail-tailed
wallaby

Black
rhinoceros

Aye-aye

AUSTRALIA

Madagascar has
lost 66% of its
tropical forest.

Much of
Australia's
range and
cropland
has turned
to desert.

Blue whale

A thinning of the ozone layer occurs
over Antarctica during summer.

ANTARCTICA

What Are Nonrenewable Resources? Resources that exist in a fixed quantity in the earth's crust and thus theoretically can be completely used up are called **nonrenewable**, or **exhaustible**, **resources**. On a time scale of millions to billions of years, such resources can be renewed by geological processes. However, on the much shorter human time scale of hundreds to thousands of years, these resources can be depleted much faster than they are formed. These exhaustible resources include *energy resources* (coal, oil, natural gas, and uranium, which cannot be recycled), *metallic mineral resources* (iron, copper, aluminum, which can be recycled), and *nonmetallic mineral resources* (salt, clay, sand, phosphates), which are usually difficult or too costly to recycle). A **mineral** is any hard, usually crystalline material that is formed naturally. Soil and most rocks consist of two or more minerals. We know how to find and extract more than 100 nonrenewable minerals from the earth's crust. We convert these raw materials into many everyday items and then we discard, reuse, or recycle them.

Figure 1-15 shows the production and exhaustion cycle of a nonrenewable energy or mineral resource. In practice, we never completely exhaust a nonrenewable mineral resource. However, such a resource becomes *economically depleted* when the costs of exploiting what is left exceed its economic value. At that point, we have five choices: recycle or reuse existing supplies (except for nonrenewable energy resources, which cannot be recycled or reused), waste less, use less, try to develop a substitute, or do without and wait millions of years for more to be produced.

Some nonrenewable material resources, such as copper and aluminum, can be recycled or reused to extend supplies. **Recycling** involves collecting and reprocessing a resource into new products. For example, glass bottles can be crushed and melted to make new bottles or other glass items. **Reuse** involves using a resource over and over in the same form. For example, glass bottles can be collected, washed, and refilled many times.

Recycling nonrenewable metallic resources requires much less energy, water, and other resources and produces much less pollution and environmental degradation than exploiting virgin metallic resources. Reuse of such resources requires even less energy and other resources than recycling, and it results in less pollution and environmental degradation.

Nonrenewable *energy* resources, such as coal, oil, and natural gas, can't be recycled or reused. Once burned, the useful energy in these fossil fuels is gone, leaving behind waste heat and polluting exhaust gases. Most of the per capita economic growth shown in Figure 1-8 has been fueled by cheap nonrenewable oil, which is expected to be economically depleted within 40–80 years.

Most published estimates of the supply of a given nonrenewable resource refer to **reserves**: known deposits from which a usable mineral can be profitably extracted at current prices. Reserves can be increased when new deposits are found or when price increases make it profitable to extract identified deposits that were previously considered too expensive to exploit.

Some environmentalists and resource experts believe that the greatest danger may not be the exhaustion of nonrenewable resources but the damage that their extraction, processing, and conversion to products do to the environment in the form of energy use, land disturbance, soil erosion, water pollution, and air pollution

1-4 POLLUTION

What Is Pollution and Where Does It Come From? Any addition to air, water, soil, or food that threatens the health, survival, or activities of humans or other living organisms is called **pollution**. Most pollutants are solid, liquid, or gaseous by-products or wastes produced when a resource is extracted, processed, made into products, or used. Pollution can also take the form of unwanted energy emissions, such as excessive heat, noise, or radiation.

Pollutants can enter the environment naturally (for example, from volcanic eruptions) or through human (anthropogenic) activities (for example, from burning coal). Most pollution from human activities occurs in or near urban and industrial areas, where

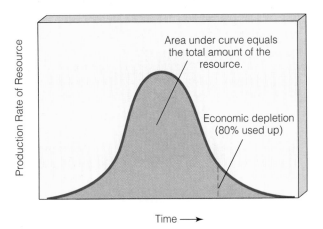

Figure 1-15 Full production and exhaustion cycle of a nonrenewable resource such as copper, iron, oil, or coal. Usually, a nonrenewable resource is considered economically depleted when 80% of its total supply has been extracted and used. Normally, it costs too much to extract and process the remaining 20%.

pollutants are concentrated. Industrialized agriculture is also a major source of pollution. Some pollutants contaminate the areas where they are produced; others are carried by winds or flowing water to other areas. Pollution does not respect local, state, or national boundaries.

Some pollutants come from single, identifiable sources, such as the smokestack of a power plant, the drainpipe of a meat-packing plant, or the exhaust pipe of an automobile. These are called **point sources**. Other pollutants come from dispersed (and often difficult to identify) **nonpoint sources**. Examples are the runoff of fertilizers and pesticides (from farmlands, golf courses, and suburban lawns and gardens) into streams and lakes and pesticides sprayed into the air or blown by the wind into the atmosphere. It is much easier and cheaper to identify and control pollution from point sources than from widely dispersed nonpoint sources.

What Types of Harm Are Caused by Pollutants?
Unwanted effects of pollutants include disruption of life-support systems for humans and other species, damage to wildlife, damage to human health, damage to property, and nuisances such as noise and unpleasant smells, tastes, and sights.

Three factors determine how severe the harmful effects of a pollutant are. One is its *chemical nature*: how active and harmful it is to living organisms. Another is its **concentration**: the amount per unit of volume or weight of air, water, soil, or body weight. Concentration is sometimes expressed in *parts per million (ppm)*, with 1 ppm corresponding to one part pollutant per one million parts of the gas, liquid, or solid mixture in which the pollutant is found. Smaller concentration units are *parts per billion (ppb)* and *parts per trillion (ppt)*.

Parts per million, billion, or trillion may seem like negligible amounts of pollution. Nevertheless, concentrations of *some* pollutants at such low levels can have serious effects on people, other animals, and plants.

One way to lower the concentration of a pollutant is to dilute it in a large volume of air or water. Until we started overwhelming the air and waterways with pollutants, dilution was *the* solution to pollution. Now it is only a partial solution.

The third factor is a pollutant's *persistence*: how long it stays in the air, water, soil, or body. **Degradable**, or **nonpersistent**, **pollutants** are broken down completely or reduced to acceptable levels by natural physical, chemical, and biological processes. Complex chemical pollutants broken down (metabolized) into simpler chemicals by living organisms (usually by specialized bacteria) are called **biodegradable pollutants**. Human sewage in a river, for example, is biodegraded fairly quickly by bacteria if the sewage is not added faster than it can be broken down.

Many of the substances we introduce into the environment take decades or longer to degrade. Examples of these **slowly degradable**, or **persistent**, **pollutants** include the insecticide DDT and most plastics.

Nondegradable pollutants cannot be broken down by natural processes. Examples include the toxic elements lead and mercury. The best ways to deal with nondegradable pollutants (and slowly degradable pollutants) are to avoid releasing them into the environment or to recycle or reuse them. Removing them from contaminated air, water, or soil is expensive, and sometimes impossible.

We know little about the possible harmful effects of 90% of the 72,000 synthetic chemicals now in commercial use and the roughly 1,000 new ones added each year. Our knowledge about the effects of the other 10% of these chemicals is limited, mostly because it is quite difficult, time-consuming, and expensive to get this information. Even if we determine the main health and other environmental risks associated with a particular chemical, we know little about its possible interactions with other chemicals or about the effects of such interactions on human health, other organisms, and life-support processes.

Solutions: What Can We Do About Pollution?
There are two basic approaches to dealing with pollution: prevent it from reaching the environment or clean it up if it does (Figure 1-16). **Pollution prevention**, or **input pollution control**, is a *throughput solution*. It slows or eliminates the production of pollutants, often by switching to less harmful chemicals or processes. Pollution can be prevented (or at least reduced) by the four Rs of resource use: *refuse* (don't use), *reduce*, *reuse*, and *recycle*.

Pollution cleanup, or **output pollution control**, involves cleaning up pollutants after they have been produced. However, environmentalists have identified three major problems with relying primarily on pollution cleanup. First, *it is often only a temporary bandage as long as population and consumption levels continue to grow without corresponding improvements in pollution control technology*. For example, adding catalytic converters to cars has reduced air pollution, but increases in the number of cars and in the total distance each travels (increased throughput) have reduced the effectiveness of this cleanup approach.

Second, *pollution cleanup often removes a pollutant from one part of the environment only to cause pollution in another*. We can collect garbage, but the garbage is then either burned (perhaps causing air pollution and leaving a toxic ash that must be put somewhere); dumped into streams, lakes, and oceans (perhaps causing water pollution); or buried (perhaps causing soil and groundwater pollution). Third, *once pollutants have*

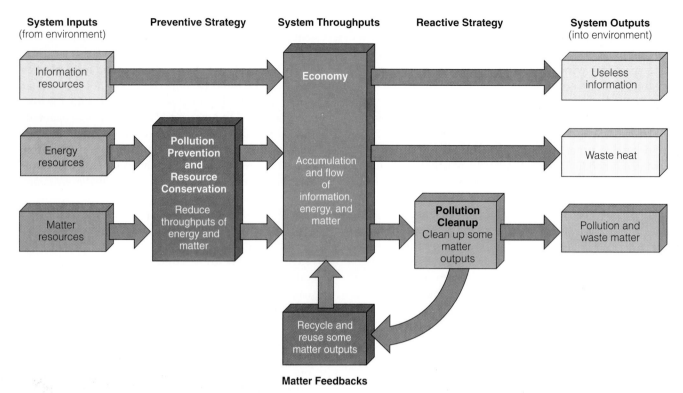

Figure 1-16 Inputs, throughputs, and outputs of a system (an economy) with two strategies for reducing pollution. Pollution prevention, or input pollution control, is based on reducing or eliminating pollution and conserving resources by reducing the throughputs (flows) of matter and energy resources through an economy. Throughput can also be reduced by recycling or reusing some or most of the output of pollution or waste matter. Pollution cleanup, or output pollution control, involves trying to reduce pollutants to acceptable levels after they have been produced. Both approaches are needed, but environmentalists and a growing number of economists believe that much greater emphasis should be placed on pollution prevention.

entered and become dispersed in the air and water (and in some cases, the soil) at harmful levels, it usually costs too much to reduce them to acceptable concentrations.

Both pollution prevention and pollution cleanup are needed, but environmentalists and some economists urge us to emphasize prevention because it works better and is cheaper than cleanup. For widely dispersed and difficult-to-identify nonpoint pollution, hazardous wastes, and slowly degradable and nondegradable pollutants, pollution prevention is the most effective (perhaps the only) approach. As Benjamin Franklin reminded us long ago, "An ounce of prevention is worth a pound of cure."

An increasing number of businesses have found that *pollution prevention pays.* So far, however, about 99% of environmental spending in the United States (and in most other countries) is devoted to pollution cleanup and only 1% to pollution prevention, a situation that environmental scientists and some economists believe must be reversed as soon as possible.

Both pollution prevention and pollution cleanup can be encouraged either by the *carrot approach* of using incentives such as various subsidies and tax write-offs or by the *stick approach* of regulations and taxes. Most analysts believe that a mix of both approaches is probably best because excessive regulation and too much taxation can incite resistance and cause a political backlash. Achieving the right balance is a difficult task.

1-5 ENVIRONMENTAL AND RESOURCE PROBLEMS: CAUSES AND CONNECTIONS

What Are Key Environmental Problems and Their Root Causes? We face a number of interconnected environmental and resource problems (Figure 1-17). The first step in dealing with these problems is to identify their underlying causes. According to environmentalists, these include the following:

- Rapid population growth

- Rapid and wasteful use of resources with too little emphasis on pollution prevention and waste reduction

Q: How long will it take to double the propulation of a country with a population growth rate of 2% per year?

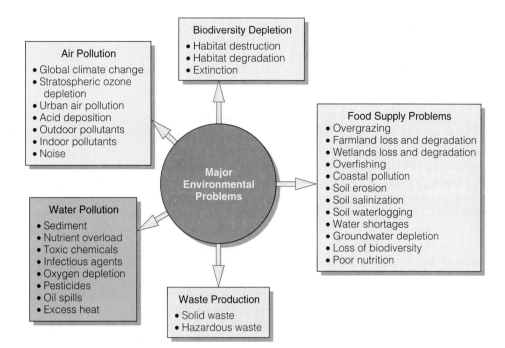

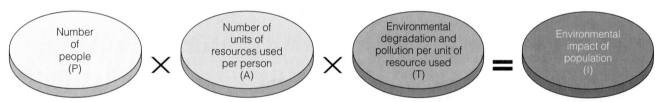

Figure 1-17 Major environmental and resource problems.

Figure 1-18 Simplified model of how three factors—population, affluence, and technology—affect the environmental impact (I) of population. According to this model, the damage we do to the earth is equal to the number of people there are, multiplied by the amount of resources each person uses, multiplied by the amount of pollution, resource waste, and environmental degradation involved in extracting, making, and using each unit of resource.

- Simplification and degradation of parts of the earth's life-support systems

- Poverty, which can drive poor people to use potentially renewable resources unsustainably for short-term survival and often exposes the poor to health risks and other environmental risks

- Failure of economic and political systems to encourage earth-sustaining forms of economic development and discourage earth-degrading forms of economic growth

- Failure of economic and political systems to have market prices include the overall environmental cost of an economic good or service

- Our urge to dominate and manage nature for our use with far too little knowledge about how nature works

How Are Environmental Problems and Their Root Causes Connected? Once we have identified environmental problems and their root causes, the next step is to understand how they are connected to one another. The three-factor model in Figure 1-18 is a good starting point.

According to this simple model, a given area's total environmental degradation and pollution—that is, the environmental impact of population—depends on three factors: the number of people (population size, P), the average number of units of resources each person uses (per capita consumption or affluence, A), and the amount of environmental degradation and pollution produced for each unit of resource used (the environmental destructiveness of the technologies used to provide and consume resources, T). This model, developed in the early 1970s by biologist Paul

Ehrlich and physicist John Holdren, can be summarized in simplified form as

Population × Affluence × Technology = Environmental impact

or

$$P \quad \times \quad A \quad \times \quad T \quad = \quad I$$

Figure 1-19 shows how the three factors depicted in Figure 1-18 can interact in developing countries and developed countries. In developing countries, population size and the resulting degradation of potentially renewable resources (as the poor struggle to stay alive) tend to be the key factors in total environmental impact.

In developed countries, high rates of per capita resource use (and the resulting high levels of pollution and environmental degradation per person) are believed to be the key factors determining overall environmental impact. For example, it is estimated that the average U.S. citizen consumes 35 times as much as the average citizen of India and 100 times as much as the average person in the world's poorest countries. *Thus, poor parents in a developing country would need 70–200 children to have the same lifetime environmental impact as two children in a typical U.S. family.* Many of the world's leading scientists warn that if we keep adding 80–90 million more people each year (all of whom want to become affluent), our life-support systems in many parts of the world will be overwhelmed.

Some forms of technology, such as polluting factories and motor vehicles and energy-wasting devices, increase environmental impact by raising the *T* factor in the equation. Other technologies, such as pollution control, solar cells, and more energy-efficient devices, can lower environmental impact by decreasing the *T* factor in the equation.

The three-factor model in Figure 1-18 can help us understand how key environmental problems and some of their causes are connected. However, the interconnected problems we face involve a number of poorly understood interactions among many more factors than those in the three-factor model (Figure 1-20).

People Overpopulation

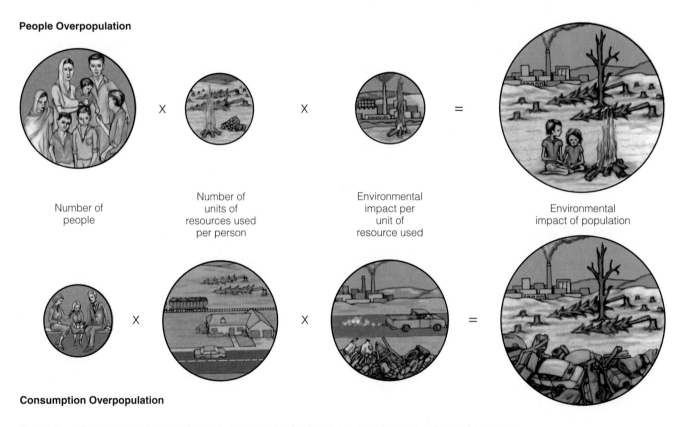

Number of people

Number of units of resources used per person

Environmental impact per unit of resource used

Environmental impact of population

Consumption Overpopulation

Figure 1-19 Environmental impact of developing countries (top) and developed countries (bottom) based on the relative importance of the factors in the model shown in Figure 1-18. Circle size shows the relative importance of each factor. The size of the T factor can be reduced by improved technology for controlling and preventing pollution, resource waste, and environmental degradation.

Q: What % of the world's projected population growth between now and 2050 is expected to take place in developing countries?

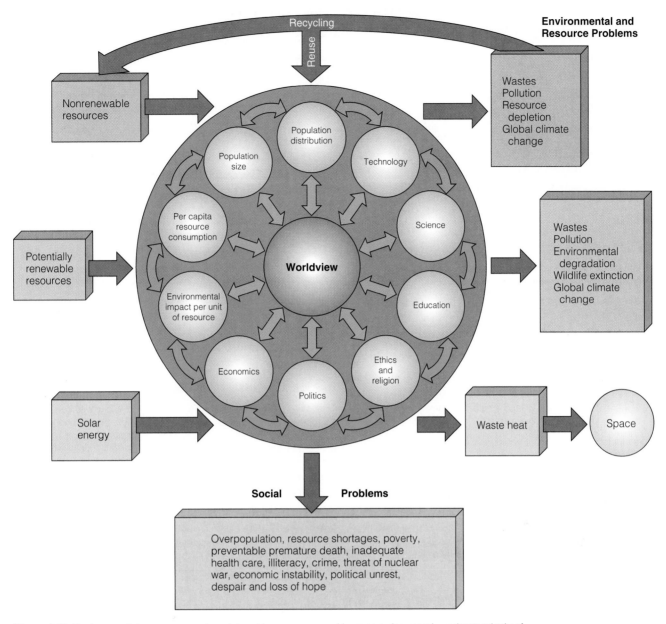

Figure 1-20 Environmental, resource, and social problems are caused by a complex, poorly understood mix of interacting factors, as illustrated by this simplified model.

✺ 1-6 IS OUR PRESENT COURSE SUSTAINABLE?

Are Things Getting Better or Worse? There are conflicting views about how serious our population and environmental problems are and what should be done about them (Table 1-1). Some analysts, mostly economists, contend that the world is not overpopulated and that people are our most important resource as consumers and producers to fuel continued economic growth and as sources of technological innovation.

They also believe that human ingenuity and technological advances will allow us to clean up pollution to acceptable levels, find substitutes for any resources that become scarce, and keep expanding the earth's ability to support more humans as we have done in the past. They don't acknowledge any insurmountable limits to population growth or economic growth. They accuse most scientists and environmentalists of exaggerating the seriousness of the problems we face and of failing to appreciate the progress (Table 1-1, left) that has been made in improving quality of life and protecting the environment.

Table 1-1 Good and Bad News About Environmental Problems

Some Good News	Some Bad but Challenging News
Annual world population growth slowed from 2.2% to 1.47% between 1960 and 1997.	The world's population is still growing rapidly and is projected to increase from 5.84 billion to 8 billion between 1997 and 2025.
Global life expectancy rose from 33 to 66 years between 1900 and 1997.	Life expectancy in developing countries is 10 years less than in developed countries.
Annual global infant mortality dropped by 65% between 1900 and 1997.	Infant mortality in developing countries is 7.1 times higher than in developed countries and could be improved in many developed countries.
Global grain production has outpaced population growth since 1978.	Since 1980 population growth has exceeded grain production in 69 developing countries and the global per capita fish catch has declined by 7.5% between 1988 and 1995. Future food production may be limited by its harmful environmental impacts.
The percentage of the world's population that are hungry has been reduced since 1960.	Because of population growth there are now more hungry people (about 1 million) than ever.
Biodiversity loss has been reduced by higher crop yields on less land.	The challenge is to increase yields while sharply reducing the growing environmental impacts of such intensive production.
Biodiversity loss has been reduced by increased urbanization, with more people living on less land	This effect is misleading because the higher resource use per person in cities must be supplied by croplands, forests, grasslands, fisheries, and mines elsewhere. This often diminishes biodiversity and increases environmental degradation in these areas that sustain otherwise unsustainable cities.
At least 5% of the world's land area has been set aside to protect wildlife and wildlife habitats.	This is half of the estimated minimum needed. Many wildlife sanctuaries exist on paper only and receive little protection and many important types of biological communities are not protected. Every hour an estimated two to eight wildlife species become extinct because of human activities; this rate is expected to increase over the next few decades.
Total forest area in temperate industrial countries increased during the 1980s.	Much of this increase came from replacing sustainable, biologically diverse old-growth forests with simplified and vulnerable tree farms. This has led to reduced forest biodiversity.
Tropical deforestation has been overblown and enough wood, paper, and food can be produced by converting much of such diverse forests to tree farms, cropland, and grazing land.	If current deforestation rates continue, within 30 to 50 years little of these forests will remain. This will lead to a massive loss of biodiversity (because such forests contain at least 50% of the world's terrestrial species).
Conservation tillage and no–till cultivation can cut soil erosion by about 75%.	These techniques are widely used in the United States but are rarely used in other countries. Topsoil is eroding faster than it forms on about one-third of the world's cropland. Deforestation and construction are increasing soil erosion in many countries.
Food industry scientists contend that the benefits of pesticides outweigh their harmful effects.	Most environmental and health scientists in this field contend that the harmful effects of pesticides outweigh their beneficial effects. Reliance mostly on chemical pesticides (all of which eventually fail because of genetic resistance by pests) should be replaced with an ecological approach called integrated pest management (IPM).

On the other hand, environmentalists, many leading scientists, and a small number of economists contend that we are depleting and degrading the earth's natural capital at an accelerating rate and that this is leading to serious environmental and economic harm (Table 1-1, right). They are encouraged by the progress that has been made but point out how much more must be done to help make the earth more sustainable for present and future human generations and for other species that support us and other forms of life.

What Is the Scientific Consensus About Environmental Problems? On November 18, 1992, some 1,680 of the world's senior scientists from 70

Q: How many people are added to the population of developing countries every four days?

Table 1-1 *(continued)*

Some Good News	Some Bad but Challenging News
A very small minority of scientists contend that possible global warming is not a serious problem.	The vast majority of scientists in this field contend that the possible global warming is one of the world's most serious environmental and economic problems.
Some economists say that reducing the threat of global warming by shifting from nonrenewable fossil fuels to renewable solar-based energy sources and reducing deforestation (especially in the tropics) over the next few decades will be too costly.	In the long run it will cost much more (ecologically and economically) if such shifts are not made. Even if global warming were not a serious problem, these shifts are needed to reduce pollution and environmental degradation and to protect biodiversity.
A very small minority of scientists say that depletion of ozone in the stratosphere (the atmosphere's second layer) is not a serious problem.	The vast majority of scientists in this field contend that ozone depletion in the stratosphere is a very serious environmental problem.
Water is recycled by a natural cycle and the earth has much more water than it needs. Dams and water transfer projects can meet projected water shortages.	Earth's water is not distributed equally and much of it is inaccessible, unnecessarily wasted, or polluted. Currently, about 1.2 billion people lack access to clean drinking water. By 2025 at least 3 billion people in 90 countries are expected to face chronic shortages of water. Large dams and water transfer projects lead to serious environmental degradation.
Since 1950 proven supplies (reserves) of virtually all nonrenewable fossil fuel and key mineral resources have increased significantly.	The exponential increase in the use of fossil fuel and mineral resources is causing massive land degradation, water pollution, and air pollution. These harmful environmental effects—not supplies or costs—may limit future resource use.
Adjusted for inflation, most nonrenewable fossil fuel and key mineral resources cost less today than in 1950.	These resource prices are low because most of the harmful environmental and health costs associated with their production are not included in their market prices. Such artificially low resource prices encourage waste, pollution, and environmental degradation and discourage waste reduction and pollution prevention.
In the United States, the recycling and composting of solid municipal waste (produced by households and businesses) increased from 7% to 25% between 1970 and 1996.	At least 60% of all municipal solid waste could be reused, recycled, or composted. Only about 1.5% of the estimated solid waste produced in the United States is municipal solid waste. The other 98.5% comes mostly from mining, fossil fuel production, agriculture, and industrial activities, with very little of it being recycled, reused, or composted.
Some scientists contend that the health risks from exposure to toxic and hazardous chemicals are overblown.	The harmful health effects of most chemicals are unknown and a growing body of evidence suggests that their harmful effects (especially to the nervous, immune, and endocrine systems) have been underestimated.
Since 1970 air and water pollution levels in most industrialized countries have dropped significantly for most pollutants.	Gains in pollution control in industrialized countries by laws enacted in the 1970s are being threatened by more people using more resources and a lack of emphasis on pollution prevention. Little progress has been made in reducing air and water pollution in most developing countries.

countries, including 102 of the 196 living scientists who are Nobel laureates, signed and sent an urgent warning to government leaders of all nations. According to this warning,

The environment is suffering critical stress. . . . Our massive tampering with the world's interdependent web of life—coupled with the environmental damage inflicted by deforestation, species loss, and climate change—could trigger widespread adverse effects, including unpredictable collapses of critical biological systems whose interactions and dynamics we only imperfectly understand. Uncertainty over the extent of these effects cannot excuse complacency or delay in facing the threats. . . . No more than one or a few decades remain before the chance to avert the threats we now confront will be lost and the prospects for humanity immeasurably diminished. . . . Whether industrialized or not, we all have but one lifeboat. No nation can escape injury when global biological systems are damaged. . . . We must recognize the earth's limited capacity to provide for us.

Also in 1992, the prestigious U.S. National Academy of Sciences and the Royal Society of London issued a joint report, their first ever, which began,

If current predictions of population growth prove accurate and patterns of human activity on the planet remain unchanged, science and technology may not be able to prevent either irreversible degradation of the environment or continued poverty for much of the world.

Some past short-term prophecies of environmental doom by a small number of scientists have not been borne out and critics have used this to urge people to ignore all warnings from scientists and environmentalists. However, the more recent warnings just cited are not the views of a small number of scientists but the consensus of the mainstream scientific community, consisting of most of the world's key researchers on environmental problems. Some analysts also point out that many of the warnings of environmental doom made in the 1960s and 1970s were averted because people realized that they might come true and worked to pass legislation and carry out individual acts to help keep them from coming true (a *prevention* result).

Whom Should We Believe?   There is no easy answer to this question. It depends mostly on how each of us believes the world works and what role the human species can and should play on this planet (Sections 2-2 and 2-3) and on our ability to use critical thinking to evaluate opposing claims. People with widely differing fundamental beliefs can take the same data, be logically consistent, and arrive at quite different conclusions (Table 1-1) because they start with different assumptions and are often seeking answers to different questions.

What we believe and do is also influenced by whether we have an optimistic (*the glass is half full*) or a pessimistic (*the glass is half empty*) outlook about the future. People with a realistic but hopeful outlook recognize that the future is represented by the top half of the glass and that what we put into this part of the glass is the key issue. They rejoice in how much has been accomplished in dealing with environmental problems since 1970 but recognize that we should also

roll up our sleeves and work together to see that such accomplishments are sustained and expanded.

Do We Need to Make a New Cultural Change?
Most environmentalists believe that there is an urgent need to begin shifting from our present array of industrialized and partially industrialized societies to a variety of more sustainable *earth-wisdom societies* throughout the world, by launching an *environmental* or *sustainability revolution* to take place over the next 50 years (Guest Essay, pp. 26–27). This new cultural change calls for us to change the ways we view and treat the earth (and thus ourselves) by using environmental science to learn more about how nature sustains itself and by mimicking such processes in human cultural systems.

Environmentalists call for us to shift our efforts from pollution cleanup to pollution prevention, from waste disposal (mostly burial and burning) to waste prevention and reduction, from species protection to habitat protection, and from increased resource use to increased resource conservation.

They urge us to use our political and economic systems to reward earth-sustaining economic activities and discourage those that harm the earth. They also believe that we should allow many parts of the world we have damaged to heal, help restore severely damaged areas, and protect remaining wild areas from destructive forms of economic development.

Others, mostly economists and business leaders, contend that environmentalists have overblown the problems we face. They believe that the solution to our environmental problems involves technological innovation and greatly increased economic growth through participation in a global economy based on mostly unregulated free-market capitalism.

How Can We Work with the Earth?   This chapter has presented an overview of both the serious problems most environmentalists and many of the world's most prominent scientists believe we face (also summarized in the yellow boxes of the Concepts and Connections diagram inside the back cover) and their root causes. It has also summarized key arguments about how serious environmental problems are. The rest of this book presents a more detailed analysis of these problems, the controversies they have created, and solutions proposed by various scientists, environmentalists, and other analysts.

The *bad news* is the environmental problems we face and their root causes, as outlined in this chapter. The *good news* is how much has been done (Table 1-1) and that it's not too late to replace our earth-degrading actions with earth-sustaining ones, as discussed throughout this book. The key is *earth wisdom*—learning as

Q: What percentage of the world's resources are used by people in developed countries?

much as we can about how the earth sustains itself and adapts to ever changing environmental conditions—and integrating such lessons from nature into the ways we think and act. The Spotlight (right) gives some guidelines various analysts have suggested for developing more sustainable societies by working with the earth.

Another key to dealing with our environmental problems and challenges is recognizing that *individuals matter.* Anthropologist Margaret Mead has summarized our potential for change: "Never doubt that a small group of thoughtful, committed citizens can change the world. Indeed, it is the only thing that ever has."

What's the use of a house if you don't have a decent planet to put it on?

HENRY DAVID THOREAU

CRITICAL THINKING

1. Is the world overpopulated? Explain. Is the United States overpopulated? Explain.

2. (a) Do you believe that the society you live in is on an unsustainable path? Explain. What about the world as a whole? **(b)** Do you believe that it is possible for the society you live in to become a sustainable society within the next 50 years? Explain. **(c)** Would you classify yourself as a technological optimist, environmental pessimist, or hopeful environmental realist? Explain.

3. Do you favor instituting policies designed to reduce population growth and stabilize **(a)** the size of the world's population as soon as possible and **(b)** the size of the U.S. population (or the population of the country where you live) as soon as possible? Explain. If you agree that population stabilization is desirable, what three major policies do you believe should be implemented to accomplish this goal?

4. Explain why you agree or disagree with the following propositions:
   a. High levels of resource use by the United States and other developed countries are more beneficial than harmful.
   b. The economic growth from high levels of resource use in developed countries provides money for more financial aid to developing countries for reducing pollution, environmental degradation, and poverty.
   c. Stabilizing population is not desirable because without more consumers, economic growth would stop.

5. Explain why you agree or disagree with the following proposition: The world will never run out of potentially renewable resources and most currently used nonrenewable resources because technological innovations will produce substitutes, reduce resource waste, or allow use

Some Guidelines for Working with the Earth

SPOTLIGHT

- Leave the earth as good as or better than we found it.
- Take no more than we need.
- Try not to harm life, air, water, or soil.
- Sustain biodiversity.
- Help maintain the earth's capacity for self-repair.
- Don't use potentially renewable resources (soil, water, forests, grasslands, and wildlife) faster than they are replenished.
- Don't waste resources.
- Don't release pollutants into the environment faster than the earth's natural processes can dilute or assimilate them.
- Emphasize pollution prevention and waste reduction.
- Slow the rate of population growth.
- Reduce poverty.

Specific things you can do to work with the earth by trying to implement such guidelines are listed in Appendix 5.

Critical Thinking

Which of these guidelines do you agree with and which do you disagree with? Why? Can you add any other guidelines?

of lower grades of scarce nonrenewable resources.

6. Suppose enforcement of government pollution control regulations meant that you would lose your job. If you had a choice, would you choose unemployment and a cleaner environment or employment and a dirtier environment? Can you think of ways to avoid this dilemma? Explain.

7. Do you believe that your current lifestyle is sustainable? If the answer is yes, explain why and include the impact of the world's other 5.84 billion people on your ability to sustain your current lifestyle. If your answer is no, explain why and indicate things you could do now to make your lifestyle more sustainable. Which of these things do you actually plan to do?

*8. What are the major resource and environmental problems in **(a)** the city, town, or rural area where you

---

*These are laboratory exercises or individual or class projects. Additional exercises and projects are found in the laboratory manual and in the *Green Lives, Green Campuses* workbook that can be used with this course.

Launching the Environmental Revolution*

Lester R. Brown

Lester R. Brown is president of the World-watch Institute, a private nonprofit research institute he founded in 1974 that is devoted to analysis of global environmental issues. Under his leadership, the institute publishes the annual State of the World Report, *considered by environmentalists and world leaders to be the best way to become informed about key environmental issues. It also publishes monographs on specific topics,* World Watch *magazine, and a series of* Environmental Alert *books. He is author of a dozen books, recipient of the McArthur Foundation Genius Award, and winner of the United Nations' 1989 environment prize. He has been described by the* Washington Post *as "one of the world's most influential thinkers."*

Our world of the mid-1990s faces potentially convulsive change. The question is, In what direction will it take us? Will the change come from strong worldwide initiatives that reverse the degradation of the planet and restore hope for the future, or will it come from continuing environmental deterioration that leads to economic decline and social instability?

Muddling through will not work. Either we will turn things around quickly or the self-reinforcing internal dynamic of the deterioration-and-decline scenario will take over. The policy decisions we make in the years immediately ahead will determine whether our children live in a world of development or decline.

There is no precedent for the rapid and substantial change we need to make. Building an environmentally sustainable future depends on restructuring the global economy, major shifts in human reproductive behavior, and dramatic changes in values and lifestyles. Doing all this quickly adds up to a revolution that is driven and defined by the need to restore and preserve the earth's environmental systems. If this *environmental revolution* succeeds, it will rank with the agricultural and industrial revolutions as one of the great economic and social transformations in human history.

Like the agricultural revolution, it will dramatically alter population trends. Although the former set the stage for enormous increases in human numbers,

*Updated excerpt from my expanded version of these ideas in "Launching the Environmental Revolution," State of the World 1992 (New York: W.W. Norton, 1992).

this revolution will succeed only if it stabilizes human population size, reestablishing a balance between people and natural systems on which they depend. In contrast to the industrial revolution, which was based on a shift to fossil fuels, this new transformation will be based on a shift away from fossil fuels.

The two earlier revolutions were driven by technological advances—the first by the discovery of farming and the second by the invention of the steam engine, which converted the energy in coal into mechanical power. The environmental revolution, though it will obviously need new technologies, will be driven primarily by the restructuring of the global economy so that it does not destroy its natural support systems.

The pace of the environmental revolution needs to be far faster than that of its predecessors. The agricultural revolution began some 10,000 years ago, and the industrial revolution has been under way for about two centuries. But if the environmental revolution is to succeed, it must be compressed into a few decades.

Progress in the agricultural revolution was measured almost exclusively in the growth in food output that eventually enabled farmers to produce a surplus that could feed city dwellers. Similarly, industrial progress was gained by success in expanding the output of raw materials and manufactured goods. The environmental revolution will be judged by whether it can shift the world economy into an environmentally sustainable development path, one that leads to greater economic security, healthier lifestyles, and a worldwide improvement in the human condition.

Many still do not see the need for such an economic and social transformation. They see the earth's deteriorating physical condition as a peripheral matter that can be dealt with by minor policy adjustments. But 25 years of effort have failed to stem the tide of environmental degradation. There is now too much evidence on too many fronts to take these issues lightly.

Already the planet's degradation is damaging human health, slowing the growth in world food production, and reversing economic progress in dozens of countries. By the age of 10, thousands of children living in southern California's Los Angeles basin have respiratory systems that are permanently impaired by polluted air. Some 300,000 people in the former Soviet Union are being treated for radiation sickness caused by the Chernobyl nuclear power plant accident. The accelerated depletion

live and **(b)** the state where you live? Which of these problems affect you directly?

*9. Make a list of the resources you truly need. Then make another list of the resources you use each day only because you want them. Finally, make a third list of resources you want and hope to use in the future. Compare your lists with those compiled by other members

of your class, and relate the overall result to the tragedy of the commons.

*10. Make a concept map of this chapter's major ideas using the section heads and subheads and the key terms (in boldface type). Look at the inside back cover and in Appendix 4 for information about concept maps.

Q: How many of the world's people attempt to survive on an annual income of about $1 per year?

of ozone in the stratosphere in the northern hemisphere will lead to an estimated additional 20,000 skin cancer fatalities over the next half century in the United States alone. Worldwide, millions of lives are at stake. These examples, and countless others, show that our health is closely linked to that of the planet.

A scarcity of new cropland and fresh water plus the negative effects of soil erosion, air pollution, and hotter summers on crop yields is slowing the growth of the world grain harvest. Combined with continuing rapid population growth, this has reversed the steady rise in grain output per person that the world had become accustomed to. Between 1950 and 1984, the historical peak year, world grain production per person climbed by nearly 40%. Since then, it has fallen roughly 1% a year, with the drop concentrated in poor countries. With food imports in these nations restricted by rising external debt, there are far more hungry people today than ever before.

On the economic front, the signs are equally ominous: Soil erosion, deforestation, and overgrazing are adversely affecting productivity in the farming, forestry, and livestock sectors, slowing overall economic growth in agriculturally based economies. The World Bank reports that after three decades of broad-based economic gains, incomes fell during the 1980s in 40 developing countries. Collectively, these nations contain more than 800 million people—almost three times the population of North America and nearly one-sixth that of the world. In Nigeria, the most populous country in the ill-fated group, the incomes of 123 million people fell a painful 29%, exceeding the fall in U.S. incomes during the depression decade of the 1930s.

Anyone who thinks these environmental, agricultural, and economic trends can easily be reversed need only look at population projections. Those of us born before the middle of this century have seen the world population more than double to 5.84 billion. We have witnessed the environmental effects of adding 3 billion people, especially in developing countries. We can see the loss of tree cover, the devastation of grasslands, the soil erosion, the crowding and poverty, the land hunger, and the air and water pollution associated with this addition of people. But what if 4.2 billion more people are added by 2050, over 90% of them in developing countries, as now projected by UN population experts?

The decline in living standards that was once predicted by some ecologists from the combination of continuing rapid population growth, spreading environmental degradation, and rising external debt has become a reality for one-sixth of humanity. Moreover, if a more comprehensive system of national economic accounting were used—one that incorporated losses of natural capital, such as topsoil and forests, the destruction of productive grasslands, the extinction of plant and animal species, and the health costs of air and water pollution, nuclear radiation, and increased ultraviolet radiation—it might well show that most of humanity suffered a decline in living conditions in the 1980s and 1990s.

Today, we study archaeological sites of civilizations that were undermined by environmental deterioration. The wheatlands that made North Africa the granary of the Roman Empire are now largely desert. The early civilizations of the Tigris–Euphrates Basin declined as the waterlogging and salting of irrigation systems slowly shrank their food supply. And the collapse of the Mayan civilization that flourished in the Guatemalan lowlands from the third century B.C. to the ninth century A.D. may have been triggered by deforestation and soil erosion.

No one knows for certain why centers of Mayan culture and art fell into neglect, or whether the population of 1 million to 3 million moved or died off, but recent progress in deciphering hieroglyphs in the area adds credence to the environmental decline hypothesis. One of those involved with the project, Linda Schele of the University of Texas, observes: "They were worried about war at the end. Ecological disasters, too. Deforestation. Starvation. I think the population rose to the limits their technology could bear. They were so close to the edge, if anything went wrong, it was all over."

Whether the Mayan economy had become environmentally unsustainable before it actually began to decline, we do not know. We do know that ours is.

Critical Thinking

1. Do you agree with the author that we need to bring about an environmental revolution within a few decades? Explain.

2. Do you believe that this can be done by making minor adjustments in the global economy or that the global economy must be restructured to put less strain on the earth's natural systems?

SUGGESTED READINGS

See Internet Sources and Further Readings at the back of the book, and InfoTrac.

2 CULTURAL CHANGES, WORLDVIEWS, ETHICS, AND SUSTAINABILITY

A.D. *2060: Green Times on Planet Earth**

Mary Wilkins sat in the living room of the passive solar earth-sheltered house she shared with her daughter Jane and her family. It was July 4, 2060: Independence Day.

She walked into the greenhouse and gazed out at the large blue and green earth flag (Figure 2-1) they flew with the American flag. Her eyes shifted to her grandchildren, Lynn and Jeffrey, running and hollering as they played tag on the neighborhood commons. She heard the hum of solar-powered pumps trickling water to rows of organically grown vegetables, and glanced at the fish in the aquaculture and waste treatment tanks.

Things began changing rapidly in 2004 when environmentalists, workers, and ordinary citizens began working together to sustain the earth's life-support systems (and thus the human species) by learning how to work with the earth. They decided to live by one of nature's most fundamental lessons: Sustainability depends on cooperating and pursuing diverse strategies adapted to local conditions, problems, and cultural beliefs.

Mary returned to the cavelike coolness of her earth-sheltered house and began putting the finishing touches on the children's costumes for this afternoon's pageant in Rachel Carson Park. It would honor earth heroes who began the Age of Ecology in the 20th century, as well as those who continued this tradition in the 21st century. She smiled when she thought of Jeffrey's delight at being chosen to play Aldo Leopold, who in the late 1940s began urging people to work with the earth. Her pride swelled when Lynn was chosen to play Rachel Carson, who in the 1960s alerted us to threats from increasing exposure to pesticides and other harmful chemicals.

Even in her most idealistic dreams, she had never guessed she would see the hemorrhaging loss of global biodiversity slowed to a trickle. There were no more wars over oil, and most air pollution gradually disappeared when energy from the sun, wind, and hydrogen (produced by using solar energy to

Figure 2-1 Some people display the earth flag as a symbol of their commitment to working with the earth at the individual, local, national, and international levels. (Courtesy of Earth Flag Co., 33 Roberts Road, Cambridge, Mass. 02138)

decompose water) replaced fossil and nuclear fuels. Now there was greatly increased emphasis on pollution prevention and waste reduction. Walking and bicycling had increased in cities and towns designed as vibrant communities for people instead of cars. Low-polluting and safe ecocars got 128 kilometers per liter (300 miles per gallon) and there was efficient mass transportation.

World population had stabilized at 8 billion in 2030 and then had begun a slow decline. The rate of global warming had slowed significantly, and international treaties enacted in the 1990s effectively banned the chemicals that had begun depleting ozone in the stratosphere.

Two hours later, she, her daughter Jane, and her son-in-law Gene watched with pride as 40 beautiful children honored the leaders of the Age of Ecology. At the end, Lynn stepped forward and said, "Today we have honored many earth heroes, but the real heroes are the ordinary people in this audience and around the world who worked to help sustain the earth. Thank you, Grandma, Mom, Dad, and everyone here for giving us such a wonderful gift. We promise to leave the earth even better for our children and grandchildren and all living creatures."

Environmentalists and many of the world's prominent scientists believe that we are at a critical turning point: a time to make crucial choices and to act on them. One option is to start on a path that could lead to a world like the one just described; another is to stay on our present path, which could result in a world like the one described on page 268. The choice is ours; to refuse to decide is to decide.

---

* Compare this hopeful scenario with the worst-case scenario that opens Chapter 10.

A continent ages quickly once we come.

Ernest Hemingway

This chapter focuses on the following questions:

- What major effects have hunter–gatherer societies, nonindustrialized agricultural societies, and industrialized societies had on the environment?

- What major human-centered environmental worldviews guide most industrial societies?

- What are some life-centered and earth-centered worldviews?

- What ethical guidelines might be used to help us work with the earth?

- How can we live more sustainably?

2-1 CULTURAL CHANGES

What Major Human Cultural Changes Have Taken Place? Fossil and anthropological evidence suggests that the current form of our species, *Homo sapiens sapiens*, has walked the earth for only about 60,000 years (some recent evidence suggests 90,000 years), an instant in the planet's estimated 4.6-billion-year existence. Until about 12,000 years ago, we were mostly hunter–gatherers who moved as needed to find enough food for survival. Since then, there have been two major cultural shifts: the *agricultural revolution*, which began 10,000–12,000 years ago, and the *industrial revolution*, which began about 275 years ago.

These cultural revolutions have given us much more energy (Figure 2-2) and new technologies with which to alter and control more of the planet to meet our basic needs and increasing desires. By expanding food supplies, lengthening life spans, and raising living standards for many people, each cultural shift contributed to the expansion of the human population. More people fed, bred, and spread (Figure 2-3). However, the results include skyrocketing resource use, pollution, and accelerating environmental degradation.

How Did Ancient Hunting-and-Gathering Societies Affect the Environment? During most of our 60,000-year existence, we were **hunter–gatherers** who survived by collecting edible wild plant parts, hunting, fishing, and scavenging meat from animals killed by other predators. Archaeological and anthropological evidence indicates that our hunter–gatherer ancestors typically lived in small bands (of fewer than 50 people) who worked together to get enough food to survive. Most groups were nomadic, picking up their few possessions and moving from place to place as needed to find enough food.

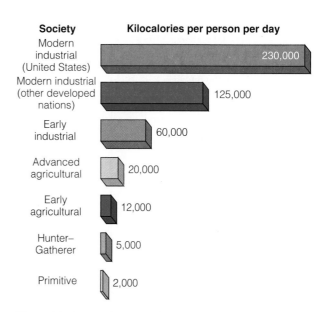

Figure 2-2 Average direct and indirect per capita daily energy use at various stages of human cultural development. A calorie is the amount of energy needed to raise the temperature of 1 gram of water 1°C (1.8°F); a kilocalorie is 1,000 calories. Food calories or calories expended during exercise are kilocalories, sometimes designated Calories (with a capital *C*).

The earliest hunter–gatherers (and those still living this way today) survived through *earth wisdom*: expert knowledge of their natural surroundings. They discovered that a variety of plants and animals could be eaten and used as medicines. They knew where to find water, how to predict the weather with reasonable accuracy, how plant availability changed throughout the year, and how some game animals migrated to get enough food.

These dwellers in nature had only three energy sources: **(1)** sunlight captured by plants (which also served as food for the animals they hunted), **(2)** fire, and **(3)** their own muscle power. Because of high infant mortality and an estimated average life expectancy of 30–40 years, hunter–gatherer populations grew very slowly (Figure 2-3).

As hunter–gatherers gradually improved their tools and hunting practices, their harmful effects on the environment increased. Some worked together to hunt herds of reindeer, mammoths, European bison, and other big game. Some advanced hunter–gatherers used fire to flush game from forest thickets and grasslands toward waiting hunters. They also stampeded herds into traps or over cliffs, killing many more animals than they needed. Some learned that burning vegetation promoted the growth of plants they could eat or that were favored by certain game animals.

Advanced hunter–gatherers had a greater impact on their environment than did early hunter–gatherers. Their use of fire, in particular, converted forests into

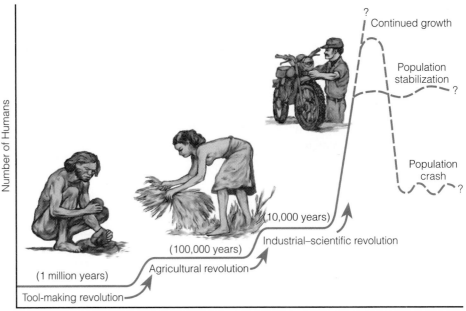

Figure 2-3 Expansion of the earth's carrying capacity for humans. Technological innovation has led to major cultural changes, and we have displaced and depleted numerous species that compete with us for—and provide us with—resources. Dashed lines represent three alternative futures: uninhibited human population growth, population stabilization, and growth followed by a crash and stabilization at a much lower level.

grasslands. There is some evidence that they contributed to—perhaps even caused—the extinction of some large animals, including the mastodon, saber-toothed tiger, giant sloth, cave bear, mammoth, and giant bison. Other researchers attribute such extinctions mostly to rapid changes in climate, especially ice ages. As these early humans moved to new areas, many probably carried plant seeds and roots with them. This altered the distribution of plants and, in some cases, the types of animals feeding on the newly introduced plants.

Hunter–gatherers exploited their environment to survive, but their environmental impact was usually limited and local. They relied on potentially renewable resources, and their use of resources was low. Most of the damage they caused was easily repaired by natural processes because of their small groups and frequent migrations.

Both early and advanced hunter–gatherers were *dwellers in nature* who survived by learning to work with nature and with one another in small groups. Some people tend to make sweeping generalizations and to romanticize the wide variety of hunter–gatherer groups, but evidence reveals that some groups were peaceful and some were aggressive, and some respected nature and some did not.

How Has the Agricultural Revolution Affected the Environment? Some 10,000–12,000 years ago, a cultural shift known as the **agricultural revolution** began in several regions of the world. It involved a gradual move from a lifestyle based on nomadic hunting and gathering to one centered on settled agri-

cultural communities in which people domesticated wild animals and cultivated wild plants.

Plant cultivation probably developed in many areas, especially in the tropical forests of southeast Asia, northeast Africa, and Mexico. People discovered that they could grow various wild food plants from roots or tubers (fleshy underground stems). Often they planted a mixture of food crops and tree crops, an ancient and sustainable form of **agroforestry**. To prepare the land for planting, they cleared small patches of tropical forests by cutting down trees and other vegetation and then burning the underbrush (Figure 2-4). The ashes fertilized the nutrient-poor soils in this **slash-and-burn cultivation**.

These early growers also used various forms of **shifting cultivation** (Figure 2-4). After a plot had been used for several years, the soil would be depleted of nutrients or reinvaded by the forest. Then the growers moved and cleared a new plot. They learned that each abandoned patch had to be left fallow (unplanted) for 10–30 years before the soil became fertile enough to grow crops again. While patches were regenerating, growers used them for tree crops, medicines, fuelwood, and other purposes. In this manner, early growers practiced sustainable cultivation within tropical forests.

These early practices involved **subsistence farming**, in which a family grew only enough food for itself. Their dependence mostly on human muscle power and crude stone or stick tools meant that growers could cultivate only small plots; thus they had little impact on their environment. However, the main reason that this form of agriculture was sustainable was that the population size and density of these early

Q: How many children under age 5 die each day in poor countries of causes that could be prevented?

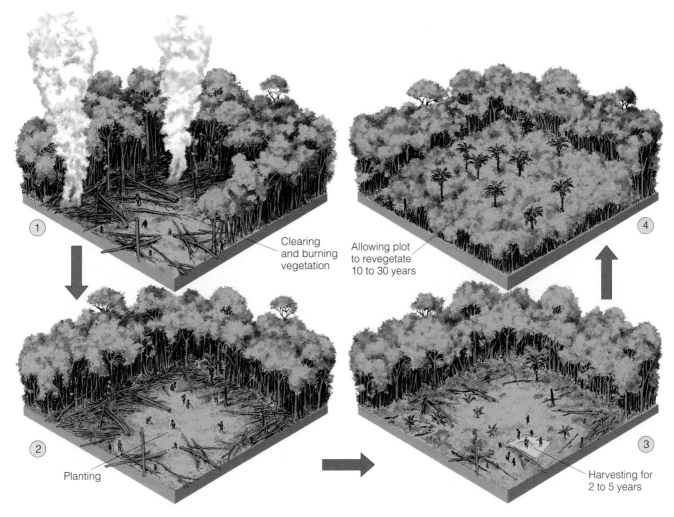

Figure 2-4 The first crop-growing technique may have been a combination of slash-and-burn and shifting cultivation in tropical forests. This method is sustainable only if small plots of the forest are cleared, cultivated for no more than 5 years, and then allowed to regenerate for 10 to 30 years to renew soil fertility. Indigenous cultures have developed many variations of this technique and have found ways to make some nondestructive uses of former plots while they are being regenerated.

farmers were low. This meant that people could move on to other areas and leave abandoned plots unplanted for the several decades needed to restore soil fertility.

The invention of the metal plow pulled by domesticated animals allowed farmers to cultivate larger plots of land and to break up fertile grassland soils, which previously could not be cultivated because of their dense root systems. In some arid regions, early farmers increased crop output by diverting water from nearby streams into hand-dug ditches and canals to irrigate crops.

The gradual shift from hunting and gathering to farming had several significant effects:

- *Using domesticated animals to plow fields, haul loads, and perform other tasks increased the average energy use*

per person and hence the ability to expand agriculture (Figure 2-2).

- *Birth rates rose faster than death rates, and thus the population increased, mostly because the larger, more reliable food supply could support more people* (Figure 2-3).

- *People cleared increasingly larger fields and built irrigation systems to transfer water from one place to another.*

- *People began accumulating material goods.* Nomadic hunter–gatherers could not carry many possessions in their travels, but farmers living in one place could acquire as much as they could afford.

- *Farmers could grow more than enough food for their families.* They could store the excess for a "rainy day" or use it to barter with craftspeople who specialized in weaving, tool making, and pottery.

- *Urbanization—the formation of villages, towns, and cities—became practical.* With fewer people needed to provide food, many people left farms and moved to villages, where they took up crafts and other occupations. Some villages grew into towns and cities, which served as centers for trade, government, and religion.

- *Conflict between societies became more common as ownership of land and water rights became crucial economic issues.* Armies and their leaders rose to power and conquered large areas of land and water supplies. These rulers forced powerless people (slaves and landless peasants) to do the hard, disagreeable work of producing food and constructing things such as irrigation systems, temples, and walled fortresses.

- *The survival of wild plants and animals, once vital to humanity, became less important.* Wild animals, which competed with livestock for grass and fed on crops, became enemies to be killed or driven from their habitats. Wild plants invading cropfields became weeds to be eliminated.

The growing populations of these emerging civilizations needed more food and more wood for fuel and building materials, so people cut down vast forests and plowed up large expanses of grassland. Such extensive land clearing degraded or destroyed the habitats of many wild plants and animals, causing or hastening their extinction.

Many of these cleared lands were poorly managed. Soil erosion, salt buildup in irrigated soils, and overgrazing of grasslands by huge herds of livestock helped turn fertile land into desert; topsoil washed into streams, lakes, and irrigation canals. The gradual degradation of the vital resource base of soil, water, forests, grazing land, and wildlife converted many once-productive landscapes into barren regions. This degradation was a factor in the downfall of many great civilizations in the Middle East, North Africa, and the Mediterranean (Guest Essay, p. 26). Historian Henry Kissinger reminds us, "As a historian, you have to be conscious of the fact that every civilization that has ever existed has ultimately collapsed."

The shift of people to towns and cities, the emergence of specialized occupations and new technologies, and the expansion of commerce and trade greatly increased the demand for metals and other nonrenewable mineral resources. The expansion of mining degraded land and water. Increased production and use of material goods created growing volumes of wastes. Towns and cities concentrated sewage and other wastes, polluted the air and water, and greatly increased the spread of diseases.

The spread of agriculture meant that most of the world's human population gradually shifted from being hunter–gatherers, working with nature in order to survive, to becoming shepherds, farmers, and urban dwellers trying to tame and manage nature to survive and prosper.

How Has the Industrial Revolution Affected the Environment? The next great cultural shift, the **industrial revolution**, began in England in the mid-1700s and spread to the United States in the 1800s. It multiplied per capita energy consumption and thus the power of humans to shape the earth to their will and fuel economic growth. Production, commerce, trade, and distribution of goods all expanded rapidly.

The industrial revolution represented a shift from dependence on *potentially renewable* wood (with supplies dwindling in some areas because of unsustainable cutting) and flowing water to dependence on *nonrenewable* fossil fuels. Coal-fired steam engines were invented to pump water and perform other tasks. Eventually an array of new machines were developed, powered by coal (and later by oil and natural gas). The new machines in turn led to a switch from small-scale, localized production of handmade goods to large-scale production of machine-made goods in centralized factories within rapidly growing industrial cities.

Factory towns grew into cities as rural people came to the factories for work. There they worked long hours under noisy, dirty, and hazardous conditions. Other workers toiled in dangerous coal mines. In these early industrial cities, coal smoke belching out of chimneys was so heavy that many people died of lung ailments. Ash and soot covered everything, and on some days the smoke was so thick that it blotted out the sun.

Fossil-fuel–powered farm machinery, commercial fertilizers, and new plant-breeding techniques increased per acre crop yields. This development helped protect biodiversity by reducing the need to expand the area of cropland to grow food. Because fewer farmers were needed, more people migrated to cities. With a larger and more reliable food supply, the size of the human population began the sharp increase in numbers still going on today.

After World War I (1914–18), more efficient machines and mass-production techniques were developed. This led to the basis of today's advanced industrial societies in places such as the United States, Canada, Japan, Australia, and western Europe.

Advanced industrial societies provide a variety of benefits to most people living in them, including mass production of many useful and economically affordable products; a sharp increase in agricultural productivity; lower infant mortality and higher life expectancy because of better sanitation, hygiene, nutrition, and medical care; a decrease in the rate of population

Q: What percentage of environmental spending in the United States is devoted to preventing pollution?

growth; better health, birth control methods, and education; methods for controlling pollution; and greater average income and old-age security.

These important benefits of industrialized societies have been accompanied by the resource and environmental problems we face today. Industrialization also isolates more people from nature and reduces understanding of the important ecological and economic services nature provides (Figure 1-2).

2-2 ENVIRONMENTAL WORLDVIEWS IN INDUSTRIAL SOCIETIES

How Shall We Live? A Clash of Cultures and Values There are conflicting views about how serious our environmental problems are and what we should do about them. These conflicts arise mostly out of differing **environmental worldviews**: how people think the world works, what they think their role in the world should be, and what they believe is right and wrong environmental behavior (**environmental ethics**).

People with widely differing environmental worldviews can take the same data, be logically consistent, and arrive at quite different conclusions (Table 1-1) because they start with different assumptions and are often seeking answers to different questions.

There are many different types of environmental worldviews, as summarized in Figure 2-5. Most can be divided into two groups according to whether they are *individual centered* (atomistic) or *earth centered* (holistic). Atomistic environmental worldviews tend to be *human centered* (anthropocentric) or *life centered* (biocentric, with the primary focus on either individual species or individual organisms). Holistic or ecocentric environmental worldviews are either *ecosystem centered* or *ecosphere* (life-support system) *centered*.

What Are the Major Human-Centered Environmental Worldviews? Most people in today's industrial consumer societies have a **planetary management worldview**, which has become increasingly accepted during the past 50 years. According to this human-centered environmental worldview, human beings, as the planet's most important and dominant species, can and should manage the planet mostly for their own benefit. Other species are seen as having only *instrumental value*; that is, their value depends on whether they are useful to us.

The basic environmental beliefs of this worldview include the following:

- *We are the planet's most important species, and we are in charge of the rest of nature.* This idea crops up when people talk about "our" planet or "our" earth and when people talk about "saving the earth."

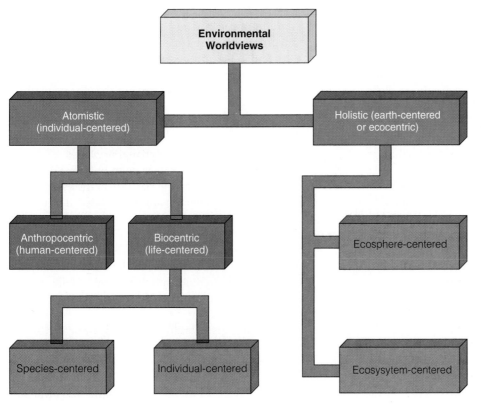

Figure 2-5 General types of environmental worldviews. (Diagram developed by Jane Heinze-Fry)

- *There is always more.* The earth has an essentially unlimited supply of resources, to which we gain access via science and technology. If we deplete a resource, we will find substitutes. To deal with pollutants, we can invent technology to clean them up, dump them into space, or move into space ourselves. If we extinguish other species, we can use genetic engineering to create new and better ones.

- *All economic growth is good, more economic growth is better, and the potential for economic growth is essentially limitless.*

- *Our success depends on how well we can understand, control, and manage the earth's life-support systems for our benefit.*

People with this (and related) environmental worldviews seek answers to questions such as, How can we keep economic growth or throughput of resources growing exponentially? How can we become better managers of the entire planet? How can we control and manage the pollutants and wastes we produce and the environmental degradation we cause? This worldview is widely supported because it is said to be the primary driving force behind the major improvements in the human condition since the beginning of the industrial revolution (Table 1-1, left).

There are several variations of this environmental worldview. Some people belong to what might be called the *no-problem school*: There are no environmental, population, or resource problems that cannot be solved by more economic growth, better management, and better technology.

Another group, the *free-market school*, believes that the best way to manage the planet for human benefit is through a free-market global economy with minimal government interference and regulations. Free-market advocates would convert all public property resources to private property resources and let the global marketplace, governed by free-market competition (pure capitalism), decide essentially everything.

Another variation, held by many people, is that we do have serious environmental, resource, and population problems but that we can deal with them by engaging in *responsible planetary management*. People holding this view follow the pragmatic principle of *enlightened self-interest*: Better earth care is better self-care. They believe that we can sustain our species with a mixture of market-based competition, better technology, and some government intervention to promote ecologically sustainable forms of economic development, protect environmental quality and private property rights, and protect and manage public and common property resources.

Still another variation of the responsible planetary management environmental worldview is the *spaceship-earth worldview.* Earth is seen as a spaceship: a complex machine that we can understand, dominate, change, and manage in order to prevent environmental overload and provide a good life for everyone.

Other people advocate that our management of the earth be guided by the principle of *stewardship*: We have an ethical responsibility to be caring and responsible managers or stewards, who tend the earth as if it were a garden. According to this view, we can and should make the world a better place for ourselves and other species through love, care, knowledge, and technology.

2-3 LIFE-CENTERED AND EARTH-CENTERED ENVIRONMENTAL WORLDVIEWS

Can We Manage the Planet? Some people believe that any human-centered worldview will eventually fail because it wrongly assumes that we now have (or can gain) enough knowledge to become effective managers or stewards of the earth. They compare our pursuit of unlimited economic growth on a finite planet to being on a treadmill that moves faster and faster. Sooner or later, they believe, we will either fall off the treadmill or damage it because our knowledge and managerial skills are so limited compared to the incredible complexity and adaptability of earth's life-support systems.

These people argue that the unregulated free-market approach won't work because it is based on mushrooming losses of earth capital (Figures 1-2, 1-5, 1-13, and 1-14) and because it focuses on short-term economic benefits regardless of the harmful long-term consequences. They also contend that the spaceship-earth and stewardship versions of planetary management will not work because such human constructs are oversimplified and misleading ways to view an incredibly complex and ever-changing planet.

For example, these critics point out that we do not even know how many species live on the earth, much less what their roles are and how they interact with one another and their nonliving environment. We have only an inkling of what goes on in a handful of soil, a meadow, a patch of forest, a pond, or any other part of the earth. Such analysts liken us to technicians who think we can build and repair automobile engines after a couple of minutes of training with a vacuum cleaner motor.

As biologist David Ehrenfeld puts it, "In no important instance have we been able to demonstrate comprehensive successful management of the world, nor do we understand it well enough to manage it even in theory." Environmental educator

Q: What percentage of the world's population lives in the United States?

David Orr (Guest Essay, p. 191) says we are losing rather than gaining the knowledge and wisdom needed to adapt creatively to continually changing environmental conditions:

On balance, I think, we are becoming more ignorant because we are losing cultural knowledge about how to inhabit our places on the planet sustainably, while impoverishing the genetic knowledge accumulated through millions of years of evolution. . . . Most research is aimed to further domination of the planet. Considerably less of it is directed at understanding the effects of domination. Less still is aimed to develop ecologically sound alternatives that enable us to live within natural limits.

To critics of the planetary management environmental worldview, our task is not to learn how to pilot spaceship earth, but instead to give up our fantasies of omnipotence and base our actions on earth wisdom—on learning to work with the earth by becoming more responsible earth citizens. Even if we had enough knowledge and wisdom to manage spaceship earth, some critics see this approach as requiring us to give up individual freedom in order to survive. Life on spaceship earth under a comprehensive system of planetary management or world government might be very much like the regimented life of astronauts in their capsule. The astronauts have virtually no individual freedom; essentially all of their actions are dictated by a central command (ground control) without

---

Deep Ecology

SPOTLIGHT

Deep ecology is an ecocentric environmental set of beliefs or worldview developed in 1972 by Norwegian philosopher Arnie Naess. In 1984 Naess, philosopher George Sessions, and sociologist Bill Devall drew up a list of eight beliefs of deep ecology. In 1993 these beliefs were modified somewhat by Naess.*

Here is the modified list:

1. The well-being and flourishing of human and nonhuman life on the earth have inherent value in themselves. These values are independent of the usefulness of the nonhuman world for human purposes.

2. The fundamental interdependence, richness, and diversity of life-forms contribute to the flourishing of human and nonhuman life on the earth.

3. Humans have no right to reduce this interdependence, richness, and diversity except to satisfy vital needs.

4. Present human interference with the nonhuman world is excessive, and the situation is worsening rapidly.

*For a set of readings about the nature of deep ecology, see George Sessions, ed., *Deep Ecology for the 21st Century* (Boston: Shambhala, 1995).

5. Because of point 4, it would be better for humans, and much better for nonhumans, if there were a substantial decrease in the human population.

6. Policies must therefore be changed. These policies affect basic economic, technological, and ideological structures. The resulting state of affairs will be deeply different from the past.

7. The ideological change is mainly that of appreciating *life quality* (involving situations of inherent value) rather than adhering to an ever-higher material standard of living.

8. Those who subscribe to these points have an obligation directly or indirectly to try to implement the necessary changes.

Naess has also described some lifestyle guidelines compatible with the basic beliefs of deep ecology. They include **(1)** appreciating all forms of life, **(2)** protecting or restoring local ecosystems, **(3)** using simple means, **(4)** consuming less, **(5)** emphasizing satisfying vital needs rather than desires, **(6)** attempting to live *in* nature and promote community, **(7)** appreciating ethnic and cultural differences, **(8)** working to improve the standard of living for the world's poor, **(9)** working to eliminate

injustice toward fellow humans or other species, and **(10)** acting nonviolently.

Since 1984, deep ecology has become a major philosophical and ethical movement, used by a growing number of people from different philosophical and religious backgrounds to bring about social and environmental change. Hundreds of articles and books have been written interpreting, analyzing, praising, and criticizing this new ecocentric environmental worldview.

Deep ecology is not an eco-religion, nor is it antireligious or antihuman, as some of its critics have claimed. Instead, it is a set of beliefs designed to have us think more deeply about the inherent value of all life on the earth and about our obligations toward both human and nonhuman life.

Critical Thinking

1. Which, if any, of the eight basic beliefs of the deep ecology environmental worldview do you agree with? Explain.

2. List five major effects on your life and lifestyle if virtually everyone lived by the beliefs of deep ecology.

---

which they cannot survive. To these analysts, managing earth as a stripped-down spaceship to prevent disaster leaves little room for human freedom, novelty, adaptability, or long-term sustainability.

Theologian Thomas Berry calls the industrial consumer society built on the human-centered, planetary management environmental worldview the "supreme pathology of all history":

> We can break the mountains apart; we can drain the rivers and flood the valleys. We can turn the most luxuriant forests into throwaway paper products. We can tear apart the great grass cover of the western plains, and pour toxic chemicals into the soil and pesticides onto the fields, until the soil is dead and blows away in the wind. We can pollute the air with acids, the rivers with sewage, the seas with oil—all this in a kind of intoxication with our power for devastation. . . . We can invent computers capable of processing ten million calculations per second. And why? To increase the volume and speed with which we move natural resources through the consumer economy to the junk pile or the waste heap. . . . If, in these activities, the topography of the planet is damaged, if the environment is made inhospitable for a multitude of living species, then so be it. We are, supposedly, creating a technological wonderworld. . . . But our supposed progress . . . is bringing us to a wasteworld instead of a wonderworld.

What Are Some Major Biocentric and Ecocentric Worldviews? Critics of human-centered environmental worldviews believe that such worldviews should be expanded to recognize the *inherent value* of all forms of life (that is, value that exists regardless of these life-forms' use to us).

Most people with a life-centered (biocentric) worldview argue that our actions should not lead to the *premature* extinction of species. Some analysts give species a hierarchy of values. For example, some believe we have greater responsibility to protect animal species than to protect plant species. But critics point out that plants are what keep most animals alive. Other analysts determine the survival rights of various species based on the harm they do to humans. For example, they see nothing inherently wrong with trying to wipe out pest species such as rats, cockroaches, mosquitoes, and indeed most insects (p. 70); species of disease-carrying bacteria; and species that we fear such as ecologically important alligators, sharks, spiders, and bats.

Some proponents go further and believe that each individual organism, not just each species, has an inherent right to survive. As in the species-centered approach, some people (the animal rights movement) place more value on the individuals of animal species than on plant species and on microorganisms. Even

then, the emphasis is usually on relatively large mammals such as dogs, cats, elephants, and bears. Animal rights advocates have not mounted major campaigns to protect individual bats, spiders, sharks, or snakes from being killed, nor have they taken a strong stand that doing so is wrong.

Trying to decide what types of species or individuals should be protected from premature extinction or death resulting from human activities is an ethical dilemma. It is hard to know where to draw the line and be ethically consistent.

Others believe that we must go beyond this biocentric worldview, which focuses on species and individual organisms. They see our primary role as limiting our actions to those that do not degrade or destroy the earth's life-support systems (and thus threaten the existence of our own species). In other words, they have an *earth-centered*, or ecocentric, environmental worldview, devoted to preserving earth's biodiversity and ecological integrity (Figure 2-5).

Their view is that we are part of, not apart from, the community of life and the ecological processes that sustain all life. Aldo Leopold summed up this idea in 1948: "All ethics rest upon a single premise: that the individual is a member of a community of interdependent parts."

There are many life-centered and earth-centered environmental worldviews, and several of them overlap in some of their beliefs. One ecocentric worldview is the **earth-wisdom worldview**. It is based on the following major beliefs, which are the opposite of those making up the planetary management worldview:

- *Nature exists for all of the earth's species, not just for us.* We need the earth, but the earth does not need us.

- *There is not always more.* The earth's limited resources should not be wasted but instead used sustainably for us and all species.

- *Some forms of economic growth are environmentally beneficial and should be encouraged, but some are environmentally harmful and should be discouraged.*

- *Our success depends on learning to cooperate with one another and with the rest of nature by learning how to work with the earth.* Because nature is so incredibly complex and always changing, we will never have enough information and understanding to even come close to managing the planet effectively. Instead, we must learn as much as we can about how the earth works, sustains itself, and adapts to changing conditions, and then use these lessons from nature to guide our actions.

People with such worldviews seek answers to the following questions: How can we design and use economic and political systems to encourage earth-

Q: What percentage of the world's mineral resources and nonrenewable energy is used by the United States?

sustaining forms of development and discourage earth-degrading ones? What is sustainability, and what do we need to do to live sustainably on the planet? How can we produce fewer pollutants and wastes and not cause so much environmental degradation? Some analysts call for us to shift our image and worldview of nature as a foe to be conquered and exploited to nature as a nurturing mother that we should work with rather than against (see cartoon, below). A related ecocentric environmental worldview is the *deep ecology worldview* (Spotlight, p. 35).

Are Biocentrists and Ecocentrists Antihuman and Antireligious? Many anti-environmentalists accuse people who promote various biocentric and ecocentric worldviews of being antihuman or against celebrating humanity's special qualities and achievements. However, those with life-centered and earth-centered worldviews consider their environmental worldviews to be profoundly prohuman. To them, recognizing the inherent value of all life and not degrading the earth's life-support systems are ways to benefit all people in this and future generations (Connections, right).

Others say that we do not need to be biocentrists or ecocentrists to value life or the earth because human-centered stewardship and responsible planetary management environmental worldviews also call for us to value individuals, species, and the earth's life-support systems as part of our responsibility as earth's caretakers.

Why Should We Care About Future Generations?

According to biologist David W. Ehrenfeld, caring about future generations enough not to degrade the earth's life-support systems is important because it gives future generations options for dealing with the problems they will face. He points out that if our ancestors had left for us the ecological devastation we are leaving our descendants, our options for enjoyment—perhaps even for survival—would be quite limited.

In response to the question, "What can future generations do for us?" Ehrenfeld gives the following answer: "They give us a reason for treating our ecological home respectfully, so that our lives as well as theirs will be enriched."

In thinking about our responsibility toward future generations, some analysts believe that we should consider the wisdom given to us in the 18th century by the Iroquois Confederation of Native Americans: *In our every deliberation, we must consider the impact of our decisions on the next seven generations.*

Critical Thinking

What obligations, if any, concerning the environment do you have to future generations? Be honest about your feelings. To how many future generations do you have responsibilities? List the most important environmental benefits and harmful conditions passed on to you by the previous two generations.

©William K. Day, Tribune Media Service. Used by permission.

A: About 25%

2-4 SOLUTIONS: LIVING SUSTAINABLY

How Should We Evaluate Sustainability Proposals? **Sustainability** has become a buzzword. Hundreds of programs have been proposed or implemented in its name, some of them useful and some questionable or harmful. Lester Brown (Guest Essay, p. 26) has this simple test for any sustainability proposal: "Does this policy or action lower carbon emissions? Does it reduce the generation of toxic wastes? Does it slow population growth? Does it increase the earth's tree cover? Does it reduce emissions of chemicals that deplete the earth's vital ozone layer or that can lead to warming of the atmosphere? Does it reduce air pollution? Does it reduce radioactive waste generation? Does it lead to less soil erosion? Does it protect the planet's biodiversity?"

Additional questions also help us evaluate sustainability proposals. Does it deplete earth capital? Does it diminish cultural diversity? Does it reduce poverty, hunger, and disease? Does it promote individual and community self-reliance? Does it prevent pollution? Does it reduce resource waste? Does it save energy? Does it transfer the most resource-efficient and environmentally benign technologies to developing countries?

The search for sustainability presumes that, for any proposed course of action, we know (or can discover) what is sustainable—that is, the levels and thresholds of environmental carrying capacity for a region or for the entire world. Unfortunately, we cannot do this very well with current knowledge. Until we can, a growing number of environmentalists believe that we should follow what is called the *precautionary principle*: playing it safe by using various prevention guidelines and strategies for developing sustainable societies.

Solutions: What Are Some Ethical Guidelines for Working with the Earth? According to Robert Cahn, "the main ingredients of an environmental ethic are caring about the planet and all of its inhabitants, allowing unselfishness to control the immediate self-interest that harms others, and living each day so as to leave the lightest possible footprints on the planet." Various ethicists and philosophers have developed a variety of ethical guidelines for living more sustainably on the earth. Such guidelines can be used by anyone, whether he or she has a human-centered stewardship environmental worldview or a biocentric or ecocentric environmental worldview.

Ecosphere and Ecosystems

- We should try to understand and work with the rest of nature to help sustain the ecological integrity, biodiversity, and adaptability of the earth's life-support systems.

- When we must alter nature to meet our needs or wants, we should carefully evaluate our proposed actions and choose methods that do the least possible short- and long-term environmental harm.

Species and Cultures

- We should work to preserve as much of the earth's genetic variety as possible because it is the raw material for all future evolution and genetic engineering.

- We have the right to defend ourselves against individuals of species that do us harm and to use individuals of species to meet our vital needs, but we should strive not to cause the premature extinction of any wild species.

- The best ways to protect species and individuals of species are to protect the ecosystems in which they live and to help restore those we have degraded.

- No human culture should become extinct because of our actions.

Individual Responsibility

- We should not inflict unnecessary suffering or pain on any animal we raise or hunt for food or use for scientific or other purposes.

- We should leave the earth as good as or better than we found it.

- We should use no more of the earth's resources than we need.

- We should work with the earth to help heal the ecological wounds we have inflicted.

Solutions: What Is Earth Education? Most environmentalists believe that learning how to live sustainably requires a foundation of earth education that relies heavily on an interdisciplinary and holistic approach to learning and a lifelong commitment to such education (Guest Essay, pp. 42–43). Among the most important goals of such an education are the following:

- *Developing respect or reverence for all life.*

- *Understanding as much as we can about how the earth works and sustains itself, and using such earth wisdom to guide our lives, communities, and societies.*

- *Understanding as much as we can about connections and interactions*—those within nature, between people and the rest of nature, among people with different cultures and beliefs, among generations, among the

Q: What percentage of the world's people live in developed countries?

problems we face, and among the solutions to these problems.

- *Becoming wisdom seekers instead of information vessels.* We need to learn how to sift through mountains of facts and ideas to find the nuggets of knowledge and wisdom that are worth knowing.

- *Understanding and evaluating one's worldview and seeing this as a lifelong process* (Individuals Matter, right).

- *Learning how to evaluate the beneficial and harmful consequences of one's lifestyle and profession on the earth, today and in the future.*

- *Using critical thinking skills to evaluate and resist advertising.* As humorist Will Rogers put it, "Too many people spend money they haven't earned to buy things they don't want, to impress people they don't like."

- *Fostering a desire to make the world a better place and to act on this desire.* As David Orr puts it, education should help students "make the leap from 'I know' to 'I care' to 'I'll do something.'"

According to environmental educator Mitchell Thomashow, four basic questions should be at the heart of environmental education:

- Where do the things I consume come from?

- What do I know about the place where I live?

- How am I connected to the earth and other living things?

- What are my purpose and responsibility as a human being?

How we answer these questions determines our *ecological identity.*

Solutions: How Can Direct Experiences Help Us Learn How to Work with the Earth? In addition to top-down formal education, some analysts believe that we need to learn by experience how to walk more lightly on the earth (Connections, p. 40). One source of wisdom is bottom-up education obtained by *listening to children.* Listen to young children and you will find that many of them believe that much of what we are doing to the earth (and thus to them) is stupid and wrong, and they do not accept the excuses we give for not changing the harmful way we act toward the earth.

A related goal is to *learn how to live more simply.* Although seeking happiness through the single-minded pursuit of material things is considered folly by virtually every major religion and philosophy, it is preached

Mindquake: Evaluating One's Environmental Worldview

INDIVIDUALS MATTER

Questioning and perhaps even changing one's environmental worldview can be difficult and threatening and can set off a cultural *mindquake* that involves examining many of one's most basic beliefs. However, once individuals change their worldview, it no longer makes sense for them to do things in the old ways. If enough people do this, then tremendous cultural change, once considered impossible, can take place rapidly.

Most environmentalists urge us to think about what our basic environmental beliefs are and why we have them. They believe that evaluating our beliefs and being open to the possibility of changing them should be one of our most important lifelong activities.

As this book emphasizes, most environmental issues are filled with controversy and uncertainty. A clearly right or wrong path is not easy to discover and is usually strongly influenced by our environmental worldview. As philosopher Hegel pointed out nearly two centuries ago, tragedy is not the conflict between right and wrong, but the conflict between right and right.

Critical Thinking

What are the basic beliefs of your current environmental worldview? At the end of this course evaluate your answer to learn which (if any) of your basic environmental beliefs have changed.

incessantly by modern advertising. Some affluent people in developed countries, however, are adopting a lifestyle of *voluntary simplicity*, doing and enjoying more with less. They agree with Edward Goldsmith that "the more we look at it, the more it is apparent that economic growth is a device for providing us with the superfluous at the cost of the indispensable."

Voluntary simplicity is based on Mahatma Gandhi's *principle of enoughness*: "The earth provides enough to satisfy every person's need but not every person's greed. . . . When we take more than we need, we are simply taking from each other, borrowing from the future, or destroying the environment and other species." It means asking oneself, "How much is enough?" This is not an easy thing to do because people in affluent societies are conditioned to want more and more, and they often think of such wants as vital needs.

Learning from the Earth

Formal earth education is important, but Aldo Leopold, Henry David Thoreau, Gary Snyder, David Brower, Stephen Jay Gould, David Orr, Mitchell Thomashow, and many others believe it's not enough.

They suggest that we reenchant our senses and kindle a sense of awe, wonder, and humility by standing under the stars, sitting in a forest, taking in the majesty and power of an ocean, or experiencing a stream, lake, or other part of untamed nature.

We might pick up a handful of soil and try to sense the teeming microscopic life in it that keep us alive. We might look at a tree, a mountain, a rock, or a bee and try to sense how they are a part of us, and we a part of them as interdependent participants in the earth's life-sustaining recycling processes.

Earth thinker Michael J. Cohen suggests that each of us recognize who we really are by saying,

I am a desire for water, air, food, love, warmth, beauty, freedom, sensations, life, community, place, and spirit in the natural world.... I have two mothers: my human mother and my planet mother, Earth. The planet is my womb of life.

Many psychologists believe that consciously or unconsciously we spend much of our lives in a search for roots—something to anchor us in a bewildering and frightening sea of change. As philosopher Simone Weil observed, "To be rooted is perhaps the most important and least recognized need of the human soul."

Earth philosophers say that to be rooted, each of us needs to find a *sense of place*—a stream, a mountain, a yard, a neighborhood lot, or any piece of the earth we feel at one

with as a place we know, experience emotionally, and love. It can be a place where we live or a place we occasionally visit and experience in our inner being. When we become part of a place, it becomes a part of us. Then we are driven to defend it from harm and to help heal its wounds. As poet–philosopher Gary Snyder puts it, "Find our place on the planet, dig in, and take responsibility from there."

Critical Thinking

Some analysts believe that learning earth wisdom by experiencing the earth and forming an emotional bond with its life-forms and processes is unscientific, mystical poppycock based on a romanticized view of nature. They believe that better scientific understanding of how the earth works and improved technology are the only ways to achieve sustainability. Do you agree or disagree? Explain.

However, voluntary simplicity by those who have more than they need should not be confused with the *forced simplicity* of the poor, who do not have enough to meet their basic needs for food, clothing, shelter, clean water and air, and good health.

After a lifetime of studying the growth and decline of the world's human civilizations, historian Arnold Toynbee summarized the true measure of a civilization's growth in what he called the *law of progressive simplification*: "True growth occurs as civilizations transfer an increasing proportion of energy and attention from the material side of life to the nonmaterial side and thereby develop their culture, capacity for compassion, sense of community, and strength of democracy."

Earth care also means *not using guilt and fear to motivate other people to work with the earth and other people.* We need to nurture, reassure, understand, and love, rather than threaten, one another. Finally, we need to *have fun and take time to enjoy life.* We shouldn't get so intense and serious that we can't laugh every day and enjoy wild nature, beauty, friendship, and love.

Solutions: How Can We Move Beyond Blame, Guilt, and Denial to Responsibility? According to many psychologists, when we first encounter an environmental problem, our initial response is often to find someone or something to blame: greedy industrialists, uncaring politicians, misguided worldviews. It is the fault of such villains, and we are the victims.

This can lead to despair, denial, and inaction because we feel powerless to stop or influence these forces. There are also so many complex and interconnected environmental problems and conflicting views about their seriousness and possible solutions that we feel overwhelmed and wonder whether there is any way out—another emotion leading to denial and inaction.

Upon closer examination we may realize that we all make some direct or indirect contributions to the environmental problems we face. We don't want to feel guilty or bad about all of the things we are not doing, so we avoid thinking about them—another path leading to denial and inaction.

Q: How much did global average life expectancy increase between 1900 and 1997?

How do we move beyond immobilizing blame, fear, and guilt to engaging in more responsible environmental actions in our daily lives? Analysts have suggested several ways to do this.

First, we need to recognize and avoid common mental traps that lead to denial, indifference, and inaction. These traps include **(1)** *gloom-and-doom pessimism* (it's hopeless), **(2)** *blind technological optimism* (science and technofixes will always save us), **(3)** *fatalism* (we have no control over our actions and the future), **(4)** *extrapolation to infinity* (if I can't change the entire world quickly, I won't try to change any of it), **(5)** *paralysis by analysis* (searching for the perfect worldview, philosophy, solutions, and scientific information before doing anything), and **(6)** *faith in simple, easy answers.*

Second, we should recognize that no one can even come close to doing all of the things people suggest (or that we know we should be doing) to work with the earth. We should focus our energy on the few things that we feel most strongly about and that we can do something about. Instead of focusing on and feeling guilty about the things we haven't done, rejoice in the good things we have done; then jump in and do more to make the earth a better place.

Third, we should base our actions on a sense of *hope*, which history has shown to be the major energizing force for bringing about change. The secret is to keep our empowering feelings of hope and joy slightly ahead of our immobilizing feelings of despair. We need to acknowledge our *grief* about the harm the human species has done to the earth, ourselves, and future generations. Then we can move through the various stages of grief and become actively engaged in working with the earth in our daily lives.

Fourth, it is important to recognize that there is no single correct or best solution to the environmental problems we face. Indeed, one of nature's most important lessons (from evolution) is that preserving diversity or a rainbow of possibilities is the best way to adapt to earth's largely unpredictable, ever-changing conditions. Each human culture and environmental worldview provides different outlooks, wisdom, and insights for helping us learn how to work with the earth and make cultural changes in response to changes in environmental conditions. We are all in this together and need to work together to find a spectrum of flexible and adaptable solutions to the problems we face.

Solutions: What Are the Major Components of an Earth-Wisdom Revolution? Many environmentalists call for us to make a new cultural change in the way we think about and use the earth's endowment of resources. Such an *earth-wisdom revolution* would

have several phases. One is an *efficiency revolution* that involves not wasting matter and energy resources, using a combination of technological advances, lifestyle changes, recycling, and reuse.

A second phase is a *pollution prevention revolution* built on the efficiency revolution. It reduces pollution and environmental degradation by reducing the waste of matter and energy resources; it does so by mimicking the earth's chemical cycling processes in which each organism's wastes serve as resource inputs for other organisms. Pollution prevention also involves keeping highly toxic substances from being released into the environment by recycling or reusing them within industrial processes, trying to find less harmful or easily biodegradable substitutes, or not producing such substances at all.

A third phase is a *sufficiency revolution*, which means trying to meet the basic needs of all people on the planet and asking how many material things we really need to have a decent and meaningful life.

A fourth phase of this new cultural change is a *demographic revolution* based on bringing the size and growth rate of the human population into balance with the earth's carrying capacity for humans and other life forms.

Opponents of such an environmental cultural change like to paint environmentalists as messengers of gloom, doom, and hopelessness. However, *the real message of environmentalism is not gloom and doom, fear, and catastrophe but hope and a positive vision of the future.* This is an exciting message of challenge and adventure as we struggle to find better and more responsible ways to live on this planet.

Envision the world as a system of all kinds of matter cycles and energy flows. See these life-sustaining processes as a beautiful and diverse web of interrelationships—a kaleidoscope of patterns, rhythms, and connections whose very complexity and multitude of possibilities remind us that cooperation, sharing, honesty, humility, and love should be the guidelines for our behavior toward one another and the earth.

When there is no dream, the people perish.
PROVERBS 29:18

CRITICAL THINKING

1. Would we be better off if agricultural practices had never been developed and we were still hunters and gatherers? Explain.

2. List the most important benefits and drawbacks of an advanced industrial society such as the United States. Do the benefits outweigh the drawbacks? Explain. What

Envisioning a Sustainable Society

Lester W. Milbrath

GUEST ESSAY

Lester W. Milbrath is director of the Research Program in Environment and Society and professor emeritus of political science and sociology at the State University of New York at Buffalo. During his distinguished career he has served as director of the Environmental Studies Center at SUNY/Buffalo (1976–87) and taught at Northwestern University, Duke University, the University of Tennessee, National Taiwan University in Taipei, and Aarhus University in Denmark. He has also been a visiting research scholar at the Australian National University and at Mannheim University in Germany. His research has focused on the relationships among science, society, and citizen participation in environmental policy decisions, with emphasis on environmental perceptions, beliefs, attitudes, and values. He has written numerous articles and books. His book Envisioning a Sustainable Society: Learning Our Way Out *(1989) summarizes a lifetime of studying our environmental predicament. It is considered one of the best analyses of what we can do to learn how to work with the earth. His most recent book is* Learning to Think Environmentally While There Is Still Time *(1995).*

The 1992 Earth Summit at Rio popularized the goal of sustainable development. Most of the heads of state meeting there believed that goal could be achieved by developing better technology and by writing better laws, agreements, and treaties, and by enforcing them. Unfortunately, their approach was flawed and will not achieve sustainability because they do not understand the nature of the crisis in our earthly home.

Try this thought experiment: Imagine that, suddenly, all the humans disappeared, but all the buildings, roads, shopping malls, factories, automobiles, and other artifacts of modern civilization were left behind. What then? After three or four centuries, buildings would have crumbled, vehicles would have rusted and fallen apart, and plants would have recolonized fields, roads, parking lots, even buildings. Water, air, and soil would gradually clear up; some endangered species would flourish. Nature would thrive splendidly without us.

That mental experiment makes it clear that we do not have an environmental crisis; we have a crisis of civilization. Heads of state meeting at the Earth Summit neither understood nor dealt with civilization's most crucial problems: Humans are reproducing at such epidemic

rates that world population is expected to almost double to 10–11 billion in 44 years; resource depletion and waste generation could easily triple or even quadruple over that period; waste discharges are already beginning to change the way the biosphere works; climate change and ozone loss will reduce the productivity of ecosystems just when hordes of new humans will be looking for sustenance, and will destroy the confidence people need in order to invest in the future.

Without intending to, we have created a civilization that is headed for destruction. Either we learn to control our growth in population and in economic activity, or nature will use death to control it for us.

Present-day society is not capable of producing a solution because it is disabled by the values our leaders constantly trumpet: economic growth, jobs, consumption, competitiveness, power, and domination. Societies pursuing these goals cannot avoid depleting their resources, degrading nature, poisoning life with wastes, and upsetting biospheric systems. *We have no choice but to change; resisting change will make us victims of change.*

But how do we transform to a sustainable society? My answer, which I believe is the only answer, is that *we must learn our way*. Nature, and the imperatives of its laws, will be our most powerful teacher as we learn our way to a new society. Most crucially, we must learn how to think about values.

Life in a viable ecosystem must become the core value of a sustainable society; that means all life, not just human life. Ecosystems function splendidly without humans (or any animals for that matter), but human society would die without viable ecosystems. Individuals seeking life quality require a well-functioning society living in well-functioning ecosystems. We must give top priority to the ecosystems that support us, and second priority to our societies.

A sustainable society would affirm love as a primary value and extend it not only to those near and dear, but to people in other lands, to future generations, and to other species. A sustainable society emphasizes partnership rather than domination, cooperation over competition, love over power. A sustainable society affirms justice and security as primary values.

A sustainable society would encourage self-realization—helping people to become all they are capable of being, rather than spending and consuming—as the key to a fulfilling life. A sustainable society would

are the alternatives?

3. Do you believe that everyone has the right to have as many children as they want? Explain. If not, how would you limit such a right, including your own?

4. Do you believe that each member of the human species

has a right to pollute and to use as many resources as they want? Explain. If not, how would you limit such a right, including your own?

5. Which (if any) of the ethical guidelines on pages 38–39 do you disagree with? Explain. Can you suggest

Q: About how long has the latest version of the human species been on the earth?

make long-lasting products to be cherished and conserved. People would learn a love of beauty and simplicity.

A sustainable society would utilize both planning and markets as basic and supplementary information systems. Markets fail us because they can neither anticipate the future nor make moral choices between objects and between policies. Markets also cannot provide public goods such as schools, parks, and environmental protection, which are just as important for life quality as private goods.

A sustainable society would continue further development of science and technology because we need practical creative solutions that are both environmentally sound and economically feasible. However, we should recognize that those who control science and technology can use them to dominate all other creatures; we must learn to develop social controls of science and technology to make our society more sustainable. We should not allow the deployment of powerful new technologies that can induce sweeping changes in economic patterns, lifestyles, governance, and social values without careful forethought regarding their long-term impacts.

Conscious social learning would become the dynamic of social change in a sustainable society—not only to deal with pressing problems, but also to realize a vision of a good society. Meaningful and lasting social change occurs when nearly everyone learns the necessity of change and the value of working toward it.

Ecological thinking is different from most thinking that guides modern society. For example, the following key maxims derived from the law of conservation and matter, the laws of thermodynamics, and the workings of ecosystems are routinely violated in contemporary thinking and discourse: (1) *Everything must go somewhere (there is no away)*; (2) *energy should not be wasted because all use of energy produces disorder in the environment*; (3) *we can never do just one thing (everything is connected)*; and (4) *we must constantly keep asking, "and then what?"* Every schoolchild and every adult should learn these simple truths; we need to reaffirm the tradition that knowledge of nature's workings and a respect for all life are basic to a true education. We should require such environmental education of all students, just as we now require every student to study history.

Ecological thinking recognizes that a proper understanding of the world requires people to learn how to think holistically, systematically, and futuristically. Because everything is connected to everything else, we must learn to anticipate second-, third-, and higher-order consequences for any contemplated major societal action. A society learning to be sustainable would redesign government to maximize its ability to learn. It would use the government learning process to promote social learning. It would require that people who govern listen to citizens, not only to keep the process open for public participation, but also to cultivate mutual learning between officials and citizens.

In the recognition that our health and welfare are vitally affected by how people, businesses, and governments in other lands behave, a sustainable society would strive for an effective system of planetary politics. It would encourage transnational social movements and political parties. It would nurture planetwide social learning.

Learning our way to a new society cannot occur, however, until enough people become aware of the need for major societal change. So long as contemporary society is working reasonably well, and leaders keep telling us that society is on the right track, the mass of people will not listen to a message urging significant change. For that reason, urgently needed change will probably be delayed, and conditions on our planet are likely to get worse before they can get better. Nature will be our most powerful teacher, especially when biospheric systems no longer work the way they used to. In times of great system turbulence, social learning can be extraordinarily swift.

Our species has a special gift: the ability to recall the past and foresee the future. Once we have a vision of the future, every decision becomes a moral decision. Even the decision not to act becomes a moral judgment. Those who understand what is happening to the only home for us and other species are not free to shrink from the responsibility to help make the transition to a sustainable society.

Critical Thinking

1. Do you agree or disagree that we can only learn our way to a sustainable society? Explain.

2. Do you think we will learn our way to a sustainable society? Explain. What role, if any, do you intend to play in this process?

any additional ethical guidelines for working with the earth?

6. Consider these hypothetical questions.
   a. Would you accept employment on a project that you knew would kill or harm people? Degrade or destroy a wild habitat? Explain.

   b. If you were granted three wishes, what would they be?

   c. If you didn't have to work for a living, what would you do with your time?

d. Could you live without an automobile? Explain.

e. Could you live without TV? Explain.

f. If you won $10 million in a lottery, how would you spend the money?

g. Do you believe that wolves have as much right to eat sheep as people do? Explain.

7. Do you believe that people have the right to do anything they want with land that they own? Explain. If not, what specific limitations would you put on such private property ownership rights?

8. Review your experience with the mental traps described on pp. 40–41. Which of these traps have you fallen into, if any? Were you aware that you had been ensnared by these mental traps? Do you plan to free yourself from these traps? How?

9. If you knew you were going to die and had an opportunity to address everyone in the world for 5 minutes, what would you say? Write out your 5-minute speech and compare it with those of other members of your class.

*10. Write an essay in which you try to identify key environmental experiences that have influenced your life and thus help form your current ecological identity. Examples may include fond childhood memories of special places where you formed some connections with the earth through emotional experiences; places that you knew and cherished that have been polluted, developed, or destroyed; key environmental events that forced you to think about your environmental values; individuals or educational experiences that influenced your understanding and concern about environmental problems and challenges; and direct experience and contemplation of wild places. Have you arrived at your current ecological identity mostly through formal education, directly experiencing nature, or both? Share you experiences with other members of your class.

*11. Make a detailed list of *everything* you own. Then write a short essay about how these possessions affect your sense of self-esteem, comfort, security, and happiness. Do you feel burdened by the need to take care of so many possessions and worrying that they may be lost, stolen, or destroyed? Examine whether you feel guilty about owning too many things, especially those that mostly fill wants, not needs. Consider the general environmental impact of your possessions by evaluating the resources used in making them and how much pollution and environmental degradation are involved. Share these feelings about your possessions with other members of your class.

*12. Make a concept map of this chapter's major ideas, using the section heads and subheads and the key terms (in boldface type). Look at the inside back cover and in Appendix 4 for information about concept maps.

SUGGESTED READINGS

See Internet Sources and Further Readings at the back of the book, and InfoTrac.

PART II

PRINCIPLES AND CONCEPTS

Animal and vegetable life is too complicated a problem for human intelligence to solve, and we can never know how wide a circle of disturbance we produce in the harmonies of nature when we throw the smallest pebble into the ocean of organic life.

GEORGE PERKINS MARSH

3 SCIENCE, SYSTEMS, MATTER, AND ENERGY

Saving Energy, Money, and Jobs in Osage, Iowa

Osage, Iowa (population about 4,000), has developed into the energy-efficiency capital of the United States. Its transformation began in 1974 when Wes Birdsall, general manager of Osage Municipal Gas and Electric Company, started urging the townspeople to save energy and reduce their natural gas and electric bills. The utility would also save money by not having to add new electrical generating facilities.

Birdsall started his crusade by telling home-owners about the importance of insulating walls and ceilings and plugging leaky windows and doors. He also advised people to replace their incandescent light bulbs with more efficient fluorescent bulbs and to turn down the temperature on water heaters and wrap them with insulation—economic boons to the local hardware and lighting stores. The utility company even gave away free water heater blankets. Birdsall also suggested saving water and fuel by installing low-flow shower heads.

Birdsall then stepped up his efforts by offering to give every building in town a free thermogram—an infrared scan that shows where heat escapes (Figure 3-1). When people could see the energy (and their money) hemorrhaging out of their buildings, they took action to plug the leaks, again helping the local economy. Birdsall then stepped up his campaign even more, announcing that no new houses could be hooked up to the company's natural gas line unless they met minimum energy-efficiency standards.

Since 1974, the town has reduced its natural gas consumption by 45%, no mean feat in a locale with frigid winter temperatures. In addition, the utility company saved enough money to prepay all its debt, accumulate a cash surplus, and cut inflation-adjusted electricity rates by a third (which attracted two new factories to the area). Furthermore, each household saves more than $1,000 per year; this money supports jobs, and most of it circulates in the local economy. Before the town's energy-efficiency revolution, about $1.2 million a year left town to buy energy. What are your local utility companies and community doing to improve energy efficiency and stimulate the local economy?

Figure 3-1 An infrared photo showing heat loss (red, white, and yellow colors) around the windows, doors, roofs, and foundations of houses and stores in Plymouth, Michigan. Wes Birdsall provided similar thermograms of houses in Osage, Iowa. The average house in the United States has heat leaks and air infiltration equivalent to leaving a window wide open during the heating season. Because of poor design, most office buildings and houses in this country waste about half the energy used to heat and cool them. Americans pay about $300 billion a year for this wasted heat—more than the entire annual military budget. (VANSCAN' Continuous Mobile Thermogram by Daedalus Enterprises, Inc.)

The laws of thermodynamics control the rise and fall of political systems, the freedom or bondage of nations, the movements of commerce and industry, the origins of wealth and poverty, and the general physical welfare of the human race.

FREDERICK SODDY (NOBEL LAUREATE, CHEMISTRY)

This chapter focuses on the following questions:

- What are science and technology? What is environmental science, and what are some of its limitations?

- What are major components and behaviors of complex systems?

- What are the basic forms of matter? What is matter made of? What makes matter useful to us as a resource?

- What are the major forms of energy? What makes energy useful to us as a resource?

- What are physical and chemical changes? What scientific law governs changes of matter from one physical or chemical form to another?

- What are the three main types of nuclear changes that matter can undergo?

- What two scientific laws govern changes of energy from one form to another?

- How are the scientific laws governing changes of matter and energy from one form to another related to resource use and environmental disruption?

3-1 SCIENCE, TECHNOLOGY, ENVIRONMENTAL SCIENCE, AND CRITICAL THINKING

What Is Science and What Do Scientists Do?
Which of the following statements is true?

- Science emphasizes logic over creativity, imagination, and intuition.

- Science establishes absolute truth or proof about nature.

- Science has a method—a how-to scheme—for learning about nature.

The answer is that all of them are false or mostly false. Let's see why.

Science is an attempt to discover order in nature and to use that knowledge to make predictions about what should happen in nature. Science is based on the assumption that there is discoverable order in nature. As Albert Einstein once said, "The whole of science is nothing more than a refinement of everyday thinking." Figure 3-2 summarizes the more systematic version of the everyday thinking process used by scientists.

The first thing scientists must do is ask a question or identify a problem to be investigated. Then scientists working on this problem collect **scientific data**, or facts, by making observations and taking measurements. The procedure a scientist uses to study some phenomenon under known conditions is called an **experiment**. Some experiments are conducted in the laboratory, but others are carried out in nature. The resulting scientific data or facts must be verified or confirmed by repeated observations and measurements, ideally by several different investigators.

The primary goal of science is not facts themselves, but a new idea, principle, or model that connects and explains certain facts and leads to useful predictions about what should happen in nature. Scientists working on a particular problem try to come up with a variety of possible explanations, or **scientific hypotheses**, of what they (or other scientists) observe in nature.

To be accepted, a scientific hypothesis not only must explain scientific data and phenomena but also should make predictions that can be used to test the validity of the hypothesis. Once a scientific hypothesis is invented, experiments are conducted (and repeated to be sure they are reproducible) to test

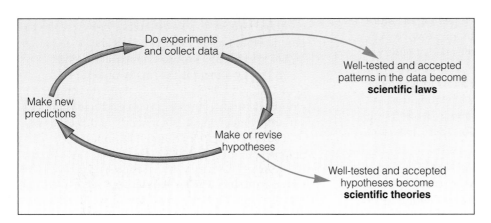

Figure 3-2 What scientists do: a summary of the scientific process, a form of critical thinking. Facts (data) are gathered and verified by repeated experiments, data are analyzed to see whether there is a consistent pattern of behavior that can be summarized as a scientific law, hypotheses are proposed to explain the data, and deductions or predictions are made and tested to evaluate each hypothesis. A hypothesis that is supported by a great deal of evidence and widely accepted by the scientific community becomes a scientific theory.

the deductions or predictions. Experiments can eliminate (disprove) various hypotheses, but they can never prove that any hypothesis is the best (most useful) or the only explanation. All scientists can say is that experiments support the validity and predictions of a particular hypothesis.

One method scientists use to test a hypothesis is to develop a **model**, an approximate representation or simulation of a system being studied. There are many types of models: mental, conceptual, graphic, physical, and mathematical.

Scientists are skeptics; they want lots of evidence before they are willing to accept the accuracy of any data or the usefulness of a particular hypothesis or model. Usually a succession of scientists working on the same problem—often working around the world—subject each other's data, hypotheses, and models to careful scrutiny. The process takes this form:

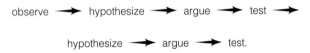

It continues until some general consensus or agreement is reached.

If many experiments by different scientists support a particular hypothesis, it becomes a **scientific theory**: an idea, principle, or model that usually ties together and explains many facts that previously appeared to be unrelated and is supported by a great deal of evidence.

Nonscientists often use the word *theory* incorrectly when they mean to refer to a scientific hypothesis, a tentative explanation that needs further evaluation. The statement, "Oh, that's just a theory," made in everyday conversation implies a lack of knowledge and careful testing—the opposite of the scientific meaning of the word. To scientists, theories are not to be taken lightly. They are ideas or principles stated with a high degree of certainty because they are supported by a great deal of evidence.

Another important result of science is a **scientific law**: a description of what we find happening in nature over and over in the same way, without known exception. For example, after making thousands of observations and measurements over many decades, scientists discovered what is called the **second law of energy** or **thermodynamics**. One simple way of stating this law is that heat always flows spontaneously from hot to cold—something you learned the first time you touched a hot object.

How Do Scientists Learn About Nature?

We often hear about *the* scientific method, but in reality there are many **scientific methods**: ways scientists gather data and formulate and test scientific hypotheses, models, theories, and laws (Figure 3-2). Instead of being a recipe, each scientific method involves trying to answer a set of questions, with no particular guidelines for answering them:

- What is the question about nature to be answered?

- What relevant facts and data are already known?

- What new data (observations and measurements) should be collected, and how should this be done?

- After the data are collected, can they be used to formulate a scientific law?

- How can a hypothesis be invented that explains the data and predicts new facts? Is this the simplest and only reasonable hypothesis?

- What new experiments can be done to test the hypothesis (and modify it if necessary) so it can become a scientific theory?

The scientific process requires not only logical reasoning, but also imagination, creativity, and intuition. According to physicist Albert Einstein, "There is no completely logical way to a new scientific idea." Intuition, imagination, and creativity are as important in science as they are in poetry, art, music, and other great adventures of the human spirit that awaken us to the wonder, mystery, and beauty of life, the earth, and the universe.

Most processes or parts of nature that scientists seek to understand are influenced by a number of *variables* or *factors*. One way scientists test a hypothesis about the effects of a particular variable is to conduct a *controlled experiment*. This is done by setting up two groups: an *experimental group*, in which the chosen variable is changed in a known way, and a *control group*, in which the chosen variable is not changed. The experiment is designed so that all components of each group are as identical as possible and experience the same conditions, except for the single factor being varied in the experimental group. If the experiment is designed properly, any difference between the two groups should result from a variable that was changed in the experimental group (Connections, right).

A basic problem is that many of nature's components and processes, especially those investigated by environmental scientists, involve a huge number of variables interacting in often poorly understood ways. In such cases, it is very difficult or impossible to carry out meaningful controlled experiments.

Can Scientists Prove Anything? In a word, no. *Scientists can disprove things, but they can never prove anything.* To win us over to their particular viewpoint,

Q: Do scientists establish absolute proof or truth?

people often say that something has or has not been "scientifically proven." Either they don't understand the nature of science, or they are trying to mislead us by falsely implying that science yields absolute proof or certainty.

Instead of certainty or absolute truth or proof, scientists speak of degrees of probability or uncertainty. Scientists might predict that if we do a certain thing, then (based on the data, hypotheses, theories, and laws underlying the processes involved) there is a high, moderate, or low probability that a given result will happen.

The goal of the rigorous scientific process is to reduce the degree of uncertainty as much as possible. However, the more complex the system being studied, the greater the degree of uncertainty or unpredictability about its behavior.

Are Scientists Always Objective? Scientists are human beings who have conscious and unconscious values, biases, and beliefs that can influence how they design and interpret data and the reliability they attach to various scientific hypotheses and theories. Part of the critical thinking process for scientists and those evaluating the results of scientific research is to understand that scientists do have biases and to try to identify them.

The idea that scientists cannot always reach the goal of complete objectivity and that the results of science are not entirely value-free should not be taken to mean that shoddy thinking is acceptable in science. The standards of evidence in science are very high, and scientific results are published so that other scientists can expose errors, challenge fuzzy or biased thinking, and discover rare but occasional cheating (faking scientific results).

Despite its limitations, science is the best way we have come up with to get reliable knowledge about how nature works. It has revealed many of nature's secrets and relationships, sparked our imagination and sense of wonder and awe, challenged old ways of thinking, and demonstrated the fuzziness of many of our worldviews.

How Does Frontier Science Differ from Consensus Science? News reports often focus on new scientific "breakthroughs" and on disputes among scientists over the validity of preliminary (untested) data, hypotheses, and models (which are by definition tentative). This aspect of science, controversial because it has not been widely tested and accepted, is called **frontier science**.

By contrast, **consensus science** consists of data, theories, and laws that are widely accepted by scientists considered experts in the field involved. This

What's Harming the Robins?

CONNECTIONS

Suppose a scientist has observed an abnormality in the growth of robin embryos in a certain area. She knows that the area has been sprayed with a pesticide and suspects that the chemical may be causing the abnormalities she has observed.

To test this hypothesis, the scientist carries out a *controlled experiment*. She maintains two groups of robin embryos of the same age in the laboratory. Each group is exposed to exactly the same conditions of light, temperature, food supply, and so on, except that the embryos in the experimental group are exposed to a known amount of the pesticide in question.

The embryos in both groups are then examined over an identical period of time for the abnormality. If there is a statistically significantly larger number of the abnormalities in the experimental group than in the control group, then the results support the idea that the pesticide is the culprit.

To be sure there were no errors in the procedure, the experiment should be repeated several times by the original researcher and ideally by one or more other scientists. Conducting such replications is a standard procedure in science.

Critical Thinking

Can you find flaws in this experiment that might lead you to question the scientist's conclusions? (*Hint*: What other factors in nature—not the laboratory—and in the embryos themselves could possibly explain the results?)

aspect of science is very reliable but is rarely considered newsworthy. One way to find out what scientists generally agree on is to seek out reports by scientific bodies such as the U.S. National Academy of Sciences and the British Royal Society that attempt to summarize consensus among experts in key areas of science.

To give a sense of balance and fairness, members of the media often quote one or more scientists in the very small number of scientists who criticize the consensus view of the vast majority of scientists in a particular field. Instead of balance, this usually leads to a biased presentation based on giving a minority view about the same weight as the vast majority of scientists holding a consensus view (see Guest Essay, pp. 52–53).

Some fundamental advances in scientific knowledge are made when a creative scientist proposes and uses experiments to establish a new hypothesis that eventually alters the consensus view about a particular topic. However, merely criticizing the consensus view without coming up with a new view and using scientific experiments to establish that view is not very useful and often misleads the public. Science thrives on new ideas but demands that they be backed up by hard evidence before they can be widely accepted as part of consensus science.

What Is Technology? Technology is the creation of new products and processes intended to improve our efficiency, our chances for survival, our comfort level, and our quality of life. In many cases, technology develops from known scientific laws and theories. Scientists invented the laser, for example, by applying knowledge about the internal structure of atoms. Applied scientific knowledge about chemistry has given us nylon, pesticides, laundry detergents, pollution control devices, and countless other products. Applications of theories in nuclear physics led to nuclear weapons and nuclear power plants.

However, some technologies arose long before anyone understood the underlying scientific principles. For example, aspirin, extracted from the bark of a willow tree, relieved pain and fever long before anyone found out how it did so. Similarly, photography was invented by people who had no inkling of its chemistry, and farmers crossbred new strains of crops and livestock long before biologists understood the principles of genetics.

Science and technology usually differ in the way the information and ideas they produce are shared. Many of the results of scientific research are published and distributed freely to be tested, challenged, verified, or modified. In contrast, many technological discoveries are kept secret until the new process or product is patented. However, the basis of some technology is published in journals and enjoys the same kind of public distribution and peer review as science.

What Is Environmental Science and What Are Its Limitations? Environmental science is the study of how we and other species interact with one another and with the nonliving environment (matter and energy). It is a *physical and social science* that integrates knowledge from a wide range of disciplines including physics, chemistry, biology (especially ecology), geology, meteorology, geography, resource technology and engineering, resource conservation and management, demography (the study of population dynamics), economics, politics, sociology, psychology, and ethics. In other words, it is a study of how the parts of nature and human societies operate and interact—a study of *connections* and *interactions* (see inside back cover of this book).

There is controversy over some of the knowledge provided by environmental science, for much of it falls into the realm of frontier science. One problem involves *arguments over the validity of data*. There is no way to measure accurately how many metric tons of soil are eroded worldwide, how many hectares of tropical forest are cut, how many species become extinct, or how many metric tons of certain pollutants are emitted into the atmosphere or aquatic systems each year.

We may legitimately argue over the numbers, but the point environmental scientists want to make is that the trends in these phenomena are significant enough to be evaluated and addressed. Such environmental data should not be dismissed because they are "only estimates" (which are all we can ever have). However, this does not relieve investigators from the responsibility of getting the best estimates possible and pointing out that they *are* estimates.

Another limitation is that *most environmental problems involve so many variables and such complex interactions that we don't have enough information or sufficiently sophisticated models to aid in understanding them very well.* Much progress has been made during the past 50 years (and especially the past 25 years), but we still know much too little about how the earth works, about its current state of environmental health, and about the effects of our activities on its life-support systems.

Reputable scientists in a field may state contradictory opinions about the meaning of the data from experiments (especially frontier scientific results) and the validity of various hypotheses. This is especially true in environmental science, where hypotheses and predictions are almost always based on inadequate data and understanding of the complex systems involved.

Because environmental problems won't go away, at some point we must evaluate the available (but always inadequate) information and make political and economic decisions. Without sufficient scientific evidence, such decisions are often based primarily on individual and societal values, which is why environmental aspects of differing worldviews are at the heart of most environmental controversies (Sections 2-2 and 2-3).

Understanding and evaluating our worldviews and trying to uncover the biases we all have is a prerequisite to critical thinking (Guest Essay, pp. 52–53). For as American psychologist William James said, "a great many people think they are thinking when they are merely rearranging their prejudices."

Q: On what three factors does life on earth depend?

📑 3-2 MODELS AND BEHAVIOR OF SYSTEMS

Why Are Models of Complex Systems Useful?

A **system** is a set of components that function and interact in some regular and theoretically predictable manner. The environment consists of a number of systems, each of which consists of various interacting living and nonliving things.

Over time, people have learned the value of using models as approximate representations or simulations of real systems to find out how systems work and to evaluate which ideas or hypotheses work.

Some of the most powerful technologies invented by humans are mathematical models, which are used to supplement our mental models. *Mathematical models* are equations that help us perceive and predict various things. Like mental models, mathematical models are imperfect approximations of reality. They tend to make predictions ranging from fairly accurate to very accurate, depending on the model and the accuracy of the data used to formulate the model.

The basic process for developing mathematical models is essentially trial and error, similar to the processes depicted in Figure 3-2. Making a mathematical model usually requires going many times through three familiar steps: **(1)** Make a guess and write down some equations, **(2)** compute the predictions implied by the equations, and **(3)** compare the predictions with observations, the predictions of mental models, and existing experimental data, and scientific hypotheses, laws, and theories.

Mathematical models are important because they can give us improved perceptions and predictions, especially concerning matters for which our mental models are weak. Research has shown that people's mental models tend to be especially unreliable **(1)** when there are many interacting variables, **(2)** when we attempt to extrapolate from too few experiences to a general case, **(3)** when consequences follow actions only after long delays, **(4)** when the consequences of actions lead to other consequences, **(5)** when responses are especially variable from one time to the next, and **(6)** when controlled experiments (Connections, p. 49) are impossible, too slow, or too expensive to conduct. Under such conditions, a good mathematical model can do better than most mental models.

After building and testing a model, scientists apply it to a useful purpose by predicting what will happen under a variety of alternative conditions. In effect, they use mathematical models to answer *if–then* questions: "*If* we do such and such, *then* what is likely to happen now and in the future?"

What Are Some Basic Components and Behaviors of System Models?

Any system being studied or modeled has one or more **inputs** of things such as matter, energy, or information.

Inputs flow through a system at a certain rate. Such **flows** or **throughputs** of matter, energy, or information through a system are represented by arrows. Forms of matter, energy, or information flowing out of a system are called **outputs** and end up in *sinks* in the environment. Examples of such sinks are the atmosphere, bodies of water, underground water, soil, and land surfaces. Figure 3-3 is a generalized diagram of inputs, throughputs, and outputs for an economic system.

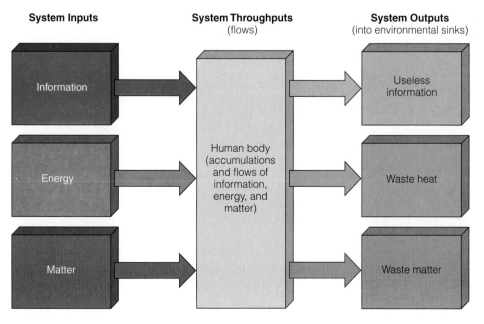

Figure 3-3 Generalized diagram of inputs, throughputs (flows), and outputs of information, matter, and energy in a system.

System Inputs — Information, Energy, Matter

System Throughputs (flows) — Human body (accumulations and flows of information, energy, and matter)

System Outputs (into environmental sinks) — Useless information, Waste heat, Waste matter

---

A: One-way flow of energy from the sun, cycling of matter, and gravity

Critical Thinking and Environmental Studies

Jane Heinze-Fry

GUEST ESSAY

Jane Heinze-Fry has a Ph.D. in environmental education and is currently the author of the instructor's manual and several other supplements that accompany this textbook. She is author of Green Lives, Green Campuses *and* Critical Thinking and Environmental Studies: A Beginner's Guide *(with G. Tyler Miller as coauthor). Previously, she taught and directed environmental studies at Sweet Briar College. She also taught biology to students at the junior high, high school, and college levels. Her interdisciplinary orientation is reflected in her concept maps, including the ones inside the back cover and in Appendix 4 of this textbook.*

Defining *critical thinking* is not easy. Critical thinking has come to mean different things to different people. In my opinion, *critical thinking* is thinking which moves learners **(1)** to more meaningful learning instead of rote learning; **(2)** to higher levels of learning (application, analysis, synthesis, and evaluation as opposed to knowledge and comprehension); **(3)** to apply concepts and principles to real-world experience and situations; **(4)** to make judgments about knowledge and values claims; and **(5)** to enhance problem-solving skills.

Learners engaged in critical thinking make efforts to

- Clarify their understanding of new concepts
- Connect new knowledge to prior knowledge and experience
- Make judgments about claims made by people in scientific articles, books, advertisements
- Develop creative thinking by learning how to exercise both right- and left- brain capabilities
- Relate what they have learned to your own life experiences
- Understand and evaluate their environmental worldview
- Take positions on issues
- Engage in problem-solving processes

Whenever we are faced with new information, we need to evaluate it by using critical thinking. Do we believe the information or not and why? Do the claims seem reasonable or exaggerated? Here are some rules for evaluating evidence and claims:

1. Gather all of the information you can.

2. Be sure that all key terms and concepts are defined and that you understand these definitions.

3. Question how the information (data) was obtained.

- Were the studies involved well-designed and carried out?
- Was there an experimental group and a control group? Were the control and experimental groups treated identically except for the variable changed in the experimental group?
- Did the investigators repeat their experiments several times and get essentially the same results?
- Were the results verified by one or more other investigators?

4. Question the conclusions derived from the data.

- Do the data support the claims, conclusions, and predictions?
- Are there other possible interpretations? Are there more reasonable interpretations?
- Do the conclusions involve a correlation or *apparent* connection between two or more variables, or do they imply a strong *cause and effect* relationship between such variables?
- Are the conclusions based on the results of original research by experts in the field involved, or are they conclusions drawn by reporters or scientists in other fields?

5. Try to determine the assumptions and biases of the investigators and then question them.

- Do the investigators have a monetary or political advantage in the outcome of the investigation or issue involved?
- What are the underlying basic assumptions or worldviews of the investigators? Would investigators with different basic assumptions or worldviews take the same data and come to different conclusions?

A **feedback loop** occurs when one change leads to some other change, which then eventually either reinforces or slows the original change. Feedback loops determine how things happen over time.

Feedback loops occur when an output of matter, energy, or information is fed back into the system as an input (Figure 3-3). For example, recycling aluminum cans involves melting aluminum and feeding it back into an economic system to make new aluminum products. This feedback loop of matter reduces the need to find, extract, and process virgin aluminum ore. It also reduces the flow of waste matter (discarded aluminum cans) into the environment.

Q: What is the sun's source of energy?

6. Expect and tolerate uncertainty. Recognize that science is a dynamic process that provides only a certain degree of probability or certainty and that the more complex the system or process being investigated the greater the degree of uncertainty.

- Are the data, claims, and conclusions based on the tentative results of *frontier science* or the more reliable and widely accepted results of *consensus science?*

7. Look at the big picture (think holistically).

- How do the results and conclusions fit into the whole system (the earth, ecosystem, economy, political system) involved?

- What additional data and experiments are needed to relate the results to the whole system?

8. Based on these steps, take a position by either rejecting or conditionally accepting the claims.

- Reject claims not based on any evidence, based on insufficient evidence, or on evidence coming from questionable sources.

- If evidence does not support a claim, reject it and state the conclusion you would draw.

- If the evidence supports the claims, conditionally accept the claims with the understanding that your support may change if new evidence disproves the claim.

In addition to these rules to enhance making reasoned judgments, there are a number of other strategies to improve your critical thinking skills. Some strategies focus on improving thinking; others on attitudes and values; others on actions.

Thinking strategies focus on learning how to use your left brain (which is good at logic and analyzing) and your right brain (which is good at visualizing and creating). They include constructing models, clarifying concepts, brainstorming, defining problems, creating alternative solutions, visualizing future possibilities, and exercising your creative brain paths.

Attitudes and values attitudes focus on establishing an ecological identity, reflecting on the interaction of lifestyle and the environment, understanding and evaluating your worldview, and using values analysis for evaluating proposed environmental policies and making personal environmental decisions. *Action strategies* focus on problem solving. They include techniques for visualizing problems, redefining problems, considering alternative solutions, creating plans of action, and developing strategies for implementing action plans.

In the environmental course you are taking, there are many opportunities to develop your critical thinking skills. Your textbook offers critical thinking questions at the end of chapters, Guest Essays, and most boxed material. If your course uses the supplement *Critical Thinking and Environmental Studies: A Beginner's Guide*, you will learn the critical thinking strategies mentioned above. Courses using the supplement *Green Lives, Green Campuses* can offer you additional opportunities to use critical thinking in evaluating your environmental lifestyle and in making an environmental audit of your campus.

Learning how to think critically is essential in helping you evaluate the validity and usefulness of what you read in newspapers, magazines, and books (such as this textbook), what you hear in lectures and speeches, and what you see and hear on the news and in advertisements.

Critical Thinking

1. Can you come up with an example where critical thinking has helped you make a major change in one or more of your beliefs or helped you make an important personal decision? Can you think of a decision you made that may have come out better if you had used critical thinking skills such as those discussed in this essay?

2. Rote learning often involves the "memorize and spit back" strategy. Meaningful learning (including critical thinking) goes far beyond memorization and requires us to evaluate the validity of what we learn. Currently, about what percentage of your learning involves rote learning and what percentage involves critical thinking as defined at the beginning of this essay?

Feedback loops can be either positive or negative. A **positive feedback loop** is a runaway cycle in which a change in a certain direction provides information that causes a system to change further in the same direction. The environment is full of positive feedback loops, some desirable and some undesirable. For example, as long as there are more human births than deaths on this planet, the human population will grow exponentially. This form of positive feedback can be viewed as desirable or undesirable depending on whether one sees population growth as an economic benefit (more workers and consumers) or an environmental drawback (more crowding, pollution, and resource depletion and degradation).

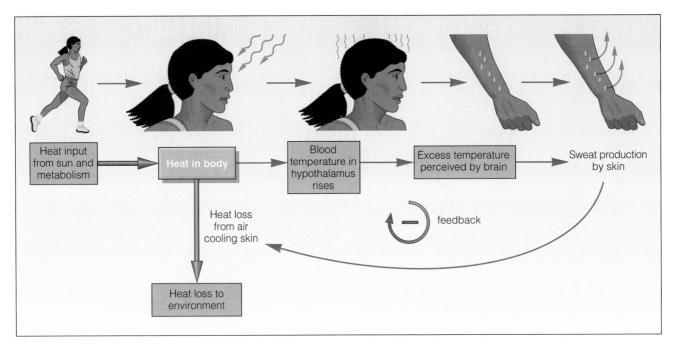

Figure 3-4 A simple negative feedback loop: keeping cool on a hot day. The temperature of the human body is controlled by a system that uses a negative feedback loop to counteract external and internal heat sources and maintain homeostasis. As the brain senses a rise in body temperature, sweating is increased. As the sweat evaporates, it removes heat from the body. Skin sensors detect the cooling and feed this information back to the brain. This negative feedback loop then stops sweat production.

In a **negative feedback loop**, one change leads to a lessening of that change. To survive, each of us must maintain our body temperature, for example, within a certain range, regardless of whether the temperature outside is steamy or freezing (Figure 3-4). This phenomenon is called **homeostasis**: the maintenance of favorable internal conditions despite fluctuations in external conditions. Homeostatic systems consist of one or more negative feedback loops that help maintain constant internal conditions when changes occur.

Most systems, such as the temperature regulating system of your body, contain one or a series of *coupled positive and negative feedback loops* (Figure 3-5). Normally a negative feedback regulates your body temperature (Figure 3-4). However, if your body temperature exceeds 42°C (108°F), your built-in temperature control system breaks down as your body's metabolism (the heat-generating chemical reactions necessary for life) produces more heat than your sweat-dampened skin can get rid of. Then a positive feedback loop caused by overloading the system (Figure 3-5) overwhelms the negative feedback loop (Figure 3-4). These conditions produce a net gain in body heat, which speeds up your metabolism even further, producing even more body heat, and so on, until death from heatstroke occurs.

3-3 MATTER: FORMS, STRUCTURE, AND QUALITY

What Are Nature's Building Blocks? Matter is anything that has mass (the amount of material in an object) and takes up space. Matter includes the solids, liquids, and gases around us and within us. Matter is found in two *chemical forms*: **elements** (the distinctive building blocks of matter that make up every material substance) and **compounds** (two or more different elements held together in fixed proportions by attractive forces called *chemical bonds*). Various elements, compounds, or both can be found together in **mixtures**.

All matter is built from the 112 known chemical elements. (Ninety-two of them occur naturally and the other 20 have been synthesized in laboratories.) Each has properties that make it unique, just as each of the 26 letters in the alphabet is different from the others. To simplify things, chemists represent each element by a one- or two-letter symbol: hydrogen (H), carbon (C), oxygen (O), nitrogen (N), phosphorus (P), sulfur (S), chlorine (Cl), fluorine (F), bromine (Br), sodium (Na), calcium (Ca), lead (Pb), mercury (Hg), and uranium (U), to mention but a few.

If we had a supermicroscope capable of looking at individual elements and compounds, we could see that they are made up of three types of building

Q: How do most producer organisms get the nutrients they need?

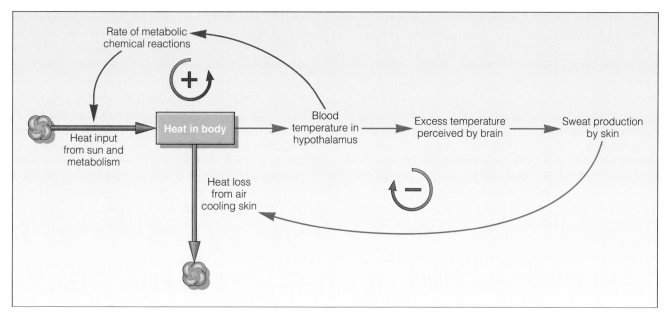

Figure 3-5 Coupled negative and positive feedback loops involved in temperature control of the human body. Homeostasis works in a limited range only: Above a certain body temperature, metabolic rates get out of control and generate large amounts of heat. This positive (runaway) feedback loop generates more heat than the negative feedback loop can get rid of and body temperature increases out of control, resulting in death.

blocks: **atoms** (the smallest unit of matter that is unique to a particular element), **ions** (electrically charged atoms or combinations of atoms), and **molecules** (combinations of two or more atoms of the same or different elements held together by chemical bonds). Because ions and molecules are formed from atoms, *atoms are the ultimate building blocks for all matter*.

Some elements are found in nature as molecules. Examples are nitrogen and oxygen, which together make up about 99% of the volume of the air we breathe. Two atoms of nitrogen (N) combine to form a nitrogen gas molecule, with the shorthand formula N_2 (read as "N-two"). The subscript after the element's symbol indicates the number of atoms of that element in a molecule. Similarly, most of the oxygen gas in the atmosphere exists as O_2 (read as "O-two") molecules. A small amount of oxygen, found mostly in the second layer of the atmosphere (stratosphere), exists as O_3 (read as "O-three") molecules; this gaseous form of oxygen is called *ozone*.

Elements can combine to form an almost limitless number of compounds, just as the letters of the alphabet can be combined to form almost a million English words. So far, chemists have identified more than 10 million compounds.

Matter is also found in three *physical states*: solid, liquid, and gas. For example, water exists as ice, liquid water, and water vapor, depending on its temperature and pressure. The three physical states of matter differ in the relative spacing and orderliness of its atoms, ions, or molecules, with solids having the most compact and orderly arrangement and gases the least compact and orderly arrangement.

What Are Atoms and Ions? If we increased the magnification of our supermicroscope, we would find that each different type of atom contains a certain number of *subatomic particles*. The main building blocks of an atom are positively charged **protons** (p), uncharged **neutrons** (n), and negatively charged **electrons** (e). Protons and neutrons are the heaviest particles, with electrons having an almost negligible mass compared to them. Other subatomic particles have been identified in recent years, but they need not concern us here.

Each atom consists of an extremely small center, or **nucleus**, containing protons and neutrons, and one or more electrons in rapid motion somewhere outside the nucleus. We can describe the locations of electrons only in terms of their probability of being at any given place outside the nucleus. This is analogous to saying that a certain number of tiny gnats are found somewhere in a cloud, without being able to identify their exact positions. Each element has its own specific **atomic number**, equal to the number of protons

A: They produce them through photosynthesis

CHAPTER 3 **55**

in the nucleus of each of its atoms. The simplest element, hydrogen (H), has only 1 proton in its nucleus, so its atomic number is 1. Carbon (C), with 6 protons, has an atomic number of 6, whereas uranium (U), a much larger atom, has 92 protons and an atomic number of 92.

Each atom has an equal number of positively charged protons and negatively charged electrons. Because these electrical charges cancel one another, *the atom as a whole has no net electrical charge.* For example, an uncharged hydrogen atom has one positively charged proton in its nucleus and one negatively charged electron outside its nucleus. Similarly, each atom of uranium has 92 protons in its nucleus and 92 electrons outside.

Because electrons have so little mass compared with the mass of a proton or a neutron, *most of an atom's mass is concentrated in its nucleus.* We describe the mass of an atom in terms of its **mass number**: the total number of neutrons and protons in its nucleus. A hydrogen atom with 1 proton and no neutrons in its nucleus has a mass number of 1, and an atom of uranium with 92 protons and 143 neutrons in its nucleus has a mass number of 235.

Although all atoms of an element have the same number of protons in their nuclei, they may have different numbers of uncharged neutrons in their nuclei, and thus may have different mass numbers. Various forms of an element having the same atomic number but a different mass number are called **isotopes** of that element. Isotopes are identified by attaching their mass numbers to the name or symbol of the element. For example, hydrogen has three isotopes: hydrogen-1 (H-1), hydrogen-2 (H-2, or deuterium), and hydrogen-3

(H-3, or tritium). A natural sample of an element contains a mixture of its isotopes in a fixed proportion or percentage abundance by weight (Figure 3-6).

Atoms of some elements can lose or gain one or more electrons to form **ions**: atoms or groups of atoms with one or more net positive (+) or negative (−) electrical charges. For example, an atom of sodium (Na) (atomic number 11) with 11 positively charged protons and 11 negatively charged electrons can lose one of its electrons. It then becomes a sodium ion with a positive charge of 1 (Na^+) because it now has 11 positive charges (protons) but only 10 negative charges (electrons). An atom of chlorine (Cl) (with an atomic number of 17) can gain an electron and become a chlorine ion with a negative charge of 1 (Cl^-), because it then has 17 positively charged protons and 18 negatively charged electrons.

The number of positive or negative charges on an ion is shown as a superscript after the symbol for an atom or a group of atoms. Examples of other positive ions are calcium ions (Ca^{2+}) and ammonium ions (NH_4^+); other common negative ions are nitrate ions (NO_3^-), sulfate ions (SO_4^{2-}), and phosphate ions(PO_4^{3-}).

What Holds the Atoms and Ions in Compounds Together? Most matter exists as compounds. Chemists use a shorthand *chemical formula* to show the number of atoms (or ions) of each type in a compound. The formula contains the symbols for each of the elements present and uses subscripts to represent the number of atoms or ions of each element in the compound's basic structural unit. Compounds made up of oppositely charged ions are called *ionic compounds*, and those made

Figure 3-6 Isotopes of hydrogen and uranium. All isotopes of hydrogen have an atomic number of 1 because each has one proton in its nucleus; similarly, all uranium isotopes have an atomic number of 92. However, each isotope of these elements has a different mass number because its nucleus contains a different number of neutrons. Figures in parentheses indicate the percentage abundance by weight of each isotope in a natural sample of the element.

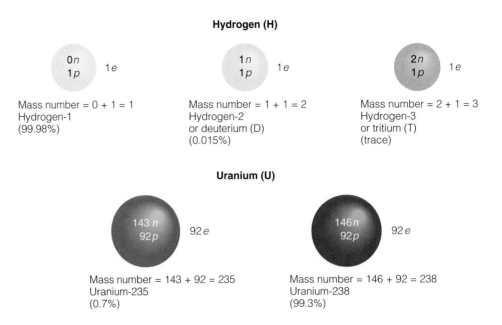

Hydrogen (H)

$0n$ $1p$ $1e$
Mass number = 0 + 1 = 1
Hydrogen-1
(99.98%)

$1n$ $1p$ $1e$
Mass number = 1 + 1 = 2
Hydrogen-2
or deuterium (D)
(0.015%)

$2n$ $1p$ $1e$
Mass number = 2 + 1 = 3
Hydrogen-3
or tritium (T)
(trace)

Uranium (U)

$143n$ $92p$ $92e$
Mass number = 143 + 92 = 235
Uranium-235
(0.7%)

$146n$ $92p$ $92e$
Mass number = 146 + 92 = 238
Uranium-238
(99.3%)

Q: How much of the high-quality energy is transferred from one trophic level to another in a food chain or web?

up of molecules of uncharged atoms are called *covalent* or *molecular compounds*.

Sodium chloride (table salt), an *ionic compound*, is represented by the formula NaCl. It consists of a network of oppositely charged ions (Na^+ and Cl^-) held together by the forces of attraction between oppositely electric charges. The strong forces of attraction between such oppositely charged ions are called *ionic bonds*.

Water, a *covalent* or *molecular compound*, consists of molecules made up of uncharged atoms of hydrogen (H) and oxygen (O). Each water molecule consists of two hydrogen atoms chemically bonded to an oxygen atom, yielding H_2O (read as "H-two-O") molecules. The bonds between the atoms in such molecules are called *covalent bonds* and are formed when the atoms in the molecule share one or more pairs of their electrons.

What Are Organic and Inorganic Compounds?

Table sugar, vitamins, plastics, aspirin, penicillin, and many other important materials have one thing in common: They are *organic compounds* containing carbon atoms combined with each other and with atoms of one or more other elements such as hydrogen, oxygen, nitrogen, sulfur, phosphorus, chlorine, and fluorine. Virtually all organic compounds are molecular compounds held together by covalent bonds. Organic compounds can be either natural or synthetic (such as plastics and many drugs made by humans).

Among the millions of known organic (carbon-based) compounds are

- *Hydrocarbons*: compounds of carbon and hydrogen atoms. An example is methane (CH_4), the main component of natural gas.

- *Chlorinated hydrocarbons*: compounds of carbon, hydrogen, and chlorine atoms. Examples are DDT ($C_{14}H_9Cl_5$, an insecticide) and toxic PCBs (such as $C_{12}H_5Cl_5$), oily compounds used as insulating materials in electric transformers.

- *Chlorofluorocarbons* (CFCs): compounds of carbon, chlorine, and fluorine atoms. An example is Freon-12 (CCl_2F_2), until recently widely used as a coolant in refrigerators and air conditioners, as an aerosol propellant, and as a foaming agent for making some plastics.

- *Simple carbohydrates* (simple sugars): certain types of compounds of carbon, hydrogen, and oxygen atoms. An example is glucose ($C_6H_{12}O_6$), which most plants and animals break down in their cells to obtain energy.

Larger and more complex organic compounds, called *polymers*, consist of a number of basic structural or molecular units (*monomers*) linked by chemical bonds, somewhat like cars linked in a freight train. The three major types of organic polymers are complex carbohydrates, proteins, and nucleic acids.

Complex carbohydrates are made by linking a number of simple carbohydrate molecules such as glucose ($C_6H_{12}O_6$). Examples are the complex starches in rice and potatoes and cellulose found in the walls around plant cells.

Proteins are produced in cells by linking different sequences of about 20 different monomers known as *alpha-amino acids*, whose number and sequence in each protein are specified by the genetic code found in DNA molecules in an organism's cells. Most animals, including humans, can make about 10 of these alpha-amino acids in their cells. Sufficient quantities of the other 10, known as *essential alpha-amino acids*, must be obtained from food to prevent protein deficiency diseases. Various protein molecules important for cell structure and energy storage act as *enzymes* to control the rate at which all of the chemical reactions in a cell (cellular metabolism) take place. From only about 20 alpha-amino acid molecules, earth's lifeforms can make tens of millions of different protein molecules.

Nucleic acids are made by linking hundreds to thousands of five different types of monomers, called *nucleotides*. Each nucleotide consists of a phosphate group, a sugar molecule containing five carbon atoms (deoxyribose in DNA molecules and ribose in RNA molecules), and one of four different nucleotide bases (represented by *A, G, C,* and *T*—the first letter in each of their names). In the cells of living organisms, these nucleotide units combine in different numbers and sequences to form nucleic acids such as various types of DNA and RNA.

Bonds formed between parts of the four nucleotides in DNA hold two DNA strands together like a spiral staircase, forming a double helix. DNA molecules can unwind and replicate themselves. They contain the hereditary instructions for assembling new cells and for assembling the proteins each cell needs to survive and reproduce. Each sequence of three nucleotides in a DNA molecule carries the code or instruction for making a specific alpha-amino acid; various alpha-amino acids are then linked in cells to form specific protein molecules. RNA molecules carry instructions (provided by DNA molecules) for producing proteins within cells.

Genes consist of specific sequences of nucleotides in a DNA molecule. Each gene carries the codes (each consisting of three nucleotides) required to make various proteins. These coded units of genetic information about specific traits are passed on from parents to offspring during reproduction. Occasionally one or more of the nucleotide bases in a gene sequence are deleted, added, or replaced. Such changes, called gene **mutations**,

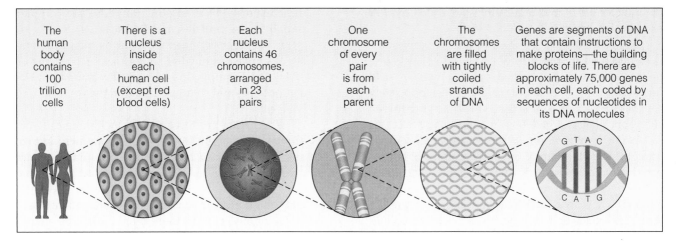

| The human body contains 100 trillion cells | There is a nucleus inside each human cell (except red blood cells) | Each nucleus contains 46 chromosomes, arranged in 23 pairs | One chromosome of every pair is from each parent | The chromosomes are filled with tightly coiled strands of DNA | Genes are segments of DNA that contain instructions to make proteins—the building blocks of life. There are approximately 75,000 genes in each cell, each coded by sequences of nucleotides in its DNA molecules |

Figure 3-7 Relationships among cells, nuclei, chromosomes, DNA, and genes.

can be helpful or harmful to an organism and its offspring, or not affect them at all.

Chromosomes are combinations of genes that make up a single DNA molecule, together with a number of proteins. Genetic information coded in your chromosomal DNA is what makes you different from an oak leaf, an alligator, or a flea, and from your parents as well. The relationships of genetic material to cells are depicted in Figure 3-7.

All other compounds are called *inorganic compounds.* Some of the inorganic compounds discussed in this book are sodium chloride (NaCl), water (H_2O), nitrous oxide (N_2O), nitric oxide (NO), carbon monoxide (CO), carbon dioxide (CO_2),* nitrogen dioxide (NO_2), sulfur dioxide (SO_2), ammonia (NH_3), hydrogen sulfide (H_2S), sulfuric acid (H_2SO_4), and nitric acid (HNO_3).

What Is Matter Quality? From a human standpoint, we can classify matter according to its quality or usefulness to us. **Matter quality** is a measure of how useful a matter resource is, based on its availability and concentration. **High-quality matter** is organized, concentrated, and usually found near the earth's surface, and has great potential for use as a matter resource; **low-quality matter** is disorganized, dilute, and often deep underground or dispersed in the ocean or the atmosphere, and usually has little potential for use as a matter resource (Figure 3-8).

An aluminum can is a more concentrated, higher-quality form of aluminum than aluminum ore containing the same amount of aluminum. That's why it takes

---

* Classifying compounds as organic or inorganic is somewhat arbitrary. All organic compounds contain one or more carbon atoms, but CO and CO_2 are classified as inorganic compounds.

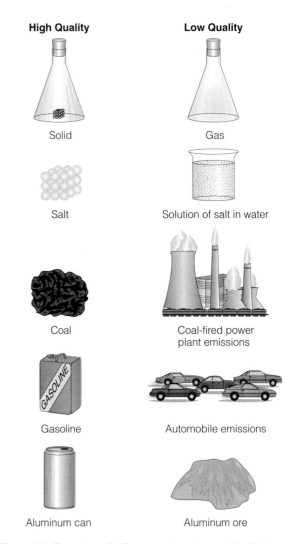

High Quality   **Low Quality**

Solid   Gas

Salt   Solution of salt in water

Coal   Coal-fired power plant emissions

Gasoline   Automobile emissions

Aluminum can   Aluminum ore

Figure 3-8 Examples of differences in matter quality. High-quality matter (left-hand column) is fairly easy to extract and is concentrated; low-quality matter (right-hand column) is more difficult to extract and is more dispersed than high-quality matter.

Q: What percentage of the world's net primary productivity on land is used, wasted, or destroyed by the human population?

less energy, water, and money to recycle an aluminum can than to make a new can from aluminum ore.

3-4 ENERGY: FORMS AND QUALITY

What Different Forms of Energy Do We Encounter? Energy is the capacity to do work and transfer heat. Work is performed when an object—be it a grain of sand, this book, or a giant boulder—is moved over some distance. Work, or matter movement, also is needed to boil water (to change it into the more dispersed and faster-moving water molecules in steam) or to burn natural gas to heat a house or cook food. Energy is also the heat that flows automatically from a hot object to a cold object when they come in contact. Touch a hot stove, and you experience this energy flow in a painful way.

Energy comes in many forms: light; heat; electricity; chemical energy stored in the chemical bonds in coal, sugar, and other materials; the mechanical energy of moving matter such as flowing water, wind (air masses), and joggers; and nuclear energy emitted from the nuclei of certain isotopes.

Scientists classify energy as either kinetic or potential. **Kinetic energy** is the energy that matter has because of its mass and its speed or velocity. It is energy in action or motion. Wind (a moving mass of air), flowing streams, falling rocks, heat flowing from a body at a high temperature to one at a lower temperature, electricity (flowing electrons), moving cars—all have kinetic energy.

Electromagnetic radiation is a form of kinetic energy consisting of a wide band or spectrum of electromagnetic waves that differ in wavelength (distance between successive peaks or troughs) and energy content (Figure 3-9). Examples are radio waves, TV waves, microwaves, infrared radiation, visible light, ultraviolet radiation, X rays, gamma rays, and cosmic rays. Infrared radiation is now being used to show us where energy is leaking out of our houses and buildings (Figure 3-1). Satellite scans use infrared radiation, ultraviolet radiation, and microwaves to show where plant life is found on land and at sea and to identify temperature changes in the upper layers of the oceans.

Cosmic rays, gamma rays, X rays, and ultraviolet radiation have enough energy to knock electrons from atoms and change them to positively charged ions. The resulting highly reactive electrons and ions can disrupt living cells, interfere with body processes, and cause many types of sickness, including various cancers. These potentially harmful forms of electromagnetic radiation are called **ionizing radiation**. The other forms of electromagnetic radiation do not contain enough energy to form ions and are called **nonionizing radiation**.

Heat is the total kinetic energy of all the moving atoms, ions, or molecules within a given substance, excluding the overall motion of the whole object. **Temperature** is a measure of the average speed of motion of the atoms, ions, or molecules in a sample of matter at a given moment. A substance can have a high heat content (much mass and many moving atoms, ions, or molecules) but a low temperature

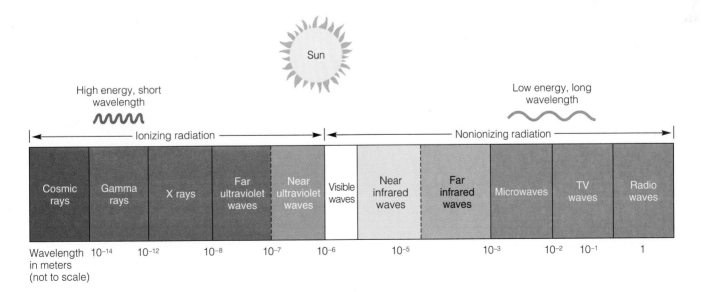

Figure 3-9 The electromagnetic spectrum: the range of electromagnetic waves, which differ in wavelength (distance between successive peaks or troughs) and energy content.

(low average molecular speed). For example, the total heat content of a lake is enormous, but its average temperature is low. Another substance can have a low heat content and a high temperature; a cup of hot coffee, for example, has a much lower heat content than a lake, but its temperature is much higher.

Potential energy is stored energy that is potentially available for use. A rock held in your hand, an unlit stick of dynamite, still water behind a dam, the gasoline in a car tank, and the nuclear energy stored in the nuclei of atoms all have potential energy because of their position or the position of their parts. Potential energy can be changed to kinetic energy. When you drop a rock, its potential energy changes into kinetic energy. When you burn gasoline in a car engine, the potential energy stored in the chemical bonds of its molecules changes into heat, light, and mechanical (kinetic) energy that propels the car.

What Is Energy Quality? From a human standpoint, the measure of an energy source's ability to do useful work is called its **energy quality** (Figure 3-10).

High-quality energy is organized or concentrated and can perform much useful work. Examples are electricity, coal, gasoline, concentrated sunlight, nuclei of uranium-235 used as fuel in nuclear power plants, and heat concentrated in small amounts of matter so that its temperature is high.

By contrast, **low-quality energy** is disorganized or dispersed and has little ability to do useful work. An example is heat dispersed in the moving molecules of a large amount of matter (such as the atmosphere or a large body of water) so that its temperature is low. Thus, even though the total amount of heat stored in the Atlantic Ocean is greater than the amount of high-quality chemical energy stored in all the oil deposits of Saudi Arabia, the ocean's heat is so widely dispersed that it can't be used to move things or to heat things to high temperatures.

We use energy to accomplish certain tasks, each requiring a certain minimum energy quality. It makes sense to match the quality of an energy source with the quality of energy needed to perform a particular task (Figure 3-10) because doing so saves energy and usually money.

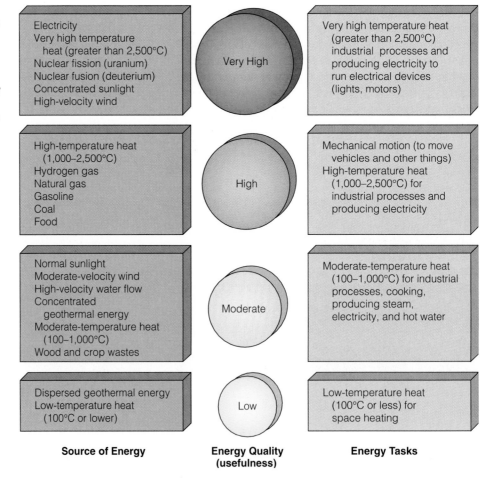

Figure 3-10 Categories of the quality (usefulness for performing various energy tasks) of different sources of energy. *High-quality energy* is concentrated and has great ability to perform useful work; *low-quality energy* is dispersed and has little ability to do useful work. To avoid unnecessary energy waste, it is best to match the quality of an energy source with the quality of energy needed to perform a task.

Electricity
Very high temperature heat (greater than 2,500°C)
Nuclear fission (uranium)
Nuclear fusion (deuterium)
Concentrated sunlight
High-velocity wind

Very High

Very high temperature heat (greater than 2,500°C) industrial processes and producing electricity to run electrical devices (lights, motors)

High-temperature heat (1,000–2,500°C)
Hydrogen gas
Natural gas
Gasoline
Coal
Food

High

Mechanical motion (to move vehicles and other things)
High-temperature heat (1,000–2,500°C) for industrial processes and producing electricity

Normal sunlight
Moderate-velocity wind
High-velocity water flow
Concentrated geothermal energy
Moderate-temperature heat (100–1,000°C)
Wood and crop wastes

Moderate

Moderate-temperature heat (100–1,000°C) for industrial processes, cooking, producing steam, electricity, and hot water

Dispersed geothermal energy
Low-temperature heat (100°C or lower)

Low

Low-temperature heat (100°C or less) for space heating

Source of Energy   **Energy Quality (usefulness)**   **Energy Tasks**

Q: What two gases make up 99% of the volume of air in the troposphere?

3-5 PHYSICAL AND CHEMICAL CHANGES AND THE LAW OF CONSERVATION OF MATTER

What Is the Difference Between a Physical and a Chemical Change? A **physical change** involves no change in chemical composition. Cutting a piece of aluminum foil into small pieces is one example. Changing a substance from one physical state to another is a second example. When solid water (ice) is melted or liquid water is boiled, none of the H_2O molecules involved are altered; instead, the molecules are organized in different spatial (physical) patterns. In a **chemical change** or **chemical reaction**, on the other hand, the chemical compositions of the elements or compounds are altered.

Chemists use shorthand chemical equations to represent what happens in a chemical reaction. A chemical equation shows the chemical formulas for the *reactants* (initial chemicals) and the *products* (chemicals produced), with an arrow placed between them. For example, when coal burns completely, the solid carbon (C) it contains combines with oxygen gas (O_2) from the atmosphere to form the gaseous compound carbon dioxide (CO_2):

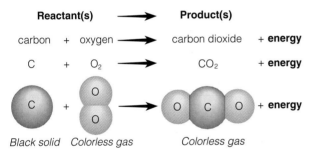

Energy is given off in this reaction, making coal a useful fuel. The reaction also shows how the complete burning of coal (or any of the carbon-containing compounds in wood, natural gas, oil, and gasoline) gives off carbon dioxide gas, which is a key gas that can lead to projected global warming of the lower atmosphere (troposphere).

Physical and chemical changes illustrate the connections between matter and energy: *All physical or chemical changes either require or give off energy.* For example, it takes energy to convert liquid water to steam. However, when the steam contacts a cooler object (such as your skin), it releases energy as it is converted back to liquid water, which is why steam can cause severe burns.

What Is the Law of Conservation of Matter? Why There Is No "Away" The earth loses some gaseous molecules to space, and it gains small amounts of matter from space, mostly in the form of occasional meteorites and cosmic dust. These losses and gains of matter are minute compared with the earth's total mass, which means that *the earth has essentially all the matter it will ever have.* In terms of matter, the earth is essentially a closed system (although in terms of energy it is an open system, receiving energy from the sun and emitting heat back into space). Over billions of years natural processes have evolved for continuously cycling key chemicals between living things and their nonliving environment (soil, air, and water).

People commonly talk about consuming or using up material resources, but the truth is that *we don't consume matter*—we only use some of the earth's resources for a while. We take materials from the earth, carry them to another part of the globe, and process them into products that are used and then discarded, burned, buried, reused, or recycled.

In so doing, *we may change various elements and compounds from one physical or chemical form to another, but in no physical and chemical change can we create or destroy any of the atoms involved.* All we can do is rearrange them into different spatial patterns (physical changes) or different combinations (chemical changes). The italicized statement above, based on many thousands of measurements, is known as the **law of conservation of matter**. In describing chemical reactions, chemists use a shorthand bookkeeping system to make sure atoms are neither created nor destroyed, as required by the law of conservation of matter (Spotlight, p. 62).

The law of conservation of matter means that there really is no "away" in "to throw away." *Everything we think we have thrown away is still here with us in one form or another.* We can collect dust and soot from the smokestacks of industrial plants, but these solid wastes must then be put somewhere. We can remove substances from polluted water at a sewage treatment plant, but the gooey sludge must either be burned (producing some air pollution), buried (possibly contaminating underground water supplies), or cleaned up and applied to the land as fertilizer (dangerous if the sludge contains nondegradable toxic metals such as lead and mercury). Banning use of the pesticide DDT in the United States but still selling it abroad means that it can return to the United States as DDT residues in imported coffee, fruit, and other foods or as fallout from air masses moved long distances by winds—something environmentalists call the *circle of poison.*

We can make the environment cleaner and convert some potentially harmful chemicals into less harmful physical or chemical forms. However, the law of conservation of matter means that we will always be faced with the problem of what to do with some quantity of wastes. By placing much greater emphasis on pollution prevention, waste reduction, and more efficient use of resources, we can greatly reduce the amount of wastes we add to the environment (Guest Essay, pp. 68–69).

Keeping Track of Atoms

In keeping with the law of conservation of matter, each side of a chemical equation must have the same number of *atoms* of each element involved. When this is the case, the equation is said to be *balanced*. The equation for the burning of carbon (C + $O_2 \longrightarrow CO_2$) is balanced because there is one atom of carbon and two atoms of oxygen on both sides of the equation.

Now consider the following chemical reaction: When electricity is passed through water (H_2O), the latter can be broken down into hydrogen (H_2) and oxygen (O_2), as represented by the following equation:

$$H_2O \longrightarrow H_2 + O_2$$

H_2O	H_2	O_2
2 H atoms	2 H atoms	2 O atoms
1 O atom		

This equation is unbalanced because there is one atom of oxygen on the left but two atoms on the right.

We can't change the subscripts of any of the formulas to balance this equation because then we would be changing the arrangements of the atoms involved. Instead, we could use different numbers of the *molecules* involved to balance the equation. For example, we could use two water molecules:

$$2 H_2O \longrightarrow H_2 + O_2$$

$2 H_2O$	H_2	O_2
4 H atoms	2 H atoms	2 O atoms
2 O atoms		

This equation is still unbalanced because even though the numbers of oxygen atoms on both sides are now equal, the numbers of hydrogen atoms are not.

We can correct this by having the reaction produce two hydrogen molecules.

$$2 H_2O \longrightarrow 2 H_2 + O_2$$

$2 H_2O$	$2 H_2$	O_2
4 H atoms	4 H atoms	2 O atoms
2 O atoms		

Now the equation is balanced, and the law of conservation of matter has not been violated. We see that for every two molecules of water through which we pass electricity, two hydrogen molecules and one oxygen molecule are produced.

See if you can balance the chemical equation for the reaction of nitrogen gas (N_2) with hydrogen gas (H_2) to form ammonia gas (NH_3).

Critical Thinking

1. Balancing equations is based on the law of conservation of matter. Do you believe that this is an ironclad law of nature or one that through new scientific discoveries could be overthrown? Explain.

2. Imagine that you have the power to revoke the law of conservation of matter. List the three major ways this would affect your life.

3-6 NUCLEAR CHANGES

What Is Natural Radioactivity? In addition to physical and chemical changes, matter can undergo a third type of change known as a **nuclear change**. This occurs when nuclei of certain isotopes spontaneously change or are made to change into one or more different isotopes. Three types of nuclear change are natural radioactive decay, nuclear fission, and nuclear fusion.

Natural radioactive decay is a nuclear change in which unstable isotopes spontaneously emit fast-moving particles, high-energy radiation, or both at a fixed rate. The unstable isotopes are called **radioactive isotopes** or **radioisotopes**. Radioactive decay into various isotopes continues until the original isotope is changed into a stable isotope that is not radioactive.

Radiation emitted by radioisotopes is damaging ionizing radiation. The most common form of ionizing energy released from radioisotopes is **gamma rays**, a form of high-energy electromagnetic radiation (Figure 3-9). High-speed ionizing particles emitted from the nuclei of radioactive isotopes are most commonly of two types: **alpha particles** (fast-moving, positively charged chunks of matter that consist of two protons and two neutrons) and **beta particles** (high-speed electrons).

Each type of radioisotope spontaneously decays at a characteristic rate into a different isotope. This rate of decay is expressed in terms of **half-life**: the time needed for *one-half* of the nuclei in a radioisotope to decay and emit their radiation to form a different isotope. The decay continues, often producing a series of different radioisotopes, until a nonradioactive isotope is formed. Each radioisotope has a characteristic half-life, which may range from a few millionths of a second to several billion years (Table 3-1).

An isotope's half-life cannot be changed by temperature, pressure, chemical reactions, or any other known factor. Half-life can be used to estimate how long a sample of a radioisotope must be stored in a safe container before it decays to what is considered a safe

Q: What gas in the stratosphere keeps 95% of the sun's harmful ultraviolet radiation from reaching the earth's surface?

Table 3-1 Half-Lives of Selected Radioisotopes

Isotope	Half-Life	Radiation Emitted
Potassium-42	12.4 hours	Alpha, beta
Iodine-131	8 days	Beta, gamma
Cobalt-60	5.27 years	Beta, gamma
Hydrogen-3 (tritium)	12.5 years	Beta
Strontium-90	28 years	Beta
Carbon-14	5,370 years	Beta
Plutonium-239	24,000 years	Alpha, gamma
Uranium-235	710 million years	Alpha, gamma
Uranium-238	4.5 billion years	Alpha, gamma

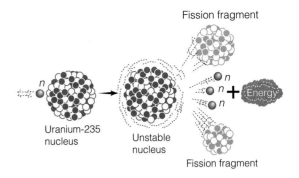

Figure 3-11 Fission of a uranium-235 nucleus by a neutron *(n)*.

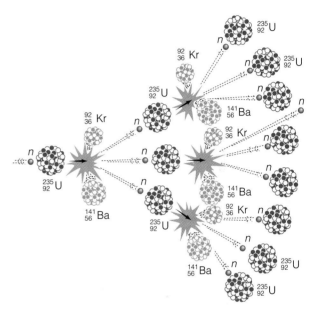

Figure 3-12 A nuclear chain reaction initiated by one neutron triggering fission in a single uranium-235 nucleus. This figure illustrates only a few of the trillions of fissions caused when a single uranium-235 nucleus is split within a critical mass of uranium-235 nuclei. The elements krypton (Kr) and barium (Ba), shown here as fission fragments, are only two of many possibilities.

level. A general rule is that such decay takes about 10 half-lives. Thus people must be protected from radioactive waste containing iodine-131 (which concentrates in the thyroid gland) for 80 days (10×8 days). In contrast, plutonium-239 (which is produced in nuclear reactors and used as the explosive in some nuclear weapons) can cause lung cancer when its particles are inhaled in minute amounts; it must be stored safely for 240,000 years ($10 \times 24,000$ years)—about four times longer than the latest version of our species has existed.

What Is Nuclear Fission? Splitting Nuclei

Nuclear fission is a nuclear change in which nuclei of certain isotopes with large mass numbers (such as uranium-235) are split apart into lighter nuclei when struck by neutrons; each fission releases two or three more neutrons and energy (Figure 3-11). Each of these neutrons, in turn, can cause an additional fission. For these multiple fissions to take place, enough fissionable nuclei must be present to provide the **critical mass** needed for efficient capture of these neutrons.

Multiple fissions within a critical mass form a **chain reaction**, which releases an enormous amount of energy (Figure 3-12). Living cells can be damaged by the ionizing radiation released by the radioactive lighter nuclei and by high-speed neutrons produced by nuclear fission.

In an atomic bomb, an enormous amount of energy is released in a fraction of a second in an uncontrolled nuclear fission chain reaction. This reaction is initiated by an explosive charge, which suddenly pushes two masses of fissionable fuel together, causing the fuel to reach the critical mass needed for a chain reaction.

In the reactor of a nuclear power plant, the rate at which the nuclear fission chain reaction takes place is controlled so that under normal operation only one of every two or three neutrons released is used to split another nucleus. In conventional nuclear fission reactors, the splitting of uranium-235 nuclei releases heat, which produces high-pressure steam to spin turbines and thus generates electricity.

What is Nuclear Fusion? Forcing Nuclei to Combine

Nuclear fusion is a nuclear change in which two isotopes of light elements, such as hydrogen, are forced together at extremely high temperatures until they fuse to form a heavier nucleus, releasing energy in the process. Temperatures of at least 100 million degrees C are needed to force the positively charged nuclei (which strongly repel one another) to fuse.

Nuclear fusion is much more difficult to initiate than nuclear fission, but once started it releases far more energy per unit of fuel than does fission. Fusion of hydrogen nuclei to form helium nuclei is the source of energy in the sun and other stars.

After World War II, the principle of *uncontrolled nuclear fusion* was used to develop extremely powerful hydrogen, or thermonuclear, weapons. These weapons use the D–T fusion reaction in which a hydrogen-2, or deuterium (D), nucleus and a hydrogen-3 (tritium, T) nucleus are fused to form a larger, helium-4 nucleus, a neutron, and energy (Figure 3-13).

Scientists have also tried to develop *controlled nuclear fusion*, in which the D–T reaction is used to produce heat that can be converted into electricity. Despite more than 40 years of research, this process is still in the laboratory stage. Even if it becomes technologically and economically feasible, it probably won't be a practical source of energy until 2050 or later.

3-7 THE TWO IRONCLAD LAWS OF ENERGY

What Is the First Law of Energy? You Can't Get Something for Nothing Scientists have observed energy being changed from one form to another in millions of physical and chemical changes, but they have never been able to detect either the creation or destruction of any energy (except in nuclear changes). The results of their experiments have been summarized in the **law of conservation of energy**, also known as the **first law of energy** or the **first law of thermodynamics**: *In all physical and chemical changes, energy is neither created nor destroyed, but it may be converted from one form to another.*

This scientific law tells us that when one form of energy is converted to another form in any physical or chemical change *energy input always equals energy output*. No matter how hard we try or how clever we are, we can't get more energy out of a system than we put in; in other words, *we can't get something for nothing in terms of energy quantity*.

What Is the Second Law of Energy? You Can't Even Break Even Because the first law of energy states that energy can be neither created nor destroyed, it's tempting to think that there will always be enough energy; yet if we fill a car's tank with gasoline and drive around or use a flashlight battery until it is dead, something has been lost. If it isn't energy, what is it? The answer is *energy quality* (Figure 3-10), the amount of energy available that can perform useful work.

Countless experiments have shown that when energy is changed from one form to another, a decrease in energy quality always occurs. The results of these experiments have been summarized in what is called the **second law of energy** or the **second law of thermodynamics**: *When energy is changed from one form to another, some of the useful energy is always degraded to lower-quality, more dispersed, less useful energy.* This degraded energy usually takes the form of heat given off at a low temperature to the surroundings (environment). There it is dispersed by the random motion of air or water molecules and becomes even more disorderly and less useful. Another way to state the second law of energy is that *heat always flows spontaneously from hot (high-quality energy) to cold (low-quality energy)*.

Basically, this law says that in any energy conversion, we always end up with *less* usable energy than we started with. So not only can we not get something

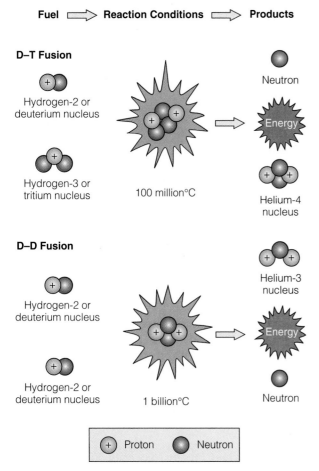

Figure 3-13 The deuterium–tritium (D–T) and deuterium–deuterium (D–D) nuclear fusion reactions, which take place at extremely high temperatures.

Q: What are the three major types of biomes?

for nothing in terms of energy quantity, *we can't even break even in terms of energy quality because energy always goes from a more useful to a less useful form.* The more energy we use, the more low-grade energy (heat) we add to the environment. No one has ever found a violation of this fundamental scientific law (see quotes at the beginning and end of this chapter).

Consider three examples of the second energy law in action. First, when a car is driven, only about 10% of the high-quality chemical energy available in its gasoline fuel is converted into mechanical energy (to propel the vehicle) and electrical energy (to run its electrical systems); the remaining 90% is degraded to low-quality heat that is released into the environment and eventually lost into space. Second, when electrical energy flows through filament wires in an incandescent light bulb, it is changed into about 5% useful light and 95% low-quality heat that flows into the environment; this so-called *light bulb* is really a *heat bulb*. Third, in living systems, solar energy is converted into chemical energy (photosynthesis and food) and then into mechanical energy (moving, thinking, and living); high-quality energy is degraded during this change of forms (Figure 3-14).

The second law of energy also means that *we can never recycle or reuse high-quality energy to perform useful work.* Once the concentrated energy in a serving of food, a liter of gasoline, a lump of coal, or a chunk of uranium is released, it is degraded to low-quality heat that is dispersed into the environment. We can heat air or water at a low temperature and upgrade it to high-quality energy, but the second law of energy tells us that it will take more high-quality energy to do this than we get in return.

Connections: How Does the Second Energy Law Affect Life? To form and maintain the highly ordered arrangement of molecules and the organized biochemical processes in your body, you must continually get and use high-quality matter and energy resources from your surroundings. As you use these resources, you add low-quality heat and low-quality waste matter to your surroundings. Your body continuously gives off heat equal to that of a 100-watt incandescent light bulb; this is why a closed room full of people gets warm. You also continuously break down solid and large molecules (such as glucose) into

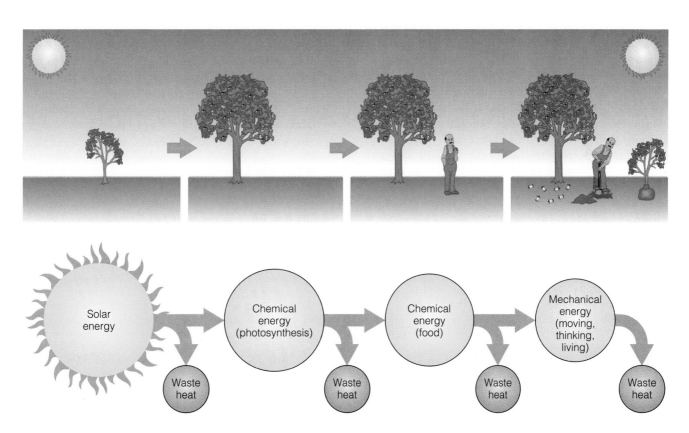

Figure 3-14 The second energy law in action in living systems. Each time energy is changed from one form to another, some of the initial input of high-quality energy is degraded, usually to low-quality heat that disperses into the environment.

A: Deserts, grasslands, and forests

smaller molecules of carbon dioxide gas and water vapor, which are dispersed in the atmosphere.

Planting, growing, processing, and cooking food all require high-quality energy and matter resources that add low-quality heat and waste materials to the environment. In addition, enormous amounts of low-quality heat and waste matter are added to the environment when concentrated deposits of minerals and fuels are extracted from the earth's crust, processed, and used.

3-8 CONNECTIONS: MATTER AND ENERGY LAWS AND ENVIRONMENTAL PROBLEMS

What Are High-Throughput or High-Waste Societies? As a result of the law of conservation of matter and the second law of energy, individual resource use automatically adds some waste heat and waste matter to the environment. Your individual use of matter and energy resources and your additions of waste heat and matter to the environment may seem small and insignificant. But there are 1.2 billion people in developed countries rapidly using large quantities of matter and energy resources. Meanwhile, the 4.7 billion people in developing countries hope to be able to use more of these resources, and each year there are about 86 million new consumers of the earth's energy and matter resources.

Most of today's advanced industrialized countries are **high-waste** or **high-throughput societies** that attempt to sustain ever-increasing economic growth by increasing the *throughput* of matter and energy resources in their economic systems (Figure 3-15). These resources flow through the economies of such societies to planetary *sinks* (air, water, soil, organisms), where pollutants and wastes end up and can accumulate to harmful levels.

However, the scientific laws of matter and energy discussed in this chapter tell us that if more and more people continue to use and waste more and more energy and matter resources at an increasing rate, eventually the capacity of the environment to dilute and degrade waste matter and absorb waste heat will be exceeded. Thus, *at some point high-waste or high-throughput societies become unsustainable.*

What Are Matter-Recycling Societies? A stopgap solution to this problem is to convert an unsustainable high-throughput society to a **matter-recycling society**. The goal of such a conversion is to allow economic growth to continue without depleting matter resources or producing excessive pollution and environmental degradation. As we have learned, however, there is no free lunch when it comes to energy and energy quality.

Even though recycling matter saves energy, the two laws of energy tell us that *recycling matter resources always requires expenditure of high-quality energy (which cannot be recycled) and adds waste heat to the environment.* For the long run, a matter-recycling society based on continuing population growth and per capita resource consumption must have an inexhaustible supply of affordable high-quality energy, and its environment must have an infinite capacity to absorb and disperse waste heat and to dilute and degrade waste matter.

There is also a limit to the number of times some materials, such as paper fiber, can be recycled before they become unusable. Changing to a matter-recycling society is an important way to buy some time, but it does not allow more and more people to use more and more resources indefinitely, even if all of them were somehow perfectly recycled.

What Are Low-Waste Societies? Learning from Nature The three scientific laws governing matter and energy changes suggest that the best long-term solution to our environmental and resource problems is

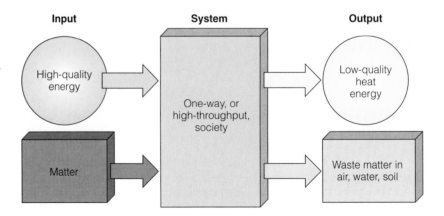

Figure 3-15 The eventually unsustainable high-waste or high-throughput societies of most developed countries are based on maximizing the rates of energy and matter flow. This process is rapidly converting the world's high-quality matter and energy resources into waste, pollution, and low-quality heat.

Input — High-quality energy — Matter

System — One-way, or high-throughput, society

Output — Low-quality heat energy — Waste matter in air, water, soil

Q: What percentage of the earth's tropical forests have been cleared or damaged?

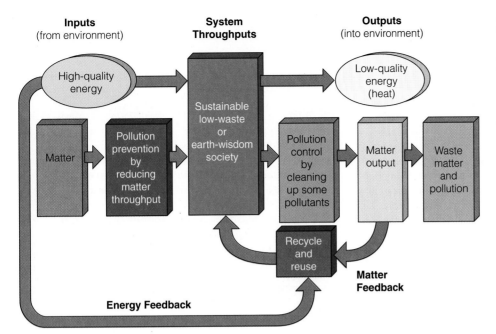

Inputs
(from environment)

System Throughputs

Outputs
(into environment)

High-quality energy

Low-quality energy (heat)

Matter

Pollution prevention by reducing matter throughput

Sustainable low-waste or earth-wisdom society

Pollution control by cleaning up some pollutants

Matter output

Waste matter and pollution

Recycle and reuse

Matter Feedback

Energy Feedback

Figure 3-16 A sustainable low-waste or low-throughput society, based on energy flow and matter recycling, works with nature to reduce throughput. This is done by **(1)** reusing and recycling most nonrenewable matter resources, **(2)** using potentially renewable resources no faster than they are replenished, **(3)** using matter and energy resources efficiently, **(4)** reducing unnecessary consumption, **(5)** emphasizing pollution prevention and waste reduction, and **(6)** controlling population growth.

to shift from a society based on maximizing matter and energy flow (throughput) to a sustainable **low-waste society** or **earth-wisdom society**. The major features of such a society are summarized in Figure 3-16.

According to many scientists, because of the three basic scientific laws of matter and energy, we all depend on one another and on the rest of nature for our survival; we are all in it together. In the next two chapters, we will apply these laws to living systems and look at some biological principles that can teach us how to work with nature.

The second law of thermodynamics holds, I think, the supreme position among laws of nature. . . . If your theory is found to be against the second law of thermodynamics, I can give you no hope.

ARTHUR S. EDDINGTON

CRITICAL THINKING

1. Do you think cheating, in the form of faking or doctoring the results of experiments, is more or less likely in science than in other areas of knowledge? Explain.

2. To what extent are scientists responsible for the applications of knowledge they discover? Should scientists abandon research because of its possible harmful uses? Explain.

3. Respond to the following statements:
   a. It has never been scientifically proven that anyone has ever died from smoking cigarettes.
   b. The greenhouse theory—that certain gases (such as water vapor and carbon dioxide) trap heat

in the atmosphere and thus influence the average temperature and climate of the earth—is not a reliable idea because it is only a scientific theory.

4. Try to find an advertisement or an article describing or using some aspect of science in which **(a)** the concept of scientific proof is misused, **(b)** the term *theory* is used when it should have been *hypothesis*, and **(c)** a consensus scientific finding is dismissed or downplayed because it is "only a theory."

5. Explain why we don't really consume anything and why we can never really throw matter away.

6. A tree grows and increases its mass. Explain why this isn't a violation of the law of conservation of matter.

7. If there is no "away," why isn't the world filled with waste matter?

8. Someone wants you to invest money in an automobile engine that will produce more energy than the energy in the fuel (such as gasoline or electricity) you use to run the motor. What is your response? Explain.

9. Use the second energy law to explain why a barrel of oil can be used only once as a fuel.

10.(a) Use the law of conservation of matter to explain why a matter-recycling society will sooner or later be necessary. **(b)** Use the first and second laws of energy to explain why, in the long run, we will need a low-waste or low-throughput society, not just a matter-recycling society.

11.(a) Imagine that you have the power to violate the law of conservation of energy (the first energy law) for one day. What are the three most important things you would do with this power? **(b)** Repeat this process, imagining that you have the power to violate the second law of energy for one day.

A New Environmentalism for the 1990s and Beyond

GUEST ESSAY

Peter Montague

Peter Montague is director of the Environmental Research Foundation in Washington, D.C., which studies and informs the public about environmental problems and the technologies and policies that might help solve them. He has served as project administrator of a hazardous waste research program at Princeton University and has taught courses in environmental impact analysis at the University of New Mexico. He is the coauthor of two books on toxic heavy metals in the natural environment and is editor of Rachel's Environment and Health Weekly, *an informative and readable newsletter on environmental problems and solutions.*

Environmentalism as we have known it for over 25 years is dead. The environmentalism of the 1970s advocated strict numerical controls on releases into the environment of *dangerous wastes* (any unwanted or uncontrolled materials that can harm living things or disrupt ecosystems). But industry's ability to create new hazards quickly outstripped government's ability to establish adequate controls and enforcement programs.

After so many years of effort by government and by concerned citizens (the environmental movement), the overwhelming majority of dangerous chemicals is still not regulated in any way. Even those few that are covered by regulations have not been adequately controlled.

In short, the *pollution management* approach to environmental protection has failed and stands discredited; *pollution prevention* is our only hope. An ounce of prevention really is worth a pound of cure.

Here, in list form, is the situation facing environmentalists today:

- *All waste disposal—landfilling, incineration, deep-well injection—is polluting because "disposal" means dispersal into the environment.* Once wastes are created, they cannot be contained or controlled because of the scientific laws of matter and energy. The old environmentalism failed to recognize this important truth and thus squandered enormous resources trying to achieve the impossible. While we in the United States currently spend about $90 billion per year on pollution control, the global environment is increasingly threatened by a buildup of heat because of heat-trapping gases we emit into the atmosphere. At least half the surface of the planet is being subjected to damaging ultraviolet radiation from the sun as a result of ozone-depleting chemicals we have discharged into the atmosphere. Vast regions of the United States, Canada, and Europe are suffering from loss of forests, crop productivity, and fish as a result of a mixture of air pollutants, produced mostly by burning fossil fuels in power plants and automobiles. Soil and water are dangerously polluted at thousands of locales where municipal garbage and industrial wastes have been (and continue to be) dumped or incinerated; thousands of such sites remain to be discovered, according to U.S. government estimates.

- *The inevitable result of our reliance upon waste treatment and disposal systems has been an unrelenting buildup of toxic synthetic materials in humans and other forms of life worldwide.* For example, breast milk of women in industrialized countries is so contaminated with pesticides and industrial hydrocarbons that, if human milk were bottled and sold com-

---

*12. Use the library or Internet to find examples of various perpetual motion machines and inventions that allegedly violate the two laws of energy (thermodynamics) by producing more high-quality energy than the high-quality energy needed to make them run. What has happened to these schemes and machines (many of them developed by scam artists to attract money from investors)?

*13. Make a concept map of this chapter's major ideas using the section heads and subheads and the key terms

(in boldface type). Look inside the back cover and in Appendix 4 for information about concept maps.

SUGGESTED READINGS

See Internet Sources and Further Readings at the back of the book, and InfoTrac.

Q: What five levels of the organization of matter are the focus of ecology?

mercially, it could be banned by the Food and Drug Administration as unsafe for human consumption. If a whale today beaches itself on U.S. shores and dies, its body must be treated as a "hazardous waste" because whales contain concentrations of PCBs (polychlorinated biphenyl, a class of industrial toxins) legally defined as hazardous.

■ *The ability of humans and other life-forms to adapt to changes in their environment is strictly limited by the genetic code each form of life inherits.* Continued contamination at a rate hundreds of times faster than we can adapt will subject humans to increasingly widespread sickness and degradation of the species, and could ultimately lead to extinction.

■ *Damage to humans (and to other life-forms) is abundantly documented.* Birds, fish, and humans in industrialized countries are enduring steadily rising levels of cancer, genetic mutations, and damage to their nervous, immune, and hormonal systems as a result of pollution.

If we will but look, the handwriting is on the wall everywhere.

To deal with these problems, industrial societies must abandon their reliance upon waste treatment and disposal and upon the regulatory system of numerical standards created to manage the damage that results from relying on waste disposal instead of waste prevention. We must—relatively quickly—move the industrialized and industrializing countries to new technical approaches accompanied by new industrial goals—namely, "clean production" or zero discharge systems.

The concept of "clean production" involves industrial systems that avoid or eliminate dangerous wastes and dangerous products and minimize the use and waste of raw materials, water, and energy. Goods manufactured in a clean production process must not damage natural ecosystems throughout their entire life cycle, including **(1)** raw materials selection, extraction, and processing; **(2)** product conceptualization, design, manufacture, and assembly; **(3)** materials transport during all phases; **(4)** industrial and household usage; and **(5)** reintroduction of the product into industrial systems or into the environment when it no longer serves a useful function.

Clean production does not rely on "end-of-pipe" pollution controls such as filters or scrubbers or chemical, physical, or biological treatment. Measures that pretend to reduce the volume of waste by incineration or concentration, that mask the hazard by dilution, or that transfer pollutants from one environmental medium to another are also excluded from the concept of "clean production."

A new industrial pattern, and a new environmentalism, is thus emerging. It insists that the long-term well-being of humans and other species must be factored into our production and consumption plans. These new requirements are not optional; human survival and life quality depend upon our willingness to make, and pay for, the necessary changes.

Critical Thinking

1. Do you agree with the author that the *pollution management* approach to environmental protection practiced during the past 25 years has failed and must be replaced with a *pollution prevention* approach? Explain.

2. List key economic, health, consumption, and lifestyle changes you might experience as a consequence of putting much greater emphasis on pollution prevention. What changes might the next generation face?

4 ECOSYSTEMS AND HOW THEY WORK: CONNECTIONS IN NATURE

Have You Thanked the Insects Today?

Insects have a bad reputation. We classify many insect species as *pests* because they compete with us for food, spread human diseases (such as malaria), and invade our lawns, gardens, and houses. Some people have "bugitis," fear all insects, and think that the only good bug is a dead bug. However, this view fails to recognize the vital roles insects play in helping sustain life on earth.

A large proportion of the earth's plant species (including many trees) depend on insects to pollinate their flowers (Figure 4-1, left). In turn, we and other land-dwelling animals depend on plants for food, either by eating them or by consuming animals that eat them.

Some insects, such as the praying mantis (Figure 4-1, right), feed on some of the insect species we classify as pests. Indeed, insects that eat other insects help control the populations of at least half the species of insects we call pests.

Suppose all insects disappeared today. Within a year most of the earth's amphibians, reptiles, birds, and mammals would become extinct because of the disappearance of so much plant life. The earth would be covered with rotting vegetation and animal carcasses being decomposed by unimaginably huge hordes of bacteria and fungi.

Fortunately, this is not a realistic scenario because insects, which have been around for at least 400 million years, are phenomenally successful forms of life. They were the first animals to invade the land and, later, the air. Today they are by far the planet's most diverse, abundant, and successful group of animals.

Insects can rapidly evolve new genetic traits, such as resistance to pesticides. They also have an exceptional ability to evolve into new species when faced with new environmental conditions, and they are extremely resistant to extinction. Because of their ability to develop genetic resistance to pesticides quickly, our efforts to eradicate the insects we don't like using chemical warfare will always fail eventually. Moreover, many of the pesticides we use reduce the populations of insect species that help keep pest insect species under control.

Some people have wondered whether insects will take over the world if the human race extinguishes itself. This is the wrong question. Insects are already in charge. Insects can thrive without newcomers such as us, but we and most other land organisms would quickly perish without them.

Learning about the roles insects play in nature requires us to understand how insects and other organisms living in a biological *community* (such as a forest or pond) interact with one another and with the nonliving environment. Ecology is the science that studies such relationships and interactions in nature, as discussed in this chapter and the two chapters that follow.

Figure 4-1 Insects play important roles in helping sustain life on earth. The bright green caterpillar moth feeding on pollen in crocus flower (left) and other insects pollinate flowering plants that serve as food for many plant eaters. The praying mantis eating a grasshopper (right) and many other insect species help control the populations of at least half of the insect species we classify as pests. (Left, Stephen Hopkins/Planet Earth Pictures; right, M. H. Sharp/Photo Researchers, Inc.)

This chapter focuses on answering the following questions about *ecosystems* (communities of species interacting with one another and with their nonliving environment of matter and energy):

- What basic processes keep us and other organisms alive?
- What are the major living and nonliving parts of an ecosystem?
- What happens to energy in an ecosystem?
- What happens to matter in an ecosystem?
- How do scientists study ecosystems?
- What roles (niches) do different types of organisms play in an ecosystem?
- How do different types of organisms interact with one another?

4-1 LIFE AND EARTH'S LIFE-SUPPORT SYSTEMS

What Is Life? The **cell** is the basic unit of life. Each cell is bounded by an outer membrane or wall and contains genetic material (DNA) and other structures needed to perform its life functions. All forms of life

- *Are made of cells that have highly organized internal structures and organizations.* Organisms may consist of a singe cell (such as bacteria) or of many cells.

- *Have characteristic types of deoxyribonucleic acid (DNA) molecules in each cell.* DNA is the stuff of which genes—the basic units of heredity—are made. These self-replicating molecules contain the instructions both for making new cells and for assembling proteins and other molecules each cell needs in order to survive and reproduce (Figure 3-7).

- *Can capture and transform matter and energy from their environment to supply their needs for survival, growth, and reproduction.* The complex set of chemical reactions that carry out this role in cells and organisms is called **metabolism**.

- *Can maintain favorable internal conditions, despite changes in their external environment, through homeostasis (Section 3-2) if not overwhelmed.*

- *Perpetuate themselves through reproduction: the production of offspring by one or more parents.* **Asexual reproduction** generally occurs by simple cell division and is common in single-cell organisms. In this case, the mother cell divides to produce two identical daughter cells that are clones of the mother cell. **Sexual reproduction** occurs in organisms that produce offspring by combining sex cells or gametes (such as ovum and sperm) from both parents. This produces offspring that have combinations of traits from their parents. Sexual reproduction usually gives the offspring a greater chance of survival under changing environmental conditions than the genetic clones produced by asexual reproduction.

- *Can adapt to changes in environmental conditions by inheriting beneficial **mutations** (random changes in the structure or number of DNA molecules) and through recombining existing genes during reproduction.* Mutations can be harmful, neutral, or beneficial. Beneficial mutations give rise to new genetic traits called *adaptive traits*. Such traits allow organisms to adapt to changing environmental conditions and thus produce more offspring than those without such traits, a process called **natural selection**. Through successive generations, these adaptive traits become more prevalent in a population. The theory explaining this change in the genetic makeup of a population through successive generations is called **evolution**. If evolution continues long enough, it can result in new species. Populations that cannot adapt decline and may become extinct. These ongoing processes of extinction and evolution, in response to environmental changes over billions of years, have led to the diversity of life-forms found on the earth today. This *biodiversity*, a vital part of earth capital, sustains life and provides the genetic raw material for adaptation to future environmental changes.

What Are the Major Parts of Earth's Life-Support Systems? We can think of the earth as being made up of several layers or concentric spheres (Figure 4-2). The **atmosphere** is a thin envelope of air around the planet. Its inner layer, the **troposphere**, extends only about 17 kilometers (11 miles) above sea level but contains most of the planet's air, mostly nitrogen (78%) and oxygen (21%). The next layer, stretching 17–48 kilometers (11–30 miles) above the earth's surface, is called the **stratosphere**. Its lower portion contains enough ozone (O_3) to filter out most of the sun's harmful ultraviolet radiation, thus allowing life on land and in the surface layers of bodies of water to exist.

The **hydrosphere** consists of the earth's liquid water (both surface and underground), ice (polar ice, icebergs, and ice in permafrost), and water vapor in the atmosphere. The **lithosphere** is the earth's crust and upper mantle; the crust contains nonrenewable fossil fuels and minerals we use as well as potentially renewable soil chemicals (nutrients) required for plant life.

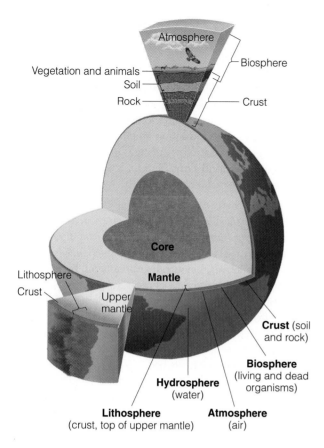

Figure 4-2 The general structure of the earth. The atmosphere (shown here) consists of several layers, including the troposphere (innermost layer) and the stratosphere (second layer).

The **ecosphere**, or **biosphere**, is the portion of the earth in which living (biotic) organisms exist and interact with one another and with their nonliving (abiotic) environment. The ecosphere includes most of the hydrosphere and parts of the lower atmosphere and upper lithosphere, reaching from the deepest ocean floor, 20 kilometers (12 miles) below sea level, to the tops of the highest mountains. If the earth were an apple, the ecosphere would be no thicker than the apple's skin. *The goal of ecology is to understand the interactions in this thin, life-supporting global skin of air, water, soil, and organisms.*

What Sustains Life on Earth? Life on the earth depends on three interconnected factors (Figure 4-3):

- The *one-way flow of high-quality (usable) energy* from the sun, first through materials and living things in their feeding interactions, then into the environment as low-quality energy (mostly heat dispersed into air or water molecules at a low temperature), and eventually back into space as infrared radiation. This is a example of *input* of high-quality energy from the sun, its flow or *throughput* through the ecosphere, and its

output of degraded energy as heat into the atmosphere and eventually back into space.

- The *cycling of matter or nutrients* (all atoms, ions, or molecules needed for survival by living organisms) through parts of the ecosphere. Each chemical cycle can be viewed as a system with *inputs* of matter and solar energy (used to drive the cycle) and their flow or *throughput* through parts of the ecosphere. In each chemical cycle, the *output* of degraded energy flows as heat into the atmosphere and eventually back into space, and the *output* of matter is returned to the system (*feedback*) to be cycled again and again.

- *Gravity*, caused mostly by the attraction between the sun and the earth, which allows the planet to hold onto its atmosphere and causes the downward movement of chemicals in the matter cycles.

Because the earth is closed to significant inputs of matter from space, essentially all of the nutrients used by organisms are already present on earth and must be recycled again and again for life to continue.

How Does the Sun Help Sustain Life on Earth? The sun is a middle-aged star that lights and warms the planet. It also supplies the energy for **photosynthesis**, the process used by green plants and some bacteria to synthesize the compounds that keep them alive and feed most other organisms. Solar energy also powers the cycling of matter and drives the climate and weather systems that distribute heat and fresh water over the earth's surface.

The sun is a gigantic fireball of hydrogen (72%) and helium (28%) gases. Temperatures and pressures in its inner core are so high that hydrogen nuclei fuse to form helium nuclei (Figure 3-13), releasing enormous amounts of energy.

The sun, then, is really a gigantic nuclear fusion (thermonuclear) reactor running on hydrogen fuel. This enormous reactor radiates energy in all directions as electromagnetic radiation (Figure 3-9). Moving at the speed of light, this radiation makes the 150-million-kilometer (93-million-mile) trip between the sun and the earth in slightly more than 8 minutes.

Because the earth is a tiny sphere in the vastness of space, it receives only about one-billionth of this output of energy. Much of this energy is either reflected away or absorbed by chemicals in its atmosphere. Most of what reaches the troposphere is visible light, infrared radiation (heat), and the small amount of ultraviolet radiation that is not absorbed by ozone in the stratosphere. About 28% of the solar energy reaching the troposphere is reflected back into space by clouds, chemicals, dust, and the earth's surface land and water (Figure 4-4).

Q: What percentage of North America's original temperate deciduous forest land has been cleared?

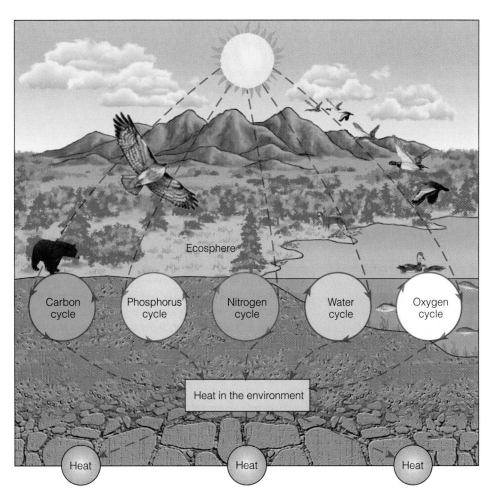

Figure 4-3 Life on the earth depends on the *one-way flow of energy* (dashed lines) from the sun through the ecosphere, the *cycling of crucial elements* (solid lines around circles), and *gravity*, which keeps atmospheric gases from escaping into space and draws chemicals downward in the matter cycles. This simplified conceptual model depicts only a few of the many cycling elements.

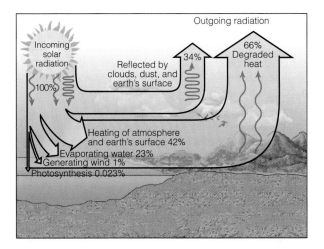

Figure 4-4 The flow of energy to and from the earth. The ultimate source of energy in most ecosystems is sunlight.

Most of the remaining 72% of solar energy warms the troposphere and land, evaporates water and cycles it through the ecosphere, and generates winds. A tiny fraction (about 0.023%) is captured by green plants and bacteria, fueling photosynthesis to make the organic compounds that most life-forms need to survive.

Most of this unreflected solar radiation is degraded into infrared radiation (which we experience as heat) as it interacts with the earth. How fast this heat flows through the atmosphere and back into space is affected by tropospheric heat-trapping (greenhouse) gases, such as water vapor, carbon dioxide, methane, nitrous oxide, and ozone. Without this atmospheric thermal blanket, known as the *natural greenhouse effect*, the earth would be nearly as cold as Mars, and life as we know it could not exist.

Connections: How Do Nutrient Cycles Sustain Life? Any atom, ion, or molecule an organism needs to live, grow, or reproduce is called a **nutrient**. Some elements (such as carbon, oxygen, hydrogen, nitrogen, phosphorus, sulfur, potassium, calcium, magnesium, and iron) are needed in fairly large amounts, whereas others (such as sodium, zinc, copper, chlorine, and iodine) are needed in small or even trace amounts.

These nutrient atoms, ions, and molecules are continuously cycled from the nonliving environment (air, water, soil, and rock) to living organisms (biota) and then back again in what are called **nutrient cycles**, or **biogeochemical cycles** (literally, life–earth–chemical cycles). These cycles, driven directly or indirectly by incoming solar energy and gravity, include the carbon, oxygen, nitrogen, phosphorus, and hydrologic (water) cycles (Figure 4-3).

Because the earth is essentially a closed system in terms of matter, the planet's chemical cycles are vital for all life, and they explain why without death there could be no life. The cycle of reproduction, growth, death, and decay of organisms keeps renewing the chemicals that support life.

The earth's chemical cycles also connect past, present, and future forms of life. Some of the carbon atoms in your skin may once have been part of a leaf, a dinosaur's skin, or a layer of limestone rock. Some of the oxygen molecules you just inhaled may have been inhaled by your grandmother, by Plato, or by a hunter–gatherer who lived 25,000 years ago.

🌐 4-2   ECOSYSTEM COMPONENTS

What Is Ecology? What organisms live in a coral reef, a field, or a pond? How do they get enough food and energy to stay alive? How do these organisms interact with one another? What changes might this coral reef, field, or pond undergo through time?

Ecology, created as a discipline about 100 years ago by MIT chemist Ellen Swallow, is the science that attempts to answer such questions. **Ecology** (from the Greek words *oikos*, "house" or "place to live," and *logos*, "study of") is the study of how organisms interact with one another and with their nonliving environment (including such factors as sunlight, temperature, moisture, and vital nutrients). The key word in this definition is *interact*. Ecologists focus on trying to understand the interactions among organisms, populations, communities, ecosystems, and the ecosphere (Figure 4-5).

An **organism** is any form of life. Organisms can be classified into **species**, groups of organisms that resemble one another in appearance, behavior, chemistry, and genetic endowment. Organisms that reproduce sexually are classified in the same species if, under natural conditions, they can actually or potentially breed with one another and produce live, fertile offspring.

We don't know how many species exist on the earth. Estimates range from 5 million to 100 million, most of them insects (p. 70) and microorganisms. So far biologists have identified and named only about 1.8 million species. Biologists know a fair amount about roughly one-third of the known species, but detailed roles and interactions of only a few.

Each identified and unknown species is the result of a long evolutionary history involving the storage of an immense amount of unique and irreplaceable genetic information about how to survive under specific environmental conditions. According to biologist Edward O. Wilson, if the genetic information encoded in the DNA found in the approximately 100,000 genes of a house mouse were translated into a printed text, it would fill all of the books in each of the 15 editions of the *Encyclopaedia Britannica* published since 1768.

Ecologists make a distinction between wild species and domesticated species. A **wild species** is one that exists as a population of individuals in a natural habitat, ideally similar to the one in which its ancestors evolved in. A **domesticated species**—such as cows, sheep, food crops, animals in zoos, plants in arboretums, and lawn grasses, flowers, and trees—is one that has been plucked from its normal ecological environment to support the needs and wants of humans and thus plays a much weaker ecological and evolutionary role than wild species. Domesticated plants and animals need to be externally fed and propagated. They are usually subjected to artificial selection and genetic selection, and thus are unlikely to leave any long-term evolutionary legacy because they contain only a portion of their evolutionary past.

A **population** consists of all members of the same species occupying a specific area at the same time (Figure 4-6). Examples are all sunfish in a pond, all white oak trees in a forest, and all people in a country. In most natural populations, individuals vary slightly in their genetic makeup, which is why they don't all look or behave exactly alike—a phenomenon called **genetic diversity** (Figure 4-7). Populations are dynamic groups that change in size, age distribution, density, and genetic composition as a result of changes in environmental conditions.

The place where a population (or an individual organism) normally lives is known as its **habitat**. It may be as large as an ocean or prairie or as small as the underside of a rotting log or the intestine of a termite. Populations of all the different species occupying a particular place make up a **community**, or **biological community**: a complex interacting network of plants, animals, and microorganisms.

An **ecosystem** is a community of different species interacting with one another and with their nonliving environment of matter and energy. The size of an ecosystem is somewhat arbitrary; it is defined by the particular system we wish to study. The unit of study may be small, such as a particular stream or field or a patch of woods, desert, or marsh. Or the units may be large, generalized types of terrestrial (land) ecosystems such as a particular type of grassland, forest, or desert. Ecosystems can be natural or artificial (human-created). Examples of human-created ecosystems are cropfields, farm ponds, and reservoirs or artificial lakes created behind dams. All of the earth's ecosystems together make up what we call the biosphere, or ecosphere.

Q: How much of earth's surface is covered by oceans?

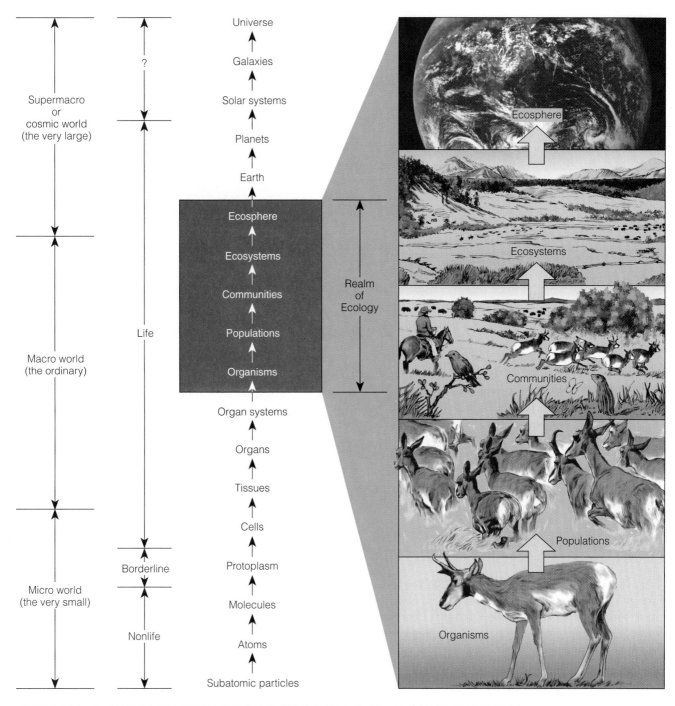

Figure 4-5 Levels of organization of matter. Note that ecology focuses on five levels of this hierarchical model.

For convenience, scientists usually consider an ecosystem under study to be an isolated unit. However, natural ecosystems rarely have distinct boundaries and are not truly self-contained, self-sustaining systems. Instead, one ecosystem tends to merge with the next in a transitional zone called an **ecotone**, a region containing a mixture of species from adjacent regions and often species not found in either of the bordering ecosystems. For example, a marsh or wetland found between dry land and the open water of a lake or ocean is an ecotone.

Climate—long-term weather—is the main factor determining what type of life, especially what plants, will thrive in a given land area. Viewed from outer space, the earth resembles an enormous jigsaw puzzle consisting of large masses of land and vast expanses of ocean (pp. 2–3 and Figure 1-1). Biologists have divided the terrestrial (land) portion of the ecosphere

Figure 4-6 A population of monarch butterflies wintering in Michoacán, Mexico. The geographic distribution of this butterfly coincides with that of the milkweed plant on which monarch larvae and caterpillars feed. (Frans Lanting/Bruce Coleman Ltd.)

Figure 4-7 The genetic diversity among individuals of one species of Caribbean snail is reflected in the variations in shell color and banding patterns. (Alan Solem)

into **biomes**, large regions (such as forests, deserts, and grasslands) characterized by a distinct climate and specific life-forms, especially vegetation adapted to it (Figure 4-8). Each biome consists of many ecosystems whose communities have adapted to differences in climate, soil, and other factors throughout the biome.

Marine and freshwater portions of the ecosphere can be divided into **aquatic life zones**, each containing numerous ecosystems. Aquatic life zones are the aquatic equivalent of biomes. Examples include freshwater life zones (such as lakes and streams) and ocean or marine life zones (such as estuaries, coastlines, coral reefs, and the deep ocean). The earth's major land biomes and aquatic life zones are discussed in more detail in Chapter 5.

Why Is Biodiversity So Important? As environmental conditions have changed over billions of years, many species have become extinct and new ones have formed. The result of these changes is **biological diversity**, or **biodiversity**: the forms of life that can best survive the variety of conditions currently found on the earth. As you learned in Chapter 1, biodiversity includes **genetic diversity** (variability in the genetic makeup among individuals in a single species, Figure 4-7), **species diversity** (the variety of species in different habitats on the earth), and **ecological diversity** (the variety of biological communities that interact with one another and with their nonliving environments).

Another term for biodiversity is **wildness**: the existence of wild gene pools, species, and ecosystems that are completely or mostly undisturbed by human activities. In the words of Henry David Thoreau, "In wildness is the preservation of the world."

We are utterly dependent on this mostly unknown biocapital. This rich variety of genes, species, and ecosystems gives us food, wood, fibers, energy, raw materials, industrial chemicals, and medicines, and it pours hundreds of billions of dollars yearly into the global economy.

The earth's life-forms and ecosystems also provide recycling, purification, and natural pest control. Every species here today contains genetic information that represents thousands to millions of years of adaptation to the earth's changing environmental conditions and is the raw material for future adaptations. Biodiversity is nature's insurance policy against disasters.

Some people also include *human cultural diversity* as part of the earth's biodiversity. The variety of human cultures represents numerous social and technological solutions that have enabled us to survive and adapt to and work with the earth.

What Are the Major Living Components of Ecosystems? The ecosphere and its ecosystems can be separated into two parts: **biotic**, or living, components (plants, animals, and microorganisms, sometimes referred to as *biota*), and **abiotic**, or nonliving, components (water, air, nutrients, and solar energy). Figures 4-9 and 4-10 are greatly simplified diagrams of some of the biotic and abiotic components in a freshwater pond ecosystem and a field.

Living organisms in ecosystems are usually classified as either *producers* or *consumers*, based on how they get food. **Producers**, sometimes called **autotrophs** (self-feeders), make their own food from compounds obtained from their environment. On land, most producers are green plants. In freshwater and marine ecosystems, algae and plants are the major producers near shorelines; in open water the dominant producers floating and drifting are *phytoplankton*, most of them microscopic. Only producers make their own

Q: What percentage of the earth's wetlands have been destroyed or polluted?

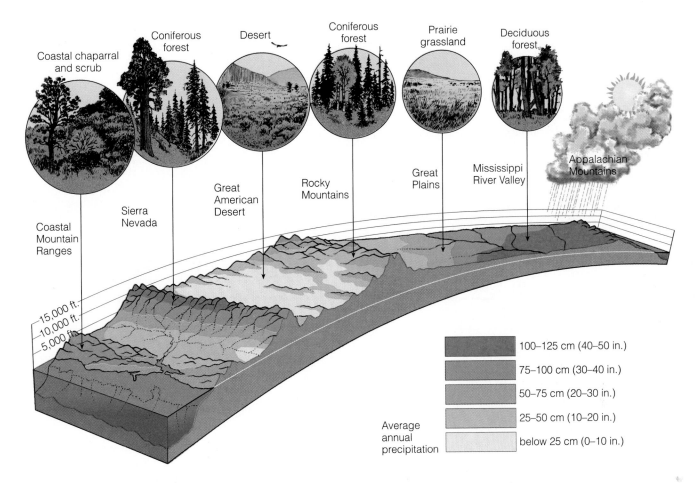

Figure 4-8 Major biomes found along the 39th parallel across the United States. The differences reflect changes in climate, mainly differences in average annual precipitation and temperature (not shown).

Average annual precipitation

- 100–125 cm (40–50 in.)
- 75–100 cm (30–40 in.)
- 50–75 cm (20–30 in.)
- 25–50 cm (10–20 in.)
- below 25 cm (0–10 in.)

food; all other organisms are consumers, which depend directly or indirectly on food provided by producers.

Most producers capture sunlight to make sugars (such as glucose, $C_6H_{12}O_6$) and other complex organic compounds from inorganic (abiotic) nutrients in the environment. This process is called **photosynthesis**. In most green plants, *chlorophyll* (a pigment molecule that gives plants their green color) traps solar energy for use in photosynthesis and converts it into chemical energy. Although a sequence of hundreds of chemical changes takes place during photosynthesis, the overall reaction can be summarized as follows:

$$\text{carbon dioxide} + \text{water} + \textbf{solar energy} \longrightarrow \text{glucose} + \text{oxygen}$$

$$6\ CO_2 + 6\ H_2O + \textbf{solar energy} \longrightarrow C_6H_{12}O_6 + 6\ O_2$$

A few producers, mostly specialized bacteria, can convert simple compounds from their environment into more complex nutrient compounds without sunlight, a process called **chemosynthesis**. In one such case, the source of energy is heat generated by the decay of radioactive elements deep in the earth's core; this heat is released at hot-water (hydrothermal) vents in the ocean's depths, where new crust is constantly being formed and reformed. In the pitch darkness around such vents, large populations of specialized producer bacteria use this geothermal energy to convert dissolved hydrogen sulfide (H_2S) and carbon dioxide into organic nutrient molecules. These bacteria in turn become food for a variety of aquatic animals, including huge tube worms and a variety of clams, crabs, mussels, and barnacles.

All other organisms in an ecosystem are **consumers** or **heterotrophs** ("other-feeders"), which get their energy and nutrients by feeding on other organisms or their remains. There are several classes of consumers, depending on their primary source of food. **Herbivores** (plant eaters) are called **primary consumers** because they feed directly on producers. **Carnivores** (meat eaters) feed on other consumers; those called **secondary consumers** feed only on primary consumers (herbivores). Most secondary consumers are

A: 25–50% (55% in the United States, 91% in California)

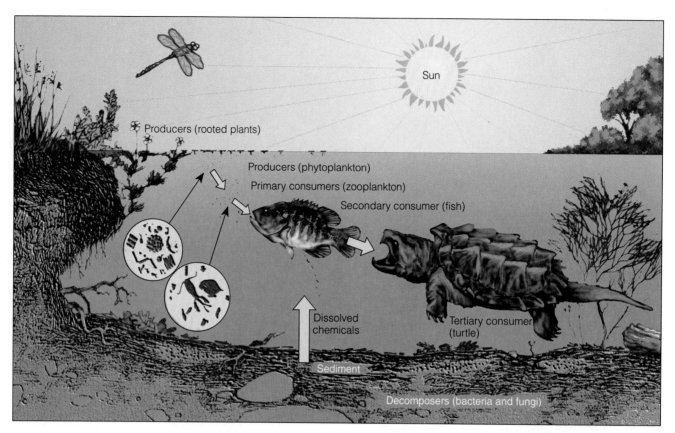

Figure 4-9 Major components of a freshwater pond ecosystem.

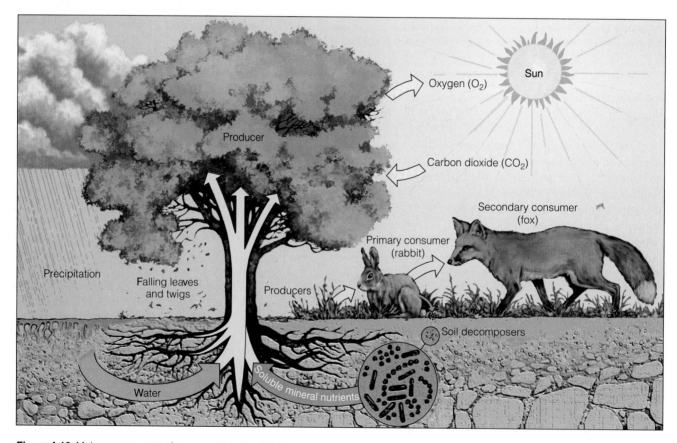

Figure 4-10 Major components of an ecosystem in a field.

Q: How old is the earth?

animals, but a few plants (such as the Venus's-flytrap plant) trap and digest insects. **Tertiary (higher-level) consumers** feed only on other carnivores. **Omnivores** are consumers that eat both plants and animals; examples are pigs, rats, foxes, bears, cockroaches, and humans.

Other consumers, called **scavengers**, feed on dead organisms that either were killed by other organisms or died naturally. Vultures, flies, crows, hyenas, and some species of sharks, beetles, and ants are examples of scavengers. **Detritivores** (detritus feeders and decomposers) live off **detritus** (pronounced di-TRI-tus), or parts of dead organisms and cast-off fragments and wastes of living organisms (Figure 4-11). **Detritus feeders**, such as crabs, carpenter ants, termites, earthworms, and wood beetles, extract nutrients from partly decomposed organic matter in leaf litter, plant debris, and animal dung.

Decomposers, mostly certain types of bacteria and fungi, are consumers that complete the final breakdown and recycling of organic materials from the remains or wastes of all organisms. They recycle organic matter in ecosystems by breaking down dead organic material (detritus) to get nutrients and releasing the resulting simpler inorganic compounds into the soil and water, where they can be taken up as nutrients by producers. In turn, decomposers are important food sources for worms and insects living in the soil and water. When we say that something is **biodegradable**, we mean that it can be broken down by decomposers. Figures 4-9 and 4-10 show various types of producers and consumers.

Both producers and consumers use the chemical energy stored in glucose and other organic compounds to fuel their life processes. In most cells, this energy is released by the process of **aerobic respiration**, which uses oxygen to convert organic nutrients back into carbon dioxide and water. The net effect of the hundreds of steps in this complex process is represented by the following reaction:

$$\text{glucose} + \text{oxygen} \longrightarrow \text{carbon dioxide} + \text{water} + \textbf{energy}$$

$$C_6H_{12}O_6 + 6\,O_2 \longrightarrow 6\,CO_2 + 6\,H_2O + \textbf{energy}$$

Although the detailed steps differ, the net chemical change for aerobic respiration is the opposite of that for photosynthesis, which takes place during the day, when sunlight is available; aerobic respiration can happen day or night.

Some decomposers get the energy they need through the breakdown of glucose (or other nutrients) in the absence of oxygen. This form of cellular respiration is called **anaerobic respiration** or **fermentation**. Instead of carbon dioxide and water, the end products of this process are compounds such as methane gas (CH_4), ethyl alcohol (C_2H_6O), acetic acid (the main component of vinegar, $C_2H_4O_2$), and hydrogen sulfide (H_2S, when sulfur compounds are broken down).

The survival of any individual organism depends on the *flow of matter and energy* through its body. However, an ecosystem as a whole survives primarily through a combination of *matter recycling* (rather than one-way flow) and *one-way energy flow* (Figure 4-12). Decomposers complete the cycle of matter by breaking down detritus into inorganic nutrients that are usable by producers. Without decomposers, the entire world would soon be knee-deep in plant litter, dead animal bodies, animal wastes, and garbage. Most life as we know it would no longer exist.

What Are the Major Nonliving Components of Ecosystems?

The nonliving, or abiotic, components of an ecosystem are the physical and chemical factors that influence living organisms. Some important physical factors affecting land ecosystems are sunlight, temperature, precipitation, wind, latitude (distance from the equator), altitude (distance above sea level), frequency of fire, and nature of the soil. For aquatic ecosystems, water currents and the amount of suspended solid material are major physical factors.

Important chemical factors affecting ecosystems are the supply of water and air in the soil and the supply of plant nutrients or toxic substances dissolved in soil moisture (or in water in aquatic habitats). In aquatic ecosystems, salinity and the level of dissolved oxygen are also major chemical factors.

How Much Change in Abiotic Factors Can Populations Tolerate?

Different species thrive under different physical conditions. Some need bright sunlight; others thrive better in shade. Some require a hot environment, others a cool or cold one. Some do best under very wet conditions; others under fairly dry conditions.

Each population in an ecosystem has a **range of tolerance** to variations in its physical and chemical environment (Figure 4-13). For example, trout thrive in colder water than do bass or perch, which in turn need colder water than catfish. Individuals within a population may also have slightly different tolerance ranges for temperature or other factors because of small differences in genetic makeup, health, and age. Thus, although a trout population may do best within a narrow band of temperatures (*optimum level or range*), a few

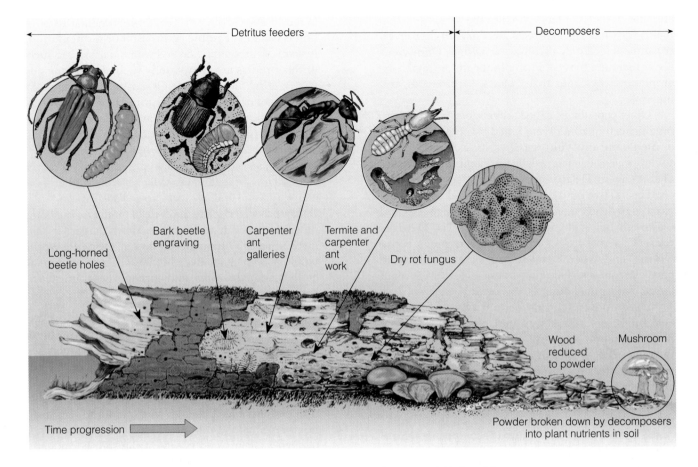

Detritus feeders — | — Decomposers

Long-horned beetle holes

Bark beetle engraving

Carpenter ant galleries

Termite and carpenter ant work

Dry rot fungus

Wood reduced to powder

Mushroom

Time progression

Powder broken down by decomposers into plant nutrients in soil

Figure 4-11 Some detritivores, called *detritus feeders*, directly consume tiny fragments of this log. Other detritivores, called *decomposers* (mostly fungi and bacteria), digest complex organic chemicals in fragments of the log into simpler inorganic nutrients. If these nutrients are not washed away or otherwise removed from the system, they can be used again by producers near this location.

individuals can survive both above and below that band. As Figure 4-13 shows, tolerance has its limits, beyond which none of the trout can survive.

These observations are summarized in the **law of tolerance**: *The existence, abundance, and distribution of a species in an ecosystem are determined by whether the levels of one or more physical or chemical factors fall within the range tolerated by that species.* In other words, there are minimum and maximum limits for physical conditions (such as temperature) and concentrations of chemical substances, called **tolerance limits**, beyond which no members of a particular species can survive.

A species may have a wide range of tolerance to some factors and a narrow range of tolerance to others. Most organisms are least tolerant during the juvenile or reproductive stages of their life cycles. Highly toler-

ant species can live in a variety of habitats with widely different conditions.

Some species can adjust their tolerance to physical or chemical factors if change is gradual, just as you can tolerate a hotter bath by slowly adding hot water. This adjustment to slowly changing new conditions, or **acclimation**, is a useful adaptation. However, acclimation has limits, and as change occurs a species comes closer to its absolute limit. Then, suddenly, without warning, the next small change triggers a **threshold effect**, a harmful or even fatal reaction as the tolerance limit is exceeded. It is like the proverbial straw that breaks an already heavily loaded camel's back.

The threshold effect explains why many environmental problems seem to arise suddenly. For example, when spruce trees suddenly begin dying in large numbers, the cause may be decades of exposure to

Q: When did the first forms of life arise on the earth?

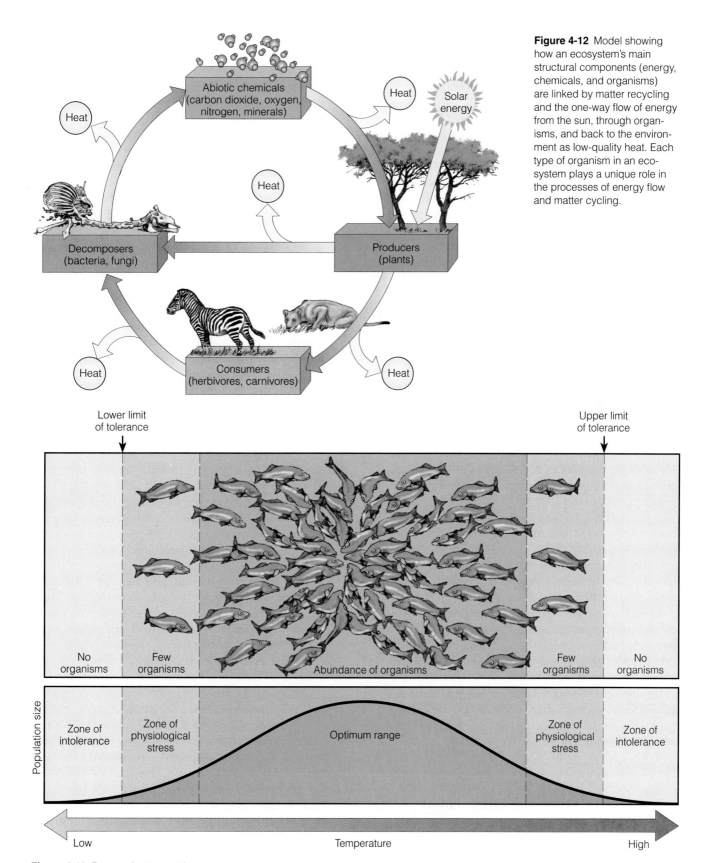

Figure 4-12 Model showing how an ecosystem's main structural components (energy, chemicals, and organisms) are linked by matter recycling and the one-way flow of energy from the sun, through organisms, and back to the environment as low-quality heat. Each type of organism in an ecosystem plays a unique role in the processes of energy flow and matter cycling.

Figure 4-13 Range of tolerance for a population of organisms to an abiotic environmental factor—in this case, temperature.

A: 3.6–3.8 billion years ago (probably bacterial cells)

numerous air pollutants, which can make the trees more vulnerable to drought, cold weather, diseases, and insects. Prevention of pollution and environmental degradation is the best way to keep thresholds from being exceeded.

What Factor Often Limits the Size of a Population? Although the organisms in a population are affected by a variety of environmental factors, one factor, known as the **limiting factor**, often turns out to be more important than others in regulating population growth. This ecological principle, related to the law of tolerance, is called the **limiting factor principle**: *Too much or too little of any abiotic factor can limit or prevent growth of a population, even if all other factors are at or near the optimum range of tolerance.*

Limiting factors in land ecosystems include temperature, water, light, and soil nutrients. On land, precipitation often is the limiting factor. Lack of water in a desert limits the growth of plants. Soil nutrients can also serve as a limiting factor on land. Suppose a farmer plants corn in phosphorus-poor soil. Even if water, nitrogen, potassium, and other nutrients are at optimum levels, the corn will stop growing when it uses up the available phosphorus. Here, the amount of phosphorus in the soil limits corn growth.

Too much of an abiotic factor can also be limiting. For example, plants can be killed by too much water or too much fertilizer, a common mistake of many beginning gardeners.

The limiting factor for a particular population can sometimes change. At the beginning of a plant's growing season, for example, temperature may be the limiting factor; later on, the supply of a particular nutrient may limit growth; and if a drought occurs, water may be the limiting factor.

One limiting factor in aquatic ecosystems is **salinity** (the amounts of various salts dissolved in a given volume of water). Aquatic ecosystems can be divided into surface, middle, and bottom layers or life zones; important limiting factors for these layers are temperature, sunlight, **dissolved oxygen (DO) content** (the amount of oxygen gas dissolved in a given volume of water at a particular temperature and pressure), and availability of nutrients.

4-3 CONNECTIONS: ENERGY FLOW IN ECOSYSTEMS

How Does Energy Flow Through Ecosystems? Food Chains and Food Webs All organisms, whether dead or alive, are potential sources of food for other organisms. A caterpillar eats a leaf, a robin eats the caterpillar, and a hawk eats the robin. When leaf, caterpillar, robin, and hawk have all died, they in turn are consumed by decomposers. As a result, *there is little waste in natural ecosystems.*

The sequence of organisms, each of which is a source of food for the next, is called a **food chain**. It determines how energy and nutrients move from one organism to another through the ecosystem (Figure 4-14). Energy enters most ecosystems as high-quality sunlight, which is converted to nutrients by photosynthesizing producers (mostly plants). The energy is then passed on to consumers and eventually to decomposers. As each organism uses the high-quality chemical energy in its food to move, grow, and reproduce, this energy is converted into low-quality heat that flows into the environment in accordance with the second energy law.

Ecologists assign each of the organisms in an ecosystem to a *feeding level*, or **trophic level** (from the Greek word *trophos*, "nourishment"), depending on whether it is a producer or a consumer and on what it eats or decomposes. Producers belong to the first trophic level, primary consumers to the second trophic level, secondary consumers to the third, and so on. Detritivores process detritus from all trophic levels.

Real ecosystems are more complex than this. Most consumers feed on more than one type of organism, and most organisms are eaten by more than one type of consumer. Because most species participate in several different food chains, the organisms in most ecosystems form a complex network of interconnected food chains called a **food web** (Figure 4-15).

How Can We Represent the Energy Flow and Storage of Biomass in an Ecosystem? Each trophic level in a food chain or web contains a certain amount of **biomass**, the combined dry weight of all organic matter contained in its organisms. Biomass is measured as the *dry* weight of organisms because the water they contain is not a source of energy or nutritional value. Biomass represents the chemical energy stored in the organic matter of a trophic level.

In a food chain or web, energy stored in biomass is transferred from one trophic level to another, with some usable energy degraded and lost to the environment as low-quality heat in each transfer. At each successive trophic level, some of the available biomass is neither eaten, digested, nor absorbed; it simply goes through the intestinal tract of the consumer and is expelled as fecal waste.

Thus, only a small portion of what is eaten and digested is actually converted into an organism's bodily material or biomass, and the amount of usable energy

Q: How many species are there on the earth?

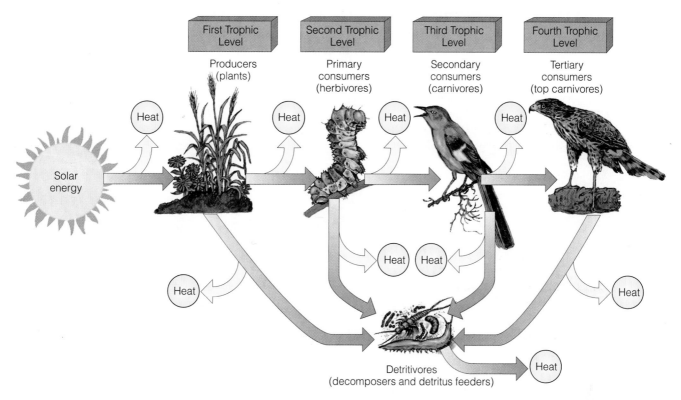

Figure 4-14 Model of a food chain. The arrows show how chemical energy in food flows through various *trophic levels* or energy transfers; most of the energy is degraded to heat in accordance with the second law of energy. Food chains rarely have more than four trophic levels.

available to each successive trophic level declines. The percentage of usable energy transferred as biomass from one trophic level to the next varies from 5% to 20% (that is, 80–95% is lost), depending on the types of species and the ecosystem involved. Assuming a 90% loss at each transfer, if green plants in an area manage to capture 10,000 units of energy from the sun, then only about 1,000 units of energy will be available to support herbivores, and only about 100 units to support carnivores.

The more trophic levels or steps in a food chain or web, the greater the cumulative loss of usable energy as energy flows through the various trophic levels. The **pyramid of energy flow*** in Figure 4-16 illustrates this energy loss for a simple food chain, assuming a 90% energy loss with each transfer. Figure 4-17 shows the pyramid of energy flow during one year for an aquatic ecosystem in Silver Springs, Florida. Pyramids of energy flow *always* have an upright pyramidal shape

---

* Because such pyramids represent energy flows, not energy storage, they should not be called a pyramid of energy (a common error in many biology and environmental science textbooks).

because of the automatic degradation of energy quality required by the second law of energy.

Energy flow pyramids explain why the earth can support more people if they eat at lower trophic levels by consuming grains, vegetables, and fruits directly (for example, grain ⟶ human), rather than passing such crops through another trophic level and eating grain eaters (grain ⟶ steer ⟶ human).

The large loss in energy between successive trophic levels also explains why food chains and webs rarely have more than four or five trophic levels. In most cases, too little energy is left after four or five transfers to support organisms feeding at these high trophic levels. This explains why top carnivores such as eagles, hawks, tigers, and white sharks are few in number and are usually the first to suffer when the ecosystems that support them are disrupted. This makes such species especially vulnerable to extinction.

The storage of biomass at various trophic levels in an ecosystem can be represented by a **pyramid of biomass** (Figure 4-18, p. 86). Ecologists estimate biomass by harvesting organisms from random patches

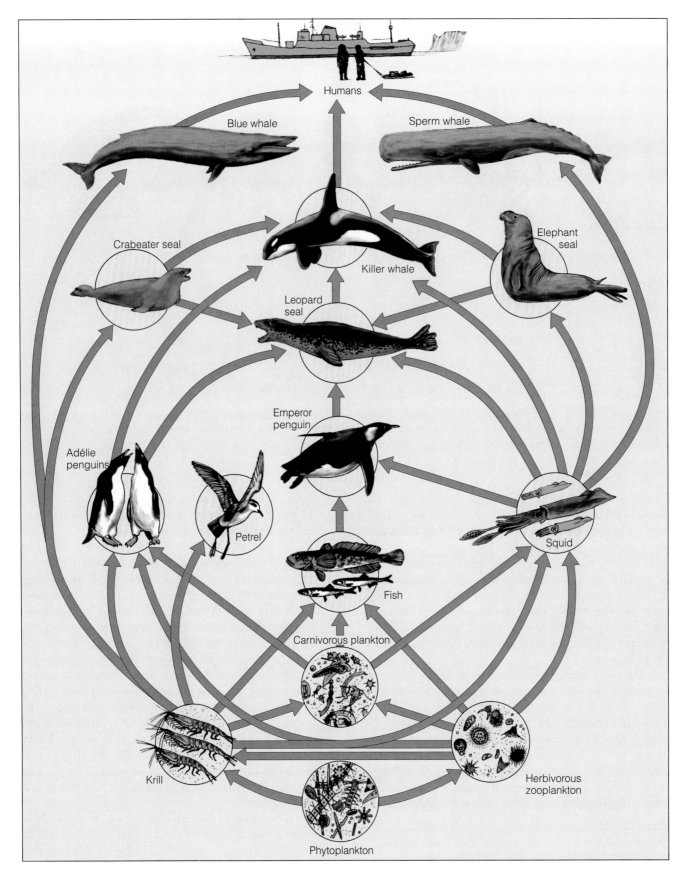

Figure 4-15 Model of a greatly simplified food web in the Antarctic. Many more participants in the web, including an array of decomposer organisms, are not depicted here.

Q: How many of earth's species have been identified?

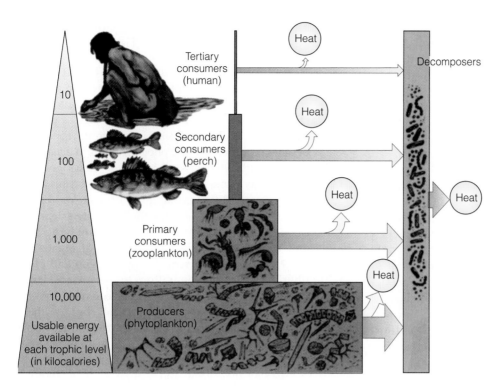

Figure 4-16 Generalized pyramid of energy flow, showing the decrease in usable energy available at each succeeding trophic level in a food chain or web. This conceptual model assumes a 90% loss in usable energy to the environment (in the form of low-quality heat) with each transfer from one trophic level to another. In nature, such losses vary from 80% to 95%. Because of the degradation of energy quality required by the second law of energy, these models always have a pyramidal shape.

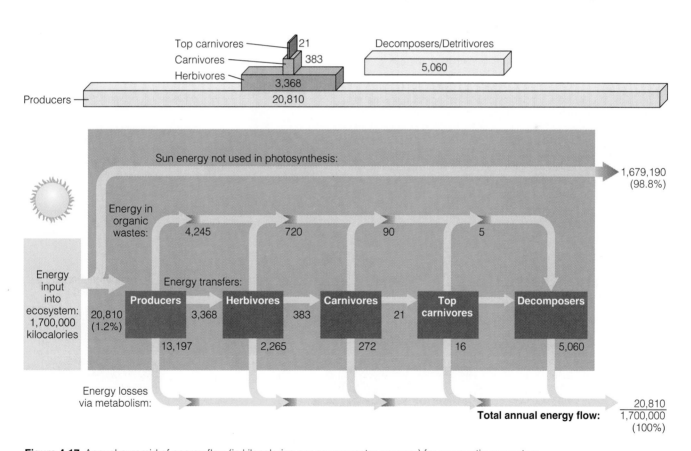

Figure 4-17 Annual pyramid of energy flow (in kilocalories per square meter per year) for an aquatic ecosystem in Silver Springs, Florida. The pyramid is constructed by using the data on energy flow through this ecosystem shown in the bottom drawing. (Used by permission from Cecie Starr, *Biology: Concepts and Applications*, 2nd ed., Belmont, Calif.: Wadsworth, 1994)

A: About 1.8 million

or narrow strips in an ecosystem. The sample organisms are then sorted according to trophic levels, dried, and weighed. These data are used to plot a pyramid of biomass. Because the typical relationship involves many producers but not so many primary consumers and just a few secondary consumers, the graph usually looks like an upright pyramid (Figure 4-18, left). However, such graphs for some ecosystems do not have the typical upright pyramid shape (Figure 4-18, right).

By estimating the number of organisms at each trophic level, ecologists can also create a graphic display called a **pyramid of numbers** for an ecosystem (Figure 4-19, left). For example, 1,000 metric tons of grass might support 27,000,000 grasshoppers, which might support 90,000 frogs, which might support 300 trout, which in turn could feed 1 person for about 30 days. By eliminating trout from their diet, 30 people could survive for 30 days by consuming 10 frogs a day. If the food situation were more desperate, frogs could be eliminated from the diet and 900 people could survive for 30 days by eating about 100 grasshoppers a day. (Fried grasshoppers are a delicacy in some countries and contain more protein per unit of weight than most conventional forms of meat and meat products.)

These graphs for some ecosystems do not have the typical upright pyramid shape (Figure 4-19, right).

How Rapidly Do Producers in Different Ecosystems Produce Biomass? The *rate* at which an ecosystem's producers convert solar energy into chemical energy as biomass is the ecosystem's **gross primary productivity (GPP)**. In effect, it is the rate at which plants or other producers can use photosynthesis to make more plant material (biomass). Figure 4-20 shows how this productivity varies in different parts of the earth.

Figure 4-20 shows that gross primary productivity is generally greatest in the shallow waters near continents; along coral reefs where abundant light, heat, and nutrients stimulate the growth of algae; and where upwelling currents bring nitrogen and phosphorus from the ocean bottom to the surface. High-nutrient, upwelling regions of the ocean make up only one-thousandth of the total ocean area, but they have high enough primary productivities to produce nearly one-half of the world's marine fish. Most of the open ocean has a low gross primary productivity.

To stay alive, grow, and reproduce, an ecosystem's producers must use some of the total biomass they pro-

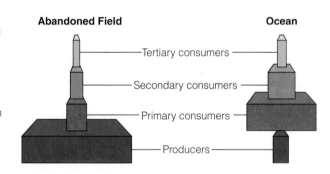

Figure 4-18 Generalized graphs of biomass of organisms in the various trophic levels for two ecosystems. The size of each tier in this conceptual model represents the dry weight per square meter of all organisms at that trophic level. For most land ecosystems, the total biomass at each successive trophic level decreases, yielding a pyramid of biomass with a large base of producers, topped by a series of increasingly smaller biomasses at higher trophic levels (left). In the open waters of aquatic ecosystems (right), the biomass of primary consumers (zooplankton) can actually exceed that of producers (which are microscopic phytoplankton that grow and reproduce rapidly, not large plants that grow and reproduce slowly). The zooplankton eat the phytoplankton almost as fast as they are produced. As a result, the producer population is never very large, and the graph is not an upright pyramid.

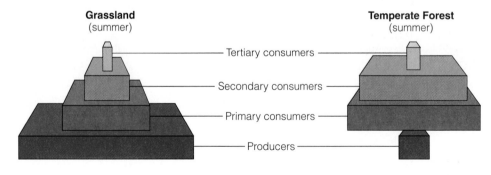

Figure 4-19 Generalized graphs of numbers of organisms in the various trophic levels for two ecosystems. Numbers of organisms for grasslands and many other ecosystems taper off from the producer level to the higher trophic levels, forming an upright pyramid (left). For other ecosystems, however, the graph can take different shapes. For example, a temperate forest (right) has a few large producers (the trees) that support a much larger number of small primary consumers (insects) that feed on the trees.

Q: What percentage of all species that have ever lived have become extinct?

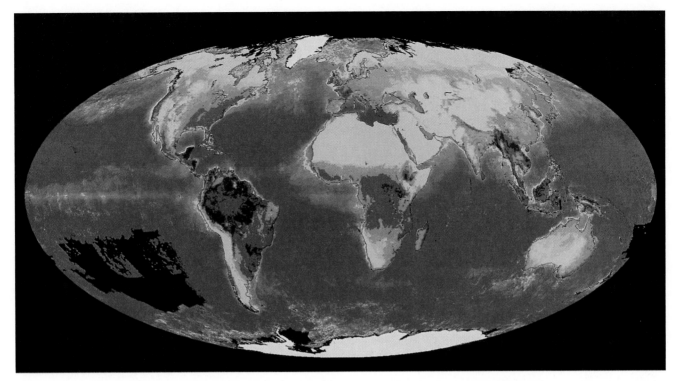

Figure 4-20 Three years of satellite data on the earth's gross primary productivity. Rain forests and other highly productive areas appear as dark green, deserts as yellow. The concentration of phytoplankton, a primary indicator of ocean productivity, ranges from red (highest) to orange, yellow, green, and blue (lowest). (Gene Carl Feldman, Compton J. Tucker-NASA/Goddard Space Flight Center)

duce for their own respiration. Only what is left, called **net primary productivity (NPP)**, is available for use as food by other organisms (consumers) in an ecosystem:

Net primary productivity = Rate at which producers store chemical energy stored as biomass (produced by photosynthesis) − Rate at which producers use chemical energy stored as biomass (through aerobic respiration)

Net primary productivity is the *rate* at which energy for use by consumers is stored in new biomass (cells, leaves, roots, and stems). It is usually measured in units of the energy or biomass available to consumers in a specified area over a given time. It is measured in kilocalories per square meter per year ($kcal/m^2/yr$) or in grams of biomass created per square meter per year ($g/m^2/yr$). Ultimately, the planet's total net primary productivity limits the number of consumers, including humans, that can survive on the earth. In other words, *the earth's total net primary productivity is the upper limit determining the planet's carrying capacity for all species.*

Various ecosystems and life zones differ in their net primary productivity (Figure 4-21). Estuaries, swamps and marshes, and tropical rain forests are highly pro-

ductive; open ocean, tundra (arctic and alpine grasslands), and desert are the least productive. It is tempting to conclude that to feed our hungry millions we should harvest plants in estuaries, swamps, and marshes, or clear tropical forests and plant crops. The grasses in estuaries, swamps, and marshes cannot be eaten by people, however, and they are vital food sources (and spawning areas) for fish, shrimp, and other aquatic life that provide us and other consumers with protein.

In tropical forests, most nutrients are stored in the vegetation rather than in the soil. When the trees are removed, the nutrient-poor soils are rapidly depleted of their nutrients by frequent rains and growing crops. Crops can be grown only for a short time without massive and expensive applications of commercial fertilizers. This explains why many ecologists urge us to protect, not clear, large areas of tropical forest to supply food.

Agricultural land is a highly modified ecosystem in which we try to increase the net primary productivity and biomass of selected crop plants by adding water (irrigation) and nutrients (fertilizers). Nitrogen as nitrate (NO_3^-) and phosphorus as phosphate (PO_4^{3-}) are the most common nutrients in fertilizers because they are most often the nutrients limiting crop growth. Despite such inputs, the net primary productivity of

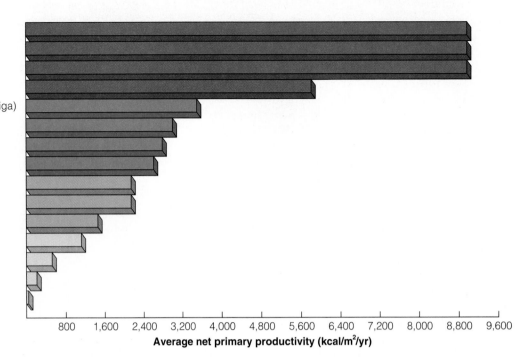

Type of Ecosystem

Estuaries
Swamps and marshes
Tropical rain forest
Temperate forest
Northern coniferous forest (taiga)
Savanna
Agricultural land
Woodland and shrubland
Temperate grassland
Lakes and streams
Continental shelf
Open ocean
Tundra (arctic and alpine)
Desert scrub
Extreme desert

800 1,600 2,400 3,200 4,000 4,800 5,600 6,400 7,200 8,000 8,800 9,600

Average net primary productivity (kcal/m²/yr)

Figure 4-21 Estimated annual average net primary productivity per unit of area in major life zones and ecosystems, expressed as kilocalories of energy produced per square meter per year (kcal/m²/yr). (Data from R. H. Whittaker, *Communities and Ecosystems*, 2d ed., New York: Macmillan, 1975)

agricultural land is not particularly high compared to that of other ecosystems (Figure 4-21).

Excluding uninhabitable areas of rock, ice, desert, and steep mountain terrain, humans have taken over, disturbed, or degraded about 73% of the earth's land surface (Figure 1-5). Ecologists have estimated that humans now use, waste, or destroy about 27% of the earth's total potential net primary productivity and 40% of the net primary productivity of the planet's terrestrial ecosystems.

This is the main reason why we are crowding out or eliminating the habitats and food supplies for a growing number of other species. If current estimates of our use of the earth's annual net primary productivity are reasonably correct, what will happen to us and to other species if the human population doubles over the next 40–50 years and per capita consumption of rises sharply?

4-4 CONNECTIONS: MATTER CYCLING IN ECOSYSTEMS

How Is Carbon Cycled in the Ecosphere? Carbon is essential to life as we know it. It is the basic building block of the carbohydrates, fats, proteins, nu-

cleic acids such as DNA and RNA, and other organic compounds necessary for life. The **carbon cycle**, a global gaseous cycle, is based on carbon dioxide gas, which makes up only 0.036% of the volume of the troposphere and is also dissolved in water.

As a heat-trapping gas, carbon dioxide is a key component of nature's thermostat. If the carbon cycle removes too much CO_2 from the atmosphere, the earth will cool; if the cycle generates too much, the earth will get warmer. Thus even slight changes in the carbon cycle can affect climate and ultimately the types of life that can exist on various parts of the planet.

Terrestrial producers remove CO_2 from the atmosphere and aquatic producers remove it from the water. They then use photosynthesis to convert CO_2 into complex carbohydrates such as glucose ($C_6H_{12}O_6$).

The cells in oxygen-consuming producers, consumers, and decomposers then carry out aerobic respiration, which breaks down glucose and other complex organic compounds and converts the carbon back to CO_2 in the atmosphere or water, for reuse by producers. This linkage between photosynthesis in producers and aerobic respiration in producers, consumers, and decomposers circulates carbon in the ecosphere and is a major part of the global carbon cycle (Figure 4-22). Oxygen and hydrogen, the other elements in carbohydrates, cycle almost in step with carbon.

Q: What two countries have the world's largest populations?

Over millions of years such buried organic matter is compressed between layers of sediment, where it forms carbon-containing fossil fuels such as coal and oil (Figure 4-22). This carbon is not released to the atmosphere as CO_2 for recycling until these fuels are extracted and burned or until long-term geological processes expose these deposits to air. In the short time period of a few hundred years, we have been extracting and burning fossil fuels that took millions of year to form from dead plant matter, which is why fossil fuels are nonrenewable resources on a human time scale.

Oceans also play a major role in regulating the level of carbon dioxide in the atmosphere. Some carbon dioxide gas, which is readily soluble in water, stays dissolved in the sea, some is removed by photosynthesizing producers, and some reacts with seawater to form carbonate ions (CO_3^{2-}) and bicarbonate ions (HCO_3^-). As water warms, more dissolved CO_2 returns to the atmosphere, just as more carbon dioxide fizzes out of a carbonated beverage when it warms.

In marine ecosystems, some organisms take up dissolved CO_2 molecules, carbonate ions, or bicarbonate ions from ocean water. These ions can then react with calcium ions (Ca^{2+}) in seawater to form slightly soluble carbonate compounds such as calcium carbonate ($CaCO_3$) to build the shells and skeletons of marine organisms. When these organisms die, tiny particles of their shells and bone drift slowly to the ocean depths and are buried for eons (as long as 400 million years) in deep bottom sediments (Figure 4-22), where under immense pressure they are converted into limestone rock. About 55 times more carbon is stored in dissolved carbon compounds and in insoluble marine sediments than is stored in the atmosphere.

Since 1800 and especially since 1950, as world population and resource use have soared, we have disturbed the carbon cycle in two ways that add more carbon dioxide to the atmosphere than oceans and plants so far have been able to remove: (1) Forest and brush removal has less vegetation to absorb CO_2 through photosynthesis, and (2) burning fossil fuels and wood produces CO_2 that flows into the atmosphere.

Computer models of the earth's climate systems suggest that increased concentration of CO_2 (and of other heat-trapping gases we're adding to the atmosphere) could enhance the planet's natural greenhouse effect. This natural greenhouse effect, caused mostly by water vapor and CO_2 in the atmosphere, is a key factor in producing livable climates on the earth over the roughly 60,000 years since the current species of human arrived on the scene.

However, measurements show that we are adding more CO_2 to the atmosphere than oceans, plants, and soil so far have been able to remove. This could alter climate patterns for hundreds to thousands of years as the carbon cycle adjusts to these rapid inputs. It could also disrupt global food production and wildlife habitats and possibly raise the average sea level, as discussed in more detail in Chapter 10.

How Is Nitrogen Cycled in the Ecosphere? Bacteria in Action Although chemically unreactive nitrogen gas (N_2) makes up 78% of the volume of the troposphere, it cannot be absorbed and used directly as a nutrient by multicellular plants or animals. Fortunately, lightning and certain bacteria convert nitrogen gas into compounds that can enter food webs as part of the **nitrogen cycle** (Figure 4-23, p. 92).

In the first step in the nitrogen cycle, called *nitrogen fixation*, specialized bacteria convert gaseous nitrogen (N_2) to ammonia (NH_3) by the reaction $N_2 + 3H_2 \longrightarrow 2NH_3$. This is done mostly by cyanobacteria in soil and water and by *Rhizobium* bacteria living in small nodules (swellings) on the root systems of legumes, a huge family of plants with about 18,000 species, including soybeans, alfalfa, and clover (Figure 4-24, p. 92).

Plants can use ammonia or ammonium ions (NH_4^+) formed when ammonia reacts with water as a source of nitrogen. However, in a two-step process called *nitrification*, most of the ammonia in soil is converted by specialized aerobic bacteria to nitrite ions (NO_2^-), which are toxic to plants and then to nitrate ions (NO_3^-), which are easily taken up by plants as a nutrient.

In a process called *assimilation*, plant roots then absorb inorganic ammonia, ammonium ions, and nitrate ions formed by nitrogen fixation and nitrification in soil water. They use these ions to make nitrogen-containing organic molecules such as DNA, amino acids, and proteins. Animals in turn get their nitrogen by eating plants or plant-eating animals.

After nitrogen has served its purpose in living organisms, vast armies of specialized decomposer bacteria convert the nitrogen-rich organic compounds, wastes, cast-off particles, and dead bodies of organisms into simpler nitrogen-containing inorganic compounds such as ammonia (NH_3) and water-soluble salts containing ammonium ions (NH_4^+). This process is known as *ammonification*.

In a process called *denitrification*, other specialized bacteria (mostly anaerobic bacteria in waterlogged soil or in the bottom sediments of lakes, oceans, swamps, and bogs) then convert NH_3 and NH_4^+ back into nitrite (NO_2^-) and nitrate (NO_3^-) ions and then into nitrogen gas (N_2) and nitrous oxide gas (N_2O). These are then released to the atmosphere to begin the cycle again.

Although the nitrogen cycle provides large amounts of ammonium and nitrate ions for use by terrestrial plants, compounds containing these ions are soluble in water and thus can easily be removed (leached) by water

Figure 4-22 Simplified conceptual model of the global carbon cycle. The left portion shows the movement of carbon through marine ecosystems, and the right portion shows its movement through terrestrial ecosystems. Carbon reservoirs are shown as boxes. (Modified by permission from Cecie Starr and Ralph Taggart, *Biology: The Unity and Diversity of Life*, 6th ed., Belmont, Calif.: Wadsworth, 1992)

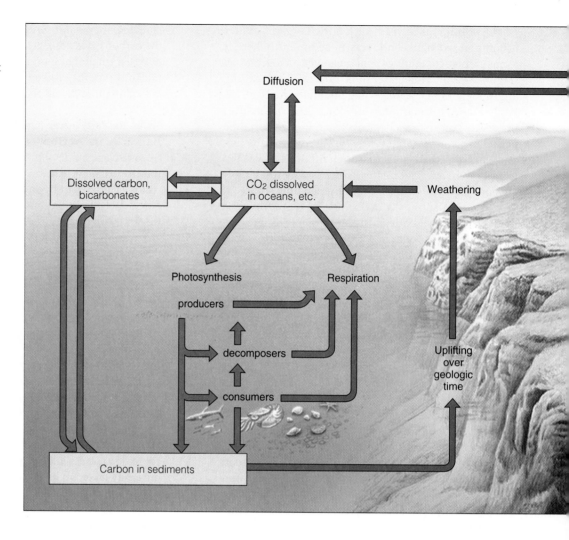

flowing across or percolating down through soil. Thus, nitrogen is often a factor limiting plant growth.

Humans intervene in the nitrogen cycle in several ways. *First*, we emit large quantities of nitric oxide (NO) into the atmosphere when we burn any fuel. Most of this NO is produced when nitrogen and oxygen molecules in the air combine at high temperatures: $N_2 + O_2 \longrightarrow 2NO$. In the atmosphere, this nitric oxide combines with oxygen to form nitrogen dioxide gas (NO_2), which can react with water vapor to form nitric acid (HNO_3). Droplets of HNO_3 dissolved in rain or snow are components of **acid deposition** (commonly called **acid rain**), which along with other air pollutants can damage and weaken trees and upset aquatic systems.

Second, humans emit heat-trapping nitrous oxide gas (N_2O) into the atmosphere through the action of anaerobic bacteria on livestock wastes and commercial inorganic fertilizers applied to the soil. *Third*, we remove nitrogen from the earth's crust when we mine nitrogen-containing mineral deposits (such as ammonium nitrate or NH_4NO_3) for fertilizers, deplete nitro-

gen from topsoil by harvesting nitrogen-rich crops, and leach water-soluble nitrate ions from soil through irrigation. *Fourth*, we remove nitrogen from topsoil when we burn grasslands and clear forests before planting crops. At the same time, we emit nitrogen oxides into the atmosphere.

Fifth, we add excess nitrogen compounds to aquatic systems in agricultural runoff and discharge of municipal sewage. This excess of plant nutrients stimulates rapid growth of photosynthesizing algae and other aquatic plants. The subsequent breakdown of dead algae by aerobic decomposers can deplete the water of dissolved oxygen and can disrupt aquatic systems.

How Is Phosphorus Cycled in the Ecosphere?

Phosphorus circulates through water, the earth's crust, and living organisms in the **phosphorus cycle** (Figure 4-25). In this cycle, phosphorus moves slowly from phosphate deposits on land and in shallow ocean sediments to living organisms, and then much more slowly back to the land and ocean. Bacteria are less important here than in the nitrogen cycle. Unlike car-

Q: At what rate was the world's population growing in 1997?

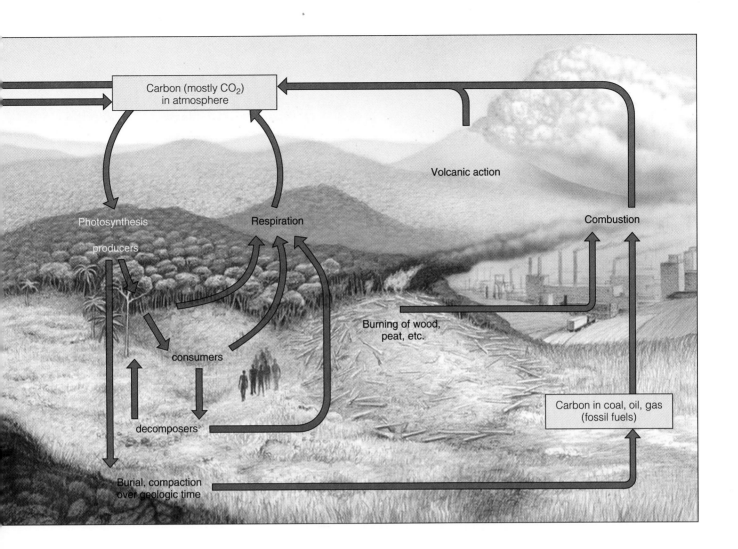

Carbon (mostly CO₂) in atmosphere

Photosynthesis

producers

Respiration

Volcanic action

Combustion

consumers

Burning of wood, peat, etc.

decomposers

Carbon in coal, oil, gas (fossil fuels)

Burial, compaction over geologic time

bon and nitrogen, very little phosphorus circulates in the atmosphere because at the earth's normal temperatures and pressures, phosphorus and its compounds are not gases. Phosphorus is found in the atmosphere only as small particles of dust. In contrast to the carbon cycle, the phosphorus cycle is slow, and on a short human time scale much phosphorus flows one way from the land to the oceans.

Phosphorous is typically found as phosphate salts containing phosphate ions (PO_4^{3-}) in terrestrial rock formations and ocean bottom sediments. Phosphorus released by the slow breakdown, or *weathering*, of terrestrial phosphate rock deposits is dissolved in soil water and then taken up by plant roots. Wind can also carry phosphate particles long distances.

Because most soils contain little phosphate, it is often the limiting factor for plant growth on land unless phosphorus (as phosphate salts mined from the earth) is applied to the soil as a fertilizer. Phosphorus also limits the growth of producer populations in many freshwater streams and lakes because phosphate salts are only slightly soluble in water.

Phosphorus cycles much more rapidly through the living components of ecosystems than it does through geological formations. Animals get phosphorus by eating producers or animals that have eaten producers. Animal wastes and the decay of dead animals and producers return much of this phosphorus as phosphate to the soil, to streams, and eventually to the ocean bottom as insoluble deposits of phosphate rock. This phosphorus remains there for millions to hundreds of millions of years before being returned to the land by geologic processes that over millions of years may push up and expose the seafloor. Weathering then slowly releases phosphorus from the exposed rocks and continues the cycle.

Some phosphate returns to the land as guano, phosphate-rich manure typically of fish-eating birds such as pelicans and cormorants. This return is small, though, compared with the phosphate transferred from the land to the oceans each year by natural processes and human activities. Severe erosion, especially by human activities, accelerates this transfer of phosphate from terrestrial ecosystems to the sea.

A: 1.47% (down from 2.2% in 1963)

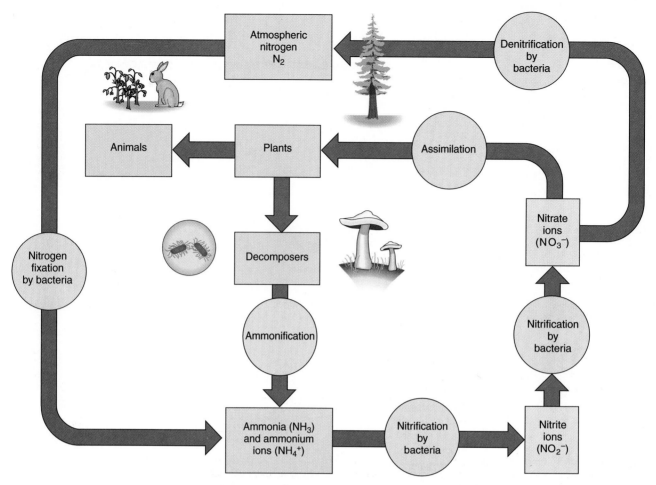

Figure 4-23 Greatly simplified conceptual model of the nitrogen cycle in a terrestrial ecosystem. Nitrogen reservoirs are shown as boxes, and processes changing one form of nitrogen to another are shown in circles.

Figure 4-24 Plants in the legume family (which includes alfalfa, clover, peas, beans, mesquite, and acacias) have root nodules where *Rhizobium* bacteria fix nitrogen; that is, they convert nitrogen gas (N_2) into ammonia (NH_3), which in soil water forms ammonium ions (NH_4^+), which are converted to nitrate ions (NO_3^-) that are taken up by the roots of plants. This process benefits both species: The bacteria convert atmospheric nitrogen into a form usable by the plants, and the plants provide the bacteria with some simple carbohydrates. (E. R. Degginger)

Humans intervene in the phosphorus cycle chiefly in three ways: *First,* we mine large quantities of phosphate rock for use in commercial inorganic fertilizers and detergents. Surface mining of phosphate leaves huge mining pits and slurry ponds that mar the landscape, can pollute nearby surface and groundwater, and are expensive to reclaim.

Second, we are sharply reducing the available phosphate and primary productivity of tropical forests by cutting them. In such ecosystems, hardly any phosphorus is found in the soil. Rather, it's in the ecosystem's plant and animal life, which is rapidly recycled from dead plants and animals by hordes of decomposers. When such forests are cut and burned, most remaining phosphorus and other nutrients are readily washed away by heavy rains, and the land becomes unproductive.

Third, we add excess phosphate to aquatic ecosystems in runoff of animal wastes from livestock feedlots, runoff of commercial phosphate fertilizers from cropland, and discharge of municipal sewage. Too much of this nutrient causes explosive growth of cyanobacteria,

Q: Worldwide, what is the average number of children per woman?

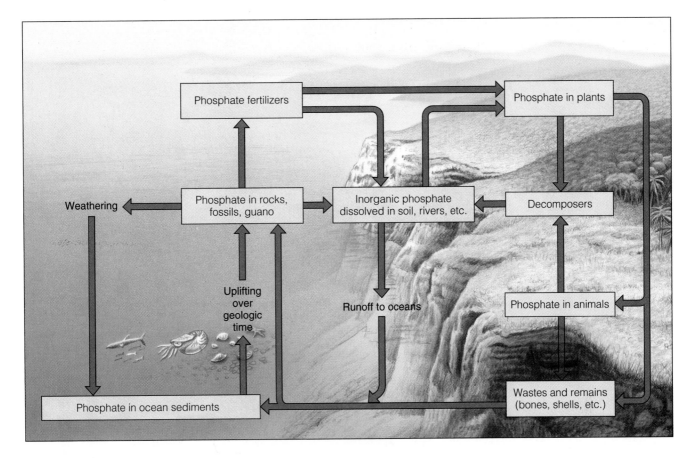

Figure 4-25 Simplified conceptual model of the phosphorus cycle. Phosphorus reservoirs are shown as boxes; the processes shown change one form of phosphorus to another. (Modified by permission from Cecie Starr and Ralph Taggart, *Biology: The Unity and Diversity of Life*, 6th ed., Belmont, Calif.: Wadsworth, 1992)

algae, and aquatic plants, with surface mats of such species blocking sunlight. As a result, plants rooted on the bottom may die because sunlight becomes the limiting factor. Huge numbers of cyanobacteria (formerly known as blue-green algae), algae, and plants age and die in an overfertilized lake. Then decomposing aerobic bacteria feeding on their dead cells use up so much dissolved oxygen that some species of fish and other aquatic animals may suffocate.

How Is Sulfur Cycled in the Ecosphere? Sulfur circulates through the ecosphere in the **sulfur cycle** (Figure 4-26). Much of the earth's sulfur is tied up underground in rocks (such as iron disulfide or pyrite) and minerals, including sulfate (SO_4^{2-}) salts (such as hydrous calcium sulfate or gypsum) buried deep under ocean sediments.

Sulfur also enters the atmosphere from several natural sources. Hydrogen sulfide (H_2S), a colorless, highly poisonous gas with a rotten-egg smell, is released from active volcanoes and by the breakdown of organic matter in swamps, bogs, and tidal flats caused

by decomposers that don't use oxygen (anaerobic decomposers). Sulfur dioxide (SO_2), a colorless, suffocating gas, also comes from volcanoes. Particles of sulfate (SO_4^{2-}) salts, such as ammonium sulfate, enter the atmosphere from sea spray.

In the atmosphere, sulfur dioxide reacts with oxygen to produce sulfur trioxide gas (SO_3). Some of the sulfur dioxide then reacts with water droplets in the atmosphere to produce tiny droplets of sulfuric acid (H_2SO_4). Sulfur dioxide also reacts with other chemicals in the atmosphere such as ammonia to produce tiny particles of sulfate salts. These droplets and particles fall to the earth as components of acid deposition, which along with other air pollutants can harm trees and aquatic life. Droplets of sulfuric acid are also produced from dimethylsulfide (DMS) emitted into the atmosphere by many species of plankton.

About a third of all sulfur (including 99% of the sulfur dioxide) that reaches the atmosphere comes from human activities. We intervene in the atmospheric phase of the sulfur cycle in several ways: **(1)** by burning sulfur-containing coal and oil to produce electric

Figure 4-26 Simplified conceptual model of the sulfur cycle.

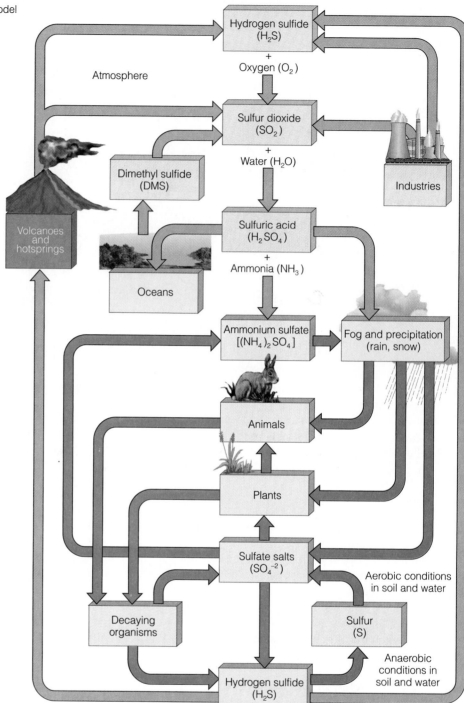

power, producing about two-thirds of the human inputs of sulfur dioxide; **(2)** by refining petroleum; **(3)** by using smelting to convert sulfur compounds of metallic minerals into free metals such as copper, lead, and zinc; and **(4)** by using other industrial processes.

How Is Water Cycled in the Ecosphere? The **hydrologic cycle**, or **water cycle**, which collects, purifies, and distributes the earth's fixed supply of water,

is shown in simplified form in Figure 4-27. The main processes in this water recycling and purifying cycle are **evaporation** (conversion of water into water vapor), **transpiration** (evaporation from leaves of water extracted from soil by roots and transported throughout the plant), **condensation** (conversion of water vapor into droplets of liquid water), **precipitation** (rain, sleet, hail, and snow), **infiltration** (movement of water into soil), **percolation** (downward flow

Q: How many teenage women (ages 15 to 19) become pregnant each year in the United States?

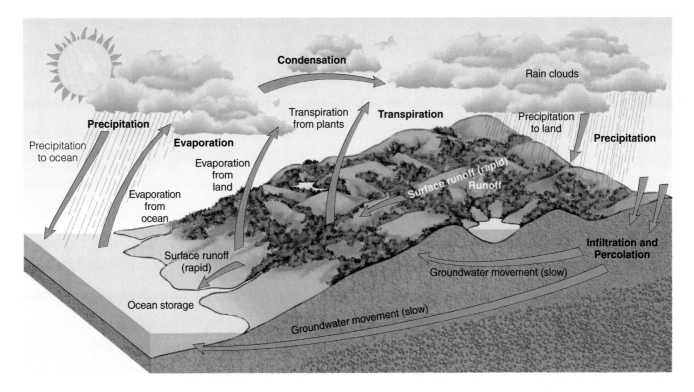

Figure 4-27 Simplified conceptual model of the hydrologic cycle.

of water through soil and permeable rock formations to groundwater storage areas called aquifers), and **runoff** (downslope surface movement back to the sea to begin the cycle again).

The water cycle is powered by energy from the sun and by gravity. Incoming solar energy evaporates water from oceans, streams, lakes, soil, and vegetation. About 84% of water vapor in the atmosphere comes from the oceans, which cover about 71% of the earth's surface; the rest comes from land. On a global scale, the amount of water vapor entering the atmosphere is equal to the amount returning to the earth's surface as precipitation.

The amount of water vapor air can hold depends on its temperature; warm air is capable of holding more water vapor than cold air. **Absolute humidity** is the amount of water vapor found in a certain mass of air; it is usually expressed as grams of water per kilogram of air. **Relative humidity** is the amount of water vapor in a certain mass of air, expressed as a percentage of the maximum amount it could hold at that temperature. For example, a relative humidity of 60% at 27°C (80°F) means that each kilogram (or other unit of mass) of air contains 60% of the maximum amount of water vapor it could hold at that temperature.

Winds and air masses transport water vapor over various parts of the earth's surface, often over long distances. Falling temperatures cause the water vapor to condense into tiny droplets that form clouds or fog.

For precipitation to occur, air must contain **condensation nuclei**: tiny particles on which droplets of water vapor can collect. Volcanic ash, soil dust, smoke, sea salts, and particulate matter emitted by factories, coal-burning power plants, and vehicles are sources of such particles. The temperature at which condensation occurs is called the **dew point**.

Some of the fresh water returning to the earth's surface as precipitation becomes locked in glaciers. Most of the precipitation falling on terrestrial ecosystems becomes *surface runoff* flowing into streams and lakes, which eventually carry water back to the oceans, where it can be evaporated to cycle again.

Besides replenishing streams and lakes, surface runoff also causes soil erosion, which moves soil and weathered rock fragments from one place to another. Water is thus the primary sculptor of the earth's landscape.

Some of the water returning to the land soaks into (infiltrates) the soil and porous rock and then percolates downward, dissolving minerals from porous rocks on the way. This water is stored as groundwater in the pores and cracks of rocks.

Where the pores are joined, a network of water channels allows water to flow through the porous rock. Such water-laden rock is called an **aquifer**, and the level of the earth's land crust to which it is filled is called the **water table**. This underground water flows slowly downhill through rock pores and seeps out into streams and lakes or comes out in springs. Eventually this water evaporates or reaches the sea to continue the cycle.

Humans intervene in the water cycle in two main ways: *First*, we withdraw large quantities of fresh water from streams, lakes, and underground sources. In heavily populated or heavily irrigated areas, withdrawals have led to groundwater depletion or intrusion of ocean salt water into underground water supplies. *Second*, we clear vegetation from land for agriculture, mining, road and building construction, and other activities. This increases runoff and reduces infiltration that recharges groundwater supplies; it also increases the risk of flooding and accelerates soil erosion and landslides.

4-5 HOW DO ECOLOGISTS LEARN ABOUT ECOSYSTEMS?

What Is Field Research? Ecologists try to unravel some of nature's secrets in three general ways: field research, laboratory research, and system analysis. Each approach has its advantages and disadvantages.

Field research, sometimes called muddy-boots biology, involves going into nature and observing and measuring the structure of ecosystems and what happens in them. Most of what we know about the structure and functioning of ecosystems and how ecosystems change with changing environmental conditions has come from such research.

However, because the parts and interactions among the components of even simple ecosystems are so complex, it is expensive, time-consuming, and difficult to get such information by carrying out field experiments. Because of the large number of interacting variables in nature, it is often difficult to set up meaningful controlled experiments in which only one variable is varied.

What Is Laboratory Research? In the past 45 years, ecologists have increasingly supplemented field research by using *laboratory research* to set up, observe, and make measurements of model ecosystems and populations under laboratory conditions. Such simplified systems have been set up in containers such as culture tubes, bottles, aquarium tanks, and greenhouses, and in chambers where temperature, light, CO_2, humidity, and other variables can be carefully controlled. In such systems, it is easier for scientists to carry out controlled experiments. Often such laboratory experiments are quicker and cheaper than similar experiments in the field.

Such laboratory research has led to enormous amounts of information and many fruitful hypotheses. Often laboratory research can isolate cause-and-effect relationships that can be tested further in the field or by system analysis using computer simulations. Indoor and outdoor chambers can also be used to study the effects of increasing levels of atmospheric CO_2 and ultraviolet light on populations, diversity, and the productivity of producers (including important agricultural crops). This can help us project the consequences of ozone depletion in the stratosphere and possible global warming. Such experiments can be carried out in the field only by waiting for such changes to occur, far too late to prevent possible harmful consequences.

Currently, about 60% of ecological research involves laboratory experiments and 25% is carried out in the field. However, there is the important question of whether what scientists observe and measure in a simplified, controlled system under laboratory conditions takes place in the same way in the more complex and dynamic conditions found in nature. Thus, the results of laboratory research must be coupled with and supported by field research.

What Is System Analysis? Since the late 1960s ecologists have made increasing use of *system analysis* (Section 3-2) to simulate ecosystems and study their structure and function. System analysis is a useful tool for ecosystem research and management because ecosystems are complex and dynamic and involve inputs, accumulations, flows (throughputs), outputs, multiple coupled feedback loops, and time delays. The advantage of system analysis is that it can help us understand large and very complex systems (such as rivers, oceans, forests, grasslands, or cities) that cannot be adequately studied and modeled by field and laboratory research. Most such analysis is carried out by a team of investigators because a single person rarely has all the knowledge and skills needed for this approach.

The simulations and predictions made using ecosystem models are no better than the data and assumptions used to develop the models. Thus, careful field and laboratory ecological research must be used to provide the baseline data and causal relationships among key variables needed to develop and test ecosystem models. Often system analysis also reveals surprising results and new questions that must be verified by a combination of field and laboratory research. Thus, meaningful ecological research involves a combination of field research, laboratory research, and system analysis.

🌐 4-6 CONNECTIONS: ROLES OF SPECIES IN ECOSYSTEMS

What Role Does a Species Play in an Ecosystem? The **ecological niche**, or simply **niche** (pronounced *nitch*), of a species is its way of life or functional role—

Q: What percentage of the world's population is under age 15?

everything it does to survive and reproduce—in an ecosystem. This includes where the species normally lives (its habitat or biological address), what it eats, what eats it, how it reproduces, what nutrients it needs, its range of tolerance for various physical and chemical factors (Figure 4-13), its range of movement, how it interacts with other living and nonliving components of the ecosystems in which it is found, and the role it plays in the flow of energy and cycling of matter in an ecosystem.

A species's **fundamental niche** is the full potential range of physical, chemical, and biological conditions and resources it could theoretically use if there were no direct competition from other species. Because species interact and often compete with other species in a particular ecosystem, however, their niches overlap. As a result, a species usually occupies only part of its fundamental niche in a particular community or ecosystem— what ecologists call its **realized niche**. By analogy, you may be capable of being president of a particular company (your *fundamental professional niche*), but competition from others may mean that you may become only a vice president (your *realized professional niche*).

Is It Better to Be a Generalist or a Specialist Species?
Species can be broadly classified as generalists or specialists, according to their niches. **Generalist species** have broad niches: They can live in many different places, eat a variety of foods, and tolerate a wide range of environmental conditions. Flies, cockroaches, mice, rats, white-tailed deer, black bears, raccoons, coyotes, bullfrogs, copperheads, robins, channel catfish, and humans are all generalist species.

Specialist species have narrow niches: They may be able to live in only one type of habitat, tolerate only a narrow range of climatic and other environmental conditions, or use only one or a few types of food. This makes them more prone to becoming endangered when environmental conditions change. Examples of specialists are tiger salamanders, which can breed only in fishless ponds so their larvae won't be eaten; red-cockaded woodpeckers, which carve nest-holes almost exclusively in old (at least 75 years) longleaf pines; spotted owls, which require old-growth forests in the Pacific Northwest for food and shelter; and China's giant panda, which feeds almost exclusively on various types of bamboo.

In a tropical rain forest, an incredibly diverse array of species survives by occupying specialized ecological niches in various distinct layers of vegetation exposed to different levels of light. The widespread clearing and degradation of such forests is dooming large numbers of such specialized species to extinction.

Is it better to be a generalist than a specialist? It depends. When environmental conditions are fairly constant, as in a tropical rain forest, specialists have an advantage because they have fewer competitors. When environments are changing rapidly, however, the adaptable generalist is usually better off than the specialist.

How Can We Classify the Roles Various Species Play in Ecosystems?
When examining ecosystems, ecologists often apply particular labels—such as *native*, *nonnative*, *indicator*, or *keystone*—to various species to clarify some of the ecological roles they play. Any given species may function as more than one of these four types in a particular ecosystem.

Species that normally live and thrive in a particular ecosystem are known as **native species**. Others that migrate into an ecosystem or are deliberately or accidentally introduced into an ecosystem by humans are called **nonnative species**, **exotic species**, or **alien species**. Some of these introduced species (such as crop species and game for sport hunting) are beneficial to humans, but some thrive and crowd out many native species by taking over all or most of their niches.

From a human standpoint, niche takeover by some nonnative species can become a nightmare. In 1957, wild African bees were imported to Brazil to help increase honey production. Instead, these bees have displaced domestic honeybees and reduced the honey supply. Since then these nonnative bee species, popularly known as "killer bees," have moved northward into Central America (killing 150 people in Mexico since 1986); by the summer of 1994 they had become established in Texas (one death in 1994), Arizona (one death in 1993), New Mexico, and Puerto Rico. They are now heading north at 240 kilometers (150 miles) per year, although they will be stopped eventually by cold winters in the central United States.

Although they are not the killer bees portrayed in some horror movies, these bees are aggressive and unpredictable. They have killed thousands of domesticated animals and an estimated 1,000 people in the western hemisphere. Fortunately, most people not allergic to bee stings can run away. Most people killed by these honeybees have died because they fell down or became trapped and could not flee.

Species that serve as early warnings that a community or an ecosystem is being damaged are called **indicator species**. For example, research indicates that a major factor in the current decline of migratory, insect-eating songbirds in North America is loss or fragmentation of habitat. The tropical forests of Latin America and the Caribbean that are the birds' winter habitats are rapidly disappearing. Their summer habitats in North America are also disappearing or are being fragmented into patches that make the birds more vulnerable to attack by predators and parasites. Birds are excellent biological indicators because they are found almost everywhere and respond quickly to environmental change. Some amphibians (frogs, toads, and salamanders), which live part of their lives in water and part on land, are also indicator species (Connections, right).

Some ecologists call species whose roles in an ecosystem are much more important than their abundance or biomass would suggest **keystone species**, although this designation is controversial.* For example, in tropical forests, various species of bees, bats, ants, and hummingbirds play keystone roles by pollinating flowering plants, dispersing seed, or both. Some keystone species—including the wolf, leopard, lion, giant anteater, great white shark, and giant armadillo—are top predators that exert a stabilizing effect on their ecosystems by feeding on and regulating the populations of certain species.

Sea otters are keystone species that keep sea urchins from depleting kelp beds in offshore waters from Alaska to southern California. If the sea otter is removed, too little kelp is left to support the diverse community of crustaceans, mollusks, fish, and marine mammals such as fish-eating harbor seals. In addition to sea urchins, sea otters feed on abalone, crabs, and mollusks. This makes them unpopular with people who make a living harvesting crabs and abalone.

Some ecologists consider large grazing animals, such as endangered elephants and rhinoceroses, as keystone species in the savanna grasslands and woodlands of Africa. By pushing over, breaking, or uprooting trees, elephants create forest openings. This promotes the growth of grasses and other forage that benefit smaller grazing species such as antelope. It also accelerates rates of nutrient cycling. Grazing pressure from rhinoceroses transforms medium-tall grasslands into patches of short and tall grassland, which improves food quality for smaller, more selective grazing animals.

The loss of a keystone species can lead to population crashes and extinctions of other species that depend on it for certain services, a ripple or domino effect that spreads throughout an ecosystem. According to biologist Edward O. Wilson, "The loss of a keystone species is like a drill accidentally striking a power line. It causes lights to go out all over."

---

* All species play some role in their ecosystems and thus are important. Whereas some scientists consider all species equally important, others consider certain species to be more important than others, at least in helping maintain the ecosystems they are a part of.

Connections: Why Should We Care About the American Alligator? Some ecologists classify the North American alligator as a keystone species because of its important ecological roles in helping maintain the structure, function, and sustainability of its natural ecosystems.

The American alligator, North America's largest reptile, has no natural predators except humans. Hunters once killed large numbers of these animals for their exotic meat and their supple belly skin, used to make shoes, belts, and pocketbooks.

Other people considered alligators to be useless, dangerous vermin and hunted them for sport or out of hatred. Between 1950 and 1960, hunters wiped out 90% of the alligators in Louisiana, and by the 1960s the alligator population in the Florida Everglades was also near extinction.

People who say "So what?" are overlooking the alligator's important ecological roles in subtropical wetland ecosystems. Alligators dig deep depressions, or gator holes, that collect fresh water during dry spells, serve as refuges for aquatic life, and supply fresh water and food for many animals. Large alligator nesting mounds provide nesting and feeding sites for species of herons and egrets. Alligators also eat large numbers of predatory gar fish and thus help maintain populations of game fish such as bass and bream.

As alligators move from gator holes to nesting mounds, they help keep areas of open water free of invading vegetation. Without these ecosystem services, freshwater ponds and coastal wetlands found in the alligator's habitat would be filled in by shrubs and trees, and dozens of species would disappear.

In 1967 the U.S. government placed the American alligator on the endangered species list. Protected from hunters, the alligator population made a strong comeback in many areas by 1975—too strong, according both to those who find alligators in their backyards and swimming pools and to duck hunters, whose retriever dogs are sometimes eaten by alligators. Large alligators have also been known to eat pigs, deer, and even cattle, dragging them under water to drown before dismembering them.

In 1977 the U.S. Fish and Wildlife Service reclassified the American alligator from an **endangered** to a **threatened species** in Florida, Louisiana, and Texas, where 90% of the animals live. In 1987 this reclassification was extended to seven other states. Alligators now number perhaps 3 million, most in Florida and Louisiana. It is generally illegal to kill members of a threatened species, but limited kills by licensed hunters are allowed in some areas of Florida, Louisiana, and South Carolina to control the population. The comeback of the American alligator is an important success story in wildlife conservation.

Q: How many legal immigrants are admitted to the United States each year?

Why are Amphibians Vanishing?

Amphibians first appeared about 350 million years ago. These cold-blooded creatures range in size from a frog that can sit on your thumb to a Japanese salamander that is about 1.5 meters (5 feet) long.

Fossil records suggest that frogs and toads, the oldest of today's amphibians, existed as long as 150 million years ago; such longevity testifies to their adaptability. Within the last decade or two, however, hundreds of the world's estimated 5,100 amphibian species (including 2,700 species of frogs and toads) have been vanishing or declining in diverse locations, even in protected wildlife reserves and parks.

Scientists have identified a variety of possible causes for such declines. For some species, such as Costa Rican golden toads, diebacks may be caused by prolonged drought, which dries up breeding pools so that few tadpoles survive. Dehydration can also weaken amphibians, making them more susceptible to fatal viruses, bacteria, and fungi.

In other cases, the culprit may be pollution. Because amphibians live part of their lives in water and part on land, they are exposed to water, soil, and air pollutants. Their insect diet guarantees them abundant food, but in farming areas it also exposes them to pesticides. The soft, permeable skin that allows them to absorb oxygen from water also makes them extremely

sensitive to waterborne pollutants. In some regions the leading culprit causing amphibian losses may be increased acidity of the water in lakes and ponds from acid deposition (acid rain).

Their eggs may also be sensitive to increases in ultraviolet radiation caused by reductions in stratospheric ozone. Especially vulnerable amphibian species may be those living at cooler and higher elevations, where the ozone layer is thinnest and where individuls must bask in the sunlight to stay warm. Recent evidence suggests that environmental pollutants that mimic estrogens and disrupt the immune and endocrine systems of amphibians may play a role in their population declines.

In Asia and in France, where frog legs are a delicacy, overhunting may play a part. In other areas, immigration or introduction of non-native predators and competitors (such as fish) can threaten amphibian populations.

Loss of habitat—or its fragmentation into pieces too small to support populations of some amphibians—is a problem in some places. Once small, isolated populations decline to a certain level, they may not be able to recover or can more easily be wiped out by a chance event.

Scientists are concerned about amphibians' decline for three reasons:

- It suggests that the world's environmental health is deteriorating

rapidly because amphibians are generally tough survivors.

- Adult amphibians play important roles in the world's ecosystems. For example, amphibians eat more insects (including mosquitoes) than do birds. In some habitats, extinction of certain amphibian species could also result in extinction of other species, such as reptiles, birds, aquatic insects, fish, mammals, and other amphibians that feed on them or their larvae.

- From a human perspective amphibians represent a genetic storehouse of pharmaceutical products waiting to be discovered. Hundreds of secretions from amphibian skin have been isolated, and some of these compounds are being used as painkillers and in treating burns and heart attacks. Others are being evaluated for their use as antiviral and antibacterial medicines.

As indicator species, amphibians may be sending us an important message. They don't need us, but we and other species need them.

Critical Thinking

On an evolutionary time scale all species eventually become extinct. Some suggest that the widespread disappearance of amphibians is the result of natural responses to changing environmental conditions. Others contend that these losses are caused mostly by human actions and that such declines are a warning of possible harm for our own species and other species. What is your position? Why?

The increased demand for alligator meat and hides has created a booming business in alligator farms, especially in Florida. By controlling diet and other conditions, alligator farm operators have quadrupled the species's reproductive rate, doubled its growth rate, and reduced mortality from 35% to 1%. Such success reduces the need for illegal hunting of wild alligators.

4-7 CONNECTIONS: HOW DO SPECIES INTERACT?

How Do Species Interact? An Overview When different species in an ecosystem have activities or resource requirements in common, they may interact with one another. Members of these species may be harmed by, benefit from, or be unaffected by the interaction.

There are three basic types of interactions among species: *interspecific competition*, *predation*, and *symbiosis*.

Interspecific competition (competition among species) occurs when two or more species compete for food, space, or any other limited resource. This competition harms the competing species to varying degrees, depending on which is the best competitor.

In **predation**, members of one species (the *predator*) feed directly on all or part of a living organism of another species (the *prey*). However, they do not live on or in the prey. The prey may or may not die from the interaction. In this interaction the predator benefits, and the individual prey is clearly harmed.

Symbiosis (from the Greek word for "living together") is a long-lasting relationship in which species live together in an intimate association. There are three general types of symbiosis: *parasitism*, *mutualism*, and *commensalism*.

Parasitism occurs when one species (the *parasite*) feeds on part of another organism (the *host*) by living on or in the host for a significant portion of the host's life. In this symbiotic relationship the parasite benefits and the host is harmed. In **mutualism** two species involved in a symbiotic relationship interact in ways that benefit both. **Commensalism** is a symbiotic interaction that benefits one species but neither harms nor helps the other species much, if at all.

These interactions play important roles in regulating the populations of species and in helping them survive changes in environmental conditions, as discussed in more detail in Section 5-4.

How Do Species Compete with One Another?

As long as commonly used resources are abundant, different species can share them. This allows each species to come closer to occupying the *fundamental niche* it could theoretically use if there were no competition from other species.

In ecosystems, however, most species face competition from one or more other species for one or more valuable limited resources (such as food, sunlight, water, soil nutrients, space, nest sites, good places to hide). Because of such interspecific competition, parts of the fundamental niches of different species often overlap significantly. With significant niche overlap, one of the competing species must either migrate to another area (if possible), shift its feeding habits or behavior through natural selection and evolution, suffer a sharp population decline, or become extinct in that area.

Species compete in two ways. In **interference competition**, one species may limit another's access to some resource, regardless of its abundance. This often takes the form of behavior in which members of a species establish a *territory* they defend against other invading species. For example, one species of hummingbird may defend patches of spring wildflowers from which it gets nectar by chasing away members of other hummingbird species. In desert and grassland habitats, many plants release chemicals into the soil that either prevent the growth of competing species or reduce their seeds' germination rates.

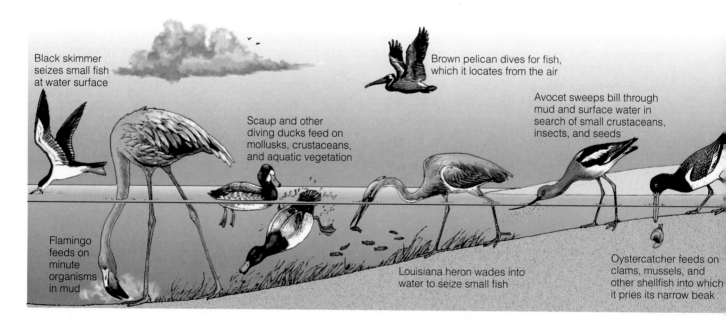

Figure 4-28 Specialized feeding niches of various species of birds in a coastal wetland. Such resource partitioning reduces competition and allows sharing of limited resources.

Q: How many illegal immigrants enter the United States each year?

In **exploitation competition**, competing species have roughly equal access to a specific resource but differ in how fast or efficiently they exploit it. As a result, one species gets more of the resource and thus hampers the growth, reproduction, or survival of the other species. In **competitive exclusion**, one species eliminates the other in a particular area through competition for limited resources.

How Have Some Species Reduced or Avoided Competition? Over a time scale long enough for evolution to occur, species that compete for the same resources may evolve adaptations that reduce or avoid competition or overlap of their fundamental niches, a process that increases biological diversity instead of leading to local extinction. One way this happens is through **resource partitioning**, the dividing up of scarce resources so that species with similar requirements use them at different times, in different ways, or in different places (Figure 4-28). In effect, negative feedback loops (Section 3-2) evolve that allow them to share the wealth, with each competing species occupying a realized niche that makes up only part of its fundamental niche.

For example, where lions and leopards live in the same area, lions take mostly larger animals as prey; leopards take smaller ones. Hawks and owls feed on similar prey, but hawks hunt during the day; owls hunt at night. Some bird species feed on the ground, whereas others seek food in trees and shrubs. Ecologist Robert H. MacArthur studied the feeding habits of five species of warblers (small insect-eating birds) that coexist in the forests of the northeastern United States and in the adjacent area of Canada. Although they appear to be competing for the same food resources, MacArthur found that the bird species reduce competition by spending at least half their time hunting for insects in different parts of trees (Figure 4-29).

How Do Predator and Prey Species Interact? Recall that in predation, members of a predator species feed on members of a prey species, but they do not live on or in the prey. Together, the two kinds of organisms, such as lions (the predator or hunter) and zebras (the prey or hunted), are said to have a **predator–prey relationship**, as depicted in Figures 4-9, 4-10, 4-14, and 4-15.

A predator–prey relationship is a positive feedback system (Figure 3-5), at least for the victor (predator), who gets more of scarce resources than does the loser (prey). This *win–lose* system is limited, however, by negative feedback when the prey population falls below the level needed to support the predator population.

We normally use the term *predator* for animals (carnivores) that feed on other animals. However, the term also applies to animals (herbivores) that feed on plants and even to carnivorous plants such as the Venus's-flytrap and aquatic bladderworts, which capture living prey (insects) in a leaf that closes like a steel trap.

At the individual level, members of the prey species are clearly harmed, but at the population level, predation can benefit the prey species because predators often kill the sick, weak, and aged members (Case Study, p. 103). Reducing the prey population gives remaining prey greater access to the available food supply. It can also improve the genetic stock of the prey population, which enhances its chances of reproductive success and long-term survival.

Some people tend to view predators with contempt. When a hawk tries to capture and feed on a rabbit, some tend to root for the rabbit. Yet the hawk (like all predators) is merely trying to get enough food to feed itself and its young; in the process, it is merely doing what it is genetically programmed to do.

What Are Parasites and Why Are They Important? Recall that parasitism is an interaction in which a member of one species (the parasite) obtains its nourishment by living on, in, or near a member of another species (its host) over an extended time. Parasitism can be viewed as a special form of predation, but unlike a conventional predator, a parasite is usually smaller than its host (prey); remains closely associated with, draws nourishment from, and gradually weakens its host over a long time; and rarely kill its host.

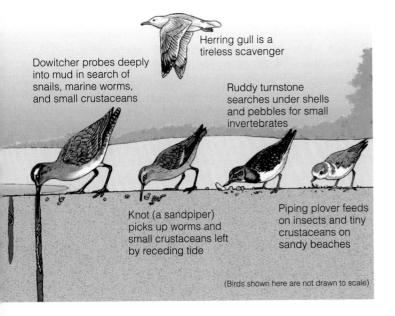

Dowitcher probes deeply into mud in search of snails, marine worms, and small crustaceans

Herring gull is a tireless scavenger

Ruddy turnstone searches under shells and pebbles for small invertebrates

Knot (a sandpiper) picks up worms and small crustaceans left by receding tide

Piping plover feeds on insects and tiny crustaceans on sandy beaches

(Birds shown here are not drawn to scale)

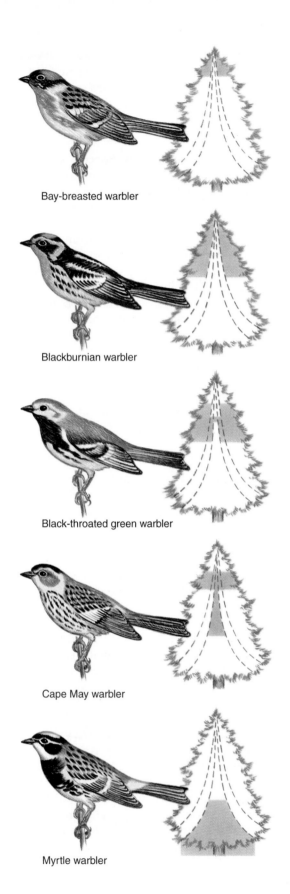

Bay-breasted warbler

Blackburnian warbler

Black-throated green warbler

Cape May warbler

Myrtle warbler

Figure 4-29 Resource partitioning of five species of common insect-eating warblers in spruce forests of Maine. Each species minimizes competition with the others for food by spending at least half its feeding time in a distinct portion (shaded areas) of the spruce trees; each also consumes somewhat different insect species from the others. (After R. H. MacArthur, "Population Ecology of Some Warblers in Northeastern Coniferous Forests," *Ecology* 36 [1958]:533–36)

Some parasites, such as tapeworms (often yards long) and plasmodium microorganisms that cause malaria, live *inside* their hosts and are called *endoparasites*. Some biologists classify disease-causing bacteria and viruses as endoparasites, but others do not. Other parasites, such as lice, ticks, mosquitoes, mistletoe plants, and fungi (that cause diseases such as athlete's foot), and lampreys (Figure 4-30), attach themselves to the *outside* of their hosts and are called *ectoparasites*. Some parasites, especially ectoparasites, move from one host to another, as fleas and ticks do; others, such as tapeworms, spend their adult lives with a single host.

From the host's point of view parasites are harmful, but parasites play important ecological roles. Collectively, the incredibly complex matrix of parasitic relationships in an ecosystem acts somewhat like a glue that helps hold the species in an ecosystem together. Parasites living within their hosts also help dampen drastic swings in population sizes and parasites promote biodiversity by helping prevent some organisms from becoming too plentiful.

How Do Species Interact So That Both Species Benefit? *Mutualism* is a symbiotic relationship in which both interacting species benefit in various ways. These include having pollen dispersed for reproduction, being supplied with food, and receiving protection. It is tempting to think of mutualism as an example of cooperation between species, but it actually involves each species benefiting by exploiting the other.

The pollination relationship between flowering plants and insects is one of the most common forms of mutualism. Butterflies and bees depend on flowers for food in the form of nectar and pollen. In turn, the flowering plants depend on bees, bats, or other pollinators (Figure 4-1, left) to carry their male reproductive cells (sperm in pollen grains) to the female flowering parts of other flowers of the same species.

Other examples are the mutualistic relationships between rhinos and oxpeckers (Figure 4-31) and between legume plants and *Rhizobium* bacteria that live in nodules on their roots (Figure 4-24).

Other important mutualistic relationships exist between animals and the vast armies of bacteria in their digestive system that break down (digest) their food. The bacteria gain a safe home with a steady food sup-

Q: What percent of the world's people live in urban areas?

Why Are Sharks Important Species?

The world's 350 shark species range in size from the dwarf dog shark, about the size of a large goldfish, to the whale shark, the world's largest fish at 18 meters (60 feet) long. Various shark species, feeding at the top of food webs, cull injured and sick animals from the ocean and thus play an important ecological role. Without such shark species the oceans would be overcrowded with dead and dying fish.

Many people, influenced by movies and popular novels, think of sharks as people-eating monsters. But the two largest species—the whale shark and the basking shark—sustain their enormous bulk by filtering out and swallowing huge quantities of plankton (small free-floating sea creatures).

Every year, members of a few species of shark—mostly great white, bull, tiger, gray reef, lemon, and blue—injure about 100 people worldwide and kill between 5 and 10. Most attacks are by great white sharks, which feed on sea lions and other marine mammals and sometimes mistake divers in wet suits

and swimmers on surfboards for their usual prey. A typical oceangoer is 150 times more likely to be killed by lightning than by a shark.

For every shark that injures a person, we kill 500,000 to 1 million sharks, for a total of 50 to 100 million sharks each year. Sharks are killed mostly for their fins, widely used in Asia as a soup ingredient and as a pharmaceutical cure-all.

Sharks are also killed for their livers, meat (especially mako and thresher), and jaws (especially great whites), or just because we fear them. Some sharks (especially blue, mako, and oceanic whitetip) die when they are trapped in nets deployed to catch swordfish, tuna, shrimp, and other commercially important species.

Sharks also help save human lives. In addition to providing people with food, they are helping us learn how to fight cancer (which sharks almost never get), bacteria, and viruses. Their highly effective immune system is being studied because it allows wounds to heal without becoming infected, and their blood is being studied in connection with AIDS research.

Chemicals extracted from shark cartilage (the elastic connective tissue attached to bones) have killed several types of cancerous tumors in laboratory animals. This research someday might help prolong your life or the life of a loved one.

Sharks have several natural traits that make them prone to population declines from overfishing. Unlike most other fish, they have only a few offspring (between 2 and 10) once every year or two. Depending on the species, sharks require 10 to 15 years (and in some cases, 24 years) to reach sexual maturity and begin reproducing. Sharks also have long gestation (pregnancy) periods—as much as 24 months for some species.

With more than 400 million years of evolution behind them, sharks have had a long time to get things right. Preserving this evolutionary genetic wisdom begins with the recognition that sharks don't need us, but we and other species need them.

Critical Thinking

Do you fear sharks? Why? How do you feel about killing sharks in large numbers? Why?

---

ply; the animal gains more efficient access to a large source of energy. In addition to helping digest your food, bacteria in your intestines synthesize vitamin K and the B-complex vitamins, which you can't make.

Research indicates that mutualism increases when resources become scarce. In other words, when the going gets tough, the tough often survive by evolving mutually beneficial relationships with other species.

How Do Species Interact So That One Benefits but the Other Is Not Harmed? Commensalism is a symbiotic association in which one species benefits while the other is neither helped nor harmed to any significant degree. An example is the intimate relationship in tropical waters between various species of clownfish and sea anemones, marine animals with stinging tenta-

Figure 4-30 Parasitism. Sea lampreys use their suckerlike mouths to attach themselves to their fish hosts. They then bore a hole in the fish with their teeth and feed on its blood. (Tom Stack)

A: 43% (74% in developed countries and 36% in developing countries) in 1997

Figure 4-31 Mutualism. Oxpeckers (or tickbirds) feed on the parasitic ticks that infest large, thick-skinned animals such as this endangered black rhinoceros in Kenya. The rhino benefits by having parasites removed from its body, and oxpeckers benefit by having a dependable source of food. The oxpecker also serves as a sentinel for the rhino by making a fierce hissing sound when alarmed. (Joe McDonald/Tom Stack & Associates)

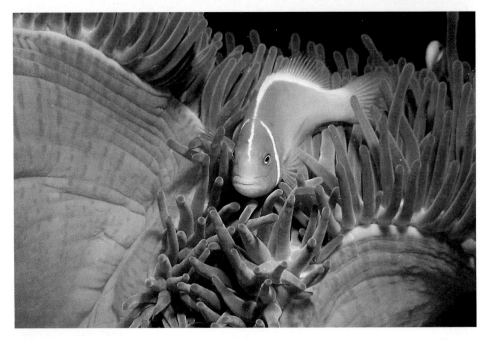

Figure 4-32 Commensalism. This clownfish in the Coral Sea, Australia, has a commensalistic relationship with deadly sea anemones, whose tentacles quickly paralyze most other fishes that touch them. The clownfish gains protection and food by feeding on scraps left over from fish killed by the sea anemones, which seem to be neither helped nor harmed by this relationship. (Carl Roessler/FPG International)

cles that paralyze most fish that touch them (Figure 4-32). The clownfish are not harmed by sea anemones. They gain protection by living among the deadly tentacles, and they feed on the detritus left from the meals of the anemone. The sea anemones seem to neither benefit nor suffer harm from this relationship.

Another example is the commensalistic relationship between various trees and other plants called *epiphytes* (such as various types of orchids and bromeliads) that attach themselves to the trunks or branches of large trees (Figure 4-33) in tropical and subtropical forests. These so-called air plants benefit by living in an elevated spot that gives them better access to sunlight. Their position in the tree allows them to get most of their water from the humid air and from rain collecting in their usually cupped leaves. They also absorb nutrient salts falling from the tree's upper leaves and limbs and from the dust in rainwater.

In this chapter we have seen that the essential features of the living and nonliving parts of individual ecosystems, and of the ecosphere as a whole, are *interdependence* and *connectedness*. Without the services performed by diverse communities of species, we would be starving, gasping for breath, and drowning in our

Q: What percentage of the world's cars are found in the United States?

Figure 4-33 Commensalism. This white orchid (an epiphyte or air plant from the tropical forests of Latin America) roots in the fork of a tree rather than the soil without penetrating or harming the tree. In this interaction, the epiphytes gain access to water, other nutrient debris, and sunlight; the tree apparently remains unharmed. (Kenneth W. Fink/Ardea London)

own wastes. The next chapter shows how this interdependence is the key to understanding the earth's major life zones and ecosystems and how populations, communities, ecosystems can change when environmental conditions change.

All things come from earth, and to earth they all return.
MENANDER (342–290 B.C.)

CRITICAL THINKING

1. (a) A bumper sticker asks, "Have you thanked a green plant today?" Give two reasons for appreciating a green plant. **(b)** Trace the sources of the materials that make up the bumper sticker and then decide whether the sticker itself is a sound application of the slogan. **(c)** Explain how decomposers help keep you alive.

2. (a) How would you set up a self-sustaining aquarium for tropical fish? **(b)** Suppose you have a balanced aquarium sealed with a clear glass top. Can life continue in the aquarium indefinitely as long as the sun shines regularly on it? **(c)** A friend cleans out your aquarium and removes all the soil and plants, leaving only the fish and water. What will happen?

3. Using the second law of energy, explain why there

is such a sharp decrease in usable energy as energy flows through a food chain or web. Doesn't an energy loss at each step violate the first law of energy? Explain.

4. Using the second law of energy, explain why many poor people in developing countries live mostly on a vegetarian diet.

5. Which cause a larger loss of energy from an ecosystem: a herbivore eating a plant or a carnivore eating an animal? Explain.

6. Carbon, nitrogen, and phosphorus are cycled in the ecosphere. Why do farmers not need to apply carbon to grow their crops but often need to supply fertilizer containing nitrogen and phosphorus?

7. Explain in ecological terms how the human practice of embalming a corpse and then entombing it in an airtight metal container could be viewed as contrary to the earth's natural processes of decay and recycling. Even though such practices are done for important cultural, religious, and health reasons, a growing number of people prefer to be cremated and then have their ashes spread on the land or at sea; they want their atoms back in action in the earth's vital chemical cycles as soon as possible after their death. What is your preference? Why?

8. Imagine that after you die you are somehow given the choice to live again as any organism you choose except a human being. What organism would you choose to be, and why?

9. Suppose that somehow all species of parasites were eliminated from the earth. How would this affect you?

*10. Use the library or Internet to find and describe two species not discussed in this textbook that are engaged in **(a)** a commensalistic interaction, **(b)** a mutualistic interaction, and **(c)** a parasite–host relationship.

*11. Write a brief scenario describing the sequence of consequences to us and to other forms of life (identify some of these organisms) if **(a)** all decomposers and detritus feeders were somehow eliminated and **(b)** all producers on land and in the upper zone of aquatic ecosystems were eliminated by drastic increases in ultraviolet radiation because of a large drop in ozone in the stratosphere.

*12. Make a concept map of this chapter's major ideas using the section heads and subheads and the key terms (in boldface type). Look at the inside back cover and in Appendix 4 for information about concept maps.

SUGGESTED READINGS

See Internet Sources and Further Readings at the back of the book, and InfoTrac.

Earth Healing: From Rice Back to Rushes

Some people's deepest fears are linked with swamps, marshes, quicksand, and the "things" that lurk there. Driven by such fears and ignorance about wetlands as well as by a desire for land and a hunger for profit, we have drained swamps and marshes relentlessly for centuries.

Belatedly, we have begun to question such campaigns against nature as we learn more about the ecological importance of marshes and other wetlands in removing pollutants from water and providing free flood control, key habitats for many wildlife species, and seafood for us and other species. Can we turn back the clock to restore or rehabilitate lost marshes?

California rancher Jim Callender decided to try. In 1982 he bought 20 hectares (50 acres) of Sacramento Valley ricefield that had been a marsh until the early 1970s. The previous owner had destroyed it, bulldoz-ing, draining, leveling, uprooting the native plants and spraying with chemicals to kill the snails and other food of the waterfowl.

Callender and his friends set out to restore the marshland. They hollowed out low areas, built up islands, replanted tules and bulrushes, reintroduced smartweed and other plants needed by birds, and planted fast-growing Peking willows.

After 6 years of care, hand-planting, and annual seeding with a mixture of watergrass, smartweed, and rice, the marsh is once again a part of the Pacific Flyway used by migratory waterfowl (Figure 5-1). Many birds pass through on their way south in autumn and north in the spring, and some mallards and wood ducks nest there.

Jim Callender and a few others have shown that at least part of the continent's wetlands heritage can be reclaimed with planning and hard work. Such earth healing is vital, but the real challenge is to protect remaining wetlands (and other undisturbed ecosystems) from harm in the first place.

Figure 5-1 Snow geese migrating along the Pacific flyway in eastern Oregon. (Pat and Tom Leeson/Photo Researchers, Inc.)

We cannot command nature except by obeying her.
SIR FRANCIS BACON

This chapter focuses on answers to the following questions:

- What are the major types of biomes, and how are they being affected by human activities?
- What are the basic types of aquatic life zones, and how are they being affected by human activities?
- How are living systems affected by stress?
- How can populations of species change their size, density, and makeup in response to environmental stress?
- What major types of life are found on the earth, and how do scientists account for the emergence of life on the earth?
- How can populations adapt to environmental changes through evolution, adaptation, and natural selection?
- How do extinction of existing species and formation of new species affect biodiversity?
- How do communities and ecosystems change as environmental conditions change?
- What impacts do human activities have on populations, communities, and ecosystems?

5-1 BIOMES: CLIMATE AND PLANT LIFE ON LAND

Why Are There Different Plants in Different Places? Why is one area of the earth's land surface a desert, another a grassland, and another a forest? Why are there different types of deserts, grasslands, and forests?

The general answer to these questions is differences in **climate**: the physical properties of the troposphere of an area based on analysis of its weather records over a long period (at least 30 years). The two main factors determining an area's climate are *temperature*, with its seasonal variations, and the amount and distribution of *precipitation*. Figure 5-2 is a generalized map of the earth's major climate zones.

Figure 5-3 and the photo on pp. 2–3 show distributions of **biomes**—terrestrial regions with characteristic types of natural, undisturbed plant communities adapted to the climate of an area. By comparing Figure 5-3 with Figure 5-2, you can see how the world's major biomes vary with climate. Figure 4-8 shows major biomes in the United States as one moves through different climates along the 39th parallel.

For plants, *precipitation is generally the limiting factor that determines whether a land area is desert, grassland, or forest*. Taken together, average annual precipitation and temperature (along with soil type) are the most important factors in producing tropical, temperate, or polar deserts, grasslands, and forests (Figure 5-4, p. 110).

On maps such as the one in Figure 5-3, biomes are presented as having sharp boundaries and being covered with the same general type of vegetation. In reality, most biomes don't have sharp boundaries and blend into one another in transitional zones or ecotones. Also, the types and numbers of plants in a biome vary from one location to another because of small variations in climate (microclimates), soil types, and natural and human-caused disturbances (Figure 1-5).

Climate and vegetation vary with **latitude** (distance from the equator) and **altitude** (elevation above sea level). If you travel from the equator toward either pole, you will generally encounter ever colder climates and zones of vegetation adapted to those climates. Similarly, as elevation above sea level increases, climate becomes colder. Thus, if you climb a tall mountain from its base to its summit, you can observe changes in plant life similar to those you would encounter in traveling from equator to poles (Figure 5-5, pp. 110–111)).

Why Do Plant Sizes, Shapes, and Survival Strategies Differ? Arctic soils are wet and nutrient rich. So why are there no trees in the Arctic, and why are the plants there so close to the ground? Why don't desert plants such as cacti have leaves? Why do trees in most forests found in both the warm tropics and in cold areas such as Canada and Sweden keep their leaves year-round, whereas most trees in temperate forests lose their leaves in winter?

Research has shown that the size and shape of a plant species (and whether it keeps its leaves year-round) represent the best strategy for gathering sunlight for photosynthesis and for not becoming too hot or too cold. Plants exposed to cold air year-round or during winter need a design that keeps them from losing too much heat and water. For example, trees or tall plants in the cold, windy arctic grasslands (tundra) would lose too much of their heat for survival.

Desert plants exposed to the sun all day must be able to lose enough heat so that they don't overheat and die. They must also conserve enough water for survival. **Succulent** (fleshy) **plants**, such as saguaro (pronounced sa-WA-ro) cacti (see photo in Table of Contents) survive in dry climates by having a sticklike shape, no leaves, and the ability to store water and synthesize food in their expandable, fleshy tissue. The plant's shape and lack of leaves give it the smallest possible surface area exposed to sunlight and the largest possible area away from the sun for radiating heat out. It also reduces water loss by opening its pores (stomata) only at night.

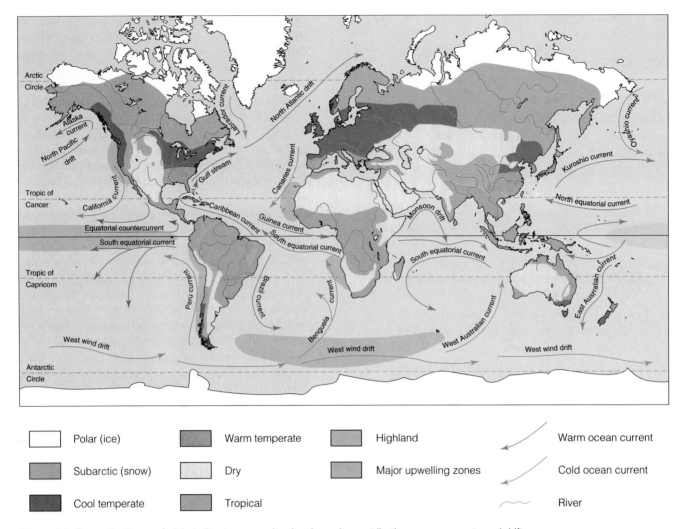

Figure 5-2 Generalized map of global climate zones, showing the major contributing ocean currents and drifts. Large variations in climate are dictated mainly by temperature (with its seasonal variations) and by the quantity and distribution of precipitation.

Legend:
- Polar (ice)
- Subarctic (snow)
- Cool temperate
- Warm temperate
- Dry
- Tropical
- Highland
- Major upwelling zones
- Warm ocean current
- Cold ocean current
- River

Trees in the canopies of wet tropical rain forests tend to be **broadleaf evergreen plants**, which keep most of their broad leaves year-round. The large surface area of the leaves allows them to collect ample sunlight for photosynthesis and radiate out heat during the hot summer.

In a climate with a cold (and sometimes dry) winter, keeping such leaves would cause plants to lose too much heat and water for survival. In such climates, **broadleaf deciduous plants**, such as oak and maple trees, survive drought and cold by shedding their leaves and becoming dormant during such periods.

If we move further north to areas such as Canada and Sweden, where summers are cool and short, this strategy is less successful. The evolutionary solution in such areas is **coniferous** (cone-bearing) **evergreen plants** (such as spruces, pines, and firs). These plants keep some of their narrow pointed leaves (needles) all year. The waxy coating, shape, and clustering of conifer needles slow down heat loss and evaporation during the long, cold winter. Additionally, by keeping their leaves all winter, such trees are ready to take advantage of the brief summer without having to take time to grow new needles.

What Are Deserts and Semideserts? A **desert** is an area where evaporation exceeds precipitation. Precipitation is typically less than 25 centimeters (10 inches) a year and is often scattered unevenly throughout the year. Low rainfall combined with different average temperatures creates tropical, temperate, and cold deserts (Figure 5-4).

In *tropical deserts*, such as the southern Sahara in Africa, temperatures are usually high year-round and there is very little rainfall. These driest places on earth typically have few plants and a hard, windblown surface strewn with rocks and some sand.

Q: What percentage of the world's population own cars?

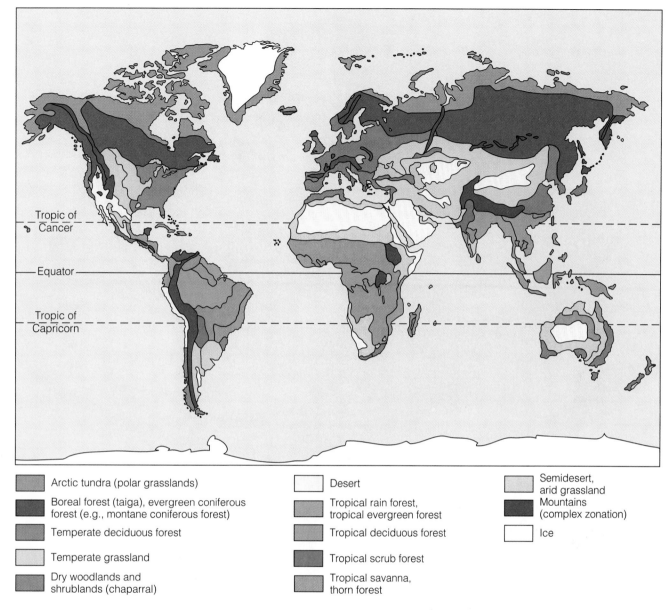

▦ Arctic tundra (polar grasslands)	▢ Desert	▦ Semidesert, arid grassland
▦ Boreal forest (taiga), evergreen coniferous forest (e.g., montane coniferous forest)	▦ Tropical rain forest, tropical evergreen forest	▦ Mountains (complex zonation)
▦ Temperate deciduous forest	▦ Tropical deciduous forest	▢ Ice
▦ Temperate grassland	▦ Tropical scrub forest	
▦ Dry woodlands and shrublands (chaparral)	▦ Tropical savanna, thorn forest	

Figure 5-3 The earth's major biomes—the main types of natural vegetation in different undisturbed land areas—result primarily from differences in climate. Each biome contains many ecosystems whose communities have adapted to differences in climate, soil, and other environmental factors. In reality, people have removed or altered much of this natural vegetation for farming, livestock grazing, harvesting lumber and fuelwood, mining, and constructing villages and cities, thereby altering the biomes.

In *temperate deserts*, such as the Mojave in southern California, daytime temperatures are hot in summer and cool in winter, and there is more precipitation than in tropical deserts. The vegetation is sparse, consisting mostly of widely dispersed, drought-resistant shrubs and cacti or other succulents and animals are adapted to the lack of water and temperature variations (Figure 5-6, p. 112).

In *cold deserts*, such as the Gobi Desert in China, winters are cold and summers are warm or hot; precipitation is low. In the semiarid zones between deserts and grasslands, we find *semidesert*. This biome is dominated by thorn trees and shrubs adapted to a long dry season followed by brief, sometimes heavy rains.

Survival in the desert has two themes: "Beat the heat" and "Every drop of water counts." Some of the plants are evergreens with wax-coated leaves (creosote bush) that minimize transpiration. Mesquite plants grow deep roots to tap into groundwater, whereas fleshy-stemmed, short (prickly pear; Figure 5-6) and tall (saguaro) cacti spread their shallow roots wide to collect scarce water for storage in their spongy tissues.

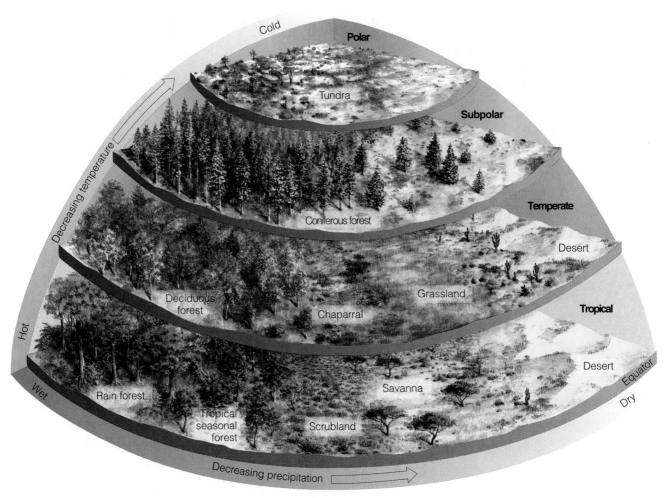

Figure 5-4 Average precipitation and average temperature, acting together as limiting factors over a period of 30 or more years, determine the type of desert, grassland, or forest biome in a particular area. Although the actual situation is much more complex, this simplified diagram explains how climate determines the types and amounts of vegetation found in an area left undisturbed by human activities. (Used by permission of Macmillan Publishing Company, from Derek Elsom, *The Earth*, New York: Macmillan, 1992. Copyright © 1992 by Marshall Editions Developments Limited)

Q: What percentage of the cars carrying people to and from work in the United States have only one passenger?

Many desert plants are annual wildflowers and grasses that store much of their biomass in seeds during dry periods and remain inactive (sometimes for years) until they receive enough water to sprout. Shortly after a rain, in a frenzy of biological activity, these plants germinate their seeds, grow, carpet the desert with a dazzling array of colorful flowers, produce new seed, and die—all in only a few weeks.

Most desert animals are small (which reduces heat gain and loss) and escape the heat by hiding in burrows or rocky crevices by day and come out at night or in the early morning hours. Some desert animals also have physical adaptations for conserving water (Spotlight, right). Insects and reptiles have thick outer coverings to minimize water loss through evaporation. Some desert animals become dormant during periods of extreme heat or drought and are active only during the cooler months of the year.

What Impacts Do Humans Have on Desert Ecosystems? When disturbed, desert ecosystems take a long time to recover because of their slow plant growth, low species diversity, slow nutrient cycling (because of little bacterial activity in their soils), and shortages of water. Already sparse vegetation destroyed by livestock overgrazing and off-road vehicles may take decades to grow back. Four-wheel drive vehicles and motorcycles can also collapse underground burrows where many desert animals live.

Figure 5-5 Generalized effects of latitude and altitude on climate and biomes. Parallel changes in vegetation type occur when we travel from the equator to the poles or from lowlands to mountaintops.

Altitude

Mountain ice and snow

Tundra (herbs, lichens, mosses)

Coniferous forests

Deciduous forests

Tropical forests

Tropical forests

Some major human impacts on deserts are as follows:

- Rapidly growing, large desert cities in countries such as Saudi Arabia and Egypt and in the southwestern United States (such as Palm Springs, California, Las Vegas, Nevada, and Phoenix, Arizona).

- Irrigating desert areas (such as parts of southern California) to grow crops. When this is done salts may accumulate in the soil (salinization) as the water evaporates, and crop productivity is limited.

- Depletion of underground water (aquifers) as desert cities and irrigation expand.

SPOTLIGHT

The Kangaroo Rat: Water Miser and Keystone Species

The kangaroo rat is a remarkable mammal superbly adapted for conserving water in its desert environment. As the desert's chief seed eater, it is also a keystone species that helps support other desert species and helps keep desert shrubland from becoming grassland.

This rodent comes out of its burrow only at night, when the air is cool and water evaporation has slowed. It seeks dry seeds that it quickly stuffs into its cheek pouches.

After a night of foraging it empties its cache of seeds into its cool burrow, where they soak up water exhaled in the rodent's breath. When the rodent eats these seeds, it gets this water back.

The kangaroo rat does not drink water; its water comes from the recycled moisture in the seeds and from water produced when sugars in the seeds undergo aerobic respiration during digestion.

Some of the water vapor in the rat's breath also condenses on the cool inside surface of its nose. This condensed water then diffuses back to its body.

Kangaroo rats have no sweat glands, so they don't lose water by perspiration. In addition, they save water by excreting hard, dry feces and thick, nearly solid urine produced by their extremely efficient kidneys.

Critical Thinking

Water is scarce in much of the southwestern United States where the kangaroo rat lives. However, this area has one of the highest rates of human population growth. As this happens, what ecological lesson can we learn from the kangaroo rat about how to survive in this area (and other water-poor areas throughout the world)?

Figure 5-6 Some components and interactions in a temperate desert biome. When these organisms die, their organic matter is broken down by decomposers into minerals used by plants. Transfers of matter and energy among producers, primary consumers (herbivores), and secondary (or higher-level) consumers (carnivores) are indicated by colored arrows.

Producer to primary consumer

Primary to secondary consumer

Secondary to higher level consumer

All producers and consumers to decomposers

Q: What percentage of Americans use public transportation to get to and from work?

- Disruption and pollution by the extraction of oil and a variety of minerals and building materials (such as road rock and sand).

- Using remote desert areas as sites for storage of toxic and radioactive wastes, underground testing of nuclear weapons, and practice exercises by heavy tanks and other military vehicles.

The largest and most untapped resource of deserts is abundant sunlight. Over the next 40–50 years many analysts expect industrialized societies to make a transition from reliance on nonrenewable fossil and nuclear fuels to greatly increased use of renewable solar energy. If such a shift takes place, large areas of many deserts near urban areas will be covered with arrays of solar collectors and solar cells used to produce electricity and hydrogen gas. Great care must be taken to do this in ways that don't seriously disrupt such desert ecosystems.

What Are the Major Types of Grasslands? A region with enough average annual precipitation to allow grass (and in some areas, a few trees) to prosper, but with precipitation so erratic that drought and fire prevent large stands of trees from growing, is called a **grassland**. Grasses in these biomes are renewable resources, if not overgrazed, because grass plants grow out from the bottom; thus, their stems can grow again after being nibbled off by grazing animals.

The three main types of grasslands—tropical, temperate, and polar (tundra)—result from combinations of low average precipitation and various average temperatures (Figure 5-4). *Tropical grasslands* are found in areas with high average temperatures, low to moderate precipitation, and a prolonged dry season.

One type of tropical grassland, called a *savanna*, usually has scattered shrubs and isolated small trees, warm temperatures year-round, two prolonged dry seasons, and abundant rain the rest of the year. African tropical savannas contain enormous herds of *grazing* (grass- and herb-eating) and *browsing* (twig- and leaf-nibbling) hoofed animals, including wildebeests (Figure 5-7), gazelles, zebras, giraffes, and antelopes. These and other large herbivores have evolved specialized eating habits that minimize interspecific competition for resources (resource partitioning): Giraffes eat leaves and shoots from the tops of trees, elephants eat leaves and branches further down, Thompson's gazelles and wildebeests prefer short grass, and zebras graze on longer grass and stems.

Temperate grasslands cover vast expanses of plains and gently rolling hills in the interiors of North and South America, Europe, and Asia (Figure 5-3). Winters there are bitterly cold and summers are hot and dry. Annual precipitation is fairly sparse and falls un-

Figure 5-7 Serengeti tropical savanna in Tanzania, Africa, an example of one type of tropical grassland. Most savannas consist of grasslands punctuated by stands of deciduous shrubs and trees, which shed their leaves during the dry season and thus avoid excessive water loss. More large, hoofed, plant-eating mammals (ungulates), such as the wildebeest shown here, live in this biome than anywhere else. (Jonathan Scott/Planet Earth Pictures)

evenly through the year. Drought, occasional fires, and intense grazing inhibit the growth of trees and bushes, except along rivers.

Types of temperate grasslands are the *tall-grass prairies* (Figure 5-8) and *short-grass prairies* of the midwestern and western United States and Canada, the South American *pampas*, the African *veldt*, and the *steppes* of central Europe and Asia. Here winds blow almost continuously and evaporation is rapid, leading to recurring fires in the summer and fall. Deep and nutrient-rich soil in such grasslands is held in place by a thick network of intertwined roots of drought-tolerant grasses, unless the topsoil is plowed up and allowed to blow away by prolonged exposure to the high winds found in these biomes. These temperate grasslands, sometimes called the world's breadbaskets, produce most of the world's wheat and other cereal grains and many of them are used for grazing cattle and sheep.

Polar grasslands, or *arctic tundra*, are found just south of the Arctic polar ice cap (Figure 5-3). During most of the year these treeless plains are bitterly cold,

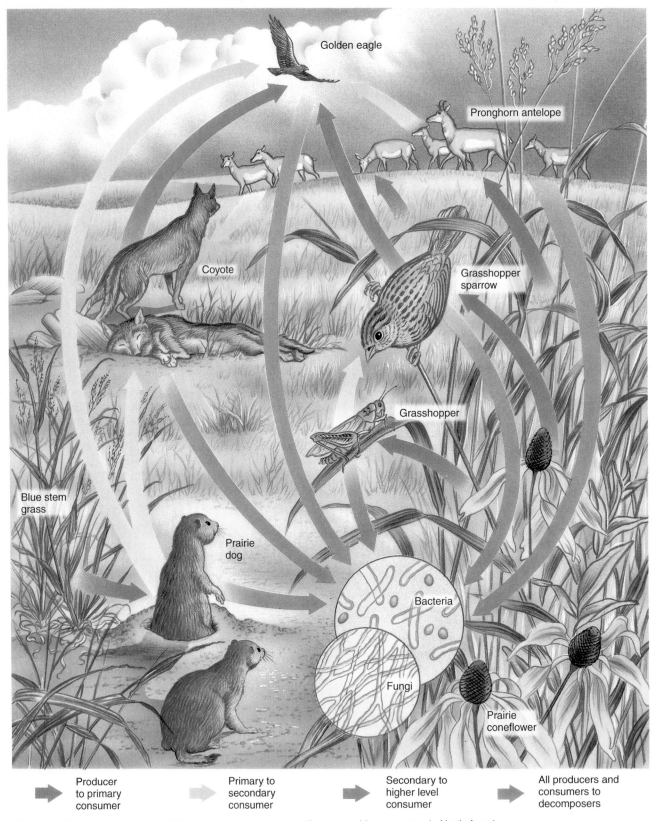

| Producer to primary consumer | Primary to secondary consumer | Secondary to higher level consumer | All producers and consumers to decomposers |

Figure 5-8 Some components and interactions in a temperate tall-grass prairie ecosystem in North America. When these organisms die, their organic matter is broken down by decomposers into minerals used by plants. Transfers of matter and energy among producers, primary consumers (herbivores), and secondary (or higher-level) consumers (carnivores) are indicated by colored arrows.

Q: What percentage of Americans walk or use a bicycle to get to and from work?

swept by frigid winds, and covered with ice and snow. Winters are long and dark, and the scant precipitation falls mostly as snow. This biome is carpeted with a thick, spongy mat of low-growing plants, primarily grasses, mosses, and dwarf woody shrubs (Figure 5-9). Most of the annual growth of these plants occurs during the 6- to 8-week summer, when sunlight shines almost around the clock.

One effect of the extreme cold is *permafrost*, a perennially frozen layer of the soil that forms when the water there freezes. In summer, water near the surface thaws, but the permafrost soil layer below stays frozen and prevents liquid water at the surface from seeping into the ground. Thus, during the brief summer the soil above the permafrost layer remains waterlogged, forming a large number of shallow lakes, marshes, bogs, ponds, and other seasonal wetlands. Hordes of mosquitoes, blackflies, and other insects thrive in these shallow surface pools. They feed large colonies of migratory birds, especially waterfowl, that return from the south to nest and breed in the bogs and ponds.

Its low rate of decomposition, shallow soil, short growing season, and slow plant growth rate make this biome especially vulnerable to disruption; vegetation destroyed by human activities can take decades to grow back. Damage by spills of oil or toxic waste may take far longer.

What Impacts Do Humans Have on Grassland Ecosystems? Some major human impacts on deserts are as follows:

- Burning, plowing up, and converting some areas of savanna into cropland. By releasing large quantities of carbon dioxide into the atmosphere, this practice may contribute to the greenhouse effect as much as (if not more than) the highly publicized clearing and burning of tropical rain forests.

- Disruption of the movement of livestock by traditional nomadic pastoralists in tropical and temperate grasslands as governments and aid agencies encourage the drilling of wells. As a result, pastures around wells have been overgrazed, trampled by thousands of hooves, and converted into less productive desert and semidesert.

- Plowing of large areas of temperate grasslands in North America, western Europe, and Ukraine and converting them to highly productive cropland (Figure 5-10). As long as temperate grasslands keep their fertile soil and the climate does not change, they can continue producing much of the world's cereal grains. However, overgrazing, mismanagement, and occasional prolonged droughts lead to severe wind erosion and loss of topsoil, which can convert temperate grasslands into desert or semidesert shrubland.

- Long-term damage to the fragile arctic tundra in Alaska and Siberia because of oil exploration and drilling, air pollution, spills or leaks of oil and toxic wastes, and disruption of soil and vegetation by vehicles.

What Are the Major Types of Forests? Undisturbed areas with moderate to high average annual precipitation tend to be covered with **forest** containing various species of trees and smaller forms of vegetation. *Tropical rain forests* are a type of evergreen broadleaf

Figure 5-9 Polar grassland (arctic tundra) in Alaska in summer. During the long, dark, cold winter, this land is covered with snow and ice. Its low-growing plants are adapted to the lack of sunlight and water, freezing temperatures, and constant high winds. Below the surface there is a frozen layer of soil called permafrost. Most of this layer stays frozen year-round, except in the brief summer, when its top portion melts. (Charlie Ott/Photo Researchers, Inc.)

Figure 5-10 Replacement of a temperate grassland with a monoculture crop near Blythe, California. When the tangled root network of natural grasses is removed, the fertile topsoil is subject to severe wind erosion unless it is covered with some type of vegetation. If global warming accelerates over the next 50 years, many of these grasslands may become too hot and dry for farming, thus threatening the world's food supply. (National Archives/EPA Documerica)

forest (Figure 5-11) found near the equator (Figure 5-3), where hot, moisture-laden air rises and dumps its moisture.

These forests have a warm annual mean temperature (which varies little, daily or seasonally), high humidity, and heavy rainfall almost daily. The almost unchanging climate means that water and temperature are not limiting factors, as in other biomes; instead, soil nutrients are the main limiting factors.

Tropical rain forests have incredible biodiversity. These diverse life-forms occupy a variety of specialized niches in distinct layers, based mostly on their need for sunlight (Figure 5-12). The stratification of specialized plant and animal niches in various layers enables species to avoid or minimize interspecific competition for resources (resource partitioning). It also results in coexistence of a great variety of species (biodiversity). Although tropical rain forests cover only about 2% of the earth's land surface, they are habitats for 50–80% of the earth's terrestrial species.

Because of the warm, moist conditions, dropped leaves and dead animals break down quickly. This rapid recycling of scarce soil nutrients is why there is little litter on the ground. Instead of being stored in the soil, most minerals released by decomposition are quickly taken up by plants. Thus, most of a tropical forest's nutrients are stored in its living organisms.

Moving a little farther from the equator, we find *tropical deciduous forests* (sometimes called tropical monsoon forests or tropical seasonal forests). They are usually located between tropical rain forests and tropical savannas (Figure 5-3). These forests are warm year-round, and most of their plentiful rainfall occurs during a wet (monsoon) season that is followed by a long dry season. Where the dry season is

especially long, we find *tropical scrub forests* (Figure 5-3) containing mostly small deciduous trees and shrubs.

Temperate deciduous forests (Figure 5-13) grow in areas with moderate average temperatures that change significantly with the season. These areas have long, warm summers, cold but not too severe winters, and abundant precipitation, often spread fairly evenly throughout the year. This biome is dominated by a few species of broadleaf deciduous trees such as oak, hickory, maple, poplar, sycamore, and beech. They survive cold winters by dropping their leaves in the fall and becoming dormant (see photo in Table of Contents). Each spring they sprout new leaves that change in the fall into a blazing array of reds and golds before dropping. Because of the fairly low rate of decomposition, these forests accumulate a thick layer of slowly decaying leaf litter that is a storehouse of nutrients.

Evergreen coniferous forests, also called *boreal forests* (meaning "northern forests") and *taigas* (pronounced "TIE-guhs"), are found just south of the arctic tundra in northern regions across North America, Asia, and Europe (Figure 5-3). In this subarctic climate, winters are long, dry, and extremely cold; in the northernmost taiga, sunlight is available only 6–8 hours a day. Summers are short, with mild to warm temperatures, and the sun typically shines 19 hours a day.

Most boreal forests are dominated by a few species of evergreen conifer trees such as spruce, fir, cedar, hemlock, and pine. The tiny, needle-shaped, waxy-coated leaves of these trees can withstand the intense cold and drought of winter when snow blankets the ground. Plant diversity is low in these forests because few species can survive the winters when soil moisture is frozen. Because trees grow slowly in the cold northern climate, these forests take a long time to recover from disruption.

Q: How many people have been killed by motor vehicles since the first automobile was built in 1885?

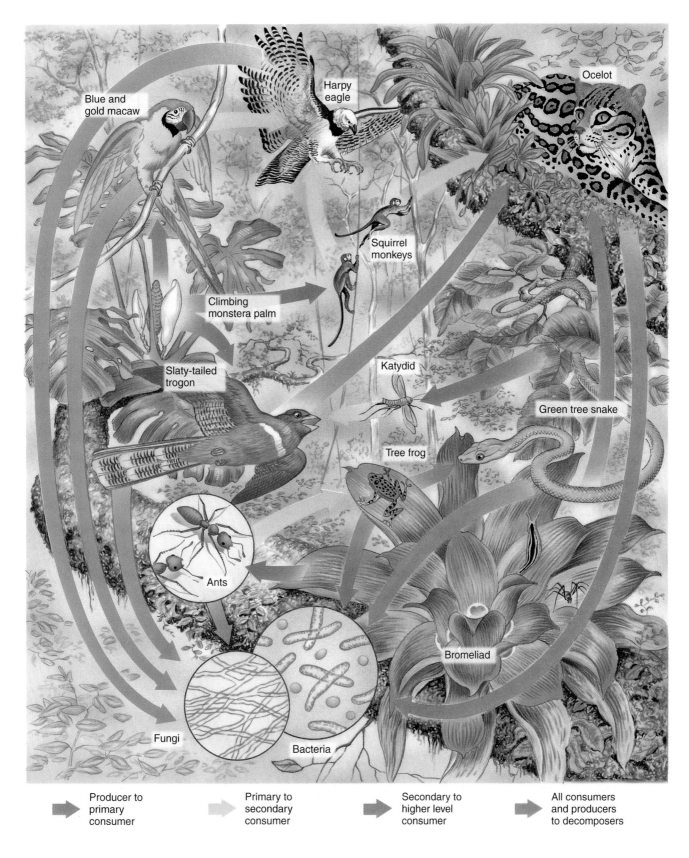

Figure 5-11 Some components and interactions in a tropical rain forest ecosystem. When these organisms die, their organic matter is broken down by decomposers into minerals used by plants. Transfers of matter and energy among producers, primary consumers (herbivores), and secondary (or higher-level) consumers (carnivores) are indicated by colored arrows.

Producer to primary consumer

Primary to secondary consumer

Secondary to higher level consumer

All consumers and producers to decomposers

A: About 18 million (3 million in the United States—twice the number of soldiers killed in all U.S. wars)   CHAPTER 5   **117**

Figure 5-12 Stratification of specialized plant and animal niches in various layers of a tropical rain forest. The presence of these specialized niches enables species to avoid or minimize competition for resources and results in the coexistence of a great variety of species (biodiversity).

Beneath the stands of trees, there is a deep layer of partially decomposed conifer needles and leaf litter. Decomposition is slow because of the low temperatures, the waxy coating of conifer needles, and the high acidity. As the conifer needles decompose, they make the thin, nutrient-poor soil acidic and prevent most other plants (except certain shrubs) from growing on the forest floor (interference competition).

These biomes contain a variety of wildlife (Figure 5-14). During the brief summer the soil becomes waterlogged, forming acidic bogs, or *muskegs*, in low-lying areas of these forests. Warblers and other insect-eating birds feed on hordes of flies, mosquitoes, and caterpillars.

In scattered coastal temperate areas with ample rainfall or moisture from dense ocean fogs, we find *coastal coniferous forests* or *temperate rain forests*. Along the coast of North America, from Canada to northern California, these biomes are dominated by dense stands of large conifers such as Sitka spruce, Douglas fir, and magnificent redwoods.

What Impacts Do Humans Have on Forest Ecosystems? Some major human impacts on forests are as follows:

- Rapid clearing and degradation of tropical rain forests and tropical seasonal forests for timber, grazing land, and agriculture, leading to severe erosion of their already nutrient-poor soils. Within 50 years only scattered fragments of these diverse forests might remain, causing a massive irreversible loss of the earth's vital biodiversity within your lifetime.

Q: Worldwide, how many people are killed each year by motor vehicles?

Broad-winged hawk

Hairy woodpecker

Gray squirrel

White oak

White-footed mouse

White-tailed deer

Metallic wood-boring beetle and larvae

Mountain winterberry

Shagbark hickory

May beetle

Racer

Long-tailed weasel

Wood frog

Fungi

Bacteria

▶ Producer to primary consumer	▶ Primary to secondary consumer	▶ Secondary to higher level consumer	▶ All consumers and producers to decomposers

Figure 5-13 Some components and interactions in a deciduous forest ecosystem. When these organisms die, their organic matter is broken down by decomposers into minerals used by plants. Transfers of matter and energy among producers, primary consumers (herbivores), and secondary (or higher-level) consumers (carnivores) are indicated by colored arrows.

Blue jay

Balsam fir

Great horned owl

Marten

Wolf

Moose

White spruce

Bebb willow

Showshoe hare

Pine sawyer beetle and larvae

Starflower

Fungi

Bacteria

Bunchberry

| ▶ | Producer to primary consumer | ▷ | Primary to secondary consumer | ▶ | Secondary to higher level consumer | ▷ | All producers and consumers to decomposers |

Figure 5-14 Some components and interactions in an evergreen coniferous (taiga) biome. When these organisms die, their organic matter is broken down by decomposers into minerals used by plants. Transfers of matter and energy among producers, primary consumers (herbivores), and secondary (or higher-level) consumers (carnivores) are indicated by colored arrows. Many of these ancient forests along the Pacific Coast from Canada to northern California have been clear-cut.

Q: Worldwide, what percentage of urban land is devoted to roads and parking?

- Extensive clearing of *temperate deciduous forests* in Europe, Asia, and North America mostly for timber, growing food, and urban development. In North America, about 99.9% of the original stands of temperate deciduous forests have been cleared for such purposes. Some have been converted to managed *tree farms* or *tree plantations*, where a single species is grown for timber, pulpwood, or Christmas trees (see photo in Table of Contents).

- Widespread clearing of large areas of evergreen coniferous forests by loggers in North America, Finland, Sweden, and Canada. Within a decade the vast boreal forests of Siberia and Russia may also disappear because of logging and mining.

Why Are Mountains Ecologically Important? Some of the world's most spectacular and important environments are mountains, which make up about 20% the earth's land surface. Mountains are places where dramatic changes in altitude, climate, soil, and vegetation take place over a very short distance (Figure 5-5). Because of the steep slopes, mountain soils are especially prone to erosion when the vegetation holding them in place is removed by natural changes (such as climate) or human activities.

Many freestanding mountains are *islands of biodiversity* surrounded by a sea of lower-elevation landscapes transformed by human activities. As a result, many mountain areas contain species found nowhere else on earth; they are also sanctuaries for animal species driven from lowland areas.

The ice and snow of mountaintops help regulate the earth's climate by reflecting solar radiation back into space. Sea levels depend on the melting of glacial ice, most of which is locked up in Antarctica, the most mountainous of all continents. Mountain regions also contain the majority of the world's forests, which contain much of the world's biodiversity.

What Impacts Do Humans Have on Mountain Ecosystems? Despite their ecological, economic, and cultural importance, the fate of mountain ecosystems has not been a high-priority item of governments or many environmental organizations. Mountain ecosystems are coming under increasing environmental pressure from several major trends:

- *Rapidly increasing population, especially in developing countries.* This is forcing landless poor people, refugees, and minority populations to migrate uphill and try to survive on less stable soils. These newcomers often use mountain soils and forests unsustainably in a desperate struggle to survive or because of a lack of knowledge about how to grow food, raise livestock, and harvest wood in these habitats.

- *Increased commercial extraction of timber and mineral resources.*

- *A growing number of hydroelectric dams and reservoirs.* Rivers in mountains are attractive sites for dams and reservoirs because the elevation and slope of mountains increase the force of flowing water. These reservoirs flood mountain slopes, and the dams alter the types and abundance of species in rivers.

- *Increased air pollution from growing urban and industrial centers and increased use of automobiles.* Trees and other vegetation at high elevations (especially conifers such as spruce) are bathed year-round in air pollutants such as ozone and acidic compounds, carried there by prevailing winds from cities, factories, and coal-burning power plants.

- *Changes in climate and levels of ultraviolet (UV) radiation brought about by human activities.* If global warming occurs as projected during the next century, many species of mountain plants could be displaced by lowland species that have higher rates of growth and reproduction. Similarly, any increase in UV radiation brought about by depletion of the ozone layer may have a pronounced effect on mountain life, which is already exposed to high levels of ultraviolet radiation because of altitude.

- *Increased warfare in mountainous areas.* In 1993, 22 of 28 major armed conflicts took place primarily in mountains.

5-2 LIFE IN AQUATIC ENVIRONMENTS

What Are the Two Major Types of Aquatic Life Zones? The aquatic equivalents of biomes are called *aquatic life zones.* The major types of organisms found in aquatic environments are determined by the water's *salinity* (the amounts of various salts such as sodium chloride or NaCl dissolved in a given volume of water). As a result, aquatic life zones are divided into two major types: *saltwater* or *marine* (such as estuaries, coastlines, coral reefs, coastal marshes, mangrove swamps, and the deep ocean) and *freshwater* (such as mostly nonflowing lakes and ponds, flowing streams, and inland wetlands).

Living in water has its advantages. Water's buoyancy provides physical support, and limited fluctuations in temperature greatly reduce the risks from drying out or becoming overheated. Required nutrients are dissolved and readily available, and potentially toxic metabolic wastes secreted by aquatic organisms are diluted and dispersed.

Most aquatic life zones can be divided into surface, middle, and bottom layers. Important factors determining the types and numbers of organisms

A: At least 33% (50% in the United States)

found in these layers are *temperature, access to sunlight for photosynthesis, dissolved oxygen content,* and *availability of nutrients* such as carbon (as dissolved CO_2 gas), nitrogen (as NO_3^-), and phosphorus (mostly as PO_4^{3-}) for producers.

Why Are the Oceans Important? A more accurate name for Earth would be *Ocean* because saltwater oceans cover about 71% of its surface (Figure 5-15). The oceans play key roles in the survival of virtually all life on earth. Because solar heat is distributed through ocean currents (Figure 5-2) and because ocean water evaporates as part of the global hydrologic cycle (Figure 4-27), oceans play a major role in regulating the earth's climate. They also participate in other important nutrient cycles.

By serving as a gigantic reservoir for carbon dioxide, oceans help regulate the temperature of the troposphere. Oceans provide habitats for about 250,000 species of marine plants and animals, which are food for many other organisms (including humans). In addition, many human-produced wastes that flow into or are dumped into the ocean are dispersed by currents (and thus are often diluted to less harmful levels).

What Is the Coastal Zone? Oceans have two major life zones: the coastal zone and the open sea (Figure 5-16). The **coastal zone** is the relatively warm, nutrient-rich, shallow water that extends from the high-tide mark on land to the gently sloping, shallow edge of the *continental shelf* (the submerged part of the continents). Although it makes up less than 10% of the ocean's area, the coastal zone contains 90% of all marine species and is the site of most of the large commercial marine fisheries. Most ecosystems found in the coastal zone have a very high primary productiv-

ity (Figure 4-20) and net primary productivity per unit of area (Figure 4-21) because of the zone's ample supplies of sunlight and plant nutrients (deposited from land and stirred up by wind and ocean currents).

One highly productive area in the coastal zone is an **estuary**, a partially enclosed area of coastal water where seawater mixes with fresh water and nutrients from rivers, streams, and runoff from land (Figure 5-17). The constant water movement stirs up the nutrient-rich silt, making it available to producers.

According to one estimate, just 0.4 hectares (1 acre) of tidal estuary provides an estimated $75,000 worth of free waste treatment and has a value of about $83,000 when recreation and fish for food are included. By comparison, 0.4 hectare (1 acre) of prime farmland in Kansas has a top value of about $1,200 and an annual production value of about $600.

Areas of coastal land that are covered all or part of the year with salt water are called **coastal wetlands**. They are breeding grounds and habitats for a variety of waterfowl and other wildlife; they also serve as popular areas for recreational activities such as boating, fishing, and hunting. They also help maintain the quality of coastal waters by diluting, filtering, and settling out sediments, excess nutrients, and pollutants. In addition, coastal wetlands protect lives and property during floods by absorbing and slowing the flow of water, and during storms they buffer shores against damage and erosion.

In temperate areas, including the United States, coastal wetlands usually consist of a mixture of *bays, lagoons, salt flats, mud flats,* and *salt marshes* (Figure 5-18), in which grasses are the dominant vegetation. These highly productive ecosystems serve as nurseries and habitats for shrimp and many other aquatic animals.

Along warm tropical coasts where there is too much silt for coral reefs to grow, we find highly productive **mangrove swamps** (see photo in Table of Contents), dominated by about 55 species of salt-tolerant trees or shrubs known as mangroves. These swamps help protect the coastline from erosion and reduce damage from typhoons and hurricanes. They also trap sediment washed off the land and provide breeding, nursery, and feeding grounds for some 2,000 species of fish, invertebrates, and plants.

Some coasts have steep *rocky shores* pounded by waves. Visitors are often surprised at the variety of life found in the numerous pools and other niches in the rocks in the intertidal zone of rocky shores (Figure 5-19, left). Other coasts have gently sloping *barrier beaches,* or *sandy shores,* with niches for different marine organisms, including crabs, lugworms, clams, ghost shrimp, sand dollars, and flounder (Figure 5-19, right). These sandy beaches and their adjoining coastal wetlands are also home to a variety of shorebirds that

Ocean hemisphere   Land–ocean hemisphere

Figure 5-15 The ocean planet. The salty oceans cover about 71% of the earth's surface. About 97% of the earth's water is in the interconnected oceans, which cover 90% of the planet's mostly ocean hemisphere (left) and 50% of its land–ocean hemisphere (right). The average depth of the world's oceans is 3.8 kilometers (2.4 miles).

Q: What is the average government subsidy per car in the United States?

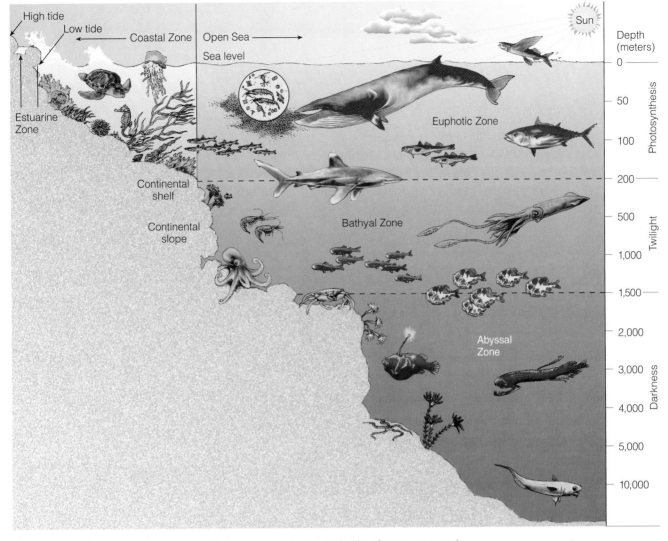

Figure 5-16 Major life zones in an ocean. (Not drawn to scale. Actual depths of zones may vary.)

Figure 5-17 View from the space shuttle of the sediment plume at the mouth of Madagascar's Betsiboka River as it flows through the estuary and into the Mozambique Channel, which separates this huge island from the African coast. Topography, heavy rainfall, and the clearing of forests for agriculture make Madagascar the world's most eroded country. (NASA)

A: $1,600-$3,200 per year

Figure 5-18 Some components and interactions in a salt marsh ecosystem. When these organisms die, their organic matter is broken down by decomposers into minerals used by plants. Transfers of matter and energy among consumers (herbivores) and secondary (or higher-level) consumers (carnivores) are indicated by colored arrows. These and other temperate coastal wetlands trap nutrients and sediment flowing in from rivers and nearby land and thus have a high net primary productivity. They also filter out and degrade some of the pollutants deposited by rivers and land runoff. Up to 65% of the coastal marshes along the Atlantic Coast of the United States have been lost to development and pollution, and much of what remains is threatened.

Q: What do many economists believe is the best way to prevent or reduce pollution?

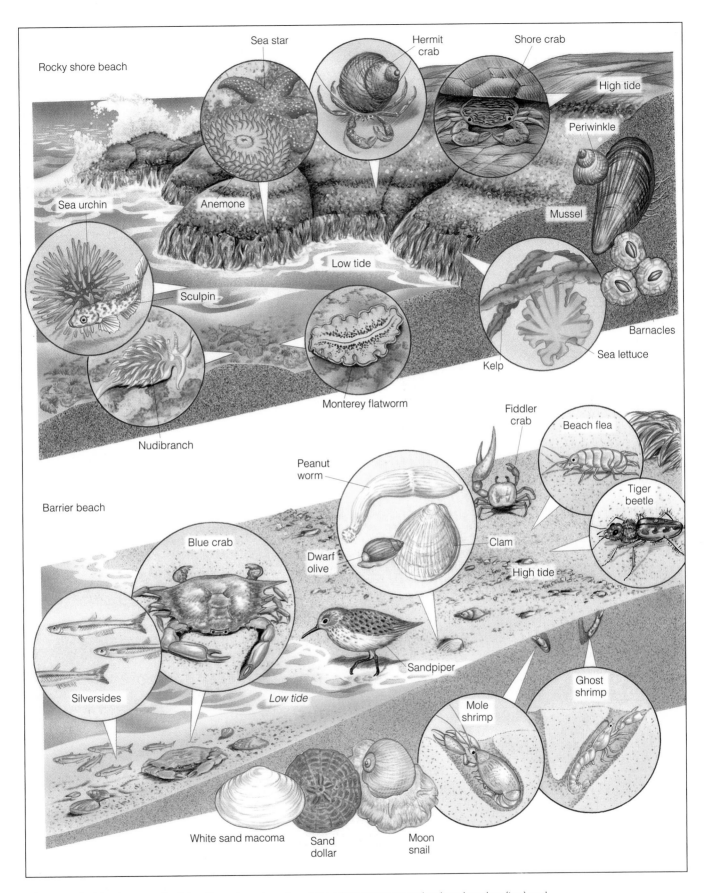

Figure 5-19 Living between the tides. Some organisms found in various zones on rocky shore beaches (top) and barrier or sandy beaches (bottom).

feed in specialized niches on crustaceans, insects, and other organisms (Figure 4-28).

If not destroyed by human activities, one or more rows of natural sand dunes on barrier beaches (with the sand held in place by the roots of grasses) serve as the first line of defense against the ravages of the sea (Figure 5-20). However, such beaches are prime sites for development. When coastal developers remove the protective dunes or build behind the first set of dunes, storms can flood and even sweep away seaside buildings and severely erode the unprotected beaches.

Along some coasts (such as most of North America's Atlantic and Gulf coasts) are **barrier islands**: long, thin, low offshore islands of sediment that generally run parallel to the shore. These islands help protect the mainland, estuaries, lagoons, and coastal wetlands by dispersing the energy of approaching storm waves. Sooner or later many of the structures humans build on low-lying barrier islands, such as Atlantic City, New Jersey, Miami Beach, Florida, and Ocean City, Maryland (Figure 5-21), are damaged or destroyed by flooding, severe beach erosion, or major storms (including hurricanes).

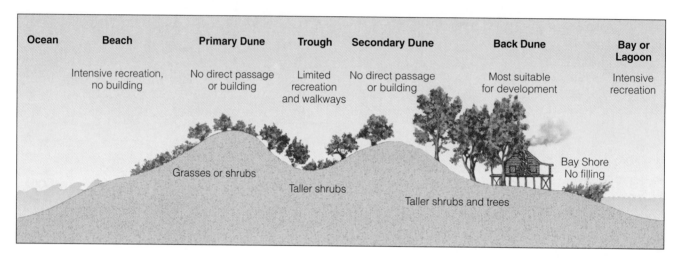

Ocean	Beach	Primary Dune	Trough	Secondary Dune	Back Dune	Bay or Lagoon
	Intensive recreation, no building	No direct passage or building	Limited recreation and walkways	No direct passage or building	Most suitable for development	Intensive recreation

Grasses or shrubs

Taller shrubs

Taller shrubs and trees

Bay Shore
No filling

Figure 5-20 Primary and secondary dunes on gently sloping sandy or barrier beaches play an important role in protecting the land from erosion by the sea. The roots of various grasses that colonize the dunes help hold the sand in place. Ideally, construction and development should be allowed only behind the second strip of dunes; walkways to the beach should be built over the dunes to keep them intact. This not only helps preserve barrier beaches, but also protects structures from being damaged and washed away by wind, high tides, beach erosion, and flooding from storm surges. This type of protection is rare, however, because the short-term economic value of oceanfront land is considered to be much higher than its long-term ecological and economic values.

Figure 5-21 A developed barrier island: Ocean City, Maryland, host to 8 million visitors a year. To keep up with shifting sands, taxpayers spend millions of dollars to pump sand onto the beaches and to rebuild natural sand dunes; they may end up spending millions more to keep buildings from sinking. Barrier islands lack effective protection against flooding and damage from severe storms; within a few hours, barrier islands may be cut in two or destroyed by a hurricane. If global warming raises average sea levels, as projected, most of these valuable pieces of real estate will be under water. (G. H. Demetrakas/O.C. Camera)

Q: How many of the world's people live in absolute poverty?

Case Study: Why Are Coral Reefs Important?

In the shallow coastal zones of warm tropical and subtropical oceans we often find **coral reefs** (Figure 5-22). These beautiful natural wonders are among the world's oldest and most diverse and productive ecosystems and are homes for one-fourth of all marine species (Figure 5-23).

Coral reefs are formed by massive colonies of tiny animals called *polyps* that are close relatives of jellyfish. They slowly build reefs by secreting a protective crust of limestone (calcium carbonate) around their soft bodies. As the polyps reproduce, the reef expands upward and outward; when polyps die, their empty crusts or outer skeleton remain as a platform for more reef growth.

Coral reefs are actually a joint venture between the polyps and tiny single-celled algae called *zooxanthellae* that live in the tissues of the polyps. In this *mutualistic relationship*, the microscopic photosynthesizing algae provide the polyp with food and oxygen. The polyps in turn provide a well-protected home for the algae and make nutrients such as nitrogen and phosphorus available that are usually scarce in tropical waters. The algae and other producers give corals most of their bright colors and provide plentiful food for a variety of marine life. Many species of reef fish maintain mutualistic or commensalistic relationships (Figure 4-32) with other reef species.

By forming limestone shells, coral polyps help remove CO_2 from the atmosphere as part of the carbon cycle. The reefs also act as natural barriers that help protect 15% of the world's coastlines from battering waves and storms. In addition, the reefs build beaches, atolls, and islands, and they provide food, jobs, and building materials for some of the world's poorest countries.

What Impacts Do Humans Have on Coastal Zones? Coastal zones are among the most densely populated and most intensely used and polluted ecosystems. Currently, nearly two-thirds of the world's population—some 3.9 billion people—live along coasts or within 160 kilometers (100 miles) of a coast. By 2025, it's estimated that 75%, or 6.2 billion people, will reside on or near coastal areas.

Since 1900 the world has lost approximately half of its coastal wetlands, primarily through coastal development. In the past 200 years nearly 55% of the area of estuaries and coastal wetlands in the United States has been destroyed or damaged, primarily because of dredging and filling and waste contamination. California alone has lost 91% of its original coastal wetlands, but Florida has lost the largest area

Figure 5-22 A healthy coral reef in the Philippines covered by colorful algae (below) and a bleached coral reef in the Bahamas that has lost most of its algae (left) because of changes in the environment (such as cloudy water or extreme temperatures). With the algae gone, the white limestone of the coral skeleton becomes visible. If the environmental stress is not removed, the corals die. These diverse and productive ecosystems are being damaged and destroyed at an alarming rate. (Left, Karl & Jill Wallin/FPG International; below, Robert Wicklund)

A: At least 1.2 billion—about one in every five people on earth

Figure 5-23 Some components and interactions in a coral reef ecosystem. When these organisms die, their organic matter is broken down by decomposers into minerals used by plants. Transfers of matter and energy among producers, primary consumers (herbivores), and secondary (or higher-level) consumers (carnivores) are indicated by colored arrows.

Q: Will improving environmental quality in the United States cause a net gain or a net loss of jobs?

of such wetlands in the United States. A 1994 study by researchers at the University of California at Berkeley estimated that the value of the long-term earth capital services provided by California's remaining 184,000 hectares (454,000 acres) of wetlands is $124.5 billion.

Mangrove swamps worldwide are under assault, with only about half of the world's original coastal mangrove swamps remaining. Large areas have been cut down for timber, fuelwood, and wood chips, to create aquaculture ponds for raising fish and shellfish, and to expand agricultural land (especially rice fields) and urban areas.

Coastal ecosystems are particularly vulnerable to toxic contamination because they trap pesticides, heavy metals, and other pollutants, concentrating them to very high levels. On any given day 37% of U.S. coastal shellfish beds are closed to commercial or sport fishing, usually because of contamination from sewage treatment plants, septic tank systems, and urban runoff.

Despite their ecological importance, coral reefs are disappearing and being degraded at an alarming rate. They are vulnerable to damage because they grow slowly, are easily disrupted, and thrive only in clear, warm, and fairly shallow water of constant high salinity.

The biggest threats to many of the world's coral reefs come from human activities, especially the deposition of eroded soil produced by deforestation, construction, agriculture, mining, dredging, and poor land management along increasingly populated coastlines. The suspended soil sediment that washes downriver to the sea or erodes from coastal areas smothers coral polyps or blocks their sunlight. Such silting is one cause of coral reef bleaching. It occurs when a reef loses its colorful algae and other producers, exposing the colorless coral animals and the underlying white skeleton of calcium carbonate (Figure 5-22, top).

Other threats to coral reefs include increased ultraviolet radiation from depletion of stratospheric ozone, runoff of toxic pesticides and industrial chemicals, commercial fishing boats, use of cyanide or dynamite to stun fish so they can be harvested, removing coral building material, oil spills, damage from tourists and recreational divers, and runoff of phosphate-rich fertilizer, raw sewage, and wastewater from sewage treatment plants.

Marine biologists estimate that humans have directly or indirectly caused the death of 10% of the world's coral reefs (especially those in Southeast Asia and the Caribbean). Another 30% of the remaining reefs are in critical condition, and 30% more are threatened; only 30% are stable. If the current rates of destruction continue, another 60% of the reefs could be gone in the next 20–40 years.

What Is the Open Sea? The sharp increase in water depth at the edge of the continental shelf separates the coastal zone from the **open sea**, which is divided into three vertical zones—euphotic, bathyal, and abyssal—based primarily on the penetration of sunlight (Figure 5-16). This vast volume of ocean contains only about 10% of all marine species.

Except at an occasional equatorial upwelling, where currents bring up nutrients from the ocean bottom, average primary productivity and net primary productivity per unit of area are quite low in the open sea (Figure 4-21). This is because sunlight cannot penetrate the lower layers and because the surface layer normally has fairly low levels of nutrients for phytoplankton, which are the main photosynthetic producers of the open ocean. However, because the open sea covers so much of the earth's surface (Figure 5-15), it makes the largest contribution to the earth's overall net primary productivity.

What Are the Major Characteristics of Freshwater Lakes? Large natural bodies of standing fresh water—formed when precipitation, runoff, or groundwater seepage fills depressions in the earth's surface—are called **lakes**. Causes of such depressions include glaciation (the Great Lakes of North America), crustal displacement accompanied by or causing earthquakes (Lake Nyasa in East Africa), and volcanic activity (Crater Lake in Oregon).

Lakes are fed by rainfall, melting snow, and the streams that drain the surrounding watershed. Lakes normally consist of distinct zones (Figure 5-24), providing habitats and niches for different species.

Ecologists classify lakes according to their nutrient content and their primary productivity. A newly formed lake generally has a small supply of plant nutrients and is called an **oligotrophic** (poorly nourished) **lake** (Figure 5-25, bottom). This type of lake is often deep, with steep banks. Because of its low net primary productivity, such a lake usually has crystal-clear blue or green water, with small populations of both phytoplankton and fish, such as smallmouth bass and trout.

Over time, sediment washes into an oligotrophic lake, and plants grow and decompose to form bottom sediments. A lake with a large or excessive supply of nutrients (mostly nitrates and phosphates) needed by producers is called a **eutrophic** (well-nourished) **lake** (Figure 5-25, top). Such lakes are typically shallow, and their water is generally a murky brown or green with very poor visibility. Because of their high levels of nutrients, these lakes have a high net primary productivity. Human inputs of nutrients from the atmosphere and from nearby urban and agricultural areas can accelerate the eutrophication of lakes—a process called *cultural eutrophication*. Many lakes fall somewhere between the two extremes of nutrient enrichment and are called

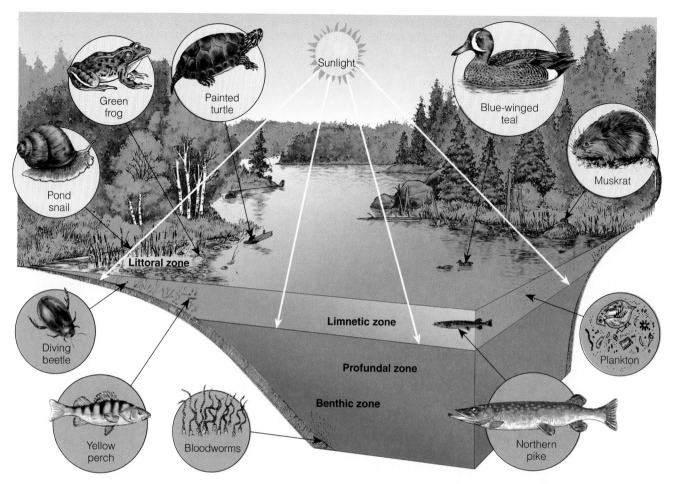

Figure 5-24 The distinct zones of life in a temperate-zone lake.

mesotrophic lakes. Human impacts on lakes are discussed in Section 11-5.

What Are the Major Characteristics of Freshwater Streams? Precipitation that doesn't sink into the ground or evaporate is **surface water**. It becomes **runoff** when it flows into streams and eventually to the ocean as part of the hydrologic cycle (Figure 4-27). This entire land area, which delivers water, sediment, and dissolved substances via small streams to a larger stream or river (and ultimately to the sea), is called a **watershed**, or a **drainage basin**.

The downward flow of surface water and groundwater from mountain highlands to the sea takes place in three zones (Figure 5-26). Because of different environmental conditions in each zone, a *river system* is actually a series of different ecosystems. In the first, narrow zone, headwater or mountain highland streams of cold, clear water rush over waterfalls and rapids. As this turbulent water flows and tumbles downward, it dissolves large amounts of oxygen from the air. Here plants such as algae and mosses are attached to rocks. The fish are cold-water fish such as trout, which need lots of dissolved oxygen.

In the second zone, the headwater streams merge to form wider, deeper streams that flow down gentler slopes with fewer obstacles. The warmer water and other conditions in this zone support more producers (phytoplankton and a variety of cool-water and warm-water fish species such as black bass) with slightly lower oxygen requirements.

In the third zone, streams join into wider and deeper rivers that meander across broad, flat valleys. Water in this zone usually has higher temperatures and less dissolved oxygen than water in the first two zones. The main channels of these slow-moving, wide, and murky rivers support distinctive varieties of fish (carp and catfish), whereas their backwaters support species similar to those present in lakes. Humans sometimes straighten, deepen, and widen meandering streams to improve navigation and to help reduce flooding and bank erosion, but such *stream channelization* is controversial.

As streams flow downhill, they become powerful shapers of land. Over millions of years the friction of moving water levels mountains and cuts deep canyons; the rock and soil the water removes are deposited as sediment in low-lying areas. Human impacts on streams are discussed in Section 11-5.

Q: What is the world's largest environmental group?

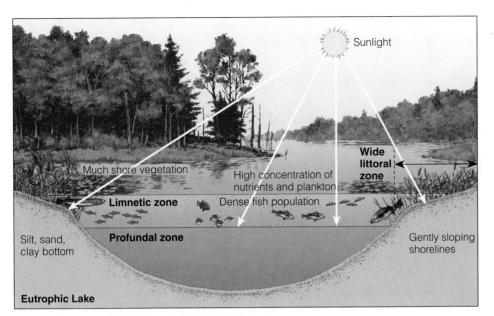

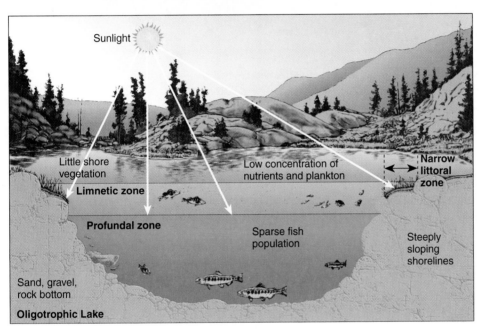

Figure 5-25 A eutrophic, or nutrient-rich, lake (top) and an oligotrophic, or nutrient-poor, lake (bottom). Mesotrophic lakes fall between these two extremes of nutrient enrichment.

Why Are Freshwater Inland Wetlands Important?

Lands covered with fresh water all or part of the time (excluding lakes, reservoirs, and streams) and located away from coastal areas are called **inland wetlands**. They include marshes, prairie potholes (depressions carved out by glaciers), swamps (dominated by trees and shrubs), mud flats, floodplains, bogs (rain-fed, peat-rich areas), wet meadows, and the wet arctic tundra in summer. Some wetlands are huge; others are small.

Some wetlands are covered with water year-round; others, such as prairie potholes, floodplain wetlands, and bottomland hardwood swamps are *seasonal wetlands*, usually underwater or soggy for only a short time each year. Some stay dry for years before filling with water again. In such cases, only the composition of the soil or the presence of plants such as cattails, bulrushes, or red maples may indicate that a given area is really a wetland.

Inland wetlands provide habitats for fish, migratory waterfowl, and other wildlife, and they improve water quality by filtering, diluting, and degrading toxic wastes, excess nutrients, sediments, and other pollutants. Floodplain wetlands near rivers reduce flooding and erosion by absorbing stormwater and releasing it slowly and by absorbing overflows from streams and lakes. According to Audubon Society estimates, if the remaining wetlands in the United States

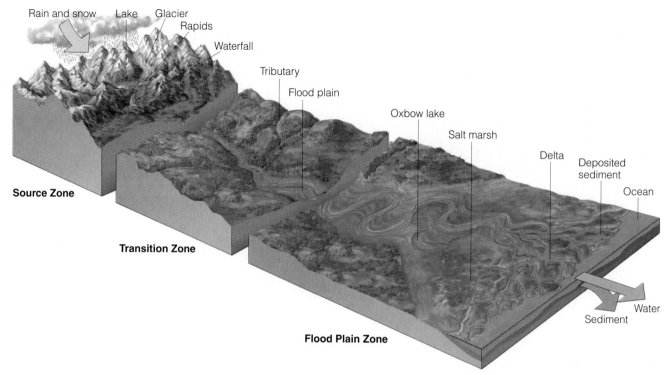

Figure 5-26 The three zones in the downhill flow of water: mountain (headwater) streams; wider, lower-elevation streams; and rivers, which empty into the ocean.

were destroyed, additional flood-control costs would be $7.7 billion to $31 billion per year.

By storing water, many seasonal and year-round wetlands allow increased infiltration, thus helping recharge groundwater supplies. Inland wetlands also help replenish groundwater supplies, a primary water source for over 50% of the U.S. population. They play significant roles in the global carbon, nitrogen, sulfur, and water cycles. Additionally, they provide recreation (especially waterfowl hunting) and are used to grow crops such as blueberries, cranberries, and rice.

What Impacts Do Humans Have on Inland Wetlands? Despite the ecological importance of year-round and seasonal inland wetlands, many are drained, dredged, filled in, or covered over. Each year some 1,200 square kilometers (470 square miles) of inland wetland in the United States are lost, about 80% to agriculture and the rest to mining, forestry, oil and gas extraction, highways, and urban development. Other countries have suffered similar losses.

A federal permit is now required in the United States to fill wetlands or deposit dredged or fill material in them. This law is poorly enforced, however, and there are continuing attempts to weaken it by using unscientific criteria to classify areas as wetlands. Only about 8% of remaining inland wetlands is under fed-

eral protection, and federal, state, and local protection of wetlands is weak.

Some damaged inland wetlands can be partially restored (p. 106) and new wetlands can be created. However, experience has also shown that at least half of the attempts to create new wetlands fail to replace lost ones, and most of those that are created bear little resemblance to natural wetlands. Restoring and creating wetlands is also expensive. To aquatic scientists and environmentalists, the best and cheapest thing to do is to protect existing wetlands from being degraded or destroyed by human activities.

5-3 RESPONSES OF LIVING SYSTEMS TO ENVIRONMENTAL STRESS

How Do Living Systems Cope with Changing Environmental Conditions? Recall from Section 3-2 that *homeostasis* is the maintenance of favorable internal conditions in a system (such as your body, a population, an economic system, or an ecosystem) despite fluctuations in external conditions. Such systems maintain favorable internal conditions through *feedback loops*, in which information fed back into a system causes the system to change (Figures 3-4 and 3-5).

All living systems, from single-celled organisms to the ecosphere, contain complex networks of negative

Q: What environmental and lifestyle factor causes the most death and suffering?

and positive feedback loops that interact to provide some degree of stability or sustainability over each system's expected life span. Because of such homeostatic processes, these systems have some ability to withstand or recover from externally imposed changes or stresses (Table 5-1), provided those stresses are not too severe. In other words, they have some degree of *stability*. However, this stability is maintained only by constant dynamic change in response to changing environmental conditions.

What Is Stability? It's useful to distinguish among three aspects of stability in living systems. **Inertia**, or **persistence**, is the ability of a living system to resist being disturbed or altered. **Constancy** is the ability of a living system such as a population to maintain a certain size or keep its numbers within the limits imposed by available resources. **Resilience** is the ability of a living system to bounce back after an external disturbance that is not too drastic.

Populations, communities, and ecosystems are so complex and variable that ecologists have little understanding of how they maintain inertia, constancy, and resilience while continually responding to changes in environmental conditions. Ecologists also find it difficult to predict which one or combination of environmental factors (Table 5-1) will stress ecosystems beyond their range of tolerance (Figure 4-13).

However, scientists have learned that the signs of ill health in stressed ecosystems include a drop in primary productivity, increased nutrient losses, decline or extinction of indicator species, larger populations of insect pests or disease organisms, a decline in species diversity, and the presence of contaminants.

How Are Homeostatic Systems Affected by Time Delays? Complex systems often show **time delays** between the input of a stimulus and the response to it. A long delay can sometimes mean that corrective negative feedback action comes too late. For example, a smoker exposed to cancer-causing chemicals in cigarette smoke may not get lung cancer for 20–30 years; by then it's too late for a negative feedback action (not smoking) to be effective.

Prolonged delays are involved in toxic dump leaks, depletion of the ozone layer, possible global warming and climate change from carbon dioxide (Connections, p. 134) and other chemicals we add to the atmosphere, destruction of forests from prolonged exposure to air pollutants, and extinction or near extinction of wildlife species.

How Can Synergy Affect Homeostatic Systems? In arithmetic, 1 plus 1 always equals 2. But in some of the complex systems found in nature, 1 plus 1 may

Table 5-1 Unfavorable Changes that Disturb Ecosystems

Natural Changes

Catastrophic	Drought
	Flood
	Fire
	Volcanic eruption
	Earthquake
	Hurricane
	Landslide
	Change in stream course
	Disease
Gradual	Changes in climate
	Immigration
	Adaptation and evolution
	Changes in plant and animal life (ecological succession)

Human-Cause Changes

Catastrophic	Deforestation
	Overgrazing
	Plowing
	Erosion
	Pesticides
	Fires
	Mining
	Toxic releases (can also be gradual)
	Urbanization (can also be gradual)
Gradual	Salt buildup in soil from irrigation (salinization)
	Waterlogging of soil from irrigation
	Compaction of soil from agriculture equipment
	Depletion of groundwater (aquifers)
	Water pollution (can also be catastrophic)
	Air pollution (can also be catastrophic)
	Loss and degradation of wildlife habitat (can also be catastrophic)
	Killing of predator and "pest" species
	Introduction of alien species
	Overhunting
	Overfishing
	Excessive tourism

A: Smoking (kills an estimated 3 million per year, including 419,000 Americans)

Controlling Carbon Dioxide Levels: Feedback and Time Delay

The atmosphere's level of carbon dioxide is, in principle, controlled by a negative feedback loop: More carbon dioxide makes many plants grow better, so the plant growth rate and population expand, absorbing more carbon dioxide (and making more oxygen). But it takes time to grow more plants, so the response is delayed.

Planting (and carefully tending) trees has been proposed as a way to help offset projected global warming caused by the increased levels of carbon dioxide we are adding to the atmosphere, mostly from burning fossil fuels and clearing and burning vegetation. Planting trees could remove some of the CO_2 we emit into the atmosphere because trees remove CO_2 from the atmosphere through photosynthesis.

However, estimates indicate that planting trees is not a complete solution to the carbon dioxide problem because trees take a long time to grow, many don't make it to maturity, and the amount of carbon dioxide we are adding to the atmosphere is enormous. Mathematical models indicate that to completely remove our excess inputs of carbon dioxide, each person on the earth would have to plant (and tend) about 1,000 trees per year indefinitely. In addition, when trees mature and die their decay releases the carbon they stored back into the atmosphere as CO_2.

Critical Thinking

Although planting large numbers of trees may not reduce CO_2 emissions enough to slow projected global, can you come up with other ecological reasons for planting large numbers of trees throughout the world?

add up to more than 2 because of synergistic interactions. A **synergistic interaction** occurs when two or more processes interact so that the combined effect is greater than the sum of their separate effects.

Synergy can result when two people work together in accomplishing a task. For example, suppose we need to move a 136-kilogram (300-pound) tree that has fallen across the road. By ourselves, each of us can lift only, say, 45 kilograms (100 pounds). However, if we cooperate and use our muscles properly, together we can move the tree out of the way. That's using synergy to solve a problem.

In effect, synergy amplifies the action of positive feedback loops and thus can be an amplifier of change

we believe is favorable. But synergy can also amplify harmful changes (Connections, right).

By identifying potentially harmful synergistic interactions and the leverage points that activate them, we can anticipate, counteract, and even prevent some environmental problems. Thus, we may be able to counter harmful synergisms, promote beneficial ones, and accelerate the improvement of life on the earth.

Does Species Diversity Increase Ecosystem Stability? In the 1960s most ecologists believed that the greater the species diversity and the accompanying web of feeding and biotic interactions in an ecosystem, the greater its stability. According to this hypothesis, an ecosystem with a diversity of species and feeding paths has more ways to respond to most environmental stresses because it does not have "all its eggs in one basket." However, most recent research indicates that there are many exceptions to this intuitively appealing idea.

Of course, there is a minimum threshold of species diversity below which ecosystems cannot function; no ecosystem can function without some plants and decomposers. (Note that we and other animal consumers are not absolutely necessary for life to continue on the earth.) Beyond this it is difficult to know whether simple ecosystems are less stable than complex ones, or to identify the threshold below which complex ecosystems fail.

In part because some species play redundant roles (niches) in ecosystems, we don't know how many or which species can be eliminated before the entire ecosystem begins to lose stability or collapse. However, research indicates that ecosystems with more species tend to have higher net primary productivities than simpler ecosystems and can also be more resilient. For example, a recent study of grasslands showed that species-rich fields had a smaller decline in net primary productivity during prolonged drought and rebounded more quickly than species-poor fields in the same area.

This supports the idea that some level of biodiversity provides insurance against catastrophe. But there is uncertainty over how much biodiversity is needed in various ecosystems. For example, some recent research suggests that average annual net primary productivity reaches a peak at 10–40 producer species. Many ecosystems contain more producer species than this, but it is difficult to distinguish between those that are essential and those that aren't.

Part of the problem is that ecologists disagree on how to define *stability* and *diversity*. Does an ecosystem need both high inertia and high resilience to be considered stable? Evidence suggests that some ecosystems have one of these properties but not the

Q: What is the projected annual global death toll from smoking by 2020?

Some Harmful Synergistic Interactions

CONNECTIONS

The consensus of atmospheric scientists is that human inputs of various chemicals into the atmosphere are causing *depletion of ozone* in the second layer of the atmosphere (the stratosphere), especially above Antarctica, and are projected to cause *warming* in the troposphere (the lowest layer of the earth's atmosphere). There is evidence that these two human-caused problems can interact synergistically to make matters worse.

As the troposphere warms, the stratosphere cools, increasing ice-cloud formation above the south pole. These ice clouds can increase stratospheric ozone depletion providing solid surfaces on which ozone-depleting chemical reactions

occur, leading to a loss of ozone or *ozone thinning* above Antarctica several months a year. Thus, projected global warming in the troposphere could worsen ozone depletion in the stratosphere.

To make matters worse, warming of the troposphere from burning fossil fuels and deforestation speeds up chemical reactions that lead to smog formation. A reduction in beneficial ozone in the stratosphere also lets more harmful ultraviolet radiation reach the earth's surface and produce more smog in the troposphere.

Thus, global warming in the troposphere and depletion of beneficial ozone in the stratosphere can interact synergistically to produce more harmful ozone in the troposphere. Furthermore, plants and

animals weakened by any one of these threats are more vulnerable to disease, heat, and other stresses as a result of further synergistic interactions.

Critical Thinking

Pollution that forms tiny particles and droplets of certain chemicals in the troposphere can reflect some incoming solar radiation. This can lead to some cooling of the troposphere and thus help counteract the warming of the troposphere from other chemicals we inject into the atmosphere—an example of a corrective negative feedback loop. Should we increase the levels of harmful pollutants into the troposphere to help counteract projected global warming? Explain.

---

other. For example, tropical rain forests have high species diversity and high inertia; that is, they are resistant to significant alteration or destruction. However, once a large tract of tropical forest is severely degraded, the ecosystem's resilience is so low that the forest may not be restored. Nutrients (which are stored primarily in the vegetation, not in the soil) and other factors needed for recovery may no longer be present. Such a large-scale loss of forest cover may so change the local or regional climate that forests can no longer be supported.

Grasslands, by contrast, are much less diverse than most forests, and because they burn easily they have low inertia. However, because most of their plant matter is stored in underground roots, these ecosystems have high resilience and recover quickly. A grassland can be destroyed only if its roots are plowed up and something else without such extensive root systems is planted in its place, or if it is severely overgrazed by livestock or other herbivores.

Another difficulty is that populations, communities, and ecosystems are rarely, if ever, at equilibrium. Instead, nature is in a continuing state of disturbance, fluctuation, and change; this means that an ecologist studying an ecosystem is trying to investigate a moving target. Indeed, ecologists recognize that disturbances of ecosystems are integral parts of the way nature works. Clearly, we have a long way to go in understanding how the factors in natural communities

and ecosystems respond to changes in environmental conditions.

What Determines the Number of Species in an Ecosystem? There are no simple answers to this question. Two factors affecting the species diversity (or richness) of an ecosystem are its size and degree of isolation. In the 1960s Robert MacArthur and Edward O. Wilson began studying communities on islands to discover why large islands tend to have more species of a certain category (such as birds or ferns) than do small islands.

To explain these differences in species diversity with island size, MacArthur and Wilson proposed what is called the **species equilibrium model** or the **theory of island biogeography**. According to this model, the number of species found on an island is determined by a balance between two factors: the *immigration rate* of species to the island from other inhabited areas, and the *extinction rate* of species established on the island. The model predicts that at some point the rates of immigration and extinction will reach an equilibrium point, which determines the island's average number of different species (species diversity).

The model also predicts that immigration and extinction rates (and thus species diversity) are affected by two important variables: the *size of the island* and its *distance from a mainland source of immigrant species*.

According to the model, a small island tends to have a lower species diversity than a large one for two reasons: First, a small island is hard for potential immigrants to find and thus should have a lower immigration rate than a larger island. Second, because a small island normally has fewer resources and less diverse habitats, it should have a higher extinction rate (because of increased competitive exclusion) than a larger island.

The model also predicts that an island's distance from a mainland source of new species is important in determining its species diversity. Assume that we have two islands of about equal size and that all other factors are roughly the same. According to the model, the island closest to a mainland source of immigrant species will have a higher immigration rate and thus a higher species diversity (assuming that extinction rates on both islands are about the same).

MacArthur and Wilson's original model or scientific hypothesis has been tested and supported by a series of field experiments. As a result, biologists have accepted it well enough to elevate it to the status of an important and useful scientific theory. In recent years, it has even been applied to the protection of wildlife on land (Connections, below).

5-4 POPULATION RESPONSES TO STRESS: POPULATION DYNAMICS

What Limits Population Growth? Populations are dynamic; they change in *size*, *density* (number of individuals in a certain space), *dispersion* (spatial pattern—such as clumping, uniform dispersion, or random dispersion—in which the members of a population are found in their habitat), and age distribution (proportion of individuals of each age in a population) in response to environmental stress or changes in environmental conditions (Table 5-2). These changes are called **population dynamics**.

Four variables—births, deaths, immigration, and emigration—govern changes in population size. A population gains individuals by birth and immigration and loses them by death and emigration:

Population change =
(Births + Immigration) – (Deaths + Emigration)

These variables in turn depend on changes in resource availability or on other environmental changes (Figure 5-27).

Using Island Biogeography Theory to Protect Mainland Communities and Species

CONNECTIONS

Ecologists and conservation biologists are now applying the species equilibrium model or theory of island biogeography to communities on the mainland. Because of widespread disruption and fragmentation by humans, most remaining wildlife sanctuaries on land are *habitat islands* surrounded by an inhospitable "sea" of disturbed and unsuitable habitat. According to the species equilibrium model, the species diversity in these terrestrial patches or islands should be determined by their size and by their distance from other patches that serve as sources of colonists.

Conservation biologists are using the model to help them locate (and try to protect) areas in the most danger of losing much of their species diversity. Researchers are also using the model to estimate what size a nature preserve should

be in a particular area to prevent it from losing species.

Of course, the size of a preserve also depends on which species we hope to protect. Members of endangered species such as tigers and grizzly bears, for example, feed over a large home range. They need a much larger area of undisturbed habitat than many other species to preserve a viable population.

For example, the average grizzly bear has a home range of 250–1,000 square kilometers (100–400 square miles). Thus, a minimum viable population of 50–90 grizzly bears needs a reserve of up to 90,000 square kilometers (35,000 square miles) for survival. This is much larger than the 53,500-square-kilometer (20,600-square-mile) Greater Yellowstone ecosystem, one of the world's largest protected habitat islands (containing a national park and six national forests).

Conservation biologists are

also using the species equilibrium model to estimate how closely a series of small wildlife reserves should be spaced to allow the possibility of immigration from one preserve to another if a species in one reserve becomes locally extinct.

In addition, they are using the model to estimate the size and number of protected corridors needed to connect various reserves and encourage the spread of protected species among them. Much more research must be done to answer such practical questions, but progress is being made through this application of ecological theory to wildlife conservation.

Critical Thinking

How would you respond to the statement that the theory of island biogeography should not be taken too seriously because it is only a theory?

Q: How many U.S. workers die prematurely from exposure to toxic substances?

Table 5-2 Ecosystem Characteristics at Immature and Mature Stages of Ecological Succession

Characteristic	Immature Ecosystem (Early Successional Stage)	Mature Ecosystem (Late Successional Stage)
Ecosystem Structure		
Plant size	Small	Large
Species diversity	Low	High
Trophic structure	Mostly producers, few decomposers	Mixture of producers, consumers, and decomposers
Ecological niches	Few, mostly generalized	Many, mostly specialized
Community organization (number of interconnecting links)	Low	High
Ecosystem Function		
Biomass	Low	High
Net primary productivity	High	Low
Food chains and webs	Simple, mostly plant → herbivore with few decomposers	Complex, dominated by decomposers
Efficiency of nutrient recycling	Low	High
Efficiency of energy use	Low	High

Populations vary in their capacity for growth. The **biotic potential**, or **reproductive potential**, of a population is the *maximum* rate (*r*) at which it could grow if it had unlimited resources. Generally, individuals in populations with a high biotic potential *reproduce early in life, have short generation times* (the time between successive generations), *can reproduce many times* (have a long reproductive life), and *produce many offspring each time they reproduce.*

No population can grow exponentially indefinitely. In the real world, a rapidly growing population reaches some size limit imposed by a shortage of one or more limiting factors, such as light, water, space, or nutrients. *There are always limits to population growth in nature.*

Environmental resistance consists of all the factors acting jointly to limit the growth of a population. They determine **carrying capacity (*K*)**, the number of individuals of a given species that can be sustained indefinitely in a given area. The population size of a species in a given place and time is determined by the interplay between its biotic potential and environmental resistance (Figure 5-27).

Any population growing exponentially starts out slowly and then goes through a rapid, mostly unrestricted exponential growth phase. If plotted, this sequence yields a *J-shaped curve* (Figure 1-1). However, because of environmental resistance its growth tends to level off once the carrying capacity of the area is reached. (In most cases, the size of such a population fluctuates slightly above and below the carrying capacity.) An idealized plot of this type of growth yields a sigmoid or *S-shaped curve* (Figure 5-28).

The populations of some species don't make such a smooth transition from a J-shaped curve to an S-shaped curve. Instead, they temporarily use up their resource base (for example, by eating more plants or animals than can be replenished). Thus, the population temporarily *overshoots* or exceeds the carrying capacity of its habitat.

This overshoot occurs because of a *reproductive time lag*, the period required for the birth rate to fall and the death rate to rise in response to resource overconsumption; in other words, the corrective negative feedback does not take effect immediately. Unless the excess individuals switch to new resources or move to an area with more favorable conditions, the population will suffer a *dieback* or *crash* (Figure 5-29).

Humans are not exempt from overshoot and dieback. Ireland, for example, experienced a population crash after a fungus destroyed the potato crop in 1845. About 1 million people died and 3 million people emigrated to other countries.

Figure 5-27 Factors that tend to increase or decrease populations. Population size at any given time is determined by the balance between growth factors (biotic potential) and decrease factors (environmental resistance).

Population Size

Growth Factors (biotic potential)

Abiotic

Favorable light

Favorable temperature

Favorable chemical environment (optimal level of critical nutrients)

Biotic

High reproductive rate

Generalized niche

Adequate food supply

Suitable habitat

Ability to compete for resources

Ability to hide from or defend against predators

Ability to resist diseases and parasites

Ability to migrate and live in other habitats

Ability to adapt to environmental change

Decrease Factors (environmental resistance)

Abiotic

Too much or too little light

Temperature too high or too low

Unfavorable chemical environment (too much or too little of critical nutrients)

Biotic

Low reproductive rate

Specialized niche

Inadequate food supply

Unsuitable or destroyed habitat

Too many competitors

Insufficient ability to hide from or defend against predators

Inability to resist diseases and parasites

Inability to migrate and live in other habitats

Inability to adapt to environmental change

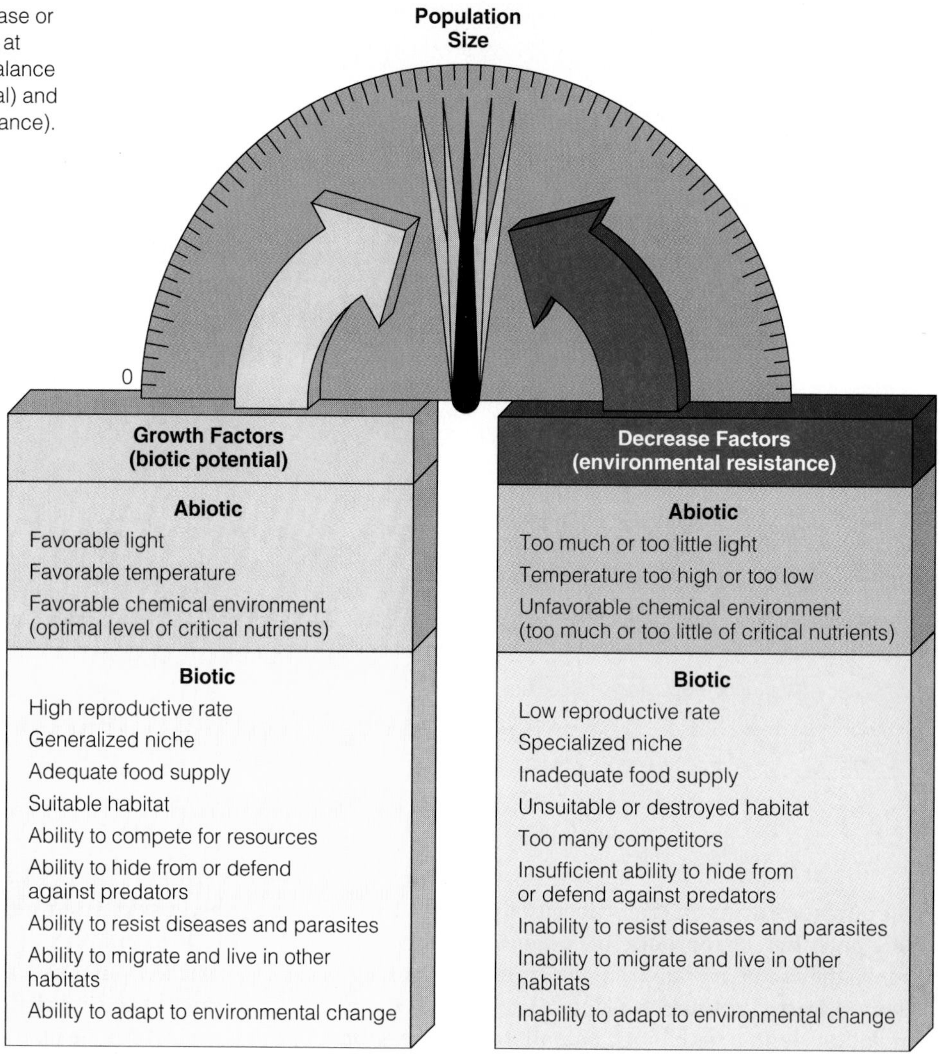

An earlier example of overshoot and collapse involved Easter Island (Rapa Nui), a small island about the size of Staten Island isolated in the great expanse of the South Pacific. It was first colonized by Polynesians about 2,500 years ago. The civilization they developed was based on the island's trees, which were used for shelter, tools, boats, fuel, food, and clothing. The people flourished, the population peaking at about 10,000. But they were unable to stop themselves from using up the precious trees—an example of the tragedy of the commons (p. 11). Each person who cut a tree reaped immediate personal benefits while helping to doom the civilization as a whole. As they started to run out of the wood that supported them, the people turned to warfare and, possibly, cannibalism. Both the population and the civilization collapsed. When Dutch explorers first reached the island on Easter Day, 1722, they found only about 2,000 inhabitants struggling under primitive conditions on a mostly barren island.

Technological, social, and other cultural changes have extended the earth's carrying capacity for the human species (Figure 2-3). We have increased food production and used large amounts of energy and matter resources to make normally uninhabitable areas of the earth habitable. However, there is growing concern about how long we will be able to keep doing this on a planet with a finite size and resources but an exponentially growing population and per capita resource use.

Carrying capacity is not a simple, fixed quantity, but rather a variable determined by many factors. Examples include competition within and among species, immigration and emigration, natural and human-caused catastrophic events, and seasonal fluctuations in the supply of food, water, hiding places,

Q: What percentage of the 72,000 chemicals in commercial use have been thoroughly screened for toxicity?

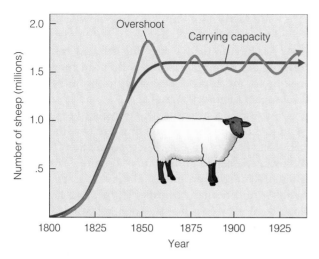

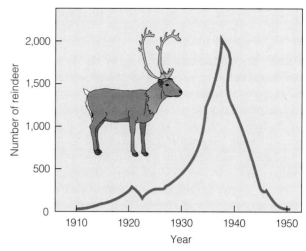

Figure 5-28 S-shaped population change curve of sheep population on the island of Tasmania between 1800 and 1925. After sheep were introduced in 1800 their population grew exponentially because of ample food. By 1855, however, they overshot the land's carrying capacity. Their numbers then stabilized and oscillated around a carrying capacity size of about 1.6 million sheep. Sheep are large and long-lived, and reproduce slowly. As a result, their population size does not vary much from year to year once they reach the carrying capacity of their habitat (assuming that other environmental factors do not lower the carrying capacity).

Figure 5-29 Exponential growth, overshoot, and population crash of reindeer introduced to a small island off the southwest coast of Alaska.

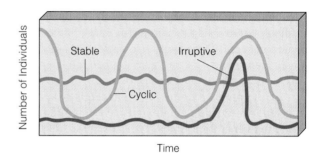

Figure 5-30 General types of idealized population change curves found in nature. A species with a *stable* population fluctuates slightly above and below its carrying capacity. A species with a *cyclic* population size undergoes boom–bust cycles and one with an *irruptive* pattern has an occasional sharp increase in size, followed by a sharp decline.

and nesting sites. Figure 5-30 shows the three generalized types of population change curves found in nature.

How Does Population Density Affect Population Growth? Some limiting factors have a greater effect as a population's density increases. Examples of such *density-dependent population controls* are competition for resources, predation, parasitism, and disease. Dense populations generally have lower birth rates, slower growth rates, and higher death rates than less dense populations.

As prey populations become more dense, their members compete more for limited resources. This can lead to a larger number of weakened individuals that are easy targets for predators. Individuals in a dense population are also more likely to be infected by parasites or contagious disease organisms.

In some species, physiological and sociological control mechanisms limit reproduction as population density rises. Overcrowding in populations of mice and rats causes hormonal changes that can inhibit sexual maturity, lower sexual activity, and reduce milk production in nursing females. Stress from crowding in these and several other species also reduces the number of offspring produced per litter through such mechanisms as spontaneous abortion. In some species

crowding may lead to population control through cannibalism and killing of the young.

Density-independent population controls affect a population's size regardless of its population density. Examples include floods, hurricanes, earthquakes, landslides, severe drought, unseasonable weather, fire, destruction of habitat (such as clearing a forest of its trees or filling in a wetland), and pesticide spraying. For example, a severe freeze in late spring can kill many individuals in a plant population, regardless of its density.

In practice, it is hard to separate the effects of density-dependent and density-independent factors because they often interact to limit population growth. For example, a severe drought would kill some individuals of a desert population of prairie dogs regardless of its

A: About 10% (only 2% have been adequately screened as carcinogens, teratogens, or mutagens) CHAPTER 5 **139**

population size, but a higher proportion would die if the population density were high.

What Reproductive Strategies Do Species Use to Survive? Each species has a characteristic mode of reproduction. At one extreme are species that reproduce early. They have many (usually small) offspring each time they reproduce, reach reproductive age rapidly, and have short generation times. They give their offspring little or no parental care or protection to help them survive. Species with this reproductive strategy overcome the massive loss of their offspring by producing so many young that a few will survive to reproduce many offspring to begin the cycle again.

Species with such a capacity for a high rate of population growth (*r*) are called **r-strategists** (Figure 5-31, left). Algae, bacteria, rodents, annual plants (such as dandelions), many bony fish, and most insects (p. 70) are examples. Such species tend to be *opportunists*, reproducing and dispersing rapidly when conditions are favorable or when a new habitat or niche becomes available for invasion. Changing or unfavorable environmental conditions can cause such populations to crash. Hence, most r-strategists go through irregular and unstable boom–bust cycles.

At the other extreme are **K-strategists**, species that tend to reproduce late and have few offspring with long generation times. Typically these offspring develop inside their mother (where they are safe).

They are fairly large and mature slowly. They are cared for and protected by one or both parents until they reach reproductive age. This strategy results in a few big and strong individuals that can compete for resources and reproduce a few young to begin the cycle again (Figure 5-31, right).

Such species are called K-strategists because they tend to maintain their population size near their habitat's carrying capacity (*K*). Their populations typically follow an *S*-shaped growth curve (Figure 5-28). Examples are most large mammals (such as elephants, whales, and humans), birds of prey, and large and long-lived plants (such as the saguaro cactus, redwood trees, and most tropical rain forest trees). Many K-strategist species, especially those with long generation times and low reproductive rates (such as elephants, rhinoceroses, and sharks), are prone to extinction.

In agriculture we raise both r-strategists (crops) and K-strategists (livestock). Many organisms have reproductive strategies between the extremes of r-strategists and K-strategists, or they change from one extreme to the other under certain environmental conditions.

The reproductive strategy of a species may give it a temporary advantage, but *the availability of niches for individuals of a population in a particular area is what determines its ultimate population size.* Regardless of how fast a species can make babies, there can be no more dandelions than there are dandelion niches and no more zebras than there are zebra niches in a particular area.

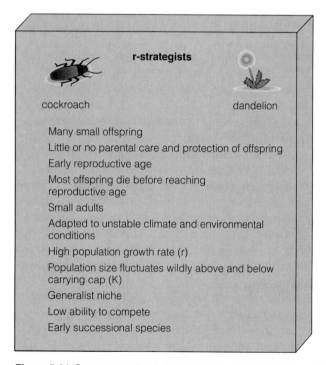

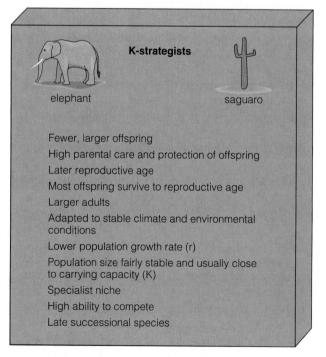

Figure 5-31 Some generalized characteristics of r-strategists and K-strategists. Many species have characteristics between these two extremes.

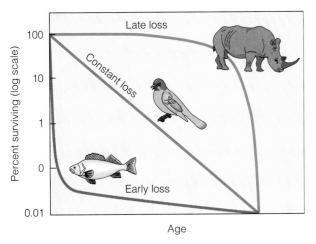

Figure 5-32 Three generalized types of survivorship curves for populations of different species, obtained by showing the percentages of the members of a population surviving at different ages. For a *late loss* population (such as elephants, rhinoceroses, and humans), there is typically high survivorship to a certain age, then high mortality. A *constant loss* population (such as many songbirds) shows a fairly constant death rate at all ages. For an *early loss* population (such as annual plants and many bony fish species), survivorship is low early in life. These generalized survivorship curves only approximate the behavior of species.

Individuals of species with different reproductive strategies tend to have different *life expectancies*. One way to represent the age structure of a population is with a **survivorship curve**, which shows the number of survivors of each age group for a particular species. There are three generalized types of survivorship curves: late loss, early loss, and constant loss (Figure 5-32).

5-5 RESPONSES TO CHANGING CONDITIONS: THE RISE OF LIFE ON EARTH

What Types of Organisms Are Found on the Earth? On the basis of their cell structure, biologists classify all organisms as either eukaryotic or prokaryotic. All organisms except bacteria are **eukaryotic**: Their cells are surrounded by a membrane and have a *nucleus* (a membrane-bounded structure containing genetic material in the form of DNA) and several other internal parts.

Bacterial cells are **prokaryotic** (which means "before nucleus"): They are surrounded by a membrane but have no distinct nucleus or other internal parts enclosed by membranes. Although most familiar organisms are eukaryotic, they could not exist without hordes of microscopic prokaryotic organisms (bacteria).

Scientists group organisms into various categories based on their common characteristics, a process called *taxonomic classification*. The largest category is the *kingdom*, which includes all organisms that have one or perhaps several common features. In this book, the earth's organisms are classified into five kingdoms: monera, protists, fungi, plants, and animals (Figure 5-33). Most bacteria, fungi, and protists are **microorganisms**: organisms that are so small that they can be seen only by using a microscope.

Monera (bacteria and cyanobacteria) are single-celled, microscopic prokaryotic organisms. Most bacteria play a vital role as decomposers, which break down the tissue of dead organisms into simpler compounds that serve as nutrients for the bacteria and that are eventually reused as nutrients by plants. A few bacteria—*Streptococcus* (which causes strep throat) and *Salmonella* (which causes food poisoning), for example—can cause diseases in humans. However, your body is inhabited by billions of beneficial bacteria that help keep you healthy by aiding in food digestion and by crowding out disease-causing bacteria.

Protists (protista) are mostly single-celled eukaryotic organisms such as diatoms, dinoflagellates, amoebas, golden brown and yellow-green algae, and protozoans. Some protists cause human diseases such as malaria, sleeping sickness, and Chagas's disease. **Fungi** are mostly many-celled (sometimes microscopic) eukaryotic organisms such as mushrooms, molds, mildews, and yeasts. Many fungi are decomposers. Other fungi kill various plants and cause huge losses of crops and valuable trees.

Plants (plantae) are mostly many-celled eukaryotic organisms such as red, brown, and green algae and mosses, ferns, conifers, and flowering plants (whose flowers produce seeds that perpetuate the species). Some plants such as corn and marigolds are **annuals**, which complete their life cycles in one growing season; others are **perennials**, which can live for more than 2 years, such as roses, grapes, elms, and magnolias.

Animals (animalia) are also many-celled, eukaryotic organisms. Most, called **invertebrates**, have no backbones. They include sponges, jellyfish, worms, arthropods (insects, shrimp, and spiders), mollusks (snails, clams, and octopuses), and echinoderms (sea urchins and sea stars). Insects play roles that are vital to our existence (p. 70). **Vertebrates** (animals with backbones and a brain protected by skull bones) include fishes (sharks and tuna), amphibians (frogs and salamanders), reptiles (crocodiles and snakes), birds (eagles and robins), and mammals (bats, elephants, whales, and humans).

How Did Life Emerge on the Earth? How did life on the earth evolve to its present system of diverse

A: Essentially none

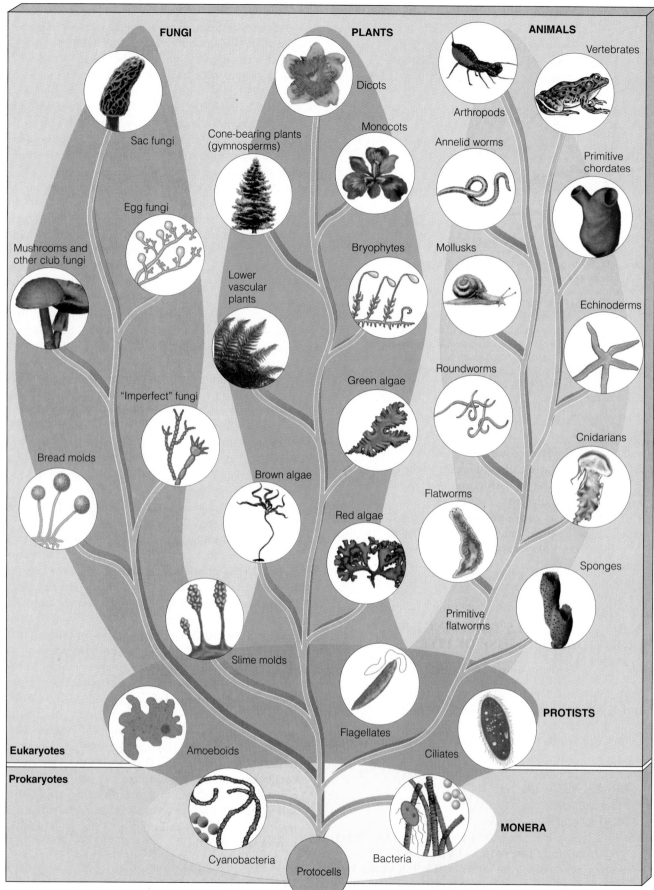

Figure 5-33 Kingdoms of the living world, one of several schemes for classifying the earth's diverse species into major groups. Most biologists believe that protocells gave rise to single-celled prokaryotes and that these in turn evolved into the more complex protists, fungi, plants, and animals that make up the earth's stunning biodiversity. This is also a greatly simplified overview of *evolution*—changes in the genetic makeup of populations through successive generations over hundreds of millions of years, in which one species leads to the appearance of many other species and groups of species.

kingdoms and species (Figure 5-34) living in an interlocking network of matter cycles, energy flows, and species interactions? We don't know the full answer to this question, but a growing body of scientific evidence suggests what might have happened.

When this diverse and continually accumulating evidence is pieced together, an important idea emerges: *The evolution of life is linked to the physical and chemical evolution of the earth*. It becomes apparent that the earth has the right physical and chemical conditions for life as we know it to exist (Connections, right).

A second conclusion from this body of evidence is that life on the earth developed in two phases over the past 4.7–4.8 billion years: *chemical evolution* of the organic molecules, biopolymers, and systems of chemical reactions needed to form the first protocells (about 1 billion years); and *biological evolution* of single-celled organisms (first prokaryotes and then eukaryotes) and then multicellular organisms (about 3.7–3.8 billion years) (Figure 5-34).

How Did Chemical Evolution Take Place? Here is an overview of how scientists believe life *may* have formed and evolved on the earth, based on the current evidence. Some 4.6–4.7 billion years ago a cloud of cosmic dust condensed into planet earth, which soon turned molten from meteorite impacts and from the heat of radioactively decaying elements in its interior. As cooling took place, the outermost portion of the molten sphere solidified to form a thin, hardened crust of rocks, devoid of atmosphere and oceans.

Volcanic eruptions and comets hitting the lifeless earth pierced its thin crust, releasing water vapor and other gases from the molten interior. Eventually the crust cooled enough for the water vapor to condense and fall to the surface as rain. This rain eroded minerals from rocks, and solutions of these minerals collected in depressions to form the early oceans that covered most of the globe.

Research suggests that most of the planet's first atmosphere had formed by 4.4 billion years ago. The exact composition of this primitive atmosphere is unknown, but atmospheric scientists believe it was dominated by carbon dioxide (CO_2), nitrogen (N_2), and water vapor (H_2O). Trace amounts of methane (CH_4), ammonia (NH_3), hydrogen sulfide (H_2S), and hydrogen chloride (HCl) were also probably present (although scientists disagree over the relative amounts of these gases).

Whatever the composition of this primitive atmosphere, scientists agree that it had no oxygen gas (O_2) because this element is so chemically reactive that it would have combined into compounds. (The only reason today's atmosphere has so much O_2 is that plants and some aerobic bacteria produce it in vast

CONNECTIONS

Earth: The Just-Right, Resilient Planet

Like Goldilocks tasting porridge at the Three Bears' house, life on the earth as we know it requires a certain temperature range: Venus is much too hot and Mars is much too cold, but the earth is *just right*. (Otherwise, you wouldn't be reading these words.)

Life as we know it depends on liquid water. Again, temperature is crucial; life on the earth requires average temperatures between the freezing and boiling points of water, between 0°C and 100°C (32°F and 212°F) at the earth's range of atmospheric pressures.

The earth's orbit around the sun is the right distance from the sun to provide these conditions. If the earth were much closer, it would be too hot—like Venus—for water vapor to condense to form rain. If it were much farther away, its surface would be so cold—like Mars—that its water would exist only as ice. The earth also spins on a tilted axis; if it didn't, the side facing the sun would be too hot and the other side too cold for water-based life to exist. So far, the temperature has been, like Baby Bear's porridge, just right.

The earth is also the right size; that is, it has enough gravitational mass to keep its iron–nickel core molten and to keep the gaseous molecules in its atmosphere from flying off into space. (A much smaller earth would be unable to hold onto an atmosphere consisting of such light molecules as N_2, O_2, CO_2, and H_2O.) The slow transfer of its internal heat (geothermal energy) to the surface also helps keep the planet at the right temperature for life. And thanks to the development of photosynthesizing bacteria over 2 billion years ago, an ozone sunscreen protects us and many other forms of life from an overdose of ultraviolet radiation.

On a time scale of millions of years, the earth is also enormously resilient and adaptive. Its average temperatures have remained between the freezing and boiling points of water even though the sun's energy output has increased by about 30% over the 3.6 billion years since life arose. In short, the earth is just right for life as we know it.

Critical Thinking

1. Which do you believe is in greater danger: the earth or the human species? Explain.

2. Do you believe that humans can learn enough about the earth to manage it (mostly for the human species) on a global scale? Explain. If not, what are the alternatives?

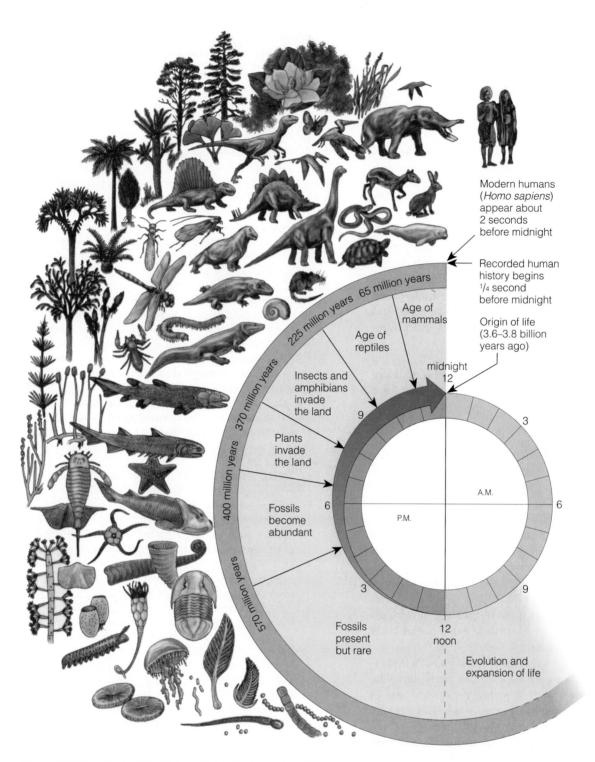

Figure 5-34 Greatly simplified history of biological evolution of life on the earth, which was preceded by about 0.5–1 billion years of chemical evolution. The early span of biological evolution on the earth, between about 3.7 billion and 570 million years ago, was dominated by microorganisms (mostly bacteria, and later, protists). Plants and animals evolved only about 570 million years ago, and humans arrived on the scene only a very short time ago. If we compress the earth's 3.7-billion-year history of life to a 24-hour time scale, the first human species (*Homo hablis*) appeared about 47–94 seconds before midnight and our species (*Homo sapiens sapiens*) appeared about 1.4 seconds before midnight. Agriculture began only 0.25 second before midnight, and the industrial revolution has been around for only 0.007 second. (Adapted from George Gaylord Simpson and William S. Beck, *Life: An Introduction to Biology,* 2d ed., New York: Harcourt Brace Jovanovich, 1965)

Q: What percentage of commercially used chemicals in the United States are regulated by federal and state governments?

quantities through photosynthesis. But this is getting ahead of the story.)

Energy from electrical discharges (lightning), heat from volcanoes, and intense ultraviolet (UV) light and other forms of solar radiation was readily available for the synthesis of biologically important organic molecules from the inorganic chemicals found in the earth's primitive atmosphere—an idea first proposed in 1923 by Russian biochemist Alexander Oparin.

In a number of experiments conducted since 1953, various mixtures of gases believed to have been in the earth's early atmosphere have been put in closed, sterilized glass containers; they have then been subjected to spark discharges to simulate lightning and heat. In these experiments, compounds necessary for life—various amino acids (the building-block molecules of proteins), simple carbohydrates, nucleic acids (the building block molecules of DNA and RNA), and other small organic compounds—formed from the inorganic gaseous molecules. These experiments supported Oparin's hypothesis, although many details are missing or hotly debated.

Another possibility is that simple organic molecules necessary for life formed on dust particles in space and reached the earth on meteorites or comets (or on countless interplanetary dust particles floating around in space when the earth was formed). Another hypothesis is that life could have arisen around the mineral-rich and very hot *hydrothermal vents*, which sit atop cracks in the ocean floor leading to subterranean chambers of molten rock. We don't know which of these processes might have produced the organic molecules necessary for life, but all of these hypotheses are reasonable explanations of how this could have happened.

Once these building-block organic molecules formed in the early atmosphere, they were removed by rain. Then they accumulated and underwent countless chemical reactions in the earth's warm, shallow waters. After several hundred million years of different chemical combinations in this hot organic soup, conglomerates of proteins, RNA, and other biopolymers may have combined to form membrane-bound *protocells*: small globules that could take up materials from their environment and grow and divide (much like living cells). Again, there are several hypotheses explaining how this could have happened. With these forerunners of living cells, the stage was set for the drama of biological evolution (Figure 5-34).

How Did Life Evolve? Over time, it is believed that the protocells developed into single-celled, bacteria-alike prokaryotes having the properties we describe as life (although the details of how this might have happened are hotly debated by scientists). These anaerobic cells probably developed either in the muddy

sediments of tidal flats or at least 10 meters (30 feet) below the ocean's surface, protected there from the intense UV radiation that bathed the earth at that time.

Scientists believe that these single-celled anaerobic bacteria multiplied and underwent genetic changes (mutated) for about a billion years in the earth's warm, shallow seas. The result was a variety of new types of prokaryotic cells.

Scientists contend that during this long early period, life could not have survived on land—a hypothesis supported by a lack of land fossils. There was no ozone layer to shield the DNA and other molecules of early life from bombardment by intense ultraviolet radiation. About 2.3–2.5 billion years ago, however, something happened in the ocean that would drastically change the earth: the development of photosynthetic prokaryotes called *cyanobacteria*. These cells could remove carbon dioxide from the water and (powered by sunlight) combine it with water to make the carbohydrates they needed. In the process, they released oxygen (O_2) into both the ocean and the atmosphere.

The resulting *oxygen revolution* took place over about half a billion years and opened the way for the evolution of a great variety of oxygen-using (aerobic) bacteria. Later came more complex organisms, first in the seas and then on land (after a protective ozone layer formed in the stratosphere) (Figure 5-34).

As oxygen accumulated in the atmosphere, some was converted by incoming solar energy into ozone (O_3), which began forming in the lower stratosphere. This shield protected life-forms from the sun's deadly UV radiation, allowing green plants to live closer to the ocean's surface. Fossil evidence suggests that about 400–500 million years ago UV levels were low enough for the first plants to exist on the land. Over the next several hundred million years a variety of land plants and animals arose, followed by mammals (and eventually) the first humans (Figure 5-34).

5-6 POPULATION RESPONSES TO STRESS: EVOLUTION, ADAPTATION, AND NATURAL SELECTION

What Is Evolution? Within limits, populations can adapt to changes in environmental conditions. The major driving force of adaptation to environmental change is believed by most biologists to be **biological evolution**, or **evolution**: the change in a population's genetic makeup through successive generations. Note that *populations, not individuals, evolve by becoming genetically different*.

According to the widely accepted **theory of evolution**, all life-forms developed from earlier life-forms.

Although this theory conflicts with the creation stories of most religions, it is the explanation the overwhelming majority of biologists accept for the way life has changed over the past 3.7–3.8 billion years and for why life is so diverse today.

How Do We Know What Organisms Lived in the Past? Most of what we know of the earth's life history comes from **fossils**: mineralized or petrified replicas of skeletons, bones, teeth, shells, leaves, and seeds, or impressions of such items.

Despite its importance, the fossil record is uneven and incomplete. Some life-forms left no fossils, some fossils have decomposed, and others are yet to be found. So far we have found fossils representing only about 1% of the species believed to have ever lived. Examining the few fossils we have in order to hypothesize about the other species believed to have lived on the earth has limitations—somewhat like a blind person trying to describe what an elephant looks like by being able to feel only the tip of its tail.

How Does Evolution Work? The raw material of evolution is the development of *genetic variability* in a population. A population's **gene pool** is the sum total of all genes possessed by the individuals of the population of a species and evolution at this level is a change in a population's gene pool over time.

Although members of a population generally have the same number and kinds of genes, a particular gene may have two or more different molecular forms, called **alleles**. Different combinations of these alleles are inherited so that different members of a population have genetic diversity (Figure 4-7).

The source of all new alleles in evolution is **mutations**: random changes in the structure or number of DNA molecules in a cell. One way mutations occur is by exposure to external agents such as radioactivity, X rays, and natural and human-made chemicals (called *mutagens*). Another source of mutations is random mistakes that are sometimes made in coded genetic instructions when DNA molecules are copied (each time a cell divides and whenever an organism reproduces). Mutations can occur in any cells, but only those in reproductive cells are passed on to offspring.

Some mutations are harmless, but many are harmful, altering traits in such a way that an individual cannot survive (lethal mutations). Every so often a mutation is beneficial. The result is new genetic traits that give their bearer and its offspring better chances for survival and reproduction, either under existing environmental conditions or when such conditions change. Any genetically controlled trait that helps an organism survive and reproduce under a given set of environmental conditions is called an **adaptation**, or **adaptive trait**. For example, a structural adaptation such as *coloration* can allow prey to hide prey from predators (Figure 5-35).

It is important to understand that mutations are random and totally unpredictable, the only source of totally new genetic raw material (alleles), and very rare. Once created by mutation, however, new alleles can be shuffled together or recombined randomly to create new combinations of genes in populations of sexually reproducing species.

What Role Does Natural Selection Play in Evolution? Because of the random shuffling or recombination of various alleles produced by mutations, certain individuals in a population may by chance have one or more beneficial adaptations that help them to survive under various environmental conditions. As a result, they are more likely to reproduce (and thus produce more offspring with the same favorable adaptations) than are individuals without such adaptations. This process is known as **differential reproduction**.

Natural selection occurs when the combined processes of adaptation and differential reproduction result in a particular beneficial gene or set of genes becoming more common in succeeding generations. It occurs when some members of a population have heritable traits that enable them to survive and produce more offspring than other members of the population.

One misconception about evolution arises from the interpretation of the expression "survival of the fittest," sometimes used to describe how natural selection works. This has often been misinterpreted as "survival of the strongest."

To biologists, however, *fitness* is a measure of reproductive success, so that the fittest individuals are those that leave the most descendants. For example, members of a population that are better at hiding from predators are more fit (can live to produce more offspring) than those that aren't as good at hiding, regardless of how strong they are. Instead of tooth-and-claw competition, evidence indicates that natural selection favors populations of species that avoid direct competition by producing offspring that can occupy niches different from those of other species (Figures 4-28 and 4-29).

What Is Coevolution? Some biologists have proposed that interactions between species can also result in evolution in each of their populations. According to this hypothesis, when populations of two different species interact over a long time, changes in the gene pool of one species can lead to changes in the gene pool of the other species. This process is called **coevolution**.

For example, individual plants in a population may evolve defenses such as camouflage, thorns, or poisons against efficient herbivores. In turn, some herbivores in

Q: How many children under age 5 die each year in developing countries from mostly preventable infectious diseases?

Figure 5-35 Two varieties of peppered moths found in England illustrate one kind of adaptation: camouflage. Before the industrial revolution in the mid-1800s, the speckled light-gray form of this moth was prevalent. When these night-flying moths rested on light-gray lichens on tree trunks during the day, their color camouflaged them from predators (left). A dark-gray form also existed but was quite rare. However, during the industrial revolution, when soot and other pollutants from factory smokestacks began killing lichens and darkening tree trunks, the dark form became the common one, especially near industrial cities. In this new environment, the dark form blended in with the blackened trees, whereas the light form was highly visible to its bird predators (right). Through natural selection, the dark form began to survive and reproduce at a greater rate than its light-colored kin. (Both varieties appear in each photo. Can you spot them?) (Left, Michael Tweedie/NHPA; right, Kim Taylor/Bruce Coleman Ltd.)

the population may have genetic characteristics that enable them to overcome these defenses and produce more offspring than those lacking them. In such coevolution, adaptation follows adaptation in something like an ongoing, long-term arms race among individuals in interacting populations of different species.

What Limits Adaptation? Shouldn't evolution lead to perfectly adapted organisms? Shouldn't adaptations to new environmental conditions allow our skin to become more resistant to the harmful effects of ultraviolet radiation, our lungs to cope with air pollutants, and our livers to become better at detoxifying pollutants? The answer to these questions is *no* because there are limits to adaptations in nature.

First, *a change in environmental conditions can lead to adaptation only for traits already present in the gene pool of a population*. Environmental change in the form of increased pollution did not produce the dark form of the peppered moth (Figure 5-35); the dark trait was already present in the population (probably as a result of a mutation). Natural selection favored individuals

that already had this trait because it made them more suited (adapted) to the changed environment.

Second, *even if a beneficial heritable trait is present in a population, that population's ability to adapt can be limited by its reproductive capacity*. If members of a population can't reproduce quickly enough to adapt to a particular environmental change, all its members can die. Populations of genetically diverse r-strategists (weeds, mosquitoes, rats, or bacteria), which can produce hordes of offspring early in their life cycle, can adapt to a change in environmental conditions in a short time, although many do go through boom–bust cycles (Figure 5-30). In contrast, populations of K-strategists (elephants, tigers, sharks, or humans) cannot produce large numbers of offspring rapidly, and they take a long time (typically thousands or even millions of years) to adapt through natural selection (Figure 5-31).

Finally, *even if a favorable genetic trait is present in a population, most of its members would have to die or become sterile so that individuals with the trait could predominate and pass the trait on*—hardly a desirable solution to the environmental problems humans face.

A: About 10 million (an average of 27,000 per day)

How Do New Species Evolve? Under certain circumstances natural selection can lead to an entirely new species. In this process, called **speciation**, two species arise from one in response to changes in environmental conditions.

The most common mechanism of speciation (especially among animals) takes place in two phases: geographic isolation and reproductive isolation. **Geographic isolation** occurs when two populations of a species or two groups of the same population become physically separated for fairly long periods into areas with different environmental conditions. For example, part of a population may migrate in search of food and then begin living in another area with different environmental conditions (Figure 5-36). Populations may also become separated by a physical barrier (such as a mountain range, stream, lake, or road) by a change such as a volcanic eruption or earthquake, or when a few individuals are carried to a new area by wind or water.

The second phase of speciation is **reproductive isolation**. It occurs as mutation and natural selection operate independently in two geographically isolated populations and change the allele frequencies in different ways—a process called *divergence*. If divergence continues long enough, members of the geographically and reproductively isolated populations may become so different in genetic makeup that they can't interbreed—or if they do, they can't produce live, fertile offspring. Then one species has become two, and *speciation* has occurred through *divergent evolution*.

In a few rapidly reproducing organisms (mostly r-strategists), this type of speciation may occur within hundreds of years; with most species (especially K-strategists), however, it takes from tens of thousands

to millions of years. Given this time scale, it is difficult to observe and document the appearance of a new species. As a result, there are many controversial hypotheses about the details of speciation.

How Do Species Become Extinct? After evolution, the second process affecting the number and types of species on the earth is **extinction**. When environmental conditions change, a species may either evolve (become better adapted) or cease to exist (become extinct).

The earth's long-term patterns of speciation and extinction have been affected by several major factors: **(1)** large-scale movements of the continents (continental drift) over millions of years (Figure 5-37), **(2)** gradual climate changes caused by continental drift and slight shifts in the earth's orbit around the sun, and **(3)** rapid climate change caused by catastrophic events (such as large volcanic eruptions and huge meteorites and asteroids crashing into the earth). Such events create dust clouds that shut down or sharply reduce photosynthesis long enough to eliminate huge numbers of producers—and shortly thereafter the consumers feeding on them.

Extinction is the ultimate fate of all species, just as death is for all individual organisms. Biologists estimate that 99.9% of all the species that have ever existed are now extinct.

Some species inevitably disappear at some low rate, called **background extinction**, as local conditions change. In contrast, **mass extinction** is an abrupt rise in extinction rates above the background level. It is a catastrophic, widespread (often global) event in which large groups of existing species (perhaps 25–70%) are wiped out.

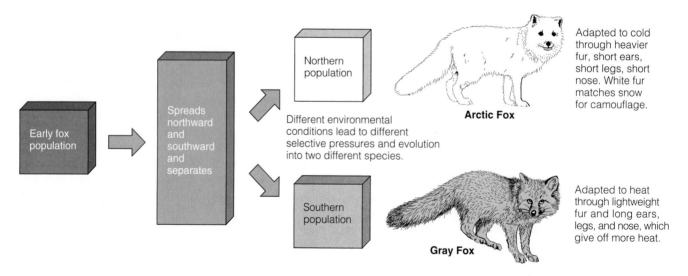

Figure 5-36 How geographic isolation can lead to reproductive isolation, divergence, and speciation.

Q: How much of the money spent on health care in the United States is used to prevent disease?

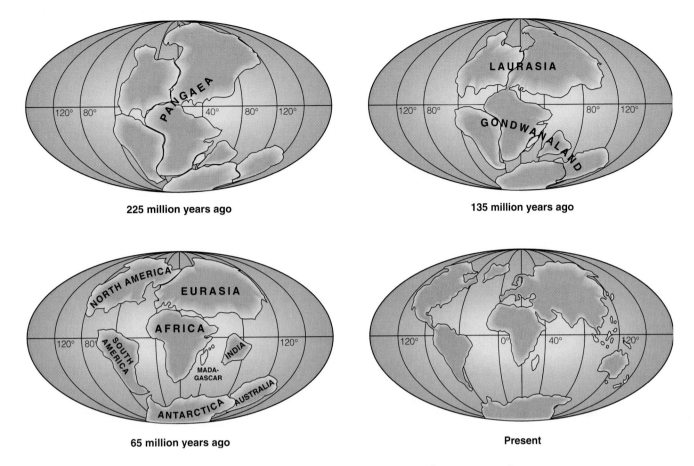

Figure 5-37 Continental drift, the extremely slow movement of continents over millions of years on several gigantic plates. This plays a role in the extinction of species and the rise of new: Populations are geographically and eventually reproductively isolated as land masses float apart and new coastal regions are created. Rock and fossil evidence indicates that about 200–250 million years ago all of the earth's present-day continents were locked together in a supercontinent called Pangaea. About 180 million years ago, Pangaea began splitting apart as the earth's huge plates separated and eventually resulted in today's locations of the continents.

Most mass extinctions are believed to result from global climate changes that kill many species and leave behind those able to adapt to the new conditions. Fossil and geological evidence indicates that the earth's species have experienced five great mass extinctions (20–60 million years apart) during the past 500 million years (Figure 5-38). Smaller extinctions (involving loss of perhaps 15–24% of all species) have come in between. The last mass extinction took place about 65 million years ago, when the dinosaurs became extinct after thriving for 140 million years.

A crisis for one species is an opportunity for another. The fact that millions of species exist today means that speciation, on average, has kept ahead of extinction. Evidence shows that the earth's mass extinctions have been followed by periods of recovery called **adaptive radiations**, in which numerous new species have evolved over several million years to fill new or vacated ecological niches in changed environments (Figure 5-38). The disappearance of dinosaurs at the end of the Mesozoic era (about 65 million years ago), for example, was followed by an evolutionary explosion for mammals (Figure 5-39). Fossil records suggest that it takes 10 million years or more for adaptive radiations to rebuild biological diversity after a mass extinction.

How Do Speciation and Extinction Affect Biodiversity? Speciation minus extinction equals *biodiversity*, the planet's genetic raw material for future evolution in response to changing environmental conditions. In this long-term give-and-take between extinction and speciation, mass extinctions temporarily reduce biodiversity. However, they also create evolutionary opportunities for surviving species to undergo

A: About 5%

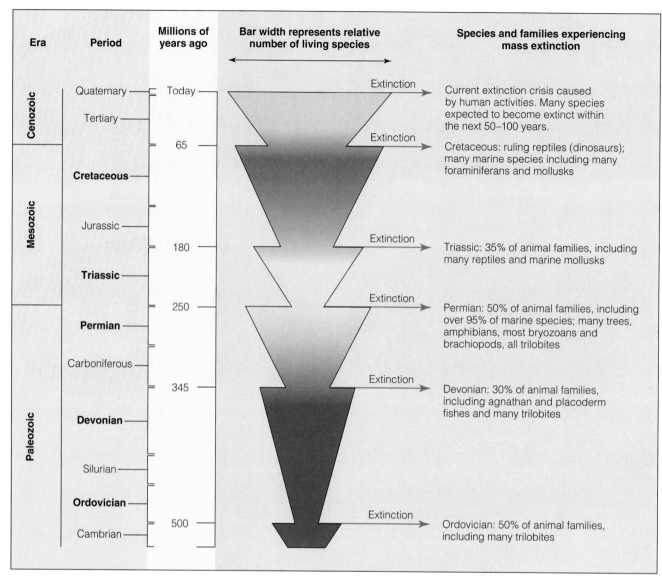

Figure 5-38 Over millions to hundreds of millions of years, there have been dramatic exits (mass extinctions) and grand entrances (speciation and radiations) of large groups of species. Fossil and radioactive dating evidence indicate that five major mass extinctions (indicated by arrows) have taken place over the past 500 million years. Mass extinctions leave large numbers of niches unoccupied and create new ones. As a result, each mass extinction has been followed by periods of recovery (represented by the wedge shapes) called *adaptive radiations* in which (over 10 million years or more) new species evolve to fill new or vacated ecological niches. Many scientists say that we are now in the midst of a sixth mass extinction, caused primarily by overhunting and the increasing elimination, degradation, and fragmentation of wildlife habitats as a result of human activities.

adaptive radiations to fill unoccupied and new niches (Figure 5-39).

Although extinction is a natural process, humans have become a major force in the premature extinction of species. As population and resource consumption increase over the next 50 years and we take over more and more of the planet's surface (Figure 1-5), we may cause the extinction of up to a quarter of the earth's current species, each the product of millions to billions of years of evolution.

On our short time scale, such catastrophic losses cannot be recouped by formation of new species; it took tens of millions of years after each of the earth's five great mass extinctions for life to recover to the previous level of biodiversity. Genetic engineering cannot stop this loss of biodiversity because genetic engineers do not create new genes. Rather, they transfer existing genes or gene fragments from one organism to another and thus rely on natural biodiversity for their raw material.

Q: What does the EPA consider to be the three most dangerous indoor air pollutants in developed countries?

🜨 5-7 COMMUNITY AND ECOSYSTEM RESPONSES TO STRESS: ECOLOGICAL SUCCESSION

How Do Communities and Ecosystems Cope with Change? One characteristic of all communities and ecosystems is that their structures, especially vegetation, are constantly changing in response to changing environmental conditions. This gradual process is called **ecological succession**, or **community development**. Change in the types of plants (and thus other species) in a community is a normal process reflecting the results of the continuing struggle among species with different adaptations to obtain food, light, space, protection, and other resources.

The various stages of succession have different patterns of species diversity, trophic structure, niches, nutrient cycling, and energy flow and efficiency (Table 5-2 and Figure 5-40).

What Is Primary Succession? Ecologists recognize two types of ecological succession, primary and secondary, depending on the conditions present at the beginning of the process. **Primary succession** involves the development of biotic communities in an essentially lifeless area where there is no soil or bottom sediment (Figure 5-41). Examples of such barren areas include bare rock exposed by a retreating glacier or severe soil erosion, newly cooled lava, an abandoned highway or parking lot, or a newly created shallow pond or reservoir.

After such a large-scale disturbance, life usually returns first in the form of a few hardy *pioneer species* (such as wind-dispersed microbes, mosses, and

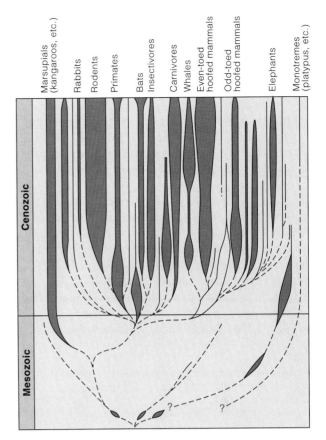

Figure 5-39 Adaptive radiation of mammals began in the first 10–12 million years of the Cenozoic era (which began about 65 million years ago) and continues today. This development of a large number of new species is thought to have resulted when huge numbers of new and vacated ecological niches became available after the mass extinction of dinosaurs near the end of the Mesozoic era. (Used by permission from Cecie Starr and Ralph Taggart, *Biology: The Unity and Diversity of Life*, Belmont, Calif.: Wadsworth, 1995)

Figure 5-40 Graphs of some generalized changes taking place during ecological succession.

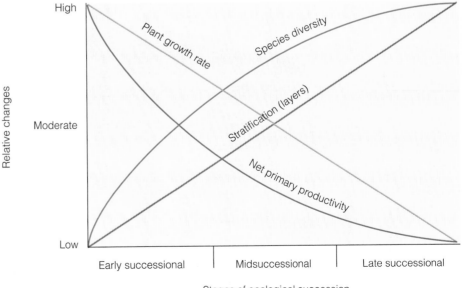

lichens). These first plants grow close to the ground, and most are r-strategists with the ability to establish large populations quickly in a new area, often under harsh conditions. Biologist Edward O. Wilson calls these species "nature's sprinters."

Sometimes pioneer species make an area suitable for *early successional plant species* with different niche requirements. Recent research indicates that this process, called *facilitation*, plays an important role only in primary succession when pioneer species first begin building soil.

After hundreds of years the soil may be deep and fertile enough to store enough moisture and nutrients to support the growth of less hardy *midsuccessional plant species* of grass and low shrubs. These in turn are replaced by trees that require lots of sunlight and are adapted to the area's climate and soil.

As these tree species grow and create shade, they are replaced by *late successional plant species* (mostly trees) that can tolerate shade. Unless fire, flooding, se-vere erosion, tree cutting, climate change, or other natural or human processes disturb the area, what was once bare rock becomes a stable and complex forest community (Figure 5-41).

What Is Secondary Succession? The more common type of succession is **secondary succession**. This begins in an area where the natural vegetation has been disturbed, removed, or destroyed but the soil or bottom sediment remains. Candidates for secondary succession include abandoned farmlands, burned or cut forests, heavily polluted streams, and land that has been dammed or flooded to produce a fairly shallow reservoir or pond. Because some soil or sediment is present, new vegetation can usually sprout within a few weeks.

In the central (Piedmont) region of North Carolina, European settlers cleared the mature native oak and hickory forests and replanted the land with crops. Some of the land was subsequently abandoned because of

Figure 5-41 Primary succession over several hundred years of plant communities on bare rock exposed by a retreating glacier on Isle Royal in northern Lake Superior.

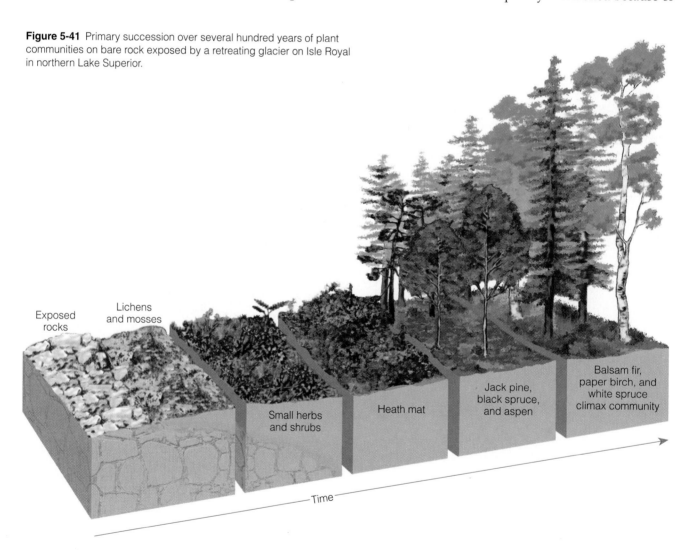

Exposed rocks

Lichens and mosses

Small herbs and shrubs

Heath mat

Jack pine, black spruce, and aspen

Balsam fir, paper birch, and white spruce climax community

Time

Q: What is the most dangerous indoor air pollutant in developing countries?

erosion and loss of soil nutrients. Figure 5-42 shows how such abandoned farmland, still covered with a thick layer of soil, has undergone secondary succession.

Descriptions of ecological succession usually focus on changes in vegetation. However, these changes in turn affect food and shelter for various types of animals. Thus, as succession proceeds the numbers and types of animals and decomposers also change.

A common process governing secondary succession (and primary succession after soil has been built up) is *inhibition*. This occurs when early species hinder the establishment and growth of other species by various means, including reducing exposure to sunlight and depositing harmful chemicals into the soil. Succession then can proceed only when a fire, bulldozer, or other disturbance removes most of the existing species.

In other cases, late successional plants are largely unaffected by plants at earlier stages of succession, a phenomenon known as *tolerance*. Tolerance may explain why late successional plants can thrive in mature communities without eliminating some early successional and midsuccessional plants.

What Is the Role of Disturbance in Ecological Succession? All forms of life face environmental conditions that change, sometimes gradually, sometimes suddenly. Disturbances such as those shown in Table 5-1 play a major role in ecological succession by converting a particular stage of succession to an earlier stage.

We tend to think of large terrestrial communities undergoing succession as being covered with a predictable blanket of vegetation. However, a close look at almost any ecosystem reveals that it consists of an ever-changing, irregular mosaic of patches at different stages of succession—the result of a variety of small and medium-sized disturbances.

The primary goal of agriculture is high productivity by selected plant species such as wheat, rice, and corn. In modern industrialized agriculture, this is achieved by replacing species-rich late successional communities (such as mature grasslands and forests) with an early successional community; such a community often consists of a single crop (monoculture) of opportunistic species that puts most of its primary productivity into edible parts instead of tall grass, trunks, thick limbs, and

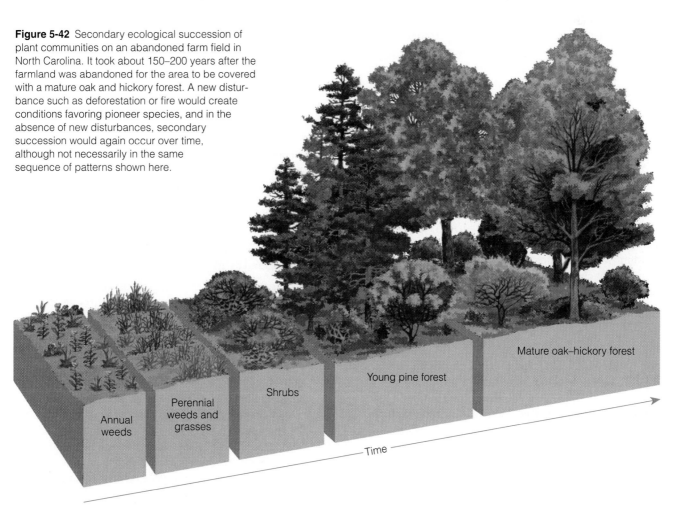

Figure 5-42 Secondary ecological succession of plant communities on an abandoned farm field in North Carolina. It took about 150–200 years after the farmland was abandoned for the area to be covered with a mature oak and hickory forest. A new disturbance such as deforestation or fire would create conditions favoring pioneer species, and in the absence of new disturbances, secondary succession would again occur over time, although not necessarily in the same sequence of patterns shown here.

Annual weeds

Perennial weeds and grasses

Shrubs

Young pine forest

Mature oak–hickory forest

Time

roots. Herbicides keep out other species (such as weeds) that would compete with the food crops for soil nutrients. Similarly, timber companies attempt to increase timber productivity by replacing diverse forests with farms or plantations of a single, fast-growing species (see photo in Table of Contents); they also sometimes use herbicides to kill competing plants.

How Predictable Is Succession? It is tempting to conclude that ecological succession is an orderly sequence in which each stage leads predictably to the next, more stable stage. According to this classic view, succession proceeds until an area is occupied by a predictable type of *climax community* dominated by a few long-lived plant species (which Edward O. Wilson calls "nature's long-distance runners").

However, research has shown that the sequences of species and community types that appear during primary or secondary succession can be highly variable and unpredictable. Research indicates that there is no ecological plan leading to ecological balance or equilibrium in an ideally adapted and stable climax community.

Rather, succession reflects the ongoing struggle by different species for enough light, nutrients, food, and space to survive and to give them a reproductive advantage over other species. In other words, ecological succession consists of a mixture of species, each "doing its own thing" by attempting to occupy as much of its fundamental niche as possible.

Even if succession were an orderly process, our knowledge of ecosystems is so limited that we could not predict the course of a given succession. We don't know the normal range for most of the variables in a given ecosystem, and we can't predict the effects of small, random, chaotic events on changes in ecosystem structure and function. This explains why a growing number of ecologists prefer terms such as *vegetation or community development* or *biotic change* instead of succession (which implies some ordered and predictable sequence of changes). Many ecologists have also have replaced the term *climax community* with terms such as *stable* or *mature community* (Table 5-2).

5-8 HUMAN IMPACT ON ECOSYSTEMS: LEARNING FROM NATURE

How Have We Modified Natural Ecosystems? To survive and support growing numbers of people, we have greatly increased the number and area of the earth's natural systems that we have modified, cultivated, built on, or degraded (Figure 1-5). We have used technology to severely alter much of the rest of nature by

- *Simplifying natural ecosystems.* We eliminate some wildlife habitats by plowing grasslands, clearing forests, and filling in wetlands, often replacing their thousands of interrelated plant and animal species with one crop or one kind of tree—called *monocultures*—or with buildings, highways, and parking lots. Then we spend a lot of time, energy, and money trying to protect such monocultures from invasion by opportunist species of plants (weeds), pests (mostly insects), and pathogens (fungi, viruses, or bacteria that harm the plants we want to grow).

- *Strengthening some populations of pest species and disease-causing bacteria by speeding up natural selection and causing genetic resistance through overuse of pesticides.*

- *Eliminating some predators.* Ranchers, who don't want bison or prairie dogs competing with their sheep for grass, want to eradicate those species. They also want to eliminate wolves, coyotes, eagles, and other predators that occasionally kill sheep. Big game hunters also push for elimination of predators that prey on game species.

- *Deliberately or accidentally introducing new species,* some beneficial and some harmful to us and other species. In the late 1800s several Chinese chestnut trees brought to the United States were infected with a fungus that spread to the American chestnut, once found throughout much of the eastern United States. Between 1910 and 1940, this accidentally introduced fungus virtually eliminated the American chestnut.

- *Overharvesting potentially renewable resources.* Ranchers and nomadic herders sometimes allow livestock to overgraze grasslands until erosion converts these ecosystems to less productive semideserts or deserts. Farmers sometimes deplete the soil of nutrients by excessive crop growing. Species of fish are overharvested. Wildlife species with economically valuable parts (such as elephant tusks, rhinoceros horns, and tiger skins) are endangered by illegal hunting (poaching).

- *Interfering with the normal chemical cycling and energy flows (throughputs) in ecosystems.* Soil nutrients can be easily eroded from monoculture crop fields, tree farms, construction sites, and other simplified ecosystems and can overload and disrupt other ecosystems such as lakes and coastal ecosystems. Chemicals such as chlorofluorocarbons (CFCs) released into the atmosphere can increase the flow of harmful ultraviolet energy reaching the earth by reducing ozone levels in the stratosphere. Emissions of carbon dioxide and other greenhouse gases—from burning fossil fuels and from clearing and burning forests and grasslands—may trigger global climate change by disrupting energy flow through the atmosphere.

Q: How much air pollution is emitted into the atmosphere each year by a typical motor vehicle in the United States?

To survive we must exploit and modify parts of nature. However, we are beginning to understand that any human intrusion into nature has multiple effects, most of them unpredictable (Connections, right).

The challenge is to maintain a balance between simplified, human-altered ecosystems and the neighboring, more complex natural ecosystems on which we and other forms of life depend, and to slow down the rates at which we are altering nature for our purposes. If we simplify and degrade too much of the planet to meet our needs and wants, what's at risk is not the earth but our own species. According to biodiversity expert E. O. Wilson, "If this planet were under surveillance by biologists from another world, I think they would look at us and say, 'Here is a species in the mid-stages of self-destruction.'" The evolutionary lesson to be learned from nature is that no species can get "too big for its britches," at least not for long.

Solutions: Learning from Nature How to Live Sustainably Scientific research indicates that living systems have six key features: *interdependence, diversity, resilience, adaptability, unpredictability,* and *limits.* Many biologists believe that the best way for us to live sustainably is to learn about and to mimic the processes and adaptations by which nature sustains itself. Here are some basic ecological lessons from nature:

- Most ecosystems use sunlight as their primary source of energy. Plants (producers) capture nonpolluting and virtually inexhaustible solar energy and convert it to chemical energy that keeps them, consumers (herbivorous and carnivorous animals), and decomposers alive.

- Ecosystems replenish nutrients and dispose of wastes by recycling chemicals. There is virtually no waste in nature. The waste outputs and decayed flesh of one organism are resource inputs for other organisms.

- Soil, water, air, plants, and animals are renewed through natural processes.

- Energy is always required to produce or maintain an energy flow or to recycle chemicals.

- Biodiversity takes various forms in different parts of the world because species diversity, genetic diversity, and ecological diversity have evolved over billions of years under different environmental conditions.

- Complex networks of positive and negative feedback loops give organisms and populations information and control mechanisms for adapting, within limits, to changing conditions.

Ecological Surprises

CONNECTIONS

Malaria once infected 9 out of 10 people in North Borneo, now known as Brunei. In 1955 the World Health Organization (WHO) began spraying the island with dieldrin (a DDT relative) to kill malaria-carrying mosquitoes. The program was so successful that the dreaded disease was virtually eliminated.

Other, unexpected things began to happen, however. The dieldrin also killed other insects, including flies and cockroaches living in houses. At first the islanders applauded this turn of events, but then small lizards that also lived in the houses died after gorging themselves on dieldrin-contaminated insects. Next, cats began dying after feeding on the lizards. Then, in the absence of cats, rats flourished and overran the villages. When the people became threatened by sylvatic plague carried by rat fleas, WHO parachuted healthy cats onto the island to help control the rats.

Then the villagers' roofs began to fall in. The dieldrin had killed wasps and other insects that fed on a type of caterpillar that either avoided or was not affected by the insecticide. With most of its predators eliminated, the caterpillar population exploded, munching its way through its favorite food: the leaves used in thatched roofs.

Ultimately, this episode ended happily: Both malaria and the unexpected effects of the spraying program were brought under control. Nevertheless, the chain of unforeseen events emphasizes the unpredictability of interfering with an ecosystem.

Critical Thinking

Do you believe that the beneficial effects of spraying pesticides on North Borneo outweighed the resulting unexpected and harmful effects? Explain.

- The population size and growth rate of all species are controlled by their interactions with other species and with their nonliving environment. In nature there are always limits to population growth.

Currently, humans are violating these principles of sustainability. No one knows how long we can continue doing this.

Biologists have formulated several important principles that can help guide us in our search for more sustainable lifestyles:

- *We are part of, not apart from, the earth's dynamic web of life.*

- *Our lives, lifestyles, and economies are totally dependent on the sun and the earth.*

- *We can never do merely one thing*—what biologist Garrett Hardin calls the **first law of human ecology**.

- *Everything is connected to everything else; we are all in it together.* We are connected to all living organisms through the long evolutionary history contained in our DNA. We are connected to the earth through our interactions with air, water, soil, and other living organisms making up the ever-changing web of life. The destiny of all species is a shared one. The primary goal of ecology is to discover which connections in nature are the strongest, most important, and most vulnerable to disruption.

We need not—and indeed cannot—stop growing food or building cities. Indeed, concentrating people in cities and increasing food supplies by raising the yields per area of cropland are both ways to help protect much of the earth's biodiversity from being destroyed or degraded, as long as the harmful environmental side effects of such activities are kept under control. The key lesson is that we need earth wisdom, care, restraint, humility, cooperation, and love as we alter the ecosphere to meet our needs and wants. Earth care is self-care.

If we love our children, we must love the earth with tender care and pass it on, diverse and beautiful, so that on a warm spring day 10,000 years hence they can feel peace in a sea of grass, can watch a bee visit a flower, can hear a sandpiper call in the sky, and can find joy in being alive.

HUGH H. ILTIS

CRITICAL THINKING

1. Why do deserts and arctic tundra support a much smaller biomass of animals than do tropical forests?

2. The deep oceans are vast and are located far away from human habitats; why not use them as the depository for our radioactive and other hazardous wastes? Give reasons for your response.

3. What factors in your lifestyle contribute to the destruction and degradation of coastal and inland wetlands?

4. Someone tries to sell you several brightly colored pieces of dry coral. Explain in biological terms why this transaction is probably a rip-off.

5. You are a defense attorney arguing in court for sparing an undeveloped old-growth tropical rain forest and a coral reef from severe degradation or destruction by development. Write out your closing statement for the defense of each of these ecosystems. If the judge decides you can only save one of the ecosystems, which one would you choose, and why?

6. Why are pest species likely to be extreme r-strategists? Why are many endangered species likely to be extreme K-strategists?

7. If after your death you could come back as a member of a particular type of species, what type of survivorship curve (Figure 5-32) would you like to have, and why?

8. Given current environmental conditions, if you had a choice would you rather be an *r-strategist* or a *K-strategist*? Explain your answer. What implications does your decision have for your current lifestyle?

9. How would you respond to someone who says that they don't believe in evolution because it is "just a theory"?

10. Someone tells you not to worry about air pollution because through natural selection the human species will develop lungs that can detoxify pollutants. How would you reply?

11. Explain why a simplified ecosystem such as a cornfield is usually much more vulnerable to harm from insects and plant diseases than a more complex, natural ecosystem such as a grassland.

*12. What type of biome do you live in or near? What effects have human activities over the past 50 years had on the characteristic vegetation and animal life normally found in the biome you live in? How is your own lifestyle affecting this biome?

*13. Visit a nearby land area or aquatic system such as a wetland, and look for signs of ecological succession. If possible, compare the types of species found on an abandoned farm field with another nearby area that is at a more mature stage of ecological succession.

*14. Use the principles of sustainability found in nature (p. 155) to evaluate the sustainability of the following parts of human systems: **(a)** transportation, **(b)** cities, **(c)** agriculture, **(d)** manufacturing, **(e)** waste disposal, and **(f)** your own lifestyle. Compare your analysis with those made by other members of your class.

*15. Make a concept map of this chapter's major ideas, using the section heads and subheads and the key terms (in boldface type). Look at the inside of the back cover and in Appendix 4 for information about concept maps.

SUGGESTED READINGS

See Internet Sources and Further Readings at the back of the book, and InfoTrac.

Q: How many people in the United States die prematurely each year because of air pollution?

We Haven't Evolved to Be Environmental Altruists, but We Can Solve Environmental Problems

Bobbi S. Low

GUEST ESSAY *Bobbi S. Low, an evolutionary and behavioral ecologist, is a professor of resource ecology at the University of Michigan. She works on issues of resource control and fertility, and on sex-related differences in resource use.*

Our Evolutionary Inheritance

We've created a series of environmental messes: acid deposition, possible global climate change, leaking toxic dumps, water pollution, and soil erosion. Often, we identify workable solutions, but we can't transcend our differences long enough to get the job done. We call these "social traps," but they are really evolutionary traps.

What strikes me, as an evolutionary biologist, is that the more we study the problem, the more it seems that precisely those behaviors we have evolved—the things that enhanced our survivorship and reproduction in past environments—are making solutions difficult! If this is true and if we understand it, we can figure out how to solve our problems.

At the heart of evolutionary history is the "selfish gene." Individuals live and die, but their genes (or their "replicates") can be immortal. This fact leads to interesting behavioral complexities. Obviously, competitive behavior can perpetuate genes in all species. But in social species like us, cooperative behaviors are often quite effective in passing our genes on—*if* those we help are our relatives, with whom we share genes, or friends who will help us in return (reciprocity).

In our evolutionary past, we mostly lived in small groups of related families. Even when societies became larger and more complex, extended families (often including a network of friends) remained central. These were the people who mattered most to us, not the rest of the world. Our main concerns were getting enough resources from the environment to meet our needs, maintaining satisfactory and stable friendships, finding mates, and raising families. Predicting the effect of our actions decades in the future was never a priority; in fact, uncertainties in the environment usually made such long-term planning futile. Most of the time, our populations and technology were sufficiently limited that we did only local damage to the environment.

As a result, we evolved to strive for resources; seldom (if ever) were we "rewarded" for conscious restraint. We evolved to be efficient, short-term, local environmental managers, not long-term regional or global conservationists. Now that we are so numerous, and have such effective technologies, our short-sighted, self-centered tendencies—which served us so well in the past—cause us difficulties.

Impediments to Solving Environmental Problems

Several factors—involving either our evolved tendencies or external conditions—interact to make solving environmental problems difficult:

- *Limited information.* It's obvious that we need information about the state of any resource, and the effect of our use of it, if we're to be efficient resource managers. The potentially usable supplies of a number of resources (coal, oil, natural gas, and water in aquifers) are still poorly known.

- *Discounting the future.* Perhaps because we had little control over changing environments in the past, we have evolved to "discount" future benefits if current costs are involved, and to discount future costs if we can get benefits now. Like Wimpy in the old Popeye cartoon, "We'll gladly pay you Tuesday, for a hamburger today." The time frame over which people are willing to pay now for future benefits is very short: about 3–5 years. Benefits any farther in the future are not considered as worth paying for.

- *Externalizing costs.* Even better than a hamburger today, paid for later, is a hamburger we get somebody else to pay for! So it's not surprising that much effort is expended in exporting ("externalizing") our harmful costs such as pollution, and having someone else pay the resulting environmental and health bills—another example of our "selfish genes" in action.

- *Common property versus private ownership.* Some resources that we call "commons" are open to everyone and can be degraded by overuse. The classic case of the resulting "tragedy of the commons," described by Garret Hardin, was the English grazing common: land on which everyone grazed their sheep and cattle. Now, if I put an extra sheep on the land, I may exceed the carrying capacity and hurt the land. But because we all share in that cost (and my part of the total cost is small) while I alone profit from the extra sheep, exploiting the commons for my individual gain is tempting. Many current environmental problems involve overexploitation of the commons: possible global warming, ozone depletion, acid deposition, whaling, toxic disposal. A frequently proposed solution to the commons dilemma is privatization, because private owners have an incentive to keep the resource workable and healthy. But not everyone shares the same interest. If I owned all the whales in the world, I might decide to exterminate them and bank the money! If private Northwest logging

(continued)

interests make the decisions on logging old growth forests, we might have no habitat for spotted owls and other species dependent on the existence of large areas of old-growth forests.

Strategies for Working with Our Evolved Behaviors

One way to solve our problems is to design strategies that work with, rather than against, our evolved tendencies. Here are some known strategies that work:

- *Getting information.*
- *Using persuasion and telling success stories.* It can help to exhort ourselves (and others) to "Do the Right Thing" partly because we have evolved in complex social groups, where the opinions of others matter. So far, relatively local problems in which we can see progress are the best candidates for this approach.
- *Accomplishing small wins.* When we "think globally and act locally" by working on local environmental issues we are more likely to see results, feel reinforced, and continue our efforts.
- *Using economic incentives.* States that require returnable deposits on beverage containers have fewer bottles and cans lying about. Because businesses measure success by profit, using economic incentives to make it possible for them to "do well by doing good" is an effective strategy that works with our "selfish genes."
- *Establishing regulations.* Economists prefer strategies that work in the marketplace but keep people from externalizing costs. For example, we can place an economic value on each unit of pollution and allow companies to trade their pollution permits in the marketplace. But that's not always possible, and regulations are often difficult to monitor and enforce. Governmental regulations can also be outstripped by new technological advances, which can have such perverse effects as making it illegal to adopt newer, cheaper, more effective technology.
- *Communicating and forming coalitions.* Participating in local recycling is a common success story: The costs

and benefits are local, and we see results quickly. Successful programs typically have both economic and social incentives. People are more likely to cooperate for the good of the group (even if it might mean a bit less for themselves) *if* they can establish communication and get to know the others in the group.

What Each of Us Can Do

All of these strategies can be useful and can be combined for greater effectiveness. Here are some things you and I can do with surprisingly little effort:

- *Continue to gather information.*
- *Look at your habits in light of this information.* What things would require little effort, now that you know they are important to change? Recycling, turning off the lights, conserving water—every little bit helps.
- *Cooperate with others to solve problems.* Join and support organizations (coalitions) whose goals you support.
- *Contact your elected representatives about things that matter to you.* You'll be appealing to their self-interest, because they rely on votes to get re-elected.
- *Vote, even (perhaps especially) in local elections.* Many decisions that affect your daily life are made locally, including the ecological issues in your own backyard.

Critical Thinking

1. Do you believe that we have "selfish" genes? If so, do you believe that we can overcome some of the environmentally harmful actions we take in the name of survival and self-interest by using the strategies listed in this essay? Which strategies do you think will work best? Which ones might work best for you?

2. Choose one small-scale and one large-scale environmental problem and propose solutions to each of them. Which evolved human traits contribute to the problems? Which proposed solutions are likely to be most and least successful, and why?

6 THE HUMAN POPULATION: GROWTH AND DISTRIBUTION

Slowing Population Growth in Thailand

Can a country sharply reduce its population growth in only 15 years? Thailand did.

In 1971 Thailand adopted a national policy to reduce its population growth. When the program began, the country's population was growing at a rate of 3.2% per year and the average Thai family had 6.4 children. Fifteen years later, the country's population growth rate had been cut in half to 1.6%. By 1997 the rate had fallen to 1.1%, and the average number of children per family was 1.9. Thailand's population is projected to grow from 60 million in 1997 to 71 million by 2025.

There are several reasons for this impressive feat: the creativity of the government-supported family-planning program, high literacy among women (90%), an increasing economic role for women, advances in women's rights, better health care for mothers and children, the openness of the Thai people to new ideas, the willingness of the government to encourage and financially support family planning and to work with the nonprofit Population and Community Development Association (PCDA), and support of family planning by the country's religious leaders (95% of Thais are Buddhist). Buddhist scripture teaches that "many children make you poor."

This remarkable transition was catalyzed by the charismatic leadership of Mechai Viravidaiya, a public relations genius and former government economist who launched the PCDA in 1974 to help make family planning a national goal. Between 1971 and 1997 the percentage of married women using modern birth control rose from 15% to 64%—higher that the 60% usage in developed countries and the 49% usage in developing countries.

Mechai also set up an economic development program that has been a factor in doubling of the country's per capita income between 1971 and 1997. A German-financed revolving loan program was established to enable people participating in family-planning programs to install toilets and drinking water systems. Low-rate loans were also offered to farmers practicing family planning. The government also offers loans to individuals from a fund that increases as their village's level of contraceptive use rises.

All is not completely rosy. Although Thailand has done well in slowing population growth and raising per capita income, it has been less successful in reducing pollution and improving public health, especially maternal health and control of AIDS and other sexually transmitted diseases.

Its capital, Bangkok, remains one of the world's most polluted and congested cities. It is plagued with notoriously high levels of traffic congestion and air pollution (Figure 6-1). The typical motorist in Bangkok spends 44 days a year sitting in traffic jams, costing $2.3 billion in lost work time.

Figure 6-1 This policeman in Bangkok, Thailand, is wearing a mask to reduce his intake of air polluted mainly by automobiles. Bangkok is one of the world's most car-clogged cities, with car commutes averaging 3 hours per day. Roughly one of every nine of its residents has respiratory ailments of some sort. (NHPA/Martin Harvey)

The problems to be faced are vast and complex, but come down to this: 5.9 billion people are breeding exponentially. The process of fulfilling their wants and needs is stripping earth of its biotic capacity to produce life; a climactic burst of consumption by a single species is overwhelming the skies, earth, waters, and fauna.

PAUL HAWKEN

This chapter is devoted to answering the following questions:

- How is population size affected by birth, death, fertility, and migration rates?
- How is population size affected by the percentage of males and females at each age level?
- How can population growth be slowed?
- What success have India and China had in slowing population growth?
- How is the world's population distributed between rural and urban areas?
- How do transportation systems shape urban areas and growth?
- How can cities be made more livable and sustainable?

🌐 6-1 FACTORS AFFECTING HUMAN POPULATION SIZE

How Is Population Size Affected by Birth and Death Rates? Populations grow or decline through the interplay of three factors: births, deaths, and migration. **Population change** is calculated by subtracting the number of people leaving a population (through death and emigration) from the number entering it (through birth and immigration) during a specified period of time (usually a year):

$$\text{Population change} = \begin{pmatrix} \text{Births} \\ + \\ \text{Immigration} \end{pmatrix} - \begin{pmatrix} \text{Deaths} \\ + \\ \text{Emigration} \end{pmatrix}$$

When births plus immigration exceed deaths plus emigration, population increases; when the reverse is true, population declines. When these factors balance out, population size remains stable, a condition known as **zero population growth (ZPG)**.

Instead of using the total numbers of births and deaths per year, demographers use two statistics: the **birth rate**, or **crude birth rate** (the number of live births per 1,000 people in a population in a given year), and the **death rate**, or **crude death rate** (the number of deaths per 1,000 people in a population in a given year). Figure 6-2 shows the crude birth and death rates for various groupings of countries in 1997.

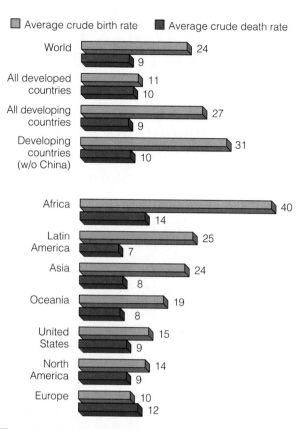

Figure 6-2 Average crude birth and death rates for various groupings of countries in 1997. (Data from Population Reference Bureau)

The rate of the world's annual population change (excluding migration) is usually expressed as a percentage*

$$\begin{aligned}
\text{Annual rate of natural population change (\%)} &= \frac{\text{Birth rate} - \text{Death rate}}{1{,}000 \text{ persons}} \times 100 \\
&= \frac{\text{Birth rate} - \text{Death rate}}{10}
\end{aligned}$$

The rate of the world's annual population growth (natural increase) dropped 26% between 1963 and 1997, from 2.2% to 1.47%. This is good news, but during the same period the population base rose by about 82%, from 3.2 billion to 5.84 billion. This 33% drop in the rate of population increase is roughly analogous to learning that the truck heading straight at you has slowed from 100 kilometers per hour to 67 while its

---

* Crude birth and death rates that have not been rounded off to the nearest whole number are often used to calculate natural change; the result is then rounded off to the nearest tenth of a percent. Consequently, use of the rounded-off crude birth and death rate figures shown in Figure 6-2 will not always produce the rounded-off percentage growth figures shown in Figure 6-3.

Q: Worldwide, how many people live in cities where outdoor air is unhealthy to breathe?

weight increased by 82%. In other words, the problem of exponential population growth has not disappeared; it's just occurring at a slower rate.

Figure 6-3 presents the annual rates of population change for major parts of the world in 1997. An annual natural increase rate of 1–3% may seem small, but such exponential growth rates lead to enormous increases in population size over a 100-year period. For example, a population growing by 3% a year will increase its population 19-fold in a century.

The current annual population increase rate of 1.47% adds about 86 million people per year (5.84 billion × 1.47% = 85.8 million)—equal to adding another Los Angeles every 2 weeks, another New York City every month, a Germany every year, and a United States every 3 years. Despite the drop in the rate of population growth, the larger base of population means that the number of people (86 million) added in 1997 was much higher than the 69 million added in 1963 when the world's population growth rate reached its peak.

In numbers of people, China (with 1.24 billion in 1997, one of every five people in the world) and India (with 970 million) dwarf all other countries. Together they make up 38% of the world's population. The United States, with 268 million people, has the world's third largest population but only 4.6% of the world's people. Figure 6-4 gives projected population growth in various regions between 1997 and 2025;

more than 95% of this growth is projected to take place in developing countries, where hunger and poverty have become a way of life for almost a billion people.

How Have Global Fertility Rates Changed?

Two types of fertility rates affect a country's population size and growth rate. The first type, **replacement-level fertility**, is the number of children a couple must bear to replace themselves. It is slightly higher than two children per couple (2.1 in developed countries and as high as 2.5 in some developing countries in 1997), mostly because some female children die before reaching their reproductive years.

Lowering fertility rates to replacement level does not mean an immediate halt in population growth (zero population growth); there are so many future parents already alive that if each had an average of 2.1 children and their children also had 2.1 children, the population would continue to grow for 50 years or more (assuming that death rates don't rise).

The second type of fertility rate, and the most useful measure of fertility for projecting future population change, is the **total fertility rate (TFR)**: an estimate of the average number of children a woman will have during her childbearing years under current age-specific birth rates. In 1997 the worldwide average TFR was 3.0 children per woman. It was 1.6 in developed countries (down from 2.5 in 1950) and 3.4 in developing countries (down from 6.5 in 1950). This

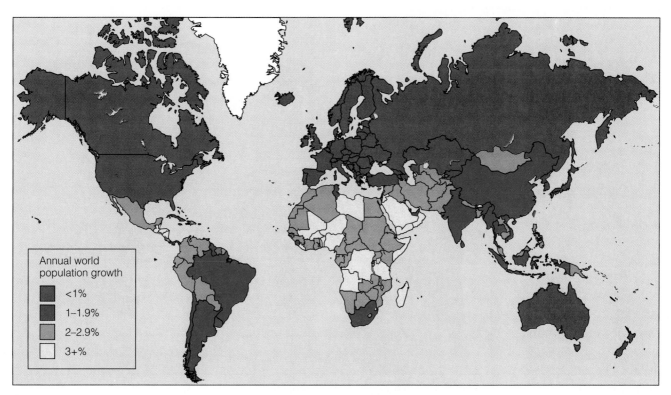

Figure 6-3 Average annual rate of population change (natural increase) in 1997. (Data from Population Reference Bureau)

A: About 1.3 billion, or one person in four

drop in the average number of children born to women in developing countries is an impressive decline, but this level of fertility is still far above the replacement level.

TFRs vary considerably throughout the world (Figure 6-5), with the highest rate by far in Africa (5.6 children per woman). If the world's TFR remains at 3.0, the human population would reach 694 billion by the year 2150—some 120 times the current population and a sobering illustration of the enormous power of exponential population growth.

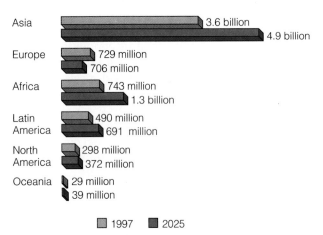

Figure 6-4 Population projections by region, 1997–2025. (Data from United Nations and Population Reference Bureau)

Population experts expect TFRs in developed countries to remain around 1.6 and those in developing countries to drop to around 2.3 by 2025; these rates are the basis of the population projections in Figure 6-4. That is good news, but it will still lead to a projected world population of around 8 billion by 2025, with more than 90% of this growth taking place in developing countries (Figure 1-7).

How Have Fertility Rates Changed in the United States? The population of the United States has grown from 76 million in 1900 to 268 million in 1997—a 2.5-fold increase—even though the country's total fertility rate has oscillated wildly (Figure 6-6). In 1957, the peak of the post–World War II baby boom, the TFR reached 3.7 children per woman. Since then it has generally declined, remaining at or below replacement level since 1972.

The drop in the total fertility rate has led to a decline in the rate of population growth in the United States. However, the country's population is still growing faster than that of most developed countries and is not even close to zero population growth. Including immigration, the U.S. population of 268 million grew by 1.17% in 1997—more than double the mean rate of the world's industrialized nations. This growth added about 3.1 million people: 1.8 million more births than deaths (accounting for about 60% of the growth),

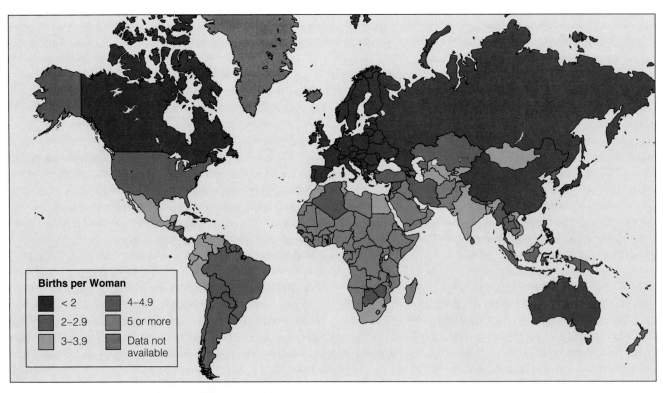

Figure 6-5 Total fertility rates (TFRs) in 1997. (Data from Population Reference Bureau)

Q: What is the greenhouse effect?

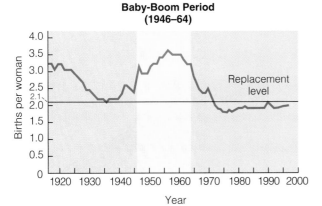

Figure 6-6 Total fertility rate for the United States between 1917 and 1997. (Data from Population Reference Bureau and U.S. Census Bureau)

935,000 legal immigrants and refugees, and an estimated 400,000 illegal immigrants. This is equivalent to adding another California every 10 years.

According to U.S. Bureau of Census projections, the U.S. population will increase from 268 million to 383 million between 1997 and 2050—a 43% increase, with no stabilization on the horizon. This is a moderate projection. A less conservative estimate projects a population of 507 million by 2050, almost double the population in 1997. Because of a high per capita rate of resource use, each addition to the U.S. population has an enormous environmental impact.

The four main reasons for this projected growth are **(1)** the large number of baby-boom women who are still in their childbearing years (even though the total fertility rate has remained at or below replacement level for 21 years, there has been a large increase in the number of potential mothers), **(2)** an increase in the number of unmarried mothers (including teenagers), **(3)** a continuation of higher fertility rates for women in some racial and ethnic groups than for Caucasian women, and **(4)** high levels of legal and illegal immigration (which accounts for about 39% of current U.S. population growth). If immigration continues at current levels, new immigrants and their descendants would account for 80 million, or 70% of the projected 115-million increase in the U.S. population between 1997 and 2050.

What Factors Affect Birth Rates and Fertility Rates? Among the most significant and interrelated factors affecting a country's average birth rate and total fertility rate are the following:

- *Average level of education and affluence.* Birth and fertility rates are usually lower in developed countries, where levels of education and affluence are higher than in developing countries.

- *Importance of children as a part of the labor force.* Rates tend to be higher in developing countries (especially in rural areas, where children begin working at an early age).

- *Urbanization.* People living in urban areas usually have better access to family planning services and tend to have fewer children than those living in rural areas, where children are needed to perform essential tasks.

- *Cost of raising and educating children.* Rates tend to be lower in developed countries, where raising children is much more costly because children don't enter the labor force until their late teens or early 20s.

- *Educational and employment opportunities for women.* Rates tend to be low when women have access to education and paid employment outside the home. Total fertility rates tend to decline as the female literacy rate increases.

- *Infant mortality rate.* In areas with low infant mortality rates, people tend to have smaller families because fewer children die at an early age.

- *Average age at marriage* (or more precisely, the average age at which women have their first child). Women normally have fewer children when their average age at marriage is 25 or older.

- *Availability of private and public pension systems.* Pensions eliminate the need of parents to have many children to help support them in old age.

- *Availability of legal abortions.* There are an estimated 30 million legal abortions and 11–22 million illegal abortions worldwide each year.

- *Availability of reliable methods of birth control* (Figure 6-7).

- *Religious beliefs, traditions, and cultural norms.* In some countries, these factors favor large families and strongly oppose abortion and some forms of birth control.

What Factors Affect Death Rates? The rapid growth of the world's population over the past 100 years is not the result of a rise in the crude birth rate; rather, it has been caused largely by a decline in crude death rates (especially in developing countries; Figure 6-8). More people started living longer (and fewer infants died) because of increased food supplies and distribution, better nutrition, improvements in medical and public health technology (such as immunizations and antibiotics), improvements in sanitation and personal hygiene, and safer water supplies (which have curtailed the spread of many infectious diseases).

Two useful indicators of overall health in a country or region are **life expectancy** (the average number

---

A: Warming of the earth's lower atmosphere (troposphere, etc.) because of the presence of heat-trapping gases   CHAPTER 6   **163**

Extremely Effective

Method	
Total abstinence	100%
Sterilization	99.6%
Hormonal implant (Norplant)	99%

Highly Effective

Method	
IUD with slow-release hormones	98%
IUD plus spermicide	98%
Vaginal pouch ("female condom")	97%
IUD	95%
Condom (good brand) plus spermicide	95%
Oral contraceptive	94%

Effective

Method	
Cervical cap	89%
Condom (good brand)	86%
Diaphragm plus spermicide	84%
Rhythm method (Billings, Sympto-Thermal)	84%
Vaginal sponge impregnated with spermicide	83%
Spermicide (foam)	82%

Moderately Effective

Method	
Spermicide (creams, jellies, suppositories)	75%
Rhythm method (daily temperature readings)	74%
Withdrawal	74%
Condom (cheap brand)	70%

Unreliable

Method	
Douche	40%
Chance (no method)	10%

Figure 6-7 Typical effectiveness of birth control methods in the United States. Percentages are based on the number of undesired pregnancies per 100 couples using a specific method as their sole form of birth control for a year. For example, a 94% effectiveness rating for oral contraceptives means that for every 100 women using the pill regularly for 1 year, 6 will get pregnant. Effectiveness rates tend to be lower in developing countries, primarily because of lack of education. (Data from Alan Guttmacher Institute)

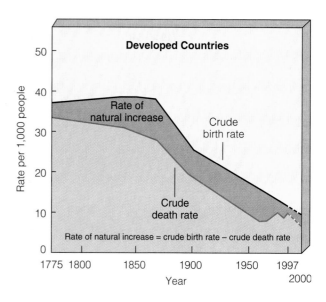

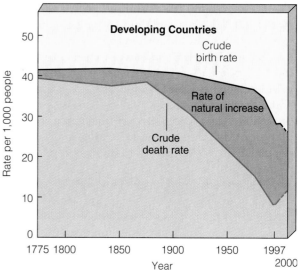

Figure 6-8 Changes in crude birth and death rates for developed and developing countries between 1775 and 1997, and projected rates (dashed lines) to 2000. (Data from Population Reference Bureau and United Nations)

of years a newborn infant can expect to live) and the **infant mortality rate** (the number of babies out of every 1,000 born each year that die within a year of birth). In most cases, a low life expectancy in an area is the result of high infant mortality. Life expectancy has increased since 1965 to an average of 71 years in developed countries and 62 years in developing countries in 1997. But in the world's poorest countries, mainly in Asia and Africa, life expectancy is only about 50 years.

Because it reflects the general level of nutrition and health care, infant mortality is probably the single most important measure of a society's quality of life (Figure 6-9). A high infant mortality rate usually indicates insufficient food (undernutrition), poor nutrition (malnutrition),

Q: Is there doubt about the validity of the greenhouse effect?

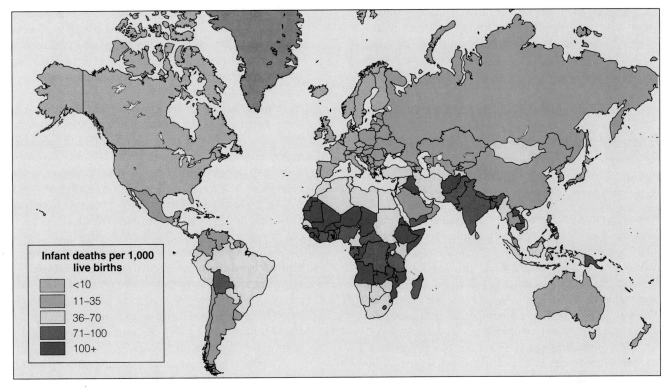

Figure 6-9 Infant mortality rates in 1997. (Data from Population Reference Bureau)

and a high incidence of infectious disease (usually from contaminated drinking water). Between 1965 and 1997, the world's infant mortality rate dropped from 20 per 1,000 live births to 9 in developed countries, and from 118 to 64 in developing countries. This is an impressive achievement, but it still means that at least 9 million infants die of preventable causes during their first year of life—an average of 24,700 unnecessary infant deaths per day.

Although the U.S. infant mortality rate of 7.3 per 1,000 in 1997 was low by world standards, 27 other countries had lower rates. Three factors that keep the U.S. infant mortality rate higher than it could be are inadequate health care (for poor women during pregnancy and for their babies after birth), drug addiction among pregnant women, and the high birth rate among teenage women. The United States has the highest teenage pregnancy rate of any industrialized country. Babies born to teenagers are more likely to have low birth weights, the most important factor in infant deaths.

How Is Migration Related to Environmental Degradation? Environmental Refugees The population of a given geographic area is also affected by movement of people into (immigration) and out of (emigration) that area. Population movement, both within and between countries, is usually desirable.

Typically, people voluntarily move from less affluent areas of low opportunity to more affluent areas of higher opportunity. The economies of many receiving countries can benefit from migrant labor. However, legal and illegal immigration of large numbers of unskilled workers into many countries are increasingly being seen as an unwanted economic burden and a source of social disruption.

Some migration is involuntary and involves refugees displaced by armed conflict, environmental degradation, or natural disaster. According to a study by the Climate Institute, by 1994 there were at least 25 million *environmental refugees*, who fled their homes because of problems such as drought, desertification, deforestation, soil erosion, and resource shortages. This study projected that the number of environmental refugees could reach 50 million by the year 2010 and 200 million in the next century if global warming projections are correct.

Dealing with today's 25 million environmental refugees is a monumental problem that affects the environmental, economic, and military security of a growing number of countries (see the Guest Essay by Norman Myers, pp. 295–296). To put this in perspective, World War II produced about 7 million homeless refugees in Europe.

Most countries influence their rates of population growth to some extent by restricting immigration; only a few countries accept large numbers of

A: No, without it, earth would be too cold for life as we know it to exist

immigrants or refugees. Only about 1% of the annual population growth in developing countries is absorbed by developed countries through international migration. Thus, population change for most countries is determined mainly by the difference between their birth rates and death rates.

Migration within countries, especially from rural to urban areas, plays an important role in the population dynamics of cities, towns, and rural areas, as discussed in Section 6-6.

6-2 POPULATION AGE STRUCTURE

What Are Age Structure Diagrams? As mentioned earlier, even if the replacement-level fertility rate of 2.1 were magically achieved globally tomorrow, the world's population would keep growing for at least another 50 years, stabilizing at about 8.6 billion (assuming no increase in death rates). The reason for this is the **age structure** of a population, or the proportion of the population (or of each sex) at each age level.

Demographers typically construct a population age structure diagram by plotting the percentages or numbers of males and females in the total population in each of three age categories: *prereproductive* (ages 0–14), *reproductive* (ages 15–44), and *postreproductive* (ages 45 and up). Figure 6-10 presents generalized age structure diagrams for countries with rapid, slow, zero, and negative population growth rates.

How Does Age Structure Affect Population Growth? Any country with many people below age 15 (represented by a wide base in Figure 6-10, left) has a powerful built-in *momentum* to increase its population size unless death rates rise sharply. The number of births rises even if women have only one or two children because of the large number of women who will soon be moving into their reproductive years.

In 1997, half the world's 2.9 billion women were in the reproductive age group and 32% of the people on the planet were under 15 years old, poised to move into their prime reproductive years. In developing countries the number is even higher: 35% compared with 20% in developed countries. This powerful force for continued population growth, mostly in developing countries (Figure 6-11), will be slowed only by an effective program to reduce birth rates or by a catastrophic rise in death rates.

How Can Age Structure Diagrams Be Used to Make Population and Economic Projections?
The 78-million-person increase that occurred in the U.S. population between 1946 and 1964, known as the *baby boom* (Figure 6-6), will continue to move up through the country's age structure as the members of this group grow older (Figure 6-12).

Baby boomers now make up nearly half of all adult Americans. As a result, they dominate the population's demand for goods and services and play an increasingly important role in deciding who gets elected and what laws are passed. Baby boomers who created the youth market in their teens and 20s are now creating the 50-something market.

Between 1996 and 2040, the proportion of the U.S. population age 65 and older is projected to increase from 13% to 21%. The economic burden of helping support so many retired baby boomers will fall on the

Figure 6-10 Population age structure diagrams for countries with rapid, slow, zero, and negative population growth rates. (Data from Population Reference Bureau)

Q: What greenhouse gas is being emitted to the atmosphere in the largest quantity from human activities?

baby-bust generation: people born since 1965 (when total fertility rates fell sharply and have remained below 2.1 since 1970; Figure 6-6). Retired baby boomers may use their political clout to force the smaller number of people in the baby-bust generation to pay higher income, health-care, and Social Security taxes. This could lead to much resentment and conflicts between the two generations.

In other respects the baby-bust generation should have an easier time than the baby-boom generation. Fewer people will be competing for educational opportunities, jobs, and services, and labor shortages may drive up their wages, at least for jobs requiring education or technical training beyond high school. On the other hand, members of the baby-bust group may find it difficult to get job promotions as they reach middle age because most upper-level positions will be occupied by members of the much larger baby-boom group. Many baby boomers may delay retirement because of improved health and the need to accumulate adequate retirement funds.

From these few projections we can see that any booms or busts in the age structure of a population create social and economic changes that ripple through a society for decades.

What Are Some Effects of Population Decline?

Populations can decline for decades after replacement level is reached if they don't have a youth-dominated age structure. When more people are in their post-reproductive years than in their reproductive and pre-reproductive years, there are more deaths than births, even at replacement-level fertility.

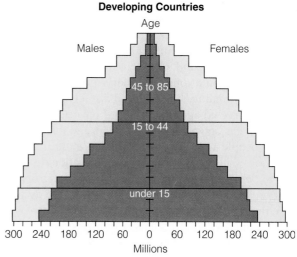

Figure 6-11 Age structure and projected population growth in developing countries and developed countries, 1985–2025. (Data from Population Reference Bureau)

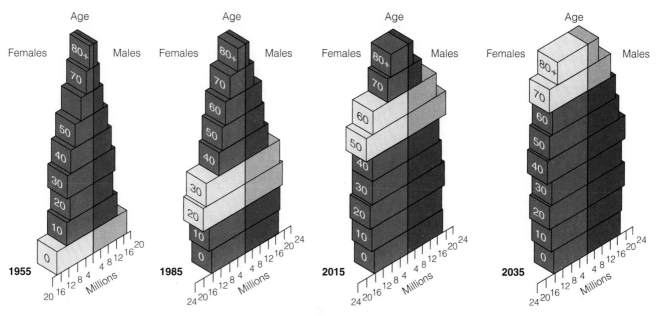

Figure 6-12 Tracking the baby-boom generation in the United States. (Data from Population Reference Bureau and U.S. Census Bureau)

A: Carbon dioxide (CO_2)

The populations of most of the world's countries are projected to grow throughout most of the 21st century. By 1997, however, 37 countries with 684 million people—12% of humanity—had roughly stable populations (annual growth rates below 0.3%) or declining populations. In other words, about one-eighth of humanity has achieved a stable population. As the projected age structure of the world's population changes between 1990 and 2150, and the percentage of people age 65 or older increases, more and more countries will begin experiencing population declines.

If population decline is gradual, its negative effects can usually be managed. But rapid population decline, like rapid population growth, can lead to severe economic and social problems. Countries undergoing rapid population decline have a sharp rise in the proportion of older people, who consume a large share of medical care, Social Security, and other costly public services. A country with a declining population can also face labor shortages unless it relies on greatly increased automation, immigration of foreign workers, or both.

Fearing that declining populations will threaten their economic well-being and national security, some European countries have offered economic incentives to encourage more births. In some cases such incentives have slowed the rate of decline, but not enough to prevent a population decrease. Massive immigration is also a solution to population decline and labor shortages, but so far most European countries have opposed this approach.

Case Study: The Graying of Japan In only 7 years, between 1949 and 1956, Japan (Figure 6-13) cut its birth, total fertility, and population growth rates in half. The main reason was widespread access to family planning implemented by the post–World War II U.S. occupation forces and the Japanese government. Since 1956 these rates have declined further, mostly because of access to family planning services and other factors: cramped housing, high land prices, late marriage ages, and high costs of education.

In 1949 Japan's total fertility rate was 4.5; in 1997 it was 1.5, one of the world's lowest. If this trend continues and immigration doesn't rise, Japan's population should begin decreasing around 2006 and could shrink to 65–96 million by 2090, depending on its fertility rate and immigration rate (which is currently negligible).

As Japan approaches zero population growth, it is beginning to face some of the problems of an aging population. Japan's universal health insurance and pension systems used about 43% of the national income in 1997. This economic burden is projected to rise to 60% or higher in 2020. Japanese economists worry that the steep taxes needed to fund these services could discourage economic growth.

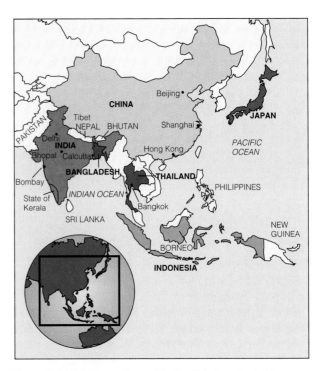

Figure 6-13 Where are Japan, Thailand, Indonesia, India, China, and Bangladesh? Some of the countries highlighted here are discussed in other chapters.

Since 1980 Japan has been feeling the effects of a declining workforce. This is one reason it has invested heavily in automation and encouraged women to work outside the home (about 43% were employed in 1997).

The population of Japan is 99% Japanese. Fearing a breakdown in its social cohesiveness, the government has been unwilling to increase immigration to provide more workers. Despite this official policy, the country is becoming increasingly dependent on illegal immigrants to keep its economic engines running. How Japan deals with these problems will be watched closely by other countries as they make the transition to zero population growth and, eventually, to population decline.

6-3 SOLUTIONS: INFLUENCING POPULATION SIZE

Case Study: Immigration in the United States
A country can increase or maintain its population size by encouraging immigration, or it can decrease or maintain its population by encouraging emigration. Only a few countries—chiefly Canada, Australia, and the United States—allow large annual increases in population from immigration.

In 1997 the United States received about 935,000 legal immigrants and refugees (Figure 6-14) and

Q: Is the possibility of global warming from human inputs of greenhouse gases a serious threat?

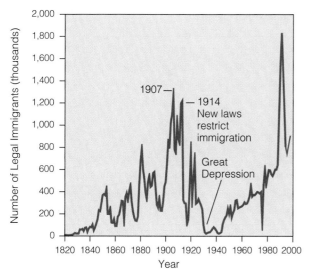

Figure 6-14 Legal immigration to the United States, 1820–1996. The large increase in immigration since 1989 resulted mostly from the Immigration Reform and Control Act of 1986, which granted legal status to illegal immigrants who could show that they had been living in the country for several years. (Data from U.S. Immigration and Naturalization Service)

400,000 illegal immigrants, together accounting for 39% of the country's population growth. The Immigration and Naturalization Service estimates that there are about 1–3 million illegal immigrants in the United States. Currently, more than 75% of all legal immigrants live in six states: California, Florida, Illinois, New York, New Jersey, and Texas. If illegal immigrants are included, this figure rises to about 90%.

Immigrants place a tax burden on residents of such states. In California, for example, the average household pays an extra $1,178 in taxes per year because of immigrants. However, according to a 1997 study by the National Academy of Sciences, the work performed and taxes paid by immigrants add as much as $10 billion per year to the overall U.S. economy. The study estimated that during his or her lifetime each immigrant will pay an average of $80,000 more in taxes than he or she costs in services.

Between 1820 and 1960, most legal immigrants to the United States came from Europe; since then, most have come from Asia and Latin America. If current trends continue, by 2050 almost half of the U.S. population will be Spanish speaking (up from 10% in 1993).

In 1995 the U.S. Commission on Immigration Reform recommended reducing the number of legal immigrants and refugees to about 700,000 per year for a transition period and then to 550,000 a year. Some demographers and environmentalists go further and call for lowering the annual ceiling for legal immigrants and refugees into the United States to 300,000–450,000, or for limiting legal immigration to about 20% of annual population growth.

Most of these analysts also support efforts to sharply reduce illegal immigration, although some are concerned that a crackdown on illegal immigrants can also lead to discrimination against legal immigrants. Proponents argue that such policies would allow the United States to stabilize its population sooner and help reduce the country's enormous environmental impact (Figure 1-19). Others oppose reducing current levels of legal immigration, arguing that it would diminish the historical role of the United States as a place of opportunity for the world's poor and oppressed.

What Are the Pros and Cons of Reducing Births? Because raising the death rate is not desirable, lowering the birth rate is the focus of most efforts to slow population growth. Today about 93% of the world's population (and 91% of the people in developing countries) live in countries with fertility reduction programs. The funding for and effectiveness of these programs vary widely; few governments spend more than 1% of the national budget on them.

The unprecedented projected doubling of the human population from 5 to 10 billion or more between 1985 and 2050 raises some important questions. Can we provide enough food, energy, water, sanitation, education, health care, and housing for twice as many people? Can we reduce already serious poverty so that people can get enough food and other basic necessities, without being forced to use potentially renewable resources unsustainably to survive? Can the world provide an adequate standard of living for twice as many people without causing massive environmental damage?

There is intense controversy concerning these questions, whether the earth is overpopulated, and what measures, if any, should be taken to slow population growth. To some the planet is already overpopulated (Figure 6-15), but others claim that if everyone existed at a minimum survival level, the earth could support 20–48 billion people. This would require that everyone exist on a diet of grain only, that all arable land be cultivated, and that much of the earth's crust be mined to a depth of 1.6 kilometers (1 mile).

Other analysts believe the planet could support 7–12 billion people at a decent standard of living by distributing land and food more equally. However, even if such optimistic estimates are technologically, politically, and environmentally feasible, many analysts doubt that our social and political structures can adapt to such a crowded and stressful world.

Others believe that asking how many people the world can support is the wrong question—equivalent

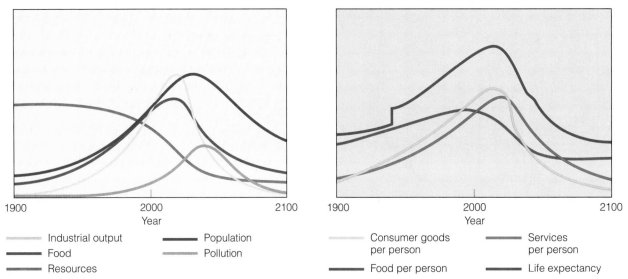

Industrial output ·········· **Population** ▬▬▬

Food ▬▬▬ **Pollution** ▬▬▬

Resources ▬▬▬

Consumer goods per person ▬▬▬ **Services per person** ▬▬▬

Food per person ▬▬▬ **Life expectancy** ▬▬▬

Figure 6-15 Plots of a computer model projecting what might happen if the world's population and economy continue growing exponentially at 1990 levels, assuming no major policy changes or technological innovations. This scenario projects that the world has already overshot some of its limits and that if current trends continue unchanged, we face global economic and environmental collapse sometime in the next century. (Used by permission from Donella Meadows et al., *Beyond the Limits: Confronting Global Collapse, Envisioning a Sustainable Future*, White River Junction, Vt.: Chelsea Green, 1992)

to asking how many cigarettes one can smoke before getting lung cancer. Instead, they say, we should be asking what the *optimum sustainable population* of the earth might be, based on the planet's *cultural carrying capacity* (Guest Essay, pp. 172–173). Such an optimum level would allow most people to live in reasonable comfort and freedom without impairing the ability of the planet to sustain future generations. No one knows what this optimum population might be. Some consider it a meaningless concept; some put it at 20 billion, others at 8 billion, and others at a level below today's population size.

Those who don't believe that the earth is overpopulated point out that the average life span of the world's 5.84 billion people is longer today than at any time in the past. Those holding this view say that talk of a population crash is alarmist, that the world can support billions more people, and that people are the world's most valuable resource for solving the problems we face. To these analysts, the primary cause of poverty is not population growth but the lack of free and productive economic systems in developing countries.

They argue that without more babies, developed countries with declining populations will face a shortage of workers, taxpayers, scientists and engineers, consumers, and soldiers needed to maintain healthy economic growth, national security, and global power

and influence. These analysts urge the governments of the United States and other developed countries to give tax breaks and other economic incentives to couples who have more than two children.

Some people view any form of population regulation as a violation of their religious beliefs, whereas others see it as an intrusion into their privacy and personal freedom. They believe that all people should be free to have as many children as they want. Some developing countries and some members of minorities in developed countries regard population control as a form of genocide to keep their numbers and power from rising.

Proponents of slowing and eventually stopping population growth point out that we fail to provide the basic necessities for one out of six people on the earth today. If we can't (or won't) do this now, how will we be able to do this for twice as many people within the next 45 years?

Proponents of slowing population growth also consider overpopulation as a threat to the earth's life-support systems. They contend that if we don't sharply lower birth rates, we are deciding by default to raise death rates for humans and greatly increase environmental harm (Figure 1-19). In 1992, for example, the highly respected U.S. National Academy of Sciences and the Royal Society of London issued the following joint statement: "If current predictions of

Q: What are the most likely effects of a warmer world?

population growth and patterns of human activity on the planet remain unchanged, science and technology may not be able to prevent either irreversible degradation of the environment or continued poverty for much of the world."

Proponents of this view recognize that population growth is not the only cause of our environmental and resource problems. However, they argue that adding several hundred million more people in developed countries and several billion more in developing countries can only intensify many environmental and social problems. They call for drastic changes to prevent accelerating environmental decline (Figure 6-16).

Those favoring slowing population growth point out that technological innovation, not sheer numbers of people, is the key to military and economic power. Otherwise, England, Germany, Japan, and Taiwan, with fairly small populations, would have little global economic and military power and China and India would rule the world.

Proponents of slowing and eventually halting population growth believe that developed countries should establish an official goal of stabilizing their populations as soon as possible. Proponents also believe that developed countries have a better chance of influencing developing countries to reduce their population growth if they officially recognize the need to stabilize their own populations and reduce their unnecessary waste of the earth's resources.

These analysts believe that people should have the freedom to produce as many children as they want. However, such freedom would apply only if it did not reduce the quality of other people's lives now and in the future, either by impairing the earth's ability to sustain life or by causing social disruption. They point out that limiting the freedom of individuals to do anything they want in order to protect the freedom of other individuals is the basis of most laws in modern societies. What is your opinion on this issue?

How Can Economic Development Help Reduce Births? Demographers have examined the birth and death rates of western European countries that industrialized during the nineteenth century, and from these data they developed a hypothesis of population change known as the **demographic transition**: As countries become industrialized, first their death rates and then their birth rates decline.

According to this hypothesis, the transition takes place in four distinct stages (Figure 6-17, p. 174). In the *preindustrial stage*, harsh living conditions lead to a high birth rate (to compensate for high infant mortality) and a high death rate. Thus there is little population growth.

State of the World

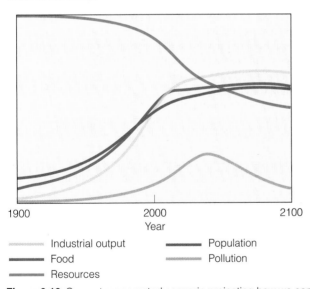

Material Standard of Living

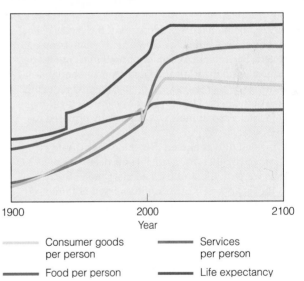

Industrial output | Population
Food | Pollution
Resources

Consumer goods per person | Services per person
Food per person | Life expectancy

Figure 6-16 Computer-generated scenario projecting how we can avoid overshoot and collapse and make a fairly smooth transition to a sustainable future. It assumes that **(1)** technology allows us to double supplies of nonrenewable resources, double crop and timber yields, cut soil erosion in half, and double the efficiency of resource use within 20 years; **(2)** 100% effective birth control was made available to everyone by 1995; **(3)** no couple has more than two children, beginning in 1995; and **(4)** per capita industrial output is stabilized at 1990 levels. Another computer run projects that waiting until 2015 to implement these changes would lead to collapse and overshoot sometime around 2075, followed by a transition to sustainability by 2100. (Used by permission from Donella Meadows et al., *Beyond the Limits: Confronting Global Collapse, Envisioning a Sustainable Future*, White River Junction, Vt.: Chelsea Green, 1992)

A: Shifts in food-growing regions and water supplies, rising sea levels, and decreased biodiversity

Moral Implications of Cultural Carrying Capacity

Garrett Hardin

GUEST ESSAY

As longtime professor of human ecology at the University of California at Santa Barbara, Garrett Hardin made important contributions to relating ethics to biology. He has raised hard ethical questions, sometimes taken unpopular stands, and forced people to think deeply about environmental problems and their possible solutions. He is best known for his 1968 essay "The Tragedy of the Commons," which has had a significant impact on the disciplines of economics and political science, and on the management of potentially renewable resources. His many books include Promethean Ethics, Filters Against Folly: How to Survive Despite Economists, Ecologists, and the Merely Eloquent, *and* Living Within Limits *(see Further Readings).*

For many years, Angel Island in San Francisco Bay was plagued with too many deer. A few animals transplanted there in the early 1900s lacked predators and rapidly increased to nearly 300 deer—far beyond the carrying capacity of the island. Scrawny, underfed animals tugged at the heartstrings of Californians, who carried extra food for them from the mainland to the island.

Such well-meaning charity worsened the plight of the deer. Excess animals trampled the soil, stripped the bark from small trees, and destroyed seedlings of all kinds. The net effect was to lower the island's carrying capacity, year by year, as the deer continued to multiply in a deteriorating habitat.

State game managers proposed that the excess deer be shot by skilled hunters. "How cruel!" some people protested. Then the managers proposed that coyotes be introduced onto the island. Though not big enough to kill adult deer, coyotes can kill fawns, thereby reducing the size of the herd. However, the Society for the Prevention of Cruelty to Animals was adamantly opposed to this proposal.

In the end, it was agreed that some deer would be transported to other areas suitable for deer. A total of 203 animals were caught and trucked many miles away. From the fate of a sample of animals fitted with radio

collars, it was estimated that 85% of the transported deer died within a year (most of them within two months) from various causes: predation by coyotes, bobcats, and domestic dogs; shooting by poachers and legal hunters; and being run over by automobiles.

The net cost (in 1982 dollars) for relocating each animal surviving for a year was $2,876. The state refused to continue financing the program, and no volunteers stepped forward to pay future bills.

Angel Island is a microcosm of the planet as a whole. Organisms reproduce exponentially, but the environment doesn't increase at all. The moral is a simple ecological commandment: *Thou shalt not transgress the carrying capacity.*

Now let's examine the situation for humans. A competent physicist has placed global human carrying capacity at 50 billion, about 10 times the current world population. Before you give in to the temptation to urge women to have more babies, consider what Robert Malthus said nearly 200 years ago: "There should be no more people in a country than could enjoy daily a glass of wine and piece of beef for dinner."

A diet of grain or bread and water is symbolic of minimum living standards; wine and beef are symbolic of higher living standards that make greater demands on the environment. When land that could produce plants for direct human consumption is used to grow grapes for wine or corn for cattle, more energy is expended to feed the human population. Because carrying capacity is defined as the *maximum* number of animals (humans) an area can support, using part of the area to support such cultural luxuries as wine and beef reduces the carrying capacity. This reduced capacity is called the *cultural carrying capacity*, and it is always smaller than simple carrying capacity.

Energy is the common "coin of the realm" for all competing demands on the environment. Energy saved by giving up a luxury can be used to produce more food staples and support more people. We could increase the simple carrying capacity of the earth by giving up any (or all) of the following "luxuries": street lighting, vacations, private cars, air conditioning, and artistic perfor-

In the *transitional stage*, industrialization begins, food production rises, and health care improves. Death rates drop and birth rates remain high, so the population grows rapidly (typically 2.5–3% a year).

In the *industrial stage*, industrialization is widespread. The birth rate drops and eventually approaches the death rate. Reasons for this convergence of rates include better access to birth control, decline in the infant mortality rate, increased job opportunities for women, and the high costs of raising children who

don't enter the workforce until after high school or college. Population growth continues, but at a slower and perhaps fluctuating rate, depending on economic conditions. Most developed countries are now in this third stage, and a few developing countries are entering this stage.

In the *postindustrial stage*, the birth rate declines even further, equaling the death rate and thus reaching zero population growth. Then the birth rate falls below the death rate and total population size slowly

Q: What are the major ways to reduce the threat of global warming?

mances of all sorts. But what we consider "luxuries" depends on our values as individuals and societies, and values are largely matters of choice. At one extreme, we could maximize the number of human beings living at the lowest possible level of comfort. Or we could try to optimize the quality of life for a much smaller human population.

What is the carrying capacity of the earth? is a scientific question. It may be possible to support 50 billion people at a "bread and water" level. Is that what we choose? The question, What is the cultural carrying capacity? requires that we debate questions of value, about which opinions differ.

An even greater difficulty must be faced. So far, we have been treating carrying capacity as a *global* issue, as if there were some global sovereignty capable of enforcing a solution on all people. But there is no global sovereignty ("one world"), nor is there any prospect of one in the foreseeable future. Thus, we must ask how some 200 nations are to coexist in a finite global environment if different sovereignties adopt different standards of living.

Consider a protected redwood forest that produces neither food for humans nor lumber for houses. Because people must travel many kilometers to visit it, the forest is a net loss in the national energy budget. However, for those fortunate enough to wander through the cathedral-like aisles beneath an evergreen vault, a redwood forest does something precious for the human spirit. But then intrudes an appeal from a distant land, where millions are starving because their population has overshot the carrying capacity; we are asked to save lives by sending food. So long as we have surpluses, we may safely indulge in the pleasures of philanthropy. But after we have run out of our surpluses, then what?

A spokesperson for the needy from that land makes a proposal: "If you would only cut down your redwood forests, you could use the lumber to build houses and then grow potatoes on the land, shipping the food to us. Since we are all passengers together on Spaceship Earth, are you not duty bound to do so? Which is more precious, trees or human beings?"

This last question may sound ethically compelling, but let's look at the consequences of assigning a preemptive and supreme value to human lives. At least 2 billion people in the world are poorer than the 39 million "legally poor" in America, and their numbers are increasing by about 40 million per year. Unless this increase is halted, sharing food and energy on the basis of need would require the sacrifice of one amenity after another in rich countries. The ultimate result of sharing would be complete poverty everywhere on the earth in order to maintain the earth's simple carrying capacity. Is that the best humanity can do?

To date, there has been overwhelmingly negative reaction to all proposals to make international philanthropy conditional on the cessation of population growth by poor, overpopulated recipient nations. Foreign aid is governed by two apparently inflexible assumptions:

- The right to produce children is a universal, irrevocable right of every nation, no matter how hard it presses against the carrying capacity of its territory.

- When lives are in danger, the moral obligation of rich countries to save human lives is absolute and undeniable.

Considered separately, each of these two well-meaning doctrines might be defensible; taken together, they constitute a fatal recipe. If humanity gives maximum carrying capacity precedence over problems of cultural carrying capacity, the result will be universal poverty and environmental ruin.

Or do you see an escape from this harsh dilemma?

Critical Thinking

1. What population size would allow the world's people to have good quality of life? What do you believe is the cultural carrying capacity of the United States? Should the United States have a national policy to establish this population size as soon as possible? Explain.

2. Do you support the two principles this essay lists as the basis of foreign aid to needy countries? If not, what changes would you make in the requirements for receiving such aid?

decreases. Emphasis shifts from unsustainable to sustainable forms of economic development. Some 37 countries (most of them in western Europe) containing about 12% of the world's population have entered this stage. To most population experts, the challenge is to help the remaining 88% of humanity reach this stage.

In most developing countries today, death rates have fallen much more than birth rates. In other words, these developing countries—mostly in Southeast Asia, Africa, and Latin America—are still in the transitional

stage, halfway up the economic ladder, with high population growth rates. Some economists believe that developing countries will make the demographic transition over the next few decades without increased family-planning efforts.

However, despite encouraging declines in fertility, some population analysts fear that the still-rapid population growth in many developing countries will outstrip economic growth and overwhelm local life-support systems. This could cause many of these

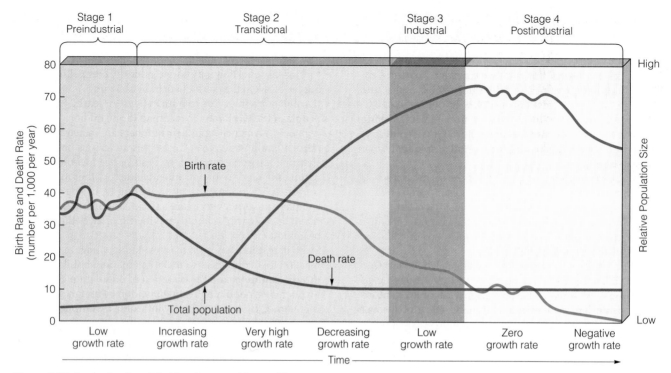

Figure 6-17 Generalized model of the demographic transition.

countries to be caught in a *demographic trap*, something that is currently happening in a number of developing countries, especially in Africa.

Analysts also point out that some of the conditions that allowed developed countries to develop are not available to many of today's developing countries. Even with large and growing populations, many developing countries do not have enough skilled workers to produce the high-tech products needed to compete in the global economy. Most low- and middle-income developing countries also lack the capital and resources needed for rapid economic development. Furthermore, the amount of economic assistance for developing countries, which are struggling under tremendous debts, has been decreasing since 1980. Indeed, since the mid-1980s developing countries have paid developed countries $40–50 billion a year more (mostly in debt interest) than they have received from these countries.

How Can Family Planning Help Reduce Births?
Family planning provides educational and clinical services that help couples choose how many children to have and when to have them. Such programs vary from culture to culture, but most provide information on birth spacing, birth control, breast-feeding, and prenatal care. Mostly on religious grounds, some people are opposed to family-planning services that provide contraceptives or perform abortions.

Family planning has been an important factor in increasing the proportion of married women in developing countries who use modern contraception. It has gone from 10% in the 1960s to 49% in 1997 (36% if China is excluded); this is close to the 60% usage by married women in developed countries. Family planning is also responsible for at least 40% of the drop in total fertility rates in developing countries, from 6 in 1960 to 3.4 in 1997 (4 if China is excluded).

Family-planning programs save a society money by reducing the need for children's social services. Proponents also argue that providing access to family planning throughout the world would bring about a sharp decline in the estimated 50 million legal and illegal abortions per year.

Family planning, coupled with improved prenatal care and health services, can also help reduce the risk of childbearing. Each year at least 600,000 women die from pregnancy-related causes, with 99% of such deaths occurring in developing countries.

The effectiveness of family planning varies with program design and funding. It has been a significant factor in reducing birth and fertility rates in populous countries such as China (Section 6-4), Indonesia, and Brazil. Family planning has also played a major role in reducing population growth in Japan, Thailand (p. 159), Mexico, South Korea, Taiwan, and several other countries with moderate to small populations. These successful programs based on committed

Q: Is ozone depletion from CFCs and other human-made chemicals in the stratosphere a serious threat?

leadership, local implementation, and wide availability of contraceptives demonstrate that population rates can be decreased significantly within 15–30 years.

Despite such successes, family planning has had moderate to poor results in more populous developing countries such as India, Egypt, Pakistan, and Nigeria. Results have also been poor in 79 less populous developing countries, especially in Africa and Latin America, with high or very high population growth rates.

According to UN studies, an estimated 300 million women in developing countries want to limit the number and determine the spacing of their children, but they lack access to services. Extending family-planning services to these women and to those who will soon be entering their reproductive years could prevent an estimated 5.8 million births a year and over 130,000 abortions a day. Other analysts call for expanding existing family-planning programs to include teenagers and sexually active unmarried women, who now are often excluded.

Some analysts urge that programs be broadened to educate males about the importance of having fewer children and taking more responsibility for raising them. Proponents also argue that much more research is needed on developing new, more effective, and acceptable methods of birth control for men. What do you think about this proposal?

Family planning could be provided in developing countries to all couples who want it for about $17 billion a year, the equivalent of less than a week's worth of worldwide military expenditures. If developed countries provided one-third of this $17 billion, each person in the developed countries would spend only $4.80 a year. This would help reduce world population by about 2.7 billion people, shrinking average family size from 3.0 to 2.1 children. However, the U.S. Congress cut international family-planning assistance funds by 87% from 1995 levels for 1996 and by 60% for 1997. According to the Alan Guttmacher Institute, the 1996 funding cuts increased abortions and led to 134,000 additional infant deaths and 8,000 deaths among women during pregnancy and childbirth.

Many analysts call for stepped up family-planning efforts in developed countries, especially the United States because of its higher growth rate than other industrialized, high-consumption countries. Elimination of unplanned births in the United States would result in about 1.8 million fewer births and about 900,000 fewer abortions per year, almost half of them among teenagers. Some analysts urge pro-choice and pro-life groups to join forces in greatly reducing unplanned births, especially among teenagers. What do you think?

How Can Economic Rewards and Penalties Be Used to Help Reduce Births? Some population experts argue that family planning, even coupled with economic development, cannot lower birth and fertility rates quickly enough to avoid a sharp rise in death rates in many developing countries. They point to studies showing that most couples in developing countries want three or four children, which is well above the replacement-level fertility required to bring about eventual population stabilization.

These analysts believe that we must go beyond family planning and offer economic rewards and penalties to help slow population growth. About 20 countries offer small payments to people who agree to use contraceptives or to be sterilized; however, such payments are most likely to attract people who already have all the children they want. Some of these countries also pay doctors and family-planning workers for each sterilization they perform and each IUD they insert.

Some countries, including China, penalize couples who have more than one or two children by raising their taxes, charging other fees, or eliminating income tax deductions for a couple's third child (as in Singapore, Hong Kong, and Ghana). Families who have more children than the prescribed limit may also lose health-care benefits, food allotments, and job options. However, programs that withhold food or increase the cost of raising children punish innocent children for the actions of their parents.

Economic rewards and penalties designed to lower birth rates work best if they encourage (rather than mandate) people to have fewer children, reinforce existing customs and trends toward smaller families, do not penalize people who produced large families before the programs were established, and increase a poor family's economic status. Once a country's population growth is out of control, however, it may be forced to use coercive methods to prevent mass starvation and hardship, as has been the case for China (Section 6-4).

How Can Empowering Women Help Reduce Births? Studies show that women tend to have fewer and healthier children and live longer when they have access to education and to paying jobs outside the home, and when they live in societies in which their individual rights are not suppressed.

Women, roughly half of the world's population, do almost all of the world's domestic work and child care and provide more health care with little or no pay than all the world's organized health services combined. They also do more than half the work associated with growing food, gathering fuelwood, and hauling water. As one Brazilian woman put it, "For poor women the only holiday is when you are asleep." Women's unpaid work, at an estimated value of

$11 trillion annually (almost half of the annual total global output), is not included in a country's GDP.

Women work two-thirds of all hours worked, but receive only one-tenth of the world's income and own a mere 0.01% of the world's property. In most developing countries, women don't have the legal right to own land or borrow money to increase agricultural productivity. Women also make up 70% of the world's poor and almost two-thirds of the more than 960 million adults who can neither read nor write.

Women are almost universally excluded from economic and political decision making. They hold only 14% of the world's administrative and managerial positions and occupy only 10% of parliamentary seats.

Most analysts believe that women everywhere should have full legal rights and the opportunity to become educated and earn income outside the home. This would not only slow population growth but also promote human rights and freedom. However, empowering women by seeking gender equality will require some major social changes that will be difficult to achieve in male-dominated societies.

6-4   CASE STUDIES: SLOWING POPULATION GROWTH IN INDIA AND CHINA

What Success Has India Had in Controlling Its Population Growth?   The world's first national family-planning program began in India (Figure 6-13) in 1952, when its population was nearly 400 million. In 1997, after 45 years of population control efforts, India was the world's second most populous country, with a population of 970 million—3.8 times as many people as the United States.

In 1952 India added 5 million people to its population; in 1997 it added 18 million—49,300 more mouths to feed each day. With 35% of its people under age 15, India's population is projected to reach 1.4 billion by 2025 and possibly 1.9 billion before leveling off early in the 22nd century. In 1997, women in India averaged 3.5 children, down from 5.3 in 1970. Despite this decline in fertility, India's population is growing exponentially at about 1.9% a year.

India's people are among the poorest in the world, with an average per capita income of about $340 a year; for 30% of the population it is less than $100 a year, or 27¢ a day. Nearly half of India's labor force is unemployed or can find only occasional work. Although India is currently self-sufficient in food-grain production, about 40% of its population today suffers from malnutrition, mostly because of poverty. Life

expectancy is only 59 years and the infant mortality rate is 75 deaths per 1,000 live births.

Some analysts fear that India's already serious malnutrition and health problems will worsen as its population continues to grow rapidly. With 16% of the world's people, India has just 2.3% of the world's land resources and 1.7% of the world's forests. Some 40% of India's cropland is degraded as a result of soil erosion, waterlogging, salinization, overgrazing, and deforestation. About 70% of India's water is seriously polluted, and sanitation services are often inadequate.

Without its long-standing family-planning program, India's population and environmental problems would be growing even faster. Still, to its supporters the results of the program have been disappointing because of poor planning, bureaucratic inefficiency, the low status of women (despite constitutional guarantees of equality), extreme poverty, and a lack of administrative and financial support.

Even though the government has provided information about the advantages of small families for years, Indian women still have an average of 3.5 children because most couples believe they need many children to do work and care for them in old age. Many social and cultural norms favor large families, including the strong preference for male children; some couples keep having children until they produce one or more boys. These factors in part explain why even though 90% of Indian couples know of at least one modern birth control method, only 36% actually use one.

What Success Has China Had in Controlling Its Population Growth?   China has the world's largest population. Although China is roughly the same size as the United States, it has about 4.6 times as many people.

Since 1970, China (Figure 6-13) has made impressive efforts to feed its people and bring its population growth under control. Between 1972 and 1997 China achieved a remarkable drop in its crude birth rate, from 32 to 17 per 1,000 people, and its total fertility rate dropped from 5.7 to 1.8 children per woman. Since 1985 its infant mortality rate has been almost one-half the rate in India. Life expectancy in China is 70 years, 11 years higher than in India. China's per capita income of $620 is almost twice that in India. Despite these achievements, with the world's largest population (1.24 billion) and a growth rate of 1.0%, China had about 12 million more mouths to feed in 1997. Its population is projected to reach 1.6 billion by 2025.

To achieve its sharp drop in fertility, China has established the most extensive, intrusive, and strict population control program in the world. Couples are strongly urged to postpone the age at which they marry and to have no more than one child. Married

Q: How long do CFCs stay in the atmosphere?

couples have ready access to free sterilization, contraceptives, and abortion. Paramedics and mobile units ensure access even in rural areas.

Couples who pledge to have no more than one child are given extra food, larger pensions, better housing, free medical care, and salary bonuses; their child will be given free school tuition and preferential treatment in employment when he or she enters the job market. Couples who break their pledge lose all the benefits. The result is that 81% of married women in China are using modern contraception, compared to 60% in developed countries and only 36% in other developing countries.

Government officials realized in the 1960s that the only alternative to strict population control was mass starvation. China is a dictatorship, and thus, unlike India, it has been able to impose a consistent population policy throughout society. Moreover, Chinese society is fairly homogeneous and has a widespread common written language, which aids in educating people about the need for family planning and in implementing policies for slowing population growth.

China's large and still growing population has an enormous environmental impact that could reduce its ability to produce enough food and threaten the health of many of its people. China has 21% of the world's population but only 7% of its fresh water and cropland, 3% of its forests, and 2% of its oil. Soil erosion in China is serious and is apparently getting worse. However, China, which for several years has had one of the world's highest rates of economic growth, has moved more people out of poverty in a shorter time than any country in history. Most countries prefer to avoid the coercive elements of China's population control program. Coercion is not only incompatible with democratic values and notions of basic human rights, but also ineffective in the long run because sooner or later people resist. However, other parts of this program could be used in many developing countries. Especially useful is the practice of localizing the program rather than asking people to go to distant centers. Perhaps the best lesson for other countries is to act to curb population growth before the choice must be made between mass starvation and coercive measures that severely restrict human freedom.

6-5 CUTTING GLOBAL POPULATION GROWTH

In 1994 the United Nations held its third once-in-a-decade Conference on Population and Development in Cairo, Egypt. One of the conference's goals was to encourage action to stabilize the world's population at 7.8 billion by 2050, instead of the projected 11–12.5 billion. The major goals of the resulting 20-year population plan, endorsed by 180 governments, are to

- Provide universal access to family-planning services and reproductive health care
- Improve the health care of infants, children, and pregnant women
- Encourage development and implementation of national population policies as part of social and economic development policies
- Bring about more equitable relationships between men and women, with emphasis on improving the status of women and expanding education and job opportunities for young women
- Increase access to education, especially for girls
- Increase the involvement of men in child-rearing responsibilities and family planning
- Take steps to eradicate poverty
- Reduce and eliminate unsustainable patterns of production and consumption

Many analysts applaud these goals, but some call them wishful thinking. Even if they wanted to, most governments could not afford to implement many of these goals.

However, the experience of Japan, Thailand (p. 159), South Korea, Taiwan, and China indicates that a country can achieve replacement-level fertility within 15–30 years. Such experience also suggests that the best way to slow population growth is a combination of investing in family planning, reducing poverty, and elevating the status of women.

Because countries differ in population growth rates, use and availability of resources, and social structure, the mix of these factors must be tailored to each country's situation. Furthermore, most analysts believe that government policy makers should devise policies that minimize the environmental impact of population growth in their effort to achieve sustainability.

6-6 POPULATION DISTRIBUTION: URBANIZATION AND URBAN GROWTH

How Fast Are Urban Areas Growing? An **urban area** is often defined as a town or city with a population of more than 2,500 people (although some countries set the minimum at 10,000–50,000 residents). A **rural area** is usually defined as an area with a population of less than 2,500 people.

A country's **degree of urbanization** is the percentage of its population living in an urban area. **Urban growth** is the rate of increase of urban populations.

Urban areas grow in two ways: by *natural increase* (more births than deaths) and by *immigration* (mostly from rural areas). Because cities are the main centers for new jobs, higher income, education, innovation, culture, better health care, and trade, people are drawn to urban areas in search of jobs, a better life, and freedom from the constraints of village cultural life. Urban growth in developing countries is also fueled by government policies that distribute most income and social services to urban dwellers (especially in capital cities, where a country's leaders live) at the expense of rural dwellers. People may also be pushed from rural areas into urban areas by factors such as poverty, lack of land, declining agricultural work, famine, and war.

Modern mechanized agriculture, for example, uses fewer farm laborers and allows large landowners to buy out subsistence farmers who cannot afford to modernize. Without jobs or land, these people are forced to move to cities. The urban poor fortunate enough to find employment usually must work long hours for low wages at jobs that may expose them to dust, hazardous chemicals, excessive noise, and dangerous machinery.

Between 1950 and 1997 the number of people living in the world's urban areas increased 12-fold, from 200 million to 2.5 billion. By 2025 it is projected to reach 5.5 billion, almost equal to the world's current population. About 90% of this urban growth will occur in developing countries. At current rates the world's population will double in 47 years, the urban population in 22 years, and the urban population of developing countries in 20 years.

Several trends are important in understanding the problems and challenges of urban growth on this rapidly urbanizing planet. *First*, the proportion of the global population living in urban areas increased between 1850 and 1997 from 2% to 43%.

During the 1990s, more than 70% of the world's population increase is expected to occur in urban areas, adding about 67 million people a year to these already overburdened areas. This is equivalent to adding about four cities the size of New York each year. If UN projections are correct, by 2025 about 63% of the world's people will be living in urban areas.

Second, the number of large cities is mushrooming. In 1960 there were 111 cities with populations of more than 1 million; today there are 293, and some analysts expect this number to increase to at least 400 by 2025. Currently, 1 person out of every 10 lives in a city with a million or more inhabitants, and many live in the world's 15 *megacities*, with 10 million or more

people. The United Nations projects that by 2000 the world will have 21 megacities, 17 of them in developing countries.

However, the move to urban areas is not limited to the world's 100 or so largest urban areas. The fastest urban growth is occurring in the 30,000 or so medium-size cities in developing countries.

As they grow outward, separate urban areas may merge to form a *megalopolis*. For example, the remaining open space between Boston and Washington, D.C., is rapidly urbanizing and coalescing. This 800-kilometer- (500-mile-) long urban area, sometimes called *Bowash* (Figure 6-18), contains almost 60 million people, more than twice Canada's entire population.

Third, developing countries, with 36% urbanization, contain 1.7 billion urban dwellers—more than the total populations of Europe, North America, Latin America, and Japan combined. The urban population in developing countries is growing at 3.5% per year and they are projected to reach at least 57% urbanization by 2025. Most of this growth will occur in large cities, which already have trouble supplying their residents with water, food, housing, jobs, sanitation, and basic services.

Fourth, in developed countries (with 74% urbanization), urban growth is less than 1% per year, much slower than in developing countries. Still, developed countries should reach 84% urbanization by 2025.

Finally, poverty is becoming increasingly urbanized as more poor people migrate from rural to urban areas. The United Nations estimates that at least 1 billion people, 17% of the world's current population, live either in the crowded *slums* of inner cities or in

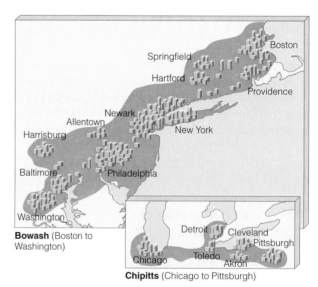

Bowash (Boston to Washington)

Chipitts (Chicago to Pittsburgh)

Figure 6-18 Two megalopolises: Bowash, consisting of urban sprawl and coalescence between Boston and Washington, D.C., and Chipitts, extending from Chicago to Pittsburgh.

Q: If all ozone-depleting substances were banned now, when might the ozone layer return to 1985 levels?

vast, mostly illegal *squatter settlements* and *shanty-towns*, where people move onto undeveloped land (usually without the owner's permission) and build shacks made of packing crates, plastic sheets, corrugated metal pipes, or whatever they can scavenge (Spotlight, below).

Case Study: Mexico City About 15.6 million people—about one of every six Mexicans—live in Mexico City (Figure 6-19), the world's fourth most populous city. Every day an additional 2,000 poverty-stricken rural peasants pour into the city, hoping to find a better life. This adds about 750,000 new people per year, equivalent to having to provide food, water, sanitation, jobs, and other services for a new city the size of Baltimore or San Francisco every year.

Mexico City suffers from severe air pollution, high unemployment (close to 50%), deafening noise, congestion, and a soaring crime rate. More than one-third of its residents live in crowded slums (called *barrios*) or squatter settlements, without running water or electricity. At least 8 million people have no sewer facilities.

This means that huge amounts of human waste are deposited in gutters and vacant lots every day, attracting armies of rats and swarms of flies. When the winds pick up dried excrement, a *fecal snow* often falls on parts of the city, leading to widespread salmonella and hepatitis infections, especially among children.

Some 4 million motor vehicles, 30,000 factories (45% of all Mexican industries), and leaking unburned liquefied petroleum gas (LPG) from stoves and heaters spew pollutants into the atmosphere. Air pollution is intensified because the city lies in a basin surrounded by mountains, and frequent thermal inversions trap pollutants at ground level. Since 1982 the amount of contamination in the city's smog-choked air has more than tripled; breathing the air is said to be roughly equivalent to smoking two packs of cigarettes a day. The city's health costs from air pollution (largely from automobiles) are estimated at $1.5 billion per year.

Pediatricians estimate that 85% of childhood illnesses in the city are related to air pollution and believe that the only way parents can improve the health of

The Urban Poor

SPOTLIGHT

Squatter settlements and shantytowns in developing countries usually lack clean water supplies, sewers, electricity, and roads. Often the land on which they are built is not suitable for human habitation because of air and water pollution and hazardous wastes from nearby factories, or because the land is especially subject to landslides, flooding, earthquakes, or volcanic eruptions.

An estimated 100 million people are homeless and sleep on the streets (Figure 1-10) or wherever they can. Half of all urban children under age 15 in developing countries live in conditions of extreme poverty, and about one-fifth of them are street children with little or no family support. In Cairo, Egypt, children of kindergarten age can be found digging through clods of ox dung, looking for undigested kernels of corn to eat.

Many cities do not provide squatter settlements and shanty-towns with adequate drinking water, sanitation facilities, electricity, food, health care, housing, schools, or jobs. Not only do these cities lack the needed money, but their officials fear that improving services will attract even more of the rural poor. Many city governments regularly bulldoze squatter shacks and send police to drive the illegal settlers out. The people then either move back in or develop another shantytown somewhere else.

Shantytowns and squatter settlements are also found in some developed countries. For example, at least 200,00 immigrants live in the *colonias* shantytowns in Texas along the southern Rio Grande. Living conditions there are as bad as in similar settlements in the cities of developing countries. In the United States, most inner cities have concentrations of poor people.

Despite joblessness, squalor, overcrowding, environmental hazards, and rampant disease, most squatter and slum residents are better off than the rural poor. With

better access to family-planning programs, they tend to have fewer children, who have better access to schools. Most residents are adaptable and resilient and have hope for a better future. Many squatter settlements provide a sense of community and a vital safety net of neighbors, friends, and relatives for the poor. A few squatter communities have organized to improve their living conditions.

Critical Thinking

1. What three things do you believe should be done to reduce the numbers and improve living conditions for the urban poor? For the rural poor?

2. Should squatters around cities of developing countries be given title to land they don't own? Explain. What are the alternatives?

Figure 6-19 The locations of Mexico, Brazil, Belize, and Costa Rica. Many other countries highlighted here are discussed in other chapters.

such children is to get them out of the city. Writer Carlos Fuentes has nicknamed this megacity "Makesicko City." Because of the air pollution, many foreign companies and governments give imported workers additional hazard pay for working in Mexico City.

The Mexican government is industrializing other parts of the country in an attempt to slow migration to Mexico City. In 1991 the government closed the city's huge state-run oil refinery and ordered many of the industrial plants in the basin to go elsewhere by 1994. Cars have been banned from a 50-block central zone. Taxis built before 1985 have been taken off the streets, and trucks can run only on liquefied petroleum gas (LPG). The government began phasing in unleaded gasoline in 1991, but it will be years before millions of older vehicles, which burn leaded gas, are eliminated. The government has also planted 25 million trees to help clean the air and has purchased some land to provide green space for the city.

The city's air and water pollution cause an estimated 100,000 premature deaths a year. These problems, already at crisis levels, will become even worse if the city grows as projected to 25.8 million people by the end of this century.

If you were in charge of Mexico City, what would you do?

How Urbanized Is the United States? In 1800 only 5% of Americans lived in cities. Since then, three major internal population shifts have taken place in the United States. As a result of the first shift, *migration to large central cities*, about 75% of Americans live in 350 *metropolitan areas* (cities and towns with at least 50,000 people). Nearly half of the country's population lives in consolidated metropolitan areas containing 1 million or more residents (Figure 6-20).

In the second shift, more people began *migrating from large central cities to suburbs and smaller cities*. Since 1970 this type of migration has followed new jobs to such areas. Today about 41% of the country's urban dwellers live in central cities, and 59% live in suburbs.

The third shift, which has been taking place for several decades, is *migration from the North and East to the South and West*. Since 1980 about 80% of the U.S. population increase has occurred in the South and West, particularly near the coasts. This shift is expected to continue.

What Are the Major Urban Problems in the United States? Here is some great news. Since 1920, many of the worst urban environmental problems in the United States (and other developed countries) have been significantly reduced. Most people have better working and housing conditions and air and water quality have improved. Better sanitation, public water supplies, and medical care have slashed death rates and the prevalence of sickness from malnutrition and transmittable diseases such as measles, diphtheria, typhoid fever, pneumonia, and tuberculosis. Furthermore, concentrating most of the population in urban areas has helped protect the country's biodiversity by reducing the destruction and degradation of wildlife habitat.

The biggest problems facing many cities in the United States (especially older ones) are deteriorating services, aging infrastructures (streets, schools, bridges, housing, sewers), budget crunches from lost tax revenues and rising costs as businesses and more

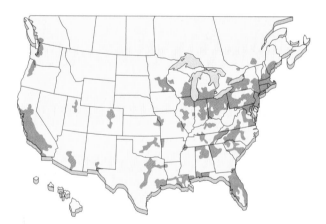

Figure 6-20 Major urban regions in the United States by the year 2000. Nearly half (48%) of Americans live in *consolidated metropolitan areas* with 1 million or more people. (Data from U.S. Census Bureau)

Q: What percentage of your body weight consists of water?

affluent people move out, and rising poverty in many central city areas. As a result, violence, drug traffic and abuse, crime, decay, and blight have increased in parts of central cities. Unemployment rates in some inner-city areas are typically 50% or higher.

6-7 URBAN RESOURCE AND ENVIRONMENTAL PROBLEMS

What Are the Environmental Pros and Cons of Urban Areas? Most of today's urban areas don't even come close to being self-sustaining; they survive only by importing food, water, energy, minerals, and other resources from farms, forests, mines, and watersheds. They also produce enormous quantities of wastes that can pollute air, water, and land within and outside their boundaries (Figure 6-21).

The 43% of the world's people currently living in urban areas occupy only about 5% of the planet's land area. However, supplying these urban dwellers with resources is a major reason that humans have disturbed 73% of the earth's land area, if uninhabitable ice and rock areas are excluded (Figure 1-5).

However, urbanization has some environmental benefits. Recycling is more economically feasible because of the large concentration of recyclable materials. The environmental pressures from population growth are reduced because birth rates in urban areas usually are three to four times lower than in rural areas. Cities provide better opportunities to educate people about environmental issues and to mobilize residents to deal with environmental problems. In addition, per capita expenditures on environmental protection are higher in urban areas.

Concentrating people in urban areas also helps preserve biodiversity by reducing the stress on wildlife habitats. This effect is important but is not as great as it first appears because of the large areas of the earth's land area that must be disturbed and degraded to provide urban dwellers with food, water, energy, minerals, and other resources (Figure 1-5). Furthermore, rural or wildlife areas downwind or downstream from urban areas are receptacles for much of the pollution produced in urban areas.

Some analysts call for seeking a more sustainable relationship between cities and the living world. To do this will require converting high-waste, unsustainable cities with a *linear metabolism* (based on an ever-increasing throughput of resources and output of wastes; Figure 3-15) to low-waste, sustainable cities with a *circular metabolism* (based on efficient use of resources, reuse, recycling, pollution prevention, and waste reduction; Figure 3-16).

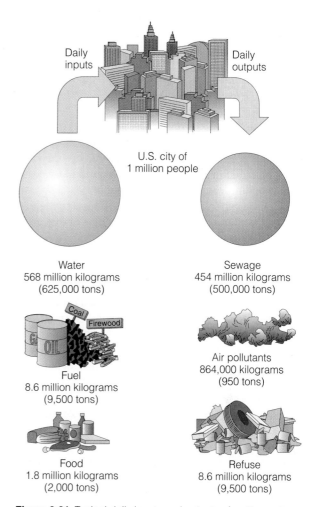

Figure 6-21 Typical daily inputs and outputs of matter and energy for a U.S. city of 1 million people.

Water
568 million kilograms
(625,000 tons)

Sewage
454 million kilograms
(500,000 tons)

Fuel
8.6 million kilograms
(9,500 tons)

Air pollutants
864,000 kilograms
(950 tons)

Food
1.8 million kilograms
(2,000 tons)

Refuse
8.6 million kilograms
(9,500 tons)

What Are the Major Resource and Environmental Problems of Urban Areas?

Major urban areas face a number of resource and environmental problems. *Most cities have few trees, shrubs, or other natural vegetation* that absorbs air pollutants, gives off oxygen, helps cool the air (as water transpires from their leaves), muffles noise, provides wildlife habitats, and gives aesthetic pleasure. As one observer remarked, "Most cities are places where they cut down the trees and then name the streets after them." This is a tragic loss. According to the American Forestry Association, one city tree provides over $57,000 worth of air conditioning, erosion and stormwater control, wildlife shelter, and air pollution control over a 50-year lifetime.

Most cities produce little of their own food. However, people can grow their food by planting community gardens in unused lots and by using window-box and balcony planters, and gardens or greenhouses built on the roofs of apartment buildings. Urban gardens currently provide 15% of the world's food supply, and this proportion could be increased. Cities can also encourage farmers' markets, which lower food prices by allowing farmers to sell directly to customers. This also helps prevent nearby farmland from being swallowed up by urban sprawl.

Cities are generally warmer, rainier, foggier, and cloudier than suburbs and nearby rural areas. The enormous amounts of heat generated by cars, factories, furnaces, lights, air conditioners, and people in cities create an **urban heat island** (Figure 6-22) surrounded by cooler suburban and rural areas. The dome of heat also traps pollutants, especially tiny solid particles (suspended particulate matter), creating a **dust dome** above urban areas. If wind speeds increase, the dust dome elongates downwind to form a **dust plume**, which can spread the city's pollutants for hundreds of kilometers. As urban areas grow and merge, individual heat islands also merge, which can affect the climate of a large area and keep polluted air from being diluted and cleansed.

Many cities have water supply and flooding problems. As cities grow and their water demands increase, expensive reservoirs and canals must be built and deeper wells drilled. The transfer of water to urban areas deprives rural and wild areas of surface water and sometimes depletes groundwater faster than it is replenished.

Covering land with buildings, asphalt, and concrete causes precipitation to run off quickly; it can overload sewers and storm drains, contributing to water pollution and flooding in cities and downstream areas. In car-dominated cities, stormwater running off roads and parking lots is contaminated with oil, road salt, and toxic liquids; in suburbs large amounts of fertilizers and pesticides run off lawns and golf courses. Large unbroken expanses of concrete or asphalt can also prevent precipitation from entering the soil to renew groundwater.

Many cities are built on floodplain areas subject to natural flooding. Floodplains are considered prime land for urbanization because they are flat, accessible, and near rivers. The poor often have little choice but to live on areas experiencing frequent flooding and landslides.

Many of the world's largest cities are in coastal areas. If an enhanced greenhouse effect increases the average atmospheric temperature as projected, a rise in average sea level of even a meter could flood many of these cities, perhaps sometime during the 21st century.

Urban residents are generally subjected to much higher concentrations of pollutants than are rural residents. According to the World Health Organization, more than 1.1 billion people—about one-fifth of humanity—live in urban areas where air pollution levels exceed healthful levels. Air pollution control in most cities in developing countries is lax because of lenient pollution laws, lack of enforcement, corrupt officials, inadequate testing equipment, and a shortage of funds.

The World Bank estimates that almost two-thirds of urban residents in developing countries don't have adequate sanitation facilities. In the developing world it is estimated that 90% of all sewage is discharged directly into rivers, lakes, and coastal waters without treatment of any kind. Air pollution is discussed in more detail in Chapters 9 and 10, water pollution in Chapter 11, and solid and hazardous wastes in Chapter 13.

Most urban dwellers also are subjected to excessive noise (Figure 6-23). Harmful effects from prolonged exposure to excessive noise include permanent hearing loss, high blood pressure (hypertension), muscle tension, migraine headaches, higher cholesterol levels, gastric ulcers, irritability, insomnia, and psychological disorders, including increased aggression.

Urban areas have beneficial and harmful effects on human health. Many aspects of urban life benefit human health, including better access to education, social services, and medical care. On the other hand, high-density city life increases the spread of infectious diseases (especially if adequate drinking water and sewage systems are not available), physical injuries (mostly from industrial and traffic accidents), and health problems caused by increased exposure to pollution and noise (Figure 6-23). The World Health Organization estimates that 600 million urban dwellers in developing countries live or work in life- and health-threatening environments. According to the World Bank, at least 220 million people in cities in developing countries don't have safe drinking water.

Another problem is the *loss of rural land, fertile soil, forests, and wildlife habitats as cities expand.* Once prime agricultural land or forestland is paved over or built on, it is lost for food production and habitat for most of its former wildlife. As land values near urban areas rise, taxes on nearby farmland increase so much that many farmers are forced to sell their land. They can

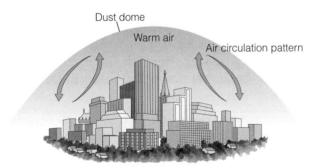

Figure 6-22 An urban heat island creates patterns of air circulation that create a dust dome over the city. Winds can elongate the dome toward downwind areas. A strong cold front can blow the dome away; this lowers local pollution levels but increases pollution in downwind areas.

Q: What percentage of the earth's enormous supply of water is available to us as usable fresh water?

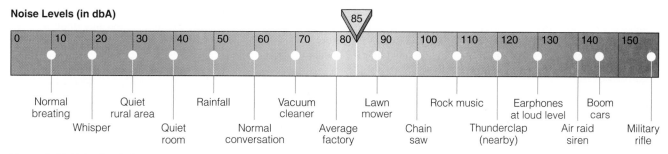

Permanent damage begins after 8-hour exposure

Noise Levels (in dbA)

| 0 | 10 | 20 | 30 | 40 | 50 | 60 | 70 | 80 | 85 | 90 | 100 | 110 | 120 | 130 | 140 | 150 |

Normal breating | Quiet rural area | Rainfall | Vacuum cleaner | Lawn mower | Rock music | Earphones at loud level | Boom cars

Whisper | Quiet room | Normal conversation | Average factory | Chain saw | Thunderclap (nearby) | Air raid siren | Military rifle

Figure 6-23 Noise levels (in decibel-A or dbA sound pressure units) of some common sounds. You are being exposed to a sound level high enough to cause permanent hearing damage if you need to raise your voice to be heard above the racket, if a noise causes your ears to ring, or if nearby speech seems muffled. Prolonged exposure to lower noise levels and occasional loud sounds may not damage your hearing but can greatly increase internal stress. There are five major ways to control noise: **(1)** Modify noisy activities and devices to produce less noise, **(2)** shield noisy devices or processes, **(3)** shield workers or other receivers from the noise, **(4)** move noisy operations or things away from people, and **(5)** use antinoise—a new technology that cancels out one noise with another.

make much more money selling to developers than raising corn or cows.

As a city expands, more energy is needed to transport food to its people; this in turn causes more pollution. In coastal areas, urban growth destroys or pollutes ecologically valuable wetlands.

6-8 TRANSPORTATION AND URBAN DEVELOPMENT

How Do Transportation Systems Affect Urban Development? If a city cannot spread outward, it must grow vertically—upward and downward (below ground)—so that it occupies a small land area with a high population density. Most people living in such compact cities walk, ride bicycles, or use energy-efficient mass transit. Residents often live in multistory apartment buildings; with few outside walls in many apartments, heating and cooling costs are reduced. Many European cities and urban areas such as Hong Kong and Tokyo are compact and tend to be more energy-efficient than the dispersed cities in the United States, Canada, and Australia, where ample land is often available for outward expansion.

A combination of cheap gasoline, plentiful land, and a network of highways produces dispersed, automobile-oriented cities with low population density, often called *urban sprawl*. Most people living in such urban areas live in single-family houses with unshared walls that lose and gain heat rapidly unless they are well insulated and airtight. Urban sprawl also gobbles up unspoiled natural habitats, paves over fertile farmland, and promotes heavy dependence on the automobile.

Who Has Most of the World's Motor Vehicles? There are two main types of ground transportation: *individual* (such as cars, motor scooters, bicycles, and walking) and *mass* (mostly buses and rail systems). Only about 8% of the world's population own cars, and only 10% can afford to. In developing countries as few as 1% of the people can afford a car; they travel mostly by foot, bicycle, or motor scooter. However, because of projected population growth and economic growth, between 1997 and 2020 the number of motor vehicles in the world is expected to increase from 496 million to almost 1 billion (with most of the increase in Asia, Latin America, and eastern Europe), greatly increasing congestion, pollution, land disruption, and use of energy and matter resources.

Despite having only 4.6% of the world's people, the United States has 35% of the world's cars and trucks. In the United States the car is used for 98% of all urban transportation and for 86% of travel to work (with 73% of Americans driving to work alone). Americans drive 3 billion kilometers (2 billion miles) each year—as far as the rest of the world combined. No wonder British author J. B. Priestley remarked, "In America, the cars have become the people." Such "automania" is spreading to developing countries. Despite their many advantages, there are serious drawbacks to relying on motor vehicles as the major form of transportation (Pro/Con, p. 184).

Are Motor Scooters the Answer? A growing number of people in developing countries who cannot afford cars are using motor scooters, which produce more air pollution than cars. Most burn a mixture of oil and kerosene in small, inefficient, and noisy engines

The automobile provides convenience and unprecedented mobility. To many people, cars are also symbols of power, sex, excitement, social status, and success. Moreover, much of the world's economy is built on producing motor vehicles and supplying roads, services, and repairs for them. In the United States, $1 of every $4 spent and one of every six nonfarm jobs is connected to the automobile.

Despite their economic and personal benefits, motor vehicles have many destructive effects on people and the environment. Since 1885, when Karl Benz built the first automobile, almost 18 million people have been killed in motor vehicle accidents. According to the World Health Organization, this global death toll increases by an estimated 885,000 people per year (an average of 2,400 deaths per day), and annually about 10 million people are injured or permanently disabled in motor vehicle accidents.

In the United States alone, 16 million motor vehicle accidents (up from 7 million in 1970) kill about 42,000 people each year and injure almost 5 million people, at least 300,000 of them severely. *More Americans have been killed by cars than in all the country's wars.* In 1994 the costs from motor vehicle accidents in the United States exceeded $150.5 billion—about $574 per citizen.

Motor vehicles are also the largest source of air pollution (including 22% of global CO_2 emissions), laying a haze of smog over the world's cities. In the United States, motor vehicles produce at least 50% of the air pollution, even though emission standards are as strict as any in the world. Gains in fuel efficiency and emission reductions have been largely offset by the

increase in cars and a more than doubling of the distance Americans traveled by car between 1970 and 1997. Two-thirds of the oil used in the United States and half of the world's total oil consumption are devoted to transportation.

Automobiles and freeways have caused social fragmentation by disrupting neighborhoods, choking cities with traffic, and increasing congestion, noise, and stress. By making long commutes and shopping trips possible, automobiles and highways have helped create urban sprawl and have reduced use of more efficient forms of transportation.

Worldwide, at least a third of urban land is devoted to roads, parking lots, gasoline stations, and other automobile-related uses. In the United States, more land is now devoted to cars than to housing. Half the land in an average U.S. city is used for cars, prompting urban expert Lewis Mumford to suggest that the U.S. national flower should be the concrete cloverleaf.

Car-culture cities have not delivered the promised convenience and speed of travel. In 1907 the average speed of horse-drawn vehicles through the borough of Manhattan was 18.5 kilometers (11.5 miles) per hour; today cars and trucks creep along Manhattan streets at an average speed of 5 kilometers (3 miles) per hour. If current trends continue, U.S. motorists will spend an average of two years of their lifetimes in traffic jams, imprisoned in metal boxes that were supposed to provide speed, freedom, and mobility. The U.S. economy loses at least $100 billion a year because of time lost in traffic delays. Even if the money is available, building more roads is not the answer because, as economist Robert Samuelson put it, "cars expand to fill available concrete."

Thus, the major hidden costs of driving include deaths and injuries

from accidents, the value of time wasted in traffic jams, air pollution, increased threats from global warming, the drop in property values near roads because of noise and congestion, and the cost of maintaining a formidable military presence in the Middle East to ensure access to oil. Two recent estimates by economists put these hidden costs of driving in the United States at roughly $300–350 billion per year—about 5% of the country's GDP.

Environmentalists and a number of economists suggest that one way to break this increasingly destructive cycle of positive feedback is to make drivers pay directly for most of the true costs of automobile use. This could be done by including the current harmful hidden costs of driving as a tax on gasoline and by phasing out government subsidies for motor vehicle owners—a user-pays approach.

Deciding to include the hidden costs in the market prices of cars, trucks, and gasoline up front makes economic and environmental sense. However, this approach faces massive political opposition from the general public (mostly because they are unaware of the huge hidden costs they are already paying) and from the powerful transportation-related industries. In addition, such tools for encouraging more use of mass transportation will not work unless fast, efficient, reliable, and affordable forms of such transportation alternatives are available.

Critical Thinking

If you own a car (or hope to own one), what conditions (if any) would encourage you to rely less on the automobile and instead encourage you to travel to school or work by bicycle or motor scooter, on foot, by mass transit, or by a car or van pool?

Q: What is the largest global use of water withdrawn from surface or groundwater sources?

that emit clouds of air pollutants. Because they are cheap, their numbers are increasing three times faster than cars and trucks in developing countries.

One solution is to replace these smog machines with zero pollution (except at power plants supplying the electricity for recharging the batteries) and quiet electric scooters. Recently Taiwan and Indonesia have introduced air pollution control legislation that may spur the use of electric scooters. The major weakness of electric cars is their limited range. However, this is less of a problem with much lighter electric scooters.

Are Riding Bicycles and Walking Alternatives to the Car? Globally, bicycles outsell cars by almost 3 to 1 because most people can afford a bicycle whereas fewer than 10% can afford a car. Besides being inexpensive to buy and maintain, bicycles produce no pollution, are rarely a serious danger to pedestrians or cyclists, take few resources to make, and are the most energy-efficient form of transportation (including walking).

In 1996 a California firm developed a simple $15 bicycle that eliminates the chain by attaching the pedals to the front wheel. This inexpensive design could greatly increase bicycle use in developing countries.

In urban traffic, cars and bicycles move at about the same average speed. Using separate bike paths or lanes running along roads, cyclists can make most trips shorter than 8 kilometers (5 miles) faster than drivers.

In China, 50–80% of urban trips are made by bicycle and the government gives subsidies to those who bicycle to work. Many cities in western Europe and Japan have taken back the streets for pedestrians, cyclists, and children by banning cars or slowing motor traffic in residential and shopping areas. In the Netherlands (with more bicycle paths than any other country), bicycle travel makes up 30% of all urban trips, and in Japan 15% of all commuters ride bicycles to work or to commuter-rail stations.

For longer trips, secure bike parking spaces can be provided at mass transit stations, and buses and trains can be equipped to carry bicycles. Such *bike-and-ride* systems are widely used in Japan, Germany, the Netherlands, and Denmark. In Seattle, Washington, all city buses are equipped with bicycle racks.

Only about 2% of commuters in the United States bicycle to work (the "no-pollute commute"), even though half of all U.S. commutes are under 8 kilometers (5 miles). However, according to recent polls, 20% of Americans say they would bicycle to work if safe bike lanes were available and if their employers provided secure bike storage and showers at work.

However, as developing countries experience economic growth, some governments are discouraging bicycle use, viewing it as a sign of backwardness. In 1993, for example, Chinese officials banned bicycles from Shanghai's main street and prestigious shopping area so they wouldn't be seen by tourists and wealthy Chinese shoppers. Similarly, officials in Jakarta, Indonesia, have confiscated thousands of cycle rickshaws to project a more modern image for visitors.

In addition, unless efficient and affordable forms of mass transportation are already in place, many urban residents in developing countries abandon bikes and walking as soon as they can afford to buy motor scooters or cars, as is happening in China, India, and Indonesia. Even in the Netherlands, which has a long-tradition of bicycle use and ample bicycle paths, more and more people are using cars instead of bicycles.

What Are the Pros and Cons of Mass Transit? In the United States mass transit accounts for only 8% of all passenger travel, compared with 15% in Germany and 47% in Japan. Only 20% of the U.S. federal gasoline tax goes to mass transit; the remaining 80% goes to highways. This encourages states and cities to invest in highways instead of mass transit.

Rapid-rail, suburban train, and trolley systems can transport large numbers of people at high speed. They are also more energy-efficient, produce less air pollution, cause fewer injuries and deaths, and take up less land than motor vehicles. However, they are efficient and cost effective only where many people live along a narrow corridor and can easily reach properly spaced stations (as in Hong Kong).

In western Europe and Japan, a new generation of streamlined, comfortable, and low-polluting high-speed rail (HSR) lines is being used for medium-distance travel between cities. These *bullet* or super-trains travel on new or upgraded tracks at speeds up to 330 kilometers (200 miles) per hour. They are ideal for trips of approximately 200–1,000 kilometers (120–620 miles). For every kilometer of travel, such trains consume only one-third as much energy per rider as a commercial airplane and one-sixth as much as a car carrying only one driver. High-speed train systems are expensive to run and maintain, however, and they must operate along heavily used transportation routes to be profitable.

The United States has lagged behind in the development of a high-speed train network. Such a system could be developed at a reasonable cost by upgrading existing intercity tracks and train systems on key routes (Figure 6-24), rather than building new and expensive rail rights-of-way. Criticism that such a system would cost too much in government subsidies ignores the fact that motor vehicle transportation receives subsidies of $300–600 billion per year in the United States. Some

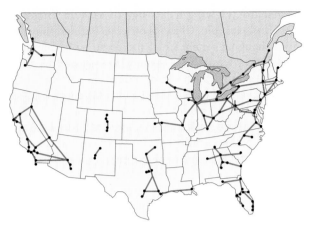

Figure 6-24 Potential routes for high-speed bullet trains in the United States and parts of Canada. Such a system would allow rapid, comfortable, safe, and affordable travel between major cities in a region. It would greatly reduce dependence on cars, buses, and airplanes for trips among these urban areas. (Data from High Speed Rail Association)

analysts believe that phasing out some of these subsidies and applying them to developing a national rail system might be a far more efficient use of increasingly limited government funds.

Bus systems are more flexible than rail systems; they can run throughout sprawling cities and be rerouted overnight if transportation patterns change. They also require less capital and have lower operating costs than heavy-rail systems. However, because they must offer low fares to attract riders, bus systems often cost more to operate than they bring in. Furthermore, unless they operate in separate express lanes, buses often get caught in traffic. Because buses are cost-effective only when full, they are sometimes supplemented by car pools, van pools, and jitneys (small vans or minibuses traveling along regular and stop-on-demand routes). Curitiba, Brazil, has developed one of the world's best bus systems and has led the way in becoming a more sustainable city (Solutions, right).

Is It Feasible to Reduce Automobile Use?

The major hidden costs of driving include deaths and injuries from accidents, the value of time wasted in traffic jams, air pollution, increased threats from global warming, the drop in property values near roads because of noise and congestion, and the cost of maintaining a formidable military presence in the Middle East to ensure access to oil. Two recent estimates by economists put these hidden costs of driving in the United States at roughly $300–350 billion per year—about 5% of the country's GDP.

Environmentalists and a number of economists suggest that one way to break this increasingly de-

structive cycle of positive feedback is to make drivers pay directly for most of the true costs of automobile use. This could be done by including the current harmful hidden costs of driving as a tax on gasoline and by phasing out government subsides for motor vehicle owners—a *user-pays* approach.

Currently, U.S. drivers pay low gasoline taxes (compared to most other developed countries) that are used to build roads and other transportation infrastructures, but these taxes cover only 60–69% of the total costs. The remainder is subsidized by federal, state, and local governments. Governments also subsidize drivers by allowing tax write-offs for business-related car mileage or other car-related costs. U.S. employers that provide parking facilities for their workers can deduct these expenses from their taxes. This gives U.S. auto commuters a tax-free benefit worth $2,400–4,800 per year.

Making heavy trucks pay for the road damage they cause would shift more freight to energy-efficient rail systems. It's estimated that heavy trucks cause 95% of all damage to U.S. highways, with one heavy truck causing as much highway wear and tear as 9,600 cars. Current U.S. government subsidies for trucks not only drain public funds but also give trucking an unfair economic advantage over more efficient and less damaging rail freight.

According to a study by the World Resources Institute, federal, state, and local government automobile subsidies in the United States amount to $300–600 billion a year (depending on the costs included), an average subsidy of $1,600–3,200 per vehicle. Taxpayers (drivers and nondrivers alike) foot this bill mostly unknowingly. If drivers had to pay these hidden costs directly in the form of a gasoline tax, the tax on each gallon would be $5–6 and a corrective negative feedback loop would take effect. Pollution emission fees, toll charges on roads (especially during peak traffic times), and higher parking fees are also ways to have motorists pay for the environmental and social costs of driving—a user-pays approach based on full-cost pricing.

In addition, such tools for encouraging more use of mass transportation will not work unless fast, efficient, reliable, and affordable mass transportation is available. Furthermore, most people who can afford cars are virtually addicted to them and most people who can't afford them hope to buy one someday.

6-9 SOME SOLUTIONS TO URBAN PROBLEMS

What Is Conventional Land-Use Planning?

Most urban areas and some rural areas use some form

SOLUTIONS

One of the world's showcase ecocities is Curitiba, Brazil, with a population of 2.1 million. It is a prosperous, booming, clean, and mostly economically self-sufficient city.

Trees are everywhere in Curitiba because city officials have given neighborhoods over 1.5 million trees to plant and care for. No tree in the city can be cut down without a permit; for every tree cut, two must be planted.

After persistent flooding in the 1950s and 1960s, city officials put certain flood-prone areas off-limits for building, passed laws protecting natural drainage systems, built artificial lakes to contain flood-waters, and converted many river-banks and floodplains into parks. This design-with-nature strategy has made costly flooding a thing of the past and has greatly increased the amount of open or green space per capita (which increased 100-fold during the city's period of rapid population growth between 1950 and 1997).

The air is clean because the city is not built around the car. There are 145 kilometers (90 miles) of bike paths, and more are being built. With the support of shopkeepers, many streets in the downtown shopping district have been converted to pedestrian zones in which no cars are allowed. Abandoned factories and buildings have been recycled into sports and other recreational facilities.

A key feature of Curitiba's success is its integration of transportation and land-use planning. City officials decided to develop a sophisticated bus system instead of a more expensive and less flexible subway or light-rail system.

The key concept was to channel the city's development along five major transportation and high-density corridors that extend outward from the central city like spokes on a bicycle wheel. Each corridor has exclusive lanes for express buses.

Curitiba probably has the world's best bus system. Each day a network of clean and efficient buses carries over 1.5 million passengers—75% of the city's commuters and shoppers—at a low cost (20–40¢ per ride with unlimited transfers) on express bus lanes. Only high-rise apartment buildings are allowed near major bus routes, and each building must devote the bottom two floors to stores, which reduces the need for residents to travel. As a result, Curitiba has one of Brazil's lowest outdoor air pollution rates.

The city recycles roughly 70% of its paper (equivalent to saving nearly 1,200 trees per day) and 60% of its metal, glass, and plastic, which is sorted by households for collection. Recovered materials are sold mostly to the city's more than 340 major industries.

The city bought a plot of land downwind of downtown as an industrial park. The city put in streets, services, housing, and schools, ran a special worker's bus line to the area, and enacted stiff air and water pollution control laws. This has attracted clean national and foreign corporations.

To equip the poor with basic technical training skills needed for jobs, the city set up old buses as roving technical training schools; courses cost the equivalent of two bus tokens. Each bus gives courses for three months in a particular area and then moves to another area.

The poor can swap sorted trash for food or bus tokens and receive free medical, dental, and child care; there are also 40 feeding centers for street children. As a result, the infant mortality rate has fallen by more than 60% since 1977.

All of these things have been accomplished despite Curitiba's enormous population growth from 300,000 in 1950 to 2.2 million in 1997, as rural poor people have flocked to the city.

Curitiba has slums, shantytowns, and most of the problems of other cities. But most of its citizens have a sense of vision, solidarity, pride, and hope, and they are committed to making their city even better.

One secret of Curitiba's success is the willingness of citizens to work together to create a better future. Another is city officials who genuinely care about providing a high quality of life for *all* of the city's inhabitants. The entire program is the brainchild of architect and former college teacher Jaime Lerner, an energetic and charismatic leader, who has served as the city's major three times since the 1970s.

City leaders and citizens have worked together to make Curitiba a living laboratory for sustainable living by favoring public transportation over the private automobile, by planning carefully, and by working with the environment instead of against it.

Critical Thinking

1. Why do you think that Curitiba has been so successful in its efforts to become an ecocity, compared to the generally unsustainable cities found in most developed and developing countries?

2. What is the city or area where you live doing to make itself more ecologically and economically sustainable?

of **land-use planning** to determine the best present and future use of each parcel of land in the area. Much land-use planning is based on the assumption that substantial future population growth and economic development should be encouraged, regardless of the environmental and other consequences. Typically this leads to uncontrolled or poorly controlled urban growth and sprawl.

A major reason for this often destructive process is that in the United States 90% of the revenue that local governments use to provide schools, police and fire protection, public water and sewer systems, and other public services comes from *property taxes* levied on all buildings and property based on their economic value.

However, the costs of providing more basic services to accommodate economic and population growth often exceed the tax revenues accompanying increased property values. Because local governments can rarely raise property taxes enough to meet expanding needs, they often try to raise more money by promoting economic growth. Typically the long-term result is a destructive positive feedback loop of economic growth leading to environmental degradation. If taxes get too high, businesses and residents move away, decreasing the tax base and reducing tax revenues. This causes further environmental and social decay as governments are forced to cut the quantity and quality of services or to raise the tax rate again (which drives more people away and further worsens the situation).

What Is Ecological Land-Use Planning? Environmentalists challenge the all-growth-is-good dogma that is the basis of most land-use planning. They urge communities to use comprehensive, regional **ecological land-use planning**, in which additional variables are integrated into a model designed to anticipate a region's present and future needs and problems. It is a complex process that takes into account geological, ecological, economic, health, and social factors (Solutions, right).

Ecological land-use planning sounds good on paper, but it is not widely used for several reasons. First, local officials seeking reelection every few years usually focus on short-term rather than long-term problems and can often be influenced by economically powerful developers. Second, officials (and often the majority of citizens) are often unwilling to pay for costly ecological land-use planning and implementation, even though a well-designed plan can prevent or ease many urban problems and save money in the long run. Finally, it's difficult to get municipalities within a region to cooperate in planning efforts. As a result, an ecologically sound development plan in one

area may be undermined by unsound development in nearby areas.

In developing countries, most cities don't have the information or funding to carry out ecological land-use planning. Urban maps are often 20–30 years old and lack descriptions of large areas of cities, especially those growing rapidly because of squatter settlements.

How Can Land Use Be Controlled? Once a plan is developed, governments control the uses of various parcels of land by legal and economic methods. The most widely used approach is **zoning**, in which various parcels of land are designated for certain uses. Zoning can be used to control growth and to protect areas from certain types of development. However, zoning can be influenced or modified by developers. Overly strict zoning can discourage innovative approaches to solving urban problems.

Currently, most zoning sets up separate areas for residential, commercial, and industrial activities (which made sense in the smokestack days when factories polluted cities and few suburbs existed). The resulting separation of homes, jobs, and shops by long distances requires increased car and energy use and promotes urban sprawl. Changing zoning laws to encourage such reintegration of homes, workplaces, and shopping areas would reduce urban sprawl, energy waste, and loss of community—an example of instituting a corrective or negative feedback process.

Local governments can also control the rate of development by limiting the number of building permits, sewer hookups, roads, and other services. They can also require an environmental impact analysis for proposed roads and development projects. Developers can be required to pay a fee that includes the full cost of additional services such as roads and water and sewer lines for all development projects.

Land can be taxed on the basis of its *actual* use as agricultural land or forestland, rather than on the basis of its most profitable *potential* use. This would keep farmers and other landowners from being forced to sell land to pay their tax bills, but would decrease tax revenues for cities. Tax breaks can also be given to landowners who agree in legally binding conservation easements to use their land only for specified purposes, such as agriculture, wildlife habitat, or nondestructive forms of recreation.

Development rights that restrict land use can be purchased and land trusts can be used to buy and protect ecologically valuable land. Such purchases can be made by private groups such as the Nature Conservancy or the Audubon Society, by local nonprofit, tax-exempt charitable organizations, and by government agencies.

Q: Worldwide, how much of the water withdrawn for irrigation is wasted?

Controlling where highways and streets are built may be far more influential than land-use planning in determining where development will take place. For example, concentrating high-density development along bus lines or other forms of mass transportation can reduce car use and urban sprawl, as has happened Portland, Oregon; Toronto, Canada; Vienna, Austria; Stockholm, Sweden; Copenhagen, Denmark; and Curitiba, Brazil (Solutions, p. 187).

How Can We Make Cities More Sustainable?

An important goal in coming decades should be to make urban areas more self-reliant, sustainable, and enjoyable places to live. In a sustainable and ecologically healthy city, called an *ecocity* or *green city*, people walk or cycle for most short trips; they walk or bike to bus, metro, or trolley stops for longer urban trips. Rapid-rail transport between cities would replace many long drives and medium-distance airplane flights.

In such cities emphasis is placed on pollution prevention, recycling and reuse (at least 60% of all municipal wastes), use of renewable energy resources, encouraging rather than assaulting biodiversity, and use composting to help create rather than destroy soil. They are based on good ecological design (Guest Essay, pp. 191–192).

Trees and plants adapted to the local climate and soils are planted throughout an ecocity to provide shade and beauty, to reduce pollution and noise, and to supply habitats for wildlife. Abandoned lots and polluted creeks are cleaned up and restored. Nearby forests, grasslands, wetlands, and farms are preserved instead of being devoured by urban sprawl. Much of an ecocity's food comes from nearby organic farms, solar greenhouses, community gardens, and small gardens on rooftops, in yards, and in window boxes.

An ecocity is people oriented, not car oriented. Its residents are able to walk or bike to most places, including work, and to take low-polluting mass transit.

Ways to make existing and new suburbs more sustainable and livable include **(1)** giving up big lawns, **(2)** building houses and apartments in small dense clusters so that more community open space is available, **(3)** developing a town center (a plaza, square, or green) that is a focus of civic life and community cohesiveness, **(4)** planting lots of new trees rather than cutting down existing ones, and **(5)** discouraging excessive dependence on the automobile and encouraging walking and bicycling.

In a few decades people may be wondering why they allowed cars to dominate their lives and degrade the environment for so long. They may take seriously the advice Lewis Mumford gave Americans over three decades ago: "Forget the damned motor car and build

Ecological Land-Use Planning

SOLUTIONS

Six basic steps are involved in ecological land-use planning:

1. *Make an environmental and social inventory.* Experts survey geological factors (soil type, floodplains, water availability), ecological factors (wildlife habitats, stream quality, pollution), economic factors (housing, transportation, industrial development), and health and social factors (disease, crime rates, poverty). A top priority is to identify and protect areas that are critical for preserving water quality, supplying drinking water, and reducing erosion and that are most likely to suffer from toxic wastes and natural hazards such as flooding.

2. *Identify and prioritize goals.* What are the primary goals? To encourage or discourage further economic development (at least some types) and population growth? To preserve prime cropland, forests, and wetlands from development? To reduce soil erosion?

3. *Develop individual and composite maps.* Data for each factor surveyed in the environmental and social inventory are plotted on separate transparent plastic maps. The transparencies are then superimposed or combined by computer into three composite maps, one each for geological, ecological, and socioeconomic factors.

4. *Develop a master composite.* The three composite maps are combined to form a master composite, which shows how the variables interact and indicates the suitability of various areas for different types of land use.

5. *Develop a master plan.* The master composite (or a series of alternative master composites) is evaluated by experts, public officials, and the general public, and a final master plan is drawn up and approved.

6. *Implement the master plan.* The plan is set in motion, monitored, updated, and revised as needed by the appropriate government, legal, environmental, and social agencies.

Critical Thinking

Why do you think so few areas use ecological land-use planning? Would you be willing to pay slightly higher local taxes to support such planning where you live? Explain.

A: 70–80%, which evaporates or seeps into the ground before reaching crops

cities for lovers and friends." Examples of cities that have attempted to become more ecologically sustainable include Curitiba, Brazil (Solutions, p. 187), and Davis, California.

Case Study: Davis, California   The ecocity is not a futuristic dream. The citizens and elected officials of Davis, California—a city of about 40,000 people about 130 kilometers (80 miles) northeast of San Francisco—committed themselves in the early 1970s to making their city ecologically sustainable.

City building codes encourage the use of solar energy for water and space heating. All new homes must meet high standards of energy efficiency, and when an existing home changes hands the buyer must bring it up to the energy conservation standards for new homes. In Davis's Village Homes development, America's first solar neighborhood, houses are heated by solar energy. They face into a common open space reserved for people and bicycles; cars are restricted to streets, which are located only on the periphery of the development. The neighborhood also has commonly shared orchards, vineyards, gardens, playgrounds, playing fields, and a solar-heated community center used for day care, meetings, and social gatherings.

Since 1975 Davis has cut its use of energy for heating and cooling in half. It has a solar power plant, and some of the electricity it produces is sold to the regional utility company. Eventually the city plans to generate all of its own electricity.

The city discourages the use of automobiles and encourages the use of bicycles by closing some streets to automobiles, building bike lanes on major streets, and building bicycle paths. Any new housing tract must have a separate bike lane, and some city employees are given bikes. As a result, 28,000 bicycles account for 40% of all in-city transportation, and less land is needed for parking spaces. This heavy dependence on the bicycle is aided by the city's warm climate and flat terrain.

Davis limits the type and rate of its growth, and it maintains a mix of homes for people with low, medium, and high incomes. Development of the fertile farmland surrounding the city for residential or commercial
use is restricted. Davis also limits the size of shopping centers to encourage smaller neighborhood shopping centers, each easily reached by foot or bicycle.

The test of the quality of life in an advanced economic society is now largely in the quality of urban life. Romance may still belong to the countryside—but the present reality of life abides in the city.

JOHN KENNETH GALBRAITH

CRITICAL THINKING

1. Why are falling birth rates not necessarily a reliable indicator of future population growth trends?

2. Why is it rational for a poor couple in India to have five or six children? What changes might induce such a couple to consider their behavior irrational?

3. Do you believe that the population of your own country is too high? Explain. What about the population of the area in which you live?

4. Evaluate the claims made by those opposing the reduction of births and those promoting a reduction in births, as discussed on pp. 169–171. Which position do you support, and why?

5. Explain why you agree or disagree with each of the following proposals:
   a. The number of legal immigrants and refugees allowed into the United States each year should be sharply reduced.
   b. Illegal immigration into the United States should be sharply decreased. If you agree, how would you go about achieving this?
   c. Families in the United States should be given financial incentives to have more children to prevent population decline.
   d. The United States should adopt an official policy to stabilize its population and reduce unnecessary resource waste and consumption as rapidly as possible.
   e. Everyone should have the right to have as many children as they want.

6. Some people have proposed that the earth could solve its population problem by shipping people off to space colonies, each containing about 10,000 people. Assuming that such large-scale, self-sustaining space stations could be built (which can't be done with existing technology), how many people would have to be shipped off each day to provide living spaces for the approximately 86 million people being added to the population each year? Current space shuttles can handle about 6 to 8 passengers. Assuming that this capacity can be increased to 100 passengers per shuttle, how many shuttles would have to be launched per day to take care of the 86 million people being added each year? According to your calculations, determine whether this proposal is a logical solution to the earth's population problem.

7. Why has China been more successful than India in reducing its rate of population growth? Do you agree with China's current population control policies? Explain. What alternatives, if any, would you suggest?

8. Congratulations—you have just been put in charge of the world. List the five most important features of your population policy.

9. Do you believe that the United States (or the country where you live) should develop a comprehensive and integrated mass transit system over the next 20 years,

Q: What percentage of the world's population lives in areas that have prolonged droughts?

The Ecological Design Arts

David W. Orr

GUEST ESSAY

David W. Orr is professor of environmental studies at Oberlin College and one of the nation's most respected environmental educators. He is author of numerous environmental articles and three books, including Ecological Literacy *and* Earth in Mind. *He is education editor for* Conservation Biology *and a member of the editorial advisory board of* Orion Nature Quarterly.

If *Homo sapiens sapiens* entered its industrial civilization in an intergalactic design competition, it would be tossed out in the qualifying round. It doesn't fit. It won't last. The scale is wrong. And even its defenders admit that it's not very pretty. The most glaring design failures of industrial/technologically driven societies are the loss of diversity of all kinds, impending climate change, pollution, and soil erosion.

Industrial civilization, of course, wasn't designed at all; it was mostly imposed by single-minded individuals, armed with one doctrine of human progress or another, each requiring a homogenization of nature and society. These individuals for the most part had no knowledge of "ecological design arts"—the set of perceptual and analytic abilities, ecological wisdom, and practical wherewithal needed to make things that fit into a world of microbes, plants, animals, and energy laws.

Good ecological design incorporates understanding about how nature works into the ways we design, build, and live. It is required in our designs of farms, houses, neighborhoods, cities, transportation systems, technologies, economies, energy policies, and just about anything that directly or indirectly requires energy or materials or governs their use.

When human artifacts and systems are well designed, they are in harmony with the ecological patterns in which they are embedded. When poorly designed, they undermine those larger patterns, creating pollution, higher costs, and social stress. Bad design is not simply an engineering problem, although better engineering would often help. Its roots go deeper.

Good ecological design has certain common characteristics, including correct scale, simplicity, efficient use of resources, a close fit between means and ends, durability, redundancy, and resilience. These characteristics are often place-specific, or, in John Todd's words, "elegant solutions predicated on the uniqueness of place." Good design also solves more than one problem at a time and promotes human competence, efficient and frugal use of resources, and sound regional economies. Where good design becomes part of the social fabric at all levels, unanticipated positive side effects multiply. When people fail to design with ecological competence, unwanted side effects and disasters multiply.

The pollution, violence, social decay, and waste all around us indicate that we have designed things badly, for, I think, three primary reasons. First, as long as land and energy were cheap and the world was relatively empty, we did not need to master the discipline of good design. The result was sprawling cities, wasteful economies, waste dumped into the environment, bigger and less efficient automobiles and buildings, and conversion of entire forests into junk mail and Kleenex—all in the name of economic growth and convenience.

Second, design intelligence fails when greed, narrow self-interest, and individualism take over. Good design is a cooperative community process requiring people who share common values and goals that bring them together and hold them together. American cities, with their extremes of poverty and opulence, are products of people who believe they have little in common with one another. Greed, suspicion, and fear undermine good community and good design alike.

Third, poor design results from poorly equipped minds. Good design can only be done by people who understand harmony, patterns, and systems. Industrial cleverness, on the contrary, is mostly evident in the minutiae of things, not in their totality or in their overall harmony. Good design requires a breadth of view that causes people to ask how human artifacts and purposes fit within a particular culture and place. It also requires ecological intelligence, by which I mean an intimate familiarity with how nature works.

An example of good ecological design is found in John Todd's "living machines," which are carefully orchestrated ensembles of plants, aquatic animals, technology, solar energy, and high-tech materials to purify wastewater, but without the expense, energy use, and chemical hazards of conventional sewage treatment technology. Todd's "living machines" resemble greenhouses filled with plants and aquatic animals [Figure 11-30]. Wastewater enters at one end and purified water leaves at the other. In between, an ensemble of organisms driven by sunlight use and remove nutrients, break down toxics, and incorporate heavy metals in plant tissues.

Ecological design standards also apply to the making of public policy. For example, the Clean Air Act of 1970 required car manufacturers to install catalytic converters to remove air pollutants. Two decades later, emissions per vehicle are down substantially, but because more cars are on the road, air quality is about the same—an example of inadequate ecological design. A sounder design approach to transportation would create better access among housing, schools, jobs, stores, and recreation areas; build better public transit systems; restore and improve railroads; and create bike trails and walkways.

An education in the ecological design arts would foster the ability to see things in their ecological context, in-

(continued)

tegrating firsthand experience and practical competence with theoretical knowledge about how nature works. It would aim to equip people to build households, institutions, farms, communities, corporations, and economies that **(1)** do not emit carbon dioxide or other heat-trapping gases, **(2)** operate on renewable energy, **(3)** preserve biological diversity, **(4)** recycle material and organic wastes, and **(5)** promote sustainable local and regional economies.

The outline of a curriculum in ecological design arts can be found in recent work in ecological restoration, ecological engineering, solar design, landscape architecture, sustainable agriculture, sustainable forestry, energy efficiency, ecological economics, and least-cost end-use analysis. A program in ecological design would weave these and simi-lar elements together around actual design objectives that aim to make students smarter about systems and about how specific things and processes fit in their ecological context. With such an education we can develop the habits of mind, analytical skills, and practical competence needed to help sustain the earth for us and other species.

Critical Thinking

1. Does your school offer courses or a curriculum in ecological design? If not, suggest some reasons why it does not.

2. Use the principles of good ecological design to evaluate how well your campus is designed. Suggest ways to improve its design.

---

including building an efficient rail network for travel within and among its major cities? How would you pay for such a system?

*10. Assume that your entire class (or manageable groups of your class) is charged with coming up with a plan for halving the world's rate of population growth within the next 20 years. Develop a detailed plan that would achieve this goal, including any differences between policies in developing countries and developed countries. Justify each part of your plan. Predict what problems you might face in implementing the plan, and devise strategies for dealing with these problems.

*11. As a class project, evaluate land use and land-use planning by your school, draw up an improved plan based on ecological principles, and submit the plan to school officials.

*12. Make a concept map of this chapter's major ideas, using the section heads and subheads and the key terms (in boldface type). Look at the inside back cover and in Appendix 4 for information about concept maps.

SUGGESTED READINGS

See Internet Sources and Further Readings at the back of the book, and InfoTrac.

7 ENVIRONMENTAL ECONOMICS AND POLITICS

To Grow or Not to Grow: Is That the Question?

Most economists, investors, and business leaders argue that we must have unlimited economic growth to create jobs, satisfy people's economic needs and desires, clean up the environment, and help reduce poverty.

These analysts see the earth as an essentially unlimited source of raw materials and the environment as an infinite sink for wastes; they believe technological innovation can overcome any resource or environmental limits. To people with this view, environmentalists put endangered species above endangered people; they threaten jobs and are against the economic growth needed for human survival and gains in the quality of life.

On the other hand, environmentalists and a small but growing number of economists and business leaders argue that economic systems depend on resources and services provided by the sun and by the earth's basic components and processes (Figures 7-1 and 1-2) and that a healthy economy ultimately depends on a healthy ecosphere. They believe that if we continue to support economic growth by consuming earth capital instead of living off sustainable earth income, such forms of growth will threaten business and impair the planet's life-support systems for humans and many other species.

If these beliefs are correct, then over the next few decades we must replace the economics of unlimited growth with the economics of sustainability. These modified economic systems would unite ecology and commerce by giving rewards (subsidies) to earth-sustaining businesses and activities and by penalizing (taxing and regulating) earth-degrading activities, with the overall goal of providing both economic and environmental security.

The question is not so much, "To grow or not to grow?" but rather, "How can we grow without plundering the planet?" or "How can we grow as if the earth matters?"

Figure 7-1 Earth, air, fire, water, and life. Most environmentalists and a growing number of economists believe that these basic components of earth capital, which sustain us and other species and all economies, are undervalued in the economic marketplace. (Greg Vaughn/Tom Stack & Associates)

When it is asked how much it will cost to protect the environment, one more question should be asked: How much will it cost our civilization if we do not?

GAYLORD NELSON

In this chapter, we seek answers to the following questions:

- What are economic goods and resources, and how are they provided?

- How should we measure economic growth?

- How can economics help improve environmental quality?

- How can we sharply reduce poverty?

- Should we gradually shift to an earth-sustaining economy, and if so how might this be done?

- How is environmental policy made in the United States?

- How can people affect environmental policy?

- What are the major goals and tactics of environmental and anti-environmental groups?

- How might global environmental policy be improved?

7-1 ECONOMIC SYSTEMS AND ENVIRONMENTAL PROBLEMS

What Supports and Drives Economies? An **economy** is a system of production, distribution, and consumption of *economic goods*: any material items or services that satisfy people's wants or needs. In an economy, individuals, businesses, and societies make **economic decisions** about what goods and services to produce, how to produce them, how much to produce, how to distribute them, and what to buy and sell.

The kinds of capital that produce material goods and services in an economy are called **economic resources**. They fall into three groups:

- **Earth capital** or **natural resources**: goods and services produced by the earth's natural processes (Figure 1-2). These include the planet's air, water, and land; nutrients and minerals in the soil and deeper in the earth's crust; wild and domesticated plants and animals (biodiversity); and nature's dilution, waste disposal, pest control, and recycling services. There are no other sources for these materials, which support all economies and lifestyles.

Figure 7-2 In a pure market economic system, economic goods and money would flow between households and businesses in a closed loop. People in households spend money to buy goods that firms produce, and firms spend money to buy factors of production (natural capital, manufactured capital, and human capital). In many economics textbooks, this and other economic systems are depicted, as here, as if they were self-contained and thus independent of the ecosphere—a model that reinforces the idea that unlimited economic growth of any kind is sustainable.

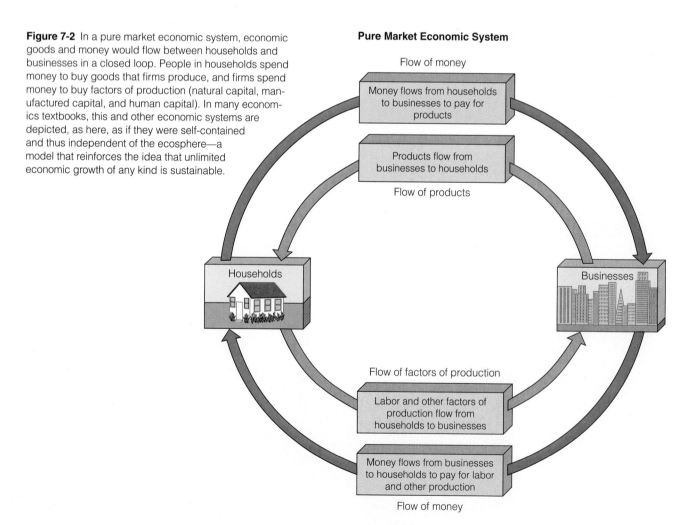

Pure Market Economic System

Flow of money

Money flows from households to businesses to pay for products

Products flow from businesses to households

Flow of products

Households

Businesses

Flow of factors of production

Labor and other factors of production flow from households to businesses

Money flows from businesses to households to pay for labor and other production

Flow of money

Q: What percentage of drinking water in the United States is withdrawn from groundwater?

- **Manufactured capital**: items made from earth capital with the help of human capital. This type of capital includes tools, machinery, equipment, factory buildings, and transportation and distribution facilities.

- **Human capital**: people's physical and mental talents. Workers sell their time and talents for wages. Managers take responsibility for combining earth capital, manufactured capital, and human capital to produce economic goods. In market-based systems, entrepreneurs and investors put up the money needed to produce an economic good with the intention of making a profit on their investment.

What Are the Major Types of Economic Systems? There are two major types of economic systems: centrally planned and market based. In a **pure command economic system**, or **centrally planned economy**, all economic decisions are made by the government. This command-and-control system assumes that government control and ownership of the means of production are the most efficient and equitable way to produce, use, and distribute goods and services.

In a **pure market economic system,** also known as **pure capitalism,** all economic decisions are made in *markets,* in which buyers (demanders) and sellers (suppliers) of economic goods freely interact without governmental or other interference. All economic resources are owned by private individuals and institutions rather than by the government. All buying and selling is based on *pure competition*, in which no seller or buyer is powerful enough to control the supply, demand, or price of a good. All sellers and buyers have full access to the market and enough information about the beneficial and harmful aspects of economic goods to make informed decisions.

Economists often depict pure capitalism as a circular flow of economic goods and money between households and businesses operating essentially independently of the ecosphere (Figure 7-2). By contrast, environmentalists and a small but growing number of economists emphasize the dependence of this or any economic system on the ecosphere (Figure 7-3).

In a pure capitalist system, a business has no legal allegiance to a particular nation, no obligation to supply any particular good or service, and no obligation

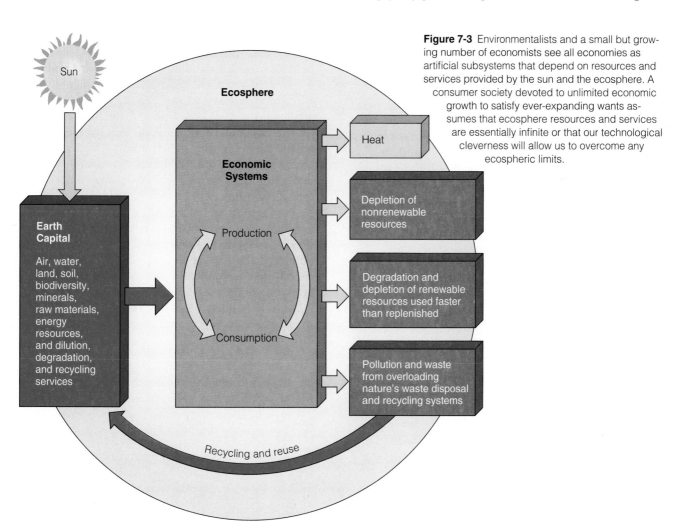

Figure 7-3 Environmentalists and a small but growing number of economists see all economies as artificial subsystems that depend on resources and services provided by the sun and the ecosphere. A consumer society devoted to unlimited economic growth to satisfy ever-expanding wants assumes that ecosphere resources and services are essentially infinite or that our technological cleverness will allow us to overcome any ecospheric limits.

A: About 50% (96% in rural areas and 20% in urban areas)

to provide jobs, safe workplaces, or environmental protection. A company's only obligation is to produce the highest possible short-term economic return (profit) for the owners or stockholders whose financial capital the company is using to do business.

Economic decisions in a pure market system are governed by interactions of *demand*, *supply*, and *price*. Buyers want to pay as little as possible for an economic good, and sellers want to set the highest price possible. **Market equilibrium** occurs when the quantity supplied equals the quantity demanded, and the price is no higher than buyers are willing to pay and no lower than sellers are willing to accept (Figure 7-4). Various factors can change the demand and supply of an economic good and establish a new market equilibrium.

In reality, all countries have **mixed economic systems** that fall somewhere between the pure market and pure command systems. The economic systems of countries such as China and North Korea fall toward the command-and-control end of the economic spectrum, whereas those of countries such as the United States and Canada fall toward the market-based end of the spectrum. Most other countries fall somewhere in between.

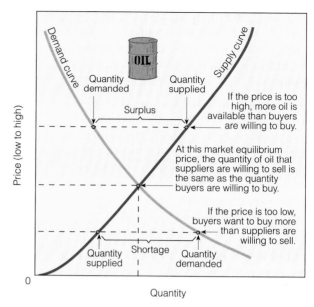

Figure 7-4 Supply, demand, and market equilibrium for a pure market system. If price, supply, and demand are the only factors involved, market equilibrium occurs at the point at which the demand and supply curves intersect.

7-2 ECONOMIC GROWTH AND EXTERNAL COSTS

How Is Economic Growth Measured? Virtually all economies seek **economic growth**: an increase in the capacity of the economy to provide goods and services for people's final use. Such growth is usually accomplished by maximizing the flow of matter and energy resources (throughput) by means of population growth (more consumers), more consumption per person, or both.

Economic growth is usually measured by the increase in a country's **gross domestic product (GDP)**—the market value (in current dollars) of all goods and services produced by an economy within its borders for final use during a year—and by its **gross national product (GNP)**—the GDP plus net income from abroad. To get a clearer picture, economists use the **real GNP** or **GDP**: the GNP or GDP adjusted for *inflation* (any increase in the average price level of final goods and services).

To show how the average person's slice of the economic pie is changing, economists use the **real per capita GNP** or **GDP**: the real GNP or GDP divided by the total population. If population expands faster than economic growth, then real per capita GNP or GDP falls.

Is Economic Growth Sustainable? Most economists believe that the capacity for economic growth is virtually unlimited because of the earth's vast amount of resources and the ability of human ingenuity to overcome resource shortages and environmental problems through science and technology (Figure 7-2). In other words, they believe that continued economic growth based on ever-increasing throughputs of matter and energy (Figure 3-15) is sustainable. Other economists accept the idea of limits to economic growth but believe that we will not reach such limits in the foreseeable future.

To environmentalists and a small but growing number of economists and business leaders, the notion of *sustainable economic growth* is nonsense because nothing that is based on the consumption of the earth capital that sustains all economies (Figure 7-3) can grow indefinitely. According to business leader Paul Hawken (Guest Essay, pp. 198–199),

We have reached a point where the value we do add to our economy is now being outweighed by the value we are removing, not only from future generations in terms of diminished resources, but from ourselves in terms of unlivable cities, deadening jobs, deteriorating health, and rising crime. In biological terms, we have become a parasite and are devouring our host.

Instead of unlimited economic growth, such critics call for **ecologically sustainable development**. This occurs when the total human population size and

Q: What percentage of the world's water use is provided by desalination?

resource use in the world (or in a region) are limited to a level that does not exceed the carrying capacity of the existing natural capital and are therefore sustainable. This is accomplished by reducing the throughput of matter and energy resources through economies (Figure 3-16).

According to system analysis expert and environmentalist Donella Meadows:

> What we need is smart development, not dumb growth. . . . When something grows it gets quantitatively bigger. When something develops it gets qualitatively better. . . . Smart development invests in insulation, efficient cars, and ever-renewed sources of energy. It ensures that forests and fields continue to produce wood, paper, and food, recharge wells, harbor wildlife, and attract tourists. Dumb growth crashes around looking for more oil. It clearcuts forests to keep loggers and sawmills going just a few more years until the trees run out. . . . It covers the landscape with the same kind of honkytonk ugliness tourists leave home to escape. . . . We need to meet dumb growth with smart questions. What really needs to grow? Who will benefit? Who will pay? What will last?

Suppose this view is correct and our collective consumption approaches or runs into some of the earth's natural limits. Then the difficult political and ethical question becomes how to share a roughly fixed economic pie instead of how to increase the size of the pie. According to Lester Brown (Guest Essay, pp. 26–27) if this happens, the growth-driven economics and politics of the last century will need to be replaced with a new economics and politics of scarcity and sustainability.

Are GNP and GDP Useful Measures of Quality of Life and Environmental Degradation? We are urged to buy and consume more and more so that the GNP and GDP will rise, making the country where one lives and the world a better place for everyone. The truth is that GNP and GDP indicators are poor measures of human welfare, environmental health, or even economic health, for the following reasons.

GNP and GDP hide the negative effects (on humans and on the rest of the ecosphere) of producing many goods and services. Pollution, crime, sickness, death, and depletion of natural resources are all counted as positive gains in the GDP or GNP. Every time an irreplaceable old-growth forest is cut down or a wetland is filled, the GDP and GNP go up. Every time that a chemical or radiation causes cancer and the victim is treated, the GNP and GDP go up. They also rise because of the funeral expenses for the 150,000–350,000 people killed prematurely each year in the United States from air

pollution. The $2.2 billion that Exxon spent partially cleaning up the oil spill from the *Exxon Valdez* tanker also raised the GDP and GNP.

Pollution is counted as a *triple positive gain* even though it decreases the quality of life for hundreds of millions of people and should be subtracted from the GNP. It is counted as a gain in GDP when it is first produced, counted again when society pays to clean it up partially, and counted as a third gain when people become sick or die from exposure to the pollution.

GNP and GDP don't include the depletion and degradation of natural resources or earth capital on which all economies depend. A country can be headed toward ecological bankruptcy—exhausting its mineral resources, eroding its soils, cutting down its forests, destroying its wetlands and estuaries, and depleting its wildlife and fisheries. At the same time it can have a rapidly rising GNP and GDP, at least for a while, until its environmental debts come due. Any business (or nation) that counted depletion of its capital (assets) as current income would always have a rosy but false picture of its true financial condition.

GNP and GDP hide or underestimate some of the positive effects of responsible behavior on society. More energy-efficient light bulbs, appliances, and cars reduce electric and gasoline bills and pollution, but these beneficial effects register as a decline in GNP and GDP. GDP and GNP indicators also exclude the labor we put into volunteer work, the health care we give loved ones, the food we grow for ourselves, and the cooking, cleaning, and repairs we do for ourselves.

GNP and GDP tell us nothing about economic justice (Guest Essay, p. 390–391). They don't reveal how resources, income, or the harmful effects of economic growth (pollution, waste dumps, land degradation) are distributed among the people in a country. UNICEF suggests that countries should be ranked not by average per capita GNP or GDP but by average or median income of the poorest 40% of their people.

Solutions: How Can Environmental Accounting Help? Economists have never claimed that GNP and GDP indicators are good measures of environmental health and human welfare, but most governments and business leaders use them that way. Environmentalists and a growing number of economists believe that GNP and GDP indicators should be replaced or supplemented with widely publicized *environmental* and *social indicators* that give a more realistic picture by subtracting from the GDP and GNP things that lead to a lower quality of life and depletion of earth capital.

In 1972 economists William Nordhaus and James Tobin developed an indicator called *net economic welfare (NEW)* to estimate the annual change in a country's quality of life. They calculate a price tag for

Natural Capital

Paul G. Hawken

Paul G. Hawken is a practical visionary who understands both business and ecology and can communicate what is needed to make the transition to an earth-sustaining, or restorative, economy. In addition to founding Smith & Hawken, a retail company known for its environmental initiatives, he has written seven widely acclaimed books including Growing a Business *(1987),* The Ecology of Commerce *(1993), and* Factor 10, The Next Industrial Revolution *(1997, with Amory and Hunter Lovins). He produced and hosted* Growing a Business, *a series for public television shown nationwide on 210 stations and now shown in 115 countries. His book,* The Ecology of Commerce, *was hailed as the best business book of 1993 and one of the most important books of this century. In 1987,* Inc. *magazine named him one of the 12 best entrepreneurs of the 1980s, and in 1995 he was named by the* Utne Reader *as one of the 100 visionaries who could change our lives. He has also been described as the poet laureate of American capitalism.*

Great ideas, in hindsight, seem obvious. The concept of natural capital [or earth capital, Figure 1-2] is such an idea. Natural capital refers to the myriad necessary and valuable resources and ecological processes that we rely upon to produce our food, products, and services. These include the obvious contributions of clean air and water, and the lesser noted functions of the environment such as processor and locus of our industrial waste.

The concept of natural capital is not a new one. Economists have long noted that natural capital is a factor in industrial production, albeit a marginal factor. A new view is emerging, the proposition that our economic systems cannot long endure without taking the flow of renewable and nonrenewable resources through economies into account. The "value" of natural capital is becoming paramount to the success of all business. While economists may still insist that its value is less than that of labor, wealth, or technology, it is doubtful that this view can be profitably supported by business in the long term.

This revision of neo-classical economics, yet to be accepted by mainstream academicians, stabilizes both the theory and practice of free-market capitalism and provides business and public policy with a powerful

new tool for growth, profitability, and the public good. It is as if we had been sitting on a two-legged stool for the past century, wondering why our economies have become increasingly unbalanced and unsteady.

The concept of natural capital, when intelligently linked to human and manufactured capital, provides the critical connection between the satisfaction of human needs, the continued prosperity of business, and the preservation (some might even suggest the restoration) of the earth's living natural systems.

Most Americans are filled with cornucopian fantasies of technological prowess, where human ingenuity bypasses natural limits and creates unimagined abundance. Optimism easily intertwines with the belief that nanotechnology, biotechnology, computers, and technologies yet to developed will eliminate hunger, disease, and want.

Dreams of alleviating human suffering are worthy, but what they usually overlook is the absolute necessity for fertile soil, ocean fisheries, a stable climate, biological diversity, and pure water, all of which we are losing, and none of which can be substantively created by any human-made technology known or imagined.

In our pursuit of dominance over the natural world, we have not taken into account the basic principle that industrialism, for all its sophistication, is enormously inefficient with respect to resources, energy, and waste. It is difficult for neo-classical economists, whose hypotheses and theories originated in a time of resource abundance, to understand that the very success of linear industrial systems [Figure 3-15] has laid the groundwork for the next stage in economic evolution. This next stage, whatever it may be called, is being brought about by powerful and much-delayed feedback from high levels of inefficiency and waste. Information from extractive and destructive activities going back a hundred years or more is being incorporated into the market and political economy. As that happens, the foundation of industrialism is giving way while the basis for the next industrial revolution is being established.

This shift is profoundly biological. It's not about the celebration of nature, although that is certainly part of it; it's about the incorporation of natural systems into our industrial life, into our way of making things, our way of processing things and deprocessing things. The reason

pollution and other "negative" goods and services included in the GNP and GDP and then subtract them to give the NEW. Economist David Pearce estimates that pollution and natural resource degradation subtract 1–5% from the GDP of developed countries (and 5–15% for developing countries), and that in general these percentages are rising. Other studies tally harmful environmental costs (excluding those from global warming and fossil-fuel depletion) at 2% of the GDP in Japan and Australia, 12–15% in China, 23% in Germany, 40% in Sweden, and 45% in the United States.

Dividing a country's NEW by its population gives the *per capita NEW*. Since 1940 the real per capita NEW in the United States has risen at about

Q: Worldwide, what percentage of the water withdrawn is unnecessarily wasted?

this shift is going to happen is because cyclical industrial systems work better than linear ones. They close the loop and reincorporate wastes as part of the production cycle [Figures 3-16 and 7-3]. There are no landfills in a cyclical society.

If there is so much inefficiency in our system, why isn't it more apparent? The inefficiencies are masked by a financial system in which money, prices, and markets give us improper information. Markets are not giving us proper information about how much our suburbs, spandex, and plastic drinking water bottles truly cost. Instead, we are getting such proper information from our beleaguered air- and watersheds, the overworked and eroded soils, the life-degrading inner cities and rural counties, the breakdown of stability worldwide, and the conflicts based on resource shortages; all these are providing the information that our prices should be giving us but don't.

Prices don't give us good information for a simple reason: improper accounting. Natural capital has never been placed on the balance sheets of companies or the countries of the world. To paraphrase G. K. Chesterton, it could fairly be said that capitalism might be a good idea except that we have never tried it yet. And try we must and will, for capitalism cannot be fully attained or practiced until, as any accounting student will tell us, we have an accurate balance sheet.

As it stands, our economic system is based on accounting principles that would bankrupt a company. Not surprisingly, it is posing problems for the world as a whole. When natural capital is placed on the balance sheet, not as a free amenity of infinite supply, but as an integral and valuable part of the production process, everything changes. The near obsessive pursuit of improvement in human productivity becomes balanced by the need for improved resource productivity. Using more and more resources to make fewer people more productive flies in the face of what we now need and require to improve our society and the environment. After all, it is people we have more of, not natural resources, so it is people we must use in order to reduce the flow (throughput) of matter and energy resources through economies.

And that is what can happen when we move from linear extractive systems [Figure 3-15] to cyclical ones [Figure 3-16]. Instead of the huge capital investments required to find, extract, and process nonrenewable oil, when we shift to renewable wind energy we spend our money on windmill maintenance instead of hydrocarbons. We get more people gainfully employed, reduce capital requirements, and produce almost no pollution.

Many people sincerely believe that an economic system based on the integrity of natural systems is unworkable. To answer that concern, we may well want to reverse the question and ask How is it that we have created an economic system that tells us it is cheaper to destroy the earth than to maintain it? We know this is not the way to take care of our cars, houses, and bridges, but somehow we have managed to overlook a pricing system that discounts the future, and sells off the past. Or to put it another way, how did we create an economic system that confuses capital with income?

Can we come up with and implement a more rational economic system? I think so. It is right before us. It requires no new theories, only commonsense. It is based on the simple but powerful proposition that *all capital must be valued*.

While there may be no *right* way to value a forest or a river, there is a *wrong* way, which is to give it no value at all. If we have doubts about how to value a 500-year-old tree, we need only ask how much it would cost to make a new one from scratch. Or a new river. Or even a new atmosphere.

The work of the future is the absorption and integration of the worth of living systems into every aspect of our culture and commerce, so that human systems mimic natural systems. Only by doing this can our cultures reflect growth and harmony rather than damage and discord.

Critical Thinking

1. If you were placed totally in charge of the world's economy, what are the three most important things you would do? Compare your answers with those of other members of your class.

2. Do you agree with the author of this essay that we must absorb and integrate the worth of living systems into every aspect of our culture and commerce? How would you go about doing this?

---

half the rate of the real per capita GNP, and since 1968 the gap between these two indicators has been widening.

Economist Robert Repetto and other researchers at the World Resources Institute have developed a *net national product (NNP)* that includes the depletion and destruction of natural resources as a factor in GNP.

They have applied this indicator to Indonesia and to Costa Rica.

Herman E. Daly, John B. Cobb, Jr., and Clifford W. Cobb have developed an *index of sustainable economic welfare (ISEW)* and applied it to the United States. This comprehensive indicator of human welfare measures per capita GNP adjusted for inequalities in income distribution, depletion of nonrenewable resources, loss of wetlands, loss of farmland from soil erosion and urbanization, the cost of air and water pollution, and estimates of long-term environmental damage from ozone depletion and possible global warming. After rising by 42% between 1950 and 1976, this indicator fell 14% between 1977 and 1990.

A more recent similar indicator is called the *Genuine Progress Indicator (GPI)*. When this indicator is applied to the United States, the GPI per person has steadily declined since 1973. Indeed, between 1973 and 1994 the GDP rose from $12,500 to $17,000 per person while the GPI fell from $6,500 to $4,000 per person (Figure 7-5).

Such indicators are far from perfect and require many crude estimates. However, they are more accurate than GNP and GDP as measures of life quality and environmental quality and even economic well-being per person. Without such indicators, we don't know much about what is happening to people, the environment, and the planet's natural resource base. We fail to see what needs to be done, and we don't have a way to measure what types of policies work. In effect, we are trying to guide national and global economies through treacherous economic and envi-

ronmental waters at ever-increasing speeds using faulty radar.

The *good news* is that such indicators exist. The *bad news* is that so far they are not widely used.

What Are Internal and External Costs?

All economic goods and services have both internal and external costs. For example, the price a consumer pays for a car reflects the costs of the factory, raw materials, labor, marketing, and shipping, as well as a markup to allow the car company and its dealers some profits. After a car is purchased, the buyer must pay for gasoline, maintenance, and repair. All these direct costs, which are paid for by the seller and the buyer of an economic good, are called **internal costs**.

Making, distributing, and using any economic good or service also involve **externalities**: social costs or benefits not included in the market price. For example, if a car dealer builds an aesthetically pleasing showroom and grounds, that is an **external benefit** to people who enjoy the sight at no cost.

On the other hand, extracting and processing raw materials to make and propel cars depletes nonrenewable energy and mineral resources, produces solid and hazardous wastes, disturbs land, pollutes the air and water, contributes to depletion of stratospheric ozone and possible global climate change, and reduces biodiversity and ecological integrity. These harmful effects are **external costs** passed on to workers, the public, and in some cases future generations. Car owners add to these external costs when they throw trash out of a car, drive a gas guzzler (which produces more air pollution than a more efficient car), disconnect or don't maintain a car's air-pollution control devices, or don't keep the engine tuned.

Because these harmful costs aren't included in the market price, people don't connect them with car ownership. Still, everyone pays these hidden costs sooner or later, in the form of higher costs for health care and health insurance and higher taxes for pollution control.

To conventional economists, external costs are minor defects in the flow of production and consumption in a self-contained economy not significantly dependent on earth capital (Figure 7-2). They assume that these defects can be rectified through technological innovations and by using the profits made from additional economic growth in a free-market economy.

To environmentalists and an increasing number of economists and business leaders, harmful externalities are a warning sign that our economic systems are stressing the ecosphere and depleting earth capital (Figures 1-2 and 7-3). They believe that these harmful external costs should be included in the market prices of goods and services—a process economists call *internalizing the external costs* (Guest Essay, pp. 198–199).

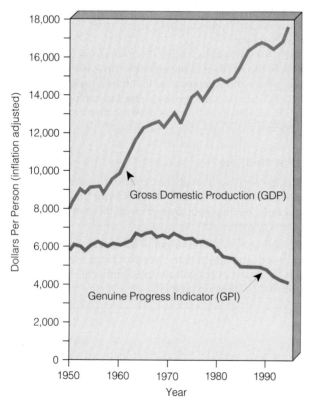

Figure 7-5 Comparison of per capita gross domestic product (GDP) and per capita genuine progress indicator (GPI) in the United States, 1950–1994. Units of per capita GDP and GPI are inflation adjusted using 1982 dollars. (Data from Clifford Cobb, Ted Halstead, and Jonathan Rowe)

Q: How many people don't have a safe supply of drinking water?

⊕ 7-3 SOLUTIONS: USING ECONOMICS TO IMPROVE ENVIRONMENTAL QUALITY

Should We Shift to Full-Cost Pricing? As long as businesses receive subsidies and tax breaks for extracting and using virgin resources and are not taxed for the pollutants they produce, few will volunteer to reduce short-term profits by becoming more environmentally responsible. Assume you own a company and believe it's wrong to subject your workers to hazardous conditions and pollute the environment beyond what natural processes can handle. Suppose you voluntarily improve safety conditions for your workers and install pollution controls, but your competitors don't. Then your product will cost more than theirs, and you will be at a competitive disadvantage. Your profits will decline; you may eventually go bankrupt and have to lay off your employees.

One way of dealing with the problem of harmful external costs is for the government to levy taxes, pass laws, provide subsidies, or use other strategies that encourage or force producers to include all or most of these costs in the market prices of economic goods and services. Then that price would be the **full cost** of these goods and services: internal costs plus short- and long-term external costs. The two major goals are (1) to close the gap between real and false prices by having prices that tell the environmental truth and (2) to have people and businesses pay the full costs of the harm they do to others and the environment.

Full-cost pricing involves *internalizing the external costs*, which requires government action because few companies will intentionally increase their cost of doing business unless their competitors must do so as well. If the market prices of economic goods reflected all or most of their full estimated cost, economic growth would be redirected. We would increase the beneficial parts of the GNP and GDP (and decrease the harmful parts), increase production of beneficial goods, and raise the net economic welfare. Preventing pollution would become more profitable than cleaning it up, and waste reduction, recycling, and reuse would be more profitable than burying or burning most of the waste we produce.

Because external costs would be internalized, the market prices for most goods and services would rise. But the total price we would pay would be about the same because the hidden external costs related to each product would already be included in its market price. Using full-cost pricing to internalize external costs provides consumers with information needed to make informed economic decisions about the effects of their lifestyles on the planet's life-support systems.

However, as external costs are internalized, economists and environmentalists warn that governments must reduce income, payroll, and other taxes and must withdraw subsidies formerly used to hide and pay for these external costs. Otherwise, consumers will face higher market prices without tax relief—an unjust policy guaranteed to fail.

Some goods and services would cost less because internalizing external costs encourages producers to find ways to cut costs (by inventing more resource-efficient methods of production); it also encourages producers to offer more earth-sustaining (or *green*) products. Jobs would be lost in earth-degrading businesses, but at least as many (some analysts say more) jobs could be created in earth-sustaining businesses. If this change in the way market prices are established took place over several decades, most current earth-degrading businesses would have time to transform themselves into profitable earth-sustaining businesses.

Full-cost pricing seems to make a lot of sense. Why isn't it more widely done? One reason is that many producers of harmful and wasteful goods would have to charge so much that they couldn't stay in business, or they would have to give up government subsidies and tax breaks that have helped hide the external costs of their goods and services. Another problem is that it is hard to put a price tag on many of the harmful environmental and health costs.

Studies estimate that governments around the globe spend about $600 billion per year to subsidize deforestation, overfishing, overgrazing, unsustainable agriculture, nonrenewable fossil fuels and nuclear energy, groundwater depletion, and other environmentally destructive activities—what environmental expert Norman Myers (Guest Essay, pp. 295–296) calls *perverse subsidies*. It is estimated that eliminating these earth-degrading subsidies that distort the global economy would allow about a 7% cut in the global tax burden of $7.5 trillion per year and encourage job creation and investment.

Despite the difficulties, proponents believe that full-cost pricing for harmful environmental and health effects deserves a serious try. They argue that doing the best we can to estimate and internalize current external costs is far better than continuing the current pricing system, which gives too little or misleading information about the environmental and health effects of goods and services. The key question is whether the problems with the *tell-the-truth full-cost* pricing system are worse than those with the current *hide-the-true-cost* pricing system.

How Useful Is Cost–Benefit Analysis? One of the chief tools corporations and governments use in making economic decisions is **cost–benefit analysis**. This approach involves comparing the estimated short-term and long-term costs (losses) with the estimated benefits (gains) for various courses of action.

Cost–benefit analyses can be useful guides and can indicate the cheapest way to go, but they can also be misused (Spotlight, below). They can even do great harm if they aren't carefully conducted and scrutinized.

Furthermore, as Worldwatch Institute researcher David Rodman reminds us, *"Environmental problems, like most important policy issues, involve more than costs and benefits; they also involve rights and wrongs, values and visions. If crime paid, cost–benefit analysis would endorse it."*

To minimize possible abuses, environmentalists and economists advocate the following guidelines for all cost–benefit analyses: **(1)** Use uniform standards; **(2)** clearly state all assumptions; **(3)** evaluate the reliability of all data inputs as high, medium, or low; **(4)** make projections using low, medium, and high discount rates; **(5)** show the estimated range of costs and benefits based on various sets of assumptions; **(6)** estimate the short- and long-term benefits and costs

---

Some Problems with Cost–Benefit Analysis

SPOTLIGHT

There are several controversies about cost–benefit analysis. One involves the **discount rate**—an estimate of a resource's future economic value compared to its present value. *The size of the discount rate (usually given as a percentage) is a primary factor affecting the outcome of any cost–benefit analysis.*

At a zero discount rate, a stand of redwood trees worth $1 million today will still be worth $1 million 50 years from now; thus there is no need to cut them down for short-term economic gain. However, at a 10% annual discount rate (normally used by most businesses and by the U.S. Office of Management and Budget), the same stand will be worth only $10,000 in 50 years. As a result, it makes short-term economic sense to cut them down now and invest the profits in something else.

Proponents of high (5–10%) discount rates argue that inflation will reduce the value of their future earnings. They also fear that innovation or changes in consumer preferences will make a product or service obsolete.

Environmentalists question this belief, pointing out that current economic systems are based on depleting the natural capital that supports them. High discount rates worsen this situation by encouraging such rapid exploitation of resources for immediate payoffs that

sustainable use of most potentially renewable natural resources is virtually impossible. They believe that unique and scarce resources should be protected by having a 0% or even a negative discount rate and that discount rates of 1–3% would make it profitable to use other resources sustainably or slowly. At its core, the choice of a discount rate is really an *ethical decision* about our responsibility to future generations.

Another problem with cost–benefit analysis is determining *who benefits and who is harmed*. In the United States, an estimated 100,000 employees die each year because of exposure to hazardous chemicals and other safety hazards at work, and an additional 400,000 are seriously injured from such exposure. In many other countries (especially developing countries), the situation is much worse. Is this a necessary or an unnecessary cost of doing business?

Another limitation of cost–benefit analysis is that *many things we value cannot easily be reduced to dollars and cents*. We can put estimated price tags on human life, good health, clean air and water, pollution and accidents that are prevented, wilderness, the northern spotted owl, and on various forms of earth capital (Figure 1-2). However, the dollar values different people assign to such things vary widely because of different assumptions, discount rates, and value judgments, leading to a wide range of projected costs and benefits.

Because these and other estimates of costs and benefits are so variable, *figures can easily be weighted to achieve the outcome desired by proponents or opponents of a proposed project or action.* For example, one industry-sponsored cost–benefit study estimated that compliance with a standard to protect U.S. workers from vinyl chloride would cost $65–90 billion; in fact, less than $1 billion was actually needed to comply with the standard.

In 1996, 11 prominent U.S. economists published a joint statement concluding, "We suggest that benefit–cost analysis has a potentially important role to play in helping inform regulatory decision-making, although it should not be the sole basis for such decision-making." Using cost–benefit analysis to make decisions is somewhat like trying to detect a car speeding at 160 kilometers per hour (100 miles per hour) with a radar device so unreliable that at best it can only tell us that the car's speed is somewhere between 80 kph (50 mph) and 8,000 kph (5,000 mph).

Critical Thinking

Do you believe that cost–benefit analysis should be used as the primary way to evaluate whether any new environmental law or regulation should be put into effect and whether any existing environmental law or regulation should be weakened or strengthened? Explain. What should be the role of cost–benefit analysis? Why?

Q: How many people die prematurely every year from drinking contaminated water?

to all affected population groups; **(7)** estimate the effectiveness of the project or form of regulation instead of assuming (as is often done) that all projects and regulations will be executed with 100% efficiency and effectiveness; and **(8)** open the evaluations to public review and discussion.

Should We Rely Mostly on Regulations or Market Forces? Most economists agree that controlling or preventing pollution and reducing resource waste require government intervention in the marketplace. Such government action can take the form of regulation, the use of market forces, or some combination of these approaches.

Regulation is a *command-and-control* approach. It involves enacting and enforcing laws, for example, that set pollution standards, regulate harmful activities, ban the release of toxic chemicals into the environment, and require that certain irreplaceable or slowly replenished resources be protected from unsustainable use (or from any use at all).

Most studies of the effects of environmental regulations in the United States have found that they do businesses very little harm. Indeed, in many cases they have led to improvements in resource use efficiency, which reduce costs, and to innovative products and industrial processes, which increase profits. A recent study by economist Robert Repetto at the World Resources Institute showed that between 1970 and 1990, the U.S. industries that spent the most on pollution control fared significantly better than average in the global marketplace.

However, business leaders and many environmentalists in the United States agree that some pollution control regulations discourage innovation by being too prescriptive and costly and must be modified. They propose using regulations to set goals but then freeing industries to meet such goals in any way that works.

Market forces can help improve environmental quality and reduce resource waste, mostly by encouraging the internalization of external costs. This is based on a fundamental principle of the marketplace in today's mixed economic systems: *What we reward—mostly by subsidies and tax breaks—we tend to get more of, and what we discourage—mostly by regulations and taxes—we tend to get less of.*

One way to put this principle into practice would be *to phase in government subsidies that encourage earth-sustaining behavior and phase out current perverse subsidies that encourage earth-degrading behavior.* The difficulty with this ecologically and economically appealing *carrot approach* is that removing or adding subsidies involves political decisions; these are easily swayed by powerful economic interests that want to preserve ecologically unsound subsidies to increase their short-term profits.

Another market approach is for the government to *grant tradable pollution and resource-use rights.* For example, a total limit on emissions of a pollutant or use of a resource could be set, and the total would then be allocated among manufacturers or users by permit. Permit holders not using their entire allocation could use it as a credit against future expansion, use it in another part of their operation, or sell it to other companies. Tradable rights could also be established among countries to preserve biodiversity and to reduce emissions of greenhouse gases, ozone-destroying chemicals, lead, and various air and water pollutants with harmful regional (or global) effects.

Some environmentalists support tradable pollution rights as an improvement over the current regulatory approach. Other environmentalists believe that allowing companies to buy and trade rights to pollute is wrong because it allows the wealthiest companies to continue polluting, thereby excluding smaller companies from the market. Others point out that whereas pollution rights charge for the right to pollute up to certain limits, the command-and-control approach gives away the right to pollute up to a certain level.

Many environmentalists also view pollution rights trading as a shell game, designed not to reduce overall pollution but to allow polluters to shift harm from one place to another. Critics also argue that the use of marketable permits creates an incentive for fraud because most pollution control regulations are based on self-reporting of pollution outputs (and government monitoring of such outputs is inadequate). Thus, potential permit sellers have a strong incentive to overstate their pollution output reductions, and potential buyers have an incentive to understate their pollution outputs.

Another market-based method is to *enact green taxes or effluent fees* that would internalize many of the harmful external costs of production and consumption. This method could include taxes on each unit of pollution discharged into the air or water, each unit of hazardous or nuclear waste produced, each unit of virgin resources used, and each unit of fossil fuel used.

Experience in the Netherlands and Germany shows that phasing in such taxes can encourage creativity in solving environmental problems and reducing costs by preventing pollution, using fewer resources, and developing new earth-sustaining technologies and products. For example, in the Netherlands fees for emissions of toxic metals (such as cadmium, lead, and mercury) into waterways have been gradually increased since 1970. According to studies, these taxes were the major factor in reducing emissions of toxic metals 86–97% between 1976 and 1994.

There are two major problems with this tax punishment, or stick, approach. *First*, because the taxes or effluent fees are set politically rather than by markets, elected officials find it easier to aim for popular approval rather than economic and ecological efficiency. As a result, the taxes are usually too low to be effective and undermine the entire concept.

Second, elected officials are likely to see such taxes as ways of raising revenue instead of improving economic and ecological efficiency. However, economists point out that *green taxes* on pollution output, resource depletion, and environmental degradation would work if they reduced or replaced income, payroll, or other taxes—and if the poor were given a basic safety net to reduce the regressive nature of consumption taxes on essentials such as food, fuel, and housing. In other words, environmental taxes must be seen as a *tax-shifting* instead of *tax-increasing* approach.

Charging user fees is another market-based method. For example, users would pay fees to cover all or most costs for grazing livestock, extracting lumber and minerals from public lands, using water provided by government-financed projects, and using public lands for recreation. In principle, this *user-pays* approach is favored by the general public. However, it is opposed by ranchers, timber harvesters, miners, and tourists who benefit from having taxpayers subsidize their low-cost use of public lands and resources.

Another market approach would *require businesses to post a pollution prevention or assurance bond* when they plan to develop a new mine, plant, incinerator, landfill, or development and before they introduce a new chemical or new technology. The size of the performance bond would be based on estimates of the worst-case consequences of each project, chemical, or technology as determined by the producer and reviewed by an in-dependent panel of risk experts. Each deposit would be kept in an interest-producing escrow account. After a set length of time, the deposit (with interest) would be returned *minus* environmental costs. If harm occurred, all or part of the bond would be used for cleanup and environmental restoration and to pay damages to those who were harmed. This approach is similar to the performance bonds contractors are now required to post for major construction projects. Understandably, this approach is opposed by businesses that would have to post such bonds.

Each of these approaches has advantages and disadvantages (Table 7-1). Currently, in the United States private industry and local, state, and federal government spend about $130 billion a year to comply with federal environmental regulations. Studies estimate that greater reliance on market-based policies could cut these expenditures by one-third to one-half.

Most analysts see a combination of command-and-control and market-based approaches as the best solutions to most environmental problems. Regulatory abuses and excessively expensive regulations should be eliminated. However, regulations play an important environmental role and sometimes benefit polluters by stimulating companies to innovate in ways that make them more competitive. Much more research must be done to determine the most effective mix of regulatory and market-based strategies for each type of environmental problem.

Should We Emphasize Pollution Control or Pollution Prevention? Shouldn't our goal be zero pollution? Ideally, yes; in the real world, not necessarily. First, natural processes can handle some of our wastes,

Table 7-1 Economic Solutions to Pollution and Resource Waste

Solution	Internalizes External Costs	Innovation	International Competitiveness	Administrative Costs	Increases Government Revenue
Regulation	Partially	Can encourage	Decreased*	High	No
Subsidies	No	Can encourage	Increased	Low	No
Withdrawing harmful subsidies	Yes	Can encourage	Decreased*	Low	Yes
Tradable rights	Yes	Encourages	Decreased*	Low	Yes
Green taxes	Yes	Encourages	Decreased*	Low	Yes
User fees	Yes	Can encourage	Decreased*	Low	Yes
Pollution-prevention bonds	Yes	Encourages	Decreased*	Low	No

*Unless more cost-effective and productive technologies are developed.

Q: What is the largest source of water pollution in the United States?

as long as we don't destroy, degrade, or overload these processes. However, environmentalists argue that harmful chemicals that either cannot be degraded by natural processes or that break down very slowly should not be released into the environment, or should be released only in small amounts and regulated by special permit.

Second, as long as we continue to rely on pollution control, we can't afford zero pollution. After we've removed a certain proportion of the pollutants in air, water, or soil, the cleanup cost per additional unit of pollutant rises sharply (Figure 7-6). Beyond a certain point, the cleanup costs exceed the harmful costs of pollution. Some businesses could then go bankrupt, and some people could lose jobs, homes, and savings. On the other hand, if we don't go far enough, dealing with the harmful external effects of pollution can cost more than pollution reduction.

To find the breakeven point, economists plot two curves: a curve of the estimated economic costs of cleaning up pollution and a curve of the estimated social (external) costs of pollution. Adding the two curves together, we get a third curve showing the total costs. The lowest point on this third curve is the point of the *optimal level of pollution* (Figure 7-7).

On a graph, this looks neat and simple, but environmentalists and business leaders often disagree in their estimates of the harmful costs of pollution. This approach assumes that we know which substances are harmful and how much each part of the environment

can handle without serious environmental harm—things we probably will never know precisely.

Most environmentalists and a growing number of economists and business leaders believe that environmental laws should emphasize pollution prevention, which avoids most of the regulatory problems and excessive costs of end-of-pipe pollution control. What do you think?

Is Encouraging Global Free Trade Environmentally Helpful or Harmful?

On April 15, 1994, representatives of 120 nations signed the Uruguay Round of the General Agreement on Tariffs and Trade (GATT). This is a revised version of the 1948 GATT convention, which attempted to lower tariff barriers to world trade among member nations. The new GATT establishes a World Trade Organization (WTO), giving it the status of a major international organization (similar to the United Nations and the World Bank) and the power to oversee and enforce the agreement.

The 1989 Free Trade Agreement (FTA) between Canada and the United States and the 1993 North American Free Trade Agreement (NAFTA) among Canada, the United States, and Mexico are also designed to remove trade barriers among participating nations.

Proponents argue that agreements to reduce global trade barriers have a number of important benefits. First, *such agreements will benefit developing countries*, whose products are often at a competitive disadvantage

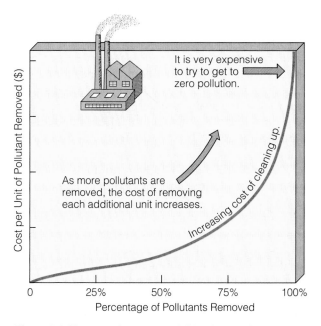

Figure 7-6 The cost of removing additional units of pollution rises exponentially, which explains why it is usually cheaper to prevent pollution than to clean it up.

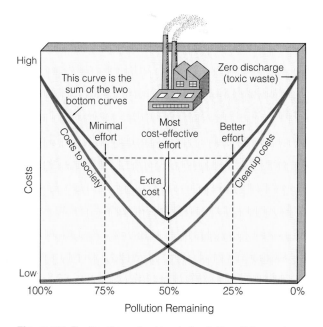

Figure 7-7 Finding the optimal level of pollution. This graph shows the optimal level at 50%, but the actual level varies depending on the pollutant.

A: Agriculture (responsible for almost two-thirds)

Improving Trade Agreements

SOLUTIONS

Critics of current trade agreements have suggested a number of ways to encourage free trade without sacrificing environmental protection or health and worker safety. They would rewrite and correct the serious weaknesses in GATT (and in FTA and NAFTA) and turn it into GAST: the *General Agreement for Sustainable Trade*. They offer the following suggestions for doing this:

- Having low or no trade tariffs on imports from nations that promote ecologically sustainable development by using environmental accounting and full-cost pricing

- Judging GATT or any trade agreement primarily on how it benefits the environment, workers, and the poorest 40% of humanity

- Setting minimum environmental, consumer protection, and worker health and safety standards for all participating countries

- Allowing any country to set more stringent versions of such standards without being penalized by fines or other means

- Requiring all panels or bodies setting and enforcing GATT standards to have environmental, labor, consumer, and health representatives from developed countries and developing countries alike

- Opening all discussions and findings of any GATT panel or other WTO body to global public scrutiny

- Allowing international environmental agreements to prevail when they conflict with GATT or any other trade agreement

Unless citizens exert intense pressure on legislators, critics warn that such safeguards will not be incorporated into international trade agreements.

Critical Thinking

1. Do you believe that the current GATT treaty will on balance help or hinder environmental protection **(a)** globally, **(b)** in the country where you live, and **(c)** in your community? Explain.

2. Explain why you agree or disagree with each of the above suggestions. If you agree with all or most of these proposals, how could they be implemented? What role could you play in making such changes?

in the global marketplace because of trade barriers erected by developed countries. Second, *they can allow consumers to buy more things at cheaper prices, thus stimulating economic growth in all countries*. Using this idea, proponents of GATT contend that it will save U.S. consumers $35 billion a year ($300 per household), boost the U.S. economy by as much as $219 billion a year after 10 years, and add millions of jobs.

Because of such potential benefits, some environmental groups generally supported FTA and NAFTA. However, most environmental groups, and those concerned with consumer protection and worker health and safety, oppose the new version of GATT for several reasons.

First, *they believe that GATT will not provide ample economic benefits for everyone*. This hypothesis is valid if only products, but not factories or workers, cross borders. Currently, 84% of the benefits of world trade flow to the richest one-fifth of the world's population, but only 0.9% to the poorest one-fifth. At best, the new version of GATT would not change this situation, and it could worsen it.

Second, they contend that *GATT will increase the economic and political power of multinational corporations and decrease the power of small businesses, citizens, and democratically elected governments*. The revised GATT treaty was developed mostly in secret by government and business officials dominated by the interests of multinational corporations based in developed countries. Under the new agreement, any nation not abiding by the ruling of a GATT panel (whose deliberations are not open to public scrutiny) could be charged heavy fines by the WTO or by a country whose complaint is upheld.

Third, *GATT will probably weaken environmental and health and safety standards in developed countries*. Under the new GATT, any country could be fined (or be subjected to tariff fees) if any portion of it adopted stricter environmental, health, worker safety, resource use, or other standards than the uniform global standards established by the WTO (dominated by multinational companies based in developed countries).

Faced with cheaper foreign products, domestic businesses operating in the international marketplace will have three choices: **(1)** go out of business, **(2)** move some or all of their operations abroad to take advantage of cheaper labor and less restrictive environmental and worker safety regulations, or **(3)** lobby to weaken domestic environmental, health, and worker and consumer safety laws. Under such pressures resulting from GATT, elected officials in the United States and other developed countries are likely to find it politically and economically necessary to weaken such laws to stem a flow of businesses, capital, tax revenue, and jobs out of their countries.

Q: What is the largest source of water pollution from oil?

Environmentalists point out that the existing FTA, NAFTA, and GATT trade agreements have lowered some environmental standards. British Columbia discontinued a government-funded tree-planting program because the United States (under pressure from U.S. lumber companies) argued that as an unfair subsidy to Canada's timber industry, it violated the 1989 FTA agreement. Canada had to relax its regulations concerning pesticides and food irradiation to bring them in line with the weaker ones in the United States. The Canadian timber industry is urging its government to challenge a U.S. law requiring the use of recycled fiber in newsprint on the grounds that it is a trade barrier.

Under the new GATT, environmentalists fear that government bans on the export or import of raw logs, intended to slow the destruction of rain forests or ancient forests, could be overturned. Germany could be forced to repeal its law requiring recycling of all beverage containers. Countries might not be able to restrict imports of hazardous wastes, banned medicines, and dangerous pesticides. International moratoriums and quotas on whale harvests, as well as bans on trading ivory to protect elephants, could be overturned. A developing country banning dirty industries could be accused of violating free trade.

The new GATT wouldn't prohibit all such activities, but countries could be fined heavily for engaging in them. Because this would reduce profits, the net effect would be to retard or discourage the implementation of these and many other national environmental laws.

Under the old GATT a country receiving an unfavorable ruling by a GATT panel could block the decision unless the ruling was unanimous. Under the new GATT, panel rulings cannot be blocked. Critics of the latest version of GATT call for it to be improved (Solutions, left).

7-4 SOLUTIONS: REDUCING POVERTY

Does the Trickle-Down Approach to Reducing Poverty Work? Poverty is usually defined as the inability to meet one's basic economic needs. Currently, an estimated 1.3 billion people in developing countries—one of every five people on the planet—have an annual income of less than $370 per year. This income of roughly $1 per day is the World Bank's definition of poverty.

Poverty causes premature deaths and preventable health problems. It also tends to increase birth rates and often pushes people to use potentially renewable resources unsustainably in order to survive.

Most economists believe that a growing economy is the best way to help the poor. This is the so-called

trickle-down hypothesis: Economic growth creates more jobs, enables more of the increased wealth to reach workers, and provides greater tax revenues that can be used to help the poor help themselves.

However, the facts suggest that either the hypothesis is wrong or it has not been applied. Instead of trickling down, most of the benefits of economic growth as measured by income have flowed up since 1960, making the top one-fifth of the world's people much richer and the bottom one-fifth poorer; most of those in between have lost or gained only slightly in real per capita GNP (Figure 1-8). This trend has accelerated in the 1980s and 1990s, with the richest fifth of the world's people receiving 82.7% of the world's income in 1991 and the poorest fifth receiving only 1.4% (Figure 7-8).

How Can Poverty Be Reduced? Analysts point out that reducing poverty requires the governments of most developing countries to make policy changes, including shifting more of the national budget to help the rural and urban poor work their way out of poverty and giving villages, villagers, and the urban poor title to common lands and to crops and trees they plant on them.

Analysts also urge developed countries and the wealthy in developing countries to help reduce

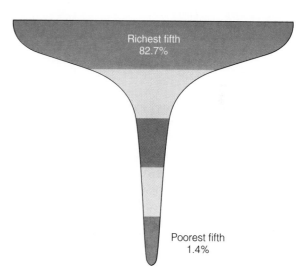

Figure 7-8 Data on the global distribution of income showing that, instead of trickling down, most of the world's income has flowed up, with the richest 20% of the world's population receiving more of the world's income than all of the remaining 80% of the world's population in 1991. Each horizontal band in this diagram represents one-fifth of the world's population. This upward flow of global income has increased since 1960 and has accelerated in the 1980s and 1990s. Do you favor or oppose this trend? Explain. If you oppose this trend, what do you believe are the most important ways to correct it? (Data from UN Development Programme)

poverty. Several controversial ways to do this have been suggested. First, *forgive at least 60% of the almost $2 trillion that developing countries owe to developed countries and international lending agencies.* Some of this debt can be forgiven in exchange for carefully monitored agreements by the governments of developing countries to increase expenditures for family planning, health care, education, land redistribution, protection and restoration of biodiversity, and more sustainable use of renewable resources.

Another proposal is to *increase the nonmilitary aid to developing countries from developed countries.* Proponents say that this aid should go directly to the poor to help them become more self-reliant. Analysts also recommend that we *shift most international aid from large-scale to small-scale projects intended to benefit local communities of the poor.*

Analysts also suggest that *international lending agencies should be required to use a standard environmental and social impact analysis to evaluate any proposed development project.* Proponents believe that no project should be supported unless its net environmental impact using full-cost accounting is favorable, most of its benefits go to the poorest 40% of the people affected, and the local people it affects are involved in planning and executing the project. They call for careful monitoring of all projects and an immediate halt in funding whenever environmental safeguards are not followed. A final proposal is to *establish policies that encourage both developed countries and developing countries to slow population growth and stabilize their populations* (Section 6-3).

⬙ 7-5 SOLUTIONS: CONVERTING TO EARTH-SUSTAINING ECONOMIES

How Can We Make Working with the Earth Profitable? Greening Business In 1994 entrepreneur and business leader Paul Hawken (Guest Essay, pp. 198–199) wrote *The Ecology of Commerce*, a widely acclaimed book describing what's wrong with business and how it can be transformed to work with the earth. Hawken and several other business leaders and economists have laid out the following principles for transforming the planet's current earth-degrading economic systems into earth-sustaining, or restorative, economies over the next several decades:

- *Reward (subsidize) earth-sustaining behavior.*

- *Discourage (tax and don't subsidize) earth-degrading behavior.*

- *Use full-cost accounting to include the ecological value of natural resources in their market prices.*

- *Use environmental and social indicators to measure progress toward environmental and economic sustainability and human well-being* (Figure 7-5).

- *Use full-cost pricing to include the external costs of goods and services in their market prices.*

- *Replace taxes on income and profits with taxes on throughput of matter and energy.*

- *Use low discount rates for evaluating future worth of irreplaceable or vulnerable resources* (Spotlight, p. 202).

- *Establish public utilities to manage and protect public lands and fisheries.*

- *Revoke the government-granted charters of environmentally and socially irresponsible businesses.*

- *Make environmental concerns a key part of all trade agreements and of all loans made by international lending agencies* (Solutions, p. 206).

- *Reduce waste of energy, water, and mineral resources.*

- *Protect biodiversity.*

- *Reduce future ecological damage and repair past ecological damage.*

- *Reduce poverty.*

- *Slow population growth.*

Paul Hawken's simple golden rule for such an economy is, *"Leave the world better than you found it, take no more than you need, try not to harm life or the environment, and make amends if you do."*

How Can We Make the Transition to an Earth-Sustaining Economy? Even if people believe that an earth-sustaining economy is desirable, is it possible to make such a drastic change in the way we think and act? Some environmentalists, economists, and business leaders say that it's not only possible but imperative and that it can be done over the next 40–50 years. They point out that *the environmental revolution is also an economic revolution* that uses a mix of regulations and market-based approaches to reward earth-sustaining behaviors by businesses and consumers and discourages earth-degrading behavior.

According to Paul Hawken, this new approach to economic thinking and actions recognizes that most business leaders are not evil, earth-degrading ogres. Instead, they are trapped in a system that by design rewards them (with the highest profits and salaries, and best chances for promotion) for maximizing short-term profits for owners and investors, regardless of the harmful short- and long-term environmental and social effects.

Hawken argues that an earth-sustaining economy would free business leaders, workers, and investors

Q: In the United States, how many underground tanks storing gasoline and other hazardous chemicals are leaking?

from this ethical dilemma and allow them to be financially compensated and respected for doing socially and ecologically responsible work, improving environmental quality, and still making hefty profits for owners and stockholders. Making this shift should also create jobs (Connections, below).

Hawken and others offer the following outline of how the shift from current unsustainable economies to sustainable ones could be made over the next 40–50 years. Over the first 20 years, *all* government subsidies encouraging resource depletion, waste, pollution, and environmental degradation would be phased out and replaced with taxes on such activities. During that same period, taxes on incomes and profits would be phased out and new government subsidies would be phased in for businesses built around recycling and reuse, reducing waste, preventing pollution, improving energy efficiency, using renewable energy, protecting biodiversity, and restoring ecosystems. This gradual and mostly predictable change would encour-

age a shift to more sustainable economies while allowing businesses to plan ahead.

This system for change represents a shift in determining which economic actions are rewarded (subsidized) and which ones are discouraged. These plans would be well publicized and would take effect over decades, giving businesses time to adjust. Because businesses go where the profits are, many of today's earth-degrading businesses might well be tomorrow's earth-sustaining businesses—a win–win solution for business, future generations, and the earth. Economic models indicate that after this first phase is completed the entire economy would be transformed within another 30–40 years.

The problem in making this shift is not economics, but politics. It involves the difficult task of convincing business leaders and elected officials to begin changing the current system of rewards and penalties, the profits from which have given them economic and political power. Without the active participation of

---

Jobs and the Environment

CONNECTIONS

Critics of environmental laws and regulations contend that they cause the loss of large numbers of jobs in the United States and in other developed countries. In fact, studies show that the opposite is true. U.S. Department of Labor statistics reveal that only 0.1% of the jobs lost in the United States between 1987 to 1990 were the result of environmental regulations.

In 1989, the Business Roundtable, an association of chief executive officers of U.S. companies, projected that a minimum of 200,000 and possibly as many as 1–2 million jobs would be lost if the Clean Air Act of 1990 were passed. However, a 1995 study by economist Eban Goodstein showed that by June 1994 only 2,363 jobs had been lost because of this act.

Instead of causing a net loss of jobs, environmental protection is a major growth industry that creates new jobs. Some 1 million jobs are expected to be added to the U.S. environmental protection industry

alone between 1995 and 2005; increasing the aluminum recycling rate in the United States to 75% would create 350,000 more jobs. Collecting and refilling reusable containers creates many more jobs per dollar of investment than using throwaway containers, and most of the new jobs are created in local communities. Reforestation, ecological restoration, sustainable agriculture, and integrated pest management are all labor-intensive activities requiring low to moderate skill levels.

A congressional study concluded that investing $115 billion per year in solar energy and in improving energy efficiency in the United States would eliminate about 1 million jobs in oil, gas, coal, and electricity production but would create 2 million other new jobs; investment of the money saved by reducing energy waste could create another 2 million jobs.

Although a host of new jobs would be created in an earth-sustaining economy, jobs would be lost in some industries, regions, and communities. Ways suggested

by various analysts to ease the transition include providing tax breaks to make it more profitable for companies to keep or hire more workers instead of replacing them with machines; using incentives to encourage location of new, emerging industries in hard-hit communities, helping such areas diversify their economic base; and providing income and retraining assistance for workers displaced from environmentally destructive businesses (a *Superfund for Workers*).

Critical Thinking

1. What major things (if any) has the federal government in the United States (or the country where you live) done to stimulate the growth of environmental jobs? What major things (if any) has the government done to discourage the growth of environmental jobs?

2. Do you believe that the government should have a significant role in stimulating environmental jobs? Explain.

A: At least 1 million

How Is Germany Investing in the Future and the Earth?

German political and business leaders see sales of environmental protection goods and services—already a roughly $408-billion-per-year business in 1994 (and expected to rise to $572 billion by 2001)—as a major source of new markets and income in the next century because environmental standards and concerns are expected to rise everywhere.

Stricter environmental standards and regulations in Germany have paid off in a cleaner environment and the development of innovative green technologies that can be sold at home and abroad in a rapidly growing market. Mostly because of stricter air pollution regulations, German companies have developed some of the world's cleanest and most efficient gas turbines, and they have invented the world's first steel mill that uses no coal to make steel. Germany sells these and other improved environmental technologies globally.

On the other hand, Germany continues to heavily subsidize its coal-mining industry, which increases environmental degradation and pollution and adds large amounts of carbon dioxide to the atmosphere. According to environmental expert Norman Myers (Guest Essay, pp. 295–296), these perverse subsidies are now so high that the German government would save money by closing all of its coal mines and sending the miners home on full pay for the rest of their lives.

In 1977 the German government started the Blue Angel product-labeling program to inform consumers about products that cause the least environmental harm. Most international companies now use the German market to test and evaluate *green* products.

Germany has also revolutionized the recycling business. German car companies are required to pick up and recycle all domestic cars they make. Bar-coded parts and disassembly plants can dismantle an auto for recycling in 20 minutes. Such *take-back* requirements are being extended to almost all products to reduce use of energy and virgin raw materials. Germany

plans to sell its newly developed recycling technologies to other countries.

The German government has also supported research and development aimed at making Germany the world's leader in solar-cell technology (Section 18-4) and hydrogen fuel (Section 18-8), which it expects to provide a rapidly increasing share of the world's energy. Finally, Germany provides about $1 billion per year in green foreign aid to developing countries. Much of the aid is designed to stimulate demand for German technologies such as solar-powered lights, solar cells, and wind-powered water pumps.

Critical Thinking

1. What major steps (if any) are the government and businesses in the country where you live taking in making the transition to a more earth-sustaining economy?

2. What three major things could you do to help promote a shift to such an economy in your country and in your local community?

---

forward-thinking business and political leaders and strong political pressure from citizens, there is little hope of making the transition to an earth-sustaining economy.

On the other hand, history shows that such changes can occur over 40–50 years with the appropriate system of carrots (subsidies) and sticks (taxes and regulations). Several countries, such as the Netherlands and Germany (Case Study, above) are beginning to take steps in this direction.

Can We Change Economic Gears in the Next Few Decades? Critics claim that a shift toward an earth-sustaining economy won't happen because it would be opposed by people whose subsidies were being eliminated and whose activities were being taxed. However, investors and businesspeople have just as much at stake in helping sustain the earth (and thus the human species and economies) as anyone else. Forward-looking investors, corporate executives, and the governments of coun-

tries such as Japan, Germany, the Netherlands, and the United States recognize that earth-sustaining businesses with good environmental management will prosper as the environmental revolution proceeds. Companies and countries that fail to invest in a green future may find that they do not have a future.

7-6 POLITICS AND ENVIRONMENTAL POLICY

How Does Social Change Occur in Representative Democracies? **Politics** is the process by which individuals and groups try to influence or control the policies and actions of governments at the local, state, national, or international levels. Politics is concerned with who has power over the distribution of resources and benefits—who gets what, when, and how.

Democracy is government "by the people" through elected officials and representatives. In a *constitutional* *democracy*, a constitution provides the basis of government authority, limits government power by mandating free elections, and guarantees freely expressed public opinion.

In passing laws, developing budgets, and formulating regulations, government decision makers must deal with pressure from many competing *special-interest groups*. Each group advocates passing laws favorable to its cause and weakening or repealing laws unfavorable to its position. This includes cheap and better access to resources, subsidies, relief from taxes, and the shaping of regulations to foster the group's goals.

Some special-interest groups are *profit-making organizations* such as corporations, and others are *nonprofit, nongovernment organizations (NGOs)*. Examples of NGOs are educational institutions, labor unions, and mainstream and grassroots environmental organizations.

Most political decisions made in democracies result from bargaining, accommodation, and compromise among leaders of competing *elites*, or power brokers. The overarching goal of government by competing elites is to maintain the overall economic and political stability of the system (status quo) by making only gradual change; this goal does *not* involve questioning or changing the rules of the game (the fundamental societal beliefs) that gave the elites their political or economic power.

One disadvantage of this deliberate design for stability is that democratic governments tend to *react* to crises instead of acting to prevent them. This means that there is a built-in bias against policies for protecting the environment because they often call for prevention of crises instead of reaction to them. Also, such policies sometimes call for fundamental changes in societal beliefs that can threaten the power of government and business elites.

Case Study: How Is Environmental Policy Made in the United States? The major function of the federal government in the United States is to develop and implement *policy* for dealing with various issues. This policy is typically composed of various *laws* passed by the legislative branch, *regulations* instituted by the executive branch to put laws into effect, and enough *funding* to implement and enforce the laws and regulations.

The first step in establishing federal environmental policy (or any other policy) is to persuade lawmakers that a problem exists and that the government has a responsibility to find solutions to it. Once over that hurdle, lawmakers try to pass laws to deal with the problem. Most environmental bills are evaluated by as many as 10 committees in both the House of Representatives and the Senate. Effective proposals are often weakened by this fragmentation and by lobbying from groups opposing the law. Nonetheless, since the 1970s a number of environmental laws have been passed in the United States (Appendix 3 and Solutions, p. 212).

Even if an environmental (or other) law is passed, Congress must appropriate enough funds to implement and enforce it. Indeed, developing and adopting a budget is the most important and controversial thing the executive and legislative branches do. Developing a budget involves answering two key questions: What resource use and distribution problems will be addressed, and how much of the limited tax revenues will be used to address each problem?

Next, regulations for implementing the law are drawn up by the appropriate government department or agency. Groups try to influence how the regulations are written and enforced; some of the affected parties may even challenge the final regulations in court. Finally, the department or agency implements and enforces the approved regulations. Proponents or affected groups may take the agency to court for failing to implement and enforce the regulations or for enforcing them too rigidly.

The net result of this fragmentation, lack of funding, and influence of the regulated on the regulators is *incremental decision making*. As a result, only small changes are usually made in existing policies and programs.

Solutions: How Can Individuals Affect Environmental Policy? A major theme of this book is that individuals matter. History shows that significant change comes from the *bottom up*, not the top down. Without grassroots political action by millions of individual citizens and organized groups, the air you breathe and the water you drink today would be much more polluted, and much more of the earth's biodiversity would have disappeared.

Individuals can influence and change government policies in constitutional democracies in several ways. They can **(1)** vote for candidates and ballot measures; **(2)** contribute money and time to candidates seeking office; **(3)** lobby, write, fax, E-mail, or call elected representatives, asking them to pass or oppose certain laws, establish certain policies, and fund various programs (Appendix 5); **(4)** use education and persuasion; **(5)** expose fraud, waste, and illegal activities in government (whistle-blowing); **(6)** file lawsuits; and **(7)** participate in grassroots activities to bring about change or enforce existing laws and regulations.

Solutions: What Are the Three Types of Environmental Leadership? There are three types of environmental leadership. One is *leading by example*, in which people use their own lifestyles to show others that change is possible and beneficial.

SOLUTIONS

Environmentalists and their supporters have persuaded the U.S. Congress to enact a number of important federal environmental and resource protection laws, as discussed throughout this text and listed in Appendix 3. These laws seek to protect the environment using the following approaches:

- *Setting standards for pollution levels or limiting emissions or effluents for various classes of pollutants* (Federal Water Pollution Control Act and Clean Air Acts)

- *Screening new substances for safety before they are widely used* (Toxic Substances Control Act)

- *Requiring comprehensive evaluation of the environmental impact of an activity before it is undertaken by a federal agency* (National Environmental Policy Act)

- *Setting aside or protecting various ecosystems, resources, and species from harm* (Wilderness Act and Endangered Species Act)

- *Encouraging resource conservation* (Resource Conservation and Recovery Act and National Energy Act)

Some environmental laws contain glowing rhetoric about goals but little guidance about how to meet them, leaving this task to regulatory agencies and the courts. In other cases, the laws or presidential executive orders specify one or more of the following general principles for setting regulations:

- *No unreasonable risk*: food regulations in the Food, Drug, and Cosmetic Act

- *No risk*: the zero-discharge goals of the Safe Drinking Water and Clean Water Acts

- *Standards based on best available technology*: the Clean Air, Clean Water, and Safe Drinking Water Acts

- *Risk–benefit balancing* (Section 8-6): pesticide regulations and a presidential executive order signed by President Clinton in 1993 that requires all government regulatory agencies to include risk assessment in all of their decisions (something similar to what anti-environmental vice president Dan Quayle tried but failed to accomplish during the Bush administration)

- *Cost–benefit balancing* (Spotlight, p. 202): the Toxic Substances Control Act and a presidential executive order (initiated by President Reagan) that gives the Office of Management and Budget the power to delay indefinitely, or even veto, any federal regulation not proven to have the least cost to society

Critical Thinking

Pick one of the U.S. environmental laws listed in Appendix 3 (or in the country where you live). Use the library or the Internet to evaluate the law's major strengths and weakness. Decide whether the law should be weakened, strengthened, or abolished and explain why. List the three most important ways you believe the law should be strengthened or weakened.

Another involves *working within existing economic and political systems to bring about environmental improvement, often in new, creative ways.* People can influence political elites by campaigning and voting for candidates and by communicating with elected officials (Appendix 5). They can also work within the system by choosing an environmental career (Individuals Matter, right).

A third type of leadership involves *challenging the system and basic societal values, as well as proposing and working for better solutions to environmental problems.* Leadership is more than being against something; it also involves coming up with better ways to accomplish various goals. All three types of leadership are needed. For an inspiring example of what can be done in a short time, see the Guest Essay on pages 418–419.

Solutions: How Can We Make Government More Responsive to All Citizens? According to most political analysts, *the biggest problem that keeps elected officials from being more responsive to the environmental and other needs and problems of all citizens is that the enormous amounts of money needed to run for office* or to get reelected come mostly from wealthy and powerful individuals and corporate interests. Most analysts and about 80% of citizens polled agree that the U.S. political system is based more on "money talks" than on citizens' needs—one reason so many Americans don't bother to vote. Even those who do vote often feel they're simply choosing between the lesser of two evils. Many analysts see drastic reform in election financing as the key to making the government more responsive to all citizens (Solutions, p. 214).

7-7 ENVIRONMENTAL AND ANTI-ENVIRONMENTAL GROUPS

What Are the Roles of Mainstream Environmental Groups? Many types of environmental groups are at work at the local, state, national, and international levels (Appendix 1). Environmental organizations range from multimillion-dollar *mainstream* groups, led by chief executive officers and staffed by experts, to *grassroots* neighborhood groups formed to do battle on local environmental issues. More than 8 million U.S. citizens belong to environmental organizations, which together received $894 million in contributions and income in 1994 (up from $630 million in 1990).

Mainstream environmental groups are active primarily at the national level and to a lesser extent at the state level; often they form coalitions to work together on issues. Some mainstream organizations (such as Greenpeace) funnel substantial funds to local activists and projects (Figure 7-9). The mainstream Sierra Club, for example, prefers grassroots action but still works to influence national environmental legislation. The Environmental Defense Fund and the Natural Resources Defense Council prefer legal action against corporations that degrade the environment and government agencies that fail to enforce environmental laws and regulations.

Some groups focus their efforts on specific issues, such as population (Zero Population Growth), protecting habitats (Wilderness Society and Nature Conservancy, Solutions, p. 472), and wildlife conservation (National Audubon Society, National Wildlife Federation, and the World Wildlife Fund). Other organizations concentrate on education and research (Worldwatch Institute, Rocky Mountain Institute, Population Reference Bureau, and World Resources Institute). Still other groups provide information, training, and assistance to localities and grassroots organizations

Figure 7-9 The Mississippi River between Baton Rouge and New Orleans, Louisiana, is lined with oil refineries and petrochemical plants. Along this corridor, known as "Cancer Alley" because of its abnormally high cancer rates, tons of carcinogenic and mutagenic chemicals leak into groundwater or are discharged into the river. In 1988 an environmental alliance of residents protested chemical dumping and groundwater pollution in their communities by marching the 137 kilometers (85 miles) from Baton Rouge to New Orleans. (Sam Kittner/ Greenpeace)

Environmental Careers

Besides committed earth citizens, the environmental movement needs dedicated professionals working to help sustain the earth. In the United States (and in other developed countries), the *green job market* is one of the fastest-growing segments of the economy. Already some 3 million people in the United States are working in the environmental field, and more than 125,000 new positions are being created each year.

Many employers are actively seeking environmentally educated graduates. They are especially interested in people with scientific and engineering backgrounds and in people with double majors (business and ecology, for example) or double minors.*

Environmental career opportunities exist in a large number of fields: environmental engineering (currently the fastest growing job market), sustainable forestry and range management, parks and recreation, air and water quality control, solid-waste and hazardous-waste management, recycling, urban and rural land-use planning, computer modeling, ecological restoration, and soil, water, fishery, and wildlife conservation and management.

Environmental careers can also be found in education, environmental planning, environmental management, environmental health, toxicology, geology, ecology, conservation biology, chemistry, climatology, population dynamics and regulation (demography), law, risk analysis, risk management, accounting, environmental journalism (Guest Essay, pp. 220–221), design and architecture, energy conservation and analysis, renewable-energy technologies, hydrology, consulting, public relations, activism and lobbying, economics, diplomacy, development and marketing, publishing (environmental magazines and books), and law enforcement (pollution detection and enforcement teams).

Critical Thinking

1. Is the green job market really one of the fastest-growing segments of the economy? Use the library or the Internet to help answer this question.

2. Have you considered an environmental career? Why or why not?

*For details, consult the Environmental Careers Organization (see Appendix 1), *The New Complete Guide to Environmental Careers* (Covelo, Calif.: Island Press, 1993); Nicholas Basta, *The Environmental Career Guide* (New York: Wiley, 1991); Joan Moody and Richard Wizansky, eds. *Earth Works: Nationwide Guide to Green Jobs* (New York: Harper-Collins-West, 1994); and the *Environmental Career Directory* (Detroit: Visible Ink Press, 1993). Environmental jobs are listed in publications such as *Earth Work, Environmental Career Opportunities, Environmental Opportunities, The Job Seeker,* and *EcoNet,* all listed in Appendix 1.

A: None

SOLUTIONS

A growing number of analysts of all political persuasions see *election finance reform* as the single most important way to reduce the influence of so-called money votes. They urge citizens not to get diverted and fragmented over other issues but to focus their efforts on this crucial issue as the key to making government more responsive to ordinary citizens.

According to former senator, presidential candidate, and archconservative leader Barry Goldwater, "The sheer cost of running for office is having a corrosive effect not only on American politics but on the quality of American government. . . . Something must be done to liberate candidates from their dependence on special-interest money."

One suggestion for reducing undue influence by powerful special interests would be to let the people (taxpayers) *alone* finance all federal, state, and local election campaigns, with low spending limits. Candidates and parties could not accept direct or indirect donations from any other individuals, groups, or parties, with *absolutely no exceptions*.

Once elected, officials could not use free mailing, staff employees, or other privileges to aid their election campaigns, nor could they accept donations or any kind of direct or indirect financial aid from any individual, corporation, political parties, or interest group for their future election campaigns, or for any other reason that could even remotely influence their votes on legislation. Violators of this new *Public Funding Elections Act* would be barred from the campaign or removed from office. Anyone making illegal donations would be subject to large fines and possible jail sentences.

Having all elections financed entirely by public funds would cost each U.S. taxpayer only about $5–10 per year (as part of their income taxes) for all federal elections (and a much smaller amount for state and local elections)—a small investment for making democracy more responsive to ordinary people.

With such a reform, elected officials could spend their time governing instead of raising money and catering to powerful special interests. Office seekers would not need to be wealthy, and public service could become a way to help the country and the earth, not a lifelong profession or route to wealth and power. Special-interest groups would be heard because of the validity of their ideas, not the size of their pocketbooks.

Proponents of this reform contend that this is an issue that ordinary people of all political persuasions could work together on. From this fundamental political reform, other democratic reforms could then flow.

The problem is getting members of Congress (and state legislators) to pass a virtually foolproof and constitutionally acceptable plan that would put them on equal financial footing with their challengers—and that might possibly decrease their chances of getting reelected. Supporters of public financing of all elections argue that the way out of this dilemma is to band together to find and elect candidates who pledge to bring about this fundamental political reform, and then vote them out of office if they don't.

Critical Thinking

1. Do you agree or disagree with this approach to election reform? Explain. What role would you take in bringing about or opposing such a reform?

2. Why might the big 10 and other mainstream environmental groups in the United States not support such a reform?

(Citizens' Clearinghouse for Hazardous Waste and the Institute for Local Self-Reliance).

Mainstream groups work within the political system; many have been major forces in persuading Congress to pass and strengthen environmental laws, as well as fighting off attempts to weaken or repeal such laws. However, these groups must continually guard both against having their efforts subverted by the political system they work to change and against losing touch with ordinary people and nature in the insulated atmosphere of national and state capitals.

All of the 10 largest U.S. mainstream environmental organizations—the "Group of 10"—rely heavily on corporate donations, and many of them have corporate executives as board members, trustees, or council members. Proponents of this corporate involvement argue that it is a way to raise much-needed funds and to influence industry; opponents believe that it is a way for corporations to unduly influence environmental organizations. Regardless of the relative merits of these arguments, the net effect has been to cause some internal conflict within these environmental organizations and to drive a wedge between them and many grassroots environmental activists.

What Are the Roles of Grassroots Environmental Groups? The base of the environmental movement in the United States consists of at least 6,000 grassroots citizens' groups organized to protect themselves against local pollution and envi-

Q: How much damage has mining activities done to federal public lands in the United States?

ronmental damage. According to political analyst Konrad von Moltke, "There isn't a government in the world that would have done anything for the environment if it weren't for the citizen groups." Environmental grassroots groups are also active on college campuses and public schools across the United States (Individuals Matter, below).

What Are the Goals and Tactics of the Anti-Environmental Movement? A small but growing number of political and business leaders in the United States and in other developed countries see improving environmental quality as a way to stimulate innovation, increase profits, and create jobs while working with the earth (Section 7-5 and Guest Essay, p. 198). However, leaders of some corporations and many people in positions of economic and political power see environmental laws and regulations as threats to their wealth and power, and they vigorously oppose such efforts.

Since 1980 anti-environmentalists in the United States have mounted a massive campaign to weaken or repeal existing environmental laws and to destroy the credibility and effectiveness of the environmental movement.

Whatever your own position, it is useful to understand the following general legal and political tactics being used against the environmental movement. Start by *establishing an enemy, to create fear and to divert people's attention and energy away from the real issues.* Prey on the fears of ordinary people by labeling environmentalists as the *green menace*—antibusiness, antipeople, antireligious, scaremongering radical extremists who are crippling the economy, hurting small business owners, costing taxpayers billions of dollars in unnecessary regulations, robbing small landowners of the right to do what they want with their land, trying to lock up the natural resources on public lands, and threatening jobs, national security, and traditional values.

When accused of using wild exaggeration and smear tactics against environmentalists to raise money and spread distrust, hate, and fear, Ron Arnold, one of the leaders of the anti-environmental Wise-Use movement (Case Study, p. 217), said, "Facts don't really matter. In politics, perception is reality."

Environmental Action on Campuses

INDIVIDUALS MATTER

Since 1988 there has been a boom in environmental awareness on college campuses and some public schools across the United States. Much of this momentum began in 1989, when the Student Environmental Action Coalition (SEAC) at the University of North Carolina at Chapel Hill held the first national student environmental conference on the UNC campus.

SEAC groups are active on 700 campuses, and the National Wildlife Federation's Campus Ecology Program has groups on 578 campuses.* Most student environmental groups work with members of the faculty and administration to bring about environmental improvements on their own campuses or schools and in their local communities.

Many of these groups focus on making an environmental audit of their own campuses or schools; they then use the data gathered to propose changes that will make their campuses or schools more ecologically sustainable, usually saving them money in the process.** In addition, students who have learned to do careful research on and develop solutions for environmental problems will be able to use these skills the rest of their lives, wherever they live.

Such audits have resulted in numerous improvements. For example, an energy management plan developed by Morris A. Pierce, a graduate student at the University of Rochester, was adopted by that school's board of trustees.

Using this plan, a capital investment of $33 million is projected to save the university $60 million over 20 years. At the University of Colorado at Boulder, students implemented a sophisticated system that recycles 35–40% of the university's solid waste.

Students at Harvard University established a Green-Cup competition among dorms to save energy and water. This program has spread to several other universities. At Dartmouth College, students have established an organic farm to supply the food service with vegetables and herbs.

Critical Thinking

What student environmental groups (if any) are active at your school? How many people actively participate in these groups? What environmental beneficial things have they done? What actions (if any) taken by such groups do you disagree with? Why?

*These efforts are described in *Ecodemia: Campus Environmental Stewardship at the Turn of the 21st Century* (Washington, D.C.: National Wildlife Federation, 1995) and in the *Campus Environmental Yearbook*, published annually by the National Wildlife Federation.

**Details for conducting such audits are found in *Campus Ecology: A Guide to Assessing Environmental Quality and Creating Strategies for Change* by April Smith and SEAC, and in *Green Lives, Green Campuses*, written by Jane Heinze-Fry as a supplement to this textbook.

A: $33–72 billion, with most cleanup costs to be paid by taxpayers

Here are some of the other basic tactics used. *Weaken and intimidate.* Infiltrate and spy on environmental groups to learn their plans and to create dissension and mistrust. Slap individual activists and small grassroots groups with nuisance lawsuits. Fire whistle-blowers who expose harmful or illegal environmental practices, or shift them to less influential jobs. Intimidate and threaten individual scientists, scientific writers, and mainstream scientific professional groups that stand in the way of corporate antienvironmental goals by lawsuits and threats of lawsuits.

Threaten or use violence. If environmental activists become too effective, harass them with phone calls threatening their lives or the lives of their family members. Try to have them fired, kill their pets, trash their homes and offices, cut phone lines, slash tires, sabotage their cars, plant drugs in their cars and notify the police, burn their houses or businesses, assault them, fire warning shots into their houses or cars—all things that have happened to environmentalists in the United States in recent years. Former Interior Secretary James Watt publicly declared in 1990, "If the troubles from the environmentalists cannot be solved in the jury box or the ballot box, perhaps the cartridge box should be used."

Influence public opinion. Get reporters with prominent newspapers to run articles that favor anti-environmental interests and sow seeds of doubt about the claims of the environmental movement. Commission books and articles, establish slick magazines, set up think tanks, and hire PR firms to deny that environmental problems exist or are serious. Use fax and letter-writing campaigns against corporations that provide financial support for TV or radio programs critical of environmentally harmful business practices or anti-environmental concerns.

Either don't collect data, or keep it secret. Then claim that nothing can be done about a problem because of a lack of information. Exert political pressure to weaken government studies that are unfavorable to your position, limit their availability to the public, and decrease funding for their agencies.

Exploit the built-in limitations of science and public ignorance about the nature of science to sow seeds of doubt. Mislead the public by saying that claims against your position have not been scientifically proven. If that doesn't work, imply that scientific findings that are contrary to your position should not be taken seriously because science can't really prove anything and can only put forth theories (Section 3-1). Ignore consensus science and instead focus on frontier science when it supports your position. Undermine the credibility of mainstream science by claiming that consensus science derived from exhaustive analysis and peer review is not sound science because it is a conspiracy by the scientific establishment to suppress minority dissenting views. When frontier science goes against your position, point out that such results are only preliminary and should not be taken seriously. These tactics should drive scientists and environmentalists crazy and undermine public support of environmental causes.

Build up your public environmental image. Set up a few showcase projects and then run print and TV ads portraying what you are doing to protect the environment. Form or donate money to well-publicized partnerships with major environmental groups to build up your image and increase profits.

Delay and wear out reformers. Set up time-consuming and costly legal and bureaucratic roadblocks to forestall change and tire reformers out. Wear out, financially drain, and tie up your adversaries in court. Use paralysis by analysis to delay action by appointing blue ribbon panels to study problems and make recommendations. Pay little attention to their advice, and then call for more research (which is usually needed but can also be an excuse for inaction).

Urge the president and Congress to require that all government environmental decisions and regulations be evaluated primarily by cost–benefit and risk–benefit analysis. Then try to load the evaluation panels with experts who support your position. If this doesn't work, challenge the analyses in courts. The built-in limitations (Spotlight, p. 202, and Section 8-6), information gaps, and vulnerable assumptions of many such analyses could take years to unravel. All of this would be achieved without trying to repeal existing environmental laws (a politically unpopular cause).

Support unenforceable legislation and regulations. This undermines confidence in government, splits up opposition groups, and helps convince the public that real change is impossible.

When unfavorable environmental laws are passed, urge legislators not to fund the laws. Passing the cost of federally unfunded environmental mandates on to financially strapped states and cities will generate intense political pressure against such laws by citizens, mayors, and governors. Because Congress is unlikely to come up with funding for these laws, they will probably be weakened, unenforced, or repealed.

Require federal and state governments to compensate property owners whenever environmental, health safety, or zoning laws either limit how the owners can use their property or decrease its financial value—something called a regulatory taking. This will make such laws and regulations too expensive to enforce and will keep new ones from being established. Environmentalists charge that such a takings law would also require taxpayers to pay

Q: In tropical and temperate areas, how long does it take to renew 2.54 centimeters (1 inch) of topsoil?

The Wise-Use Movement

Since 1988, several hundred local and regional grass-roots groups in the United States have formed a national anti-environmental coalition called the *Wise-Use movement*. Much of their money comes from real estate developers and from timber, mining, oil, coal, and ranching interests and traditional supporters of right-wing causes.

According to Ron Arnold, who played a key role in setting up this movement, the specific goals of the Wise-Use movement are as follows:

- *Cut all old-growth forests in the national forests and replace them with tree plantations.*

- *Modify the Endangered Species Act so that economic factors override preservation of endangered and threatened species.*

- *Eliminate government restrictions on wetlands development.*

- *Open all national parks, national wildlife refuges, and wilderness areas to oil drilling, mining, off-road vehicles (ORVs), and commercial development.*

- *Do away with the National Park Service and launch a 20-year construction program of new concessions run by private firms in the national parks.*

- *Continue mining on public lands under the provisions of the 1872 Mining Law (p. 338), which allows mining interests not only to pay no royalties to taxpayers for hard-rock minerals they remove, but also to buy public lands for a pittance.*

- *Recognize private property rights to mining claims, water, grazing permits, and timber contracts on public lands; do not raise fees for these activities.*

Ideally, sell resource-rich public lands to private enterprise.

- *Provide civil penalties against anyone who legally challenges economic action or development on federal lands.*

- *Allow pro-industry (Wise-Use) groups or individuals to sue as "harmed parties" on behalf of industries threatened by environmentalists.*

Most anti-environmental groups either affiliated with or generally supportive of the Wise-Use movement and its offshoot, the *Alliance for America* (formed in 1991), use environmentally friendly names.* Here are a few examples: the *U.S. Council for Energy Awareness* (nuclear power industry), *America the Beautiful* (packaging industry), *Partnership for Plastics Progress* (plastics industry), *American Council on Science and Health* (food and pesticide industries), *National Wetlands Coalition* (real estate developers and oil and gas companies), *Global Climate Coalition* (50 corporations and trade associations opposed to reducing fossil-fuel use to slow the rate of global warming), *Blue Ribbon Coalition* (off-road vehicle users and manufacturers who want all public lands opened up to ORVs), *American Forest Resource Alliance* (timber and logging companies), *American Farm Bureau Federation* (agricultural chemicals industries), and *People for the West* (mining industry).

*For lists and information about these organizations, see *The Greenpeace Guide to Anti-Environmental Organizations* (1993, Odonian Press, Box 7776, Berkeley, CA 94707); *Masks of Deception: Corporate Front Groups in America* by Mark Megalli and Andy Friedman (1993, Essential Information, P.O. Box 19367, Washington, DC 20036); *Let the People Judge: A Reader on the Wise-Use Movement* (Covelo, Calif.: Island Press, 1993); and *The War Against the Greens* (San Francisco: Sierra Club Books, 1994).

The Wise-Use movement also includes the *Sahara Club*, which advocates violence against environmentalists in defense of its members' "right" to dirt-bike in wilderness areas. One of its leaders tells audiences to "Throw environmentalists off the bridge. Water optional," and that "You can't reason with eco-freaks, but you sure can scare them." Another group is *Citizens for the Environment* (an industry-backed education group that counters environmental ideas with such slogans as "Recycling doesn't save forests," "Packaging prevents waste," and "Global warming and ozone depletion are hoaxes").

According to Jay D. Hair, former president of the National Wildlife Federation, the Wise-Use movement is "merely a wise disguise for a well-financed, industry-backed campaign that preys on the economic woes and fears of U.S. citizens. . . . These organizations with benign-sounding names are not grass-roots—they are astro-turf laid down with big corporate money. And they are out to . . . ensure that certain special interests will be allowed to continue to pollute and exploit public resources for private profit."

According to environmentalist B. J. Bergman, the basic principles of the Wise-Use movement are, "If it grows, cut it; if it flows, dam it; if it's underground, extract it; if it's swampy, fill it; if it moves, kill it."

Critical Thinking

Do you agree or disagree with each of the major goals of the Wise-Use movement? Explain. What positive effects has the Wise-Use movement had on the environmental movement?

people not to do something they shouldn't (or don't really plan to) be doing in the first place: *pay me and I won't pollute, develop this land, build an incinerator or landfill here, or fill in this wetland.* Indeed, some people may go into the no-pollution, no-development business by buying up land at a low price and then make a profit by getting taxpayers to pay them not to develop the land.

Environmentalists recognize the need to respect private property rights. Indeed, the desire to protect health and property values is the main driving force behind the opposition of grassroots environmentalists to incinerators, landfills, and toxic waste dumps, and some forms of development in their communities. However, environmentalists don't believe that the solution is to replace the *polluter- or degrader-pays* principle found in most environmental laws with a *taxpayer-pays* approach.

Some government officials and environmentalists are looking for solutions to this difficult and highly controversial problem without having to spend large sums of money or being forced not to enforce most environmental, safety, and health regulations. One possibility is to give tax deductions to landowners who agree not to cut timber, farm, develop, drain wetlands, or harm endangered species on their land. In other words, the government could reward people for not carrying out earth-degrading activities on property they own. Another possibility is to eliminate or modify environmental regulations that go too far or that are unjustly enforced (Solutions, right).

A final anti-environmental tactic is *divide and conquer.* Keep people and interest groups fighting with one another so they can't get together on vital issues that threaten the status quo (such as election campaign financing (Solutions, p. 214). Give corporate money to mainstream environmental organizations to influence them and to create dissension between mainstream and grassroots organizations. At the same time, donate funds to elected officials to influence them to weaken or eliminate environmental, health, and worker safety laws and by funding mainstream and grassroots anti-environmental groups (Case Study, p. 217).

Whom Should We Believe? We need to ask tough questions of people on both sides of environmental issues and gather and carefully evaluate the evidence for each position by using the techniques of critical thinking (Guest Essay, p. 52). In addition, we need to distinguish between frontier and consensus science when claims are made (Section 3-1).

To evaluate statements by environmentalists and anti-environmentalists alike, citizens, elected officials, and environmental reporters are urged to identify the consensus of most of the scientists in the particular field involved and not give equal weight to a small minority of scientists in such fields who disagree with the current consensus view. Such reporting under the guise of balance may make a story more interesting, but it represents an unbalanced approach that can mislead the public.

One problem is that as the *easy work* has been done, the focus of environmental issues has shifted to more complex and controversial environmental problems that are harder to understand and solve. Examples are global climate change, ozone depletion, biodiversity protection, nonpoint water pollution (such as runoff from farms and lawns), and protection of unseen groundwater. Explaining such complex issues to the public and mobilizing support for the often controversial solutions is quite difficult (Guest Essay, pp. 220–221).

7-8 GLOBAL ENVIRONMENTAL POLICY

Should We Expand the Concept of Security?
Countries have legitimate *national security* interests, but without adequate soil, water, clean air, and biodiversity, no nation can be secure. Thus, to a growing number of policy analysts, national security ultimately depends on *global environmental security* based on sustainable use of the ecosphere (Guest Essay, p. 198).

Countries are also legitimately concerned with *economic security.* However, because all economies are supported by the ecosphere (Figure 7-3 and Guest Essay, p. 198), economic security also depends on environmental security.

Proponents of emphasizing environmental security propose that national governments have a council of advisers made up of highly qualified experts in environmental, economic, and military security. Any major decision would require integrating all three security concerns. These analysts also call for all countries to make environmental security a major focus of diplomacy and government policy at all levels. What do you think?

What Progress Has Been Made in Developing International Environmental Cooperation and Policy? Since the 1972 UN Conference on the Human Environment was held in Stockholm, Sweden, some progress has been made in addressing environmental issues at the global level. Today, 115 nations have environmental protection agencies, and more than 170 international environmental treaties have been signed concerning issues such as endangered species, ozone depletion, ocean pollution, global warming, biodiversity, and export of hazardous waste. The 1972 conference also created the UN Environment Programme (UNEP) to negotiate environmental treaties and to help implement them.

In June 1992 the second UN Conference on the Human Environment, known as the *Rio Earth Summit,*

Q: Worldwide, how much topsoil is eroded each year?

Environmentalists agree that some government laws and regulations go too far and that bureaucrats sometimes develop and impose ridiculous and excessively costly regulations. But they argue that the solution is to stop regulatory abuse, not throw out or seriously weaken the body of laws and regulations that help protect the public good.

According to environmentalist William Ashworth,

If we wish to make progress, it will do us no good to replace one failed system with another that failed just as badly. Government regulation, after all, didn't fall out of the sky; it was erected, piece by piece, as an attempt to deal with the damage caused by unrestrained property rights and the unregulated free-market system. . . . We do not need to deconstruct regulation, but to reconstruct it.

To accomplish this, a growing number of analysts urge environmentalists to take a hard look at existing environmental laws and regulations. Which laws (or parts of laws) have worked and why? Which ones have failed, and why? Which government bureaucracies concerned with developing environmental and resource regulations have either abused their power or have not been responsive enough to the needs of the people? How can such abuses be corrected? What existing environmental laws (or parts of laws) should be repealed or modified?

How can a balanced program of regulation and market-based approaches (Table 7-1) be used to achieve environmental goals? What environmental problems lend themselves to market-based approaches (free-market environmentalism), and which ones do not? How can the principles of pollution prevention and waste reduction become the guiding principles for environmental legislation and regulation? What is the minimum amount of environmental legislation and regulation needed?

These are important issues that environmentalists, business leaders, elected officials, and government regulators need to address with a cooperative, problem-solving spirit. Too much is at stake in terms of economic and environmental health to fall into confrontational patterns that inhibit change and a creative search for win–win solutions.

Critical Thinking

Analyze a particular environmental law in the United States (Appendix 3) or in the country where you live to come up with ways in which it could be improved. Develop a strategy for bringing about such changes.

---

was held in Rio de Janeiro, Brazil. More than 100 heads of state, thousands of officials, and more than 1,400 accredited nongovernment organizations (NGOs) from 178 nations met to develop plans for addressing environmental issues.

The major official results included **(1)** an *Earth Charter*, a nonbinding statement of broad principles for guiding environmental policy that commits countries that sign it to pursue sustainable development and work toward eradicating poverty; **(2)** *Agenda 21*, a nonbinding detailed action plan to guide countries toward sustainable development and protection of the global environment during the 21st century; **(3)** a *forestry agreement* that is a broad, nonbinding statement of principles of forest management and protection; **(4)** a *convention on climate change* that requires countries to use their best efforts to reduce their emissions of greenhouse gases; **(5)** a *convention on protecting biodiversity* that calls for countries to develop strategies for the conservation and sustainable use of biological diversity; and **(6)** establishment of the *UN Commission on Sustainable Development*, composed of high-level government representatives charged with carrying out and overseeing the implementation of these agreements.

Many environmentalists were disappointed because these accomplishments consist of nonbinding agreements without sufficient incentives for their implementation. The convention on climate change, for example, lacks the targets and timetables for stabilizing carbon dioxide emissions favored by all major industrial countries except the United States (which signed the treaty only after such items were eliminated). The forest-protection statement was so watered down that most environmentalists consider it virtually useless. In addition, countries did not commit even the minimum amount of money that conference organizers said was needed to begin implementing Agenda 21.

In 1997, 5 years after the conference, there was little improvement in the major environmental problems discussed at the Rio summit. Does this mean that the conference was a failure? No, for two reasons. *First*, it gave the world a forum for discussing and seeking solutions to environmental problems. This led to general agreement on some key principles that, with enough political pressure from the bottom up, could be implemented or improved.

Second, paralleling the official meeting was a Global Forum that brought together 18,000 people

---

A: About 24 billion metric tons (26 billion tons)

GUEST ESSAY

Covering the Environment

Andrew C. Revkin

Andrew (Andy) C. Revkin, currently an environmental reporter for The New York Times, *has written about the environment and science since 1982. He is author of two widely acclaimed environmental books:* The Burning Season: The Murder of Chico Mendes and the Fight for the Amazon Basin Forest *(Plume Paperback, 1994), which was the basis of the Emmy-winning HBO film of the same name, and* Global Warming: Understanding the Forecast *(Abbeville Press, 1992). He is formerly a senior editor of* Discover *magazine and senior writer at* Science Digest. *He has won more than a half-dozen national writing awards, including the Robert F. Kennedy award for* The Burning Season *and the American Association for the Advancement of Science/Westinghouse Science Journalism Award. He has also taught a course on environment and energy reporting at the Columbia University Graduate School of Journalism.*

When I traveled to Brazil in 1989 to write a biography of Chico Mendes [Figure 16-12], the slain leader of the movement to save the Amazon rain forest, Chico's friends were at first very suspicious. "Why should we talk to you?" they asked me. "How do we know you are going to tell the truth?"

They said they were skeptical because, shortly after the murder, dozens of journalists had flown in for a day or two and then left. As a result, many of the articles or television reports were simplistic or inaccurate. Only after I had stayed for weeks, and then months, did people open up. And only then did the complexities of the story of the invasion of the rain forest and the struggle to defend it start to make a little sense.

But I was very lucky. For the first time in my career, I had the luxury of being able to spend substantial time on the story. Most often, a reporter has a few hours, or maybe a few days, or—rarely—a week or two to figure out how to explain a complicated subject in an interesting and accurate way.

The lack of time affects journalists covering everything from war to sports. Something that is news today doesn't stay news for very long. But for a writer covering the environment, other factors combine with deadline pressure to make the challenge of effective communication particularly difficult.

First there is the complexity and unfamiliarity of science. A baseball writer or a political writer can assume at least some basic knowledge on the reader's part of the rules of the game. But when writing about the deterioration of a diffuse layer of ozone high in the atmosphere that helps block harmful ultraviolet radiation, almost every step must be explained. It is no easy task to fit in all the essential ideas while still telling a story that does not cause readers to flip to the comics or movie reviews.

Making the task more difficult is the widespread lack of understanding of some of the most basic scientific concepts—not just among readers, but also among the editors who decide what stories make front pages or evening broadcasts. Recently, a biology professor at Oberlin College, Michael Zimmerman, designed a survey to gauge the basic scientific literacy of Americans. He decided to send a little scientific quiz to the managing editors of the nation's 1,563 daily newspapers. He was not trying to pick on the editors. He just figured that they represented a decent cross section of educated America.

The results of the survey were disquieting, to say the least. When asked if it were true or false that "dinosaurs and humans lived contemporaneously," only 51% of the editors disagreed strongly with the statement. Thirty-seven percent either agreed or had no opinion on the matter. That means a big chunk of our newspapers are being edited by people who subscribe, at least tacitly, to what amounts to the "Flintstones Theory" of evolution.

Another impediment to good environmental reporting is something one of my journalism professors used to call "the MEGO factor"—with the acronym standing for "my eyes glaze over." Editors are bombarded with so much information every day that they become numb to its significance. If a reporter asks to write an update on toxic chemicals in a river or the latest findings on climate change at many papers, magazines, and television networks, the response is likely to be along the lines of, "Haven't we done that story?" Nothing causes an editor's eyes to glaze faster than a complicated, subtle topic.

Another impediment to effective environmental journalism is the endless appetite for the sound bite or

from more than 1,400 NGOs in 178 countries. These NGOs worked behind the scenes to influence official policy, formulated their own agendas and treaties for helping sustain the earth, learned from one another, and developed a series of new global networks, alliances, and projects. In the long run, these newly formed networks and alliances may play the greatest role in helping to monitor, support, and implement the commitments and plans developed by the formal conference.

Bringing about environmental improvement at local and national levels is difficult enough, but doing

Q: How much land has become desertified during the past 50 years?

snappy quote. Journalists often try to create balance in a story by quoting a yea-sayer and a nay-sayer. As one journalist put it, echoing a famous law of physics, "For every Ph.D., there is an equal and opposite Ph.D." This is a quick and easy way of establishing that the reporter has no bias. The problem is that the loudest voices on the issue—whether it's abortion or pollution—are often the most suspect. Scientists who can deliver well-honed sound bites may have been spending more time in front of microphones than microscopes. Too often, such stories confuse instead of inform.

Finally, there is the need for articles to be timely, to have a "news peg" to hang on, in journalistic parlance. In stark contrast to traditional news events, most environmental problems develop in a creeping, almost imperceptible fashion. Few writers reported on the dangerous conditions of chemical plants overseas, the risks from human errors at nuclear power plants, or the narrow channel in Prince William Sound, Alaska, before the chemical catastrophe at a pesticide plant in Bhopal, India [Case Study, p. 426], the accident at Three Mile Island nuclear power plant in Pennsylvania [Section 19-4], or the grounding of the *Exxon Valdez* oil tanker in Alaska's Prince William Sound [Section 11-6].

The tendency is to wait for a press release before covering a topic. In the 1980s, networks and many national publications gave top billing to a report on risks to children from Alar, a pesticide used on apple crops. Apple growers in the Northwest were nearly bankrupted as apple juice sales plummeted. Soon it became clear that the threat was drastically overstated, and most apples were not even treated with the compound. This kind of reactive reporting can end up confusing readers and creating an air of cynical skepticism about environmental threats.

One result, according to many scientists and health experts, is that the public has become overly fearful of such things as toxic dumps, and does not take seriously such threats to humanity's future as the continuing growth in concentrations of heat-trapping gases in the atmosphere. We fret about parts per billion of certain synthetic chemicals in food, while happily heading to the beach for a new dose of ultraviolet radiation—and increasingly getting there in fleets of gas-guzzling Blazers

and Broncos and thus contributing to global warming.

Many journalists have worked long and hard to overcome some of these impediments to coverage of environmental issues. Editors are becoming more informed. Reporters have created a Society of Environmental Journalists, which is fostering a daily debate on the Internet and its newsletter on ways to do a better job. The result is a steady improvement in the national debate on such issues as how to make regulations work more effectively to cut pollution, how to use land in ways that mesh the needs of people and threatened species, and how to mesh perception and reality when it comes to the risks of chemicals and food in the environment.

Ultimately, though, part of the responsibility rests with readers as well as writers. Consumers of news on television or in the print media become better informed when they treat reports with a skeptical eye and seek a variety of sources. Just as good journalism results when a reporter seeks a multiplicity of sources for a story, good citizenship results when people seek to understand an issue by relying on more than one medium for their information—consuming magazines, newspapers, television reports, books, using the Internet, and more.

Critical Thinking

1. Do your eyes generally glaze over when you read a newspaper or magazine story covering some complex environmental issue such as ozone depletion, greenhouse warming, biodiversity, or environmental economics? How does this relate to the quality of information you are receiving about issues that prominent scientists and environmentalists say are vital?

2. How well do you meet your responsibility as a consumer of information to seek a variety of sources on key environmental issues? Have you become skeptical of environmental stories? Why? Have you become too skeptical of such stories, assuming that virtually all of them are suspect?

this at the international level is like trying to change the direction of a huge floating iceberg. However, history shows that in the long run persistent grassroots pressure can bring about important social and environmental changes. The situation is difficult and challenging, but not hopeless.

Can We Develop Earth-Sustaining Political and Economic Systems Over the Next Few Decades? Environmentalists say we can if we care enough to make the necessary commitment. Despite the claims of anti-environmentalists, environmentalists are not people-hating, antibusiness people (although any diverse movement has a very small number of people who take such radical positions). Instead, environmentalists have an exciting and positive vision of a new earth-sustaining, people-sustaining, and people-empowering society.

Environmentalists believe that making this new cultural transition over the next 40–50 years should be our most important and urgent goal. Making this new cultural change will be controversial, and like all significant change it will not be predictable, orderly, or painless.

In making this new cultural change, many environmentalists urge us not to fall into the trap of *thinking that we can save the earth*. We are absolutely dependent on the earth, but the earth's existence and functioning does not depend on us. We can impair some of the planet's natural processes at local, regional, and perhaps at global levels, and we can cause the premature extinction of species. However, fossils and other evidence show that on a time scale of hundreds to millions of years the earth is resilient and adaptable and will continue functioning and changing with or without us. What is at stake is not the earth, but the future quality of life for most human beings and—if we go too far—the existence of our species and many other species.

As the wagon driver said when they came to a long, hard hill, "Them that's going on with us, get out and push. Them that ain't, get out of the way."

ROBERT FULGHUM

CRITICAL THINKING

1. The primary goal of all current economic systems is to maximize economic growth by producing and consuming more and more economic goods. Do you agree with that goal? Explain. What are the alternatives?

2. Do you believe that cost–benefit analysis should be the primary method for making all decisions about how limited federal, state, and local government funds are used? Explain. If not, what are the alternatives?

3. Do you favor internalizing the external costs of pollution and unnecessary resource waste? Explain. How might it affect your lifestyle? Wildlife? Any children you might have?

4. a. Do you believe that we should establish optimal levels or zero-discharge levels for most of the chemicals we release into the environment? Explain. What effects would adopting zero-discharge levels have on your life?

   b. Should we assume that all chemicals we release or propose to release into the environment are potentially harmful until proven otherwise? Explain. What effects would adopting this principle have on your life?

5. Do you agree or disagree with the proposals various analysts have made for sharply reducing poverty as discussed on pp. 207–208? Explain.

6. Do you agree or disagree with the guidelines for an earth-sustaining economy discussed on pp. 208–210? Explain.

7. Suppose that a presidential candidate ran on a platform calling for the federal government to phase in a tax on gasoline so that, over 5–10 years, the price of gasoline would rise to $3–5 a gallon (as is the case in Japan and most western European nations). The candidate argues that this tax increase is necessary to encourage conservation of oil and gasoline, to reduce air pollution, and to enhance future economic, environmental, and military security. Some of the tax revenue would be used to provide tax relief or other aid to people with incomes below a certain level (the poor and lower middle class), who would be hardest hit by such a consumption tax. Income taxes on the poor and middle class would be reduced by an amount roughly equal to the increase in gasoline taxes. Would you vote for this candidate who wants to triple the price of gasoline? Explain.

8. Do you believe that landowners should be compensated by the federal, state, or local government when any government regulation leads to loss in the market value of their property? Explain. What effects might such action have on environmental quality and public health and safety? What are the alternatives?

9. Which of the tactics, if any, used by the anti-environmental movement (pp. 215–218) do you think are appropriate? Explain. Which, if any, of these same tactics have been used by the environmental movement?

*10. List all the economic goods you use; then identify those that meet your basic needs and those that satisfy your wants. Identify any economic wants you would be willing to give up, those you believe you should give up but are unwilling to give up, and those you hope to give up in the future. List what is likely to make you happy and improve the quality of your life. Relate the results of this analysis to your personal impact on the environment. Compare your results with those of your classmates.

*11. Make a concept map of this chapter's major ideas, using the section heads and subheads and the key terms (in boldface type). Look at the inside back cover and in Appendix 4 for information about concept maps.

SUGGESTED READINGS

See Internet Sources and Further Readings at the back of the book, and InfoTrac.

8 RISK, TOXICOLOGY, AND HUMAN HEALTH

The Big Killer

What is roughly the diameter of a 30-caliber bullet, can be bought almost anywhere, is highly addictive, and kills about 8,200 people every day? It's a cigarette.

Cigarette smoking is the single most preventable major cause of death and suffering among adults. The World Health Organization (WHO) estimates that each year tobacco contributes to the premature deaths of at least 3 million people from heart disease, lung cancer, other cancers, bronchitis, emphysema, and stroke. The annual death toll from smoking-related diseases is projected by WHO to reach 10 million by 2020 and 12 million (primarily in developing countries) by 2050—an average of almost 33,000 preventable deaths per day.

In 1993 smoking killed about 419,000 Americans (up from 390,000 in 1985), an average of 1,150 deaths per day (Figure 8-1). This death toll is equivalent to three fully loaded jumbo jets crashing every day with no survivors. Smoking causes more deaths each year in the United States than do all illegal drugs, alcohol (the second most harmful drug after nicotine), automobile accidents, suicide, and homicide combined (Figure 8-1).

The overwhelming consensus in the scientific community is that the nicotine (and probably the acetaldehyde) in inhaled tobacco smoke is highly addictive. Only 1 in 10 people who try to quit smoking succeed—about the same relapse rate as for recovering alcoholics and those addicted to heroin or crack cocaine.

A British government study showed that adolescents who smoke more than one cigarette have an 85% chance of becoming smokers. About 80% of all smokers say they wish they had never started smoking and have tried to quit.

Government agencies and independent economists estimate that smoking costs the United States up to $100 billion a year ($50 billion in medical care alone in 1993). This is an average of $4.10 per pack of cigarettes sold in expenses related to premature death, disability, medical treatment, increased insurance costs, and lost earnings and productivity because of illness.

Many health experts urge that a $2–4 federal tax be added to the price of a pack of cigarettes. This would mean that the users of cigarettes (and other tobacco products), not the rest of society, would pay a much greater share of the health, economic, and social costs associated with their smoking—a *user-pays* approach.

Other suggestions for reducing the death toll and health effects of smoking in the United States include (1) banning all cigarette advertising (as has been done in France), (2) forbidding the sale of cigarettes and other tobacco products to anyone under 21 (with strict penalties for violators), (3) banning all cigarette vending machines, (4) classifying nicotine as an addictive and dangerous drug (and placing its use in tobacco or other products under the jurisdiction of the Food and Drug Administration), and (5) eliminating all federal subsidies and tax breaks to U.S. tobacco farmers and tobacco companies.

All of us face hazards to our health and well-being. Some of them, such as smoking, are avoidable, but others are not. When we evaluate the hazards we face, we must decide whether the risks of damage from each hazard outweigh the short- and long-term benefits and how we can reduce the hazards and minimize the risks.

The discussion in this chapter answers several general questions:

- What types of hazards do people face?
- What is toxicology, and how do scientists measure toxicity?
- What chemical hazards do people face, and how can they be measured?
- What physical hazards do people face from earthquakes and volcanic eruptions?
- What types of disease (biological hazards) threaten people in developing countries and developed countries?
- How can risks be estimated, managed, and reduced?

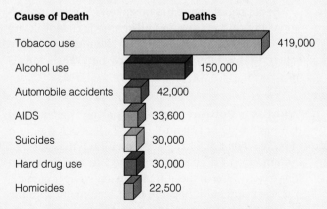

Cause of Death	Deaths
Tobacco use	419,000
Alcohol use	150,000
Automobile accidents	42,000
AIDS	33,600
Suicides	30,000
Hard drug use	30,000
Homicides	22,500

Figure 8-1 Deaths in the United States from tobacco use and other causes in 1993. Smoking is by far the nation's leading cause of preventable death, causing more premature deaths each year than all the other categories in this figure combined. Cardiovascular disease causes about 180,000 of the 419,000 smoking-related deaths per year, followed by 120,000 from lung cancer, and 9,000 from inhaling secondhand smoke. (Data from National Center for Health Statistics)

For the first time in the history of the world, every human being is now subjected to dangerous chemicals from the moment of conception until death.

RACHEL CARSON

8-1 RISK AND HAZARDS

What Is Risk? Risk is the possibility of suffering harm from a *hazard* that can cause injury, disease, economic loss, or environmental damage. Risk is expressed in terms of **probability**: a mathematical statement about how likely it is that some event or effect will occur.

The probability of a risk is expressed as a value ranging from 0 (absolute certainty that there is no risk, which can never be shown) to 1.0 (absolute certainty that there is a risk). For example, the lifetime cancer risk from exposure to a particular chemical may have a probability of 0.000001, or 1 in 1 million. This means that one of every 1 million people exposed to the chemical at a specified average daily dose will develop cancer over a typical lifetime (usually considered to be 70 years).

Risk assessment involves using data, hypotheses, and models to estimate the probability of harm to human health, to society, or to the environment that may result from exposure to specific hazards.

What Are the Major Types of Hazards?
The various kinds of hazards we face can be categorized as follows:

- *Cultural hazards* such as unsafe working conditions, smoking (p. 223), poor diet, drugs, drinking, driving, criminal assault, unsafe sex, and poverty

- *Chemical hazards* from harmful chemicals in the air (Chapter 9), water (Chapter 11), soil (Chapter 12), and food (Chapter 14)

- *Physical hazards* such as ionizing radiation, noise, fire, tornadoes, hurricanes, earthquakes, volcanic eruptions, and floods

- *Biological hazards* from pathogens (bacteria, viruses, and parasites), pollen and other allergens, and animals such as bees and poisonous snakes

8-2 TOXICOLOGY

What Determines Whether a Chemical Is Harmful? Dose and Response
The study of the adverse effects of chemicals on health is called **toxicology**. **Toxicity** is a measure of how harmful a substance is.

The amount of a potentially harmful substance that a person has ingested, inhaled, or absorbed through the skin is called the **dose**, and the amount of resulting type and amount of damage to health are called the **response**.

There are two major types of responses to a harmful dose of a substance. An *acute effect* is an immediate or rapid harmful reaction to an exposure; it can range from dizziness or a rash to death. A *chronic effect* is a permanent or long-lasting consequence (kidney or liver damage, for example) of exposure to a harmful substance.

Whether a chemical is harmful depends on the size of the dose over a certain time, how often an exposure occurs, who is exposed (adult or child, for example), and how well the body's detoxification systems (liver, lungs, and kidneys) work.

Two other factors affecting dose and response are *bioaccumulation* and *biomagnification*. **Bioaccumulation** is an increase in the concentration of a chemical in specific organs or tissues at a level higher than would normally be expected.

The levels of some toxins in the environment can also be magnified as they pass through food chains and webs by a process called **biomagnification** (Figure 8-2). Examples of chemicals that can be biomagnified include long-lived, fat-soluble organic compounds such as the pesticide DDT and PCBs (oily chemicals used in electrical transformers) and some radioactive isotopes (such as strontium-90).

The detection of trace amounts of a chemical in air, water, or food does not necessarily mean that it is there at a level harmful to most people. In some cases, all we may be doing is finding trace levels we could not detect before. Indeed, practically any synthetic or natural chemical, even water, can be harmful if ingested in a large enough quantity. The critical question is how much exposure to a particular toxic chemical causes a harmful response.

Some people have the mistaken idea that all natural chemicals are safe and all synthetic chemicals are harmful. In fact, many synthetic chemicals are quite safe if used as intended, and many natural chemicals are deadly.

What Is a Poison? Legally, a **poison** is a chemical that has an LD_{50} of 50 milligrams or less per kilogram of body weight. An LD_{50} is the **median lethal dose,** or the amount of a chemical received in one dose that kills exactly 50% of the animals (usually rats and mice) in a test population (usually 60–200 animals) within a 14-day period. *LD* stands for lethal dose, and the subscript *50* refers to the percentage of test organisms for which the dose was lethal.

Q: How much of the planet's land is threatened by desertification?

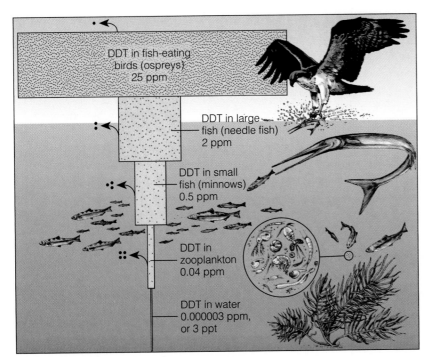

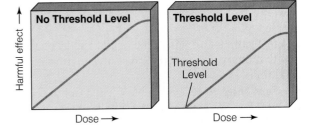

Figure 8-2 Bioaccumulation and biomagnification. DDT is a fat-soluble chemical that can bioaccumulate in the fatty tissues of animals. In a food chain or food web, the accumulated concentrations of DDT can be biologically magnified in the bodies of animals at each higher trophic level. This diagram shows that the concentration of DDT in the fatty tissues of organisms was biomagnified about 10 million times in this food chain in an estuary near Long Island Sound. If each phytoplankton organism in such a food chain takes up from the water and retains one unit of DDT, a small fish eating thousands of zooplankton (which feed on the phytoplankton) will store thousands of units of DDT in its fatty tissue. Then each large fish that eats 10 of the smaller fish will ingest and store tens of thousands of units, and each bird (or human) that eats several large fish will ingest hundreds of thousands of units. Dots represent DDT, and arrows show small losses of DDT through respiration and excretion.

Chemicals vary widely in their toxicity (Table 8-1); some poisons can cause serious harm or death after a single acute exposure at extremely low doses, whereas others cause such harm only at such huge doses that it is nearly impossible to get enough into the body. Most chemicals fall between these two extremes.

How Do Scientists Determine Toxicity?

Three methods are used to determine the level at which a substance poses a health threat; each has certain limitations. One is *case reports* (usually made by physicians) about people suffering some adverse health effect or death after exposure to a chemical. Such information often involves accidental poisonings, drug overdoses, homicides, or suicide attempts.

A second method relies on *laboratory investigations* (usually on test animals) to determine toxicity, residence time, what parts of the body are affected, and (sometimes) how the harm takes place. The third method is *epidemiology*, which involves studies of populations of humans exposed to certain chemicals or diseases.

Case reports are usually the least valuable source for determining toxicity because the actual dose and the exposed person's health status are often not known. However, such reports can provide clues about environmental hazards and suggest the need for laboratory investigations.

Acute toxicity and chronic toxicity are usually determined by tests on live laboratory animals (especially mice and rats, which are small and prolific and can be housed inexpensively in large numbers) and on

Figure 8-3 Two hypothetical dose–response curves. The curve on the left represents harmful effects that occur with increasing doses of a chemical or ionizing radiation; no dose is considered to be safe. The curve on the right shows the response of exposure to a chemical or ionizing radiation in which harmful effects appear only when the dose exceeds a certain threshold level. There is considerable uncertainty and controversy over which of these models applies to various harmful agents because of the difficulty in estimating the response to very low doses. Nonlinear threshold and nonthreshold dose–response curves, in which the graphs are curved instead of being straight lines, have also been observed.

bacteria. Such tests are also made on cell and tissue cultures and chicken egg membranes.

Acute toxicity tests are run to develop a **dose–response curve**, which shows the effects of various doses of a toxic agent on a group of test organisms (Figure 8-3). Such tests are *controlled experiments* in which the effects of the chemical on a *test group* are compared with the responses of a *control group* of organisms not exposed to the chemical. Care is taken to ensure that organisms in each group are as identical

Table 8-1 Toxicity Ratings and Average Lethal Doses for Humans

Toxicity Rating	LD$_{50}$ (milligrams per kilogram of body weight)*	Average Lethal Dose[†]	Examples
Supertoxic	Less than 0.01	Less than 1 drop	Nerve gases, botulism toxin, mushroom toxins, dioxin (TCDD)
Extremely toxic	Less than 5	Less than 7 drops	Potassium cyanide, heroin, atropine, parathion, nicotine
Very toxic	5–50	7 drops to 1 teaspoon	Mercury salts, morphine, codeine
Toxic	50–500	1 teaspoon to 1 ounce	Lead salts, DDT, sodium hydroxide fluoride, sulfuric acid, caffeine, carbon tetrachloride
Moderately toxic	500–5,000	1 ounce to 1 pint	Methyl (wood) alcohol, ether, phenobarbital, amphetamines (speed), kerosene, aspirin
Slightly toxic	5,000–15,000	1 pint to 1 quart	Ethyl alcohol, Lysol, soaps
Essentially nontoxic	15,000 or greater	More than 1 quart	Water, glycerin, table sugar

*Dosage that kills 50% of individuals exposed.

[†]Amounts of substances that are liquids at room temperature when given to a 70.4-kilogram (155-pound) human.

in age, health status, and genetic makeup as possible and that they are exposed to the same environmental conditions.

Fairly high dose levels are used in order to reduce the number of test animals needed, obtain results quickly, and lower costs. Otherwise, tests would have to be run on millions of laboratory animals for many years, and manufacturers couldn't afford to test most chemicals. For the same reasons, the results of high-dose exposures are usually extrapolated to low-dose levels using mathematical models. Then the extrapolated low-dose results on the test organisms are extrapolated to humans to estimate LD$_{50}$ values for acute toxicity (Table 8-1).

According to the *linear dose–response model*, any dose of a toxic chemical or ionizing radiation has a certain risk of causing harm (Figure 8-3, left). With the *threshold dose–response model* (Figure 8-3, right) there is a threshold dose below which no detectable harmful effects occur, presumably because the body can repair the damage caused by low doses of some substances. It is extremely difficult to establish which of these models applies at low doses. To err on the side of safety, the linear or nonthreshold dose–response model is often assumed.

Some scientists challenge the validity of extrapolating data from test animals to humans because human physiology and metabolism are often different from those of the test animals. Others counter that such tests and models work fairly well (especially for

revealing cancer risks) when the correct experimental animal is chosen or when a chemical is toxic to several different species of test animals.

Animal tests take 2–5 years and cost $200,000 to $2 million per substance tested. They are opposed by animal rights groups that want to ban all use of test animals (or to ensure that experimental animals are treated in the most humane manner possible). Scientists are looking for substitute methods, but they point out that some animal testing is needed because the known alternatives cannot adequately mimic the complex biochemical interactions of a live animal.

Another approach to testing for toxicity and identifying the agents causing diseases is **epidemiology**: the study of the patterns of disease or toxicity to find out why some people get sick and others do not. Typically, the health of people exposed to a particular toxic agent or disease organism (the experimental group) is compared with the health of another group of statistically similar people not exposed to these conditions (the control group).

Epidemiology has limitations. For many toxic agents, too few people have been exposed to sufficiently high levels to allow detection of statistically significant differences. Because people are exposed to many different toxic agents and disease-causing factors throughout their lives, it is usually impossible to link conclusively an observed epidemiological effect with exposure to a particular toxic agent. All an epidemiological study can do is to establish strong,

Q: What major food-producing country is doing the most to reduce soil erosion?

moderate, weak, or no statistical associations between a hazard and a health problem. Because epidemiology can be used to evaluate only the hazards to which people have already been exposed, it is rarely useful for predicting the effects of new technologies, substances, or diseases.

Thus, all methods for estimating toxicity levels and risks have serious limitations, but they are all we have. To take this uncertainty into account and minimize harm, standards for allowed exposure to toxic substances and radiation are typically set at levels 100 or even 1,000 times lower than the estimated harmful levels.

Despite their many limitations, carefully conducted and evaluated toxicity studies are important sources of information used to help us understand dose and response effects and to estimate and set exposure standards. But citizens, lawmakers, and regulatory officials must recognize the huge uncertainties and guesswork involved in all such studies.

✦ 8-3 CHEMICAL HAZARDS

What Are Toxic and Hazardous Chemicals?
Toxic chemicals are generally defined as substances that are fatal to over 50% of test animals (LD_{50}) at given concentrations. **Hazardous chemicals** cause harm by **(1)** being flammable or explosive, **(2)** irritating or damaging the skin or lungs (strong acidic or alkaline substances such as oven cleaners), **(3)** interfering with or preventing oxygen uptake and distribution (asphyxiants such as carbon monoxide and hydrogen sulfide), or **(4)** inducing allergic reactions of the immune system (allergens).

What Are Mutagens, Teratogens, and Carcinogens?
Mutagens are agents, such as chemicals and radiation, that cause random *mutations*, or changes in the DNA molecules found in cells. Mutations in a sperm or egg cell may be passed on to future generations and cause diseases such as manic depression, cystic fibrosis, hemophilia, sickle-cell anemia, Down's syndrome, and some types of cancer. Those in other cells are not inherited but may cause harmful effects such as tumors. Although some mutations are harmful, most are of no consequence, probably because all organisms have biochemical repair mechanisms that can find and correct mistakes or changes in the DNA code.

Teratogens are chemicals, radiation, or viruses that cause birth defects while the human embryo is growing and developing during pregnancy, especially during the first 3 months. Chemicals known to cause birth defects in laboratory animals include PCBs, thalidomide, steroid hormones, and heavy metals such as arsenic, cadmium, lead, and mercury.

Carcinogens are chemicals, radiation, or viruses that cause or promote the growth of a malignant (cancerous) tumor, in which certain cells multiply uncontrollably. Many cancerous tumors spread by **metastasis** when malignant cells break off from tumors and travel in body fluids to other parts of the body. There, they start new tumors, making treatment much more difficult.

Because there are more than 100 types of cancer (depending on the types of cells involved), there are many different causes. These include genetic predisposition, viral infections, and exposure to various mutagens and carcinogens.

According to the World Health Organization, environmental and lifestyle factors play a key role in causing or promoting up to 80% of all cancers. Major sources of carcinogens are cigarette smoke (30–40% of cancers), diet (20–30%), occupational exposure (5–15%), and environmental pollutants (1–10%). About 10–20% of cancers are believed to be caused by inherited genetic factors or by certain viruses.

Typically, 10–40 years may elapse between the initial exposure to a carcinogen and the appearance of detectable symptoms. Partly because of this time lag, many healthy teenagers and young adults have trouble believing that their smoking (p. 223), drinking, eating, and other lifestyle habits today could lead to some form of cancer before they reach age 50.

How Can Chemicals Harm the Immune, Nervous, and Endocrine Systems?
Since the 1970s a growing body of research on wildlife and laboratory animals and epidemiological studies of humans indicates that long-term (often low-level) exposure to various toxic chemicals in the environment can cause damage by disrupting the body's immune, nervous, and endocrine systems.

The *immune system* consists of myriad specialized cells and tissues that protect the body against disease and harmful substances by forming antibodies to invading agents and rendering them harmless. Synthetic chemicals, viruses such as HIV, and ionizing radiation that weaken the human immune system leave the body wide open to attacks by allergens and infectious bacteria, viruses, and protozoans. Recent studies of laboratory animals and wildlife as well as epidemiological studies of humans (especially in developing countries) have linked suppression of the immune system to several widely used pesticides.

The human *nervous system* (brain, spinal cord, and peripheral nerves) is also being threatened by synthetic chemicals in the environment. Many poisons are *neurotoxins*, which attack nerve cells (neurons). Examples are **(1)** chlorinated hydrocarbons (DDT, PCBs, dioxins); **(2)** organophosphate pesticides; **(3)** formaldehyde;

(4) various compounds of arsenic, mercury, lead, and cadmium; and **(5)** widely used industrial solvents such as trichloroethylene (TCE), toluene, and xylene.

The *endocrine system* is a complex set of organs and tissues whose actions are coordinated by chemical messengers called *hormones*. These hormones (which are secreted in extremely low levels into the bloodstream) control sexual reproduction, growth, development, and behavior in humans and other animals.

Each type of hormone has a specific molecular shape that allows it to attach only to certain cell receptors; they fit together somewhat like a molecular key fitting into a specially shaped keyhole on a receptor molecule (Figure 8-4, left). Once bonded together, the hormone and its receptor molecule move onto the cell's nucleus to execute the chemical message carried by the hormone.

Case Study: Do Hormone Disrupters Threaten the Health of Wildlife and Humans?

Over the last few years experts from a number of disciplines have been piecing together field studies on wildlife, studies on laboratory animals, and epidemiological studies of human populations indicating that a variety of human-made chemicals can act as *hormone disrupters*.

Some, called *hormone mimics*, are estrogenlike chemicals that disrupt the endocrine system by attaching to estrogen receptor molecules (Figure 8-4, center). Others, called *hormone blockers*, disrupt the endocrine system by preventing natural hormones such as

androgens (male sex hormones) from attaching to their receptors (Figure 8-4, right). There is also growing concern about pollutants that can act as *thyroid disrupters* and cause growth, weight, brain, and behavioral disorders.

So far 51 chemicals, many of them widely used, have been shown to act at extremely low levels as hormone disrupters in wildlife, laboratory animals, and some populations of humans. Examples include **(1)** dioxins (chlorinated hydrocarbons that are unwanted byproducts when chlorine-containing compounds are incinerated, Section 13-7), **(2)** certain PCBs (a group of chlorinated hydrocarbons widely distributed in the environment and capable of being biologically magnified in food chains), **(3)** various chemicals in certain plastics, **(4)** some pesticides, and **(5)** lead and mercury.

Numerous wildlife and laboratory studies reveal various possible effects of estrogen mimics and hormone blockers. Here are a few of many examples: **(1)** ranch minks fed Lake Michigan fish contaminated with endocrine disrupters such as DDT and PCBs failed to reproduce, **(2)** alligators in a Florida lake polluted by a nearby hazardous-waste site had such abnormally small penises and low testosterone levels that they couldn't reproduce, **(3)** rats exposed to PCBs in the womb and infancy tend to be hyperactive, and **(4)** estrogenlike p-nonylphenol (added to PVC plastic) can inhibit the growth of testicles in laboratory animals.

Most natural hormones are broken down or excreted. However, many of the synthetic hormone

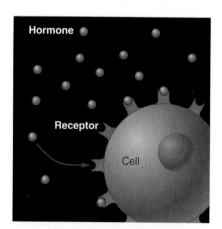 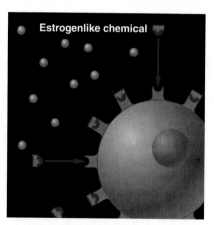 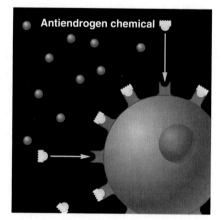

Normal Hormone Process   **Hormone Mimic**   **Hormone Blocker**

Figure 8-4 Hormones are molecules that act as messengers in the endocrine system to regulate various bodily processes, including reproduction, growth, and development. Each type of hormone has a unique molecular shape that allows it to attach to specially shaped receptors on the surface of or inside cells and transmit its chemical message (left). Molecules of certain pesticides and other molecules have shapes similar to those of natural hormones. Some of these hormone impostors, called *hormone mimics*, disrupt the endocrine system by attaching to estrogen receptor molecules (center). Others, called *hormone blockers*, prevent natural hormones such as androgens (male sex hormones) from attaching to their receptors (right). Some pollutants called *thyroid disrupters* may disrupt hormones released by thyroid glands and cause growth and weight disorders and brain and behavioral disorders.

Q: What is the largest source of solid waste in the United States?

impostors are stable, fat-soluble compounds whose concentrations can be biomagnified as they move through food chains and webs (Figure 8-2). Thus, they can pose a special threat to humans and other carnivores dining at the top of food webs.

Here are the results of a few of the many frontier science studies suggesting that hormone disrupters can affect the human endocrine system.*

■ Between the 1940s and early several 1970s, several million pregnant women were given a synthetic estrogen called diethylstilbestrol (DES) to help prevent miscarriage. Later it was found that daughters of women taking DES had a much higher risk of vaginal cancer and their sons had a higher incidence of undescended testicles and testicular cancer.

■ In 1973 estrogen mimics called PBBs accidentally got into cattle feed in Michigan, and from there into beef. Pregnant women who ate the beef (and whose breast milk had high levels of PBBs) had sons with undersized penises and malformed testicles.

■ A higher than normal number of the children of women in Taiwan who ate rice oil contaminated with PCBs in 1979 have suffered growth retardation and slightly lower IQs, and many of the boys have abnormally small penises.

■ During the past 50 years there have been dramatic increases in testicular and prostate cancer in humans almost everywhere.

■ There has been a sharp rise in endometriosis in increasingly younger women in the United States (from a handful of cases 50 years ago to 5 million cases today). This painful inflammation of the uterine lining often causes infertility. A recent study by German scientists found that women with endometriosis were more likely than others to have high levels of PCBs in their blood.

Much more research is need to verify such frontier science findings and to determine whether low levels of most hormone-disrupting chemicals in the environment pose a threat to the human population. In 1996 the National Academy of Sciences began an evaluation of endocrine disrupters and the EPA made evaluation of this potentially serious problem a top research priority.

Why Do We Know So Little About the Harmful Effects of Chemicals? According to risk assessment expert Joseph V. Rodricks, "Toxicologists know a great deal about a few chemicals, a little about many, and next to nothing about most." The U.S. National Academy of Sciences estimates that only about 10% of the 72,000 chemicals in commercial use have been thoroughly screened for toxicity and only 2% have been adequately tested to determine whether they are carcinogens, teratogens, or mutagens. Hardly any of these chemicals have been screened for damage to the nervous, endocrine, and immune systems.

Each year we introduce into the marketplace about 1,000 new chemicals, about whose potentially harmful effects we have little knowledge. Currently, about 99.5% of the commercially used chemicals in the United States are entirely unregulated by federal and state governments.

There are three major reasons for this lack of information. *First*, under existing laws most chemicals are considered innocent until proven guilty. No one is required to investigate whether they are harmful. *Second*, there are not enough funds, personnel, facilities, and test animals to provide such information for more than a small fraction of the many chemicals we encounter in our daily lives.

Third, even if we could make a reasonable estimate of the biggest risks associated with particular technologies or chemicals (a very difficult and expensive thing to do), we know little about their possible interactions with other technologies and chemicals or about the effects of such interactions on human health and ecosystems. For example, just to study the possible three-chemical interactions among the top 500 most widely used industrial chemicals would require 20.7 million experiments—a physical and financial impossibility.

The difficulty and expense of getting information about the harmful effects of chemicals is one reason an increasing number of environmentalists and health officials are pushing for much greater emphasis on *pollution prevention*. This strategy greatly reduces the need for statistically uncertain and controversial toxicity studies and exposure standards. It also reduces the risk posed by potentially hazardous chemicals and products and their possible but poorly understood multiple interactions.

🌐 8-4 PHYSICAL HAZARDS: EARTHQUAKES AND VOLCANIC ERUPTIONS†

What Are Earthquakes? Stress in the earth's crust can cause solid rock to deform until it suddenly fractures and shifts along the fracture, producing a *fault*.

---

* For a readable summary of this research, see Theo Coburn, *Our Stolen Future* (New York: Dutton, 1996).

† **Kenneth J. Van Dellen**, professor of geology and environmental science, Macomb Community College, is the primary author of this section, with assistance from G. Tyler Miller, Jr.

The faulting or a later abrupt movement on an existing fault causes an **earthquake**. An earthquake has certain features and effects (Figure 8-5). When the stressed parts of the earth suddenly fracture or shift, energy is released as shock waves, which move outward from the earthquake's focus like ripples in a pool of water.

One way of measuring the severity of an earthquake is by its *magnitude* on a modified version of the Richter scale. The magnitude is a measure of the amount of energy released in the earthquake, as indicated by the amplitude (size) of the vibrations when they reach a recording instrument (seismograph). Using this approach, seismologists rate earthquakes as *insignificant* (less than 4.0 on the Richter scale), *minor* (4.0–4.9), *damaging* (5.0–5.9), *destructive* (6.0–6.9), *major* (7.0–7.9), and *great* (over 8.0). Each unit on the Richter scale represents an amplitude that is 10 times greater than the next smaller unit, so a magnitude 5.0 earthquake is 10 times greater than a magnitude 4.0, and a magnitude 6.0 quake is 100 times greater than a magnitude 4.0 quake.

The northern California earthquake of 1989 had a Richter magnitude of 7.1 and caused damage within a radius of 97 kilometers (60 miles) from its epicenter (Figure 8-6). Earthquakes often have *aftershocks* that gradually decrease in frequency over a period of up to several months, and some have *foreshocks* from seconds to weeks before the main shock.

The primary effects of earthquakes include shaking and sometimes a permanent vertical or horizontal displacement of the ground. These effects may have serious consequences for people and for buildings, bridges, freeway overpasses, dams, and pipelines. Secondary effects of earthquakes include various types of mass wasting (such as rockfalls and rockslides), urban fires, and flooding caused by subsidence of land. Coastal areas can also be severely damaged by large, earthquake-generated water waves, called *tsunamis* (misnamed "tidal waves," even though they have nothing to do with tides), that travel as fast as 950 kilometers (590 miles) per hour.

Solutions: How Can We Reduce Earthquake Hazards? Loss of life and property from earthquakes can be reduced. To do this we **(1)** examine historical records and make geologic measurements to locate active fault zones, **(2)** make maps showing areas in which ground conditions are more subject to shaking and thus are high-risk areas (Figure 8-7), **(3)** establish building codes that regulate the placement and design of buildings in areas of high risk, and **(4)** ideally, learn to predict when and where earthquakes will occur.

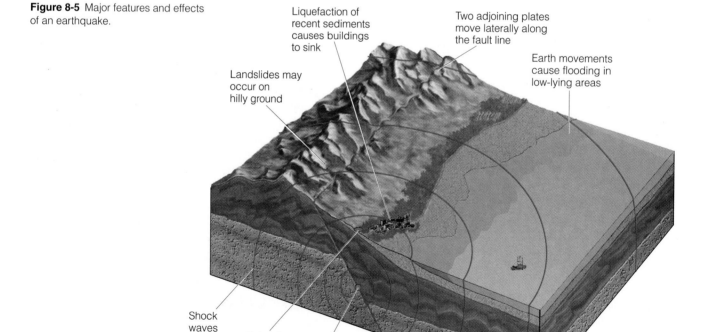

Figure 8-5 Major features and effects of an earthquake.

Liquefaction of recent sediments causes buildings to sink

Two adjoining plates move laterally along the fault line

Earth movements cause flooding in low-lying areas

Landslides may occur on hilly ground

Shock waves

Epicenter

Focus

Q: What percentage of the solid waste produced in the United States is municipal solid waste (garbage)?

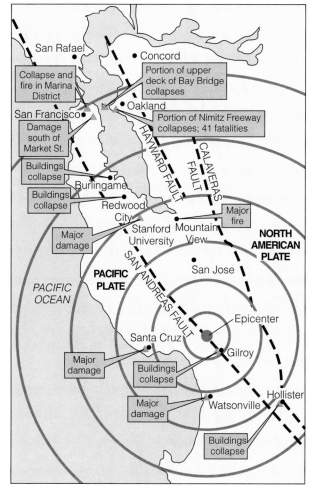

Figure 8-6 At 5:04 P.M. on October 17, 1989, a magnitude 7.1 earthquake occurred in northern California along the San Andreas Fault. It was the largest earthquake in northern California since 1906, when the great San Francisco quake and the resulting fires destroyed much of the city. In the 1989 quake, the most extensive damage was within a radius of 32 kilometers (20 miles) from the epicenter, but seismic waves caused major damage as far away as San Francisco and Oakland. Sixty-seven people were killed, and official damage estimates were as high as $10 billion. This was North America's costliest natural disaster until Hurricane Andrew devastated Florida in 1992.

What Are Volcanoes? An active **volcano** occurs where magma (molten rock) reaches the earth's surface through a central vent or a long crack (fissure). Volcanic activity can release *ejecta* (debris ranging from large chunks of lava rock to ash that may be glowing hot), liquid lava, and gases (water vapor, carbon dioxide, sulfur dioxide, nitrogen oxides, and others) into the environment.

Some volcanoes, such as those at Mount St. Helens in Washington (which erupted in 1980) and Mount

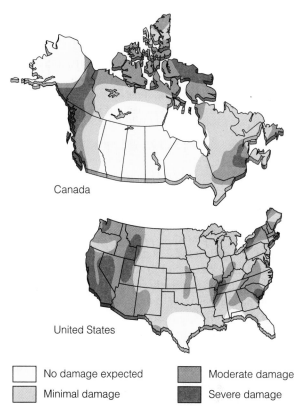

No damage expected	Moderate damage
Minimal damage	Severe damage

Figure 8-7 Expected damage from earthquakes in Canada and the contiguous United States. Except for a few regions along the Atlantic and the Gulf coasts, virtually every part of the continental United States is subject to some risk from earthquakes. Several areas have a risk of moderate to major damage. This map is based on earthquake records. (Data from U.S. Geological Survey)

Pinatubo in the Philippines (which erupted in 1991), have a steep, flaring cone shape and usually erupt explosively. Others, such as those in the islands of Iceland and Hawaii (Figure 7-1), typically erupt more quietly. Most produce lava flows, which can cover roads and villages and ignite brush, trees, and homes.

Land-use planning, better prediction of volcanic eruptions, and development of effective evacuation plans can reduce the loss of human life (and sometimes property) from volcanic eruptions. Prediction of volcanic activity has improved considerably during this century, but every volcano has its own personality. Like earthquake maps, the eruptive history of a volcano or volcanic center can provide some indication of where the risks are. Volcanologists are now studying precursor phenomena such as tilting or swelling of the cone, changes in magnetic and thermal properties of the volcano, changes in gas composition, and increased seismic activity.

We tend to think negatively of volcanic activity, but it also provides some benefits. One is outstanding scenery in the form of majestic mountains, some lakes (such as Crater Lake in Oregon; see photo in the Table of Contents), and other landforms. Perhaps the most important benefit of volcanism is the highly fertile soils produced by the weathering of lava.

On May 18, 1980, Mount St. Helens, in the Cascade Range near the Washington–Oregon border, erupted in what has been called the worst volcanic disaster in U.S. history.

Fifty-seven people died in the eruption, and several hundred cabins and homes were destroyed or severely damaged. Tens of thousand of hectares of forest were obliterated, along with campgrounds and bridges. An estimated 7,000 big-game animals (bear, deer, elk, mountain lions) died, as did millions of smaller animals and birds and some 11 million fish. Salmon hatcheries were damaged, and crops (including alfalfa, apples, potatoes, and wheat) were lost. Many people living in the area also lost their jobs.

On the plus side, trace elements from ash that were added to the soil may benefit agriculture in the long run. Furthermore, increased tourism to the area brought new jobs and income. By 1990 many biologists were surprised at how fast various forms of life had begun colonizing many of the most devastated areas. This rapid recovery has taught biologists important and often surprising lessons about nature's ability to recover from what seems to be almost total devastation.

8-5 BIOLOGICAL HAZARDS: DISEASE IN DEVELOPED AND DEVELOPING COUNTRIES

What Are the Major Types of Disease? A transmissible disease is caused by a living organism (such as a bacterium, virus, protozoa, or parasite) and can be spread from one person to another. The infectious agents, called *pathogens*, are spread by air, water, food, body fluids, some insects, and other nonhuman carriers (called *vectors*).

Typically, a *bacterium* is a one-celled microorganism capable of replicating itself by simple division. A *virus* is a microscopic, noncellular infectious agent. Its DNA contains instructions for making more viruses, but it has no apparatus to do so. In order to replicate, a virus must invade a host cell and take over the cell's DNA to create a factory for producing more viruses.

The world's seven deadliest infectious diseases are *acute respiratory infections,* mostly pneumonia (caused by bacteria and viruses and killing about 4.7 million people per year), *diarrheal diseases* (caused by bacteria and viruses that kill about 3.1 million people per year), *tuberculosis* (a bacterial disease killing about 3.1 million people per year), *malaria* (caused by parasitic protozoa and killing at least 2.1 million people per year), *AIDS* (a viral disease with a death toll of about 1.7 million in 1996), *hepatitis B* (a viral disease killing about 1 million people per year), and *measles* (a viral disease with a death toll of about 1 million per year).

In developing countries infectious diseases account for about 42% of all deaths each year, compared to only 1.2% in developed countries. According to the World Health Organization and UNICEF, every year in developing countries at least 10 million children under the age of 5 die of mostly preventable infectious diseases—an average of at least 27,000 premature deaths per day. About 80% of all illnesses in developing countries are caused by waterborne infectious diseases (such as diarrhea, hepatitis, typhoid fever, and cholera), mainly from unsafe drinking water and inadequate sanitation systems (Spotlight, p. 234).

Diseases such as cardiovascular (heart and blood vessel) disorders, most cancers, diabetes, bronchitis, emphysema, and malnutrition typically have multiple (and often unknown) causes. They also tend to develop slowly and progressively over time. Because they are not caused by living organisms and do not spread from one person to another, they are classified as **nontransmissible diseases**.

As a country industrializes, it usually makes an *epidemiological transition*: The infectious diseases of childhood become less important and the chronic diseases of adulthood (heart disease and stroke, cancer, and respiratory conditions) become more important in causing mortality.

Case Study: Tuberculosis, a Bacterial Disease
One of the world's most underreported stories in the 1990s has been the rapid spread of tuberculosis (TB), a highly infectious bacterial disease that currently kills about 3.1 million people per year (Figure 8-8). The current TB epidemic is so severe that in 1993 the World Health Organization (WHO) declared a global state of emergency. The bacterium causing TB Infection moves from person to person, mainly in airborne droplets produced by coughing, sneezing, singing, or even talking. Whereas TB kills about 3.1 million people per year, highly publicized Ebola viruses, which are hard to transmit, have killed about 650 people over the past 20 years—an average of 33 people per year.

Until recently the incidence of TB had fallen sharply (except among the poor), mostly because of

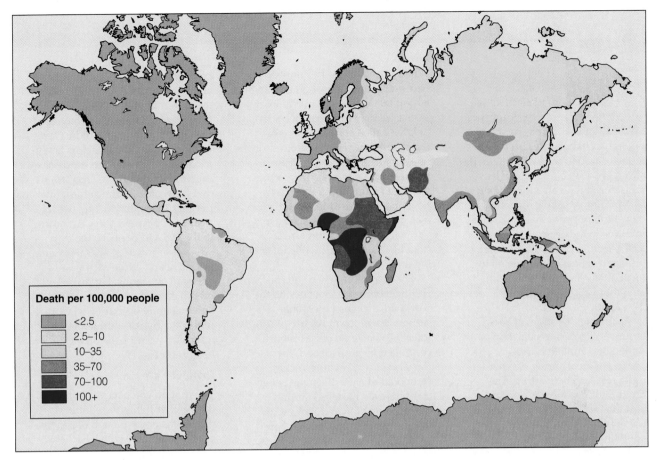

Figure 8-8 The current global tuberculosis (TB) epidemic. This easily transmitted disease is spreading rapidly and now kills about 3.1 million people a year. Without global effort to identify and fully treat people with active TB, the death toll is projected to rise to about 4 million a year by 2005. (Data from World Health Organization)

prevention programs (such as X-ray screening and detection of people with active TB) and treatment with antibiotics, which began in the 1940s.

Major reasons for the recent epidemic of TB are **(1)** poor TB screening and control programs (especially in developing countries, where about 95% of the new cases occur), **(2)** development of strains of the tuberculosis bacterium that are genetically resistant to virtually all effective antibiotics (typically leading to mortality rates of over 50%, Spotlight, p. 234), **(3)** population growth and increased urbanization (which increase contacts among people), and **(4)** the spread of AIDS, which greatly weakens the immune system and allows TB bacteria to multiply.

Slowing the spread of the disease involves early identification and treatment of people with active TB, usually those with a chronic cough. Treatment with a combination of four inexpensive drugs can cure 90% of those with active TB. To be successful, however, the patients must take the drugs *every* day for 6 to 8 months. Because the symptoms disappear after a few weeks, many patients think they are cured and stop taking the drugs. This allows the disease to recur in a hard-to-treat form, it spreads to other people, and drug-resistant strains of TB bacteria develop.

How Rapidly Are Viral Diseases Spreading?

Viral diseases include *influenza* or *flu* (transmitted by the bodily fluids or airborne emissions of an infected person), *Ebola* (transmitted by the blood or other body fluids of an infected person), *rabies* (transmitted by dogs, coyotes, racoons, skunks, and bats), and *AIDS* (Connections, page 236). Viruses, like bacteria, can genetically adapt rapidly to different conditions.

Flu viruses move through the air and are highly contagious. In 1918–19 a flu epidemic infected almost half the world's population and killed 20–30 million people. Although health officials worry about the emergence of new viral diseases (such those caused by Ebola viruses), they recognize that the greatest virus

A: Enough to rebuild the country's airline fleet every 3 months

There is growing and alarming evidence that we may be losing our war against infectious bacterial diseases because bacteria are among the earth's ultimate survivors. When a colony of bacteria is dosed with an antibiotic such as penicillin, most die, but a few have mutant genes that make them immune to the drug. Through natural selection (Section 5-6), a single mutant can pass such traits on to most of its offspring, which can amount to 16,777,216 in only 24 hours!

Each time this strain of bacterium is exposed to penicillin or some other antibiotic, a larger proportion of its offspring is genetically resistant to the drug. The rapid multiplication of resistant bacteria in a victim is made easier because the antibiotics also wipe out their bacterial competitors.

Even worse, bacteria can become genetically resistant to antibiotics they have never been exposed to.

When a resistant and a nonresistant bacterium touch one another (say in a hospital bedsheet or in a human stomach), they can exchange a loop of DNA called a plasmid, thereby transferring genetic resistance from one organism to another. This process allows bacteria to become resistant to antibiotics they have never been exposed to. Bacteria can also pick up genetic resistance to various antibiotics from viruses (that have acquired it while infecting other bacteria).

The incredible genetic adaptability of bacteria is one reason the world now faces a potentially serious rise in the incidence of some infectious bacterial diseases once controlled by antibiotics. Other factors also play a key role, including (1) spread of bacteria (some beneficial and some harmful) around the globe by human travel and the trade of goods; (2) overuse of antibiotics by doctors, often at the insistence of their patients; (3) failure of many patients to take all of their prescribed antibiotics, which pro-

motes bacterial resistance; (4) availability of antibiotics in many countries without prescriptions; and (5) widespread use of antibiotics in the livestock and dairy industries.

The result of these factors acting together is that every major disease-causing bacterium now has strains that resist at least one of the roughly 160 antibiotics we use to treat bacterial infections. According to the U.S. Office of Technology Assessment, at least 19,000 patients in U.S. hospitals die each year of bacterial infections that resist antibiotics doctors use for treatment, and this number could rise. Currently, the risk of contracting an infection during a stay in a U.S. hospital is 1 in 15.

Critical Thinking

1. What do you believe are the three major things we should do to reduce the development of genetically resistant infectious bacteria?

2. Use the library or the Internet to determine what progress is being made in dealing with this problem.

health threat to humans is the emergence of new, very virulent stains of influenza.

There is a growing threat from the spread of *acquired immune deficiency syndrome (AIDS)*, which is caused by the human immunodeficiency virus (HIV). The HIV virus itself isn't deadly, but it kills immune cells and leaves the body defenseless against all sorts of other infectious bacteria and viruses. The HIV virus can be transmitted during unprotected sexual activity, from one intravenous drug user to another through shared needles, from an infected mother to an infant during birth, and by exposure to infected blood.

According to the Global AIDS Policy Coalition at the Harvard School of Health, by the end of 1996 at least 36.2 million people (90% of them living in developing nations in Asia and Africa, with 23 million in sub-Saharan Africa) were infected with HIV. In 1996 a record 5.6 million people contracted HIV. Currently, India has more HIV-positive people than any other country.

Within about 7–10 years, 95% of those with HIV develop AIDS. This long incubation period means

that infected people often spread the virus for several years before they learn that they are infected. There is as yet no cure for AIDS, although drugs may help some infected people live longer. By the end of 1996, 12.5 million of the 36.2 million people infected with HIV had acquired full-blown AIDS, with a record number (1.7 million) dying from AIDS in 1996.

Once a viral infection starts it is much harder to fight than infections by bacteria and protozoans. Only a few antiviral drugs exist because most drugs that will kill a virus also harm the cells of its host. Treating viral infections with antibiotics is useless and merely increases genetic resistance in disease-causing bacteria. Medicine's only effective weapon against viruses are preventive vaccines that provide antibodies to ward off viral infections. Immunization with vaccines has helped tame viral diseases such as smallpox, polio, rabies, influenza, and measles.

Case Study: Malaria, a Protozoal Disease

About 45% (or 2.6 billion people) of the world's population live in tropical and subtropical regions in which

Q: What percentage of glass beverage bottles in the United States are refillable?

malaria is present (Figure 8-9). Currently, some 250–300 million people are infected with malaria parasites, and at least 110 million new cases occur each year (97 million of them in Africa). Malaria's symptoms come and go and include fever and chills, anemia, an enlarged spleen, severe abdominal pain and headaches, extreme weakness, and greater susceptibility to other diseases. The disease kills at least 2.1 million people each year, more than half of them children under age 5.

Malaria is caused by four species of protozoa of the genus *Plasmodium*. Most cases of the disease are transmitted when an uninfected female of any one of 60 species of *Anopheles* mosquito bites an infected person, ingests blood that contains the parasite, and later bites an uninfected person (Figure 8-10). When this happens, *Plasmodium* parasites move out of the mosquito and into the human's bloodstream, multiply in the liver, and then enter blood cells to continue multiplying. Malaria can also be transmitted by blood transfusions or by sharing needles. This cycle repeats itself until immunity develops, treatment is given, or the victim dies.

During the 1950s and 1960s the spread of malaria was sharply curtailed by draining swamplands and marshes, spraying breeding areas with insecticides, and using drugs to kill the parasites in the bloodstream. Since 1970, however, malaria has come roaring back. Most species of the malaria-carrying *Anopheles* mosquitoes have become genetically resistant to most of the insecticides used. Worse, the *Plasmodium* parasites have become genetically resistant to the common antimalarial drugs.

Researchers are working to develop new antimalarial drugs, vaccines, and biological controls for *Anopheles* mosquitoes. Such approaches are underfunded and have proved more difficult than originally thought. Researchers are also studying the feasibility of altering the mosquitoes' genetic makeup so that they cannot carry and transmit the parasite to humans.

Prevention is the best approach to slowing the spread of malaria. Methods include increasing water flow in irrigation systems to prevent mosquito larvae from developing (an expensive and wasteful use of water), using mosquito nets dipped in a nontoxic insecticide (permethrin) in windows and doors of homes, cultivating fish that feed on mosquito larvae (biological control), clearing vegetation around houses, and planting trees that soak up water in low-lying marsh areas where mosquitoes thrive (a method that can degrade or destroy ecologically important wetlands).

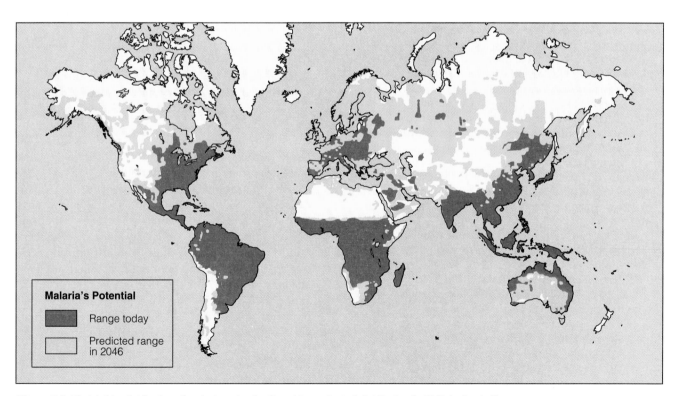

Figure 8-9 Worldwide distribution of malaria today (red) and its projected distribution in 2046 (yellow). About 45% of the world's current population lives in areas in which malaria is present, with the disease killing about 2.1 million people a year. If the world becomes warmer, as projected by current climate models, by 2046 malaria could affect 60% of the world's population. (Data from the World Health Organization)

A: About 11% (compared to 95% in Finland and 73% in Germany)

Sex Can Be Hazardous to Your Health

Sexually transmitted diseases (STDs) are typically passed on during sexual activity; many STDs can also be transmitted from mother to infant during birth, from one intravenous (IV) drug user to another through shared needles, and by exposure to infected blood.

Worldwide there are about 250 million new infections and 750,000 deaths from STDs each year. By the year 2000 the number of deaths from STDs is expected to reach 1.5 million—an average of 4,100 per day. In the United States, STDs strike at least 8 million people (3 million of them teenagers) per year. The number of new reported cases of most STDs has risen every year since 1981.

Major STDs caused by bacteria are *chlamydia*, *gonorrhea*, and *syphilis*. These bacterial diseases can be treated with antibiotics if caught in time. However, some of the bacteria causing these diseases are becoming genetically resistant to an increasing number of the antibiotics used to control them.

Major STDs caused by viruses include *genital warts*, *genital herpes*, *hepatitis B*, and *acquired immune deficiency syndrome* or *AIDS* (which is caused by the human immunodeficiency virus, HIV). The AIDS virus itself isn't deadly, but it kills off immune cells and leaves the body defenseless against all sorts of other infectious bacteria and viruses. These viral diseases—with the possible exception of genital warts—are currently incurable.

According to the World Health Organization, by mid-1995 at least 19 million people (12 million of them in sub-Saharan Africa) were infected with HIV. By 2000, an estimated 40 million people are expected to be infected with the virus—80% of them in Asia and Africa. Heterosexual transmission accounts for about 90% of the new infections worldwide and for about 8% in the United States (up from 2% in 1985).

Within about 7–10 years, 95% of those with HIV develop AIDS. This long incubation period means that infected people often spread the virus for several years before they learn that they are infected. There is as yet no cure for AIDS, although drugs may help some infected people live longer.

Recent research on AIDS patients with lymphoma suggests that the AIDS virus (HIV) can invade cells and activate normally dormant cancer-causing genes. If this research is confirmed, attempts to develop an AIDS vaccine (a weakened strain of HIV) may be dangerous. The inoculation may provide protection against AIDS but cause a cancer.

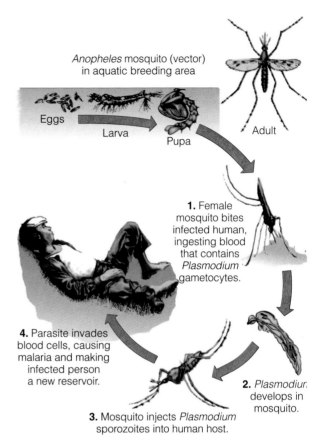

Anopheles mosquito (vector) in aquatic breeding area

Eggs
Larva
Pupa
Adult

1. Female mosquito bites infected human, ingesting blood that contains *Plasmodium* gametocytes.

2. *Plasmodium* develops in mosquito.

3. Mosquito injects *Plasmodium* sporozoites into human host.

4. Parasite invades blood cells, causing malaria and making infected person a new reservoir.

Figure 8-10 The life cycle of malaria.

What Are the Major Diseases in Developed Countries? As a country industrializes, it usually makes an *epidemiological transition*: The infectious diseases of childhood become less important, and the chronic diseases of adulthood (heart disease and stroke, cancer, and respiratory conditions) become more important in causing mortality.

In 1994 about 2.3 million people died of all causes in the United States: 39% from heart attacks and strokes, 24% from cancer, 5% from infectious diseases (mostly pneumonia, influenza, and AIDS), and 4% from accidents (half from automobile accidents). Almost two-thirds of these deaths result from chronic diseases (such as heart attack, stroke, and cancer) that take a long time to develop and that have multiple causes. They are largely related to location (urban or rural), work environment, diet, smoking (p. 223), amount of exercise, sexual habits (Connections, left), and the level of use of alcohol or other harmful drugs.

According to the Department of Health and Human Services, about 95% of the money spent on health care in the United States each year is used to *treat* rather than to *prevent* disease—one reason why health care costs are so high and continue to climb. Health experts estimate that changing harmful lifestyle factors could prevent 40–70% of all premature deaths, one-third of all cases of acute disability, and two-thirds of all cases of chronic disability in the United States.

The U.S. death rate from infectious diseases increased 58% between 1980 and 1992 (36% from AIDS and 22% from other infectious diseases). In 1980, infectious diseases ranked as America's fifth leading killer. By 1992 they were the third leading cause of death, representing a partial reversal of the epidemiological transition in developed countries.

Solutions: How Can We Reduce Infectious and Other Diseases? According to health scientists and public health officials, disease in developing and developed countries could be greatly reduced by

- Greatly increasing research on tropical diseases. The World Health Organization estimates that only 3% of the money spent worldwide each year for such research is devoted to malaria and other tropical diseases, even though more people suffer and die worldwide from these diseases than from all others combined.

- Mounting a global campaign to reduce overcrowding, unsafe drinking water, poor sanitation, and malnutrition.

- Increasing funding for monitoring and responding to disease outbreaks to reduce the chances that a small outbreak can turn into an epidemic.

- Not using powerful antibiotics to treat minor infections or undiagnosed symptoms.

- Educating the public to understand the dangers of overuse of antibiotics and the need to take all of the antibiotics in any prescription.

- Not using antibiotics that have caused widespread genetic resistance, coupled with rotating from one antibiotic to another.

- Not selling antibiotics without a prescription (allowed in many countries).

- Sharply reducing the use of antibiotics in livestock.

- Insisting that doctors, nurses, and orderlies strictly maintain hygienic standards at all times.

- Putting much more money into the development of vaccines to prevent infections by bacteria and viruses responsible for most disease and death.

- Emphasizing preventive health care, especially in developing countries (Solutions, p. 240).

⊛ 8-6 RISK ANALYSIS

How Can We Estimate Risks? Risk analysis involves identifying hazards and evaluating their associated risks (*risk assessment*), ranking risks (*comparative risk analysis*), determining options and making decisions about reducing or eliminating risks (*risk management*), and informing decision makers and the public about risks (*risk communication*).

Risk assessment involves determining the types of hazards involved, estimating the probability that each hazard will occur, and estimating how many people are likely to be exposed to it and how many may suffer serious harm. Statistical probabilities based on past experience, animal testing and other tests, and epidemiological studies are used to estimate risks from older technologies and products (Section 8-1). For new technologies and products, much more uncertain statistical probabilities, based on models rather than actual experience, must be calculated.

The left side of Figure 8-11 is an example of *comparative risk analysis*, summarizing the greatest ecological and health risks identified by a panel of scientists acting as advisers to the U.S. Environmental Protection Agency. Note the considerable difference between the comparison of risk by scientists (Figure 8-11, left) and the general public (Figure 8-11, right). These differences result largely from failure of professional risk evaluators to communicate the nature of risks and their relative importance to the public, teachers, and members of the media. Much of our risk education is based on often misleading media reports on the latest risk scare of the week (based mainly on frontier science) that do not put such risks in perspective.

The key question is whether the estimated short- and long-term risks of using a particular technology or product outweigh the estimated short- and long-term benefits of other alternatives. One method for making such evaluations is **risk–benefit analysis**, which involves estimating such benefits and risks involved.

What Are the Greatest Risks People Face? The greatest risks most people face today are rarely dramatic enough to make the daily news. In terms of reduced life span from malnutrition, exposure to disease-causing organisms and dangerous chemicals, and lack of basic health care, *the greatest risk by far is poverty* (Figure 8-12).

After the health risks associated with poverty, the greatest risks of premature death are mostly the result of voluntary—and thus correctable—choices people make about their lifestyles (Figures 8-1 and 8-12). By far the best ways to reduce one's risk of premature death and serious health risks are to not smoke, avoid

Figure 8-11 Comparative risk analysis of the most serious ecological and health problems according to scientists acting as advisers to the U.S. Environmental Protection Agency (left column). Risks in each of these categories are not listed in rank order. The right side of this figure represents polls showing how U.S. citizens rank the ecological and health risks they perceive as being the most serious. Why do you think there is such a great difference between the ranking by risk experts and by the general public? (Data from Science Advisory Board, *Reducing Risks*, Washington, D.C.: Environmental Protection Agency, 1990)

Scientists
(Not in rank order
in each category)

Citizens
(In rank order)

High-Risk Health Problems
- Indoor air pollution
- Outdoor air pollution
- Worker exposure to industrial or farm chemicals
- Pollutants in drinking water
- Pesticide residues on food
- Toxic chemicals in consumer products

High-Risk Ecological Problems
- Global climate change
- Stratospheric ozone depletion
- Wildlife habitat alteration and destruction
- Species extinction and loss of biodiversity

High-Risk Problems
- Hazardous waste sites
- Industrial water pollution
- Occupational exposure to chemicals
- Oil spills
- Stratospheric ozone depletion
- Nuclear power-plant accidents
- Industrial accidents releasing pollutants
- Radioactive wastes
- Air pollution from factories
- Leaking underground tanks

Medium-Risk Ecological Problems
- Acid deposition
- Pesticides
- Airborne toxic chemicals
- Toxic chemicals, nutrients, and sediment in surface waters

Medium-Risk Problems
- Coastal water contamination
- Solid waste and litter
- Pesticide risks to farm workers
- Water pollution from sewage plants

Low-Risk Ecological Problems
- Oil spills
- Groundwater pollution
- Radioactive isotopes
- Acid runoff to surface waters
- Thermal pollution

Low-Risk Problems
- Air pollution from vehicles
- Pesticide residues in foods
- Global climate change
- Drinking water contamination

excess sunlight (which ages skin and causes skin cancer), not drink alcohol or drink only in moderation (no more than two drinks per day), reduce consumption of foods containing cholesterol and saturated fats, eat a variety of fruits and vegetables, exercise regularly, lose excess weight, and (for those who can afford a car) drive as safely as possible in a vehicle with the best available safety equipment.

How Can We Estimate Risks for Technological Systems? The more complex a technological system and the more people needed to design and run it, the more difficult it is to estimate the risks. The overall reliability of any technological system (expressed as a percentage) is the product of two factors (multiplied by 100):

$$\text{System reliability \%} = \text{Technology reliability} \times \text{Human reliability} \times 100$$

With careful design, quality control, maintenance, and monitoring, a highly complex system such as a nuclear power plant or a space shuttle can achieve a high degree of technology reliability. However, human reliability is almost always much lower than technology reliability and is virtually impossible to predict; to err is human. Suppose, for example, that the technology reliability of a nuclear power plant is 95% (0.95) and that human reliability is 75% (0.75). Then the overall system reliability is only 71% (0.95 × 0.75 × 100 = 71%). Even if we could make the technology 100% reliable (1.0), the overall system reliability would still be only 75% (1.0 × 0.75 × 100 = 75%). The crucial dependence of even the most carefully designed systems on unpredictable human reliability helps explain essentially "impossible" tragedies such as the Chernobyl (p. 538) nuclear power-plant accident and the explosion of the space shuttle *Challenger*.

Q: What do scientists believe are the least desirable ways to deal with solid waste?

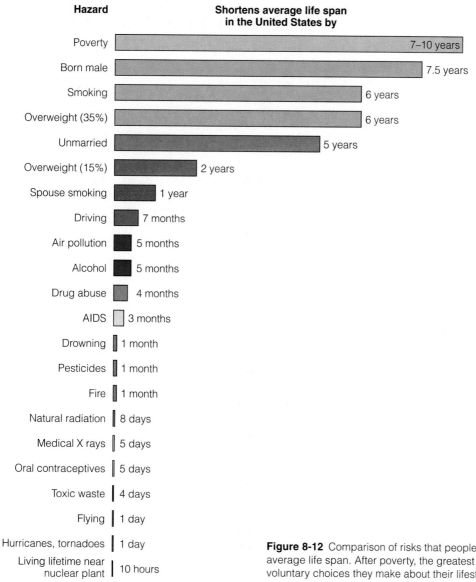

Hazard	Shortens average life span in the United States by
Poverty	7–10 years
Born male	7.5 years
Smoking	6 years
Overweight (35%)	6 years
Unmarried	5 years
Overweight (15%)	2 years
Spouse smoking	1 year
Driving	7 months
Air pollution	5 months
Alcohol	5 months
Drug abuse	4 months
AIDS	3 months
Drowning	1 month
Pesticides	1 month
Fire	1 month
Natural radiation	8 days
Medical X rays	5 days
Oral contraceptives	5 days
Toxic waste	4 days
Flying	1 day
Hurricanes, tornadoes	1 day
Living lifetime near nuclear plant	10 hours

Figure 8-12 Comparison of risks that people face expressed in terms of shorter average life span. After poverty, the greatest risks people face are mostly from voluntary choices they make about their lifestyles. (Data from Bernard L. Cohen)

One way to make a system more foolproof or fail-safe is to move more of the potentially fallible elements from the human side to the technical side. However, chance events such as a lightning bolt can knock out automatic control system. No machine or computer program can completely replace human judgment. Of course, the parts in any automated control system are manufactured, assembled, tested, certified, and maintained by fallible human beings. Computer software programs used to monitor and control complex systems can also contain human errors, or they can be deliberately modified by computer viruses to malfunction. The pros and cons of genetic engineering reveal the difficulty in evaluating a new technology (Pro/Con, p. 241).

What Are the Limitations of Risk Assessment and Risk–Benefit Analysis? Risk assessment is a young science that has many built-in uncertainties and limitations. It depends on toxicology assessments that have scientific and economic limitations (Section 8-3). Each additional step in risk assessment (and related risk–benefit analysis) also has uncertainties and economic limitations.

Here are some of the key questions involved in risk assessment:

- How reliable are risk assessment data and models?

- Who profits from allowing certain levels of harmful chemicals into the environment, and who suffers? Who decides this?

With adequate funding, the health of people in developing countries (and the poor in developed countries) can be improved dramatically, quickly, and cheaply by providing the following forms of mostly preventive health care:

- Family-planning counseling.

- Better nutrition, prenatal care, and birth assistance for pregnant women. At least 600,000 women in developing countries die each year of mostly preventable pregnancy-related causes, compared with about 6,000 in developed countries.

- Better nutrition for children.

- Greatly improved postnatal care (including promotion of breast-feeding) to reduce infant mortality. Breast-fed babies get natural immunity to many diseases from antibodies in their mothers' milk.

- Immunization against the world's five largest preventable infectious diseases: tetanus, measles, diphtheria, typhoid fever, and polio. Between 1971 and 1992, the percentage of children in developing countries immunized against these diseases rose from 10% to 80%, saving about 9 million lives a year

- Oral rehydration therapy for victims of diarrheal diseases, which cause about one-fourth of all deaths of children under the age of 5. A simple solution of boiled water, salt, and sugar or rice, at a cost of only a few cents per person, can prevent death from dehydration. According to the British medical journal *Lancet*, this simple treatment is "the most important medical advance of the century."

- Careful and selective use of antibiotics for infections.

- Clean drinking water and sanitation facilities for the one-third of the world's population that lacks them.

According to the World Health Organization, extending such primary health care to all the world's people would cost an additional $10 billion per year, a mere 4% of what the world spends every year on cigarettes or devotes every 4 days to military spending. The cost of this program is about $1 per child. Experience shows that the resulting drop in infant mortality would also help reduce population growth.

Critical Thinking

1. Do you believe that developed countries should foot at least half of the bill implementing such proposals? What major economic and environmental advantages would this have for developed countries?

2. How many dollars per year of your taxes would you be willing to spend for such a preventive health program in developing countries?

- Should estimates emphasize short-term risks, or should more weight be put on long-term risks? Who should make this decision?

- Should the primary goal of risk analysis be to determine how much risk is acceptable (the current approach) or to figure out how to do the least damage (a prevention approach)?

- Who should do a particular risk–benefit analysis or risk assessment, and who should review the results? A government agency? Independent scientists? The general public?

- Should cumulative effects of various risks be considered, or should risks be considered separately, as is usually done? Suppose a pesticide is found to have an annual risk of killing one person in a million from cancer, the current EPA limit. Cumulatively, however, effects from 40 such pesticides might kill 40, or even 400, of every 1 million people. Is this acceptable?

- Should risk levels be higher for workers (as is almost always the case) than for the general public? What say should workers and their families have in

this decision? According to government estimates, the exposure of workers to toxic chemicals in the United States causes 50,000–70,000 deaths, at least half from cancer, and 350,000 new cases of illness per year. The situation is much worse in developing countries. Is this a necessary cost of doing business?

Some see risk analysis as a useful and much-needed tool. Others see it as a way to justify premeditated murder in the name of profit. According to the National Academy of Sciences, exposure to toxic chemicals is responsible for 2–4% of the 521,000 cancer deaths in the United States; this amounts to 10,400–20,800 premature cancer deaths per year. According to hazardous waste expert Peter Montague (Guest Essay, pp. 68–69), "The explicit aim of risk assessment is to convince people that some number of citizens *must* be killed each year to maintain a national lifestyle based on necessities like Saran Wrap, throwaway cameras, and lawns without dandelions."

These critics also accuse industries of favoring risk analysis because so little is known about health risks from pollutants and because the data that do exist are

Q: What country recycles the largest amount of its mu1nicipal solid waste?

Genetic Engineering: Savior or a Potential Monster?

Genetic engineers have learned how to splice genes and recombine sequences of existing DNA molecules in organisms to produce DNA with new genetic characteristics (recombinant DNA). Thus, they can transfer traits from one species to another (Figure 8-13) without waiting for new genetic combinations to evolve through the process of natural selection.

This rapidly developing technology excites many scientists and investors, who see it as a way to produce, patent, and sell high-yield crops and livestock with more protein and greater resistance to diseases, pests, frost, and drought. They also hope to create bacteria that can destroy oil spills, degrade toxic wastes, concentrate metals found in low-grade ores, and serve as biological factories for new vaccines, drugs, and therapeutic hormones. Gene therapy, its proponents argue, could eliminate certain genetic diseases.

When the U.S. Supreme Court

Figure 8-13 An example of genetic engineering. The six-month-old mouse on the left is normal; the other mouse of the same age contains a human growth hormone gene in the chromosomes of all its cells. Mice with the human growth hormone gene normally grow two to three times faster and reach a size twice that of mice without the gene. (R. L. Brinster and R. E. Hammer, School of Veterinary Medicine, University of Pennsylvania)

ruled in 1980 that genetically engineered organisms could be patented, investors began pouring billions of dollars into the fledgling biotech industry. Genetic engineering has produced a drug that reduces heart attack damage and agents that fight diabetes, hemophilia, and some forms of cancer; it has also been used to diagnose AIDS and cancer.

Genetically altered bacteria have been used to manufacture more effective vaccines and hormones that can stimulate human growth. In agriculture, gene transfer has been used to develop strawberries that resist frost, tomatoes that stay fresh and tasty longer, and smaller cows that produce more milk.

Some people worry that biotechnology may run amok. They argue that it should be kept under strict control because we don't understand well enough how nature works to allow unregulated genetic alteration of humans and other species. They also point to serious and widespread ecological problems that have resulted from the accidental or deliberate introduction nonnative organisms into biological communities (Section 17-3).

Genetically engineered organisms might also mutate, change their form and behavior, migrate to other areas, and alter the genetic traits of existing wild species; unlike defective cars and other products, they couldn't be recalled. Proponents argue that the risk of biotech-caused ecological catastrophes is small.

Critics are also concerned that one of the most serious effects of widespread use of biotechnology is reduction of the world's vital genetic diversity and thus biodiversity. Already the world's 20 major food crops have become 70% less genetically diverse because a wide range of wild strains have been replaced with only a few varieties developed by crossbreeding; using bioengineered crop strains could hasten this loss of vital biodiversity.

Development of all new varieties by crossbreeding or genetic engineering is based on mixing certain traits found in wild varieties. Thus, reducing the natural genetic diversity of wild plants and animals by replacing them with a few stains of genetically engineered alternatives undermines the genetic engineers' ability to produce new combinations in the laboratory.

Genetically engineered plants can also affect human health. Use of the genetically engineered bovine growth hormone (recently approved by the FDA to increase milk production in cows) raises the incidence of udder infections. This requires giving cows higher doses of antibiotics, which can accelerate the problem of genetic resistance to widely used antibiotics.

In 1989, a committee of prominent ecologists appointed by the Ecological Society of America warned that the ecological effects of new combinations of genetic traits from different species would be difficult to predict. Their report called for a case-by-case review of any proposed environmental releases, as well as carefully regulated, small-scale field tests before any bioengineered organism is put into commercial use.

This controversy illustrates the difficulty of balancing the actual and potential benefits of a technology with its actual and potential risks of harm.

Critical Thinking

1. What government controls, if any, do you believe should be applied to the development and use of genetic engineering? How would you enforce any restrictions?

2. Use the library or the Internet to find out what controls now exist on biotechnology in the United States (or the country where you live) and how well such controls are enforced.

How Useful Are Risk Assessment and Risk–Benefit Analysis?

SPOTLIGHT

When we add new chemicals or technologies, we are mostly flying blind about their possible harmful effects. For example, a recent study has documented the significant uncertainties involved in even relatively simple risk assessment. Eleven European governments established 11 different teams of their best scientists and engineers (including those from private companies) to assess the hazards and risks from a small plant storing only one hazardous chemical (ammonia). The 11 teams, consisting of world-class experts analyzing this very simple system, disagreed with one another on fundamental points and varied in their assessments of the hazards by a factor of 25,000!

Thus, the current built-in uncertainty in risk assessment and risk–benefit analysis (and cost–benefit analysis, Section 7-3) is analogous to a radar device that can detect a car speeding at 160 kilometers (100 miles) per hour but can tell us only that the car is traveling somewhere between 0.16 kilometers (0.1 miles) per hour and 160,000 kilometers (100,000 miles) per hour. Such inherent uncertainty explains

why regulators setting human exposure levels for toxic substances usually divide the best results by 100 to 1,000 to provide the public with a margin of safety.

Despite the inevitable uncertainties involved, risk assessment and risk–benefit analysis are useful ways to **(1)** organize available information, **(2)** identify significant hazards, **(3)** focus on areas that need more research, **(4)** help regulators decide how money for reducing risks should be allocated, and **(5)** stimulate people to make more informed decisions about health and environmental goals and priorities.

However, at best risk assessments and risk–benefit analyses yield only a range of probabilities and uncertainties based on different assumptions—not the precise numbers that decision makers and the general public wish they had. Also politics, economics, and value judgments are involved at every step of the risk analysis process. Environmentalists and health officials believe that all such analyses should contain the wide range of assumptions, probabilities, and uncertainties involved and be open to full public review.

Critics of risk assessment and risk–benefit analysis argue that the

main decision-making tools we should rely on are *looking at the available alternatives to a particular chemical or technology and having a full public discussion of the major advantages (benefits) and disadvantages (costs) of those alternatives.*

The goal of this approach is not to find out how much risk is acceptable but to find out the least damaging reasonable alternatives by asking, "Which alternative will bring sufficient benefits and minimize damage to humans and to the earth?" They argue that if the alternatives are fairly examined, emphasis will shift from trying to determine "acceptable" risk levels to trying to reduce the risks as much as possible by *pollution prevention.*

Critical Thinking

Do you believe that **(a)** risk assessment, risk–benefit analysis, and cost–benefit analysis (Section 7-3) should be used as the *primary* tool for establishing any federal health, safety, or environmental regulation? or **(b)** that the emphasis should be placed on fair evaluation of alternatives with the goal of finding the least harmful (and most affordable) alternative? Explain.

controversial. The result is that risk assessment and risk–benefit analysis can be crafted to support almost any conclusion and then called scientific decision making. The huge uncertainties in risk assessment and risk–benefit analysis (Spotlight, above) also allow industries to delay regulatory decisions for decades by challenging data in the courts.

How Should Risks Be Managed? Once an assessment of risk is made, decisions must be made about what to do about the risk. **Risk management** includes the administrative, political, and economic actions taken to decide whether and how to reduce a particular societal risk to a certain level and at what cost.

Risk management involves deciding **(1)** which of the vast number of risks facing society should be

evaluated and managed and in what order or priority with the limited funds available, **(2)** how reliable the risk–benefit analysis or risk assessment performed for each risk is, **(3)** how much risk is acceptable, **(4)** how much money it will take to reduce each risk to an acceptable level, **(5)** how much each risk will be reduced if available funds are limited (as is almost always the case), **(6)** and how the risk management plan will be communicated to the public, monitored, and enforced. Each step in this process involves making value judgments and weighing trade-offs to find some reasonable compromise among conflicting political, economic, health, and environmental interests.

How Well Do We Perceive Risks? Most of us do poorly in assessing the relative risks from the hazards

Q: What percentage of U.S. municipal solid waste could be recycled, composted, or reused?

that surround us (Figures 8-11 and 8-12). Many people deny or shrug off the high-risk chances of dying (or injury) from voluntary activities they enjoy, such as motorcycling (1 in 50 participating), smoking (1 in 300 participants by age 65 for a pack-a-day smoker), hang-gliding (1 in 1,250), and driving (1 in 2,500 without a seat belt and 1 death in 5,000 with a seat belt).

Some of these same people may be terrified about the possibility of dying from a commercial airplane crash (1 in 4.6 million), exposure to asbestos in schools (1 in 11 million), train crash (1 in 20 million), snakebite (1 in 36 million), shark attack (1 in 300 million), or exposure to trichloroethylene (TCE) in drinking water at the trace levels allowed by the EPA (1 in 2 billion).

Being bombarded with news about people killed or harmed by various hazards distorts our sense of risk. However, *the most important good news each year is that about 99.1% of the people on the earth didn't die.** But that's not what we see on TV, hear, or read about every day. Despite the greatly increased use of synthetic chemicals in food production and processing, the general health and average life expectancy of people in the United States (and most developed countries) have increased during the past 50 years.

Our perceptions of risk and our responses to perceived risks often have little to do with how risky the experts say something is (Figures 8-11 and 8-12). The public generally sees a technology or a product as being riskier than experts do when

- *It is new or complex rather than familiar.* Examples include genetic engineering or nuclear power, as opposed to large dams or coal-fired power plants.

- *It is perceived as being mostly involuntary.* Examples include nuclear power plants or food additives, as opposed to driving or smoking.

- *It is viewed as unnecessary rather than as beneficial or necessary.* Examples might include using chlorofluorocarbon (CFC) propellants in aerosol spray cans or using food additives that increase sales appeal, as opposed to cars or aspirin.

- *Its use involves a large, well-publicized death toll from a single catastrophic accident rather than the same or an even larger death toll spread out over a longer time.* Examples might include a severe nuclear power plant accident, an industrial explosion, or a plane crash, as opposed to coal-burning power plants, automobiles, or smoking.

- *Its use involves unfair distribution of the risks.* Citizens are outraged when government officials decide to put a hazardous-waste landfill or incinerator in or near their neighborhood, even when the decision is based on risk–benefit analysis. This is usually seen as politics, not science. Residents will not be satisfied by estimates that the lifetime risks of cancer death from the facility are no greater than, say, 1 in 100,000. Living near the facility means that they, not the 99,999 people living farther away, have a much higher risk of dying from cancer by having this risk involuntarily imposed on them.

- *The people affected are not involved in the decision-making process from start to finish.*

- *Its use does not involve a sincere search for and evaluation of alternatives.* People who believe that their lives and the lives of their children are being threatened want to know what the alternatives are and which alternative provides the least harm to them and the earth.

Better education and communication about the nature of risks will help bring the public's perceptions of various risks closer to those of professional risk evaluators. However, such education will not eliminate the emotional, cultural, and ethical factors that decision makers must take into account in determining the acceptability of a particular risk and in evaluating the possible alternatives.

Not all waste and pollution can be eliminated. . . . What is absolutely crucial, however, is to recognize that pollution prevention should be the first choice and the option against which all other options are judged. The burden of proof imposed on individuals, companies, and institutions should be to show that pollution prevention options have been thoroughly examined, evaluated, and used before lesser options are chosen.

JOEL HIRSCHORN

CRITICAL THINKING

1. Explain why you agree or disagree with the proposals made by health officials for reducing the death toll and other harmful effects from smoking given on p. 223.

2. In 1997 U.S. tobacco company officials and attorneys general of states suing tobacco companies for state-related Medicaid costs met to negotiate a solution. One possibility was for the tobacco industry to pay $375 billion to states over 25 years to cover the states' Medicaid expenses related to harm from smoking. Tobacco companies could fund this payment by raising the price of a pack of cigarettes by about 50¢, an increase critics say would do little to reduce sales. In return for this payment, tobacco companies want the U.S. Congress to guarantee that any individual who sues them for any

---

* The world's crude death rate in 1997 was 9 per 1,000 people. With 5.84 billion people this amounts to about 52.6 million deaths ($9/1,000 \times 5.84$ billion = 52.6 million). These deaths made up 0.9% of the world's population in 1997 (52.2 million/5.8 billion $\times$ 100 = 0.9%).

A: 60–80% by weight

health-related illness could collect no more than $250,000 in damages. Values of most tobacco stocks went up in response to this possible settlement. Are you for or against such an agreement? Explain. What has happened with this issue?

3. Should we have zero pollution levels for all hazardous chemicals? Explain.

4. Do you believe that health and safety standards in the workplace should be strengthened and enforced more vigorously, even if this causes a loss of jobs when companies transfer operations to countries with weaker standards? Explain.

5. Evaluate the following statements:
   a. Because almost any chemical can cause some harm in a large enough dose, we shouldn't get so worked up about exposure to toxic chemicals.
   b. We shouldn't worry so much about exposure to toxic chemicals because through genetic adaptation we can develop immunity to such chemicals.
   c. We shouldn't worry so much about exposure to toxic chemicals because we can use genetic engineering to reduce or eliminate such problems.

6. Before you read this chapter were you aware of the serious global TB epidemic that prematurely kills an estimated 3.1 million people each year? Why do you think that this important story has gotten so little media attention compared to much less serious health issues?

7. List the three most important things you believe should be done to reduce the overall threat to human health from AIDS.

8. What are the five major risks that you as an individual face from your lifestyle, where you live, and what you do for a living? Which of these risks are voluntary and which are involuntary? List the five most important things you can do to reduce these risks. Which of these things do you actually plan to do?

9. How would you answer each of the questions raised about risk assessment and risk management on pp. 238–239? Explain each of your answers.

10. Do you believe that we should shift the emphasis from risk assessment and risk–benefit analysis to pollution prevention and risk reduction? Explain. What beneficial and harmful effects might such a shift have on your life? On the life of any child you might have?

*11. Conduct a survey to determine the major physical, chemical, and biological hazards to human health at your school. Develop a plan for preventing or reducing these hazards. Present your findings to school officials.

*12. Make a concept map of this chapter's major ideas, using the section heads and subheads and the key terms (in boldface type). Look at the inside back cover and in Appendix 4 for information about concept maps.

SUGGESTED READINGS

See Internet Sources and Further Readings at the back of the book, and InfoTrac.

PART III

RESOURCES: AIR, WATER, SOIL, MINERALS, AND WASTES

In our every deliberation, we must consider the impact of our decisions on the next seven generations.
IROQUOIS CONFEDERATION, 18TH CENTURY.

9 AIR AND AIR POLLUTION

When Is a Lichen Like a Canary?

Nineteenth-century coal miners took canaries with them into the mines—not for their songs, but for the moment when they stopped singing. Then the miners knew the air was bad and it was time to get out of the mine.

Today we use sophisticated equipment to monitor air quality, but living things such as lichens (Figure 9-1) still have a role in warning us of bad air. A lichen consists of a fungus and an alga living together, usually in a mutually beneficial (mutualistic) partnership.

With more than 20,000 known species, lichens can live almost anywhere: on rocks, trees, bare soil, buildings, gravestones, and even sun-bleached bones. Some lichens survive for more than 4,000 years.

These hearty pioneer species are good air pollution detectors because they are always absorbing air as a source of nourishment. Certain lichen species are sensitive to specific air-polluting chemicals. Old man's beard (*Usnea trichodea*) (Figure 9-1, right) and yellow *Evernia* lichens, for example, sicken or die in the presence of too much sulfur dioxide.

Because lichens are widespread, long-lived, and reliably anchored in place, they can also help track pollution to its source. The scientist who discovered sulfur dioxide pollution on Isle Royale in Lake Superior, where no car or smokestack had ever intruded, used *Evernia* lichens to point the finger northward to coal-burning facilities at Thunder Bay, Canada.

Radioactive particles spewed into the atmosphere by the Chernobyl nuclear power-plant disaster (p. 538) fell to the ground over much of northern Scandinavia, spread, and were absorbed by lichens that carpet much of Lapland. The area's Saami people depend on reindeer meat for food, and the reindeer feed on lichens. After Chernobyl more than 70,000 reindeer had to be killed and the meat discarded because it was too radioactive to eat. However, scientists helped the Saami identify which of the remaining reindeer to move by analyzing lichens (which absorbed some of the radioactive fallout) to pinpoint the most contaminated areas.

Last but not least, lichens can replace electronic monitoring stations that would cost more than $100,000 each. This is not so much a triumph of nature over technology as a partnership between the two, for technicians use highly sophisticated methods to analyze lichens for pollution and measure their rates of photosynthesis.

Because we all must breathe air from a global atmospheric commons, air pollution anywhere is a potential threat elsewhere, and in some cases, everywhere. Lichens can alert us to the danger, but as with all forms of pollution, the only real solution is prevention.

Figure 9-1 Red and yellow crustose lichens growing on slate rock in the foothills of the Sierra Nevada near Merced, California (below), and *Usnea trichodea* lichen growing on a branch of a larch tree in Gifford Pinchot National Park, Washington (right). The vulnerability of various species of lichens to specific pollutants can help detect levels of specific air pollutants and track down their sources. (Left, Kenneth W. Fink/Ardea London; right, Milton Rand/Tom Stack & Associates)

I thought I saw a blue jay this morning. But the smog was so bad that it turned out to be a cardinal holding its breath.

MICHAEL J. COHEN

This chapter is devoted to answering the following questions:

- What layers are found in the atmosphere?
- What are air pollutants, and where do they come from?
- What is smog?
- What is acid deposition?
- What are the harmful effects of air pollutants?
- How can we prevent and control air pollution?

9-1 THE ATMOSPHERE

What Is the Troposphere? Weather Breeder
We live at the bottom of a "sea" of air called the **atmosphere**. This thin envelope of life-sustaining gases surrounding the earth is divided into several spherical layers characterized by abrupt changes in temperature, the result of differences in the absorption of incoming solar energy (Figure 9-2).

About 75% of the mass of the earth's air is found in the atmosphere's innermost layer, the **troposphere**, which extends only about 17 kilometers (11 miles) above sea level at the equator and about 8 kilometers (5 miles) over the poles. If the earth were an apple, this lower layer containing the air we breathe would be no thicker than the apple's skin. This thin and turbulent layer of rising and falling air currents and winds is the planet's weather breeder.

Throughout the earth's long history the composition of the troposphere has varied considerably (Section 5-5). Today, about 99% of the volume of clean, dry air in the troposphere consists of two gases: nitrogen (78%) and oxygen (21%). The remainder has slightly less than 1% argon (Ar), 0.036% carbon dioxide (CO_2), and trace amounts of several other gases. Air in the troposphere also holds water vapor in amounts varying from 0.01% by volume at the frigid poles to 5% in the humid tropics.

The average pressure exerted by the gases in the atmosphere decreases with altitude because the average density (mass of gases per unit volume) decreases with altitude. Temperature also declines with altitude in the troposphere but abruptly begins to rise at the top of this zone, called the *tropopause*. This temperature change limits mixing between the troposphere and upper layers of the atmosphere.

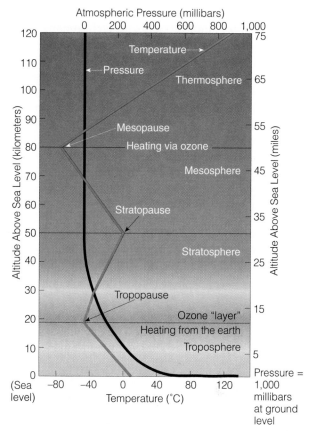

Figure 9-2 The earth's current atmosphere consists of several layers. Most ultraviolet radiation from the sun is absorbed by ozone (O_3) in the stratosphere, which is found primarily in the *ozone layer* between 17 and 26 kilometers (11–16 miles) above sea level.

What Is the Stratosphere? Earth's Global Sunscreen The tropopause marks the end of the troposphere and the beginning of the **stratosphere**, the atmosphere's second layer, which extends from about 17 to 48 kilometers (11–30 miles) above the earth's surface (Figure 9-2). Although the stratosphere contains less matter than the troposphere, its composition is similar, with two notable exceptions: Its volume of water vapor is about 1,000 times less, and its volume of ozone (O_3) is about 1,000 times greater.

Stratospheric ozone is produced when some of the oxygen molecules there interact with lightning and UV radiation emitted by the sun. Ozone is continuously being formed and destroyed, but as long as the rates of these two reversible processes are equal, the average concentration of ozone in the stratosphere remains constant (although the concentration varies at different altitudes and at different places).

The "global sunscreen" of ozone in the stratosphere keeps about 99% of the sun's harmful ultraviolet radiation from reaching the earth's surface (Section 10-6).

This UV filter (1) allows humans and other forms of life to exist on land; (2) helps protect humans from sunburn, skin and eye cancer, cataracts, and damage to the immune system; and (3) prevents much of the oxygen in the troposphere from being converted to ozone, a harmful air pollutant. The trace amounts of ozone that do form in the troposphere are a component of urban smog and they damage plants, the respiratory systems of humans and other animals, and materials such as rubber.

Thus, our good health, and that of many other species, depends on having enough ozone in the stratosphere and as little ozone as possible in the troposphere. There is considerable evidence that some human activities are both increasing the amount of harmful ozone in the tropospheric air we must breathe and decreasing the amount of beneficial ozone in the stratosphere (Figure 8-11, left, and Section 10-5).

Unlike air in the troposphere, that in the stratosphere is calm, with little vertical mixing. Pilots like to fly in this layer because it has so little turbulence and such excellent visibility (due to the almost complete absence of clouds). Flying in the stratosphere also improves fuel efficiency because the thin air offers little resistance to the forward thrust of the plane. Temperature rises with altitude in the stratosphere until there is another temperature reversal at the *stratopause*. This change marks the end of the stratosphere and the beginning of the atmosphere's next layer: the *mesosphere* (Figure 9-2).

How Are We Disrupting the Earth's Gaseous Nutrient Cycles? The earth's biogeochemical cycles (Section 4-4) work fine as long as we don't disrupt them by overloading them at certain points or removing too many vital chemicals at other points. However, our activities are disrupting the gaseous parts of some of these cycles.

We add about one-fourth as much CO_2 to the troposphere as the rest of nature does by burning up nature's one-time-only deposit of fossil fuels and clearing forests. This massive human intrusion into the carbon cycle (Figure 4-22) has the potential to warm the earth and alter global climate and food-producing regions (Section 10-4). We also disrupt natural energy flows (Figure 4-4) by producing massive heat islands and dust domes over urban areas (Figure 6-22). The high-risk ecological problems (Figure 8-11) of possible global warming and depletion of ozone in the stratosphere are discussed in detail in Chapter 10.

By burning fossil fuels and using nitrogen fertilizers, we are releasing three times more nitrogen oxides (NO, NO_2, and N_2O) and gaseous ammonia (NH_3) into the troposphere than do natural processes in the nitro-gen cycle (Figure 4-23). In the troposphere, most of these nitrogen oxides are converted to nitric acid vapor (HNO_3) and acid-forming nitrate salts that dissolve in water and return to the earth, where they can increase the acidity of soils, streams, and lakes and harm plant and animal life.

Our inputs of sulfur dioxide (SO_2) into the troposphere (mostly from petroleum refining and burning of coal and oil; see photo in the Table of Contents) are about twice as high as natural inputs from the sulfur cycle (Figure 4-26). Much of this sulfur dioxide is converted to sulfuric acid (H_2SO_4) and sulfate salts that return to the earth's surface as components of acid deposition (Section 9-2).

These are only a few of the chemicals we are spewing into the troposphere. Small amounts of toxic metals such as arsenic, cadmium, and lead also circulate in the ecosphere in chemical cycles. Scientists estimate that we now inject into the troposphere about twice as much arsenic as nature does, 7 times as much cadmium, and 17 times as much lead.

9-2 OUTDOOR AIR POLLUTION: POLLUTANTS, SMOG, AND ACID DEPOSITION

What Are the Major Types and Sources of Air Pollution? **Air pollution** is the presence of one or more chemicals in the atmosphere in quantities and duration that cause harm to humans, other forms of life, and materials. As clean air in the troposphere moves across the earth's surface, it collects the products of natural events (dust storms and volcanic eruptions) and human activities (emissions from cars and smokestacks). These potential pollutants, called **primary pollutants**, are mixed vertically and horizontally and are dispersed and diluted by the churning air in the troposphere. While in the troposphere, some of these primary pollutants may react with one another or with the basic components of air to form new pollutants, called **secondary pollutants** (Figure 9-3).

Long-lived primary and secondary pollutants can travel great distances before they return to the earth's surface as solid particles, droplets, or chemicals dissolved in precipitation. Table 9-1 lists the major classes of pollutants commonly found in outdoor air.

Most pollutants in urban areas enter the atmosphere from the burning of fossil fuels in both power plants and factories (*stationary sources*) and in motor vehicles (*mobile sources*). In car-clogged cities such as Los Angeles, California, São Paulo, Brazil, Bangkok, Thailand (Figure 6-1), Rome, Italy, and Mexico City, Mexico (Case Study, p. 179), motor vehicles are

Q: According to the EPA, how many landfills in the United States will eventually leak?

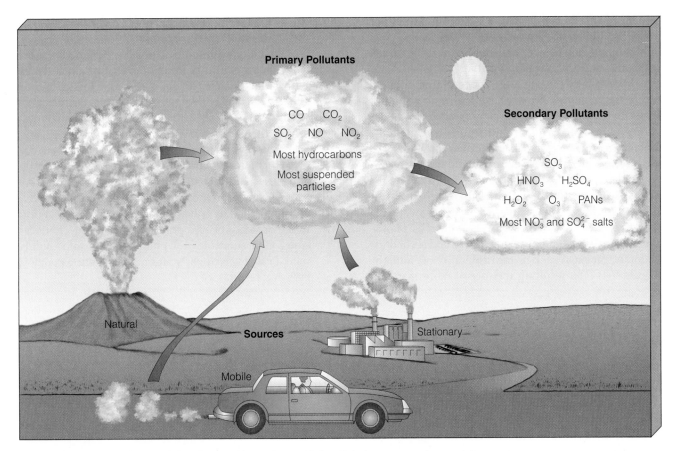

Figure 9-3 Sources and types of air pollutants. Human inputs of air pollutants may come from mobile sources (cars) and stationary sources (industrial and power plants). Some primary air pollutants may react with one another or with other chemicals in the air to form secondary air pollutants.

Table 9-1 Major Classes of Air Pollutants	
Class	**Examples**
Carbon oxides	Carbon monoxide (CO), carbon dioxide (CO_2)
Sulfur oxides	Sulfur dioxide (SO_2), sulfur trioxide (SO_3)
Nitrogen oxides	Nitric oxide (NO), nitrogen dioxide (NO_2), nitrous oxide (N_2O) (NO and NO_2 are often lumped together and labeled NO_x)
Volatile organic compounds	Methane (CH_4), propane (C_3H_8), benzene (C_6H_6), chlorofluorocarbons (CFCs)
Suspended particles	Solid particles (dust, soot, asbestos, lead, nitrate and sulfate salts), liquid droplets (sulfuric acid, PCBs, dioxins, pesticides)
Photochemical oxidants	Ozone (O_3), peroxyacyl nitrates (PANs), hydrogen peroxide (H_2O_2), aldehydes
Radioactive substances	Radon-222, iodine-131, strontium-90, plutonium-239 (Table 3-1)
Toxic compounds	Trace amounts of at least 600 toxic substances (many of them volatile organic compounds), 60 of them known to cause cancer in test animals

A: All of them

responsible for 80–88% of the air pollution. According to the World Health Organization, more than 1.1 billion people—one of every five—live in urban areas where the air is unhealthy to breathe.

Because they contain large concentrations of cars and factories, cities normally have higher air pollution levels than rural areas. However, prevailing winds can spread long-lived primary and secondary air pollutants from emissions in urban and industrial areas to the countryside and to other downwind urban areas.

What Is Photochemical Smog? Brown-Air Smog

Any chemical reaction activated by light is called a *photochemical reaction*. Air pollution known as **photochemical smog** is a mixture of primary and secondary pollutants formed under the influence of sunlight (Figure 9-4). The resulting mixture of more than 100

chemicals is dominated by ozone, a highly reactive gas that harms most living organisms.

Virtually all modern cities have photochemical smog, However, it is much more common in cities with sunny, warm, dry climates and lots of motor vehicles such as Los Angeles, California, Denver, Colorado, and Salt Lake City, Utah, in the United States, as well as Sydney, Australia, Mexico City, Mexico, and São Paulo and Buenos Aires, Brazil.

The hotter the day, the higher the levels of ozone and other components of photochemical smog. As traffic increases in the morning, levels of NO and NO_2 and unburned hydrocarbons rise and begin reacting in the presence of sunlight to produce photochemical smog. On a sunny day the photochemical smog builds up to peak levels by early afternoon, irritating people's eyes and respiratory tracts.

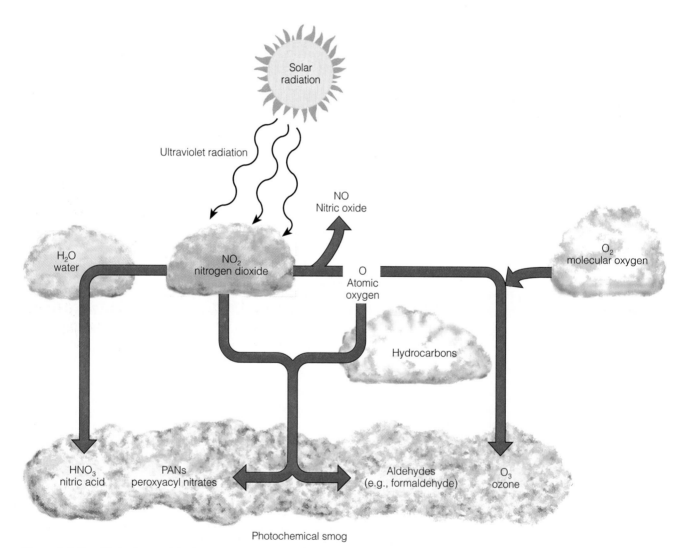

Figure 9-4 Simplified scheme of the formation of photochemical smog. The severity of smog is generally associated with atmospheric concentrations of ozone at ground level.

Q: What percentage of the hazardous waste produced in the United States is regulated by federal laws?

What Is Industrial Smog? Gray-Air Smog

Thirty years ago cities such as London, England, and Chicago and Pittsburgh in the United States burned large amounts of coal and heavy oil (which contain sulfur impurities) in power plants and factories and for space heating. During winter, people in such cities were exposed to **industrial smog** consisting mostly of sulfur dioxide, suspended droplets of sulfuric acid (formed from some of the sulfur dioxide, Figure 9-3), and a variety of suspended solid particles and droplets (called aerosols).

Urban industrial smog is rarely a problem today in most developed countries because coal and heavy oil are burned only in large boilers with reasonably good pollution control or with tall smokestacks. However, industrial smog is a problem in industrialized urban areas of China, India, Ukraine, and some eastern European countries, where large quantities of coal are burned with inadequate pollution controls.

What Factors Influence the Formation of Photochemical and Industrial Smog?

The frequency and severity of smog in an area depend on several things: the local climate and topography, the population density, the amount of industry, and the fuels used in industry, heating, and transportation. In areas with high average annual precipitation, rain and snow help cleanse the air of pollutants. Winds help sweep pollutants away and bring in fresh air, but they may also transfer some pollutants to downwind areas.

Hills and mountains tend to reduce the flow of air in valleys below them and allow pollutant levels to build up at ground level. Buildings in cities generally slow wind speed, thereby reducing dilution and removal of pollutants.

During the day the sun warms the air near the earth's surface. Normally this heated air expands and rises, carrying low-lying pollutants higher into the troposphere (Figure 9-5, left). Colder, denser air from surrounding high-pressure areas then sinks into the low-pressure area created when the hot air rises. This continual mixing of the air helps keep pollutants from reaching dangerous concentrations near the ground.

Sometimes, however, a layer of dense, cool air can be trapped beneath a layer of less dense, warm air in an urban basin or valley, causing a phenomenon known as a **temperature inversion**, or a **thermal inversion** (Figure 9-5, right). The changing temperature (temperature gradient) in the warm air above the pool of cool air prevents ascending air currents (that would disperse and dilute pollutants) from developing. These inversions usually last for only a few hours, but when a high-pressure air mass stalls over an area, they can last for several days, allowing air pollutants at ground level to build up to harmful and even lethal concentrations.

The United States suffered its first major air pollution disaster in Donora, Pennsylvania, a small industrial town southeast of Pittsburgh. The town is surrounded by mountains and nestled in the Monongahela Valley, which is subject to frequent thermal inversions (Figure 9-5, right). In 1948 Donora experienced a thermal inversion that trapped pollutants from steel mills, a zinc smelter, a sulfuric acid plant, and several other industrial plants for 5 days. About 6,000 of the town's 14,000 inhabitants fell ill, and 20 of them died.

A city with several million people and motor vehicles in an area with a sunny climate, light winds, mountains on three sides, and the ocean on the other has ideal conditions for photochemical smog worsened by frequent thermal inversions. This describes California's Los Angeles basin, which has 14 million people, 23 million motor vehicles, thousands of factories, and thermal inversions at least half of the year. Despite having the world's toughest air-pollution control program, Los Angeles is the air pollution capital of the United States. Other cities with frequent thermal inversions

 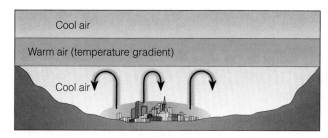

Figure 9-5 Thermal inversion. The change in temperature (temperature gradient) in the warm air (right) prevents ascending air currents from rising and dispersing and traps pollutants in the cool pool of air near the ground. Because of their topography, Los Angeles in the United States and Mexico City in Mexico have frequent thermal inversions, many of them prolonged during the summer months.

are Denver in the United States, Mexico City, Mexico (Case Study, p. 179), Rio de Janeiro and São Paulo in Brazil, and Beijing and Shenyang in China.

What Is Acid Deposition? To reduce local air pollution (and meet government standards without having to add expensive air-pollution control devices), most coal-burning power plants, ore smelters, and other industrial plants in developed countries use tall smokestacks to emit sulfur dioxide, suspended particles, and nitrogen oxides above the inversion layer. As this practice spread in the 1960s and 1970s, pollution in downwind areas began to increase. In addition to smokestack emissions, large quantities of nitrogen oxides are also released by motor vehicles.

This "dilution solution" reduces local air pollution. However, it increases pollution downwind because what goes up must come down—another example of connections or unintended consequences. As the primary pollutants sulfur dioxide and nitrogen oxides are transported as much as 1,000 kilometers (600 miles) by prevailing winds, they form secondary pollutants such as nitric acid vapor, droplets of sulfuric acid, and particles of acid-forming sulfate and nitrate salts (Figure 9-3).

These chemicals descend to the earth's surface in two forms: *wet* (as acidic rain, snow, fog, and cloud vapor) and *dry* (as acidic particles). The resulting mixture is called **acid deposition** (Figure 9-6). Although this form of pollution is commonly called *acid rain*, *acid deposition* is a better term because the acidity can reach the earth's surface not only in rain but also as gases and as solid particles.

Acidity of substances in water is commonly expressed in terms of *pH*. Solutions with pH values less than 7 are acidic, and those with pH values greater than 7 are alkaline or basic (Figure 9-7). Natural precipitation is slightly acidic, with a pH of 5.0–5.6. However, primarily because of acid deposition, typical rain in the eastern United States is now about 10 times more acidic, with a pH of 4.3. In some areas it is 100 times more acidic, with a pH of 3—as acidic as vinegar. Some cities and mountaintops are bathed in a fog as acidic as lemon juice, with a pH of 2.3—about 1,000 times the acidity of normal precipitation.

What Areas Are Most Affected by Acid Deposition? Acid deposition occurs on a regional rather than global basis because the acidic components remain in the atmosphere only for a few days. However, acid deposition is a serious regional problem (Figure 9-8) in many areas downwind from coal-burning power plants, smelters, factories, and large urban areas. How seriously vegetation and aquatic life in nearby lakes are affected by an area receiving acid deposition depends mostly on whether its soils are acidic or basic.

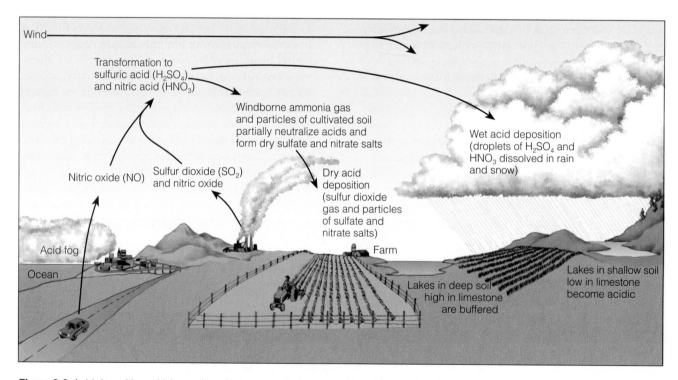

Figure 9-6 Acid deposition, which consists of rain, snow, dust, or gas with a pH lower than 5.6, is commonly called acid rain. Soils and lakes vary in their ability to buffer or remove excess acidity.

Q: What do scientists believe are the most desirable ways to deal with hazardous waste?

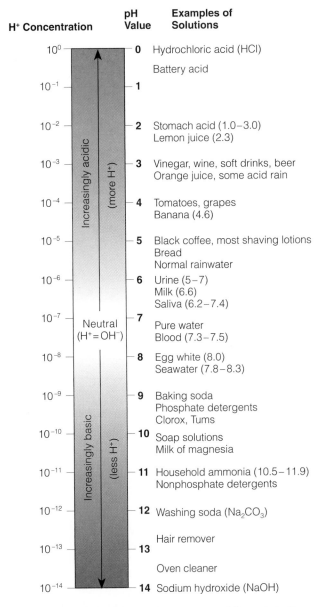

H⁺ Concentration	pH Value	Examples of Solutions

10^0 — **0** Hydrochloric acid (HCl)

Battery acid

10^{-1} — **1**

10^{-2} — **2** Stomach acid (1.0–3.0)
Lemon juice (2.3)

10^{-3} — **3** Vinegar, wine, soft drinks, beer
Orange juice, some acid rain

10^{-4} — **4** Tomatoes, grapes
Banana (4.6)

10^{-5} — **5** Black coffee, most shaving lotions
Bread
Normal rainwater

10^{-6} — **6** Urine (5–7)
Milk (6.6)
Saliva (6.2–7.4)

10^{-7} — **7** Pure water
Blood (7.3–7.5)

10^{-8} — **8** Egg white (8.0)
Seawater (7.8–8.3)

10^{-9} — **9** Baking soda
Phosphate detergents
Clorox, Tums

10^{-10} — **10** Soap solutions
Milk of magnesia

10^{-11} — **11** Household ammonia (10.5–11.9)
Nonphosphate detergents

10^{-12} — **12** Washing soda (Na_2CO_3)

Hair remover

10^{-13} — **13**

Oven cleaner

10^{-14} — **14** Sodium hydroxide (NaOH)

Increasingly acidic (more H⁺)

Neutral (H⁺ = OH⁻)

Increasingly basic (less H⁺)

Figure 9-7 The pH scale, used to measure acidity and alkalinity of water solutions. Values shown are approximate. A solution with a pH less than 7 is acidic, a neutral solution has a pH of 7, and one with a pH greater than 7 is basic. The lower the pH, the more acidic the solution. Each whole-number drop in pH represents a 10-fold increase in acidity.

In some areas soils are basic enough to neutralize or buffer some inputs of acids. The ecosystems most harmed by acid deposition are those containing thin, acidic soils without such natural buffering (Figure 9-8, green areas) and those where the buffering capacity of soils has been depleted because of decades of exposure to acid deposition.

Many acid-producing chemicals generated by power plants, factories, smelters, and cars in one coun-try may be exported to others by prevailing winds. For example, more than three-fourths of the acid deposition in Norway, Switzerland, Austria, Sweden, the Netherlands, and Finland is blown to those countries from industrialized areas of western Europe (especially the United Kingdom and Germany) and eastern Europe.

Chemical detective work indicates that more than half the acid deposition in southeastern Canada and the eastern United States originates from coal- and oil-burning power plants and factories in the states of Ohio, Indiana, Pennsylvania, Illinois, Missouri, West Virginia, and Tennessee. In areas near and downwind from large urban areas, emissions of NO and NO_2 (mostly from cars) leading to the formation of nitric acid may be the main culprit.

Within a few decades, NO_x and SO_2 emissions from developing countries are expected to outstrip those from developed countries, leading to greatly increased damage from acid deposition over a much wider area, especially where soils are sensitive to acidification.

What Are the Effects of Acid Deposition? Risk analysis experts rate acid deposition as a medium-risk ecological problem and a high risk to human health (Figure 8-11, left). Acid deposition has many harmful ecological effects, especially when the pH falls below 5.1 for terrestrial systems and below 5.5 for aquatic systems. It also contributes to human respiratory diseases such as bronchitis and asthma (which can cause premature death), and it damages statues, buildings, metals, and car finishes.

Acid deposition and other air pollutants such as ozone (O_3) can damage tree foliage directly, but the most serious effect is weakening trees so they become more susceptible to other types of damage (Figure 9-9). The areas hardest hit by acid deposition are mountain-top forests, which tend to have thin soils without much buffering capacity. Trees on mountaintops, especially conifers such as red spruce that keep their leaves year-round, are bathed almost continuously in very acidic fog and clouds.

A combination of acid deposition and other air pollutants (especially ozone) can make trees more susceptible to stresses such as cold temperatures, diseases, insects, drought, and fungi (which thrive under acidic conditions), and to a drop in net primary productivity from loss of soil plant nutrients. Although the final cause of tree damage or death may be mosses, insect attacks, diseases, and lack of plant nutrients, the underlying cause is often years of exposure to an atmospheric cocktail of air pollutants and soil overloaded with acids. Drops in forest productivity because of depletion of soil nutrients and acid-buffering chemicals can reduce biodiversity and have significant economic implications for timber companies.

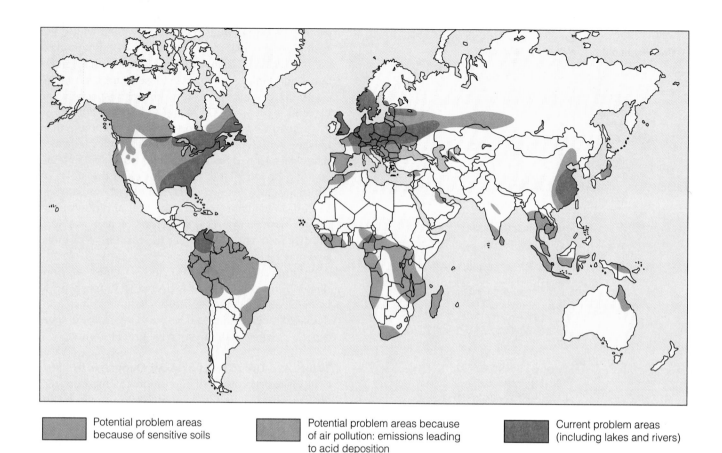

| | Potential problem areas because of sensitive soils | | Potential problem areas because of air pollution: emissions leading to acid deposition | | Current problem areas (including lakes and rivers) |

Figure 9-8 Regions where acid deposition is now a problem and regions with the potential to develop this problem, either because of increased air pollution (mostly from power plants, industrial plants, and ore smelters) or because of soils that cannot neutralize inputs of acidic compounds (green areas and most red areas). (Data from World Resource Institute and U.S. Environmental Protection Agency)

Because it can take decades to hundreds of years for soil to replenish nutrients leached out by acid deposition, losses in plant productivity in damaged areas could continue for decades even if emissions of sulfur dioxide and nitrogen oxides are reduced by air-pollution control programs. Until recently scientists expected some of the lost forest productivity to be offset by increased productivity from the larger input of nitric acid and nitrate salts from acid deposition in areas where this nutrient is the limiting factor. However, recent research revealed that much of the nitrate raining down on forest areas in parts of Germany and Norway damaged by air pollution is not being taken up by the trees or by nitrogen-using microbes in the soil.

Acid deposition can also release aluminum ions (Al^{3+}) attached to soil minerals. Once released from soil particles, these water-soluble ions can damage tree roots. When washed into lakes, aluminum ions can also kill many kinds of fish by stimulating excessive mucus formation, which asphyxiates the fish by clogging their gills.

Excess acidity can contaminate fish in some lakes with highly toxic methylmercury. Increased acidity of lakes apparently converts moderately toxic inorganic mercury compounds in lake-bottom sediments into highly toxic methylmercury, which is more soluble in the fatty tissue of animals and can be biomagnified to higher concentrations in aquatic food chains and webs (Figure 8-2). However, acid runoff into lakes and streams is rated by risk analysis experts as a low-risk ecological problem (Figure 8-11, left).

How Serious Is Acid Deposition in the United States? A large-scale government-sponsored research study on the *ecological effects* of acid deposition in the United States in the 1980s concluded that the problem was serious but not yet at a crisis stage. However, numerous health studies have shown that the effects from exposure to the chemical components of acid deposition are a *serious health problem* (Figure 8-11 and Section 9-4) and also damages materials (Table 9-2).

Representatives of coal companies and industries that burn coal and oil claim that adding expensive air pollution control equipment or burning low-sulfur coal or oil costs more than the resulting health and

Q: What do scientists believe are the least desirable ways of handling hazardous waste?

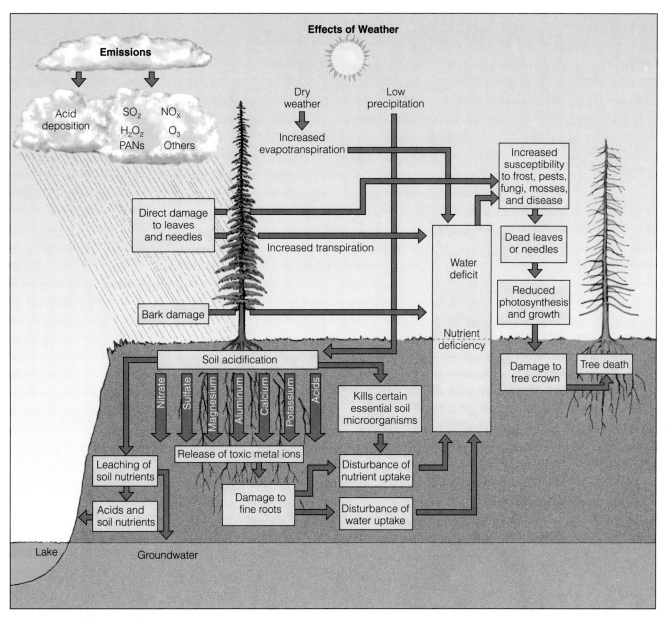

Figure 9-9 Possible or suspected harmful effects of prolonged exposure to an atmospheric cocktail of air pollutants on trees.

environmental benefits are worth. However, a 1990 government-sponsored economic study indicated that the benefits of controlling acid deposition in the United States are worth at least $5 billion (some say $10 billion) per year, about 50% more than the claimed costs of controlling acid deposition. The benefits may be much greater because according to the EPA the actual cleanup costs of SO_2 in 1994 were about one-tenth of the estimate given by industry when they opposed the new standards set by the Clean Air Act of 1990.

Many scientists support greatly reducing emissions from coal- and oil-burning facilities to reduce their harmful effects on human health and materials and to *prevent* acidic compounds in soil and aquatic systems from exceeding the tolerance levels of various species and eventually serious and costly ecological and economic damage.

Progress is being made. A 1993 study by the U.S. Geological Survey found that the concentration of sulfate ions, a key component of acid deposition, declined at 26 out of 33 U.S. rainwater collection sites between 1980 and 1991. In addition, U.S. sulfur dioxide emissions dropped 30% between 1970 and 1993 and are expected to fall further by 2000 because of the requirements of the Clean Air Act of 1990.

Table 9-2 Harmful Effects of Air Pollution on Materials

Material	Effects	Principal Air Pollutants
Stone and concrete	Surface erosion, discoloration, soiling	Sulfur dioxide, sulfuric acid, nitric acid, particulate matter
Metals	Corrosion, tarnishing, loss of strength	Sulfur dioxide, sulfuric acid, nitric acid, particulate matter, hydrogen sulfide
Ceramics and glass	Surface erosion	Hydrogen fluoride, particulate matter
Paints	Surface erosion, discoloration, soiling	Sulfur dioxide, hydrogen sulfide, ozone, particulate matter
Paper	Embrittlement, discoloration	Sulfur dioxide
Rubber	Cracking, loss of strength	Ozone
Leather	Surface deterioration, loss of strength	Sulfur dioxide
Textiles	Deterioration, fading, soiling	Sulfur dioxide, nitrogen dioxide, ozone, particulate matter

Solutions: What Can Be Done to Reduce Acid Deposition? According to most scientists studying acid deposition, the best solutions are *prevention approaches*. They include **(1)** reducing energy use and thus air pollution by improving energy efficiency (Section 18-3); **(2)** switching from coal to cleaner-burning natural gas (Section 19-2) and renewable energy resources (Chapter 18); **(3)** removing sulfur from coal before it is burned; **(4)** burning low-sulfur coal; **(5)** removing SO_2, particulates, and nitrogen oxides from smokestack gases; and **(6)** removing nitrogen oxides from motor vehicle exhaust.

Reducing coal use, the major culprit, will be economically and politically difficult. For example, China (the world's largest user of coal) and India (the fourth largest user) are using their own coal reserves to fuel rapid industrial growth and so far have not put much money into pollution control.

Some *cleanup approaches* can be used, but they are expensive and merely mask some of the symptoms temporarily without treating underlying causes. Acidified lakes can be neutralized by treating them or the surrounding soil with large amounts of limestone or lime. However, such liming is an expensive and temporary remedy that usually must be repeated on a yearly basis. Using lime to reduce excess acidity in U.S. lakes would cost at least $8 billion per year.

Liming can also kill some types of plankton and aquatic plants and can harm wetland plants that need acidic water. It is also difficult to know how much of the lime to put where (in the water or at selected places on the ground). In addition, some recent research suggests that liming can increase populations of microbes that deplete carbon stored in the slowly decaying soil matter (humus) and thus reduce forest productivity. Recently, however, researchers in England found that adding a small amount of phosphate fertilizer can neutralize excess acidity in a lake.

✥ 9-3 INDOOR AIR POLLUTION

What Are the Types and Sources of Indoor Air Pollution? If you are reading this book indoors, you may be inhaling more air pollutants with each breath than if you were outside (Figure 9-10). According to EPA studies, in the United States levels of 11 common pollutants are generally 2–5 times higher inside homes and commercial buildings than outdoors, and as much as 70 times higher in some cases. A 1993 study found that pollution levels inside cars in traffic-clogged U.S. urban areas can be up to 18 times higher than those outside the vehicles.

The health risks from exposure to such chemicals are magnified because people spend 70–98% of their time indoors. In 1990 the EPA placed indoor air pollution at the top of the list of 18 sources of cancer risk, and it is rated by risk analysis scientists as a high-risk health problem for humans (Figure 8-11, left). At greatest risk are smokers, infants and children under age 5, the old, the sick, pregnant women, people with respiratory or heart problems, and factory workers.

Danish and U.S. EPA studies have linked pollutants found in buildings to dizziness, headaches, coughing, sneezing, nausea, burning eyes, chronic fatigue, and flulike symptoms, known as the *sick building syndrome*. A building is considered "sick" when at least 20% of its occupants suffer persistent symptoms that disappear when they go outside.

Q: How is legally regulated hazardous waste in the United States dealt with?

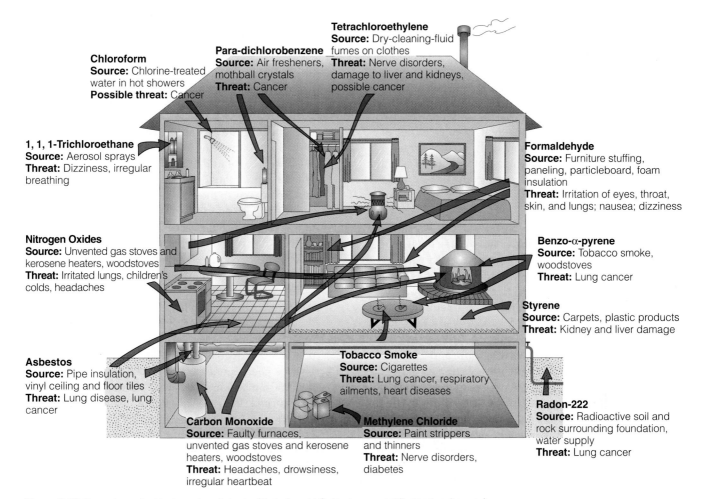

Chloroform
Source: Chlorine-treated water in hot showers
Possible threat: Cancer

Para-dichlorobenzene
Source: Air fresheners, mothball crystals
Threat: Cancer

Tetrachloroethylene
Source: Dry-cleaning-fluid fumes on clothes
Threat: Nerve disorders, damage to liver and kidneys, possible cancer

1, 1, 1-Trichloroethane
Source: Aerosol sprays
Threat: Dizziness, irregular breathing

Formaldehyde
Source: Furniture stuffing, paneling, particleboard, foam insulation
Threat: Irritation of eyes, throat, skin, and lungs; nausea; dizziness

Nitrogen Oxides
Source: Unvented gas stoves and kerosene heaters, woodstoves
Threat: Irritated lungs, children's colds, headaches

Benzo-α-pyrene
Source: Tobacco smoke, woodstoves
Threat: Lung cancer

Styrene
Source: Carpets, plastic products
Threat: Kidney and liver damage

Asbestos
Source: Pipe insulation, vinyl ceiling and floor tiles
Threat: Lung disease, lung cancer

Tobacco Smoke
Source: Cigarettes
Threat: Lung cancer, respiratory ailments, heart diseases

Carbon Monoxide
Source: Faulty furnaces, unvented gas stoves and kerosene heaters, woodstoves
Threat: Headaches, drowsiness, irregular heartbeat

Methylene Chloride
Source: Paint strippers and thinners
Threat: Nerve disorders, diabetes

Radon-222
Source: Radioactive soil and rock surrounding foundation, water supply
Threat: Lung cancer

Figure 9-10 Some important indoor air pollutants. (Data from U.S. Environmental Protection Agency)

New buildings are more commonly "sick" than old ones because of reduced air exchange (to save energy) and chemicals released from new carpeting and furniture. According to the EPA, at least 17% of the 4 million commercial buildings in the United States are considered "sick" (including EPA's headquarters). Indoor air pollution in the United States costs an estimated $100 billion per year in absenteeism, reduced productivity, and health costs.

A 1994 study by Cornell University researchers suggested that the primary culprits causing "sick building" symptoms may be mineral fibers falling from ceiling tiles and blowing in from the lining of air conditioning ducts. The study also suggests why workers who spend a lot of time in front of computer terminals may suffer more than coworkers who don't: Electrostatic fields generated by computer monitors attract the fibers, thus exposing people sitting in front of the screens to more of them.

According to the EPA and public health officials, cigarette smoke (p. 223), formaldehyde, asbestos, and radioactive radon-222 gas are the four most danger-

ous indoor air pollutants. A number of research studies on laboratory animals have also identified tiny fibers of *fiberglass* as a widespread and potentially potent carcinogen in indoor air.

The chemical that causes most people difficulty is *formaldehyde*, an extremely irritating gas. As many as 20 million Americans suffer from chronic breathing problems, dizziness, rash, headaches, sore throat, sinus and eye irritation, and nausea caused by daily exposure to low levels of formaldehyde emitted (outgassed) from common building materials (such as plywood, particleboard, and paneling), furniture, drapes, upholstery, and adhesives in carpeting and wallpaper (Figure 9-10). The EPA estimates that as many as 1 out of every 5,000 people who live in manufactured homes for more than 10 years will develop cancer from formaldehyde exposure.

In developing countries, the burning of wood, dung, and crop residues in open fires or in unvented or poorly vented stoves for cooking and heating exposes inhabitants, especially women and young children, to very high levels of particulate air pollution.

Partly as a result, respiratory illnesses are a major cause of death and illness among the poor in most developing countries.

Case Study: What Should Be Done About Asbestos? There is intense controversy over what to do about possible exposure to tiny fibers of asbestos, a name given to several different fibrous forms of silicate minerals. Unless completely sealed within a product, asbestos easily crumbles into a dust of fibers tiny enough to become suspended in the air and to be inhaled deep into the lungs, where they remain for many years.

Prolonged exposure to asbestos fibers can cause *asbestosis* (a chronic, sometimes fatal disease that eventually makes breathing nearly impossible and was recognized as a hazard among asbestos workers as early as 1924), *lung cancer*, and *mesothelioma* (an inoperable cancer of the chest cavity lining). Epidemiological studies have shown that lung cancer death rates for nonsmoking asbestos workers were 5 times higher than for nonsmokers in a control group, and 53 times higher for asbestos workers who smoked.

Most of these diseases occur in people exposed for years to high levels of asbestos fibers. This group includes asbestos miners, insulators, pipefitters, shipyard employees, and workers in asbestos-producing factories. By the year 2000 it is estimated that 300,000 American workers will have died prematurely because of exposure to asbestos fibers. After being swamped with health claims from workers, most U.S. asbestos manufacturing companies have either declared bankruptcy or have moved their operations to other countries (such as Mexico and Brazil) with weaker environmental laws and lax enforcement.

In recent years the focus has shifted from asbestos workers to concern over possible health effects of inhalation of low levels of asbestos fibers by the public in buildings. In the United States between 1900 and 1986, asbestos was sprayed on ceilings and walls of schools and other public and private buildings for fireproofing, soundproofing, insulation of heaters and pipes, and wall and ceiling decoration. The EPA banned those uses in 1974.

In 1989 the EPA ordered a ban on almost all remaining uses of asbestos (such as brake linings, roofing shingles, and water pipes) in the United States by 1997. Representatives of the asbestos industry in the United States and Canada (which now produces most of the asbestos used in the United States) challenged the ban in court, contending that with proper precautions asbestos products can be safely used and that the costs of the ban outweigh the benefits. Industry officials also pointed to some controversial evidence that most of the harm from asbestos comes from inhalation of needle-shaped amphibole fibers, which are rarely found in buildings. In 1991 a federal appeals court overturned the 1989 EPA ban.

In 1988 the EPA estimated that more than 760,000 buildings—one of every seven commercial and public buildings in the United States (including 30,000 schools)—contained asbestos that had crumbled or could crumble and release fibers. Removal of asbestos from such buildings could cost $50–200 billion, with about $10 billion being spent by 1993.

Critics contend that the health benefits of asbestos removal from many schools, homes, and other buildings are not worth the costs, unless measurements (not just visual inspection) indicate that the buildings have high levels of airborne asbestos fibers, especially amphibole fibers.* They call for sealing, wrapping, and other forms of containment instead of removal of most asbestos, and they point out that improper or unnecessary removal can release more asbestos fibers than sealing off asbestos that is not crumbling.

After much controversy and huge expenditures of money on asbestos removal, there is now general agreement that the degree of risk from low-level exposure to asbestos fibers is unclear and asbestos should not be removed from buildings where it has not been damaged or disturbed. Instead it should be sealed or wrapped, with removal only as a last, carefully conducted resort.

Critics of environmentalists charge that much of the government-required removal of asbestos from schools and other public buildings was unnecessary and wasted billions of dollars—an example of environmental and regulatory overkill.

Case Study: Is Your Home Contaminated with Radon Gas? Radon-222 is a colorless, odorless, tasteless, naturally occurring radioactive gas produced by the radioactive decay of uranium-238. Small amounts of uranium-238 are found in most soil and rock, but this isotope is much more concentrated in underground deposits of minerals such as uranium, phosphate, granite, and shale.

When radon gas from such deposits seeps upward through the soil and is released outdoors, it disperses quickly in the atmosphere and decays to harmless levels. However, when the gas is drawn into buildings through cracks, drains, and hollow concrete blocks (Figure 9-11), or seeps into groundwater in underground wells over such deposits, it can build up to high levels.

Radon-222 gas quickly decays into solid particles of other radioactive elements that, if inhaled, expose

---

* If you plan to buy or live in a house built before 1980, you may want to have its air tested for asbestos fibers. To get a free list of certified asbestos laboratories that charge $25–50 to test a sample, send a self-addressed stamped envelope to NIST/NVLAP, Building 411, Room A124, Gaithersburg, MD 20899, or call the EPA's Toxic Substances Control Hotline at 202-554-1404.

Q: How many sites in the United States contain potentially hazardous wastes?

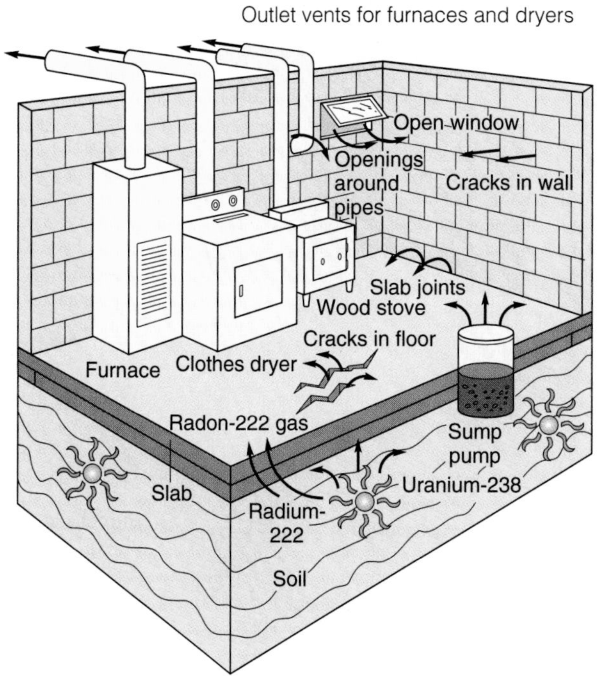

Figure 9-11 Sources and paths of entry for indoor radon-222 gas. (Data from U.S. Environmental Protection Agency)

lung tissue to a large amount of ionizing radiation from alpha particles. Assuming that there is no threshold for harm caused by inhaled asbestos fibers (Figure 8-3, left), scientists have extrapolated the harmful effects of high-level exposure to radon on uranium miners to low levels of exposure in homes. Using this approach, they estimate that prolonged exposure (defined as 75% of one's time spent in the same home for a lifetime of 70 years) to low levels of radon or radon acting together with smoking is responsible for 6,000–36,000 of the 130,000 lung cancer deaths each year in the United States (13,600 deaths is the best estimate).

Scientists who believe there is a threshold dose before radon is harmful say that these estimates are too high. The results of several epidemiological studies in various countries are mixed and provide no clear evidence to support or refute a connection between lung cancer deaths and inhaled radon in homes.

According to the EPA, prolonged exposure to average radon levels above 4 picocuries* per liter of air in a closed house is considered unsafe. Other researchers cite evidence suggesting that radon becomes dangerous only if indoor levels exceed 20 picocuries per liter—the level accepted in Canada, Sweden, and Norway. Such controversy over acceptable radon levels demonstrates the uncertainties and

problems inherent in risk assessment and risk management (Section 8-6).

EPA indoor radon surveys suggest that 4–5 million U.S. homes may have annual radon levels above 4 picocuries per liter of air and that 50,000–100,000 homes may have levels above 20 picocuries per liter. If the 4 picocuries per liter standard is adopted (as proposed by the EPA), the cost of testing and correcting the problem could run about $50 billion, with a 15–20% reduction in radon-related deaths. Some researchers argue that it makes more sense to spend perhaps only $500 million to find and fix homes and buildings with radon levels above 20 picocuries per liter until more reliable data are available on the threat from exposure to lower levels of radon.

Because radon "hot spots" can occur almost anywhere, it's impossible to know which buildings have unsafe levels of radon without conducting tests. In 1988 the EPA and the U.S. Surgeon General's Office recommended that everyone living in a detached house, a town house, a mobile home, or on the first three floors of an apartment building test for radon.[†] Ideally, radon levels should be continuously monitored in the main living areas (not basements or crawl spaces) for 2 months to a year. By 1996 only about 6% of U.S. households had conducted radon tests (most lasting only 2 to 7 days and costing $20–100 per home).

If testing reveals an unacceptable level, homeowners can consult the free EPA publication *Radon Reduction Methods* for ways to reduce radon levels and health risks. According to the EPA, radon control could add $350–500 to the cost of a new home, and correcting a radon problem in an existing house could run $800–2,500.

9-4 EFFECTS OF AIR POLLUTION ON LIVING ORGANISMS AND MATERIALS

How Is Human Health Harmed by Air Pollutants? Your respiratory system (Figure 9-12) has a number of mechanisms that help protect you from air pollution. Hairs in your nose filter out large particles. Sticky mucus in the lining of your upper respiratory tract captures smaller (but not the smallest) particles and dissolves some gaseous pollutants. Sneezing and coughing expel contaminated air and mucus when your respiratory system is irritated by pollutants. The cells of your upper respiratory tract are also lined with hundreds of thousands of tiny, mucus-coated hairlike structures called *cilia* that continually wave back and forth, transporting mucus and the

---

* A picocurie is a trillionth of a curie, which is the amount of radioactivity emitted by a gram of radium.

† For information, see "Radon Detectors: How to Find Out if Your House Has a Radon Problem," *Consumer Reports*, July 1987.

A: About 38,000 (perhaps as many as 425,000)

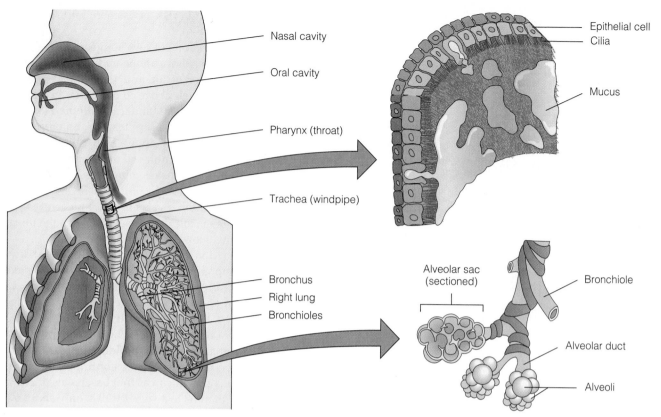

Figure 9-12 Components of the human respiratory system.

pollutants they trap to your throat (where they are either swallowed or expelled).

Years of smoking and exposure to air pollutants can overload or break down these natural defenses, causing or contributing to respiratory diseases. Examples are **(1)** *lung cancer,* **(2)** *asthma* (typically an allergic reaction causing sudden episodes of muscle spasms in the bronchial walls, resulting in acute shortness of breath), **(3)** *chronic bronchitis* (persistent inflammation and damage to the cells lining the bronchi and bronchioles, causing mucus buildup, painful coughing, and shortness of breath), and **(4)** *emphysema* (irreversible damage to air sacs or alveoli leading to abnormal dilation of air spaces, loss of lung elasticity, and acute shortness of breath; Figure 9-13). Elderly people, infants, pregnant women, and people with heart disease, asthma, or other respiratory diseases are especially vulnerable to air pollution.

About 90% of the *carbon monoxide* (CO)—a colorless, odorless, poisonous gas—in the troposphere comes from natural sources. Most of this is produced by reaction in the upper troposphere between methane (emitted mostly by the anaerobic decay of organic matter in swamps, bogs, and marshes) and oxygen. Because this CO is diluted by the turbulent air flows in the troposphere, it does not build up to harmful levels.

However, the remaining 10% of the CO added to the atmosphere comes from the incomplete burning of carbon-containing chemicals (primarily fossil fuels). Cigarette smoking (p. 223) is responsible for the largest human exposure to CO, but this gas is also released by motor vehicles, kerosene heaters, woodstoves, fireplaces, and faulty heating systems.

CO reacts with hemoglobin in red blood cells and thus reduces the ability of blood to carry oxygen. This impairs perception and thinking, slows reflexes, and causes headaches, drowsiness, dizziness, and nausea. CO can also trigger heart attacks and angina attacks in people with heart disease. It can damage the development of fetuses and young children and aggravate chronic bronchitis, emphysema, and anemia. Exposure to high levels of CO causes collapse, coma, irreversible damage to brain cells, and even death. Carbon monoxide detectors (similar in size to smoke detectors) are now available for about $50 and are a good safety investment for homeowners.

Inhaling *suspended particulate matter* aggravates bronchitis and asthma, and long-term exposure can contribute to development of chronic respiratory disease and cancer. Invisible fine particles, 1/28 as thin as a human hair, emitted by incinerators, motor vehicles, radial tires, wind erosion, wood-burning fireplaces, and power and industrial plants are especially hazardous.

Q: How many of the world's estimated 30,000 edible plants feed most of the world's people?

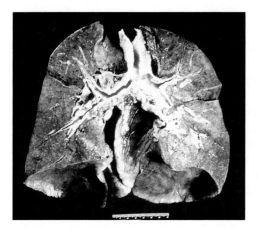
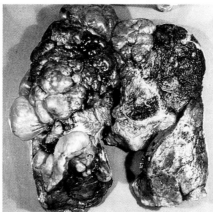

Figure 9-13 Normal human lungs (left) and the lungs of a person who died from emphysema (right). Prolonged smoking and exposure to air pollutants can cause emphysema in anyone, but about 2% of emphysema cases result from a defective gene that reduces the elasticity of the air sacs in the lungs. Anyone with this hereditary condition, for which testing is available, should not smoke and should not live or work in a highly polluted area. (O. Auerbach/Visuals Unlimited)

Such tiny particles are not effectively captured by modern air-pollution control equipment, and they are small enough to penetrate the respiratory system's natural defenses against air pollution. They can also bring with them droplets or other particles of toxic or cancer-causing pollutants that become attached to their surfaces. Once lodged deep within the lungs, these fine particles can cause chronic irritation that can trigger asthma attacks, aggravate other lung diseases, cause lung cancer, and interfere with the blood's ability to take in oxygen and release CO_2. This strains the heart, increasing the risk of death from heart disease.

Several recent studies of air pollution in U.S. cities have indicated that fine particles at levels below current standards prematurely kill 65,000–150,000 Americans each year. To date there is no known threshold level below which the harmful effects of fine particles disappear. The World Bank estimates that if particulate levels were reduced globally to WHO guidelines, 300,00–700,000 premature deaths per year could be prevented.

Sulfur dioxide causes some constriction of the airways in healthy people and severe restriction in people with asthma. Chronic exposure causes a condition similar to bronchitis. Sulfur dioxide and suspended particles react to form far more hazardous acid sulfate particles, which are inhaled more deeply into the lungs than SO_2 and remain there for long periods. According to the World Health Organization, at least 625 million people are exposed to unhealthy levels of sulfur dioxide from fossil-fuel burning.

Nitrogen oxides (especially NO_2) can irritate the lungs, aggravate asthma or chronic bronchitis, cause conditions similar to chronic bronchitis and emphysema, and increase susceptibility to respiratory infections such as the flu and common colds (especially in young children and elderly people).

Research indicates that many *volatile organic compounds* (such as benzene and formaldehyde) and *toxic particulates* (such as lead, cadmium, PCBs, and dioxins) can cause mutations, reproductive problems, or cancer.

Evidence also shows that inhaling *ozone*, a component of photochemical smog (Figure 9-4), causes coughing, chest pain, shortness of breath, and eye, nose, and throat irritation. It also aggravates chronic diseases such as asthma, bronchitis, emphysema, and heart trouble and reduces resistance to colds and pneumonia. Many U.S. cities often exceed safe levels, especially during warm weather. A 1987 study also showed that after factoring out smoking, long-time Los Angeles residents exposed to ozone had double the risk of cancer compared to residents of cleaner cities.

How Many People Die Prematurely from Air Pollution? No one knows how many people die prematurely from respiratory or cardiac problems caused or aggravated by air pollution. Such figures are difficult to estimate by risk analysis because people are exposed to so many different pollutants over their lifetimes.

In the United States, estimates of annual deaths related to outdoor air pollution range from 65,000 to 200,000 (most from exposure to fine particles). If indoor air pollution is included, estimated annual deaths from air pollution in the United States range from 150,000 to 350,000 people—equivalent to 1–2 fully loaded 400-passenger jumbo jets crashing *each day* with no survivors.

Millions more become ill and lose work time. According to the EPA and the American Lung Association, air pollution in the United States costs at least $150 billion annually in health care and lost work productivity, with $100 billion of that caused by indoor air pollution.

The World Health Organization estimates that worldwide about 1.3 billion people, mostly in developing countries, live in areas where outdoor air is unhealthy to breathe, causing 300,000–700,000 premature deaths each year. These estimates may be much

A: 15 (with most provided by wheat, rice, corn, and potato)

Figure 9-14 The dead zone around a nickel smelter in Sudbury, Ontario, Canada. Sulfur dioxide and other fumes released from the smelter over several decades killed the forest once found on this land. (A. J. Copley/Visuals Unlimited)

too low. According to a 1993 Resources for the Future study, more than 900,000 people in China alone may be dying prematurely each year from pollution-related lung disease.

How Are Plants Damaged by Air Pollutants?

Some gaseous pollutants (especially ozone) damage leaves of crop plants and trees directly when they enter leaf pores. Chronic exposure of leaves and needles to air pollutants can break down the waxy coating that helps prevent excessive water loss and damage from diseases, pests, drought, and frost. Such exposure also interferes with photosynthesis and plant growth, reduces nutrient uptake, and causes leaves or needles to turn yellow or brown and drop off (Figure 9-9). Spruce, fir, and other conifers, especially at high elevations, are most vulnerable to air pollution because of their long life spans and the year-round exposure of their needles to a mixture of air pollutants.

Prolonged exposure to high levels of several air pollutants from smelters (plants that extract metal from ores) can kill trees and most other vegetation in an area (Figure 9-14). Lengthy exposure to a mixture of air pollutants from coal-burning power and industrial plants and from cars can also damage trees and many other plants. However, the effects may not become visible for several decades, when large numbers of trees suddenly begin dying off because of depletion of soil nutrients and increased susceptibility to pests, diseases, fungi, and drought. This phenomenon, known as *Waldsterben* (forest death), has turned whole forests of spruce, fir, and beech into stump-studded meadows and mountainsides.

Trees at high elevations in mountain biomes (Section 5-1) (especially in Europe), with their thin and easily erodible soils and almost year-round exposure to air pollutants, have suffered the most damage. The diebacks of trees in such areas lead to extensive soil erosion, which in turn leads to increased flooding and avalanches (particularly in the heavily developed Alps).

Surveys in 1993 and 1994 revealed that almost one out of four trees in Europe had been damaged by air pollution, resulting in a loss of more than 25% of their leaves. Such damage occurs in virtually all European countries but is highest in the Czech Republic, where 57% of all trees suffer moderate or severe defoliation or have died (photo on p. 245). It is estimated that air pollution has been a key factor in reducing the overall productivity of European forests by about 16% and causing damage valued at roughly $30 billion per year.

Forest diebacks have also occurred in the United States, mostly on high-elevation slopes that face moving air masses and are dominated by red spruce. The most seriously affected areas are in the Appalachian Mountains, where research reported in 1994 found a rapid die-off of more than 70 tree species and standing tree skeletons of more than 20 species. Ozone and other air pollutants are believed to be responsible for this forest degradation (Figure 9-9) and loss of biodiversity.

Air pollution, mostly by ozone, also threatens some crops—especially corn, wheat, and soybeans, the three most important U.S. crops—and is reducing U.S. food production by 5–10%. In the United States, estimates of economic losses to agriculture as a result of air pollution range from $1.9–5.4 billion per year.

How Can Air Pollutants Damage Aquatic Life?

High acidity (low pH) can severely harm the aquatic life in freshwater lakes, both where the surrounding soils have little acid-neutralizing capacity and in the northern hemisphere, where there is significant winter snowfall. Much of the damage to aquatic life in such areas is a result of *acid shock* caused by the sudden runoff of large amounts of highly acidic water and aluminum ions into lakes and streams, when snow melts in the spring or after unusually heavy rains. The aluminum ions leached from the soil and lake sediment by this sudden input of acid can kill fish and inhibit their reproduction.

As the acidity of a lake increases and its food chain is disrupted, there is a decline in net primary productivity. This can turn a moderately eutrophic lake (Figure 5-25, top) into a clear blue oligotrophic lake (Figure 5-25, bottom).

Because of excess acidity, at least 16,000 lakes in Norway and Sweden contain no fish, and 52,000 more

Q: What percentage of the earth's land area is suitable for cultivation?

lakes have lost most of their acid-neutralizing capacity. In Canada some 14,000 acidified lakes are almost fish graveyards, and 150,000 more are in peril.

In the United States about 9,000 lakes are threatened with excess acidity, one-third of them seriously. Most of them are concentrated in the Northeast and the upper Midwest—especially Minnesota, Wisconsin, and the upper Great Lakes—where 80% of the lakes and streams are threatened by excess acidity. Over 10% of some 200 lakes in New York's Adirondack Mountains are too acidic to support fish.

What Are the Harmful Effects of Air Pollutants on Materials? Each year air pollutants cause billions of dollars in damage to various materials we use (Table 9-2). The fallout of soot and grit on buildings, cars, and clothing requires costly cleaning. Air pollutants break down exterior paint on cars and houses, and they deteriorate roofing materials. Irreplaceable marble statues, historic buildings, and stained glass windows around the world have been pitted, gouged, and discolored by air pollutants. For example, the famous Greek ruins on the Acropolis in Athens have deteriorated more during the past 50 years than during the previous 2,000 years. Damage to buildings in the United States from acid deposition alone is estimated at $5 billion per year.

✥ 9-5 SOLUTIONS: PREVENTING AND REDUCING AIR POLLUTION

How Have Laws Been Used to Reduce Air Pollution in the United States? The U.S. Congress passed Clean Air Acts in 1970, 1977, and 1990, providing federal air pollution regulations that are enforced by each state. These laws required the EPA to establish *national ambient air quality standards (NAAQS)* for seven outdoor pollutants: suspended particulate matter, sulfur oxides, carbon monoxide, nitrogen oxides, ozone, volatile organic compounds, and lead. Each standard specifies the maximum allowable level, averaged over a specific period, for a certain pollutant in outdoor (ambient) air.

The Clean Air Act has worked. Between 1970 and 1995 levels of major air pollutants decreased nationally almost 30%, despite significant increases in population size and a doubling of economic growth. According to EPA data, between 1986 and 1995 lead levels in U.S. air decreased by 78% (98% since 1970), carbon monoxide 36%, sulfur dioxide 37%, suspended particulate matter 10 micrometers or less in diameter 17%, nitrogen dioxide 14%, and ground level ozone 6%.

Between 1990 and 1995 ozone levels in U.S. urban areas fell by 50%. Result: 50 million people breathe cleaner air. Nitrogen dioxide levels have not dropped much since 1980 because of a combination of inadequate automobile emission standards and more vehicles traveling longer average distances.

Without the 1970 standards for emissions of pollutants, air pollution levels would be much higher today. Even so, today the EPA estimates that 80 million Americans still live in areas that exceed at least one air pollution standard.

A 1996 study by the EPA found that the benefits of the Clean Air Act greatly exceed costs. Between 1970 and 1990, the U.S. spent about $436 billion (in 1990 dollars) to comply with clean air regulations. Total human health and ecological benefits during the same 20-year period were estimated at $2.7–14.6 trillion (in 1990 dollars)—6–33 times higher than the costs.

How Can U.S. Air Pollution Laws Be Improved? The Clean Air Act of 1990 was an important step in the right direction, but most environmentalists point to the following major deficiencies in this law:

■ *Continuing to rely almost entirely on pollution cleanup rather than pollution prevention.* In the United States, the air pollutant with the largest drop (98% between 1970 and 1995) in its atmospheric level was lead, which was virtually banned in gasoline. This shows the effectiveness of the pollution prevention approach.

■ *Failing to sharply increase the fuel efficiency standards for cars and light trucks.* According to environmental scientists, this would reduce oil imports and air pollution more quickly and effectively than any other method and would save consumers enormous amounts of money (Section 18-3).

■ *Not requiring stricter emission standards for fine particulates.* In 1997 the EPA proposed stricter standards for fine particles that would prevent an estimated 20,000 premature deaths, 500,000 asthma attacks, and 9,000 hospital admissions per year and save $1 billion worth of crop losses. The EPA estimates the cost of implementing the standards at $7 billion per year, with estimated health and other benefits of $120 billion per year. Industries opposing these regulations put the cost of implementing the regulations at $200 billion per year.*

---

* In 1990 industry lobbyists predicted that new air pollution standards to reduce acid deposition would cost $1,500 for each 1-ton reduction in sulfur dioxide emissions, compared to a $500-per-ton estimate by the EPA. Because the regulations spurred companies to innovate and look at alternatives, the real cost turned out to be less than $100 per ton.

- *Giving municipal trash incinerators 30-year permits,* which locks the nation into hazardous air pollution emissions and toxic waste from incinerators well into the 21st century. This also undermines pollution prevention, recycling, and reuse (Chapter 13).

- *Setting weak standards for air pollution emissions from incinerators,* thus allowing unnecessary emissions of mercury, lead, dioxins, and other toxic pollutants (Section 13-7).

- *Doing too little to reduce emissions of carbon dioxide and other greenhouse gases* (Section 10-5).

Should We Use the Marketplace to Reduce Pollution? To help reduce SO_2 emissions, the Clean Air Act of 1990 allows an *emissions trading policy,* which enables the 110 most polluting power plants in 21 states (primarily in the Midwest and East) to buy and sell SO_2 pollution rights.

Each year a power plant is given a certain number of pollution credits or rights that allow it to emit a certain amount of SO_2. A utility that emits less SO_2 than its limit receives more pollution credits. It can use these credits to avoid reductions in SO_2 emissions from some of its other facilities, bank them for future plant expansions, or sell them to other utilities, private citizens, or environmental groups.

Proponents of this system argue that it allows the marketplace to determine the cheapest, most efficient way to get the job done instead of having the government dictate how to control pollution. If this market-based approach works for reducing SO_2 emissions, it could be applied to other air and water pollutants.

Some environmentalists see this market approach as an improvement over the current regulatory approach, as long as it achieves net reduction in SO_2 pollution. This would be done by limiting the total number of credits and gradually lowering the annual number of credits (to encourage pollution prevention and the development of better pollution control technologies), something that is not required by the 1990 Clean Air Act. Without such reductions, critics contend that the system of tradable pollution rights is essentially an economic shell game, with no continuing progress in reducing overall SO_2 emissions.

Some environmentalists also contend that marketing pollution rights allows utilities with older, dirtier power plants to buy their way out and keep on emitting unacceptable levels of SO_2. They also warn that this approach creates incentives to cheat. Because air quality regulation is based largely on self-reporting of emissions and because pollution monitoring is incomplete and imprecise, sellers of permits will benefit by understating their reductions (to get more permits), and permit buyers will benefit by underreporting emissions (to reduce their permit purchases). What do you think?

How Can We Reduce Outdoor Air Pollution? Figure 9-15 summarizes ways to reduce emissions of sulfur oxides, nitrogen oxides, and particulate matter from stationary sources (such as electric power plants and industrial plants that burn coal). Until recently, emphasis has been primarily on dispersing and diluting the pollutants by using tall smokestacks or adding equipment that removes some of the particulate pollutants after they are produced (Figure 9-16). Under the sulfur-reduction requirements of the Clean Air Act of 1990, more utilities are switching to low-sulfur coal to reduce SO_2 emissions. Environmentalists call for taxes on air pollutant emissions and greater emphasis on prevention methods.

Figure 9-17 lists ways to reduce emissions from motor vehicles, the primary culprits in producing photochemical smog. Ford Motor Company and Englehard Corporation (which develops and makes catalysts) are working together to develop a catalyst-coated car radiator capable of destroying up to 90% of the ozone and carbon monoxide in the atmosphere. Use of alternative vehicle fuels to reduce air pollution is evaluated in Table 9-3.

Recently, a University of Colorado professor developed a 1-second highway test for auto emissions. A beam of infrared light is sent across a highway to an air pollution detector that measures levels of carbon monoxide, hydrocarbons, and nitrogen oxides in a car's exhaust as it passes by. The information is fed into a computer and a video camera captures the car's license plate number and stores it in the computer. Drivers can be sent notices directing them to correct faulty emissions and have the vehicle rechecked at an official testing station or face large fines.

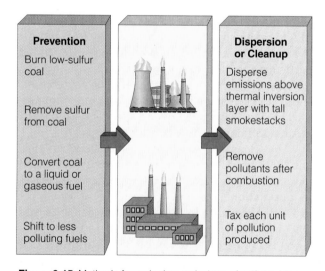

Figure 9-15 Methods for reducing emissions of sulfur oxides, nitrogen oxides, and particulate matter from stationary sources such as coal-burning electric power plants and industrial plants.

Q: What percentage of the world's food is produced on irrigated cropland?

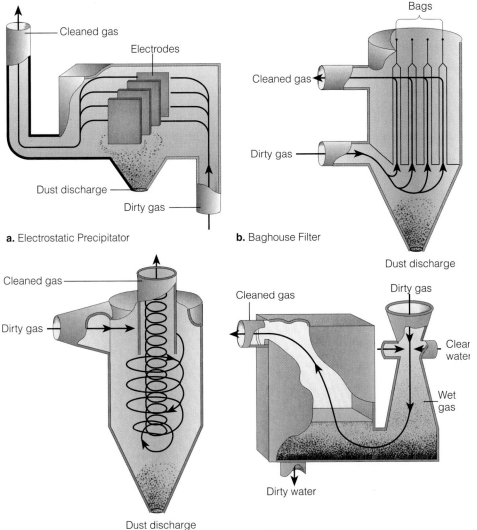

Figure 9-16 Four commonly used methods for removing particulates from the exhaust gases of electric power and industrial plants. Of these, only baghouse filters remove many of the more hazardous fine particles. All these methods produce hazardous materials that must be disposed of safely, and except for cyclone separators, all of them are expensive. The wet scrubber can also reduce sulfur dioxide emissions.

a. Electrostatic Precipitator

b. Baghouse Filter

c. Cyclone Separator

d. Wet Scrubber

An important way to make significant reductions in air pollution is to get older, high-polluting vehicles off the road. A badly maintained older vehicle can emit 100 times more pollutants than a properly maintained modern vehicle. According to EPA estimates, 10% of the vehicles on the road in the United States emit 50–60% of the pollutants. A problem is that many old cars are owned by people who can't afford to buy a newer car. One suggestion would be to pay people to take their old cars off the road, which would result in huge savings in health and air pollution control costs. Companies could buy up old cars (as oil companies in the Los Angeles area are now doing) to get emissions credits that enable them to extend the lives of their polluting facilities temporarily while new technology is installed.

There is some good news. From 1982 to 1993, overall U.S. smog levels dropped by 8% even as population,

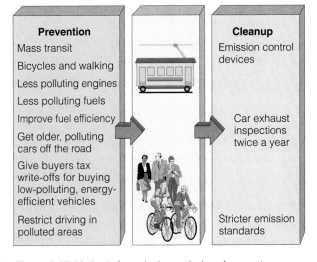

Figure 9-17 Methods for reducing emissions from motor vehicles.

Table 9-3 Evaluation of Alternatives to Gasoline

Advantages	Disadvantages
Compressed Natural Gas	
Fairly abundant, inexpensive domestic and global supplies	Large fuel tank required; one-fourth the range
Low emissions of hydrocarbons, CO, and CO_2	Expensive engine modification required ($2,000)
Vehicle development advanced; well suited for fleet vehicles	New filling stations required
Reduced engine maintenance	Nonrenewable resource
Electricity	
Renewable if not generated from fossil fuels or nuclear power	Limited range and power
Zero vehicle emissions	Batteries expensive
Electric grid in place	Slow refueling (6–8 hours)
Efficient and quiet	Power-plant emissions if generated from coal or oil
Reformulated Gasoline (Oxygenated Fuel)	
No new filling stations required	Nonrenewable resource
Low to moderate emission reduction of CO	Dependence on imported oil perpetuated
No engine modification required	No emission reduction of CO_2
	Higher cost
	Water resources contaminated by leakage and spills
Methanol	
High octane	Large fuel tank required; one-half the range
Emission reduction of CO_2 (total amount depends on method of production)	Corrosive to metal, rubber, plastic
Reduced total air pollution (30–40%)	Increased emissions of potentially carcinogenic formaldehyde
	High CO_2 emissions if generated by coal
	High capital cost to produce
	Hard to start in cold weather
Ethanol	
High octane	Large fuel tank required; lower range
Emission reduction of CO_2 (total amount depends on distillation process and efficiency of crop growing)	Much higher cost
Emission reduction of CO	Corn supply limited
Potentially renewable	Competition with food growing for cropland
	Smog formation possible
	Corrosive
	Hard to start in cold weather
Solar–Hydrogen	
Renewable if produced using solar energy	Nonrenewable if generated by fossil fuels or nuclear power
Lower flammability	Large fuel tank required
Virtually emission-free	No distribution system in place
Zero emissions of CO_2	Engine redesign required
Nontoxic	Currently expensive

traffic, and industry rose. In 1993 the number of Americans living in areas where ozone concentrations exceeded Clean Air Act standards was 54 million, the lowest total in 20 years and down from the 100 million people exposed to excessive ozone in 1992. In 1995 Los Angeles had 103 unsafe air days, compared to 239 in 1988.

Despite such successes, smog levels in the Los Angeles basin (which is subject to frequent thermal inversions) are too high much of the year (30% of the days in 1995) and could rise as population and consumption increase. The 1990 Clean Air Act gives Los Angeles until 2010 to meet federal air pollution standards. The EPA estimates that meeting the standards will cost California $4–6 billion annually from 1995 through 2010. However, a 1989 study estimated that implementing the program will save California an estimated $9.4 billion a year in health costs, reduced crop yields, and lowered productivity, leading to a net economic gain of $4–5 billion per year.

Q: How many people on average does one U.S. farmer feed?

California's South Coast Air Quality Management District Council developed a drastic program to produce a fivefold reduction in ozone, photochemical smog, and other major air pollutants in the Los Angeles area by 2009. This plan would **(1)** sharply reduce use of gasoline-burning engines over two decades by converting cars, trucks, buses, chain saws, outboard motors, and lawn mowers to run on electricity or on alternative fuels; **(2)** outlaw drive-through facilities to keep vehicles from idling in lines; **(3)** substantially raise parking fees and assess high fees for families owning more than one car; **(4)** require gas stations to use a hydrocarbon-vapor recovery system on gas pumps and to sell alternative fuels (Table 9-3); **(5)** strictly control or relocate industrial plants and businesses that release large quantities of hydrocarbons and other pollutants; and **(6)** find substitutes for or ban consumer products that release hydrocarbons, including aerosol propellants, paints, household cleaners, and barbecue starter fluids. Such measures are a glimpse of what many other cities may have to do if people, cars, and industries continue proliferating.

Here is some more good news. Since the 1960s, Tokyo, Japan (with a current population of about 26 million), has implemented a strict air-pollution control program that has sharply reduced levels of sulfur dioxide, carbon monoxide, and ozone. During the past 25 years outdoor air quality in most western European cities has also improved. However, outdoor air quality has remained about the same or has gotten worse in most rapidly growing urban areas in developing countries.

How Can We Reduce Indoor Air Pollution?
In the United States indoor air pollution poses a greater threat to health for many people than outdoor air pollution. Yet the EPA spends about $500 million per year fighting outdoor air pollution and only about $13 million a year on indoor air pollution.

To reduce indoor air pollution, it's not necessary to impose indoor air quality standards and monitor the more than 100 million homes and buildings in the United States. Instead, air pollution experts suggest that indoor air pollution reduction can be achieved by several means (Figure 9-18). Another possibility for cleaner indoor air in high-rise buildings is rooftop greenhouses through which building air can be circulated. Some actions you can take to reduce your exposure to indoor air pollutants are listed in Appendix 5.

In developing countries, indoor air pollution from open fires and leaky and inefficient wood- or charcoal-burning stoves (and the resulting high levels of respiratory illnesses) could be reduced if governments gave people simple stoves that burn biofuels more efficiently (which would also reduce deforestation) and

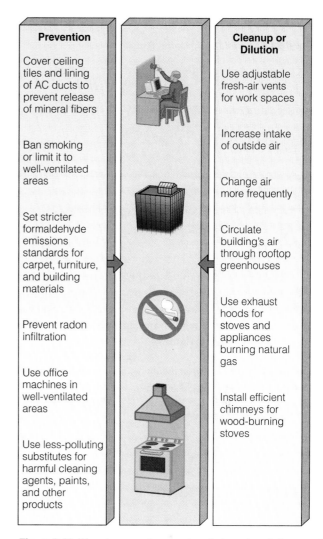

Figure 9-18 Ways to prevent and reduce indoor air pollution.

that are vented outside, or provided them with simple solar cookers (Figure 18-21d).

How Can We Protect the Atmosphere? An Integrated Approach
As population and consumption rise, we can generate new air pollution faster than we can clean up the old, even in developed countries with strict air-pollution control laws. As a result, environmentalists believe that protecting the atmosphere, and thus the health of people and many other organisms, will require a global approach that integrates many different strategies. Suggestions for doing this over the next 40–50 years include the following:

- *Putting more emphasis on pollution prevention*

- *Improving energy efficiency*

- *Reducing use of fossil fuels (especially coal and oil)*

- *Increasing use of renewable energy*

- *Slowing population growth*

- *Integrating air pollution, water pollution, energy, land-use, population, economic, and trade policies*

- *Regulating air quality for an entire region or airshed*

- *Phasing in full-cost pricing, mostly by taxing the production of air pollutants*

- *Distributing cheap and efficient cookstoves and solar cookstoves in developing countries*

- *Transferring the latest energy-efficiency, renewable-energy, pollution prevention, and pollution control technologies to developing countries*

Making such changes will be controversial and expensive. However, proponents argue that not implementing such an integrated approach will cost far more in money, poor human health and premature death, and ecological damage.

Turning the corner on air pollution requires moving beyond patchwork, end-of-pipe approaches to confront pollution at its sources. This will mean reorienting energy, transportation, and industrial structures toward prevention.

HILARY F. FRENCH

CRITICAL THINKING

1. Evaluate the pros and cons of the following statement: "Because we have not proven absolutely that anyone has died or suffered serious disease from nitrogen oxides, current federal emission standards for this pollutant should be relaxed."

2. What climate and topographical factors in your local community intensify of air pollution, and which of these factors help reduce air pollution?

3. Should all tall smokestacks be banned? Explain.

4. Should annual government-held auctions of marketable trading permits be used as a primary way of controlling and reducing air pollution? Explain. What conditions, if any, would you put on this approach?

5. Do you agree or disagree with the possible weaknesses of the U.S. Clean Air Act listed on pp. 263–264? Defend each of your choices. Can you identify other weaknesses?

*6. Evaluate your exposure to some or all of the indoor air pollutants in Figure 9-10 where you work and live. Come up with a plan for reducing your exposure to these pollutants.

*7. Do buildings in your school contain asbestos? If so, what are the indoor levels? Should this asbestos be removed? If indoor asbestos testing has not been done, talk with school officials about having it done.

*8. Have dormitories and other buildings on your campus been tested for radon? If so, what were the results? What has been done about areas with unacceptable levels? If this testing has not been done, talk with school officials about having it done.

*9. Use the library or the Internet to evaluate the effectiveness of the Clean Air Act of 1990. List its major accomplishments, disappointments, and weaknesses.

*10. Make a concept map of this chapter's major ideas, using the section heads and subheads and the key terms (in boldface type). Look at the inside back cover and in Appendix 4 for information about concept maps.

SUGGESTED READINGS

See Internet Sources and Further Readings at the back of the book, and InfoTrac.

10 CLIMATE, GLOBAL WARMING, AND OZONE LOSS

A.D. 2060: Hard Times on Planet Earth*

Mary Wilkins sat in the living room of her underground home in Illinois (Figure 10-1), which she shared with her daughter Jane and her family. It was July 4, 2060: Independence Day. There would be no parade or barbecue or fireworks today; people didn't stay outside for long because of the searing heat from global warming. With the food riots and martial law in place since 2040, people stayed home.

Many of her friends and millions of other Americans had long ago migrated to Canada, to find a cooler climate and more plentiful food supply after America's Midwestern breadbasket and central and southern California had been mostly abandoned because of a lack of water and food. Her friend June had recently written, wondering where to go now that Canadian farmland was drying up.

Just then a door opened. Her daughter Jane came out, ready to go to work. Behind her, shouts and squeals erupted from her grandchildren, Lynn and Jeffrey, who were playing Refugees and Border Patrol. It had been a popular game since 2020, when the United States had built a "Great Wall" with armed guards along its border with Mexico in a mostly vain attempt to keep out millions of Latin Americans trying to find food and work in the north.

---

* Compare this fictional worst-case scenario with the hopeful scenario that opens Chapter 2.

Mary sighed as she went to corral her pale and undernourished grandchildren. She felt sorry for them and began telling them about the old days, before the Warming. There were green parks to play in and green trees to climb, swimming pools full of water, and lakes and rivers everywhere. In winter, cold white stuff called snow fell from the sky and could be gathered up in balls to throw at each other.

She also told them that almost everyone had a car. "What's a car?" asked Jeffrey, the younger child. "Is it like the bus that Mommy rides to work?" "Yes, only much smaller—just for one person or one family," Mary answered. "It could go fast and ran on a fuel called gasoline—much too rare to be used anymore." "Why did people let things get so bad?" asked Lynn, the older child.

Mary's eyes filled with tears as she took the child in her arms. "Why didn't we listen to the warnings of scientists in the 1980s and 1990s?" she asked herself. Then she looked at Lynn and admitted, "Because we didn't want to believe anything bad could happen."

Although our species has existed for only an eyeblink of the earth's history, evidence indicates that we may be altering its atmosphere 10–100 times faster than the natural rate of change over the past 10,000 years.

Many scientists believe that global warming (from our binge of fossil fuel burning and deforestation) and depletion of stratospheric ozone (from our use of chlorofluorocarbons and other chemicals) will threaten life as we know it in the next century if we don't take action now.

Other analysts believe that these problems are overblown or that we need more research before acting to reduce such trends by phasing out fossil fuels and halting deforestation. Most environmental scientists argue that even if projected global warming doesn't happen, these are things we should be doing anyway.

Figure 10-1 An earth-sheltered house in Will County, Illinois, in the United States. About 13,000 families across the United States have built such houses. Mary Wilkins's fictional house in 2060 could be similar to this one. (Pat Armstrong/Visuals Unlimited)

We, humanity, have finally done it: disturbed the environment on a global scale.

THOMAS E. LOVEJOY

This chapter is devoted to answering the following questions:

- What key factors determine variations in the earth's climate?
- Can we really make the earth warmer, and if so, what will a few degrees matter?
- What can we do about possible global warming?
- Are we depleting ozone in the stratosphere, and why should we care?
- What can we do to slow ozone depletion?

10-1 WEATHER AND CLIMATE: A BRIEF INTRODUCTION

How Does Weather Differ from Climate?

At every moment at any spot on the earth, the troposphere (the inner layer of the atmosphere containing most of the earth's air, Figure 9-2) has a particular set of physical properties. Examples are temperature, pressure, humidity, precipitation (type and rate), sunshine, cloud cover (types, amounts, and heights), and wind (direction and speed). These short-term properties of the troposphere at a given place and time are what we call **weather**.

Climate is the physical properties of the troposphere of an area based on analysis of its weather records over a long period (at least 30 years). The two main factors determining an area's climate are *temperature*, with its seasonal variations, and the amount and distribution of *precipitation*. Figure 5-2 is a generalized map of the earth's major climate zones.

How Does the Global Circulation of Air Affect Regional Climates?

The temperature and precipitation patterns that lead to different climates (Figure 5-2) are caused primarily by the way air circulates over the earth's surface. Several factors determine global air circulation patterns.

One is *long-term variations in the amount of solar energy striking the earth*. Such variation occurs because of occasional changes in solar output, slight planetary shifts in which the earth's axis wobbles (22,000-year cycle) and tilts (44,000-year cycle) as it revolves around the sun, and minute changes in the shape of its orbit around the sun (100,000-year cycle).

A second factor is the *uneven heating of the earth's surface*. Air is heated much more at the equator (where the sun's rays strike directly throughout the year) than at the poles (where sunlight strikes at an angle and is thus spread out over a much greater area). These differences help explain why tropical regions near the equator are hot, polar regions are cold, and temperate regions in between generally have intermediate average temperatures (Figure 5-2).

Third, *seasonal changes occur because the earth's axis* (an imaginary line connecting the north and south poles) is *tilted*; as a result, various regions are tipped toward or away from the sun as the earth makes its annual revolution (Figure 10-2). This creates opposite seasons in the northern and southern hemispheres.

Fourth, *the earth rotates on its axis*, which prevents air currents from moving due north and south from the equator. Forces in the atmosphere created by this rotation deflect winds (moving air masses) to the right in the northern hemisphere and to the left in the southern hemisphere in what is called the *Coriolis effect*. The result is six huge convection cells of swirling air masses—three north and three south of the equator—that convey or move heat and water from one area to another (Figure 10-3).

Finally, climate and global air circulation are affected by the *properties of air and water*. When heated by the sun, ocean water evaporates and removes heat from the oceans to the atmosphere, especially near the hot equator. This moist, hot air expands, becomes less dense (weighs less per unit of volume), and rises in fairly narrow vortices. These upward spirals create an area of low pressure at the earth's surface.

As this moisture-laden air rises, it cools and releases moisture as condensation (because cold air can hold less water vapor than warm air). When water vapor condenses it releases heat, which radiates into space. The resulting cooler, drier air becomes denser,

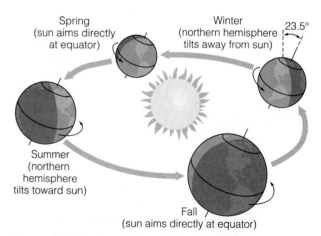

Figure 10-2 The effects of the earth's tilted axis. As the planet makes its annual revolution around the sun on an axis tilted about 23.5°, various regions are tipped toward or away from the sun. This produces the variations in the amount of solar energy reaching the earth and causes the seasons.

Q: How many units of energy are required to put one unit of food energy on the table in the United States?

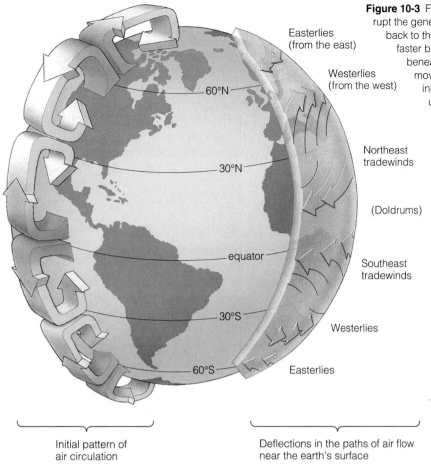

Figure 10-3 Formation of prevailing surface winds, which disrupt the general flow of air from the equator to the poles and back to the equator. As the earth rotates, its surface turns faster beneath air masses at the equator and slower beneath those at the poles. This deflects air masses moving north and south to the west or east, creating six huge convection cells in which air swirls upward and then descends toward the earth's surface at different latitudes. The direction of air movement in these cells sets up belts of prevailing winds that distribute air and moisture over the earth's surface. These winds affect the general types of climate found in different areas and drive the circulation of ocean currents. (Used by permission from Cecie Starr and Ralph Taggart, *Biology: The Unity and Diversity of Life*, 7th ed., Belmont, Calif.: Wadsworth, 1995)

Easterlies (from the east)

Westerlies (from the west)

Northeast tradewinds

(Doldrums)

Southeast tradewinds

Westerlies

Easterlies

60°N

30°N

equator

30°S

60°S

Initial pattern of air circulation

Deflections in the paths of air flow near the earth's surface

sinks (subsides), and creates an area of high pressure. As this air mass flows across the earth's surface, it picks up heat and moisture and begins to rise again. The resulting small and giant convection cells circulate air, heat, and moisture both vertically and from place to place in the troposphere, leading to different climates and patterns of vegetation (Figure 10-4).

How Do Ocean Currents Affect Regional Climates? The factors just listed, plus differences in water density, cause warm and cold ocean currents (Figure 5-2). Ocean currents, like air currents, redistribute heat and thus influence climate and vegetation, especially near coastal areas. For example, without the warm Gulf Stream, which transports 25 times more water than all the world's rivers, the climate of northwestern Europe would be subarctic. Currents also help mix ocean waters and distribute nutrients and dissolved oxygen needed by aquatic organisms.

Along some steep, western coasts of continents, almost constant trade winds blow offshore, pushing surface water away from the land. This outgoing surface water is replaced by an **upwelling** of cold, nutrient-rich bottom water. Upwellings, whether far from shore or near shore (Figure 5-2), bring plant

nutrients from the deeper parts of the ocean to the surface and support large populations of phytoplankton, zooplankton, fish, and fish-eating seabirds. Changes in prevailing winds can change the temperature of surface waters, weaken or alter ocean currents, suppress upwellings, and trigger weather changes over at least two-thirds of the globe (Figure 10-5).

How Does the Chemical Makeup of the Atmosphere Affect Climate? The Greenhouse Effect and the Ozone Layer Small amounts of heat-trapping gases such as water vapor (H_2O), carbon dioxide (CO_2), ozone (O_3), methane (CH_4), nitrous oxide (N_2O), and chlorofluorocarbons (CFCs) play a key role in determining the earth's average temperatures and thus its climates.

Together, these gases, known as **greenhouse gases**, act somewhat like the glass panes of a greenhouse: They allow light, infrared radiation, and some ultraviolet radiation from the sun (Figure 3-9) to pass through the troposphere. The earth's surface then absorbs much of this solar energy and degrades it to longer-wave, infrared radiation (that is, heat), which then rises into the troposphere (Figure 4-4). Some of this heat escapes into space; some is absorbed

A: About 10 (a loss of 9 units of energy)

by molecules of greenhouse gases, warming the air; and some radiates back toward the earth's surface. This natural trapping of heat in the troposphere is called the **greenhouse effect** (Figure 10-6).

The amount of heat trapped in the troposphere depends primarily on the concentrations of greenhouse gases and the length of time they stay in the atmosphere. The two predominant greenhouse gases in the troposphere are water vapor, controlled by the hydrologic cycle (Figure 4-27), and carbon dioxide, controlled by the global carbon cycle (Figure 4-22).

The primary heat-trapping gas is water vapor; because its concentration in the atmosphere is high (1–5%), inputs of water vapor from human activities have little effect on this chemical's greenhouse effects. By contrast, the concentration of carbon dioxide in the atmosphere is so small (0.036%) that fairly large input of CO_2 from human activities can significantly affect the amount of heat trapped in the atmosphere.

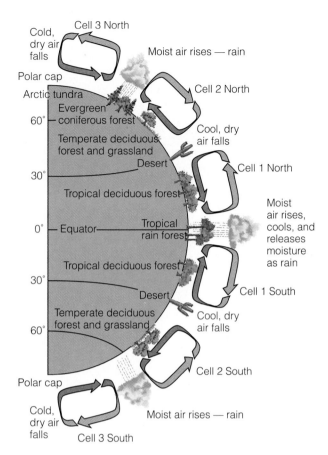

Figure 10-4 Model of global air circulation and biomes. Heat and moisture are distributed over the earth's surface by vertical convection currents that form into six large convection cells (called Hadley cells) at different latitudes. The direction of air flow and the ascent and descent of air masses in these convection cells determine the earth's general climatic zones. The uneven distribution of heat and moisture over the planet's surface leads to the forests, grasslands, and deserts that make up the earth's biomes.

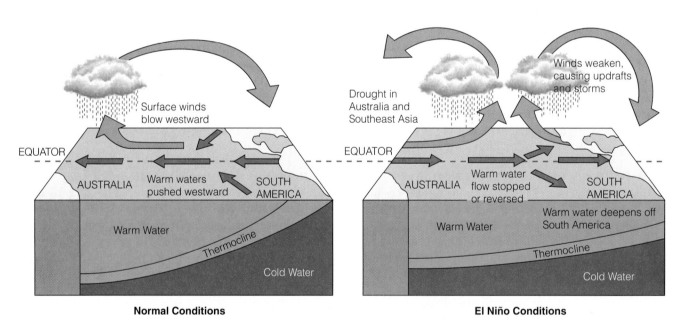

Figure 10-5 Normal surface winds blowing westward cause shore upwellings of cold, nutrient-rich bottom water in the tropical Pacific Ocean near the coast of Peru (left). The warm and cold water are separated by a zone of gradual temperature change called the thermocline. Every few years a climate shift known as the *El Niño–Southern Oscillation (ENSO)* disrupts this pattern. Westward surface winds weaken, which depresses the coastal upwellings and warms the surface waters off South America (right). ENSOs typically last for several months to over a year and occur every 3 or 4 years, although the interval has been as long as 7 years. When an ENSO lasts 12 months or longer, it severely disrupts populations of plankton, fish, and seabirds in upwelling areas and can trigger extreme weather changes over much of the globe.

Q: How much of the world's cropland is used to grow livestock feed?

Figure 10-6 The greenhouse effect. Without the atmospheric warming provided by this natural effect, the earth would be a cold and mostly lifeless planet. According to the widely accepted greenhouse theory, when concentrations of greenhouse gases in the atmosphere rise, the average temperature of the troposphere also rises. (Used by permission from Cecie Starr and Ralph Taggart, *Biology: The Unity and Diversity of Life*, 7th ed., Belmont, Calif.: Wadsworth, 1995)

The greenhouse effect, first proposed by Swedish chemist Svante Arrhenius in 1896, has been confirmed by numerous laboratory experiments and atmospheric measurements. It is one of the most widely accepted theories in the atmospheric sciences. Indeed, without its current greenhouse gases (especially water vapor), the earth would be a cold and lifeless planet with an average surface temperature of $-18°C$ ($0°F$) instead of its current $15°C$ ($59°F$).

Measured atmospheric levels of certain greenhouse gases—CO_2, CFCs, methane, and nitrous oxide—have risen substantially in recent decades (Figure 10-7). Although molecules of CFCs, methane, and nitrous oxide trap much more heat per molecule than CO_2, the much larger input of CO_2 makes it the most important greenhouse gas produced by human activities. Most of the increased levels of these greenhouse gases since 1958 have been caused by human activities: burning fossil fuels, agriculture, deforestation, and use of CFCs.

The developed countries produce about 70% of these CO_2 emissions (mostly from burning fossil fuels). The United States alone accounts for about 23% of global CO_2 emissions from human activities, followed by China (14%), Russia (7%), and Japan (5%).

Ice core analysis reveals that at the beginning of the industrial revolution the atmospheric concentration of CO_2 was about 280 parts per million (referred to as the *preindustrial level*). Between 1860 and 1995 the concentration of CO_2 in the atmosphere grew exponentially to 360 parts per million (Figure 10-7a), higher than at any time in the past 150,000 years. The atmospheric concentrations of CO_2 and other greenhouse gases are projected to double from preindustrial (1860) levels sometime during the next century—probably by 2050—and then continue to rise.

We and other species currently benefit from a comfortable level of greenhouse gases that typically undergo only minor, slow fluctuations over hundreds

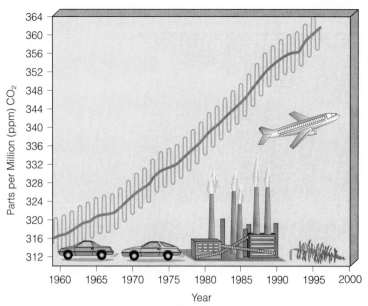

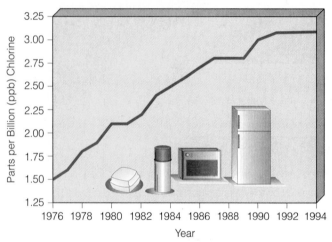

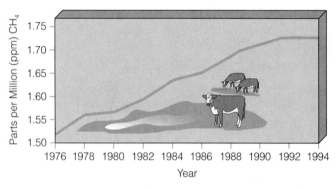

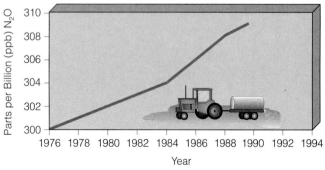

Figure 10-7 Increases in average concentrations of major greenhouse gases in the troposphere, mostly as a result of human activities. (Data from Electric Power Research Institute. Adapted and updated by permission from Cecie Starr and Ralph Taggart, *Biology: The Unity and Diversity of Life*, 6th ed., Belmont, Calif.: Wadsworth, 1992)

a. Carbon dioxide (CO$_2$) is responsible for 50–60% of the global warming from greenhouse gases produced by human activities since preindustrial times. The main sources are fossil fuel burning (75%) and land clearing and burning (25%). Most of the CO$_2$ comes from burning coal, but an increasing fraction is released from motor vehicle exhaust. CO$_2$ remains in the atmosphere for 50–200 years. The annual rise and fall of CO$_2$ levels shown in the graph result from less photosynthesis during winter and more during summer.

b. Chlorofluorocarbons (CFCs) contribute to global warming in the troposphere and deplete ozone in the stratosphere. The main sources are leaking air conditioners and refrigerators, evaporation of industrial solvents, production of plastic foams, and aerosol propellants. CFCs take 10–20 years to reach the stratosphere and generally trap 1,500–7,000 times as much heat per molecule as CO$_2$ while they are in the troposphere. This heating effect in the troposphere may be partially offset by the cooling caused when CFCs deplete ozone during their 65- to 135-year stay in the stratosphere. Their use is being phased out.

c. Methane (CH$_4$) is produced when anaerobic bacteria break down dead organic matter in moist places that lack oxygen. These areas include swamps and other natural wetlands, rice paddies, and landfills, and the intestinal tracts of cattle, sheep, and termites. Production and use of oil and natural gas (especially from leaks in natural gas pipelines) and incomplete burning of organic materials (including biomass burning in the tropics) also are significant sources. CH$_4$ stays in the troposphere for 9–15 years. Each CH$_4$ molecule traps about 25 times as much heat as a CO$_2$ molecule. Methane levels have stopped growing since 1991, possibly because of slightly better control of massive leaks in Russia's natural gas system.

d. Nitrous oxide (N$_2$O) can trap heat in the troposphere and deplete ozone in the stratosphere. It is released from nylon production, burning of biomass and nitrogen-rich fuels (especially coal), and the breakdown of nitrogen fertilizers in soil, livestock wastes, and nitrate-contaminated groundwater. Its life span in the troposphere is about 120 years, and it traps about 230 times as much heat per molecule as CO$_2$.

Q: What percentage of U.S. cropland is used to produce fruits and vegetables?

to thousands of years. However, crude but rapidly improving mathematical models of the earth's climate indicate that natural or human-induced global warming (or cooling) taking place over a few decades could be disastrous for human societies and many forms of life.

In a band of the stratosphere 17–26 kilometers (11–16 miles) above the earth's surface (Figure 9-2), oxygen (O_2) is continuously converted to ozone (O_3) and back to oxygen by a sequence of reactions initiated by ultraviolet radiation from the sun ($3O_2 + UV \rightarrow 2O_3$). The result is a thin veil of protective ozone at very low concentrations (up to 10 parts per million) that absorbs at least 95% of the harmful incoming ultraviolet radiation from the sun and prevents it from reaching the earth's surface. Normally, the average levels of ozone in this life-saving layer don't change much because the rate of ozone destruction is equal to its rate of formation.

Ultraviolet (UV) radiation reaching the stratosphere is composed of three bands: A, B, and C. The ozone layer blocks out nearly all the highest-energy, shortest-wavelength radiation (UV-C), approximately half of the next highest band (UV-B), a small part of the lowest-energy radiation (UV-A). Besides preventing most of the sun's harmful ultraviolet radiation from reaching the earth's surface, stratospheric ozone creates warm layers of air that prevent churning gases in the troposphere from entering the stratosphere. This *thermal cap* is important in determining the average temperature of the troposphere and thus the earth's current climates.

✸ 10-2 PAST AND FUTURE CHANGES IN THE EARTH'S CLIMATE

What Is the Scientific Consensus About the Earth's Past Temperatures? Between 1990 and 1995 the Intergovernmental Panel on Climate Change (IPCC), a network of about 2,500 of the world's lead-

ing climate experts from 70 nations, published several reports evaluating the best available evidence concerning past changes in global temperatures and models projecting future changes in global temperatures and climate.

Layers of ancient ice in Antarctic glaciers provide a time capsule whose contents can be analyzed to provide information about the temperature and contents of the atmosphere in the ancient past. Such analyses and other data show that the earth's average surface temperature has fluctuated considerably over geologic time. These data show that during the past 800,000 years several ice ages have covered much of the planet with thick ice. Each glacial period lasted about 100,000 years and was followed by a warmer interglacial period of 10,000–12,500 years (Figure 10-8).

For the past 10,000 years we have enjoyed the warmth of the latest interglacial period (called the Holocene). This climatic stability has prevented drastic changes in the nature of soils and vegetation patterns throughout the world, allowing large increases in food production and thus in population (Figure 2-3). However, even small temperature changes during this period have led to large migrations of peoples in response to changed agricultural and grazing conditions.

Analysis of gases in bubbles trapped in ancient ice show that over the past 160,000 years tropospheric water vapor levels (the dominant greenhouse gas) have remained fairly constant. During most of this period levels of CO_2 have fluctuated between 190 and 290 parts per million. Estimated changes in the levels of tropospheric CO_2 correlate fairly closely with estimated variations in the earth's mean surface temperature during the past 160,000 years (Figure 10-9).

Since 1860 (when measurements began), mean global temperature (after correcting for urban heat island effects; see Figure 6-22) has risen 0.3–0.6°C (0.5–1.1°F). The temperature rose about 0.3°C between 1946 and 1995 (Figure 10-10). Since 1860 the twelve

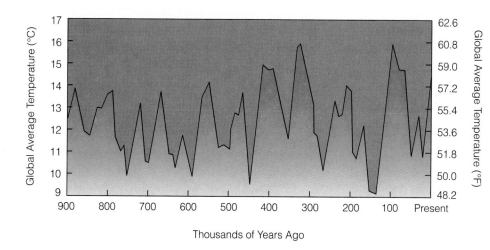

Figure 10-8 During the past 900,000 years the earth has experienced cycles of ice ages, each lasting about 100,000 years and then followed by warmer interglacial periods lasting 10,000–12,500 years. The warm interglacial period during the past 10,000 years has been a major factor in the development of agriculture, human civilizations, and population growth.

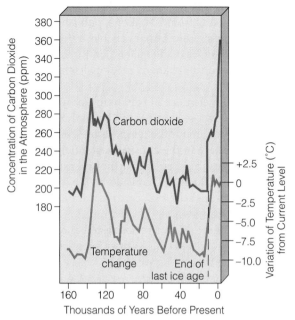

Figure 10-9 Estimated long-term variations in mean global surface temperature and average tropospheric carbon dioxide levels over the past 160,000 years. These CO_2 levels were obtained by inserting metal tubes deep into Antarctic glaciers, removing the ice, and analyzing bubbles of ancient air trapped in ice at various depths throughout the past. Such analyses reveal that since the last great ice age ended about 10,000 years ago, we have enjoyed a warm interglacial period. The rough correlation between tropospheric CO_2 levels and temperature shown in these estimates based on ice core data suggests a connection.

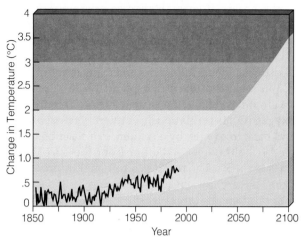

Figure 10-10 Recorded changes in the earth's mean surface temperature between 1860 and 1996 (dark line). The curved yellow region shows global warming projected by various computer models of the earth's climate systems. Note that the climate model projections roughly match the 0.3–0.6°C (0.5–1.1°F) recorded temperature increase between 1860 and 1996. Current models indicate that the average global temperature will rise by 1–3.5°C (1.8–6.3°F) during the next century. However, this projection assumes that air pollution from sulfate aerosols will continue to increase and exert a slight cooling effect, despite the fact that sulfur emissions have leveled off since 1960. The projected warming shown here could be *overestimated* or *underestimated* by a factor of two. (Data from U.S. National Academy of Sciences and National Center for Atmospheric Research)

warmest years occurred between 1979 and 1997; 1997 was the hottest of all, followed closely by 1995 and 1990.

Many uncertainties remain. Some or even all of the roughly 0.5°C rise since 1860 could result from normal fluctuations in the mean global temperature. On balance, however, the Intergovernmental Panel on Climate Change concluded in its 1995 report that "the observed increase over the last century is unlikely to be entirely due to natural causes" and "the balance of evidence suggests that there is a discernible human influence on global climate."

How Do Scientists Model Greenhouse Warming? Computer Models as Crystal Balls

To project the effects of increases in greenhouse gases on average global temperature and changes in the earth's climate, scientists develop mathematical models of such systems and run them on supercomputers. The most sophisticated of these climate models are called *general circulation models* (GCMs). Recently they have been coupled with models of ocean circulation and inputs of aerosols (tiny particles and droplets) into the atmosphere by volcanoes and pollution from human activities.

Current models simulate the large-scale features of the current climate fairly well and have reproduced some climate features of the distant past fairly accurately. The greatest uncertainties in current models involve incorporating the effects of clouds and the ecosphere on climate.

These models can provide us with scenarios of what *could* happen based on various assumptions and data fed into each model. How well the results correspond to the real world depends on (1) the design and assumptions of each model, (2) the accuracy of the data used, (3) magnification of tiny errors over time, (4) factors in the earth's climate system that amplify (positive feedback) or dampen (negative feedback) changes in average global temperatures, and (5) the effects of totally unexpected or unpredictable events (chaos). The models can also help us evaluate the possible effects of various ways of slowing or even halting possible global warming.

Figure 10-11 shows the series of models involved in making predictions about the effects of increased levels of greenhouse gases (model 1), on average global temperature (model 2), changes in regional

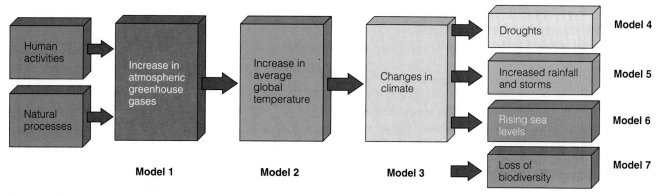

Figure 10-11 Generalized schematic of the greenhouse effect and models of its possible consequences. The greenhouse effect itself (model 1) is well established. The increase in average global temperature resulting from higher concentrations of greenhouse gases (model 2) is projected by crude climate models. Current models cannot consistently project either the specific climate changes (model 3) or their consequences (models 4–7) in various parts of the world.

climate in various parts of the world (model 3), and possible effects of such changes in these different areas (models 4 through 7). Current versions of models 1 and 2 project that the earth's average global temperature will increase within a certain range as atmospheric levels of greenhouse gases rise. However, most models disagree on how projected rises in average global temperature (global warming) might affect the climate in different areas (model 3) and the effects of such changes in these areas (models 4 through 7).

Despite their limitations, mathematical models are the most useful tools we have for understanding and projecting the behavior of complex systems such as global climate. Climate models are cloudy crystal balls, but without them we are groping around in total darkness trying to understand and deal with serious issues.

What Is the Scientific Consensus About Future Global Warming and Its Effects? According to the latest climate models, the IPCC projects that the earth's mean surface temperature will rise 1–3.5°C (1.8–6.3°F) between 1990 and 2100 (Figure 10-10). This may not seem like much, but even at the lowest projected increase of 1.0°C, the earth would be warmer than it has been for 10,000 years. Current models project that climate change, once begun, will continue for hundreds of years.

According to the models, the northern hemisphere should warm more and faster than the southern hemisphere because the latter has more heat-absorbing ocean than land (Figure 5-15) and because water cools more slowly than land. Current climate models project a more pronounced warming at the earth's poles. Measurements reveal that the surface temperatures at nine stations north of the Arctic circle have risen by about 5.5°C (9.9°F) since 1968. Since 1947 the average

summer temperature at Antarctica has risen by almost 2°C (3.6°F) and the thick, massive ice shelves surrounding the continent are beginning to break up. In January 1995, an iceberg almost as big as Rhode Island or Luxembourg broke off, and ominous cracks are appearing in many parts of the ice shelf.

Climate models also project that as the earth's atmosphere warms, the rate of water evaporation will increase and global average precipitation will rise. At the mid- to high latitudes (where most of North America, Europe, and Japan are located) average precipitation is projected to increase (especially during the cold season). Much of it will come in heavy showers or thunderstorms (thus increasing flooding) rather than in gentler, longer-lasting rainfalls. Such changes are already being observed in many parts of the northern hemisphere.

With a warmer climate global sea levels will rise, mainly because water expands slightly when heated. Satellite data indicate that ocean surface temperatures have been rising at a rate of about 0.1°C (0.18°F) per year since the 1980s. Current climate models of global warming project that global sea levels will rise by 15–95 centimeters (6–37 inches) between 1990 and 2100, with a best estimate of 48 centimeters (19 inches).

The models also indicate that if atmospheric warming at the poles caused ice sheets and glaciers to melt even partially, global sea level would rise much more. However, this is expected to take hundreds of years to occur.

IPCC scientists warn that warming or cooling by more than 1°C (1.8°F) over a few decades (instead of over many centuries, as has been the pattern during the last 10,000 years) will cause serious disruptions of the current structure and functioning of earth's ecosystems and of human economic and social systems.

A: About 840 million in 1995 (down from 940 million in 1970)

10-3 GLOBAL WARMING OR A LOT OF HOT AIR? HOW SERIOUS IS THE THREAT?

Will the Earth Really Get Warmer? There is much controversy over whether we are already experiencing global warming, how warm temperatures might be in the future, and the effects of such temperature increases. One problem is that many of the past measurements and estimates of the earth's average temperature are imprecise, and we have only about 100 years of fairly accurate data. With such limited data, it is difficult to separate out the normal short-term ups and downs of global temperatures (called *climate noise*) from an overall rise in average global temperature.

What Factors Might Amplify or Dampen Global Warming? Scientists have also identified a number of factors that might amplify (positive feedback) or dampen (negative feedback) a rise in average atmospheric temperature. These factors influence both how fast temperatures might climb and what the effects might be on various areas. Let's look more closely at the possible effects of such factors.

One factor is *changes in the amount of solar energy reaching the earth*. Solar output varies by about 0.1% over the 11-year and 22-year sunspot cycles—and over 80-year and other much longer cycles. These up-and-down changes in solar output can temporarily warm or cool the earth and thus affect the projections of climate models. Two 1992 studies concluded that the projected warming power of greenhouse gases should outweigh the climatic influence of the sun over at least the next 50 years.

Another problem is understanding the *effects of oceans on climate*. If the oceans warm up enough, some of the dissolved CO_2 will bubble out into the atmosphere (just as in a glass of carbonated ginger ale left out in the sun), amplifying and accelerating global warming (positive feedback).

We know that the oceans currently help moderate tropospheric temperature by removing about 29% of the excess CO_2 we pump into the atmosphere (negative feedback), but we don't know whether they can absorb more. Global warming could be dampened (negative feedback) if the oceans absorbed more heat, but this depends on how long the heat takes to reach deeper layers. Recent measurements indicate that deep vertical mixing in the ocean occurs extremely slowly (taking hundreds of years) in most places because water density increases with depth, inhibiting mixing of different layers.

There is also the possibility that deep ocean currents could be disrupted. At present, these currents (driven largely by differences in water density and winds) act like a gigantic conveyor belt, transferring heat from one place to another and storing carbon dioxide in the deep sea (Figure 10-12). There is concern that global warming could halt this thermal conveyor belt by reducing the density and salinity of water in the North Atlantic. If this loop stalls out, evidence from past climate changes indicates that this could trigger atmosphere temperature changes of more than 5°C (9°F) over periods as short as 40 years (positive feedback).

Changes in the *water vapor content and the amount and types of cloud cover* also affect climate. Warmer temperatures would increase evaporation and the water-holding capacity of the air and create more clouds. Significant increases in water vapor, a potent greenhouse gas, could enhance warming (positive feedback).

However, it is difficult to predict the net effect of additional clouds on climate. They could have a

Figure 10-12 This loop of ocean water stores carbon dioxide in the deep sea and brings warmth to Europe. It occurs when ocean water in the North Atlantic is dense enough (because of its salt content and cold temperature) to sink to the ocean bottom and well up in the warmer Pacific (helping cool that part of the world). Then a shallower return current aided by winds brings warmer and less salty—and thus less dense—water to the Atlantic, which can then cool and sink to begin the cycle again. If this heat conveyor belt or loop stalls out because of a drop in density of ocean water in the North Atlantic (possibly from changes caused by global warming), massive climate changes over much of the earth's surface could occur within only a few decades. Models indicate that this oceanic heat conveyor belt would return, but only after hundreds or thousands of years.

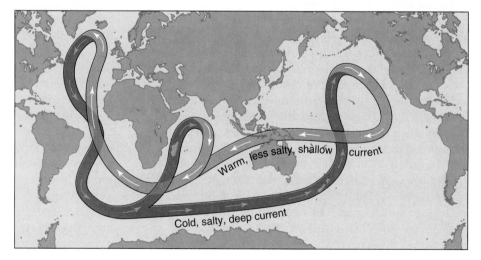

Warm, less salty, shallow current

Cold, salty, deep current

Q: How many people die each year from hunger-related causes?

warming effect by trapping heat (positive feedback) or a cooling (negative feedback) effect by reflecting sunlight back into space. The net result of these two opposing effects depends on whether it is day or night and on the type (thin or thick) and altitude of clouds. Scientists don't know which of these factors might predominate or how cloud types and heights might vary in different parts of the world as a result of global warming.

Climate is also influenced by *changes in polar ice*. The light-colored Greenland and Antarctic ice sheets act like enormous mirrors, reflecting sunlight back into space. If warmer temperatures melted some of this ice and exposed darker ground or ocean, more sunlight would be absorbed and warming would be accelerated. Then more ice would melt, further accelerating the rise in atmospheric temperature (positive feedback).

On the other hand, the early stages of global warming might increase the amount of the earth's water stored as ice. Warmer air would carry more water vapor, which could drop more snow on some polar glaciers, especially the gigantic Antarctic ice sheet. If snow accumulated faster than ice was lost, the ice sheet would grow, reflect more sunlight, and help cool the atmosphere, perhaps leading to a new ice age within a thousand years.

Climate can also be affected by *air pollution*. Projected global warming might be partially offset by *aerosols* (tiny droplets and solid particles) of various air pollutants released or formed in the atmosphere by volcanic eruptions and human activities. This occurs because during daytime they reflect some of the incoming sunlight back into space. However, nights would be warmer because the clouds would still be there and prevent some of the heat stored in the earth's surface (land and water) during the day from being radiated back into space. These pollutants may explain why most recent warming in the northern hemisphere occurs at night.

However, these interactions are complex. Pollutants in the lower troposphere can either warm or cool the air, depending on the reflectivity of the underlying surface. We also know little about the effects of aerosols on the properties of clouds. These contradictory and patchy effects and uncertainties, plus improved air pollution control, make it unlikely that air pollutants will counteract projected global warming very much in the next half century. Aerosols also fall back to the earth or are washed out of the atmosphere within weeks or months, whereas CO_2 and other greenhouse gases remain in the atmosphere for decades to several hundred years.

We could maintain or increase levels of aerosol air pollutants to offset possible global warming. However, because these pollutants already kill hundreds of thousands of people a year and damage vegetation (including food crops), they are being reduced.

Another uncertainty is the *effect of increased CO_2 on photosynthesis*. Some studies suggest that more CO_2 in the atmosphere is likely to increase the rate of photosynthesis in areas with adequate amounts of water and other soil nutrients. This would remove more CO_2 from the atmosphere and help slow global warming (negative feedback). Other studies suggest that this effect varies with different types of plants and in different climate zones. Also, much of any increased plant growth could also be offset by plant-eating insects that breed more rapidly and year-round in warmer temperatures. Weeds may also grow more rapidly at the expense of food crops.

Another factor that can affect CO_2 levels is *forest turnover*: how fast trees grow and die in a forest. A 1994 study indicated that the turnover in tropical forests worldwide is accelerating. Besides reducing the biodiversity of tree species in tropical forests, this change could enhance global warming (positive feedback) by reducing removal of CO_2 from the atmosphere because less dense, faster-growing trees require less CO_2 for growth.

Global warming could be accelerated by *increased release of methane, a potent greenhouse gas, from wetlands*. A 1994 study indicated that increased uptake of CO_2 by wetland plants could boost emissions of methane by providing more organic matter for methane-producing anaerobic bacteria to decompose (positive feedback). Some scientists also speculate that in a warmer world huge amounts of methane now tied up in arctic tundra soils and in muds on the bottom of the Arctic Ocean might be released if the blanket of permafrost in tundra soils melts and the oceans warm considerably. Conversely, some scientists believe that bacteria in tundra soils would rapidly oxidize the escaping methane to CO_2, a less potent but still important greenhouse gas.

There is also concern over *how rapidly the earth's climate might change*. If moderate change takes place gradually over several hundred years, people in areas with unfavorable climate changes may be able to adapt to the new conditions. However, if the projected global temperature change takes place over several decades or all within the next century, we may not be able to switch food-growing regions and relocate the large portion of the world's population living near coastal areas fast enough. The result would be large numbers of premature deaths from lack of food, as well as social and economic chaos. Such rapid changes would also reduce the earth's biodiversity because many species couldn't move or adapt.

Recent data from analysis of ice cores suggest that the earth's climate has shifted often and more drastically and quickly than previously thought. This new evidence indicates that average temperatures during the warm interglacial period that began about

A: At least 10 million (some say 20 million)

125,000 years ago (Figure 10-8) varied as much as 10°C (18°F) in only a decade or two and that such warming and cooling periods each lasted 1,000 years or more.

If these findings are correct and also apply to the current interglacial period, fairly small rises in greenhouse gas concentrations could trigger rapid up-and-down shifts in average global temperatures. Such rapid shifts would be disastrous for humans and many other forms of life on the earth.

As a result of the factors discussed in this section, climate scientists estimate that their projections about *global warming and rises in average sea levels during the next 50–100 years could be half the current projections (the best-case scenario) or double them (the worst-case scenario)*. In any event, possible climate change and its effects are likely to be erratic and mostly unpredictable.

✿ 10-4 SOME POSSIBLE EFFECTS OF A WARMER WORLD

Why Should We Worry if the Earth's Temperature Rises a Few Degrees? So what's the big deal? Why should we worry about a possible rise of only a few degrees in the earth's average temperature? We often have that much change between May and July, or even between yesterday and today.

This is a common critical thinking trap that many people fall into. The key point is that we are not talking about normal swings in *local weather*, but about a projected *global* change in *climate*.

A warmer global climate could have a number of possible effects. One is *changes in food production*, which could increase in some areas and drop in others. Archeological evidence and computer models indicate that climate belts would shift northward by 100–150 kilometers (60–90 miles) or upward 150 meters (500 feet) in altitude for each 1°C (1.8°F) rise in global temperature.

Whether such poleward shifts lead to increased crop productivity in new crop-growing areas depends mainly on two factors: the fertility of the soil in such regions and the availability of enormous amounts of money to build a new agricultural infrastructure (for irrigation and food storage and distribution). In parts of Asia, food production in more northern areas could increase because of favorable soils. However, in North America the northward expansion of crop-growing regions from the midwestern United States into Canada would be limited by the thinner and less fertile soils there. Siberia also lacks rich soils and couldn't make up for lower crop production in Ukraine.

Current climate models project 10–70% declines in the global yield of key food crops and a loss in current cropland area of 10–50%, especially in most poor countries. Currently, we can't predict where changes in crop-growing capacity might occur or how long such changes might last. However, we do know that drops in global crop yields of only 10% would cause large increases in hunger and starvation, especially in poor countries, and cause economic and social chaos.

Global warming would also *reduce water supplies* in some areas. Lakes, streams, and aquifers in some areas that have provided water to ecosystems, croplands, and cities for centuries could shrink or dry up altogether. This would force entire populations to migrate to areas with adequate water supplies—if they could. So far we can't say with much certainty where this might happen.

Global warming will also lead to a *change in the makeup and location of many of the world's forests*. Forests in temperate and subarctic regions would move toward the poles or to higher altitudes, leaving more grassland and shrubland in their wake. However, tree species move slowly through the growth of new trees along forest edges—typically about 0.9 kilometer (0.5 mile) per year or 9 kilometers (5 miles) per decade.

According to the 1995 report of the IPCC, midlatitude climate zones are projected to shift northward by 550 kilometers (340 miles) over the next century. At that rate some tree species such as beech (Figure 10-13) might not be able to migrate fast enough and would die out. Also, species that already live at high latitudes and on mountaintops would have nowhere to go, and many would become extinct. According to the IPCC, over the next century "entire forest types might disappear, including half of the world's dry tropical forests." Such forest diebacks would release carbon stored in their biomass and in surrounding soils and accelerate global warming (positive feedback).

Oregon State University scientists project that drying from global warming could cause massive *wildfires* in up to 90% of North American forests. If widespread fires occurred, large numbers of homes and large areas of wildlife habitats would be destroyed. Huge amounts of CO_2 would be injected into the atmosphere, which would accelerate global warming.

Climate change would lead to *reductions in biodiversity* in many areas. Large-scale forest diebacks would cause mass extinction of plant and animal species that couldn't migrate to new areas. Fish would die as temperatures soared in streams and lakes and as lowered water levels concentrated pesticides. Any

Q: What human activity has the most harmful overall environmental impact?

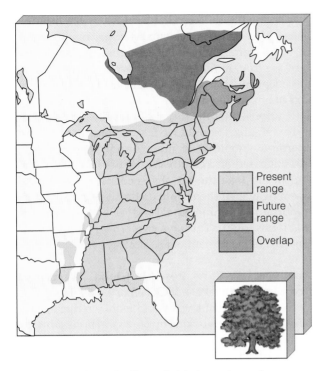

Figure 10-13 Possible effects of global warming on the geographic range of beech trees. According to one projection, if CO_2 emissions doubled between 1990 and 2050, beech trees (now common throughout the eastern United States) would survive only in a greatly reduced range in northern Maine and southeastern Canada. This is only one of a number of tree species whose geographic ranges would be drastically changed by global warming. (Data from Margaret B. Davis and Catherine Zabinski, University of Minnesota)

Legend:
- Present range
- Future range
- Overlap

shifts in regional climate would threaten many parks, wildlife reserves, wilderness areas, wetlands, and coral reefs, wiping out many current efforts to stem the loss of biodiversity.

In a warmer world, water in the world's oceans would expand and lead to a *rise in sea level*. Even the modest rise of 48 centimeters (19 inches) projected to occur by 2100 would flood coastal regions—where about one-third of the world's people and economic infrastructure are concentrated—as well as lowlands and deltas where crops are grown. It would also destroy most coral reefs, move barrier islands farther inland, accelerate coastal erosion, contaminate coastal aquifers with salt water, reduce already declining global fish catches, and flood tanks storing oil and other hazardous chemicals in coastal areas.

If warming at the poles caused ice sheets and glaciers to melt even partially, the global sea level would rise far more. One comedian jokes that he plans to buy land in Kansas because it will probably become

valuable beach-front property; another boasts that she isn't worried because she lives in a houseboat—the "Noah strategy."

In a warmer world, *weather extremes* are expected to increase in number and severity. Prolonged heat waves and droughts could become the norm in many areas, taking a huge toll on many humans and ecosystems. As the upper layers of seawater warm, damaging hurricanes, typhoons, tornadoes, and violent storms will increase in intensity and occur more frequently.

Any major increase in weather extremes could bankrupt the insurance and banking industries. Between 1990 and 1995, the global insurance industry paid out $57 billion in weather-related claims, compared to $17 billion in the 1980s.

Alarmed by an unprecedented series of hurricanes, floods, droughts, and wildfires in recent years, a growing number of the world's insurance companies are sharply raising their premiums for coverage of damages from such events and eliminating coverage in high-risk areas. Executives of some insurance companies are also pressuring government leaders to get more serious about slowing possible global warming and are spurring improvements in energy efficiency and increased use of renewable energy through their procurement and investment policies.

Global warming also poses *threats to human health*. According to the 1995 IPCC report, global warming would bring more heat waves. This would double or triple heat-related deaths among the elderly and people with heart disease.

A warmer world would also disrupt supplies of food and fresh water, displacing millions of people and altering disease patterns in unpredictable ways. The spread of warmer and wetter tropical climates from the equator would bring malaria (Figure 8-9), encephalitis, yellow fever, dengue fever, and other insect-borne diseases to formerly temperate zones.

Atmospheric warming also affects the respiratory tract by increasing air pollution in winter months and increasing exposure to dusts, pollens, and smog in summer months. Sea-level rise could spread infectious disease by flooding coastal sewage and sanitation systems.

Climate change would lead to *a growing number of environmental refugees*. According to Norman Myers (Guest Essay, pp. 295–296), by 2050 global warming could produce as many as 50–150 million environmental refugees (compared to 7 million war refugees in Europe after World War II). Most of these refugees would illegally migrate to other countries, causing much social disorder and political instability. Thus, global warming has serious implications for the foreign, military, and economic security policies of nations.

A: Agriculture

10-5 SOLUTIONS: DEALING WITH THE THREAT OF GLOBAL WARMING

Should We Do More Research or Act Now?

There are three schools of thought concerning global warming. A *very small* group of scientists (many of them not experts in climate research or heavily funded by the oil and coal industries) contend that global warming is not a threat; some popular press commentators and writers even claim that it is a hoax. Widespread reporting of this *no-problem* minority view in the media has clouded the issue, cooled public support for action, and slowed international negotiations to deal with this threat.

A second group of scientists and economists believe we should wait until we have more information about the global climate system, possible global warming, and its effects before we take any action. Proponents of this *waiting strategy* question whether we should spend hundreds of billions of dollars phasing out fossil fuels and replacing deforestation with reforestation (and in the process risk disrupting national and global economies) to help ward off something that might not happen. They call for more research before making such far-reaching decisions.

A third group of scientists point out that greatly increased spending on research about the possibility and effects of global warming will not provide the certainty decision makers want because the global climate system is so complex. These scientists urge us to adopt a *precautionary strategy*. They believe that when dealing with risky and far-reaching environmental problems such as possible global warming, the safest course is to take informed action *before* there is overwhelming scientific knowledge to justify acting.

Those who favor doing nothing or waiting before acting point out that there is a 50% chance that we are *overestimating* the impact of rising greenhouse gases. However, those urging action point out that there is also a 50% chance that we are *underestimating* such effects.

If global warming does occur as projected, it will take place gradually over many decades until it crosses thresholds and triggers obvious and serious effects. By then it will be too late to take corrective action. In the early stages of projected global warming it will be easy for people to deny that anything serious is happening. Psychologist Robert Ornstein calls this denial the *boiled frog syndrome*. He describes it as like trying to alert a frog to danger as it sits in a pan of water very slowly being heated on the stove. If the frog could talk it would say, "I'm a little warmer, but I'm doing fine." As the water gets hotter, we would warn the frog that it will die, but it might reply, "The temperature has been increasing for a long time, and I'm still alive."

Eventually the frog dies because it has no evolutionary experience of the lethal effects of boiling water and thus cannot perceive its situation as dangerous. Like the frog, humans face a possible future without precedent, and our senses are unable to pick up warnings of impending danger.

Suppose that the threat of global warming does not materialize. Should we take actions that will cost enormous amounts of money and create political turmoil based on a cloudy crystal ball? Some say that we should take the actions needed to slow global warming even if there were no threat because of their important environmental and economic benefits (Solutions, right). This *no-regrets strategy* is an important part of the precautionary strategy.

How Can We Slow Possible Global Warming?

According to the 1994 IPCC report, stabilizing CO_2 levels at the current level would require reducing current global CO_2 emissions by 66–83%. This is a highly unlikely and politically charged change. The International Energy Agency projects that CO_2 emissions will increase by nearly 50% between 1990 and 2010, with most of the increase coming from developing countries. According to climate models, even stabilizing CO_2 concentrations at 450 ppm requires cutting CO_2 emission by more than half.

Figure 10-14 presents a variety of solutions analysts have suggested to slow possible global warming; none of these solutions is being vigorously pursued. The quickest, cheapest, and most effective way to reduce emissions of CO_2 and other air pollutants over the next two to three decades is to use energy more efficiently (Section 18-2 and Solutions, right).

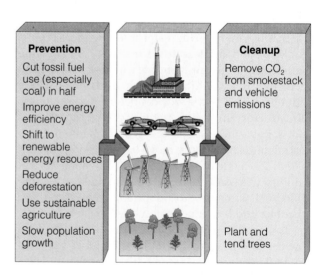

Figure 10-14 Methods for slowing possible global warming.

Q: How many of the world's 17 major marine fishery zones have been overfished?

Some analysts call for increased use of nuclear power (Section 19-4) because it produces only about one-sixth as much CO_2 per unit of electricity as coal. Other analysts argue that the danger of large-scale releases of highly radioactive materials from nuclear power-plant accidents and the very high cost of nuclear power (Section 19-4) make it a much less desirable option than improving energy efficiency and relying more on renewable energy resources (Chapter 18).

Using natural gas (Section 19-2) could help us make the 40- to 50-year transition to an age of energy efficiency and renewable energy. When burned, natural gas emits only half as much CO_2 per unit of energy as coal, and it emits far smaller amounts of most other air pollutants. Shifting from high-carbon fuels such as coal to low-carbon fuels such as natural gas could reduce CO_2 emissions by as much as 40%. However, without effective maintenance, more reliance on natural gas can increase inputs of methane (a potent greenhouse gas) from leaking tanks and pipelines, and thus increase global warming.

One method for reducing CO_2 emissions would be to phase out government subsidies for fossil fuels over a decade and gradually phase in *carbon taxes* on fossil fuels (especially coal and gasoline) based on their emissions of CO_2 and other air pollutants. To be politically feasible, analysts warn that these consumption tax increases should be matched by declines in taxes on income, labor, or capital. In 1997 more than 2,000 economists, including six Nobel laureates, signed a statement calling for carbon taxes as part of an international system of tradable permits for greenhouse gas emissions.

Reducing deforestation (Section 16-4) and switching to more sustainable agriculture (Section 14-4) would reduce CO_2 emissions and help preserve biodiversity. According to most analysts, slowing population growth is also crucial. If we cut per capita greenhouse gas emissions in half but world population doubles, we're back where we started. Some analysts argue that it is vital to global environmental security (Guest Essay, pp. 295–296) that developed countries transfer energy efficiency, renewable energy, pollution prevention, and waste reduction technologies to developing countries as soon as possible.

It has also been suggested that we remove CO_2 from the exhaust gases of fossil fuel–burning vehicles, furnaces, and industrial boilers. However, available methods can remove only about 30% of the CO_2, and using them would double the cost of electricity.

Some call for a massive global reforestation program as a strategy for slowing global warming. However, studies suggest that such a program (requiring each person in the world to plant and tend an average of 1,000 trees every year) would offset only about 3 years of our current CO_2 emissions from burning

Energy Efficiency to the Rescue

SOLUTIONS

According to energy expert Amory Lovins (Guest Essay, p. 536), *the major remedies for slowing possible global warming are things we should be doing already even if there were no threat of global warming.* He argues that if we waste less energy, reduce air pollution by cutting down on our use of fossil fuels and switching to renewable forms of energy, and harvest trees sustainably, we and other forms of life—in this and future generations—would be better off even if these actions had nothing to do with global climate.

According to Lovins, improving energy efficiency (Sections 18-2 and 18-3) would be the fastest, cheapest, and surest way to slash emissions of CO_2 and most other air pollutants within two decades, using existing technology. He estimates that increased energy efficiency would also save the world up to $1 trillion per year in reduced energy costs—as much as the annual global military budget.

Using energy more efficiently would also reduce pollution, help protect biodiversity, and deter arguments among governments about how CO_2 reductions should be divided up and enforced. This approach would also make the world's supplies of fossil fuel last longer, reduce international tensions over who gets the dwindling oil supplies, and allow more time to phase in renewable energy.

According to a 1990 government report, controlling emissions of greenhouse gases in the United States will cost about $10 billion per year over the next century, for a total of $1 trillion. Cutting annual U.S. oil imports by 20% by wasting less energy would cover the annual $10 billion projected costs of reducing greenhouse gas emissions.

To Lovins and most environmentalists, greatly improving worldwide energy efficiency *now* is a money-saving, life-saving, biodiversity-saving, win–win, no-regrets proposition that we should not refuse, even if climate change were not an issue. Atmospheric scientist Fred Singer, a strong critic of global warming projections (as well as of ozone depletion), agrees.

Critical Thinking

1. Do you agree that improving energy efficiency **(a)** is an important way to reduce the input of CO_2 into the atmosphere and **(b)** should be done regardless of its impact on the threat of global warming?

2. Why do you think there has been little emphasis on improving energy efficiency? Explain. (See Sections 18-2 and 18-3.)

fossil fuels. However, a global program for planting and tending trees would help restore deforested and degraded land and reduce soil erosion, water pollution, and loss of biodiversity. Furthermore, a recent study suggests that forests store much more carbon than previously thought, primarily in peat and other organic matter in soils.

Some scientists have suggested various technofixes for dealing with possible global warming, including adding iron to the oceans to stimulate the growth of marine algae (which could remove more CO_2 through photosynthesis), unfurling gigantic foil-surfaced sun mirrors in space to reduce solar input, and injecting sunlight-reflecting sulfate particulates into the stratosphere to cool the earth's surface.

Many of these costly schemes might not work, and most would probably produce unpredictable short- and long-term harmful environmental effects. Moreover, once started, those that work could never be stopped without a renewed rise in CO_2 levels. Instead of spending huge sums of money on such schemes, many scientists believe it would be much more effective and cheaper to improve energy efficiency and shift to renewable forms of energy that don't produce carbon dioxide (Chapter 18).

What Has Been Done to Reduce Greenhouse Gas Emissions? At the 1992 Earth Summit in Rio de Janeiro, 106 nations approved a Convention on Climate Change, in which developed countries committed themselves to reducing their emissions of CO_2 and other greenhouse gases to 1990 levels by the year 2000. However, the convention does not *require* countries to reach this goal.

Environmentalists call the treaty, which officially went into effect in 1994, an important beginning. However, most environmentalists contend that it does not go far enough toward reaching the at least 60% cut in 1995 global emissions of greenhouse gases IPCC scientists say will be needed to reduce projected climate change to manageable levels before the end of the next century.

According to 1995 UN projections, only about half of the 35 industrial countries committing themselves to this goal will actually reduce their greenhouse gas emissions to 1990 levels by the year 2000. Instead of falling, energy-related CO_2 emissions are projected to rise 30–93% by 2010.

Currently, developing countries contribute only about one-third of global CO_2 emissions. However, their emissions are increasing by about 5% per year, for a doubling time of only 14 years. If coal burning continues as expected, by 2025 China will emit three times more CO_2 than the current total of the United States, Japan, and Canada combined (even though

China's per capita CO_2 emissions would still be less than those of America).

In December 1997 representatives of 160 nations met in Kyoto, Japan to negotiate a new treaty to help slow global warming. The resulting treaty would

- require developed countries to cut greenhouse emissions by an average of 5.2% below 1990 levels between 2008 and 2012.

- require European industrialized nations to lower greenhouse emissions to 8% below 1990 levels. The United States would have to cut greenhouse emissions by 7% below 1990 levels, and Japan would have to reduce their emissions by 6% below 1990 levels between 2008 and 2012. In 1996, U.S. greenhouse gas emissions were 9% above 1990 levels and by 2012 are projected to be 30% higher than they were in 1990.

- not require developing countries to make any cuts in their greenhouse gas emissions because of an earlier treaty unless they choose to do so.

- allow emissions trading, in which a country that beats its target goal for reducing greenhouse gas emissions can sell its excess reductions to countries that failed to meet their reduction goals.

- allow forested countries to get a break in their quotas because trees absorb carbon dioxide.

- allow penalties for countries that violate the treaty, to be determined later.

Some analysts praise the agreement as a very modest but important step in dealing with the problem of potential global warming. However, there is strong opposition in the U.S. Senate against ratifying the treaty. In addition, climate scientists estimate that it would take a 60% reduction in emissions of greenhouse gases below 1990 levels to slow projected global warming to an acceptable level—compared to the only 5.2% reduction goal set by the Kyoto treaty.

How Can We Prepare for Possible Global Warming? It seems clear that many (perhaps most) of the things climate experts have recommended (Figure 10-14) either will not be done or will be done too slowly. As a result, some analysts suggest that we should also begin to prepare for the effects of long-term global warming. Figure 10-15 shows some ways to do this.

Implementing the key measures for slowing or responding to climate change listed in Figures 10-14 and 10-15 will cost a lot of money. However, studies indicate that in the long run the savings would greatly exceed the costs. Some actions you can take to reduce the threat of global warming are given in Appendix 5.

Q: What percentage of U.S. crops are grown using organic methods (no pesticides or commercial fertilizers)?

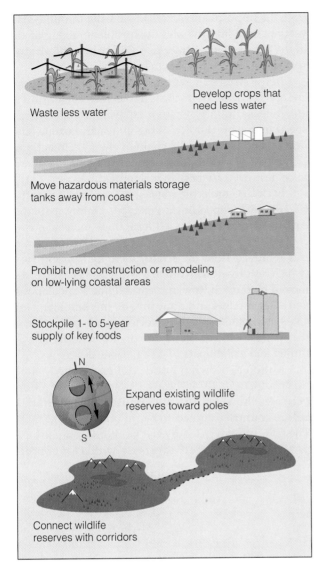

Figure 10-15 Ways to prepare for the possible long-term effects of global warming.

⊕ 10-6 OZONE DEPLETION: IS IT A SERIOUS THREAT?

What Is the Threat from Ozone Depletion?

Evolution of photosynthetic, oxygen-producing bacteria has produced a stratospheric global sunscreen: the ozone layer. Its presence for the past 450 million years has allowed life to develop and expand on land and in the surface layers of aquatic systems.

A handful of scientists dismiss the threat of ozone depletion from human-produced chemicals. Based on measurements and models (Figure 10-16), however, the overwhelming consensus of researchers in this field is that ozone depletion by certain chlorine- and bromine-containing chemicals emitted into the atmosphere by human activities is a serious long-term threat to human health, animal life, and the sunlight-driven primary producers (mostly plants) that support the earth's food chains and webs.

Ozone concentrations in the stratosphere have been measured since the mid-1960s at more than 30 locations around the world and also by satellites since 1970. These measurements show that during the 1980s normal ozone levels dropped 5–15% in winter above the temperate and tropical zones of both hemispheres—three times the losses measured in the 1970s. Globally, the earth lost an average of about 4% of its stratospheric ozone between 1979 and 1994. According to a 1995 report by prominent atmospheric scientists, average global ozone levels are projected to drop 7–13% during the 1990s.

What Causes Ozone Depletion? From Dream Chemicals to Nightmare Chemicals

This situation started when Thomas Midgley, Jr., a General Motors chemist, discovered the first chlorofluorocarbon (CFC) in 1930, and chemists then made similar compounds to create a family of highly useful CFCs.

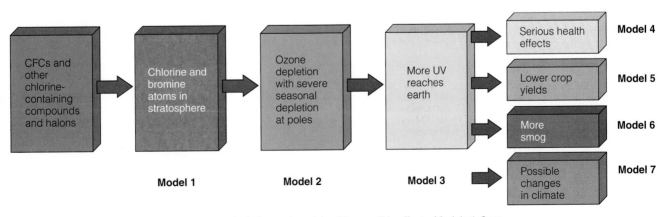

Figure 10-16 Generalized schematic of ozone depletion and models of its possible effects. Models 1–3 are widely accepted. The nature and extent of the effects in models 4–7 are still being studied.

The two most widely used are CFC-11 (trichlorofluoromethane, CCl_3F) and CFC-12 (dichlorofluoromethane, CCl_2F_2), known by their trade name as Freons.

These chemically stable, odorless, nonflammable, nontoxic, and noncorrosive compounds seemed to be dream chemicals. Cheap to make, they became popular as coolants in air conditioners and refrigerators (replacing toxic sulfur dioxide and ammonia), propellants in aerosol spray cans, cleaners for electronic parts such as computer chips, sterilants for hospital instruments, fumigants for granaries and ship cargo holds, and bubbles in plastic foam used for insulation and packaging.

But CFCs were too good to be true. In 1974 calculations by chemists Sherwood Rowland and Mario Molina (building on earlier work by Paul Crutzen) indicated that CFCs were creating a global chemical time bomb by lowering the average concentration of ozone in the stratosphere. They shocked both the scientific community and the $28-billion-per-year CFC industry by calling for an immediate ban of CFCs in spray cans (for which substitutes were readily available).

Here's what Rowland and Molina found: Spray cans, discarded or leaky refrigeration and air conditioning equipment, and the production and burning of plastic foam products release CFCs into the atmosphere. Because these molecules are insoluble in water and are chemically unreactive, they are not removed from the troposphere. As result—mostly through convection, random drift, and the turbulent mixing of air in the troposphere—they rise slowly into the stratosphere, taking 10–20 years to make the journey.

In the stratosphere, under the influence of high-energy UV radiation, these molecules break down and release highly reactive chlorine atoms, which speed up the breakdown of highly reactive ozone (O_3) into O_2 and O in a cyclic chain of chemical reactions (Figure 10-17). This causes ozone in the stratosphere to be destroyed faster than it is formed.

Each CFC molecule can last in the stratosphere for 65–110 years (depending on its type). During that time each chlorine atom released from these molecules can convert as many as 100,000 molecules of O_3 to O_2 before it is removed from the stratosphere by forming HCl (which diffuses downward to the troposphere and is removed by rain). If Rowland and Molina's calculations and later models and atmospheric measurements of CFCs in the stratosphere are correct (as almost all scientists in this field believe), these dream molecules have turned into a nightmare of global ozone destroyers.

Although Rowland and Molina warned us of this problem in 1974, it took 15 years of interaction between the scientific and political communities before countries agreed to begin phasing out CFCs. The CFC industry (led by the Du Pont Company), a powerful, well-funded adversary with a lot of prof-

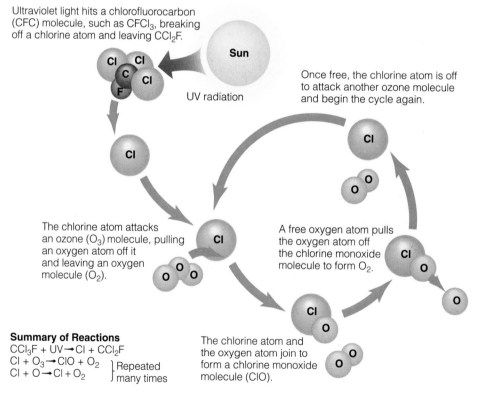

Figure 10-17 A simplified summary of how chlorofluorocarbons (CFCs) and other chlorine-containing compounds destroy ozone in the stratosphere. Note that chlorine atoms are continuously regenerated as they react with ozone. Thus, they act as *catalysts*, chemicals that speed up chemical reactions without themselves being used up by the reaction. Bromine atoms released from bromine-containing compounds that reach the stratosphere also destroy ozone by a similar mechanism.

Ultraviolet light hits a chlorofluorocarbon (CFC) molecule, such as $CFCl_3$, breaking off a chlorine atom and leaving CCl_2F.

Sun

UV radiation

Once free, the chlorine atom is off to attack another ozone molecule and begin the cycle again.

The chlorine atom attacks an ozone (O_3) molecule, pulling an oxygen atom off it and leaving an oxygen molecule (O_2).

A free oxygen atom pulls the oxygen atom off the chlorine monoxide molecule to form O_2.

The chlorine atom and the oxygen atom join to form a chlorine monoxide molecule (ClO).

Summary of Reactions
$CCl_3F + UV \rightarrow Cl + CCl_2F$
$Cl + O_3 \rightarrow ClO + O_2$ } Repeated
$Cl + O \rightarrow Cl + O_2$ } many times

Q: What percentage of the food produced in the United States is wasted?

its and jobs at stake, attacked Rowland and Molina. However, they held their ground, expanded their research, and explained the meaning of their calculations to other scientists, elected officials, and the media. It was not until 1988—14 years after Rowland and Molina's study—that Du Pont officials acknowledged that CFCs were depleting the ozone layer and agreed to stop producing them once they found substitutes.* In 1995 Rowland and Molina (along with Paul Crutzen) received the Nobel prize in chemistry for their work.

What Other Chemicals Deplete Stratospheric Ozone?

CFCs are not the only ozone-eaters; other chemicals can release highly reactive chlorine and bromine atoms if they reach the stratosphere and are exposed to intense UV radiation. Collectively, all ozone-depleting compounds are called *ODCs*.

One group consists of long-lived bromine-containing compounds such as *halons* and *HBFCs*, both used in fire extinguishers. Another is *methyl bromide* (CH_3Br), a widely used fumigant. Another group consists of chlorine-containing compounds such as *carbon tetrachloride* (CCl_4), a cheap, highly toxic solvent, and toxic *methyl chloroform*, or 1,1,1-trichloroethane ($C_2H_3Cl_3$), used as a cleaning solvent

---

* For a fascinating account of how corporate stalling, politics, economics, and science can interact, see Sharon Roan's *Ozone Crisis: The 15-Year Evolution of a Sudden Global Emergency* (New York: Wiley, 1989).

for clothes and metals and as a propellant in more than 160 consumer products, such as correction fluid, dry-cleaning sprays, spray adhesives, and other aerosols. Another source of ozone depletion is the emission of hydrogen chloride (HCl) into the stratosphere by the U.S. space shuttles.

Why Is There Seasonal Thinning of Ozone over the Poles?

Sometimes the news about ozone loss has taken scientists by surprise. The first surprise came in 1984, when researchers analyzing satellite data discovered that 40–50% of the ozone in the upper stratosphere over Antarctica was being destroyed during the Antarctic spring and early summer (September–December), when sunlight returned after the dark Antarctic winter. Since then, this seasonal Antarctic *ozone thinning* (incorrectly called an ozone hole) has expanded in most years (Figure 10-18), with a typical seasonal loss of about 50% (but as high as 100% in some spots). In 1992 and 1993, seasonal ozone thinning above Antarctica covered an area three times the size of the continental United States.

This pronounced seasonal loss of stratospheric ozone had not been predicted by computer models or detected by earlier satellite measurements. However, when NASA scientists looked at earlier satellite measurements they found that their computers had been programmed to regard very low ozone readings as errors. When they went back and looked at the earlier rejected data, they found that average seasonal ozone levels over Antarctica had been dropping since 1974

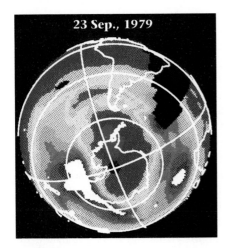 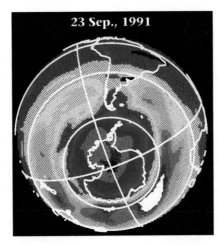 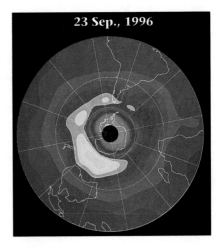

Figure 10-18 Seasonal thinning or loss of ozone (shown by shades of pink) in the upper stratosphere over Antarctica as measured by satellite on September 23 of 1979, 1991, and 1996. Since 1987 this area of seasonal thinning (in which the concentration of ozone is cut at least in half) has spanned an area over Antarctica larger than the continental United States or Europe. Some good news is that a recent computer model suggests that this huge area of seasonal ozone thinning should not get any larger. It should also be decreasing in size over the next 50 years as the ozone-depleting chemicals are phased out and those already in the atmosphere are eventually destroyed. (Source: NASA/GSFC and NOAA)

A: About 25%

and that the problem had been getting worse every year (Figure 10-18).

The "smoking gun" considered by most atmospheric scientists to establish the link between ozone depletion and CFCs was found in 1987, when a research plane flew into the stratosphere above Antarctica and measured chlorine (as ClO derived primarily from CFCs, Figure 10-17) and ozone concentrations. As ClO concentrations rose, ozone concentrations dropped.

Measurements have indicated that CFCs are the primary culprits. Each sunless winter, steady winds blow in a circular pattern over the earth's poles, creating *polar vortices*: huge swirling masses of very cold air that are trapped above the poles until the sun returns a few months later. When water droplets in clouds enter these large circling streams of frigid air, they form tiny ice crystals. The surfaces of these ice crystals collect CFCs and other ozone-depleting chemicals in the stratosphere and speed up (catalyze) the chemical reactions that release Cl atoms and ClO. Instead of entering a chain reaction of ozone destruction (Figure 10-17), the ClO atoms combine with one another to form Cl_2O_2 molecules. In the dark of winter the Cl_2O_2 molecules cannot react with ozone, so they accumulate in the polar vortex.

When sunlight and the antarctic spring return 2–3 months later (in October), the light breaks up the stored Cl_2O_2 molecules, releasing large numbers of Cl atoms and initiating the catalyzed chlorine cycle (Figure 10-17). Within weeks, this destroys 40–50% of the ozone above Antarctica (and 100% in some places). The returning sunlight gradually melts the ice crystals, breaks up the vortex of trapped polar air, and allows it to begin mixing again with the rest of the atmosphere. Then new ozone forms over Antarctica until the next dark winter.

When the vortex breaks up, huge masses of ozone-depleted air above Antarctica flow northward and linger for a few weeks over parts of Australia, New Zealand, South America, and South Africa. This raises biologically damaging UV-B levels in these areas by 3–10%, and in some years by as much as 20%.

In 1988 scientists discovered that similar but less severe ozone thinning occurs over the north pole during the arctic spring and early summer (February–June), with a seasonal ozone loss of 10–38% (compared to a typical 50% loss above Antarctica). When this mass of air above the Arctic breaks up each spring, large masses of ozone-depleted air flow south to linger over parts of Europe, North America, and Asia.

So far, seasonal ozone loss over the north pole is much lower than that over the south pole. One reason is that air masses flowing toward and away from the Arctic pass alternately over both land and water instead of mostly water, as in the Antarctic (Figure 5-15). This produces intense atmospheric disturbances that cause the arctic vortex to wander, lurch, and be warmer and less stable than the antarctic vortex.

However, the situation there could change. In 1992 atmospheric scientists warned that if rising levels of greenhouse gases warm the troposphere, the resulting insulating blanket could cool the stratosphere. This could increase the size and duration of seasonal ozone-depleting vortexes over both poles, perhaps leading to large and regular arctic ozone thinning. As a result, ozone levels over parts of the northern hemisphere (including the United States) would decline sharply. According to measurements by the U.S. National Oceanic and Atmospheric Administration (NOAA), atmospheric ozone over the northern hemisphere was depleted by 10–25% between 1979 and 1996.

Is Ozone Depletion Really a Serious Problem?
Political talk show commentator Rush Limbaugh, zoologist and former head of the Atomic Energy Commission Dixy Lee Ray (now deceased), physicist and climate scientist S. Fred Singer, and articles and books in the popular press have claimed that ozone depletion by CFCs is a hoax, or at least a vastly overblown problem. The Pro/Con box on p. 290 summarizes some of their charges and the answers by prominent research scientists studying ozone depletion.

Why Should We Be Worried About Ozone Depletion? Life in the Ultraviolet Zone Why should we care about ozone loss? From a human standpoint the answer is that with less ozone in the stratosphere, more biologically damaging UV-B radiation will reach the earth's surface and give humans worse sunburns, more cataracts (a clouding of the lens that reduces vision and can cause blindness if not corrected) and more skin cancers (Connections, p. 291).

According to UN Environment Programme estimates, the additional UV-B radiation reaching the earth's surface resulting from an annual 10% loss of global ozone (already a likely possibility within a few years) could lead to 300,000 additional cases of squamous cell cancer (Figure 10-19, left) and basal cell cancer (Figure 10-19, center) worldwide each year, 4,500–9,000 additional cases of potentially fatal malignant melanoma (Figure 10-19, right) each year, and 1.5 million new cases of cataracts (which account for over half of the world's 25–35 million cases of blindness) each year.

Cases of skin cancer and cataracts are increasing in Australia, New Zealand, South Africa, Argentina, and Chile, where the ozone layer is very thin for sev-

Q: What percentage of the world's potential food supply is lost to pests?

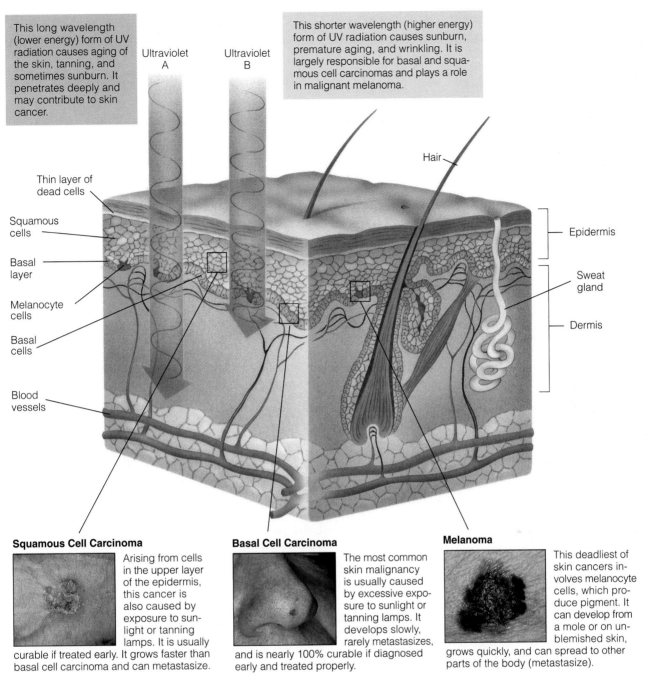

This long wavelength (lower energy) form of UV radiation causes aging of the skin, tanning, and sometimes sunburn. It penetrates deeply and may contribute to skin cancer.

This shorter wavelength (higher energy) form of UV radiation causes sunburn, premature aging, and wrinkling. It is largely responsible for basal and squamous cell carcinomas and plays a role in malignant melanoma.

Ultraviolet A

Ultraviolet B

Hair

Thin layer of dead cells

Squamous cells

Basal layer

Melanocyte cells

Basal cells

Blood vessels

Epidermis

Sweat gland

Dermis

Squamous Cell Carcinoma

Arising from cells in the upper layer of the epidermis, this cancer is also caused by exposure to sunlight or tanning lamps. It is usually curable if treated early. It grows faster than basal cell carcinoma and can metastasize.

Basal Cell Carcinoma

The most common skin malignancy is usually caused by excessive exposure to sunlight or tanning lamps. It develops slowly, rarely metastasizes, and is nearly 100% curable if diagnosed early and treated properly.

Melanoma

This deadliest of skin cancers involves melanocyte cells, which produce pigment. It can develop from a mole or on unblemished skin, grows quickly, and can spread to other parts of the body (metastasize).

Figure 10-19 Structure of the human skin and the relationships between ultraviolet (UV-A and UV-B) radiation and the three types of skin cancer. The incidence of these types of cancer is rising, mostly because more fair-skinned people have increased their exposure to sunlight by moving to areas with sunnier climates and by spending more of their leisure time exposed to sunlight. If ozone-destroying chemicals continue to reduce stratospheric ozone levels, incidence of these types of cancers is expected to rise. (Source: The Skin Cancer Foundation)

eral months each year. Australian television stations now broadcast daily UV levels, warning Australians (with the world's highest rate of skin cancer) to stay inside during bad spells and to protect themselves from the sun's rays with hats, clothing, and sunscreens when they go out during daytime. These precautions are required by law for schoolchildren. Some newspapers and TV weather forecasts in the United States report the EPA and National Weather Service daily ultraviolet index.

Assuming that we phase out *all* ozone-destroying chemicals over the next three decades, the EPA estimates that projected ozone thinning during the 1980s and 1990s will lead to 12 million new cases of

A: About 55% (35% before harvest and 20% after harvest)

Charge: There is no ozone hole, and the whole idea is merely a scientific theory.

Response: Technically, it is not correct to call this phenomenon an ozone *hole*. Instead, it is a thinning or partial depletion of normal ozone levels in the stratosphere. A scientific theory represents a widely accepted idea or principle that has a very high degree of certainty because it has been supported by a great deal of evidence. Nothing in science can be proven absolutely, but there is overwhelming scientific evidence of ozone thinning in the stratosphere.

Charge: There is no scientific proof that an ozone hole exists.

Response: Ground-based and satellite measurements clearly show seasonal ozone thinning above Antarctica and the Arctic, and other measurements reveal a lower overall thinning everywhere except over the tropics.

Charge: CFC molecules can't reach the stratosphere because they are heavier than air, and no measurements have detected them in the stratosphere.

Response: The atmosphere is like a turbulent fluid in which gases, both lighter and heavier than air, are mixed thoroughly by the churning movements of large air masses. Since 1975 measuring instruments on balloons and satellites have clearly shown that CFCs are transported high into the stratosphere.

Charge: Sodium chloride (NaCl) from the evaporation of sea spray contributes more chlorine to the stratosphere than do CFCs.

Response: Particles of NaCl from sea spray (unlike CFCs) are soluble in water and are washed out of the lower atmosphere. Measurements have detected no sodium in the lower stratosphere.

Charge: Volcanic eruptions and biomass burning have added much more chlorine (Cl) to the stratosphere than have human-caused CFC emissions.

Response: Most of the water-soluble HCl injected into the troposphere from occasional large-scale volcanic eruptions is washed out by rain before it reaches the stratosphere. A serious calculation error by one of the skeptics (Dixy Lee Ray) greatly overestimated the amount of HCl injected into the stratosphere by recent volcanic eruptions. Measurements indicate that no more than 20% of the chlorine from biomass burning (in the form of methyl chloride, CH_3Cl) reaches the stratosphere; this amount of chlorine is about five times less than the current contribution from CFCs. Even Fred Singer (who is highly skeptical about some aspects of ozone depletion models) stated in 1993 that "CFCs make the major contribution to stratospheric chlorine." However, recent research indicates that aerosol particles of sulfate salts produced from sulfur dioxide emitted by volcanic eruptions can destroy stratospheric ozone for several years after an eruption. These volcanic aerosols, plus the human-related inputs of ozone-destroying chemicals, can seriously deplete ozone in the stratosphere.

Charge: Seasonal ozone thinning over Antarctica is a natural phenomenon because it appeared long before CFCs were in wide use.

Response: In the late 1950s, British scientist Gordon Dobson discovered that the antarctic polar vortex leads to some seasonal ozone loss through natural causes. Measurements indicate that between 1956 and 1976 the natural pattern of slight ozone loss above Antarctica did not change significantly. Since 1976, however, seasonal losses of ozone above Antarctica have increased dramatically and have been linked by measurements and models to rising levels of CFCs in the stratosphere. A 1994 review of Dobson's 1958 data concluded that in 1958 there was no credible evidence for a significant ozone thinning (such as that observed since 1976) over Antarctica.

Charge: The best models of ozone production and destruction estimate that global ozone levels should be 10% less than we actually observe.

Response: Scientists recognize that their models need to be improved. However, they point out that their models of ozone depletion by chlorine- and bromine-containing chemicals produced by human activities still account for about 90% of the observed ozone depletion.

Charge: The expected increase in UV-B radiation caused by stratospheric ozone loss has not been detected in urban areas in the United States and most other developed countries.

Response: Increased UV levels have been detected in urban areas in countries in the southern hemisphere such as Australia, New Zealand, South Africa, Argentina, and Chile (as well as in Toronto, Canada) that are exposed to ozone-depleted air drifting away from Antarctica after the seasonal polar vortex breaks up each year. Some ozone depletion experts hypothesize that significant ground-level increases in UV-B have not been observed in many urban areas (especially in industrialized countries in the northern hemisphere) because ozone-laden smog over most cities may filter out some of the UV-B. They argue that polluting our way out of increased UV-B levels in such cities is unacceptable; it will continue to cause premature death and widespread health problems for humans and damage trees and crops (Section 9-4).

Critical Thinking

Do you believe that ozone depletion in the stratosphere is a serious problem or one that has been greatly overblown? Explain.

Q: What percentage of U.S. crops are lost to pests?

Considerable research indicates that years of exposure to UV-B ionizing radiation in sunlight is the primary cause of *squamous cell* (Figure 10-19, left) and *basal cell* (Figure 10-19, center) *skin cancers*, which together make up 95% of all skin cancers. Typically there is a 15–40 year lag between excessive exposure to UV-B and development of these cancers.

Caucasian children and adolescents who get only a single severe sunburn double their chances of getting these two types of cancers. Some 90–95% of these types of skin cancer can be cured if detected early enough, although their removal may leave disfiguring scars. These cancers kill only 1–2% of their victims, but this still amounts to about 2,300 deaths in the United States each year.*

A third type of skin cancer, *malignant melanoma* (Figure 10-19, right), occurs in pigmented areas like moles anywhere on the body's surface. This type of cancer can spread rapidly (within a few months) to other organs. It kills about one-fourth of its victims (most under age 40) within 5 years, despite surgery, chemotherapy,

*I have had six basal cell cancers on my face and neck because of "catching too many rays" in my younger years. I wish I had known then what I know now.

and radiation treatments. Each year it kills about 100,000 people (including 6,900 Americans in 1994), mostly Caucasians, but it can be cured if detected early enough. The lag time between first substantial exposure to UV radiation (apparently UV-A and UV-B) and the occurrence of melanoma is 15–25 years.

Evidence indicates that people (especially Caucasians) who get three or more blistering sunburns before age 20 are five times more likely to develop malignant melanoma than those who have never had severe sunburns. About 10% of those who get malignant melanoma have an inherited gene that makes them especially susceptible to the disease.

To protect yourself, the safest course is to stay out of the sun (especially between 10 A.M. and 3 P.M., when UV levels are highest) and avoid tanning parlors. When you are in the sun, wear tightly woven protective clothing, a wide-brimmed hat, and sunglasses that protect against UV-A and UV-B radiation (ordinary sunglasses may actually harm your eyes by dilating your pupils so that more UV radiation strikes the retina). Because UV rays can penetrate clouds, overcast does not protect you; neither does shade because UV rays can reflect off sand, snow, water, or patio floors. People who take antibiotics and women who take birth

control pills are more susceptible to UV damage.

Apply sunscreen with a protection factor of 15 or more (25 if you have light skin) to all exposed skin, and reapply it after swimming or excessive perspiration. Most people don't realize that the protection factors for sunscreens are based on using one full ounce of the product—far more than most people apply. Children who use a sunscreen with a protection factor of 15 every time they are in the sun from birth to age 18 decrease their chance of getting skin cancer by 80%; babies under a year old should not be exposed to the sun at all.

Become familiar with your moles and examine your skin at least once a month. The warning signs of skin cancer are a change in the size, shape, or color of a mole or wart (the major sign of malignant melanoma, which must be treated quickly); sudden appearance of dark spots on the skin; or a sore that keeps oozing, bleeding, and crusting over but does not heal. Be alert for precancerous growths (reddish-brown spots with a scaly crust). If you observe any of these signs, consult a doctor immediately.

Critical Thinking

What precautions, if any, do you take to reduce your chances of getting skin cancer from exposure to sunlight? Explain why you do or don't take such precautions.

---

skin cancer and 200,000 additional skin cancer deaths in the United States alone over the next 50 years.

Other effects from increased UV exposure are

■ Suppression of the immune system. This makes the body more susceptible to infectious diseases and some forms of cancer.

■ An increase in eye-burning, highly damaging acid deposition (Figure 9-6), and ozone in smog in the troposphere (Figure 9-4).

■ Lower yields of key crops such as corn, rice, cotton, soybeans, beans, peas, sorghum, and wheat, with estimated losses totaling $2.5 billion per year in the United States before the middle of the 21st century.

■ A serious decline in forest productivity of the many tree species sensitive to UV-B radiation. This could reduce CO_2 uptake and enhance global warming.

■ Increased breakdown and degradation of materials such as various types of paints, plastics, and outdoor

---

A: About 37% (7% higher than in the 1940s despite a 33-fold increase in pesticide use since then)

materials. Such damage could cost billions of dollars per year.

- Reduction in the productivity of surface-dwelling phytoplankton, which could upset aquatic food webs, decrease yields of seafood eaten by humans, and possibly accelerate global warming by decreasing the oceanic uptake of CO_2 by phytoplankton.

- Damage to the ecological structure and function of lakes because of deeper penetration of UV light caused by a synergistic interaction between ozone depletion, global warming, and acid deposition (Section 9-2).

In a worst-case ozone depletion scenario, most people would have to avoid the sun altogether (see cartoon below). Even cattle could graze only at dusk, and farmers and other outdoor workers might need to limit their exposure to the sun to minutes. Fortunately, we are acting to prevent such a scenario.

Some critics who believe that the threat of ozone depletion has been overblown argue that we should not worry about a 10%, 20%, or even 50% increase in UV-B radiation reaching the earth's surface. According to Fred Singer, moving just 97 kilometers (60 miles) closer to the equator involves potential exposure to 10% more UV-B radiation.

Many scientists contend that this argument ignores two crucial points. First, the southward movement of the U.S. population (and people in many other countries) is one reason skin cancer rates have risen. Any increased exposure to UV-B radiation because of ozone depletion would add to this already serious health threat. Second, we are talking about the likely possibility of rapid (within decades), widespread, long-lasting, and essentially unpredictable ecological disruptions in species adapted to existing levels of background UV-B radiation.

"I MISS THE OZONE LAYER...."

Humans can quickly make cultural adaptations to increased UV-B radiation by staying out of the sun, protecting their skin with clothing, and applying sunscreens. However, plants and other animals that help support us and other forms of life can't make such changes except through the long process of biological evolution.

10-7 SOLUTIONS: PROTECTING THE OZONE LAYER

How Can We Protect the Ozone Layer?

The scientific consensus of researchers in this field is that we should immediately stop producing all ozone-depleting chemicals. Even with immediate action, the models indicate that it will take 50–60 years for the ozone layer to return to 1975 levels and another 100–200 years for full recovery to pre-1950 levels.

Substitutes are already available for most uses of CFCs (Table 10-1), and others are being developed (Individuals Matter, p. 294). One substitute for CFCs is *hydrochlorofluorocarbons* or *HCFCs* (such as $CHClF_2$, containing fewer chlorine atoms per molecule than CFCs). Because of their shorter lifetimes in the stratosphere, these compounds should have only about 2.5% of the ozone-depleting potential of CFCs. If used in massive quantities, however, HCFCs would still cause ozone depletion and act as potent greenhouse gases.

Another substitute for CFCs is *hydrofluorocarbons* or *HFCs* (such as CF_3, containing fluorine but no chlorine or bromine). Recent research indicates that HFCs have a negligible effect on ozone depletion. However, they may need to be restricted or phased out because they are powerful greenhouse gases that remain in the atmosphere much longer than CFCs or CO_2. In addition, a 1996 astudy indicated that one widely used HCFC (HCFC-123) may be causing acute hepatitis and other living abnormalities.

To a growing number of scientists, hydrocarbons (HCs) such as propane and butane are a better way to reduce ozone depletion while doing little to enhance global warming. HCs are especially useful as coolants and insulating foam in refrigerators recently developed by German scientists.

Developing countries can use HC refrigerator technology to leapfrog ahead of industrialized countries without having to invest in costly HFC and HCFC technologies that will have to be phased out within a few decades. This approach is also less costly because HCs cannot be patented and can be manufactured locally, reducing the need to import expensive HFCs and HCFCs.

Q: What percentage of insecticides applied to crops in the United States reach the target pests?

Table 10-1 CFC Substitutes

Types	Pros	Cons
HCFCs (hydrochlorofluorocarbons)	Break down faster (2–20 years). Pose about 90% less danger to ozone layer. Can be used in aerosol sprays, refrigeration, air conditioning, foam, and cleaning agents.	Are greenhouse gases. Will still deplete ozone, especially if used in large quantities. Health effects largely unknown. HCFC-123 causes benign tumors in the pancreas and testes of male rats and may be banned for use in aerosol sprays, foam, and cleaning agents. May lower energy efficiency of appliances. Can be degraded to trifluoroacetate (TFA), which can inhibit plant growth in wetlands.
HFCs (hydrofluorocarbons)	Break down faster (2–20 years). Do not contain ozone-destroying chlorine. Can be used in aerosol sprays, refrigeration, air conditioning, and insulating foam.	Are greenhouse gases. Safety questions about flammability and toxicity still unresolved. Can be degraded to trifluoroacetate (TFA), which can inhibit plant growth in wetlands. May lower energy efficiency of appliances. Production of HFC-134a, a refrigerant substitute, yields an equal amount of methyl chloroform, a serious ozone depleter.
Hydrocarbons (HCs) (such as propane and butane)	Cheap and readily available. Can be used in aerosol sprays, refrigeration, foam, and cleaning agents.	Can be flammable and poisonous. Some increase in ground-level air pollution.
Ammonia	Simple alternative for refrigerators; widely used before CFCs.	Toxic if inhaled. Must be handled carefully.
Water and Steam	Effective for some cleaning operations and for sterilizing medical instruments.	Creates polluted water that must be treated. Wastes water unless the used water is cleaned up and reused.
Terpenes (from the rinds of lemons and other citrus fruits)	Effective for cleaning electronic parts.	None.
Helium	Effective coolant for refrigerators, freezers, and air conditioners.	This rare gas may become scarce if use is widespread, but very little coolant is needed per appliance.

Can Technofixes Save Us? What about a quick fix from technology so that we can keep on using CFCs? Physicist Alfred Wong has proposed that each year we launch a fleet of 20–30 football-field-long, radio-controlled blimps into the stratosphere above Antarctica. Hanging from each blimp would be a huge curtain of electrical wires that would inject negatively charged electrons into the stratosphere when exposed to high voltages (produced by electricity from huge panels of solar cells). Based on 4 years of laboratory experiments, Wong believes that ozone-destroying chlorine atoms (Cl) in the stratosphere would each pick up an electron and be converted to chloride ions ($Cl + e \longrightarrow Cl^-$) that would not react with ozone. A second suspended sheet of positively charged wires could be used to attract the negatively charged ions and remove them from the stratosphere. Wong esti-mates that it would cost about $400 million a year to remove 10–30% of the chlorine atoms formed in the stratosphere each year.

However, atmospheric chemist Ralph Ciecerone believes that this plan won't work because other chemical species in the stratosphere snatch electrons more readily than does chlorine. This scheme could also have unpredictable side effects on atmospheric chemistry.

Others have suggested using tens of thousands of lasers to blast CFCs out of the atmosphere before they can reach the stratosphere. However, the energy required to do this would be enormous and expensive, and decades of research would be needed to perfect the types of lasers needed. Moreover, we can't predict the possible effects of such powerful laser blasts on climate, birds, or planes.

A: 1–2% (and often less than 0.1%)

Ray Turner and His Refrigerator

Ray Turner, an aerospace manager at Hughes Aircraft in California, made an important low-tech, ozone-saving discovery by using his head—and his refrigerator. His concern for the environment led him to look for a cheap and simple substitute for the CFCs used as cleaning agents for removing films of oxidation from the electronic circuit boards manufactured at his plant and elsewhere.

He started his search by looking in his refrigerator for a better circuit board cleaner. He decided to put drops of various substances on a corroded penny to see whether any of them would remove the film of oxidation. Then he used his soldering gun to see whether solder would stick to the surface of the penny, indicating that the film had been cleaned off.

First, he tried vinegar. No luck. Then he tried some ground-up lemon peel, also a failure. Next he tried a drop of lemon juice and watched as the solder took hold. The rest, as they say, is history.

Today, Hughes Aircraft uses inexpensive citrus-based solvents that are CFC-free to clean circuit boards. This new cleaning technique has reduced circuit board defects by about 75% at Hughes. And Turner got a hefty bonus. Now other companies, such as AT&T, clean computer boards and chips using acidic chemicals extracted from cantaloupes, peaches, and plums. Maybe you can find a solution to an environmental problem in your refrigerator, grocery or drugstore, or backyard.

Solutions: Some Hopeful Progress In 1987, 36 nations meeting in Montreal developed a treaty, commonly known as the *Montreal Protocol*, to cut emissions of CFCs (but not other ozone depleters) into the atmosphere by about 35% between 1989 and 2000. After hearing more bad news about ozone depletion, representatives of 93 countries met in London in 1990 and in Copenhagen in 1992 and adopted a protocol accelerating the phaseout of key ozone-depleting chemicals, with some phaseout schedules accelerated in 1995 and 1997.

The agreements reached so far are important examples of global cooperation in response to serious threats to global environmental security (Guest Essay, pp. 295–296). Because of these agreements, CFC production fell by 76% between 1988 (its peak production year) and 1995. Global production of halons, car-

bon tetrachloride, and methyl chloroform has also dropped sharply, but that of methyl bromide and HCFCs continues to rise.

Developed countries have set up a $250-million fund to help developing countries make the switch away from CFCs, but much more money will be needed. It is encouraging that 58 developing countries have committed themselves to phasing out CFCs early. However, the world's two most populous countries, China and India, have refused to sign the ozone agreements. Even if the 1992 agreements are upheld, scientists estimate that the ozone layer will continue to be depleted until around 2080 and will cause ozone losses of 10–30% over the northern hemisphere (where most of the world's people live). However, without the 1992 international agreement, ozone depletion would be a much more serious threat.

Will the International Treaty to Slow Ozone Depletion Work? There is growing concern that the requirements of the Copenhagen agreement might not be met. By 1995 there was a political and economic backlash in the United States against this treaty. This was caused by widely publicized (but scientifically refuted; see Pro/Con on p. 290) attacks on the overwhelming evidence and scientific consensus that ozone depletion is a very serious problem.

The effectiveness of the treaty is also being undermined by a rapidly growing black market in CFCs (apparently being smuggled into the United States and other developed countries from Russia and China). This black market is being stimulated by dwindling supplies of CFCs (phased out by 1996), the high prices of some CFC substitutes, and the costly conversion of some older air conditioners and refrigerators to use replacement chemicals. There are also signs that some countries are cheating and not living up to the requirements of the ozone treaty.

Even if the ozone treaty is only partially implemented, it has set an important precedent for global cooperation and action when faced with potential global disaster. However, international cooperation in dealing with projected global warming is much more difficult because the evidence for global warming is less clear-cut. Moreover, lowering our inputs of greenhouse gases by greatly reducing use of fossil fuels and greatly slowing deforestation is economically and politically difficult to do.

Some people have wondered whether there is intelligent life in other parts of the universe; others wonder whether there is intelligent life on the earth. To them, if we can seriously deal with the interconnected global problems of loss of biodiversity (Chapters 5, 14, 16, and 17), possible climate change, and depletion of stratospheric ozone from human activities,

Q: Since 1945, how many premature deaths from insect-transmitted diseases have been saved by using insecticides?

Environmental Security

GUEST ESSAY

Norman Myers

Norman Myers is an international consultant in environment and development, with emphasis on conservation of wildlife species and tropical forests, and is one of the world's leading environmental experts. His research and consulting have taken him to 80 countries. He has served as a consultant for many development agencies and research organizations, including the U.S. National Academy of Sciences, the World Bank, the Organization for Economic Cooperation and Development, various UN agencies, and the World Resources Institute. Among his many publications (see Further Readings) are Conversion of Tropical Moist Forests *(1980),* A Wealth of Wild Species *(1983),* The Primary Source: Tropical Forests and Our Future *(1984),* The Gaia Atlas of Planet Management *(1992),* The Gaia Atlas of Future Worlds *(1990),* Ultimate Security: The Environmental Basis of Political Security *(1993), and* Scarcity or Abundance: A Debate on Environment *(1994).*

> *There is a new and different threat to our national security emerging—the destruction of our environments. I believe that one of our key national security objectives must be to reverse the accelerating pace of environmental destruction around the globe.*
>
> —*Senator Sam Nunn, former chairman of the U.S. Senate Armed Services Committee*

We are already engaged in World War III. It is a war against nature, and it is simply no contest. As a result, the threat from the skies is no longer missiles but ozone-layer depletion and global warming, and the threat on land is from soil erosion.

These and other environmental threats, including overpopulation and widespread poverty in developing countries, have been described by a growing number of political leaders and military planners as the greatest threat we face short of nuclear war. So a new concept is emerging in councils of foreign policy makers and military strategists: the environmental dimension to security issues.

According to this idea, we should move beyond traditional thinking about security concepts, and consider a series of environmental factors underpinning our material welfare. These factors include such natural resources as a nation's soil, water, forests, grasslands, and fisheries, and the climatic patterns and biogeochemical cycles that maintain the life-support systems of all nations [Figures 1-2 and 7-3].

If a nation's environmental foundations are degraded or depleted, its economy may well decline, its social fabric deteriorate, and its political structure become destabilized as growing numbers of people seek to sustain themselves from declining resource stocks. The likely outcome is tension and conflict within a nation and possibly with other nations.

Thus, national security is no longer about fighting forces and weaponry alone. It relates increasingly to watersheds, croplands, forests, genetic resources, climate, and other factors that, taken altogether, are as crucial to a nation's security as are military factors.

Consider Ethiopia and its agricultural decline as a source of conflict. The country's traditional farming area in the highlands was losing an estimated 1 billion tons of soil a year by the early 1970s (for comparison, the United States loses an estimated 2.8 billion tons a year, from a cropland area 20 times as large). The resulting drop in agricultural production led to food shortages in Ethiopian cities. Ensuing disorders precipitated the overthrow of Emperor Haile Selassie in 1974.

The new regime did not move fast enough, despite some efforts to restore agriculture, so throngs of impoverished peasants streamed into the country's lowlands, including an area that straddles the border with Somalia and is of long-standing dispute between the two countries. This resulted in an outbreak of hostilities in 1977, which threatened oil tankers heading from the Persian Gulf to the industrialized nations of the Western world.

Between 1976 and 1980, Ethiopia spent an average of $225 million a year on military activities. If this money had been used to safeguard topsoil, tree cover, and associated factors of the natural-resource base in traditional farmlands of Ethiopia, the migration to the lowlands would have been far less likely. Ironically, the amount that the United Nations has budgeted (but not spent) for Ethiopia for anti-erosion, reforestation, and related measures under its Anti-Desertification Plan suggests that no more than $50 million a year would have been needed to counter much of the problem, had this investment in environmental security been undertaken in due time. By comparison, the amount required to counter Ethiopia's famine during 1985 alone amounted to $500 million for relief measures alone.

In many countries, decline of the natural-resource base that underpins agriculture has led to increased imports, rising prices, and ultimately to outright shortages of food. In turn, these shortages have helped trigger civil disorders, military outbreaks, and downfalls of governments.

Food-supply disputes have also occurred among developed countries. For example, in the North Atlantic, Great Britain and Iceland have come to the edge of hostilities over declining cod stocks. At least 16 similar major clashes over declining fishery stocks have occurred in other parts of the world. Such disputes

(continued)

are likely to increase in the future in light of the failure of fisheries around the world to maintain sustainable yields.

Another source of conflict is water supplies. In the Middle East, competition for scarce water is a major factor in political confrontations, and shortages are projected to grow even more acute [p. 298 and Figure 11-1]. Israel went to war in 1967 in part because the Arabs were trying to divert the headwaters of the River Jordan. And currently, Israel—the world's most water-efficient country—is beginning to suffer from critical water deficits.

Of 200 major river systems, almost 150 are shared by two nations, and more than 50 by three to ten nations; all in all, these rivers support almost 40% of the world's population. As many as 80 countries, with 40% of the world's population, already experience water shortages.

Other conflicts occur because of deforestation. The Ganges River system is dependent on tree cover in its catchment zone in the Himalayas. Because of deforestation, flooding during the annual monsoon has become so widespread that it regularly imposes damages to crops, livestock, and property worth $1 billion a year among downstream communities in India, even though most of the deforestation occurs in Nepal.

In every category of environmental decline there are major implications for international relations. They all act, to a varying degree, as sources of economic disruption and political tension. Although they may not trigger outright confrontation, they help to destabilize societies in an already increasingly unstable world.

In the instances cited here, the environmental issues are readily apparent. In other cases, the impact is more deferred and diffuse, as in the case of species extinctions and gene depletion, with all this ultimately implies for future genetic contributions to agriculture, medicine, industry, and energy. Probably the most deferred and diffuse impact of all—although altogether the most significant—will prove to be climatic change.

Buildup of carbon dioxide and other greenhouse gases in the global atmosphere will, if it persists as projected, cause far-reaching disruptions for temperature and rainfall patterns.

We cannot dispatch battalions to turn back the deserts, we cannot launch a flotilla to resist the rising seas, we cannot send fighter planes to counter the greenhouse effect. Instead, we can achieve more enduring, widespread, and true security by safeguarding our environments than we can by engaging in military buildups. Increasingly, it is becoming a case of trees now or tanks later.

Major environmental problems recognize no geographic boundaries; the winds carry no passports. The new security will be security for us all, or for none. This means working toward our new security *together*—another big break from the past. Peace with the earth means peace with each other, and vice versa. Hopefully, the prospect of catastrophic breakdown can motivate us to achieve ultimate peace on the earth, by learning how to cooperate with one another and with the rest of nature.

Critical Thinking

1. Do you agree or disagree with the position that we need to place much more emphasis on environmental security? Should we treat it with the same degree of seriousness, analysis, and funding as we do economic security and military security, or should environmental security have higher or lower priority? Defend your answers.

2. Some 300 million couples in developing countries want to reduce their family size but lack the family planning services to do so. If we were to meet this need, we would reduce the ultimate global population by at least 2 billion. The cost would be $6 billion per year, equivalent to two and one-half days of military spending worldwide. Why do you think it is not being done? What might be the environmental and economic security benefits?

then the answer is a hopeful yes. This means recognizing that *prevention* is the best (and in the long run, the least costly) way to deal with global environmental problems. Otherwise, they believe the answer is a tragic no.

The atmosphere is the key symbol of global interdependence. If we can't solve some of our problems in the face of threats to this global commons, then I can't be very optimistic about the future of the world.

MARGARET MEAD

Q: Worldwide, how many people are poisoned each year by pesticides?

CRITICAL THINKING

1. Do you believe that possible global warming from an enhanced greenhouse effect caused at least partly by human activities is a serious problem or one that has been greatly overblown? Explain.

2. Explain why you agree or disagree with each of the proposals listed in Figure 10-14 (for slowing down emissions of greenhouse gases into the atmosphere) and those given in Figure 10-15 (for preparing for the effects of global warming). What effects would carrying out these proposals have on your own lifestyle and on those of any grandchildren you might have? What might be the effects of *not* carrying out these actions?

3. What consumption patterns and other features of your lifestyle directly add greenhouse gases to the atmosphere? Which, if any, of these things would you be willing to give up to slow projected global warming and reduce other forms of air pollution?

4. Do people have the right to use the atmosphere as a dumping ground for pollutants? Explain. If not, how would you restrict such activities?

5. Use the information in Table 10-1 to determine which chemicals should be given priority as substitutes for CFCs.

6. In 1995 and 1996, some members of the U.S. House of Representatives introduced legislation that would cancel U.S. participation in the international ozone treaty on the basis that there is no scientific proof of ozone thinning. Criticize this position on the basis of the nature of science (Section 3-1) and the information in the Pro/Con box on p. 290.

7. In preparation for the 1992 UN Conference on the Human Environment in Rio de Janeiro, President Bush's top White House economic adviser gave an address in Williamsburg, Virginia, to representatives of governments from a number of countries. He told his audience not to worry about global warming because the average temperature increases scientists are predicting were much less than the temperature increase he experienced in coming from Washington, D.C., to Williamsburg. What is the fundamental flaw in this reasoning?

*8. How has the climate changed in the area where you live during the past 50 years? Investigate the beneficial and harmful effects of these changes. How have these changes benefited or harmed you personally?

*9. As a class, conduct a poll of students at your school to determine whether they believe that possible global warming from an enhanced greenhouse effect is a very serious problem, a moderately serious problem, or of little concern. Tally the results to see whether there are differences related to year in school, political leaning (liberal, conservative, independent), or sex of poll participants.

*10. As a class, conduct a poll of students at your school to determine whether they believe that stratospheric ozone depletion is a very serious problem, a moderately serious problem, or of little concern. Tally the results to see whether there are differences related to year in school, political leaning (liberal, conservative, independent), or sex of poll participants.

*11. Make a concept map of this chapter's major ideas, using the section heads and subheads and the key terms (in boldface type). Look at the inside back cover and in Appendix 4 for information about concept maps.

SUGGESTED READINGS

See Internet Sources and Further Readings at the back of the book, and InfoTrac.

11 WATER RESOURCES AND WATER POLLUTION

Water Wars in the Middle East

Because of differences in climate, some parts of the earth have an abundance of water, whereas other parts have a shortage. As population, agriculture, and industrialization grow, there is increasing competition for water, especially in dry regions such as the Middle East.

The next war in the Middle East may well be fought over water, not oil. Most water in this dry region comes from three shared river basins: the Jordan, the Tigris–Euphrates, and the Nile (Figure 11-1). Water in much of this arid region is already in short supply, and the human population in this region is projected to double within only 25 years.

Ethiopia, which controls the headwaters that feed 86% of the Nile's flow, has plans to divert more of this water; so does Sudan. This could reduce the amount of water available to desperately water-short Egypt, whose terrain is desert except for a thin green strip of irrigated cropland running down its middle along the Nile and its delta.

Between 1997 and 2025, Egypt's population is expected to increase from 65 million to 98 million, greatly increasing the demand for already scarce water. Egypt's options are to (1) go to war with Sudan and Ethiopia to obtain more water, (2) slash population growth, (3) improve irrigation efficiency, (4) spend $2 billion to build the world's longest concrete canal and a massive pumping station to pump water out of Lake Nasser (the reservoir created from the Nile by the Aswan High Dam, Figure 11-1) and create a lush new valley of irrigated cropland in the middle of the desert, (5) import more grain to reduce the need for irrigation water, (6) work out water-sharing agreements with other countries, or (7) suffer the harsh human and economic consequences.

There is also fierce competition for water among Jordan, Syria, and Israel, which get most of their water from the Jordan River basin (Figure 11-1). Israel irrigates two-thirds of its croplands and uses water more efficiently than any other country in the world. However, within the next few years, its supply is projected to fall by up to 30% short of demand because of increased immigration.

Some 90 million people currently live in the water-short basins of the Tigris and Euphrates rivers (Figure 11-1), and by the year 2020 the population there is projected to almost double to 170 million. Turkey, located at the headwaters of these two rivers, has abundant water, and it plans to build 22 dams along the upper Tigris and Euphrates and to construct pipelines to transport and sell water to parched Saudi Arabia and Kuwait and perhaps to Syria, Israel, and Jordan. The greatest threat to Iraq is a cutoff of its water supply by Turkey and Syria.

Clearly, distribution of water resources will be a key issue in any future peace talks in this region. Resolving these problems will require a combination of regional cooperation in allocating water supplies, slowed population growth, and improved efficiency in the use of water resources.

By the middle of the next century, almost twice as many people will be trying to share the same amount of fresh water the earth has today. Already, 1.2 billion people lack access to clean drinking water, 2.2 billion live without sewage systems, and two-thirds of the world's households don't have running water. As fresh water becomes scarcer, access to water resources will be a major factor in determining the economic, environmental, and military security of a growing number of countries around the globe (Guest Essay, pp. 295–296).

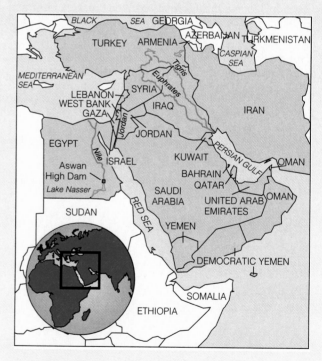

Figure 11-1 The Middle East, whose countries have some of the highest population growth rates in the world. Because of the dry climate, food production depends heavily on irrigation. Existing conflicts among countries in this region over access to water may soon overshadow both long-standing religious and ethnic clashes and attempts to take over valuable oil supplies.

Our liquid planet glows like a soft blue sapphire in the hard-edged darkness of space. There is nothing else like it in the solar system. It is because of water.

JOHN TODD

In this chapter we answer the following questions:

- What are water's unique physical properties?

- How much fresh water is available to us, and how much of it are we using?

- What areas face water shortages, and what can be done about this problem?

- What areas suffer from flooding, and what can be done to reduce the risk of flooding and flood damage?

- What pollutes water, where do the pollutants come from, and what effects do they have?

- What are the water pollution problems in streams, lakes, oceans, and groundwater?

- How can we prevent and reduce water pollution?

11-1 WATER'S IMPORTANCE AND UNIQUE PROPERTIES

Why Is Water So Important? We live on the water planet, with a precious film of water—most of it salt water—covering about 71% of the earth's surface (Figure 5-15). The earth's organisms are made up mostly of water; a tree is about 60% water by weight, and most animals (including humans) are about 50–65% water.

Each of us needs only a dozen or so cupfuls of water per day to survive, but huge amounts of water are needed to supply us with food, shelter, and our other needs and wants. Water also plays a key role in sculpting the earth's surface, moderating climate, and diluting pollutants. In fact, without water the earth would have no oceans, no life as we know it, and no people.

What Are Some Important Properties of Water? Water is a remarkable substance with a unique combination of properties:

- *There are strong forces of attraction (called hydrogen bonds) between molecules of water.* These attractive forces between water molecules are the major factor determining water's unique properties.

- *Water exists as a liquid over a wide temperature range because of the strong forces of attraction between water molecules.* Its high boiling point of 100°C (212°F) and

low freezing point of 0°C (32°F) mean that water remains a liquid in most climates on the earth.

- *Liquid water changes temperature very slowly because it can store a large amount of heat without a large change in temperature.* This high heat capacity helps protect living organisms from the shock of abrupt temperature changes; it also moderates the earth's climate and makes water an excellent coolant for car engines, power plants, and heat-producing industrial processes.

- *It takes a lot of heat to evaporate liquid water because of the strong forces of attraction between its molecules.* Water's ability to absorb large amounts of heat as it changes into water vapor—and to release this heat as the vapor condenses back to liquid water—is a primary factor in distributing heat throughout the world (Figure 10-4). This property makes evaporation of water an effective cooling process, which is why you feel cooler when perspiration or bathwater evaporates from your skin.

- *Liquid water can dissolve a variety of compounds.* This enables it to carry dissolved nutrients into the tissues of living organisms, to flush waste products out of those tissues, to serve as an all-purpose cleanser, and to help remove and dilute the water-soluble wastes of civilization. Water's superiority as a solvent also means that it is easily polluted by water-soluble wastes.

- *The strong attractive forces between the molecules of liquid water cause its surface to contract (high surface tension) and also to adhere to and coat a solid (high wetting ability).* Together these properties allow water to rise through a plant from the roots to the leaves (capillary action).

- *Unlike most liquids, water expands when it freezes and becomes ice.* This means that ice has a lower density (mass per unit of volume) than liquid water. Thus ice floats on water, and as air temperatures fall below freezing, bodies of water freeze from the top down instead of from the bottom up. Without this property, lakes and streams in cold climates would freeze solid, and most of their current forms of aquatic life could not exist. Because water expands upon freezing, it can also break pipes, crack engine blocks (which is why we use antifreeze), and break up streets and fracture rocks (thus forming soil).

Water, the lifeblood of the ecosphere, is truly a wondrous substance; it connects us to one another, to other forms of life, and to the entire planet. Despite its importance, water is one of our most poorly managed resources. We waste it and pollute it, and we also charge too little for making it available, thus encouraging greater waste and pollution of this potentially renewable resource, for which there is no substitute.

11-2 SUPPLY, RENEWAL, AND USE OF WATER RESOURCES

How Much Fresh Water Is Available? Only a tiny fraction of the planet's abundant water is available to us as fresh water. About 97% by volume is found in the oceans and is too salty for drinking, irrigation, or industry (except as a coolant).

The remaining 3% is fresh water. About 2.997% of it is locked up in ice caps or glaciers or is buried so deep that it costs too much to extract. Only about 0.003% of the earth's total volume of water is easily available to us as soil moisture, usable groundwater, water vapor, and lakes and streams. If the world's water supply were only 100 liters (26 gallons), our usable supply of fresh water would be only about 0.003 liter (one-half teaspoon) (Figure 11-2).

Fortunately, the available fresh water amounts to a generous supply that is continuously collected, purified, recycled, and distributed in the solar-powered *hydrologic cycle* (Figure 4-27), as long as we don't overload it with slowly degradable and nondegradable wastes or withdraw it from underground supplies faster than it is replenished. Unfortunately, we are doing both.

Differences in average annual precipitation divide the world's countries and people into water haves and have-nots. For example, Canada, with only 0.5% of the world's population, has 20% of the world's fresh water supply. By contrast, China, with 21% of the world's people, has only 7% of the world's freshwater supply.

As population, irrigation, and industrialization increase, water shortages in already water-short regions will intensify, and wars over water may erupt (p. 298). Projected global warming also might cause changes in rainfall patterns and disrupt water supplies in unpredictable ways (Section 10-4).

What Is Surface Water? The fresh water we use first arrives as the result of precipitation. Precipitation that does not infiltrate the ground or return to the atmosphere by evaporation (including transpiration) is called **surface runoff** that flows into streams, lakes, wetlands, and reservoirs.

A **watershed**, also called a **drainage basin**, is a region from which water drains into a stream, stream system, lake, reservoir, or other water body.

What Is Groundwater? Some precipitation infiltrates the ground and percolates downward through voids (pores, fractures, crevices, and other spaces) in soil and rock (Figure 11-3). The water in these voids is called **groundwater**.

Close to the surface, the voids have little moisture in them. However, below some depth, in what is called the **zone of saturation**, they are filled with water except for an occasional air bubble. The surface of the zone of saturation, at the boundary with the unsaturated zone above, is the **water table**. The water table falls in dry weather and rises in wet weather.

Porous, water-saturated layers of sand, gravel, or bedrock through which groundwater flows are called

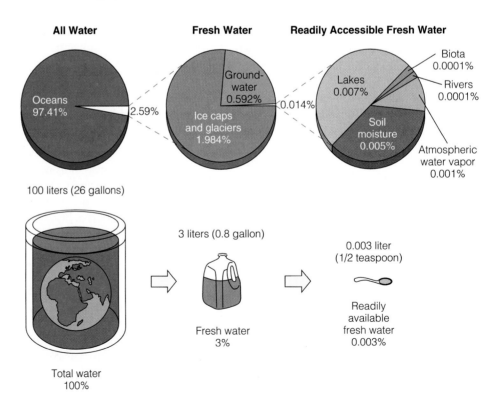

Figure 11-2 The planet's water budget. Only a tiny fraction by volume of the world's water supply is fresh water available for human use.

Q: What is the most serious drawback to using chemicals to control pests (especially insects)?

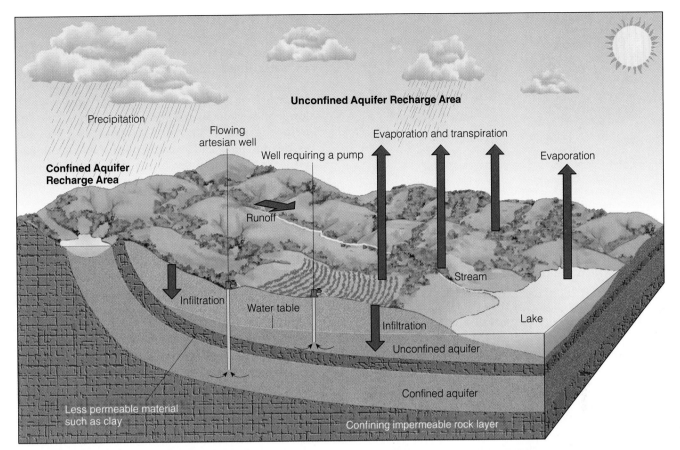

Figure 11-3 The groundwater system. An *unconfined aquifer* is an aquifer with a water table. A *confined aquifer* is bounded above and below by less permeable beds of rock. Groundwater in this type of aquifer is confined under pressure.

aquifers (Figure 11-3). Any area of land through which water passes downward or laterally into an aquifer is called a **recharge area**. Aquifers are replenished naturally by precipitation that percolates downward through soil and rock in what is called **natural recharge**, but some are recharged from the side by *lateral recharge*.

Groundwater moves from the recharge area through an aquifer and out to a discharge area (well, spring, lake, geyser, stream, or ocean) as part of the hydrologic cycle. Groundwater normally moves from points of high elevation and pressure to points of lower elevation and pressure. This movement is quite slow, typically only a meter or so (about 3 feet) per year and rarely more than 0.3 meter (1 foot) per day.

Some aquifers get very little (if any) recharge. Often found fairly deep underground and formed tens of thousands of years ago, they are (on a human time scale) nonrenewable resources. Withdrawals from such aquifers amount to *water mining* that, if kept up, will deplete these ancient deposits of liquid earth capital.

How Do We Use the World's Fresh Water? Since 1950 the global rate of water withdrawal from surface and groundwater sources has increased almost five-fold and per capita use has tripled. According to a 1996 study, humans currently use about 54% of the global surface runoff that is realistically available from the hydrologic cycle. Because of increased population growth and economic development, global withdrawal rates of surface water are projected to at least double in the next two decades and exceed the available surface runoff in a growing number of areas.

Uses of withdrawn water vary from one region to another and from one country to another (Figure 11-4). Averaged globally, about 65% of all water withdrawn each year from rivers, lakes, and aquifers is used to irrigate 16% of the world's cropland. Some 60–80% of this water either evaporates or seeps into the ground before reaching crops.

Worldwide, about 25% of the water withdrawn is used for energy production (oil and gas production and power-plant cooling) and industrial processing, cleaning, and waste removal. Agricultural and manufactured products require large amounts of water,

A: Development of genetic resistance to the chemicals by target pests

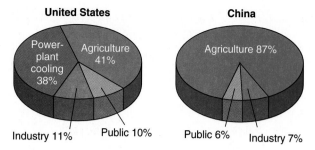

United States

Power-plant cooling 38%

Agriculture 41%

Industry 11%   Public 10%

China

Agriculture 87%

Public 6%   Industry 7%

Figure 11-4 Use of water in the United States and China. The United States has the world's highest per capita use of water, amounting to an average of 6,000 liters (1,600 gallons) per person every day. About half of the water used in the United States is unnecessarily wasted. (Data from Worldwatch Institute and World Resources Institute)

much of which could be used more efficiently and reused. It takes about 380,000 liters (100,000 gallons) to make an automobile, 3,800 liters (1,000 gallons) to produce 454 grams (1 pound) of aluminum, 3,000 liters (800 gallons) to produce 454 grams (1 pound) of grain-fed beef in a feedlot (where large numbers of cattle are confined to a fairly small area), and 100 liters (26 gallons) to produce 1 kilogram (2.2 pounds) of paper.

Domestic and municipal use accounts for about 10% of worldwide water withdrawals and about 13–16% of withdrawals in developed countries. As population, urbanization, and industrialization grow, the volume of wastewater needing treatment will increase enormously.

Case Study: Fresh Water Resources in the United States Although the United States has plenty of fresh water, much of it is in the wrong place at the wrong time or is contaminated by agricultural and industrial practices. The eastern states usually have ample precipitation, whereas many of the western states have too little. In the East, the largest uses for water are for energy production, cooling, and manufacturing; in the West, the largest use by far is for irrigation (which accounts for about 85% of all water use).

In many parts of the eastern United States the most serious water problems are flooding, occasional urban shortages, and pollution. For example, the 3 million residents of Long Island, New York, get most of their water from an aquifer that is becoming severely contaminated. The major water problem in the arid and semiarid areas of the western half of the country is a shortage of runoff caused by low precipitation, high evaporation, and recurring prolonged drought. Water tables in many areas are dropping rapidly, as farmers and cities deplete groundwater aquifers faster than they are recharged.

⬥ 11-3 TOO LITTLE WATER

What Causes Fresh Water Shortages? According to water expert Malin Falkenmark, there are four causes of water scarcity: **(1)** a *dry climate* (Figure 5-2), **(2)** *drought* (a period in which precipitation is much lower and evaporation is higher than normal), **(3)** *desiccation* (drying of the soil because of such activities as deforestation and overgrazing by livestock), and **(4)** *water stress* (low per capita availability of water due to increasing numbers of people relying on fixed levels of runoff).

Since the 1970s water scarcity intensified by prolonged drought (mostly in areas with dry climates, where 40% of the world's people live) has killed more than 24,000 people per year and created many environmental refugees. In water-short areas, many women and children must walk long distances each day, carrying heavy jars or cans, to get a meager supply of sometimes contaminated water. Millions of poor people in developing countries have no choice but to try to survive on drought-prone land. If global warming occurs as projected, severe droughts may become more common in some areas of the world.

According to a 1995 World Bank study, 30 countries containing 40% of the world's population (2.3 billion people) now experience chronic water shortages that threaten their agriculture and industry and the health of their people (Figure 11-5). By 2025 at least 3 billion people in 90 countries are expected to face severe water stress. In most of these countries the problem is not a shortage of water but the wasteful and unsustainable use of normally available supplies.

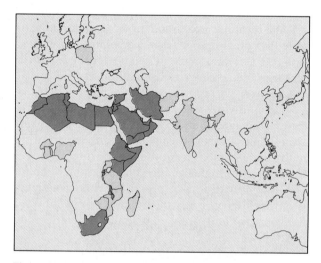

Figure 11-5 Water-short countries. Countries suffering extreme water stress (red) have fewer than 500 cubic meters (650 cubic yards) of fresh water per person; those suffering from severe water stress (yellow) have fewer than 1,000 cubic meters (1,300 cubic yards) of fresh water per person. (Data from Population Action International, Malin Falkenmark, and Peter Gleick).

Q: How many premature cancer deaths in the United States are caused by exposure to pesticide residues in foods?

In the United States, many major urban centers (especially those in the West and Midwest) are located in areas that don't have enough water or are projected to have water shortages by 2000 (Figure 11-6). Experts project that current shortages and conflicts over water supplies will get much worse as more industries and people migrate west and compete with farmers for scarce water. These shortages could worsen even more if climate warms as a result of an enhanced greenhouse effect.

A number of analysts believe that *access to water resources, a key foreign policy and environmental security issue for water-short countries in the 1990s, will become even more important early in the next century* (Guest Essay, pp. 295–296). Almost 150 of the world's 214 major river systems (57 of them in Africa) are shared by 2 countries, and another 50 are shared by 3 to 10 nations. Some 40% of the world's population already clashes over water, especially in the Middle East (p. 298).

Some areas have lots of water, but the largest rivers (which carry most of the runoff) are far from agricultural and population centers where the water is needed. For example, South America has the largest annual water runoff of any continent, but 60% of the runoff flows through the Amazon River in remote areas where few people live.

How Can Water Supplies Be Increased? There are five ways to increase supply of fresh water in a particular area: **(1)** build dams and reservoirs to store runoff, **(2)** bring in surface water from another area, **(3)** withdraw groundwater, **(4)** convert salt water to fresh water (desalination), and **(5)** improve the efficiency of water use.

In developed countries, people tend to live where the climate is favorable and then bring in water from another watershed. In developing countries, most people (especially the rural poor) must settle where the water is and try to capture and use as much precipitation as they can.

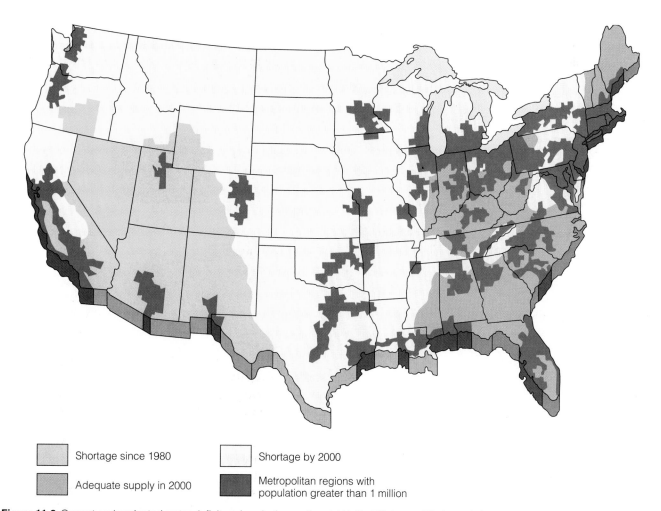

Shortage since 1980

Shortage by 2000

Adequate supply in 2000

Metropolitan regions with population greater than 1 million

Figure 11-6 Current and projected water-deficit regions in the continental United States and their proximity to metropolitan areas having populations greater than 1 million. (Data from U.S. Water Resources Council and U.S. Geological Survey)

A: 4,000–20,000 per year, according to the National Academy of Sciences

What Are the Pros and Cons of Large Dams and Reservoirs? Large dams and reservoirs have benefits and drawbacks (Figure 11-7). Large reservoirs created by damming streams can capture and store water from rain and melting snow. This water can then be released as desired to produce hydroelectric power at the dam site, irrigate land below the dam, control flooding of land below the reservoir, and provide water carried to towns and cities by aqueducts. Reservoirs also provide recreational activities such as swimming, fishing, and boating. Between 25% and 50% of the total runoff on every continent is now captured and controlled by dams and reservoirs, and many more large projects are planned.

When completed, China's Three Gorges project on the mountainous upper reaches of the Yangtze River will be the world's largest hydroelectric dam and reservoir. This superdam will supply power to industries and to 150 million Chinese. The project will also help China reduce its dependence on coal and help hold back the Yangtze River's floodwaters (which have claimed at least 500,000 lives in this century).

However, the project will also flood large areas of some of the best farmland in the region and 800 existing factories; it will also displace 1.25 million people, including 100 towns and 2 cities (each with about 100,000 people), and flood some of China's most scenic land. Critics charge that the dam will transform the Yangtze into a sewer for industrial wastes and that the immense pressure from the huge volume of water in the reservoir could trigger landslides and earthquakes. Furthermore, as the reservoir fills up with sediment, the dam could overflow and expose half a million people to severe flooding (especially if the reservoir is kept filled at a high level, as planned, to provide maximum hydroelectric power).

A series of dams on a river, especially in arid areas, can reduce downstream flow to a trickle and prevent it from reaching the sea (Figures 5-17 and 5-26) as a part of the hydrologic cycle (Figure 4-27). Since 1972 dams have prevented China's great Huang He (Yellow River) from reaching the sea. Since then, the lower part of the river has run dry every year for progressively longer periods. In the United States, the heavily dammed Colorado River disappears into the Arizona desert and rarely reaches the Gulf of California. This threatens the survival of species that spawn in such rivers, destroys estuaries that serve as breeding grounds for numerous aquatic species, and increases contamination of aquifers near coasts with saltwater (Figures 11-11 and 11-13, p. 308).

What Are the Pros and Cons of Watershed Transfers? The California Experience Tunnels, aqueducts, and underground pipes can transfer stream runoff collected by dams and reservoirs from water-

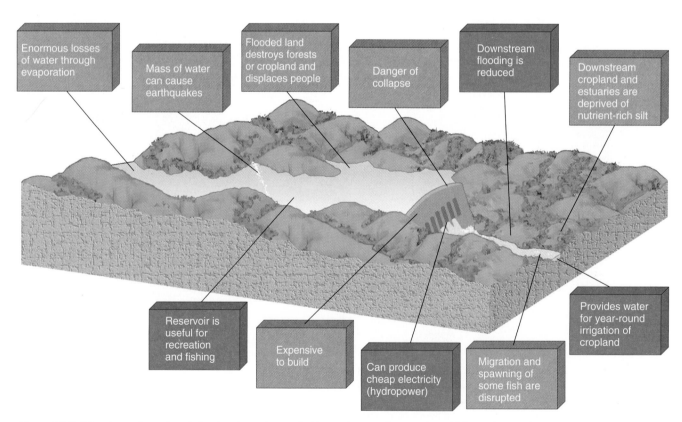

Figure 11-7 Advantages (green) and disadvantages (orange) of large dams and reservoirs, which can be used to produce electricity.

Q: What percentage of U.S. lawns are treated with pesticides?

rich watersheds to water-poor areas. Although such transfers have benefits, they also create environmental problems. Indeed, most of the world's large-scale water transfers illustrate the important ecological principle that *you can't do just one thing*.

One of the world's largest watershed transfer projects is the California Water Project. In California, the basic water problem is that 75% of the population lives south of Sacramento but 75% of the rain falls north of it. The California Water Project uses a maze of giant dams, pumps, and aqueducts to transport water from water-rich northern California to heavily populated areas and to arid and semiarid agricultural regions, mostly in southern California (Figure 11-8).

For decades, northern and southern Californians have been feuding over how the state's water should be allocated under this project. Southern Californians say they need more water from the north to support Los Angeles, San Diego, and other growing urban areas and to grow more crops. Although agriculture uses 82% of the water withdrawn in California, it accounts for only 2.5% of the state's economic output. Irrigation for just two crops, alfalfa (to feed cattle) and cotton, uses as much water as the residential needs of all 32 million Californians.

Opponents in the north say that sending more water south would degrade the Sacramento River, threaten fisheries, and reduce the flushing action that helps clean San Francisco Bay of pollutants. They also argue that much of the water already sent south is unnecessarily wasted and that making irrigation just 10% more efficient would provide enough

water for domestic and industrial uses in southern California.

If water supplies in California were to drop sharply because of projected global warming, the amount of water delivered by the huge distribution system would plummet. Most irrigated agriculture in California would have to be abandoned and much of the population of southern California might have to move to areas with more water. The 6-year drought that California experienced between 1986 and 1992 was perhaps just a small taste of the future.

Pumping out more groundwater is not the answer. Throughout much of California, groundwater is already being withdrawn faster than it is replenished. Most analysts see improving irrigation efficiency and allowing farmers to sell their legal rights to withdraw certain amounts of water from the river as much quicker and cheaper solutions.

Case Study: The James Bay Watershed Transfer Project Another major watershed transfer project is the James Bay project, a $60-billion, 50-year scheme to harness the wild rivers that flow into Quebec's James and Hudson Bays to produce electric power for Canadian and U.S. consumers (Figure 11-9). If completed,

Figure 11-9 If completed, the James Bay project in northern Quebec will alter or reverse the flow of 19 major rivers and flood an area the size of Washington State to produce hydropower for consumers in Quebec and the United States, especially in New York State. Phase I of this 50-year project is completed; phase II was postponed indefinitely in 1994 because of a surplus of electricity and opposition by environmentalists and by the indigenous Cree, whose ancestral hunting grounds would have been flooded.

Figure 11-8 The California Water Project and the Central Arizona Project involve large-scale transfers of water from one watershed to another. Arrows show the general direction of water flow.

A: About 40%

this megaproject would **(1)** construct 600 dams and dikes that will reverse or alter the flow of 19 giant rivers covering a watershed three times the size of New York State, **(2)** flood an area of boreal forest and tundra equal in area to Washington State or Germany, and **(3)** displace thousands of indigenous Cree and Inuit who for 5,000 years have lived off James Bay by subsistence hunting, fishing, and trapping.

After 20 years and $16 billion, Phase I has been completed. The second and much larger phase was postponed indefinitely in 1994 because of an excess of power generated, opposition by the Cree (whose ancestral hunting grounds would have been flooded) and Canadian and U.S. environmentalists, and New York State's cancellation of two contacts to buy electricity produced by Phase II.

Case Study: The Aral Sea Watershed Transfer Disaster The shrinking of the Aral Sea (Figure 11-10) is a result of a large-scale water-transfer project in an area of the former Soviet Union with the driest climate in Central Asia. Since 1960, enormous amounts of irrigation water have been diverted from the inland Aral Sea and its two feeder rivers to irrigate cotton, vegetable, fruit, and rice crops to supply much of the region's needs. The irrigation canal, the world's longest, stretches over 1,300 kilometers (800 miles).

Figure 11-10 Once the world's fourth-largest freshwater lake, the Aral Sea has been shrinking and getting saltier since 1960 because most of the water from the rivers that replenish it has been diverted to grow cotton and food crops. As the lake shrinks, it leaves behind a salty desert, economic ruin, increasing health problems, and severe ecological disruption.

This water diversion (coupled with droughts) has caused a regional ecological, economic, and health disaster, described by one former Soviet official as "ten times worse than the 1986 Chernobyl nuclear power-plant accident." The sea's salinity has tripled, its surface area has shrunk by 54%, and its volume has decreased by almost 75%; the two supply rivers are now mere trickles. About 30,000 square kilometers (11,600 square miles) of former lake bottom has become a human-made salt desert. The process continues, and within a few decades the once enormous Aral Sea may be reduced to a few small brine lakes.

Some 20 of the 24 native fish species have become extinct, devastating the area's fishing industry, which once provided work for more than 60,000 people. Two major fishing towns are surrounded by a desert containing stranded fishing boats and rusting commercial ships that now lie more than 70 kilometers (44 miles) from the lake's receded shore. Roughly half of the area's bird and mammal species have also disappeared.

Winds pick up the salty dust that encrusts the lake's now-exposed bed and blow it onto fields as far away as 300 kilometers (190 miles) away. As the salt spreads, it kills crops, trees, and wildlife and destroys pastureland. This phenomenon has added a new term to our list of environmental ills: *salt rain.*

These changes have also affected the area's already semiarid climate. The once-huge Aral Sea acted as a thermal buffer, moderating the heat of summer and the extreme cold of winter. Now there is less rain, summers are hotter, winters are colder, and the growing season is shorter. Cotton and crop yields have dropped dramatically.

To raise yields, farmers have increased inputs of herbicides, insecticides, and fertilizers on some crops. Many of these chemicals have percolated downward and accumulated to dangerous levels in the groundwater, from which most of the region's drinking water comes. The low river flows have also concentrated salts, pesticides, and other toxic chemicals, making surface water supplies hazardous to drink.

During the mid-1980s, the area near the shrunken Aral Sea had the highest levels of infant mortality and maternal mortality in the former Soviet Union. Between 1980 and 1993, kidney and liver diseases (especially cancers) increased 30- to 40-fold, arthritic diseases 60-fold, chronic bronchitis 30-fold, typhoid fever 30-fold, and hepatitis 7-fold.

Ways to deal with the ecological, economic, and health problems caused by this disaster include **(1)** charging farmers more for irrigation water to reduce waste and encourage a shift to less water-intensive crops, **(2)** decreasing irrigation water quotas, **(3)** introducing water-saving technologies, at a cost of at least $50 million, **(4)** developing a regional integrated water management plan, **(5)** planting protective forest belts, **(6)** using

Q: What do most scientists believe is the best way to control pests?

underground water to supplement irrigation water and to lower the water table to reduce waterlogging and salinization, **(7)** improving health services, and **(8)** slowing the area's rapid population growth (2.5% per year).

Even with help from foreign countries, the United Nations, and agencies such as the World Bank, the money needed to save the Aral Sea and restore the ecological and economic services may not be available to this extremely poor area. What has happened to the Aral Sea basin is a stark reminder that everything in nature is connected and that preventing an ecological problem is much cheaper than trying to deal with its harmful consequences.

What Are the Pros and Cons of Withdrawing Groundwater? In the United States, about half of the drinking water (96% in rural areas and 20% in urban areas) and 40% of irrigation water is pumped from aquifers. In Florida, Hawaii, Idaho, Mississippi, Nebraska, and New Mexico, more than 90% of the population depends on groundwater for drinking water.

Overuse of groundwater can cause or intensify several problems: *aquifer depletion*, *aquifer subsidence* (sinking of land when groundwater is withdrawn), and *intrusion of salt water into aquifers* (Figure 11-11). Groundwater can also become contaminated from industrial and agricultural activities, septic tanks, and

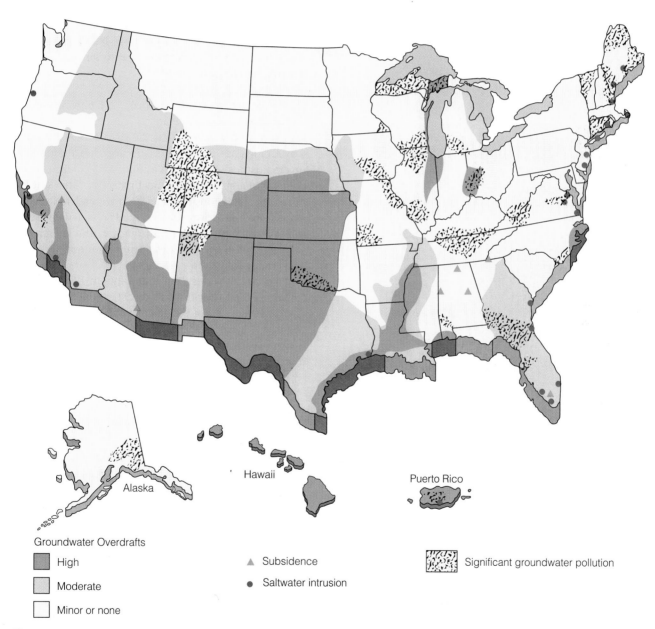

Figure 11-11 Areas of greatest aquifer depletion, subsidence, saltwater intrusion, and groundwater contamination in the United States. (Data from U.S. Water Resources Council and U.S. Geological Survey)

A: Integrated pest management (IPM)

other sources, as discussed in Section 11-5. Because groundwater is the source of about 40% of the stream flow in the United States, groundwater depletion also robs streams of water.

Currently, *groundwater in the United States is being withdrawn at four times its replacement rate.* The most serious overdrafts are occurring in parts of the huge Ogallala Aquifer, extending from southern South Dakota to central Texas (Figure 11-12 and Case Study, right) and in parts of the arid southwestern United States (Figure 11-6), especially California's Central Valley, which is the country's vegetable basket. Aquifer depletion is also a problem in Saudi Arabia, northern China, northern Africa (especially Libya and Tunisia, where the rate of withdrawal is estimated to be 10 times the rate of discharge), southern Europe, the Middle East, and parts of Mexico, Thailand, and India.

When fresh water from an aquifer near a coast is withdrawn faster than it is recharged, salt water intrudes into the aquifer (Figure 11-13). Such intrusion threatens to irreversibly contaminate the drinking water of many towns and cities along the Atlantic and Gulf coasts (Figure 11-11) and in the coastal areas of Israel, Syria, and the Arabian Gulf states. Inland movement of salt water into aquifers also occurs when rivers that normally empty into the ocean have so much water diverted and managed by dams and irrigation projects that the rivers no longer empty into the sea.

Ways to slow groundwater depletion include controlling population growth, not planting water-thirsty crops in dry areas, developing crop strains that require less water and are more resistant to heat stress, and wasting less irrigation water.

How Useful Is Desalination? The removal of dissolved salts from ocean water or from brackish (slightly salty) groundwater—called **desalination**—is another way to increase fresh water supplies. Distillation and reverse osmosis are the two most widely used methods. *Distillation* involves heating salt water until it evaporates and condenses as fresh water, leaving salts behind in solid form. In *reverse osmosis*, salt water is pumped at high pressure through a thin membrane whose pores allow water molecules, but not dissolved salts, to pass through.

About 7,500 desalination plants in 120 countries (especially in the arid Middle East and parts of North Africa) provide about 0.1% of the fresh water used by humans. However, desalination has a downside. Because it uses vast amounts of electricity, water produced in this way costs three to five times more than

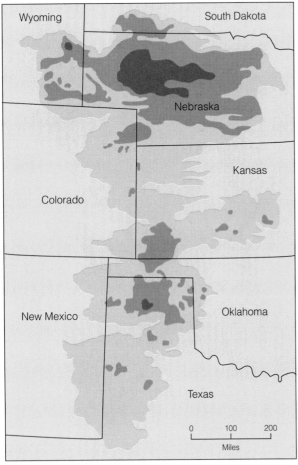

Figure 11-12 The Ogallala, the world's largest known aquifer. If the water in this aquifer were above ground, it could cover all 50 states with 0.5 meter (1.5 feet) of water. This aquifer, which is renewed very slowly, is being depleted (especially at its thin southern end in parts of Texas, New Mexico, Oklahoma, and Kansas) to grow crops, raise cattle, and provide urban dwellers and industries with water. (Data from U.S. Geological Survey).

Saturated thickness of Ogallala Aquifer:

- Less than 61 meters (200 ft)
- 61–183 meters (200–600 ft)
- More than 183 meters (600 ft) (as much as 370 meters or 1,200 ft in places)

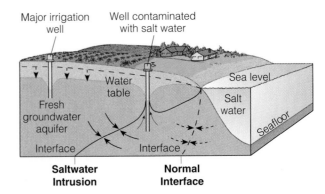

Figure 11-13 Saltwater intrusion along a coastal region. When the water table is lowered, the normal interface (dotted line) between fresh and saline groundwater moves inland (solid line).

Q: What percentage of the USDA's research and education budget is spent on integrated pest management?

Mining Groundwater: The Shrinking Ogallala Aquifer

Vast amounts of water pumped from the Ogallala, the world's largest known aquifer (Figure 11-12), have helped transform vast areas of arid high plains prairie land into one of the largest and most productive agricultural regions in the United States.

With yields two to three times those of dryland farming, irrigated farming is a profitable venture. As a result, this region now produces 20% of U.S. agricultural output (including 40% of its feedlot beef), valued at $32 billion. This production has brought prosperity to many farmers and merchants in this region, but the hidden environmental (and economic) cost has been increasing depletion of the aquifer in some areas.

Although this aquifer is gigantic, it is essentially a nonrenewable aquifer (stored during the retreat of the last ice age about 15,000–30,000 years ago) with an extremely slow recharge rate. In some areas, water is being pumped out of the aquifer 8–10 times faster than the natural recharge rate. The northernmost states (Wyoming, North Dakota, South Dakota, and parts of Colorado) still have ample supplies. However, supplies in parts of the southern states, where the aquifer is thinner (Figure 11-12), are being depleted rapidly.

Water experts project that at the current rate of withdrawal, one-fourth of the aquifer's original supply will be depleted by 2020, and much sooner in areas where it is shallow. It will take thousands of years to replenish the aquifer. Government farm policies encourage depletion of the aquifer by subsidizing water-thirsty crops that require irrigation and by providing crop-disaster payments. The greater the use of groundwater, the greater this tax break—an example of positive feedback in action.

Depletion of this essentially nonrenewable water resource can be delayed. Farmers can use more efficient forms of irrigation; in some areas they may have to switch to crops that require less water, or to dryland farming. Total irrigated area is already declining in five of the seven states using this aquifer because of rising drilling and pumping costs as the water table drops.

Cities using this groundwater can improve their water management by wasting less water and using less for lawns and golf courses. Individuals enjoying the benefits of this aquifer can pitch in by installing water-saving toilets and showerheads and converting their lawns to plants that can survive in an arid climate with little watering.

Critical Thinking

Do you believe that farmers and cattle raisers using water withdrawn from the Ogallala should receive federal subsidies to grow crops and raise livestock that require large amounts of irrigation water? Explain. What are the alternatives?

---

water from conventional sources. Distributing the water from coastal desalination plants costs even more in terms of the energy needed to pump the water uphill and inland.

Desalination also produces large quantities of brine containing high levels of salt and other minerals. Dumping the concentrated brine into the ocean near the plants might seem to be the logical solution, but this would increase the local salt concentration and threaten food resources in estuary waters. If these wastes were dumped on the land, they could contaminate groundwater and surface water.

Desalination can provide fresh water for coastal cities in arid countries (such as sparsely populated Saudi Arabia), where the cost of getting fresh water by any method is high. However, desalinated water will probably never be cheap enough to irrigate conventional crops or to meet much of the world's demand for fresh water unless affordable, efficient solar-powered distillation plants can be developed and someone can figure out what to do with the resulting mountains of salt. Instead of spending less money to import more wheat, Saudi Arabia uses desalinated water to cultivate wheat in the desert at about seven times the world price per bushel.

Can Cloud Seeding and Towing Icebergs Improve Water Supplies? For years several countries, particularly the United States, have been experimenting with seeding clouds with tiny particles of chemicals (such as silver iodide) to form water condensation nuclei and thus produce more rain over dry regions and more snow over mountains. However, cloud seeding is not useful in very dry areas where it is most needed because rain clouds are rarely available there. Furthermore, widespread cloud seeding would introduce large amounts of the cloud-seeding chemicals into soil and water systems, possibly harming people, wildlife, and agricultural productivity.

Another obstacle to cloud seeding is legal disputes over the ownership of water in clouds. During the 1977 drought in the western United States, the attorney general of Idaho accused officials in neighboring Washington of "cloud rustling" and threatened to file suit in federal court.

There also have been proposals to tow massive icebergs to arid coastal areas (such as Saudi Arabia and southern California) and then to pump the fresh water from the melting bergs ashore. However, the technology for doing this is not available, and the costs may be too high, especially for water-short developing countries.

Solutions: Why Is Reducing Water Waste So Important? Increasing the water supply in some areas is important, but soaring population, food needs, and industrialization (and unpredictable shifts in water supplies) will eventually outstrip this approach. It makes much more sense, economically and environmentally, to use water more efficiently.

Mohamed El-Ashry of the World Resources Institute estimates that *65–70% of the water people use throughout the world is wasted through evaporation, leaks, and other losses*. The United States, the world's largest user of water, does slightly better but still loses about 50% of the water it withdraws. El-Ashry believes that it is economically and technically feasible to reduce water waste to 15%, thereby meeting most of the world's water needs for the foreseeable future.

Conserving water would have many other benefits, including reducing the burden on wastewater plants and septic systems, decreasing pollution of surface water and groundwater, reducing the number of expensive dams and water-transfer projects that destroy wildlife habitats and displace people, slowing depletion of groundwater aquifers, and saving energy and money needed to treat and distribute water.

Why Is So Much Water Wasted? A prime cause of water waste in the United States (and in most countries) is artificially low water prices resulting from government subsidies. Cheap water is the only reason farmers in Arizona and southern California can grow water-thirsty crops such as alfalfa in the middle of the desert. It also enables affluent people in desert areas to keep their lawns and golf courses green.

Water subsidies are paid for by all taxpayers through higher taxes. Because the harmful environmental and groundwater-depletion costs caused by these perverse subsidies don't show up on monthly water bills, consumers have little incentive to use less water or to adopt water-conserving devices and processes. Some analysts believe that sharply raising the price of federally subsidized water would encourage investments in improving water efficiency, and many of the West's water supply problems could be eased. However, so far Western members of the U.S. Congress have prevented such increases.

Another reason for the water waste in the United States (and many other countries) is that the responsibility for water resource management in a particular watershed may be divided among many state and local governments rather than being handled by one authority. The Chicago metropolitan area, for example, has 349 water-supply systems divided among some 2,000 local units of government over a six-county area.

In sharp contrast is the regional approach to water management used in England and Wales. The British Water Act of 1973 replaced more than 1,600 agencies with 10 regional water authorities based on natural watershed boundaries. Each water authority owns, finances, and manages all water supply and waste treatment facilities in its region. The responsibilities of each authority include water pollution control, water-based recreation, land drainage and flood control, inland navigation, and inland fisheries. Each water authority is managed by a group of elected local officials and a smaller number of officials appointed by the national government.

Solutions: How Can We Waste Less Water in Irrigation? Worldwide, about 65% of the water that is diverted from rivers or pumped out of aquifers is used for irrigating about 16% of the world's cropland. Because only about 40% of this water reaches crops, more efficient use of even a small amount of irrigation water would free large amounts of water for other uses.

Most irrigation systems distribute water from a groundwater well or a surface canal by downslope or gravity flow through unlined ditches in cropfields so that the water can be absorbed by crops (Figure 11-14, left). However, this flood irrigation method delivers far more water than needed for crop growth, and because of evaporation, deep percolation (seepage), and runoff, only 50–60% of the water reaches crops.

Seepage can be reduced by placing plastic, concrete, or tile liners in the large irrigation canals that distribute water to unlined irrigation ditches. Lasers can be used to make sure that fields are level so that water flowing into unlined irrigation ditches is distributed more evenly. Small check dams of earth and stone can capture runoff from hillsides and channel it to fields. Holding ponds can store rainfall or capture irrigation water for recycling to crops.

Many U.S. farmers served by the dwindling Ogallala Aquifer now use center-pivot sprinkler systems (Figure 11-14, right), with which 70–80% of the water reaches crops. Some of these farmers are switching to low-energy precision-application (LEPA) sprinklers. These systems bring 75–85% (some claim 85–95%) of the water to crops by spraying it closer to the ground and in larger droplets than the center-pivot system; they also reduce energy use and costs by 20–30%. However, because of the high initial costs, such sprinklers are used on only about 1% of the world's irrigated cropland.

Q: What percentage of earth's land area is covered by forests?

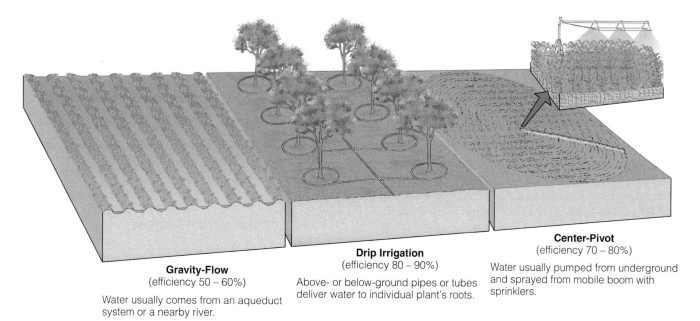

Gravity-Flow
(efficiency 50 – 60%)
Water usually comes from an aqueduct system or a nearby river.

Drip Irrigation
(efficiency 80 – 90%)
Above- or below-ground pipes or tubes deliver water to individual plant's roots.

Center-Pivot
(efficiency 70 – 80%)
Water usually pumped from underground and sprayed from mobile boom with sprinklers.

Figure 11-14 Major irrigation systems.

In the 1960s, highly efficient trickle or drip irrigation systems were developed in arid Israel. A network of perforated piping, installed at or below the ground surface, releases a trickle of water close to plant roots (Figure 11-14, center), minimizing evaporation and seepage and bringing 80–90% of the water to crops. Drip systems tend to be too expensive for most poor farmers and for use on low-value row crops. However, they are economically feasible for high-profit fruit, vegetable, and orchard crops, and for home gardens. Researchers are testing drip irrigation systems that cost one-third to one-tenth as much as current drip systems.

Irrigation efficiency can also be improved by computer-controlled systems that monitor soil moisture and provide water only when necessary. Farmers can also switch to more water-efficient, drought-resistant, and salt-tolerant crop varieties and use nighttime irrigation to reduce loss to evaporation. In addition, farmers can use organic farming techniques, which produce higher crop yields per hectare and require only one-fourth of the water and commercial fertilizer of conventional farming.

Since 1950 Israel has used many of these techniques to slash irrigation water waste by about 84% while irrigating 44% more land. The government also gradually removed most government subsidies to raise the price of irrigation water. Israel also imports most of its wheat and meat, concentrating on fruits, vegetables, and flowers that require less water. However, in other countries where government-funded water schemes and other subsidies provide artificially cheap water, farmers have little incentive to invest in water-saving techniques.

As fresh water becomes scarce and cities consume water once used for irrigation, carefully treated urban wastewater (which is rich in nitrate and phosphate plant nutrients) could be used for irrigation. Israel now treats and reuses 55% of its municipal sewage water, mostly for irrigating inedible crops such as cotton and flax, and plans to reuse 80% of this flow within the next few years.

Solutions: How Can We Waste Less Water in Industry, Homes, and Businesses? Manufacturing processes can use recycled water or be redesigned to save water. Japan and Israel lead the world in conserving and recycling water in industry. One paper mill in Hadera, Israel, uses one-tenth as much water as most other paper mills. Manufacturing aluminum from recycled scrap rather than from virgin ores can reduce water needs by 97%.

Nearly half of the water supplied by municipal water systems in the United States is used to flush toilets and water lawns, and another 15% is lost through leaky pipes. In a typical U.S. home, flushing toilets, washing hands, and bathing account for about 78% of the water used. In the arid western United States and in dry Australia, lawn and garden watering can take 80% of a household's daily usage, and much of this water is unnecessarily wasted.

Green lawns in an arid or semiarid area can be replaced with vegetation adapted to a dry climate—a form of landscaping called *xeriscaping*, from the

Greek word *xeros*, meaning "dry." A xeriscaped yard typically uses 30–80% less water than a conventional one. Drip irrigation can water gardens and other vegetation around homes.

More than half the water supply in Cairo, Lima, Mexico City, and Jakarta disappears before it can be used, mostly from leaks. In Cairo people often have to wade ankle deep across streets because of leaky water pipes. Leaky pipes, water mains, toilets, bathtubs, and faucets waste 20–35% of water withdrawn from public supplies in the United States and the United Kingdom.

Many cities offer no incentive to reduce leaks and waste. About one-fifth of all U.S. public water systems don't have water meters and charge a single low rate (often less than $100 a year for an average family) for virtually unlimited use of high-quality water. Many apartment dwellers have little incentive to conserve water because their water use is included in their rent.

In Boulder, Colorado, the introduction of water meters reduced water use by more than one-third. Tucson, Arizona, a desert city with ordinances that require conserving and reusing water, now consumes half as much water per capita as Las Vegas, a desert city where water conservation is still voluntary.

A California water utility gives rebates for water-saving toilets; it also distributed some 35,000 water-saving showerheads, cutting per capita water use 40% in only 1 year. On January 1, 1994, the Comprehensive Energy Act of 1992 required that all new toilets sold in the United States use no more than 6 liters (1.6 gallons) per flush. Similar laws have been passed in Mexico and in Ontario, Canada.

A low-flow showerhead costing about $20 saves about $34–$56 per year in water heating costs. Audits conducted in Brown University's environmental studies program showed that the school could save $44,000 a year by using low-flow showerheads in dormitories.

Two decades of experience with droughts in northern California has shown that water demand can be cut by more than 50% for homes, 60% for parks, and 20% for businesses without economic hardships. Savings in water use—called *negaliters* or *negagallons*—can also save money. For instance, a 1994 New York City study showed that providing a $240 rebate to customers who convert to water-saving toilets and showerheads would cost only one-third of the $8- to $12-billion price tag for developing new water supplies for the city's residents.

Used water (gray water) from bathtubs, showers, bathroom sinks, and clothes washers can be collected, stored, treated, and reused for irrigation and other purposes. California has become the first state to legalize reuse of gray water to irrigate landscapes. An estimated 50–75% of the water used by a typical house could be reused as gray water.

In some parts of the United States, people can lease systems that purify and completely recycle wastewater from houses, apartments, or office buildings. Such a system can be installed in a small outside shed and serviced for a monthly fee roughly equal to that charged by most city water-and-sewer-systems. In Tokyo, all the water used in Mitsubishi's 60-story office building is purified for reuse by an automated recycling system. Some actions you can take to waste less water are listed in Appendix 5.

11-4 TOO MUCH WATER

What Are the Causes and Effects of Flooding? Natural flooding by streams, the most common type of flooding, is caused primarily by heavy rain or rapid melting of snow; this causes water in the stream to overflow its normal channel and to cover the adjacent area, called a **floodplain** (Figure 11-15).

People have settled on floodplains since the beginnings of agriculture. The soil is fertile, and ample water is available for irrigation. Communities can use the water for transportation of people or goods, and floodplains (being flat) are suitable for cropland, buildings, highways, and railroads. People may decide that all of these benefits outweigh the risk of flooding (if they are even aware of the risk).

On marine coasts, flooding results most often from wind-driven storm surges and rain-swollen streams associated with tropical cyclones (typhoons and hurricanes). Flooding can also occur on the shorelines of large inland lakes.

Floods are a natural phenomenon and have several benefits. They provide the world's most productive farmland because they are regularly covered with nutrient-rich silt left after floodwaters recede. Floods also recharge groundwater under plains and refill wetlands, which help keep rivers flowing during droughts and provide important breeding and feeding grounds for fish and waterfowl.

Each year, flooding (Figure 11-15) kills thousands of people and causes tens of billions of dollars in property damage, including massive flooding in the United States in 1993 along the flood plains of the Mississippi, Missouri, and Illinois rivers (Figure 11-16).

Floods, like droughts, are usually considered natural disasters, but since the 1960s human activities have contributed to the sharp rise in flood deaths and damages. Indeed, floods account for about 39% of all deaths from natural disasters—more than any other type of such disasters.

One way humans increase the likelihood of flooding and the resulting damage is by removing

Q: At current loss rates, how long will it take to destroy most remaining tropical forests?

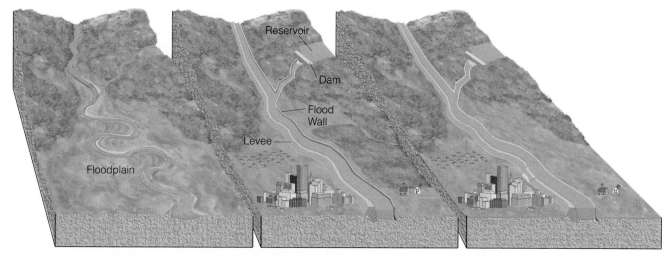

Figure 11-15 Land in a natural floodplain (left) is often flooded after prolonged rains. When the floodwaters recede, alluvial deposits of silt are left behind, creating a nutrient-rich soil. To reduce the threat of flooding (and thus allow people to live in floodplains), rivers have been dammed to create reservoirs, to store and release water as needed. They have also been narrowed, straightened, and equipped with protective levees and walls (middle). However, these alterations can give a false sense of security to floodplain dwellers, who actually live in high-risk areas. In the long run, such measures can greatly increase flood damage. Although dams, levees, and walls prevent flooding in most years, they can be overwhelmed by prolonged rains (right), as happened in the midwestern United States during the summer of 1993 (Figure 11-16).

water-absorbing vegetation, especially on hillsides (Figure 11-17); another is by living on floodplains. In many developing countries, the poor have little choice but to try to survive in flood-prone areas. In developed countries, however, people deliberately settle on floodplains and then expect dams, levees, and other devices to protect them from floodwaters—solutions that don't work when heavier-than-normal rains come (Figure 11-16).

Urbanization also increases flooding (even with only moderate rainfall) by replacing vegetation and soil with highways, parking lots, and buildings, which leads to rapid runoff of rainwater. If sea levels rise during the next century, as projected, many low-lying coastal cities, wetlands, and croplands will be under water.

Case Study: Living Dangerously in Bangladesh
Bangladesh (Figure 6-13) is one of the world's most densely populated countries, with 122 million people packed into an area roughly the size of Wisconsin. Its population is projected to reach 180 million by 2025. Bangladesh is also one of the world's poorest countries, with an average per capita GNP of about $240, or 66¢ per day.

More than 80% of the country consists of floodplains and shifting islands of silt formed by a delta at the mouths of three major rivers. Runoff from annual monsoon rains in the Himalaya Mountains of India, Nepal, Bhutan, and China flows down the rivers through Bangladesh into the Bay of Bengal.

The people of Bangladesh are used to moderate annual flooding during the summer monsoon season, and they depend on the floodwaters to grow rice, their primary source of food. The annual deposit of eroded Himalayan soil in the delta basin also helps maintain soil fertility.

In the past, great floods occurred every 50 years or so, but during the 1970s and 1980s they came about every 4 years. Bangladesh's flood problems begin in the Himalayan watershed, where a combination of rapid population growth, deforestation, overgrazing, and unsustainable farming on steep, easily erodible mountain slopes has greatly diminished the ability of the soil to absorb water. Instead of being absorbed and released slowly, water from the monsoon rains runs off the denuded Himalayan foothills, carrying vital topsoil with it (Figure 11-17). This runoff, combined with heavier-than-normal monsoon rains, has increased the severity of flooding along Himalayan rivers and in Bangladesh—another example of connections.

In 1988 a disastrous flood covered two-thirds of Bangladesh's land area for 3 weeks and leveled 2 million homes after the heaviest monsoon rains in 70 years. At least 2,000 people drowned, and 30 million people—one of every four—were left homeless. More than a quarter of the country's crops were destroyed, costing at least $1.5 billion and causing thousands of people to die of starvation.

In their struggle to survive, the poor in Bangladesh have cleared many of the country's coastal

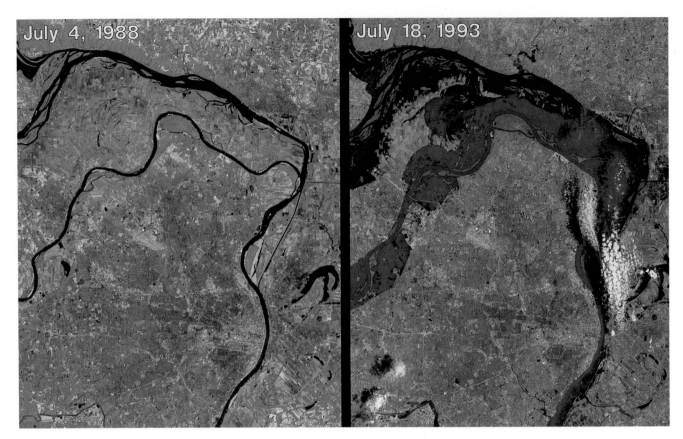

Figure 11-16 Satellite images of the area around St. Louis before flooding on July 4, 1988 (left), and on July 18, 1993, after severe flooding from prolonged rains. (Earth Satellite Corporation)

mangrove forests for fuelwood and for cultivation of crops. This has led to more severe flooding because these coastal wetlands shelter the low-lying coastal areas from storm surges and cyclones.

Sixteen devastating cyclones have slammed into Bangladesh since 1961; in 1970, as many as 1 million people drowned in one storm. Another surge killed an estimated 140,000 people in 1991. Damages and deaths from cyclones in areas still protected by mangrove forests are much lower than in areas where the forests have been cleared.

The severity of this problem can be reduced if Bangladesh, Bhutan, China, India, and Nepal cooperate in reforestation and flood control measures and reduce their population growth.

Solutions: How Can We Reduce Flooding Risks?

One controversial way of reducing flooding is *channelization*, in which a section of a stream is deepened, widened, or straightened to allow more rapid runoff (Figure 11-15, middle). Although channelization can reduce upstream flooding, the increased flow of water can also increase upstream bank erosion and downstream flooding and deposition of sediment.

Artificial levees and embankments along stream banks can reduce the chances of water overflowing into nearby floodplains. They may be permanent or temporary (such as sandbags placed when a flood is imminent). Levees, like channelization, contain and speed up stream flow and increase the water's capacity for doing damage downstream. If a levee breaks or the water spills over it, floodwater may persist long after the stream discharge has decreased (Figure 11-15, right).

Many people in the midwestern United States learned this harsh lesson during massive flooding from heavy rains during the summer of 1993 (Figure 11-16). After the floodwaters receded, a toxic sludge laced with pesticides, industrial wastes, and raw sewage was left on large areas of land along the floodplains of the Mississippi, Missouri, and Illinois rivers. Severe erosion from the 1993 flooding reduced crop yields on already heavily eroded fields.

A levee can contribute to increased destruction when major floods occur, but the destruction happens *downriver* from each levee. Thus, downriver landowners need levees to protect against upriver levees, which are needed to protect against flood crests caused by levees further upriver. The result is a spiraling *levee*

Q: Worldwide, how many trees are planted for every ten cut?

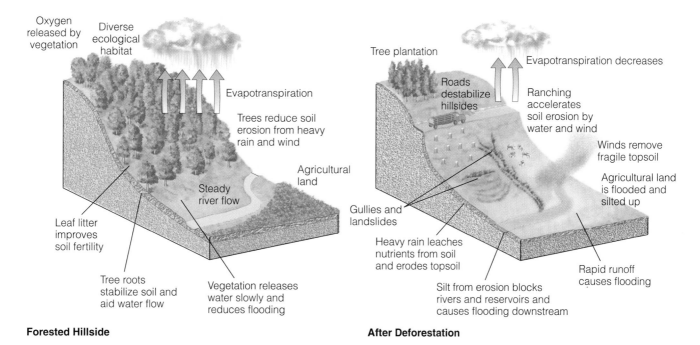

Oxygen released by vegetation

Diverse ecological habitat

Evapotranspiration

Trees reduce soil erosion from heavy rain and wind

Agricultural land

Steady river flow

Leaf litter improves soil fertility

Tree roots stabilize soil and aid water flow

Vegetation releases water slowly and reduces flooding

Forested Hillside

Tree plantation

Roads destabilize hillsides

Evapotranspiration decreases

Ranching accelerates soil erosion by water and wind

Winds remove fragile topsoil

Agricultural land is flooded and silted up

Gullies and landslides

Heavy rain leaches nutrients from soil and erodes topsoil

Silt from erosion blocks rivers and reservoirs and causes flooding downstream

Rapid runoff causes flooding

After Deforestation

Figure 11-17 A hillside before and after deforestation. Once a hillside has been deforested—for timber and fuelwood, grazing livestock, or unsustainable farming—water from precipitation rushes down the denuded slopes, eroding precious topsoil and flooding downstream areas. A 3,000-year-old Chinese proverb says, "To protect your rivers, protect your mountains."

race that can eventually cause destruction for almost everyone when the incredible force of a hugely flood-swollen river inevitably spills over or takes out levees.

A *flood control dam* built across a stream can hold back, store, and release water more gradually (Figure 11-15, middle). The dam and its reservoir may also provide such secondary benefits as hydroelectric power, water for irrigation, and recreational facilities. A reservoir can reduce floods only if its water level is kept low. However, it is much more profitable for dam operators to keep water levels high (for producing electricity and supplying irrigation water). As a result, after prolonged rains the reservoir can overflow, or operators may release large volumes of water to prevent overflow, thereby worsening the severity of flooding downstream. The reservoir gradually fills with sediment (that once fertilized downstream floodplains) until it is useless, and it also has other drawbacks (Figure 11-7).

Some flood control dams have failed for one reason or another, causing sudden, catastrophic flooding and threatening lives, property, and wildlife. According to the Federal Emergency Management Agency, the United States has about 1,300 unsafe dams in popu-lated areas, and the dam safety programs of most states are inadequate because of weak laws and budget cuts.

Over the past 65 years the U.S. Army Corps of Engineers has spent $25 billion on channelization, dams, and levees. Even so, the financial losses from flood damage have increased steadily during this period. The massive floods in the midwestern United States in 1993 caused an estimated $12 billion in damage to property and crops; billions more will have to be spent to repair the more than 800 levees and embankments that were topped or breached.

This disaster also demonstrated that dams, levees, and channelization can give a false sense of security and encourage people to settle on floodplains and thus worsen the severity of flood damage. Some countries are realizing that the risks of managing natural water flow can outweigh the benefits; Germany, for example, plans to bulldoze through dikes and allow parts of floodplains to flood regularly.

From an environmental viewpoint, *floodplain management* is the best approach. The first step is to construct a *flood-frequency curve* (based on historical records and an examination of vegetation) to determine how often *on average* a flood of a certain size occurs in a particular area. This doesn't tell us exactly when floods will occur, but it provides a general idea of how often they might occur, based on past history.

Using these data, a plan is developed **(1)** to prohibit certain types of buildings or activities in high-risk zones, **(2)** to elevate or otherwise floodproof buildings that are allowed on the legally defined floodplain, and **(3)** to construct a floodway that allows floodwater to flow through the community with minimal damage. Floodplain management based on thousands of years of experience can be summed up in one idea: *Sooner or later the river (or the ocean) always wins.*

In the United States, the Federal Flood Disaster Protection Act of 1973 requires local governments to adopt floodplain development regulations in order to be eligible for federal flood insurance. It also denies federal funding to proposed construction projects in designated flood hazard areas. The federal flood insurance program underwrites $185 billion in policies because private insurance companies are unwilling to fully insure people who live in flood-prone areas. This federal program thus can actually encourage people to build on floodplains and low-lying coastal areas.

Some economists argue that it would make more economic sense if the government would buy the 2% of the country's land that repeatedly floods, instead of continuing to make disaster payments. Others believe that people should be free to live in flood-prone areas, but that those choosing to do so should accept the risk without the assistance of federal flood insurance. Attempts to reduce or eliminate federal flood insurance coverage (or to restrict development on floodplains) are usually defeated politically by intense protests from property owners who already live in these risky areas.

11-5 POLLUTION OF STREAMS, LAKES, AND GROUNDWATER

What Are the Major Water Pollutants? **Water pollution** is any chemical, biological, or physical change in water quality that has a harmful effect on living organisms or makes water unsuitable for desired uses. There are several classes of water pollutants. One is *disease-causing agents (pathogens)*, which include bacteria, viruses, protozoa, and parasitic worms that enter water from domestic sewage and untreated human and animal wastes (Table 11-1). According to a 1995 World Bank study, contaminated water (lack of clean drinking water and lack of sanitation) causes about 80% of the diseases in developing countries and kills about 10 million people annually—an average of 27,000 premature deaths per day, more than half of them children under age 5.

A good indicator of the quality of water for drinking or swimming is the number of colonies of *coliform bacteria* present in a 100-milliliter (0.1-quart) sample of water. The World Health Organization recommends a coliform bacteria count of 0 colonies per 100 milliliters for drinking water, and the EPA recommends a maximum level for swimming water of 200 colonies per 100 milliliters. Because the average human excretes about 2 billion such organisms a day, we can see how easily untreated sewage can contaminate water.

A second category of water pollutants is *oxygen-demanding wastes*, organic wastes that can be decomposed by aerobic (oxygen-requiring) bacteria. Large populations of bacteria decomposing these wastes can degrade water quality by depleting water of dissolved oxygen (Figure 11-18), causing fish and other forms of oxygen-consuming aquatic life to die. The

Table 11-1 Common Diseases Transmitted to Humans Through Contaminated Drinking Water

Type of Organism	Disease	Effects
Bacteria	Typhoid fever	Diarrhea, severe vomiting, enlarged spleen, inflamed intestine; often fatal if untreated
	Cholera	Diarrhea, severe vomiting, dehydration; often fatal if untreated
	Bacterial dysentery	Diarrhea; rarely fatal except in infants without proper treatment
	Enteritis	Severe stomach pain, nausea, vomiting; rarely fatal
Viruses	Infectious hepatitis	Fever, severe headache, loss of appetite, abdominal pain, jaundice, enlarged liver; rarely fatal but may cause permanent liver damage
Parasitic protozoa	Amoebic dysentery	Severe diarrhea, headache, abdominal pain, chills, fever; if not treated can cause liver abscess, bowel perforation, and death
	Giardia	Diarrhea, abdominal cramps, flatulence, belching, fatigue
Parasitic worms	Schistosomiasis	Abdominal pain, skin rash, anemia, chronic fatigue, and chronic general ill health

Q: What percentage of the earth's species live in tropical forests?

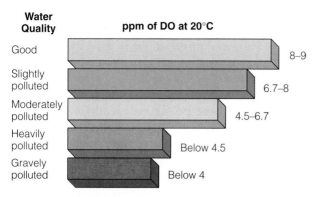

Figure 11-18 Water quality and dissolved oxygen (DO) content in parts per million (ppm) at 20°C (68°F). The solubility of oxygen decreases as the water temperature increases. Only a few species of fish can survive in water with fewer than 4 ppm of dissolved oxygen.

quantity of oxygen-demanding wastes in water can be determined by measuring the **biological oxygen demand (BOD)**: the amount of dissolved oxygen needed by aerobic decomposers to break down the organic materials in a certain volume of water over a 5-day incubation period at 20°C (68°F).

A third class of water pollutants is *water-soluble inorganic chemicals*, which include acids, salts, and compounds of toxic metals such as mercury and lead. High levels of these chemicals can make water unfit to drink, harm fish and other aquatic life, lower crop yields, and accelerate corrosion of metals exposed to such water.

Inorganic plant nutrients are another class of water pollutants. They are water-soluble nitrates and phosphates that can cause excessive growth of algae and other aquatic plants, which then die and decay, depleting water of dissolved oxygen and killing fish. Drinking water with excessive levels of nitrates lowers the oxygen-carrying capacity of the blood; this can kill unborn children and infants, especially those under 1 year old.

Water can also be polluted by a variety of *organic chemicals*, which include oil, gasoline, plastics, pesticides, cleaning solvents, detergents, and many other chemicals. They threaten human health and harm fish and other aquatic life.

By far the biggest class of water pollutants by weight is *sediment*, or *suspended matter*—insoluble particles of soil and other solids that become suspended in water, mostly when soil is eroded from the land. Sediment clouds water and reduces photosynthesis; it also disrupts aquatic food webs and carries pesticides, bacteria, and other harmful substances. Sediment that settles out destroys feeding and spawning grounds of fish. It also clogs and fills lakes, artificial reservoirs, stream channels, and harbors.

Water can also be polluted by *water-soluble radioactive isotopes*, some of which are concentrated or biologically magnified in various tissues and organs as they pass through food chains and webs (Figure 8-2). Ionizing radiation emitted by such isotopes can cause birth defects, cancer, and genetic damage.

Heat absorbed by water used to cool industrial and power plants can lower water quality. The resulting rise in water temperature, called *thermal pollution*, lowers dissolved oxygen levels and makes aquatic organisms more vulnerable to disease, parasites, and toxic chemicals.

Another form of water pollution, *genetic pollution*, occurs when aquatic systems are disrupted by the deliberate or accidental introduction of *nonnative species*. Some of these species can crowd out native species, reduce biodiversity, and cause economic losses.

What Are Point and Nonpoint Sources of Water Pollution? **Point sources** discharge pollutants at specific locations through pipes, ditches, or sewers into bodies of surface water. Examples include factories, sewage treatment plants (which remove some but not all pollutants), active and abandoned underground mines, offshore oil wells, and oil tankers. Because point sources are at specific places, they are fairly easy to identify, monitor, and regulate. In developed countries many industrial discharges are strictly controlled, whereas in most developing countries such discharges are largely uncontrolled.

Nonpoint sources are sources that cannot be traced to any single site of discharge. They are usually large land areas or airsheds that pollute water by runoff, subsurface flow, or deposition from the atmosphere. Examples include acid deposition (Figure 9-6), runoff of chemicals into surface water (including stormwater), and seepage into the ground from croplands, livestock feedlots, logged forests, streets, lawns, and parking lots.

In the United States, nonpoint pollution from agriculture—mostly in the form of sediment, inorganic fertilizers, manure, salts dissolved in irrigation water, and pesticides—is responsible for an estimated 64% of the total mass of pollutants entering streams and 57% of those entering lakes. According to the EPA, nonpoint runoff of stormwater causes 33% of all contamination in U.S. lakes and estuaries, and 10% of all stream contamination. Little progress has been made in controlling nonpoint water pollution because of the difficulty and expense of identifying and controlling discharges from so many diffuse sources.

What Are the Pollution Problems of Streams? Flowing streams, including large ones called *rivers*, can recover rapidly from degradable, oxygen-demanding wastes and excess heat by a combination of dilution and bacterial decay. This natural recovery process works as long as streams are not overloaded with these pollutants and as long as their flow is not reduced by

A: At least 50% (some say 90%)

drought, damming, or diversion for agriculture and industry. However, these natural dilution and biodegradation processes do not eliminate slowly degradable and nondegradable pollutants.

The breakdown of degradable wastes by bacteria depletes dissolved oxygen, which reduces or eliminates populations of organisms with high oxygen requirements until the stream is cleansed of wastes. The depth and width of the resulting *oxygen sag curve* (Figure 11-19) (and thus the time and distance required for a stream to recover) depend on the stream's volume, flow rate, temperature, pH level, and the volume of incoming degradable wastes. Similar oxygen sag curves can be plotted when heated water from industrial and power plants is discharged into streams.

What Progress Has Been Made in Reducing Stream Pollution?

Requiring cities to withdraw their drinking water downstream rather than upstream (as is done now) would dramatically improve water

quality because each city would be forced to clean up its own waste outputs rather than passing them downstream. However, upstream users, who already have the use of fairly clean water without high cleanup costs, fight this pollution prevention approach.

Here is some good news. Water pollution control laws enacted in the 1970s have greatly increased the number and quality of wastewater treatment plants in the United States and many other developed countries; laws have also required industries to reduce or eliminate point-source discharges into surface waters. These efforts, spurred by individuals (Individuals Matter, right), have enabled the United States to hold the line against increased pollution of most of its streams by disease-causing agents and oxygen-demanding wastes. This is an impressive accomplishment considering the rise in economic activity and population since the laws were passed.

One success story is the cleanup of Ohio's Cuyahoga River, which was so polluted that in 1959 and

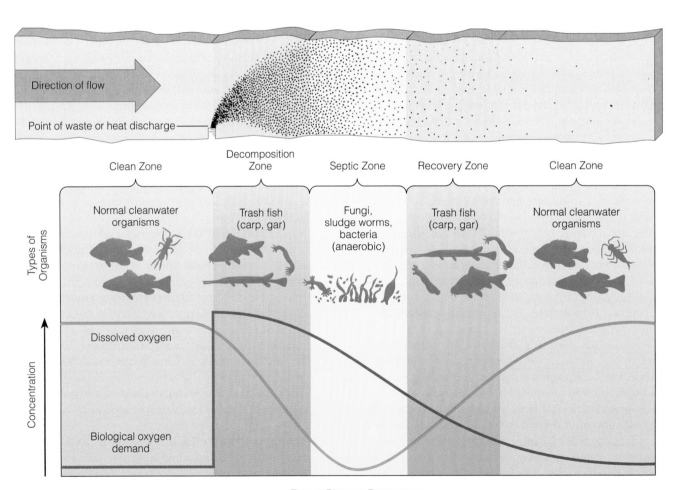

Figure 11-19 Dilution and decay of degradable, oxygen-demanding wastes and heat, showing the oxygen sag curve (orange) and the curve of oxygen demand (blue). Depending on flow rates and the amount of pollutants, streams recover from oxygen-demanding wastes and heat if they are given enough time and are not overloaded.

Q: What percentage of tropical forest plants have been studied for their possible use as human resources?

Rescuing a River

In the 1960s, Marion Stoddart (Figure 11-20) moved to Groton, Massachusetts, on the Nashua River, then considered one of the nation's filthiest rivers. Dead fish bobbed on its waves, and at times the water was red, green, or blue from pigments discharged by paper mills.

Instead of thinking nothing could be done, Marion Stoddart committed herself to restoring the Nashua and establishing public parklands along its banks. She didn't start by filing lawsuits or organizing demonstrations; instead she created a careful cleanup plan and approached state officials with it in 1962. They laughed, but she was not deterred and began practicing the most time-honored skill of politics: one-on-one persuasion. She identified the power brokers in the riverside communities and began to educate them, to win them over, and she got them to cooperate in cleaning up the river.

She got the state to ban open dumping in the river. When promised federal matching funds

for building the treatment plant failed to materialize, Stoddart gathered 13,000 signatures on a petition sent to President Nixon. The funds arrived in a hurry.

Stoddart's next success was getting a federal grant to beautify the river. She hired high school dropouts to clear away mounds of debris. When the river cleanup was completed, she persuaded communities along the river to create some 2,400 hectares (6,000 acres) of riverside park and woodlands along both banks.

Now, almost three decades later, the Nashua is still clean. Several new water treatment plants have been built, and a citizens' group founded by Stoddart keeps watch on water quality. The river supports many kinds of fish and other wildlife, and its waters are used for canoeing and other kinds of recreation.

For her efforts, Stoddart has been named by the UN Environment Programme as an outstanding worldwide worker for the environment. However, she might say that the blue and canoeable Nashua itself is her best reward.

Figure 11-20 Earth citizen Marion Stoddart canoeing down the Nashua River near Groton, Massachusetts. She spent over two decades spearheading successful efforts to have this river cleaned up.

again in 1969 it caught fire and burned for several days as it flowed through Cleveland, Ohio. The highly publicized image of this burning river prompted city, state, and federal officials to enact laws limiting the discharge of industrial wastes into the river and sewage systems and to appropriate funds to upgrade sewage treatment facilities. Today the river has made a comeback and is widely used by boaters and anglers.

Pollution control laws passed since 1970 have also led to improvements in dissolved oxygen content in many streams in Canada, Japan, and most western European countries. A spectacular cleanup has occurred in Great Britain. In the 1950s the Thames River was little more than a flowing anaerobic sewer, but after more than 30 years of effort, $250 million of British taxpayers' money, and millions more spent by industry, the Thames has made a remarkable recovery. Commercial fishing is thriving, and many species of waterfowl and wading birds have returned to their former feeding grounds.

What Is the Bad News About Stream Pollution?

Despite progress in improving stream quality in most developed countries, large fish kills and contamination of drinking water still occur. Most of these disasters are caused by accidental or deliberate releases of toxic inorganic and organic chemicals by industries, malfunctioning sewage treatment plants, and nonpoint runoff of pesticides and nutrients (eroded soil, fertilizer, and animal waste) from cropland, urban development, and animal feedlots (Individuals Matter, p. 320).

Available data indicate that stream pollution from huge discharges of sewage and industrial wastes is a serious and growing problem in most developing countries, where waste treatment is practically nonexistent. Numerous streams in the former Soviet Union and in eastern European countries are severely polluted. Currently, more than two-thirds of India's water resources are polluted with industrial wastes and sewage. Of the 78 streams monitored in China, 54 are seriously polluted with untreated sewage and

Dr. JoAnn M. Burkholder is an aquatic ecologist at North Carolina State University. She knows what it is to be crippled by a dangerous new type of microbe and to experience political heat for going public with your research to alert people to a serious health threat.

In 1986 she investigated why a colleague's laboratory research fish were dying mysteriously and discovered that the culprit was a new microbe she named *Pfiesteria* (pronounced fee-STEER-ee-ah) *piscida*. Since then she has been the world's leading researcher on this bizarre and deadly microbe.

She discovered that this microscopic organism can assume at least 24 forms in its lifetime as it moves from river bottoms to midwaters to feed on fish. Without suitable prey, the microbe can masquerade as a plant or lie dormant for years. Under certain conditions, it can change and release toxins that stun fish and shellfish in rivers and coastal estuaries, allowing the microbe to attack and often kill such species.

In 1993 Dr. Burkholder and her chief research aide experienced nausea, burning eyes, cramps, weakness, and severe loss of memory and mental powers from breathing toxic fumes released in tanks of fish dying from *Pfiesteria* attacks. They eventually recovered, but still cannot exercise strenuously without severe shortness of breath and respiratory illness.

She has found that the organism can live in waters as far north as the Delaware Bay and as far south as the Gulf of Mexico. With the proper nutrient environment and other conditions, it may be able to thrive throughout much of the world.

Dr. Burkholder believes that the increased nutrient flow from fertilized croplands, industrial development, and the recent huge growth in hog farms along some major North Carolina rivers flowing into coastal estuaries has led to outbreaks of *Pfiesteria*. The result has been repeated kills of millions of fish in the state's coastal estuaries, with the bodies of the dead fish covered with bleeding sores torn open by *Pfiesteria*.

She went public with her findings and urged state legislators to put curbs on hog farming and enact much tougher laws to reduce the flow of nutrients and other pollutants into the state's rivers. Hog farmers, developers, farming interests, fishing industry officials, tourism industry officials, and some state officials reacted negatively to her political activism. Some challenged her character and competence and accused her of using the results of preliminary research to push for questionable policies. She also received anonymous death threats.

Dr. Burkholder has not backed down and continues to criticize state health officials and legislators for not taking her concerns about public health seriously enough. In light of state and national publicity,* state officials have softened their public criticism.

With the help of increased state and federal research funds, she and her colleagues are trying to find out the chemical makeup of the toxin the organism releases, what causes it to transform itself into a fish killer, how it can affect people, and whether it is safe for people to eat fish and shellfish or swim in waters contaminated with these microbes.

Critical Thinking

1. What do you think should be done about this problem?

2. If it turns out to be a serious problem, should activities such as hog farming, commercial fishing, and ocean swimming be curtailed until bacterial levels are under control?

* For a popularized description of her research and political battle to alert the public and elected officials to the dangers posed by this microbe, see Rodney Barker's *And the Waters Turned to Blood: The Ultimate Biological Threat* (New York: Simon & Schuster, 1997).

industrial wastes, and 20% of China's rivers are too polluted to use for irrigation. In Latin America and Africa, most streams passing through urban or industrial areas are severely polluted.

What Are the Pollution Problems of Lakes?
In lakes, reservoirs, and ponds, dilution is often less effective than in streams. Lakes and reservoirs often contain stratified layers that undergo little vertical mixing, and ponds contain small volumes of water. Stratification also reduces levels of dissolved oxygen, especially in the bottom layer. In addition, lakes, reservoirs, and ponds have little flow, further reducing dilution and replenishment of dissolved oxygen. The flushing and changing of water in lakes and large artificial reservoirs can take from 1 to 100 years, compared with several days to several weeks for streams.

Thus, lakes, reservoirs, and ponds are more vulnerable than streams to contamination by plant nutrients, oil, pesticides, and toxic substances such as lead, mercury, and selenium. These contaminants can destroy both bottom life and fish and birds that feed on contaminated aquatic organisms. For example, selenium-contaminated water flowing from irrigated croplands into ponds and lakes in and near the Kesterson National Wildlife Refuge in California's San Joaquin Valley has killed thousands of waterfowl and fish and has affected populations of other forms of wildlife that feed on such species. Atmospheric fallout and runoff of acids are a serious problem in lakes vulnerable to acid deposition (Figure 9-6).

Concentrations of some chemicals, such as DDT (Figure 8-2), PCBs (Figure 11-21), some radioactive isotopes, and some mercury compounds can be biologically magnified as they pass through food webs in lakes. Many toxic chemicals also enter lakes and reservoirs from the atmosphere.

Lakes receive inputs of nutrients and silt from the surrounding land basin as a result of natural erosion and runoff. This natural nutrient enrichment of lakes is called **eutrophication**. Over time, some of these lakes become more eutrophic (Figure 5-25, top), but others don't because of differences in the surrounding water basin. Near urban or agricultural areas, human activities can greatly accelerate the input of nutrients to a lake, which results in a process known as **cultural**

eutrophication. Such a change is caused mostly by nitrate- and phosphate-containing effluents from sewage treatment plants, runoff of fertilizers and animal wastes, and accelerated erosion of nutrient-rich topsoil (Figure 11-22).

During hot weather or drought, this nutrient overload produces dense growths of organisms such as algae, cyanobacteria, water hyacinths, and duckweed. Dissolved oxygen (in both the surface layer of water near the shore and in the bottom layer) is depleted when large masses of algae die, fall to the bottom, and are decomposed by aerobic bacteria. This oxygen depletion can kill fish and other aquatic animals. If excess nutrients continue to flow into a lake, anaerobic bacteria take over and produce gaseous decomposition products such as smelly, highly toxic hydrogen sulfide and flammable methane.

About one-third of the 100,000 medium-to-large lakes and about 85% of the large lakes near major population centers in the United States suffer from some

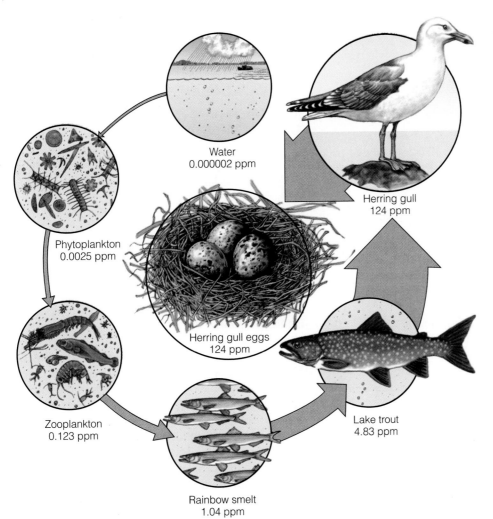

Figure 11-21 Biological magnification of PCBs (polychlorinated biphenyls) in an aquatic food chain in the Great Lakes. Most of the 209 different PCBs are insoluble in water, soluble in fats, and resistant to biological and chemical degradation—properties that result in their accumulation in the tissues of organisms and their biological amplification in food chains and webs. Although the long-term health effects on people exposed to low levels of PCBs are unknown, high doses of PCBs in laboratory animals produce liver and kidney damage, gastric disorders, birth defects, skin lesions, hormonal changes, smaller penis size, and tumors. Boys in Taiwan exposed to PCBs while in their mothers' wombs developed smaller penises than other Taiwanese boys. In the United States, manufacture and use of PCBs have been banned since 1976; before then, millions of metric tons of these long-lived chemicals were released into the environment, many of them ending up in bottom sediments of lakes, streams, and oceans.

Water
0.000002 ppm

Herring gull
124 ppm

Phytoplankton
0.0025 ppm

Herring gull eggs
124 ppm

Lake trout
4.83 ppm

Zooplankton
0.123 ppm

Rainbow smelt
1.04 ppm

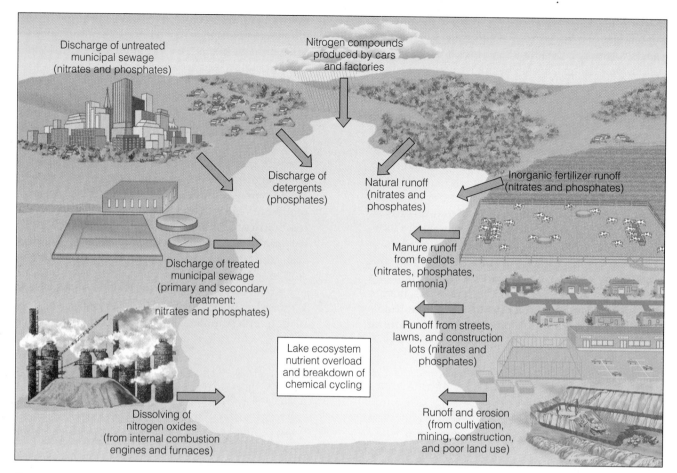

Figure 11-22 Principal sources of nutrient overload causing cultural eutrophication in lakes. The amount of nutrients from each source varies according to the types and amounts of human activities occurring in each airshed and watershed. Levels of dissolved oxygen (Figure 11-18) drop when enlarged populations of algae and plants (stimulated by increased nutrient input) die and are decomposed by aerobic bacteria. Lowered oxygen levels can kill fish and other aquatic life and reduce the aesthetic and recreational values of the lake.

degree of cultural eutrophication. One-fourth of China's lakes are classified as eutrophic.

Ways to *prevent* or reduce cultural eutrophication include advanced waste treatment (Section 11-7), bans or limits on phosphates in household detergents and other cleaning agents, and soil conservation and land-use control to reduce nutrient runoff.

Major *cleanup methods* are dredging bottom sediments to remove excess nutrient buildup, removing excess weeds, controlling undesirable plant growth with herbicides and algicides, and pumping air through lakes and reservoirs to avoid oxygen depletion (an expensive and energy-intensive method).

As usual, pollution prevention is more effective and usually cheaper in the long run than pollution control. If excessive inputs of limiting plant nutrients stop, a lake can usually return to its previous state.

Seattle's Lake Washington is a success story of recovery from severe eutrophication after decades of use as a sewage repository. The recovery took place after

the sewage was diverted into Puget Sound. This worked for three reasons. First, a large body of water (Puget Sound) was available to receive the sewage wastes. Second, the lake had not yet filled with weeds and sediment because of its large size and depth. Third, preventive corrective action was taken before the lake had become a shallow, highly eutrophic lake (Figure 5-25, top).

Case Study: Chemical and Genetic Pollution in the Great Lakes The five interconnected Great Lakes (Figure 11-23) contain at least 95% of the surface fresh water in the United States and 20% of the world's fresh surface water. The Great Lakes basin is home for about 38 million people, about 30% of the Canadian population and 13% of the U.S. population.

Despite their enormous size, these lakes are vulnerable to pollution from point and nonpoint sources because less than 1% of the water entering the Great Lakes flows out to the St. Lawrence River each year. In

Q: How much of the world's area of tropical forests is managed sustainably?

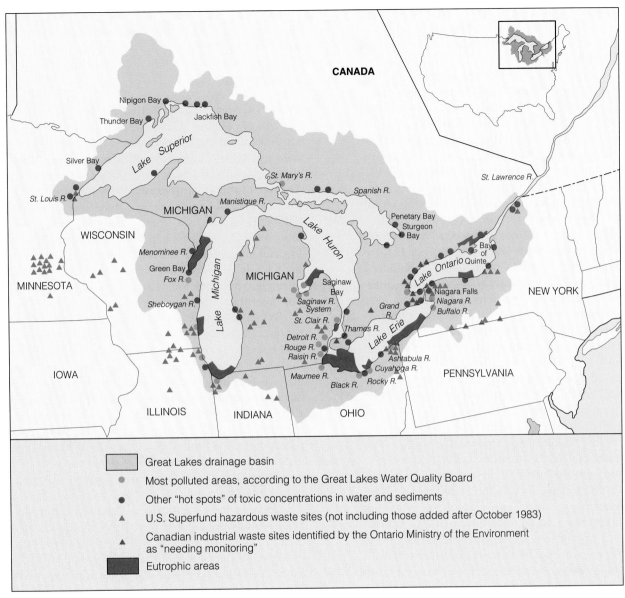

Figure 11-23 The Great Lakes basin and the locations of some of its water quality problems. (Data from Environmental Protection Agency)

Legend:
- Great Lakes drainage basin
- Most polluted areas, according to the Great Lakes Water Quality Board
- Other "hot spots" of toxic concentrations in water and sediments
- U.S. Superfund hazardous waste sites (not including those added after October 1983)
- Canadian industrial waste sites identified by the Ontario Ministry of the Environment as "needing monitoring"
- Eutrophic areas

addition to land runoff, these lakes receive large quantities of acids, pesticides, and other toxic chemicals by deposition from the atmosphere (often blown in from hundreds or thousands of kilometers away).

By the 1960s many areas of the Great Lakes were suffering from severe cultural eutrophication, huge fish kills, and contamination from bacteria and other wastes. The impact on Lake Erie was particularly intense because it is the shallowest of the Great Lakes. Many bathing beaches had to be closed, and by 1970 the lake had lost nearly all its native fish.

Since 1972, a $20-billion pollution-control program, carried out jointly by Canada and the United States, has significantly decreased levels of phosphates, coliform bacteria, and many toxic industrial chemicals in the Great Lakes. Algae blooms have also decreased, dissolved oxygen levels and sport and commercial fishing have increased, and most swimming beaches have reopened. These improvements were brought about mainly by new or upgraded sewage treatment plants, better treatment of industrial wastes, and banning of phosphate detergents, household cleaners, and water conditioners. Even so, less than 3% of the lakes' shoreline is clean enough for swimming or for supplying drinking water.

Levels of several toxic chlorinated hydrocarbon pollutants such as DDT and PCBs in Great Lakes water have dropped to their lowest levels in two decades. Despite this progress, contamination from toxic wastes flowing into the lakes (especially Lakes

A: About 0.1%

Erie and Ontario) from land runoff, streams, and atmospheric deposition (which accounts for an estimated 50% of the input of toxic compounds) is still a serious problem.

Toxic chemicals such as PCBs have built up in food chains and webs (Figure 11-21), contaminated many types of sport fish, and depleted populations of birds, river otters, and other animals feeding on contaminated fish. There is growing evidence and concern about possible effects of exposure to small amounts of many of these substances on the hormone systems of wildlife and humans (Case Study, p. 238). A survey by Wisconsin biologists revealed that one fish in four taken from the Great Lakes is unsafe for human consumption.

In 1991 the U.S. government passed a law requiring accelerated cleanup of the lakes, especially of 42 toxic hot spots, and an immediate reduction in air pollutant emissions in the region. However, meeting these goals may be delayed by a lack of federal and state funds.

Some environmentalists call for a ban on the use of chlorine as a bleach in the pulp and paper industry around the Great Lakes, a ban on all new incinerators in the area, and an immediate ban on discharge into the lakes of 70 toxic chemicals that threaten human health and wildlife. Understandably, officials of these industries strongly oppose such bans.

Great Lakes fisheries also face threats from *genetic pollution*. In 1986 larvae of a nonnative species, the *zebra mussel*, arrived in water discharged from a European ship near Detroit. With no known natural enemies, these tiny mussels have run amok; they deplete the food supply for other lake species, clog irrigation pipes, shut down water intake systems for power plants and city water supplies, foul beaches, and grow in huge masses on boat hulls, piers, and other surfaces.

Zebra mussels cost the Great Lakes basin at least $500 million per year, and the annual costs could reach $5 billion within a few years. The zebra mussel is expected to spread unchecked and dramatically alter most freshwater communities in parts of the United States and southern Canada within a few years, with damage costing tens of billions of dollars.

However, zebra mussels may be good news for a number of aquatic plants. By consuming algae and other microorganisms, the mussels increase water clarity. Clearer waters permit deeper penetration of sunlight and more photosynthesis, allowing some native plants to thrive and return the plant composition of Lake Erie (and presumably other lakes) closer to what it was 100 years ago. Because the plants provide food and increase dissolved oxygen, their comeback may benefit certain aquatic animals (including the mussels).

There is more bad news, however. In 1991 a larger and potentially more destructive species, the *quagga mussel*, invaded the Great Lakes, probably brought in by a Russian freighter. It can survive at greater depths and tolerate more extreme temperatures than the zebra mussel. There is concern that it may eventually colonize areas such as the Chesapeake Bay and waterways in parts of Florida.

Why Is Groundwater Pollution Such a Serious Problem? Highly visible oil spills get a lot of media attention, but a much greater threat to human health is the out-of-sight pollution of groundwater (Figure 11-24), a prime source of water for drinking and irrigation. This vital form of earth capital is easy to deplete and pollute because much of it is renewed so slowly. Although experts rate groundwater pollution as a low-risk ecological problem, they consider pollutants in drinking water (much of it from groundwater) a high-risk health problem (Figure 8-11, left). Laws protecting groundwater are weak in the United States and nonexistent in most countries.

When groundwater becomes contaminated, it cannot cleanse itself of degradable wastes, as surface water can if it is not overloaded. Because groundwater flows are slow and not turbulent, contaminants are not effectively diluted and dispersed. Groundwater also has much smaller populations of decomposing bacteria than surface-water systems, and its cold temperature slows down decomposition reactions. Thus, it can take hundreds to thousands of years for contaminated groundwater to cleanse itself of degradable wastes; nondegradable wastes are there permanently on a human time scale.

Crude estimates indicate that up to 25% of the usable groundwater in the United States is contaminated (and in some areas as much as 75%). In New Jersey, for example, every major aquifer is contaminated. In California, pesticides contaminate the drinking water of more than 1 million people. In Florida, where 92% of the residents rely on groundwater for drinking, over 1,000 wells have been closed. The EPA has documented groundwater contamination by 74 pesticides in 38 states.

Groundwater can be contaminated from a number of sources, including underground storage tanks, landfills, abandoned hazardous-waste dumps, deep wells used to dispose of liquid hazardous wastes, and industrial waste storage lagoons located above or near aquifers (Figure 11-24). An EPA survey found that one-third of 26,000 industrial waste ponds and lagoons have no liners to prevent toxic liquid wastes from seeping into aquifers, and one-third of those sites are within 1.6 kilometers (1 mile) of a drinking water well.

The EPA estimates that at least 1 million underground tanks storing gasoline, diesel fuel, and toxic solvents are leaking their contents into groundwater. A slow gasoline leak of just 4 liters (1 gallon) per day

Q: How many people cannot find or buy enough fuelwood to meet their basic needs?

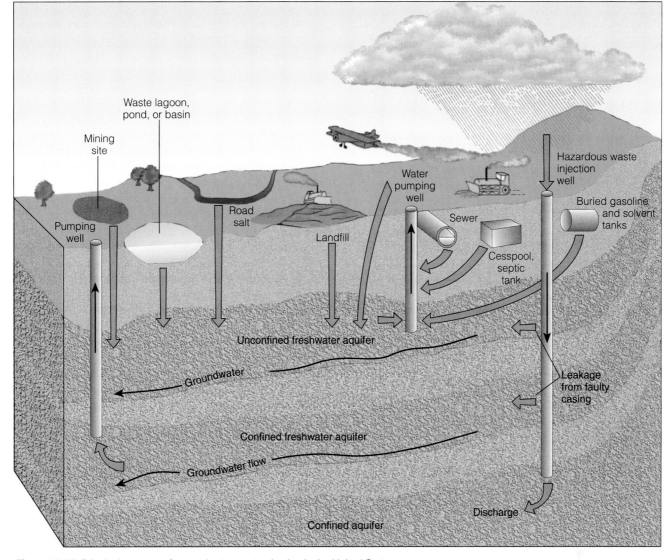

Figure 11-24 Principal sources of groundwater contamination in the United States.

can seriously contaminate the water supply for 50,000 people. Such slow leaks usually remain undetected until someone discovers that a well is contaminated.

Determining the extent of a leak can cost $25,000–250,000. Cleanup costs range from $10,000 for a small spill to $250,000 or more if the chemical reaches an aquifer, and complete cleanup is rarely possible. Replacing a leaking tank adds an additional $10,000–60,000. Legal fees and damages to injured parties can run into the millions. Current regulations should reduce leakage from new tanks in the United States but would do little about the millions of older tanks that are toxic time bombs.

Solutions: How Can We Protect Groundwater?
Pumping polluted groundwater to the surface, clean-

ing it up, and returning it to the aquifer is usually too expensive (for a single aquifer, $5 million or more). Recent attempts to pump and treat contaminated aquifers indicate that it may take 50–1,000 years of continuous pumping before all the contamination is forced to the surface and drinking-water quality is achieved. Thus, *preventing contamination by various means is considered the only effective way to protect groundwater resources.* Ways to do this include

- *Monitoring aquifers near landfills and underground tanks*

- *Requiring leak detection systems for existing and new underground tanks used to store hazardous liquids*

- *Requiring liability insurance for old and new underground tanks used to store hazardous liquids*

- *Banning or more strictly regulating disposal of hazardous wastes in deep injection wells and in landfills*

- *Storing hazardous liquids above ground in tanks with systems for detecting and collecting any leaks*

11-6 OCEAN POLLUTION

How Much Pollution Can the Oceans Tolerate?

The oceans are the ultimate sink for much of the waste matter we produce, as summarized in the African proverb, "Water may flow in a thousand channels, but it all returns to the sea."

Oceans can dilute, disperse, and degrade large amounts of raw sewage, sewage sludge, oil, and some types of industrial waste, especially in deep-water areas. Marine life has also proved to be much more resilient than some scientists had expected, leading some to suggest that it is generally safer to dump sewage sludge and most other hazardous wastes into the deep ocean than to bury them on land or burn them in incinerators.

Other scientists dispute this idea, pointing out that we know less about the deep ocean than we do about outer space. They add that dumping waste in the ocean would delay urgently needed pollution prevention and promote further degradation of this vital part of the earth's life-support system.

How Do Pollutants Affect Coastal Areas?

Coastal areas—especially wetlands and estuaries, coral reefs, and mangrove swamps—bear the brunt of our enormous inputs of wastes into the ocean. This is not surprising because half the world's population lives on or within 100 kilometers (160 miles) of the coast, and coastal populations are growing at a more rapid rate than global population. Fourteen of the world's 15 largest metropolitan areas, each with 10 million people or more, are near coastal waters.

In most coastal developing countries (and in some coastal developed countries), municipal sewage and industrial wastes are often dumped into the sea without treatment. The most polluted seas lie off the densely populated coasts of Bangladesh, India, Pakistan, Indonesia, Malaysia, Thailand, and the Philippines. About 85% of the sewage from large cities along the Mediterranean Sea, which has a coastal population of 200 million people during tourist season, is discharged into the sea untreated, causing widespread beach pollution and shellfish contamination.

In the United States about 35% of all municipal sewage ends up virtually untreated in marine waters. Most U.S. harbors and bays are badly polluted from municipal sewage, industrial wastes, and oil.

Each year at least one-third of the area of U.S. coastal waters around the lower 48 states is closed to shellfish harvesters because of pollution and habitat disruption.

California's Santa Monica Bay is the filming site for the widely watched television show *Baywatch*, which gives an illusion of a clean California beach lifestyle. What viewers don't know is that the bay is so polluted that the actors get extra pay each time they enter the water and are chemically cleaned afterward.

Runoff of sewage and agricultural wastes into coastal waters introduces large quantities of nitrogen and phosphorus, which can cause explosive growth of algae. These algal blooms—called red, brown, or green tides, depending on their color—damage fisheries, reduce tourism, and poison seafood. When the algae die and decompose, coastal waters are depleted of oxygen and a variety of marine species die.

Case Study: The Chesapeake Bay The Chesapeake Bay, the largest estuary in the United States, is in trouble because of human activities. Between 1940 and 1996, the number of people living in the Chesapeake Bay area grew from 3.7 million to 15.5 million, and within a few years its population may reach 18 million.

The estuary receives wastes from point and non-point sources scattered throughout a huge drainage basin that includes 9 large rivers and 141 smaller streams and creeks in parts of six states (Figure 11-25). The bay has become a huge pollution sink because it is quite shallow and because only 1% of the waste entering it is flushed into the Atlantic Ocean.

Levels of phosphates and nitrates have risen sharply in many parts of the bay, causing algae blooms and oxygen depletion (Figure 11-25). Studies have shown that point sources, primarily sewage treatment plants, contribute about 60% by weight of the phosphates. Nonpoint sources—mostly runoff from urban, suburban, and agricultural land and deposition from the atmosphere—are the origins of about 60% by weight of the nitrates.

Air pollutants account for nearly 35% of the nitrogen entering the estuary. In addition, large quantities of pesticides run off cropland and urban lawns, and industries discharge large amounts of toxic wastes, often in violation of their discharge permits. Commercial harvests of oysters, crabs, and several important fish have fallen sharply since 1960 because of a combination of overfishing, pollution, and disease.

In the 1980s the Chesapeake Bay Program, the country's most ambitious attempt at integrated coastal management, was implemented. Results have been impressive. Since 1983 more than $700 million in

Q: By 2000, how many people may not be able to get enough fuelwood?

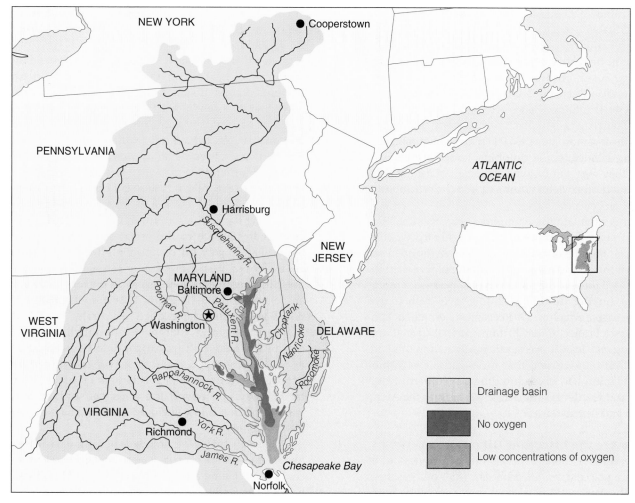

Figure 11-25 Chesapeake Bay, the largest estuary in the United States, is severely degraded as a result of water pollution from point and nonpoint sources in six states and from deposition of air pollutants.

state and federal funds have been spent on a Chesapeake Bay cleanup program that will ultimately cost several billion dollars. Between 1987 and 1993, phosphorus levels declined 16% through a combination of bans on phosphorus-containing detergents, upgrades of municipal sewage treatment plants, and soil erosion controls and nutrient management on agricultural land. During this same period nitrogen levels dropped 7%—a significant achievement given the increasing population in the watershed and the fact that more than a third of the nitrogen inputs come from the atmosphere.

However, a greater effort will be required to reach the goals of a 40% reduction in nutrient levels and a significant improvement in habitat water quality throughout the bay. This will be especially difficult because the area's population is expected to grow by 25% between 1995 and 2020. Moreover, the bay will

soon be invaded by zebra and quagga mussels. So far, however, the Chesapeake Bay Program shows what can be done when interested parties work together to achieve goals that benefit both wildlife and people.

What Pollutants Are Dumped into the Ocean?
Dumping of industrial waste off U.S. coasts has stopped, although it still occurs in a number of other developed countries and some developing countries. However, barges and ships still legally dump large quantities of **dredge spoils** (materials, often laden with toxic metals, scraped from the bottoms of harbors and rivers to maintain shipping channels) at 110 sites off the Atlantic, Pacific, and Gulf coasts.

In addition, many countries dump into the ocean large quantities of **sewage sludge**: a gooey mixture of toxic chemicals, infectious agents, and settled solids removed from wastewater at sewage treatment plants. Since 1992 this practice has been banned in the United States.

Fifty countries with at least 80% of the world's merchant fleet have agreed not to dump sewage and garbage at sea, but this agreement is difficult to enforce and is often violated. Most ship owners save money by dumping wastes at sea and risk only small fines if they are caught. Each year as many as 2 million seabirds and more than 100,000 marine mammals (including whales, seals, dolphins, and sea lions) die when they ingest or become entangled in fishing nets, ropes, and other debris dumped into the sea and discarded on beaches.

Under the London Dumping Convention of 1972, 100 countries agreed not to dump highly toxic pollutants and high-level radioactive wastes in the open sea beyond the boundaries of their national jurisdiction. Since 1983 these same nations have observed a moratorium on the dumping of low-level radioactive wastes at sea, which in 1994 became a permanent ban. However, France, Great Britain, Russia, China, and Belgium may legally exempt themselves from this ban. In 1992 it was learned that for decades the former Soviet Union had been dumping large quantities of high- and low-level radioactive wastes into the Arctic Ocean and its tributaries.

What Are the Effects of Oil on Ocean Ecosystems?

Crude petroleum (oil as it comes out of the ground) and *refined petroleum* (fuel oil, gasoline, and other processed petroleum products) are accidentally or deliberately released into the environment from a number of sources.

Although tanker accidents and blowouts at offshore drilling rigs (when oil escapes under high pressure from a borehole in the ocean floor) get most of the publicity, more oil is released during normal operation of offshore wells, from washing tankers and releasing the oily water, and from pipeline and storage tank leaks. A 1993 Friends of the Earth study estimated that each year U.S. oil companies unnecessarily spill, leak, or waste an amount of oil equal to that shipped by 1,000 huge *Exxon Valdez* tankers—more oil than Australia uses. Oil pollution from shipping in the Mediterranean Sea is equivalent to 17 huge *Exxon Valdez* tankers emptying their tanks per year.

Natural oil seeps also release large amounts of oil into the ocean at some sites, but most ocean oil pollution comes from activities on land. Almost half (some experts estimate 90%) of the oil reaching the oceans is waste oil dumped, spilled, or leaked onto the land or into sewers by cities, individuals, and industries. Each year, a volume of oil equal to 20 times the amount spilled by the *Exxon Valdez* is improperly disposed of by about 50 million U.S. motorists changing their own motor oil. Worldwide, about 10% of the oil that reaches the ocean comes from the atmosphere, mostly from smoke emitted by oil fires.

The effects of oil on ocean ecosystems depend on a number of factors: type of oil (crude or refined), amount released, distance of release from shore, time of year, weather conditions, average water temperature, and ocean currents. Volatile organic hydrocarbons in oil immediately kill a number of aquatic organisms, especially in their vulnerable larval forms. Some other chemicals form tarlike globs that float on the surface.

This floating oil coats the feathers of birds (especially diving birds) and the fur of marine mammals, destroying the animals' natural insulation and buoyancy; many drown or die of exposure from loss of body heat. Heavy oil components that sink to the ocean floor or wash into estuaries can smother bottom-dwelling organisms such as crabs, oysters, mussels, and clams or make them unfit for human consumption. Some oil spills have killed reef corals.

Research shows that most (but not all) forms of marine life recover from exposure to large amounts of crude oil within 3 years. However, recovery from exposure to refined oil, especially in estuaries, may take 10 years or longer. The effects of spills in cold waters and in shallow enclosed gulfs and bays generally last longer.

Oil slicks that wash onto beaches can have a serious economic impact on coastal residents, who lose income from fishing and tourist activities. Oil-polluted beaches washed by strong waves or currents become clean after about a year, but beaches in sheltered areas remain contaminated for several years. Estuaries and salt marshes suffer the most and longest-lasting damage. Despite their localized harmful effects, oil spills are rated by experts as a low-risk ecological problem (Figure 8-11, left).

Case Study: The *Exxon Valdez* Oil Spill

Crude oil from Alaska's North Slope fields near Prudhoe Bay is carried by pipeline to the port of Valdez and then shipped by tanker to the West Coast. On March 24, 1989, the *Exxon Valdez*, a tanker more than three football fields long, went off course in a 16-kilometer-wide (11-mile-wide) channel in Prince William Sound near Valdez, Alaska. It hit submerged rocks, creating the worst oil spill ever in U.S. waters (Figure 11-26).

The rapidly spreading oil slick coated more than 1,600 kilometers (1,000 miles) of shoreline, almost the length of the shoreline between New Jersey and South Carolina. The full loss of wildlife will never be known because most of the dead animals sank and decomposed without being counted.

In 1994 a jury awarded members of the fishing industry, landowners, and other Alaska residents $5 bil-

Q: What percentage of the original old-growth forests in the United States and Canada have been cut?

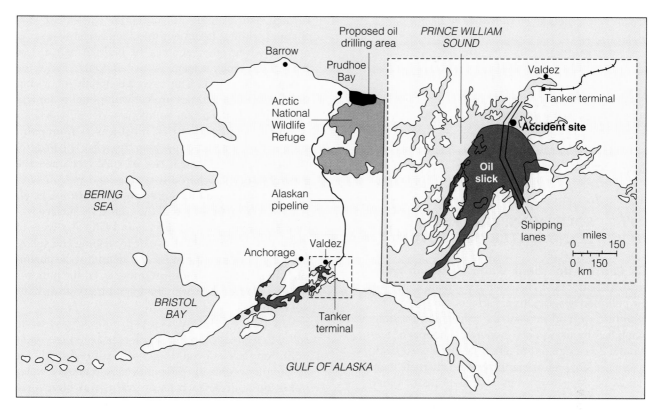

Figure 11-26 The location of Alaska's Prince William Sound, the site of the oil spill from the tanker *Exxon Valdez* on March 24, 1989.

lion in damages and penalties. However, Exxon has appealed the decision and it may be tied up in the courts for decades.

This roughly $8.5-billion accident might have been prevented if Exxon had spent only $22.5 million to fit the tanker with a double hull (which it still has not done). In the early 1970s interior secretary Rogers Morton told Congress that all oil tankers using Alaskan waters would have double hulls, but under pressure from oil companies the requirement was dropped.

Today virtually all merchant ships have double hulls, but only 15% of oil tankers have such hulls. Legislation passed since the spill requires all new tankers to have double hulls and all existing large single-hulled oil tankers to be phased out between 1995 and 2015. However, the oil industry is working to weaken these and other stricter requirements enacted since the spill as the public memory of the accident fades. Oil company officials also hope to return the repaired but still single-hulled *Exxon Valdez* tanker (under a new name) to operation in Prince William Sound.

This spill highlighted the importance of pollution prevention and the advantages of shifting to improved energy efficiency and renewable energy (Chapter 18) to reduce dependence on oil. Even with the best technology and a fast response by well-trained people, sci-

entists estimate that no more than 11–15% of the oil from a major spill can be recovered.

Solutions: How Can We Protect Coastal Waters?
The key to protecting oceans is to reduce the flow of pollution from the land and from streams emptying into the ocean. Such efforts must also be integrated with efforts to prevent and control air pollution because an estimated 33% of all pollutants entering the ocean worldwide comes from air emissions from land-based sources.

Some ways various analysts have suggested to prevent and reduce excessive pollution of coastal waters are

Prevention

■ *Encourage or require separate sewage and storm runoff lines in coastal urban areas*

■ *Discourage ocean dumping of sludge and hazardous dredged materials*

■ *Protect sensitive and ecologically valuable coastal areas from development, oil drilling, and oil shipping*

■ *Use ecological land-use planning to control and regulate coastal development*

- *Require double hulls for all oil tankers by 2002*
- *Recycle used oil*

Cleanup

- *Improve oil-spill cleanup capabilities*
- *Require at least secondary treatment of coastal sewage, or use wetlands, solar aquatic, or other environmentally acceptable methods*

11-7 SOLUTIONS: PREVENTING AND REDUCING WATER POLLUTION

What Can We Do About Water Pollution from Nonpoint Sources? The leading nonpoint source of water pollution is agriculture. Farmers can sharply reduce fertilizer runoff into surface waters and leaching into aquifers by using only moderate amounts of fertilizer and by using none at all on steeply sloped land. They can use slow-release fertilizers and alternate their plantings between row crops and soybeans or other nitrogen-fixing plants to reduce the need for fertilizer. Farmers can also plant buffer zones of permanent vegetation between cultivated fields and nearby surface water.

Applying pesticides only when needed can reduce pesticide runoff and leaching. Farmers can also reduce the need for pesticides by using biological control or integrated pest management (Section 15-5). Nonfarm uses of inorganic fertilizers and pesticides—on golf courses, lawns, and public lands, for example—can also be sharply reduced and replaced with organic methods.

Livestock growers can control runoff and infiltration of manure from feedlots and barnyards by managing animal density, planting buffers, and not locating feedlots on land near surface water when that land slopes steeply toward the water. Diverting the runoff into well-designed detention basins would allow this nutrient-rich water to be pumped out and applied as fertilizer to cropland or forestland.

Another way to reduce nonpoint water pollution, especially from eroded soil, is to reforest critical watersheds. Besides reducing water pollution from sediments, reforestation would reduce soil erosion and the severity of flooding; it would also help slow projected global warming and loss of wildlife habitat.

What Can We Do About Water Pollution from Point Sources? The Legal Approach In many developing countries and in some developed countries,

sewage and waterborne industrial wastes are discharged without treatment into the nearest waterway or into wastewater lagoons. In Latin America, less than 2% of urban sewage is treated. Only 15% of the urban wastewater in China receives treatment, and in India treatment facilities protect water quality for less than a third of the urban population.

In developed countries, most wastes from point sources are purified to varying degrees. The Federal Water Pollution Control Act of 1972 (renamed the Clean Water Act when it was amended in 1977) and the 1987 Water Quality Act form the basis of U.S. efforts to control pollution of the country's surface waters. The main goals of the Clean Water Act were to make all U.S. surface waters safe for fishing and swimming by 1983 and to restore and maintain the chemical, physical, and biological integrity of the nation's waters. Progress has been made, but these goals have not been met.

In 1995 the EPA developed a *discharge trading policy* designed to use market forces to reduce water pollution (as has been done with sulfur dioxide for air pollution control). The policy would allow a water pollution source, such as an industrial plant or a sewage treatment plant, to sell credits for its excess reductions to another facility that can't reduce its discharges as cheaply.

The Clean Water Act of 1972 has led to significant improvements in U.S. water quality between 1972 and 1992, The percentage of U.S. rivers and lakes tested that have become fishable and swimmable increased from 36% to 62%.

On the other hand, about 44% of lakes, 37% of rivers, and 32% of estuaries in the United States are still unsafe for fishing, swimming, and other recreational uses. Fish caught in more than 1,400 different waterways are unsafe to eat because of high levels of pesticides and other toxic substances.

Some environmentalists call for the Clean Water Act to be strengthened by **(1)** increasing funding and the authority to control nonpoint sources of pollution; **(2)** strengthening programs to prevent and control toxic water pollution, including phasing out certain toxic discharges (such as many organic chemicals containing chlorine); **(3)** providing more funding and authority for integrated watershed and airshed planning to protect groundwater and surface water from contamination; **(4)** permitting states with good records of environmental stewardship to take over parts of the clean water program, under looser federal control; and **(5)** expanding the ability of citizens to bring lawsuits to ensure that water pollution laws are enforced.

Many industries, state and local officials, and farmers and developers oppose these proposals, contending

Q: What percentage of all U.S. land consists of public lands?

that the Clean Water Act's regulations are already too restrictive and costly. Farmers and developers also see the law as a curb on their rights as property owners to fill in wetlands; they believe that they should be compensated for property value losses because of federal wetland protection regulations. State and local officials want more discretion in testing for and meeting water quality standards. They argue that in many communities it is too expensive and unnecessary to test for all the water pollutants required by federal law.

What Can We Do About Water Pollution from Point Sources? The Technological Approach

In rural and suburban areas with suitable soils, sewage from each house is usually discharged into a **septic tank** (Figure 11-27). About 25% of all homes in the United States are served by septic tanks, which should be cleaned out every 3–5 years by a reputable contractor so that they won't contribute to groundwater pollution.

In urban areas, most waterborne wastes from homes, businesses, factories, and storm runoff flow through a network of sewer pipes to wastewater treatment plants. Some cities have separate lines for stormwater runoff, but in 1,200 U.S. cities the lines for these two systems are combined because it is cheaper. When rains cause combined sewer systems to overflow, they discharge untreated sewage directly into surface waters.

When sewage reaches a treatment plant, it can undergo up to three levels of purification, depending on the type of plant and the degree of purity desired.

Primary sewage treatment is a mechanical process that uses screens to filter out debris such as sticks, stones, and rags; suspended solids settle out as sludge in a settling tank (Figure 11-28). Improved primary treatment uses chemically treated polymers to remove suspended solids more thoroughly.

Secondary sewage treatment is a biological process in which aerobic bacteria are used to remove up to 90% of biodegradable, oxygen-demanding organic wastes (Figure 11-28). Some plants use *trickling filters*, in which aerobic bacteria degrade sewage as it seeps through a bed of crushed stones covered with bacteria and protozoa. Others use an *activated sludge process*, in which the sewage is pumped into a large tank and mixed for several hours with bacteria-rich sludge and air bubbles to facilitate degradation by microorganisms. The water then goes to a sedimentation tank, where most of the suspended solids and microorganisms settle out as sludge. The sludge produced from primary or secondary treatment is broken down in an anaerobic digester and either incinerated, dumped into the ocean or a landfill, or applied to land as fertilizer.

Even after secondary treatment, wastewater still contains about 3–5% by weight of the original oxygen-demanding wastes, 3% of the suspended solids, 50% of the nitrogen (mostly as nitrates), 70% of the phosphorus (mostly as phosphates), and 30% of most toxic metal compounds and synthetic organic chemicals. Virtually no long-lived radioactive isotopes or persistent organic substances such as pesticides are removed.

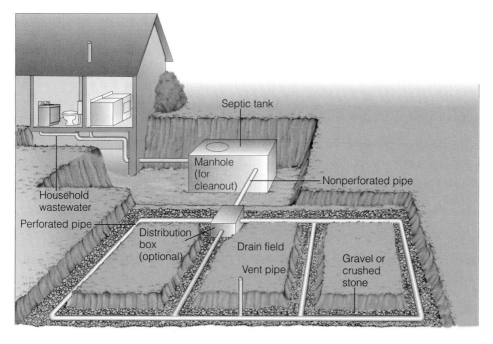

Figure 11-27 Septic tank system used for disposal of domestic sewage and wastewater in rural and suburban areas. This system traps greases and large solids and discharges the remaining wastes over a large drainage field. As these wastes percolate downward, the soil filters out some potential pollutants, and soil bacteria decompose biodegradable materials. To be effective, septic tank systems must be properly installed in soils with adequate drainage, not placed too close together or too near well sites, and pumped out when the settling tank becomes full.

A: About 42% (35% managed by the federal government)

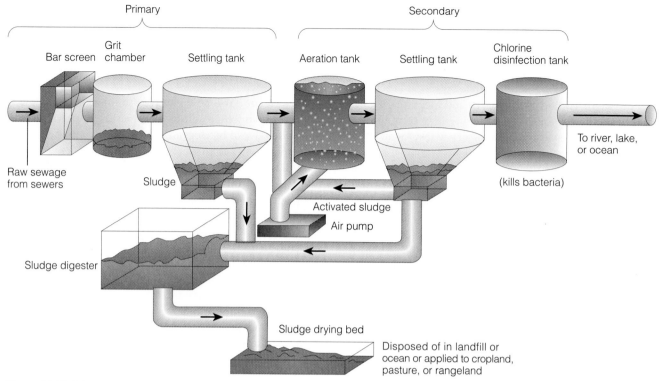

Primary

Secondary

Bar screen | Grit chamber | Settling tank | Aeration tank | Settling tank | Chlorine disinfection tank

Raw sewage from sewers

Sludge

To river, lake, or ocean

(kills bacteria)

Activated sludge

Air pump

Sludge digester

Sludge drying bed

Disposed of in landfill or ocean or applied to cropland, pasture, or rangeland

Figure 11-28 Primary and secondary sewage treatment.

As a result of the Clean Water Act, most U.S. cities have secondary sewage treatment plants. In 1989, however, the EPA found that more than 66% of sewage treatment plants have either water-quality or public-health problems, and studies by the General Accounting Office have shown that most industries have violated regulations. Moreover, 500 cities have failed to meet federal standards for sewage treatment plants, and 34 East Coast cities simply screen out large floating objects from their sewage before discharging it into coastal waters.

Advanced sewage treatment is a series of specialized chemical and physical processes that remove specific pollutants left in the water after primary and secondary treatment (Figure 11-29). Types of advanced treatment vary according to the specific contaminants to be removed. Without advanced treatment, sewage treatment plant effluents contain enough nitrates and phosphates to contribute to accelerated eutrophication of lakes, slow-moving streams, and coastal waters. Advanced treatment is rarely used because such plants typically cost twice as much to build and four times as much to operate as secondary plants.

Before water is discharged after primary, secondary, or advanced treatment, it is bleached (to remove water coloration) and disinfected (to kill disease-carrying bacteria and some but not all viruses). The usual method for doing this is *chlorination*. However, chlorine can react with organic materials in water to form small amounts of chlorinated hydrocarbons, some of which cause cancers in test animals. According to some preliminary research in 1992, chlorinated drinking water may cause 7–10% of all cancers in the United States. There is also growing evidence that some chlorinated hydrocarbons may damage the human nervous, immune, and endocrine systems (Section 8-3). Other disinfectants such as ozone and ultraviolet light are used in some places, but they cost more than chlorination and are not as long-lasting.

Sewage treatment produces a toxic, gooey sludge that must be disposed of or recycled as fertilizer. About 54% by weight of all municipal sludge produced in the United States is applied to farmland, forests, highway medians, and degraded land as fertilizer, and 9% is composted. The rest is dumped in conventional landfills (where it can contaminate groundwater) or incinerated (which can pollute the air with traces of toxic chemicals, and the resulting toxic ash is usually buried in landfills that EPA experts say will eventually leak).

Q: Where is most of the federally owned and managed land in the United States?

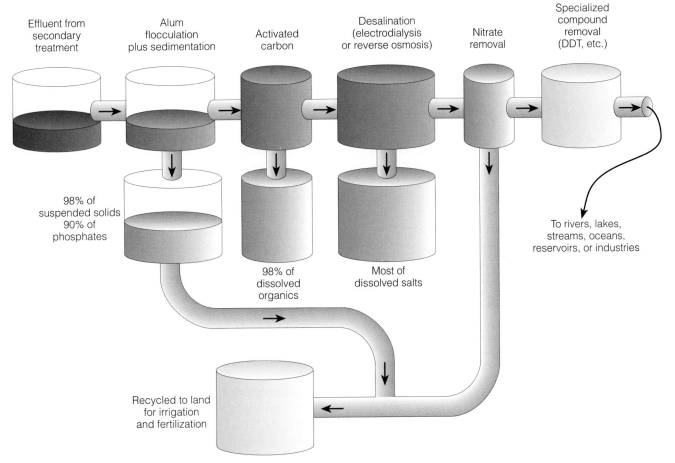

Figure 11-29 Advanced sewage treatment. Often only one (or two) of these processes is used to remove specific pollutants in a particular area. This expensive method is not widely used.

Before it is applied to land, sewage sludge can be heated to kill harmful bacteria, as is done in Switzerland and parts of Germany; it can also be treated to remove toxic metals and organic chemicals before application, but such treatment can be expensive. (The best and cheapest solution is to prevent these toxics from reaching sewage treatment plants.) Untreated sludge can be applied to land not used for crops or livestock, such as forests, surface-mined land, golf courses, lawns, cemeteries, and highway medians.

Since 1992, when Congress banned ocean dumping of sludge. its use as a fertilizer has increased sharply. However, most sewage sludge used as fertilizer in the United States is not adequately treated to kill harmful bacteria and remove toxic metals and organic chemicals, which may end up in food products or leach into groundwater. A growing number of health problems and lawsuits have resulted from use of sludge to fertilize crops.

Solutions: Working with Nature to Purify Sewage Some communities and individuals are seeking better ways to purify contaminated water by working with nature. A low-tech, low-cost alternative to expensive waste treatment plants is to create an artificial wetland, as the residents of Arcata, California, did.

In this coastal town of 17,000, some 63 hectares (155 acres) of wetlands has been created between the town and the adjacent Humboldt Bay. The marshes, developed on land that was once a dump, act as an inexpensive, natural waste treatment plant. The project was completed in 1974 for $3 million less than the estimated cost of a conventional treatment plant.

Here's how it works: First, sewage is held in sedimentation tanks, where the solids settle out as sludge that is removed and processed for use as fertilizer. The liquid is pumped into oxidation ponds, where remaining wastes are broken down by bacteria. After a month or so, the water is released into the artificial marshes,

A: Alaska (73%) and the western states (22%)

CHAPTER 11　　333

Figure 11-30 At the Providence, Rhode Island, Solar Sewage Plant, biologist John Todd demonstrates how ecological waste engineering in a greenhouse can be used to purify wastewater. Todd and others are conducting research to perfect such solar–aquatic systems based on working with nature. (Ocean Arks International)

where it is further filtered and cleansed by plants and bacteria. Although the water is clean enough to be discharged directly into the bay, state law requires that it first be chlorinated. So the town chlorinates the water and then dechlorinates it before sending it into the bay, where oyster beds thrive. Some water from the marshes is piped into the city's salmon hatchery.

The marshes and lagoons are an Audubon Society bird sanctuary and provide habitats for thousands of otters, seabirds, and marine animals; the treatment center is a city park and attracts many tourists. The town even celebrates its natural sewage treatment system with an annual "Flush with Pride" festival. Over 150 cities and towns in the United States now use natural and artificial wetlands for treating sewage.

Is it possible to use natural processes for treating wastewater if there isn't a wetland available or enough land on which to create one? According to ecologist John Todd, it is: Just set up a greenhouse lagoon and use sunshine the way nature does (Figure 11-30). The process begins when sewage flows into a greenhouse containing rows of large tanks full of aquatic plants such as water hyacinths, cattails, and bulrushes. In these tanks, algae and microorganisms decompose wastes into nutrients absorbed by the plants.

The decomposition is speeded up by sunlight streaming into the greenhouse. Then the water passes through an artificial marsh of sand, gravel, and bulrush plants, which filters out algae and organic waste. Next the water flows into aquarium tanks, where snails and zooplankton consume microorganisms and are in turn consumed by crayfish, tilapia, and other fish that can be eaten or sold as bait. After 10 days, the now-clear water flows into a second artificial marsh for final filtering and cleansing. When working properly, such solar–aquatic treatment systems have produced water fit for drinking.

These more natural alternatives to building expensive treatment plants may not solve the waste problems of large cities, but they can help and are an attractive alternative for small towns, the edges of urban areas, and rural areas.

Is the Water Safe to Drink? Many rivers in eastern Europe, Latin America, and Asia are used as sources of drinking water but are severely polluted, as are some rivers in developed countries. Aquifers used as sources of drinking water in many developed countries and developing countries are becoming contaminated with pesticides, fertilizers, and hazardous organic chemicals. In China, 41 large cities get their drinking water from polluted groundwater. In Russia, half of all tap water is unfit to drink, and a third of the aquifers are too contaminated for drinking purposes.

In 1980 the UN recommended spending $300 billion to supply all the world's people with clean drinking water and adequate sanitation by 1990. The $30 billion annual cost of this program is about what the world spends every 10 days for military purposes. Only about $1.5 billion per year was actually spent.

How Is Drinking Water Purified? Treatment of water for drinking by city dwellers is much like wastewater treatment. Areas that depend on surface water usually store it in a reservoir for several days to improve clarity and taste by allowing the dissolved oxygen content to increase and suspended matter to settle out. The water is then pumped to a purification plant, where it is treated to meet government drinking water standards. Usually the water is run through sand filters and activated charcoal before it is disinfected. In areas with very pure sources of groundwater, little treatment is necessary.

How Is the Quality of Drinking Water Protected? About 54 countries, most of them in North America and Europe, have safe drinking water

Q: What percentage of timber in the United States is harvested by clear-cutting?

standards. The U.S. Safe Drinking Water Act of 1974 requires the EPA to establish national drinking water standards, called *maximum contaminant levels*, for any pollutants that *may* have adverse effects on human health. This act has helped improve drinking water in much of the United States, but attempts to weaken this law continue. At least 700 potential pollutants have been found in municipal drinking water supplies. Maximum contaminant levels have not been set for potentially dangerous water pollutants such as certain synthetic organic compounds, radioactive materials, toxic metals, and pathogens.

Privately owned wells are not required to meet federal drinking water standards, primarily because of the costs of testing each well regularly (at least $1,000) and ideological opposition to mandatory testing and compliance by some homeowners.

According to a 1994 study by the Natural Resources Defense Council (NRDC), in 1992 the drinking water of 50 million people—one American in five—violated one or more EPA pollutant standards. In most cases people were not notified (as required by the 1974 law) when their drinking water was contaminated. This study also estimated that contaminated drinking water is responsible for 7 million illnesses and 1,200 deaths per year in the United States.

According to the Natural Resources Defense Council, U.S. drinking water supplies could be made safer at a cost of only about $30 a year per household. One might expect that such information would lead to public pressure to upgrade drinking water in the United States. Instead, Congress is being pressured by water-polluting industries to weaken the Safe Drinking Water Act by **(1)** eliminating national tests of drinking water, **(2)** eliminating the requirement that the media be advised of emergency water health violations and that water system officials notify their customers of such violations, **(3)** allowing states to give drinking water systems a permanent right to violate the standard for a given contaminant if the provider claims it can't afford to comply, and **(4)** eliminating the requirement that water systems use affordable, feasible technology to remove cancer-causing contaminants.

Environmentalists call for the U.S. Safe Drinking Water Act to be strengthened by **(1)** reducing testing and administrative costs and improving treatment by combining at least half of the 50,000 water systems that serve less than 3,300 people each with larger ones nearby, **(2)** strengthening and enforcing public notification requirements about violations of drinking water standards, and **(3)** banning all lead in new plumbing pipes, faucets, and fixtures (current law allows fixtures with up to 10% lead to be sold as lead free).

Is Bottled Water the Answer? The United States has one of the world's best drinking water supply systems. Yet about half of all Americans worry about getting sick from tap water contaminants and many drink bottled water or install expensive water-purification systems. Studies indicate that many of these consumers are being ripped off and in some cases may end up drinking water that is dirtier than they can get from their taps.

To be safe, consumers purchasing bottled water should determine whether the bottler belongs to the International Bottled Water Association (IBWA) and adheres to its testing requirements.* The IBWA requires its members to test for 181 contaminants, and annually it sends an inspector to bottling plants to check all pertinent records and ensure that the plant is run cleanly. Some companies pay $2,500 annually to obtain more stringent certification by the National Sanitation Foundation, an independent agency that tests for 200 chemical and biological contaminants.

Before buying expensive home water purifiers, health officials suggest that consumers have their water tested by local health authorities or private labs† to identify what contaminants, if any, must be removed; then they should buy a unit that does the required job. Independent experts contend that unless tests show otherwise, for most urban and suburban Americans served by large municipal drinking water systems, home water treatment systems aren't worth the expense and maintenance hassles.

Buyers should be suspicious of door-to-door salespeople, telephone appeals, scare tactics, and companies offering free water tests (which often are neither accurate nor carried out by certified labs). They should carefully check out companies selling such equipment and demand a copy of purifying claims by EPA-certified laboratories. Buyers should also be wary of claims that a treatment device has been *approved* by the EPA. Although the EPA does *register* such devices, it neither tests nor approves them.

Solutions: How Can We Use Water Resources More Sustainably? According to water resource expert Sandra Postel, none of the technological solutions

---

* Check for the IBWA seal of approval on the bottle or contact the International Bottled Water Association (113 North Henry Street, Alexandria, VA 22314; phone: 703-683-5213) for a member list.

† Contact the state health or other appropriate department for help in finding laboratories that are certified to do tests. You can also call EPA's Safe Drinking Water Hotline from 8:30 A.M. to 4:30 P.M. EST at 800-426-4791 or 202-382-5533.

for supplying more water—dams, watershed transfers, tapping groundwater, and using water more efficiently —deal with the three underlying forces that can lead humans to use such a potentially renewable resource in unsustainable ways: **(1)** depletion or degradation of a shared resource, which shrinks the resource pie to be shared (the tragedy of the commons); **(2)** population growth, which forces the resource pie to be divided into smaller slices; and **(3)** unequal distribution or access (primarily because of poverty), which means that some countries and people get larger slices than others.

Sustainable water use is based on the commonsense principle stated in an old Inca proverb: "The frog does not drink up the pond in which it lives."

Sustainable use of potentially renewable groundwater means that the rate of extraction should not exceed the rate of recharge. Determining what constitutes sustainable use of shared rivers is much more complex because available water from a river varies with the time of year and with conditions such as drought and higher-than-normal precipitation. Sustainable water use also requires an integrated plan governing water use, sewage treatment, and water pollution among all users of a water basin (as is done in Great Britain).

Many analysts believe that a key to reducing water waste and providing more equitable access to water resources is for governments to phase out subsidies that reduce the price of water for those benefiting from such subsidies (often large farmers and industries).

According to environmentalists, a sustainable approach to dealing with water pollution requires that we shift our emphasis from pollution cleanup to pollution prevention. This involves **(1)** *source reduction* to reduce the toxicity or volume of pollutants (for example, replacing organic solvent-based inks and paints with water-based materials), **(2)** *reuse* of wastewater instead of discharging it (for example, reusing treated wastewater for irrigation), and **(3)** *recycling* pollutants (for example, cleaning up and recycling contaminated solvents for reuse) instead of discharging them.

To make such a shift, we need to accept that the environment—air, water, soil, life—is an interconnected whole. Without an integrated approach to all forms of pollution, environmentalists argue, we will continue to shift environmental problems from one part of the environment to another. Some actions you can take to help reduce water pollution are listed in Appendix 5.

It is not until the well runs dry that we know the worth of water.

BENJAMIN FRANKLIN

CRITICAL THINKING

1. How do human activities increase the harmful effects of prolonged drought? How can these effects be reduced?

2. Explain how dams and reservoirs can cause more flood damage than they prevent. Should all proposed large dam and reservoir projects be scrapped? Explain.

3. Should federal subsidies of irrigation projects in the western United States (or in the country where you live) be gradually phased out? Explain.

4. Should prices of water for all uses be raised sharply to include more of its environmental costs and to encourage water conservation? Explain. What harmful and beneficial effects might this have on business and jobs, on your lifestyle and the lifestyles of any children or grandchildren you might have, on the poor, and on the environment?

5. List five major ways to conserve water on a personal level (see Appendix 5). Which, if any, of these practices do you now use or intend to use?

6. How do human activities contribute to flooding and flood damage? How can these effects be reduced?

7. Some analysts argue that the U.S. Federal Flood Disaster Protection Act should be repealed because it encourages people to build on areas with a high risk of flooding by providing them with government-backed insurance (paid for mostly by people who not do live in high flood-risk areas) to rebuild after a flood. Instead, they would replace it with an act that provides emergency relief to victims of a disastrous flood but does not help flood victims rebuild in such risky areas. Do you agree or disagree with this proposal? Explain. Would your position be the same if you lived in an area prone to flooding?

8. Why is dilution not always the solution to water pollution? Give examples and conditions for which this solution is, and is not, applicable.

9. How can a stream cleanse itself of oxygen-demanding wastes? Under what conditions will this natural cleansing system fail?

10. Should all dumping of wastes in the ocean be banned? Explain. If so, where would you put the wastes instead? What exceptions would you permit, and why?

11. Your town (Town B) is located on a river between towns A and C. What are the rights and responsibilities of upstream communities to downstream communities? Should sewage and industrial wastes be dumped at the upstream end of a community that generates them?

*12. In your community,
   a. What are the major sources of the water supply?
   b. How is water use divided among agricultural, industrial, power-plant cooling, and public uses? Who are the biggest consumers of water?

Q: What percentage of the U.S. Forest Service budget is devoted to timber sales?

c. What has happened to water prices during the past 20 years? Are they too low to encourage water conservation and reuse?

d. What water supply problems are projected?

e. How is water being wasted?

*13. In your community,

a. What are the principal nonpoint sources of contamination of surface water and groundwater?

b. What is the source of drinking water?

c. How is drinking water treated?

d. What contaminants are tested for?

e. How many times during each of the past 5 years have levels of tested contaminants violated federal standards? Was the public notified about the violations?

f. Is groundwater contamination a problem? If so, where, and what has been done about the problem?

14. Make a concept map of this chapter's major ideas, using the section heads and subheads and the key terms (in boldface type). Look at the inside back cover and in Appendix 4 for information about concept maps.

SUGGESTED READINGS

See Internet Sources and Further Readings at the back of the book, and InfoTrac.

12 MINERALS AND SOIL

The Great Terrain Robbery

Want to get rich at the taxpayers' expense? You can if you know how to make use of a little known mining law passed in 1872 to encourage mining of gold, silver, lead, copper, uranium, and other hard-rock minerals on U.S. public lands.

Under this 1872 law, a person or corporation can assume legal ownership of any public land not classified as wilderness or park simply by *patenting* it. This involves declaring their belief that the land contains valuable hard-rock minerals, spending $500 to improve the land for mineral development, filing a claim, and then paying the federal government

Figure 12-1 Bear Trap Creek in Montana is one example of how gold mining can contaminate water with highly toxic cyanide or mercury used to extract gold from its ore. Air and water also convert the sulfur in gold ore to sulfuric acid, which releases toxic metals such as cadmium and copper into streams and groundwater. (Bryan Peterson)

$6–12 per hectare ($2.50–5.00 an acre) for the land. So far public lands containing at least $285 billion of the public's mineral resources have been transferred to private interests at these bargain-basement prices.

In 1993 the Manville Corporation paid $10,000 for federal land in Montana that contains an estimated $32 billion of platinum and palladium. Secretary of Interior Bruce Babbitt was forced by the courts in 1994 to sell to a Canadian corporation for only $10,000 federal land in Nevada containing an estimated $10–15 billion worth of gold. In 1995 he had to sign over federal land in Idaho containing $1 billion worth of minerals to a Danish company that paid just $275 for the bonanza.

Mining companies operating under this law— *almost half of them controlled by foreign corporations*— annually remove mineral resources worth about $3.6 billion on land they have bought at absurdly low prices. They pay no royalties to the U.S. Treasury.

The 1872 law does not even require that patented property be mined. Land speculators have often purchased such property at 1872 prices and then sold it or leased it for thousands of times what they paid. In 1986 a mining company paid the government $42,500 for oil shale land in Colorado and a month later sold it to Shell Oil for $37 million. Other owners have developed sites as casinos, ski resorts, golf courses, and vacation-home developments. According to Senator Dale Bumpers, "The 1872 mining law is a license to steal and the biggest scam in America."

There is also no provision in the 1872 law requiring reclamation of damaged land. Acids and toxic metal residues continue to leach from thousands of the almost 558,000 abandoned hard-rock mines and open pits, mostly in the West. One result is some 19,000 kilometers (12,000 miles) of polluted streams (Figure 12-1); another result is that 56 abandoned mine sites are now listed on the EPA's Superfund list of the nation's worst hazardous waste dump sites.

It is estimated that the taxpayers' cleanup costs for all land damaged by hard-rock mining on existing or sold public lands will be $33–72 billion, depending on whether groundwater and toxic waste cleanup are included. It's common for a company to mine a site, abandon it, file for bankruptcy, and leave the public with the cleanup bill. Environmentalists have been trying, without success, to have this law revised to protect taxpayers and the environment.

Below that thin layer comprising the delicate organism known as the soil is a planet as lifeless as the moon.

G. Y. JACKS AND R. O. WHYTE

In this chapter we seek answers to the following questions:

- What are the major geologic processes occurring on and in the earth?

- How does the rock cycle recycle earth materials and concentrate mineral resources?

- How fast are nonfuel mineral supplies being used up, and how can we increase supplies of key minerals?

- What are the environmental effects of extracting and using mineral resources?

- What is soil, and what types are best for growing crops?

- Why should we worry about soil erosion?

- How can we control erosion and reduce the loss of nutrients from topsoil?

12-1 GEOLOGIC PROCESSES*

What Is the Earth's Structure? As the primitive earth cooled over eons, its interior separated into three major concentric zones, which geologists identify as the core, the mantle, and the crust (Figure 4-2). Various indirect measurements indicate that the earth's innermost zone, the **core**, is made mostly of iron (with perhaps some nickel). The core has a solid inner part, surrounded by a liquid core of molten material.

The earth's core is surrounded by a thick, solid zone called the **mantle**. This largest zone of the earth's interior is rich in the elements iron (its major constituent), silicon, oxygen, and magnesium. Most of the mantle is solid rock, but under its rigid outermost part there is a zone of very hot, partly melted rock that flows like soft plastic. This plastic region of the mantle is called the *asthenosphere.*

The outermost and thinnest zone of the earth is called the **crust**. It consists of the *continental crust,* which underlies the continents (including the continental shelves extending into the oceans), and the *oceanic crust,* which underlies the ocean basins and covers 71% of the earth's surface (Figure 12-2).

** **Kenneth J. Van Dellen,** professor of geology and environmental science, Macomb Community College, is the primary author of this and the next section, with assistance from G. Tyler Miller, Jr.*

What Is Plate Tectonics? Internal Geologic Processes We tend to think of the earth's crust, mantle, and core as fairly static and unchanging. However, these parts of the earth are constantly changing because of geologic processes taking place within the earth and on the earth's surface, most over thousands to millions of years.

A map of the earth's earthquakes and volcanoes shows that most of these phenomena occur along certain lines or belts on the earth's surface (Figure 12-3a). The areas of the earth outlined by these major belts are called **plates** (Figure 12-3b). They are about 100 kilometers (60 miles) thick and are composed of the crust and the rigid, outermost part of the mantle (above the asthenosphere)—a combination called the **lithosphere**. These plates move constantly, supported by the slowly flowing asthenosphere like large pieces of ice floating on the surface of a lake during the spring breakup. Some plates move faster than others, but a typical speed is about the rate at which fingernails grow.

The theory explaining the movements of the plates and the processes that occur at their boundaries is called **plate tectonics**. The concept, which became widely accepted by geologists in the 1960s, was developed from an earlier idea called *continental drift.* Throughout the earth's history, continents have split and joined as plates have drifted thousands of kilometers back and forth across the planet's surface (Figure 5-37).

Plate motion produces mountains (including volcanoes), the oceanic ridge system, trenches, and other features of the earth's surface (Figure 12-2), certain natural hazards are likely to be found at plate boundaries (Figure 12-3a), and plate movements and interactions also concentrate many of the minerals we extract and use.

The theory of plate tectonics also helps explain how certain patterns of biological evolution occurred. By reconstructing the course of continental drift over millions of years (Figure 5-37), we can trace how lifeforms migrated from one area to another when continents that are now far apart were still joined together. As the continents separated, speciation occurred as populations became geographically and reproductively isolated (Figure 5-36).

Lithospheric plates have three types of boundaries (Figure 12-4, p. 342). At a **divergent plate boundary** the plates move apart in opposite directions (Figure 12-4, top), and at a **convergent plate boundary** they are pushed together by internal forces (Figures 12-3b and 12-4, middle). At most convergent plate boundaries, oceanic lithosphere is carried downward (subducted) under the island arc or the continent at a **subduction zone**. A trench ordinarily forms at the boundary between the two converging plates (Figure 12-4, middle).

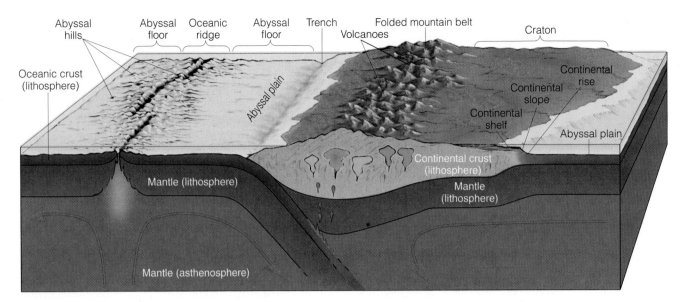

Figure 12-2 Major features of the earth's crust and upper mantle. The lithosphere, composed of the crust and outermost mantle, is rigid and brittle. The asthenosphere, a zone in the mantle, can be deformed by heat and pressure (that is, it is plastic).

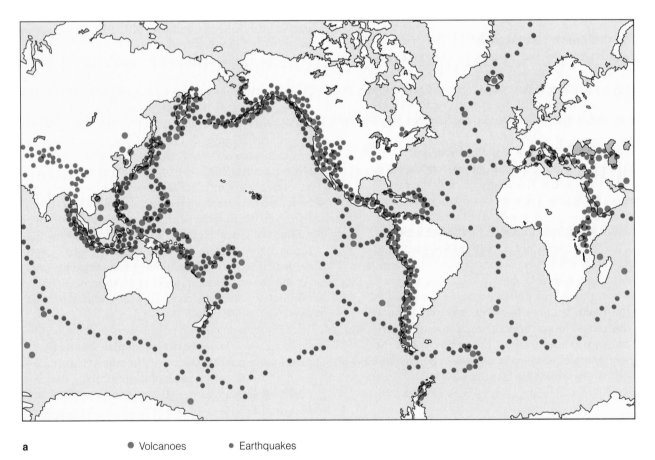

a   ● Volcanoes   ● Earthquakes

Figure 12-3 Earthquake and volcano sites are distributed mostly in bands along the planet's surface **(a)**. These bands correspond to the patterns for the types of lithospheric plate boundaries **(b)** shown in Figure 12-4.

Q: Between 1978 and 1993, how much money did the Forest Service lose on timber sales?

Stresses in the plate undergoing subduction cause earthquakes at convergent plate boundaries.

The third type of plate boundary, called a **transform fault**, occurs where plates move in opposite but parallel directions along a fracture (fault) in the lithosphere (Figure 12-4, bottom). In other words, the plates slide past one another. Like the other types of plate boundaries, most transform faults are on the ocean floor.

What Geologic Processes Occur on the Earth's Surface? Geological changes based directly or indirectly on energy from the sun and on gravity (rather than on heat in the earth's interior) are called *external processes*. Whereas internal processes generally build up the earth's surface, external processes tend to wear it down.

A major external process is **erosion.** It is the process by which loosened material (as well as material not yet separated) is dissolved, loosened, or worn away from one part of the earth's surface and deposited in other places. Streams, the most important agent of erosion, operate everywhere on the earth except in the polar regions. They produce ordinary valleys and canyons, and they may form deltas where streams flow into lakes and oceans (Figure 12-5). Some erosion is also caused when wind blows particles of soil from one area to another.

Loosened material that can be eroded is usually produced by **weathering**, which can occur as a result of mechanical processes, chemical processes, or both. In *mechanical weathering*, a large rock mass is broken into smaller fragments of the original material, similar to the results you would get by using a hammer to break a rock into small fragments. The most important agent of mechanical weathering is *frost wedging*, in which water collects in pores and cracks of rock, expands upon freezing, and splits off pieces of the rock.

In *chemical weathering*, a mass of rock is decomposed by one or more chemical reactions, resulting in products that are chemically different from the original material. The products usually include both solid and dissolved components. Most chemical weathering involves a reaction of rock material with oxygen, carbon dioxide, and moisture in the atmosphere and the ground.

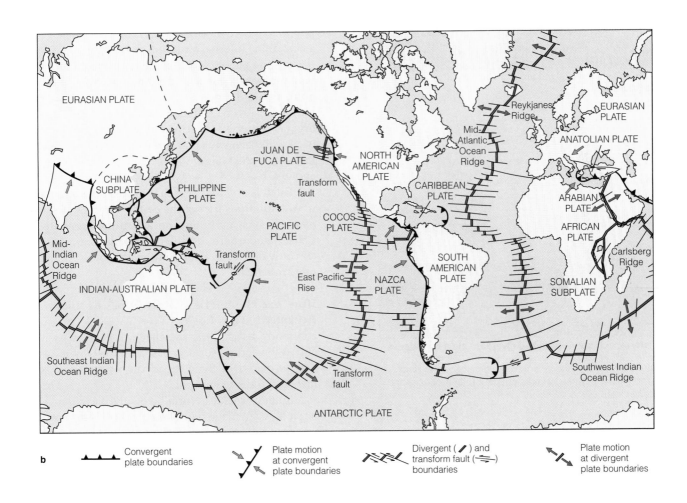

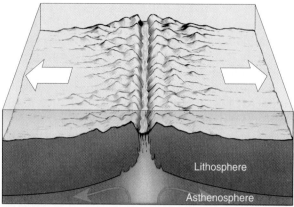

Oceanic ridge at a divergent plate boundary

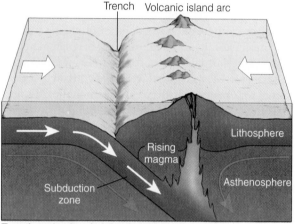

Trench and volcanic island arc at a convergent plate boundary

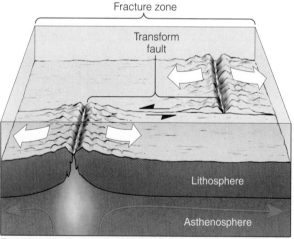

Transform fault connecting two divergent plate boundaries

Figure 12-4 Types of boundaries between the earth's lithospheric plates. All three boundaries occur both in oceans and on continents.

Disintegration of rock by mechanical weathering accelerates chemical weathering by increasing the surface area that can be attacked by chemical weathering agents. This is similar to the way granulated sugar dissolves much faster than a large chunk of sugar. Because chemical weathering is also aided by higher temperatures and precipitation, it occurs most rapidly in the tropics and next most rapidly in temperate climates. Weathering is responsible for the development of soil, as discussed in Section 12-5. Human activities, particularly those that destroy vegetation, accelerate erosion, as discussed in Section 12-6.

12-2 MINERAL RESOURCES AND THE ROCK CYCLE

What Are Minerals and Rocks? The earth's crust, which is still forming in various places, is composed of minerals and rocks. It is the source of virtually all the nonrenewable resources we use: fossil fuels, metallic minerals, and nonmetallic minerals (Figure 1-11). It is also the source of soil and of the elements that make up our bodies and those of other living organisms.

A **mineral** is an element or an inorganic compound that occurs naturally and is solid. It usually has a crystalline internal structure made up of an orderly, three-dimensional arrangement of atoms or ions. Some minerals consist of a single element, such as gold, silver, diamond (carbon), or sulfur. However, most of the over 2,000 identified minerals occur as inorganic compounds formed by various combinations of the eight elements that make up 98.5% by weight of the earth's crust (Figure 12-6). Examples are salt, mica, and quartz.

Rock is any material that makes up a large, natural, continuous part of the earth's crust. Some kinds of rock, such as limestone (calcium carbonate, or $CaCO_3$) and quartzite (silicon dioxide, or SiO_2), contain only one mineral, but most rocks consist of two or more minerals.

How Are the Earth's Three Types of Rock Recycled? The Rock Cycle Geologic processes constantly redistribute the chemical elements within and at the earth's surface. Based on the way it forms, rock is placed in three broad classes: igneous, sedimentary, and metamorphic.

Igneous rock can form below the earth's surface, or on it, when magma (molten rock) wells up from the upper mantle or deep crust, cools, and hardens into rock. Examples are granite (formed underground) and lava lock (formed aboveground when molten lava cools and hardens). Although often covered by sedi-

Q: How much of U.S. public rangeland is in unsatisfactory (fair or poor) condition?

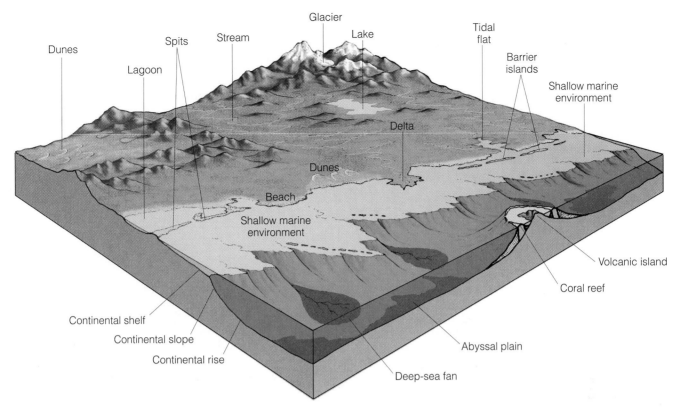

Figure 12-5 The variety of landforms and sedimentary environments depicted here is mainly the result of *external processes*, powered primarily by solar energy (as it drives the hydrologic cycle and wind) and gravity, with some assistance from organisms such as reef-building corals.

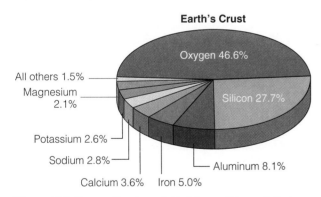

Figure 12-6 Composition by weight of the earth's crust. Various combinations of only eight elements make up the bulk of most minerals.

mentary rocks or soil, igneous rocks form the bulk of the earth's crust. They also are the main source of many nonfuel mineral resources.

Sedimentary rock forms from sediment in several ways. Most such rocks are formed when preexisting rocks are weathered and eroded into small pieces, transported from their sources, and deposited in a body of surface water. As these deposited layers become buried and compacted, the resulting pressure causes their particles to bond together to form sedimentary rocks such as sandstone and shale. Some sedimentary rocks, such as dolomite and limestone, are formed from the compacted shells, skeletons, and other remains of dead organisms. Two types of coal—lignite and bituminous coal—are sedimentary rocks derived from plant remains.

Metamorphic rock is produced when a preexisting rock is subjected to high temperatures (which may cause it to melt partially), high pressures, chemically active fluids, or a combination of those agents. Examples are anthracite (a form of coal), slate, and marble.

Rocks are constantly exposed to various physical and chemical conditions that can change them over time. The interaction of processes that change rocks from one type to another is called the **rock cycle** (Figure 12-7). Recycling material over millions of years, this slowest of the earth's cyclic processes is responsible for concentrating the planet's nonrenewable mineral resources on which humans depend.

Figure 12-7 The rock cycle, the slowest of the earth's cyclic processes. The earth's materials are recycled over millions of years by three processes: melting, erosion, and metamorphism, which produce igneous, sedimentary, and metamorphic rocks. Rock of any of the three classes can be converted to rock of either of the other two classes (or can even be recycled within its own class).

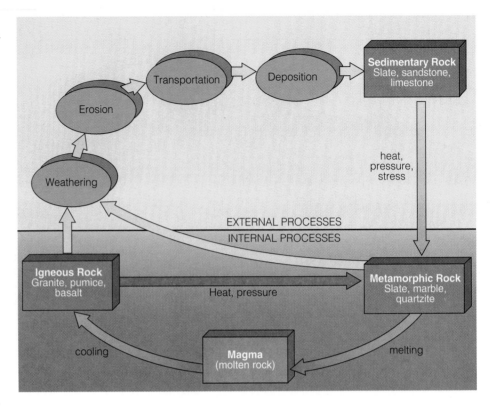

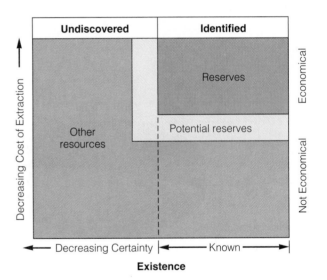

Figure 12-8 General classification of mineral resources. (The area shown for each class does not represent its relative abundance.) In theory, all mineral resources classified as *other resources* could become reserves because of rising mineral prices, improved mineral location and extraction technology, or both. In practice, geologists expect only a fraction of other resources to become reserves. The area labeled *potential reserves* shows the way reserves normally increase.

What Are Mineral Resources? A mineral resource is a concentration of naturally occurring solid, liquid, or gaseous material in or on the earth's crust that can be extracted and processed into useful materials at an affordable cost. The earth's internal and external processes have produced numerous mineral resources, which on a human time scale are essentially nonrenewable because of the slowness of the rock cycle. Mineral resources include *energy resources* (coal, oil, natural gas, uranium, geothermal energy), *metallic mineral resources* (iron, copper, aluminum), and *nonmetallic mineral resources* (salt, gypsum, clay, sand, phosphates, water, and soil).

An **ore** is a metal-yielding material that can be economically extracted at a given time. To be profitable, copper in copper ore must be concentrated to 86 times its crustal average, gold 1,000 times, and mercury an astonishing 100,000 times.

We know how to find and extract more than 100 nonrenewable minerals from the earth's crust. We con-

vert these minerals into many everyday items that we either use and discard (Figure 3-15) or learn to reuse, recycle, or use less wastefully (Figure 3-16).

The U.S. Geological Survey (USGS) divides mineral resources into two broad categories, *identified* and *undiscovered* (Figure 12-8). **Identified resources** are deposits of a particular mineral resource that have a *known* location, quantity, and quality, or deposits for which these parameters are estimated from direct geological evidence and measurements. **Undiscovered**

Q: How much do 29,000 U.S. ranchers with permits to graze on public lands get in federal subsidies?

Figure 12-9 This open-pit copper mine in Bingham, Utah, is the largest human-made hole in the world—4.0 kilometers (2.5 miles) in diameter and 0.8 kilometer (0.5 mile) deep. The amount of material removed from this mine is seven times the amount moved to build the Panama Canal. (Don Green/Kennecott Copper Corporation, now owned by British Petroleum)

Figure 12-10 Spoil banks, the results of unrestored area strip mining of coal near Mulla, Colorado. Restoration of newly strip-mined areas is now required in the United States, but many previously mined areas have not been restored. (National Archives/EPA Documerica)

resources are potential supplies of a particular mineral resource that are assumed to exist on the basis of geologic knowledge and theory (although specific locations, quality, and amounts are unknown).

Reserves are identified resources that can be extracted economically at current prices using current mining technology. **Other resources** are identified and undiscovered resources not classified as reserves.

Most published estimates of particular mineral resources refer only to reserves. Reserves can increase when exploration finds previously undiscovered economic-grade mineral resources, or when identified subeconomic-grade mineral resources become economically viable because of new technology or higher prices. Figure 12-8 shows a region labeled *potential reserves*, indicating how resources can be expanded. Theoretically, all of the *other resources* could eventually be converted to reserves, but this is highly unlikely.

How Do We Find and Remove Mineral Deposits?

Mining companies use several methods to find promising mineral deposits. Geological information about plate tectonics and mineral formation suggests areas worthy of closer study. Aerial photos and satellite images sometimes reveal rock formations associated with certain minerals. Other instruments on aircraft and satellites can detect mineral deposits by their effects on the earth's magnetic or gravitational fields.

After profitable deposits of minerals are located, deep deposits are removed by **subsurface mining**, and shallow deposits by **surface mining**. In surface mining, mechanized equipment strips away the **overbur-**

den of soil and rock and usually discards it; such waste material is called **spoil**. Surface mining extracts about 90% of the mineral and rock resources, and more than 60% of the coal by weight in the United States.

The type of surface mining used depends on the resource being sought and on local topography. In **open-pit mining** (Figure 12-9), machines dig holes and remove ores such as iron and copper. This method is also used for sand and gravel, and for building stone such as limestone, sandstone, slate, granite, and marble. Another form of surface mining is **dredging**, in which chain buckets and draglines scrape up underwater mineral deposits.

Strip mining is surface mining in which bulldozers, power shovels, or stripping wheels remove the overburden in strips. It is used mostly for extracting coal and some phosphate rock. Most surface-mined coal is removed by area strip mining or contour strip mining, depending on the terrain.

Area strip mining is used where the terrain is fairly flat. An earthmover strips away the overburden, and then a power shovel digs a cut to remove the mineral deposit, such as coal. After the mineral is removed, the trench is filled with overburden, and a new cut is made parallel to the previous one. The process is repeated over the entire site. If the land is not restored, area strip mining leaves a wavy series of highly erodible hills of rubble called *spoil banks* (Figure 12-10).

Contour strip mining is used in hilly or mountainous terrain (Figure 12-11). A power shovel cuts a series of terraces into the side of a hill. An earthmover removes the overburden, and a power shovel extracts

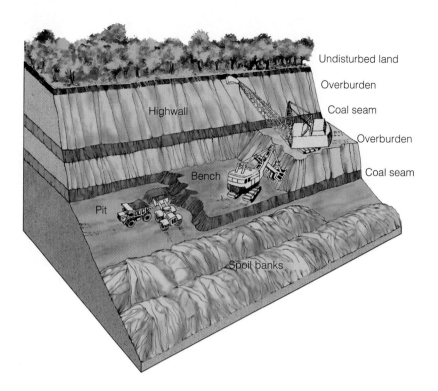

Figure 12-11 Contour strip mining of coal.

the coal; the overburden from each new terrace is dumped onto the one below. Unless the land is restored, a wall of dirt is left in front of a highly erodible bank of soil and rock called a *highwall*. Sometimes giant augers are used to drill horizontally into a hillside to extract underground coal.

In the United States, contour strip mining for coal is used mostly in the mountainous Appalachian region. Restoring land on surface mining sites can reverse mining's devastating effects. However, much of the land previously strip-mined for coal in the Appalachian region has not been restored. In addition, about three-fourths of the coal that can be surface mined in the United States is in arid and semiarid regions in the West, where the soil and climate usually prevent full restoration.

Subsurface mining is used to remove coal and various metal ores that are too deep to be extracted by surface mining. Miners dig a deep vertical shaft, blast subsurface tunnels and chambers to get to the deposit, and haul the coal or ore to the surface. Subsurface mining disturbs less than one-tenth as much land as surface mining and usually produces less waste material. However, it leaves much of the resource in the ground and is more dangerous and expensive than surface mining. Roofs and walls of underground mines collapse, trapping and killing miners, explosions of dust and natural gas injure or kill them, and prolonged inhalation of mining dust causes lung diseases.

12-3 ESTIMATING SUPPLIES OF NONRENEWABLE MINERAL RESOURCES

Will There Be Enough Mineral Resources? The future supply of nonrenewable minerals depends on two factors: the actual or potential supply and the rate at which that supply is being used. We never completely run out of any mineral. However, a mineral becomes *economically depleted* when the costs of finding, extracting, transporting, and processing the remaining deposits exceed the returns. At that point we have five choices: recycle or reuse existing supplies, waste less, use less, find a substitute, or do without.

As mentioned earlier, most published estimates of the supplies of a given resource refer to *reserves*: known deposits from which a usable mineral can be extracted profitably at current prices (Figure 12-8). **Depletion time** is the time it takes to use up a certain proportion (usually 80%) of the reserves of a mineral at a given rate of use. When experts disagree about depletion times, they are often using different assumptions about supply and rate of use (Figure 12-12). A traditional measure of the projected availability of nonrenewable resources is the **reserve-to-production ratio**: the number of years that proven reserves of a particular nonrenewable mineral will last at current annual production rates.

Estimates of reserves are continually changing because new deposits are often discovered, and new

Q: What % of the income of private concessionaires in U.S. national parks is paid in user fees to the government?

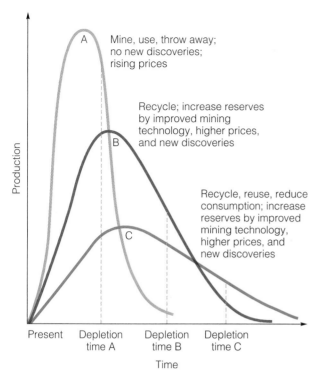

Figure 12-12 Depletion curves for a nonrenewable resource (such as aluminum or copper) using three sets of assumptions. Dashed vertical lines represent times when 80% depletion occurs.

Curve A label: Mine, use, throw away; no new discoveries; rising prices

Curve B label: Recycle; increase reserves by improved mining technology, higher prices, and new discoveries

Curve C label: Recycle, reuse, reduce consumption; increase reserves by improved mining technology, higher prices, and new discoveries

mining and processing can allow some of the minerals classified as other resources (Figure 12-8) to be converted to reserves. Under these circumstances, the reserve-to-production ratio is the best available projection of the current estimated supply and its estimated depletion time.

The shortest depletion time assumes no recycling or reuse and no increase in reserves (curve A, Figure 12-12). A longer depletion time assumes that recycling will stretch existing reserves and that better mining technology, higher prices, and new discoveries will increase reserves (curve B, Figure 12-12). An even longer depletion time assumes that new discoveries will further expand reserves and that recycling, reuse, and reduced consumption will extend supplies (curve C, Figure 12-12). Finding a substitute for a resource leads to a new set of depletion curves for the new resource.

While world population doubled between 1950 and 1993, global production of six key metals (aluminum, copper, lead, nickel, tin, and zinc) increased more than eightfold. During this same period, world reserves of copper increased almost fivefold, lead almost threefold, zinc fourfold, and aluminum almost ninefold. Furthermore, the prices of most metals today have changed little in constant dollars over the last 150 years, mostly because of government subsidies and

failure to include the harmful environmental effects of metal mining and processing in their market prices.

According to resource experts, the United States will never again be self-sufficient in oil or many key metal resources. Currently, the United States imports 50% or more of 24 of its 42 most important nonfuel minerals. Some are imported because they are used faster than they can be produced from domestic supplies; others are imported because foreign ore deposits are of a higher grade and cheaper to extract than remaining U.S. reserves.

How Does Economics Affect Resource Supplies? Geologic processes determine the quantity and location of a mineral resource in the earth's crust; economics determines what part of the known supply will be used.

According to standard economic theory, in a competitive free market a plentiful resource is cheap because supply exceeds demand (Figure 7-4); when a resource becomes scarce its price rises, stimulating exploration and development of better mining technology. Rising prices also make it profitable to mine ores of lower grades and encourage the search for substitutes. However, this theory may no longer apply to most developed countries because in the economic systems of such countries, industry and government control supply, demand, and prices of minerals to such a large extent that a truly competitive free market does not exist.

Most mineral prices are artificially low because countries subsidize development of their domestic mineral resources to help promote economic growth and national security. In the United States, for instance, mining companies get depletion allowances amounting to 5–22% of their gross income, depending on the mineral. In addition, the companies can deduct much of their costs for finding and developing mineral deposits. Moreover, hard-rock mining companies operating in the United States get public land and the minerals they extract essentially for free (p. 338). Between 1982 and 1995, these mining subsidies cost U.S. taxpayers about $5.5 billion.

Another problem is that the cost of nonfuel mineral resources is only a small part of the final cost of goods. Thus, because scarcity of minerals does not raise the market prices of products very much, industries and consumers have no incentive to reduce demand for products in time to avoid economic depletion of the minerals.

Most mineral prices are low because most of the harmful environmental costs of mining and processing are not included in their prices. As a result, mining companies and manufacturers have little incentive to reduce resource waste and pollution as long as they can pass many of the harmful environmental costs of

Mining with Microbes

SOLUTIONS

One way to improve mining technology is the use of microorganisms for in-place (in situ) mining, which would remove desired metals from ores while leaving the surrounding environment undisturbed. This biological approach to mining would also reduce both the air pollution associated with the smelting of metal ores and the water pollution associated with using hazardous chemicals such as cyanides to extract gold.

Once an ore deposit has been identified and deemed economically viable, wells are drilled into it and the ore is fractured. Then the ore is inoculated—with either natural or genetically engineered bacteria—to extract the desired metal. Next the ore is flooded with water, which is then pumped to the surface, where the desired metal is removed.

This technique permits economical extraction from low-grade ores, which are increasingly being used as high-grade ores are depleted. Since 1958, the copper industry has been using natural strains of the bacterium *Thiobacillus ferroxidans* to remove copper from low-grade copper ore. Currently, at least 25% of all copper produced worldwide, worth more than $1 billion a year, comes from such biomining.

Microbiological processing of ores is slow, however: It can take decades to remove the same amount of material that conventional methods can remove within months or years. So far, biological methods are economically feasible only with low-grade ore (such as gold and copper), for which conventional techniques are too expensive.

Critical Thinking

If you had a large sum of money to invest, would you invest it in the microbiological processing of aluminum ore? Explain.

their production on to society. Environmentalists and some economists argue that taxing rather than subsidizing the extraction of nonfuel mineral resources would provide governments with revenue, create incentives for more efficient resource use, promote waste reduction and pollution prevention, and encourage recycling and reuse. So far, leaders of politically powerful resource extraction industries have been able to prevent significant taxation of the resources they extract in most countries.

Can We Find Enough New Land-Based Mineral Deposits? Geologic exploration guided by better knowledge, satellite surveys, and other new techniques will increase current reserves of most minerals. Although most of the easily accessible, high-grade deposits are already known, new deposits will be found, mostly in unexplored areas of developing countries.

However, exploring for new resources is an expensive and risky financial venture. Typically, if geologists identify 10,000 possible deposits of a given resource, only 1,000 sites are worth exploring; only 100 justify drilling, trenching, or tunneling; and only 1 becomes a producing mine or well. Even if large new supplies are found, no nonrenewable mineral supply can stand up to continued exponential growth in its use.

One factor limiting production of nonfuel minerals is lack of investment capital. With today's fluctuating mineral markets and rising costs, investors are wary of tying up large sums for long periods with no assurance of a reasonable return.

Can We Get Enough Minerals by Mining Lower-Grade Ores? Some analysts contend that all we need to do to increase supplies of any mineral is to extract lower grades of ore. They point to new earth-moving equipment, improved techniques for removing impurities, and other technological advances in mineral extraction and processing during the past few decades.

In 1900 the average copper ore mined in the United States was about 5% copper by weight; today it is 0.5%, and copper costs less (adjusted for inflation). New methods of mineral extraction may allow even lower-grade ores of some metals to be used (Solutions, left).

Several factors limit the mining of lower-grade ores, however. As poorer grade ores are mined, a point is reached at which it costs more to mine and process most such resources than they are worth (unless we have a virtually inexhaustible source of cheap energy). Availability of fresh water also may limit the supply of some mineral resources. To extract and process most minerals by conventional means requires large amounts of water and many mineral-rich areas lack fresh water. Finally, exploitation of lower-grade ores may be limited by the environmental impact of waste material produced during mining and processing. At some point, the costs of land restoration and pollution control exceed the current value of the minerals, unless we continue to pass these harmful costs on to society and to future generations.

Can We Get Enough Minerals by Mining the Oceans? Ocean mineral resources are found in three areas: seawater, sediments and deposits on the shallow continental shelf (Figure 12-5), and sediments and

Q: How much of all U.S. land area is protected as wilderness?

nodules on the deep-ocean floor. Most of the chemical elements found in seawater occur in such low concentrations that recovering them takes more energy and money than they are worth. Only magnesium, bromine, and sodium chloride are abundant enough to be extracted profitably at current prices with existing technology.

Deposits of minerals (mostly sediments) along the continental shelf and near shorelines are already significant sources of sand, gravel, phosphates, sulfur, tin, copper, iron, tungsten, silver, titanium, platinum, and even diamonds.

The deep-ocean floor at various sites may be a future source of manganese and other metals. Manganese-rich nodules may cover 20% of the world's ocean floors and have been found in large quantities at a few sites. These cherry- to potato-size rocks are 30–40% manganese by weight; they also contain small amounts of other important metals such as nickel, copper, and cobalt. They might be sucked up from the ocean floor by giant vacuum pipes or scooped up by buckets on a continuous cable operated by a mining ship.

Nodule beds in international waters are not being developed today because of squabbles over who owns them and how any profits should be distributed among the world's nations. An international Law of the Sea Treaty (signed by the United States in 1994) may resolve some of these issues.

Environmentalists recognize that seabed mining would probably cause less harm than mining on land. However, they are concerned that removing seabed mineral deposits and dumping back unwanted material will stir up ocean sediments, which could destroy seafloor organisms and have unknown effects on poorly understood ocean food webs. Surface waters might also be polluted by the discharge of sediments from mining ships and rigs.

Can We Find Substitutes for Scarce Nonrenewable Mineral Resources?

The Materials Revolution Some analysts believe that even if supplies of key minerals become very expensive or scarce, human ingenuity will find substitutes. They point to the current materials revolution in which silicon and new materials, particularly ceramics and plastics, are being developed and used as replacements for metals.

Ceramics have many advantages over conventional metals. They are harder, stronger, lighter, and longer-lasting than many metals. They withstand intense heat and do not corrode. Within a few decades we may have high-temperature ceramic superconductors in which electricity flows without resistance. Such a development may lead to faster computers,

more efficient power transmission, and affordable electromagnets for propelling magnetic levitation trains.

Plastics also have advantages over many metals. High-strength plastics and composite materials strengthened by lightweight carbon and glass fibers are likely to transform the automobile and aerospace industries. They cost less to produce than metals because they require less energy, don't need painting, and can be molded into any shape. New plastics and gels are also being developed to provide superinsulation without taking up much space. One new plastic can withstand extremely high temperatures and is not even affected by exposure to the most intense laser beams.

Substitutes can undoubtedly be found for many scarce mineral resources, but the search is costly, and phasing a substitute into a complex manufacturing process takes time. While a vanishing mineral is being replaced, people and businesses dependent on it may suffer economic hardships. Moreover, finding substitutes for some key materials may be difficult or impossible. Examples are helium, phosphorus for phosphate fertilizers, manganese for making steel, and copper for wiring motors and generators. Finally, some substitutes are inferior to the minerals they replace. For example, even though aluminum could replace copper in electrical wiring, producing aluminum takes much more energy than producing copper, and aluminum wiring presents a greater fire hazard than copper wiring.

⚙ 12-4 ENVIRONMENTAL EFFECTS OF EXTRACTING AND USING MINERAL RESOURCES

What Are the Environmental Impacts of Using Mineral Resources? The mining, processing, and use of crustal resources requires enormous amounts of energy, and often causes land disturbance, erosion, and air and water pollution (Figure 12-13).

Mining can affect the environment in several ways. Most noticeable are scarring and disruption of the land surface (Figures 12-9, 12-10, and 12-11). Underground fires in coal mines cannot always be put out. Land above underground mines collapses or subsides, causing houses to tilt, sewer lines to crack, gas mains to break, and groundwater systems to be disrupted. In addition, spoil heaps and tailings can be eroded by wind and water. The air can be contaminated with dust and toxic substances, and water pollution is a serious concern.

Past and present mining operations for metallic and nonmetallic minerals occupy only a small percentage of the total land area in any country (0.25% of the

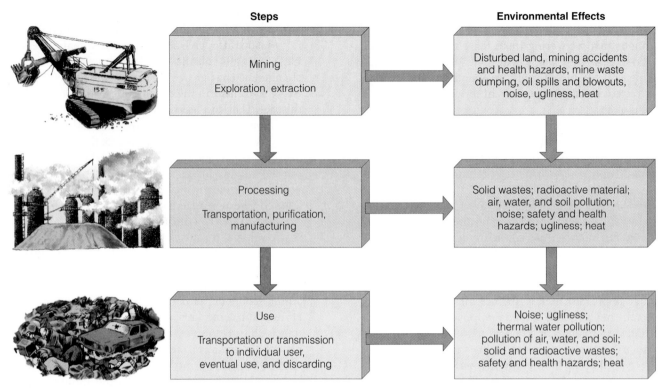

Steps	Environmental Effects
Mining Exploration, extraction	Disturbed land, mining accidents and health hazards, mine waste dumping, oil spills and blowouts, noise, ugliness, heat
Processing Transportation, purification, manufacturing	Solid wastes; radioactive material; air, water, and soil pollution; noise; safety and health hazards; ugliness; heat
Use Transportation or transmission to individual user, eventual use, and discarding	Noise; ugliness; thermal water pollution; pollution of air, water, and soil; solid and radioactive wastes; safety and health hazards; heat

Figure 12-13 Some harmful environmental effects of mineral extraction, processing, and use. The energy used to carry out each step causes additional pollution and environmental degradation. Harm could be minimized by requiring mining, processing, and manufacturing companies to include the full costs of the pollution and environmental degradation in the prices of their products. Many of these *external* costs are now passed on to society in the form of poorer health, increased health and insurance costs, and increased taxes to deal with pollution and environmental degradation.

United States). However, the scars from mining are long lasting (Figure 12-9), and the resulting air and water pollution can extend beyond the limits of ground disturbance.

Rainwater seeping through a mine or mine wastes can carry sulfuric acid (H_2SO_4, produced when aerobic bacteria act on iron sulfide minerals in spoil) to nearby streams and groundwater (Figure 12-14). Such *acid mine drainage* can destroy aquatic life and contaminate water supplies. Other harmful materials that either run off or are dissolved from underground mines or aboveground mining wastes include radioactive uranium compounds and compounds of toxic metals such as lead, mercury, arsenic, or cadmium (Figure 12-1).

After extraction from the ground, many resources must be separated from other matter, a process that can pollute the air and water. Ore, for example, typically contains two parts: the ore mineral, which contains the desired metal, and the **gangue**, which is the waste mineral material. **Beneficiation**, or separation in a mill of the ore mineral from the gangue, produces waste called **tailings.** Piles of tailings are ugly, and toxic metals blown or leached from them by rainfall can contaminate surface and groundwater.

Most ore minerals do not consist of pure metal, so **smelting** is done to separate the metal from the other elements in the ore mineral. Without effective pollution control equipment, smelters emit enormous quantities of air pollutants, which damage vegetation and soils in the surrounding area. Pollutants include sulfur dioxide, soot, and tiny particles of arsenic, cadmium, lead, and other toxic elements and compounds found in many ores.

Decades of uncontrolled sulfur dioxide emissions from copper-smelting operations near Copperhill and Ducktown, Tennessee, killed all the vegetation over a wide area around the smelter; another dead vegetation zone was created around the Sudbury, Ontario, nickel smelter (Figure 9-14). New dead vegetation zones have formed in parts of eastern Europe, the former Soviet Union, and Chile. Smelters also cause water pollution and produce liquid and solid hazardous wastes that must be disposed of safely.

Some companies are using improved technology to reduce pollution from smelting, reduce production costs, and save costly cleanup bills and liability for damages. For example, the new $880-million Kennecott smelter refinery at Bingham, Utah, is expected to be

Q: What percentage of the world's population relies on plants or plant extracts for medicines?

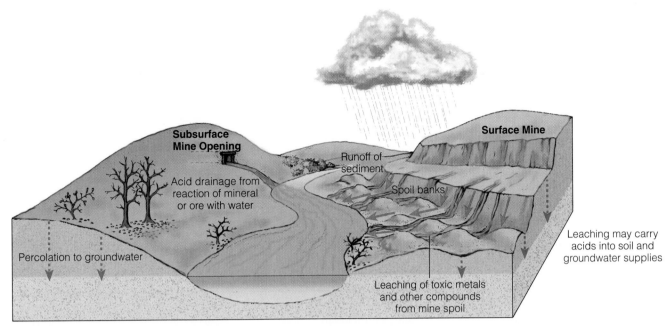

Figure 12-14 Pollution and degradation of a stream and groundwater by runoff of acids—called acid mine drainage—and by toxic chemicals from surface and subsurface mining. These substances can kill fish and other aquatic life. Acid mine drainage has damaged over 26,000 kilometers (16,000 miles) of streams in the United States, mostly in Appalachia and the west.

among the cleanest in the world. It also is projected to reduce production costs by 53%, making Kennecott one of the world's cheapest and cleanest copper producers.

Are There Environmental Limits to Resource Extraction and Use? Some environmentalists and resource experts believe that the greatest danger from high levels of resource consumption may not be the exhaustion of resources but the damage that their extraction, processing, and conversion to products do to the environment (Figure 12-13).

The minerals industry accounts for 5–10% of world energy use, making it a major contributor to air and water pollution and emissions of greenhouse gases. As more remote, deeper deposits are mined, even more energy will be needed to dig bigger holes and to transport the metal ores over greater distances.

The *grade* of an ore—its percentage of metal content—largely determines the environmental impact of mining it (Case Study, p. 352); more accessible and higher-grade ores are generally exploited first. As they are depleted, it takes more money, energy, water, and other materials to exploit lower-grade ores, and environmental effects increase accordingly.

Should the U.S. 1872 Mining Law Be Reformed to Reduce Environmental Harm and Save Taxpayers Money? Environmentalists point out that mining is the only natural resource industry in the United States that by law can buy public lands; it is also the only resource industry that pays no rents or royalties for resource extraction. Environmentalists and a growing number of citizens support a drastic reform of the 1872 mining law (p. 338), including the following changes:

- *Prohibiting buying of public land for mining but allowing such land to be leased for mining for up to 20 years.*

- *Requiring a full environmental impact assessment of the proposed mining activities before a mining lease is approved.*

- *Setting strict environmental standards for preventing and controlling pollution and environmental degradation resulting from mining activities during the period of the lease.*

- *Requiring companies leasing public land to post an environmental performance bond to cover estimated environmental damage and ecological restoration costs based on the environmental impact study of the project.* After such costs are deducted as needed and restoration is complete, remaining funds plus interest would be returned to the mining company.

- *Requiring mining companies to pay rent to cover all government (taxpayer) costs in evaluating and monitoring any leased mining site.*

- *Requiring mining companies to pay a 12.5% royalty on the gross (not net) values of all minerals removed.*

- *Making mining companies legally and financially responsible for environmental cleanup and restoration of each site.*

A: About 80%

The Environment and the New Gold Rush

Miners must extract and process massive quantities of soil and rock to end up with small quantities of gold. Gold miners typically remove ore equal to the weight of 50 automobiles to extract an amount of gold that would fit inside your clenched fist.

The mountains of solid waste remaining after gold is extracted from its ore are left piled near the mine sites and can pollute the air, surface water (Figure 12-1), and groundwater. Gold-bearing rock tends to contain large quantities of sulfur, which forms sulfuric acid when exposed to air and water (Figure 12-14). In addition to killing aquatic life, the acid puts highly toxic metals such as cadmium and copper into solution.

In Australia and North America, a new mining technology, called *cyanide heap leaching*, has been cheap enough to allow mining companies to level entire mountains containing very low-grade gold ore. To extract the gold, miners spray a cyanide solution (which reacts with gold) onto huge open-air piles of crushed ore. They then collect the solution in leach beds and overflow ponds, recirculate it a number of times, and extract gold from it. Unfortunately, cyanide is extremely toxic and can be harmful or lethal to people,

plants, and wildlife, especially to birds and mammals drawn to cyanide collection ponds as a source of water.

Cyanide leach pads and collection ponds can also leak or overflow, posing threats to underground drinking water supplies and to wildlife (especially fish) in lakes and streams. Special liners beneath the ore heaps and in the collection ponds are supposed to prevent leaks, but some have failed; according to the EPA all such liners will eventually leak.

In the United States, companies have used the 1872 mining law to buy public land for practically nothing, mine a site, abandon it, file for bankruptcy, and leave the public with the cleanup bill (p. 338). A glaring example is the Summitville gold mine site in the San Juan Mountains of southern Colorado. A Canadian company bought the land from the federal government for a pittance, spent $1 million developing the site, and then removed $98 million worth of gold. Shoddy construction allowed acids and toxic metals to leak from the site and poison a 27-kilometer (17-mile) stretch of the Alamosa River, the source of irrigation water for farms and ranches in the San Luis Valley.

The company then declared bankruptcy and abandoned the property, but only after being

allowed to retrieve $2.5 million of the $7.5 million reclamation bond it had posted with the state. Summitville is now a Superfund site; the EPA spends $40,000 a day just to contain the site's toxic wastes. Ultimately, the EPA expects to spend about $120 million to finish the cleanup.

The gold rush of the 1980s and 1990s has also caused millions of miners—many of them landless poor—in various Latin American, Asian, and African developing countries to stream into tropical forests and other areas in search of gold. These small-scale miners use destructive mining techniques such as digging large pits by hand, river dredging, and hydraulic mining (a technique, outlawed in the United States, in which water jets wash entire hillsides into sluice boxes). Highly toxic mercury is usually used to extract the gold from the other materials. In the process, much of the mercury ends up contaminating water supplies and fish consumed by people.

Critical Thinking

Do you believe that the harmful environmental impacts of gold mining should be much more strictly regulated in **(a)** the United States and **(b)** other countries? If so, what major regulations would you like to see? How would you see that such controls are implemented and enforced?

Mining companies claim that charging royalties for minerals taken from public lands and requiring them to pay for cleanup will force them to do their mining in other countries, which would cost American jobs and reduce tax revenues. They also argue that their average cost for patenting public land under the 1872 law is about $42,000 per hectare ($17,000 per acre) when their mining development costs are included.

Environmentalists counter that mining companies would still make a reasonable profit on the high-value minerals such as gold and platinum they get from public lands and that threats to move operations elsewhere are a rarely implemented scare tactic (green-

mail). For example, gold costs miners about $30 per ounce to extract, but in recent years it has been sold for $320–395 per ounce. Even with a 12.5% royalty and responsibility for cleanup costs—as required for oil, gas, and coal companies—hard-rock mining companies can turn a hefty profit on high-price minerals such as gold and platinum.

Environmentalists also point out that Canada, Australia, South Africa, and other countries that are major extractors of hard-rock minerals don't sell public lands to mining companies, and they require the companies to pay rent on any public land they lease and royalties on the minerals they extract.

Q: What percentage of the medicines sold in the world have active ingredients extracted from wildlife (mostly plants)?

Mostly because of the political influence of mining companies and their congressional allies, this law stands little chance of serious reform in the near future without intense pressure from citizens.

12-5 SOIL: THE BASE OF LIFE

What Are the Major Layers Found in Mature Soils? The material we call **soil** is a complex mixture of eroded rock, mineral nutrients, decaying organic matter, water, air, and billions of living organisms, most of them microscopic decomposers (Figure 12-15). Although soil is a potentially renewable resource, it is produced very slowly by the weathering of rock, deposit of sediments by erosion, and decomposition of organic matter in dead organisms.

Mature soils are arranged in a series of zones called **soil horizons**, each with a distinct texture and composition that varies with different types of soils. A cross-sectional view of the horizons in a soil is called a **soil profile**. Most mature soils have at least three of the possible horizons (Figure 12-15).

The top layer, the *surface-litter layer*, or *O horizon*, consists mostly of freshly fallen and partially decomposed leaves, twigs, animal waste, fungi, and other organic materials. Normally, it is brown or black. The *topsoil layer*, or *A horizon*, is a porous mixture of partially decomposed organic matter, called **humus**, and some inorganic mineral particles. It is usually darker and looser than deeper layers. The roots of most plants and most of a soil's organic matter are concentrated in these two upper layers. As long as these layers are anchored by vegetation, soil stores water and releases it in a nourishing trickle instead of a devastating flood.

The two top layers of most well-developed soils teem with bacteria, fungi, earthworms, and small insects that interact in complex food webs (Figure 12-16). Bacteria and other decomposer microorganisms found

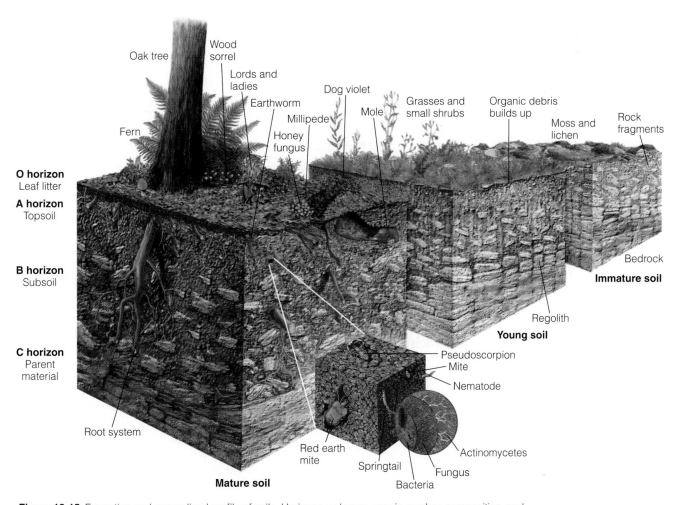

O horizon
Leaf litter

A horizon
Topsoil

B horizon
Subsoil

C horizon
Parent material

Oak tree Wood sorrel Lords and ladies Dog violet Grasses and small shrubs Organic debris builds up Moss and lichen Rock fragments

Fern Earthworm Mole

Honey fungus Millipede

Root system

Red earth mite Springtail Bacteria Fungus Actinomycetes Nematode Mite Pseudoscorpion

Bedrock

Immature soil

Regolith

Young soil

Mature soil

Figure 12-15 Formation and generalized profile of soils. Horizons, or layers, vary in number, composition, and thickness, depending on the type of soil. (Used by permission of Macmillan Publishing Company from Derek Elsom, *Earth*, New York: Macmillan, 1992. Copyright ©1992 by Marshall Editions Developments Limited)

A: At least 45% (25% in the United States), worth $100 billion per year

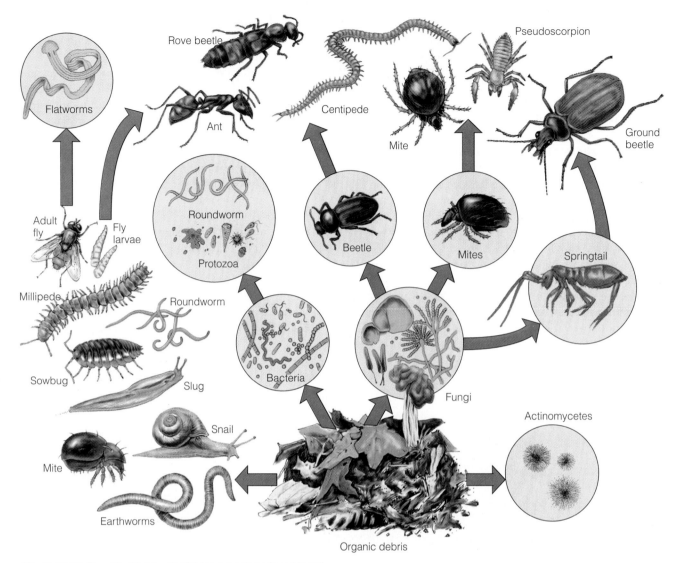

Figure 12-16 Simplified food web of living organisms found in soil.

by the billions in every handful of topsoil recycle the nutrients we and other land organisms need (Figure 12-17). They break down some complex organic compounds into simpler inorganic compounds soluble in water. Soil moisture carrying these dissolved nutrients is drawn up by the roots of plants and transported through stems and into leaves.

Some organic litter in the two top layers is broken down into a sticky, brown residue of partially decomposed organic material (humus). Because this humus is only slightly soluble in water, most of it stays in the topsoil layer. A fertile soil that produces high crop yields has a thick topsoil layer with lots of humus, which helps topsoil hold water and nutrients taken up by plant roots.

The color of its topsoil tells us a lot about how useful a soil is for growing crops. For example, dark-brown or black topsoil is nitrogen-rich and high in

organic matter. Gray, bright yellow, or red topsoils are low in organic matter and will need nitrogen enrichment to support most crops.

The *B horizon (subsoil)* and the *C horizon (parent material)* contain most of a soil's inorganic matter, mostly broken-down rock consisting of varying mixtures of sand, silt, clay, and gravel. The C horizon lies on a base of unweathered parent rock called *bedrock*.

The spaces, or pores, between the solid organic and inorganic particles in the upper and lower soil layers contain varying amounts of air (mostly nitrogen and oxygen gas) and water. Plant roots need oxygen for cellular respiration.

Some of the precipitation that reaches the soil percolates through the soil layers and occupies many of the soil's open spaces or pores. This downward movement of water through soil is called **infiltration**. As the water seeps down, it dissolves various soil compo-

Q: What percentage of the world's estimated plant species have been evaluated for their medical uses?

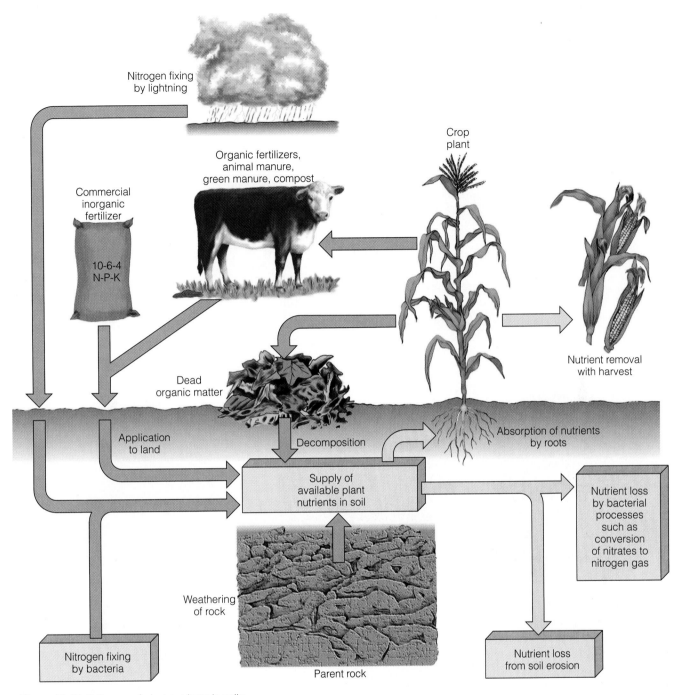

Figure 12-17 Pathways of plant nutrients in soils.

nents in upper layers and carries them to lower layers in a process called **leaching**.

Soils develop and mature slowly. The earth's current mature soils vary widely from biome to biome in color, content, pore space, acidity, and depth. Five important soil types, each with a distinct profile, are shown in Figure 12-18. Most of the world's crops are grown on soils exposed when grasslands and deciduous forests are cleared.

How Do Soils Differ in Texture, Porosity, and Acidity? Soils vary in their content of *clay* (very fine particles), *silt* (fine particles), *sand* (medium-size particles), and *gravel* (coarse to very coarse particles). The relative amounts of the different sizes and types of mineral particles determine **soil texture**, as depicted in Figure 12-19. Soils with roughly equal mixtures of clay, sand, silt, and humus are called **loams**.

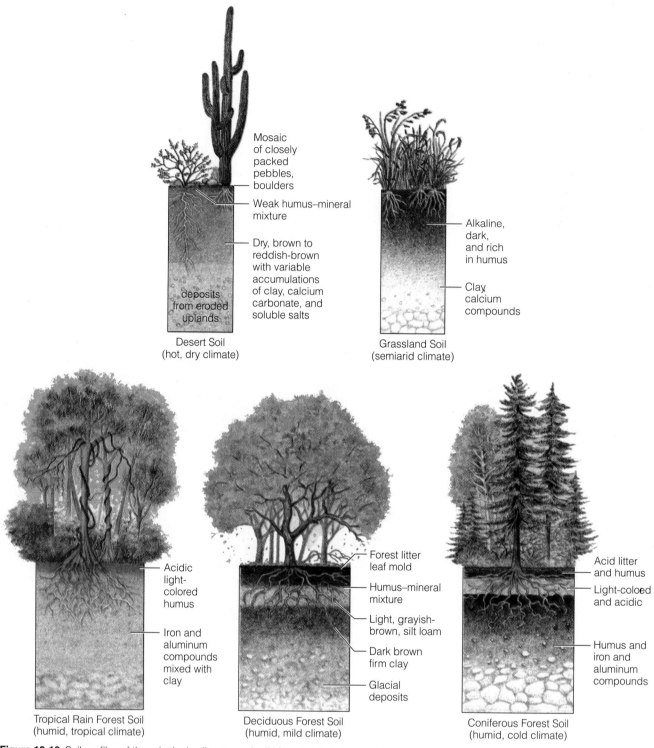

Figure 12-18 Soil profiles of the principal soil types typically found in five different biomes.

Labels in figure:

Desert Soil (hot, dry climate)
- Mosaic of closely packed pebbles, boulders
- Weak humus–mineral mixture
- Dry, brown to reddish-brown with variable accumulations of clay, calcium carbonate, and soluble salts
- deposits from eroded uplands

Grassland Soil (semiarid climate)
- Alkaline, dark, and rich in humus
- Clay, calcium compounds

Tropical Rain Forest Soil (humid, tropical climate)
- Acidic light-colored humus
- Iron and aluminum compounds mixed with clay

Deciduous Forest Soil (humid, mild climate)
- Forest litter leaf mold
- Humus–mineral mixture
- Light, grayish-brown, silt loam
- Dark brown firm clay
- Glacial deposits

Coniferous Forest Soil (humid, cold climate)
- Acid litter and humus
- Light-colored and acidic
- Humus and iron and aluminum compounds

To get an idea of a soil's texture, take a small amount of topsoil, moisten it, and rub it between your fingers and thumb. A gritty feel means that it contains a lot of sand. A sticky feel means a high clay content, and you should be able to roll it into a clump. Silt-laden soil feels smooth, like flour. A loam topsoil, which is best suited for plant growth, has a texture between these extremes—a crumbly, spongy feeling—with many of its particles clumped loosely together.

Soil texture helps determine **soil porosity**, a measure of the volume of pores or spaces per volume of soil and of the average distances between those spaces.

Q: How many of earth's species are believed to become extinct each year because of human activities?

A porous soil has many pores and can hold more water and air than a less porous soil. The average size of the spaces or pores in a soil determines **soil permeability**: the rate at which water and air move from upper to lower soil layers. Soil porosity is also influenced by **soil structure**: how soil particles are organized and clumped together.

Soil texture, porosity, and permeability determine a soil's *water-holding capacity*, *aeration* or *oxygen content* (the ability of air to move through the soil), and *workability* (how easily it can be cultivated). Table 12-1 compares the main physical and chemical properties of sand, clay, silt, and loam soils.

Loams are the best soils for growing most crops because they hold lots of water, but not too tightly for

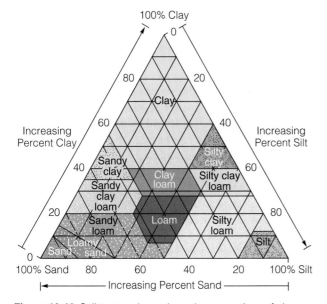

Figure 12-19 Soil texture depends on the proportions of clay, silt, and sand particles in the soil. Soil texture affects soil porosity, the average number and spacing of pores in a given volume of soil. Loams—roughly equal mixtures of clay, sand, silt, and humus—are the best soils for growing most crops. (Data from Soil Conservation Service)

plant roots to absorb. Sandy soils are easy to work, but water flows rapidly through them. They are useful for growing irrigated crops or those with low water requirements, such as peanuts and strawberries.

The particles in clay soils are very small and easily compacted. When these soils get wet, they form large, dense clumps, which is why wet clay can be molded into bricks and pottery. Clay soils are more porous and have a greater water-holding capacity than sandy soils, but the pore spaces are so small that these soils have a low permeability. Because little water can infiltrate to lower levels, the upper layers can easily become too waterlogged for most crops.

A numerical scale of *pH* values is used to compare the acidity and alkalinity in water solutions (Figure 9-7). The pH of a soil influences the uptake of soil nutrients by plants, which vary in the pH ranges they can tolerate.

When soils are too acidic, the acids can be partially neutralized by an alkaline substance such as lime. Because lime speeds up the decomposition of organic matter in the soil, however, manure or another organic fertilizer should also be added to maintain soil fertility.

In dry regions such as much of the western and southwestern United States, rain does not leach away calcium and other alkaline compounds, so soils in such areas may be too alkaline (pH above 7.5) for some crops. Adding sulfur, which is gradually converted into sulfuric acid by soil bacteria, reduces soil alkalinity.

🌐 12-6 SOIL EROSION

What Causes Soil Erosion? **Soil erosion** is the movement of soil components, especially surface litter and topsoil, from one place to another. The two main agents of erosion are flowing water and wind. Some soil erosion is natural—the long-term wearing down of mountains and building up of plains and deltas

Table 12-1 Useful Properties of Soils with Different Textures					
Soil Texture	Nutrient-Holding Capacity	Water-Infiltration Capacity	Water-Holding Capacity	Aeration	Workability
Clay	Good	Poor	Good	Poor	Poor
Silt	Medium	Medium	Medium	Medium	Medium
Sand	Poor	Good	Poor	Good	Good
Loam	Medium	Medium	Medium	Medium	Medium

Figure 12-20 Rill and gully erosion of vital topsoil from irrigated cropland in Arizona. (Soil Conservation Service)

due to the combined action of physical, chemical, and biological forces (Figure 12-5). In undisturbed vegetated ecosystems, the roots of plants help anchor the soil, and usually soil is not lost faster than it forms.

However, farming, logging, construction, overgrazing by livestock, off-road vehicles, deliberate burning of vegetation, and other activities that destroy plant cover leave soil vulnerable to erosion. Such human activities can speed up erosion and destroy in a few decades what nature took hundreds to thousands of years to produce. In 1937, U.S. President Franklin D. Roosevelt sent a letter to the governors of the states in which he said, *"The nation that destroys its soil destroys itself."*

Most soil erosion is caused by moving water. Soil scientists distinguish among three types of water erosion. *Sheet erosion* occurs when surface water moves down a slope or across a field in a wide flow and peels off fairly uniform sheets or layers of soil. Because the topsoil disappears evenly, sheet erosion may not be noticeable until much damage has been done. In *rill erosion* the surface water forms fast-flowing rivulets that cut small channels in the soil (Figure 12-20). In *gully erosion*, rivulets of fast-flowing water join together and with each succeeding rain cut the channels wider and deeper until they become ditches or gullies (Figure 12-20). Gully erosion usually happens on steep slopes where all or most vegetation has been removed.

Losing topsoil makes a soil less fertile and less able to hold water. The resulting sediment, the largest source of water pollution, clogs irrigation ditches, boat channels, reservoirs, and lakes. The sediment-laden water is cloudy and tastes bad, fish die, and flood risk increases. Rivers running brown with silt from human-accelerated soil erosion contain earth capital that is hemorrhaging from the land (Figure 5-17).

Soil, especially topsoil, is classified as a potentially renewable resource because it is continuously regenerated by natural processes. However, in tropical and temperate areas it takes 200–1,000 years (depending on climate and soil type) for 2.54 centimeters (1 inch) of new topsoil to form. If topsoil erodes faster than it forms on a piece of land, the soil becomes a nonrenewable resource. Annual erosion rates for farmlands throughout the world are 7–100 times the natural renewal rate (Guest Essay, pp. 366–367). Soil erosion is milder on forestland and rangeland than on cropland, but forest soil takes two to three times longer to restore itself than does cropland. Construction sites usually have the highest erosion rates by far.

How Serious Is Global Soil Erosion? According to a UN Environment Programme survey, topsoil is eroding faster than it forms on about one-third of the world's cropland, causing an estimated 85% of the world's land degradation from human activities.

A 1992 study by the World Resources Institute and the UN Environment Programme found that soil on an area equal to the size of China and India combined had been seriously eroded since 1945 (Figure 12-21). The study also found that about 15% of land scattered across the globe was too eroded to grow crops because of a combination of overgrazing (35%), deforestation (30%), and unsustainable farming (28%). Two-thirds of the seriously degraded lands are in Asia and Africa.

Each year we must feed about 87 million more people with an estimated 24 billion metric tons (26 billion tons) less topsoil, eroded as a result of human activities. The topsoil that washes and blows into the world's streams, lakes, and oceans each year would fill a train of freight cars long enough to encircle the planet 150 times. At that rate, the world is losing about 7% of its topsoil from actual or potential cropland each decade.

This situation is worsening as many poor farmers in developing countries plow up marginal (easily erodible) lands to survive. According to a 1990 UN study, agricultural mismanagement, overgrazing, deforestation, and overharvesting of fuelwood, mostly by the poor, account for about 70% of the damage done to the world's soil. In 1995, soil expert David Pimentel (Guest Essay, pp. 366–367) estimated that soil erosion causes nearly $400 billion per year worldwide in direct damage to agricultural lands and indirect damage to waterways, infrastructure, and human health, an average of $46 million in damages per hour.

Some critics say that there is no accurate way to measure global soil erosion and that the estimates of erosion and the resulting environmental and health costs are overblown. Soil erosion experts agree that all we can ever have is rough estimates, but they contend

Q: What is the greatest threat to wild species?

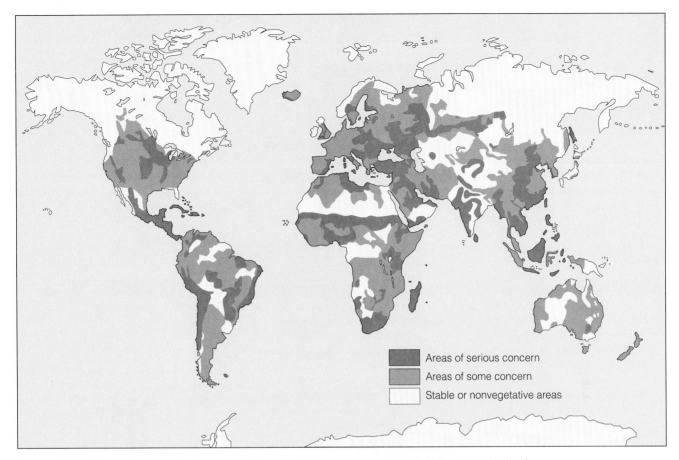

Figure 12-21 Global soil erosion. (Data from UN Environmental Programme and the World Resources Institute)

that such estimates from numerous sources show a growing and alarming increase in soil erosion, regardless of the exact numbers involved.

How Serious Is Soil Erosion in the United States? According to the Soil Conservation Service, about one-third of the nation's original prime topsoil has been washed or blown into streams, lakes, and oceans, mostly as a result of overcultivation, overgrazing, and deforestation (Case Study, p. 360).

Today, soil on cultivated land in the United States is eroding about 16 times faster than it can form. Erosion rates are even higher in heavily farmed regions, including the Great Plains, which has lost one-third or more of its topsoil in the 150 years since it was first plowed. Some of the country's most productive agricultural lands, such as those in Iowa, have lost about half their topsoil. California's soil is eroding about 80 times faster than it can be formed.

The estimated amount of topsoil that erodes away each day in the United States would fill a line of dump trucks 5,600 kilometers (3,500 miles) long. In 1995, soil expert David Pimentel (Guest Essay, pp. 366–367) estimated that the direct and indirect costs of soil erosion

and runoff in the United States exceed $44 billion per year—an average loss of $5 million per hour.

Critics say that estimates of soil erosion and damages from such erosion are overblown. They point to studies by several soil scientists concluding that if current rates of cropland erosion in the United States continue for 100 years, crop yields will be only 3–10% less than they would be without such erosion.

What Is Desertification, and How Serious Is This Problem? Desertification is a process whereby the productive potential of arid or semiarid land falls by 10% or more; this phenomenon results mostly from human activities. *Moderate desertification* is a 10–25% drop in productivity, *severe desertification* is a 25–50% drop, and *very severe desertification* is a drop of 50% or more, usually creating huge gullies and sand dunes. Desertification is a serious and growing problem in many parts of the world (Figure 12-23).

Practices that leave topsoil vulnerable to desertification include **(1)** overgrazing on fragile arid and semiarid rangelands, **(2)** deforestation without reforestation, **(3)** surface mining without land reclamation, **(4)** irrigation techniques that lead to increased erosion,

The Dust Bowl

CASE STUDY

In the 1930s, Americans learned a harsh environmental lesson when much of the topsoil in several midwestern states was lost through a combination of poor cultivation practices and prolonged drought.

Windy and dry, the vast grasslands of the Great Plains stretch across 10 states, from Texas through Montana and the Dakotas. Before settlers began grazing livestock and planting crops there in the 1870s, the deep and tangled root systems of native prairie grasses anchored the fertile topsoil firmly in place (Figure 12-18). Plowing the prairie tore up these roots, and the agricultural crops the settlers planted annually in their place had less extensive root systems.

After each harvest, the land was plowed and left bare for several months, exposing it to the plains winds. Overgrazing also destroyed large expanses of grass, denuding the ground. The stage was set for severe wind erosion and crop failures; all that was needed was a long drought.

Such a drought occurred between 1926 and 1934. In the 1930s, dust clouds created by hot, dry windstorms darkened the sky at midday in some areas; rabbits and birds choked to death on the dust. During May 1934, the entire eastern United States was blanketed by a

cloud of topsoil blown off the Great Plains, some 2,400 kilometers (1,500 miles) away. Journalists gave the Great Plains a new name: the *Dust Bowl* (Figure 12-22).

During the so-called Dirty Thirties, cropland equal in area to Connecticut and Maryland combined was stripped of topsoil, and an area the size of New Mexico was severely eroded. Thousands of displaced farm families from Oklahoma, Texas, Kansas, and Colorado migrated to California or to the industrial cities of the Midwest and East. Most found no jobs because the country was in the midst of the Great Depression.

In May 1934 Hugh Bennett of the U.S. Department of Agriculture (USDA) went before a congressional hearing in Washington to plead for new programs to protect the country's topsoil. Lawmakers took action when Great Plains dust began seeping into the hearing room.

In 1935, the United States passed the Soil Erosion Act, which established the Soil Conservation Service (SCS) as part of the USDA. With Bennett as its first head, the SCS began promoting sound conservation practices, first in the Great Plains states and later elsewhere. Soil conservation districts were formed throughout the country, and farmers and ranchers were given technical assistance in setting up soil conservation programs.

Unfortunately, these heroic efforts have not yet stopped human-accelerated erosion in the Great Plains. The basic problem is that much of the region is better suited for moderate grazing than for farming. If the earth warms as projected, the region could become even drier, and farming might have to be abandoned.

Critical Thinking

Do you think Americans learned a lesson about protecting soil as a result of the Dust Bowl in the 1930s? Explain.

Figure 12-22 The Dust Bowl of the Great Plains, where a combination of extreme drought and poor soil conservation practices led to severe wind erosion of topsoil in the 1930s. Note the connection of this area with the Ogallala aquifer in Figure 11-12.

(5) salt buildup and waterlogged soil, (6) farming on land with unsuitable terrain or soils, and (7) soil compaction by farm machinery and cattle hoofs. The consequences of desertification include worsening drought, famine, declining living standards, and swelling numbers of environmental refugees whose land is too eroded to grow crops or feed livestock.

An estimated 8.1 million square kilometers (3.1 million square miles)—an area the size of Brazil and 12 times the size of Texas—have become desertified in the past 50 years. This threatens the livelihoods of at least 900 million people in 100 countries. If current trends continue, within a few years desertification could threaten the livelihoods of 1.2 billion people.

Every year, low to moderate new desertification occurs on an estimated 60,000 square kilometers (23,000 square miles, an area the size of West Virginia); another 210,000 square kilometers (81,000 square miles, an area the size of Kansas) undergo severe desertification and lose so much soil and fertility that they are no longer economically valuable for farming or grazing.

Q: What do most wildlife scientists believe is the best way to prevent wildlife extinction from human activities?

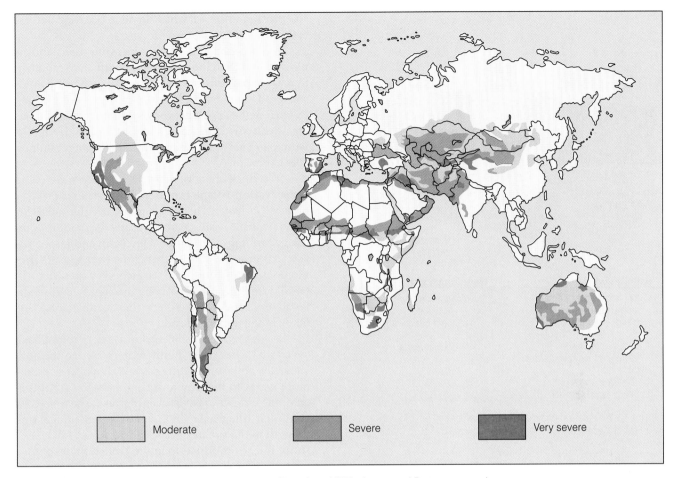

| Moderate | Severe | Very severe |

Figure 12-23 Desertification of arid and semiarid lands. (Data from UN Environmental Programme and Harold E. Dregnue)

Solutions: How Can We Slow Desertification?

The most effective way to slow desertification is to drastically reduce overgrazing, deforestation, and the destructive forms of planting, irrigation, and mining that are to blame. In addition, planting trees and grasses will anchor soil and hold water while slowing desertification and reducing the threat of global warming (Section 10-5).

Such prevention and rehabilitation would cost $10–22 billion annually for the next 20 years. This expenditure is considerably less than the estimated $42 billion annual loss in agricultural productivity from desertified land; once this potential productivity is restored, the cost of the program could be quickly recouped. So far, less than $1 billion per year is spent globally to halt this form of land degradation.

How Do Excess Salts and Water Degrade Soils?

The approximately 16% of the world's cropland that is now irrigated by various methods (Figure 11-14) produces about one-third of the world's food. Irrigated land can produce crop yields that are two to three times greater than those from rain watering, but irrigation has its downside. Most irrigation water is a dilute solution of various salts, picked up as the water flows over or through soil and rocks.

Irrigation water not absorbed into the soil evaporates, leaving behind a thin crust of dissolved salts (such as sodium chloride) in the topsoil. The accumulation of these salts, called **salinization** (Figure 12-24), stunts crop growth, lowers yields, and eventually kills plants and ruins the land. According to a 1995 study, severe salinization has reduced yields on 20% of the world's irrigated cropland, and another 30% has been moderately salinized. The most severe salinization occurs in Asia, especially in China, India, and Pakistan.

Precipitation can desalinate soil, but this takes thousands of years in arid and semiarid areas where irrigation is used. Salts can be flushed out of soil by applying much more irrigation water than is needed for crop growth, but this practice increases pumping and crop-production costs, wastes enormous amounts of water, and waterlogs plants if the water table rises close to the surface.

A: Establish a worldwide network of protected habitat areas

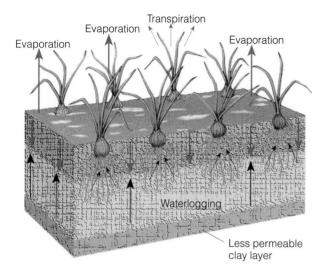

Evaporation   Evaporation   Transpiration   Evaporation

Waterlogging

Less permeable
clay layer

Salinization

1. Irrigation water contains
   small amounts of
   dissolved salts.

2. Evaporation and
   transpiration leave salts
   behind.

3. Salt builds up in soil.

Waterlogging

1. Precipitation and
   irrigation water
   percolate downward.

2. Water table rises.

Figure 12-24 Salinization and waterlogging of soil on irrigated
land without adequate drainage lead to decreased crop yields.

Heavily salinized soil can also be renewed by taking the land out of production for 2 to 5 years, installing
an underground network of perforated drainage pipes,
and flushing the soil with large quantities of low-salt
water. However, this costly scheme only slows the
salt buildup; it does not stop the process. Flushing salts
from the soil also makes downstream irrigation water
saltier unless the saline water can be drained into evaporation ponds rather than returned to the stream or canal.

Another problem with irrigation is **waterlogging**
(Figure 12-24). Farmers often apply large amounts of
irrigation water to leach salts deeper into the soil.
Without adequate drainage, however, water accumulates underground, gradually raising the water table.
Saline water then envelops the deep roots of plants,
lowering their productivity and killing them after prolonged exposure. At least one-tenth of all irrigated
land worldwide suffers from waterlogging, and the
problem is getting worse.

12-7 SOLUTIONS: SOIL CONSERVATION

**How Can Conservation Tillage Reduce Soil
Erosion?** **Soil conservation** involves reducing soil
erosion and restoring soil fertility. For hundreds of
years, farmers have used various methods to reduce

soil erosion, most of which involve keeping the soil
covered with vegetation.

In **conventional-tillage farming** the land is
plowed and then the soil is broken up and smoothed
to make a planting surface. In areas such as the midwestern United States, harsh winters prevent plowing
just before the spring growing season. Thus, cropfields
are often plowed in the fall, baring the soil during the
winter and early spring months and leaving it vulnerable to erosion.

To reduce erosion, many U.S. farmers are using
conservation-tillage farming (either *minimum-tillage*
or *no-till farming*). The idea is to disturb the soil as little as possible while planting crops. With minimum-
tillage farming, special tillers break up and loosen the
subsurface soil without turning over the topsoil, previous crop residues, and any cover vegetation. In
no-till farming, special planting machines inject seeds,
fertilizers, and weed-killers (herbicides) into slits
made in the unplowed soil.

Besides reducing soil erosion, conservation tillage
saves fuel, cuts costs, holds more water in the soil,
keeps the soil from getting packed down, and allows
more crops to be grown during a season (multiple
cropping). Yields are at least as high as those from conventional tillage. At first, conservation tillage was
thought to require more herbicides, but a 1990 USDA
study of corn production in the United States found
no real difference in levels of herbicide use between
conventional and conservation tillage systems. However, no-till cultivation of corn does leave stalks, which
can serve as habitats for the corn borer; this can potentially increase the use of pesticides.

By 1994 conservation tillage was used on about
one-third of U.S. croplands and is projected to be used
on over half of it by 2005. The USDA estimates that
using conservation tillage on 80% of U.S. cropland
would reduce soil erosion by at least half. So far, the
practice is not widely used in other parts of the world.

**How Can Terracing, Contour Farming, Strip
Cropping, and Alley Cropping Reduce Soil
Erosion?** **Terracing** can reduce soil erosion on steep
slopes, each of which is converted into a series of broad,
nearly level terraces that run across the land contour
(Figure 12-25a). Terracing retains water for crops at each
level and reduces soil erosion by controlling runoff.

In mountainous areas such as the Himalayas and
the Andes, farmers traditionally built elaborate systems of terraces to grow crops. Today, however, some
of these slopes are being farmed without terraces, leaving the land too nutrient poor to grow crops or generate new forest after only 10–40 years. Although most
poor farmers know the risk of not terracing, many have
too little time and too few workers to build terraces;
they must plant crops or starve. The resultant loss of

Q: What percentage of the earth's land surface has been set aside to protect wildlife?

a

b

c

d

Figure 12-25 Soil conservation methods: **(a)** terracing in Bali, **(b)** contour planting and strip cropping in Illinois, **(c)** alley cropping in Peru, and **(d)** windbreaks in South Dakota. (Clockwise from top left, Prato/Bruce Coleman Ltd.; U.S. Soil Conservation Service; P. A. Sanchez/North Carolina State University; U.S. Soil Conservation Service)

protective vegetation and topsoil (Figure 11-17) also greatly intensifies flooding below these watersheds.

Soil erosion can be reduced by 30–50% on gently sloping land by means of **contour farming**: plowing and planting crops in rows across, rather than up and down, the sloped contour of the land (Figure 12-25b). Each row planted along the contour of the land acts as a small dam to help hold soil and slow water runoff.

In **strip cropping**, a row crop such as corn alternates in strips with another crop (such as a grass or a grass–legume mixture) that completely covers the soil and thus reduces erosion (Figure 12-25b). The strips of the cover crop trap soil that erodes from the row crop. They also catch and reduce water runoff and help prevent the spread of pests and plant diseases. Nitrogen-fixing legumes such as soybeans or alfalfa planted in some of the strips help restore soil fertility.

Erosion can also be reduced by **alley cropping**, or **agroforestry**, a form of *intercropping* in which several crops are planted together in strips or alleys between trees and shrubs that can provide fruit or fuelwood (Figure 12-25c). The trees provide shade (which reduces water loss by evaporation) and help to retain and slowly release soil moisture. The tree and shrub trimmings can be used as mulch (green manure) for the crops and as fodder for livestock.

How Can Gully Reclamation, Windbreaks, Land Classification, and PAM Reduce Soil Erosion?
Gully reclamation can restore sloping bare land on which water runoff quickly creates gullies. Small gullies can be seeded with quick-growing plants such as oats, barley, and wheat for the first season, whereas deeper gullies can be dammed to collect silt and

A: About 6%, compared to a minimum of 10% wildlife scientists say is needed

gradually fill in the channels. Fast-growing shrubs, vines, and trees can also be planted to stabilize the soil, and channels can be built to divert water from the gully and prevent further erosion.

Windbreaks, or **shelterbelts**, can reduce wind erosion. Long rows of trees are planted to partially block the wind (Figure 12-25d). Windbreaks, which are especially effective if uncultivated land is kept covered with vegetation, also help retain soil moisture, supply some wood for fuel, and provide habitats for birds, pest-eating and pollinating insects, and other animals. Unfortunately, many of the windbreaks planted in the upper Great Plains after the 1930s Dust Bowl disaster have been cut down to make way for large irrigation systems and modern farm machinery.

Land can be evaluated with the goal of identifying easily erodible (marginal) land that should neither be planted in crops nor cleared of vegetation. In the United States, the Soil Conservation Service (SCS) has set up a classification system to identify types of land that are suitable or unsuitable for cultivation. The SCS basically relies on voluntary compliance with its guidelines in the almost 3,000 local and state soil and water conservation districts it has established, and it provides technical and economic assistance through local district offices.

A chemical called polyacrylamide (PAM) has been used recently to sharply reduce erosion of some irrigated fields, at moderate cost in most soils (except desert soils). Tests show that adding just 10 parts per million (ppm) of this white crystal to water during the first hour of irrigation can reduce soil erosion by 70–99%. After that, irrigation can continue for 19–24 hours without further treatment. The negatively charged PAM particles may work by binding to the positively charged clay particles in soils, thereby reducing erosion by increasing the cohesiveness of surface soil particles.

Case Study: Slowing Soil Erosion in the United States

Of the world's major food-producing countries, only the United States is reducing some of its soil losses through conservation tillage and government-sponsored soil conservation programs.

The 1985 Farm Act established a strategy for reducing soil erosion in the United States. In the first phase of this program, farmers are given a subsidy for highly erodible land they take out of production and replant with soil-saving grass or trees for 10 years. The land in such a *conservation reserve* cannot be farmed, grazed, or cut for hay. Farmers who violate their contracts must pay back all subsidies plus interest.

By 1994, the authorized limit of 15 million hectares (37 million acres) of highly erodible land had been placed in conservation reserves, cutting soil erosion

on U.S. cropland by almost one-third. Between 1985 and 1995, this program has cut soil losses on cropland in the United States by about 60%—a shining example of good news—and could eventually cut such losses as much as 80%. In 1996 Congress reauthorized the Conservation Reserve Program until 2002.

The second phase of the program required all farmers with highly erodible land to develop SCS-approved 5-year soil conservation plans for their entire farms by the end of 1990. A third provision of the Farm Act authorizes the government to forgive all or part of farmers' debts to the Farmers Home Administration if they agree not to farm highly erodible cropland or wetlands for 50 years. The farmers must plant trees or grass on this land or restore it to wetland.

In 1987, however, the SCS eased the standards that farmers' soil conservation plans must meet to keep them eligible for other subsidies. Environmentalists have also accused the SCS of laxity in enforcing the Farm Act's "swampbuster" provisions, which deny federal funds to farmers who drain or destroy wetlands on their property. In 1996 Congress weakened the act by allowing farmers to end their contracts without USDA approval. Despite some weaknesses, the 1985 Farm Act makes the United States the first major food-producing country to make soil conservation a national priority.

Even though these efforts to slow soil erosion are an important step, effective soil conservation is practiced on only about half of all U.S. agricultural land and on less than half of the country's most erodible cropland.

How Can We Maintain and Restore Soil Fertility?

Fertilizers partially restore plant nutrients lost by erosion, crop harvesting, and leaching. Farmers can use either **organic fertilizer** from plant and animal materials or **commercial inorganic fertilizer** produced from various minerals.

Three basic types of *organic fertilizer* are animal manure, green manure, and compost. **Animal manure** includes the dung and urine of cattle, horses, poultry, and other farm animals. It improves soil structure, adds organic nitrogen, and stimulates beneficial soil bacteria and fungi. Despite its effectiveness, the use of animal manure in the United States has decreased. One reason is that separate farms for growing crops and raising animals have replaced most mixed animal-raising and crop-farming operations. Animal manure is available at feedlots near urban areas, but transporting it to distant rural crop-growing areas usually costs too much. Thus, much of this valuable resource is wasted and can end up polluting nearby bodies of water. In addition, tractors and other motorized farm machinery have replaced horses and other draft animals that naturally added manure to the soil.

Q: What percentage of the commercial energy used in the United States is wasted?

Green manure is fresh or growing green vegetation plowed into the soil to increase the organic matter and humus available to the next crop. **Compost** is a rich natural fertilizer and soil conditioner that aerates soil, improves its ability to retain water and nutrients, helps prevent erosion, and prevents nutrients from being wasted in landfills. Farmers, homeowners, and communities produce compost by piling up alternating layers of nitrogen-rich wastes (such as grass clippings, weeds, animal manure, and vegetable kitchen scraps), carbon-rich plant wastes (dead leaves, hay, straw, sawdust), and topsoil (Individuals Matter, right). This mixture provides a home for microorganisms that aid the decomposition of the plant and manure layers. Composting also reduces the amount of waste taken to landfills and incinerators.

Another method for conserving soil nutrients is **crop rotation**. Corn, tobacco, and cotton can deplete the topsoil of nutrients (especially nitrogen) if planted on the same land several years in a row. Farmers using crop rotation may plant areas or strips with such nutrient-depleting crops one year; the next year, however, they plant the same areas with legumes—whose root nodules (Figure 4-24) add nitrogen to the soil—or with crops such as soybeans, oats, barley, rye, or sorghum. This method helps restore soil nutrients and reduces erosion by keeping the soil covered with vegetation. It also helps reduce crop losses to insects by presenting them with a changing target.

Will Inorganic Fertilizers Save the Soil? Today, many farmers (especially in developed countries) rely on *commercial inorganic fertilizers* containing nitrogen (as ammonium ions, nitrate ions, or urea), phosphorus (as phosphate ions), and potassium (as potassium ions). Other plant nutrients may also be present in low or trace amounts.

Inorganic commercial fertilizers are easily transported, stored, and applied. Worldwide, their use increased about tenfold between 1950 and 1989 but declined by 12% between 1990 and 1996. Today, the additional food they help produce feeds one of every three people in the world; without them, world food output would plummet an estimated 40%.

Commercial inorganic fertilizers have some disadvantages, however. They do not add humus to the soil. Unless animal manure and green manure are also added, the soil's content of organic matter, and thus its ability to hold water, will decrease and the soil will become compacted and less suitable for crop growth. By decreasing the soil's porosity, inorganic fertilizers also lower its oxygen content and keep added fertilizer from being taken up as efficiently. In addition, most commercial fertilizers supply only 2 or 3 of the 20-odd nutrients needed by plants. Moreover, produc-

INDIVIDUALS MATTER

Fast-Track Composting

Compost piles must be turned over every few days for aeration to speed up decomposition. Ruth Beckner has invented a special drill bit that attaches to a cordless electric drill to inject air into a compost pile with very little physical exertion. Using her COMPOST AIR ™[*] device for about three minutes per day can create good compost in about three weeks, instead of waiting 12–18 months for the pile to decompose naturally.

Worms, the planet's champion recyclers, can also help create compost efficiently and without any odors. Recycling coordinators in Seattle, Washington, and Toronto, Canada, have given out thousands of worm bins to homeowners.

* For information contact Beckner & Beckner, 15 Portola Avenue, San Rafael, CA 94903 or call 1-800-58COMPOST.

ing, transporting, and applying inorganic fertilizers require large amounts of energy and release a greenhouse gas (nitrous oxide, N_2O; Figure 10-7d).

The widespread use of commercial inorganic fertilizers, especially on sloped land near streams and lakes, also causes water pollution as some fertilizer nutrients are washed into nearby bodies of water; the resulting plant-nutrient enrichment (cultural eutrophication; Figure 11-22) causes algae blooms that use up oxygen dissolved in the water, thereby killing fish. Rainwater seeping through the soil can also leach nitrates in commercial fertilizers into groundwater. Drinking water drawn from wells containing high levels of nitrate ions can be toxic, especially for infants.

Environmental historian Donald Worster reminds us that fertilizers are not a substitute for fertile soil:

We can no more manufacture a soil with a tank of chemicals than we can invent a rain forest or produce a single bird. We may enhance the soil by helping its processes along, but we can never recreate what we destroy. The soil is a resource for which there is no substitute.

According to soil scientists, responsibility for reducing soil erosion should not be limited to farmers. Timber cutting, overgrazing, mining, and urban development carried out without proper regard for soil conservation cause at least 40% of soil erosion in the United States. Each of us has a role in seeing that these vital soil resources are used sustainably. Some things you can do to reduce soil erosion are listed in Appendix 5.

A: About 84%; 43% of this energy is wasted unnecessarily

Land Degradation and Environmental Resources

GUEST ESSAY

David Pimentel

David Pimentel is professor of insect ecology and agricultural sciences in the College of Agriculture and Life Sciences at Cornell University. He has published over 440 scientific papers and 20 books on environmental topics, including land degradation, agricultural pollution and energy use, biomass energy, and pesticides. He was one of the first ecologists to use an interdisciplinary, holistic approach in investigating complex environmental problems.

At a time when the world's human population is rapidly expanding and its need for more land to produce food, fiber, and fuelwood is also escalating, valuable land is being degraded through erosion and other means at an alarming rate. Soil degradation is of great concern because soil reformation is extremely slow. Worldwide annual erosion rates for agricultural land are about 20–100 times the average of 500 years (with a range of 220 to 1,000 years) required to renew 2.5 centimeters (1 inch) of soil in tropical and temperate areas—a renewal rate of about 1 metric ton of topsoil per hectare of land per year.

Erosion rates vary in different regions because of topography, rainfall, wind intensity, and the type of agricultural practices used. In China, for example, the average annual soil loss is reported to be about 40 metric tons per hectare (18 tons per acre), whereas the U.S. average is 18 metric tons per hectare (8 tons per acre). In states such as Iowa and Missouri, however, annual soil erosion averages more than 35 metric tons per hectare (16 tons per acre).

Worldwide, about 10 million hectares (25 million acres) of land—an area about the size of Virginia—are abandoned for crop production each year because of high erosion rates plus waterlogging, salinization, and other forms of soil degradation. In addition, according to the UN Environment Programme, crop production becomes un-economical on about 20 million hectares (49 million acres) each year because soil quality has been severely degraded.

Soil erosion also occurs in forestlands, but it is not as severe as that in the more exposed soil of agricultural land. Soil erosion in managed forests is a primary concern because the soil reformation rate in forests is about two to three times slower than that in agricultural land. To compound this erosion problem, at least 24 million hectares (59 million acres) of forest are being cleared each year throughout the world; most of this land is used to grow food and graze cattle.

The effects of agriculture and forestry are interrelated in other ways. Deforestation reduces fuelwood supplies and forces the poor in developing countries to substitute crop residue and manure for fuelwood. When these plant and animal wastes are burned instead of being returned to the land as ground cover and organic fertilizer, erosion is intensified and productivity of the land is decreased. These factors, in turn, increase pressure to convert more forestland into agricultural land, further intensifying soil erosion.

One reason that soil erosion is not a high-priority concern among many governments and farmers is that it usually occurs so slowly that its cumulative effects may take decades to become apparent. For example, the removal of 1 millimeter (1/25 inch) of soil is so small that it goes undetected. But over a 25-year period the loss would be 25 millimeters (1 inch)—which would take about 500 years to replace by natural processes.

Besides reduced soil depth, soil erosion leads to reduced crop productivity because of losses of water, organic matter, and soil nutrients. A 50% reduction of soil organic matter on a plot of land has been found to reduce corn yields as much as 25%.

When soil erodes, vital plant nutrients such as nitrogen, phosphorus, potassium, and calcium are also lost. With U.S. annual cropland erosion rates of about

---

At some point, either the loss of topsoil from the world's croplands will have to be checked by effective soil conservation practices, or the growth in the world's population will be checked by hunger and malnutrition.

LESTER R. BROWN

CRITICAL THINKING

1. Explain what would happen if plate tectonics stopped. Explain what would happen if erosion stopped.

2. Use the second law of energy (Section 3-7) to analyze the feasibility of each of the following processes. Which, if any, could be profitable without subsidies?

a. Extracting most minerals dissolved in seawater
b. Recycling minerals that are widely dispersed
c. Mining increasingly lower-grade deposits of minerals
d. Using inexhaustible solar energy to mine minerals
e. Continuing to mine, use, and recycle minerals at increasing rates

3. Explain why you support or oppose each of the proposals listed on p. 351 for reforming the U.S. 1872 Mining Law.

4. Why should everyone, not just farmers, be concerned with soil conservation?

5. Do you believe that soil erosion in the country where you live is a serious problem? Explain. What about in the

Q: What percentage of the energy input of an incandescent light bulb is converted to light?

18 metric tons per hectare (8 tons per acre), an estimated $18 billion of plant nutrients are lost annually. Using fertilizers to replace these nutrients substantially adds to the cost of crop production.

Some analysts who are unaware of the numerous and complex effects of soil erosion have falsely concluded that the damages are relatively minor. For example, they report that soil loss causes an annual reduction in crop productivity of only 0.1–0.5% in the United States. However, we must consider all the ecological effects caused by erosion, including reductions in soil depth, in availability of water for crops, and in soil organic matter and nutrients. When this is done, agronomists and ecologists report a 15–30% reduction in crop productivity, leading to increased use of costly fertilizer. Because fertilizers are not a substitute for fertile soil, they can be applied only up to certain levels before crop yields begin to decline.

Reduced agricultural productivity is only one of the effects of soil erosion. In the United States, water runoff is responsible for transporting about 3 billion metric tons (3.3 billion tons) of sediment (about 60% from agricultural land) each year to waterways in the lower 48 states. Off-site damages to U.S. water storage capacity, wildlife, and navigable waterways from these sediments cost an estimated $6 billion each year. About 25% of new water storage capacity in U.S. reservoirs is built solely to compensate for sediment buildup.

When soil sediments that include pesticides and other agricultural chemicals are carried into streams, lakes, and reservoirs, fish production is adversely affected. These contaminated sediments interfere with fish spawning, increase predation on fish, and destroy fisheries in estuarine and coastal areas.

Increased erosion and water runoff on mountain slopes flood agricultural land in the valleys below, further decreasing agricultural productivity. Eroded land also does not hold water very well, again decreasing

crop productivity. This effect is magnified in the 80 countries (with nearly 40% of the world's population) that experience frequent droughts.

Thus soil erosion is one of the world's critical problems, and if not slowed, it will seriously reduce agricultural and forestry production and degrade the quality of aquatic ecosystems. Solutions that are not particularly difficult are often not implemented because erosion occurs so gradually that we fail to acknowledge its cumulative effects until damage is irreversible. Many farmers have also been conditioned to believe that losses in soil fertility can be remedied by applying more fertilizer or by using more fossil-fuel energy.

The principal way to control soil erosion and its accompanying runoff of sediment is to maintain adequate vegetative coverage on soils [by using various methods discussed in Section 12-5]. These methods are also cost-effective, especially when off-site costs of erosion are included. Scientists, policymakers, and agriculturists need to work together to implement soil and water conservation practices before world soils lose most of their productivity.

Critical Thinking

1. Some analysts contend that average soil erosion rates around the world are low and that the soil erosion problem can easily be solved with improved agricultural technology such as no-till cultivation and increased use of commercial inorganic fertilizers. Do you agree or disagree with this position? Explain.

2. What three major things do you believe elected officials should do to decrease soil erosion and the resulting water pollution by sediment in the United States or in the country where you live?

area where you live?

6. What are the main advantages and disadvantages of commercial inorganic fertilizers? Why should both inorganic and organic fertilizers be used?

*7. Use the library or Internet to find out what key mineral resources are found in the country where you live. How long could the estimated reserves of each of these meet the current needs of your country? How long will the estimated reserves last if the use of each of these resources increases by 2% a year?

*8. As a class project, evaluate soil erosion on your school grounds. Use this information to develop a soil conservation plan for your school and then present it to school officials.

*9. What mineral resources are extracted in your local area? What mining methods are used? Do local, state, or federal laws require restoration of the landscape after mining is completed? If so, how stringently are those laws enforced?

*10. Make a concept map of this chapter's major ideas using the section heads and subheads and the key terms (in boldface type). Look at the inside back cover and in Appendix 4 for information about concept maps.

SUGGESTED READINGS

See Internet Sources and Further Readings at the back of the book, and InfoTrac.

13 SOLID AND HAZARDOUS WASTE

Figure 13-1 The Love Canal housing development near Niagara Falls, New York, was built near a hazardous-waste dump site. The photo shows the area when it was abandoned in 1980. In 1990 the EPA allowed people to buy some of the remaining houses and move back into the area, despite protests from environmentalists. (NY State Department of Environmental Conservation)

There Is No "Away": The Love Canal Tragedy

Between 1942 and 1953, Hooker Chemicals and Plastics (owned by OxyChem since 1968) sealed chemical wastes containing at least 200 different chemicals into steel drums and dumped them into an old canal excavation (called Love Canal after its builder, William Love) near Niagara Falls, New York.

In 1953 Hooker Chemicals filled the canal, covered it with clay and topsoil, and sold it to the Niagara Falls school board for $1. The company inserted in the deed a disclaimer denying legal liability for any injury caused by the wastes. In 1957 Hooker warned the school board not to disturb the clay cap because of the possible danger from toxic wastes.

By 1959 an elementary school, playing fields, and 949 homes had been built in the 10-square-block Love Canal area (Figure 13-1). Roads and sewer lines crisscrossed the dump site, some of them disrupting the clay cap covering the wastes. An expressway built at one end of the dump in the 1960s blocked groundwater from migrating to the Niagara River. This created a "bathtub effect" that allowed contaminated groundwater and rainwater to build up and overflow the disrupted cap.

Residents began complaining to city officials in 1976 about chemical smells and chemical burns their children received playing in the canal area, but these complaints were ignored. In 1977 chemicals began leaking from the badly corroded steel drums into storm sewers, gardens, basements of homes next to the canal, and the school playground.

In 1978, after considerable media publicity and pressure from residents led by Lois Gibbs (a mother galvanized into action as she watched her children come down with one illness after another; Guest Essay, pp. 372–373), the state acted. It closed the school and arranged for the 239 homes closest to the dump to be evacuated, purchased, and destroyed.

Two years later, after protests from families still living fairly close to the landfill, President Jimmy Carter declared Love Canal a federal disaster area, had the remaining families relocated, and offered federal funds to buy 564 more homes. Residents of all but 72 of the homes moved out.

The dump site was covered with a new clay cap and surrounded by a drainage system that pumps leaking wastes to a new treatment plant. Some of the residents who remained claim that the entire problem was overblown by other residents, environmentalists, and the media.

After more than 15 years of court cases, OxyChem agreed in 1994 to a $98 million settlement with New York State and agreed to be responsible for all future treatment of wastes and wastewater at the Love Canal site. In 1996 the company agreed to reimburse the federal government $129 million for the Love Canal cleanup.

Because of the difficulty in linking exposure to a variety of chemicals to specific health effects (Section 8-2), the long-term health effects of exposure to hazardous chemicals on Love Canal residents remain unknown and controversial. However, the psychological damage to the evacuated families is enormous: For the rest of their lives they will worry about the possible effects of the chemicals on themselves and their children and grandchildren.

In June 1990 the EPA declared the area (renamed Black Creek Village) safe and allowed state officials to begin selling 234 of the remaining houses at 10–20% below market value. Still, the dump has not yet been cleaned up but only fitted with a drainage system. The EPA acknowledges that the dump site will leak again, sooner or later. Buyers must sign an agreement stating that New York State and the federal government make no guarantees or representations about the safety of living in these homes.

The Love Canal incident is a vivid reminder that we can never really throw anything away, that wastes don't stay put, and that preventing pollution is much safer and cheaper than trying to clean it up.

Solid wastes are only raw materials we're too stupid to use.
ARTHUR C. CLARKE

This chapter is devoted to answering the following questions:

- What are solid waste and hazardous waste, and how much of each type is produced?

- What can we do to reduce, reuse, and recycle solid waste and hazardous waste?

- What is being done to recycle aluminum, paper, and plastics?

- What are the advantages and disadvantages of burning or burying wastes?

- What can we do to reduce exposure to lead, dioxins, and hazardous chlorine compounds?

- How can we make the transition to a low-waste society?

13-1 WASTING RESOURCES: THE HIGH-WASTE APPROACH

What Is Solid Waste, and How Much Is Produced?

The United States, with only 4.6% of the world's population, produces about 33% of the world's **solid waste**: any unwanted or discarded material that is not a liquid or a gas. The United States generates about 10 billion metric tons (11 billion tons) of solid waste per year—an average of 40 metric tons (44 tons) per person. Although garbage produced directly by households and businesses is a significant problem, *about 98.5% of the solid waste in the United States comes from mining, oil and natural gas production, agriculture, and industrial activities used to produce goods and services for consumers* (Figure 13-2). Although mining waste is the single largest category of U.S. solid waste, the EPA has done little to regulate its disposal, mostly because the U.S. Congress has exempted it from regulation as a hazardous waste.

The remaining 1.5% of solid waste produced in the United States is **municipal solid waste** (MSW) from homes and businesses in or near urban areas. The amount of municipal solid waste, often called *garbage*, produced in the United States in 1996 was enough to fill a bumper-to-bumper convoy of garbage trucks encircling the globe almost eight times. This amounted to an average of 680 kilograms (1,500 pounds) per person in the United States—two to three times that in most other developed countries and many times that in developing countries.

About 25% of the resources in MSW produced in the United States in 1996 was recycled or composted (up

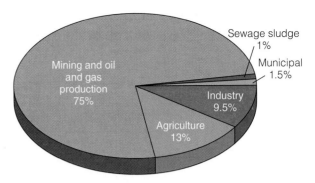

Figure 13-2 Sources of the estimated 10 billion metric tons (11 billion tons) of solid waste produced each year in the United States. Mining and industrial activities produce 65 times as much solid waste as household activities. (Data from U.S. Environmental Protection Agency and U.S. Bureau of Mines)

from 7% in 1960); the other 75% was hauled away and either dumped in landfills (60%) or burned in incinerators and waste-to-energy plants (15%) at a cost of about $40 billion (projected to rise to $75 billion by 2005).

What Does It Mean to Live in a High-Waste Society?

U.S. consumers throw away astounding amounts of solid waste, including the following quantities:

- Enough aluminum to rebuild the country's entire commercial airline fleet every 3 months

- Enough tires each year to encircle the planet almost three times

- About 18 billion disposable diapers per year, which if linked end-to-end would reach to the moon and back seven times

- About 2 billion disposable razors, 10 million computers, and 8 million television sets each year

- About 2.5 million nonreturnable plastic bottles each hour

- Some 14 billion catalogs (an average of 54 per American) and 38 billion pieces of junk mail each year

This is only part of the 1.5% of all solid waste labeled "municipal" in Figure 13-2.

What Is Hazardous Waste, and How Much Is Produced?

In the United States, **hazardous waste** is legally defined as any discarded solid or liquid material that **(1)** contains one or more of 39 toxic, carcinogenic, mutagenic, or teratogenic compounds at levels that exceed established limits (including many solvents, pesticides, and paint strippers); **(2)** catches fire easily (gasoline, paints, and solvents); **(3)** is reactive or unstable enough to explode or release toxic fumes (acids, bases, ammonia, chlorine bleach); or **(4)** is capable of corroding metal containers such as tanks, drums, and barrels (industrial cleaning agents and oven and drain cleaners).

This narrow official definition of hazardous wastes (mandated by Congress) does *not* include the following materials: **(1)** radioactive wastes (Section 19-4), **(2)** hazardous and toxic materials discarded by households (Table 13-1), **(3)** mining wastes, **(4)** oil- and gas-drilling wastes (routinely discharged into surface waters or dumped into unlined pits and landfills), **(5)** liquid waste containing organic hydrocarbon compounds (80% of all liquid hazardous waste), **(6)** cement kiln dust produced when liquid hazardous wastes are burned in a cement kiln (a practice classified as recycling by the EPA but considered dangerous *sham recycling* by environmentalists), and **(7)** wastes from the thousands of small businesses and factories that generate less than 100 kilograms (220 pounds) of hazardous waste per month.

Environmentalists call these omissions from the list of hazardous wastes *linguistic detoxification* designed to save industries and the government money and to mislead the public. Designating these excluded categories as hazardous waste would shift efforts from waste management and pollution control to waste reduction and pollution prevention. Industry representatives say that having to manage these wastes would bankrupt them or force them to move their operations to other countries with less-strict waste regulations.

The EPA estimates that at least 5.5 billion metric tons (6 billion tons) of hazardous waste are produced each year in the United States—an average of 21 metric tons (23 tons) per person. However, because only about 6% of the total is *legally* defined as hazardous waste, *94% of the country's hazardous waste is not regulated by hazardous-waste laws*. In most other countries, especially developing countries, even less of the hazardous waste is regulated.

13-2 PRODUCING LESS WASTE AND POLLUTION: REDUCING THROUGHPUT

What Are Our Options? There are two ways to deal with the solid and hazardous waste we create: *waste management* and *pollution (waste) prevention*. Waste management is a *high-waste approach* that views waste production as an unavoidable product of economic growth. It attempts to manage the resulting wastes in ways that reduce environmental harm, mostly by burying, burning, or shipping them off to another state or country. The goal is to move increasing amounts of matter and energy resources through the economy to enhance economic growth (Figure 3-15).

Preventing pollution and waste is a *low-waste approach* that views most solid and hazardous waste either as potential resources (that we should be recycling, composting, or reusing) or as harmful substances that we should not be using in the first place (Figures 3-16, 13-3, and 13-4). With this approach, taxes and subsidies (Table 7-1) are used to discourage waste production and encourage waste reduction (Guest Essays, pp. 68–69 and 372–373).

Table 13-1 Common Household Toxic and Hazardous Materials

Cleaning Products

Disinfectants

Drain, toilet, and window cleaners

Oven cleaners

Bleach and ammonia

Cleaning solvents and spot removers

Septic tank cleaners

Paint and Building Products

Latex and oil-based paints

Paint thinners, solvents, and strippers

Stains, varnishes, and lacquers

Wood preservatives

Acids for etching and rust removal

Asphalt and roof tar

Gardening and Pest Control Products

Pesticide sprays and dusts

Weed killers

Ant and rodent killers

Flea powder

Automotive Products

Gasoline

Used motor oil

Antifreeze

Battery acid

Solvents

Brake and transmission fluid

Rust inhibitor and rust remover

General Products

Dry-cell batteries (mercury and cadmium)

Artists' paints and inks

Glues and cements

Q: What percentage of the energy input of a screw-in fluorescent light bulb produces light?

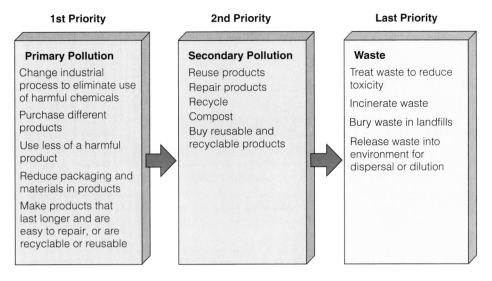

1st Priority

Primary Pollution

Change industrial process to eliminate use of harmful chemicals

Purchase different products

Use less of a harmful product

Reduce packaging and materials in products

Make products that last longer and are easy to repair, or are recyclable or reusable

2nd Priority

Secondary Pollution

Reuse products
Repair products
Recycle
Compost
Buy reusable and recyclable products

Last Priority

Waste

Treat waste to reduce toxicity

Incinerate waste

Bury waste in landfills

Release waste into environment for dispersal or dilution

Figure 13-3 Solutions: priorities suggested by prominent scientists for dealing with material use and solid waste. To date, these priorities have not been followed in the United States (and in most countries); most efforts are devoted to waste management (bury it or burn it). (U.S. Environmental Protection Agency and U.S. National Academy of Sciences)

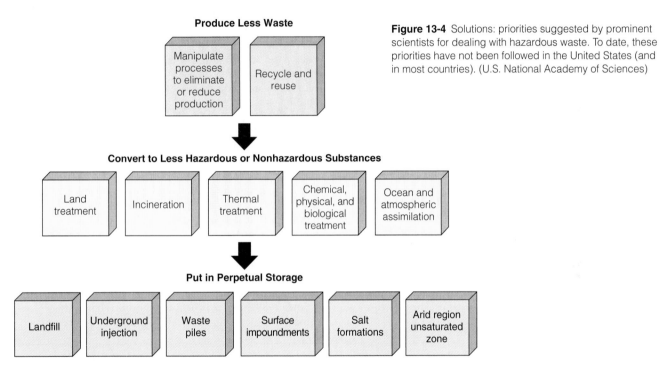

Produce Less Waste

Manipulate processes to eliminate or reduce production

Recycle and reuse

Figure 13-4 Solutions: priorities suggested by prominent scientists for dealing with hazardous waste. To date, these priorities have not been followed in the United States (and in most countries). (U.S. National Academy of Sciences)

Convert to Less Hazardous or Nonhazardous Substances

Land treatment

Incineration

Thermal treatment

Chemical, physical, and biological treatment

Ocean and atmospheric assimilation

Put in Perpetual Storage

Landfill

Underground injection

Waste piles

Surface impoundments

Salt formations

Arid region unsaturated zone

According to the U.S. National Academy of Sciences (Figures 13-3 and 13-4), the low-waste approach should have the following hierarchy of goals: **(1)** *reduce* waste and pollution, **(2)** *reuse* as many things as possible, **(3)** *recycle and compost* as much waste as possible, **(4)** *chemically or biologically treat or incinerate* waste that can't be reduced, reused, recycled, or composted, and **(5)** *bury* what is left in state-of-the-art landfills or aboveground vaults after the first four goals have been met.

Scientists estimate that in a low-waste society 60–80% of the solid and hazardous waste produced could be eliminated through reduction, reuse, and recycling (including composting). The remaining 20–40%

of such wastes would then be treated to reduce their toxicity, and what's left would be burned or buried under carefully regulated conditions. Currently, the order of priorities shown in Figures 13-3 and 13-4 for dealing with solid and hazardous wastes is reversed in the United States (and in most other countries), mostly because the costs of producing and dealing with these wastes are not included in the market prices of products.

Why Is Producing Less Waste and Pollution the Best Choice? A small but increasing number of companies are learning that reducing waste and pollution can be good for corporate profits, worker health

We Have Been Asking the Wrong Questions About Wastes

Lois Marie Gibbs

In 1977, Lois Marie Gibbs was a young housewife with two children living near the Love Canal toxic dump site. She had never engaged in any sort of political action until her children began experiencing unexplained illnesses and she learned that toxic chemicals were oozing from the dump site into many of the area's yards and basements. Then she organized her neighborhood and became the president and major strategist for the Love Canal Homeowners Association. This dedicated grassroots political action by "amateurs" brought hazardous-waste issues to national prominence and spurred passage of the federal Superfund legislation to help clean up abandoned hazardous-waste sites. Lois Gibbs then moved to Washington, D.C., and formed Citizens' Clearinghouse for Hazardous Wastes, an organization that has helped over 7,000 community organizations protect themselves from hazardous wastes. Her story is told in her autobiography, Love Canal: My Story *(State University of New York Press, 1982); she was also the subject of a CBS movie,* Lois Gibbs: The Love Canal, *which aired in 1982. Her latest book is* Dying from Dioxin *(Boston: South End Press, 1995). She is an inspiring example of what a dedicated citizen can do to change the world.*

Just about everyone knows our environment is in danger. One of the most serious threats is the massive amount of waste we put into the air, water, and ground every year. All across the United States and around the world are thousands of places that have been, and continue to be, polluted by toxic chemicals, radioactive waste, and just plain garbage.

For generations, the main question people have asked is, "Where do we put all this waste? It's got to go somewhere." That is the wrong question, as has been shown by a series of experiments in waste disposal and by the simple fact that there is no "away" in "throwaway."

We tried dumping our waste in the oceans. That was wrong. We tried injecting it into deep, underground wells. That was wrong. We've been trying to build landfills that don't leak, but that doesn't work. We've been trying to get rid of waste by burning it in high-tech incinerators; that only produces different types of pollution, such as air pollution and toxic ash. We've tried a broad range of "pollution" controls, but all that does is allow legalized, high-tech pollution. Even recycling, which is a very good thing to do, suffers from the same problem as all the other methods: It addresses waste *after* it has been produced.

For many years, people have been assuming that "it's got to go somewhere," but now many people, especially young people, are starting to ask why. Why do we produce so much waste? Why do we need products and services that have so many toxic by-products? Why can't industry change the way it makes things so that it stops producing so much waste?

These are the *right* questions. When you start asking them, you start getting answers that lead to *pollution prevention* and *waste reduction* instead of *pollution control* and *waste management*. People, young and old, who care about pollution prevention are challenging companies to stop making products with gases that reduce ozone in the ozone layer [Section 10-6] and contribute to the threatening possibility of global warming [Section 10-3]. They are asking why so many goods are wrapped in excessive, throwaway packaging. They are challenging companies that sell pesticides, cleaning fluids, batteries, and other hazardous products to either remove the toxins from those products or take them back for recovery or recycling rather than disposing of them in the environment. They are demanding alternatives to throwaway materials in general.

Since 1988, hundreds of student groups have contacted my organization to get help and advice in taking

and safety, the local community, consumers, and the environment as a whole. Such methods **(1)** save energy and virgin resources by keeping the quality of matter resources high (Figure 3-8) with a lower input of high-quality energy; **(2)** reduce the environmental effects of extracting, processing, and using resources (Figure 12-13); **(3)** improve worker health and safety by reducing exposure to toxic and hazardous materials; **(4)** decrease pollution control and waste management costs and future liability for toxic and hazardous materials; and **(5)** are usually less costly on a life cycle basis than trying to clean up pollutants and manage wastes once they are produced.

A 1992 study of 181 waste reduction initiatives in 27 U.S. firms found that two-thirds of the initiatives took 6 months or less to implement. One-fourth required no capital investment, two-thirds paid back their capital investments in 6 months or less, and 93% got them back within 3 years.

In 1975 the Minnesota Mining and Manufacturing Company (3M), which makes 60,000 different products in 100 manufacturing plants, began a Pollution Prevention Pays (3P) program. It redesigned equipment and processes, used fewer hazardous raw materials, identified hazardous chemical outputs (and recycled or sold them as raw materials to other companies), and began

Q: What percentage of the energy in gasoline is used to move a motor vehicle powered by an internal combustion engine?

these effective types of actions. Many of these groups begin by working to get polystyrene food packaging out of their school cafeterias and local fast-food restaurants.

Oregon students even took legal action to get rid of cups and plates made from bleached paper because the paper contains the deadly poison dioxin. They were asking the right questions, and they got the right answer when the school systems switched to reusable cups, plates, and utensils.

Dozens of student groups have joined with local grass-roots organizations to get toxic-waste sites cleaned up or to stop construction of new toxic-waste sites, radio-active-waste sites, or waste incinerators.

Waste issues are not simply environmental issues; they are also tied up with our economy, which is geared to producing and then disposing of waste. *Somebody* is making money from every scrap of waste and has a vested interest in keeping things the way they are. Environmentalists and industry officials constantly argue about "cost–benefit analysis," which, simply stated, poses the question of whether the benefit of controlling pollution or waste will be greater than the cost. But this is another example of asking the wrong question. The right questions are, "Who will benefit, and who will pay the cost?"

Waste issues are also issues of *justice* and *fairness*. Again, there's a lot of debate between industry officials and environmentalists, especially those in federal and state environmental agencies, about so-called "acceptable risk" [Section 8-5]. These industry officials and environmentalists decide the degree of people's exposure to toxic chemicals, but don't ask the people who will actually be exposed how they feel about it. Again, asking what is an "acceptable risk" of exposure is the wrong question. It's simply not just to expose people to chemical poisons without their consent.

Risk analysts often say, "But there's only a one in a million chance of increased death from this toxic chemi-cal." That may be true. But suppose I took a pistol and went to the edge of your neighborhood and began shooting into it. There's probably only a one in a million chance that I'd hit somebody, but would you issue me a license to do that? As long as we don't stand up for our rights and demand that "bullets" in the form of hazardous chemicals not be "fired" in our neighborhoods, we are giving environmental regulators and waste producers a license to kill a certain number of us without our even being consulted.

From my personal experience, I know that decisions made to dump wastes at Love Canal and in thousands of other places were not made purely on the basis of the best available scientific knowledge. The same holds true for decisions about how to manage the wastes we produce today and to produce less waste.

Instead, the world we live in is shaped by decisions based on money and power. If you really want to understand what's behind any given environmental issue, the first question you should ask is, "Who stands to profit from this?" Then ask, "Who is going to pay the price?" You can then identify both sides of the issue and decide whether you want to be part of the problem or part of the solution.

Critical Thinking

1. Do you believe that we should put primary emphasis on pollution prevention and waste reduction? Explain. What changes would you be willing to make in your own lifestyle to prevent pollution and reduce waste?

2. What political and economic changes, if any, do you believe must be made so that we shift from primary emphasis on waste production and waste management to primary emphasis on pollution prevention and waste reduction? What actions are you taking to bring about such changes?

making more nonpolluting products. By 1995 3M's overall waste production was down by one-third, emissions of air pollutants were reduced by 70%, and the company had saved over $750 million in waste disposal costs.

Solutions: How Can We Reduce Waste and Pollution? There are several ways to reduce waste and pollution or resource throughput (Figure 3-16). One is to *decrease consumption*, which begins by asking whether we really need a particular product (sometimes called *precycling*).

Another way to reduce throughput is to *redesign manufacturing processes and products to use less material* (Figure 13-5). For example, cars made today are lighter because an increasing proportion of their steel parts are being replaced with aluminum (80% from recycled metal) and plastic parts. Another strategy is to *design products that produce less pollution and waste fewer resources when used* (Figure 13-5). Examples include paints (which are responsible for about 8% of the volatile organic compounds emitted into the atmosphere) that use fewer volatile solvents and using more energy-efficient cars, lights, and appliances (Section 18-3).

Manufacturing processes can also be redesigned to produce less waste and pollution (Figure 13-5). Most toxic organic solvents can be recycled within plants or

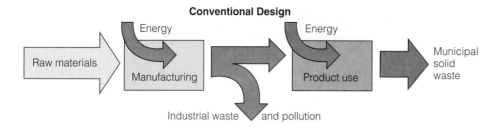

Conventional Design

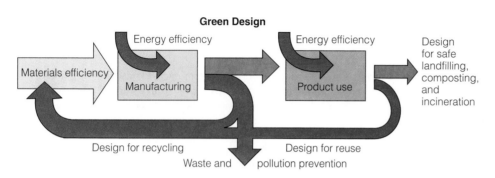

Green Design

Figure 13-5 How product design affects throughputs of matter and energy resources and outputs of pollution and solid waste. Most conventional manufacturing processes involve a one-way flow of materials from raw materials to waste and pollution. Green design reduces the overall impact by emphasizing efficient use of matter and energy resources, reduction of outputs of solid and hazardous waste, and reuse and recycling of materials. (Adapted from U.S. Congress Office of Technology Assessment, *Green Products by Design: Choices for a Cleaner U.S. Environment*, Washington, D.C.: Government Printing Office, 1992)

replaced with water-based or citrus-based solvents (Individuals Matter, p. 294). Some processes can be redesigned to eliminate washing altogether. Hydrogen peroxide can be used instead of toxic chlorine to bleach paper and other materials. For clothes, wet cleaning (with water and steam) and microwave drying can replace dry cleaning with toxic organic solvents. In effect, over the next two decades we need to change the ways we make things (Solutions, right).

People can use less hazardous (and usually cheaper) cleaning products (Table 13-2). Three inexpensive chemicals—baking soda (which can also be used as a deodorant and a toothpaste), vinegar, and borax—can be used for most cleaning and clothes bleaching. People can also use pesticides and other hazardous chemicals only when absolutely necessary and in the smallest amount possible.

Green design and life cycle assessment can help develop products that last longer and are easy to repair, reuse, remanufacture, compost, or recycle. Several European auto manufacturers now design their cars for easy disassembly and for reuse and recycling of up to 80% of their parts (75% in the United States), and they are trying to minimize the use of nonrecyclable or hazardous materials.

Products can be designed to last longer. For example, although tires are now being produced with an average life of 97,000 kilometers (60,000 miles), researchers believe this could be extended to at least 160,000 kilometers (100,000 miles).

Eliminating or reducing unnecessary packaging is another important strategy. Here are some key questions that environmentalists believe manufacturers and con-

sumers should ask about packaging: Is it necessary? Can it use fewer materials? Can it be reused? Are the resources that went into it nonrenewable or potentially renewable? Does it contain the highest feasible amount of consumer-discarded (postconsumer) recycled material? Is it designed to be easily recycled? Can it be incinerated without producing harmful air pollutants or a toxic ash? Can it be buried and decomposed in a landfill without producing chemicals that can contaminate groundwater?

Trash taxes can also be used to reduce waste. For example, in 1992 Victoria, British Columbia, instituted a trash tax of $1.20–2.10 per bag, along with a strong recycling program. Within a year, household waste fell by 18%. After Denmark imposed a solid waste charge, the recycling rate for demolition waste rose from 12% to 82% between 1987 and 1994.

13-3 REUSE

What Are the Advantages of Refillable Containers?
Reuse, a form of waste reduction, extends resource supplies, keeps high-quality matter resources from being reduced to low-quality matter waste (Figure 3-8), and reduces energy use and pollution even more than recycling. Two examples of reuse are refillable glass beverage bottles and refillable soft-drink bottles made of polyethylene terephthelate (PET) plastic. Unlike throwaway and recycled cans and bottles, refillable beverage bottles create local jobs related to their collection and refilling. Moreover, studies by Coca-Cola and PepsiCo of Canada show

Q: What is the most inefficient and costly way to produce electricity for heating an interior space or water?

An Ecoindustrial Revolution: Cleaner Production

Some analysts urge us to bring about an *ecoindustrial revolution* over the next 50 years as a way to help achieve industrial, economic, and environmental sustainability. All industrial products and processes would be redesigned and integrated into an essentially closed system of cyclical material flows.

The goals of this emerging concept of *cleaner production* (Guest Essay, pp. 68–69), or *industrial ecology*, are to reduce resource throughput, waste, and pollution (Figure 3-16) by **(1)** minimizing the input of energy and matter resources and the output of wastes, **(2)** using less material and energy per unit of output, **(3)** substituting less toxic or nontoxic chemicals in manufacturing processes or products, **(4)** recycling or reusing toxic chemicals used or produced in manufacturing processes, and **(5)** developing materials exchange systems in which one company's wastes become another company's resources and companies take back packaging and used products from consumers for reuse, recycling, repair, or remanufacturing.

In effect, companies would mimic natural chemical cycles (Section 4-4) and interact in complex *resource exchange webs* similar to food webs in natural ecosystems. With such an exchange and chemical cycling network, producing a large amount of easily reusable waste might sometimes be preferable to designing a more efficient process that produces a small amount of unusable waste.

A prototype of this concept exists in Kalunborg, Denmark, where a coal-fired power plant, an oil refinery, a municipal heating authority, a producer of sulfuric acid, a sheet rock plant, a biotechnology company, local farms, some greenhouses, a fish farm, and several other businesses are working together to exchange and convert their wastes into resources. Ecoindustrial parks are in the planning stages in areas such as Baltimore, Rochester (New York), Chattanooga, Halifax (Nova Scotia), and the Brownsville/Matamoros region along the Texas/Mexico border.

Such an ecoindustrial revolution, carried out over the next 50 years, will challenge designers, engineers, and scientists to revolutionize the way we design and make things (Guest Essay, p. 191). This will reduce pollution, improve human health, and help preserve biodiversity and ecological integrity.

This important form of learning how to work with nature will also provide economic benefits to businesses by **(1)** reducing the costs of controlling pollution and complying with pollution regulations, **(2)** improving the health of workers (thus reducing company health-care insurance expenses), **(3)** reducing legal liability for toxic and hazardous wastes, **(4)** stimulating companies to come up with new, environmentally friendly processes and products that can be sold worldwide, and **(5)** giving companies a better image among consumers, based on results rather than on public relations campaigns. Such a revolution would be a win–win situation for everyone, and for the earth.

Critical Thinking

What short- and long-term disadvantages (if any) might there be with an ecoindustrial revolution? Do you believe that it will be possible to phase in such a revolution in the country where you live over the next 2–3 decades? Explain. What are the three most important strategies for doing this?

---

that 0.5-liter (16-ounce) servings of their soft drinks cost one-third less in refillable bottles.

In 1964, 89% of all soft drinks and 50% of all beer in the United States were sold in refillable glass bottles. Today such bottles make up only about 7% of the beer and soft drink market, and only 10 states even have refillable glass bottles. The disappearance of most local bottling companies has led to a loss of local jobs, income, and tax revenues. Some call for reinstatement of this bottling reuse system in the United States; others say it isn't practical because the system of collections and returns has been dismantled.

Denmark has led the way by banning all beverage containers that can't be reused. To encourage use of refillable glass bottles, Ecuador has a refundable beverage container deposit fee that is 50% of the cost of the drink. In Finland, 95% of the soft drink, beer, wine, and spirits containers are refillable, and in Germany 73% are refillable.

Another reusable container is the metal or plastic lunchbox that most workers and schoolchildren once used. Sandwiches and refrigerator leftovers can be put in small reusable plastic containers instead of in plastic wrap and aluminum foil (most of which is not recycled). This practice and most forms of reuse save money and reduce resource waste. Another way to reduce resource waste and pollution is by the use of reusable grocery and shopping bags (Solutions, p. 377).

Table 13-2 Alternatives to Some Common Household Chemicals

Chemical	Alternative	Chemical	Alternative
Deodorant	Sprinkle baking soda on a damp wash cloth and wipe skin; dab on vanilla extract.	General surface cleaner	Mixture of vinegar, salt, and water, or mixture of borax, soap, lemon juice, and water.
Oven cleaner	Baking soda and water paste, scouring pad.	Bleach	Baking soda or borax.
Toothpaste	Baking soda.	Mildew remover	Mix ½ cup vinegar, ½ cup borax, and warm water.
Drain cleaner	Pour ½ cup salt down drain, followed by boiling water; or pour 1 handful baking soda and ½ cup white vinegar and cover lightly for one minute.	Disinfectant and general cleaner	Mix ½ cup borax in 1 gallon hot water.
Window cleaner	Add 2 teaspoons white vinegar to 1 quart warm water.	Furniture or floor polish	Mix ½ cup lemon juice and 1 cup vegetable or olive oil.
Toilet bowl, tub, and tile cleaner	Mix a paste of borax and water; rub on and let set one hour before scrubbing. Can also scrub with baking soda and a brush.	Carpet and rug shampoos	Sprinkle on cornstarch, baking soda, or borax and vacuum.
Floor cleaner	Add ½ cup vinegar to a bucket of hot water; sprinkle a sponge with borax for tough spots.	Detergents and detergent boosters	Washing soda or borax and soap powder.
Shoe polish	Polish with inside of a banana peel, then buff.	Spray starch	In a spray bottle, mix 1 tablespoon cornstarch in a pint of water.
Silver polish	Clean with baking soda and warm water.	Fabric softener	Add 1 cup white vinegar or ¼ cup baking soda to final rinse.
Air freshener	Set vinegar out in an open dish. Use an opened box of baking soda in closed areas such as refrigerators and closets. To scent the air, use pine boughs or make sachets of herbs and flowers.	Dishwasher soap	1 part borax and 1 part washing soda.
		Pesticides (indoor and outdoor)	Use natural biological controls (Section 15-5).

13-4 RECYCLING

How Can We Recycle Organic Solid Wastes? Community Composting Compost is a sweet-smelling, dark-brown, humuslike material that is rich in organic matter and soil nutrients. It is produced when microorganisms (mostly fungi and aerobic bacteria) in soil break down organic matter such as leaves, food wastes, paper, and wood in the presence of oxygen.

Biodegradable wastes make up about 35% by weight of the municipal solid waste output in the United States. Such wastes can be composted by consumers in backyard bins or indoor containers or collected and composted in centralized community facilities (as is done in many western European countries).

To compost such wastes, we mix them with soil, put the mixture into a pile or container, stir it occasionally, and let the stuff rot for several months. Heat generated by microbial decomposition rises inside the pile. The pile is periodically turned and mixed to ensure that the temperature is high enough to kill pathogens and weed seeds (but not hot enough to kill the decomposing microbes). There are also ways to speed up the composting process (Individuals Matter, p. 365).

The resulting compost can be used as an organic soil fertilizer or conditioner, as topsoil, or as a landfill cover. Compost can also be used to help restore eroded soil on hillsides and along highways, strip-mined land, overgrazed areas, and eroded cropland.

To be successful, a large-scale composting program must overcome siting problems (few people want to live near a giant compost pile or plant), control odors, and exclude toxic materials that can contaminate the compost and make it unsuitable for use as a fertilizer on crops and lawns. Three ways to control or reduce odors for large-scale composting operations are **(1)** enclosing the facilities and filtering the air inside (but residents near large composting plants still complain of unacceptable odors), **(2)** creating municipal compost operations near existing landfills or at other isolated sites, and **(3)** decomposing biodegradable wastes in a closed metal container in which air is recirculated to give precise control of available oxygen and temperature (a technique that has been used successfully in the Netherlands for 20 years).

Q: What are the two most energy-efficient ways to heat interior space?

What Are the Two Types of Recycling? There are two types of recycling for materials such as glass, metals, paper, and plastics: primary and secondary. The most desirable type is *primary* or *closed-loop recycling*, in which wastes discarded by consumers (*postconsumer wastes*) are recycled to produce new products of the same type (such as newspaper into newspaper and aluminum cans into aluminum cans).

A still useful but less desirable type is *secondary*, or *open-loop*, *recycling*, in which waste materials are converted into different products. Primary recycling reduces the amount of virgin materials in a product by 20–90%, whereas secondary recycling reduces virgin material by 25% at most.

Environmentalists urge us not to be misled by labels claiming that paper and plastic bags or other items are recyclable. Just about anything is in theory recyclable. What counts is whether an item is actually recycled and whether we complete the recycling loop by buying products using the maximum feasible content of postconsumer recycled materials.

Case Study: Recycling Municipal Solid Waste in the United States In 1996 about 25% of U.S. municipal solid waste was recycled or composted; 16 states have adopted goals of recycling or composting at least half of their municipal solid waste by 2000. By 1996 the United States had more than 7,300 municipal curbside recycling programs serving 112 million people—about 48% of the population. Recent pilot studies in several U.S. communities show that a 60–80% recycling and composting rate is possible.

The pacesetter is Seattle, with a recycling rate of 45%. Its *pay-as-you-throw* program bases garbage collection charges on the amount of waste a household generates for disposal; materials sorted out for recycling are hauled away free. Currently, more than 2,700 communities in North America have curbside pay-as-you-throw systems.

Recycling also benefits communities by creating more jobs than burying or burning wastes. Recycling 1 million tons of solid waste in the United States requires about 1,800 workers, compared to 600 workers for landfilling these wastes and only 80 jobs for incinerating them.

Is Centralized Recycling of Mixed Solid Waste the Answer? Large-scale recycling can be accomplished by collecting mixed urban waste and transporting it to centralized *materials-recovery facilities (MRFs)*. Machines shred and automatically separate the mixed waste to recover glass (which can be melted and converted to new bottles or to fiberglass insulation), iron, aluminum, and other valuable materials (Figure 13-6); these are then sold to manufac-

What Kind of Grocery Bags Should We Use?

SOLUTIONS

When you're offered a choice between plastic or paper bags for your groceries, which should you choose? The answer is *neither*. Both are environmentally harmful, and the question of which is the more damaging has no clear-cut answer.

On one hand, plastic bags degrade slowly in landfills and can harm wildlife if swallowed, and producing them pollutes the environment. On the other hand, producing the brown paper bags used in most supermarkets uses trees and pollutes the air and water.

Instead of having to choose between paper and plastic bags, you can bring your own *reusable* canvas or string containers to the store and save and reuse any paper or plastic bags you get. To encourage people to bring their own reusable bags, stores in the Netherlands charge for paper or plastic bags.

Critical Thinking

1. Apply similar reasoning to determine what kind of cup (plastic, paper, or reusable) you should use whenever possible. How could you solve the problem of getting coffee or other beverages at fast-food places and at workplaces?

2. Do you believe that grocery stores should charge for paper or plastic bags and sell reusable bags to encourage reuse? Explain. How would you go about implementing such a policy in all major grocery stores to provide an even economic playing field for all consumers?

turers as raw materials. The remaining paper, plastics, and other combustible wastes are recycled or burned. The resulting heat produces steam or electricity to run the recovery plant or to sell to nearby industries or homes.

By 1996 the United States had more than 220 materials-recovery facilities, and at least 50 more were in the planning stages. However, such plants are expensive to build and maintain, and once trash is mixed it takes a lot of money and energy to separate it, which is why some MRFs have shut down. MRFs must have a large input of garbage to make them financially successful. Thus, their owners have a vested interest in increased *throughput* of matter and energy resource to produce more trash—the reverse of what prominent scientists believe we should be doing (Figure 13-3).

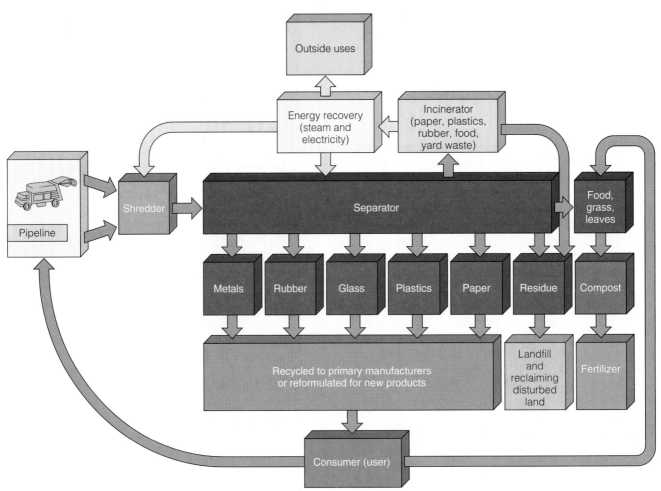

Figure 13-6 Schematic of a generalized materials-recovery facility used to sort mixed wastes for recycling and burning to produce energy. Because such plants require high volumes of trash to be economical, they discourage reuse and waste reduction.

These facilities also can emit toxic air pollutants and produce a toxic ash that must be disposed of safely. Other problems include health and accident threats to workers in poorly designed and managed plants and increased truck traffic, odor, and noise. Most communities collecting mixed solid waste can't afford to build or operate high-tech resource recovery facilities. Instead, to meet government-required recycling goals, many communities hire workers to sort the trash by hand. This is a hazardous, low-pay job.

Is Separating Solid Wastes for Recycling the Answer? Many solid-waste experts argue that it makes more sense economically and environmentally for households and businesses to separate trash into recyclable and reusable categories (such as glass, paper, metals, certain types of plastics, and compostable materials) before it is picked up. Compartmentalized city collection trucks, private haulers, or volunteer recycling organizations then pick up the segregated

wastes and sell them to scrap dealers, compost plants, and manufacturers. Another alternative (especially in less populated areas) is the establishment of a network of drop-off centers, buyback centers, and deposit-refund programs in which people deliver and either sell or donate their separated recyclable materials.

The source-separation approach produces little air and water pollution, reduces litter, and has low start-up costs and moderate operating costs. It also saves more energy and provides more jobs for unskilled workers than centralized MRFs, and it creates three to six times more jobs per unit of material than landfilling or incineration. In addition, separated recyclables are cleaner and can usually be sold for a higher price. Source separation also educates people about the need for waste reduction, reuse, and recycling (Individuals Matter, right).

In the United States, many small- and medium-scale source-separation operations—pioneers in the

Q: What is the most energy-efficient fuel for powering a motor vehicle?

recycling business—are now being squeezed out by large waste management companies operating MRFs, which need large inputs of trash to be profitable. In some communities, elected officials have signed long-term (20- to 30-year) contracts giving companies exclusive ownership of all garbage. Now owners of some small-scale recycling businesses are being sued for recycling materials that large waste management companies claim they own. The fight used to be over what to do with garbage; now, as garbage is becoming more valuable, the fight is over who owns it.

Aluminum and paper separated out for recycling are worth a lot of money. As a result, in a growing number of cities people steal these materials—from curbside containers set out by residents and from unprotected recycling drop-off centers—and sell them. This undermines municipal recycling programs by lowering the income available from selling these high-value materials.

Does Recycling Make Economic Sense? The answer is yes and no, depending on different ways of looking at the economic and environmental costs and benefits of recycling. Critics contend that recycling **(1)** has become almost a religion that is above criticism regardless of how much it costs communities, **(2)** does not make sense if it costs more to recycle materials than to send them to a landfill or incinerator (as is the case in some areas), **(3)** is often not needed to save landfill space because most areas in the United States are not running out of landfill space, and **(4)** may make economic sense for valuable and easy-to-recycle materials (such as aluminum, mercury, paper, and steel) but not for cheap or plentiful resources (such as glass from silica) and most plastics (which are expensive to recycle).

Many communities established recycling programs with the idea that they should pay for themselves. But recycling proponents argue that *recycling programs should not be judged on whether they pay for themselves any more than are conventional garbage disposal systems based on land burial or incineration.*

Moreover, recycling proponents contend that with full-cost accounting (Section 7-3), the net economic, health, and environmental benefits of recycling far outweigh the costs. A recent study by MIT economist Robert F. Stone revealed that recycling in Massachusetts yields a net social benefit of about of $254 per metric ton ($231 per ton) if both the roughly $146 per metric ton ($133 per ton) environmental benefits of recycling and the avoided solid-waste disposal costs are taken into account.

An analysis of various recycling programs revealed that cities and private waste collection firms tend to make money if they have high recycling rates

Source Separation Recycling in Some Georgia Schools

INDIVIDUALS MATTER

In Rome, Georgia, environmental educator Steve Cordle has set up and exciting project that combines the important concept of source separation recycling with environmental education.

He has designed and implemented a source separation recycling program for 17 Floyd County schools in Georgia. Before his program, which began in 1992, the schools were a major contributor to the county landfill and hardly any of their waste output was recycled.

During a 7-month period in 1996, approximately 114 metric tons (250,000 pounds) of paper, corrugated cardboard, and aluminum and steel cans were collected in separate containers. The school system stored and sold the items separated out for recycling. During this period the project added $10,139.69 to school funds, although school officials consider any money made incidental compared to the educational value of the program.

This important experiment in environmental education teaches students (and teachers) from kindergarten through high school about the need for recycling and engages them in hands-on participation in source separation. Mixing and throwing waste materials in garbage cans (which should be called *resource containers*) and hauling them off to a landfill, incinerator, or mixed-resource recycling plant is an out-of-sight, out-of-mind approach that does little to further environmental education.

Cordle is working hard to expand this idea to a larger area and hopefully to Georgia's entire school system. With proper backing and support, this program could become a model for use throughout the United States and other parts of the world.

Critical Thinking

Use the second law of energy (thermodynamics) to explain why a properly designed source separation recycling program takes less energy and produces less pollution than a centralized program that collects mixed waste over a large area and hauls it to a centralized facility where workers or machinery separate the wastes for recycling.

and a single pickup system (for both materials to be recycled and garbage that can't be recycled). Cities that lose money on recycling often have expensive dual collection systems and have not designed programs to encourage more cost-effective collection and high rates of recycling.

Finally, environmentalists point out that the primary benefit from recycling is *not* a reduction in the use of landfills and incinerators. Instead, *the major reasons for recycling are reduced use of virgin resources, reduced throughput of matter and energy resources, and reduced pollution and environmental degradation* (Figure 3-16).

Why Don't We Have More Reuse and Recycling?

Three factors that hinder recycling (and reuse) are failure to include the environmental and health costs of raw materials in the market prices of consumer items, more tax breaks and subsidies for resource-extracting industries than for recycling industries, and lack of large, steady markets for recycled materials.

Studies suggest that within 10 years the United States could recycle, reuse, and compost 60–80% by weight of the municipal solid waste resources it now throws away. Analysts have suggested various ways to overcome the obstacles to recycling, including the following:

- Taxing virgin resources and phasing out subsidies for extraction of virgin resources

- Lowering or eliminating taxes on recycled materials based on postconsumer waste content

- Providing subsidies for reuse and postconsumer waste recycling

- Requiring households and businesses to pay directly for garbage collection based on how much they throw away, with lower charges or no charges for materials separated for recycling and composting (a *pay-as-you throw* system)

- Encouraging or requiring government purchases of recycled products to help increase demand and lower prices

- Viewing landfilling and incineration of solid wastes as last resorts to be used only for wastes that can't be reused, composted, or recycled (Figure 13-3)

- Requiring ecolabels on all products, evaluating life-cycle environmental costs and listing preconsumer and postconsumer recycled content

Countries such as Germany are leading the way in recycling, reuse, and waste reduction (Case Study, right).

⚙ 13-5 CASE STUDIES: RECYCLING ALUMINUM, WASTEPAPER, AND PLASTICS

How Much Aluminum Is Being Recycled?

Worldwide, the recycling rate for aluminum in 1996 was about 35% (34% in the United States). Recycling aluminum produces 95% less air pollution and 97% less water pollution, and it requires 95% less energy than mining and processing aluminum ore.

In 1994, 62% (compared to 15% in 1973) of aluminum beverage cans in the United States were recycled. Despite this progress, about 38% of the 95 billion aluminum cans produced in 1994 in the United States were still thrown away. Laid end to end, these cans would wrap around the planet more than 120 times.

Recycling aluminum cans is great, but many environmentalists believe that these cans are an example of an unnecessary item that could be replaced by more energy-efficient and less polluting refillable glass or PET plastic bottles—a switch from recycling to reuse that also creates local jobs. One way to encourage this change would be to place a heavy tax on nonrefillable containers and no tax on reusable beverage containers, as is done in at least nine countries. What do you think?

How Much Wastepaper Is Being Recycled?

Paper (especially newspaper and cardboard) is one of the easiest materials to recycle. To make paper, a chemical such as caustic soda (sodium hydroxide) is used to convert wood chips into a soft mush (pulp) that is pressed into a thin sheet and then dried. Recycling paper involves removing its ink, glue, and coating and reconverting it to pulp that is pressed again into new paper. This process breaks down some of the paper fibers, requiring addition of some new pulp to maintain paper strength.

In 1996 the United States recycled about 40% of its wastepaper (up from 25% in 1989) and is projected to recycle 50% by the year 2000. At least 10 other countries recycle 50–98% of their wastepaper.

Recycling the Sunday newspapers in the United States alone would save 500,000 trees per week. In addition to saving trees, recycling paper **(1)** saves energy because it takes 30–64% less energy to produce the same weight of recycled paper as to make the paper from trees, **(2)** reduces air pollution from pulp mills by 74–95%, **(3)** lowers water pollution by 35%, **(4)** helps prevent groundwater contamination by toxic ink left after paper rots in landfills over a 30- to 60-year period, **(5)** conserves large quantities of water, **(6)** can save landfill space, **(7)** creates five times more jobs than harvesting trees for pulp, and **(8)** can save money.

Buying recycled paper products can save trees and energy and reduce pollution, but it does not necessarily reduce solid waste. Only products made from *postconsumer waste*—waste intercepted on its way from consumer to the landfill or incinerator—do that.

Most recycled paper is actually made from *preconsumer waste*—scraps and cuttings recovered from paper and printing plants. Because paper manufacturers have always recycled this waste, it has never contributed to landfill problems. Now this paper is labeled "recycled"

Q: How much money could be saved if the world got serious about improving energy efficiency?

Recycling, Reuse, and Waste Reduction in Germany

In 1991, Germany enacted the world's toughest packaging law, which was designed to (1) reduce the amount of waste being landfilled or incinerated, (2) reduce waste production, and (3) recycle or reuse 65% of the nation's packaging by 1995. Product distributors must take back their boxes and other containers for reuse or recycling, and incineration of packaging is not allowed.

To implement the system, over 600 German manufacturers and distributors pay fees to a nonprofit company they formed to collect, sort, and reprocess the packaging discards (coded with a green dot) of member firms.

This system was able to find markets for recycled cardboard and glass, but it soon resulted in a glut of plastics and composite materials (made of several types of plastics or other materials) that are difficult and expensive to recycle. Despite such problems, about 52% of all plastic packaging used in Germany in 1994 was recycled, compared to 29% in 1993. To deal with this problem the government recently added a sliding fee (based on the types rather than just weight of packaging) that charges manufacturers more for plastic and composite packaging than for glass and cardboard packaging.

The system is now fairly close to breaking even financially. Plastic packaging has lost one-third of its market share to glass and cardboard, and four out of five German manufacturers have reduced their use of packaging.

Critical Thinking

What might be some long-term disadvantages of Germany's program? Do you believe that a similar program should be put into place in the country where you live? Explain.

---

as a marketing ploy, giving the false impression that people who buy such products (often at higher prices) are helping the solid-waste problem. Most "recycled" paper has no more than 50% recycled fibers, with only 10% from postconsumer waste. Environmentalists propose that the government require companies to report the amount of postconsumer recycled materials in paper and other products and reserve the term *recycled* only for items using postconsumer recycled materials.

Is It Feasible to Recycle Plastics? Plastics are various types of polymer molecules made by chemically linking monomer molecules (called *petrochemicals*) produced mostly from oil and natural gas. The plastics industry is among the leading producers of hazardous waste.

Plastics now account for about 8% by weight and 20% by volume of municipal solid wastes in the United States and about 60% of the debris found on U.S. beaches. In landfills, toxic cadmium and lead compounds used as binders, colorants, and heat stabilizers can leach out of plastics and ooze into groundwater and surface water. Most plastics used today are nondegradable or take 200–400 years to degrade. Even biodegradable plastics take decades to partially decompose in landfills because conditions there lack sufficient oxygen and moisture.

Currently, only about 5% by weight of all plastic wastes and 6% of plastic packaging used in the United States are recycled because there are so many different types of plastic resin. Before they can be recycled, plastics in trash must either be separated into different types of resins by consumers or separated from mixed trash (a costly, labor-intensive procedure unless affordable automated separating technologies can be developed). Another problem is that the current price of oil is so low that the price of virgin plastic resins (except for PET, used mostly in plastic drink bottles) is about 40% lower than that of recycled resins.

Environmentalists recognize the beneficial qualities of plastics: durability (in products such as car and machine parts, carpeting, toys, furniture, reusable tubs and containers, and refillable bottles), light weight, unbreakability (compared to glass), and in some cases reusability as containers. But many environmentalists believe that some widespread uses of plastics—especially excessive and often unnecessary single-use packaging and throwaway beverage and food containers—should be sharply reduced and replaced with less harmful and wasteful alternatives. What do you think?

13-6 DETOXIFYING, BURNING, BURYING, AND EXPORTING WASTES

How Can Hazardous Waste Be Detoxified?

Denmark has the most comprehensive and effective hazardous-waste detoxification program. Each Danish municipality has at least one facility that accepts paints, solvents, and other hazardous wastes from households. Hazardous and toxic waste from industries is delivered to 21 transfer stations throughout the country. All waste is then transferred to a large treatment facility, where about 75% of it is detoxified; the rest is buried in a carefully designed and monitored landfill.

Some consider biological treatment of hazardous waste, or *bioremediation*, to be the wave of the future for cleaning up some types of toxic and hazardous waste. In this process, microorganisms, usually bacteria, are used either to destroy toxic or hazardous substances or to convert them to harmless forms. This approach mimics nature by using decomposers to recycle matter.

If toxin-degrading bacteria or fungi can be found or engineered for specific hazardous chemicals, they could clean up contaminated sites at less than half the cost of disposal in landfills, and at only about one-third the cost of on-site incineration. Bioremediation might also clean up contaminated groundwater at an affordable cost by pumping it to the surface, treating it with microorganisms, and then returning it to its aquifer.

Preliminary testing reveals that bioremediation is effective for a number of specific organic wastes, but it does not appear to work very well for toxic metals, highly concentrated chemical wastes, or complete digestion of some complex mixtures of toxic chemicals.

Is Burning Solid and Hazardous Waste the Answer? In 1996 about 15% of the municipal solid waste and 7% of the officially regulated hazardous waste in the United States was burned in incinerators. Most of this mixed municipal solid waste was burned in 150 *mass-burn incinerators* (Figure 13-7), which burn mixed trash without separating out hazardous materials (such as batteries and polyvinylchloride or PVC plastic materials) or noncombustible materials that can interfere with combustion and pollute the air.

Incinerators are costly to build, operate, and maintain and they create very few long-term jobs. Without continual maintenance and good operator

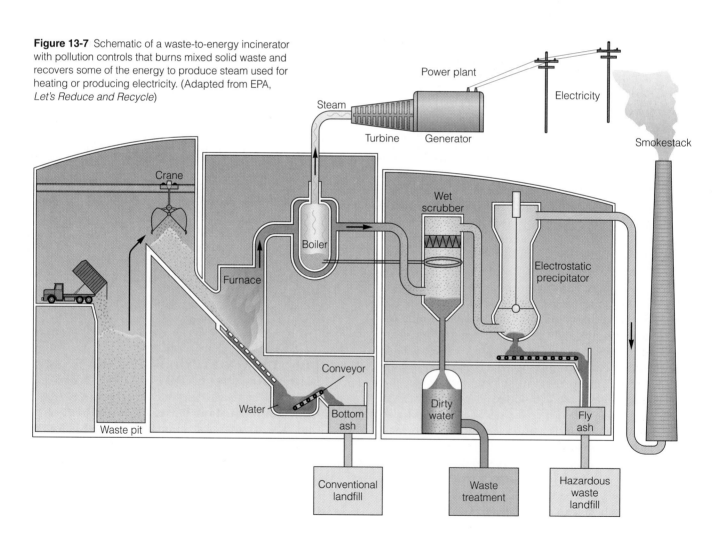

Figure 13-7 Schematic of a waste-to-energy incinerator with pollution controls that burns mixed solid waste and recovers some of the energy to produce steam used for heating or producing electricity. (Adapted from EPA, *Let's Reduce and Recycle*)

Q: What country has the highest industrial energy efficiency?

training and supervision, the air pollution control equipment on incinerators often fails, and emission standards are exceeded.

Hazardous-waste experts Peter Montague (Guest Essay, pp. 68–69), Ellen and Paul Connett, and some EPA scientists point out that even with advanced air pollution control devices (mostly a scrubber followed by a baghouse filter or electrostatic precipitators; Figures 9-16 and 13-7), *all incinerators burning hazardous or solid waste release toxic air pollutants.* Examples are very harmful fine particles of toxic lead and mercury (which cannot be removed by scrubbers) and dioxins (Section 13-7). They also leave a highly toxic ash to be disposed of in landfills that even the EPA concedes will eventually leak.

According to EPA hazardous-waste expert William Sanjour, EPA incinerator regulations don't work because

> *the regulations require no monitoring of the outside air in the vicinity of the incinerator. Because operators maintain the records, they can easily cheat. . . . Government inspectors are poorly trained and have low morale and high turnover. . . . Government inspectors typically work from nine to five Monday through Friday. So if there is anything particularly nasty to burn, it will be done at night or on weekends. When complaints come in, . . . the inspector may visit the plant but rarely finds anything. The enforcement officials tend to view the incinerator operator as their client and the public as a nuisance. . . . There is no reward to inspectors for finding serious violations.*

In 1993 the *Wall Street Journal* warned that using incinerators to burn municipal trash spells financial disaster for local governments in the United States. Since 1992 Rhode Island and West Virginia have banned solid-waste incineration because of its health threats and high cost. Sweden banned the construction of new incinerators in 1985.

Between 1985 and 1997, several solid-waste incinerators in the United States were shut down because of excessive costs and pollution, and over 280 new incinerator projects were blocked, delayed, or canceled because of intense public opposition and high costs. Most of the plants still in the planning stage may not be built as communities discover that recycling, reuse, composting, and waste reduction are cheaper, safer, and less environmentally harmful.

Japan depends on incinerators more than any country, burning about 75% of its municipal solid waste in more than 1,850 government-operated incinerators and more than 3,300 privately owned industrial incinerators. Many of these incinerators are built near populous areas.

Recently there has growing concern in Japan about rising infant deaths and health problems in areas downwind of many incinerators from emissions of toxic dioxins (Section 13-7). The head of a government advisory committee on dioxins found that the concentration of dioxins in the air in Japan is three times that in the United States and some European countries. Dioxins are also found at high levels in the fatty tissue of fish, which plays an important role in the Japanese diet.

According to the Japanese government, the amount of dioxins ingested by residents is somewhat higher than in other developed countries but still generally within levels considered to be safe. Critics contend that Japan is far behind other countries in regulating dioxin emissions from incinerators.

Is Land Disposal of Solid Waste the Answer?
Currently, about 60% by weight of the municipal solid waste in the United States is buried in sanitary landfills. A **sanitary landfill** is a garbage graveyard in which solid wastes are spread out in thin layers, compacted, and covered daily with a fresh layer of clay or plastic foam.

Modern state-of-the-art landfills on geologically suitable sites are lined with clay and plastic before being filled with garbage (Figure 13-8). The bottom is covered with a second impermeable liner, usually made of several layers of clay, thick plastic, and sand. This liner collects *leachate* (rainwater contaminated as it percolates through the solid waste) and is intended to prevent its leakage into groundwater. Collected leachate ("garbage juice") is pumped from the bottom of the landfill, stored in tanks, and sent to a regular sewage treatment plant or an on-site treatment plant. When full, the landfill is covered with clay, sand, gravel, and topsoil to prevent water from seeping in. Several wells are drilled around the landfill to monitor any leakage of leachate into nearby groundwater.

Sanitary landfills for solid wastes offer certain benefits. Air-polluting open burning is avoided. Odor is seldom a problem, and rodents and insects cannot thrive. If located properly they can reduce water pollution from leaching, but proper siting is not always achieved. A sanitary landfill can be put into operation fairly quickly, has low operating costs, and can handle a huge amount of solid waste. After a landfill has been filled, the land can be graded, planted, and used as a park, golf course, ski hill, athletic field, wildlife area, or for some other recreational purpose.

Solid-waste landfills also have drawbacks. While in operation, they cause traffic, noise, and dust; most also emit toxic gases. Paper and other biodegradable wastes break down very slowly in today's compacted and water- and oxygen-deficient landfills. Newspapers dug up from some landfills are still readable after 30 or 40 years; hot dogs, carrots, and chickens that

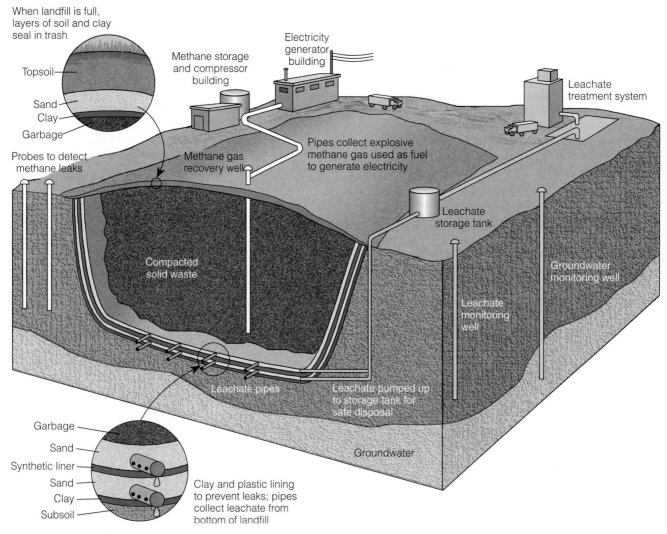

When landfill is full, layers of soil and clay seal in trash

Topsoil

Sand

Clay

Garbage

Methane storage and compressor building

Electricity generator building

Leachate treatment system

Probes to detect methane leaks

Methane gas recovery well

Pipes collect explosive methane gas used as fuel to generate electricity

Compacted solid waste

Leachate storage tank

Groundwater monitoring well

Leachate monitoring well

Leachate pipes

Leachate pumped up to storage tank for safe disposal

Groundwater

Garbage

Sand

Synthetic liner

Sand

Clay

Subsoil

Clay and plastic lining to prevent leaks; pipes collect leachate from bottom of landfill

Figure 13-8 State-of-the-art sanitary landfills are designed to eliminate or minimize environmental problems that plague older landfills. Only a few of the 7,500 municipal and industrial landfills in the United States have such a state-of-the-art design, and 85% of U.S. landfills are unlined. Even state-of-the-art landfills will eventually leak, passing both the effects of contamination and cleanup costs on to future generations.

have been dug up after 10 years have not decomposed. Landfills also deprive present and future generations of valuable reusable and recyclable resources and encourage waste production instead of pollution prevention and waste reduction.

The underground anaerobic decomposition of organic wastes at landfills produces explosive methane (a greenhouse gas), toxic hydrogen sulfide gas, and smog-forming volatile organic compounds that escape into the air. Large landfills can be equipped with vent pipes to collect these gases, and the collected methane can be burned in small power plants or in fuel cells to produce steam or electricity (Figure 13-8). Methane must be collected in all new landfills in the United States, but thousands of older and abandoned landfills don't have such systems and will emit methane for decades. In 1996, 140 U.S. landfills collected gas for use

by industry, power plants, or buildings (Individuals Matter, right).

Contamination of groundwater and nearby surface water by leachate from both unlined and lined landfills is another serious problem. Some 86% of the U.S. landfills studied have contaminated groundwater, and a fifth of all Superfund hazardous-waste sites are former municipal landfills that will cost billions of dollars to clean up. Once groundwater is contaminated, it is extremely expensive and difficult—often impossible—to clean up.

Modern double-lined landfills (Figure 13-8), required in the United States since 1996, delay the release of toxic leachate into groundwater below landfills, but they do not prevent it. These landfills are designed to accept waste for 10–40 years, and current EPA regulations require owners to maintain and

Q: How much of the heat in U.S. homes and other buildings escapes through closed windows?

monitor landfills for at least 30 years after they are closed. However, they could begin to leak after this period, passing the health risks and costs of contamination to future generations.

According to G. Fred Lee, an experienced landfill consultant, the best solution to the leachate problem is to apply clean water to landfills continuously and then collect and treat the resulting leachate in carefully designed and monitored facilities. He contends that after 10–20 years of such washing, little potential for groundwater pollution should remain. This wetting would also hasten the breakdown of wastes and thus allow old landfills to be dug out and used again.

Since 1979 there has been a sharp drop in the number of solid-waste landfills in the United States as existing landfills reached their capacity or closed; since 1997 only more expensive, state-of-the art landfills (Figure 13-8) are allowed to operate. Although a few U.S. cities, notably Philadelphia and New York, are having trouble finding nearby landfills, there is no national shortage of landfill volume because most of the smaller local landfills are being replaced by larger local and regional landfills.

Is Land Disposal of Hazardous Waste the Answer? Most hazardous waste in the United States is disposed of by deep-well injections, surface impoundments, and state-of-the-art landfills. In *deep-well disposal*, liquid hazardous wastes are pumped under pressure through a pipe into dry, porous geologic formations or into fracture zones of rock far beneath aquifers tapped for drinking and irrigation water (Figure 11-3). In theory, these liquids soak into the porous rock material and are isolated from overlying groundwater by essentially impermeable layers of rock.

Because this method is simple and cheap, its use is increasing rapidly as other methods are legally restricted or become too expensive. If sites are chosen according to the best geological and seismic data, a number of scientists believe deep wells may be a reasonably safe way to dispose of fairly dilute solutions of organic and inorganic waste; with proper site selection and care, they may be safer than incineration. Furthermore, if some use eventually is found for the waste, it could be pumped back to the surface.

However, many scientists believe that current regulations for geologic evaluation, long-term monitoring, and long-term liability if wells contaminate groundwater are inadequate. Until deep-well disposal is more carefully evaluated and regulated, many scientists and environmentalists believe that it should not be allowed to proliferate. What do you think?

Surface impoundments such as ponds, pits, or lagoons (Figure 11-24) used to store hazardous waste are supposed to be sealed with a plastic liner on the

Using Landfill Methane to Heat a School

INDIVIDUALS MATTER

In January 1997, Pattonville High School in Maryland Heights, Missouri (near St. Louis), became the first public school to use methane gas produced by a nearby landfill as a source of heat for the school's 117 classrooms and two gymnasiums.

Most students at the high school looked at the nearby landfill and saw garbage. In 1993, however, members of the school's Ecology Club saw it as an opportunity to reduce resource waste and pollution.

After doing their homework, they convinced school officials to build a pipeline to transfer waste methane gas from the landfill to the school and to rebuild two basement heating system boilers to burn the methane to heat the school. The school is expected to recoup its investment in about five years and save money thereafter.

This project reduces resource waste and pollution because before the landfill was burning off the methane gas, which emitted carbon dioxide—a greenhouse gas (Figure 10-7a)—into the atmosphere. In addition, students at the school are getting a lesson in applied environmental science by studying the operation and design of the system.

Critical Thinking

This project may not have wide application because most schools are not near landfills. Are there any renewable energy resources (Chapter 18) available at your school site that could be used to provide all or part of the school's heat or electricity? If so, consider evaluating such resources and presenting a plan to school officials.

bottom. Solid wastes settle to the bottom and accumulate, whereas water and other volatile compounds evaporate into the atmosphere. According to the EPA, however, 70% of these storage basins have no liners, and as many as 90% may threaten groundwater. Eventually all liners will leak, and waste will percolate into groundwater. Moreover, volatile compounds (such as hazardous organic solvents) can evaporate into the atmosphere, promote smog formation, and return to the ground to contaminate surface water and groundwater in other locations.

About 5% by weight of the legally regulated hazardous waste produced in the United States is concentrated, put into drums, and buried, either in one of 21 specially designed and monitored commercial hazardous-waste landfills or in one of 35 such landfills run by companies to handle their own waste. Sweden goes further and buries its concentrated hazardous wastes in underground vaults (Figure 13-9). In the United Kingdom, most hazardous wastes are mixed with household garbage and stored in hundreds of conventional landfills all over the country.

When current and future commercial hazardous-waste landfills in the United States eventually leak and threaten water supplies, many of their operators will declare bankruptcy. Then the EPA may have to put the landfills on the Superfund list, and taxpayers will pick up the tab for cleaning them up. As EPA hazardous-waste expert William Sanjour points out,

> *The real cost of dumping is not borne by the producer of the waste or the disposer, but by the people whose health and property values are destroyed when the wastes migrate onto their property and by the taxpayers who pay to clean it up. . . . It is better for liners to leak sooner rather than later, because then there will be responsible parties that they can get to clean it up. Liners don't protect communities. They protect the people who put the waste there and the politicians who let them put the waste there, because they are long since gone when the problem comes up.*

Some engineers and scientists have proposed storing hazardous wastes above ground in large, two-story, reinforced-concrete buildings until better technologies are developed (Figure 13-10). The first floor would contain no wastes but would have inspection walkways so people could easily check for leaks from the upper story. Any leakage would be collected, treated, solidified, and returned to the storage building. The buildings would have negative air pressure—caused by fans blowing in air and then filtering it—to prevent the release of toxic gases. Proponents believe that such an *in-sight* approach would be cheaper and safer than *out-of-sight* landfills, deep wells, or incinerators for many hazardous wastes. What do you think?

There is also growing concern about accidents during some of the more than 500,000 shipments of hazardous wastes (mostly to landfills and incinerators) in the United States each year. Between 1988 and 1992, approximately 34,500 toxic-chemical accidents caused 100 deaths, over 11,000 injuries, and evacuation of over 500,000 people. Few communities have the equipment and trained personnel to deal with hazardous-waste spills.

Is Exporting Waste the Answer? Some U.S. cities that lack sufficient landfill space are shipping some of their solid waste to other states or other countries, especially developing countries. However, a few of these

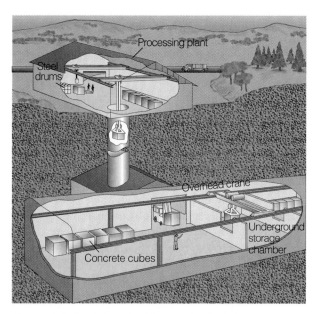

Figure 13-9 Swedish method for handling hazardous waste. Hazardous materials are placed in drums, which are embedded in concrete cubes and stored in an underground vault.

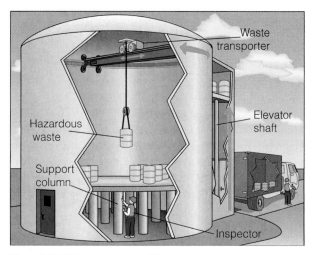

Figure 13-10 An aboveground hazardous-waste storage facility would help keep wastes from contaminating groundwater supplies.

states and some developing countries are refusing to accept delivery.

To save money and to escape regulations and local opposition, some U.S. cities and waste disposal companies legally ship large quantities of hazardous waste to other countries. U.S. exports of hazardous wastes can take place without EPA approval because U.S. laws allow exports for recycling; some exported wastes designated for recycling are instead dumped after reaching their destination.

Q: What do scientists believe are the two best ways to save oil, slow global warming, and reduce air pollution?

Waste disposal firms can charge high prices for picking up hazardous wastes. If they can then dispose of them (legally or illegally) at low costs, they pocket huge profits. Often officials of developing countries find it difficult to resist the income (often bribes) derived from receiving these wastes.

In 1994, 64 nations met in Geneva, Switzerland, and approved an immediate ban on all exports of hazardous wastes for disposal from developed countries to developing countries (including eastern Europe and the former Soviet Union). The United States was the only participating nation that failed to support the ban. By 1996 some 97 countries had imposed national bans on hazardous-waste imports.

In 1994 the Clinton administration recommended that Congress ban exports of all hazardous wastes outside North America (with some exceptions, including materials containing toxic metals such as lead exported for recycling). This ban would do little to change U.S. policy because it would still allow export of hazardous waste to Canada and Mexico, which now receive almost 98% of U.S. hazardous waste exports.

An effective worldwide ban on all hazardous-waste exports, including those slated to be recycled (often a sham), would help reduce this transfer of risk and encourage developed countries to reduce their production of such wastes. But this would not end illegal trade in these wastes because the potential profits are much too great. To most environmentalists, the only real solution to the hazardous-waste problem is not to produce it in the first place (pollution prevention).

🌐 13-7 CASE STUDIES: LEAD, DIOXINS, AND CHLORINE

How Can We Reduce Exposure to Lead? Each year 12,000–16,000 American children under age 9 are treated for acute lead poisoning, and about 200 die. About 30% of the survivors suffer from palsy, partial paralysis, blindness, and mental retardation.

Lead can also cause damage at levels far below those that cause acute lead poisoning, especially in children and unborn fetuses. In the United States, the current maximum legal level in the blood is 10 micrograms (μg) of lead per deciliter (dL, or about half a cup) of blood, or 10 μg/dL. Research indicates that children under age 6 and unborn fetuses with blood levels greater than this standard are especially vulnerable to nervous system impairment, a lowered IQ (by 4–6 points), a shortened attention span, hyperactivity, hearing damage, and various behavior disorders.

It is great news that between 1976 and 1992 the percentage of U.S. children ages 1–5 with lead levels above 10 μg/dL dropped from 85% to 6% for white children and from 98% to 21% for black children, preventing at least 9 million childhood lead poisonings. The primary reason is government regulations that phased out leaded gasoline and lead-based solder, a pollution prevention approach instituted primarily because of the pioneering research on lead toxicity in children published in 1979 by Dr. Herbert Needleman.

Even with the encouraging drop in average blood levels of lead, a large number of U.S. children and unborn fetuses still have unsafe levels of lead from exposure to a number of sources (Figure 13-11). Health scientists have proposed a number of ways to help protect children from lead poisoning:

- *Testing all children for lead by age one.*

- *Banning incineration of solid and hazardous waste or greatly increasing current pollution control standards for old and new incinerators.*

- *Phasing out leaded gasoline worldwide over the next decade.*

- *Testing older housing and buildings for leaded paint and lead dust and removing this hazard.* According to government estimates, 74% of U.S. homes contain lead-based paint.

- *Banning all lead solder in plumbing pipes and fixtures and in food cans.*

- *Removing lead from municipal drinking water systems within 10 years.*

- *Washing fresh fruits and vegetables and hands thoroughly.*

- *Testing ceramicware used to serve food for lead glazing.*

- *Reevaluating the proposed increase in electric cars propelled by lead batteries.* A 1995 study estimated that the manufacture, handling, disposal, and recycling of lead batteries for widespread use in electric vehicles could create up to 60 times the lead pollution caused by vehicles burning leaded gasoline.

Doing most of these things will cost an estimated $50 billion in the United States. But health officials say the alternative is to keep poisoning and mentally handicapping large numbers of children. What do you think?

How Dangerous Are Dioxins? Dioxins are a family of 75 chlorinated hydrocarbon compounds formed as unwanted by-products in chemical reactions (usually at high temperatures, especially in incinerators) involving chlorine and hydrocarbons. One of these compounds, TCDD—sometimes simply called dioxin—

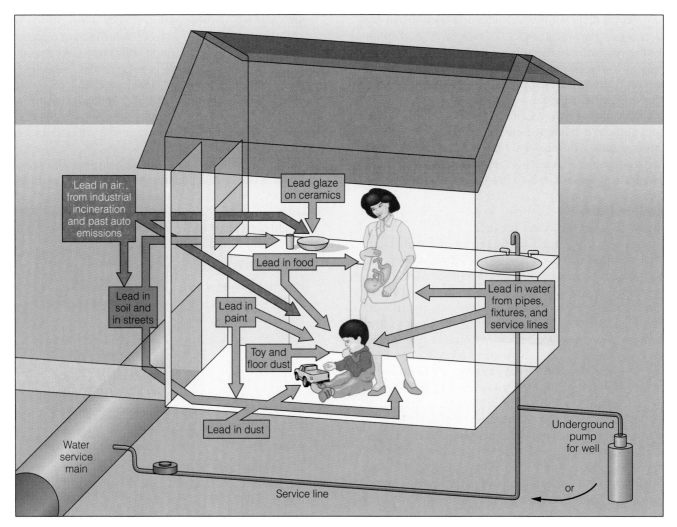

Figure 13-11 Sources of lead exposure for children and fetuses.

is the most harmful and the most widely studied. Dioxins such as TCDD are persistent chemicals that linger in the environment for decades, especially in soil and human fat tissue.

In 1990 representatives of the paper and chlorine industries claimed that the inconsistent results of health studies exonerated TCDD and other dioxins, and they pushed the EPA for a reassessment of the health risks of dioxins. This strategy may have backfired when EPA's 1994 reevaluation indicated that dioxins are an even greater health threat than previously thought. This new review concluded that dioxin is "likely to present a cancer threat to humans." Unlike most carcinogens, TCDD does not damage DNA; it apparently promotes cancer by activating DNA already damaged by other carcinogens.

This finding explains why researchers found a variety of cancers rather than a single type, as is usu-

ally the case with most carcinogens. According to the EPA review, TCDD and several other dioxins in food and air may already be a major cancer hazard; existing levels of dioxins in the U.S. population are now believed to cause an estimated 2.5–25% of all new cancers, or 26,000–260,000 new cancers each year.

This comprehensive, 3-year, $4-million review by more than 100 scientists around the world also revealed that the most powerful effects of exposure to various dioxins at or near levels already occurring in the U.S. population are seen in the reproductive system, the endocrine (hormone) system, and the immune system. These effects may pose an even greater threat to human health (especially for fetuses and newborns) than the cancer-promoting ability of various dioxins (Section 8-2). With profits at stake, there is intense industry pressure to soften or discredit this new health reassessment of dioxins.

Q: What percentage of the world's energy is provided by renewable resources?

What Should We Do About Chlorine? Modern society depends heavily on chlorine (Cl_2) and chlorine-containing compounds. This highly reactive gas is used to purify water, to bleach paper and wood pulp, and to produce household bleaching agents; it also combines with various organic compounds to form about 11,000 different chlorinated organic compounds, many of them widely used. The problem is that some of the chlorine-containing chemicals we produce and use in fairly large amounts are persistent, accumulate in body fat, and (according to animal and other toxicity studies, Section 8-2) are harmful to human health.

In 1993 the Governing Council of the American Public Health Association (APHA), one of the premier U.S. scientific and medical associations, unanimously approved a statement urging the American chemical industry to phase out use of chlorine except for producing some pharmaceuticals and for disinfection of public water supplies. The goal would be to replace most chlorine compounds with environmentally and economically acceptable substitutes over the next 10–20 years.

The three largest single uses, which together make up 60% of all chlorine use, are plastics (mostly to make PVC), solvents, and paper and pulp bleaching. Most of these uses have substitutes (although some are more expensive as long as full-cost pricing is not used) and could be phased out, as a small but growing number of companies are doing. Nonchlorine plastics and other materials could replace most uses of PVC, and incineration of PVCs could be forbidden (which would also reduce production of dioxins).

Studies estimate that 60% of the chlorinated organic solvents currently used could be phased out over a decade and replaced with less harmful and affordable substitutes, such as soap and water, steam cleaning, citrus-based solvents (Individuals Matter, p. 294), and physical cleaning (including elbow grease and blasting with plastic beads and a pressurized solution of water and baking soda). Such a phaseout has already begun, not just for environmental reasons but also to reduce legal liability for possible effects of such chemicals on workers and consumers.

Using chlorine to bleach wood pulp and paper could be replaced with less harmful processes that rely on oxygen, ozone, chlorine dioxide, and hydrogen peroxide, as several paper companies are doing. By the early 1990s the German paper industry had achieved totally chlorine-free paper production (which is 30% cheaper than processes using chlorine); the rest of Europe is following Germany's lead.

About 5% of the chlorine produced in the United States is used to purify drinking water (1%) and wastewater from sewage treatment plants (4%). Much of this could be gradually replaced with ozone and other nonchlorine purification processes. Drinking water may still need to be purified by chlorination because it continues to kill harmful microorganisms in water lines.

The remaining 35% of chlorine is used to manufacture a variety of organic and inorganic chemicals, many of them for small and specialized uses. Scrutiny of each of these chemicals is required to determine which are essential (many pharmaceuticals) and which might be replaced by affordable and environmentally acceptable substitutes.

Understandably, producers of chlorine chemicals strongly oppose such threats to their profits. However, if given proper financial incentives and 10–20 years to develop and phase in substitutes, chemical producers could profit from new markets resulting from this important part of the ecoindustrial revolution based on pollution prevention (Solutions, p. 375); they would also protect themselves from adverse publicity, possible future lawsuits, and cleanup costs.

13-8 HAZARDOUS-WASTE REGULATION IN THE UNITED STATES

What Is the Resource Conservation and Recovery Act? In 1976 the U.S. Congress passed the Resource Conservation and Recovery Act (RCRA, pronounced "RICK-ra"), which was amended in 1984. This law requires the EPA to identify hazardous wastes and set standards for their management, and it provides guidelines and financial aid for states to establish waste management programs. The law also requires all firms that store, treat, or dispose of more than 100 kilograms (220 pounds) of hazardous wastes per month to have a permit stating how such wastes are to be managed.

To reduce illegal dumping, hazardous-waste producers that are granted disposal permits by the EPA must use a cradle-to-grave system to keep track of waste transferred from point of origin to approved off-site disposal facilities. However, the EPA and state regulatory agencies don't have sufficient personnel or money to review the documentation of more than 750,000 hazardous-waste generators and 15,000 haulers each year, let alone detect and prosecute offenders. Environmentalists argue that fines are too small and that culprits don't get jail time for serious violations that can cause severe health problems and even death, sending violators the clear message that pollution pays.

What Is the Superfund Act? In 1980 the U.S. Congress passed the Comprehensive Environmental Response, Compensation, and Liability Act, commonly known as the Superfund program. This law (plus amendments in 1986 and 1990) established a $16.3-billion Superfund financed jointly by federal and state governments and by special taxes on chemical and petrochemical industries (which provide 86% of the funding). The purpose of the Superfund is to identify and clean up abandoned hazardous-waste dump sites such as Love Canal (p. 366) and leaking underground tanks that threaten human health and the environment.

To keep taxpayers from footing most of the bill, cleanups are based on the *polluter-pays principle*. The EPA is charged with locating dangerous dump sites, finding the potentially liable culprits, ordering them to pay for the entire cleanup, and suing them if they don't. When the EPA can find no responsible party, it draws money out of the Superfund for cleanup.

To implement the polluter-pays principle, the Superfund legislation considers all polluters of a site to be subject to strict, joint, and several liability. This controversial strategy means that each individual polluter (no matter what its contribution) can be held liable for the entire cost of cleaning up a site if the other parties can't be found or have gone bankrupt. Once the EPA sues any parties they consider liable, these parties typically try to reduce their cleanup costs by suing any other contributors to the site that they can find. Once stuck with the cleanup bill, responsible parties file claims with their casualty and property insurers, who usually fight such claims in court. Although this process holds identified polluters (or their insurance companies) responsible for the cleanup, the resulting lengthy legal battles slow cleanup.

Currently about 1,300 sites are on a National Priority List for cleanup because they pose a real or potential threat to nearby populations; the number may eventually grow to 2,000–3,000. By June 1997 emergency cleanup had been carried out at virtually all sites, and more than 400 sites had been cleaned up by containing the wastes to prevent leakage. The EPA has plans to stabilize an additional 500 sites by 2000. Once a site is stabilized, final cleanup (which often involves pumping up and cleaning contaminated groundwater) can take decades. Attempts are being made to find ways to improve the Superfund Act without seriously weakening it (Solutions, right).

The U.S. Office of Technology Assessment and the Waste Management Research Institute estimate that the Superfund list could eventually include at least 10,000 priority sites, with cleanup costs of up to $1 trillion, not counting legal fees. Other studies project

only 2,000 sites with estimated cleanup costs of $100–165 billion.

Cleaning up toxic military dumps will cost another $100–200 billion and take at least 30 years. Cleaning up contaminated Department of Energy sites used to make nuclear weapons will cost an additional $200–400 billion and take 30–50 years. In addition, the Department of Interior will need to spend at least $100 billion to clean up 300,000 active and abandoned mine sites, 200,000 oil and gas leases, 3,000 landfills, and 4,200 leaking underground storage tanks on public lands under its jurisdiction.

These estimated costs are only for cleanup to prevent future damage; they don't include the health and ecological costs already associated with such wastes. To environmental scientists and some economists, it is difficult to imagine a more convincing reason for emphasizing waste reduction and pollution prevention (Figures 13-3 and 13-4).

13-9 SOLUTIONS: ACHIEVING A LOW-WASTE SOCIETY

What Is the Role of Grassroots Action? Bottom-Up Change In the United States, local citizens have been stirred into action by proposals to bring incinerators, landfills, or treatment plants for hazardous and radioactive wastes into their communities. Outrage has grown as numerous studies have shown that such facilities have traditionally been located in communities populated by African Americans, Asian Americans, Hispanics, and poor whites (Guest Essay, pp. 392–393).

This loose coalition of grassroots organizations offers the following guidelines for helping achieve environmental justice for all:

- *Don't compromise our children's futures by cutting deals with polluters and regulators.*

- *Don't be bulldozed by scientific and risk analysis experts.* Risks from incinerators and landfills, when averaged over the entire country, are quite low, but the risks for the people near these facilities are much higher. These people, not the rest of the population, are the ones whose health, lives, and property values are being threatened.

- *Hold polluters—and elected officials who go along with them—personally accountable because what they are doing is wrong.*

- *Don't fall for the argument that protesters against hazardous-waste landfills, incinerators, and injection wells are holding up progress in dealing with hazardous wastes.*

Q: What is the largest untapped energy source in the United States?

Even though the Superfund was created with noble goals, there is widespread agreement that the program can and should be improved. Since the program began, polluters and their insurance companies have been working hard to do away with the polluter-pays principle at the heart of the program and make it mostly a *public-pays* approach.

This strategy, which is working, has three components: **(1)** Deny responsibility (stonewall) in order to tie up the EPA in expensive legal suits for years, **(2)** sue local governments and small businesses to make them responsible for cleanup, both as a delay tactic and to turn local governments and small businesses into opponents of Superfund's strict liability requirements, and **(3)** mount a public relations campaign declaring that toxic dumps pose little threat, that cleanup is too expensive compared to the risks involved, and that Superfund is unfair to polluters and is wasteful and ineffective.

The strict joint and several liability of the Superfund law may seem unfair to polluters, but the authors of the legislation argued that any other liability scheme wouldn't work. If the EPA had to identify and bring an enforcement action against every party liable for a dump site, the agency would be overwhelmed

with lawsuits, and the pace of cleanup would be still slower. Administrative costs would also rise sharply, causing the program to become more of a public-pays program favored by the polluters.

The EPA also points out that the strict polluter-pays principle in the Superfund Act has been effective in making illegal dump sites virtually a thing of the past—an important form of pollution prevention for the future. It has also forced waste producers who are fearful of future liability claims to reduce their production of such waste and to recycle or reuse much more of what they do generate.

A generally accepted proposal for reducing lawsuits and speeding cleanup is to set up an $8 billion Environmental Insurance Resolution Fund funded by insurers for a 10-year period. Companies found liable for cleanup of wastes they disposed of before 1986 would be able to use money from this fund rather than going to their individual insurance companies.

Although most experts agree that the Superfund's ultimate goal—permanent cleanup—should be maintained, many argue that sites should be ranked in three general categories: **(1)** sites requiring immediate full cleanup, **(2)** sites considered to pose a serious hazard but located sufficiently far from concentrations of people or endan-

gered ecosystems (these sites would receive emergency cleanup and then be isolated by barriers and signs, with more complete cleanup to come later), and **(3)** lower-risk sites requiring only stabilization (capping and containment) and monitoring.

There is also considerable pressure to involve people and local governments in communities with contaminated sites in cleanup decisions. Too often the people affected by risks from Superfund sites are not consulted until the EPA and the polluters have worked out a plan, which may or may not be in the best interests of local citizens.

Critical Thinking

1. Do you support the *polluter-pays principle* used in the original Superfund legislation? Defend your position. Use the library or Internet to find out whether there have been any recent changes in the Superfund law. If so, evaluate these changes and decide which of them you support. Defend your position.

2. Explain why you agree or disagree with the other suggestions listed above for improving the Superfund law. Use the library or the Internet to determine what changes (if any) have been made in the Superfund law and explain why you agree or disagree with these changes.

Instead, recognize that the best way to deal with waste and pollution is to produce much less of it. For example, after a charge for each unit of toxic waste produced in Germany was imposed in 1991, hazardous waste production fell more than 15% in 3 years.

■ *Oppose all hazardous-waste landfills, deep-disposal wells, and incinerators.* This will sharply raise the cost of dealing with hazardous materials, discourage location of such facilities in poor neighborhoods often populated by minorities, and encourage waste pro-

ducers and elected officials to get serious about pollution prevention.

■ *Recognize that there is no such thing as safe disposal of a toxic or hazardous waste.* For such materials, the goal should be "Not in Anyone's Backyard" (NIABY) or "Not on Planet Earth" (NOPE).

■ *Pressure elected officials to pass legislation requiring that unwanted industries and waste facilities be distributed more widely instead of being concentrated in poor and working-class neighborhoods, many populated mostly by minorities.*

GUEST ESSAY

Robert D. Bullard

Robert D. Bullard is a professor of sociology at the University of California, Riverside. For more than a decade, he has conducted research in the areas of urban land use, housing, community development, industrial facility siting, and environmental justice. His scholarship and activism have made him one of the leading experts on environmental racism—the systematic selection of communities of color as sites for waste facilities and polluting industries. He is the author of four books and more than three dozen articles, monographs, and scholarly papers that address equity concerns. His book, Dumping in Dixie: Race, Class, and Environmental Quality, 2d ed. *(Westview Press, 1994), has become a standard text in the field. Other books are* Confronting Environmental Racism *(South End Press, 1993) and* Unequal Protection: Environmental Justice and Communities of Color *(Sierra Club Books, 1994).*

Despite widespread media coverage and volumes written on the U.S. environmental movement, environmentalism and social justice have seldom been linked. Nevertheless, an environmental revolution is now taking shape in the United States that combines the environmental and social justice movements into one framework.

People of color (African Americans, Latinos, Asians, Pacific Islanders, and Native Americans), working-class people, and poor people in the United States suffer disproportionately from industrial toxins, dirty air and drinking water, unsafe work conditions, and the location of noxious facilities such as municipal landfills, incinerators, and toxic-waste dumps. Despite the government's attempts to level the playing field, all communities are not created equal.

The environmental justice movement attempts to dismantle exclusionary zoning ordinances, discriminatory land-use practices, differential enforcement of environmental regulations, disparate siting of risky technologies, and the dumping of toxic waste on the poor and people of color in the United States and in developing countries.

All communities are not treated as equals when it comes to resolving environmental and public health concerns, either. Over 300,000 farm workers (over 90% of whom are people of color) and their children are poisoned by pesticides sprayed on crops in the United States. Some 3–4 million children (many of them African Americans or Latinos living in the inner city) are poisoned by lead-based paint in old buildings, lead-soldered pipes and water mains, lead-tainted soil contaminated by industry, and air pollutants from smelters.

All communities do not bear the same burden or reap the same benefits from industrial expansion. This is true in the case of the mostly African American Emelle, Alabama (home of the nation's largest hazardous-waste landfill); Navajo lands in Arizona where uranium is mined; and the 2,000 factories known as *maquiladores*, located just across the U.S. border in Mexico.

Communities, states, and regions that contain hazardous-waste disposal facilities (importers) receive far fewer economic benefits (jobs) than the geographic locations that generate the wastes (exporters). Nationally, 60% of African Americans and 50% of Latinos live in communities with at least one uncontrolled toxic-waste site. Three of the five largest hazardous-waste landfills

- *Ban all hazardous-waste exports from one country to another.*

It is not surprising that representatives of the increasingly profitable solid and hazardous waste management industries oppose these tactics. So far they have been able to persuade U.S. federal and state legislators not to put such ideas into practice.

How Can We Make the Transition to a Low-Waste Society? According to physicist Albert Einstein, "A clever person solves a problem, a wise person avoids it." To prevent pollution and reduce waste, many environmental scientists urge us to understand and live by four key principles: **(1)** Everything is connected, **(2)** there is no "away" for the wastes we produce, **(3)** dilution is not the solution to most pollution, and **(4)** the best and cheapest way to deal with waste and pollution is to produce less of it and then reuse and recycle most of the materials we use (Figures 13-3 and 13-4). Some actions you can take to reduce your production of solid waste and hazardous waste are listed in Appendix 5.

Unless we learn to depend on fewer virgin resources— by development of secondary materials industries . . . and by redesigning goods, services, and communities—we will continue propelling ourselves toward economic and ecological disaster.

JOHN E. YOUNG AND AARON SACHS

Q: What percentage of U.S. and world energy needs could be provided by renewable resources by 2030 or sooner?

are located in communities that are predominantly African American or Latino.

The marginal status of many people of color in the United States makes them prime actors in the movement for environmental and social justice. For example, the organizing theme of the 1991 First National People of Color Environmental Summit, held in Washington, D.C., was justice, fairness, and equity. More than 650 delegates from all 50 states, as well as from Puerto Rico, Mexico, Chile, Colombia, and the Marshall Islands, participated in this historic 4-day gathering.

Environmental justice does not stop at the U.S. border. Environmental injustices exist from the *favelas* of Rio de Janeiro, Brazil, to the shantytowns of Johannesburg, South Africa. Members of the environmental justice movement are also questioning the wasteful and nonsustainable development models being exported to the developing world.

It is no mystery why grass-roots environmental justice groups in Louisiana's "Cancer Alley," Chicago's southside, and Los Angeles's East and South Central neighborhoods are attacking the institutions they blame for their underdevelopment, disenfranchisement, and poisoning. Some people see these threats to their communities as a form of genocide.

Grass-roots leaders are demanding justice. Residents of communities such as West Dallas and Texarkana (Texas), West Harlem (New York), Rosebud (South Dakota), Kettleman City (California), and Sunrise, Lions, and Wallace (Louisiana) see their struggle for environmental justice as a life-and-death matter. Unfortunately, their stories of environmental racism are not broadcast into the nation's living rooms during the nightly news, nor are they splashed across the front pages of national newspapers and magazines. To a large extent, the communities that are the victims of environmental injustice remain "invisible" to the larger society.

The environmental justice movement is led, planned, and to a large extent funded by individuals who are not part of the established environmental community or the "Big 10" environmental organizations. Most environmental justice groups are small and operate with resources generated from the local community.

For too long these groups and their leaders have been "invisible" and their stories muted. This is changing as these grass-roots groups are forcing their issues onto the nation's environmental agenda.

The United States has a long way to go in achieving environmental justice for all its citizens. The membership of decision-making boards and commissions still does not reflect the racial, ethnic, and cultural diversity of the country, and token inclusion of people of color on boards and commissions does not necessarily mean that their voices will be heard or their cultures respected. The ultimate goal of any inclusion strategy should be to democratize the decision-making process and empower disenfranchised people to speak and do for themselves.

Critical Thinking

1. Do your lifestyle and political involvement help promote or reduce environmental racism in your community and in society as a whole?

2. How would you go about helping prevent polluting factories and hazardous-waste facilities from being located in or near communities made up largely of people of color, working-class people, and poor people?

CRITICAL THINKING

1. Explain why you support or oppose the following:
   a. Requiring that all beverage containers be reusable
   b. Requiring all households and businesses to sort recyclable materials into separate containers for curbside pickup

2. Would you oppose having a hazardous-waste landfill, a waste treatment plant, a deep-injection well, or an incinerator in your community? Explain. If you oppose these disposal facilities, how do you believe that hazardous waste generated in your community and your state should be managed?

3. Give your reasons for agreeing or disagreeing with each of the following proposals for dealing with hazardous waste:

   a. Reducing the production of hazardous waste and encouraging recycling and reuse of hazardous materials by levying on producers a tax or fee for each unit of waste generated
   b. Banning all land disposal and incineration of hazardous waste to encourage recycling, reuse, and treatment and to protect air, water, and soil from contamination
   c. Providing low-interest loans, tax breaks, and other financial incentives to encourage industries producing hazardous waste to recycle, reuse, treat, destroy, and reduce generation of such waste
   d. Banning the shipment of hazardous waste from the United States to any other country
   e. Banning the shipment of hazardous waste from one state to another

4. Explain why you agree or disagree with each of the suggestions on pp. 390–392 made by a coalition

of grassroots organizations in the United States seeking environmental justice for all.

*5. For one week, keep a list of the solid waste you throw away. What percentage is materials that could be recycled, reused, or burned for energy? What percentage of the items could you have done without in the first place? Tally and compare the results for your entire class.

*6. What hazardous and solid wastes do you create? What specific things could you do to reduce, reuse, or recycle as much of these wastes as possible? Which of these things do you actually plan to do? Do you use any of the alternative chemicals shown in Table 13-2?

*7. Determine whether (a) your school and your city have recycling programs, (b) your school sells soft drinks in throwaway cans or bottles, and (c) your school bans release of throwaway helium-filled balloons at sporting events and other activities.

*8. What happens to solid waste in your community? How much is landfilled? Incinerated? Composted? Recycled? What technology is used in local landfills and incinerators? What leakage and pollution problems have local landfills or incinerators had? Does your community have a recycling program? Is it voluntary or mandatory? Does it have curbside collection? Drop-off centers? Buy-back centers?

*9. What hazardous wastes are produced at your school? In your community? What happens to these wastes?

*10. Make a concept map of this chapter's major ideas, using the section heads and subheads and the key terms (in boldface type). Look at the inside back cover and in Appendix 4 for information about concept maps.

SUGGESTED READINGS

See Internet Sources and Further Readings at the back of the book, and InfoTrac.

PART IV

SUSTAINING BIODIVERSITY AND ECOLOGICAL INTEGRITY

It is the responsibility of all who are alive today to accept the trusteeship of wildlife and to hand on to posterity, as a source of wonder and interest, knowledge, and enjoyment, the entire wealth of diverse animals and plants. This generation has no right by selfishness, wanton or intentional destruction, or neglect, to rob future generations of this rich heritage. Extermination of other creatures is a disgrace to humankind.

WORLD WILDLIFE CHARTER

14 FOOD RESOURCES

Perennial Crops on the Kansas Prairie

When you think about farms in Kansas, you probably picture seemingly endless fields of wheat or corn plowed up and planted each year. By 2040 the picture might change, thanks to pioneering work at the nonprofit Land Institute near Salina, Kansas (Figure 14-1).

The Institute, headed by plant geneticist Wes Jackson, is experimenting with an ecological approach to agriculture on the Midwestern prairie. The goal is to grow food crops by planting a mix of *perennial* grasses, legumes (a source of nitrogen fertilizer), sunflowers, grain crops, and plants that provide natural insecticides in the same field (polyculture). Because these plants are perennials, the soil doesn't have to be plowed up and prepared each year to replant them. The Institute's goal is to raise food by mimicking many of the natural conditions of the prairie without losing fertile grassland soil (Figure 12-18). Perennial polyculture is especially suitable for marginal land, leaving prime, flat land available for raising annual crops.

By eliminating yearly plowing and planting, perennial polyculture requires much less labor than conventional monoculture or diversified organic farms that grow annual crops. It reduces soil erosion because the unplowed soil is not exposed to wind and rain and also reduces pollution caused by chemical fertilizers and pesticides.

If the Institute and similar groups doing such earth-sustaining research succeed, within a few decades many people may be eating food made from perennials such as *Maximilian sunflower* (which produces seeds with as much protein as soybeans), *eastern gamma grass* (a relative of corn with three times as much protein as corn and twice as much as wheat), *Illinois bundleflower* (a wild nitrogen-producing legume that can enrich the soil and whose seeds may serve as livestock feed), and *wheatgrass* (a wild ancestor of wheat that produces a highly nutritious seed that has about two-thirds more protein than ordinary wheat).

These discoveries will come none too soon. To feed the 8 billion people projected by 2025, we must produce and distribute about as much food during the next 26 years as was produced since agriculture began about 10,000 years ago.

Figure 14-1 The Land Institute in Salina, Kansas, is a farm, a prairie laboratory, and a school dedicated to changing the way we grow food. It advocates growing a diverse mixture of edible perennial plants to supplement traditional annual monoculture crops. (Terry Evans)

There are two spiritual dangers in not owning a farm. One is the danger of supposing that breakfast comes from the grocery, and the other that heat comes from the furnace.

ALDO LEOPOLD

The discussion in this chapter answers several general questions:

- How is the world's food produced?
- What are the world's food problems?
- What can we do to help solve the world's food problems?
- How can we design and shift to more sustainable agricultural systems?

14-1   HOW IS FOOD PRODUCED?

What Plants and Animals Feed the World?   The multitude of species of plants and animals (species diversity) and the varieties of plants and animals (genetic diversity) that provide us with food are an important part of the planet's biodiversity. Biologists estimate that even though the earth has perhaps 30,000 plant species with parts that people can eat, only 15 plant and 8 animal species supply 90% of our food.

Four crops—wheat, rice, corn, and potato—make up more of the world's total food production than all other crops combined. These four, and most of our other food crops, are *annuals*, whose seeds must be replanted each year.

Two out of three or the world's people survive primarily on grains (mainly rice, wheat, and corn), mostly because they can't afford meat. As incomes rise people consume even more grain, but indirectly in the form of meat (mostly beef, pork, and chicken), eggs, milk, cheese, and other products of grain-eating domesticated livestock.

What Are the Major Types of Food Production?   There are two major types of agricultural systems: industrialized and traditional. **Industrialized agriculture**, or **high-input agriculture**, uses large amounts of fossil fuel energy, water, commercial fertilizers, and pesticides to produce huge quantities of single crops (monocultures) or livestock animals for sale. Practiced on about 25% of all cropland, mostly in developed countries (Figure 14-2), industrialized agriculture has spread since the mid-1960s to some developing countries. **Plantation agriculture**, a form of industrialized agriculture primarily in tropical developing countries, grows cash crops such as bananas, coffee, and cacao, mostly for sale in developed countries.

Traditional agriculture consists of two main types, which together are practiced by about 2.7 billion people in developing countries—almost half the people on the earth. **Traditional subsistence agriculture** typically produces only enough crops or livestock for a farm family's survival; in good years there may be a surplus to sell or to put aside for hard times. Subsistence farmers primarily use human labor and draft animals. Examples of this type of agriculture include numerous forms of shifting cultivation in tropical forests (Figure 2-4) and nomadic livestock herding.

In **traditional intensive agriculture**, farmers increase their inputs of human and draft labor, fertilizer, and water to get a higher yield per area of cultivated land and produce enough food to feed their families and to sell for income. Figure 14-3 shows the inputs of land, human and animal labor, fossil fuel energy, and capital needed to produce one unit of food energy in various types of food production.

How Have Green Revolutions Increased Food Production?   Farmers can produce more food either by farming more land or by getting higher yields per unit of area from existing cropland. Since 1950 most of the increase in global food production has resulted from increased yields per unit of cropland in an agricultural system called the **green revolution**.

This process involves three steps: **(1)** developing and planting monocultures (Figure 5-10) of selectively bred or genetically engineered high-yield varieties of key crops such as rice, wheat, and corn; **(2)** lavishing fertilizer, pesticides, and water on crops to produce high yields; and **(3)** often increasing the intensity and frequency of cropping. This approach dramatically increased crop yields in most developed countries between 1950 and 1970 in what is considered the *first green revolution* (Figure 14-4, p. 400).

A *second green revolution* has been taking place since 1967 (Figure 14-4), when fast-growing dwarf varieties of rice and wheat, specially bred for tropical and subtropical climates, were introduced into several developing countries. With sufficient fertile soil and enough fertilizer, water, and pesticides, yields of these new plants (Figure 14-5, p. 400) can be two to five times those of traditional wheat and rice varieties. The fast growth also allows farmers to grow two or even three crops a year (multiple cropping) on the same land.

Between 1970 and 1992 India doubled its total food production (primarily by use of high-yield varieties of grain) and increased per capita food production by 18%—an impressive achievement considering its large population increase during this period. Without the green revolution, India would have faced massive famines in the 1970s and 1980s.

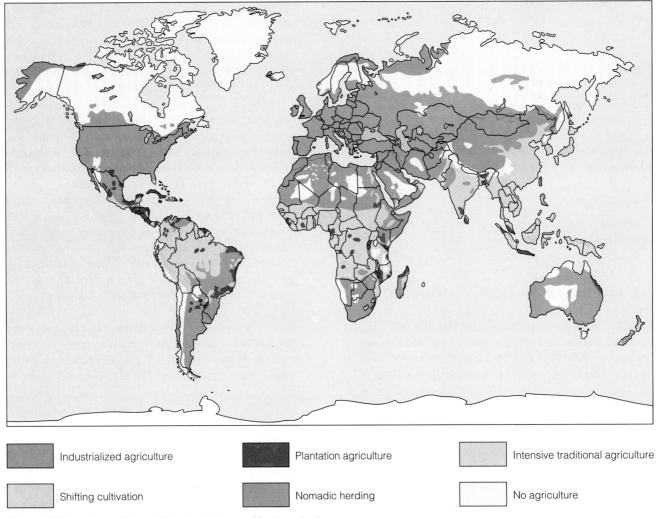

 appears here with legend:

Industrialized agriculture

Plantation agriculture

Intensive traditional agriculture

Shifting cultivation

Nomadic herding

No agriculture

Figure 14-2 Locations of the world's principal types of food production.

Producing more food on less land is also an important way to protect biodiversity by saving large areas of forests, grasslands, wetlands, and mountain terrain from being used to grow food. Since 1950 high-yield agriculture has saved an estimated 9–26 million square kilometers (3.5–10 million square miles) of such land from destruction or degradation by farming.

These yield increases depend not only on having fertile soil and ample water, but also on extensive use of fossil fuels to run machinery, produce and apply inorganic fertilizers and pesticides, and pump water for irrigation. Between 1950 and 1990, agricultural use of fossil fuels increased 4-fold, irrigated area expanded 2.5-fold, use of commercial fertilizer rose 10-fold, and use of pesticides increased 30-fold. All told, green-revolution agriculture now uses about 8% of the world's oil output.

These high inputs of energy, water, fertilizer, and pesticides on high-yield crop varieties have yielded dramatic results. At some point, however, additional inputs become useless because no more output can be squeezed from the soil and the crop varieties—the principle of diminishing returns in action. In fact, yields may even start dropping for a number of reasons: The soil erodes, loses fertility, and becomes salty and waterlogged (Figure 12-24); underground and surface water supplies become depleted and polluted with pesticides and nitrates from fertilizers; and populations of rapidly breeding pests develop genetic immunity to widely used pesticides.

For example, 5–8% of the world's irrigated cropland depends to some degree on overpumping and depletion of aquifers, especially in Libya, Iran, the Arabian peninsula, northern China, northern India, and California and the High Plains in the United States (Case Study, p. 309). With less water for irrigation, food production drops and food imports can rise. Importing 1 metric ton of wheat is equivalent to importing 1,000 metric tons of water.

Q: What percentage of the world's electricity could solar cells supply by 2050?

Case Study: Food Production in the United States Since 1940 U.S. farmers have more than doubled crop production without cultivating more land, a result of industrialized agriculture using green-revolution techniques in a favorable climate on some of the world's most fertile and productive soils. This has also kept large areas of forests, grasslands, wetlands, and easily erodible land from being converted to farmland.

Farming has become *agribusiness* as big companies and larger family-owned farms have taken control of most U.S. food production. Only about 650,000 Americans are full-time farmers. However, about 9% of the population is involved in the U.S. agricultural system, from growing and processing food to distributing it and selling it at the supermarket. In terms of total annual sales, agriculture is the biggest industry in the United States, bigger than the automotive, steel, and housing industries combined. It generates about 18% of the country's gross national product and 19% of all jobs in the private sector, employing more people than any other industry.

The U.S. agricultural system is highly productive. Currently each U.S. farmer feeds and clothes about 140 people (105 at home and 35 abroad), up from 58 in 1976. Today U.S. farms, with only 0.3% of the world's farm labor force, produce about 25% of the world's food and nearly half of the world's grain exports. U.S. residents spend an average of only 10–12% of their income on food (down from 21% in 1940), compared to 18% in Japan and 40–70% in most developing countries.

This industrialization of agriculture was made possible by the availability of cheap energy, most of it from oil. Agriculture consumes about 17% of all commercial energy in the United States each year (Figure 14-6). On average, a piece of food eaten in the United States has traveled 2,100 kilometers (1,300 miles). Processing food also requires large amounts of energy.

Most plant crops in the United States provide more food energy than the energy used to grow them. However, if we include livestock as well as crops, the U.S. food production system currently uses about three units of fossil fuel energy to produce one unit of food energy.

Energy efficiency is much lower if we look at the whole U.S. food system. Considering the energy used to grow, store, process, package, transport, refrigerate, and cook all plant and animal food, *an average of about 10 units of nonrenewable fossil fuel energy are needed to put 1 unit of food energy on the table.* By comparison, every unit of energy from human labor in subsistence farming provides at least 1 unit of food energy; with traditional intensive farming, each unit of energy provides up to 10 units of food energy.

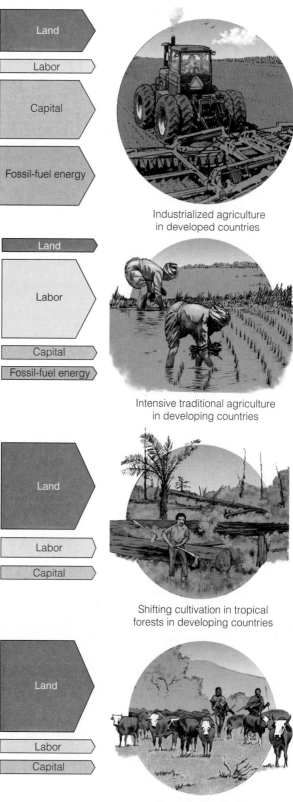

Industrialized agriculture in developed countries

Intensive traditional agriculture in developing countries

Shifting cultivation in tropical forests in developing countries

Nomadic herding in developing countries

Figure 14-3 Relative inputs of land, labor, capital, and fossil-fuel energy in four agricultural systems. An average of 60% of the people in developing countries are directly involved in producing food, compared with only 8% in developed countries (and 2% in the United States).

A: About 25% (35% in the United States) with an aggressive program

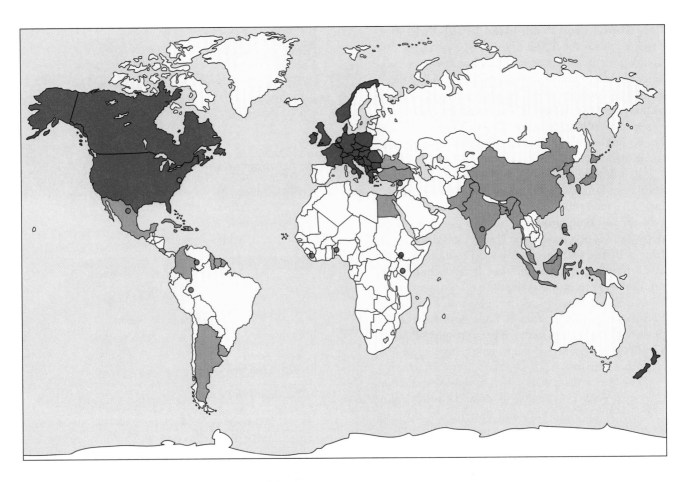

First green revolution
(developed countries)

Second green revolution
(developing countries)

• Major international agricultural
research centers and seed banks

Figure 14-4 Countries whose crop yields per unit of land area increased during the two green revolutions. The first took place in developed countries between 1950 and 1970; the second has occurred since 1967 in developing countries with enough rainfall or irrigation capacity. Several agricultural research centers and gene or seed banks play a key role in developing high-yield crop varieties.

Figure 14-5 A high-yield, semidwarf variety of rice called IR-8 (left), part of the second green revolution, was produced by crossbreeding two parent strains of rice: PETA from Indonesia (center) and DGWG from China (right). The shorter and stiffer stalks of the new variety allow the plants to support larger heads of grain without toppling over. (International Rice Research Institute, Manila)

How Are Livestock Produced and What Are the Environmental Consequences? The world's rangelands make up the second land-based food system. For thousands of years domesticated animals such as cattle, horses, oxen, sheep, chickens, and pigs have played important roles in the human economy by providing food, fertilizer, fuel, clothing, and transport.

Whereas only about 10% of the world's land is suitable for producing crops, about 20% is used for grazing cattle and sheep. Grasses on this land produce most of the world's beef and mutton.

Meat and meat products are good sources of high-quality protein. Traditionally, when both crops and livestock are grown on diversified farms (such as those found in the United States before the recent shift to industrial monocultures), the livestock returned nutrients to the soil as manure, provided draft power, and grazed on fallow fields.

Q: What percentage of the world's electricity could wind power supply by 2050?

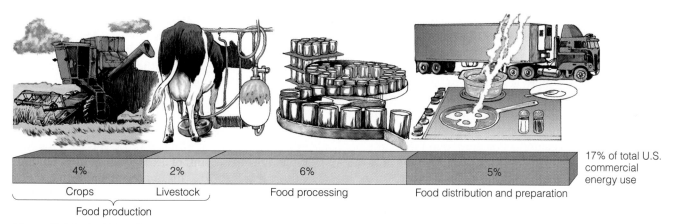

4%	2%	6%	5%	17% of total U.S. commercial energy use
Crops	Livestock	Food processing	Food distribution and preparation	

Food production

Figure 14-6 In the United States, industrialized agriculture uses about 17% of all commercial energy.

During the past 50 years the global livestock population has exploded as increased affluence has led to rising production and consumption of meat, mainly in developed countries and more recently in middle-income developing countries. Between 1950 and 1996, world meat production increased fourfold and per capita meat production rose by 29%.

Currently about 1.2 billion people in developed countries live high on the food chain by having a diet based on high consumption of meat and meat-based products. Roughly another 1 billion people living in the poorest developing countries (especially in most of Africa) where incomes are not rising consume mostly grain and live low on the food chain. The remaining 3.7 billion people live in low- or middle-income countries (especially in Asia) where rising incomes allow them to move further up the food chain by consuming more meat and meat products.

An increasing amount of livestock production in developed countries is industrialized; large numbers of cattle are typically brought to crowded feedlots, where they are fattened up for about 4 months before slaughter. Most pigs and chickens in developed countries spend their entire lives in densely populated pens and cages and are mostly fed grain grown on cropland. The meat-based diet of affluent people in developed countries and developing countries has enormous effects on resource use, environmental degradation, and pollution.

Any expansion in the supply of animal protein from livestock increasingly depends on producing feed grain and using water for irrigation and washing animal wastes away. More than half of the world's cropland (19% in the United States) is used to produce livestock feed grain (mostly field corn, sorghum, and soybeans). In 1995 livestock and fish raised for food consumed about 37% of the world's grain production (70% in the United States). Livestock use more than

half the water withdrawn each year in the United States; most of this water irrigates crops fed to livestock and washes manure from crowded livestock pens away.

About 14% of U.S. topsoil loss is directly associated with livestock grazing. Globally, overgrazing of sparse vegetation and trampling of the soil by too many livestock is the major cause of desertification in arid and semiarid areas (Figure 12-23). If everyone in the world became vegetarians and all other factors stayed the same, the world's current oil reserves would last about 260 years instead of 40–80 years.

Cattle belch out 12–15% of all the methane released into the atmosphere (Figure 10-7c). Some of the nitrogen in commercial inorganic fertilizer used to grow livestock feed is converted to nitrous oxide, another greenhouse gas (Figure 10-7d). Nitrogen in manure escapes into the atmosphere as gaseous ammonia (NH_3), a pollutant that contributes to acid deposition (Figure 9-6). Livestock in the United States produce 21 times more waste (manure) than that produced by the country's human population. Only about half of this nutrient-rich livestock waste is recycled to the soil.

Some environmentalists have called for reducing livestock production (especially cattle) as a way to feed more humans; other analysts say this won't work. Reducing livestock production would decrease its environmental impact, but it would not free up much land or grain to feed more people. Cattle and sheep that graze on rangeland use a resource (grass) that humans can't eat, and most of this land is not suitable for growing crops. Moreover, because of poverty, insufficient economic aid, and the nature of global economic and food distribution systems, very little if any additional grain grown on land used to raise livestock or livestock feed would reach the world's hungry people.

A: About 10% (10–25% in the United States)

What Is Traditional Agriculture? Agrodiversity in Action Traditional farmers in developing countries today grow about 20% of the world's food on about 75% of its cultivated land. Many traditional farmers simultaneously grow several crops on the same plot, a practice known as **interplanting**. Such crop diversity reduces the chance of losing most or all of their year's food supply to pests and other misfortunes.

Common interplanting strategies found throughout the world, mostly in developing countries, include the following:

- **Polyvarietal cultivation**, in which a plot is planted with several varieties of the same crop.

- **Intercropping**, in which two or more different crops are grown at the same time on a plot (for example, a carbohydrate-rich grain that uses soil nitrogen alongside a protein-rich legume that puts it back).

- **Agroforestry**, or **alley cropping**, in which crops and trees are planted together (Figure 12-25c). For example, a grain or legume crop can be planted around fruit-bearing orchard trees or in rows between fast-growing trees or shrubs that can be used for fuelwood or for adding nitrogen to the soil.

- **Polyculture**, a more complex form of intercropping in which many different plants maturing at various times are planted together. If cultivated properly, these plots can provide food, medicines, fuel, and natural pesticides and fertilizers on a sustainable basis.

With polyculture, root systems at different depths in the soil capture nutrients and moisture efficiently and minimize the need for fertilizer or irrigation water. Year-round plant coverage also protects the soil from wind and water erosion. The presence of various habitats for insects' natural predators means that crops need not be sprayed with insecticides. In addition,

weeds have trouble competing with the multitude of crop plants and thus can be removed fairly easily by hand, without herbicides. Crop diversity also provides insurance against bad weather: If one crop fails because of too much or too little rain, another crop may survive or even thrive.

Recent ecological research on crop yields of 14 artificial ecosystems found that on average polyculture (with four or five different crop species) produces higher yields per unit of area than high-input monoculture (Figure 5-10). This important finding has major implications for development of high-yield sustainable agriculture in developing countries by combining the techniques of traditional high-yield interplanting with modern inputs of fertilizer and irrigation.

14-2 WORLD FOOD PROBLEMS AND CHALLENGES

How Much Has Food Production Increased? Figure 14-7 shows the success story of global agriculture. Between 1950 and 1990, world grain production almost tripled (Figure 14-7, left), and per capita production rose by about 36% (Figure 14-7, right), helping reduce global hunger and malnutrition. During the same period, average food prices adjusted for inflation dropped by 25% and the amount of food traded in the world market quadrupled.

Despite these impressive achievements in food production, population growth is outstripping food production and distribution in areas that support about 2 billion people. Since 1978 grain production has lagged behind population growth in 88 developing countries. Food production in Africa has been rising steadily since 1961, but not fast enough to keep up with population growth; between 1974 and 1994 per capita production fell by 20%.

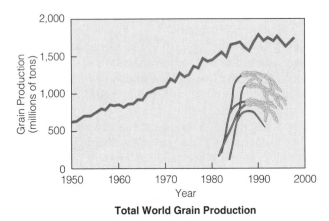

Total World Grain Production

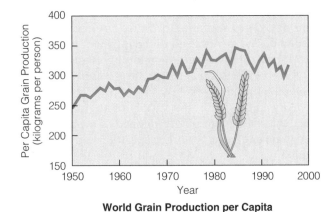

World Grain Production per Capita

Figure 14-7 Total worldwide grain production of wheat, corn, and rice, and per capita grain production, 1950–96. (Data from U.S. Department of Agriculture and Worldwatch Institute)

Q: What do many scientists believe is the best fuel to replace oil and other fossil fuels over the next 40 years?

Since 1950 global grain production has been rising but its rate of growth has slowed (Figure 14-7, left) and per capita grain production has declined slightly (Figure 14-7, right). More than 100 countries regularly import food from the United States, Canada, Australia, Argentina, western Europe, Australia, New Zealand, Thailand, and a few other countries.

Major factors leading to this slowdown in grain production (from an average of 2.6% a year between 1950 and 1990 to 0.7% a year between 1990 and 1996) are **(1)** population growth, **(2)** increasing affluence, which increases the demand for food (especially meat produced by feeding livestock grain), **(3)** degradation and loss of cropland, mostly because of erosion and to a lesser extent because of industrialization and urbanization (especially in Asia), **(4)** little growth in irrigation since 1980 (shrinking the irrigated area per person by 6% between 1980 and 1996), and **(5)** a 12% decline in global fertilizer use since 1989 (amounting to a 21% drop in per capita fertilizer use). Any drop in annual grain production by major food-exporting countries (especially the United States) has severe economic consequences for food importing nations (Connections, below).

Unless death rates rise sharply, we seem destined to have a population of around 8 billion people by 2025. To provide this many people with even a meatless subsistence diet will require doubling food production and distribution between 1995 and 2025.

How Serious Are Undernutrition, Malnutrition, and Overnutrition? People who cannot grow or buy enough food to meet their basic energy needs suffer from **undernutrition**. People getting less than 80% of their minimum calorie intake are considered to be *seriously undernourished*. Children in this category are likely to suffer from mental retardation and stunted growth. They are also much more susceptible to infectious diseases such as measles and diarrhea, which kill one child in four in developing countries.

To maintain good health and to resist disease, people need not only a certain number of calories, but also the proper amounts of protein (from animal or plant sources), carbohydrates, fats, vitamins, and minerals. People who are forced to live on a low-protein, high-carbohydrate diet consisting only of grains such as wheat, rice, or corn often suffer from **malnutrition**—deficiencies of protein and other key nutrients.

What Happens If Food Demand Exceeds the Supply?

CONNECTIONS

The demand for grain is rising. In most developed countries and rapidly industrializing middle-income countries this increasing demand is fueled by rising affluence. In the poorest developing counties (most in Africa), population growth is the major cause of the rising demand for grain.

So far the supply of food for export (primarily by the United States) has kept up with the demand. However, if bad weather or changes in climate lower food production the demand for grain is likely to exceed the supply. Then grain prices will rise.

This will greatly increase the economic and political power of a small number of grain-exporting countries, especially the United States, which controls almost half of such exports—a larger share of the world's grain exports than

the share of oil exports controlled by Saudi Arabia. Any drops in available food exports from the United States because of bad weather or long-term changes in climate can cause economic chaos in the increasing number of countries dependent on such imports.

If grain production fails to keep up with demand, China, Japan, Taiwan, and most other rapidly industrializing countries of Asia will be losers because they will have to divert an increasing amount of their income to pay for food imports. The biggest loser will be most of Africa, with low rates of economic growth and the fastest population growth rate of any continent (Figure 6-3). If food prices rise, the economic power of rural crop-growing areas should increase, whereas cities dependent on such areas for food will be losers.

If there are unfavorable climate changes in major U.S. food-growing

areas as a result of projected global warming (Figure 10-10 and Sections 10-3 and 10-4), the entire global food production system could be thrown into severe economic and political chaos.

In today's interconnected world, political and economic instability almost anywhere can affect global and national economies and the profits and losses of multinational corporations. It can also lead to hordes of hungry people and environmental refugees illegally migrating into other countries in a desperate search for food and work.

Critical Thinking

The United States is by far the world's largest food exporter. What harmful economic and environmental effects might the United States experience if its ability to supply enough food for export exceeds the demand?

Saving Children

Officials of the United Nations Children's Fund (UNICEF) estimate that between one-half and two-thirds of childhood deaths from nutrition-related causes could be prevented at an average annual cost of only $5–10 per child—only 10–19¢ per week. This life-saving program would involve the following simple measures:

- Immunizing children against childhood diseases such as measles

- Encouraging breast-feeding

- Preventing dehydration from diarrhea by giving infants a mixture of sugar and salt in a glass of water

- Preventing blindness by giving people a vitamin A capsule twice a year at a cost of about 75¢ per person

- Providing family-planning services to help mothers space births at least two years apart

- Increasing education for women, with emphasis on nutrition, sterilization of drinking water, and child care

Critical Thinking

How much money (if any) would you be willing to spend each year to help implement such a program for saving children? Why has little money been allocated for such a program?

Many of the world's desperately poor people, especially children, suffer from both undernutrition and malnutrition.

The two most common nutritional-deficiency diseases are marasmus and kwashiorkor. *Marasmus* (from the Greek word *marasmos*, "to waste away") occurs when a diet is low in both calories and protein (Figure 1-9). Most victims are either nursing infants of malnourished mothers or children who do not get enough food after being weaned from breastfeeding. If the child is treated in time with a balanced diet, most of these effects can be reversed. In practice, however, relief efforts are often too little to late to save many children from death.

Kwashiorkor (meaning "displaced child" in a West African dialect) is a severe protein deficiency occurring in infants and children ages 1–3, usually after the arrival of a new baby deprives them of breast milk. The displaced child's diet changes to grain or sweet potatoes, which provide enough calories but not enough protein. If caught soon enough, most of the harmful effects can be cured with a balanced diet; otherwise, if the child survives, stunted growth and mental retardation result.

Here's some good news. *Between 1970 and 1995 the worldwide proportion of people suffering from chronic undernutrition fell from 36% to 14%.* Also, despite population growth, *the estimated number of chronically malnourished people fell from 940 million in 1970 to 840 million in 1995.*

Despite this progress, about one of every five people in developing countries (including one of every three children below age 5) is chronically undernourished or malnourished; 87% of them live in Asia and Africa. Such people are disease prone, and adults are too weak to work productively or think clearly. As a result, their children also tend to be underfed and malnourished. If these children survive to adulthood, many are locked in a tragic malnutrition–poverty cycle (Figure 14-8) that can be perpetuated for generations.

It is estimated that each year at least 10 million people, half of them children under age 5, die prematurely from undernutrition, malnutrition, or normally nonfatal diseases such as measles and diarrhea worsened by malnutrition. Some put this annual death toll at 20 million. However, children don't have to die prematurely because of undernutrition and malnutrition (Solutions, left).

Each of us must have a small daily intake of vitamins that cannot be made in the human body. Although balanced diets, vitamin-fortified foods, and vitamin supplements have slashed the number of vitamin-deficiency diseases in developed countries, millions of cases occur each year in developing countries. Each year, for example, more than 250,000 children go blind because their diet lacks vitamin A.

Other nutritional-deficiency diseases are caused by the lack of certain minerals. For example, too little iron (a component of hemoglobin that transports oxygen in the blood) causes anemia. This mineral deficiency causes fatigue, makes infection more likely, increases a woman's chances of dying in childbirth, and increases an infant's chances of dying from infection during its first year of life. In tropical regions of Asia, Africa, and Latin America, iron-deficiency anemia affects about 350 million people.

According to the World Health Organization, some 1.6 billion people—almost one of every four—get too little iodine in their diet. Lack of iodine can cause goiter, an abnormal enlargement of the thyroid gland in the neck, which leads to deafness if untreated. Worldwide, an estimated 100 million children born to mothers with severe iodine deficiency suffer from

Q: What percentage of the world's commercial energy comes from nonrenewable resources?

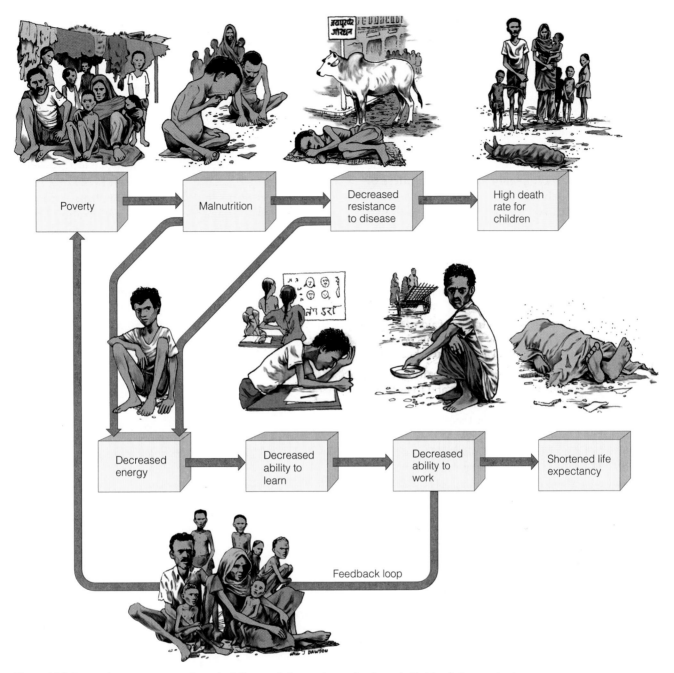

Figure 14-8 Interactions among poverty, malnutrition, and disease form a tragic cycle that tends to perpetuate such conditions in succeeding generations of families.

stunted physical growth and severe mental retardation; another 20 million have some degree of brain damage from too little iodine.

Whereas an estimated 17% of the people in developing countries suffer from undernutrition and malnutrition, about 15% of the people in developed countries suffer from **overnutrition**—an excessive intake of food, especially fats, that can cause obesity (excess body fat).

Overnutrition is associated with at least two-thirds of the deaths in the United States each year. A

study of thousands of Chinese villagers indicates that the healthiest diet for humans is largely vegetarian, with only 10–15% of calories coming from fat—in contrast to the typical meat-based diet, in which 40% of the calories come from fat.

Do We Produce Enough Food to Feed the World's People? The *good news* is that we produce more than enough food to meet the basic nutritional needs of every person on the earth today. Indeed, if distributed equally, the grain currently produced worldwide would be enough to give everyone a meat-less subsistence diet.

The *bad news* for those not getting enough food is that food is not distributed equally among the world's people because of differences in soil, climate, political and economic power, and average per capita income throughout the world. Most agricultural experts agree that *the principal cause of hunger is and will continue to be poverty,* which prevents poor people from growing or buying enough food regardless of how much is available.

If everyone ate the diet typical of a person in a developed country, with 30–40% of the calories coming from animal products, estimates suggest that the current world agricultural system would support only 2.5 billion people—less than half the current population and only one-fourth of the 10 billion people projected sometime in the next century.

Moreover, increases in global and per capita food production often hide large differences in food supply and quality among and within countries. For example, despite impressive gains in total and per capita food production since 1970, roughly 40% of India's population suffers from malnutrition because they are too poor to buy or grow enough food to meet their basic needs.

Developed countries also have pockets of poverty, hunger, and malnutrition. Tufts University researchers found in 1991 that at least 20 million people (12 million children and 8 million adults) in the United States were suffering from preventable chronic hunger and malnutrition.

What Are the Environmental Effects of Producing Food? *Agriculture has a greater harmful impact on air, soil, water, and biodiversity resources than any other human activity* (Figure 14-9). A dramatic casualty of intensive, industrialized agriculture is the drastically shrinking Aral Sea in Uzbekistan (Figure 11-10). The results of this ecodisaster are a sharp drop in the agricultural and aquatic productivity of the region, increased poverty and health problems, and economic losses of an estimated $110 billion per year.

Food prices in the United States are deceptively low because they do not include the harmful environmental and health costs associated with U.S. food production. David Pimentel (Guest Essay, pp. 366–367) has estimated that the harmful costs not included in the prices of food in the United States are $150–200 billion per year. According to a 1990 UN study, degradation of irrigated cropland, rain-fed cropland, and rangeland now costs the world more than $42 billion a year in lost crop and livestock output; this loss is roughly equal to the annual value of the entire U.S. grain harvest.

Many analysts believe that it is possible to produce enough food to feed the 8 billion people projected by 2025 through new advances in agricultural technology and by spreading the use of existing green-revolution or high-yield techniques. Other analysts disagree. They have serious doubts about the ability of new food production technologies and food distribution systems to keep up with current levels of population growth, mostly because the harmful environmental effects of agriculture will reduce yields (Case Study, pp. 408–409).

According to Worldwatch Institute estimates, between 1945 and 1990 erosion, salinization, waterlogging, and other forms of environmental degradation eliminated an area of land from food production equal to the cropland of two Canadas. This trend is expected to accelerate as modern industrialized farming and environmentally unsound subsistence farming increase in coming decades.

According to agricultural and environmental expert Lester Brown (Guest Essay, pp. 26–27) the major constraints on food production are environmental, not economic:

> *Historically, the size of the fish catch was determined by the investment in fishing trawlers, but today the sustainable yield of fisheries is the controlling factor. Until recently, the amount of water pumped was determined by the number of wells drilled, but now it is the sustainable yield of the aquifer. And, increasingly it is not the amount of fertilizer that a farmer can afford but the amount of nutrients that plants can absorb that dictates grain production levels.*

14-3 INCREASING WORLD FOOD PRODUCTION

Is Increasing Crop Yields the Answer?

Agricultural experts expect most future increases in food yields per hectare on existing cropland to result from improved strains of plants and from expansion of green-revolution technology to new parts of the world. For example, in 1994 crop scientists announced they had developed new strains of corn that can increase crop yields up to 40% in regions plagued by droughts and acidic soils. In addition, a new strain of rice developed by the International Rice Research Institute and expected to be commercially available by 2000 could increase rice yields by 20–25%, reducing the projected gap between production and demand by 2025 by one-third.

Scientists are working to create new green revo-

Q: What percentage of the commercial energy used in the United States comes from nonrenewable resources?

Biodiversity Loss

Loss and degradation of habitat from clearing grasslands and forests and draining wetlands

Fish kills from pesticide runoff

Killing of wild predators to protect livestock

Loss of genetic diversity from replacing thousands of wild crop strains with a few monoculture strains

Soil

Erosion

Loss of fertility

Salinization

Waterlogging

Desertification

Air Pollution

Greenhouse gas emissions from fossil fuel use

Other air pollutants from fossil fuel use

Pollution from pesticide sprays

Water

Aquifer depletion

Increased runoff and flooding from land cleared to grow crops

Sediment pollution from erosion

Fish kills from pesticide runoff

Surface and groundwater pollution from pesticides and fertilizers

Overfertilization of lakes and slow-moving rivers from runoff of nitrates and phosphates from fertilizers, livestock wastes, and food processing wastes

Human Health

Nitrates in drinking water

Pesticide residues in drinking water, food, and air

Contamination of drinking and swimming water with disease organisms from livestock wastes

Bacterial contamination of meat

Figure 14-9 Major environmental effects of food production.

lutions—actually *gene revolutions*—by using genetic engineering and other forms of biotechnology. Over the next 20–40 years they hope to breed high-yield plant strains that are more resistant to insects and disease, thrive on less fertilizer, make their own nitrogen fertilizer (as do legumes, Figure 4-24), do well in slightly salty soils, can withstand drought, and can use solar energy more efficiently during photosynthesis. But according to Donald Duvick, former director of research at Pioneer HiBred International (one of the world's largest seed producers), "No breakthroughs are in sight. Biotechnology, while essential to progress, will not produce sharp upward swings in yield potential except for isolated crops in certain situations."

Several factors have limited the success of the green and gene revolutions to date and may continue to do so. Without huge amounts of fertilizer and water, most green-revolution crop varieties produce yields that are no higher (and are sometimes lower) than those from traditional strains; this is why the second green revolution has not spread to many arid and semiarid areas such as much of Africa and Australia (Figure 14-4). Without ample water, good soil, and favorable weather, new genetically engineered crop strains could fail. Furthermore, the cost of genetically

Since 1970 China has made significant progress in feeding its people and in slowing its rate of population growth (Section 6-4). In addition, since 1980 the Chinese economy has quadrupled. During the 1990s China has lifted many of its people out of poverty by becoming the world's second largest economy (after the United States). With its economy growing very rapidly (typically at 9–14% a year), China could become the world's largest economy by 2010.

But with such a rapidly increasing demand for food and other resources and 14 million more people each year, a shortage of resources may slow China's ability to feed its people and sustain its rapid economic growth.

There is growing concern that crop yields may not be able to keep up with demand. A basic problem is that with 21% of the world's people, China has only 7% of the world's cropland and fresh water, 3% of its forests, and 2% of its oil. This concern was highlighted when China shifted from being a net exporter of grain in 1994 to being a net importer to being the world's second largest importer of grain (after Japan) in 1995.

Despite the country's huge area, much of western China is desert and unfit for agriculture. Thus, most of its cropland is concentrated in the eastern part of the country (Figure 14-10). Most of China's people—five times the population of the United States—live on the country's southern and eastern coasts in an area about the size of the United States east of the Mississippi River.

China irrigates 60% of its cropland, but water tables from aquifer depletion are dropping under about 10% of its cultivated area. Since 1950 China has lost an area of cropland roughly equal to the area of Argentina, mostly because of population growth, industrialization, and urbanization—trends that are expected to get worse. Government officials talk of the need to build 600 *new* cities by 2010 to accommodate the country's booming urban population and industrialization. This could lead to another 5% loss of the country's cropland by 2010.

According to Worldwatch Institute projections, China's grain production is likely to fall by at least 20% between 1990 and 2030. Even if China's booming economy resulted in no increases in meat consumption, this 20% drop would mean that by 2030 China would need to import more than the world's entire 200 million tons of grain exports in 1993 (roughly half from the United States).

However, if the increased demand for meat led to a rise in per capita grain consumption equal to the current level in Taiwan (one-half the current U.S. level), by 2030 China would have to import more grain the entire current grain output of the United States.

The Worldwatch Institute warns that if either of these scenarios is correct, no country or combination of countries has the potential to supply even a small fraction of China's potential food supply deficit. This is not even taking into account the huge food deficits that are projected in other parts of the world by 2030, especially Africa and India.

Serious and rapidly growing environmental problems may also limit China's economic growth and its ability to feed its people. China now consumes more grain, red meat, and fertilizer, and produces more steel than the United States—all leading to significant and growing pollution and environmental degradation.

With its limited oil reserves and huge supplies of coal China gets 75% of its energy from burning highly polluting coal (compared to 22% in the United States). As much as 30% of the acid deposition falling on western Japan can be traced to sulfur dioxide emissions from coal burning in China. According to one estimate, by 2035 China's SO_2 emissions will exceed those of all other industrialized countries.

China ranks second in the world (after the United States) in emissions of CO_2, the world's major greenhouse gas (Figure 10-7a). Within the next 25 years China is expected to become the world's largest emitter of CO_2, making it the major contributor to projected global warming (Figure 10-10).

An estimated 80% of the country's industrial and

engineered crop strains is too high for most of the world's subsistence farmers in developing countries.

Continuing to increase inputs of fertilizer, water, and pesticides eventually produces no additional increase in crop yields; the *J*-shaped curve of crop productivity slows down, reaches its limits, levels off, and becomes an *S*-shaped curve. Grain yields per hectare are still increasing in almost every country, but at a much slower rate. Since 1985 yields for the major grains in the three countries with the highest yields per hectare—the United States (corn), Great Britain (wheat), and Japan (rice)—have leveled off and are a key factor in the slowdown in grain production and the decline in per capita grain production since 1990 (Figure 14-7).

According to Lester Brown, president of the Worldwatch Institute,

> The old formula of combining more and more fertilizer with ever-higher yielding that helped almost triple the world grain harvest from 1950 to 1990 is no longer working very well.

Q: What percentage of earth's proven reserves of oil are in OPEC countries?

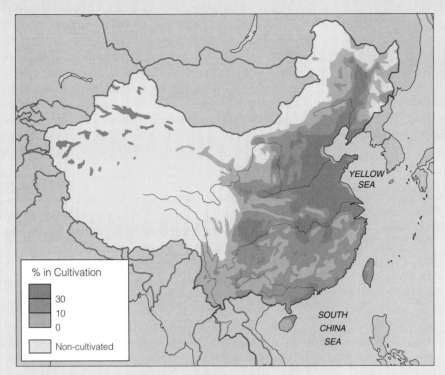

Figure 14-10 Most of China's arable land suitable for growing crops is found in the eastern part of the country. Because of population growth and environmental degradation, the amount of land available for agriculture has declined and is expected to decline more because of China's rapid economic growth, increasing affluence, growing urbanization, and increasing pollution and environmental degradation.

domestic waste is discharged untreated into rivers. As a result, 25% of its rivers are too polluted to use for irrigation. According to a top Chinese environmental official, all but 5 of China's more than 500 cities suffer from severe air pollution, which helps to explain why respiratory disease is the leading cause of death in China's urban areas.

So far China has concentrated on rapid industrialization and devoted little attention to sustainable use of its resources and to reducing pollution and environmental degradation. However, some Chinese officials are beginning to realize that without such policies its economic growth cannot be sustained.

If China begins acting now to chart a new course, it has a unique opportunity to leapfrog over the traditional Western forms of economic development and show the world how to build an environmentally sustainable economy over the next few decades.

Critical Thinking

If the scenarios about China's growing dependence on food imports are valid, how might this affect **(a)** world food prices, **(b)** your life, and **(c)** the harmful environmental impacts of food production (Figure 14-9). What ways, if any, do you suggest for dealing with this potential problem?

There has also been a tendency to use the highest yields achieved for grains such as rice, wheat, and corn as the basis for estimating yield increases elsewhere. Lester Brown argues that this is unrealistic because maximum attainable yields vary with variations in temperature, precipitation, day length (based on latitude), solar intensity, and inherent soil fertility.

Moreover, Indian economist Vandana Shiva (Guest Essay, p. 506) contends that overall gains in crop yields from new green- and gene-revolution varieties may be much lower than claimed. The reason is that the yields are based on comparisons between the output per hectare of old and new *monoculture* varieties, rather than between the even higher yields per hectare for *polyculture* cropping systems and the new monoculture varieties that often replace them.

Connections: Will Loss of Genetic Diversity Limit Crop Yields? Some agricultural scientists think that new genetically engineered or crossbred varieties will enable yields of key crops to continue rising. Other scientists question whether this is possible, primarily

Figure 14-11 The winged bean, a protein-rich annual legume from the Philippines, is one of many currently unfamiliar plants that could become important sources of food and fuel. Its edible winged pods, spinachlike leaves, tendrils, and seeds contain as much protein as soybeans, and its edible roots contain more than four times the protein of potatoes. Its seeds can be ground into flour or used to make a caffeine-free beverage that tastes like coffee. Indeed, this plant produces so many different edible parts that it has been called a supermarket on a stalk. Because of nitrogen-fixing nodules in its roots, this fast-growing plant needs little fertilizer. (Larry Mellichamp/Visuals Unlimited)

because of the environmental impacts of current forms of industrialized agriculture and the accelerating loss of biodiversity, which can limit the genetic raw material needed for future green and gene revolutions.

In India, which once had 30,000 varieties of rice, more than 75% of the rice production now comes from 10 varieties. In the United States, about 97% of the food plant varieties that are available to farmers in the 1940s no longer exist, except perhaps as a handful of seeds in a seed bank or in the backyards of a few gardeners. In other words, we are rapidly shrinking the world's genetic "library" just when we need it more than ever.

Scientists can crossbreed varieties of animal and plant life, and genetic engineers can move genes from one organism to another, but they need the genetic materials in the earth's existing species to work with. We are losing much of this genetic diversity as a small number of specially bred monoculture varieties of key crops have replaced thousands of strains of various crops and natural areas have been cleared.

The UN Food and Agriculture Organization estimates that by the year 2000, two-thirds of all seed planted in developing countries will be of uniform strains. Such genetic uniformity increases the vulnerability of food crops to pests, diseases, and harsh weather. Many biologists argue that this decreased variability, plus growing species extinction, can severely limit the potential of future green and gene revolutions.

In the mid-1970s a valuable wild corn species, the only known perennial strain of corn, was barely saved from extinction. When this strain was discovered in south central Mexico, only a few thousand stalks survived in three tiny patches that were about to be cleared by squatter cultivators and commercial loggers.

Crossbreeding this perennial strain with commercial varieties could reduce the need for yearly plowing and sowing, which would reduce soil erosion and save water and energy. Even more important, this wild corn has a built-in genetic resistance to four of the eight major corn viruses, and it grows in cooler and damper habitats than established commercial strains. Overall, the economic benefits of cultivating this barely rescued wild plant could total several billion dollars per year.

Wild varieties of the world's most important plants can be collected and stored in gene or seed banks, agricultural research centers, and botanical gardens. However, space and money severely limit the number of species that can be preserved. Many plants (such as potatoes) cannot be stored successfully as seed in gene banks. Power failures, fires, or unintentional disposal of seeds can also cause irreversible losses.

In addition, because stored seeds don't remain alive indefinitely, periodically they must be planted (germinated) and new seeds collected for storage. Unless this is done, seed banks become seed morgues. Moreover, stored plant species stop evolving; thus they may be difficult to reintroduce into their native habitats, which may have changed in the meantime.

Because of these limitations, ecologists and plant scientists warn that the only effective way to preserve the genetic diversity of most plant and animal species is to protect representative ecosystems throughout the world from agriculture and other forms of development (Chapter 16).

Will People Try New Foods? Some analysts recommend greatly increased cultivation of less widely known plants to supplement or replace such staples as wheat, rice, and corn. One of many possibilities is the winged bean, a protein-rich legume now common only in New Guinea and Southeast Asia (Figure 14-11). Insects are also important potential sources of protein, vitamins, and minerals in many parts of the world (Figure 14-12). Two basic problems are getting farmers to take the financial risk of cultivating new types of food crops and convincing consumers to try new foods.

David Pimentel (Guest Essay, pp. 366–367) and plant scientists at the Land Institute in Salina, Kansas (p. 396), believe that we could rely more on polycultures of perennial crops, which are better adapted to regional soil and climate conditions than most annual food crops. Using perennials would also eliminate the need

Q: What percentage of earth's proven reserves of oil are in the United States?

Figure 14-12 Insects are important food items in many parts of the world. *Mopani* (emperor moth caterpillars) are among several insects eaten in South Africa. However, this food is so popular that the caterpillars (known as mopane worms) are being over-harvested. Kalahari Desert dwellers eat cockroaches; lightly toasted butterflies are a favorite food in Bali; and French-fried ants are sold on the streets of Bogota, Colombia. Most of these insects are 58–78% protein by weight—three to four times as protein-rich as beef, fish, or eggs. (Anthony Bannister/Natural History Photographic Agency)

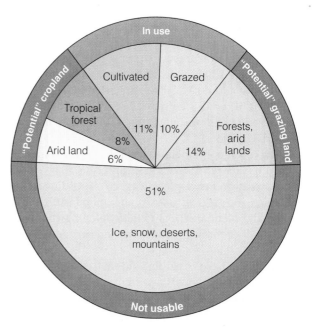

Figure 14-13 Classification of the earth's land. Theoretically, we could double the amount of cropland by clearing tropical forests and irrigating arid lands. However, converting these lands into cropland would destroy valuable forest resources, reduce earth's biodiversity, affect water quality and quantity, and cause other serious environmental problems, usually without being cost-effective.

to till soil and replant seeds each year; it would greatly reduce energy, save water, and reduce soil erosion and sediment water pollution. Of course, widespread use of perennials would reduce the profits of agribusinesses selling annual seeds, fertilizers, and pesticides, which explains why they don't favor this approach.

Is Cultivating More Land the Answer? Today nearly all of the world's best agricultural land is in use. Between 1980 and 1990 the area of the world's cropland expanded by only 2%.

Theoretically, the world's cropland could be more than doubled by clearing tropical forests and irrigating arid land, mostly in Africa and Latin America (Figure 14-13). However, many analysts believe that this potential for agricultural expansion is often overestimated because much of the land is marginal land, where cultivation is unlikely to be sustainable.

Clearing rain forests to grow crops and graze livestock, for example, can have disastrous ecological consequences, as discussed in Section 16-5. In addition, potential cropland in savanna and other semiarid land in Africa cannot be used for farming or livestock grazing because of the presence of 22 species of the tsetse fly.

Some researchers hope to develop new methods of intensive cultivation in tropical areas. But other scientists argue that it makes more ecological and economic sense to combine various ancient methods of shifting cultivation (Figure 2-4), followed by fallow periods long enough to restore soil fertility with various forms of polyculture.

Much of the world's potentially cultivable land lies in dry areas, especially in Australia and Africa. Large-scale irrigation in these areas would require large, expensive dam projects with a mixture of beneficial and harmful impacts (Figure 11-7) and large inputs of fossil fuel to pump water long distances. Large-scale irrigation could also deplete groundwater supplies by removing water faster than it is replenished. The land would need constant and expensive maintenance to prevent erosion, groundwater contamination, salinization, and waterlogging. Expanding wetland production of rice could also accelerate projected global warming by increasing atmospheric methane emissions (Figure 10-7c).

A: About 2–3%; the United States uses about 30% of global oil production each year

Thus, much of the new cropland that could be developed would be on land that is marginal for raising crops, requiring expensive inputs of fertilizer, water, and energy. Furthermore, these potential increases in cropland would not offset the projected loss of almost one-third of today's cultivated cropland due to erosion, overgrazing, waterlogging, salinization, mining, and urbanization.

Even if it is financially feasible, such expansion would reduce wildlife habitats and thus the world's biodiversity and ecological integrity. According to the UN Food and Agriculture Organization (FAO), cultivating all potential cropland in developing countries would reduce forests, woodlands, and permanent pasture by 47%.

In addition to providing wildlife habitats and conserving water and soils, these forests store 20–50 times more carbon as biomass than crops and pasture do. Clearing these forests would release a huge amount of carbon dioxide into the atmosphere and accelerate possible global warming. For these reasons, many analysts believe that a major economically profitable and environmentally sustainable expansion of cropland is unlikely.

Can We Grow More Food in Urban Areas?

Food experts project that people in urban areas could live more sustainably and save money by growing food in empty lots, on rooftops, and in their own backyards. Currently, urban gardens provide about 15% of the world's food, but this output could be increased. A study by the UN Center for Human Settlements estimated that up to 50% of the total area in many cities in developing countries is vacant public land that could be used to produce food.

According to Worldwatch Institute estimates, at least 200 million people grow some of their food and provide 800 million people with at least some of their food. Farmers in Accra, Ghana, provide the city with about 90% of its vegetables. In Singapore urban farmers supply 80% of the city's poultry and 25% of its vegetables. Farmers in or near 18 of China's largest cities provide urban dwellers with 85% of their vegetables and more than half of their meat and poultry.

Recycling nutrient-rich animal and human wastes to grow food in urban areas can also greatly reduce water pollution from cultural eutrophication (Figure 11-22). This mimicking of nature helps convert a one-way flow of nutrients from crops through humans and livestock and into overfertilized aquatic systems into a cyclical system that uses the plant nutrients in animal waste to grow more food.

In China human waste is treated and sold to farmers as fertilizer. For more than 50 years sewage-fed lagoons (ponds) in Calcutta, India, have provided the city's people with one-tenth of the fish they consume. This wastewater-fed aquaculture system also is a cost-effective way to treat sewage wastes that are usually discharged into nearby rivers and lakes, thus helping close the food nutrient loop.

Can We Harvest More Fish and Shellfish?

In addition to cropland and grazing land, the third major food producing system is **fisheries**—concentrations of particular aquatic species suitable for commercial harvesting in a given ocean area or inland body of water. About 71% of the annual commercial catch of fish and shellfish comes from the ocean; 99% of this catch is taken from plankton-rich coastal waters. However, this vital coastal zone is being disrupted and polluted at an alarming rate (Sections 5-2 and 11-6). The remainder of the annual catch comes from using aquaculture to raise fish in ponds and underwater cages (19%) and from inland freshwater fishing from lakes and rivers (10%).

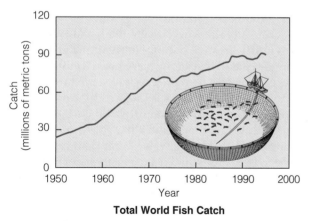

Total World Fish Catch

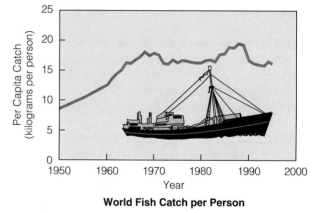

World Fish Catch per Person

Figure 14-14 Worldwide fish catch, total and per person, 1950–95. Worldwide per capita fish catch did not rise much between 1968 and 1989 and has dropped since then. (Data from UN Food and Agriculture Organization and Worldwatch Institute)

Q: What percentage of the oil used in the United States is imported?

Here is some *good news*. Between 1950 and 1995 the annual commercial fish catch from all sources increased more than fourfold, with most of this coming from an almost fivefold increase in the marine catch (Figure 14-14, left). Between 1950 and 1980 the per capita seafood catch more than doubled (Figure 14-14, right).

However, because of population growth and a slowdown in the marine catch (Figure 14-14, left), the per capita catch fell by 7.6% between 1989 and 1995 (Figure 14-14, right). The drop would have been more without the almost threefold growth of aquaculture since 1985. Because of overfishing, pollution, and population growth, the world's marine catch is not expected to increase significantly; it may even decline, and the per capita world catch *is* projected to continue declining.

Connections: How Are Overfishing and Habitat Degradation Affecting Fish Harvests?

Fish are potentially renewable resources as long as the annual harvest leaves enough breeding stock to renew the species for the next year. Ideally, an annual **sustainable yield**—the size of the annual catch that could be harvested indefinitely without a decline in the population of a species—should be established for each species to avoid depleting the stock.

However, determining sustainable yields is difficult. Estimating mobile aquatic populations isn't easy, and sustainable yields shift from year to year because of changes in climate, pollution, and other factors. Furthermore, sustainably harvesting the entire annual surplus of one species may severely reduce the population of other species that rely on it for food—another example of connections in nature.

Overfishing is the taking of so many fish that too little breeding stock is left to maintain numbers; that is, overfishing is a harvest in excess of the estimated sustainable yield. Prolonged overfishing leads to **commercial extinction**—reduction of a species to the point at which it's no longer profitable to hunt for them. Fishing fleets then move to a new species or a new region, hoping that the overfished species will eventually recover (Connections, right).

According to the UN Food and Agriculture Organization, since 1993, *15 of the world's 17 major oceanic fisheries have been fished at or beyond their estimated maximum sustainable yield for commercially valuable species and 13 have been in a state of decline*. As a result, 70% of the world's commercial fish stocks are fully exploited, overfished, or rebuilding from past overfishing and pollution.

According to the U.S. National Fish and Wildlife Foundation, 14 major commercial fish species in U.S. waters (accounting for one-fifth of the world's annual catch and half of all U.S. stocks) are so depleted that even

Commercial Fishing and the Tragedy of the Commons

CONNECTIONS

Today the commercial fishing industry is dominated by industrial fishing fleets with factory trawlers the size of football fields. They are equipped with satellite positioning equipment, sonar, massive nets, and spotter planes. Most of these industrial fishing fleets can remain at sea for a long time because they have factory ships that can process their catches.

Between 1975 and 1996, the size of the industrial fishing fleet expanded twice as fast as the rise in catches. As a result, there are now too many boats fishing for a declining number of fish. This leads to overfishing—an example of the tragedy of the commons (p. 11).

Because of the overcapacity of the fishing fleet and overfishing, it costs the global fishing industry about $125 billion a year to catch $70 billion worth of fish. Most of the $54 billion dollar annual deficit of the industry is made up by government subsidies such as fuel-tax exemptions, price controls, low-interest loans, and grants for fishing gear.

Critics contend that such subsidies accelerate overfishing. They argue that eliminating these subsidies would reduce the size of the fishing fleet by encouraging free-market competition and would allow some of the economically and biologically depleted stocks to recover.

Critical Thinking

Do you believe that government subsidies for the fishing industry should be eliminated? Explain. How would you feel about eliminating such subsidies if your livelihood depended on fishing?

if all fishing stopped immediately it would take up to 20 years for stocks to recover. Ways to manage fisheries to reduce overfishing are discussed in Section 17-6.

Is Aquaculture the Answer? **Aquaculture**, in which fish and shellfish are raised for food, supplies about 19% of the world's commercial fish harvest. Aquaculture production almost tripled between 1984 and 1996; by 2005 it may account for one-third of the world's fish production. China is the world leader in aquaculture (producing almost half of the world's output), followed by India and Japan.

There are two basic types of aquaculture. **Fish farming** involves cultivating fish in a controlled environment, often a pond or tank, and harvesting them when they reach the desired size. **Fish ranching** involves holding anadromous species such as salmon (which live part of their lives in fresh water and part in salt water) in captivity for the first few years of their lives (usually in fenced-in areas or floating cages in coastal lagoons and estuaries), releasing them, and then harvesting the adults when they return to spawn (Figure 14-15).

Species cultivated in developing countries (mostly by inland aquaculture) include carp, tilapia, milkfish, clams, and oysters, all of which feed on phytoplankton and other aquatic plants. In developed countries and some rapidly developing countries in Asia, aquaculture is used mostly to stock lakes and streams with game fish or to raise expensive fish and shellfish such as oysters, catfish, crayfish, rainbow trout, shrimp, and

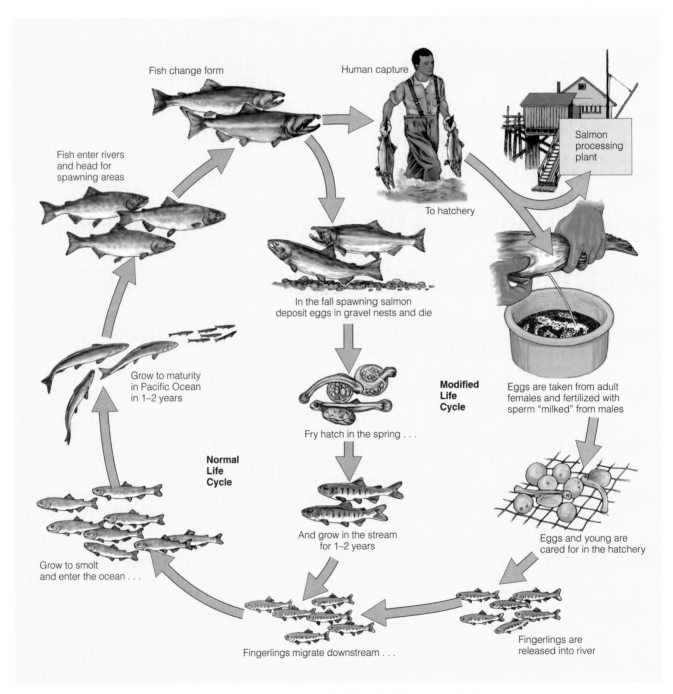

Figure 14-15 Normal life cycle of wild salmon (left) and human-modified life cycle of hatchery-raised salmon (right). Salmon spend part of their lives in fresh water and part in salt water.

Q: Will the United States ever again be self-sufficient in oil?

salmon. Aquaculture now produces 90% of all oysters, one-third of all salmon (75% in the United States), and one-quarter of the shrimp and prawns (50% in the United States) sold in the global marketplace.

Aquaculture has several advantages. It is highly efficient and can produce high yields in a small volume of water. Because little fuel is needed, yields and profits are not closely tied to the price of oil (as they are in commercial marine fishing). Crossbreeding and genetic engineering can help increase yields.

Some people in the aquaculture industry talk of a *blue revolution* that would do for fish farmers and fish ranchers what the green revolution did for grain growers. They project that freshwater and saltwater aquaculture production could double in the next 10 years.

There are problems, however. Fish farms are essentially *aquatic feedlots,* requiring large inputs of land, feed, water, and energy and producing large outputs of wastes. Large-scale aquaculture also requires considerable capital and scientific knowledge, which are in short supply in most developing countries; for another, scooping out huge ponds for fish and shrimp farming in some developing countries has destroyed ecologically important mangrove forests (see photo in Table of Contents). Pesticide runoff from nearby croplands can kill fish in aquaculture ponds, and dense populations make fish more vulnerable to bacterial and viral infections.

Chemicals used to keep nets and cages free of unwanted marine life can be toxic to nearby free-ranging marine animals. Escaped farm-raised fish may also breed with wild fish and degrade the genetic stock of such species. Moreover, without adequate pollution control, waste outputs from shrimp farming and other large-scale aquaculture operations can contaminate nearby estuaries, surface water, and groundwater and eliminate some native aquatic species. A typical salmon farm with 75,000 fish produces as much organic waste as a city of 20,000 people.

How Do Government Agricultural Policies Affect Food Production? Agriculture is a financially risky business. Whether farmers have a good year or a bad year is determined by factors over which they have little control: weather, crop prices, crop pests and diseases, interest rates, and the global market. Because of the need for reliable food supplies despite fluctuations in these factors, most governments provide various forms of assistance to farmers and consumers.

One approach is to *keep food prices artificially low.* This makes consumers happy but means that farmers may not be able to make a living. Many governments in developing countries keep food prices in cities lower than in the countryside to prevent political unrest. With food prices lower in the cities, more rural people migrate to urban areas, aggravating urban problems and unemployment and increasing the chances of political unrest—another example of harmful positive feedback in action.

A second approach is to *give farmers subsidies to keep them in business and to encourage them to increase food production.* In developed countries government price supports and other subsidies for agriculture total more than $300 billion per year. If government subsidies are too generous and the weather is good, farmers may produce more food than can be sold; food prices and profits then drop because of the surplus. Large amounts of food then become available for export or food aid to developing countries, depressing world food prices; the low prices reduce the financial incentive for farmers in developing countries to increase domestic food production. Moreover, the taxes citizens in developed countries pay to provide agricultural subsidies more than offset the lower food prices they enjoy.

A third policy is to *eliminate most or all price controls and subsidies*, allowing market competition to be the primary factor determining food prices and thus the amount of food produced. Some analysts call for phasing out all government price controls and subsidies over, say, 5–10 years and letting farmers respond to market demand. However, these analysts urge that any phaseout of farm subsidies, which in effect subsidize manufacturers of farm chemicals and machinery, should be coupled with increased aid for the poor and the lower middle class, who would suffer the most from any increase in food prices.

Many environmentalists believe that instead of eliminating all subsidies, they should be used only to reward farmers and ranchers who protect the soil, conserve water, reforest degraded land, protect and restore wetlands, and conserve wildlife. What do you think?

Another way governments and private organizations deal with a lack of food production and hunger is through food aid (Pro/Con, p. 416).

14-4 SOLUTIONS: SUSTAINABLE AGRICULTURE

What Is Sustainable Agriculture? Many agricultural scientists and experts believe that a key to reducing world hunger, poverty, and the harmful environmental effects of agriculture (Figure 14-9) is to develop systems of **sustainable agriculture**, or **low-input agriculture**, and phase them in over the next three decades.

How Useful Is International Food Aid?

Most people view international food aid as a humanitarian effort to prevent people from dying prematurely. However, some analysts contend that giving food to starving people in countries with high population growth rates does more harm than good in the long run. By not helping people grow their own food, they argue, food relief can condemn even greater numbers to premature death from starvation and disease in the future—another example of a harmful positive feedback process.

Biologist Garrett Hardin (Guest Essay, p. 172) has suggested that we use the concept of *lifeboat ethics* to decide which countries get food relief. His basic premise is that there are already too many people in the lifeboat we call earth. Thus, if food relief is given to countries that are not reducing their populations, the effect is to add more people to an already overcrowded lifeboat. Sooner or later the overloaded boat will sink, and most of the passengers will drown.

Large amounts of food relief can also depress local food prices, decrease food production, and stimulate mass migration from farms to already overburdened cities. In addition, food relief discourages local and national governments from investing in the rural agricultural development needed to enable farmers to grow enough food for the population.

Another problem is that much food relief does not reach hunger victims. Transportation networks and storage facilities are often inadequate, so some of the food rots or is devoured by pests before it can reach the hungry. Moreover, officials often steal some of the food and sell it for personal profit; some must also usually be given to officials as bribes for approving the unloading and transporting of the remaining food to the hungry.

A combination of rapid population growth and environmental degradation in developing countries can lead to social disintegration, mass migration by hordes of environmental refugees, and political strife and warfare between tribal, religious, and ethnic groups. These interacting factors lead to even less food production and can result in mass famine and killing.

Providing food and other forms of aid under such circumstances can also be risky and costly. For example, supplying starving people in Somalia in 1992 required a UN Peacekeeping force that probably cost at least 10 times more than the food that was distributed; in 1995 the UN force had to be evacuated. This illustrates the strong linkages between environmental security and economic and military security (Guest Essay, pp. 295–296).

Most critics are not against providing aid, but they believe that such aid should help countries control population growth and grow enough food to feed their population by using sustainable agricultural methods (Section 14-4). Temporary food relief, they believe, should be given only when there is a complete breakdown of an area's food supply because of natural disaster.

Critical Thinking

Is sending food to famine victims helpful or harmful in the short run? in the long run? Explain. Are there any conditions you would attach to providing such aid? Explain.

Low-input farming reduces waste of irrigation water and uses less pesticides and inorganic fertilizer. To maintain and restore soil fertility, farmers rely on good soil conservation practices (Section 12-7) and use manure, compost, and other forms of organic matter. They also emphasize biological and physical methods for controlling pests and use chemical pesticides only as a last resort (and in the smallest amounts possible). A growing number of farmers are discovering that low-input farming is often more profitable than high-input farming because they spend less money on irrigation water, fertilizer, and pesticides.

Most proponents of more sustainable agriculture are not opposed to high-yield agriculture; indeed, they see it as vital for protecting the earth's biodiversity and ecological integrity by reducing the need to cultivate new and often marginal land. Instead, they believe that current research and economic incentives should be redirected to encourage increases in yield per hectare without depleting or degrading soil, water, and biodiversity.

General guidelines for sustainable agricultural systems suggested by various analysts include the following:

- *Combine traditional high-yield polyculture and modern monoculture methods for growing crops.*

- *Grow more perennial crops (p. 396).*

- *Minimize soil erosion, salinization, and waterlogging.*

- *Reduce destruction of forests, grasslands, and wetlands for producing foods by emphasizing increased yields per area of cropland using sustainable methods.*

- *Stabilize aquifers by reducing the rate of water removal to the rate of recharge.*

- *Reduce water waste in irrigation (Section 11-3).*

Q: How does the inflation-adjusted price of gasoline in the United States in 1997 compare with its price in 1950? in 1973?

- *Reduce overfishing by reducing the catch of commercially important species to below their estimated sustainable yields.*

- *Reduce use and waste of fossil fuels (Section 18-3) and shift to an energy efficient, solar–hydrogen economy (Chapter 18).*

- *Reduce loss of agricultural land to car-based transportation systems and pollution by encouraging a shift to transportation systems based on rail, bus, electric scooters, bicycles, and walking (Section 6-8).*

- *Increase use of organic fertilizers, solar, wind, and biomass energy to grow and process crops.*

- *Emphasize biological pest control and integrated pest management (Section 15-5).*

- *Protect existing prime cropland from environmental degradation and conversion to urban or industrial uses (Section 6-9).*

- *Subsidize sustainable farming and phase out subsidies for unsustainable farming.*

- *Shift to full-cost pricing (Section 7-3) that includes the harmful environmental effects of agriculture (Figure 14-9) in food prices.*

- *Educate the public about the hidden environmental and health costs they are paying for food and the need to gradually incorporate these costs into market prices.*

- *Greatly increase research on sustainable agriculture.*

- *Set up demonstration projects throughout each country so that farmers can see how sustainable agricultural systems work.*

- *Establish training programs in sustainable agriculture for farmers and government agricultural officials and encourage the creation of college curricula in sustainable agriculture.*

- *Reduce poverty.*

- *Provide poor women with access to credit, markets, and technical food-growing advice, as well as education and health care.*

- *Slow population growth to help all of the world's countries reach the more sustainable postindustrial stage of the demographic transition (Figure 6-17).*

- *Integrate agriculture, population, urban and rural, energy, health, climate, water resource, soil resource, land use, pollution, and biodiversity protection policies.*

Can We Make the Transition to Sustainable Agriculture? A growing number agricultural analysts believe that over the next 30 years we must make a transition from unsustainable and environmentally harmful agriculture (Figure 14-9) to more sustainable forms of agriculture.

In developed countries, including the United States, even a partial shift to more environmentally sustainable food production will not be easy. It will be opposed by agribusiness, successful farmers with large investments in unsustainable forms of industrialized agriculture, and specialized farmers unwilling to learn the demanding art of farming sustainably. It might also be resisted by many consumers unwilling or unable to pay higher prices for food because full-cost accounting would include agriculture's harmful environmental and health costs in the market prices of food.

Despite such difficulties, many environmentalists believe that a new *eco-agricultural revolution* could take place throughout most of the world over the next 30 years. Whether it does occur is primarily a political and ethical issue. Some actions you can take to help promote sustainable agriculture are listed in Appendix 5.

The need to bring birth rates well below death rates, increase food production while protecting the environment, and distribute food to all who need it is the greatest challenge our species has ever faced.

PAUL AND ANNE EHRLICH

CRITICAL THINKING

1. What are the biggest advantages and disadvantages of **(a)** labor-intensive subsistence agriculture, **(b)** energy-intensive industrialized agriculture, and **(c)** sustainable agriculture?

2. Summarize the advantages and limitations of each of the following proposals for increasing world food supplies and reducing hunger over the next 30 years: **(a)** cultivating more land by clearing tropical forests and irrigating arid lands, **(b)** catching more fish in the open sea, **(c)** producing more fish and shellfish with aquaculture, and **(d)** increasing the yield per area of cropland.

3. What are the three most important things that you believe should be done to reduce hunger in the country where you live? In the world?

4. Some people argue that starving people could get enough food by eating nonconventional plants and insects; others point out that most starving people don't know what plants and insects are safe to eat and can't take a chance on experimenting when even the slightest illness could kill them. If you had no money to grow or buy food, would you collect and eat protein-rich grasshoppers, moths, beetles, or other insects?

5. If the demand for imported food increases as projected, few African countries will be able compete with more affluent Asian countries in a bidding contest for grain. Africa uses grain more efficiently by using it mostly for direct human consumption. Asians feed much of their food imports to livestock to supply a growing demand for meat and meat products as their affluence increases.

A: About the same in both cases

Alberto Ruz Buenfil

GUEST ESSAY

Alberto Ruz Buenfil, an international environmental activist, writer, and performer, is the founder of Huehuecoyotl, a land-based ecovillage in the mountains of Mexico. His articles on ecology and alternative living have appeared in publications in the United States, Canada, Mexico, Japan, and Europe. His book, Rainbow Nation Without Borders: Toward an Ecotopian Millennium *(New Mexico: Bear, 1991), has been published in English, Italian, and Spanish.*

The world's eight known species of sea turtles are all officially listed as endangered or threatened. Seven of these species nest on Mexico's Pacific and Atlantic coastlines, making Mexico the world's most important turtle nesting country. The Pacific coasts of southern Mexico, especially the shores of the state of Oaxaca, are the main sites for turtle nesting, reproduction, and conservation; they also contain some of Mexico's last reserves of wetlands.

Only recently have we begun recognizing the ecological values of wetlands, deltas, and coastal ecosystems [Section 5-2]. They provide habitats for a rich diversity of wildlife, maintain water supplies, protect shorelines from erosion, and play a role in regulating global climate.

Swamps, marshes, and bogs were once considered wastelands to be drained, filled, and turned into "productive land," especially for developing and constructing urban and tourist areas. During the past few decades, the coasts of Oaxaca have not escaped such exploitation, especially after business interests discovered the beaches of Mazunte, Zipolite, San Agustinilo, Puerto Angel, and Puerto Escondido, and the magnificent bays of Hautulco.

The villages of Mazunte and San Agustinilo were founded in the late 1960s, basically to bring in cheap labor from neighboring indigenous villages to provide workers for a slaughterhouse making products from various species of turtles nesting in the area. Members of nearly 200 indigenous farming families became fishermen and employees of the new turtle meat factory, which was fully operational in the 1970s and 1980s. According to some of these workers, nearly 2,000 turtles were killed and quartered every day during those years. At night, dozens of poachers came to collect turtle eggs from their nests.

In the 1980s, this situation came to the attention of two of the first environmental organizations to speak up in defense of species, forests, and natural resources in Mexico. They began denouncing the massacre of the area's turtles, and they pointed out the danger that some of the species might be exterminated. With support from other international organizations, they campaigned for almost 10 years, until a 1990 decree by the president of Mexico made it illegal to exploit the turtles and led to the closing of San Agustinilo's turtle slaughterhouse.

The Mexican government provided some funds, boats, and freezers to compensate for the loss of jobs. However, only about 5% of the indigenous population benefited from this compensation. Since then, most of the people have been living on the verge of starvation, and some have illegally killed protected turtles to survive. What had seemed to be an important environmental victory turned into a nightmare for a large population of indigenous people. Understandably, these people had no use for ecologists or environmentalists.

In 1990, a group called ECOSOLAR A.C. began efforts to change this situation by implementing a plan for sustainable development of the coast of Oaxaca. They were successful in obtaining funding for this project from different national and international institutions. By 1992, the members of this small but effective group had succeeded in:

- Making a detailed study of the bioregion, which, with the participation of the local people, is being used to define the possible uses of different areas

---

a. Does Africa have a greater moral claim on surplus food than Asia because of its more efficient use of grain? Explain.

b. Does the United States, with one of the world's highest levels of meat consumption, have a moral responsibility to cut its grain consumption to make more grain available for export to countries whose people get most of their food by direct consumption of grain? Explain.

6. Should all price supports and other government subsidies paid to farmers be eliminated? Explain. Try to consult one or more farmers in answering this question.

7. Should governments phase in agricultural tax breaks and subsidies to encourage farmers to switch to more sustainable farming? Explain. At the same time, should governments phase in higher taxes and reduce subsidies to discourage farmers from using unsustainable, earth-degrading forms of farming, and then use the resulting revenue to encourage earth-sustaining farming? Explain.

*8. If possible, visit both a conventional industrialized farm and an organic or low-input farm. Compare soil erosion and other forms of land degradation, use and costs of energy, use and costs of pesticides and inorganic fertilizer, use and costs of natural pest control and organic fertilizer, yields per hectare for the same crops, and overall profit per hectare for the same crops.

Q: How long will the world's oil reserves last at the current consumption rate?

- Creating a system of credits to help native inhabitants build better houses, establish small family-run restaurants, and manufacture hammocks for rent or sale to visitors

- Beginning the construction of systems for drainage, water collection, and latrines using low-impact technology and local materials and workers, as well as nurseries for local seeds and facilities for reforestation, and wildlife preservation projects

- Working with the community to promote Mazunte as a center for ecotourism—a place where visitors can experience unique ecosystems containing alligators, turtles, and hundreds of species of birds and fishes

In only two years, the native inhabitants of Mazunte and other neighboring communities completely changed their opinion about and perspective on ecology and environmentalists. In May 1992, Mazunte hosted the second annual gathering of "Earth Keepers," involving nearly 150 representatives of 35 organizations from 20 different countries.

For 1 week, these specialists shared their practical knowledge with the local people. They used their health skills to help the community set up an alternative clinic for healing arts, their skills in permaculture (a form of sustainable agriculture) and organic agriculture to improve local hatcheries and home-based plant nurseries, and their skills in ecotechnology to build both biodigestors for producing natural fertilizers from biomass and a village recycling center located at the school. In addition, artistic and cultural activities took place every night in the Center for Biological Investigations, which the people of Mazunte want to turn into Mexico's first Marine Turtles Museum. Run by local people, it will attract and educate visitors from around the world.

However, despite the legal protection of sea turtles in Mexico, illegal killing of turtles and removal of their eggs continues. A few days after the event concluded, the village of Mazunte called a general meeting attended by 150 heads of families to discuss ways to get community members to protect turtle nesting areas instead of illegally killing the turtles for food. Out of that meeting came a "Declaration of Mazunte" requesting that competent higher authorities and the president of Mexico put an immediate end to such destruction, which violates the earlier presidential decree forbidding the annihilation of turtles in Mexico.

The community went on to declare their village and neighboring environments to be Mexico's first *Farming and Fishing Reserve*. Its goals would be to protect the area's forests, water sources, wetlands, wildlife, shores, beaches, and scenic places, and to "establish new forms of relationship between humans and nature, for the well-being of today's and tomorrow's generations." This declaration has been presented to the government of Mexico and to many national and international organizations.

Mazunte is taking the lead in demonstrating that cooperation between local people and environmental experts can lead to ecologically sustainable communities that benefit local people and wildlife alike. This model can show farmers and indigenous communities everywhere how they can live sustainably on Earth and turn things around in a short time. It is a message of hope and empowerment for people seeking a better world for themselves and others.

Critical Thinking

1. What lessons that could be applied to your own life have you learned from this essay?

2. Could the rapid change toward sustainability brought about by environmentalists and local people in Mazunte be accomplished in your own community? If so, how? If not, why not?

---

*9. Use health and other local government records to determine how many people in your community suffer from undernutrition or malnutrition. Has this problem increased or decreased since 1980? What are the basic causes of this hunger problem, and what is being done to alleviate it? Share the results of your study with local officials and then present your own plan for improving efforts to reduce hunger in your community.

*10. Make a survey in the nearest urban area to estimate what percentage of the food is grown by urban dwellers. Survey unused land and use it to estimate how much it could contribute to urban food production. Use these data to draw up a plan for increasing urban food production and present it to city officials.

*11. Make a concept map of this chapter's major ideas, using the section heads and subheads and the key terms (in boldface type). Look at the inside back cover and in Appendix 4 for information about concept maps.

SUGGESTED READINGS

See Internet Sources and Further Readings at the back of the book, and InfoTrac.

15 PROTECTING FOOD RESOURCES: PESTICIDES AND PEST CONTROL

Along Came a Spider

The longest war in human history is our ongoing war against insect pests. This war was declared about 10,000 years ago, when humans first got serious about agriculture. Today we are not much closer to winning this war than we were then.

Chinese farmers recently decided that it's time to change strategies. Instead of spraying their rice and cotton fields with poisons, they build little straw huts around the fields in the fall.

If this sounds crazy, it's crazy like a fox. These farmers are giving aid and comfort to insects' worst enemy, one that has hunted them for millions of years: spiders (Figure 15-1). The little huts are for hibernating spiders. Protected from the worst of the cold by the huts, far more of the hibernating spiders become active in the spring. Ravenous after their winter fast, they scuttle off into the fields to stalk their insect prey.

Even without human help, the world's 30,000 known species of spiders kill far more insects every year than insecticides do. A typical acre of meadow or woods contains an estimated 50,000 to 2 million spiders, each devouring hundreds of insects per year.

Entomologist Willard H. Whitcomb found that leaving strips of weeds around cotton and soybean fields provides the kind of undergrowth favored by insect-eating wolf spiders (Figure 15-1, right). He also sings the praises of one type of banana spider, which lives in warm climates and can keep a house clear of cockroaches.

The idea of encouraging populations of spiders in fields, forests, and even houses scares some people because spiders have bad reputations. Perhaps because of their eight legs and numerous eyes, spiders look like dangerous space aliens instead of the helpful and mostly harmless creatures they are.

A few species of spiders, such as the black widow, the brown recluse, and eastern Australia's Sydney funnel web, are dangerous to people. However, the vast majority of spider species, including the ferocious-looking wolf spider, are harmless to humans. Even the giant tarantula rarely bites people, and its venom is too weak to harm us or other large mammals.

As biologist Thomas Eisner puts it, "Bugs are not going to inherit the earth. They own it now. So we might as well make peace with the landlord." As we seek new ways to coexist with the insect rulers of the planet, we would do well to be sure that spiders are in our corner.

Figure 15-1 Spiders are insects' worst enemies. Most spiders, like the crab spider (left) and the wolf spider (right), found in many parts of the world, are harmless to humans. (Left, James C. Cokendolpher; right, Dan Kline/Visuals Unlimited)

A weed is a plant whose virtues have not yet been discovered.

RALPH WALDO EMERSON

This chapter answers the following questions:

- What types of pesticides are used?
- What are the pros and cons of using chemicals to kill insects and weeds?
- How well is pesticide use regulated in the United States?
- What alternatives are there to using pesticides?

15-1 PESTICIDES: TYPES AND USES

How Does Nature Keep Pest Populations Under Control? A **pest** is any species that competes with us for food, invades lawns and gardens, destroys wood in houses, spreads disease, or is simply a nuisance. In natural ecosystems and many polyculture agroecosystems, natural enemies (predators, parasites, and disease organisms) control the populations of 50–90% of pest species, thus constituting a crucial type of earth capital.

When we simplify natural ecosystems, we upset these natural checks and balances, which keep any one species from taking over for very long. Then we must devise ways to protect our monoculture crops, tree farms, and lawns from insects and other pests that nature once controlled at no charge.

We have done this primarily by developing a variety of **pesticides** (or *biocides*): chemicals to kill organisms we consider undesirable. Common types of pesticides are **insecticides** (insect killers), *herbicides* (weed killers), *fungicides* (fungus killers), *nematocides* (roundworm killers), and *rodenticides* (rat and mouse killers).

Humans didn't invent the use of chemicals to repel or kill other species; plants have been producing chemicals to ward off or poison herbivores that feed on them for about 225 million years. This is a never-ending, ever-changing process: Herbivores overcome various plant defenses through natural selection, then the plants use natural selection to develop new defenses. The result of these dynamic interactions between predator and prey species is what biologists call *coevolution* (Section 5-6).

What Was the First Generation of Pesticides and Repellents? As the human population grew and agriculture spread, people began looking for ways to protect their crops, mostly by using chemicals to kill or repel insect pests. Sulfur was used as an insecticide well before 500 B.C.; by the 1400s toxic compounds of arsenic, lead, and mercury were being applied to crops as insecticides. Farmers abandoned this approach in the late 1920s, when the increasing number of human poisonings and fatalities encouraged a search for less toxic substitutes. However, traces of these nondegradable toxic metal compounds are still being taken up by tobacco, vegetables, and other crops grown on soil dosed with them long ago.

In the 1600s nicotine sulfate, extracted from tobacco leaves, came into use as an insecticide. In the mid-1800s two more natural pesticides were introduced: pyrethrum, obtained from the heads of chrysanthemum flowers, and rotenone, from the root of the derris plant and other tropical forest legumes. These *first-generation pesticides* were mainly natural substances, weapons borrowed from plants that had been at war with insects for eons.

In addition to protecting crops, people have used chemicals (mostly produced by plants) to repel or kill insects in their households, yards, and gardens; they also save money and reduce health hazards associated with using commercial insecticides.

If *ants* come indoors, they can usually be persuaded to leave within about 4 days by sprinkling repellents such as red or cayenne pepper, crushed mint leaves, or boric acid (with an anticaking agent) along their trails inside a house, and by wiping off countertops with vinegar. (However, boric acid is poisonous and should not be placed in areas accessible to small children and pets.)

Mosquitoes can be repelled by planting basil outside windows and doors and rubbing a bit of vinegar, basil oil, lime juice, or mugwort oil on exposed skin. Mosquito attacks can also be reduced by not using scented soaps or wearing perfumes, colognes, and other scented products outdoors during mosquito season. Researchers have found that the $30 million U.S. consumers spend each year on electric bug zappers to kill mosquitoes is mostly wasted because only about 3% of the insects they kill on an average night are female mosquitoes—the kind that bite.

Cockroaches (Spotlight, p. 422) can be killed by sprinkling boric acid under sinks and ranges, behind refrigerators, and in cabinets, closets, and other dark, warm places (but not in areas accessible to children and pets) or by establishing populations of banana spiders. Roaches can also be trapped by greasing the inner neck of a bottle or large jar with petroleum jelly, filling much of it with raw potato, stale beer, banana skins, or other food scraps (especially fruits), and placing a small ramp leading to it. Placing fresh or dried bay leaves in and around cupboards repels cockroaches.

Flies can be repelled by planting sweet basil and tansy (a common herb) near doorways and patios and by hanging a series of polyethylene strips in front

Cockroaches: Nature's Ultimate Survivors

SPOTLIGHT

Cockroaches, the bugs many people love to hate, have been around for about 350 million years and are one of the great success stories of evolution. The major reason they are so successful is that they are *generalists*, able to eat almost anything (including algae, dead insects, salts in tennis shoes, electrical cords, glue, paper, soap, and—when times are bad—other, weaker cockroaches). The 4,000 known cockroach species can live and breed almost anywhere except polar regions.

Some species can go for months without food, last a month without water, and withstand massive doses of radiation. One species can survive being frozen for 48 hours. The antennae of most cockroach species (which can detect minute movements of air), the vibration sensors in their knee joints, and their lightning-fast response times (faster than you can blink) allow them to evade predators and a human foot in hot pursuit. Some even have wings.

High reproductive rates also aid the survival of cockroaches. In only a year, a single Asian cockroach (especially prevalent in Florida) and its young can add about 10 million new cockroaches to the world. Their high reproductive rate also helps them quickly develop genetic resistance to almost any poison we throw at them. Most cockroaches also sample food before it enters their mouths and learn to shun foul-tasting poisons.

Only about 25 species of cockroach live in homes, but such species can carry viruses and bacteria that cause such diseases as hepatitis, polio, typhoid fever, plague, and salmonella. They can cause people to have allergic reactions ranging from watery eyes to severe wheezing. Indeed, about 60% of the 11.5 million Americans suffering from asthma are allergic to dead or live cockroaches.

Critical Thinking

How do you feel about cockroaches? If you could, would you exterminate them? What might be some ecological consequences of doing this?

of entry doors (like the ones on some grocery-store coolers). Nontoxic flypaper can be made by applying honey to strips of yellow paper (their favorite color) and hanging it from the ceiling in the center of rooms. You can also hang clusters of cloves.

Fleas can be kept off pets by using green dye or flea-repellent soaps; feeding them brewer's yeast or vitamin B; using flea powders made from eucalyptus, sage, tobacco, wormwood, or vetiver; or dipping or shampooing pets in a mixture of water and essential oils such as citronella, cedarwood, eucalyptus, fleabane, sassafras, geranium, clove, or mint. Researchers recently invented a trap that uses green-yellow light to attract fleas to an adhesive-coated surface. Another solution is to put a light over a shallow pan of water before going to bed at night and turn out all other lights, empty the water every morning, and continue for a month. (Fleas are attracted to heat and light but they can't swim.) Desiccant powders, such as Dri-Die, Perma-Guard, or SG-67, can also rid a house of flea infestations by "drying-up" the insects. Diatomaceous earth (or diatom powder) can also be sprinkled on carpets and pets to kill fleas.* This powder consists of the skeletal remains of microscopic algae and kills fleas and many other harmful insects by cutting their outer shell so they lose their body fluids.

Lawn weeds can be controlled by raising the cutting level of your lawn mower so the grass can grow 8–10 centimeters (3–4 inches) high. This gives it a strong root system that can hinder weed growth; the higher grass also provides habitats for spiders and other insects that eat insect pests. Pull weeds and douse the hole with soap solution or human urine (which is high enough in nitrogen to burn the weed).

What Was the Second Generation of Pesticides?

A major pest control revolution began in 1939, when entomologist Paul Mueller discovered that DDT (dichlorodiphenyltrichloroethane), a chemical known since 1874, was a potent insecticide. DDT, the first of the so-called *second-generation pesticides*, soon became the world's most-used pesticide, and Mueller received the Nobel prize in 1948. Since 1945 chemists have developed hundreds of synthetic organic chemicals for use as pesticides.

Worldwide, about 2.3 million metric tons (2.5 million tons) of such pesticides are used yearly—0.45 kilogram (1 pound) for each person on the earth. About 75% of these chemicals are used in developed countries, but use in developing countries is soaring. In 1995 worldwide sales of pesticides were $29 billion ($10.4 billion in the United States alone).

In the United States, about 630 different biologically active (pest-killing) ingredients and about 1,820 inert (inactive) ingredients are mixed to make some 25,000 different pesticide products. Cultivation of two crops—cotton (55%) and corn (35%)—used about

---

* Diatom powder can be purchased in bulk at stores that sell it for use in swimming pool filters. It is also found in some gardening stores under the name Permatex. Because it contains fine particles of silicate, a dust mask should be worn when applying this powder to avoid inhaling tiny particles of silicate.

Q: How long could *all* known and projected U.S. oil deposits supply the world and the United States at current consumption?

90% of the insecticides and 80% of the herbicides applied to crops in the United States in 1995.

Manufacturers add pesticides to products as diverse as paints, shampoos, carpets, mattresses, wax on produce, and contact lenses. About 25% of pesticide use in the United States is for ridding houses, gardens, lawns, parks, playing fields, swimming pools, and golf courses of unwanted pests. According to the EPA, the average lawn in the United States is doused with more than 10 times more synthetic pesticides per hectare than U.S. cropland. Each year, more than 250,000 U.S. residents become ill because of household use of pesticides, and such pesticides are a major source of accidental poisonings and deaths for children under age 5.

Some pesticides, called *broad-spectrum* agents, are toxic to many species; others, called *selective* or *narrow-spectrum* agents, are effective against a narrowly defined group of organisms. Pesticides vary in their *persistence*, the length of time they remain deadly in the environment (Table 15-1). Most organophosphates (except malathion) are highly toxic to humans and other animals, and they account for most human pesticide poisonings and deaths. In 1962 biologist Rachel Carson warned against relying on synthetic chemicals to kill insects and other species we deem pests (Individuals Matter, p. 424).

15-2 THE CASE FOR PESTICIDES

Proponents of pesticides contend that their benefits outweigh their harmful effects.

Pesticides save human lives. Since 1945 DDT and other chlorinated hydrocarbon and organophosphate insecticides have probably prevented the premature deaths of at least 7 million people from

Table 15-1 Major Types of Pesticides

Type	Examples	Persistence	Biologically Amplified?
Insecticides			
Chlorinated hydrocarbons	DDT, aldrin, dieldrin, toxaphene, lindane, chlordane, methoxychlor, mirex	High (2–15 years)	Yes
Organophosphates	Malathion, parathion, diazinon, TEPP, DDVP, mevingphos	Low to moderate (1–12 weeks), but some can last several years	No
Carbamates	Aldicarb, carbaryl (Sevin), propoxur, maneb, zineb	Low (days to weeks)	No
Botanicals	Rotenone, pyrethrum, and camphor extracted from plants, synthetic pyrethroids (variations of pyrethrum) and rotenoids (variations of rotenone)	Low (days to weeks)	No
Microbotanicals	Various bacteria, fungi, protozoa	Low (days to weeks)	No
Herbicides			
Contact chemicals	Atrazine, simazine, paraquat	Low (days to weeks)	No
Systemic chemicals	2,4-D, 2,4,5-T, Silvex, diruon, daminozide (Alar), alachlor (Lasso), glyphosate (Roundup)	Mostly low (days to weeks)	No
Soil sterilants	Trifualin, diphenamid, dalapon, butylate	Low (days)	No
Fungicides			
Various chemicals	Captan, pentachlorphenol, zeneb, methyl bromide, carbon bisulfide	Mostly low (days)	No
Fumigants			
Various chemicals	Carbon tetrachloride, ethylene dibromide, methyl bromide	Mostly high	Yes (for most)

A: World, about 1.7 years; United States, about 10 years

INDIVIDUALS MATTER

Rachel Carson

Rachel Carson (Figure 15-2) began her professional career as a biologist for the Bureau of U.S. Fisheries (later to become the U.S. Fish and Wildlife Service). In that capacity, she carried out research on oceanography and marine biology, wrote articles about the oceans and topics related to the environment, and became editor-in-chief of the bureau's publications in 1949.

In 1951, she wrote *The Sea Around Us*, which described in easily understandable terms the natural history of oceans and the harm that humans were doing them. The book was on the best-seller list for 86 weeks, sold more than 2 million copies, was translated into 32 languages, and won a National Book Award.

During the late 1940s and throughout the 1950s, the use of DDT and related compounds—to kill insects that ate food crops, attacked trees, bothered people, and transmitted diseases such as malaria—rapidly expanded.

In 1958, DDT was sprayed to control mosquitoes near the home and private bird sanctuary of Olga Huckins, a good friend of Carson. After the spraying, Huckins witnessed the agonizing deaths of several of her birds, and in distress she asked Carson if she could find someone to investigate the effects of pesticides on birds and other wildlife.

Carson decided to look into the issue herself and quickly found that almost no independent critical research on the environmental effects of pesticides existed. As a well-trained scientist, Carson surveyed the scientific literature and methodically built her case against the widespread use of pesticides.

In 1962, she published her findings in popular form in *Silent Spring*, an allusion to the silencing of "robins, catbirds, doves, jays, wrens, and scores of other bird voices" because of their exposure to pesticides. She pointed out that "for the first time in the history of the world, every human being is now subjected to dangerous chemicals, from the moment of conception until death."

Carson's book was read by many scientists, politicians, and policy makers and was embraced by the general public. However, the chemical industry viewed the book as a serious threat to booming pesticide sales and mounted a $250,000 campaign to discredit Carson. A parade of critical reviewers and industry scientists claimed that her book was full of inaccuracies, made selective use of research findings, and failed to give a balanced account of the benefits of pesticides.

Some critics even claimed that, as a woman, she was incapable of understanding the highly scientific and technical subject of pesticides. Others charged that she was a hysterical woman and a radical nature lover trying to scare the American public in order to sell books.

During this period of intense controversy Carson was suffering from terminal cancer, but she was able to defend her research and strongly counter her critics. She died in 1964—18 months after the publication of *Silent Spring*—without knowing that her efforts

Figure 15-2 Biologist Rachel Carson (1907–1964) was a pioneer in increasing public awareness of the importance of nature and the threat of pollution. She died without knowing that her efforts were a key in starting today's environmental movement. (Eric Hartmann/ Magnum)

were a driving force in the birth of what is now known as the environmental movement in the United States. To environmentalists and most citizens, Rachel Carson is an outstanding example of a dedicated scientist and effective communicator of complex scientific information.

---

insect-transmitted diseases such as malaria (carried by the *Anopheles* mosquito; Figure 8-10), bubonic plague (rat fleas), typhus (body lice and fleas), and sleeping sickness (tsetse fly).

Pesticides increase food supplies and lower food costs. About 55% of the world's potential human food supply is lost to pests before (35%) or after (20%) harvest. An estimated 37% of the potential U.S. food supply is destroyed by pests before and after harvest; insects cause 13% of these losses, plant pathogens 12%, and weeds 12%. Without pesticides, these losses would be worse, and food prices would rise (by 30–50% in the United States, according to pesticide company officials).

Q: How much new oil must be discovered and developed to continue using oil at the current rate?

Pesticides increase profits for farmers. Pesticide companies estimate that every $1 spent on pesticides leads to an increase in U.S. crop yields worth approximately $4 (but studies have shown that this benefit drops to about $2 if the harmful effects of pesticides are included).

Pesticides work faster and better than alternatives. Pesticides can control most pests quickly and at a reasonable cost. They have a long shelf life, are easily shipped and applied, and are safe when handled properly. When genetic resistance occurs, farmers can use stronger doses or switch to other pesticides.

The health risks of pesticides are insignificant compared with their benefits. According to Elizabeth Whelan, director of the American Council on Science and Health (ACSH), which presents the position of the pesticide industry, "The reality is that pesticides, when used in the approved regulatory manner, pose no risk to either farm workers or consumers." Pesticide proponents consider media reports describing pesticide health scares to be distorted science and irresponsible reporting. They also point out that about 99.99% of the pesticides we consume in our food are natural chemicals produced by plants.

Proponents point out some beneficial changes in pesticides and their use. For example, *safer and more effective pesticides are being developed.* Industry scientists are developing pesticides, such as botanicals and microbotanicals (Table 15-1), that are safer to users and less damaging to the environment. Genetic engineering also holds promise in developing pest-resistant crop strains. However, total research and development and government approval costs for a new pesticide have risen from $6 million in 1976 to $80–120 million today. As a result, new chemicals are being developed only for crops such as wheat, corn, and soybeans with large markets.

Many new pesticides are used at very low rates per unit area compared to older products. For example, application amounts per hectare for many new herbicides are 100 times lower than for older ones.

Scientists continue to search for the ideal pest-killing chemical, which would

- *Kill only the target pest*

- *Harm no other species*

- *Disappear or break down into something harmless after doing its job*

- *Not cause genetic resistance in target organisms*

- *Be cheaper than doing nothing*

Unfortunately, no known pesticide meets all these criteria, and most don't even come close.

Figure 15-3 A boll weevil, just one example of an insect capable of rapid breeding. In the cotton fields of the southern United States, these insects lay thousands of eggs, producing a new generation every 21 days and as many as six generations in a single growing season. Attempts to control the cotton boll weevil account for at least 25% of insecticide use in the United States. However, farmers are now increasing their use of natural predators and other biological methods to control this major pest. (U.S. Department of Agriculture)

15-3 THE CASE AGAINST PESTICIDES

How Do Pesticides Cause Genetic Resistance and Kill Natural Pest Enemies? Opponents of widespread use of pesticides believe that their harmful effects outweigh their benefits. The biggest problem is the development of *genetic resistance* by pest organisms. Insects breed rapidly (Figure 15-3), and within 5–10 years (much sooner in tropical areas) they can develop immunity to pesticides through natural selection (Section 5-6) and come back stronger than before (see cartoon, p. 426). Weeds and plant-disease organisms also become resistant, but more slowly.

Since 1950 at least 520 species of insects and mites, 273 weed species, 150 plant diseases, and 10 species of rodents (mostly rats) have developed genetic resistance to one or more pesticides. At least 17 species of insect pests are resistant to all major classes of insecticides, and several fungal plant diseases are now immune to most widely used fungicides. Because of genetic resistance, most widely used insecticides no longer protect people from insect-transmitted diseases in many parts of the world, leading to resurgences of diseases such as malaria (Figure 8-9).

Another problem is that *broad-spectrum insecticides kill natural predators and parasites that may have been maintaining the population of a pest species at a reasonable level.* With wolf spiders (Figure 15-1, right), wasps, predatory beetles, and other natural enemies out of the way, a rapidly reproducing insect pest species can make a strong comeback only days or weeks after initially being controlled. Although natural predators can also

Well, people succeeded in reducing biodiversity down to one species. And they always thought it would be them.

A pity we're too primitive to appreciate the irony

develop genetic resistance to pesticides, most predators can't reproduce as quickly as their insect prey.

Wiping out natural predators can also unleash new pests whose populations the predators had previously held in check, causing other unexpected effects (Connections, p. 155). Currently 100 of the 300 most destructive insect pests in the United States were secondary pests that became major pests through the effects of insecticides.

What Is the Pesticide Treadmill? When genetic resistance develops, pesticide sales representatives usually recommend more frequent applications, larger doses, or a switch to new (usually more expensive) chemicals to keep the resistant species under control. This can put farmers on a **pesticide treadmill**, whereby they pay more and more for a pest control program that often becomes less and less effective. Between the mid-1950s and late 1970s, for example, cotton growers in Central America increased the frequency of insecticide applications from 10 to 40 times per growing season; still, declining yields and falling profits forced many of the farmers into bankruptcy.

In 1989 David Pimentel (Guest Essay, pp. 366–367), an expert in insect ecology, evaluated data from more than 300 agricultural scientists and economists and came to the following conclusions:

- Although the use of synthetic pesticides has increased 33-fold since 1942, it is estimated that more of the U.S. food supply is lost to pests today (37%) than in the 1940s (31%). Losses attributed to insects almost doubled (from 7% to 13%) despite a tenfold increase in the use of synthetic insecticides.

- The estimated environmental, health, and social costs of pesticide use in the United States range from $4 billion to $10 billion per year. The International Food Policy Research Institute puts the estimate

much higher, at $100–200 billion per year, or $5–10 in damages for every dollar spent on pesticides.

- Alternative pest control practices (Section 15-5) could halve the use of chemical pesticides on 40 major U.S. crops without reducing crop yields.

- A 50% cut in U.S. pesticide use would cause retail food prices to rise by only about 0.2% but would raise average income for farmers about 9%.

Although pesticides can't be completely eliminated, numerous studies and experience show that they can be sharply reduced without decreasing yields, and in some cases yields would increase. Over the past few years, Sweden has cut pesticide use in half with virtually no decrease in the harvest. Campbell Soup uses no pesticides on tomatoes it grows in Mexico, and yields have not dropped. After a 65% cut in pesticide use on rice in Indonesia, yields increased by 15%.

Where Do Pesticides Go, and How Can They Harm Wildlife? *Pesticides don't stay put.* According to the U.S. Department of Agriculture, no more than 2% (and often less than 0.1%) of the insecticides applied to crops by aerial spraying (Figure 15-4) or by ground spraying actually reaches the target pests; less than 5% of herbicides applied to crops reaches the target weeds.

Pesticides that miss their target pests end up in the air, surface water, groundwater, bottom sediments, food, and nontarget organisms, including humans and wildlife. Pesticide waste and mobility can be reduced by using recirculating sprayers, covering spray booms to reduce drift, and using rope-wick applicators (which deliver herbicides directly to weeds and reduce herbicide use by 90%).

Some pesticides can harm wildlife. During the 1950s and 1960s, populations of fish-eating birds such as the osprey, cormorant, brown pelican, and bald eagle plummeted. Research indicated that a chemical derived from DDT, when biologically magnified in food webs (Figure 8-2), made the birds' eggshells so fragile that they could not successfully reproduce. Also hard-hit were such predatory birds as the prairie falcon, sparrow hawk, and peregrine falcon (Figure 15-5), which help control rabbits, ground squirrels, and other crop-eaters.

Since the U.S. ban on DDT in 1972, most of these species have made a comeback. In 1980, however, DDT levels were again rising in the peregrine falcon and the osprey. Scientists believe that the birds may be picking up DDT and other banned pesticides in Latin America, where they winter and where use of such chemicals is still legal. Illegal use of banned pesticides in the United States, as well as airborne drift of long-lived pesticides from developing countries onto U.S. land and surface water, may also play a role. Recently, the EPA found DDT in 99% of the freshwater fish it tested.

Q: What would happen if oil's harmful effects were included in its price and government subsidies were removed?

Figure 15-4 A crop duster spraying an insecticide on grapevines south of Fresno, California. Aircraft apply about 25% of the pesticides used on U.S. cropland, but only 0.1–2% of these insecticides actually reach the target pests. To compensate for the drift of pesticides from target to nontarget areas, aircraft apply up to 30% more pesticide than ground-based application does. (National Archives/EPA Documerica)

Figure 15-5 Until 1994 the peregrine falcon was listed as an endangered species in the United States, mostly because DDT caused their young to die before hatching because the eggshells were too thin to protect them. Only about 120 peregrine falcons were left in the lower 48 states in 1975; today there are about 1,990, most of them bred in captivity and then released into the wild in a $2.7-million-per-year recovery program. Because of the success of this program, the status of the peregrine falcon in the United States has been changed from endangered to threatened. (Hans Reinhard/Bruce Coleman Ltd.)

Each year 20% of U.S. honeybee colonies are wiped out by pesticides, and another 15% are damaged, costing farmers at least $200 million per year from reduced pollination of vital crops. Pesticide runoff from cropland, a leading cause of fish kills worldwide, kills 6–14 million fish each year in the United States. According to the U.S. Fish and Wildlife Service, pesticides menace about 20% of the endangered and threatened species in the United States.

How Can Pesticides Threaten Human Health?
According to the World Health Organization and the UN Environment Programme, an estimated 25 million agricultural workers in developing countries are seriously poisoned by pesticides each year, resulting in an estimated 220,000 deaths. Health officials believe that the actual number of pesticide-related illnesses and deaths among the world's farm workers is probably greatly underestimated because of poor records, lack of doctors and disease reporting in rural areas, and faulty diagnoses.

Farm workers in developing countries are especially vulnerable to pesticide poisoning because three-quarters of all pesticides are applied by hand. In addition, educational levels are low, warnings are few,

pesticide regulations are lax or nonexistent, and use of protective equipment is rare (especially in the hot and humid tropics). Workers' clothing also spreads pesticide contamination to family members, and it is common in developing countries for children to work or play in fields treated with pesticides. To make matters worse, some families reuse pesticide containers to store food and drinking water.

Some of these same conditions exist for farm workers in parts of the United States and in other developed countries. In the United States, for example, it is estimated that at least 300,000 farm workers suffer from pesticide-related illnesses each year. At least 30,000 of these pesticide poisonings are acute, and typically 25 U.S. farm workers die every year from such poisonings. Every year about 20,000 Americans, mostly children, get sick from home misuse or unsafe storage of pesticides.

DDT and related persistent chlorinated hydrocarbon pesticides have been banned in the United States and most developed countries. However, they were largely replaced by organophosphate pesticides (Table 15-1) that, although considerably less persistent, are much more toxic to humans and have led to increased poisonings and deaths among farm workers.

A: It would be too expensive to use and most of it would be phased out

CHAPTER 15 **427**

In 1995 the Environmental Working Group estimated that 13% of vegetables and fruits consumed in the United States may contain illegal pesticides and levels of approved pesticides above their legally allowed limits. These estimates may be too low because U.S. Food and Drug Administration (FDA) inspectors check less than 1% (about 12,000 samples) of domestic and imported food for pesticide contamination and measuring techniques by FDA laboratories can detect only about half of the legal and illegal pesticides in use. Moreover, the FDA's turnaround time for food analysis is so long that the food has often been sold and eaten before contamination is detected. Even when contaminated food is found, the growers and importers are rarely penalized.

In 1987 the EPA ranked pesticide residues in foods as the third most serious environmental health threat in the United States in terms of cancer risk. At least 75 of the active ingredients approved for use in U.S. pesticide products cause cancer in test animals. A 1993 study of Missouri children revealed a statistically significant correlation between childhood brain cancer and use of various pesticides in the home, including flea and tick collars, no-pest strips, and chemicals used to control pests such as roaches, ants, spiders, mosquitoes, and termites.

A 1995 study also found that children whose yards were treated with pesticides (mostly herbicides) were four times more likely to suffer from cancers of muscle and connective tissues than children whose yards were not treated. The study also found that children whose homes contained pest strips (with dichlorovos) faced 2.5–3 times the risk of leukemia as children whose homes didn't contain such strips.

Some scientists are becoming increasingly concerned about possible genetic mutations, birth defects, nervous system disorders (especially behavioral disorders), and effects on the immune and endocrine systems from long-term exposure to low levels of various pesticides (Section 8-3 and Case Study, p. 228). Very little research has been conducted on these potentially serious threats to human health.

Accidents and unsafe practices in pesticide manufacturing plants can expose workers, their families, and sometimes the general public to harmful levels of pesticides or toxic chemicals used in their manufacture (Case Study, below).

A Black Day in Bhopal

CASE STUDY

December 2, 1984, will long be a black day on the Indian calendar because on that date *the world's worst industrial accident* occurred at a Union Carbide pesticide plant in Bhopal, India.

Some 36 metric tons (40 tons) of highly toxic methyl isocyanate (MIC) gas, used in the manufacture of carbamate pesticides, leaked from an underground storage tank. When water accidentally entered the tank, its cooling system failed, which caused the reaction mixture to overheat and explode. Once in the atmosphere, some of the toxic MIC was converted to even more deadly hydrogen cyanide gas.

The toxic cloud of gas settled over about 78 square kilometers (30 square miles), exposing up to 600,000 people, many of them illegal squatters living near the plant because they had no other place to go.

According to Indian officials, at least 5,100 people (some say

7,000–15,000, based on the sale of shrouds and cremation wood) were killed; according to a 1996 report by the International Medical Commission on Bhopal another 50,000 to 60,000 sustained permanent injuries such as blindness or lung damage.

Indian officials claim that the accident was caused by negligence, whereas Union Carbide officials claim that it was caused by sabotage (but has presented no evidence in court to back up this charge).

The Indian Supreme Court ordered Union Carbide to pay a $470 million settlement. However, the Indian government challenged the ruling, arguing that the settlement was inadequate. In 1991, the court upheld the settlement amount.

Leaving aside fair compensation to the victims, the economic damage from the accident was estimated at $4.1 billion, so Union Carbide got off extremely lightly, explaining why the company's

stock price rose when the settlement amount was announced.

After the accident, Union Carbide reduced the corporation's liability risks for compensating victims by selling off a portion of its assets and giving much of the profits to its shareholders in the form of special dividends. In 1994, Union Carbide sold its holdings in India.

Union Carbide could probably have prevented this tragedy, which cost billions of dollars (not including the tragic loss of life and serious health effects) by spending no more than $1 million to improve plant safety.

Critical Thinking

Did Union Carbide behave responsibly or irresponsibly in this matter with respect to **(a)** those killed and injured and **(b)** its stockholders, who saw the return on their investment rise because of the accident? Explain. How could future disasters like this be prevented?

Q: How long will proven reserves of natural gas last for the world and the United States at current consumption rates?

15-4 PESTICIDE REGULATION IN THE UNITED STATES

Is the Public Adequately Protected? The Federal Insecticide, Fungicide, and Rodenticide Act (FIFRA) requires that all commercial pesticides be approved by the EPA for general or restricted use. Pesticide companies first evaluate the biologically active ingredients in their products. EPA officials then review these data. In the late 1970s the EPA discovered that data used to support registrations for more than 200 pesticide active ingredients had been falsified by a now-defunct laboratory.

When a pesticide is legally approved for use on fruits or vegetables, the EPA sets a *tolerance level* specifying the amount of toxic pesticide residue that can legally remain on the crop when the consumer eats it. According to the National Academy of Sciences, "Tolerance levels are not based primarily on health considerations." In 1993 the National Academy of Sciences recommended that the legally allowed tolerance levels for all active ingredients in pesticides be reduced 10-fold to help protect infants and children, who are more vulnerable to such chemicals than adults. By 1997 this had not been done.

Traditionally, pesticide makers have also been required to evaluate the effect of active pesticide ingredients on wildlife. Recently, however, the EPA has stopped requiring them to conduct field tests on birds and fish for newly developed pesticides.

Between 1972 and 1996 the EPA canceled or severely restricted the use of 55 active pesticide ingredients. The banned chemicals include most chlorinated hydrocarbon insecticides, several carbamates and organophosphates, and the systemic herbicides 2,4,5-T and Silvex (Table 15-1). However, banned or unregistered pesticides may be manufactured in the United States and exported to other countries (Connections, right).

FIFRA required the EPA to reevaluate the more than 600 active ingredients approved for use in pre-1972 pesticide products to determine whether any of them caused cancer, birth defects, or other health risks. However, by 1996—25 years after Congress ordered EPA to evaluate these chemicals—less than 10% of these active ingredients had been fully tested and evaluated. The EPA contends that it hasn't been able to complete this evaluation because of the difficulty and expense of evaluating the health effects of chemicals (Section 8-2).

According to a National Academy of Sciences study, federal laws regulating the use of pesticides in the United States are inadequate and poorly enforced by the EPA, FDA, and USDA. The study also concluded that up to 98% of the potential risk of developing cancer from pesticide residues on food grown in the United States

The Circle of Poison

CONNECTIONS

U.S. pesticide companies can make and export to other countries pesticides that have been banned or severely restricted—or never even approved—in the United States. But what goes around comes around.

In what environmentalists call a *circle of poison*, residues of some of these banned or unapproved chemicals can return to the United States in or on imported items such as coffee, cocoa, pineapples, and out-of-season melons, tomatoes, and grapes. More than one-fourth of the produce (fruits and vegetables) consumed in the United States is grown overseas. Persistent pesticides such as DDT can also be carried by winds from other countries to the United States.

Environmentalists have urged Congress—without success—to break this deadly circle. Supporters of pesticide exports argue that such sales increase economic growth and provide jobs. They also contend that if the United States didn't export pesticides, other countries would.

Critical Thinking

Should U.S. companies be allowed to export pesticides to other countries that have been banned, severely restricted, or not registered for use in the United States? Explain.

would be eliminated if EPA standards were as strict for pre-1972 pesticides as they are for later ones.

According to a 1987 study by the National Academy of Sciences, exposure to pesticide residues in food causes 4,000–20,000 cases of cancer per year in the United States. Because roughly 50% of those getting cancer die prematurely, this amounts to about 2,000–10,000 premature deaths per year in the United States from exposure to legally allowed pesticide residues in foods. Representatives from pesticide companies dispute these findings. Indeed, the food industry denies that anyone in the United States has ever been harmed by eating food that has been grown using pesticides for the past 50 years.

The 1,820 so-called *inert* ingredients (such as chlorinated hydrocarbon solvents) in pesticide products normally don't kill pest organisms. However, many of them may be active chemically or biologically in or on other organisms, including humans and various wildlife species. In fact, because inert ingredients make up 80–99% by weight of a pesticide product, they can pose a higher health risk than some active ingredients.

Since 1987 the EPA has been evaluating these inert ingredients; thus far it has labeled 100 of them "of known or potential toxicological concern" but has not yet banned their use. By 1997 the EPA still had no toxicity information on more than 75% of the inert ingredients used in pesticides.

FIFRA allows the EPA to leave inadequately tested pesticides on the market and to license new chemicals without full health and safety data. It also gives the EPA unlimited time to remove a chemical, even when its health and environmental risks are shown to outweigh its economic benefits. The built-in appeals and other procedures often keep a dangerous chemical on the market for up to 10 years.

The EPA can ban a chemical immediately in an emergency. Until 1990, however, the law required the EPA to use its already severely limited funds to compensate pesticide manufacturers for their remaining inventory and for all storage and disposal costs. Because compensation costs for a single chemical could exceed the agency's annual pesticide budget, the only economically feasible solution was for the EPA to allow existing stocks of a dangerous chemical to be sold. One of the 1988 amendments to FIFRA shifted some of the costs of storage and disposal of banned pesticides from the EPA to the manufacturers. This law is the only major environmental statute that does not allow citizens to sue the EPA for not enforcing the law, an essential tool to ensure government compliance.

According to a 1996 study by Charles M. Benbrook, former executive director of the U.S. National Academy of Sciences' Board of Agriculture, despite the expenditure of more than $1 billion per year of taxpayers' money to regulate pesticides, the public health hazards and ecological damage created by pesticides have not diminished in the past 30 years.

A 1993 study of pesticide safety by the U.S. National Academy of Sciences urged the government to do the following things:

- Make human health the primary consideration for setting limits on pesticide levels allowed in food.

- Collect more and better data on exposure to pesticides for different groups, including farm workers, adults, and children.

- Develop new and better test procedures for evaluating the toxicity of pesticides, especially for children.

- Consider cumulative exposures of all pesticides in food and water, especially for children, instead of basing regulations on exposure to a single pesticide. Currently, tiny quantities of nearly 100 pesticides are legally allowed in milk, which typically makes up one-fifth of a toddler's diet. A child may also

be exposed to trace amounts of 20 different pesticides on grapes, 20 on oranges, and 13 on apples. Are the cumulative levels of these pesticides safe for children?

Some progress was made with the passage of the 1996 Food Quality Protection Act:

- It requires new standards for pesticide tolerance levels in foods, based on a reasonable certainty of no harm to human health (defined for cancer as producing no more than 1 additional cancer per million people exposed to a certain pesticide over a lifetime).

- It requires manufacturers to demonstrate that the active ingredients in their pesticide products are safe for infants and children.

- It allows the EPA to apply an additional 10-fold safety factor to pesticide tolerance levels to protect infants and children.

- It requires the EPA to consider exposure to more than one pesticide when setting pesticide tolerance levels.

- It requires the EPA to develop rules for a program to screen all active and inactive ingredients for their estrogenic and endocrine effects by 1999.

15-5 SOLUTIONS: OTHER WAYS TO CONTROL PESTS

How Can Cultivation Practices Control Pests? A number of *cultivation practices* can help reduce pest damage. The type of crop planted in a field each year can be changed (crop rotation). Rows of hedges or trees can be planted around fields to hinder insect invasions and provide habitats for their natural enemies (with the added benefit of reduced soil erosion). Planting times can be adjusted so that major insect pests either starve or get eaten by their natural predators. Recent research by USDA scientists suggests that plowing at night reduces the growth of certain types of weeds by 50–60%, by preventing their buried seed from being sprouted by exposure to a short flash of light. Crops can also be grown in areas where their major pests do not exist.

Trap crops can be planted to lure pests away from the main crop. For example, in Nicaraguan cotton fields several rows of cotton are planted several months ahead of the regular crop to attract boll weevils, which can then be destroyed by hand or with small doses of pesticides.

Growers can switch from vulnerable monocultures to intercropping, agroforestry, and polyculture, which use plant diversity to reduce losses to pests

Q: How long will the world's proven reserves of coal last at current consumption rates?

Figure 15-6 The results of one example of genetic engineering against pest damage. Both tomato plants were exposed to destructive caterpillars. The normal plant's leaves are almost gone (left), whereas the genetically altered plant (right) shows little damage. (Monsanto)

Figure 15-7 Biological control of pests: An adult convergent ladybug (right) is consuming an aphid (left). (Peter J. Bryant/Biological Photo Service)

(Section 14-1). Diseased or infected plants and stalks and other crop residues that harbor pests can be removed from cropfields. Photodegradable plastic can be used to keep weeds from sprouting between crop rows. Vacuum machines can be used to gently remove harmful bugs from plants.

With the rise of industrial agriculture and the greatly increased use of synthetic pesticides over the last 30 years, many farmers in developed countries no longer use many of these traditional cultivation methods for controlling pest populations. However, as the problems and expense of using pesticides have risen since 1980, more farmers are returning to these methods.

How Can Genetically Resistant Plants and Crops Help Lower Pest Losses? Plants and animals that are genetically resistant to certain pest insects, fungi, and diseases can be developed. However, resistant varieties usually take a long time (10–20 years) and lots of money to develop by conventional crossbreeding methods. Moreover, insects and plant diseases can develop new strains that attack the once-resistant varieties, forcing scientists to develop new resistant strains continually. Genetic engineering is now helping to speed the process (Figure 15-6 and Pro/Con, p. 240).

There has been much talk about using genetic engineering to build pest resistance into crops and thus reduce the need for pesticides. However, some biotech companies have been focusing much of their genetic engineering research on developing crop strains that are resistant to herbicides—a strategy designed to increase the use and sale of herbicides.

How Can Natural Enemies Help Control Pests? Biological control using predators (Figures 15-1 and 15-7), parasites, and pathogens (disease-causing bac-

teria and viruses) can be encouraged or imported to regulate pest populations. More than 300 biological pest control projects worldwide have been successful, especially in China (p. 420). In Nigeria, crop-duster planes release parasitic wasps instead of pesticides to fight the cassava mealybug; farmers get a $178 return for every $1 they spend on the wasps. In the United States, natural enemies have been used to control more than 70 insect pests.

Biological control has several advantages. It focuses on selected target species and is nontoxic to other species, including people. Once a population of natural predators or parasites is established, biological pest control can often be self-perpetuating. Development of genetic resistance is minimized because pest and predator species interact and change together (coevolution). In the United States, biological control has saved farmers an average of $25 for every $1 invested.

However, years of research may be needed first, both to understand how a particular pest interacts with its various enemies and to choose the best biological control agent. Biological agents can't always be mass-produced, and farmers find them slower acting and more difficult to apply than pesticides. Biological agents must also be protected from pesticides sprayed in nearby fields.

Once released into the environment, biological control agents can multiply, cause unpredictable harmful ecological effects, and be impossible to recapture. Some may even become pests themselves; others (such as praying mantises) devour beneficial and pest insects alike. Indeed, species introduced for biological pest control have been strongly implicated in the extinction of nearly 100 insect species worldwide.

Figure 15-8 Infestation of a steer by screwworm fly larvae in Texas (United States). An adult steer can be killed in 10 days by thousands of maggots feeding on a single wound. (U.S. Department of Agriculture)

Figure 15-9 Pheromones can help control populations of pests, such as the red scale mites that have infested this lemon grown in Florida (United States). (Agricultural Research Service/USDA)

How Can Biopesticides Help Control Pests?

Botanicals such as synthetic pyrethroids (Table 15-1) are an increasingly popular method of pest control, and scientists are looking for new plant toxins to synthesize for mass production. Microbes are also being drafted for insect wars, especially by organic farmers. For example, *Bacillus thuringensis (Bt)* toxin is a registered pesticide sold commercially as a dry powder. Each of the thousands of strains of this common soil bacterium will kill a specific pest. Various strains of *Bt* are used by almost all organic farmers as a nonchemical pesticide.

Recently Monsanto transferred a *Bacillus thuringensis* gene to cotton and other plants. The *Bt* gene produce a protein that disrupts the digestive system of pests; insects that bite the plant die within hours. However, preliminary trials indicate that it is not very effective because of the low doses of *Bt* produced by the cotton. More bad news is that genetic resistance is already developing to some *Bt* toxins.

In addition, environmentalists have charged Monsanto with deliberately sabotaging the growing use of *Bt*, the main nonchemical pesticide used by organic farmers, to eliminate this competition. As more crops containing *Bt* genes are used, insect pests will become genetically resistant to the various forms of *Bt* and they will no longer be useful to farmers. Insect experts say this will probably happen in 3–10 years.

How Can Insect Birth Control, Sex Attractants, and Hormones Help Control Pests?

Males of some insect pest species can be raised in the laboratory, sterilized by radiation or chemicals, and then released into an infested area to mate unsuccessfully with fertile wild females. This technique works best if the females mate only once, the infested area is isolated (so that it can't be repopulated with nonsterilized males), and the insect pest population has already been reduced to a fairly low level by weather, pesticides, or other factors. Problems with this approach include high costs, the difficulties in knowing the mating times and behaviors of each target insect, the large number of sterile males needed, and the few species for which this strategy works.

The U.S. Department of Agriculture (USDA) used the sterile-male approach to essentially eliminate the screwworm fly, a major livestock pest (Figure 15-8), from the southeastern states between 1962 and 1971. To prevent resurgence of this pest, new strains of sterile male flies must be released every few years. Currently, the USDA has a cooperative male release program with the Mexican government in southern Mexico.

In many insect species, a female that is ready to mate releases a minute amount (typically about one-millionth of a gram) of a chemical sex attractant called a *pheromone*. Whether extracted from insects or synthesized in the laboratory, pheromones can lure pests into traps or attract their natural predators into cropfields (usually the more effective approach). More than 50 companies worldwide sell about 250 pheromones to control pests (Figure 15-9). These chemicals attract only one species, work in trace amounts, have little chance of causing genetic resistance, and are not harmful to nontarget species. However, it is costly and time-consuming to identify, isolate, and produce the specific sex attractant for each pest or predator species.

Each step in the insect life cycle is regulated by the timely natural release of juvenile hormones (JH) and molting hormones (MH) (Figure 15-10). These chemicals, which can be extracted from insects or synthesized in the laboratory, can disrupt an insect's normal life cycle, causing the insect to fail to reach maturity

Q: How long will proven reserves of coal in the United States last at current consumption rates?

and reproduce (Figure 15-11). Insect hormones have the same advantages as sex attractants, but they take weeks to kill an insect, are often ineffective with large infestations of insects, and sometimes break down before they can act. They must also be applied at exactly the right time in the target insect's life cycle. Moreover, they sometimes affect the target's predators and other nonpest species, and they can kill crustaceans if they get into aquatic ecosystems. Finally, like sex attractants, they are difficult and costly to produce.

How Can Hot Water Be Used to Zap Pests?
Some farmers have begun using the *Aqua Heat* machine, which sprays boiling water on crops to kill weeds and insects. Water is boiled and drawn from a large stainless steel tank mounted on a tractor and sprayed on crops using a long boom.

So far, the system has worked well on cotton, alfalfa, and potato fields and in citrus groves in Florida, where the machine was invented. The costs are roughly equal to those of using chemical pesticides.

How Can Zapping Foods with Radiation Help Control Pests?
Certain foods can be exposed after harvest to gamma rays emitted by radioactive isotopes. Such food irradiation extends food shelf life and kills insects, parasitic worms (such as trichinae in pork), and bacteria (such as salmonellae, which infect 51,000 Americans and kill 2,000 each year). A food does not become radioactive when it is irradiated, just as being exposed to X rays does not make the body radioactive.

According to the U.S. Food and Drug Administration and the World Health Organization, over 2,000 studies show that foods exposed to low doses of ionizing radiation are safe for human consumption. Currently, 37 countries (8 of them in western Europe) allow irradiation of one or more food items. Proponents of this technology contend that irradiation of food is likely to lower health hazards to people and reduce pesticide use and that its potential benefits greatly exceed the risks.

Critics contend that irradiating food destroys some of its vitamins and other nutrients and that the irradiation process may form trace amounts of toxic, possibly carcinogenic compounds in the food. They point to some studies indicating that research animals fed irradiated foods have shown increases in kidney damage, tumors, and reproductive failures. These opponents contend that more studies are needed to ensure the safety and wholesomeness of irradiated food and that it is too soon to see possible long-term effects, which might not show up for 30–40 years. Critics also argue that Americans want fresh, wholesome food, not old, possibly less nutritious food made to appear fresh and healthy by irradiation. And people prefer food

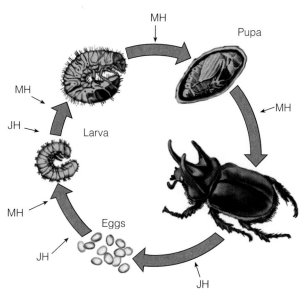

Figure 15-10 For normal insect growth, development, and reproduction to occur, certain juvenile hormones (JH) and molting hormones (MH) must be present at genetically determined stages in the insect's life cycle. If applied at the proper time, synthetic hormones disrupt the life cycles of insect pests and control their populations.

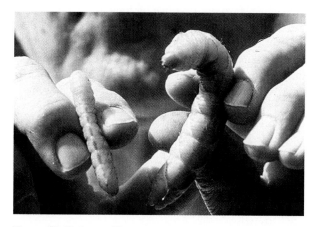

Figure 15-11 A use of hormones to prevent insects from maturing completely, making it impossible for them to reproduce. The stunted tobacco hornworm (left) was fed a compound that prevents production of molting hormones; a normal hornworm is shown on the right. (Agricultural Research Service/USDA)

that is really clean to food contaminated by fecal matter even if it is germ-free. Irradiating food will also raise the price of foods such as hamburgers by 3¢ to 7¢ per pound. It will also endanger the safety and health of food-irradiation workers.

Irradiation could also increase deaths from botulism poisoning. Current levels of irradiation do not destroy botulinum spores, but they do destroy the bacteria that give off the rotten odor warning us of their presence. Furthermore, some microorganisms could mutate when exposed to radiation, possibly creating new and more dangerous forms of botulism.

The FDA has approved the use of irradiation on poultry, beef, spices, and all domestic fruits and vegetables to delay sprouting, kill insect pests, and slow ripening. However, New York, New Jersey, and Maine have prohibited the sale and distribution of irradiated food, as have Germany, Austria, Denmark, Sweden, Switzerland, Sudan, Singapore, Australia, and New Zealand.

How Can Integrated Pest Management Help Control Pests? An increasing number of pest control experts and farmers believe that the best way to control crop pests is a carefully designed **integrated pest management (IPM)** program. In this approach, each crop and its pests are evaluated as parts of an ecological system. Then a control program is developed that includes a mix of cultivation and biological and chemical methods applied in proper sequence and with the proper timing.

The overall aim of IPM is not eradication of pest populations, but reduction of crop damage to an economically tolerable level (Figure 15-12). Fields are carefully monitored; when a damaging level of pests is reached, farmers first use biological methods (natural predators, parasites, and disease organisms) and cultivation controls, including vacuuming up harmful bugs. Small amounts of insecticides (mostly botanicals or microbotanicals) are applied only as a last resort, and different chemicals are used to slow development of genetic resistance and to avoid killing predators of pest species.

In 1986 the Indonesian government banned the use of 57 of 66 pesticides used on rice, phased out pesticide subsidies over a 2-year period, and used some of the money to help launch a nationwide program to switch to IPM, including a major farmer education program. The results were dramatic: Between 1987 and 1992, pesticide use dropped by 65%, rice production rose by 15%, and more than 250,000 farmers were trained in IPM techniques. By 1993 the program had saved the Indonesian government over $1.2 billion (most from elimination of $120-million annual pesticide subsidies)—more than enough to fund its IPM program.

In Australia use of IPM on Queensland citrus crops saves growers about $1 million per year for each 1,000 hectares (2,500 acres). Overall pesticide use fell 90% between 1980 and 1990. In 1994 farmers using IPM in Bangladesh used 80–90% less pesticide and still got yields 12–13% higher than those of non-IPM fields. In the Philippines nearly 4,000 farmers trained to use IPM used 60–98% less pesticide in 1993 and 1994 and had yield increases of 5–15%. In 1997 Mexico unveiled a program to phase out all uses of the pesticides DDT and chlordane (Table 15-1) within 10 years.

The experiences of countries such as China, Brazil, Indonesia, Australia, and the United States show that a well-designed IPM program can reduce pesticide use and pest control costs by 50–90%. IPM can also reduce preharvest pest-induced crop losses by 50%. It can improve crop yields, reduce inputs of fertilizer and irrigation water, and slow the development of genetic resistance because pests are assaulted less often and with lower doses of pesticides. Thus IPM is an important form of pollution prevention that reduces risks to wildlife and human health.

However, IPM requires expert knowledge about each pest situation, and it is slower acting than conventional pesticides. Methods developed for a crop in one area might not apply to areas with even slightly different growing conditions. Although long-term costs are typically lower than the costs of using conventional pesticides, initial costs may be higher.

Despite its promise and growth, widespread use of IPM is hindered by government subsidies of conventional chemical pesticides and by opposition from agricultural chemical companies, whose sales would drop sharply. In addition, farmers get most of their information about pest control from pesticide salespeople (and in the United States from USDA county farm agents), few of whom have adequate training in IPM.

In 1996 a study by the National Academy of Sciences recommended that the United States shift from chemically based approaches to ecologically based approaches to pest management. A growing number of scientists urge the USDA to promote IPM in the United States by **(1)** adding a 2% sales tax on pesticides and using the revenue to fund IPM research and education; **(2)** setting up a federally supported IPM

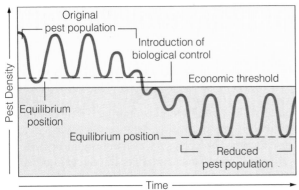

Figure 15-12 The goal of integrated pest management is to keep each pest population just below the size at which it causes economic loss.

Q: How many people in the United States die prematurely each year from coal burning?

demonstration project on at least one farm in every county; **(3)** training USDA field personnel and county farm agents in IPM so that they can help farmers use this alternative; **(4)** providing federal and state subsidies, and perhaps government-backed crop insurance, to farmers who use IPM or other approved alternatives to pesticides; and **(5)** gradually phasing out subsidies to farmers who depend almost entirely on pesticides, once effective IPM methods have been developed for major pest species. Some actions you can take to reduce your use of and exposure to pesticides are listed in Appendix 5.

We need to recognize that pest control is basically an ecological, not a chemical problem.

ROBERT L. RUDD

CRITICAL THINKING

1. Should DDT and other pesticides be banned from use in malaria control efforts throughout the world? Explain. What are the alternatives?

2. Do you believe that because essentially all pesticides eventually fail, their use should be phased out or sharply reduced and that farmers should be given economic incentives for switching to integrated pest management? Explain your position.

3. Do you agree or disagree that because DDT and the other banned chlorinated hydrocarbon pesticides pose no demonstrable threat to human health and have saved millions of lives, they should again be approved for use in the United States? Explain.

4. Should certain types of foods used in the United States be irradiated? Explain. If so, should such foods be required to carry a clear label stating that they have been irradiated? Explain.

5. What changes, if any, do you believe should be made in the Federal Insecticide, Fungicide, and Rodenticide Act regulating pesticide use in the United States?

*6. How are bugs and weeds controlled in your yard and garden? On the grounds of your school? On public school grounds, parks, and playgrounds near your home? Consider organizing an effort to have integrated pest management and organic fertilizers used on school and public grounds. Do the same thing for your yard and garden.

*7. Make a concept map of this chapter's major ideas, using the section heads and subheads and the key terms (in boldface type). Look at the inside back cover and in Appendix 4 for information about concept maps.

SUGGESTED READINGS

See Internet Sources and Further Readings at the back of the book, and InfoTrac.

16 SUSTAINING ECOSYSTEMS: FORESTS, RANGELANDS, PARKS, AND WILDERNESS

How Farmers and Loud Monkeys Saved a Forest

It's early morning in a tropical forest in the Central American country Belize (Figure 6-19). Suddenly, loud roars that trail off into wheezing moans—territorial calls of black howler monkeys (Figure 16-1)—wake up everyone in or near the wildlife sanctuary by the Belize River.

This species is the centerpiece of an experiment recruiting peasant farmers to preserve tropical forests and wildlife. The project is the brainchild of an American biologist, Robert Horwich, who in 1985 met with villagers and suggested that they establish a sanctuary that would benefit the local black howlers and themselves. He proposed that the farmers leave thin strips of forest along the edges of their fields to provide the howlers with food and a travel route through the sanctuary's patchwork of active garden plots and young and mature forest.

To date, more than 100 farmers have participated, and the 47-square-kilometer (18-square-mile) sanctuary is home for an estimated 1,100 black howlers. The idea has spread to seven other villages.

As many as 6,000 ecotourists visit the sanctuary each year to see its loud monkeys and other wildlife. Villagers serve as tour guides, cook meals for the visitors, and lodge tourists overnight in their spare rooms.

As long as this ecotourism doesn't disturb the sanctuary ecosystem, this experiment can demonstrate the benefits of integrating ecology and economics, allowing villagers to make money by helping sustain the forest and its wildlife.

Forests, rangelands, parks, wilderness, and other biodiversity sanctuaries are crucial resources that are coming under increasing pressure from population growth and economic development. Conservation biologists believe that learning how to use these potentially renewable forms of earth capital sustainably, by protecting some from exploitation and helping to heal those we have degraded, is an urgent priority.

Figure 16-1 A black howler monkey, shown in a tropical forest in Belize. The populations of six howler monkey species have declined as tropical forests have been cleared. The black howler, like more than two-thirds of the world's 150 known species of primates, is threatened with extinction. (Carol Farnetti/Planet Earth Pictures)

Forests precede civilizations, deserts follow them.

François–Auguste–René de Chateaubriand

This chapter is devoted to answering the following questions:

- What is conservation biology, and why is it important?

- What are the types of public lands in the United States, and how are they used?

- How should forest resources be managed and conserved?

- Why are tropical deforestation and fuelwood shortages serious problems, and what can be done about them?

- Why are rangelands important, and how should they be managed?

- What problems do parks face, and how should they be managed?

- Why is wilderness important, and how much wilderness and other biodiversity sanctuaries should be preserved?

✿ 16-1 BIODIVERSITY, CONSERVATION BIOLOGY, AND ECOLOGICAL INTEGRITY

Why Is Biodiversity Loss Considered the Key Environmental Problem? Every country and region has three forms of wealth: material, cultural, and biological (biodiversity). The forests, rangelands, parks, wilderness, and aquatic systems, where the genes, species, and ecosystems making up biodiversity are found, are crucial sources of biological wealth that are coming under increasing pressure from population growth and economic development.

Since 1980 there has been considerable growth in **(1)** the scientific understanding of biological wealth or biodiversity that is so vital to all life and economies, **(2)** the ecological processes such as matter cycling, energy flow, and species interactions (Chapters 4 and 5) that sustain biodiversity, and **(3)** the biological consequences of loss of biodiversity through environmental degradation.

As a result, most ecologists and other environmental scientists consider the loss of biodiversity the key environmental problem. This revolution in awareness has led to a much deeper understanding and appreciation of *interrelationships or connections in nature*—a major theme of this book.

What Are Conservation Biology and Ecological Integrity? The science-based study of biodiversity has led to the development of **conservation biology**. It is a multidisciplinary science created in the late 1970s to deal with the crisis of maintaining the genes, species, communities, and ecosystems that make up the earth's *biological diversity*. Its goals are to investigate human impacts on biodiversity and develop practical approaches to preserving biodiversity.

Conservation biology uses scientific data and concepts to find practical ways to protect critical ecosystems and biodiversity-rich areas and prevent the premature extinctions of species. Conservation biology differs from *wildlife management*, which is devoted primarily to the manipulation of the population sizes of various animal species, especially game species prized by hunters and fishers (Section 17-4).

Since the mid-1980s conservation biologists and ecologists have increasingly emphasized the preservation of **ecological integrity**, the conditions and natural processes (such as the energy flow and matter cycling in ecosystems and species interactions) that generate and maintain biodiversity and allow evolutionary change (Section 5-6) as a key mechanism for adapting to changes in environmental conditions. The **ecological health** of an area can be described in terms of the degree to which its biodiversity and ecological integrity remain intact.

Conservation biology rests on the following principles: **(1)** Biodiversity and ecological integrity are necessary to all life on earth and should not be reduced by human actions, **(2)** humans should not cause or hasten the premature extinction of populations and species by disrupting evolutionary processes and critical ecological processes, **(3)** the best way to preserve biodiversity and ecological integrity is to preserve habitats, niches, and ecological interactions, and **(4)** goals and strategies for preserving biodiversity and ecological integrity of an area must be based on a deep understanding of the ecological properties and processes of that system.

This *scientific approach* recognizes that saving wildlife means saving their habitats and not disrupting the complex interactions among species in an ecosystem (ecological integrity). It is also based on Aldo Leopold's ethical principle that something is right when it tends to maintain the earth's life-support systems for us and other species, and wrong when it doesn't.

✿ 16-2 PUBLIC LANDS IN THE UNITED STATES

What Are the Major Types of U.S. Public Lands? No nation has set aside as much of its land—about 42%—for public use, enjoyment, and wildlife as has the United States. Almost one-third of the country's land belongs to every American and is managed by the federal government; 73% of this public land is in Alaska and another 22% is in the western states (where 60% of all land is public land). These public lands are classified as multiple-use lands, moderately restricted-use lands, and restricted-use lands.

Multiple-Use Lands

■ The 156 forests (Figure 16-2) and 20 grasslands of the *National Forest System* are managed by the U.S. Forest Service. These forests are used for logging (the dominant use in most cases), mining, livestock grazing, farming, oil and gas extraction, recreation, sport hunting, sport and commercial fishing, and conservation of watershed, soil, and wildlife resources. Off-road vehicles are usually restricted to designated routes.

■ *National Resource Lands* in the western states and Alaska are managed by the Bureau of Land Management. Emphasis is on providing a secure domestic supply of energy and strategic minerals and on preserving rangelands for livestock grazing under a permit system.

Moderately Restricted-Use Lands

■ The 508 *National Wildlife Refuges* (Figure 16-2) are managed by the U.S. Fish and Wildlife Service. Most refuges protect habitats and breeding areas for waterfowl and big game to provide a harvestable supply for hunters; a few protect endangered species from extinction. Sport hunting, trapping, sport and commercial fishing, oil and gas development, mining, logging, grazing, some military activities, and farming are permitted as long as the Department of the Interior finds such uses compatible with the purposes of each unit.

Restricted-Use Lands

■ The 375 units of the *National Park System* include 54 major parks (mostly in the West) and 321 national recreation areas, monuments, memorials, battlefields, historic sites, parkways, trails, rivers, seashores, and lakeshores (Figure 16-2) managed by the National Park Service. National parks may be used only for camping, hiking, sport fishing, and boating. Motor vehicles are permitted only on roads.

■ The 630 roadless areas of the *National Wilderness Preservation System*, which lie within the national parks, national wildlife refuges, and national forests, are managed by the National Park Service (42%), Forest Service (33%), Fish and Wildlife Service (20%), and Bureau of Land Management (5%). These areas are open only for recreational activities such as hiking, sport fishing, camping, nonmotorized boating, and, in some areas, sport hunting and horseback riding. Roads, logging, livestock grazing, mining, commercial activities, and buildings are banned, except when they predate the wilderness designation.

How Should Public Lands Be Managed?

Because of the resources they contain, there has been intense controversy over how public lands should be used and managed (Spotlight, p. 440). Economists, developers, and resource extractors tend to view areas of the earth's surface in terms of their usefulness in providing mineral and other resources, their potential for development, and their ability to increase short-term economic growth.

Ecologists and conservation biologists take a scientific approach and view the earth's ecosystems in terms of their usefulness in maintaining the planet's biodiversity and life-sustaining processes. To conservation biologists and ecologists, destroying or degrading the earth's remaining intact ecosystems for short-term economic gain will make them less useful to humans and other forms of life.

16-3 MANAGING AND SUSTAINING FORESTS

What Are the Major Types of Forests?

There are three general types of forests, depending primarily on climate: tropical, temperate, and polar (Figure 5-3). Since agriculture began about 10,000 years ago, human activities have reduced the earth's forest cover by about one-quarter—from about 34% to 26% of the world's land area—and only 12% consists of intact forest ecosystems.

If the rate of cutting and degradation does not exceed the rate of regrowth, and if protecting biodiversity is emphasized, forests are renewable resources. However, forests are disappearing or are being fragmented and degraded almost everywhere, especially in tropical countries (Figure 16-3).

Old-growth forests are uncut forests and regenerated forests that have not been seriously disturbed for several hundred or thousands of years. Old-growth forests provide ecological niches for a multitude of wildlife species (Figure 5-12). These forests also have large numbers of standing dead trees (snags) and fallen logs, which are habitats for a variety of species.

Q: What would happen if coal's harmful effects were included in its market price and government subsidies were removed?

Figure 16-2 National forests, national parks, and wildlife refuges managed by the U.S. federal government. (Data from U.S. Geological Survey)

National parks and preserves

National forests

National wildlife refuges

How Should U.S. Public Lands Be Managed?

SPOTLIGHT

In recent years, the total cost of subsidies (in public funds spent and taxes and user fees not collected) given to mining (p. 338), logging, and grazing interests using public lands has exceeded $1 billion per year. In addition to losing potential revenue for resource extraction on lands they own, taxpayers must also pay for dealing with most of the resulting environmental damage.

Most environmentalists and many free-market economists believe that the following principles should govern use of public land:

■ Protection of biodiversity and the ecological integrity of vital public lands should be the primary goal.

■ No one should be given subsidies or tax breaks for using or extracting resources on public lands.

■ The American people deserve fair compensation for the use of their public lands.

■ All users or extractors of resources on public lands should be fully responsible for any environmental damage they cause.

According to U.S. Representative George Miller, "It is time for the resource industries to grow up and get off the federal bottle."

Since 1995, however, there have been increasing attempts by the U.S. Congress (influenced by ranching, timber, mining, and development interests) to pass laws that would **(1)** sell public lands or their resources to corporations or individuals, usually at less than market value; **(2)** give public lands to the states (where these interests often have more influ-

ence); **(3)** slash federal funding for administration of public lands so that unlawful exploiters have less chance of getting caught; **(4)** weaken, waive, or eliminate federal laws that regulate use of public lands; **(5)** require the government to reimburse private landowners for regulatory takings (p. 216); **(6)** allow highways to be built through national parks and wilderness areas; **(7)** redefine protected wetlands so that about half of them would no longer be protected; and **(8)** repeal or seriously weaken the Endangered Species Act.

In 1995 and 1996, most of these laws were not passed or were vetoed by President Clinton (except for the forest salvage law, Spotlight, p. 447).

Critical Thinking

Explain why you agree or disagree with the four principles listed above.

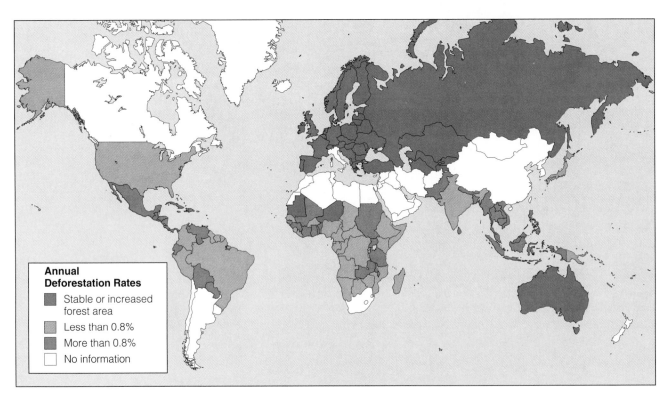

Figure 16-3 Estimated annual rates of deforestation, 1981–90. During this period the area of the world's tropical forest decreased by 8% (an average of 0.8% per year). (Data from UN Food and Agriculture Organization)

Annual Deforestation Rates
- Stable or increased forest area
- Less than 0.8%
- More than 0.8%
- No information

Q: How much of the world's electricity is supplied by nuclear power?

Decay of this dead vegetation returns plant nutrients to the soil and helps build fertile soil.

Second-growth forests are stands of trees resulting from secondary ecological succession after cutting (Figure 5-42). Most forests in the United States and other temperate areas are second-growth forests that grew back after virgin forests were cleared for timber or to create farms that were later abandoned. About 40% of tropical forests are second-growth forests.

Some second-growth stands have remained undisturbed long enough to become old-growth forests, but many are not diverse forests but rather **tree farms** or **plantations** (see photo in the Table of Contents). They are managed tracts with uniformly aged trees of one species that are harvested by clear-cutting as soon as they become commercially valuable. Then they are replanted and clear-cut again on regular cycles.

What Are the Commercial and Ecological Importance of Forests? Forests provide lumber for housing, biomass for fuelwood, pulp for paper, medicines, and many other products (Figure 16-4) worth more than $300 billion a year. The global timber trade alone amounts to more than $100 billion per year, most of which comes from forests in temperate countries. Many forestlands are also used for mining, grazing livestock, and recreation.

Forested watersheds act as giant sponges, slowing down runoff and holding water that recharges springs, streams, and groundwater. Thus, they regulate the

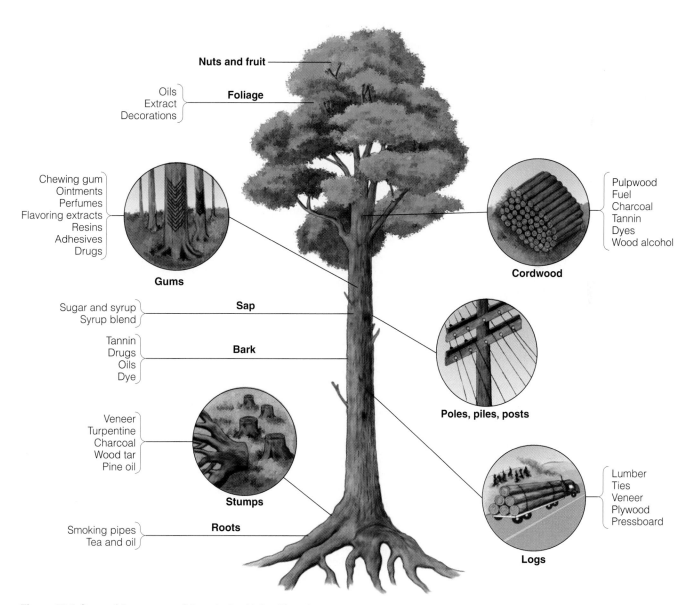

Figure 16-4 Some of the many useful products obtained from trees.

Nuts and fruit

Oils
Extract
Decorations | **Foliage**

Chewing gum
Ointments
Perfumes
Flavoring extracts
Resins
Adhesives
Drugs
Gums

Pulpwood
Fuel
Charcoal
Tannin
Dyes
Wood alcohol
Cordwood

Sugar and syrup
Syrup blend | **Sap**

Tannin
Drugs
Oils
Dye | **Bark**

Poles, piles, posts

Veneer
Turpentine
Charcoal
Wood tar
Pine oil
Stumps

Lumber
Ties
Veneer
Plywood
Pressboard
Logs

Smoking pipes
Tea and oil | **Roots**

flow of water from mountain highlands to croplands and urban areas, and they reduce the amount of sediment washing into streams, lakes, and reservoirs by reducing soil erosion.

Forests also influence climate. For example, 50–80% of the moisture in the air above tropical forests comes from trees via transpiration and evaporation. If large areas of these forests are cleared, average annual precipitation drops and the region's climate gets hotter and drier; then soils become depleted of already-scarce nutrients, and they are baked and washed away. This process can eventually convert a diverse tropical forest into a sparse grassland or even a desert.

Forests are vital to the global carbon cycle (Figure 4-22). They take up about 90% of the carbon removed from the atmosphere as CO_2 by terrestrial ecosystems and store 20–100 times more carbon per hectare than agricultural lands. They provide habitats for more wildlife species than any other biome. They also buffer us against noise, absorb air pollutants, and nourish the human spirit.

According to one calculation, a typical tree provides $196,250 worth of ecological benefits during its lifetime in the form of oxygen, air purification, soil fertility and erosion control, water recycling and humidity control, and wildlife habitats. Sold as timber, the same tree is worth only about $590. Even if such estimates are off by a factor of 100, the long-term ecological benefits of a tree still clearly exceed its short-term economic benefits.

How Much Is a Forest Worth?

SPOTLIGHT

In 1997 a group of 13 ecologists, economists, and geographers from 12 prestigious laboratories and universities attempted to estimate how much nature's ecological services (Figure 1-2) are worth. According to this ambitious and crude appraisal led by ecological economist Robert Costanza of the University of Maryland, the economic value of Mother Nature's bounty ranges from $16 trillion to $54 trillion. In other words, each year nature provides free goods and services worth somewhere between $2,800 and $9,000 per person.

Their medium estimate for nature's life-support services for humans is $33 trillion a year. This is nearly twice the $18-trillion-per-year economic output of the earth's 194 nations and almost five times the $6.9-trillion U.S. GDP in 1996.

To make these estimates, the researchers divided the earth's surface into 16 biomes (Figure 5-3) and aquatic life zones such as various types of forests, grasslands, cropland, open oceans, estuaries, wetlands, and lakes and rivers (but omitted deserts and tundra because of a lack of data). Then they agreed on a list of 17 goods and services provided by nature (Figure 1-2) and sifted through more than 100 studies that attempted to put a dollar value on such services in the 16 different types of ecosystems.

Some analysts believe that such estimates are misleading and dangerous because they amount to putting a dollar value on ecosystem services with an infinite value because they are irreplaceable. In the 1970s economist E. F. Schumacher warned, "To undertake to measure the immeasurable is absurd" and is a "pretense that everything has a price." Biologist and conservationist David Ehrenfeld's response to this evaluation was, "I am afraid that I don't see much hope for a civilization so stupid that it demands a quantitative estimate of the value of its own umbilical cord." Those who cringe at the thought of putting a dollar value on nature's ecosystem services believe that they should be protected on *moral* grounds instead of being reduced to easily manipulated cost–benefit analyses (Spotlight, p. 202).

The researchers admit that their estimates are full of assumptions and omissions and could easily be too low by a factor of 10, 1 million, or more. For example, their calculations include only estimates of the ecosystem services themselves, not the natural or earth capital that generates them, and omit the value of nonrenewable minerals and fuels. They also recognize that as the supply of ecosystem services declines their value will rise sharply and that such services can be viewed as having an infinite value.

However, they contend that their estimates are much more accurate than the *zero* value the market assigns to these ecosystem services. They hope such estimates will call people's attention to the fact that the earth's ecosystem services are absolutely essential for all humans and their economies (Figure 7-3) and that their economic value is huge. They believe that failure to estimate economic values for ecosystem services helps ensure that such vital life-support services will continue to be degraded for short-term economic gain.

Critical Thinking

Do you agree or disagree with the idea of trying to estimate the economic value of the earth's ecosystem services? Explain. What are the alternatives?

Q: Is there a scientifically and politically acceptable method for the long-term disposal of high-level radioactive waste?

The ecological benefits of complex and diverse forests are undervalued in the marketplace (Section 7-2). Until this situation is changed, most analysts believe that we will continue to sacrifice these forests and their long-term ecological services for short-term economic gain (Spotlight, left).

What Are the Major Types of Forest Management? The total volume of wood produced by a particular stand of forest varies as it goes through different stages of growth and ecological succession (Figure 16-5). If the goal is to produce fuelwood or fiber for paper production in the shortest time, the forest is usually harvested on a short rotation cycle, well before the volume of wood produced peaks (point A in Figure 16-5). Harvesting at point B in Figure 16-5 gives the maximum yield of wood per unit of time. If the goal is high-quality wood for fine furniture or veneer, managers use longer rotations to develop larger, older-growth trees (point C in Figure 16-5), whose rate of growth has leveled off and is much lower than that of young trees.

There are two basic forest management systems: even-aged and uneven-aged. With **even-aged management**, trees in a given stand are maintained at about the same age and size. A major goal of even-aged management, sometimes called *industrial forestry*, is to grow and harvest trees using monoculture techniques. This is achieved by replacing a biologically diverse natural forest with a simplified tree farm of one or two fast-growing and economically desirable species that can be harvested every 10–100 years, depending on the species. Crossbreeding and genetic engineering can improve the quality and the quantity of tree-farm wood.

Even-aged management begins with one or two cuttings of all or most trees from an area; then the site is usually replanted with seedlings of one or more species. In a natural forest, dead and fallen trees are seen as vital wildlife habitats and integral parts of a natural cycle of decay and forest renewal. In most industrial forests, they are viewed as debris to be removed and burned to make way for the growth of planted tree seedlings.

Even-aged management of an area leads to more even-aged management of that area. Once a diverse forest has been cleared and replaced with even-aged stands, the only economical thing for timber companies to do is to keep repeating the process of cutting down a stand of trees before it turns into a true forest. In even-aged management, forests are viewed primarily as lumber and fiber factories, and old-growth forests are seen as timber going to waste rather than essential centers of the earth's biodiversity and ecological integrity.

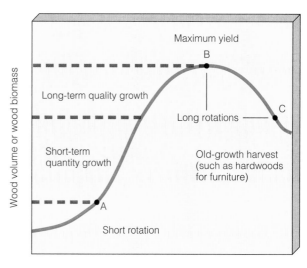

Figure 16-5 Changes in wood volume over various growth–harvest cycles in forest management. Forest management occurs over a cycle of decisions and events called a *rotation*. The most important steps in a rotation include taking an inventory of the site, developing a forest management plan, building roads into the site, preparing the site for harvest, harvesting timber, and regenerating and managing the site until the next harvest.

In **uneven-aged management**, a variety of tree species in a given stand are maintained at many ages and sizes to foster natural regeneration. Here the goals are biological diversity, long-term production of high-quality timber, a reasonable economic return, and multiple use. Mature trees are selectively cut; the removal of all trees is used only on small patches of species that benefit from such a practice.

How Are Trees Harvested? Before larger amounts of timber can be harvested, roads must be built for access and removal of timber. Even with careful design, logging roads have a number of harmful impacts (Figure 16-6). They cause erosion and increase sedimentation of waterways and cause severe habitat fragmentation and loss of biodiversity. They can expose forests to invasion by exotic pests, diseases, and introduced wildlife species. They can also open up once-inaccessible forests to farmers, miners, ranchers, hunters, and off-road vehicle users. In addition, logging roads on U.S. public lands disqualify the land for protection as wilderness.

Once loggers can reach a forest, they use various methods for harvesting the trees (Figure 16-7). With **selective cutting**, intermediate-aged or mature trees in an uneven-aged forest are cut singly or in small groups (group selection), creating gaps no larger than the height of the standing trees (Figure 16-7a). Selective cutting reduces crowding, encourages the growth of younger trees, and maintains an uneven-aged stand

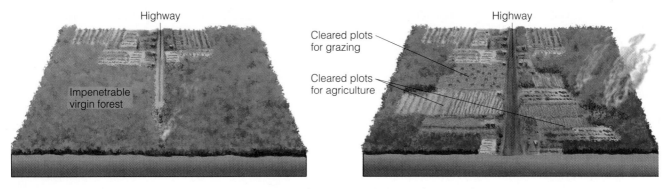

Figure 16-6 Building roads into previously inaccessible tropical forests paves the way to fragmentation, destruction, and degradation.

of trees of different species. It also allows natural regeneration from the surrounding trees, thereby avoiding the financial and environmental costs of site preparation (usually by bulldozer or herbicides) and planting of trees after clear-cutting. If done properly, selective cutting also helps protect the site from soil erosion and wind damage.

Many industry and government foresters contend that selective cutting is not profitable and that many of the trees that bring the best prices don't get enough sunlight for regeneration with selective cutting. But numerous private foresters and timber companies have been making satisfactory profits by using selective cutting of small and large areas of forests containing such species.

An unsound type of selective cutting is *high grading*, or *creaming*, which removes the most valuable trees. This practice, common in many tropical forests, ends up injuring one-third to two-thirds of the remaining trees when they are knocked over by logging equipment or by the large target trees when they are felled and removed.

Some tree species grow best in full or moderate sunlight in moderate to large clearings. Such sunloving species are usually harvested by shelterwood cutting, seed-tree cutting, or clear-cutting. **Shelterwood cutting** removes all mature trees in two or three cuttings over a period of about 10 years (Figure 16-7b). The first cut removes most mature canopy trees, unwanted tree species, and diseased, defective, and dying trees. This opens the forest floor to light but leaves enough mature trees to cast seed and to shelter growing seedlings. Some years later, after enough seedlings take hold, a second cut removes more canopy trees, although some of the best mature trees are left to shelter the young trees. After these are well established, a third cut removes the remaining mature trees, and the even-aged stand of young trees then grows to maturity.

This method allows natural seeding from the best seed trees and keeps seedlings from being crowded out. It leaves a fairly natural-looking forest that can serve a variety of purposes; it also helps reduce soil erosion and provides a good habitat for wildlife. The danger is that loggers may take too many trees in the first cut.

Seed-tree cutting harvests nearly all of a stand's trees in one cutting, leaving a few uniformly distributed seed-producing trees to regenerate the stand (Figure 16-7c). After the new trees become established, the seed trees may be harvested. By allowing several species to grow at once, seed-tree cutting leaves an aesthetically pleasing forest that is useful for recreation, deer hunting, erosion control, and wildlife conservation. Leaving the best trees for seed can also lead to genetic improvement in the new stand.

Clear-cutting is the removal of all trees from an area in a single cutting. The clear-cut area may be a whole stand (Figure 16-7d), a strip, or a series of patches. After all trees are cut, the site is usually reforested; seed is released naturally, by the harvest, or artificially as foresters broadcast seed over the site or plant seedlings raised in a nursery. If clear-cut areas are small enough, seeding may occur from nearby uncut trees.

On the positive side, clear-cutting increases timber yield per hectare, permits reforesting with genetically improved stocks of fast-growing trees, and shortens the time needed to establish a new stand of trees. It also requires much less skill and planning than other harvesting methods, and it usually provides timber companies the maximum economic return in the shortest time.

However, clear-cutting leaves ugly, unnatural forest openings and eliminates any potential recreational value for several decades. It also destroys and fragments some wildlife habitats and thus reduces biodiversity and disrupts ecological integrity. Environmental degradation from clear-cutting above ground is obvious, but equally serious damage occurs underground from the loss of fungi, worms, bacteria, and other

Q: How many people in the United States die prematurely each year from U.S. nuclear power plants?

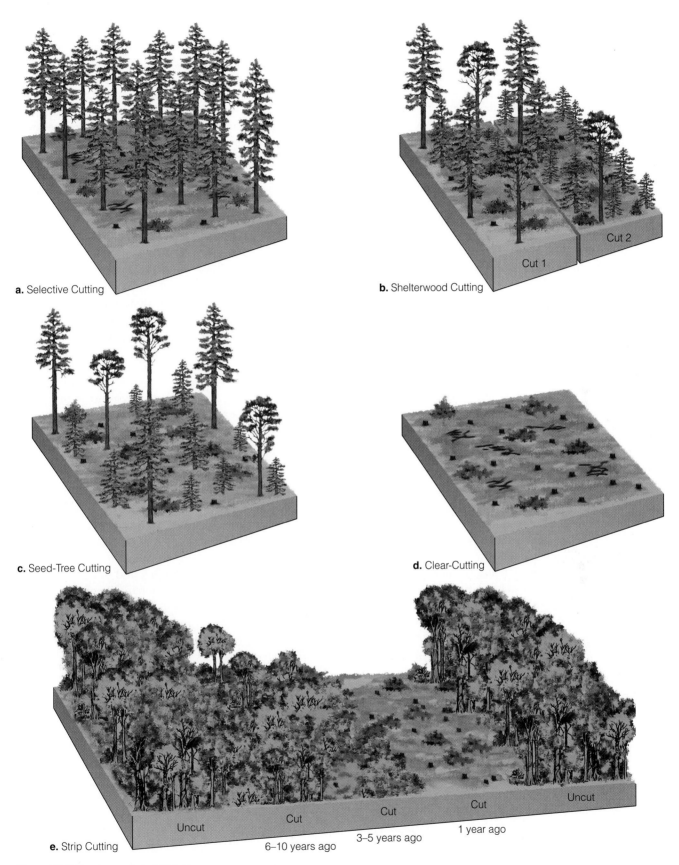

a. Selective Cutting

b. Shelterwood Cutting

Cut 1

Cut 2

c. Seed-Tree Cutting

d. Clear-Cutting

e. Strip Cutting

Uncut

Cut
6–10 years ago

Cut
3–5 years ago

Cut
1 year ago

Uncut

Figure 16-7 Tree harvesting methods.

microbes that condition soil and help protect plants from disease (Figures 12-15 and 12-16). Furthermore, trees in stands bordering clear-cut areas are more vulnerable to being blown down by windstorms. Large-scale clear-cutting on steep slopes leads to severe soil erosion, sediment water pollution, and flooding (Figure 11-17).

If done carefully and responsibly, clear-cutting is often the best way to harvest tree farms and stands of some tree species that require full or moderate sunlight for growth. However, for economic reasons it is often done irresponsibly and used on species that don't require this method.

A variation of clear-cutting that can allow a sustainable timber yield without widespread destruction is **strip cutting**. A strip of trees is clear-cut along the contour of the land, with the corridor narrow enough to allow natural regeneration within a few years (Figure 16-7e). After regeneration, another strip is cut above the first, and so on. This allows a forest to be clear-cut in narrow strips over several decades with minimal damage.

How Can Forests Be Protected from Pathogens and Insects? In a healthy and diverse forest, tree diseases and insect populations are usually controlled by interactions with other species. Thus, they rarely get out of control and seldom destroy many trees. However, a biologically simplified tree farm is vulnerable to attack by pathogens, especially parasitic fungi (such as chestnut blight and Dutch elm disease) and insects (such as bark beetles, the gypsy moth, and spruce budworm).

To conservation biologists preserving biodiversity is the best and cheapest defense against tree diseases and insects. Other methods include banning imported timber that might introduce harmful new pathogens, removing infected trees or clear-cutting infected areas and burning all debris, treating diseased trees with antibiotics, developing disease-resistant tree species, applying pesticides (Section 15-1), and using integrated pest management (Section 15-5).

How Do Fires Affect Forest Ecosystems?
Intermittent natural fires set by lightning are an important part of the ecological cycle of many forests. Some species actually need occasional fires. The seeds of the giant sequoia, the lodgepole pine, and the jack pine, for instance, are released from cones and germinate only after being exposed to intense heat.

Forest ecosystems can be affected by different types of fires. Some, called *surface fires*, usually burn only undergrowth and leaf litter on the forest floor. These fires kill seedlings and small trees but spare most mature trees, and most wild animals can escape.

In forests in which ground litter accumulates rapidly, a surface fire every 5 years or so burns away flammable material and helps prevent more destructive fires. Surface fires also release valuable mineral nutrients tied up in slowly decomposing litter and undergrowth. They increase the activity of underground nitrogen-fixing bacteria, stimulate the germination of certain tree seeds, and help control pathogens and insects. Some wildlife species—deer, moose, elk, muskrat, woodcock, and quail, for example—depend on occasional surface fires to maintain their habitats and to provide food in the form of vegetation that sprouts after fires.

Some extremely hot fires, called *crown fires*, may start on the ground but eventually burn whole trees and leap from treetop to treetop. They usually occur in forests in which no surface fires have occurred for several decades, allowing dead wood, leaves, and other flammable ground litter to build up. These rapidly burning fires can destroy most vegetation, kill wildlife, and lead to accelerated soil erosion.

Sometimes surface fires go underground and burn partially decayed leaves or peat. Such *ground fires* are most common in northern peat bogs. They may smolder for days or weeks before being detected and are difficult to extinguish.

Protecting forest resources from fire can involve four approaches: *prevention*, *prescribed burning* (setting controlled ground fires to prevent buildup of flammable material), *presuppression* (early detection and control of fires), and *suppression* (fighting fires once they have started). Means of forest-fire prevention include requiring burning permits, closing all or parts of a forest to travel and camping during periods of drought and high fire danger, and public education. The Smokey the Bear educational campaign of the Forest Service and the National Advertising Council, for example, has prevented countless forest fires in the United States, saving many lives and avoiding billions of dollars in losses.

However, ecologists contend that because it allows litter to accumulate in some forests, prevention increases the likelihood of highly destructive crown fires. Since 1972 U.S. Park Service policy has been to allow most lightning-caused fires to burn themselves out as long as they don't threaten human lives, park facilities, private property, or endangered wildlife.

After fires ravaged parts of Yellowstone National Park during the hot, dry summer of 1988 and after numerous fires broke out in a number of national forests in the West during the summer of 1994, some people called for a reversal of this policy. However, biologists contend that these fires caused more damage than they should have because the policy of fighting all fires in the park before 1972 had allowed the buildup of

Q: How many sites in the United States are contaminated with radioactive materials?

flammable ground litter and small plants. They believe that many fires in national parks, national forests, and wilderness areas should be allowed to burn as part of the natural ecological cycle of succession and regeneration (Figure 5-42). Exceptions would include fires that posed a serious threat to human-built structures.

Critics of putting out all fires in national parks also point out that in 1988 firefighters caused more damage in Yellowstone National Park by bulldozing fire lines and felling trees than if they had let the fires burn. Moreover, it would have been much cheaper for the government to rebuild burned buildings in Yellowstone than to spend $300 million to protect them. Timber companies argue that they can help prevent fires and tree disease by being allowed to make extensive salvage cuts in national forests and national parks (Spotlight, below).

How Are Forests Threatened by Air Pollution and Climate Change? Forests at high elevations and those downwind from urban and industrial centers are exposed to a variety of air pollutants that can harm trees, especially conifers. Besides doing direct harm, prolonged exposure to multiple air pollutants makes trees much more vulnerable to drought, diseases, and insects (Figure 9-9). The solution is to reduce emissions of the offending pollutants from coal-burning power plants, industrial plants, and motor vehicles (Section 9-5).

In coming decades, an even greater threat to forests (especially temperate and boreal forests) may come from regional climate changes brought about by projected global warming. Ways to deal with projected global warming are discussed in Section 10-5.

Is Industrial Forestry Sustainable? To timber companies, sustainable forestry means producing a maximum sustainable yield of commercial timber in as short a time as possible. This usually means clearing diverse, uneven-aged forests by using clear-cutting, seed-tree cutting, or shelterwood cutting (Figure 16-7b, c, and d) and replacing them with intensively managed, even-aged tree plantations that can be clear-cut again as quickly as possible.

How Much Salvage Tree Cutting Should Be Allowed on Public Lands?

SPOTLIGHT

In 1995 logging companies persuaded Congress to pass emergency logging laws (written with the aid of timber company representatives) that would allow them to reduce the threat of fire and tree diseases in national parks and national forests. This law, the Federal Lands Forest Health and Protection and Restoration Act, allows timber companies to make salvage cuts to remove dead and diseased trees, thin out stands, and remove limbs, stumps, and other flammable debris from national forests and national parks.

Environmentalists consider the law a sham that allows additional commercial logging and road building in national forests (including old-growth forests) under the guise of emergency fire and disease protection. They agree that some salvage cutting is needed in selected areas under careful guidelines, but point out that existing policy already allows regulated salvage logging.

To environmentalists this 1995 salvage law, which they call the *logging without laws* act, gives timber companies a free hand to do what they want in the national forests and national parks without having to follow any existing forestry or wildlife protection laws. President Clinton vetoed the law once but then reluctantly approved it when it was attached to an act including emergency relief for victims of the 1995 bombing of a federal building in Oklahoma.

The 1995 salvage law allows timber companies to cut large areas of healthy trees if they find only one or a few diseased or burned trees (which opens up almost any tract to cutting). Timber companies don't have to file environmental impact statements or abide by the Endangered Species Act, and they don't have to abide by the requirement that national forests be cut on a sustainable-yield basis.

They can also ignore the 16-hectare (40-acre) limit on clearcuts. Moreover, the law prohibits citizens from challenging any salvage cuts (many of which are aimed at old-growth forests that laws had previously protected by law or court orders) in the courts.

The salvage cut law also requires the Forest Service to make the cuts available within 2 years. This has allowed timber companies to make large cuts in national forests before the public gets wind of the nature of the law and demands that it be repealed. Companies also can cut valuable trees at bargain basement prices, some for less than $1 a tree. Attempts to repeal the law in 1995 and 1996 failed, and there are efforts in Congress to make the law permanent.

Critical Thinking

What are some (if any) benefits of the 1995 emergency tree salvage law? Do you agree or disagree with this law? Explain.

To conservation biologists and some foresters such *maximum sustained yield forestry* does not result in long-term *sustainable forestry*. Research reveals that industrial or monoculture forestry has lead to a decrease in the biodiversity of forests, elimination of competing species, suppression of ecologically important natural fires, draining of wetlands, and increased soil erosion and loss of soil nutrients.

For example, decades of experience with intensive even-aged management of German forests has shown that in a forest in which virtually all of the trees are repeatedly cut and removed, the soil is depleted of nutrients. Trees then have little resilience for coping with stresses such as drought, diseases, and pests (Figure 9-9). In other words, using industrial forestry to increase the maximum sustainable yields of trees (at least in the short run of a hundred years or so) leads to a decline in forest quality by reducing overall biodiversity and disrupting the ecological integrity of forests.

Ecologists point out that in nature's economy nothing goes to waste. What timber company officials call "wasted trees" are the ecological raw materials for regenerating and maintaining a forest as decomposition of dead and rotting trees returns nutrients to the soil slowly enough so that growing trees can take them up again to renew life (Figure 4-11). While they decay, dead trees and moss-covered logs also help sustain biodiversity by providing habitats and food for a variety of species.

Recently, many government and industry foresters have been advocating *New Forestry*, a variation of clear-cutting on public lands based on the principle of *ecosystem management* with the basic goal of protecting ecosystem integrity. A few dead trees (snags) are left, and sometimes a few live seed trees are left as well. Instead of piling up and burning all the unmarketable logs and debris, some of this so-called *slash* is left scattered around to create wildlife habitats and encourage biodiversity. Sometimes the sizes and shapes of the clear-cut patches are varied. Many foresters see this approach as an important step in the right direction, provided it is widely implemented and not used primarily as a smoke screen for continuing unsustainable, even-aged industrial forestry.

However, other foresters contend that this practice is not sustainable forestry. Instead of relying primarily on selective cutting, it replaces one-step clear-cutting with two-step shelterwood and seed-tree cutting, which still leads to a drastic loss of habitat, biodiversity, and ecological integrity compared to a natural forest or a selectively cut forest. Most of the woody material is still removed from the ecosystem instead of decomposing to supply nutrients for new growth. Most of the soil is exposed to the sun and storms and is compacted by heavy machinery. Herbicides are often used to kill plants that compete with the new crop of plantation trees, further reducing biodiversity and disrupting ecological integrity. According to John Dennington, head of the Sportsman's Association, "Ecosystem Management is about how to grow the most pine and fool the most people."

Solutions: What Is Sustainable Forestry?
Instead of conventional industrial forestry and New Forestry, conservation biologists and a growing number of foresters call for sustainable forest management that emphasizes the following:

- *Recycling more paper to reduce the harvest of pulpwood trees* (Section 13-5)

- *Using fibers from fast-growing plants such as kenaf to make tree-free paper, and reducing wood waste* (p. 451)

- *Growing more timber on long rotations,* generally about 100–200 years (point C, Figure 16-5), depending on the species and soil quality

- *Practicing selective cutting of individual trees or small groups of most tree species* (Figure 16-7a)

- *Minimizing fragmentation of remaining larger blocks of forest*

- *Using road building and logging methods that minimize soil erosion and compaction*

- *Practicing strip cutting* (Figure 16-7e), *instead of conventional clear-cutting, and banning clear-cutting (including seed-tree and shelterwood cutting) on land that slopes more than 15–20°*

- *Leaving most standing dead trees (snags) and fallen timber to maintain diverse wildlife habitats and to be recycled as nutrients* (Figure 4-11)

- *Including the ecological and recreational services provided by trees and a forest in evaluating their economic value.*

16-4 FOREST RESOURCES AND MANAGEMENT IN THE UNITED STATES AND CANADA

What Is the Status of Forests in the United States? Forests cover about one-third of the lower 48 United States and provide habitats for more than 80% of the country's wildlife species. Nearly two-thirds of this forestland is productive enough to grow commercially valuable trees.

Today American forests are generally bigger and often healthier than they were in 1900, when the country's population was below 100 million. This is an important accomplishment, but it can be misleading.

Q: How much will it cost U.S. taxpayers to clean up contaminated nuclear weapon production facilities?

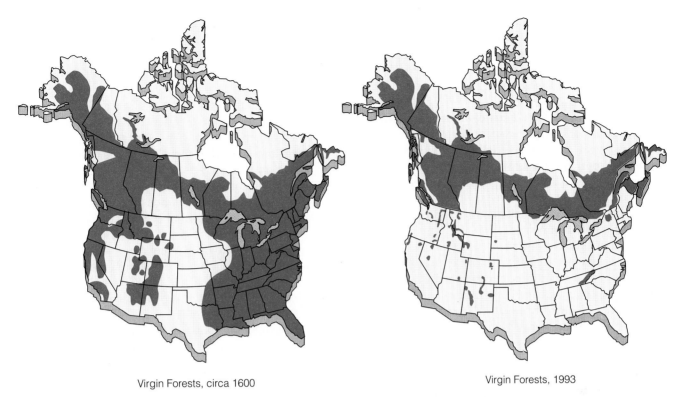

Virgin Forests, circa 1600

Virgin Forests, 1993

Figure 16-8 Vanishing old-growth forests in the United States and Canada. Since about 1600 most of the virgin forests that once covered much of the lower 48 states have been cleared away; most of the remaining old-growth forests in those states are on public lands (especially in the Pacific Northwest). About 60% of old-growth forests in western Canada have been cleared, and much of what remains is slated for cutting. (Data from the Wilderness Society, the U.S. Forest Service, and *Atlas Historique du Canada*, vol. 1)

Most of the virgin and old-growth forests in the lower 48 states have been cut (Figure 16-8), and much of what remains on private and public lands is threatened.

Most of the virgin forests that were cleared or partially cleared between 1800 and 1960 were allowed to grow back as fairly diverse second-growth (and in some cases third-growth) forests. However, since the mid-1960s a rapidly increasing area of the nation's remaining old-growth and fairly diverse second-growth forests has been clear-cut and replaced with tree plantations. Although this doesn't reduce the nation's overall tree cover, conservation biologists argue that it does reduce the amount of eventually sustainable forestland and forest biodiversity and ecological integrity.

Case Study: How Are U.S. National Forests Managed? About 22% of the commercial forest acreage in the United States is located within the 156 national forests (Figure 16-2) managed by the U.S. Forest Service. These forests serve as grazing lands for more than 3 million cattle and sheep each year. They support multibillion-dollar mining operations (p. 338), supply about 13% of the nation's timber, contain a network of roads equal in area to the entire U.S. interstate highway

system, and receive more recreational visits than any other federal public lands. The resulting networks of roads and clear-cuts severely fragment wildlife habitats, isolating many wildlife populations and making them vulnerable to genetic deterioration and predation.

The Forest Service is required by law to manage national forests according to the principles of *sustained yield* (which states that potentially renewable tree resources should not be harvested faster than they are replenished) and *multiple use* (which says that these forests should be managed simultaneously for a variety of uses such as sustainable timber harvesting, recreation, and wildlife conservation).

Environmentalists charge that especially since 1980 the Forest Service has allowed timber harvesting to become the dominant use in most national forests. To back up this charge, they point out that 70% of the Forest Service budget is devoted directly or indirectly to the sale of timber.

A major reason for this emphasis on timber cutting is that Congress allows the agency to keep most of the money it makes on timber sales, whereas any losses are passed on to taxpayers. Because logging increases its budget, the Forest Service has a powerful

A: At least $400–900 billion over the next 75 years

built-in incentive to encourage timber sales, another example of positive feedback in action. Local county officials also lobby Congress and encourage Forest Service officials to keep timber harvests high because counties get 25% of the gross receipts from national forests within their boundaries.

A drive through the 120 national forests open to logging, or a boat ride along streams in such forests, does not reveal the emphasis on timber cutting because the Forest Service leaves thin buffers of uncut trees, called *beauty strips*, along roads and streams. A flight over these forests, however, reveals that many of the trees have been clear-cut and most of the forests have been highly fragmented.

Environmentalists and the U.S. General Accounting Office have accused the Forest Service of poor financial management of public forests. By law, the Forest Service must sell timber for no less than the cost of reforesting the land from which it was harvested. However, the cost of access roads is not included in this price but is provided as a subsidy to logging com-

panies. Usually the companies also get the timber itself for less than they would pay a private landowner for an equivalent amount of timber.

Studies have shown that between 1978 and 1993, national forests lost at least $5.6 billion (some sources say $7 billion) from below-cost timber sales. Each year typically fewer than 25 of the 120 national forests open to logging make money. In effect, publicly owned national forests provide about 13% of the nation's timber at a *net loss* to taxpayers of $200–500 million a year.

Timber company officials argue that being able to get timber from federal lands fairly cheaply benefits taxpayers by keeping lumber and paper prices down. Environmentalists respond that below-cost timber prices discourage investments to produce additional timber on private lands and put more pressure for timber cutting on national and state forests.

Instead of providing economic stability for local industries and communities, logging in national forests tends to have the opposite effect. Communities that rely on proceeds from national forest timber sales experi-

How Can U.S. Federal Forest Management Be Improved?

SOLUTIONS

Foresters and conservation biologists have proposed a number of management reforms that would allow national forests in the United States to be used sustainably without seriously impairing their important ecological functions. These suggestions include the following:

- *Make sustaining biodiversity and ecological integrity the first priority of national forest management policy.*

- *Use full-cost accounting to include the ecological services provided by old-growth forests in the market price of timber harvested from national forests.*

- *Prohibit timber harvesting on at least half of remaining old-growth forests on public lands.*

- *Ban all timber cutting in national forests and fund the Forest Service completely from recreational user fees.* The Forest Service estimates that recreational user fees (as low as $3 per day) would generate three times what it earns from timber sales.

- *Sharply reduce all new road building in national forests.*

- *Allow individuals or groups to buy conservation easements that prevent timber harvesting on designated areas of public old-growth forests.* In such *conservation-for-tax-relief swaps,* purchasers would be allowed tax breaks for the funds they put up.

- *Require that timber from national forests be sold at a price that includes the costs of road building, site preparation, and site regeneration, and that all timber sales in national forests yield a profit for taxpayers.*

- *Do not use money from timber sales in national forests to supplement the Forest Service budget* because it encourages overexploitation of timber resources.

- *Eliminate the provision that returns 25% of gross receipts from national forests to counties containing the forests, or base such returns on recreational user-fee receipts only.*

- *Strictly enforce and close loopholes in the current federal ban on exporting raw logs harvested from public lands.*

- *Tax exports of raw logs but not finished wood products* (which provide and create domestic milling and other jobs).

- *Establish federally funded reforestation and ecological restoration projects on degraded public lands.* In addition to the ecological benefits, this will provide jobs for displaced forest harvesting and wood products workers.

- *Provide increased aid and job retraining for displaced workers in the forest harvesting and products industry.*

- *Require use of sustainable-forestry methods* (p. 448).

Understandably, timber company officials, most county officials, and many Forest Service officials oppose such policies and have prevented such reforms.

Critical Thinking

Explain why you agree or disagree with each of these proposals for reforming federal forest management in the United States.

Q: Where is the world's largest number of sites contaminated with radioactive materials?

ence boom-and-bust cycles as the timber is depleted and timber companies move to other areas. Analysts have proposed a number of reforms in Forest Service policy that would lead to more sustainable use and management of the national forests (Solutions, left).

Solutions: How Can We Reduce Wood Waste and Use Other Fibers to Make Paper? According to the Worldwatch Institute and forestry analysts, up to 60% of the wood now being consumed in the United States is unnecessarily wasted. Because only 13% of the total U.S. timber production comes from the national forests, reducing the waste of wood and paper products by only 13% could eliminate the need to remove any timber from the national forests. Then these public lands could be used primarily for recreation and protection of biodiversity and ecological integrity.

Ways to reduce such waste include **(1)** reducing construction waste, **(2)** using laminated boards instead of solid wood, **(3)** using less packaging, **(4)** reducing junk mail, **(5)** recycling more paper with the highest possible content of postconsumer waste, and **(6)** using paper made from fibers obtained from rapidly growing plants other than trees.

Because of deforestation and the rising costs of wood pulp, an increasing number of countries use other sources of fiber for making paper. In China, for example, 60% of the paper is made with nonwood pulp such as rice straw and other agricultural wastes left after harvest.

In Canada, Al Wong is running a pilot mill to convert wheat straw or rice straw left over after harvesting into paper. According to Wong, North America produces enough rice and wheat straw and other agricultural wastes to more than meet all of its paper needs without having to ever cut another tree. The problem is convincing the $130-billion-a-year paper industry to invest in the technology needed to produce such ecopaper or Ag-paper.

Currently, most of the small amount of treeless paper produced in the United States is made from the fibers of a woody annual plant called *kenaf* (pronounced kuh-NAHF; Figure 16-9). The tall, bamboolike stalks of this annual woody plant grow to heights of 4 meters (12 feet) in only 4–5 months, compared to 7–60 years for pine trees. Large-scale commercial production of Kenaf has occurred in India and Asia since 1900.

Kenaf thrives in tropical and temperate climates and herbicides are rarely needed because it grows faster than most weeds; few insecticides are needed because its outer fibrous covering is nearly insect-proof. Because kenaf is a nitrogen fixer, cultivating it does not deplete soil nitrogen. It also takes fewer chemicals and less energy to break kenaf down into fibers used for making paper, so less toxic waste-

Figure 16-9 Kenaf growing in a field in south Texas. This fast-growing plant can be harvested annually and its fibers used to make paper. Its widespread cultivation would decrease the need to grow and harvest trees. (U.S. Department of Agriculture)

water is produced and water treatment costs are lower than when pulpwood is used. The U.S. Department of Agriculture considers kenaf the best renewable fiber to replace trees in paper making and predicts that the market for kenaf should expand rapidly in coming years.

There are some drawbacks, but they can be overcome. Because the moisture content of kenaf is much higher than that of wood, the plants must be dried to prevent spoilage. Kenaf paper currently costs three to five times more than virgin or recycled paper stock, which is why this book is not printed on tree-free kenaf paper. However, as demand increases and producers cut production costs, the price of kenaf should come down. The basic problem is getting investors to gamble money on converting or constructing mills to produce rice, wheat, or kenaf paper without an established market.

Case Study: Should Remaining Old-Growth Forests on U.S. Public Lands Be Cut or Preserved? To timber companies, the giant living trees and rotting dead trees in the Pacific Northwest's old-growth forests are valuable resources that could

Figure 16-10 The threatened northern spotted owl, which lives almost exclusively in old-growth forests of the Pacific Northwest. Environmentalists have used the owl's threatened status both to save it and to halt or decrease logging in the national forests in which it lives, thereby advancing their wider goal of preserving biodiversity. (Pat & Tom Leeson/Photo Researchers, Inc.)

provide jobs and should be harvested for profit, not left alone to please environmentalists.

These officials also point out that the timber industry annually pumps millions of dollars into the region's economy and provides jobs for about 100,000 loggers and millworkers. Timber officials claim that protecting large areas of remaining old-growth forests on public lands will cost as many as 53,000 jobs and hurt the economies of logging and milling towns in such areas.

To conservation biologists, the remaining ancient forests on the nation's public lands are a treasure whose ecological, scientific, aesthetic, and recreational values far exceed the economic value of cutting them down for short-term gain. They argue that the fate of these forests is a national issue because these forests are owned by all U.S. citizens. They also view this as a global issue because these forests are important reservoirs of the earth's biodiversity and ecological integrity.

Conservation biologists and economists point out that the major causes of job losses in the timber industry in this region are automation, export of logs overseas (depriving U.S. millworkers of jobs while providing jobs for millworkers in other countries), and timber imports from Canada. Loggers, millworkers,

and store owners living in logging communities are caught in the middle of a highly controversial situation.

The threatened northern spotted owl (Figure 16-10) lives almost exclusively in 200-year-old Douglas fir forests in western Oregon, Washington, and northern California (mostly in 17 national forests and 5 Bureau of Land Management parcels). The species is vulnerable to extinction because of its low reproductive rate and the low survival rate of juveniles through their first 5 years. Only 2,000–3,600 pairs remain. However, logging company officials contend that the owls do not require old-growth forest and can instead adapt to younger second-growth forests.

In July 1990 the U.S. Fish and Wildlife Service added the northern spotted owl to the federal list of threatened species. This requires that its habitat be protected from logging or other practices that would decrease its chances of survival. The timber industry hopes to persuade Congress to revise the U.S. Endangered Species Act to allow for economic considerations that will reverse this protection.

The owl and other threatened species are symbols of the broader clash between timber companies who want to clear-cut most remaining old-growth stands in the national forests and conservation biologists who want to protect their priceless biodiversity and ecological integrity, or at least allow only sustainable harvesting in some areas. The question is, "Should the biodiversity and ecological integrity of the last 5–8% of remaining ancient forests on public lands in the Northwest be preserved to protect their biodiversity and ecological integrity or cut to save a final decade or so of logging jobs?" What do you think?

Case Study: Destruction of Old-Growth Forests in Western Canada With nearly 50% of its land forested (much of it in second-growth forest), Canada has 10% of the world's forests. One of every 10 jobs in Canada is directly related to forestry.

Even though provincial governments own and manage 80% of Canada's forests, the country has lost an estimated 60% of its old-growth forests (most of them publicly owned) to logging (Figure 16-8); less than 20% of what remains is in protected areas. Ninety percent of logging in Canada involves clear-cutting, a practice expected to double over the next few years.

Critics charge that the Canadian government is giving massive subsidies to logging companies (many of them Japanese owned) without providing many jobs for Canadians. Most of the profit is not even made in Canada. For example, Canada's western province of Alberta collects only $0.90 for the 16 aspen trees that will make $590 worth of pulp shipped to Japan, where a Japanese mill and workers will turn the pulp into about $1,300 worth of paper.

Q: Does using nuclear power add carbon dioxide to the atmosphere?

Environmentalists and a growing number of Canadians are outraged that the provincial government spent $50 million to buy a 4% share in MacMillan Bloedel, British Columbia's largest international forest products company. Moreover, it did this only a few weeks before it awarded the company a huge contract to clear-cut the magnificent ancient forest of Clayoquot (pronounced "clack-what") Sound on the spectacular west coast of Vancouver Island. Under intense pressure from environmental groups and subject to the glare of unfavorable international publicity, the British Columbia government announced in 1995 that it would end clear-cutting in Clayoquot Sound.

To environmentalists, the situation is grim, but there is some hope. In the 1990s citizens' groups throughout Canada formed a powerful grassroots network of environmentalists, labor union members, and citizens called Canada's Future Forest Alliance. Their goal is to direct the glare of international publicity onto Canada's rapid deforestation and to pressure the provincial and federal governments to restructure forest policies to emphasize truly sustainable forestry and social justice for native peoples.

They also encourage people to boycott companies and governments buying timber from Canada. Several companies in Europe and in the United States (including the *New York Times*) have announced that they will no longer buy wood products and paper from MacMillan Bloedel.

As Canadian forestry activist Colleen McCrory (recently awarded the Goldman Environmental Prize for her efforts to protect Canadian forests and founder of Canada's Future Forest Alliance) puts it:

> *We've got the Brazil of the North happening right here. . . . In the Amazon region, one acre of forest is lost every nine seconds to logging and forest clearing. In Canada, one acre every twelve seconds is lost to logging or burning. . . . In Brazil, the desperation of poverty and rapid population growth are the central driving factors in deforestation, whereas in Canada, the central cause of deforestation is the greed of multinational corporations, and their ability to dominate debate about forest policy because of their wealth and political influence.*

16-5 TROPICAL DEFORESTATION AND THE FUELWOOD CRISIS

How Fast Are Tropical Forests Being Cleared and Degraded? Tropical forests cover about 6% of the earth's land area (roughly the area of the lower 48 states of the United States) and grow in equatorial Latin America, Africa, and Asia (Figure 5-3).

Tropical forests come in several varieties (Figure 5-4): rain forests (which receive rainfall almost daily), tropical deciduous forests (with one or two dry seasons each year), dry and very dry deciduous forests, and forests on hills and mountains.

Climatic and biological data suggest that mature tropical forests once covered at least twice as much area as they do today, with most of the destruction occurring since 1950. Satellite scans and ground-level surveys used to estimate forest destruction indicate that large areas of tropical forests are being cut (Figure 16-3). It's estimated that an equivalent area of these forests is seriously degraded and fragmented without being destroyed outright each year.

There is considerable debate over the current rates of tropical deforestation and degradation because of difficulties in interpreting satellite images, different ways of defining deforestation and forest degradation, and political and economic factors that cause countries to hide or exaggerate deforestation. The lowest estimated rate of loss and degradation of remaining tropical forests is 62,000 square kilometers (24,000 square miles) per year. This is equivalent to a loss and degradation of *14 city blocks of tropical forest per minute*. The highest estimated rate of destruction and degradation is 308,000 square kilometers (118,000 square miles) per year—equivalent to a loss and degradation of *68 city blocks of tropical forest per minute*.

Whether the best estimate of tropical deforestation and degradation is 14 or 68 city blocks per minute (or somewhere in between), it is clear that these reservoirs of biodiversity and ecological integrity are being lost and degraded at a high rate. Scientists estimate that this annual rate of destruction and degradation could well double within another decade. Indeed, recent data indicate that the rate of clearing and burning tropical forests in the Amazon Basin increased by about 34% between 1991 and 1994.

About 40% of current tropical deforestation is taking place in South America (especially in the vast Amazon Basin). However, the *rates* of such deforestation in Southeast Asia and in Central America are about 2.7 times higher than those in South America. Haiti has lost 98% of its original forest cover, the Philippines 97%, and Madagascar 84%.

Why Should We Care About Tropical Forests? Biologists consider the plight of tropical forests to be one of the world's most serious environmental problems, primarily because these forests are home to 50–90% of the earth's terrestrial species (Figure 1-12), most of them still unknown and unnamed. Biologist Edward O. Wilson estimates that by 2022 at least 20% of tropical forest species could be gone, and as many

Figure 16-11 Rubber tapping is a potentially sustainable use of tropical forests. This 1987 photo shows the activist Chico Mendez (murdered in 1988 for his efforts to protect these forests) making a cut in a rubber tree. The cut allows milky liquid latex to trickle into a collecting cup. The latex is processed to make rubber, and the scars heal without killing the tree. During the seasons when latex is not flowing, rubber tappers make a living by gathering and selling nuts. (Randall Hyman)

Figure 16-12 Rosy periwinkle in the threatened tropical forests of Madagascar. Two compounds extracted from this plant have been used to cure most victims of two deadly cancers: lympho-cytic leukemia (which was once almost always fatal for children) and Hodgkin's disease (mostly affecting young adults). World-wide sales of these two drugs are about $180 million per year, none of which is returned to Madagascar. Only a tiny fraction of the world's tropical plants has been studied for such potential uses, and many will become extinct before we can study them. (Heather Angel)

as 50% by 2042, if current rates of tropical deforestation and degradation continue. If these estimates are correct, no extinction of this size has occurred for 65 million years.

Tropical forests touch the daily lives of everyone on the earth through the products and ecological services they provide. These forests supply half of the world's annual harvest of hardwood, hundreds of food products (including coffee, tea, cocoa, spices, nuts, chocolate, and tropical fruits), and materials such as natural latex rubber (Figure 16-11), resins, dyes, and essential oils that can be harvested sustainably. A 1988 study showed that sustainable harvesting of such non-wood products as nuts, fruits, herbs, spices, oils, medicines, and latex rubber in Amazon rain forests over 50 years would generate twice as much revenue per hectare as timber production and three times as much as cutting them and converting them to grassland for cattle ranching.

The active ingredients for 25% of the world's prescription drugs are derived from plants, most of which grow in tropical rain forests. Commercial sales of drugs with active ingredients derived from tropical forests total at least $100 billion per year worldwide and $15.5 billion per year in the United States. Some analysts put the economic value of plant-based drugs from all types of forests at $200 billion to $1.2 trillion annually.

Of the 3,000 plants identified by the National Cancer Institute (NCI) as sources of cancer-fighting chemicals, 70% come from tropical forests (Figure 16-12). Some tropical tree species can be used for many purposes (Solutions, right).

Most of the original strains of rice, wheat, and corn that supply more than half of the world's food were developed from wild tropical plants. Botanists believe that tens of thousands of plant strains with potential food value await discovery in tropical forests.

Despite this immense potential, fewer than 1% of the estimated 125,000 flowering plant species in tropical forests (and less than 3% of all the world's 240,000 such species) have been examined closely for possible use as human resources. Biologist Edward O. Wilson warns that destroying these forests and the species they contain for short-term economic gain is like burning down an ancient library before reading the books.

Case Study: Cultural Extinction in Tropical Forests An estimated 250 million people—1 person out of every 20 on the earth—belong to indigenous cultures found in about 70 countries. Many of these peoples, such as the Kuana of Panama, the Kenyah in Indonesia, the Yánesha in Peru, and the Yanomami in Brazil, have been living in and using forests sustainably for centuries. They get most of their food from hunting and gathering, trapping, and sustainable slash-and-burn and shifting cultivation (Figure 2-4).

Q: Nuclear weapons existing today could kill everyone in the world how many times over?

The Incredible Neem Tree

Wouldn't it be nice if there were a single plant that could quickly reforest bare land, provide fuelwood and lumber in dry areas, provide alternatives to toxic pesticides, be used to treat numerous diseases, and help control population growth? There is: the *neem tree*, a broadleaf evergreen member of the mahogany family.

This remarkable tropical species, native to India and Burma, is ideal for reforestation because it can grow 9 meters (30 feet) tall in only 5–7 years! It grows well in poor soil in semiarid lands such as those in Africa, providing an abundance of fuelwood, lumber, and lamp oil in the process.

It also contains various natural pesticides. Chemicals from its leaves and seeds can repel or kill over 200 insect species, including termites, gypsy moths, locusts, boll weevils, and cockroaches.

Extracts from neem seeds and leaves can be used to fight bacterial, viral, and fungal infections. Indeed, the tree's chemicals have allegedly relieved so many different afflictions that the tree has been called a "village pharmacy." Its twigs are used as an antiseptic toothbrush, and oil from its seeds is used to make toothpaste and soap.

But that's not all. Neem-seed oil evidently acts as a strong spermicide that might be used in producing a much-needed male birth control pill. According to a study by the U.S. National Academy of Sciences, the neem tree "may eventually benefit every person on the planet."

Since 1985 many U.S. patents have been awarded to U.S. and Japanese firms for neem-based products. This has raised a legal controversy. Scientists in India have been researching the neem tree since the 1920s, and farmers in India have freely shared processes they developed over decades for extracting neem-seed oil.

Indian officials and farmers claim that no company should be able to receive exclusive patents on such widespread and general knowledge based on the genetic resources of a particular country.

And if such patents *are* granted, they contend that the courts should award the country where the genetic resources were discovered and first developed a certain percentage of all profits. Vandana Shiva (Guest Essay, p. 506) calls the granting of patents for neem tree products economic, intellectual, and biological "piracy of the third world."

Patent holders say that they have taken neem material and improved it through innovative extraction and formulation processes. They believe that their investment of capital and the resulting unique products entitle them to a patent and to any profits made from such patents.

Critical Thinking

Do you believe that exclusive patents should be awarded to companies that have developed neem-based products for sale? Explain. Should Indian farmers and scientists, who originally developed and shared their knowledge of the neem tree's many benefits, receive a percentage of the profits from such products? Explain.

Many of the earth's remaining tribal peoples, representing 5,000 cultures, are vanishing as their lands are taken over for economic development. They are seeing their homelands bulldozed, cut, burned, mined, flooded, and contaminated. They are forced to adopt new ways while reeling from shock and hunger, and they are often killed by introduced diseases to which they have no immunity. Those who resist development are often killed by ranchers, miners, and settlers.

Many analysts believe that eliminating indigenous peoples, besides being wrong or even genocidal, causes a tragic loss of earth wisdom and cultural diversity. People in these cultures know more about living sustainably in tropical forests (and in other biomes) and about what plants can be used as foods and medicines than anyone.

Others believe that economic development of these mostly unused and almost empty areas should not be held back by small numbers of primitive tribal peoples who don't hold legal titles to the land they live on.

Environmentalists and others, led by organizations such as Survival International and Cultural Survival, disagree. Instead of being "unused" and "empty," they see such areas as important centers of cultural diversity, biodiversity, and ecological integrity that are be used sustainably by indigenous people as a vital part of the planet's earth capital (Figure 1-2).

They call on governments to protect the rights and the earth wisdom of remaining indigenous peoples by (1) establishing a UN Declaration on the Rights of Indigenous Peoples enforceable by international law, (2) mapping their homelands and giving them full ownership of their land and all mineral rights, (3) protecting their lands from intrusion and illegal resource extraction, (4) giving them legal control over and just compensation for marketable drugs and other products derived from their lands, and (5) creating an international organization to fight for their legal rights. What do you think?

What Causes Tropical Deforestation? Tropical deforestation results from a number of interconnected causes, all of which are related to population growth, poverty, and certain government policies that encourage deforestation (Figure 16-13). Population growth and poverty combine to drive subsistence farmers and the landless poor to tropical forests, where they try to grow enough food to survive.

Government subsidies can accelerate deforestation by making timber or other resources cheap relative to their full ecological value and by encouraging the poor to colonize tropical forests by giving them title to land they clear (as is done in Brazil, Indonesia, and Mexico). International lending agencies encourage developing countries to borrow huge sums of money from developed countries to finance projects such as roads, mines, logging operations, oil drilling, and dams in tropical forests. To stimulate economic development (and in some cases to pay the interest on loans from developed countries), these countries often sell off some of their timber and other natural resources, mostly to developed countries.

The process of degrading a tropical forest begins with a road (Figure 16-6), usually cut by logging companies. Once the forest becomes accessible, it can be cleared and increasingly degraded by a number of factors. One is unsustainable forms of small-scale farming. Hordes of poor people follow logging roads into the forest to build homes and plant crops on small cleared plots. Instead of practicing various methods of traditional and potentially sustainable shifting cultivation, these inexperienced *shifted cultivators* often practice unsustainable farming that depletes soils and destroys large tracts of forests.

According to environmental scientist Norman Myers (Guest Essay, pp. 295–296), these forest newcomers are responsible for well over half of all tropical forest destruction. Driven by rapidly growing population and poverty, their numbers and impact are rising rapidly.

Cattle ranching also degrades tropical forests. Cattle ranches, often aided by government subsidies, are often established on cropland that has been exhausted by small-scale farmers. Some farmers simply abandon their plots to ranchers; others seed their plots with grass and then sell them to ranchers. When torrential rains and overgrazing turn the usually thin and nutrient-poor tropical forest soils (Figure 12-16) into eroded wastelands, ranchers move to another area and repeat the destructive process known as *shifting ranching*.

In Brazil's Amazon Basin, the grazing of cattle allows ranchers (supported by government subsidies averaging $5.6 million per ranch) to claim large tracts of land and the mineral rights below it. Some then sell both land and mineral rights for quick profits, a process called *ghost ranching*. At least half of the 600 large ranches in the Brazilian Amazon Basin have never sent a single head of cattle to market, and about 30% are now abandoned. Although this program has been a financial success for wealthy ranchers, it's been an economic and ecological disaster for the national economy, costing the government at least $2.5 billion in lost revenue.

Clearing large areas of tropical forest for raising cash crops severely degrades these reservoirs of biodiversity. Immense plantations grow crops such as sugarcane, bananas, pineapples, peppers, strawberries, cotton, tea, and coffee, mostly for export to developed countries.

Mining and oil drilling also degrade tropical forests. Widespread gold mining is particularly damaging because it erodes soils and pollutes streams and soils with toxic mercury (Case Study, p. 352). Dams built on rivers also flood large areas of tropical forests.

Commercial logging also degrades tropical forests. Since 1950 the consumption of tropical timber has risen 15-fold. Japan alone accounts for 53% of the world's tropical timber imports, followed by Europe (32%) and the United States (15%). Currently, almost three-fourths of exported tropical timber comes from Southeast Asia. As tropical timber in Asia is depleted in the 1990s, cutting will shift to Latin America and Africa.

Although timber exports to developed countries contribute significantly to tropical forest depletion and degradation, over 80% of the trees cut in developing countries are used at home, mostly for firewood and construction. Loggers typically take only the best large and medium trees, but they also topple up to 17 other trees for each one they remove. The falling trees damage many others, with up to 70% of them eventually dying from their injuries.

Logging is directly responsible for only a small portion of tropical deforestation and degradation when compared to small-scale agriculture. However, the construction of logging and other resource access roads initiates and helps accelerate the destruction by opening up these forests to small-scale farmers, ranchers, and miners.

Solutions: How Can We Reduce Tropical Deforestation and Degradation? A number of analysts have suggested ways to protect tropical forests, to use them more sustainably, and to restore degraded areas of such forests. An important first step is to make a detailed survey to determine how much of the world is covered with tropical (and other) forests, how much has been deforested or degraded, and where. This could be done by a combination of remote-sensing satellites and ground-level evaluations, at a cost roughly equivalent to what the world spends for military purposes *every three minutes*. Conservation

Q: Can switching to increased use of nuclear power in the United States save much oil?

Primary Causes:
Rapid population growth
Poverty
Exploitive government policies
Exports to developed countries
Failure to include ecological services
in evaluating forest resources

Secondary Causes:
Roads
Logging
Unsustainable peasant farming
Cash crops
Cattle ranching
Tree plantations
Flooding from dams
Mining
Oil drilling

Bromeliad

Toucan

Scarlet macaw

Golden lion marmoset

Orchid

Blue morpho butterfly

Figure 16-13 Major interconnected causes of the destruction and degradation of tropical forests, all of which are ultimately related to population growth, poverty, and government policies that encourage deforestation.

A: No, in 1993, burning oil supplied only 3% of the country's electricity

Sustainable Agriculture and Forestry in Tropical Forests

A combination of the knowledge of indigenous peoples, ecological research, and modern technology can show people how to grow crops and harvest timber in tropical forests on a more sustainable basis.

One approach would be to show newcomers how to grow crops using forms of more sustainable agroforestry developed by indigenous peoples throughout the world. The Lacandon Maya Indians of Chiapas, Mexico, for example, have developed a multilayered system of agroforestry that allows them to cultivate up to 75 crop species on 1-hectare (2.5-acre) plots for up to 7 consecutive years. After that another plot must be planted to allow regeneration of the soil in the original plot.

Another approach being used by Yánesha Indians in the lush rain forests of Peru's Palcazú Valley, is strip-shelterbelt harvesting (Figure 16-7e) of tropical trees for lumber. They use chain saws to cut widely spaced thin strips—barely visible from an airplane—through the forest. Oxen haul the cut logs to a processing plant, where other tribe members convert them to charcoal fuel, boards, and fence posts. Tribe members market their ecologically harvested wood products in the global marketplace and act as consultants to help other forest dwellers set up similar systems.

Critical Thinking

What applications (if any) might these practices have for growing food and harvesting timber in the United States and other developed counties? How would you encourage use of such approaches for more sustainable use of forests?

Otherwise, they fear that most money and efforts will go into protecting endangered areas of tropical forests, while equally important endangered and unique ecosystems elsewhere are neglected.

Environmentalists urge governments to reduce the flow of the landless poor to tropical forests by slowing population growth and discouraging the poor from migrating to undisturbed tropical forests. They also call for a global effort to sharply reduce the poverty that leads the poor to use forests (and other resources) unsustainably for short-term survival. These analysts also call for establishing programs to help new settlers in tropical forests learn how to practice small-scale sustainable agriculture and forestry (Solutions, left).

Another suggestion is to use economic policies to help protect and sustain tropical forests. One approach is to phase in full-cost pricing over the next 2–3 decades (Section 7-3). Governments can also phase out subsidies that encourage unsustainable deforestation and phase in tariffs, use taxes, user fees, and subsidies that favor more sustainable forestry and protection of biodiversity. Another approach is to pressure banks and international lending agencies (controlled by developed countries) not to lend money for environmentally destructive projects, especially road building and dams, involving old-growth tropical forests.

Debt-for-nature swaps and conservation easements can be used to encourage countries to protect tropical forests (Figure 16-14). In a debt-for-nature swap, participating countries act as custodians for protected forest reserves in return for foreign aid or debt relief (Case Study, right); in conservation easements, a private organization, country, or group of countries compensates other countries for protecting selected forest areas. Currently less than 5% of the world's tropical forests are part of parks and preserves, and many of these are protected in name only.

Tropical timber-cutting regulations and practices can be reformed. New logging contracts could charge more for timber-cutting concessions and require companies to post adequate bonds for reforestation and restoration. Additionally, gentler methods could be used for harvesting trees. Cutting canopy vines (lianas) before felling a tree can reduce damage to neighboring trees by 20–40%, and using the least-obstructed paths to remove the logs can halve the damage to other trees.

Pressure for clearing old-growth tropical forests can be reduced by concentrating peasant farming, tree and crop plantations, and ranching activities on already-cleared tropical forest areas that are in various stages of secondary ecological succession. Finally, global efforts can be mounted to reforest and rehabilitate degraded tropical forests and watersheds.

biologists urge us to move rapidly to protect areas of tropical forests (and other biomes) that are rich in unique species and in imminent danger—so called *hot spots* or *critical ecosystems.*

Unfortunately, some of these hot spots have very high concentrations of people living near them—one of the reasons they need immediate protection. Many governments oppose such protection because of political opposition by local people who are in a desperate struggle to survive.

Some biologists believe that hot spots should be identified for all of the earth's major biomes and areas.

Q: What if nuclear power's harmful effects were included in its price and all subsidies were removed?

A Debt-for-Nature Swap in Bolivia

In 1984, biologist Thomas Lovejoy suggested that debtor nations should be rewarded for protecting their natural resources. In such **debt-for-nature swaps**, Lovejoy proposed that a certain amount of foreign debt be canceled in exchange for spending a certain sum on better natural resource management.

Typically, a conservation organization buys a specified amount of a country's debt from a bank at a discounted rate and negotiates the swap. A government or private agency must agree to enact and supervise the conservation program.

In 1987, Conservation International, a private U.S. banking consortium, purchased $650,000 of Bolivia's $5.7-billion national debt from a Citibank affiliate in Bolivia for $100,000. In exchange for not having to pay back this part of its debt, the Bolivian government agreed to expand and protect from harmful forms of development 1.5 million hectares (3.7 million acres) of tropical forest around its existing Beni Biosphere Reserve in the Amazon Basin, which contains some of the world's largest reserves of mahogany and cedar. The government was to establish maximum legal protection for the reserve and create a $250,000 fund, with the interest to be used to manage the reserve.

The plan was intended to be a model of how conservation of forest and wildlife resources could be compatible with sustainable economic development. Central to the plan was a virgin tropical forest to be set aside as a biological reserve, surrounded by a protective buffer of savanna used for sustainable grazing of livestock (Figure 16-14). Controlled commercial logging, as well as hunting and fishing by local inhabitants, would be permitted in some parts of the forest but not in the area above the tract, which was to be set aside to protect the area's watershed and to prevent erosion.

In 1997, 10 years after the agreement was signed, the Bolivian government still had not provided legal protection for the reserve. It also waited until April 1989 before contributing only $100,000 to the reserve management fund.

Meanwhile, timber companies (with government approval) have cut thousands of mahogany trees from the area; most of this lumber has been exported to the United States. The area's 5,000 native inhabitants were not consulted about the swap plan, even though they are involved in a land-ownership dispute with logging companies.

One lesson from this first debt-for-nature swap is that *legislative and budget requirements must be met before the swap is executed.* Another is that such *swaps must be carefully monitored by environmental organizations* to ensure that proposals for sustainable development are not disguises for unsustainable development.

Critical Thinking

Do you agree or disagree with the charge that many of the debt-for-nature swaps support international (mostly corporate) control over debtor countries' development while giving banks in developed countries a way to make good on part of essentially bad loans? Explain. Overall, do you believe that debt-for-nature swaps are a good idea? Explain.

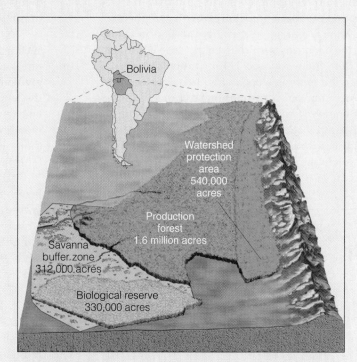

Figure 16-14 Blending economic development and conservation in a 1.5-million-hectare (3.7-million-acre) tract in Bolivia. A U.S. conservation organization arranged a debt-for-nature swap to help protect this land from destructive development.

How Serious Is the Fuelwood Crisis in Developing Countries? In 1997 about 2 billion people in 63 developing countries either could not get enough fuelwood to meet their basic needs or were forced to meet their needs by using wood faster than it was being replenished. The UN Food and Agriculture Organization projects that by the end of this century this already serious shortage of fuelwood will affect 2.7 billion people in 77 developing countries (Figure 16-15). As burning wood (or charcoal derived from wood) to boil water becomes an unaffordable luxury, waterborne infectious diseases and death will spread.

Besides deforestation and accelerated soil erosion, fuelwood scarcity has other harmful effects. It places an additional burden on the rural poor, especially women and children, who often must walk long distances searching for firewood. Buying fuelwood or charcoal can take 40% of a poor family's meager income.

An estimated 800 million poor people who can't get enough fuelwood burn dried animal dung and crop residues for cooking and heating. These natural fertilizers thus never reach the soil, cropland productivity is reduced, the land is degraded still further, and hunger and malnutrition increase—another example of harmful positive feedback in action.

Fuelwood shortages, like most environmental problems in developing countries, are driven by a combination of rapid population growth and poverty. Urbanization is another factor. City dwellers in many developing countries burn charcoal because it is much lighter than fuelwood and is thus much cheaper to transport. But burning wood in traditional earthen pits to produce charcoal consumes more than half of the wood's energy. Thus, each city dweller who burns charcoal uses twice as much wood for a given amount of energy as a rural dweller who burns firewood. This helps explain the expanding rings of deforested land that surround many cities that use charcoal as the major source of fuel.

Solutions: What Can Be Done About the Fuelwood Crisis? Developing countries can reduce the severity of the fuelwood crisis by planting more fast-growing fuelwood trees or shrubs, burning wood more efficiently, and switching to other fuels. However, fast-growing tree and shrub species used to establish fuelwood plantations must be selected carefully to prevent harm to local ecosystems.

Eucalyptus trees, for example, are being used to reforest areas threatened by desertification and to establish fuelwood plantations in some parts of the world. Because these species grow fast even on poor soils, this might seem like a good idea, but environmentalists and local villagers see it as an ecological disaster. In their native Australia these trees thrive in areas with good rainfall; when planted in arid areas, however, the trees suck up so much of the scarce water that most other plants can't grow. Then farmers don't have fodder to feed their livestock, and groundwater is not replenished. The eucalyptus trees also deplete the soil of nutrients and produce toxic compounds that accumulate in the soil because of low rainfall. In Karnata, India, villagers became so enraged over a government-sponsored project to plant these trees that they uprooted the saplings.

Experience shows that planting projects are most successful when local people participate in their planning and implementation. Programs work best that give village farmers incentives, such as ownership of the land or ownership of any trees they grow on village land. Emphasis should be placed on community woodlots, which are easy to tend and harvest, rather than on creating large fuelwood plantations located far from where the wood is needed.

Another promising method is to encourage villagers to use the sun-dried roots of various gourds and squashes as cooking fuel. These rootfuel plants, which regenerate themselves each year, produce large quantities of burnable biomass per unit area on dry deforested lands. They also help reduce soil erosion and produce an edible seed with a high protein content.

The traditional three-stone fire is a very inefficient way to burn wood, typically wasting about 94% of the wood's energy content. New, cheap, more efficient, less polluting stoves can provide both heat and light while reducing indoor air pollution, a major health threat. However, such stoves won't be widely used if they don't take into account local needs, resources, and cultural practices. The stoves must be easy and cheap to build, and the materials used to make them must be easily accessible locally (Individuals Matter, right).

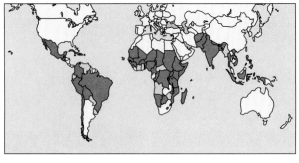

■ Acute scarcity and depletion in 1985

□ Deficits and scarcity by 2000

Figure 16-15 Scarcity of fuelwood, 1985 and 2000 (projected). (Data from UN Food and Agriculture Organization)

Q: Is nuclear fusion the answer to our energy problems?

Cheap and easily made solar ovens (Figure 18-21d) that capture sunlight can also be used to reduce wood use and air pollution in sunny and warm areas; they are often not accepted, though, by cultures that also use fires for light and heat at night and as centers for social interaction.

Despite encouraging success in some countries (such as China, Nepal, Senegal, and South Korea), most developing countries suffering from fuelwood shortages have inadequate forestry policies and budgets, and they lack trained foresters. Such countries are cutting trees for fuelwood and forest products 10–20 times faster than new trees are being planted.

16-6 MANAGING AND SUSTAINING RANGELANDS

What Is Rangeland, and Why Is It Important?
Almost half of the earth's ice-free land is **rangeland**: land that supplies forage or vegetation for grazing (grass-eating) and browsing (shrub-eating) wild and domesticated animals and is not intensively managed. Most rangelands are grasslands in arid and semiarid areas too dry for nonirrigated crops.

About 42% of the world's rangeland is used for grazing livestock; much of the rest is too dry, cold, or remote from population centers to be grazed by large numbers of livestock. About 34% of the total U.S. land area is rangeland, most of it short-grass prairie in the arid and semiarid western half of the country (Figure 5-3). About 2% of the cattle and 10% of the sheep in the United States graze on public rangelands.

Most rangeland grasses have deep and complex root systems (Figure 12-18) that help anchor the plants. Blades of rangeland grass grow from the base, not the tip. Thus, as long as only its upper half is eaten and its lower half remains, rangeland grass is a renewable resource that can be grazed again and again.

Rangeland has a number of important ecological functions. It provides forage for large numbers of wild herbivores and essential habitats for a variety of wild plant and animal species. Rangelands also act as crucial watersheds that help replenish surface and groundwater resources by absorbing and slowly releasing rainfall.

Rangeland is also a valuable resource for recreation. Since 1975 use of U.S. rangeland for activities such as hiking, camping, and hunting has risen dramatically. Some have suggested that some rangelands could be used to raise wild grazing animals for meat instead of conventional livestock (Solutions, p. 462).

INDIVIDUALS MATTER

New Stoves Help Save India's Forests and Improve Women's Health

Since 1984 Lalita Balakrishnan—supported by the All-India Women's Conference (formed in 1926 with its early mission to educate Indian women)—has put together an army of 300,000 women dedicated to spreading efficient and smokeless stoves throughout much of India. These women have helped design, build, and install 300,000 smokeless wood stoves, called *chulhas*, across the country.

These stoves cost about $5 and are made of cow-dung, mud, and hay slapped together by hand, with a metal pipe leading to the roof. The Indian government pays one-third of the stoves' cost, and all materials (except the pipe) are provided by the stove buyer.

The program now has 60 different models, with designs modified for local cultures and diets. Local women are hired to sell, install, and maintain the stoves.

About half of India's annual loss of 11,800 square kilometers (4,600 square miles) of forest is caused by use of fuelwood. During the past decade these cheap, fuel-efficient stoves have helped reduce deforestation and saved India more than 182,000 metric tons (200,000 tons) of fuelwood.

Replacement of smoky indoor fires with such stoves has also led to improved health for an estimated 1 million Indian women who previously suffered from chronic bronchitis, asthma, and almost certain premature death from inhalation of smoke from indoor open fires used for cooking with wood. This program has also raised the self-esteem of the women participating in this project.

In recognition of her efforts, the UN Environment Programme added the name of Lalita Balakrishnan to its 1996 Global 500 Roll of Honor.

In 1996 the Indian government developed a plan to distribute 2.7 million *chulhas*, mostly through state governments.

What Is Overgrazing?
Overgrazing occurs when too many animals graze for too long and exceed the carrying capacity of a grassland area. It lowers the productivity of vegetation and changes the number and types of plants in an area. Large populations of wild ruminants can overgraze rangeland during prolonged dry periods, but most overgrazing is caused by excessive numbers of domestic livestock feeding for too long in a particular area.

Wild Game Ranching

Some ecologists have suggested that wild herbivores such as eland, oryx, and Grant's gazelle could be raised on ranches on tropical savanna (Figure 5-7). Because many wild herbivores have a more diversified diet than cattle, they can make more efficient use of available vegetation and thus reduce the potential for overgrazing.

In addition, they need less water than cattle and are more resistant than cattle to animal diseases native to savanna grasslands. These animals also achieve much of their own predator control because of their long evolutionary experience with lions, leopards, cheetahs, and wild dogs.

Since 1978 David Hopcraft has carried out a successful game-ranching experiment on the Athi Plains near Nairobi, Kenya. The ranch is stocked with various native grazers and browsers, including antelope, zebras, giraffes, and ostriches.

Cattle once raised as a comparison group are being phased out and may be replaced with Cape buffalo. The yield of meat from the native herbivores has been rising steadily, the condition of the range has improved, and costs are much lower than those for raising cattle in the same region.

Critical Thinking

Do you believe that wild game ranching on public and private rangeland would be useful in the western United States? Explain. If so, what animals might be raised successfully? What policies might support such wild game ranching?

Figure 16-16 Rangeland: overgrazed (left) and lightly grazed (right). (Soil Conservation Service)

Heavy overgrazing compacts the soil, which diminishes its capacity to hold water and regenerate itself. Overgrazing also converts continuous grass cover into patches of grass and thus exposes the soil to erosion, especially by wind; then woody shrubs such as mesquite and prickly pear cactus take over. Overgrazing is the major cause of desertification in arid and semiarid lands (Figure 12-23).

Figure 16-16 shows lightly grazed and overgrazed grassland side by side. A UN assessment of earth's dryland regions showed that livestock production worth $23.2 billion ($15 billion of this in Africa and Asia) was lost in 1990 as a result of rangeland degradation.

What Is the Condition of the World's Rangelands?

Range condition is usually classified as excellent (containing more than 75% of its potential forage production), good (51–75%), fair (26–50%), and poor (0–25%). Limited data from surveys in various countries indicate that most of the world's rangelands have been degraded to some degree, mostly by desertification (Figure 12-23).

In 1990, 68% of nonarctic U.S. public rangeland was rated by the Bureau of Land Management and the General Accounting Office as being in unsatisfactory (fair or poor) condition, compared with 84% in 1936. Since 1936 U.S. public range area in excellent or good condition has doubled while the area rated as poor has shrunk by half. This is a considerable improvement, but there's still a long way to go. Conservation biologists and some range experts also point out that surveys of U.S. rangeland neglect the damage to vital riparian zones by livestock (Spotlight, right).

How Can Rangelands Be Managed?

The primary goal of rangeland management is to maximize livestock productivity without overgrazing rangeland vegetation. The most widely used method to prevent overgrazing is controlling the *stocking rate*—the number of each kind of animal placed in a given area—so that it doesn't exceed an area's carrying capacity. But determining the carrying capacity of a range site is difficult and costly, and carrying capacity varies with factors such as climatic conditions (especially drought), past grazing use, soil type, invasions by new species, kinds of grazing animals, and intensity of grazing.

Both the numbers and distribution of livestock on a rangeland must be controlled to prevent overgrazing. Ranchers can control livestock distribution by fencing off damaged rangeland and riparian zones, rotating livestock from one grazing area to another, providing supplemental feed at selected sites, and situating water holes and salt blocks in strategic places.

A more expensive and less widely used method of rangeland management involves suppressing the

Q: What are our best energy options?

growth of unwanted plants (mostly invaders) by herbicide spraying, mechanical removal, or controlled burning. A cheaper and less harmful way to discourage unwanted vegetation is controlled, short-term trampling by large numbers of livestock.

Seeding and applying fertilizer can increase growth of desirable vegetation. Even though this method usually costs too much, reseeding is an excellent way to restore severely degraded rangeland.

Case Study: Livestock and Coyotes Another aspect of rangeland management for livestock is *predator control*. For decades, hundreds of thousands of predators have been shot, trapped, and poisoned by U.S. ranchers, farmers, hunters, and federal officials in a little-known agency called Animal Damage Control (ADC).

Federally subsidized predator control programs have so reduced gray wolf (Case Study, p. 466) and grizzly bear populations that they are now endangered species. Today the U.S. Fish and Wildlife Service spends additional funds to protect and help revive them.

Coyotes (*Canis latrans*), the main target of livestock predator control programs, are smaller relatives of wolves. The effect of coyotes on livestock, especially sheep, is a controversial issue in the western United States. Sheep ranchers claim that coyotes kill large numbers of sheep on the open range and should be exterminated. However, many wildlife experts maintain that although coyotes kill some sheep, their net effect is to increase rangeland vegetation for livestock and wild herbivores by reducing grazing pressure from rodents.

Because coyotes are so prolific and adaptable, environmentalists contend that any attempt to control them is doomed to failure. They also argue that it would cost taxpayers much less to pay ranchers for each head of livestock that was killed by a coyote than to spend millions of dollars each year for predator control.

Environmentalists suggest using a combination of fences, repellents, and trained guard dogs to keep predators away. Such methods have given sheep producers in Kansas one of the country's lowest rates of livestock losses to coyotes.

In 1986 USDA researchers reported that predation can be sharply reduced by penning young lambs and cattle together for 30 days and then allowing them to graze together on the same range. When predators attack, cattle butt and kick them, protecting both themselves and the sheep. Llamas and donkeys are also tough fighters against predators and act in the same way to protect sheep. Livestock raisers have little incentive to use such methods, however, when a local ADC or state predator control official is only a phone call away and takes care of any predator problems at little or no charge.

Endangered Riparian Zones

SPOTLIGHT

According to some conservation biologists and rangeland experts, overall estimates of rangeland condition obscure severe damage to certain heavily grazed areas, especially vital **riparian zones**: thin strips of lush vegetation along streams. These zones help prevent floods (and help keep streams from drying out during droughts) by storing and releasing water slowly from spring runoff and summer storms. These centers of biodiversity also provide habitats, food, water, and shade for wildlife in the arid and semiarid western lands. Studies indicate that 65–75% of the wildlife in the western United States is totally dependent on riparian habitats.

Because cattle need lots of water, they congregate near riparian zones and feed there until the grass and shrubs are gone. As a result, riparian vegetation is destroyed by trampling and overgrazing. A 1988 U.S. General Accounting Office report concluded that "poorly managed livestock grazing is the major cause of degraded riparian habitat on federal rangelands."

Riparian areas can be restored by using fencing to restrict access to degraded areas and by developing off-stream watering sites for livestock. Sometimes protected areas can recover in a few years. Despite the threat to soil and wildlife, little has been done to protect or repair riparian zones on U.S. public land, mostly because of political opposition by ranchers and lack of awareness of the problem by the general public.

Critical Thinking

Do you believe that riparian zones on public rangelands in the United States should receive stronger protection? Explain. If so, how would you see that such protection is provided?

Case Study: Livestock and U.S. Public Rangeland About 29,000 U.S. ranchers hold what are essentially lifetime permits to graze about 4 million livestock (3 million of them cattle) on Bureau of Land Management (BLM) and Forest Service rangelands in 16 western states. About 10% of these permits are held by small livestock operators; the other 90% belong to ranchers with large livestock operations, including four U.S. billionaires, and to corporations such as Metropolitan Life Insurance, Union Oil, Getty Oil, and the Vail Ski Corporation.

Permit holders pay the federal government a grazing fee for this privilege. Since 1981 grazing fees on

---

A: Energy efficiency and renewable energy (especially solar-produced hydrogen gas)

The Eco-Rancher

Wyoming rancher Jack Turnell is one of a new breed of cowpuncher who gets along with environmentalists. He talks about riparian ecology and biodiversity as fluently as he talks about cattle: "I guess I have learned how to bridge the gap between the environmentalists, the bureaucracies, and the ranching industry."

Turnell grazes cattle on his 32,000-hectare (80,000-acre) ranch south of Cody, Wyoming, and on 16,000 hectares (40,000 acres) of Forest Service land on which he has grazing rights. For the first decade after he took over the ranch, he raised cows the conventional way. Since then, he's made some changes.

Turnell disagrees with the proposals by environmentalists to raise grazing fees and remove sheep and cattle from public rangeland. He believes that if ranchers are kicked off the public range, ranches like his will be sold to developers and chopped up into vacation sites, irreversibly destroying the range for wildlife and livestock alike.

At the same time, he believes that ranches can be operated in more ecologically sustainable ways. To demonstrate this, Turnell began systematically rotating his cows away from the riparian areas, gave up most uses of fertilizers and pesticides, and crossed his Hereford and Angus cows with a French breed that tends to congregate less around water. Most of his ranching decisions are made in consultation with range and wildlife scientists, and changes in range condition are carefully monitored with photographs.

The results have been impressive. Riparian areas on the ranch and Forest Service land are lined with willows and other plant life, providing lush habitat for an expanding population of wildlife, including pronghorn antelope, deer, moose, elk, bear, and mountain lions. And this eco-rancher now makes more money because the higher-quality grass puts more meat on his cattle.

public rangeland have been set by Congress at only one-fourth to one-eighth the going rate for comparable private land. This means that taxpayers give the roughly 1% of U.S. ranchers with federal grazing permits subsidies amounting to $60–200 million a year, depending on how the costs of federal rangeland management are calculated.

The economic value of a permit is a part of the overall worth of a ranch. It can serve as collateral for a loan and is usually automatically renewed every 10 years, allowing permit holders to treat federal land as part of their ranches. It's not surprising that politically influential permit holders have fought so hard to block any change in this system.

The public subsidy does not end with low grazing fees, however. When other costs such as water pipelines, stock ponds, weed control, livestock predator control, clearing of undesirable vegetation, planting of grass for livestock, erosion, and loss of biodiversity are added, it's estimated that taxpayers are providing 29,000 ranchers with an annual subsidy of about $2 billion (an average of $69,000 per rancher) to produce only 2% of the country's beef. In 1991 one rancher alone received subsidies worth almost $950,000, which explains why many critics charge that the public lands grazing programs are little more than rancher welfare, mostly for the well-to-do ranchers who hold 90% of the permits.

However, ranchers with permits say that grazing fees on public lands should be low because most private rangeland is more productive than public rangeland. Environmentalists contend that this varies with the land involved and that even when this is true, current grazing fees on public rangelands are much too low.

Economic studies show that federal expenditures on grazing programs exceed not only grazing fee receipts but also the profits ranchers make from grazing their livestock on public lands. Thus, some analysts suggest that the government could save money by paying ranchers with permits not to graze their livestock on public lands.

Solutions: How Can U.S. Public Rangeland Be Managed More Sustainably? Some environmentalists believe that all commercial grazing of livestock on western public lands should be phased out over the next 10–15 years. They contend that the water-poor western range (Figure 11-6) is not a very good place to raise cattle and sheep, which require a lot of water.

However, some ranchers have demonstrated that western rangeland can be grazed sustainably; they make the case that such practices should not be prohibited on public rangeland (Individuals Matter, left). Some environmentalists agree that with proper management, ranching on western rangeland is a potentially sustainable operation and that encouraging sustainable ranching practices keeps the land from being developed and destroyed. They believe that the following measures can promote sustainable use of public rangeland:

- *Allow no or only limited grazing on riparian areas.*

- *Ban grazing on rangeland in poor condition.*

- *Use competitive bidding for all grazing permits.* The existing noncompetitive system gives ranchers with essentially lifetime permits an unfair economic advantage over ranchers who can't get such permits.

Q: How much money does switching from an incandescent bulb to a fluorescent bulb save?

- *Allow individuals and environmental groups to purchase grazing permits and not use the land for grazing.*

- *Until a competitive bidding system is implemented, raise grazing fees to fair market value.*

- *Give ranchers with small holdings grazing fee discounts to reduce the financial pressure for them to sell their ranch lands to developers.*

- *Abolish rancher-dominated grazing advisory boards.*

The problem is that despite their small numbers, western ranchers with grazing permits wield enough political power to see that measures that promote sustainability are not enacted by elected officials. There are efforts in the U.S. Congress to turn management of public rangelands over almost completely to ranchers and exempt them from most environmental laws and any legal challenges.

16-7 MANAGING AND SUSTAINING NATIONAL PARKS

How Popular Is the Idea of National Parks?

Today over 1,100 national parks each larger than 1,000 hectares (2,500 acres) are located in more than 120 countries, together covering an area equal to that of Alaska, Texas, and California combined. This important achievement in global conservation was spurred by the creation of the first national park system in the United States in 1912.

The U.S. National Park System is dominated by 54 national parks, most of them in the West (Figure 16-2). These repositories of majestic beauty and biodiversity, sometimes called *America's crown jewels*, are supplemented by state, county, and city parks. Most state parks are located near urban areas and thus are more heavily used than national parks. Nature walks, guided tours, and other educational services offered by U.S. Park Service employees have given many visitors a better appreciation for and understanding of nature, but recently these services have diminished because of budget cutbacks.

How Are Parks Being Threatened?

Parks everywhere are under siege. In developing countries parks are often invaded by local people who desperately need wood, cropland, and other resources. Poachers kill animals to obtain and sell rhino horns, elephant tusks, and furs. Park services in developing countries typically have too little money and too few personnel to fight these invasions, either by force or by education. In addition, most of the world's national parks are too small to sustain many of the larger animal species.

Popularity is one of the biggest problems of national and state parks in developed countries. Because of increased numbers of roads, cars, and affluent people, annual recreational visits to major U.S. national parks increased fourfold (and visits to state parks sevenfold) between 1950 and 1996; visits are projected to increase from 266 million in 1996 to 500 million by 2010.

During the summer, the most popular U.S. national and state parks are often choked with cars and trailers and hour-long entrance backups. They are also plagued by noise, traffic jams, litter, vandalism, poaching, deteriorating trails, polluted water, rundown visitor centers, garbage piles, and crime. Many visitors to heavily used parks leave the city to commune with nature, only to find the parks as noisy, congested, and stressful as where they came from.

U.S. Park Service rangers now spend an increasing amount of their time on law enforcement instead of conservation, management, and education. Since 1976 the number of federal park rangers has not changed while the number of visitors to park units has risen by 85 million; the number of visitors is expected to rise by another 162 million in the next 20 years. Currently, there is one ranger for every 84,200 visitors to the major national parks. Because overworked rangers earn an average salary of less than $28,000, many leave for better-paying jobs.

Wolves (Case Study, p. 466), bears, and other large predators in and near various parks have all but vanished because of excessive hunting, poisoning by ranchers and federal officials, and the limited size of most parks. As a result, populations of species these predators once helped control have exploded, destroying vegetation and crowding out other species.

Nonnative species have moved into or been introduced into many parks. Wild boars (imported to North Carolina in 1912 for hunting) are threatening vegetation in part of the Great Smoky Mountains National Park. The Brazilian pepper tree has invaded Florida's Everglades National Park. Mountain goats in Washington's Olympic National Park trample native vegetation and accelerate soil erosion. At the same time that some species have moved into parks, other native species of animals and plants (including many threatened or endangered species) are being killed or removed illegally in almost half of U.S. national parks.

Some say that the greatest threat to many U.S. parks today is posed by nearby human activities. Wildlife and recreational values are threatened by mining, logging, livestock grazing, coal-burning power plants, water diversion, and urban development. Polluted air drifts hundreds of kilometers, killing ancient trees in California's Sequoia National Park and blurring the awesome vistas at Arizona's Grand Canyon. According to the National Park

Who's Afraid of the Big Gray Wolf?

CASE STUDY

At one time, the gray wolf, *Canis lupis* (Figure 16-17), ranged over most of North America, but between 1850 and 1900 some 2 million wolves were systematically shot, trapped, and poisoned by ranchers, hunters, and government employees. The idea was to make the West and the Great Plains safe for livestock and for big-game animals prized by hunters.

This strategy worked. The species is now listed as endangered in all 48 lower states except Minnesota (which has an estimated 2,000 wolves), where it is listed as threatened.

Ecologists now recognize the important role these predators once played in parts of the West and the Great Plains by culling herds of bison, elk, caribou, and mule deer of weaker animals, thereby strengthening the genetic pool of the survivors. In recent years, these herds have proliferated, devastating some of the area's vegetation, increasing erosion, and threatening the niches of other wildlife species.

In 1987, the U.S. Fish and Wildlife Service proposed that gray wolves be reintroduced into the Yellowstone ecosystem. This proposal to reestablish ecological connections that humans had eliminated brought outraged protests.

Some objections came from ranchers who feared the wolves would attack their cattle and sheep; one enraged rancher said that the idea was "like reintroducing smallpox." Other protests came from hunters who feared that the wolves would kill too many big game animals and from miners and loggers who worried that the government would force them to cease operations on wolf-populated federal lands.

National Park Service officials promised to trap or shoot any wolves that killed livestock outside

Figure 16-17 Efforts to return the gray wolf to its former habitat in Yellowstone National Park and central Idaho are vigorously opposed by ranchers, hunters, miners, and loggers, but biologists have been reintroducing gray wolves in these areas since 1994. By the end of 1996 there were 52 gray wolves in Yellowstone and 27 in Idaho. (Tom J. Ulrich/Visuals Unlimited)

designated areas and to reimburse ranchers from a private fund established by Defenders of Wildlife for any livestock verified to have been killed by wolves. However, these promises fell on deaf ears, and many ranchers and hunters worked hard and unsuccessfully to delay or defeat the plan in Congress.

Since 1995 federal wildlife officials have been catching about 30–40 gray wolves per year in Canada and relocating them in Yellowstone National Park and northern Idaho. The goal is to allow the wolf populations to increase to sustainable levels by 2002. By 1996 there were 31 gray wolves in Yellowstone and 24 in northern Idaho.

Some ranchers and hunters say that either way they'll take care of the wolves quietly—what they call the "shoot, shovel, and shut up" so-

lution. Meanwhile, there are continuing efforts in Congress to kill the program or its funding.

In 1997 the recovery program received a serious setback when a Wyoming federal appeals court judge found the wolf recovery program illegal. He ordered that the roughly 160 in Yellowstone National Park and central Idaho be removed.

Critical Thinking

Do you agree or disagree with the program to reestablish populations of the gray wolf in the Yellowstone ecosystem? Explain. What are the major drawbacks of this program? Could the money be better spent on other wildlife programs? If so, what programs would you suggest?

Q: What percentage of the oil used in the United States is for transportation?

Service, air pollution affects scenic views in national parks more than 90% of the time.

That's not all. Mountains of trash wash ashore daily at Padre Island National Seashore in Texas. Water use in Las Vegas threatens to shut down geysers in the Death Valley National Monument. Visitors to Sequoia National Park complain of raw sewage flowing through a parking lot; similar contamination has occurred for years at Kentucky's Mammoth Cave.

Unless a massive ecological restoration project is successful, Florida's Everglades National Park may dry up, making it the country's most endangered national park. Much of the water needed to sustain the park's wildlife has been diverted for crops and cities by network of canals, levees, spillways, and pumping stations to the north. Numbers of nesting birds in the park declined by 90%, from about 90,000 in 1931 to fewer than 9,000 in 1994. All other vertebrates, from deer to turtles, are down 75–95%.

By the 1970s state and federal officials recognized that Florida's massive plumbing project had been a serious ecological blunder. After over 20 years of political haggling, in 1990 state and federal governments agreed on a massive restoration project to undo some of the damage. Whether this 20-year project will work and be funded adequately remains to be seen. However, the high cost of this project is another example of a fundamental lesson from nature: Prevention is the cheapest and best way to go.

Case Study: Why Is Yellowstone's Ecosystem Unraveling? To a growing number of ecologists, river experts, and foresters the gravest danger to many U.S. national parks is coming from *within*. For example, they point to Yellowstone National Park as an ecosystem that is unraveling as stands of aspens and willows die, woody shrubs disappear, populations of once abundant animal species dwindle or disappear, and stream banks erode.

These scientists point to the explosion of the elk population as the major culprit. They have collected data indicating that between 1968 and 1996, the National Park Service's hands-off management policy of *natural regulation* has been the key factor in allowing the park's elk population to increase more than sixfold from 3,100 to about 20,000.

According to Charles Kay, a wildlife biologist at Utah State University, the rising elk population is consuming tree shoots and other forms of vegetation and changing the once diverse park into a simplified lawn by disrupting a series of important ecological connections. He has compared the vegetation in an area called the Mammoth Exclosure, which has been fenced off for 40 years, with vegetation outside the enclosure.

Inside the enclosed area there is a dense woody understory that provides habitats and food for deer, mice, and many other small mammals; outside the enclosure there is little left of the woody understory. Inside the fenced area there are dense stands of aspen, willows, and woody shrubs; outside such species are almost nonexistent. Outside the enclosed area, centuries-old stands of aspen, which reproduce by sending up shoots or suckers, have died out as their shoots have been eaten by elk.

This has resulted in an interconnected cascade of other ecological changes. Loss of these tree species has eliminated food and dam-building materials for beavers, whose numbers have plummeted in the last 60 years. As the beaver populations have dropped, so have their dams and ponds, which trapped eroded silt and slowly built up the park's streambeds. With a stronger flow of water and without aspens, cottonwoods, and willows to hold streambed soil in place, stream banks have eroded and collapsed.

Since adopting their policy of natural regulation in the early 1970s, park service officials have claimed that the large increase in elk represents the park ecosystem returning to its natural equilibrium state. Many ecologists say this is an outdated idea because recent ecological research has revealed that ecosystems are in a constant state of change and rarely attain an equilibrium point (Section 5-7). Park officials also attribute the decline in woody streamside vegetation to climate change or the long-standing policy of suppressing fires within the park.

Kay challenges these conclusions with his data on changes inside the park and the status of vegetation in many areas outside the park. For example, Eagle Creek on national forest land just north of the park near the town of Gardiner, Montana, has the same climate, no cattle grazing, and the same fire history as Yellowstone.

In this area about 3,000 elk are shot by hunters each year. Stands of aspen have regenerated because their shoots have not been devoured by elk. There are also thick stands of shrubby undergrowth, rarely found inside the park. In private and national forest land North of Gardiner along the Yellowstone River, the river is lined with a thick band of aspens, willows, and cottonwoods, in sharp contrast to the almost barren river banks in most parts of the park.

Under such mounting criticism, the National Park Service has hinted that it may reexamine its long-standing natural regulation policy. However, this may not be politically feasible because polls indicate that the public (mostly unaware of the ecological decline of the park) favors natural regulation of the park over the park's old policy of allowing hunters to kill thousands of elk each winter. What do you think should be done?

A: About 66% (Up from 53% in 1973)

Solutions: How Can Park Management in Developing Countries Be Improved? Some park managers, especially in developing countries, are developing integrated management plans that combine conservation practices with sustainable development of resources in the park and surrounding areas. In such plans, the inner core and vulnerable areas of each park are to be protected from development and treated as wilderness. Restricted numbers of people are allowed to use these areas for hiking, nature study, ecological research, and other nondestructive recreational and educational activities.

In buffer areas surrounding the core, controlled commercial logging, sustainable grazing by livestock, and sustainable hunting and fishing by local people are allowed. Money spent by park visitors adds to local income. By involving nearby residents in developing park management and restoration plans, managers and conservation biologists seek to help them see the park as a vital resource to protect and sustain rather than ruin. Costa Rica has led the way in protecting much of its parklands and other biodiversity sanctuaries (Solutions, right).

Integrated park management plans look good on paper, but they cannot be carried out without adequate funding and the support of nearby landowners and users. Moreover, in some cases the protected inner core may be too small to sustain the park's larger animal species.

Solutions: How Can Management of U.S. National Parks Be Improved? The 54 major U.S. national parks are managed under the principle of *natural regulation*, as if they were wilderness ecosystems that would sustain themselves if left alone. Many ecologists consider this a misguided policy. Most parks are far too small to even come close to sustaining themselves. Even the biggest ones, such as Yellowstone, cannot be isolated from the harmful effects caused by activities in nearby areas and from destruction from within by exploding populations of some plant-eating species (Case Study, p. 466).

The U.S. National Park Service has two goals that increasingly conflict: to preserve nature in parks and to make nature more available to the public. The Park Service must accomplish these goals with a $1.5-billion annual budget and a $6-billion backlog of maintenance, repairs, and high-priority construction projects at a time when park usage and external threats to the parks are increasing. Private developers also continually pressure Congress and the Park Service to turn national parks into high-tech, heavily developed theme or amusement parks instead of the repositories of natural ecosystems and centers for learning about nature they were intended to be.

Currently the U.S. National Park Service spends about 92% of its budget on visitor services; only 7% is spent on protecting natural resources, and a mere 1% funds environmental research that could help park officials implement better ecological management of parklands. Doubling the Park Service's current budget would cost as much as one B-2 bomber and cost each American only $5. This would be a good investment because according to a Park Service model, national park units pump about $10 billion annually into local areas.

In 1988 the Wilderness Society and the National Parks and Conservation Association made a number of suggestions for sustaining and expanding the national park system, including the following:

- *Have all user and entrance fees for national park units be used for the management, upkeep, and repair of the national parks. Currently, such fees go into the general treasury.*

- *Require integrated management plans for parks and other nearby federal lands.*

- *Increase the budget for adding new parkland near the most threatened parks.*

- *Increase the budget for buying private lands inside parks.*

- *Locate all new and some existing commercial facilities and visitor parking areas outside parks.*

- *Require private concessionaires who provide campgrounds, restaurants, hotels, and other services for park visitors to compete for contracts, raise their fees to 22% of their gross (not net) receipts, and pour receipts back into parks. In 1995, for instance, private concessionaires in national parks reported gross receipts of $700 million but paid only 2.7% ($19 million) in franchise fees to the government. Many large concessionaires have long-term contracts by which they pay the government as little as 0.75% of their gross receipts.*

- *Allow concessionaires to lease but not own facilities inside parks.*

- *Provide more funds for park system maintenance and repairs.*

- *Raise fees for park visitors and pour receipts back into parks.*

- *Restrict the number of visitors to crowded park areas.*

- *Increase the number and pay of park rangers.*

- *Survey the condition and types of wildlife species in parks.*

- *Encourage volunteers to give tours and lectures to visitors.*

- *Encourage individuals and corporations to donate money for park maintenance and repair.*

Q: What fossil fuel produces the least amount of CO_2 per unit of energy?

SOLUTIONS

Costa Rica (Figure 6-19), smaller in area than West Virginia, was once almost completely covered with tropical forests. Between 1963 and 1983, politically powerful ranching families cleared much of the country's forests to graze cattle, and they exported most of the beef to the United States and western Europe. By 1983, only 17% of the country's original tropical forest remained, and soil erosion was rampant.

Despite widespread forest loss, tiny Costa Rica is a superpower of biodiversity, with an estimated 500,000 species of plants and animals. A single park in Costa Rica is home for more bird species than all of North America.

In the mid-1970s, Costa Rica established a system of national parks and conservation reserves that by 1993 included 12% of its land (6% of it in reserves for indigenous peoples), compared to only 1.8% set aside as reserves in the lower 48 United States. As a result Costa Rica now has larger proportion of land devoted to biodiversity conservation than any other country.

One reason for this outstanding accomplishment in biodiversity protection was the establishment in 1963 of the Organization of Tropical Studies, a consortium of 44 U.S. and Costa Rican universities with the goal of promoting research and education in tropical ecology. The resulting infusion of several thousand scientists has helped Costa Ricans appreciate their country's great biodiversity. It also led to the

establishment in 1989 of the National Biodiversity Institute (INBio), a private nonprofit organization set up to survey and catalog the country's biodiversity.

This biodiversity conservation strategy has paid off. Today, revenue from tourism (almost two-thirds of it from ecotourists) is the country's largest source of outside income. If Costa Rica's attempts to protect and use biodiversity in sustainable ways are successful, it could be a model for the entire world.

However, legal and illegal deforestation threatens this plan because of the country's population growth and poverty, which still affects 10% of its people. Currently, this small country still loses roughly 400 square kilometers (150 square miles) of primary forest per year—four times the rate of loss in Brazil.

Costa Rica is also the site of one of the world's largest and most innovative ecological restoration projects. In the lowlands of the country's Guanacaste National Park, a small tropical seasonal forest is being restored and relinked to the rain forest on adjacent mountain slopes.

Daniel Janzen, professor of biology at the University of Pennsylvania and a leader in the growing field of rehabilitation and restoration of degraded ecosystems, has helped galvanize international support and has raised more than $10 million for this restoration project.

Janzen's vision is to make the nearly 40,000 people who live near the park an essential part of the restoration of the degraded forest—a concept he calls *biocultural restora-*

tion. By actively participating in the project, local residents will reap enormous educational, economic, and environmental benefits. Local farmers have been hired to sow large areas with tree seeds and to plant seedlings started in Janzen's lab.

Students in grade schools, high schools, and universities study the ecology of the park in the classroom and go on field trips to the park itself. Educational programs for civic groups and tourists from Costa Rica and elsewhere will stimulate the local economy.

The project will also serve as a training ground in tropical forest restoration for scientists from all over the world. Research scientists working on the project give guest classroom lectures and lead some of the field trips.

Janzen recognizes that in 20–40 years today's children will be running the park and the local political system. If they understand the importance of their environment, they are more likely to protect and sustain its biological resources. He believes that education, awareness, and involvement—not guards and fences—are the best ways to protect ecosystems from unsustainable use.

Critical Thinking

Why do you think that Costa Rica has been able to set aside a much larger percentage of its land for national parks and conservation reserves than any other country? Do you agree or disagree that such a plan should be implemented in the country where you live? If so, how would you go about doing this?

16-8 PROTECTING AND MANAGING WILDERNESS AND OTHER BIODIVERSITY SANCTUARIES

How Much of the Ecosphere Should We Protect from Exploitation? Most wildlife and conservation biologists believe that the best way to preserve biodiversity and ecological integrity—or wildness—

is through a worldwide network of reserves, parks, wildlife sanctuaries, wilderness, and other protected areas. Currently, about 6% of the world's land area is either strictly or partially protected in more than 20,000 nature reserves, parks, wildlife refuges, and other areas around the globe.

This is an important beginning, but conservation biologists say that to keep biodiversity and ecological integrity from being depleted, a minimum of 10–12% of the globe's land area should be protected. Moreover, many existing reserves are too small to provide any real protection for the wild species that live on them. Many existing preserves receive little protection. Currently, less than 2% of the world's protected areas are at least 10,000 square kilometers (3,900 square miles) in size, a conservative estimate of the minimum habitat needed to maintain viable populations of the largest mammals.

In 1981 the UN Educational, Scientific, and Cultural Organization (UNESCO) proposed that at least one (and ideally five or more) *biosphere reserves* be set up in each of the earth's 193 biogeographical zones. Each reserve should be large enough to prevent gradual species loss and should combine conservation and sustainable use of natural resources. To date, more than 300 biosphere reserves have been established in 76 countries.

A well-designed biosphere reserve has three zones (Figure 16-18): a *core area* containing an important ecosystem that has had little or no disturbance from human activities, a *buffer zone* where activities and uses are managed in ways that help protect the core, and a second buffer or *transition zone*, which combines

conservation and sustainable forestry, grazing, agriculture, and recreation. Buffer zones can also be used for education and research.

Proponents of large reserves say they that are the only way to maintain viable populations of large, wide-ranging species (such as panthers, elephants, and grizzly bears). Also, large reserves can sustain more species, minimize edge effects, and provide greater habitat diversity than small preserves. These advantages of large parks or reserves are based on the theory of island biogeography developed by Robert MacArthur and Edward O. Wilson (Connections, p. 136), in which the reserves are assumed to be habitat islands surrounded by expanses of often inhospitable development or other terrain.

In reality, few countries are physically or politically able to set aside and protect large biosphere reserves or other protected areas. Moreover, some research indicates that in some locales several well-placed medium-sized, isolated reserves may better protect a greater variety of habitat types and more populations of rare species than a single large one. If the population of a species is wiped out in one reserve because of fire, epidemic, or some other disaster, the species might still survive in the other reserves.

Conservation biologists also suggest establishing protected habitat corridors running between reserves to help support more species, allow migrations when environmental conditions in a reserve deteriorate, and reduce loss of genetic diversity from inbreeding. Such corridors might also help preserve animals that must make seasonal migrations to obtain food.

However, there are some drawbacks to corridors. They can allow the movement of pest species, disease, and exotic species between reserves. Also, animals moving along corridors may be exposed to greater risks of predation by both natural predators and human hunters. Although most conservation biologists support the development of wildlife corridors between reserves, they recognize the need to evaluate them on a park-by-park basis.

So far most biosphere reserves fall short of the ideal (Figure 16-18) and too little funding has been provided for their protection and management. An international fund to help developing countries protect and manage bioregions and biosphere reserves would cost $100 million per year—about what the world's nations spend on arms every 90 minutes.

Because there won't be enough money to protect most of the world's biodiversity, environmentalists believe that efforts should be focused on the most biodiverse countries and threatened species-rich areas within these countries (Figure 16-19). In the United States, conservation biologists urge Congress to pass an Endangered Ecosystems or Biodiversity Protection

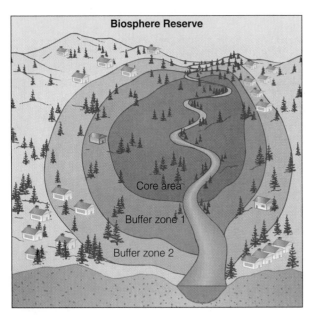

Biosphere Reserve

Core area

Buffer zone 1

Buffer zone 2

Human settlements

Research station

Tourism and education center

Figure 16-18 Design of a model biosphere reserve. In traditional parks and wildlife reserves, well-defined boundaries keep people out and wildlife in. By contrast, biosphere reserves recognize people's needs for access to sustainable use of various resources in parts of the reserve.

Q: What are the three major types of coal?

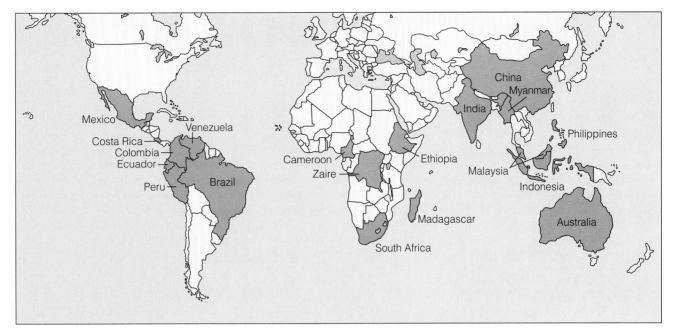

Figure 16-19 The earth's most biodiverse countries. Conservation biologists believe that efforts to preserve repositories of biodiversity as protected wilderness areas should be concentrated in these species-rich countries. (Data from Conservation International and World Wildlife Fund)

Act as an important step toward surveying and preserving the country's biodiversity.

In 1996 the World Wildlife Fund conducted a survey identifying 217 terrestrial, marine, and freshwater ecoregions throughout the world that were in the greatest need of protection. These ecoregions were identified on the basis of rareness of their habitat types, species found only in them (endemic species), total number of species (species biodiversity), and unusual ecological phenomena (such as the mass migration of caribou in the Canadian tundra). In addition to government action, private groups have helped establish biodiversity and wildlife sanctuaries (Solutions, p. 472).

Most economists, developers, and resource extractors disagree with protecting even the current 6% of the earth's remaining undisturbed ecosystems. They contend that such areas contain valuable resources that would add to economic growth. They view protected areas as "useless" and "empty" because they are not being used to provide mineral and other resources and land for development for humans.

What Is Wilderness and How Much Is Left?

Protecting biodiversity and ecological integrity means protecting *wildness*. One way to do this is to protect undeveloped lands from exploitation by setting them aside as wilderness. According to the U.S. Wilderness Act of 1964, **wilderness** consists of areas "where the earth and its community of life are untrammeled by man, where man himself is a visitor who does not remain." President Theodore Roosevelt summarized what we should do with wilderness: "Leave it as it is. You cannot improve it."

The U.S. Wilderness Society estimates that a wilderness area should contain at least 4,000 square kilometers (1,500 square miles); otherwise it can be affected by air, water, and noise pollution from nearby human activities. Conservation biologists urge that remaining wilderness be protected by law everywhere, focusing first on the most endangered spots in wilderness- and species-rich countries (Figure 16-19). Conservation biologists urge that more wilderness areas be protected in the United States (Case Study, p. 473).

A 1987 survey sponsored by the Sierra Club revealed that only about 34% of the earth's land area is undeveloped wilderness in blocks of at least 4,000 square kilometers (1,500 square miles). About 30% of these remaining wildlands are forests, many of them threatened tropical forests; tundra, ice-covered land, and desert make up most of the rest. Only about 20% of the wildlands identified in this survey are protected.

Why Preserve Wilderness? According to wilderness supporters, we need wild places where people can experience the beauty of nature and observe natural biological diversity and where they can enhance their mental and physical health by getting away from

The Nature Conservancy

Private groups play an important role in establishing wildlife refuges and other protected areas as a major strategy to protect biological diversity and ecological integrity. For example, since 1951 the Nature Conservancy has preserved over 38,000 square kilometers (15,800 square miles) of areas in the United States with unique ecological or aesthetic significance. Outside the United States, it has helped protect 80,000 square kilometers (31,000 square miles) of wildlife habitat. The Nature Conservancy now has over 800,000 members and programs in all 50 states, Latin America, and the Caribbean.

For several years, the Conservancy has been rated among the top conservation organizations, with one of the lowest overhead rates of any nonprofit organization: 85% of all contributions go directly to conservation programs.

The Nature Conservancy began in 1951 when an association of professional ecologists wanted to use their scientific knowledge of nature to conserve natural areas. Since its beginning this science-based organization has used the most sophisticated scientific knowledge available to identify and rank sites that are unique and ecologically significant and whose biodiversity or existence is threatened by development or other human activities.

Once sites are identified, a variety of techniques are used to see that they receive legal protection. Then science-based management plans are drawn up to maintain or restore the ecological health of each site and to provide long-term stewardship.

This conservation organization uses private and corporate donations to maintain a fund of over $150 million, which can be used to buy ecologically important pieces of land or wetlands threatened by development when no other option is available. Sometimes this land is then sold or donated to federal or state agencies.

If it cannot buy land for the protection of habitat, the Conservancy helps landowners obtain tax benefits in exchange for accepting legal restrictions or conservation easements preventing development. Other techniques include long-term management agreements and debt-for-nature swaps. Landowners have also received sizable tax deductions by donating their land to the Nature Conservancy in exchange for lifetime occupancy rights.

Through such efforts, this organization has created the world's largest system of private natural areas and wildlife sanctuaries, using the guiding principle of land conservation through private action.

Critical Thinking

Why do you think the antienvironmental Wise Use movement (Case Study, p. 217) has singled out the Nature Conservancy as the most important environmental group to weaken or destroy? Do you agree or disagree with this strategy? Explain.

---

noise, stress, and large numbers of people. Wilderness preservationist John Muir advised us to

Climb the mountains and get their good tidings. Nature's peace will flow into you as the sunshine into the trees. The winds will blow their freshness into you, and the storms their energy, while cares will drop off like autumn leaves.

Even those who never use the wilderness may want to know it is there, a feeling expressed by novelist Wallace Stegner:

Save a piece of country . . . and it does not matter in the slightest that only a few people every year will go into it. This is precisely its value. . . . We simply need that wild country available to us, even if we never do more than drive to its edge and look in. For it can be a means of reassuring ourselves of our sanity as creatures, a part of the geography of hope.

Recently some critics have argued that protecting wilderness for its scenic and recreational value for a small number of people is an outmoded concept that keeps some areas of the planet from being economically useful to humans. They also contend that wilderness protection diverts money, energy, and talent from learning how to better manage urban and other landscapes disturbed by human activities.

Scientists argue that this idea is outmoded because it fails to take into account the research that has revealed the ecological importance of wilderness areas for all people and all life. To most biologists, *the most important reason for protecting wilderness and other areas from exploitation and degradation is to preserve the biodiversity and ecological integrity they contribute as a vital part of earth capital* (Figure 1-2).

Wilderness areas provide mostly undisturbed habitats for wild plants and animals, protect diverse biomes from damage, and provide a laboratory in which we can discover more about how nature works. In other words, wilderness is a biodiversity and wildness bank and an ecoinsurance policy.

Instead of decreasing the ecologically inadequate 5–6% of the world's wilderness areas currently under some degree of protection, most ecologists and conser-

Q: What is the most desirable type of coal?

In the United States, preservationists have been trying to save wild areas from development since 1900. On the whole, they have fought a losing battle. Not until 1964 did Congress pass the Wilderness Act, which allowed the government to protect undeveloped tracts of public land from development as part of the National Wilderness Preservation System.

Only about 4% of U.S. land area is protected as wilderness, with almost three-fourths of it in Alaska. Only 1.8% of the land area of the lower 48 states is protected, most of it in the West.

Of the 413 wilderness areas in the lower 48 states, only four are larger than 4,000 square kilometers. Furthermore, the present wilderness preservation system includes only 81 of the country's 233 distinct

ecosystems. Like the national parks, most wilderness areas in the lower 48 states are habitat islands in a sea of development.

There remain almost 400,000 square kilometers (150,000 square miles) of scattered blocks of public lands that could qualify for designation as wilderness. Conservation biologists would like to see all this land protected as wilderness. Such efforts are strongly opposed by timber, mining, ranching, energy, and other interests who want either to extract resources from these and most other public lands or convert them to private ownership (Case Study, p. 217).

Some conservation biologists also urge that *wilderness recovery areas* be created by closing and obliterating nonessential roads in large areas of public lands, restoring habitats, allowing natural fires to burn, and reintroducing species that have been driven from such

areas. However, resource developers lobby elected officials and government agencies to build roads into national forests (and other areas being evaluated for inclusion in the wilderness system) so that they cannot be designated as wilderness; they also strongly oppose the idea of wilderness recovery areas.

Critical Thinking

1. Do you believe that all or most of the remaining potential wilderness area in the United States (or in the country where you live) should be protected? Explain.

2. Should a large portion of U.S. Forest Service and Bureau of Land Management land be reclassified as wilderness recovery areas, closed off for many uses, and allowed to undergo natural restoration as wildlife habitat? Explain. If so, how would you go about doing this?

vation biologists believe there is an urgent need to protect at least 10–12% of the earth's surface as wilderness and to allow some degraded areas to return to a more wild state. They agree that we need to learn much more about how to manage and restore land disturbed by human activities. However, they don't believe that this should be done by decreasing or unraveling wilderness protection, which will lead to even more areas disturbed by human activities.

To protect more of the earth's biodiversity and ecological integrity, conservation biologists and other environmental scientists believe the designation of future wilderness (and other protected) areas should focus on prairies, lowland forests, and wetlands that are largely absent from the current system. This will bring wilderness advocates into increasing conflict with developers, farmers, ranchers, and resource extractors who view remaining lands in these categories as prime sites for human use.

Some analysts believe that wilderness should be preserved because the wild species it contains have a right to exist (or struggle to exist) and to play their roles

in the earth's ongoing saga of biological evolution and ecological processes, without human interference.

How Can Wilderness Be Managed? To protect the most popular areas from damage, wilderness managers increasingly designate sites where camping is allowed and limit the number of people using these sites at any one time. Managers have also increased the number of wilderness rangers to patrol vulnerable areas, and they have enlisted volunteers to pick up trash discarded by thoughtless users.

Environmental historian and wilderness expert Roderick Nash suggests that wilderness areas be divided into three categories. The easily accessible, popular areas would be intensively managed and have trails, bridges, hiker's huts, outhouses, assigned campsites, and extensive ranger patrols. Large, remote wilderness areas would be used only by people who get a permit by demonstrating their wilderness skills. The third category, biologically unique areas, would be left undisturbed as gene pools of plant and animal species, with no human entry allowed.

Figure 16-20 Wangari Maathai, the first Kenyan woman to earn a Ph.D. (in anatomy) and to head an academic department (veterinary medicine) at the University of Nairobi, organized the internationally acclaimed Green Belt Movement in 1977. The goals of this highly regarded women's self-help community are to establish tree nurseries, raise seedlings, and plant a tree for each of Kenya's 27 million people. Each of the 50,000 members of the group receives a small fee for every tree that survives. By 1990 more than 10 million trees had been planted. The success of this project has sparked the creation of similar programs in more than a dozen other African countries. (William Campbell/*TIME* magazine)

What Is the U.S. National Wild and Scenic Rivers System?

In 1968 Congress passed the National Wild and Scenic Rivers Act, which allows rivers and river segments with outstanding scenic, recreational, geological, wildlife, historical, or cultural values to be protected in the National Wild and Scenic Rivers System. These waterways are to be kept free of development; they may not be widened, straightened, dredged, filled, or dammed along the designated lengths. The only activities allowed are camping, swimming, nonmotorized boating, sport hunting, and sport and commercial fishing. New mining claims are permitted in some areas, however.

Currently, only 0.2% of the country's 6 million kilometers (3.5 million miles) of waterways along some 150 stretches are protected by the Wild and Scenic River System. In contrast, 17% of the country's wild river length has been tamed and land flooded by dams and reservoirs. Environmentalists have urged Congress to add 1,500 additional river segments to the system—a goal that is vigorously opposed by some local communities and anti-environmental groups. If that goal is achieved, about 2% of the country's river systems would be protected. Environmentalists also urge that a permanent federal administrative body be established to manage the Wild and Scenic Rivers System and that states develop their own wild and scenic river programs.

How Can Change Be Brought About from the Bottom Up?

People the world over are working to protect forests. Penan tribespeople in Sarawak, Malaysia, have joined forces with environmentalists to protest destructive logging, the effects of which have reduced their population from 10,000 to less than 500. In Brazil, 500 conservation organizations have formed a coalition to preserve the country's remaining tropical forests. In the United States, members of Earth First! have perched in the tops of giant Douglas firs and lain in front of logging trucks and bulldozers to prevent the felling of ancient trees in national forests.

In 1985 the Rainforest Action Network was virtually alone in the United States in opposing tropical deforestation. Today there are nearly 200 Rainforest Action groups around the United States, most on college campuses, and a new one is formed about every 10 days. In 1991 Find Fensen, a first-year student at Princeton University, became concerned about the world's tropical forests and decided to buy and protect some rain forest. His dorm room operation grew into the Rainforest Conservancy. By 1993 such efforts to preserve tropical forests had spread to 45 campuses, attracted contributions from individuals and corporations, and resulted in the purchase and protection of hundreds of acres of rainforest in Belize.

In Kenya, Wangari Maathai (Figure 16-20) started the Green Belt Movement, a national self-help community action effort by 50,000 women and half a million schoolchildren to plant trees for firewood and to help hold the soil in place. This inspiring leader has said:

> I don't really know why I care so much. I just have something inside me that tells me that there is a problem and I have got to do something about it. And I'm sure it's the same voice that is speaking to everyone on this planet, at least everybody who seems to be concerned about the fate of the world, the fate of this planet.

In response to her efforts to sustain the earth and fight for human rights, the Kenyan government closed down her Green Belt Movement offices (she moved the headquarters into her home) and jailed her twice. In 1992 she was severely beaten by police while leading a peaceful protest against government imprisonment of several environmental and political activists.

The inspiring actions of such *earth citizens* show us that working with the earth to preserve biodiversity and ecological integrity will come mostly from the bottom up, through the actions of ordinary people, not from the top down. Some actions you can take to help sustain the earth's biodiversity and ecological integrity are listed in Appendix 5.

Q: What percentage of global emissions of carbon dioxide come from burning coal?

We abuse land because we regard it as a commodity belonging to us. When we see land as a community to which we belong, we may begin to use it with love and respect.

ALDO LEOPOLD

CRITICAL THINKING

1. Should owners of private property be able to do anything they want with their land? Explain. If not, what specific restrictions on use should be imposed?

2. Should there be limits to private property rights when biodiversity and ecological integrity are at risk? Explain. If so, what are these limits and how should they be enforced?

3. Should private companies that harvest timber from U.S. national forests continue to be subsidized by federal payments for reforestation and for building and maintaining access roads? Explain.

4. Explain why you agree or disagree with each of the proposals given in the Spotlight on p. 450 for reforming federal forest management in the United States.

5. Explain why you agree or disagree with the principles of sustainable forestry listed on p. 448.

6. What difference could the loss of essentially all the remaining old-growth tropical forests and old-growth forests in North America have on your life and on the lives of your descendants?

7. Explain why you agree or disagree with each of the proposals given on p. 458 concerning protection of the world's tropical forests.

8. Explain why you agree or disagree with the proposals given on pp. 464–465 for providing more sustainable use of public rangeland in the United States.

9. Explain why you agree or disagree with each of the proposals given on p. 468 concerning the U.S. national park system.

10. Is a protected forest or other area "useless" because it does not add significantly to economic growth? Explain. Describe the usefulness of an area in ecological terms.

*11. Evaluate timber harvesting on private and public lands in your local area. What are the most widely used harvesting methods? Try to document any harmful environmental impacts. Have the economic benefits to the community outweighed any harmful environmental effects?

*12. Evaluate cattle grazing on private and public rangeland and pastures in your local area. Try to document any harmful environmental impacts. Have the economic benefits to the community outweighed any harmful environmental effects?

*13. Make a survey of the national, state, and local parks within a 97-kilometer (60-mile) radius of your local community. How widely are they used by local residents and by outside visitors? What is their condition? Develop a plan for their improved management.

*14. Make a concept map of this chapter's major ideas, using the section heads and subheads and the key terms (in boldface type). Look at the inside back cover and in Appendix 4 for information about concept maps.

SUGGESTED READINGS

See Internet Sources and Further Readings at the back of the book, and InfoTrac.

A: About 42%

17 SUSTAINING WILD SPECIES

The Passenger Pigeon: Gone Forever

In the early 1800s, bird expert Alexander Wilson watched a single migrating flock of passenger pigeons darken the sky for over 4 hours. He estimated that this flock was more than 2 billion birds strong, some 386 kilometers (240 miles) long, and 1.6 kilometers (1 mile) wide.

By 1914 the passenger pigeon (Figure 17-1) had disappeared forever. How could a species that was once the most common bird in North America become extinct in only a few decades?

The answer is humans. The main reasons for the extinction of this species were uncontrolled commercial hunting and loss of the bird's habitat and food supply as forests were cleared to make room for farms and cities.

Passenger pigeons were good to eat, their feathers made good pillows, and their bones were widely used for fertilizer. They were easy to kill because they flew in gigantic flocks and nested in long, narrow colonies.

Beginning in 1858, passenger pigeon hunting became a big business. Shotguns, traps, artillery, and even dynamite were used. Birds were sometimes suffocated by burning grass or sulfur below their roosts. Live birds were even used as targets in shooting galleries. In 1878 one professional pigeon trapper made $60,000 by killing 3 million birds at their nesting grounds near Petoskey, Michigan. By the early 1880s commercial hunting had ceased because only a few thousand birds were left. At that point, recovery of the species was doomed because the females laid only one egg per nest. On March 24, 1900, a young boy in Ohio shot the last known wild passenger pigeon. The last passenger pigeon on earth, a hen named Martha after Martha Washington, died in the Cincinnati Zoo in 1914. Her stuffed body is now on view at the National Museum of Natural History in Washington, D.C.

Eventually, all species become extinct or evolve into new species, but humans have become the primary factor in the premature extinction of more and more species. Conservation biologists estimate that every day at least 50 (and perhaps as many as 200) species become extinct because of human activities, and the rate of this loss of biodiversity is increasing rapidly.

Figure 17-1 Passenger pigeons, extinct in the wild since 1900. The last known passenger pigeon died in the Cincinnati Zoo in 1914. (John James Audubon/The New York Historical Society)

The last word in ignorance is the person who says of an animal or plant: "What good is it?" . . . If the land mechanism as a whole is good, then every part of it is good, whether we understand it or not. . . . Harmony with land is like harmony with a friend; you cannot cherish his right hand and chop off his left.

ALDO LEOPOLD

This chapter will be devoted to answering the following questions:

- Why should we care about wildlife?
- What activities and traits of humans endanger wildlife?
- How can we prevent premature extinction of species?
- Can game animals be managed sustainably?
- Can freshwater and marine fish be managed sustainably?

🦋 17-1 WHY PRESERVE WILD SPECIES?

Why Not Let Wild Species Die? If all species eventually become extinct, why should we worry about losing a few more because of our activities? Does it matter that the passenger pigeon (Figure 17-1), the black rhinoceros (Figure 4-31), the northern spotted owl (Figure 16-10), the green sea turtle (p. 145), the 70 remaining Florida panthers (p. 395), or some unknown plant or insect in a tropical forest becomes prematurely extinct because of human activities?

Biologists contend that the answer is yes because of the economic, medical, scientific, ecological, aesthetic, and recreational value of all species. Some environmentalists go further and contend that each species has an inherent right to play its role in the ongoing evolution of life on earth until it becomes extinct without interference by humans.

What Is the Economic and Medical Importance of Wild Species? Some 90% of today's food crops were domesticated from wild tropical plants (Figure 14-5). Moreover, agricultural scientists and genetic engineers need existing wild plant species to derive today's crop strains and to develop the new crop strains of tomorrow.

Wild plants and plants domesticated from wild species supply rubber, oils, dyes, fiber, paper, lumber, and other useful products (Figure 16-4). Nitrogen-fixing microbes in the soil and in plants' root nodules (Figure 4-24) supply nitrogen to grow food crops worth almost $50 billion per year worldwide ($7 billion in the United States alone). Pollination by birds and insects is essential to many food crops, including 40 U.S. crops valued at approximately $30 billion per year.

About 80% of the world's population relies on plants or plant extracts for medicines. At least 40% of all pharmaceuticals, worth at least $100 billion per year, owe their existence to the genetic resources of wild plants (Figure 16-12), mostly from tropical developing countries. Plant-derived anticancer drugs save an estimated 30,000 lives per year in the United States. Over 3,000 antibiotics, including penicillin and tetracycline, are derived from microorganisms. Only about 5,000 of the 250,000 known plant species have been studied thoroughly for their possible medical uses.

What Is the Scientific and Ecological Importance of Wild Species? Every species can help scientists understand how life has evolved and functions, and how it will continue to evolve on this planet. Wild species also provide many of the ecological services that make up earth capital (Figure 1-2) and thus are key factors in sustaining the earth's biodiversity and ecological integrity.

They supply us (and other species) with food, recycle nutrients essential to agriculture, and help generate and maintain soils. They also produce oxygen and other gases in the atmosphere, absorb pollution, moderate the earth's climate, help regulate local climates and water supplies, reduce erosion and flooding, and store solar energy. Moreover, they detoxify poisonous substances, break down organic wastes, control potential crop pests and disease carriers, and make up a vast gene pool for future evolutionary processes.

What Is the Aesthetic and Recreational Importance of Wildlife? Wild plants and animals are a source of beauty, wonder, joy, and recreational pleasure for many people (Figure 17-2). Americans spend about $18.2 billion a year to watch wildlife— over three times the $5.8 billion they spend each year on movie tickets and the $5.9 billion they spend annually on professional sporting events.

Wildlife tourism, sometimes called *ecotourism*, is the fastest growing segment of the global travel industry and generates an estimated $30 billion in revenues each year. Conservation biologist Michael Soulé estimates that one male lion living to age 7 generates $515,000 in tourist dollars in Kenya; by contrast, if killed for its skin the lion would bring only about $1,000. Similarly, over a lifetime of 60 years, a Kenyan elephant is worth close to $1 million in ecotourist revenue.

However, according to Emily Young, professor of geography at the University of Arizona, "ecotourism is fast becoming tourism without the eco." New hotels are devouring wildlife habitat. Beachfront hotels on Mexico's Pacific coast lure visitors to watch endangered sea turtles lay their eggs, but their bright lights disorient the turtles as they lumber ashore and many fail to lay eggs.

Figure 17-2 Many species of wildlife, such as this bird of paradise in a New Guinea tropical forest, are a source of beauty and pleasure. This species is vulnerable to extinction because its mating ritual requires a throng of males to assemble and "strut their stuff" in front of females. When population numbers decline too much, mating cannot occur. (A. J. Deane/Bruce Coleman Ltd.)

Why Is It Ethically Important to Preserve Wild Species? Bioethics Some people believe that each wild species has an inherent right to exist, or to struggle to exist. This ethical stance is based on the view that each species has *intrinsic value* unrelated to its usefulness to humans.

According to this view, we have an ethical responsibility to protect species from becoming prematurely extinct as a result of human activities. Biologist Edward O. Wilson believes that deep within, most people feel obligated to protect other species and the earth's biodiversity because of the natural affinity for nature built into our genes (Spotlight, right).

Some people distinguish between the survival rights of plants and those of animals, mostly for what they consider practical reasons. Poet Alan Watts once said that he was a vegetarian "because cows scream louder than carrots."

Other people distinguish among various types of species. They might think little about killing a fly, mosquito, cockroach (Spotlight, p. 422), or rat or ridding the world of disease-causing bacteria. Unless they are strict vegetarians, they might also see no harm in having others kill domesticated animals in slaughterhouses to provide them with meat, leather, and other products. However, these same people might deplore the killing of wild animals such as deer, squirrels, or rabbits.

Some proponents go further and assert that each individual organism, not just each species, has a right to survive without human interference, just as each human being has the right to survive. Others emphasize the importance of preserving the whole spectrum of biodiversity by protecting entire ecosystems rather than individual species or organisms, as discussed in Chapter 16. This view is based on the principle that humans have an ethical obligation to prevent premature extinction of wildlife by saving their habitats and not disrupting the complex ecological interactions that sustain all life.

17-2 THE RISE AND FALL OF SPECIES

How Does Background Extinction Differ from Mass Extinction? Extinction is a natural process and eventually all species become extinct. David Raup and several other evolutionary biologists estimate that more than 99.9% of all the species that have ever existed are now extinct because of a combination of background and mass extinctions.

Each year, a small number of species become extinct naturally at a low rate, a phenomenon called the *natural, or background, rate of extinction*. Based mostly on the fossil record, evolutionary biologists estimate that the current average background rate of extinction is 3 species per year if there are about 10 million species and 30 species per year if there are 100 million species.

In contrast, **mass extinction** is an abrupt rise in extinction rates above the background level. It is a catastrophic, widespread (often global) event in which large groups of existing species (perhaps 25–70%) are wiped out. Most mass extinctions are believed to result from one or a combination of global climate changes that kill many species and leave behind those able to adapt to the new conditions.

Fossils and geological evidence indicate that the earth's species have experienced five great mass extinctions (20–60 million years apart) during the past 500 million years in which large numbers of species became extinct each year for tens of thousands to millions of years (Figure 5-38). Evidence also shows that these mass extinctions were followed by other periods, called *adaptive radiations*, when the diversity of

Q: How many new nuclear power plants have been ordered in the United States since 1978?

life increased and spread for 10 million years or more (Figures 5-38 and 5-39). The last mass extinction took place about 65 million years ago, when the dinosaurs became extinct for reasons that are hotly debated, after thriving for 140 million years.

Why Do Conservation Biologists Believe There Is a New Mass Extinction Crisis? Imagine that you have built a two-story house using wood as the basic structural material and that termites are slowly destroying various parts of the structure. How long will it be before they destroy enough of the structure for parts or even all of the house to collapse?

Conservation biologists believe that this urgent question, when applied to the earth, is one that we as a species should be asking ourselves. As we tinker with the only home for us and other species, we are rapidly removing parts of the earth's natural biodiversity and ecological integrity on which we and other species depend in ways we know little about. We are not heeding Aldo Leopold's warning: "To keep every cog and wheel is the first precaution of intelligent tinkering."

So far we have little understanding of the ecological roles of the world's identified 1.75 million species, much less of the 100 million more species that may exist. Until we have such information, most biologists believe that we should use the *precautionary principle* to prevent the premature extinction of species as a result of our activities.

It's difficult to document extinctions, for most go unrecorded. However, fossil and other evidence indicate that since agriculture began about 10,000 years ago, human activities have caused a growing number of species to become extinct (Figure 17-3). Some conservation biologists estimate that currently 18,000–73,000 species become extinct each year (on average, 50–200 species per day)—thousands of times the estimated natural background extinction rate of 3–30 species per year. Even if this estimate of the erosion of biodiversity is 100 times too high, the current extinction rate is many times the estimated rate of background extinction.

Scientists expect this rate of extinction to accelerate as the human population grows and takes over more of the planet's wildlife habitats and net primary productivity. They warn that within the next few decades we could easily lose at least 1 million of earth's species, most of them in tropical regions. According to biodiversity expert Edward O. Wilson, "Clearly, we are in the midst of one of the great extinction spasms of geological history."

Mass extinctions occurred long before humans evolved (Figure 5-38), but there are three important differences between the current mass extinction most biologists believe we are bringing about and those of

Biophilia

SPOTLIGHT

Biologist Edward O. Wilson contends that because of the billions of years of biological connections leading to the evolution of the human species (Figure 5-34), we have an inherent affinity for the natural world—a phenomenon he calls *biophilia* (love of life). He points out that we cannot erase this evolutionary imprint in our genes because of a few generations of urban living.

Evidence of this natural affinity for life is seen in the preference people have for almost any natural scene over one from an urban environment. Given a choice, people prefer to live in an area where they can see water, grassland, or a forest. More people visit zoos and aquariums than attend all professional sporting events combined.

In the 1970s I was touring the space center at Cape Canaveral in Florida. During our bus ride the tour guide pointed out each of the abandoned multimillion dollar launch sites and gave a brief history of each launch. Most of us were utterly bored. All of a sudden people started rushing to the front of the bus and staring out the window with great excitement. What they were looking at was a baby alligator—a dramatic example of how *biophilia* can triumph over *technophilia*.

Critical Thinking

Do you have a built-in affinity for wildlife and wild ecosystems (biophilia)? If so, how do you display this inherent love of wildlife in your daily actions? What patterns of your consumption help destroy and degrade wildlife?

the past. First, *the current extinction crisis is the first to be caused by a single species: our own.* By using or wasting approximately 40% of the earth's terrestrial net primary productivity (25% of the world's total net primary productivity), we are crowding out other terrestrial species. What will happen to wildlife and the services they provide for humans and other species if our population doubles in the next 40 years?

Second, *the current mass wildlife extinction is taking place in only a few decades, rather than over thousands to millions of years.* Such rapid extinction cannot be balanced by speciation because it takes 2,000–100,000 generations for new species to evolve. Fossil and other evidence related to past mass extinctions indicates that it takes millions of years to recover biodiversity through adaptive radiations (Figures 5-38 and 5-39). Thus, repercussions for humans and other species from the current human-caused mass extinction will affect the future

Figure 17-3 Some species that have become extinct largely because of human activities, mostly habitat destruction and overhunting.

Passenger pigeon   Great Auk   Dodo   Dusky seaside sparrow   *Aepyornis* (Madagascar)

course of evolution for 5–10 million years and could lead to the premature extinction of the human species.

Third, *besides killing off species, we are eliminating many biologically diverse environments such as tropical forests, tropical coral reefs, wetlands, and estuaries that in the past have served as evolutionary centers for the recovery of biodiversity after a mass extinction.* In the words of Norman Myers (Guest Essay, p. 295–296), "Within just a few human generations, we shall—in the absence of greatly expanded conservation efforts—impoverish the biosphere to an extent that will persist for at least 200,000 human generations or twenty times longer than the period since humans emerged as a species."

Mathematical models recently developed by ecologists indicate a time lag of several generations between habitat loss and extinction, primarily because habitat loss also removes potential colonization sites. If this model is correct, biologists may be grossly underestimating the magnitude of the current biodiversity meltdown.

Even without extinctions, ecosystems may lose their ability to support many forms of life because of the disappearance of local populations of key organisms. According to a 1997 study by biologist Jennifer Hughes and her colleagues at Stanford University, every year an estimated 16 million local plant and animal populations— an average of 1,800 populations every hour—disappear from the world's tropical forests.

Is There Really an Extinction Crisis? Some social scientists and a few biologists question the existence of an extinction crisis caused by our activities. They point to several problems in estimating species loss. First, we don't know how many species there are; estimates range between 5 million and 100 million. Second, it is difficult to observe species extinction, especially for species we know little or nothing about.

The annual loss of tropical forest habitat is estimated at about 1.8% per year. Edward O. Wilson and several tropical biologists who have counted species in patches of tropical forest before and after destruc-

tion or degradation estimate that this 1.8% loss in habitat results in roughly a 0.5% loss of species. However, biomes vary in the number of species they contain (species diversity), and the ratio of habitat loss to species loss varies in different biomes.

Do such estimates add up to an extinction crisis? Let us assume, as Wilson and many other biologists do, that a loss of 1 million species over several decades represents an extinction crisis. If we assume the global decline in species to be 0.5% per year, then we will lose 25,000 species per year if there are 5 million species, 100,000 per year if there are 20 million species, and 500,000 per year if there are 100 million species. If these assumptions are correct, we will lose 1 million species in 40 years if there are 5 million species, in 10 years with 20 million species, and in only 2 years with 100 million species.

Let's assume, however, that the estimate of 0.5% species loss per year is too high for the earth as a whole. If it is 0.25% per year, then we will lose 1 million species in 80 years with 5 million species, in 20 years with 20 million species, and in 4 years with 100 million species. Even if we halve the estimated species loss again, to 0.125% per year, we can still lose 1 million species within 8–160 years, enough to easily qualify the situation as an extinction crisis.

These biologists don't contend that their estimates are precise enough to make a firm prediction. Instead, they argue that there is ample evidence that we are destroying and degrading wildlife habitats at an increasing rate and that our actions certainly lead to a significant loss of species, even though the number and rate vary in different parts of the world.

What Are Endangered and Threatened Species?
Biologists distinguish among three levels of extinction: **(1)** *Local extinction* occurs when a species is no longer found in an area it once inhabited but is still found elsewhere in the world, **(2)** *ecological extinction* occurs when there are so few members of a species left that it can no longer play its ecological roles in the biological communities where it is found, and **(3)** *biological extinc-*

Q: What percentage of current U.S. nuclear power generation will be retired by 2015?

tion occurs when a species is no longer found anywhere on the earth. Biological extinction is forever.

Species heading toward biological extinction are classified as either *endangered* or *threatened* (Figure 17-4). An **endangered species** has so few individual survivors that the species could soon become extinct over all or most of its natural range. Examples are the California condor in the United States (39 in the wild), the whooping crane in North America (about 288 left), the giant panda in central China (about 1,000 left), the snow leopard in central Asia (about 2,500 left), and the black rhinoceros in Africa (about 2,400 left).

A **threatened species** is still abundant in its natural range but is declining in numbers and is likely to become endangered. Examples are the bald eagle (Figure 8-2), the grizzly bear, and the American alligator (p. 98). Endangered and threatened species are ecological smoke alarms.

Some species have characteristics that make them more vulnerable than others to premature extinction (Table 17-1). In general, species more vulnerable to extinction are found in a limited area and have a small population size, a low population density, large body size, a specialized niche (Section 4-6), and a low reproductive rate (K-strategists; Figure 5-31). Such species also tend to undergo fairly fixed migrations, feed atop long food chains or webs, and have high economic values to people.

According to a 1996 joint study by the International Union for the Conservation and Conservation International, more than 5,200 known animal species are at risk of extinction. They include 34% of the world's fish, 25% of amphibians, 25% of mammals (Figure 17-5, p. 484), 20% of reptiles, and 11% of the bird species.

Each animal species has a *critical population density* and a *minimum viable population size*, below which survival may be jeopardized because males and females have a difficult time finding each other. Once a population falls below its critical size, it continues to decline, even if the species is protected, because its death rate exceeds its birth rate. The remaining small population can easily be wiped out by fire, flood, landslide, disease, or some other single catastrophe. As population size drops, genetic diversity also decreases because the resulting smaller gene pool and inbreeding reduce a population's ability to respond to environmental changes through natural selection.

According to a 1995 study of the genetics of small populations by population geneticist Russell Lande, an endangered species must number at least 10,000, and often more, to maintain its evolutionary potential for survival. On average, only about 1,000 individuals of an animal species and 120 of a plant species remain when they are listed as endangered, according to a 1993 study by the Environmental Defense Fund.

The populations of many wild species that are not yet in danger of extinction have diminished locally or regionally to the point where they are locally or ecologically extinct. Such species may be a better indicator of the condition of entire ecosystems than endangered and threatened species. These *indicator species* can serve as early warnings so that we can prevent

Table 17-1 Characteristics of Extinction-Prone Species

Characteristic	Examples
Low reproductive rate	Blue whale, polar bear, California condor, Andean condor, passenger pigeon, giant panda, whooping crane
Specialized feeding habits	Everglades kite (eats apple snail of southern Florida), blue whale (krill in polar upwellings), black-footed ferret (prairie dogs and pocket gophers), giant panda (bamboo), koala (certain eucalyptus leaves)
Feed at high trophic levels	Bengal tiger, bald eagle, Andean condor, timber wolf
Large size	Bengal tiger, lion, elephant, Javan rhinoceros, American bison, giant panda, grizzly bear
Limited or specialized nesting or breeding areas	Kirtland's warbler (6- to 15-year-old jack pine trees), whooping crane (marshes), orangutan (only on Sumatra and Borneo), green sea turtle (lays eggs on only a few beaches), bald eagle (prefers forested shorelines), nightingale wren (only on Barro Colorado Island, Panama)
Found in only one place or region	Woodland caribou, elephant seal, Cooke's kokio, many unique island species
Fixed migratory patterns	Blue whale, Kirtland's warbler, Bachman's warbler, whooping crane
Preys on livestock or people	Timber wolf, some crocodiles
Behavioral patterns	Passenger pigeon and white-crowned pigeon (nest in large colonies), redheaded woodpecker (flies in front of cars), Carolina parakeet (when one bird is shot, rest of flock hovers over body), Key deer (forages for cigarette butts along highways—it's a "nicotine addict")

A: At least 40% and the rest by 2030

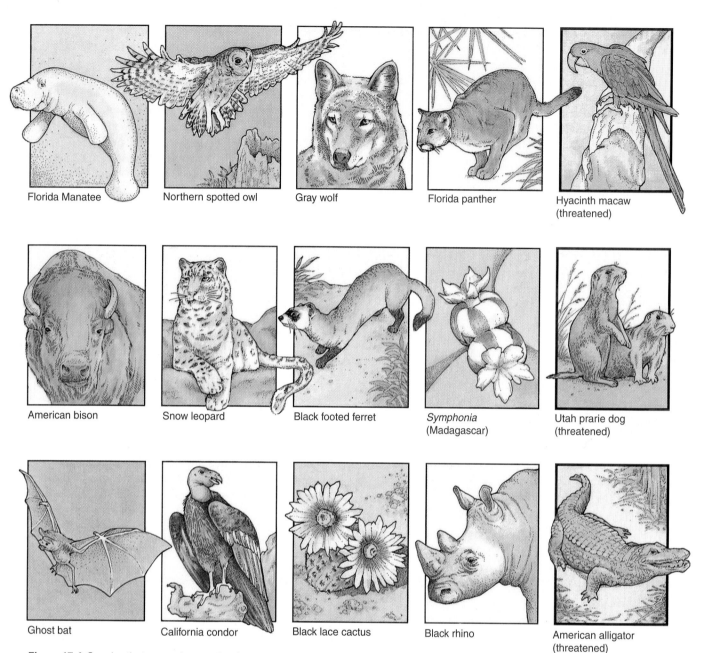

Figure 17-4 Species that are endangered or threatened largely because of human activities.

Florida Manatee	Northern spotted owl	Gray wolf	Florida panther	Hyacinth macaw (threatened)
American bison	Snow leopard	Black footed ferret	*Symphonia* (Madagascar)	Utah prarie dog (threatened)
Ghost bat	California condor	Black lace cactus	Black rhino	American alligator (threatened)

species extinction rather than respond to emergencies, often with little chance of success.

What Is the Status of Wild Species and Ecosystems in the United States? According to a 1995 study by the Nature Conservancy, about 32% of the 20,500 U.S. species studied by scientists are vulnerable to premature extinction, mostly because of human activities. The animals that are most at risk are those that depend on aquatic ecosystems, such as freshwater fish and amphibians; a large number of flowering plants are also in trouble. Approximately one-third of all plants and animals that are endangered or threatened in North America make their homes in wetlands, which are under growing pressure for draining and development and stress from pollution (Section 5-2).

Another survey published in 1995 by Defenders of the Wildlife focused on the status of ecosystems in the United States. According to the report, ecosystem health is especially threatened in 10 states: California, Hawaii, Texas, and the Southeastern states, particularly Florida. However, the report noted that "natural areas are at a biological breaking point all across the country."

Case Study: Bats Are Getting a Bad Rap
Despite their variety (950 known species) and worldwide distribution, bats—the only mammals that can fly—have certain traits that place them at risk because of human activities. Bats reproduce slowly, and many

Q: How much did global average life expectancy increase between 1900 and 1997?

Grizzly bear
(threatened)

Arabian oryx

Bald eagle
(threatened)

Kirtland's warbler

African elephant

Mojave desert tortoise
(threatened)

Swallow butterfly

Humpback chub

Golden lion tamarin

Siberian tiger

Peregrine falcon
(threatened)

Giant panda

Knowlton cactus

Whooping crane

Blue whale

bat species that live in huge colonies in caves and abandoned mines become vulnerable to destruction when people block the passageways or disturb their hibernation.

Because of unwarranted fears of bats and lack of knowledge about their vital ecological roles, several species have been driven to extinction. Currently, 26% of the world's bat species, including the ghost bat (Figure 17-6), are listed as endangered or threatened.

Conservation biologists urge us not to kill bats but to protect them because of their important ecological roles. About 70% of all bat species feed on crop-damaging nocturnal insects and other pest species such as mosquitoes, making them the primary control agents for such insects.

In some tropical forests and on many tropical islands, pollen-eating bats pollinate flowers and fruit-eating bats distribute plants throughout tropical forests by excreting undigested seeds. As keystone species, they are vital for maintaining plant biodiversity and re-generating large areas of tropical forest cleared by human activities. If you enjoy bananas, cashews, dates, figs, avocados, or mangos, you can thank bats.

Many people mistakenly view bats as fearsome, filthy, aggressive, rabies-carrying bloodsuckers. But most bat species are harmless to people, livestock, and crops. In the United States, only 10 people have died of bat-transmitted disease in four decades of record-keeping; more Americans die each year from falling coconuts.

A: From 33 to 66

Percentage of mammal species classified as threatened, by country

| ■ Less than 10 | ■ 10 to 14.9 | ■ 15 to 19.9 | ■ 20 or more | □ No data |

Figure 17-5 Percentage of mammals at risk of becoming extinct. Of the 4,327 mammal species, 1,096 (about 25%) are at risk, and 169 are critically endangered with an extremely high risk of becoming extinct in the wild in the very near future. The highest percentage of mammal species at risk are apes and monkeys (46%), moles and shrews (36%), antelopes and cattle (33%), bats (26%), wild dogs, bears, and cats (26%), and rodents (17%). Countries with the highest numbers of mammal species threatened with extinction are Indonesia (128), China (75), India (75), Brazil (71), Mexico (64), and Australia (58). (Data from International Union for the Conservation of Nature and Conservation International)

One way to protect bats from human disturbance and killing is to prevent human access to caves and mines that are major bat roosts and hibernation sites. *Bat gates*—vandalproof metal grids that allow bats to enter and leave but keep out people—are one option. Another is to educate people about the importance of bats to humans and other species. We need to see bats as valuable allies, not as enemies.

🌐 17-3 CAUSES OF DEPLETION AND PREMATURE EXTINCTION OF WILD SPECIES

What Are the Root Causes of Wildlife Depletion and Extinction? Three underlying causes of population reduction and extinction of wildlife are

(1) human population growth (Figures 1-1 and 1-7); **(2)** economic systems and policies that fail to value the environment and its ecological services (Figure 1-2), thereby promoting unsustainable exploitation; and **(3)** greater per capita resource use as a result of increasing affluence and economic growth, which is a prime factor in the exploitation and degradation of wildlife habitats for human uses (Figures 1-18 and 1-19). In developing countries, the combination of rapid population growth (Figure 6-3) and poverty push the poor to cut forests, grow crops on marginal land, overgraze grasslands, deplete fish species, and kill endangered animals for their valuable furs, tusks, or other parts in order to survive.

These underlying causes lead to other more direct causes of the endangerment and extinction of wild species, such as **(1)** habitat loss and degradation, **(2)** habi-

Q: What percentage of the world's pollution is produced by the United States?

Figure 17-6 An endangered ghost bat carrying a mouse in tropical northern Australia. This nocturnal carnivore is harmless to people. Bats are considered keystone species in many ecosystems because of their roles in pollinating plants, dispersing seeds, and controlling insect and rodent populations. (G. B. Barker/A.N.T. Photo Library)

tat fragmentation, **(3)** commercial hunting and poaching, **(4)** overfishing, **(5)** predator and pest control, **(6)** sale of exotic pets and decorative plants, **(7)** climate change and pollution, and **(8)** deliberate or accidental introduction of nonnative (exotic or alien) species into ecosystems.

What Is the Role of Habitat Loss and Fragmentation? The greatest threat to all types of wild species is reduction of habitats as we increasingly occupy or degrade more of the planet (Figures 1-5 and 17-7). According to conservation biologists, tropical deforestation (Figure 16-3) is the greatest eliminator of species, followed by destruction of coral reefs and wetlands (Section 5-2), plowing of grasslands (Figure 5-10), and pollution of freshwater and marine habitats (Sections 11-6 and 11-11).

In the lower 48 states of the United States, 98% of the tall-grass prairies have been plowed, half of the wetlands drained, and 85–95% of old-growth forests cut (Figure 16-8). Overall forest cover has been reduced by 33%. At least 500 native species have been driven to extinction, and others to near-extinction (Figure 17-4), mostly because of habitat loss and fragmentation.

Island species, many of them *endemic species* found nowhere else on earth, are especially vulnerable to extinction. Most have no other place to go and few have evolved defenses against predators or diseases accidentally or deliberately introduced onto islands. Roughly half of the plants and animals known to have become extinct since 1600 were island species, even though islands make up only a small fraction of the earth's surface. Almost 900 bird species—10% of the world's known bird species—exist on only one island. The endangered *Symphonia* (Figure 17-4) clings

to life on the large island of Madagascar, a megadiversity country where 90% of the original vegetation has been destroyed; hundreds of this island's endemic species are also threatened with extinction.

The theory of island biogeography developed by Robert MacArthur and Edward O. Wilson has been used to predict the number and percentage of species that would become extinct when habitats are destroyed or seriously degraded. This model has been extended from islands to national parks, tropical rain forests, lakes, and nature reserves (Connections, p. 136), which can be viewed as *habitat islands* in an inhospitable sea of unsuitable habitat. Such models predict that when half of the habitat on an island or a habitat island is destroyed, approximately 10% of the native species will become extinct. When 90% of the habitat is destroyed, about 50% of the species will be lost. Such calculations are one tool conservation biologists use to estimate the number of current and future extinctions in various areas.

Most national parks and other protected areas are habitat islands, many of them surrounded by potentially damaging logging, mining, energy extraction, and industrial activities. Freshwater lakes are also habitat islands that are especially vulnerable when nonnative species are introduced, as has happened in the Great Lakes (Case Study, p. 322).

Migrating species face a double habitat problem. Nearly half of the 700 U.S. bird species spend two-thirds of the year in the tropical forests of Central or South America or on Caribbean islands, returning to North America during the summer to breed. A U.S. Fish and Wildlife study showed that between 1978 and 1987, populations of 44 of the 62 surveyed species of insect-eating, migratory songbirds in North America declined; 20 species experienced drops of 25–45%.

The main culprits are logging of tropical forests in the birds' winter habitats and fragmentation of their summer forest habitats in North America. The intrusion of farms, freeways, and suburbs break forests into patches. This makes it easier for opossums, skunks, squirrels, raccoons, and blue jays to feast on the eggs and the young of migrant songbirds. Another serious threat to songbirds from such habitat fragmentation comes from parasitic cowbirds that lay their eggs in the nests of various songbird species and have those birds raise their young for them.

Approximately 68% of the world's 9,600 known bird species are declining in numbers (58%) or are threatened with extinction (11%), mostly because of habitat loss and fragmentation. Conservation biologists view this loss and decline of bird species as an early warning of the greater loss of biodiversity and ecological integrity to come. Birds are excellent environmental indicators because they live in every climate and biome, respond quickly to environmental changes in their habitats, and are easy to track and count.

A: About 25%

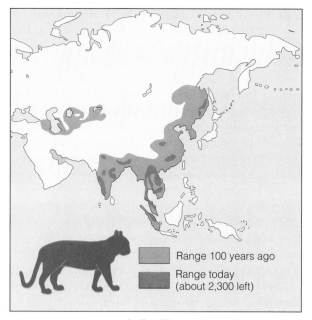

Indian Tiger

Range 100 years ago

Range today
(about 2,300 left)

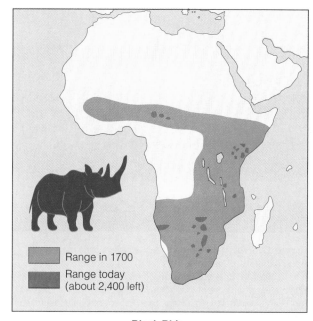

Black Rhino

Range in 1700

Range today
(about 2,400 left)

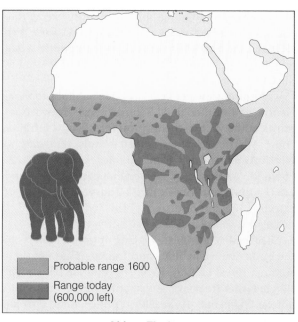

African Elephant

Probable range 1600

Range today
(600,000 left)

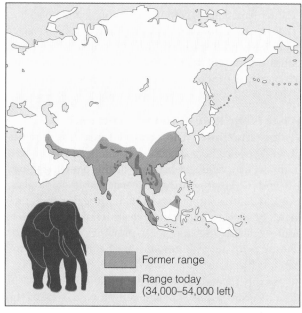

Asian or Indian Elephant

Former range

Range today
(34,000–54,000 left)

Figure 17-7 Reductions in the ranges of four wildlife species, mostly the result of a combination of habitat loss and hunting. What will happen to these and millions of other species when the global human population doubles in the next few decades? (Data from International Union for the Conservation of Nature and World Wildlife Fund)

In addition to serving as indicator species, birds play important ecological roles: They help control populations of insects (including the spruce budworm, gypsy moth, and tent caterpillar, which decimate many tree species) and rodents, pollinate a wide variety of flowering plants, and spread plants throughout their habitats by consuming plant seeds and excreting them in their droppings.

Habitat fragmentation reduces biodiversity in several ways. *First*, it increases the exposed surface area (edge), which makes many species vulnerable to predators, fires, changes in microclimate, winds, and outbreaks of disease. *Second*, many of the patches are too small to support the minimum breeding populations of some species. *Third*, fragmentation can create barriers that limit the ability of some species to disperse and colonize new areas, find enough to eat, and find mates.

What Is the Role of Commercial Hunting and Poaching? The international trade in wild plants and

Q: What percentage of the world's population live in urban areas?

animals is big business, bringing in up to $12 billion a year worldwide, with almost half of this trade involving the illegal sale of endangered and threatened species or their parts. Organized crime has moved into wildlife smuggling because of the huge profits involved (second only to drug smuggling). An estimated 60–80% of all live animals smuggled around the world die in transit.

Worldwide, some 622 species of animals and plants face extinction, mostly because of illegal trade. A live mountain gorilla is worth $150,000, a live chimpanzee $50,000, and a live Imperial Amazon macaw $30,000. Bengal tigers are at risk because a tiger fur sells for $100,000 in Tokyo. The skin of a snow leopard (Figure 17-4) can bring $14,000. Rhinoceros horn (Figure 17-8) sells for as much as $28,600 per kilogram. The pulverized bones of a rare Siberian tiger (only 450 left) used as a medicinal powder are worth more than $500,000.

Despite international protection, the world's total wild tiger population has dwindled to 4,600–8,000. Without emergency action, there may be few or no tigers left in the wild within 20 years. The main reasons for the rapid decline in the world's tigers are habitat loss (Figure 17-7) and poaching for their skins, bones (ground into a powder used to treat arthritis and rheumatism in parts of Asia), and penises (which are ground into a powder, cooked and made into "tiger wine" and a soup, which are consumed as an alleged aphrodisiac).

As more species become endangered, the demand for them on the black market soars, hastening their chances of extinction—another example of runaway positive feedback. Poaching of endangered or threatened species (many for markets in Asia) is increasing in the United States, especially in western national parks and wilderness areas covered by only 200 federal wildlife protection officers.

According to the U.S. Fish and Wildlife Service, a poached gyrfalcon sells for $120,000, a bighorn sheep head for $10,000–60,000, an average 87-gram (3-ounce) bear gallbladder (used in Asia for medicinal purposes) for $22,000, a large saguaro cactus for $5,000–15,000, a peregrine falcon (Figure 17-4) for $10,000, a polar bear for $6,000, a grizzly bear for $5,000, and a bald eagle for $2,500. Most poachers are not caught, and the money to be made far outweighs the risk of fines and the much smaller risk of imprisonment.

Case Study: Near Extinction of the American Bison

In 1500, before Europeans settled North America, 60–125 million North American bison grazed the plains, prairies, and woodlands over much of the continent. A single herd on the move might thunder past for hours. Several Native American tribes depended heavily on bison, and typically they killed only the animals they needed for food, clothing, and

Figure 17-8 Rhinoceros horns are carved into ornate dagger handles that sell for $500–12,000 in Yemen and other parts of the Middle East. In China and other parts of Asia, powdered rhino horn is used for medicinal purposes, particularly as a proven fever reducer, and occasionally as an alleged aphrodisiac. All five species of rhinoceros are threatened with extinction because of poachers (who kill them for their horns) and loss of habitat. Between 1973 and 1990 the population of African black rhinos (Figure 17-7) dropped from approximately 63,000 to about 2,400. (R. F. Porter/Ardea London)

shelter. By 1906, however, the once-vast range of the bison had shrunk to a tiny area, and the species had been driven nearly to extinction (Figure 17-9).

How did this happen? First, as settlers moved west after the Civil War, the sustainable balance between Native Americans and bison was upset. The Sioux, Cheyenne, Comanches, and other plains tribes traded bison skins to settlers for steel knives and firearms, so they began killing more bison.

The most relentless slaughter, however, was caused by the new settlers. As railroads spread westward in the late 1860s, railroad companies hired professional bison hunters, including Buffalo Bill Cody, to supply construction crews with meat. Passengers also gunned down bison from train windows for sport, leaving the carcasses to rot. Commercial hunters shot millions of bison for their hides and tongues (considered a delicacy), leaving most of the meat to rot. "Bone pickers" collected the bleached bones that whitened the prairies and shipped them east to be ground up as fertilizer.

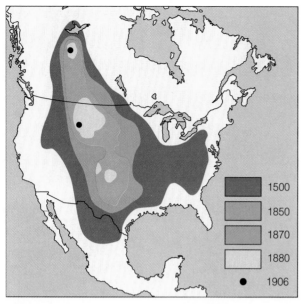

Figure 17-9 The historical range of the bison shrank severely between 1500 and 1906.

Figure 17-10 Vultures are feeding on this elephant carcass, poached in Tanzania for its ivory tusks, which are used to make jewelry, piano keys, ornamental carvings, and art objects. (E. R. Degginger)

Farmers shot bison because they damaged crops, fences, telegraph poles, and sod houses. Ranchers killed them because they competed with cattle and sheep for grass. The U.S. Army killed at least 2.5 million bison each year between 1870 and 1875 as part of their campaign to subdue the plains tribes by killing off their primary source of food.

By 1892 only 85 bison were left. They were given refuge in Yellowstone National Park and protected by an 1893 law against the killing of wild animals in national parks. In 1905, 16 people formed the American Bison Society to protect and rebuild the captive population. Soon thereafter, the federal government established the National Bison Range near Missoula, Montana. Today there are an estimated 200,000 bison, about 97% of them on privately owned ranches.

Some wildlife conservationists have suggested restoring large herds of bison on public lands in the North American plains. Not surprisingly, this idea has been strongly opposed by ranchers with permits to graze cattle and sheep on federally managed lands.

Case Study: How Should We Protect Elephants from Extinction? For decades poachers have slaughtered elephants for their valuable ivory tusks (Figure 17-10). Habitat loss (Figure 17-7) and legal and illegal trade in elephant ivory have reduced African elephant numbers from 2.5 million in 1970 to about 600,000 today.

Since 1989 this decline has been slowed by an international ban on the sale of ivory from African elephants, but things are not quite that simple. Increases in elephant populations in areas where their habitat has shrunk has resulted in widespread destruction of vegetation by these animals. This in turn reduces the niches available for other wild species. Some analysts also argue that sustainable legal harvesting of elephants for their ivory, meat, and hides, with the profits going mostly to local people, will be more successful in the long run than banning all ivory sales.

In several southern African nations in which elephant populations were high, the governments allowed a certain number of elephants to be killed each year. Income from this sustainable harvesting of elephants encouraged local people to protect elephants from unsustainable and illegal poaching, and it also provided governments with funds to help pay for conservation of elephants and other wildlife species. With the ban on ivory trade, such income dried up.

To restore these important sources of income, some wildlife conservationists and leaders of several southern African nations call for a partial lifting of the elephant ivory ban in areas in which the elephant populations are not endangered. Before being sold in the international marketplace, ivory from the sustainable culling of elephants in these areas would be marketed in a way that certifies that it was obtained legally; most of the profits would go to local people. There would also be strict measures to help prevent poaching, largely enforced by local people who would have an economic stake in preventing illegal killing of elephants.

In Africa, governments have a stockpile of at least 450 metric tons (500 tons) of ivory, most of which was confiscated from poachers. Government officials in these countries ask why they should be forced by an international ban to sit on a stockpile of ivory worth hundreds of million of dollars while their people beg and starve. Selling this ivory could provide much needed revenue for rural development, wildlife habitat acquisition, and wildlife protection.

Q: How much of the world's land area has been set aside to protect wildlife?

However, some analysts worry that most of the funds from sale of ivory stocks would end up in the pockets of corrupt bureaucrats. One proposal being considered by the World Wildlife Fund is a *debt-for-ivory swap* in which developed countries would write off part of an African country's debt in return for destruction of its stockpile of ivory tusks.

In most Western countries elephants are viewed as benign. However, according to David Western, head of the Kenya Wildlife Service, in some countries (such as Kenya) elephants are highly destructive of ecosystems and some wildlife preserves (such as Kenya's Amboseli National Park) where elephant populations are too high relative to available vegetation and space. Elephants also destroy fences, crops, and houses and sometimes kill people.

The ban on elephant ivory sales has increased the killing of other species with ivory parts such as bull walruses (for their ivory tusks, worth $500–1,500 a pair) and hippos (for their ivory teeth, worth $70 per kilogram)—another example of the principle that we can never do only one thing. Eastern Zaire's hippo population, once one of the world's largest, has dropped from about 23,000 to 11,000 since the 1989 elephant ban, mostly because of poaching for hippo teeth.

Others argue that opening up the legal elephant ivory market again, even partially, would lead to a renewal of massive poaching of elephants to supply a new ivory black market driven by increased consumption of ivory. They see continuing the ban as a way to help prevent African elephants from becoming threatened or endangered by keeping the price of ivory down.

In 1997 the Convention on International Trade in Endangered Species voted to let Zimbabwe, Botswana, and Namibia sell almost 55 metric tons (60 tons) of stockpiled ivory under strict monitoring. Profits from the sale will be used for elephant conservation programs.

What Is the Role of Overfishing in Reducing Aquatic Biodiversity? Various methods for harvesting fish are shown in Figure 17-11. Some fishing boats, called *trawlers*, catch demersal fish by dragging a funnel-shaped net held open at the neck along the ocean bottom. Newer trawling nets are large enough to swallow 12 jumbo jets in a single gulp, and even larger ones are on the way. The large mesh of the net allows most small fish to escape but

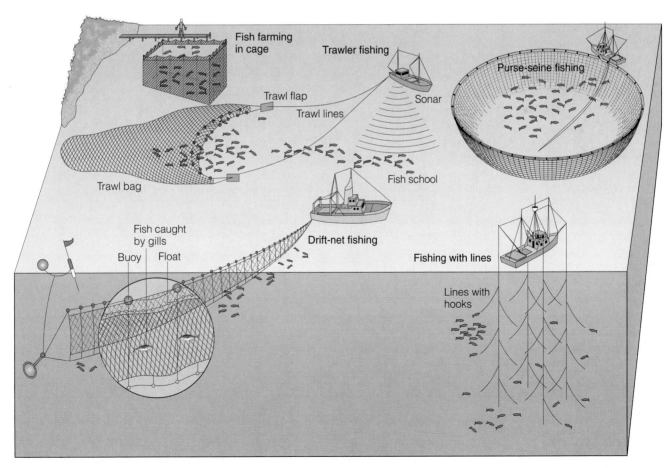

Figure 17-11 Major commercial fishing methods used to harvest various marine species.

can capture and kill other species such as seals and endangered and threatened sea turtles. Only the large fish are kept, with most of the fish and other aquatic species thrown back into the ocean either dead or dying.

Surface-dwelling species such as tuna, which feed in schools near the surface or in shallow areas, are often caught by *purse-seine fishing* (Figure 17-11). After a school is found, it's surrounded by a large purse-seine net. The net is then closed like a drawstring purse to trap the fish, and the catch is hauled aboard. Nets used to capture yellowfin tuna in the eastern tropical Pacific Ocean have also killed large numbers of dolphins, which swim on the surface above schools of the tuna. Another increasingly used method for catching fish is *longlining*, in which fishing vessels put out lines up to 120 kilometers (75 miles) long, hung with thousands of hooks.

One of the greatest threats to marine biodiversity is *drift-net fishing* (Figure 17-11). These monster nets drift in the water and catch fish when their gills become entangled in the nylon mesh. Each net descends as much as 15 meters (50 feet) below the surface and is up to 65 kilometers (40 miles) long. Almost anything that comes in contact with these nearly invisible "curtains of death" becomes entangled.

Q: Over the last 50 years what has happened to the total area of forest and woodlands in the United States?

In 1990 the UN General Assembly declared a moratorium on the use of drift nets longer than 2.5 kilometers (1.6 miles) in international waters after December 31, 1992, to help reduce overfishing and reduction of marine biodiversity. However, compliance is voluntary and there is no effective mechanism for monitoring fleets' activities over vast ocean areas and no structure for enforcement or punishment of violators. The financial rewards of illegal harvesting using drift nets far outweigh the very slim chances of being caught and having to pay small fines.

Even if illegal drift-netting could somehow be effectively banned, thousands of kilometers of *ghost nets*—stretches of netting that have been lost or abandoned and are adrift at sea—will continue to kill marine animals until the nets sink with the weight of decomposing bodies.

How Serious Are the Threats to the Biodiversity of Freshwater and Marine Systems? According to a number of aquatic scientists, the economic pressures to catch more and more fish pose a serious threat to freshwater and marine diversity. Most people are unaware how vulnerable freshwater environments are to environmental degradation and depletion.

Figure 17-12 Major species of sea turtles that have roamed the seas for 150 million years, showing their maximum adult lengths. These species are becoming endangered mostly because of loss or degradation of beach habitat (where they come ashore to lay their eggs), legal and illegal taking of their eggs (Guest Essay, pp. 418–419), and unintentional capture by commercial fishing boats.

A: It has remained about the same

Currently, about 34% of the known fish species are at risk of becoming extinct. According to a recent study by the World Resources Institute, coastal developments threaten marine biodiversity along roughly half of the world's coasts. The most obvious signs of this degradation of marine life include (1) dramatic declines in many of the world's fish stocks, mostly because of overfishing (Section 14-3), (2) destruction and bleaching of coral reefs (Figure 5-22, right), (3) massive die-offs of dolphins and seals, (4) loss of mangrove forests, (5) increasing incidences of huge blooms of toxic red tides, and (6) pollution and sedimentation.

Overfishing leads in most cases to *commercial extinction*, which is usually only a temporary depletion of fish stocks, as long as depleted areas are allowed to recover. However, large-scale fish harvesting can lead to serious depletion and extinction of species such as sea turtles (Figure 17-12, p. 490) and other marine mammals that are unintentionally caught.

What Is the Role of Predator and Pest Control?

People try to exterminate species that compete with them for food and game animals. For example, U.S. fruit farmers exterminated the Carolina parakeet around 1914 because it fed on fruit crops. The species was easy prey because when one member of a flock was shot, the rest of the birds hovered over its body, making themselves easy targets (Table 17-1).

African farmers kill large numbers of elephants to keep them from trampling and eating food crops. Ranchers, farmers, and hunters in the United States support the killing of coyotes, wolves, and other species that can prey on livestock and on species prized by game hunters. Since 1929 U.S. ranchers and government agencies have poisoned 99% of North America's prairie dogs because horses and cattle sometimes step into the burrows and break their legs. This has also nearly wiped out the endangered black-footed ferret (Figure 17-4), which preyed on the prairie dog. However, as a result of captive breeding started in 1981, 316 of these ferrets have been released into the wild in Montana, Wyoming, and South Dakota.

What Is the Role of the Market for Exotic Pets and Decorative Plants?

The global legal and illegal trade in wild species for use as pets is a huge and very lucrative business. However, for every live animal captured and sold in the pet market an estimated 50 other animals are killed.

Worldwide, over 5 million live wild birds are captured and sold legally each year and an estimated 2.5 million more are captured and sold illegally, primarily in Europe, Japan, and the United States. Over 40 species, mostly parrots, are endangered or threatened because of this wild bird trade. Collectors of exotic birds may pay up to $10,000 for a threatened hyacinth macaw (Figure 17-4) smuggled out of Brazil; however, during its lifetime, a single macaw left in the wild might yield as much as $165,000 in tourist income.

About 25 million U.S. households have exotic birds as pets, 85% of them imported. For every wild bird that reaches a pet shop legally or illegally, as many as 10 others die during capture or transport. Keeping pet birds may also be hazardous to human health. A 1992 study suggested that keeping indoor pet birds for more than 10 years doubles a person's chances of getting lung cancer.

Some exotic plants, especially orchids and cacti such as the black lace cactus (Figure 17-4), are endangered because they are gathered (often illegally) and sold to collectors to decorate houses, offices, and landscapes. A collector may pay $5,000 for a single rare orchid. A single rare mature crested saguaro cactus can earn cactus rustlers as much as $15,000. To thwart cactus rustlers, Arizona has put 222 species under state protection, with penalties of up to $1,000 and jail sentences of up to 1 year. However, only seven people are assigned to enforce this law over the entire state, and the fines are too small to discourage poaching.

What Are the Roles of Climate Change and Pollution?

A potential problem for many species is the possibility of fairly rapid (50–100 years) changes in climate, accelerated by deforestation and emissions of heat-trapping gases into the atmosphere (Sections 10-2 and 10-4). Wildlife in even the best-protected and best-managed reserves could be depleted in a few decades if such changes in climate take place.

Toxic chemicals degrade wildlife habitats, including wildlife refuges, and kill some terrestrial plants and animals and aquatic species. Slowly degradable pesticides, such as DDT, can be biologically magnified to very high concentrations in food chains and webs (Figure 8-2). The resulting high concentrations of DDT or other slowly biodegraded, fat-soluble organic chemicals can kill some organisms, reduce their ability to reproduce, or make them more vulnerable to diseases, parasites, and predators.

What Is the Role of Deliberately Introduced Species?

Travelers sometimes collect plants and animals intentionally and introduce them into new geographic regions. Many of these introduced species ultimately provide food, game, and aesthetic beauty and may even help control pests in their new environments.

However, some introduced species have no natural predators, competitors, parasites, or pathogens to control their numbers in their new habitats. This can allow them to dominate their new ecosystem and reduce or wipe out the populations of many native species, mak-

Q: How much of the municipal solid waste produced in the United States is recycled?

ing them a form of *biological pollution* (Table 17-2). The kudzu vine is an example of the effects of a deliberately introduced species (Connections, p. 494).

According to a 1993 Office of Technology Assessment study, more than 4,500 plant and animal species have been accidentally or intentionally introduced into the United States. Annual pest control costs for such species approach $100 billion and future losses are expected to be much higher. About 30% of the species in the United States on the official list of endangered and threatened species are there in part because of population declines caused by exotic species.

According to Chris Bright of the Worldwatch Institute, "To date, the study of exotics can be summed up in three negative statements: It's impossible to predict where an exotic will establish itself, or what it will do afterwards, or when it will do it." This is another lesson from nature indicating the need to use the *precautionary principle*, which recognizes that deliberately introducing any exotic species into an

---

Table 17-2 Damage Caused by Plants and Animals Imported into the United States

Name	Origin	Mode of Transport	Type of Damage
Mammals			
European wild boar	Russia	Intentionally imported (1912), escaped captivity	Destroys habitat by rooting; damages crops
Nutria (cat-sized rodent)	Argentina	Intentionally imported, escaped captivity (1940)	Alters marsh ecology; damages levees and earth dams; destroys crops
Birds			
European starling	Europe	Intentionally released (1890)	Competes with native songbirds; damages crops; transmits swine diseases; causes airport nuisance
House sparrow	England	Intentionally released by Brooklyn Institute (1853)	Damages crops; displaces native songbirds
Fish			
Carp	Germany	Intentionally released (1877)	Displaces native fish; uproots water plants; lowers waterfowl populations
Sea lamprey	North Atlantic Ocean	Entered Great Lakes via Welland Canal (1829)	Wiped out lake trout, lake whitefish, and sturgeon in Great Lakes
Walking catfish	Thailand	Imported into Florida	Destroys bass, bluegill, and other fish
Insects			
Argentine fire ant	Argentina	Probably entered via coffee shipments from Brazil (1918)	Damages crops; destroys native ant species
Camphor scale insect	Japan	Accidentally imported on nursery stock (1920s)	Damaged nearly 200 plant species in Louisiana, Texas, and Alabama
Japanese beetle	Japan	Accidentally imported on irises or azaleas (1911)	Defoliates more than 250 species of trees and other plants, including many of commercial importance
Plants			
Water hyacinth	Central America	Accidentally introduced (1884)	Clogs waterways; shades out other aquatic vegetation
Chestnut blight (fungus)	Asia	Accidentally imported on nursery plants (1900)	Killed nearly all eastern U.S. chestnut trees; disturbed forest ecology
Dutch elm disease (fungus)	Europe	Accidentally imported on infected elm timber used for veneers (1930)	Killed millions of elms; disturbed forest ecology

From *Biological Conservation* by David W. Ehrenfeld. Copyright © 1970 by Holt, Rinehart & Winston, Inc. Modified and reprinted by permission.

A: About 24% (up from 7% in 1970)

Deliberate Introduction of the Kudzu Vine

CONNECTIONS

In the 1930s, the *kudzu vine* was imported from Japan and planted in the southeastern United States to help control soil erosion. It does control erosion, but it is so prolific and difficult to kill that it engulfs hillsides, trees, houses, stream banks, patches of forest, and anything else in its path (Figure 17-13). It has spread throughout much of the southern United States and is moving north into Illinois and Pennsylvania. Kudzu could spread as far north as the Great Lakes by 2040 if projected global warming occurs.

Although kudzu is considered a menace in the United States, Asians use a powdered kudzu starch in beverages, gourmet confections, and herbal remedies for a range of diseases. A Japanese firm is building a large kudzu farm and processing plant in Alabama and will ship the extracted starch to Japan, where demand exceeds supply.

U.S. researchers are currently testing a kudzu-based drug made for reducing an alcoholic's craving for alcohol and may seek FDA approval for use in a few years. In an ironic twist, kudzu—which can engulf and kill trees—could eventually help save trees from loggers. Research at Georgia Institute of

Technology indicates that kudzu may join kenaf (Figure 16-10) as a source of tree-free paper.

Critical Thinking

On balance, do you think the potential beneficial uses of the kudzu vine outweigh its harmful ecological effects? Explain.

Figure 17-13 Kudzu taking over a field and trees near Lyman, South Carolina. This vine can grow 0.3 meter (1 foot) per day and is found from East Texas to Florida and as far north as southeastern Pennsylvania and Illinois. Kudzu was deliberately introduced into the United States for erosion control but cannot be stopped by being dug up or burned. Grazing by goats and repeated doses of herbicides can destroy it, but goats and herbicides also destroy other plants, and herbicides can contaminate water supplies. (Angelina Lax/Photo researchers, Inc.)

ecosystem is a risky thing to do. Once an introduced species has become established in an ecosystem its wholesale removal is almost impossible—somewhat like trying to get smoke back into a chimney or unscramble an egg.

What Is the Role of Accidentally Introduced Species? One example of the effects of the accidental introduction of a nonnative species was the invasion of the Great Lakes by zebra and quagga mussels (Case Study, p. 322).

Another example involves the accidental introduction of fire ants into the United States. In the late 1930s, extremely aggressive red fire ants were accidentally introduced into the United States in Mobile, Alabama, perhaps arriving on shiploads of lumber imported from South America. Without natural predators, these ants have spread rapidly by land and water (they can float) throughout the South from Texas to Florida and as far north as Tennessee and Virginia.

They have also hitched rides on truckloads of produce going to Arizona, California, New Mexico, Ore-

Q: What percentage of the municipal solid waste in the United States could be recycled?

gon, and Washington. So far, the ants' aversion to frost has kept them out of the Midwest and Northwest, but this might not last long because of genetic changes in some of their populations.

Wherever the fire ant has gone, up to 90% of native ant populations have been sharply reduced or wiped out. Interference competition by fire ants has also reduced populations of many other species, including ladybugs and many species of spiders, ticks, and cockroaches.

Their extremely painful stings have also killed deer fawn, lizards, birds, livestock, pets, and at least 100 people allergic to their venom. In the South, they have made small children afraid to play in their backyards. They have invaded cars and caused accidents by attacking the driver. They have made cropfields unplowable, disrupted phone service and electrical power, and caused some fires by chewing through underground cables.

Mass spraying of pesticides in the 1950s and 1960s temporarily reduced populations of fire ants. In the end, however, this chemical warfare hastened the advance of the fire ant by reducing populations of many native ant species. The rapidly multiplying fire ants also developed genetic resistance to heavily used pesticides.

Researchers at the U.S. Department of Agriculture are evaluating the use of a tiny parasitic fly that lays its eggs on the fire ant's body. After the eggs hatch, the larvae eat their way through the ant's head. Before releasing these biological control agents, however, researchers must be sure they will not cause problems for native ant species.

17-4 SOLUTIONS: PROTECTING WILD SPECIES FROM DEPLETION AND EXTINCTION

How Can We Protect Wildlife and Biodiversity? There are three basic approaches to managing wildlife and protecting biodiversity. The *ecosystem approach*, discussed in Chapter 16, aims to preserve balanced populations of species in their native habitats, establish legally protected wilderness areas and wildlife reserves, and eliminate or reduce the populations of nonnative species.

Biologists consider protecting ecosystems to be the best way to preserve biological diversity and ecological integrity. The basic problem is that fully or partially protected wildlife sanctuaries make up only about 6% of the world's land area, and the human population is expected to double in 40 years.

The *species approach* is based on protecting endangered species by identifying them, giving them legal protection, preserving and managing their crucial habitats, propagating them in captivity, and re-

introducing them into suitable habitats. The *wildlife management approach* manages game species for sustained yield by using laws to regulate hunting, establishing harvest quotas, developing population management plans, and using international treaties to protect migrating game species such as waterfowl. These two approaches are discussed in the remainder of this chapter.

How Can International Treaties Help Protect Endangered Species? Several internationaltreaties and conventions help protect endangered or threatened wild species. One of the most far-reaching is the 1975 Convention on International Trade in Endangered Species (CITES). This treaty, now signed by 124 countries, lists 675 species that cannot be commecially traded as live specimens or wildlife products because they are endangered or threatened.

However, enforcement of this treaty is spotty, convicted violators often pay only small fines, and member countries can exempt themselves from protecting any listed species. Furthermore, much of the estimated $6 billion annual illegal trade in wildlife and wildlife products goes on in countries that have not signed the treaty. Centers of illegal animal trade are Singapore, Argentina, Indonesia, Spain, Taiwan, and Thailand.

How Can Laws Help Protect Endangered Species? The United States controls imports and exports of endangered wildlife and wildlife products through two important laws. The Lacey Act of 1900 prohibits transporting live or dead wild animals or their parts across state borders without a federal permit. The Endangered Species Act of 1973 (amended in 1982 and 1988) makes it illegal for Americans to import or trade in any product made from an endangered or threatened species unless it is used for an approved scientific purpose or to enhance the survival of the species.

The Endangered Species Act of 1973 is one of the world's toughest environmental laws. It authorizes the National Marine Fisheries Service (NMFS) to identify and list endangered and threatened ocean species; the U.S. Fish and Wildlife Service (USFWS) identifies and lists all other endangered and threatened species. These species cannot be hunted, killed, collected, or injured in the United States.

Any decision by either agency to add or remove a species from the list must be based on biology only, not on economic or political considerations. However, economic factors can be used in deciding whether and how to protect endangered habitat and in developing recovery plans for listed species. The act also forbids federal agencies to carry out, fund, or authorize projects that would either jeopardize an endangered or threatened species or destroy or modify the critical habitat it needs to survive. On private lands, fines and even jail sentences can be imposed to ensure protection of the habitats of endangered species.

Between 1973 and 1996, the number of U.S. species included in the official endangered and threatened list increased from 92 to 959 (including an additional 566 species found elsewhere). By 1996, 3,700 more species were being evaluated as candidates for listing.

Getting listed is only half the battle. Next, the USFWS or the NMFS is supposed to prepare a plan to help the species recover. However, because of a lack of funds, final recovery plans have been developed and approved for only about two-thirds of the endangered or threatened U.S. species, and half of those plans exist only on paper.

The act requires that all commercial shipments of wildlife and wildlife products enter or leave the country through one of nine designated ports. Few illegal shipments are confiscated (Figure 17-14) because the 60 USFWS inspectors can examine only about one-fourth of the 90,000 shipments that enter and leave the United States each year. Even if caught, many violators are not prosecuted, and convicted violators often pay only a small fine.

According to a 1995 study by a panel of scientists for the U.S. National Academy of Sciences, the information needed to designate the critical habitat of an endangered or threatened species is often not available. The study also points out that just because a species is found in a particular habitat does not necessarily mean that the species requires that habitat.

This lack of important ecological data often makes identification of a species' critical habitat time-consuming and controversial, thus delaying or preventing protection. To avoid this problem the panel of scientists recommended that when a species is listed as endangered, a core amount of *survival habitat* should be set aside immediately as an emergency measure to help ensure short-term survival of the species for 25–50 years. If the critical habitat is later identified, the survival habitat area would automatically expire.

Should the Endangered Species Act Be Weakened? Since 1995 there have been intense efforts to seriously weaken the Endangered Species Act by **(1)** making the protection of endangered species on private land voluntary, **(2)** having the government pay landowners if it forces them to stop using part of their land to protect endangered species (the issue of regulatory takings, p. 216), **(3)** making it harder and more expensive to list newly endangered species by requiring government wildlife officials to navigate through a series of hearings and peer-review panels, and **(4)** giving the secretary of interior the power to permit a listed species to become extinct without trying to save it, **(5)** allowing the secretary of the interior to give any state, county, or landowner permanent exceptoin from the law, with no requirement for public notification or comment, **(6)** allowing landowners to lock in long-term endangered species management plans—known as *habitat conservation* plans (HCPs)—that exempt the owners from further obligations for 100 years or more, and **(7)** prohibiting the public from commenting on or bringing lawsuits on any changes in habitat conservation plans for endangered species.

Those who favor weakening or eliminating the Endangered Species Act argue that it has been a failure

Figure 17-14 Confiscated products made from endangered species. Because of a scarcity of funds and inspectors, probably no more than one-tenth of the illegal wildlife trade in the United States is discovered. The situation is even worse in most other countries. (Steve Hillebrand/U.S. Fish and Wildlife Service)

Q: What percentage of the rivers and lakes in the United States are fishable and swimmable?

because only seven species have been removed from the list and only 20 species have recovered enough to be reclassified from endangered to threatened. Some of these critics have spread horror stories about how environmentalists have used endangered species (many of them small and unfamiliar) to block development and resource extraction, violate public property rights, and waste tax dollars trying to save useless creatures that are on the verge of extinction anyway.

The late California Representative Sonny Bono joked that the best way to deal with endangered species is to "give them all a designated area and then blow it up." Washington Senator Slade Gordon suggests that all endangered species be removed from the wild and bred in zoos as "a way to preserve animals without blocking economic development."

Should the Endangered Species Act Be Strengthened?

Most conservation biologists and wildlife scientists contend that the Endangered Species Act has not been a failure (Spotlight, below). They also refute the charge that the act has cause severe economic losses.

The truth is that the Endangered Species Act has had virtually no impact in the nation's overall eco-nomic development. For example, between 1989 and 1992 only 55 of more than 118,000 projects evaluated by the USFWS were blocked or withdrawn as a result of the Endangered Species Act.

Moreover, *the act does allow for economic concerns.* By law, a decision to list a species must be based solely on science. But once a species is listed, economic considerations can be weighed against species protection in protecting critical habitat and in designing and implementing recovery plans. The act also allows a special Cabinet-level panel, called the "God Squad," to exempt any federal project from having to comply with the act if the economic costs are too high. In addition, the act allows the government to issue permits and exemptions to landowners with listed species living on their property.

Most biologists and wildlife conservationists believe that we should develop a new system to protect and sustain biological diversity and ecological integrity based on three principles: **(1)** Find out what species and ecosystems we have, **(2)** locate and protect the most endangered ecosystems and species, and **(3)** give private landowners financial carrots (tax breaks and write-offs) for helping protect endangered species and ecosystems.

Has the Endangered Species Act Been a Failure?

SPOTLIGHT

Critics of the Endangered Species Act call it an expensive failure because only a few species have been removed from the endangered list. Most of these critics are ranchers, developers, and officials of timber and mining companies who want more access to resources on public lands (Spotlight, p. 217).

Most biologists strongly disagree that the act has been a failure, for several reasons. *First*, species are listed only when they are already in serious danger of extinction. This is like setting up a poorly funded hospital emergency room that takes only the most desperate cases, often with little hope for recovery, and then saying it should be shut down because it has not saved enough patients.

Second, it takes decades for most species to become endangered or threatened. Thus, it should not be surprising that it usually takes decades to bring a species in critical condition back to the point where it can be removed from the list. Expecting the Endangered Species Act (which has been in existence only since 1973) to quickly repair the biological depletion of centuries is unrealistic. The most important measure of the law's success is that the conditions of almost 40% of the listed species are stable or improving. A hospital emergency room taking only the most desperate cases yet stabilizing or improving the condition of 40% of its patients would be considered an astounding success.

Third, the federal endangered species budget was only $93 million in 1996 (up from $23 million in 1993). This funding is equal to about what it takes to build 3.2 kilometers (1.5 miles) of urban interstate highway, or about 27¢ a year per U.S. citizen—a pittance to help save some of the country's irreplaceable biodiversity. One C-17 transport airplane costs $300 million—almost six times what was spent on protecting endangered species in the United States during 1996.

To most biologists it's amazing that so much has been accomplished in stabilizing or improving the condition of almost 40% of nearly terminal species on a shoestring budget. However, critics call the act an expensive failure.

Critical Thinking

1. Do you believe that the Endangered Species act should be weakened or strengthened? Explain.

2. Do you agree or disagree with each of the proposals made to **(a)** weaken the act given on p. 496, and **(b)** strengthen the act listed on p. 497. Defend each of your answers.

A step toward implementing the first principle was made in 1993 when Interior Secretary Bruce Babbitt launched a *national biological survey* to undertake a complete census of U.S. wild species, along with an inventory of the biological communities and ecosystems in which these species live. This is designed to help identify centers of biodiversity and the species and ecosystems most in need of protection. However, its budget has been severely limited by Congress, mostly because of pressure by economic interests favoring weakening or elimination of the Endangered Species Act.

Conservation biologists suggest that the second principle be implemented by greatly increasing currently minuscule funding for implementing the Endangered Species Act. They also urge Congress to pass a *National Biodiversity Protection Act* to identify, protect, and in some cases restore the country's most threatened wildlife habitats and ecosystems. The act would emphasize emergency protection of the most threatened hot spots of biodiversity. This prevention or ecosystem management approach would protect large numbers of species in functioning ecosystems and thus help reduce the number of species needing emergency care under the Endangered Species Act.

Should We Try to Protect All Endangered and Threatened Species? Because of limited funds and trained personnel, only a few endangered and threatened species can be saved. Many wildlife experts suggest that the limited funds available for preserving threatened and endangered wildlife be concentrated on species that **(1)** have the best chance for survival, **(2)** have the most ecological value to an ecosystem, and **(3)** are potentially useful for agriculture, medicine, or industry.

Others oppose such ideas on ethical grounds or contend that currently we don't have enough biological information to make such evaluations. Proponents argue that, in effect, we are already deciding by default which species to save and that despite limited knowledge, this approach is more effective and a better use of limited funds than the current one. What do you think?

How Can Wildlife Refuges and Other Protected Areas Help Protect Endangered Species? In 1903 President Theodore Roosevelt established the first U.S. federal wildlife refuge at Pelican Island off Florida's Atlantic coast to protect the brown pelican from overhunting and loss of habitat. Before the end of Roosevelt's first term he had created 49 more refuges. Since then the National Wildlife Refuge System has grown to 508 refuges (Figure 16-2). About 85% of the area included in these refuges is in Alaska.

Over three-fourths of the refuges are wetlands for protection of migratory waterfowl. About 20% of the species on the U.S. endangered and threatened

list have habitats in the refuge system, and some refuges have been set aside for specific endangered species. These have helped Florida's Key deer, the brown pelican, and the trumpeter swan to recover. Conservation biologists urge the establishment of more refuges for endangered plants. They are also urging Congress and state legislatures to allow abandoned military lands that contain significant wildlife habitat to become national or state wildlife refuges.

So far Congress has not established guidelines (such as multiple use or sustained yield) for management within the National Wildlife Refuge System, as it has for other public lands. A 1990 report by the General Accounting Office found that activities considered harmful to wildlife occur in nearly 60% of the nation's wildlife refuges. In addition, a 1986 USFWS study estimated that one federal refuge in five is contaminated with chemicals from old toxic-waste dump sites (including military bases) and runoff from nearby agricultural land. Private groups have also helped establish biodiversity and wildlife sanctuaries (Solutions, p. 372)

Case Study: Oil and Gas Development in the Arctic National Wildlife Refuge The Arctic - National Wildlife Refuge on Alaska's North Slope (Figure 17-15) contains more than one-fifth of all the land in the National Wildlife Refuge System and has been called the crown jewel of the system. The refuge's coastal plain, its most biologically productive part, is the only stretch of Alaska's arctic coastline not open to oil and gas development.

For years, U.S. oil companies have been working to change this because they believe that this coastal area *might* contain oil and natural gas deposits. They argue that such exploration is needed to increase U.S. oil and natural gas supplies and to reduce dependence

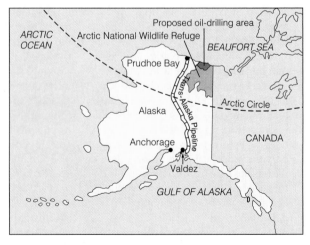

Figure 17-15 Proposed oil-drilling area in Alaska's Arctic National Wildlife Refuge. (Courtesy U.S. Fish and Wildlife Service)

Q: What percentage of the wetlands in the United States have been lost or seriously degraded?

on oil imports. However, oil companies plan to export most of any oil they find in the reserve to Japan.

Conservation biologists oppose this proposal and urge Congress to designate the entire coastal plain as wilderness. They cite Department of Interior estimates that there is only a 19% chance of finding as much oil there as the United States consumes every 6 months. Even if the oil does exist, environmentalists do not believe the potential degradation of any portion of this irreplaceable wilderness area would be worth it, especially when improvements in energy efficiency would save far more oil at a much lower cost (Sections 18-2 and 18-3).

Oil company officials claim they have developed Alaska's Prudhoe Bay oil fields without significant harm to wildlife; they also contend that the area they seek to open to oil and gas development is less than 1.5% of the entire coastal plain region. However, the huge 1989 oil spill from the tanker *Exxon Valdez* in Alaska's Prince William Sound (Figure 11-26) casts serious doubt on such claims.

Moreover, a study leaked from the U.S. Fish and Wildlife Service in 1988 revealed that oil drilling at Prudhoe Bay has caused much more air and water pollution than was anticipated before drilling began in 1972. According to this study, oil development in the coastal plain could cause the loss of 20–40% of the area's 180,000-head caribou herd, 25–50% of the remaining musk oxen, 50% or more of the wolverines, and 50% of the snow geese that live there part of the year. A 1988 EPA study also found that "violations of state and federal environmental regulations and laws are occurring at an unacceptable rate" in portions of the Prudhoe Bay area in which oil fields and facilities have been developed.

Do you believe that possible oil and gas reserves should be developed in this wildlife refuge? Why?

Can Gene Banks and Botanical Gardens Help Save Most Endangered Species? Botanists preserve genetic information and endangered plant species by storing their seeds in gene or seed banks—refrigerated, low-humidity environments (p. 410). Scientists urge that many more such banks be established, especially in developing countries. But some species can't be preserved in gene banks, and maintaining the banks is very expensive.

The world's 1,500 botanical gardens and arboreta maintain at least 40,000 plant species. However, these sanctuaries have too little storage capacity and too little funding to preserve most of the world's rare and threatened plants.

Can Zoos Help Protect Most Endangered Species? Worldwide, 1,000 zoos house about 540,000 terrestrial vertebrate animals from 3,000 species of mam-

mals, amphibians, birds, and reptiles. Many of them are neither threatened nor endangered.

Traditionally, zoos have focused on large, charismatic invertebrates because they are of the greatest interest to the public and thus provide the most income. However, zoos and animal research centers are increasingly being used to preserve some individuals of critically endangered animal species, with the long-term goal of reintroducing the species into protected wild habitats.

Two techniques for preserving such species are egg pulling and captive breeding. *Egg pulling* involves collecting wild eggs laid by critically endangered bird species and then hatching them in zoos or research centers. In *captive breeding*, some or all wild individuals of a critically endangered species are captured for breeding in captivity, with the aim of reintroducing the offspring into the wild. Other techniques for increasing the populations of captive species include artificial insemination, surgical implantation of eggs of one species into a surrogate mother of another species (embryo transfer), use of incubators, and having the young of a rare species raised by parents of a similar species (cross-fostering).

Captive breeding programs at zoos in Phoenix, San Diego, and Los Angeles saved the nearly extinct Arabian oryx (Figure 17-4), a large antelope that once lived throughout the Middle East. By the early 1970s it had been hunted nearly to extinction in the wild by people riding in Jeeps and helicopters and wielding rifles and machine guns. Since 1980 a few oryx bred in captivity have been returned to the wild in protected habitats in the Middle East; the wild population is now about 120. Endangered U.S. species now being bred in captivity and returned to the wild include the peregrine falcon (Figure 17-4, changed from endangered to threatened in 1995) and the black-footed ferret (Figure 17-4). Endangered golden lion tamarins (Figure 17-4) bred at the National Zoo in Washington, D.C., have been released in Brazilian rain forests.

For every successful reintroduction, however, there are many more species that can't go home again because their habitats no longer exist or have been too degraded to support the minimum population size needed for survival. Thus many zoos are becoming arks that cannot set free their valuable and endangered passengers.

Efforts to maintain populations of endangered species in zoos and research centers are limited by lack of space and money. The captive population of each species must number 100 to 500 individuals to avoid extinction through accident, disease, or loss of genetic diversity through inbreeding. Recent genetic research indicates that 10,000 or more individuals are needed for an endangered species to maintain its capacity for biological evolution.

The world's zoos now house only 27 endangered animal species that have populations of 100 or more individuals. It is estimated that today's zoos and research centers have space to preserve healthy and sustainable populations of only 925 of the 2,000 large vertebrate species that could vanish from the planet. It is doubtful that the more than $6 billion needed to care for these animals for 20 years will become available.

Some critics see zoos as prisons for once wild animals. They also contend that zoos foster the notion that we don't need to preserve large numbers of wild species in their natural habitats.

Whether one agrees or disagrees with this position, it is clear that zoos and botanical gardens are not a biologically or economically feasible solution for most of the world's current endangered species and the much larger number expected to become endangered over the next few decades.

17-5 WILDLIFE MANAGEMENT

How Can Wildlife Populations Be Managed?
Wildlife management entails manipulating wildlife populations (especially game species) and their habitats for their welfare and for human benefit. It includes preserving endangered and threatened wild species and enforcing wildlife laws.

In the United States, funds for state game management programs come from the sale of hunting and fishing licenses and from federal taxes on hunting and fishing equipment. Two-thirds of the states also have provisions on state income tax returns that allow individuals to contribute money to state wildlife programs. Only 10% of all government wildlife dollars are spent to study or benefit nongame species, which make up nearly 90% of the country's wildlife species. Since the passage of the Wildlife Restoration Act in 1937 there has been a spectacular increase in the number of game animals (such as white-tailed deer, wild turkeys, Rocky Mountain elk, and pronghorn antelope) sought by many sport hunters.

The first step in wildlife management is to decide which species are to be managed in a particular area, a decision that is a source of much controversy. Ecologists and conservation biologists emphasize preserving biodiversity and ecological integrity, wildlife conservationists are concerned about endangered species, birdwatchers want the greatest diversity of bird species, and hunters want sufficiently large populations of game species for harvest during hunting season. In the United States and other developed countries, most wildlife management is devoted to producing surpluses of game animals and game birds for hunters.

After goals have been set, the wildlife manager must develop a management plan. Ideally, this is based on the principles ecological succession (Figures 5-41 and 5-42, wildlife population dynamics (Figures 5-27, 5-28, 5-29, and 5-30), and an understanding of the cover, food, water, space, and other habitat requirements of each species to be managed. The manager must also consider the number of potential hunters, their likely success rates, and the regulations for preventing excessive harvesting.

This information is difficult, expensive, and time-consuming to obtain. It involves much educated guesswork and trial and error, which is why wildlife management is as much an art as a science. Management plans must also be sensitive to political pressures from conflicting groups and to budget constraints.

How Can Vegetation and Water Supplies Be Manipulated to Manage Wildlife? Wildlife managers can encourage the growth of plant species that are the preferred food and cover for a particular animal species in a given area by controlling the ecological succession of the vegetation in that area.

Wildlife species can be classified into four types, according to the stage of ecological succession at which they are most likely to be found: early successional, midsuccessional, late successional, and wilderness (Figure 17-16). *Early successional species* find food and cover in weedy pioneer plants that invade an area that has been disturbed, whether by human activities or natural phenomena (fires, volcanic activity, or glaciation).

Midsuccessional species are found around abandoned croplands and partially open areas created by logging of small stands of timber, controlled burning, and clearing of vegetation for roads, firebreaks, oil and gas pipelines, and electrical transmission lines. Such openings of the forest canopy promote the growth of vegetation favored by midsuccessional mammal and bird species; they also increase the amount of edge habitat, where two communities such as a forest and a field come together. This transition zone allows animals such as deer to feed on vegetation in clearings and then quickly escape to cover in the nearby forest.

Late successional species rely on old-growth and mature forests to produce the food and cover on which they depend. These animals require the protection of moderate-sized, old-growth forest refuges.

Wilderness species flourish only in undisturbed, mature vegetation communities, such as large areas of old-growth forests, tundra, grasslands, and deserts. They can survive only in large wilderness areas and wildlife refuges.

Various types of habitat management can be used to attract a desired species and encourage its population growth. Examples are planting seeds, transplanting certain types of vegetation, building artificial nests, and deliberately setting controlled, low-level ground

Q: What percentage of the world' species are found in tropical forests?

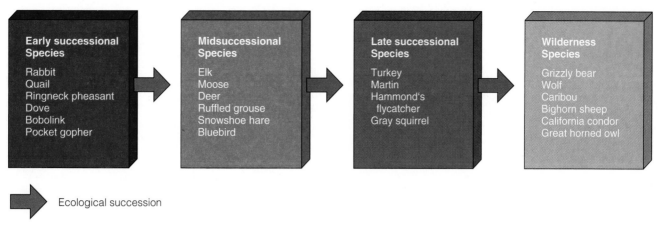

Ecological succession

Figure 17-16 Examples of wildlife species typically found in areas at different stages of ecological succession in areas of the United States with a temperate climate.

fires to help control vegetation. Wildlife managers often create or improve ponds and lakes in wildlife refuges to provide water, food, and habitat for waterfowl and other wild animals.

How Useful Is Sport Hunting in Managing Wildlife Populations? Most developed countries use sport hunting laws to manage populations of game animals. Licensed hunters are allowed to hunt only during certain portions of the year so as to protect animals during their mating season. Limits are set on the size, number, and sex of animals that can be killed, as well as on the number of hunters allowed in a given area.

Close control of sport hunting is difficult. Accurate data on game populations may not exist and may cost too much to get. People in communities near hunting areas, who benefit from money spent by hunters, may seek to have hunting quotas raised.

How Can Populations of Migratory Waterfowl Be Managed? Migratory birds, including ducks, geese, swans, and many songbirds, make north–south journeys from one habitat to another each year, usually to find food, suitable climate, and other conditions necessary for reproduction. Such bird species use many different north–south routes, called **flyways**, but only about 15 are considered major routes (Figure 17-17). Some countries along such flyways have entered into agreements and treaties to protect crucial habitats needed by such species, both along their migration routes and at each end of their journeys.

In North and South America, migratory waterfowl (ducks, geese, and swans) nest in Canada during the summer. Then during the fall they migrate to the United States and Central America along generally fixed flyways.

Canada, the United States, and Mexico have signed agreements to prevent habitat destruction and overhunting of migratory waterfowl. However, between 1972 and 1992 the estimated breeding population of ducks in North America dropped 38%, mostly because of prolonged drought in key breeding areas and degradation and destruction of wetland and grassland nesting sites by farmers (p. 106). What wetlands remain are used by flocks of ducks and geese, whose high densities in shrinking wetlands make them more vulnerable to diseases and to predators such as skunks, foxes, coyotes, raccoons, and hunters. Waterfowl in wetlands near croplands are also exposed to pollution from pesticides and other toxic chemicals such as selenium in irrigation runoff.

Wildlife officials manage waterfowl by regulating hunting, protecting existing habitats, and developing new habitats, including artificial nesting sites, ponds, and nesting islands. More than 75% of the federal wildlife refuges in the United States are wetlands used by migratory birds. Local and state agencies and private conservation groups such as Ducks Unlimited, the Audubon Society, and the Nature Conservancy have also established waterfowl refuges.

In 1986 the United States and Canada agreed to spend $1.5 billion over a 24-year period, with the goal of almost doubling the continental duck-breeding population, mostly through purchase, improvement, and protection of additional waterfowl habitats in five priority areas. According to Ducks Unlimited, the number of migratory waterfowl in North America increased by 20% between 1986 and 1994. Conservationists fear that this gain may be undermined by congressional efforts to eliminate funding for this program.

Since 1934 the Migratory Bird Hunting and Conservation Stamp Act has required waterfowl hunters to buy a duck stamp each season they hunt. Revenue

A: At least 50% (some biologists say 90%)

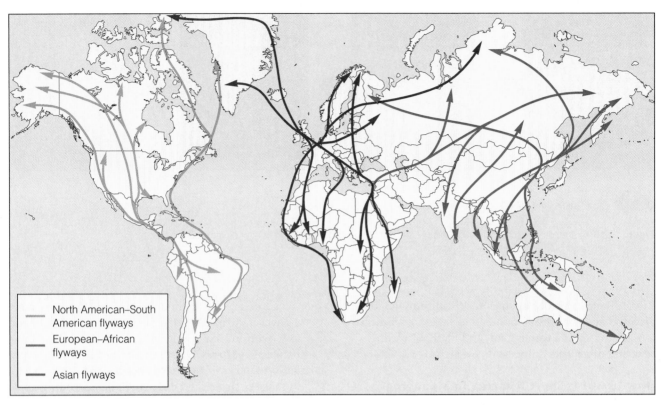

Figure 17-17 Major flyways used by migratory birds, mostly waterfowl. Each route has a number of subroutes.

from these sales goes into a fund to buy land and easements for the benefit of waterfowl.

17-6 FISHERY MANAGEMENT AND PROTECTING MARINE BIODIVERSITY

How Can Freshwater Fisheries Be Managed and Sustained? Managing freshwater fish involves encouraging populations of commercial and sport fish species and reducing or eliminating populations of less desirable species. Several techniques are used. This is usually done by regulating time and length of fishing seasons and the number and size of fish that can be taken.

Other techniques include building reservoirs and farm ponds and stocking them with fish, fertilizing nutrient-poor lakes and ponds, and protecting and creating spawning sites. Habitats can also be protected from buildup of sediment and other forms of pollution, and debris can be removed. Excessive growth of aquatic plants from cultural eutrophication (Figure 11-22) can be prevented, and small dams can be built to control water flow.

Predators, parasites, and diseases can be controlled by improving habitats, breeding genetically resistant fish varieties, and using antibiotics and disinfectants judiciously. Hatcheries can be used to restock ponds, lakes, and streams with prized species such as trout and salmon, and entire river basins can be managed to protect such valued species as salmon (Figure 14-16).

How Can Marine Fisheries Be Managed? By international law, a country's offshore fishing zone extends to 370 kilometers (200 nautical miles, or 230 statute miles) from its shores. Foreign fishing vessels can take certain quotas of fish within such zones, called *exclusive economic zones*, but only with a government's permission. Ocean areas beyond the legal jurisdiction of any country are known as the *high seas*; any limits on the use of the living and mineral common-property resources in these areas are set by international maritime law and international treaties.

Managers of marine fisheries use several techniques to help prevent overfishing and commercial extinction and allow depleted stocks to recover. Fishery commissions, councils, and advisory bodies (with representatives from countries or states using a fishery) can set annual quotas and establish rules for dividing the allowable catch among the participating countries or states, limiting fishing seasons, and regulating the type of fishing gear that can be used to harvest a particular species.

Q: How many wild species are driven to premature extinction each hour by human activities?

As voluntary associations, however, fishery commissions don't have any legal authority to compel their members to follow their rules. Furthermore, it is very difficult to estimate the sustainable yields of various marine species. Experience has shown that many fishery commissions, most of them dominated by fishing industry representatives, have not prevented overfishing.

Various suggestions have been made to reduce overfishing in U.S. waters. *First*, gradually phase out government subsidies of the fishing industry, such as low-interest loans and direct subsidies for boats and fishing operations.

Second, impose fees for harvesting fish and shellfish from publicly owned and managed offshore waters, similar to fees now imposed for grazing and logging on U.S. public lands. Currently, the U.S. fishing industry, like the hard-rock mining industry (p. 338), pays no fees or royalties on the catch it harvests from this publicly owned resource. Fishing harvest fees could be adjusted to changes in the estimated size of fish stocks, with fees rising as stocks become more depleted. Australia imposes fish harvesting fees ranging from 11% to 60% of the value of the gross catch, with an average of about 30%. Understandably, the U.S. fishing industry strongly opposes such changes. *Third*, give each fishing vessel owner a marketable quota for removing a certain amount of fish (Spotlight, right).

Why Is It Difficult to Protect Marine Biodiversity? One reason why it is difficult to protect marine biodiversity is that shore-hugging species are adversely affected by coastal development and the accompanying massive inputs of sediment and other wastes from land (Figure 5-17). This poses a severe threat to biologically diverse and highly productive coastal ecosystems such as coral reefs (Figure 5-22), marshes (Figure 5-18), and mangrove swamps (see photo in the Table of Contents).

Protecting marine biodiversity is also difficult because much of the damage is not visible to most people. In addition, the seas are viewed by many as an inexhaustible resource, capable of absorbing an almost infinite amount of waste and pollution. Finally, most of the world's ocean area lies outside the legal jurisdiction of any country and is thus an open-access resource, subject to overexploitation because of the tragedy of the commons.

Protecting marine biodiversity requires countries to enact and enforce tough regulations to protect coral reefs, mangrove swamps, and other coastal ecosystems from unsustainable use and abuse. Furthermore, much more effective international agreements are needed to protect biodiversity in the open seas.

SPOTLIGHT

Is Using Transferable Quotas to Privatize Fishing Rights a Good Idea?

In 1994 the National Marine Fisheries Service (NMFS) claimed that it could no longer manage and conserve U.S. public marine fisheries under an open-access policy. Instead in 1996 it initiated a controversial system of Individual Transferable Quotas (ITQs), in which the government gives each fishing vessel owner a specified percentage of the total allowable catch for each fishery. The largest quotas go to those who have historically taken the most fish, and ITQ owners are permitted to buy, sell, or lease their quotas like private property.

On the surface, this example of free-market environmentalism sounds like a good idea, but most environmentalists generally oppose this approach for several reasons. *First*, it would in effect transfer ownership of publicly owned fisheries to the private commercial fishing industry but still use taxpayer dollars to pay for NMFS oversight of this newly privatized resource.

Second, the scheme would shift fishing quotas from small-scale operators to large-scale operators (whose unsustainable harvesting caused most of the problem in the first place), as has happened with the ITQ system now used in New Zealand. Small fishers would be squeezed out because they don't have the capital to buy the ITQs from others.

Third, the New Zealand experience also shows that the ITQ system increases poaching and sales of illegally caught fish on the black market by small-scale fishers who received no quota or too small a quota to make a living.

Critical Thinking

Do you believe that privatizing fishing in publicly owned waters by using transferable quotas is a good or a bad idea? Explain.

17-7 CASE STUDY: THE WHALING INDUSTRY

Why Is Whaling an Example of the Tragedy of the Commons? *Cetaceans* are an order of mostly marine mammals ranging in size from the 0.9-meter (3-foot) porpoise to the giant 15- to 30-meter (50- to 100-foot) blue whale. They are divided into two major groups: toothed whales and baleen whales (Figure 17-18).

Toothed whales, such as the porpoise, sperm whale, and killer whale (orca), bite and chew their food; they

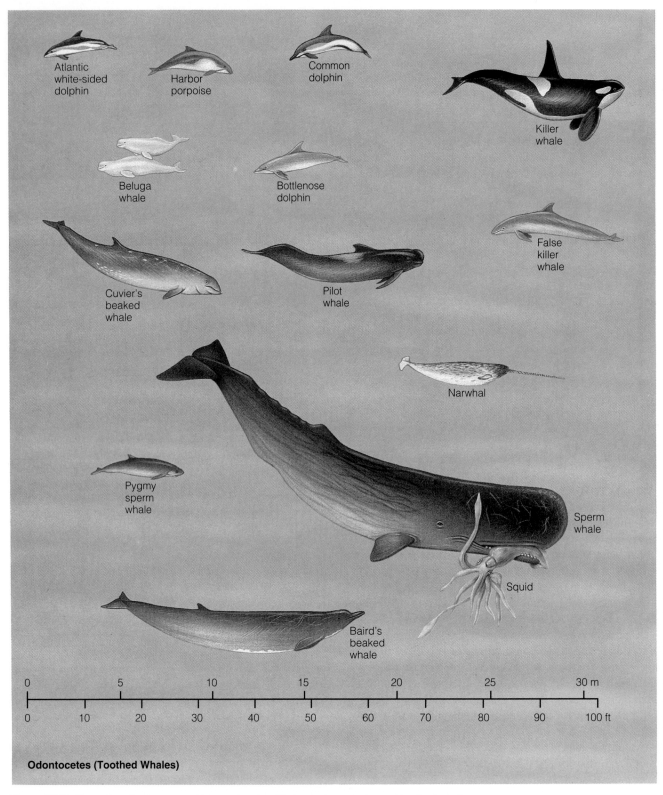

Odontocetes (Toothed Whales)

Figure 17-18 Examples of cetaceans, which can be classified as baleen whales and toothed whales.

Q: What percentage of the world's land has been partially or totally modified by human activities?

Humpback whale

Bowhead whale

Right whale

Minke whale

Blue whale

Fin whale

Feeding on krill

Sei whale

Gray whale

Mysticetes (Baleen Whales)

GUEST ESSAY

Vandana Shiva

Vandana Shiva has a background in physics and the philosophy of science, and she is director of the Research Foundation for Science, Technology, and Natural Resource Policy in Dehra Dun, India. She is also a consultant to the UN Food and Agriculture Organization and in recent years has served as a visiting professor at Oslo University in Norway, Schumacher College in England, and Mount Holyoke College in the United States. She has published 11 books and 74 articles concerned with environmental research and analysis. In 1993 she received several international environmental awards: the Global 500 Roll of Honor, the Earth Day International Award, the VIDA SANA International Award, and the Alternative Nobel Peace Prize (Right Livelihood Award).

Biodiversity, the richness of the living diversity, is severely threatened because of a worldview that perceives the value of all species only in terms of human exploitation. Because we are ignorant of the role that most species play in maintaining the intricate web of life, and because most species are "useless" from the perspective of the dominant economy, such a worldview leads to extinction of a growing number of species.

The main threat to biodiversity comes from the habit of thinking in terms of monocultures; from what I have called "monocultures of the mind." The resulting disappearance of diversity is also a disappearance of alternatives and gives rise to the TINA (there is no alternative) syndrome. Alternatives do exist, but are often excluded. Their inclusion requires an emphasis on preserving all kinds of diversity, biological and cultural, to allow the possibility of multiple choices.

This mindset underlies industrial forestry, in which tree species are useful only if they can provide raw material for construction and to make pulp and paper. Diverse natural forests are thus clear-cut and converted to monoculture plantations.

This worldview also permeates industrial agriculture. Thousands of crop varieties have disappeared because they are declared "uneconomic" and "unproductive," even though they enrich ecosystems and are nutritionally productive. Millets and legumes that require very low resource inputs and produce high nutrient value are declared "inferior" and "marginal" and pushed to extinction. In their place are introduced large-scale monocultures of globally traded crops like wheat, rice, soybeans, and maize.

In our drive to dominate nature for our purposes, tree crop and animal varieties are engineered as if they were machines, not living organisms. Reducing living organisms to machines removes their ecological resilience. Thus, crop varieties engineered to be dwarfs in order to take up high doses of chemical fertilizers to produce high yields are especially vulnerable to pests and disease. Animals engineered to be either "milk machines" or "meat factories" are also highly vulnerable to disease

---

feed mostly on squid, octopus, and other marine animals. *Baleen whales*, such as the blue, gray, humpback, and finback, are filter feeders. Instead of teeth, they have several hundred horny plates made of baleen, or whalebone, that hang down from their upper jaw. These plates filter plankton from the seawater, especially shrimplike krill, which are smaller than your thumb (Figure 4-15). Baleen whales are the most abundant group of cetaceans.

Whales are fairly easy to kill because of their large size and their need to come to the surface to breathe. Mass slaughter has become especially efficient since the advent of fast propeller-driven ships, harpoon guns, and inflation lances (which pump dead whales full of air and make them float).

Whale harvesting—mostly in international waters, a commons open to all resource extractors—has followed the classic pattern of a tragedy of the commons. Between 1925 and 1975, whalers killed an estimated 1.5 million whales; this overharvesting drove the populations of 8 of the 11 major species to commercial extinction, the point where it no longer paid to hunt and kill them.

Case Study: Near Extinction of the Blue Whale

In addition to commercial extinction, the populations of some prized species were reduced to the brink of biological extinction. A prime example is the endangered blue whale (Figure 17-18), the world's largest animal. Fully grown, it's more than 30 meters (100 feet) long—longer than three train boxcars—and weighs more than 25 elephants. The adult has a heart the size of a Volkswagen Beetle car, and some of its arteries are so big that a child could swim through them.

Blue whales spend about 8 months of the year in antarctic waters. There they find an abundant supply of krill, which they filter daily by the trillions from seawater. During the winter months, they migrate to warmer waters, where their young are born.

Before commercial whaling began, an estimated 200,000 blue whales roamed the Antarctic Ocean. Today, the species has been hunted to near biological extinction for its oil, meat, and bone. Its decline was caused by a combination of prolonged overharvesting and certain natural characteristics of blue whales. Their huge size made them easy to spot. They were caught in large numbers because they grouped to-

Q: What percentage of the world's cropland is suffering from soil erosion?

and need heavy doses of antibiotics to survive in factory farming environments.

It is in this context of the production of uniformity that conservation of biodiversity must be understood. Conservation of biodiversity is, above all, the preservation of alternatives, of keeping alive alternative forms of production for the biotechnology industry, and more important for future biological evolution on the earth.

The rapidly growing threat to biodiversity emerges from the myths that monocultures are essential for solving problems of scarcity and that there is no option but to destroy biodiversity to increase productivity. It is not true that without monoculture tree plantations there will be shortages of fuel wood, and without monocultures in agriculture there will be famines of food. Monocultures are in fact a source of scarcity and poverty, both because they destroy diversity and alternatives and because they destroy decentralized control of production and consumption systems.

Humans' escalating war against and control of other species for maximization of profits and production also influence concepts of ownership and control. People have always owned their farm animals and pets. However, the new trend toward "patenting of life forms" has changed the notion of ownership. A patent is a property right derived from an innovation; it is also called an "intellectual property right." However, when applied to the domain of life forms, this concept has far-reaching implications. A patent term is usually 20 years. If a company has genetically modified a sheep or pig or cow and gotten a patent for it, by implication the patent holder has "ownership" of future generations of the modified animal. The machine view of a living organisms thus leads to the ethically outrageous and biologically absurd idea that future generations of genetically engineered plants and animals are solely the products of human innovation and are somehow divorced from the billions of years of evolution that produced all of their genes.

If diverse species on this planet are to survive, we must shift our worldview from a monoculture paradigm that views life as a machine to one based on biodiversity and on allowing the earth's life forms to play their roles in evolution under ever-changing conditions.

It is up to citizens and activists around the world, who see reverence for life as an essential part of the ecology movement, to ensure that technology and law serve the objective of conserving the rich biodiversity of our planet, and not merely serve the objective of protecting profits through unrestrained manipulation and monopoly control of life.

Critical Thinking

1. Do you believe that monocultures of crops and trees are essential for our survival? Explain.

2. Do you believe that a company or individual should be awarded a patent for a form of life that has been genetically engineered from existing forms of life? Explain.

---

gether in their Antarctic feeding grounds. They also take 25 years to mature sexually and have only one offspring every 2–5 years, a reproductive rate that makes it difficult for the species to recover once its population falls beneath a certain threshold.

Blue whales haven't been hunted commercially since 1964 and have been classified as an endangered species since 1975. Despite this protection, some marine biologists believe that too few blue whales—an estimated 3,000–10,000—remain for the species to recover and avoid extinction.

How Have Whale Populations Been Managed? In 1946 the International Whaling Commission (IWC) was established to regulate the whaling industry by setting annual quotas to prevent overharvesting and commercial extinction. However, these quotas often were based on inadequate data or were ignored by whaling countries. Without any powers of enforcement, the IWC has been unable to stop the decline of most whale species.

In 1970 the United States stopped all commercial whaling and banned all imports of whale products. Under intense pressure from environmentalists and governments of many countries, led by the United States, the IWC has imposed a moratorium on commercial whaling since 1986. As a result, the estimated number of whales killed commercially worldwide dropped from 42,480 in 1970 to 741 in 1994.

Some whaling has continued because Japan, Norway, Peru, and the former Soviet Union have exempted themselves from the moratorium. Since 1985 Norway, Iceland, and Japan have each killed several hundred whales a year by exploiting a loophole in the IWC treaty that allows whales to be killed for scientific research. Opponents of whaling call this a sham. With whale meat fetching more than $600 per kilogram ($273 per pound) in Japan, there is a great incentive to cheat.

Should the International Ban on Whaling Be Lifted? Whaling is a traditional part of the economies and cultures of countries such as Japan, Iceland, and Norway. To many people in these cultures hunting whales is no more immoral than hunting deer or elk.

These and other whaling nations believe that the international ban on commercial whaling should be lifted, arguing that the moratorium is based on emotion, not science.

Most environmentalists disagree. Some argue that whales are peaceful, intelligent, sensitive, and highly social mammals that pose no threat to humans and should not be killed; others fear that opening the door to any commercial whaling may eventually lead to widespread harvests of whales by weakening current international disapproval and economic sanctions against commercial whaling. They cite the earlier failure of the IWC to enforce quotas that prevent commercial extinction of most whale species.

In 1994 the IWC voted 23 to 1 to establish a permanent whale sanctuary in the Antarctic Ocean. However, it will be almost impossible to monitor and prevent illegal whaling in this huge sanctuary. Moreover, Japan and Russia filed objections to the IWC decision and Japan announced its intentions to continue commercial whaling in the sanctuary.

During our short time on this planet we have gained immense power over which species, including our own, live or die. We named ourselves the wise (*sapiens*) species. Most conservation biologists believe that in the next few decades, we will learn whether we are indeed a wise species—whether we have the wisdom to learn from and work with nature to protect ourselves and other species. Some actions you can take to help protect wildlife and preserve biodiversity are listed in Appendix 5.

A greening of the human mind must precede the greening of the earth. A green mind is one that cares, saves, and shares. These are the qualities essential for conserving biological diversity now and forever.

M. S. SWAMINATHAN

CRITICAL THINKING

1. Discuss your gut-level reaction to the following statement: "It doesn't really matter that the passenger pigeon is extinct and that the blue whale, the whooping crane, the California condor, and the world's remaining species of rhinoceros and tigers are endangered mostly because of human activities, because eventually all species become extinct anyway." Be honest about your reaction and give arguments for your position.

2. **a.** Do you accept the ethical position that each *species* has the inherent right to survive without human interference, regardless of whether it serves any useful purpose for humans? Explain.

   b. Do you believe that each *individual* of an animal species has an inherent right to survive? Explain.

Would you extend such rights to individual plants and microorganisms? Explain.

3. Your lawn and house are invaded by fire ants, which can cause painful bites. What would you do?

4. Should U.S. energy companies be allowed to drill for oil and natural gas in the Arctic National Wildlife Refuge? Explain your position.

5. Should whaling nations be allowed by the International Whaling Commission to resume a limited annual harvest of minke whales as long as this species' estimated numbers don't decline significantly? Explain. Would you extend this to other commercially valuable whale species if their populations increase significantly? Explain. How would you monitor and enforce annual commercial whaling quotas?

6. Which of the following statements best describes your feelings toward wildlife: **(a)** As long as it stays in its space, wildlife is OK; **(b)** as long as I don't need its space, wildlife is OK; **(c)** I have the right to use wildlife habitat to meet my own needs; **(d)** when you've seen one redwood tree, fox, elephant, or some other form of wildlife you've seen them all, so lock up a few of each species in a zoo or wildlife park and don't worry about protecting the rest; **(e)** wildlife should be protected.

7. List your three favorite species. Examine why they are your favorites. Are they cute and cuddly-looking, like the giant panda and the koala? Do they have humanlike qualities, like apes or penguins that walk upright? Are they large, like elephants or blue whales? Are they beautiful, like tigers and monarch butterflies? Are any of them plants? Are any of them species such as bats, sharks, snakes, or spiders that most people are afraid of? Are any of them microorganisms that help keep you alive? Reflect on what your choice of favorite species tells you about your attitudes toward most wildlife.

*8. Make a log of your own consumption of all products for a single day. Relate your level and types of consumption to the decline of wildlife species and the increased destruction and degradation of wildlife habitats in the United States, in tropical forests, and in aquatic ecosystems.

*9. Identify examples of habitat destruction or degradation in your community that have had harmful effects on the populations of various wild plant and animal species. Develop a management plan for the rehabilitation of these habitats and wildlife.

*10. Make a concept map of this chapter's major ideas, using the section heads and subheads and the key terms (in boldface type). Look at the inside back cover and in Appendix 4 for information about concept maps.

SUGGESTED READINGS

See Internet Sources and Further Readings at the back of the book, and InfoTrac.

PART V

ENERGY RESOURCES

A country that runs on energy cannot afford to waste it.
BRUCE HANNON

18 ENERGY EFFICIENCY AND RENEWABLE ENERGY

Houses That Save Energy and Money

Energy experts Hunter and Amory Lovins (Guest Essay, pp. 538–539) have built a large, passively heated, superinsulated, partially earth-sheltered home in Snowmass, Colorado (Figure 18-1), where winter temperatures can drop to –40°C (–40°F).

This structure, which also houses the research center for the Rocky Mountain Institute, an office used by 40 people, gets 99% of its space and water heating and 95% of its daytime lighting from the sun. It uses one-tenth the usual amount of electricity for a structure of its size. Total energy savings repaid the cost of its energy-saving features after 10 months and are projected to pay for the entire facility in 40 years.

Superinsulating windows, already here, mean that a house can have large numbers of windows without much heat loss in cold weather or heat gain in hot weather. Thinner insulation material now being developed will allow roofs and walls to be insulated far better than in today's best superinsulated houses.

In energy-efficient houses of the near future, sensors will monitor indoor temperatures, sunlight angles, and the occupants' locations. Then small microprocessors will send heat or cooled air where it is needed and open or close window shutters to let solar energy in or reduce heat loss from windows at night and on cloudy days.

A small but growing number of people in developed and developing countries are getting their electricity from *solar cells* that convert sunlight directly into electricity. They can be attached like shingles to a roof or applied to window glass as a coating. Currently, solar cell prices are high but are expected to fall rapidly within a decade.

These are only a few of the components of the exciting *energy-efficiency and renewable energy revolution* that many analysts believe will help us make the transition to more sustainable societies over the next 40–50 years.

Figure 18-1 The Rocky Mountain Institute in Colorado. This facility is a home and a center for the study of energy efficiency and sustainable use of energy and other resources. It is also an example of energy-efficient passive solar design. (Robert Millman/Rocky Mountain Institute)

If the United States wants to save a lot of oil and money and increase national security, there are two simple ways to do it: stop driving Petropigs and stop living in energy sieves.

AMORY B. LOVINS

These are the questions answered in this chapter:

- How should we evaluate energy alternatives?
- What are the benefits and drawbacks of improving energy efficiency; using solar energy to heat buildings and water and to produce electricity; using flowing water and solar energy stored as heat in water to produce electricity; using wind to produce electricity; burning plants and organic waste (biomass) for heating buildings and water, for producing electricity, and for transportation (biofuels); producing hydrogen gas and using it to make electricity, heat buildings and water, and propel vehicles; and extracting heat from the earth's interior (geothermal energy)?

18-1 EVALUATING ENERGY RESOURCES

What Types of Energy Do We Use? *Some 99% of the energy used to heat the earth and all of our buildings comes directly from the sun.* Without this input of essentially inexhaustible solar energy, the earth's average temperature would be –240°C (–400°F) and life as we know it would not exist. Solar energy also helps recycle the carbon, oxygen, water, and other chemicals we and other organisms need to stay alive and healthy. This direct input of solar energy also produces several forms of renewable energy: wind, falling and flowing water (hydropower), and biomass (solar energy converted to chemical energy stored in chemical bonds of organic compounds in trees and other plants).

The remaining 1%, the portion we generate to supplement the solar input, is *commercial energy* sold in the marketplace. Most commercial energy comes from extracting and burning mineral resources obtained from the earth's crust, primarily nonrenewable fossil fuels (Figure 18-2).

Developed countries and developing countries differ greatly in their sources of energy (Figure 18-3) and average per capita energy use. The most important supplemental source of energy for developing countries is potentially renewable biomass, especially fuelwood, the main source of energy for heating and cooking for roughly half the world's population. Within a few decades one-fourth of the world's population in developed countries may face an oil shortage, but half the world's population in developing countries already faces a fuelwood shortage (Figure 16-15).

The United States is the world's largest user (and waster) of energy. With only 4.6% of the population, it uses 25% of the world's commercial energy, 93% from *nonrenewable* fossil fuels (85%) and nuclear energy (8%). In contrast, India, with 16.6% of the world's people, uses only about 3% of the world's commercial energy.

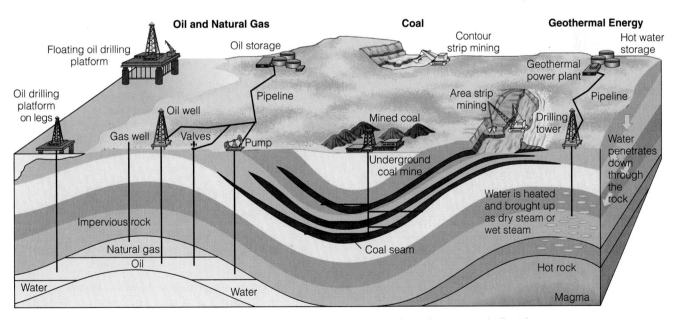

Figure 18-2 Important commercial energy resources from the earth's crust are geothermal energy, coal, oil, and natural gas. Uranium ore is also extracted from the crust and then processed to increase its concentration of uranium-235, which can be used as a fuel in nuclear reactors to produce electricity.

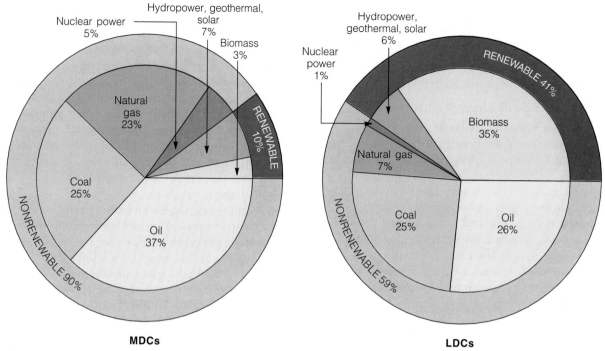

MDCs **LDCs**

Figure 18-3 Commercial energy use by source for developed countries and developing countries. Commercial energy amounts to only 1% of the energy used in the world; the other 99% comes from the sun and is not sold in the marketplace. (Data from U.S. Department of Energy, British Petroleum, and Worldwatch Institute)

How Should We Evaluate Energy Resources?

The types of energy we use and how we use them are major factors determining our quality of life and our harmful environmental effects. Our current dependence on nonrenewable fossil fuels is the primary cause of air and water pollution, land disruption, and projected global warming. Moreover, affordable oil, the most widely used energy resource in developed countries, will probably be depleted within 40–80 years and will need to be replaced by other energy resources.

What is our best immediate energy option? The general consensus is to cut out unnecessary energy waste by improving energy efficiency, as discussed in this chapter. What is our next best energy option? There is disagreement about that. Some say we should get much more of the energy we need from the sun, wind, flowing water, biomass, heat stored in the earth's interior, and hydrogen gas by making the transition to a *renewable energy* or *solar age*. These renewable energy options are evaluated in this chapter.

Others say we should burn more coal and synthetic liquid and gaseous fuels made from coal. Some believe natural gas is the answer, at least as a transition fuel to a new solar age built around improved energy efficiency and renewable energy. Others think nuclear power is the answer. These nonrenewable energy options are evaluated in Chapter 19.

Experience shows that it usually takes at least 50 years and huge investments to phase in new energy alternatives (Figure 18-4), with the exception of nuclear power, which after almost 50 years still provides only a small proportion of the world's commercial energy. Thus, we must plan for and begin the shift to a new mix of energy resources now. To do so involves answering the following questions for *each* energy alternative:

- How much of the energy source will be available in the near future (the next 15 years), intermediate future (the next 30 years), and for the long term (the next 50 years)?

- What is this source's net energy yield?

- How much will it cost to develop, phase in, and use this energy resource?

- How will extracting, transporting, and using the energy resource affect the environment?

- What will using this energy source do to help sustain the earth for us, for future generations, and for the other species living on this planet?

Q: How many people lack access to clean drinking water?

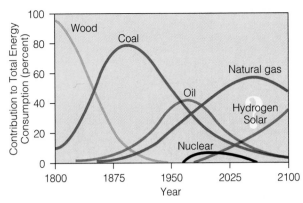

Figure 18-4 Shifts in the use of commercial energy resources in the United States since 1850, with projected changes to 2100. Shifts from wood to coal and then from coal to oil and natural gas have each taken about 50 years. Affordable oil is expected to be running out within 40–50 years, and burning fossil fuels is the primary cause of air pollution and projected warming of the atmosphere. For these reasons, most analysts believe we must make a new shift in energy resources over the next 50 years. Some believe that this shift should involve improved energy efficiency and greatly increased use of solar energy and hydrogen. (Data from U.S. Department of Energy)

What Is Net Energy? The Only Energy That Really Counts It takes energy to get energy. For example, oil must be found, pumped up from beneath the ground, transported to a refinery and converted to useful fuels (such as gasoline, diesel fuel, and heating oil), then transported to users, and finally burned in furnaces and cars before it is useful to us. Each of these steps uses energy, and the second law of energy (Section 3-7) tells us that each time we use energy to perform a task, some of it is always wasted and is degraded to low-quality energy.

The usable amount of high-quality energy available from a given quantity of an energy resource is its **net energy**: the total useful energy available from the resource over its lifetime minus the amount of energy used (the first law of energy), automatically wasted (the second law of energy), and unnecessarily wasted in finding, processing, concentrating, and transporting it to users. Net energy is like your net spendable income (your wages minus taxes and other deductions). For example, suppose that for each 10 units of energy in oil in the ground we have to use and waste 8 units of energy to find, extract, process, and transport the oil to users. Then we have only 2 units of useful energy available from each 10 units of energy in the oil.

We can look at this concept as the ratio of useful energy produced to the useful energy used to produce it. In the example just given, the *net energy ratio* would

be 10/8, or approximately 1.25. The higher the ratio, the greater the net energy yield. When the ratio is less than 1, there is a net energy loss.

Figure 18-5 shows estimated net energy ratios for various types of space heating, high-temperature heat for industrial processes, and transportation. Currently, oil has a high net energy ratio because much of it comes from large, accessible deposits such as those in the Middle East. When those sources are depleted, the net energy ratio of oil will decline and prices will rise. Then more money and more high-quality fossil fuel energy will be needed to find, process, and deliver new oil from widely dispersed small deposits, deposits buried deep in the earth's crust, or deposits located in remote areas.

Conventional nuclear energy has a low net energy ratio because large amounts of energy are required to extract and process uranium ore, convert it into a usable nuclear fuel, and build and operate power plants. In addition, more energy is needed to dismantle the plants after their 15–40 years of useful life and to store the resulting highly radioactive wastes for thousands of years.

18-2 THE IMPORTANCE OF IMPROVING ENERGY EFFICIENCY

What Is Energy Efficiency? Doing More with Less You may be surprised to learn that *84% of all commercial energy used in the United States is wasted* (Figure 18-6). About 41% of this energy is wasted automatically because of the degradation of energy quality imposed by the second law of energy. More important, about 43% is wasted unnecessarily, mostly by using fuel-wasting motor vehicles, furnaces, and other devices, and by living and working in leaky, poorly insulated, and poorly designed buildings.

People in the United States unnecessarily waste as much energy as two-thirds of the world's population consumes. Indeed, according to energy expert Amory Lovins (Guest Essay, pp. 538–539), unnecessary energy waste in the United States amounts to about $300 billion per year—an average of $570,000 per minute. In 1995 this waste was much more than the roughly $200 billion annual federal budget deficit and the entire $262 billion military budget. To Lovins the single most important way for the United States (and other countries) to improve its military, economic, and environmental security is to reduce energy waste.

According to Lovins, the easiest, fastest, and cheapest way to get more energy with the least environmental impact is to eliminate much of this energy waste by

Space Heating

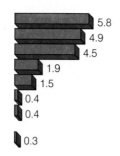

Passive solar	5.8
Natural gas	4.9
Oil	4.5
Active solar	1.9
Coal gasification	1.5
Electric resistance heating (coal-fired plant)	0.4
Electric resistance heating (natural-gas-fired plant)	0.4
Electric resistance heating (nuclear plant)	0.3

High-Temperature Industrial Heat

Surface-mined coal	28.2
Underground-mined coal	25.8
Natural gas	4.9
Oil	4.7
Coal gasification	1.5
Direct solar (highly concentrated by mirrors, heliostats, or other devices)	0.9

Transportation

Natural gas	4.9
Gasoline (refined crude oil)	4.1
Biofuel (ethyl alcohol)	1.9
Coal liquefaction	1.4
Oil shale	1.2

Figure 18-5 Net energy ratios for various energy systems over their estimated lifetimes. (Data from Colorado Energy Research Institute, *Net Energy Analysis*, 1976; and Howard T. Odum and Elisabeth C. Odum, *Energy Basis for Man and Nature*, 3d ed., New York: McGraw-Hill, 1981)

making lifestyle changes that reduce energy consumption: walking or biking for short trips, using mass transit, putting on a sweater instead of turning up the thermostat, and turning off unneeded lights.

Another equally important way is to increase the efficiency of the energy conversion devices we use. **Energy efficiency** is the percentage of total energy input that does useful work (is not converted to low-quality, essentially useless heat) in an energy conversion system. The energy conversion devices we use vary in their energy efficiencies (Figure 18-7). We can save energy and money by buying the most energy-efficient home heating systems, water heaters, cars, air conditioners, refrigerators, computers, and other appliances that are available, and by supporting research to invent even more energy-efficient devices. Some energy-efficient models may cost more initially, but in the long run they usually save money by having a lower **life cycle cost**: initial cost plus lifetime operating costs.

The net efficiency of the entire energy delivery process for a space heater, water heater, or car is determined by the efficiency of each step in the energy conversion process. For example, the sequence of energy-using (and energy-wasting) steps involved in using electricity produced from fossil or nuclear fuels is

Figure 18-8 shows the net energy efficiency for heating two well-insulated homes: one with electricity produced at a nuclear power plant, transported by wire to the home, and converted to heat (electric resistance heating); the other heated passively, with an input of direct solar energy through high-efficiency windows facing the sun, with heat stored in rocks or water for slow release. This analysis shows that the process of converting the high-quality energy in nuclear fuel to high-quality heat at several thousand degrees in the power plant, converting this heat to high-quality electricity, and then using the electricity to provide low-quality heat for warming a house to only about 20°C (68°F) is extremely wasteful of high-quality energy. Burning coal or any fossil fuel at a power plant to supply electricity for space heating is also inefficient. It is much less wasteful to collect solar energy from the

Q: What percentage of the world's households don't have running water?

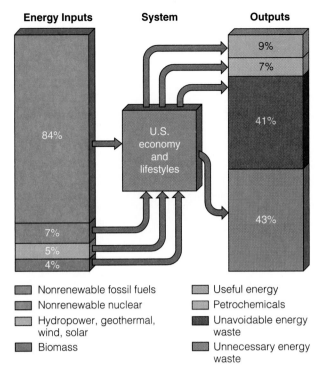

Figure 18-6 Flow of commercial energy through the U.S. economy. Note that only 16% of all commercial energy used in the United States ends up performing useful tasks or being converted to petrochemicals; the rest is either automatically and unavoidably wasted because of the second law of energy (41%) or wasted unnecessarily (43%).

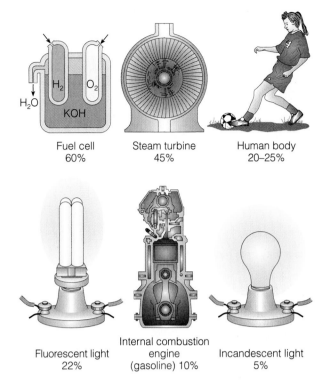

Figure 18-7 Energy efficiency of some common energy conversion devices.

environment, store the resulting heat in stone or water, and, if necessary, raise its temperature slightly to provide space heating or household hot water.

Physicist and energy expert Amory Lovins (Guest Essay, pp. 538–539) points out that using high-quality electrical energy to provide low-quality heating for living space or household water is like using a chain saw to cut butter or a sledgehammer to kill a fly. As a general rule, he suggests that we not use high-quality energy to do a job that can be done with lower-quality energy (Figure 3-10).

The logic of this point is illustrated by looking at the costs of providing heat using various fuels. In 1991 the average price of obtaining 250,000 kilocalories (1 million Btu) for heating either space or water in the United States was $6.05 using natural gas, $7.56 using kerosene, $9.30 using oil, $9.74 using propane, and $24.15 using electricity. As these numbers suggest, if you don't mind throwing away hard-earned dollars, then use electricity to heat your house and bathwater. Yet almost one of every four homes in the United States is heated by electricity.

Perhaps the three least efficient energy-using devices in widespread use today are incandescent light bulbs (which waste 95% of the energy input), vehicles with internal combustion engines (which waste 86–90% of the energy in their fuel), and nuclear power plants producing electricity for space heating or water heating (which waste 86% of the energy in their nuclear fuel, and probably 92% when the energy needed to deal with radioactive wastes and retired nuclear plants is included). Energy experts call for us to replace these devices or greatly improve their energy efficiency over the next few decades.

Why Is Reducing Energy Waste So Important?
According to Amory Lovins and other energy analysts, reducing energy waste is one of the planet's best and most important economic and environmental bargains. It

- *Makes nonrenewable fossil fuels last longer.*

- *Gives us more time to phase in renewable energy resources.*

- *Decreases dependence on oil imports (52% in 1996)*

- *Lessens the need for military intervention in the oil-rich but politically unstable Middle East.*

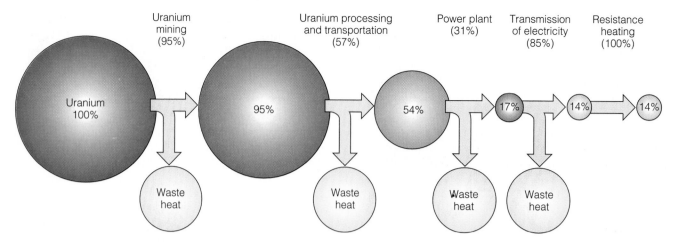

Uranium
mining
(95%)

Uranium processing
and transportation
(57%)

Power plant
(31%)

Transmission
of electricity
(85%)

Resistance
heating
(100%)

Uranium
100%

95%

54%

17%

14%

14%

Waste
heat

Waste
heat

Waste
heat

Waste
heat

Electricity from Nuclear Power Plant

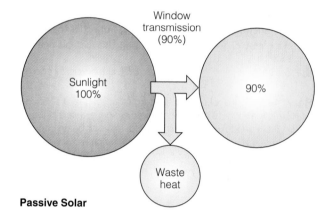

Window
transmission
(90%)

Sunlight
100%

90%

Waste
heat

Passive Solar

Figure 18-8 Comparison of net energy efficiency for two types of space heating. The cumulative net efficiency is obtained by multiplying the percentage shown inside the circle for each step by the energy efficiency for that step (shown in parentheses). Because of the second law of thermodynamics, in most cases the greater the number of steps in an energy conversion process, the lower its net energy efficiency. About 86% of the energy used to provide space heating by electricity produced at a nuclear power plant is wasted. If the additional energy needed to deal with nuclear wastes and to retire highly radioactive nuclear plants after their useful life is included, then the net energy yield for a nuclear plant is only about 8% (or 92% waste). By contrast, with passive solar heating, only about 10% of incoming solar energy is wasted.

- *Reduces local and global environmental damage* because less of each energy resource would provide the same amount of useful energy.

- *Is the cheapest and quickest way to slow projected global warming* (Solutions, p. 283).

- *Saves more money, provides more jobs, improves productivity, and promotes more economic growth per unit of energy than other alternatives.*

- *Improves competitiveness in the international marketplace.*

According to Amory Lovins, if the world really got serious about improving energy efficiency, we could save $1 trillion per year. A 1993 study by economists estimated that a full-fledged energy-efficiency program could produce 1.3 million jobs in the United States by 2010.

Why isn't there more emphasis on improving energy efficiency? The primary reason is a glut of low-cost, underpriced fossil fuels. As long as such energy is artificially cheap because its harmful environmental

costs are not included in its market prices, people are more likely to waste it and not make investments in improving energy efficiency.

18-3 WAYS TO IMPROVE ENERGY EFFICIENCY

How Can We Use Waste Heat? Could we save energy by recycling energy? No. The second law of energy tells us that we cannot recycle energy. However, we can slow the rate at which waste heat flows into the environment when high-quality energy is degraded. For a house, the best way to do this is to insulate it thoroughly, eliminate air leaks, and equip it with an air-to-air heat exchanger to prevent buildup of indoor air pollutants. Many homes in the United States are so full of leaks that their heat loss in cold weather and heat gain in hot weather are equivalent to having a large window-size hole in the wall of the house (Figure 3-1).

Q: What are the three components of biodiversity?

In office buildings and stores, waste heat from lights, computers, and other machines can be collected and distributed to reduce heating bills during cold weather; during hot weather, this heat can be collected and vented outdoors to reduce cooling bills. Waste heat from industrial plants and electrical power plants can be used to produce electricity (cogeneration); it can also be distributed through insulated pipes to heat nearby buildings, greenhouses, and fish ponds, as is done in some parts of Europe.

How Can We Save Energy in Industry? There are a number of ways to save energy and money in industry. One is **cogeneration**, the production of two useful forms of energy (such as steam and electricity) from the same fuel source. Waste heat from coal-fired and other industrial boilers can produce steam that spins turbines and generates electricity at roughly half the cost of buying it from a utility company. By using the electricity or selling it to the local power company for general use, a plant can save energy and money. Cogeneration has been widely used in western Europe for years, and its use in the United States is growing. Within 8 years cogeneration could produce more electricity than all U.S. nuclear power plants, and much more cheaply.

In Germany, small cogeneration units that run on natural gas or liquefied petroleum gas (LPG) supply restaurants, apartment buildings, and houses with all their energy. In 4–5 years they pay for themselves in saved fuel and electricity.

Replacing energy-wasting electric motors is another strategy. About 60–70% of the electricity used in U.S. industry drives electric motors, most of which are run at full speed with their output throttled to match their task—somewhat like driving with the gas pedal to the floor and the brake engaged. Each year a heavily used electric motor consumes 10 times its purchase cost in electricity—equivalent to using $120,000 worth of gasoline each year to fuel a $12,000 car. According to Amory Lovins, it would be cost-effective to scrap virtually all such motors and replace them with adjustable-speed drives. The costs would be paid back in 1–3 years, depending on how the motors were used.

Energy can also be saved by switching to high-efficiency lighting. Additionally, computer-controlled energy management systems can turn off lighting and equipment in nonproduction areas and make adjustments during periods of low production. Recycling and reuse and making products that last longer and that are easy to repair and recycle also saves energy compared to using virgin resources.

How Can We Save Energy in Producing Electricity? The Negawatt Revolution Traditionally, utilities make more money by increasing the demand for electricity. This process encourages the building of often unnecessary power plants to send electricity to inefficient appliances, heating and cooling systems, and industrial plants. To make more money, many utilities encourage their customers to use (and thus waste) even more electricity—a classic example of harmful positive feedback in action.

A small but growing number of utility companies in the United States and elsewhere are trying to reverse this wasteful process by reducing the demand for electricity. By helping customers use electricity more efficiently, these 20 utility companies do not need to finance and build expensive new power plants.

This new approach is known as *demand-side management* or the *negawatt revolution*. To reduce demand, utilities give customers cash rebates for buying efficient lights and appliances, free home-energy audits, low-interest loans for home weatherization or industrial retrofits, and lower rates to households or industries meeting certain energy-efficiency standards. To make demand-side management feasible, state and other utility regulators must allow utility investors to make a reasonable return on their money, based on the amount of energy the utilities save. Such a policy allows utility companies to shift their emphasis from producing megawatts to energy-saving negawatts.

How Can We Save Energy in Transportation? The most important way to save energy (especially oil) and money in transportation is to *increase the fuel efficiency of motor vehicles*. Between 1973 and 1985, the average fuel efficiency doubled for new American cars and rose 37% for all passenger cars on the road, but it has risen only slightly since then.

According to the U.S. Office of Technology Assessment, existing technology could raise current fuel efficiency of the entire U.S. automotive fleet to 15 kilometers per liter (35 miles per gallon) by 2010. Doing this would save U.S. consumers about $65 billion a year (about $576 per household), cause a sharp drop in emissions of CO_2 and other air pollutants, cut oil imports in half (which would significantly lower the annual U.S. trade deficit), and create about 244,000 new jobs throughout the U.S. economy.

Currently, the U.S. market includes more than 15 car models with fuel efficiencies of at least 17 kilometers per liter (40 miles per gallon), but they make up only 5% of U.S. car sales. Since 1985 at least 10 automobile companies have made nimble and peppy prototype cars that meet or exceed current safety and pollution standards, with fuel efficiencies of 29–59 kilometers per liter (67–138 miles per gallon) (Figure 18-9). If such cars were mass-produced, their slightly higher costs would be more

Figure 18-9 The Ultralite, a prototype car built by General Motors. It is fast and much safer than existing cars. When driven at 81 kph (50 mph), it gets 43 kpl (100 mpg). Replacing its engine with a 50% more efficient diesel engine (already available) would raise overall fuel efficiency to 81 kpl (190 mpg). It has a 218-kph (135-mph) top speed, is equipped with four air bags, and easily carries four adults. Its body is made from lightweight carbon-fiber composite plastics that won't dent, scratch, or corrode. With 200 fewer parts than a conventional engine, it costs about $400 less to make and costs consumers less in maintenance expenses. (General Motors)

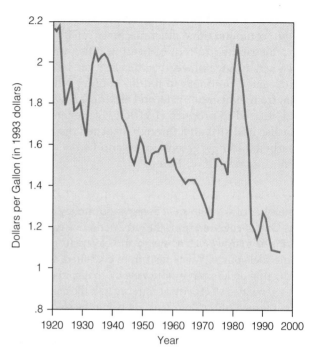

Figure 18-10 Real price of gasoline (in 1993 dollars), 1920–96. When adjusted for inflation, U.S. consumers are paying less for gasoline than at any time since 1920. This underpricing of gasoline encourages energy waste and pollution by failing to include in its market price the many harmful social and environmental costs of gasoline, which consumers ultimately end up paying anyway.

than offset by their fuel savings. The problem is that there is little consumer interest in fuel-efficient cars when the inflation-adjusted price of gasoline today is the lowest it has been since 1920 (Figure 18-10). This underpricing of gasoline encourages energy waste and pollution by failing to include in its market price the many harmful social and environmental costs of gasoline, which consumers ultimately end up paying anyway.

To encourage consumers to buy energy-efficient vehicles, Amory Lovins has suggested that a system of revenue-neutral rebates and "freebates" be established for motor vehicles. Buyers of a new vehicle would pay a fee or receive a rebate depending on its fuel efficiency. The fees on inefficient vehicles would be used to pay for the rebates on efficient ones.

Conventional battery-powered *electric cars* might help reduce dependence on oil, especially for urban commuting and short trips. Electric vehicles are extremely quiet, need little maintenance, and can accelerate rapidly. The cars themselves, called zero-emissions vehicles, produce no air pollution. However, using coal and nuclear power plants to produce the electricity needed to recharge their batteries daily does produce air pollution and nuclear wastes—something called *elsewhere pollution*.

Greatly increased manufacture, disposal, and recycling of lead batteries for widespread use of electric vehicles could also greatly increase the input of toxic lead into the environment (Section 13-7). If solar cells

or wind turbines could be used for recharging the car's batteries, CO_2 and other air pollution emissions would be virtually eliminated.

On the negative side, today's electric cars are not very efficient; they are equivalent to gasoline-powered cars that get about 7–11 kilometers per liter (16–25 miles per gallon). Current electric cars can travel only 81–161 kilometers (50–100 miles). Their batteries must be recharged for 3–8 hours (although a new device may reduce recharge time to 10–20 minutes) and must be replaced about every 48,000 kilometers (30,000 miles) at a cost of at least $2,000. This requirement, plus the electricity costs for daily recharging and buying a charger ($700–3,500), means that today's electric cars have twice the operating cost of gasoline-powered cars. A growing number of analysts believe that the future lies rather in the development of *ecocars* (Spotlight, right).

Another way to save energy is to *shift to more energy-efficient ways to move people* (Figure 18-11) *and freight*. More freight could be shifted from trucks and planes to more energy-efficient trains and ships. The fuel efficiency of other means of transport can also be increased. With improved aerodynamic design, turbocharged diesel engines, and radial tires, new

Q: What factor is most important in determining whether a land area is a desert, a grassland, or a forest?

Ecocars

There is growing interest in developing hybrid electric–internal combustion *ecocars*, or *supercars* getting 64–128+ kilometers per liter (150–300+ miles per gallon). They get most of their power from a small hybrid electric/fuel engine that makes its own electricity and uses a small battery or flywheel to provide the extra energy needed for acceleration and hill climbing.

In 1997, Toyota developed a hybrid electric/gasoline vehicle that is twice as fuel-efficient as an equivalent gasoline-powered car, while significantly reducing air pollution emissions. Toyota plans to begin marketing the car in 1998.

Such supercars will be lighter and safer than conventional cars because their bodies and many other parts will be made of low-weight composite materials (Figure 18-9) that absorb much more crash energy per unit of weight than the steel in today's cars. Today most race car bodies are made of carbon-fiber plastics, which is why race car drivers can usually walk away after hitting a wall at 320 kilometers (200 mph).

Car bodies and other parts made from composite materials won't rust and can be recycled. They also don't need to be painted (currently the most costly and polluting step in car manufacture) because the desired color can be added to the molds that shape the composite materials. Switching to hybrid ecocars with composite bodies and far fewer parts would also sharply reduce this use of minerals and the resulting pollution and environmental degradation.

Amory Lovins (Guest Essay, pp. 538–539) and other researchers project that with proper financial incentives, such ecocars could be on the market within a decade. With incentives such as tax credits or energy rebates, such cars could replace much of the existing car fleet within 10–12 years (the average road life of existing cars).

Another environmentally appealing type of ecocar is an electric vehicle that uses fuel cells powered by solar-produced hydrogen (Section 18-8). Fuel cells consist of two electrodes immersed in a solution (electrolyte) that conducts electricity. They produce electricity by combining hydrogen and oxygen ions, typically from hydrogen and oxygen gas supplied as fuel for the cell (Figure 18-7) or from gasoline. In 1997, researchers announced the development of a new type of fuel-cell system that burns hydrogen produced from gasoline. Such cars could be on the road by 2005–2010.

The cells are compact and safe, make no noise, and require little maintenance. They are also very efficient, converting 50% or more of the fuel energy to power, compared to 10–14% efficiency for gasoline-powered vehicles. More important, fuel-cell cars running on hydrogen come close to being true zero-emission vehicles because they emit only water vapor and trace amounts of nitrogen oxides, easily controlled with existing technology.

Critical Thinking

Do you believe that it is economically and politically important to build and phase in ecocars over the next 20 years? Explain. Can you think of any drawbacks from doing this?

Energy Use (thousands of Btu per passenger mile)

General aviation	11
Commercial aviation	5.7
Personal truck	4.8
Automobile	3.5
Motorcycle	2.3
AMTRAK railroad	1.8
Intercity high-speed train	1.2
Intercity bus	0.9

Figure 18-11 Energy use of various types of domestic passenger transportation: the lower the energy use, the greater the efficiency.

transport trucks can be 50% more fuel efficient than today's conventional trucks.

How Can We Save Energy in Buildings?

Heating, cooling, and lighting buildings consume about one-third of the energy used by modern societies, with much of this energy unnecessarily wasted. The 110-story, twin-towered World Trade Center in Manhattan is a monument to energy waste: It uses as much electricity as a city of 100,000 people for only about 53,000 employees.

In contrast, Atlanta's 13-story Georgia Power Company building uses 60% less energy than conventional office buildings of the same size. The largest surface of the building faces south to capture solar energy. Each floor extends out over the one below it, blocking out the higher summer sun to reduce air conditioning costs but allowing warming by the

---

A: Climate, based mostly on average temperature and average precipitation lower winter sun. Energy-efficient lights focus on desks rather than illuminating entire rooms. The Georgia Power model and other existing cost-effective commercial building technologies could reduce energy use by 75% in U.S. buildings, cut carbon dioxide emissions in half, and save more than $130 billion per year in energy bills by 2010.

In 1987, the Internationale Nederlanden (ING) Bank in Amsterdam built a new headquarters that uses 90% less energy than its predecessor. This new building saves the bank $2.4 million per year in energy costs.

There are a number of ways to improve the energy efficiency of buildings, many of them discussed in the opening of this chapter (p. 510). One is to build more *superinsulated houses* (Figure 18-12). Although such houses typically cost 5% more to build than conventional houses of the same size, this extra cost is paid back by energy savings within 5 years and can save a homeowner $50,000–100,000 over a 40-year period.

Since the mid-1980s there has been growing interest in building long-lasting, affordable, and easily constructed superinsulated houses called *straw-bale houses* with walls consisting of compacted bales of certain types of straw (available at a low cost almost everywhere) covered with plaster or adobe (Figure 18-13).* By 1995 there were 500 such homes built or under construction in the United States. Using straw, an *annually* renewable agricultural residue often burned as a waste product, for the walls reduces the need for wood and thus slows deforestation. There is little ecological impact on land because the straw comes from land already converted to pasture or cropland. The main problem is getting banks and other money lenders to recognize the potential of this type of housing and provide homeowners with construction loans.

Another way to save energy is to *use the most energy-efficient ways to heat houses* (Figure 18-14). The most energy-efficient ways to heat space are to build a superinsulated house, use passive solar heating, and use high-efficiency (85–98% efficient) natural gas furnaces. The most wasteful and expensive way is to use electric resistance heating with the electricity produced by a coal-fired or nuclear power plant.

Heat pumps can save energy and money for space heating in warm climates (where they aren't needed much), but not in cold climates; at low temperatures they automatically switch to wasteful, costly electric resistance heating (which is why utili-

---

* For information on straw-bale houses, see Steen et al., *The Straw Bale House* (White River Junction, Vt.: Chelsea Green, 1994); and GreenFire Institute, 1509 Queen Anne Ave. North, #606, Seattle, Wash. 98109, (206) 284-7470.

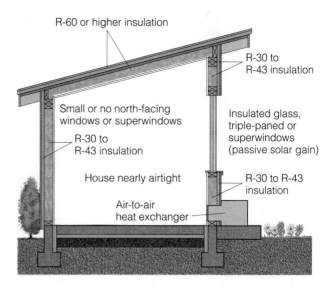

Figure 18-12 Major features of a superinsulated house. Such a house is heavily insulated and nearly airtight. Heat from direct solar gain, appliances, and human bodies warms the house, which requires little or no auxiliary heating. An air-to-air heat exchanger prevents buildup of indoor air pollution without wasting much heat.

ties wanting to sell more electricity advocate their use). Some heat pumps in their air conditioning mode are also much less efficient than many individual air conditioning units. Most heat pumps also require expensive repair every few years. However, manufacturers have developed some improved models that produce warmer air and are more efficient than older models.

The energy efficiency of existing houses can be improved significantly by adding insulation, plugging leaks, and installing energy-saving windows. About one-third of heated air in U.S. homes and buildings escapes through closed windows and holes and cracks (Figure 3-1)—equal to the energy in all the oil flowing through the Alaska pipeline every year. During hot weather these windows also let heat in, increasing the use of air conditioning.

We can also *use the most energy-efficient ways to heat household water.* An efficient method is to use tankless instant water heaters (about the size of bookcase loudspeakers) fired by natural gas or liquefied petroleum gas (LPG). These devices, widely used in many parts of Europe, heat the water instantly as it flows through a small burner chamber and provide hot water only when (and as long as) it is needed. A well-insulated, conventional natural gas or LPG water heater is also fairly efficient (although all conventional natural gas and electric resistance heaters waste energy by keeping a large tank of water hot all day and night and can run out after a long shower or two).

Q: What percentage of the world's people live on or near coastal areas?

Figure 18-13 An energy-efficient straw-bale house during construction (left) and a completed one (right). Depending on the thickness of the bales, plastered straw-bale walls have an insulating value of R-35 to R-60, compared to R-12 to R-19 in conventional house. (The R value is a measure of resistance to heat flow.) (Out on Bale, Tucson, Arizona)

Net Energy Efficiency

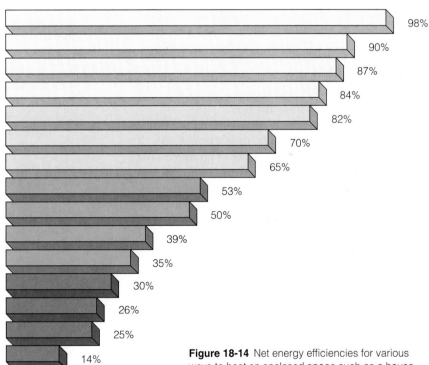

Heating method	Efficiency
Superinsulated house (100% of heat)(R-43)	98%
Passive solar (100% of heat)	90%
Passive solar (50% of heat) plus high-efficiency natural gas furnace (50% of heat)	87%
Natural gas with high-efficiency furnace	84%
Electric resistance heating (electricity from hydroelectric power plant)	82%
Natural gas with typical furnace	70%
Passsive solar (50% of heat) plus high-efficiency wood stove (50% of heat)	65%
Oil furnace	53%
Electric heat pump (electricity from coal-fired power plant)	50%
High-efficiency wood stove	39%
Active solar	35%
Electric heat pump (electricity from nuclear plant)	30%
Typical wood stove	26%
Electric resistance heating (electricity from coal-fired power plant)	25%
Electric resistance heating (electricity from nuclear plant)	14%

Figure 18-14 Net energy efficiencies for various ways to heat an enclosed space such as a house.

The most inefficient and expensive way to heat water for washing and bathing is to use electricity produced by any type of power plant. A $425 electric hot water heater can cost $5,900 in energy costs over its 15-year life, compared to about $2,900 for a comparable natural gas water heater over the same period.

Setting higher energy-efficiency standards for new buildings would also save energy. Building codes could require that all new houses use 60–80% less energy than conventional houses of the same size, as has been done in Davis, California (p. 190). Because of tough energy-efficiency standards, the average Swedish home

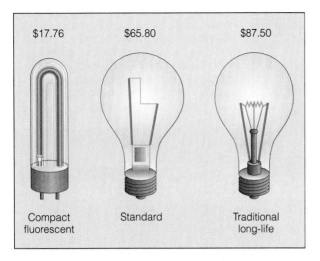

$17.76 $65.80 $87.50

Compact   Standard   Traditional
fluorescent   long-life

Figure 18-15 Cost of electricity for comparable light bulbs used for 10,000 hours. Because conventional incandescent bulbs are only 5% efficient and last only 750–1,500 hours, they waste enormous amounts of energy and money, and they add to the heat load of houses during hot weather. Socket-type fluorescent bulbs use one-fourth as much electricity as conventional bulbs. Although these new bulbs cost about $10–20 per bulb, they last 10–20 times longer than conventional incandescent bulbs used the same number of hours, and they save considerable money (compared with incandescent bulbs) over their long life. In 1998, compact fluorescent bulbs that can be dimmed went on sale. (Data from Electric Power Research Institute)

consumes about one-third as much energy as the average American home of the same size.

Another way to save energy is to *buy the most energy-efficient appliances and lights* (Figure 18-15).* If the most energy-efficient appliances and lights now available were installed in all U.S. homes over the next 20 years, the savings in energy would equal the estimated energy content of Alaska's entire North Slope oil fields. Replacing a standard incandescent bulb with an energy-efficient compact fluorescent saves about $48–70 per bulb over its 10-year life and saves enough electricity to avoid burning 180 kilograms (400 pounds) of coal. Thus, replacing 25 incandescent bulbs in a house with energy-efficient fluorescent bulbs saves $1,250–1,750.

Energy-efficient lighting could save U.S. businesses $15–20 billion per year in electricity bills. Students in Brown University's environmental studies program showed that the school could save more than

---

* Each year the American Council for an Energy-Efficient Economy (ACEEE) publishes a list of the most energy-efficient major appliances mass-produced for the U.S. market. A copy can be obtained from the council at 1001 Connecticut Ave. N.W., Suite 801, Washington, D.C. 20036. Each year they also publish *A Consumer Guide to Home Energy Savings*, available in bookstores or from the ACEEE.

$40,000 per year just by replacing the incandescent light bulbs in exit signs with fluorescents.

If all U.S. households used the most efficient frost-free refrigerator now available, 18 large (1,000 megawatt) power plants could close. Microwave ovens can cut electricity use for cooking by 25–50% (but not if used for defrosting food). Clothes dryers with moisture sensors cut energy use by 15%, and front-loading washers use 50% less energy than top-loading models yet cost no more. New microwave clothes dryers use 15% less energy than conventional electric dryers and 28% less energy than gas units while virtually eliminating wrinkles.

Improvements in energy efficiency could be encouraged by giving rebates or tax credits for building energy-efficient buildings, for improving the energy efficiency of existing buildings, and for buying high-efficiency appliances and equipment.

18-4 DIRECT USE OF SOLAR ENERGY FOR HEAT AND ELECTRICITY

What Are the Advantages of Solar Energy? The Coming Renewable-Energy Age About 92% of the known reserves and potentially available energy resources in the United States are renewable energy from the sun, wind, flowing water, biomass, and earth's internal heat. The other 8% of potentially available domestic energy resources are coal (5%), oil (2.5%), and uranium (0.5%).

Developing these mostly untapped renewable energy resources could meet 50–80% of projected U.S. energy needs by 2040 or sooner, and it could meet virtually all energy needs if coupled with improvements in energy efficiency. In 1994 Shell International Petroleum in London, which forecast the steep rise in oil prices in the 1970s, projected that renewable energy, especially solar, will dominate world energy production by 2050.

Developing renewable energy resources would save money, create two to five times more jobs per unit of electricity produced than coal and nuclear power plants, eliminate the need for oil imports, cause much less pollution and environmental damage per unit of energy used, and increase military, economic, and environmental security. In the United States, wind farms and power plants using geothermal, wood (biomass), hydropower, and high-temperature solar energy (with natural gas backup) can already produce electricity more cheaply than can new nuclear power plants, and with far fewer federal subsidies (Figure 18-16).

How Can Solar Energy Be Used to Heat Houses and Water? Buildings and water can be heated by

Q: What is carrying capacity?

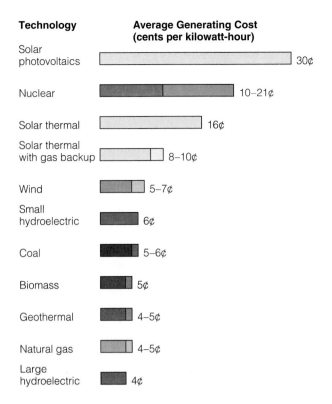

Technology	Average Generating Cost (cents per kilowatt-hour)
Solar photovoltaics	30¢
Nuclear	10–21¢
Solar thermal	16¢
Solar thermal with gas backup	8–10¢
Wind	5–7¢
Small hydroelectric	6¢
Coal	5–6¢
Biomass	5¢
Geothermal	4–5¢
Natural gas	4–5¢
Large hydroelectric	4¢

Figure 18-16 Generating costs of electricity per kilowatt-hour by various technologies in 1993. By 2000, costs per kilowatt-hour for wind are expected to decrease to 4–5¢, solar thermal with gas assistance 5–6¢, solar photovoltaic 6–12¢, natural gas 3–4¢, coal 4–5¢, and nuclear power still at 10–21¢. Costs for the other technologies shown in this figure are expected to remain about the same. (Data from U.S. Department of Energy, Council for Renewable Energy Education, Investor Responsibility Research Center, California Energy Commission, and Worldwatch Institute)

solar energy using two methods: passive and active (Figure 18-17). A **passive solar heating system** captures sunlight directly within a structure (Figures 18-1 and 18-18) and converts it into low-temperature heat for space heating. Energy-efficient windows, greenhouses, and sunspaces face the sun to collect solar energy by direct gain.

Thermal mass (heat-storing capacity), in the forms of walls and floors of concrete, adobe, brick, stone, salt-treated timber, or walls made of used tires packed with dirt, stores much of the collected solar energy as heat and releases it slowly throughout the day and night. Buildup of moisture and indoor air pollutants is minimized by an air-to-air heat exchanger, which supplies fresh air without much heat loss or gain. A small backup heating system may be used but is not necessary in many climates.

Engineer and builder Michael Sykes has designed a solar envelope house that is heated and cooled passively by solar energy and the slow storage and release of energy by massive timbers and the earth beneath the house (Figure 18-19). The front and back of this house are double walls of heavy timber impregnated with salt to increase the ability of the wood to store heat. The space between these two walls, plus the basement, forms a convection loop or envelope around the inner shell of the house. In summer, roof vents release heated air from the convection loop throughout the day; at night, these roof vents, with the aid of a fan, draw air into the loop, passively cooling the house. The interior temperature of the house typically stays within two degrees of 21°C (70°F) year-round, without any conventional cooling or heating

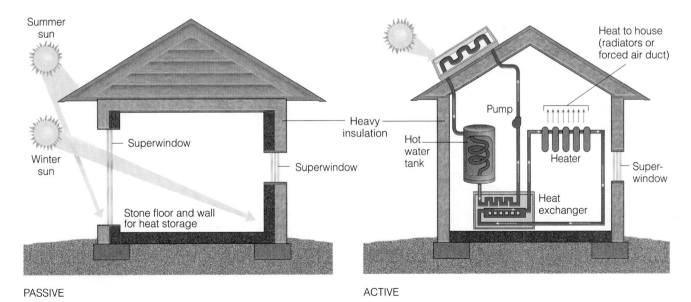

PASSIVE

ACTIVE

Figure 18-17 Passive and active solar heating for a home.

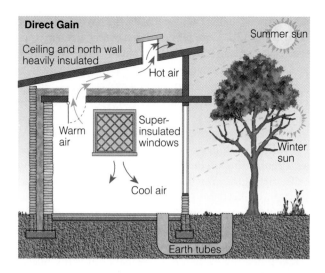

Direct Gain

Ceiling and north wall heavily insulated

Summer sun

Hot air

Warm air

Super-insulated windows

Winter sun

Cool air

Earth tubes

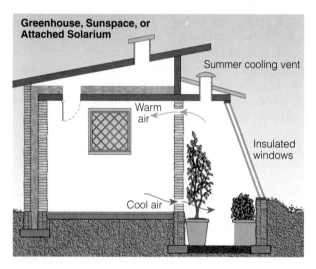

Greenhouse, Sunspace, or Attached Solarium

Summer cooling vent

Warm air

Insulated windows

Cool air

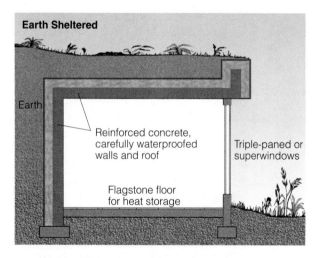

Earth Sheltered

Earth

Reinforced concrete, carefully waterproofed walls and roof

Triple-paned or superwindows

Flagstone floor for heat storage

Figure 18-18 Three examples of passive solar design for houses.

system. In cold or cloudy climates, a small wood stove or vented natural gas heater in the basement can be used as a backup to heat the air in the convection loop.

With available and developing technologies, passive solar designs can provide at least 70% of a residential building's heating needs and at least 60% of its cooling needs, and up to 60% of a *commercial* building's energy needs. In North America there are an estimated 250,000 fully passive solar homes and more than 1 million buildings that include some aspect of passive solar design. Roof-mounted passive solar water heaters, such as the thermosiphoning Copper Cricket, can supply all or most of the hot water for a typical house.

On a life cycle cost basis, good passive solar and superinsulated design is the cheapest way to heat a home or a small building in regions where ample sunlight is available more than 60% of daylight hours (Figure 18-20). Such a system usually adds 5–10% to the construction cost, but the life cycle cost of operating such a house is 30–40% lower. The typical payback time for passive solar features is 3–7 years. In 1994 the Esperanza del Sol subdivision of low-income, energy-efficient, and solar-oriented houses was built in Texas. The houses add only $13 per year to mortgage payments but save residents $480 per year in heating and cooling costs.

In an **active solar heating system**, specially designed collectors absorb solar energy and a fan or a pump supplies part of a building's space-heating or water-heating needs (Figure 18-17, right). Several connected collectors are usually mounted on a portion of the roof with an unobstructed exposure to the sun. Some of the heat can be used directly, and the rest can be stored in insulated tanks containing rocks, water, or a heat-absorbing chemical for release later as needed.

Active solar collectors can also supply hot water. In Cyprus and Jordan, active solar water heaters supply 25–65% of the hot water for homes. About 12% of houses in Japan, 37% in Australia, and 83% in Israel also use such systems. Currently there are about 1.3 million active solar hot-water systems in U.S. homes. At an average cost of $2,500, active solar water heating systems have been too expensive for most homeowners. However, in 1992 Solar Development in Riveria Beach, Florida, developed a system that costs less than $700.

Solar energy for low-temperature heating of buildings, whether collected actively or passively, is free, and the net energy yield is moderate (active) to high (passive). Both active and passive technology are well developed and can be installed quickly. No heat-trapping carbon dioxide is added to the atmosphere and environmental effects from air and water pollution are low. Land disturbance is also minimal because passive

Q: What percentage of all species that have ever lived on earth have become extinct?

Figure 18-19 Solar envelope house that is heated and cooled passively by solar energy and the earth's thermal energy. This patented Enertia design developed by Michael Sykes needs no conventional heating or cooling system in most areas. It comes in a precut kit engineered and tailored to the buyer's design goals. Sykes plants 50 trees for each one used in his house kits and his design has received both the Department of Energy's Innovation Award and the North Carolina Governor's Energy Achievement Award. (Enertia Building Systems, Rt. 1, Box 67, Wake Forest, N.C. 27587)

systems are built into structures, and active solar collectors are usually placed on rooftops. However, owners of passive and active solar systems need *solar legal rights* laws to prevent others from building structures that block their access to sunlight.

With current technology, active solar systems usually cost too much for heating most homes and small buildings because they use more materials to build, need more maintenance, and eventually deteriorate and must be replaced. In addition, some people consider active solar collectors sitting on rooftops or in yards to be ugly. However, retrofitting existing buildings is often easier with an active solar system.

How Can We Cool Houses Naturally?

Superinsulation and superinsulated windows help make buildings cooler. Passive cooling can also be provided by blocking the high summer sun with deciduous trees, window overhangs, or awnings (Figure 18-18, top). Windows and fans can be used to take advantage of breezes and keep air moving. A reflective insulating foil sheet suspended in the attic can block heat from radiating down into the house. Solar-powered air conditioners have been developed but so far are too expensive for residential use.

Earth tubes can also be used for indoor cooling (Figure 18-18, top). At a depth of 3–6 meters (10–20 feet), the soil temperature stays at about 5–13°C (41–55°F) all year long in cold northern climates, and about 19°C (67°F) in warm southern climates. Several earth tubes, simple plastic (PVC) plumbing pipes with a diameter of 10–15 centimeters (4–6 inches),

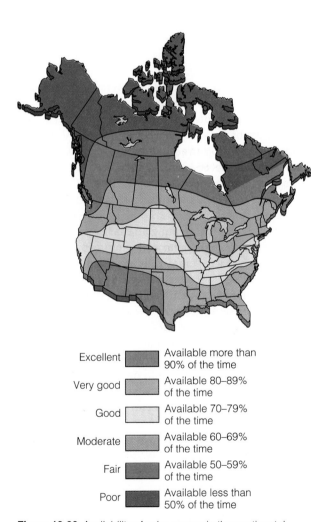

Excellent		Available more than 90% of the time
Very good		Available 80–89% of the time
Good		Available 70–79% of the time
Moderate		Available 60–69% of the time
Fair		Available 50–59% of the time
Poor		Available less than 50% of the time

Figure 18-20 Availability of solar energy in the continental United States and Canada. (Data from U.S. Department of Energy and National Wildlife Federation)

A: About 99.9%

buried about 0.6 meter (2 feet) apart and 3–6 meters deep can pipe cool and partially dehumidified air into an energy-efficient house at a cost of a few dollars per summer.* People allergic to pollen and molds should add an air purification system, but this is also necessary with a conventional cooling system.

How Can Solar Energy Be Used to Generate High-Temperature Heat and Electricity? Several so-called *solar thermal* systems collect and transform radiant energy from the sun into thermal energy (heat) at high temperatures, which can be used directly or converted to electricity (Figure 18-21). In one such *central receiver system*, called a *power tower*, huge arrays of computer-controlled mirrors called *heliostats* track the sun and focus sunlight on a central heat-collection tower (Figure 18-21a). Several government-subsidized experimental power towers are being tested in the United States (at the 10-megawatt Solar Two pilot plant in the desert near Barstow, California, which opened in 1996), Japan, Italy, France, and Spain.

In a *solar thermal plant* or *distributed receiver system*, sunlight is collected and focused on oil-filled pipes running through the middle of curved solar collectors (Figure 18-21b). This concentrated sunlight can generate temperatures high enough for industrial processes or for producing steam to run turbines and generate electricity. At night or on cloudy days, high-efficiency natural gas turbines supply backup electricity as needed. In California's Mojave Desert, such a system with a natural gas turbine backup system has produced power for 8–10¢ per kilowatt-hour—much cheaper than nuclear power plants.

Another type of distributed receiver system uses *parabolic dish collectors* (that look somewhat like TV satellite dishes) instead of parabolic troughs. These collectors can track the sun along two axes and are generally more efficient than troughs. A pilot plant is now being built in northern Australia. The U.S. Department of Energy projects that early in the 21st century parabolic dishes with gas turbine backup should be able to produce electrical power for 6¢ per kilowatt-hour.

Another promising approach to intensifying solar energy is a *nonimaging optical solar concentrator*. With this technology, the sun's rays are allowed to scramble instead of being focused on a particular point (Figure 18-21c). Experiments show that a nonimaging parabolic concentrator can intensify sunlight striking the earth 80,000 times. Because of its high efficiency and ability to generate extremely high temperatures,

nonimaging concentrators may make solar energy practical for widespread industrial and commercial use within a decade.

Inexpensive solar cookers can focus and concentrate sunlight and cook food, especially in rural villages in sunny developing countries. They can be made by fitting an insulated box big enough to hold three or four pots inside with a transparent, removable top (Figure 18-21d). Solar cookers reduce deforestation for fuelwood, the time and labor needed to collect firewood, and indoor air pollution from smoky fires.

The impact of solar thermal power plants on air and water is low. They can be built as large or as small as needed in 1–2 years, compared to 5–7 years for a coal-fired plant and 12–15 years for a nuclear power plant. This saves investors millions in interest on construction loans. Distributed system solar thermal power plants with small turbines burning natural gas as backup can produce electricity at about half the cost of a nuclear power plant (Figure 18-16). According to the Worldwatch Institute, solar power plants installed on a total land area equal to that of Panama or South Carolina could provide as much electricity as the entire world now uses.

How Can We Produce Electricity from Solar Cells? The PV Revolution In addition to the current personal computer (PC) revolution, analysts project that we will soon be in the midst of a PV or solar cell revolution. Solar energy can be converted directly into electrical energy by **photovoltaic (PV) cells**, commonly called **solar cells** (Figure 18-22). Sunlight falling on a solar cell, a transparent silicon wafer thinner than a sheet of paper, releases a flow of electrons when it strikes silicon atoms, creating an electrical current.

Because a single solar cell produces only a tiny amount of electricity, many cells are wired together in a panel providing 30–100 watts. Several panels are in turn wired together and mounted on a roof or on a rack that tracks the sun and produces electricity for a home or a building. The direct current (DC) electricity produced can be stored in batteries and used directly or converted to conventional alternating current (AC) electricity.

Solar cells are reliable and quiet, have no moving parts, and should last 20–30 years or more if encased in glass or plastic. The cells can be installed quickly and easily and expanded or moved as needed; maintenance consists of occasional washing to keep dirt from blocking the sun's rays. Arrays of cells can be located in deserts and marginal lands, along interstate highways, in yards, and on rooftops. Expandable banks of cells can also produce electricity at a small power plant, using efficient turbines burning natural gas to provide backup power when the sun isn't shining.

---

* They work. I used them in a passively heated office for 15 years.

Q: What are the six key features of living systems?

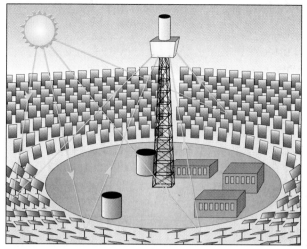

a. Solar Tower Power

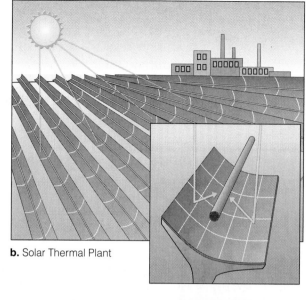

b. Solar Thermal Plant

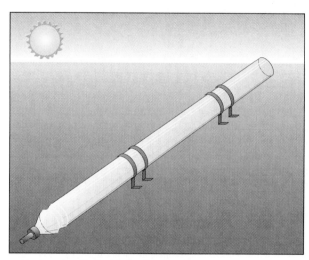

c. Nonimaging Optical Solar Concentrator

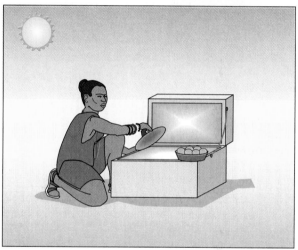

d. Solar Cooker

Figure 18-21 Several ways to collect and concentrate solar energy to produce high-temperature heat and electricity are in use. Today, solar plants are used mainly to supply reserve power for daytime peak electricity loads, especially in sunny areas with a high demand for air conditioning.

Solar cells produce no heat-trapping CO_2. Air and water pollution during operation is extremely low, air pollution from manufacture is low, land disturbance is very low for roof-mounted systems, and producing the materials doesn't require strip mining. The net energy yield is fairly high and is increasing with new designs. Whereas solar thermal systems require direct sunlight, PV cells work in cloudy weather. Traditional-looking solar-cell roof shingles and photovoltaic panels that resemble metal roofs are now available, reducing the cost of solar cell installations by saving on roof costs.

Solar cells are an ideal technology for providing electricity to 2 billion people in rural areas in most de-veloping countries, which have not committed to expensive large-scale centralized nuclear and fossil power plants and electric-line transmission systems. With financing from the World Bank, India is installing solar cell systems in 38,000 villages, and Zimbabwe is bringing solar electricity to 2,500 villages.

With an aggressive program starting now, analysts project that solar cells could supply 17% of the world's electricity by 2010—as much as nuclear power does today—at a lower cost and much lower risk. With a strong push from governments and private investors, by 2050 solar cells could provide as much as 25% of the world's electricity (at least 35% in the United States).

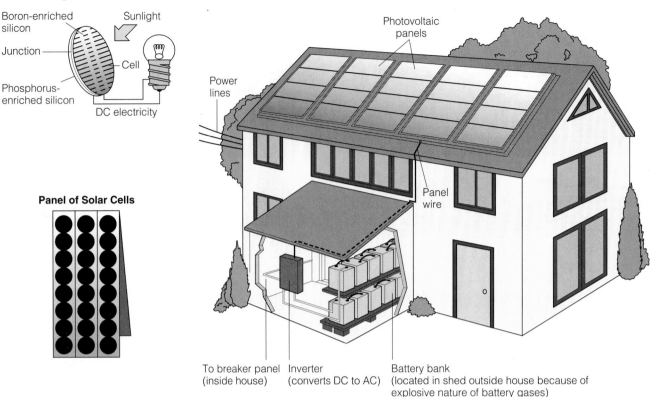

Single Solar Cell

Boron-enriched silicon

Sunlight

Junction

Cell

Phosphorus-enriched silicon

DC electricity

Panel of Solar Cells

Array of Solar Cell Panels on a Roof

Photovoltaic panels

Power lines

Panel wire

To breaker panel (inside house)

Inverter (converts DC to AC)

Battery bank (located in shed outside house because of explosive nature of battery gases)

Figure 18-22 Photovoltaic (solar) cells can provide electricity for a house. Small and easily expandable arrays of such cells can provide electricity for urban villages throughout the world without large power plants or power lines. Massive banks of such cells can also produce electricity at a small power plant. Today, at least two dozen U.S. utility companies are using photovoltaic cells in their operations; as the price of such electricity drops, usage will increase dramatically. In 1990 a Florida builder began selling tract houses that get all of their electricity from roof-mounted solar cells. Although the solar-cell systems account for about one-third of the cost of each house, the savings in electric bills will pay this off over the term of a 30-year mortgage. Soon the racks of solar cell panels shown in this diagram will be incorporated in the roof, in the form of solar-cell roof shingles or PV panel roof systems that will look like metal roofs.

There are some drawbacks, however. The current costs of solar-cell systems are high (Figure 18-16), but they should become competitive in 5–15 years and are already cost-competitive in some situations. The manufacture of solar cells produces moderate levels of water pollution, which can be eliminated by effective pollution controls. Some people find racks of solar cells on rooftops or in yards to be unsightly, but new thin and flexible rolls of cells (and solar shingles and solar-cell metal roofs) should eliminate this problem.

According to some analysts, unless federal and private research efforts on photovoltaics are increased sharply, the United States may lose out on a huge global market ($1.4 billion in 1995 and at least $5 billion per year by 2010) and may have to import solar cells from Japan, Germany, Italy, and other countries that have invested heavily in this promising technology since 1980.

🐝 18-5 PRODUCING ELECTRICITY FROM MOVING WATER AND FROM HEAT STORED IN WATER

What Is Hydroelectric Power? In a *large-scale hydropower project*, a high dam is built across a large river to create a reservoir. Some of the water stored in the reservoir is allowed to flow through huge pipes at controlled rates, spinning turbines and producing electricity. Such projects have advantages and disadvantages (Figure 11-7). In a *small-scale hydropower project*, a low dam with no reservoir (or only a small one) is built across a small stream. Output in small systems can vary with seasonal changes in stream flow.

Falling water can also produce electricity in *pumped-storage hydropower systems*, which supply extra power mainly during times of peak electrical demand. When demand is low (usually at night), pumps using

Q: What percentage of cancers are caused or promoted by environmental and lifestyle factors?

surplus electricity from a conventional power plant pump water from a lake or a reservoir to another reservoir at a higher elevation. When a power company temporarily needs more electricity than its other plants can produce, water in the upper reservoir is released, flows through turbines, and generates electricity on its return to the lower reservoir. Another possibility may be to use solar-powered pumps to raise water to the upper reservoir.

Hydroelectric power supplies about 18% of the world's electricity (95% in Norway, 50% in developing countries, and 9–10% in the United States) and 6% of its total commercial energy. In 1996 the world's three largest producers of hydroelectric power were Canada, the United States, and Brazil. Any new large supplies of hydroelectric power in the United States will be imported from Canada, which gets more than 70% of its electricity from hydropower. Within a few years, China, with 10% of the world's hydropower potential, may become the largest producer of hydroelectricity.

Hydropower has a moderate to high net useful energy yield and fairly low operating and maintenance costs. Hydroelectric plants rarely need to be shut down, and they emit no heat-trapping carbon dioxide or other air pollutants during operation. They have life spans 2–10 times those of coal and nuclear plants. Large dams also help control flooding and supply a regulated flow of irrigation water to areas below the dam.

However, hydropower, especially huge dams and reservoirs, has adverse effects on the environment (Figure 11-7). Because of increasing concern about the environmental and social consequences of large dams, there has been growing pressure on the World Bank and other development agencies to stop funding new, large-scale hydropower projects. Small-scale projects eliminate most of the harmful environmental effects of large-scale projects, but they can threaten recreational activities and aquatic life, disrupt the flow of wild and scenic rivers, and destroy wetlands.

Is Producing Electricity from Tides and Waves a Useful Option? Twice a day in high and low tides, water that flows into and out of coastal bays and estuaries can spin turbines to produce electricity (Figure 18-23). Two large tidal energy facilities are currently operating, one at La Rance in France and the other in Canada's Bay of Fundy. However, most analysts expect tidal power to make only a tiny contribution to world electricity supplies. There are few suitable sites, and construction costs are high.

The kinetic energy in ocean waves, created primarily by wind, is another potential source of electricity (Figure 18-23). Most analysts expect wave power to make little contribution to world electricity production, except in a few coastal areas with the right conditions (such as western England). Construction costs are moderate to high and the net energy yield is moderate, but equipment could be damaged or destroyed by saltwater corrosion and severe storms.

How Can We Produce Electricity from Heat Stored in Water? Japan and the United States have been evaluating the use of the large temperature differences (between the cold, deep waters and the sun-warmed surface waters) of tropical oceans for producing electricity. If economically feasible, this would be done in *ocean thermal energy conversion* (OTEC) plants anchored to the bottom of tropical oceans in suitable sites (Figure 18-23). However, most energy analysts believe that the large-scale extraction of energy from ocean thermal gradients may never compete economically with other energy alternatives. Despite 50 years of work, the technology is still in the research-and-development stage.

Saline solar ponds, usually located near inland saline seas or lakes in areas with ample sunlight, can be used to produce electricity (Figure 18-23). Heat accumulated during the daytime in the denser bottom layer can be used to produce steam that spins turbines, generating electricity. A small experimental saline solar pond power plant on the shore of the Israeli side of the Dead Sea operated for several years but was closed in 1989 because of high operating costs; smaller experimental systems have been built in California and Australia.

Freshwater solar ponds can be used to heat water and space (Figure 18-23). A shallow hole is dug and lined with concrete. A number of large, black plastic bags, each filled with several centimeters of water, are placed in the hole and then covered with fiberglass insulation panels. The panels let sunlight in but keep most of the heat stored in the water during the daytime from being lost to the atmosphere. When the water in the bags has reached its peak temperature in the afternoon, a computer turns on pumps to transfer hot water from the bags to large, insulated tanks for distribution.

Saline and freshwater solar ponds require no energy storage and backup systems. They emit no air pollution and have a moderate net energy yield. Freshwater solar ponds can be built in almost any sunny area and have moderate construction and operating costs. With adequate research and development support, proponents believe that freshwater solar ponds could supply 3–4% of U.S. electricity needs within 10 years.

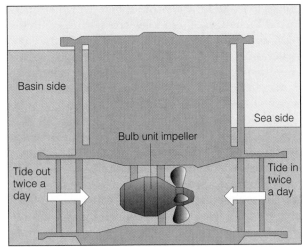

Tidal Power Plant

Basin side

Sea side

Bulb unit impeller

Tide out twice a day

Tide in twice a day

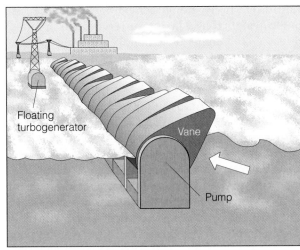

Wave Power Plant

Floating turbogenerator

Vane

Pump

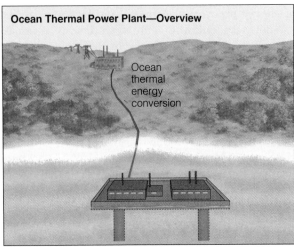

Ocean Thermal Power Plant—Overview

Ocean thermal energy conversion

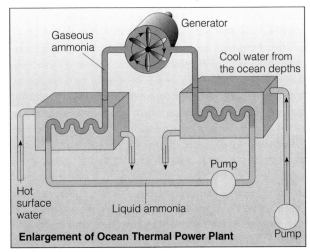

Gaseous ammonia

Generator

Cool water from the ocean depths

Hot surface water

Liquid ammonia

Pump

Pump

Enlargement of Ocean Thermal Power Plant

Ocean Thermal Electric Plant

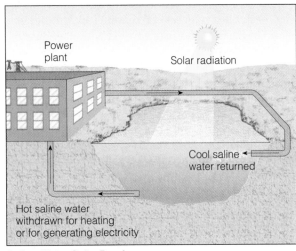

Power plant

Solar radiation

Cool saline water returned

Hot saline water withdrawn for heating or for generating electricity

Saline Water Solar Pond

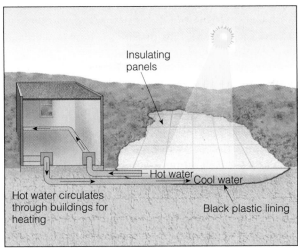

Insulating panels

Hot water circulates through buildings for heating

Hot water

Cool water

Black plastic lining

Freshwater Solar Pond

Figure 18-23 Ways to produce electricity from moving water and to tap into solar energy stored in water as heat. None of these systems are expected to be significant sources of energy in the near future.

Q: In terms of reduced life span, what is the world's greatest risk?

⊕ 18-6 PRODUCING ELECTRICITY FROM WIND

Since 1980 the use of wind to produce electricity has grown rapidly and in 1996 was the world's fastest-growing energy resource. In 1996 there were more than 25,000 wind turbines worldwide, collectively producing about 6,100 megawatts of electricity (about 17% of the world's electricity).

Wind farms (clusters of 20–100 wind turbines), most of which are highly automated, now provide about 1% of California's electricity—enough to power 280,000 homes. Sizable wind-farm projects are being planned in 12 other states. In principle, all the power needs of the United States could be provided by exploiting the wind potential of just three states: North Dakota, South Dakota, and Texas.

Wind power is a virtually unlimited source of energy at favorable sites; the global potential of wind power is about five times current world electricity use. Wind farms can be built in 6 months to a year and then easily expanded as needed. With a moderate to fairly high net energy yield, these systems emit no heat-trapping carbon dioxide or other air pollutants and need no water for cooling; manufacturing them produces little water pollution. The land under wind turbines can be used for grazing cattle or farming.

Wind power (with much lower government subsidies) has a significant cost advantage over nuclear power (Figure 18-16) and has become competitive with coal fired power plants in many places. With new technological advances and mass production, projected cost declines should make wind power one of the world's cheapest ways to produce electricity within the next decade. In the long run, electricity from large wind farms in remote areas might be used to make hydrogen gas from water during off-peak periods. The hydrogen could then be fed into a pipeline and storage system.

One drawback of wind power is that it's economical primarily in areas with steady winds. When the wind dies down, backup electricity from a utility company or from an energy storage system becomes necessary. Backup power could also come from linking wind farms with a solar-cell or hydropower plant, from efficient natural-gas–burning turbines, or from flywheels.

Other drawbacks to wind farms include visual pollution and noise, although these can be overcome by improving their design and locating them in isolated areas. Large wind farms might also interfere with the flight patterns of migratory birds in certain areas, and they have killed large birds of prey (especially hawks, falcons, and eagles) that prefer to hunt along the same ridge lines that are ideal for wind turbines. Researchers are evaluating how serious and widespread this problem is and are trying to find ways to deal with it.

Wind power experts project that by the middle of the 21st century wind power could supply more than 10% of the world's electricity and 10–25% of the electricity used in the United States. European governments are currently spending 10 times more for wind energy research and development than the U.S. government. Analysts warn that if this trend continues, U.S. consumers will be buying wind turbines from European countries and eventually from China and India, and American companies will lose out on a rapidly growing global business.

18-7 PRODUCING ENERGY FROM BIOMASS

⊕ **How Can We Produce Energy from Plants? Plant Power** Many forms of *biomass*—organic matter in plants produced through photosynthesis—can be burned directly as a solid fuel or converted into gaseous or liquid *biofuels* (Figure 18-24). The burning of wood and manure to heat buildings and cook food supplies about 13% of the world's energy (4–5% in Canada and the United States) and about 35% of the energy used in developing countries.

Biomass is a potentially renewable energy resource as long as trees and plants are not harvested faster than they grow back, a requirement that is not being met in many places. No net increase in atmospheric levels of heat-trapping carbon dioxide occurs as long as the rate of removal and burning of trees and plants and the rate of loss of below-ground organic matter do not exceed the rate of replenishment. Burning biomass fuels adds much less air pollution to the atmosphere per unit of energy produced than does the uncontrolled burning of coal.

According to a 1992 UN study, by 2050 biomass could supply as much as 55% of the total energy used globally today. Whether such projections become reality depends on the availability of large areas of productive land and adequate water and fertilizer—resources that may be in short supply in coming decades—and the ability to minimize the harmful environmental effects of such large-scale production of biomass. Currently, potentially renewable biomass is being exploited in nonrenewable and unsustainable ways, primarily because of deforestation, soil erosion, and the inefficient burning of wood in open fires and energy-wasting stoves.

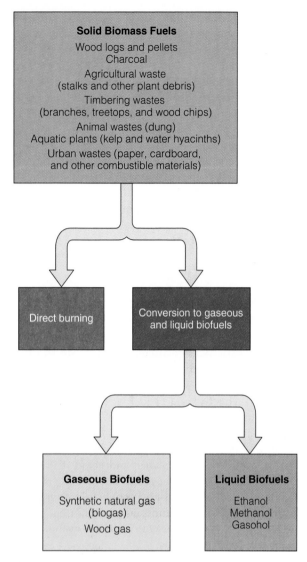

Solid Biomass Fuels

Wood logs and pellets
Charcoal
Agricultural waste
(stalks and other plant debris)
Timbering wastes
(branches, treetops, and wood chips)
Animal wastes (dung)
Aquatic plants (kelp and water hyacinths)
Urban wastes (paper, cardboard,
and other combustible materials)

Direct burning

Conversion to gaseous
and liquid biofuels

Gaseous Biofuels

Synthetic natural gas
(biogas)

Wood gas

Liquid Biofuels

Ethanol
Methanol
Gasohol

Figure 18-24 Principal types of biomass fuel.

Without effective land-use controls and replanting, widespread removal of trees and plants can deplete soil nutrients and cause excessive soil erosion, water pollution, flooding, and loss of wildlife habitat. Biomass resources also have a high moisture content (15–95%). The added weight of moisture makes collecting and hauling wood and other plant material fairly expensive, and the moisture also reduces the net energy yield.

Is Producing Gaseous and Liquid Fuels From Solid Biomass a Useful Option? Various forms of solid biomass can be converted by bacteria and various chemical processes into gaseous and liquid biofuels (Figure 18-24). Examples include *biogas*, a mixture of 60% methane (the principal component of natural gas)

and 40% carbon dioxide); *liquid ethanol* (ethyl, or grain, alcohol); and *liquid methanol* (methyl, or wood alcohol).

In China, anaerobic bacteria in more than 6 million *biogas digesters* convert organic plant and animal wastes into methane fuel for heating and cooking. These simple devices can be built for about $50 including labor. After the biogas has been separated, the solid residue is used as fertilizer on food crops or, if contaminated, on trees. When they work, biogas digesters are very efficient. However, they are also slow and unpredictable, a problem that could be corrected with development of more reliable models.

Some analysts believe that liquid ethanol and methanol produced from biomass could replace gasoline and diesel fuel when oil becomes too scarce and expensive. *Ethanol* can be made from sugar and grain crops (sugarcane, sugar beets, sorghum, and corn) by fermentation and distillation. Gasoline mixed with 10–23% pure ethanol makes *gasohol*, which can be burned in conventional gasoline engines and is sold as super unleaded or ethanol-enriched gasoline.

Another alcohol, *methanol*, is made mostly from natural gas but can also be produced at a higher cost from wood, wood wastes, agricultural wastes (such as corncobs), sewage sludge, garbage, and coal. The advantages and disadvantages of using ethanol, methanol, and several other fuels as alternatives to gasoline were summarized in Table 9-3 (p. 266). Recently, a water-based gasoline fuel has entered the picture and is currently undergoing extensive testing by state and federal agencies (Individuals Matter, right).

Is Producing Energy from Biomass Plantations a Useful Option? One way to produce biomass fuel is to plant large numbers of fast-growing trees (especially cottonwoods, poplars, sycamores, and leucaenas), shrubs, and water hyacinths in *biomass plantations*. After harvest, these "Btu bushes" can be burned directly, converted into burnable gas, or fermented into a liquid alcohol fuel. Such plantations can be located on semiarid land not needed for crops (although lack of water can limit productivity) and can be planted to reduce soil erosion and help restore degraded lands.

However, this industrialized approach to biomass production requires large areas of land (about 81 times as much land as solar cells to provide the same amount of electricity) as well as heavy use of pesticides and fertilizers, which can pollute drinking water supplies and harm wildlife. In some areas the plantations might compete with food crops for prime farmland. Conversion of large forested areas or natural grasslands into single-species biomass plantations also reduces biodiversity and ecological integrity.

Q: How many offspring can a singe bacterium have in 24 hours?

INDIVIDUALS MATTER

It seems too good to be true, but a breakthrough fuel that is more than half water could power cars, trucks, train locomotives, generators, and aircraft running on gasoline and diesel fuel within a few years.

This new, milky-looking fuel, called A-21, was developed by Reno, Nevada, inventor Rudolf Gunnerman. It consists of 55% water and 45% naphtha (a cheap-to-produce by-product of petroleum distillation). In addition, it contains lubricants, antirust and antifreeze agents, and small amounts of a blending agent developed by Gunnermann (which attracts the normally repellent water and oily naphtha to one another). In 1994 Gunnerman and diesel equipment giant Caterpillar formed Advanced Fuels, a joint venture to test and market the new fuel.

So far in every test the new fuel tops conventional gasoline and diesel fuel as a clean, safe, and cheap fuel that can be used in almost any combustion engine. In November 1995, Nevada certi-

fied A-21 as a clean alternative fuel. That means it can meet federal and Nevada laws requiring clean fuels in fleets and other vehicles. A-21 is especially attractive to the trucking industry, facing federal crackdowns on dirty diesel engine emissions.

In 1996 A-21 was also being tested in California and Illinois. If it passes these hurdles and even more stringent testing by the Department of Energy, A-21 could be used in all states. By early 1996 the new fuel was successfully being used in a city bus and a power generator in Reno, in generators at Caterpillar's Illinois plants, and in public and private vehicles in Sacramento, California.

Tests so far show that using A-21 leads to a 60% drop in EPA-monitored emissions of air pollutants such as carbon monoxide, nitrous oxide, and hydrocarbons. The vapor pressure of the fuel is about one-fifth that of gasoline. That means that vapor recovery systems won't be needed at the pump. Because gasoline vapor is what catches fire, A-21 is virtually immune to fire and explosions.

Because it doesn't burn outside engine combustion chambers, the fuel can be stored in aboveground tanks without risking explosions.

Converting vehicles would be as easy as changing spark plugs. Once that's done, the vehicle could use either the water-based fuel or conventional gasoline or diesel fuels.

Car and diesel truck drivers could see their fuel costs cut in half. That's because the refining process for producing naphtha (currently used mostly as a hardener in road tars) is faster and cheaper than producing gasoline and diesel fuel. In addition to cutting costs, using naphtha instead of gasoline and diesel fuel would eliminate up to 90% of the air pollutants produced by oil refineries. Stay tuned to see whether this fuel passes its final tests.

Critical Thinking

Can you think of some economic and political reasons why this potentially promising fuel may not be used widely? How could such hurdles be overcome?

Is Burning Wood a Useful Option? Almost 70% of the people living in developing countries heat their homes and cook their food by burning wood or charcoal. However, about 2 billion people in developing countries cannot find (or are too poor to buy) enough fuelwood to meet their needs (Figure 16-15).

Sweden leads the world in using wood as an energy source, mostly for district heating plants. Wood provides 8% of all energy used by U.S. industry, with the forest products industry (mostly paper companies and lumber mills) consuming almost two-thirds of the fuelwood used in the United States. About 20% of U.S. homes get some heat from burning wood, and about 4% burn wood as their main heating fuel (down from 7.5% in 1984).

Wood has a moderate to high net useful energy yield when collected and burned directly and efficiently near its source. However, in urban areas, to

which wood must be hauled from long distances, wood can cost homeowners more per unit of energy produced than oil or electricity. Harvesting wood can cause accidents (mostly from chain saws), and burning wood in poorly maintained or operated wood stoves can cause house fires. According to the EPA, air pollution from wood burning causes as many as 820 cancer deaths a year in the United States. Since 1990 the EPA has required all new wood stoves sold in the United States to emit at least 70% less particulate matter than earlier models.

Fireplaces can also be used for heating, but they usually result in a net loss of energy from a house. The draft of heat and gases rising up the fireplace chimney pushes out warm inside air and pulls in cold outside air through cracks and crevices throughout a house. Fireplace inserts with glass doors and blowers help, but they still waste energy compared with efficient wood-burning stoves. Energy loss can be reduced by closing the room off and cracking a window so that the fireplace won't draw much heated air from other rooms. A better solution is to run a small pipe from outside into the front of the fireplace so that it gets the air it needs during combustion.

Is Burning Agricultural and Urban Wastes a Useful Option? In agricultural areas, crop residues (such as sugarcane residues, rice husks, cotton stalks, and coconut shells) and animal manure can be collected and burned or converted into biofuels. The ash from biomass power plants and burners can sometimes be used as fertilizer.

This approach makes sense when crop residues are burned in small power plants or burners located near areas where the residues are produced. Otherwise, it takes too much energy to collect, dry, and transport the residues to power plants. Some ecologists argue that it makes more sense to use animal manure as a fertilizer and to use crop residues to feed livestock, retard soil erosion, and fertilize the soil.

An increasing number of cities in Japan, western Europe, and the United States have built incinerators that burn trash and use the energy released to produce electricity or to heat nearby buildings (Figure 13-7). However, this approach has been limited by opposition from citizens concerned about emissions of toxic gases and disposal of the resulting toxic ash. Some analysts argue that more energy is saved by composting or recycling paper and other organic wastes than by burning them. For example, recycling paper saves at least twice as much energy as is produced by burning the paper.

18-8 THE SOLAR–HYDROGEN REVOLUTION

What Can We Use to Replace Oil? Good-Bye Oil and Smog, Hello Hydrogen When oil is gone (or when what's left costs too much to use) what will we use to fuel vehicles, industry, and buildings? Some scientists say the fuel of the future is hydrogen gas (H_2) (Table 9-3, p. 266).

There is very little hydrogen gas around, but we can get it from something we have plenty of: water. Water can be split by electricity into gaseous hydrogen and oxygen (Figure 18-25). If we can make the transition to an *energy-efficient solar–hydrogen age*, we could say goodbye to smog, oil spills, acid rain, and nuclear energy, and perhaps to the threat of global warming. The reason is simple: When hydrogen burns in air, it combines with oxygen gas in the air and produces nonpolluting water vapor and some nitrogen oxides, thus eliminating most of the air pollution problems we face today.

What's the Catch? If you think using hydrogen as an energy source sounds too good to be true, you're right. Several problems must be solved to make hydrogen one of our primary energy resources, but scientists are making rapid progress in finding solutions.

One problem is that it takes energy to produce this marvelous fuel. We could use electricity from coal-burning and nuclear power plants to split water and produce hydrogen, but this subjects us to the harmful environmental effects associated with using these fuels, and it costs more than the hydrogen fuel is worth.

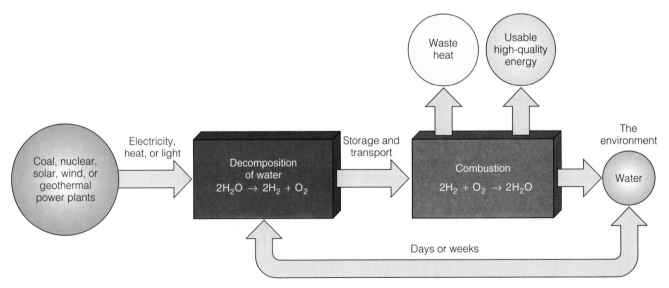

Figure 18-25 The hydrogen energy cycle. The production of hydrogen gas requires electricity, heat, or solar energy to decompose water, thus leading to a negative net energy yield. However, hydrogen is a clean-burning fuel that can replace oil, other fossil fuels, and nuclear energy. Using solar energy to produce hydrogen from water could also eliminate most air pollution and greatly reduce the threat of global warming.

Q: In terms of deaths, what is the world's most dangerous viral disease?

Most proponents of hydrogen gas believe that to get its very low pollution benefits, the energy to produce the gas from water must come from the sun, in the form of electricity generated by sources such as hydropower, solar thermal and solar cell power plants, and wind farms. If scientists and engineers can learn how to use sunlight to decompose water cheaply enough, they will set in motion a *solar–hydrogen revolution* over the next 50 years and change the world as much as the agricultural and industrial revolutions did.

Such a revolution would eliminate the air and water pollution caused by extracting, transporting, and burning fossil fuels. It would reduce the threat of global warming by sharply reducing CO_2 emissions and would also decrease the threat of wars over dwindling oil supplies. It would allow nuclear and coal-fired power plants to be phased out and let individuals produce most or all of their own energy.

Currently, using solar energy to produce hydrogen gas is too costly, but the costs of using solar energy to produce electricity are coming down. Moreover, if the health and environmental costs of using gasoline were included in its market prices (through gasoline taxes), it would cost about $1 per liter ($4 per gallon)—roughly the price in Japan and in many European countries. At this price, hydrogen produced by any form of solar energy would be competitive.

Hydrogen for vehicles could initially be produced from natural gas. Blends of natural gas and hydrogen produced from solar sources could then be phased in gradually as reserves of natural gas fuel become depleted. Hydrogen produced by solar cell power plants (in sunny, mostly desert, areas) could be carried by pipeline wherever it was needed. However, such facilities would have to be designed and managed to reduce the disruption of easily harmed desert ecosystems.

Hydrogen gas is much easier to store than electricity. It can be stored in a pressurized tank, metal powders, and activated carbon that absorb gaseous hydrogen and release it when heated for use as a fuel. If properly handled, hydrogen is a safer fuel than gasoline and natural gas. Unlike gasoline, compounds and powders containing absorbed hydrogen will not explode or burn if a vehicle's tank is ruptured in an accident. However, it's difficult to store enough hydrogen gas in a car as a compressed gas or as a metal powder for it to run very far—a problem similar to the one electric cars face. Scientists and engineers are seeking solutions to this problem.

Another possibility is to power a car with a *fuel cell* (Figure 18-7) in which hydrogen and oxygen gas combine to produce electrical current. Fuel cells produce no air pollution, have no moving parts, and have high energy efficiencies of up to 65%—several times the efficiency of conventional gasoline-powered engines and electric cars. Fuel cells could be resupplied with hydrogen in a matter of minutes, compared to the several hours needed to recharge the batteries in electric cars.

A number of prototype fuel-cell systems for cars, buses, homes, and buildings are being tested and evaluated. Currently, fuel cells are expensive and heavy, and they produce energy at a cost equivalent to gasoline at $2.50 per gallon, but this could change with more research and mass production.

The solar–hydrogen revolution has already started. Soon a German firm plans to market solar–hydrogen systems that would meet all the heating, cooling, cooking, refrigeration, and electrical needs of a home, as well as providing hydrogen fuel for one or more cars. Germany and Saudi Arabia have each built a large solar–hydrogen plant, and Germany and Russia are jointly developing a prototype commercial airliner fueled by hydrogen. (Hydrogen has about 2.5 times the energy by weight of gasoline, making it an especially attractive aviation fuel.)

We don't need to invent hydrogen-powered vehicles. Mercedes, BMW, and Mazda already have prototypes being tested on the roads. In the United States, a *hydrogen corridor* of research and production facilities is beginning to develop in the desert east of Los Angeles. It is anchored by a $2.5-million facility built by Xerox for using solar energy to convert water into hydrogen.

What's Holding Up the Solar–Hydrogen Revolution? Politics and economics, not a lack of promising technology, are the main factors holding up a more rapid transition to a solar–hydrogen age. Phasing out dependence on fossil fuels and phasing in new solar–hydrogen technologies over the next 40–50 years involves convincing investors and energy companies with strong vested interests in fossil fuels to risk a lot of capital on hydrogen. It also involves convincing governments to put up some of the money for developing hydrogen energy (as they have done for decades for fossil fuels and nuclear energy).

In the United States, large-scale government funding of hydrogen research is generally opposed by powerful oil companies, electric utilities, and automobile manufacturers, which see it as a serious threat to their profits. In contrast, the Japanese and German governments have been spending seven to eight times more on hydrogen research and development than the United States. Some analysts warn that without greatly increased government and private research and development, Americans may be buying solar–hydrogen equipment and fuel cells from Germany and Japan and may lose out on a huge global market and source of domestic jobs.

How Can We Tap the Earth's Internal Heat? Going Underground Heat contained in underground rocks and fluids is an important source of energy. Over millions of years, this **geothermal energy** from earth's mantle (Figure 12-2) has been transferred to underground concentrations of *dry steam* (steam with no water droplets), *wet steam* (a mixture of steam and water droplets), and *hot water* trapped in fractured or porous rock at various places in earth's crust. All three types are currently being used to heat space and water, and in some cases to produce electricity (Figure 18-2).

If such geothermal sites are close to the surface, wells can be drilled to extract the dry steam, wet steam (Figure 18-26), or hot water. This thermal energy can be used for space heating and to produce electricity or high-temperature heat for industrial processes.

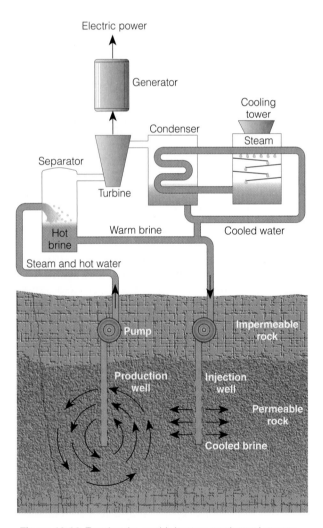

Figure 18-26 Tapping the earth's heat or geothermal energy in the form of wet steam to produce electricity.

Geothermal reservoirs containing dry steam, wet steam, or hot water can be depleted if heat is removed faster than it is renewed by natural processes. Thus, geothermal resources are nonrenewable on a human time scale, but the potential supply is so vast that it is usually classified as a potentially renewable energy resource. However, easily accessible concentrations of geothermal energy are fairly scarce.

Currently, about 22 countries (most of them in the developing world) are extracting energy from geothermal sites to produce less than 1% of the world's electricity. The United States accounts for 44% of the geothermal electricity generated worldwide, with most of the favorable sites in California, Hawaii, Nevada, and Utah.

Three other virtually nondepletable or potentially renewable sources of geothermal energy are *molten rock* (magma); *hot dry-rock zones*, where molten rock that has penetrated the earth's crust heats subsurface rock to high temperatures; and low- to moderate-temperature *warm-rock reservoir deposits*, which could be used to preheat water and run heat pumps for space heating and air conditioning.

Research is being carried out to see whether hot-dry rock zones, which can be found almost anywhere if one can drill deep enough, can provide affordable geothermal energy. According to the National Academy of Sciences, the energy potentially recoverable from such reservoirs would meet U.S. energy needs at current consumption levels for 600–700 years, but the projected cost is high.

What Are the Pros and Cons of Geothermal Energy? The biggest advantages of geothermal energy include a vast, reliable, and sometimes renewable supply of energy for areas near reservoir sites, moderate net energy yields for large and easily accessible reservoir sites, about 96% fewer CO_2 emissions per unit of energy than fossil fuels, and a competitive cost of producing electricity (Figure 18-16).

A serious limitation of geothermal energy is the scarcity of easily accessible reservoir sites. Dry steam, wet steam, and hot water geothermal reservoirs must be carefully managed or they can be depleted within a few decades. Furthermore, geothermal development in some areas can destroy or degrade forests or other ecosystems. Without pollution control, conventional single-loop heat transfer geothermal energy production causes moderate to high local air and water pollution. Noise, odor, and local climate changes can also be problems. With proper controls, however, most experts consider the environmental effects of geothermal energy to be less (or no greater) than those of fossil fuel and nuclear power plants.

Q: At current rates, how long will it take for the world's urban population to double?

Economics is the main barrier to greater use of geothermal energy. Currently, the cost of tapping geothermal energy is too high for all but the most concentrated and accessible sources.

18-10 MAKING THE TRANSITION TO A SUSTAINABLE ENERGY FUTURE

Industry, governments (national, state, and local), and individuals all have important roles to play in the development of a sustainable energy future. Brazil and Norway get more than half of their energy from hydropower, wood, and alcohol fuel. Israel, Japan, the Philippines, and Sweden plan to rely on renewable sources for most of their energy. India is greatly expanding its use of wind energy and offers full income tax deductions for renewable energy investments.

California has become the world's showcase for solar and wind power. Communities such as Davis, California (p. 190), and Osage, Iowa (p. 46), and individuals everywhere are taking energy matters into their own hands (Figures 18-1, 18-13, and 18-19). Some actions you can take to waste less energy (and thus reduce pollution and environmental degradation) and make more use of renewable energy are listed in Appendix 5.

The chief feature of the sustainable energy revolution over the next 40–50 years is likely to be significant decentralization, analogous to the computer industry's shift from large, centralized mainframes to small, widely dispersed PCs. As the rapidly growing forces for technological change in the energy industry gain momentum in the early part of the next century, such a shift is likely to take place very rapidly.

By the middle of the next century the decentralized global energy system is likely to be based mostly on improved energy efficiency and much greater use of renewable forms of energy such as wind power, solar cells, and solar-produced hydrogen. Companies and investors trying to preserve large, centralized coal-burning and nuclear power plants may find themselves saddled with expensive technological dinosaurs. The nation that becomes the leader in the development of energy efficiency and solar technologies is likely to capture the huge future global market for this industry.

In the long run, humanity has no choice but to rely on renewable energy. No matter how abundant they seem today, eventually coal and uranium will run out.

DANIEL DEUDNEY AND CHRISTOPHER FLAVIN

CRITICAL THINKING

1. A homebuilder installs electric baseboard heat and claims that "it's the cheapest and cleanest way to go." Apply your understanding of the second law of energy (thermodynamics) to evaluate his claim.

2. What are the five most important things an individual can do to save energy at home and in transportation (see Appendix 5)? Which, if any, of these do you currently do? Which, if any, do you plan to do?

3. Should the United States (or the country where you live) institute a crash program to develop solar photovoltaic cells and solar-produced hydrogen fuel? Explain.

4. Someone tells you that we can save energy by recycling it. How would you respond?

5. Congratulations. You have just won $100,000 to build a house of your choice anywhere in the country. What type of house would you build? Where would you locate it? What types of materials would you use? What types of materials would you be sure *not* to use? How would you heat and cool your house? How would you heat your water? Considering fuel and energy efficiency, what sort of lighting, stove, refrigerator, washer, and dryer would you use? Which of these appliances could you do without? Would you use a dishwasher? A garbage disposal? A trash compactor? Suppose you learn that it will cost $140,000 to build the type of energy-efficient, passive solar house you want. Considering lifetime energy costs, would it make sense for you to borrow $40,000 to complete your house or to save the $40,000 by cutting back on energy-saving and solar design features?

*6. Make an energy use study of your school or dorm, and use the findings to develop an energy-efficiency improvement program. Present your plan to school officials.

*7. Make a concept map of this chapter's major ideas, using the section heads and subheads and the key terms (in boldface type). Look at the inside back cover and in Appendix 4 for information about concept maps.

SUGGESTED READINGS

See Internet Sources and Further Readings at the back of the book, and InfoTrac.

Technology Is the Answer (But What Was the Question?)

Amory B. Lovins

GUEST ESSAY

Physicist and energy consultant Amory B. Lovins is one of the world's most recognized and articulate experts on energy strategy. In 1989 he received the Delphi Prize for environmental work; in 1990 the Wall Street Journal *named him one of the 39 people most likely to change the course of business in the 1990s. He is research director at Rocky Mountain Institute, a nonprofit resource policy center that he and his wife, Hunter, founded in Snowmass, Colorado, in 1982. He has served as a consultant to over 200 utilities, private industries, and international organizations, and to many national, state, and local governments. He is active in energy affairs in more than 35 countries and has published several hundred papers and a dozen books on energy strategies and policies.*

The answers you get depend on the questions you ask. It is fashionable to suppose that we're running out of energy and that the solution is obviously to get lots more of it. But asking how to get more energy begs the question of how much we need and what are the cheapest and least environmentally harmful ways to meet these needs.

How much energy it takes to make steel, run a sewing machine, or keep ourselves comfortable in our houses depends on how cleverly we use energy, and the more it costs, the smarter we seem to get. It is now cheaper, for example, to double the efficiency of most industrial electric motor drive systems than to fuel existing power plants to make electricity. *(Just this one saving can more than replace the entire U.S. nuclear power program.)* We know how to make lights five times as efficient as those currently in use and how to make household appliances that give us the same work as now, but use one-fifth as much energy (saving money in the process).

Ten automakers have made good-sized, peppy, safe prototype cars averaging 29–59 kilometers per liter (67–138 miles per gallon) and within a decade automakers could have cars getting 64–128 kpl (150–300 mpg) on the road, if consumers demanded such cars. We know today how to make new buildings (and many old ones) so heat-tight (but still well ventilated) that they need essentially no outside energy to maintain comfort year-round, even in severe climates. In fact, I live and work in one [Figure 18-1].

These energy-saving measures are uniformly cheaper than going out and getting more energy. However, the old view of the energy problem included a worse mistake than forgetting to ask how much energy we needed: It sought more energy, in any form, from any source, at any price—as if all kinds of energy were alike.

Just as there are different kinds of food, so there are many different forms of energy, whose different prices and qualities suit them to different uses [Figure 3-10]. There is, after all, no demand for energy as such; nobody

wants raw kilowatt-hours or barrels of sticky black goo. People instead want energy services: comfort, light, mobility, hot showers, cold beverages, and the ability to bake bread and make cement. We ought therefore to start at that part of the energy problem and ask, "What tasks do we want energy for, and what amount, type, and source of energy will do each task most cheaply?"

Electricity is a particularly high-quality, expensive form of energy. An average kilowatt-hour delivered in the United States in 1994 was priced at about 9.3¢, equivalent to buying the heat content of oil costing $154 per barrel—over eight times oil's average world price in 1994. The average cost of electricity from nuclear plants (including construction, fuel, and operating expenses) that began operation in 1988 was 13.5¢ per kilowatt-hour in 1994, equivalent on a heat basis to buying oil at about $216 per barrel.

Such costly energy might be worthwhile if it were used only for the premium tasks that require it, such as lights, motors, electronics, and smelters. But those special uses—only 8% of all delivered U.S. energy needs—are already met twice over by today's power stations. Two-fifths of our electricity is already spilling over into uneconomic, low-grade uses such as water heating, space heating, and air conditioning; yet no matter how efficiently we use electricity (even with heat pumps), we can never get our money's worth on these applications.

Thus, *supplying more electricity is irrelevant to the energy problem we have.* Even though electricity accounts for almost all the federal energy research-and-development budget and for at least half the national energy investment, it is the wrong kind of energy to meet our needs economically. Arguing about what kind of new power station to build—coal, nuclear, solar—is like shopping for the best buy in antique Chippendale chairs to burn in your stove, or for brandy to put in your car's gas tank. *It is the wrong approach.*

Indeed, *any kind of new power station is so uneconomical that if you have just built one, you will save the country money by writing it off and never operating it.* Why? Because its additional electricity can be used only for low-temperature heating and cooling (the premium "electricity-specific" uses being already filled up) and is the most expensive way of supplying those services. Saving electricity (creating negawatts) is much cheaper than making it.

The real question is, What is the cheapest way to do low-temperature heating and cooling? The answer is weather-stripping, insulation, heat exchangers, greenhouses, superwindows (which have as much insulating value as the outside wall of a typical house), window shades and overhangs, trees, and so on. These measures generally cost about half a penny to 2¢ per kilowatt-hour; the running costs *alone* for a new nuclear plant will be nearly 4¢ per kilowatt-hour, so it's cheaper not to

Q: During the 1990s, what percentage of the world's population growth is expected to occur in urban areas?

run it. In fact, under the crazy U.S. tax laws, the extra saving from not having to pay the plant's future subsidies is probably so big that by shutting the plant down society can also recover the capital cost of having built it!

If we want more electricity, we should get it from the cheapest sources first. In approximate order of increasing price, these include:

- Converting to efficient lighting equipment. This would save the United States electricity equal to the output of 120 large power plants, plus $30 billion a year in fuel and maintenance costs.

- Using more efficient electric motors to save up to half the energy used by motor systems. This would save electricity equal to the output of another 150 large power plants and repay the cost in about a year.

- Eliminating pure waste of electricity, such as lighting empty offices.

- Displacing with good architecture, weatherization, insulation, and passive and some active solar techniques the electricity now used for water heating and for space heating and cooling.

- Making appliances, smelters, and the like cost-effectively efficient.

Just these five measures can quadruple U.S. electrical efficiency, making it possible to run today's economy with no changes in lifestyles and using no power plants, whether old or new or fueled with oil, gas, coal, or uranium. We would need only the present hydroelectric capacity, readily available small-scale hydroelectric projects, and a modest amount of wind power.

If we still wanted more electricity, the next cheapest sources would include industrial cogeneration, combined heat-and-power plants, low-temperature heat engines run by industrial waste heat or by solar ponds, filling empty turbine bays and upgrading equipment in existing big dams, modern wind machines or small-scale hydroelectric turbines in good sites, steam-injected natural-gas turbines, and perhaps recent developments in solar cells with waste heat recovery.

It is only after we had clearly exhausted all these cheaper opportunities that we would even consider building a new central power station of any kind—the slowest and costliest known way to get more electricity (or to save oil).

To emphasize the importance of starting with energy end uses rather than energy sources, consider a sad little story from France involving a "spaghetti chart" (or energy flowchart)—a device energy planners often use to show how energy flows from primary sources via conversion processes to final forms and uses. In the mid-1970s, energy conservation planners in the French government started, wisely, on the right-hand side of the spaghetti chart. They found that their biggest need

for energy was to heat buildings, and that even with good heat pumps, electricity would be the costliest way to do this. So they had a fight with their nationalized utility; they won, and electric heating was supposed to be discouraged or even phased out because it was so wasteful of money and fuel.

Meanwhile, down the street, the energy supply planners (who were far more numerous and influential in the French government) were starting on the left-hand side of the spaghetti chart. They said: "Look at all that nasty imported oil coming into our country! We must replace that oil. Oil is energy. We need some other source of energy. Voilà! Reactors can give us energy; we'll build nuclear reactors all over the country." But they paid little attention to who would use that extra energy and no attention to relative prices.

Thus these two groups of the French energy establishment went on with their respective solutions to two different, indeed contradictory, French energy problems: *more energy of any kind* versus *the right kind to do each task in the most inexpensive way*. It was only in 1979 that these conflicting perceptions collided. The supply-side planners suddenly realized that the only thing they would be able to *sell* all that nuclear electricity for would be electric heating, which they had just agreed not to do.

Every industrial country is in this embarrassing position (especially if we include as "heating" air conditioning, which just means heating the outdoors instead of the indoors). Which end of the spaghetti chart we start on, or *what we think the energy problem is*, is not an academic abstraction; it *determines what we buy*. It is the most fundamental source of disagreement about energy policy.

People starting on the left side of the spaghetti chart think the problem boils down to whether to build coal or nuclear power stations (or both). People starting on the right realize that *no* kind of new power station can be an economic way to meet the needs for using electricity to provide low- and high-temperature heat and for the vehicular liquid fuels that are 92% of our energy problem.

So if we want to provide our energy services at a price we can afford, let's get straight what question our technologies are supposed to answer. Before we argue about the meatballs, let's untangle the strands of spaghetti, see where they're supposed to lead, and find out what we really need the energy *for*!

Critical Thinking

1. The author argues that building more nuclear, coal, or other electrical power plants to supply electricity for the United States is unnecessary and wasteful. Summarize the reasons for this conclusion, and give your reasons for agreeing or disagreeing with this viewpoint.

2. Do you agree or disagree that increasing the supply of energy, instead of improving energy efficiency, is the wrong answer to U.S. energy problems? Explain.

19 NONRENEWABLE ENERGY RESOURCES

Bitter Lessons from Chernobyl

Chernobyl is a chilling word recognized around the globe as the site of a horrendous nuclear disaster (Figure 19-1). On April 26, 1986, a series of explosions in a nuclear power plant in Ukraine (then in the Soviet Union) blew the massive roof off the reactor building and flung radioactive debris and dust high into the atmosphere to encircle the planet. Here are some consequences of this disaster caused by poor reactor design and human error:

- Although the official death toll from the accident is 45, Greenpeace Ukraine estimates that by 1995 the total death toll from the accident was about 32,000.

- According to the United Nations, by the end of 1995 about 375,000 people were forced to leave their homes, probably never to return. Most were not evacuated until 10 days or more after the accident.

- According to a recent UN report, some 160,000 square kilometers (62,000 square miles) of the former Soviet Union—almost the size of the state of Florida or the combined areas of Denmark and Greece—remain contaminated with radioactivity.

- The 9 million people (about one-third of them children) still living in this area (and especially the 375,000 evacuated environmental refugees) live under constant stress.

- In 1994 analysis of samples taken from the reactor core revealed that the accident released the highly radioactive contents of 80% of the core, not the 3% announced to the world.

- Over a half-million people were exposed to dangerous radioactivity, and some may suffer from cancers, thyroid tumors, and eye cataracts. Between 1986 and 1995 the rate of thyroid cancer (mostly among children) in Ukraine and nearby Belarus was more than 10 times the normal rate and 30 times higher among those evacuated from the Chernobyl area.

- Government officials say that the total cost of the accident will reach at least $358 billion—many times greater than the value of all the nuclear electricity that has ever been generated in the former Soviet Union.

The environmental refugees evacuated from the Chernobyl region had to leave their possessions behind and say good-bye to lush, green wheat fields and blossoming apple trees, to land their families had farmed for generations, to cows and goats that would be shot because the grass they ate was radioactive, and to their radioactivity-poisoned cats and dogs. They will not be able to return.

Chernobyl taught us that *a major nuclear accident anywhere is a nuclear accident everywhere.*

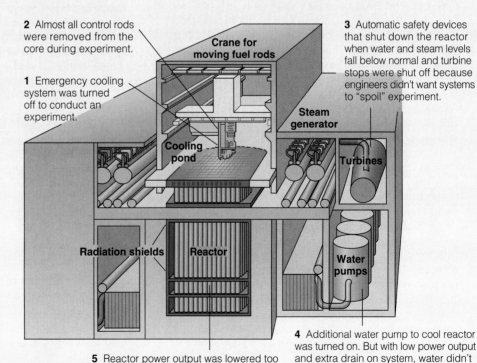

Figure 19-1 Major events leading to the Chernobyl nuclear power-plant accident on April 26, 1986, in the former Soviet Union. The accident happened because engineers turned off most of the reactor's automatic safety and warning systems (to keep them from interfering with an unauthorized safety experiment) and because of inadequate safety design of the reactor (no secondary containment shell as in Western-style reactors, no emergency core cooling system to prevent reactor core meltdown, and a design flaw that leads to unstable operation at low power). After the reactor exploded, crews exposed themselves to lethal levels of radiation to put out fires and encase the shattered reactor in a hastily constructed concrete tomb. This tomb is now sagging and full of holes that allow water to seep in and radioactive dust to drift out. Building a new tomb for the reactor will cost at least $1.5 billion—money the Ukrainian government doesn't have.

2 Almost all control rods were removed from the core during experiment.

1 Emergency cooling system was turned off to conduct an experiment.

Crane for moving fuel rods

3 Automatic safety devices that shut down the reactor when water and steam levels fall below normal and turbine stops were shut off because engineers didn't want systems to "spoil" experiment.

Steam generator

Cooling pond

Turbines

Radiation shields

Reactor

Water pumps

4 Additional water pump to cool reactor was turned on. But with low power output and extra drain on system, water didn't actually reach reactor.

5 Reactor power output was lowered too much, making it too difficult to control.

We are an interdependent world and if we ever needed a lesson in that, we got it in the oil crisis of the 1970s.

ROBERT S. MCNAMARA

In this chapter we seek answers to the following questions:

- What are the benefits and drawbacks of oil, natural gas, coal, conventional nuclear fission, breeder nuclear fission, and nuclear fusion?
- What are the best energy options?

19-1 OIL

What Is Crude Oil, and How Is It Extracted and Processed? Petroleum, or **crude oil** (oil as it comes out of the ground) is a fossil fuel produced by the decomposition of deeply buried dead organic matter from plants and animals under high temperatures and pressures over millions of years. Typically this gooey, smelly liquid consists mostly of hydrocarbons, with small amounts of sulfur, oxygen, and nitrogen impurities.

Crude oil and natural gas are often trapped together under a dome deep within the earth's crust (Figure 18-2). The crude oil is dispersed in pores and cracks in underground rock formations, like water in a sponge.

Primary oil recovery involves drilling a well and pumping out the oil that flows by gravity into the bottom of the well. After the flowing oil has been removed, water can be injected into nearby wells to force some of the remaining heavy oil to the surface, a process known as **secondary oil recovery**.

For each barrel of crude oil removed by primary and secondary recovery, two barrels of *heavy oil* are usually left. As oil prices rise, it may become economical to remove about 10% of this heavy oil by **enhanced**, or **tertiary**, **oil recovery**. Steam or CO_2 gas can be used to force some of the heavy oil into the well cavity for pumping to the surface. However, enhanced oil recovery is expensive: It takes the energy in one-third of a barrel of oil to retrieve each barrel of heavy oil. Researchers hope to use bacteria to increase tertiary recovery by 10–25% at a lower cost.

Most crude oil travels by pipeline to a *refinery* (Figure 19-2), where it is heated and distilled in gigantic columns to separate it into liquid components with different boiling points, such as naphtha (Individuals Matter, p. 533), diesel oil, heating oil, aviation fuel, and gasoline and solids such as grease, wax, and asphalt.

Some of the products of oil distillation, called **petrochemicals**, are used as raw materials in industrial organic chemicals, pesticides, plastics, synthetic fibers, paints, medicines, and many other products.

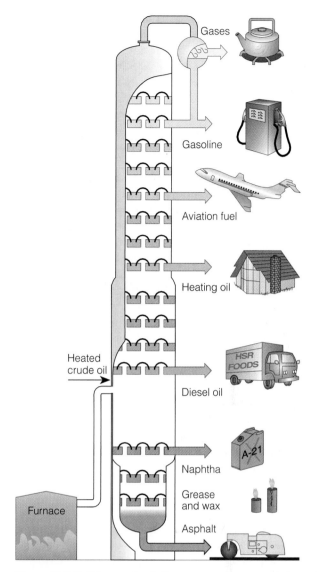

Figure 19-2 Refining crude oil. Components are removed at various levels, depending on their boiling points, in a giant distillation column. The most volatile components with the lowest boiling points are removed at the top of the column.

Who Has the World's Oil Supplies? Oil *reserves* are identified deposits from which oil can be extracted profitably at current prices with current technology. The 13 countries that make up the Organization of Petroleum Exporting Countries (OPEC)* have 67% of the world's reserves, which explains why OPEC is expected to have long-term control over world oil supplies and prices. Saudi Arabia, with 26%, has the world's largest known crude oil reserves, followed by Iraq with 10%.

---

* OPEC was formed in 1960 so that developing countries with much of the world's known and projected oil supplies could get a higher price for this resource. Today its members are Algeria, Ecuador, Gabon, Indonesia, Iran, Iraq, Kuwait, Libya, Nigeria, Qatar, Saudi Arabia, United Arab Emirates, and Venezuela.

Currently, the United States has only 2.3% of the world's oil reserves but uses nearly 30% of the oil extracted worldwide each year, 65% of it for transportation, mostly because oil is cheap (Figure 19-3). Figure 19-4 shows the locations of the largest crude oil fields in the United States and Canada. Despite an upsurge in exploration and test drilling, U.S. oil extraction has declined since 1985 and the net energy yield for small new domestic oil supplies is low and falling.

Mostly because of declining oil reserves and increased oil use, the United States imported 52% of the oil it used in 1996 (up from 36% in 1973); by 2010 it could be importing 70% or more of the oil it uses. This dependence on imported oil (about half of it from OPEC countries) and the likelihood of much higher oil prices within 10–20 years could drain the United States and other major oil-importing nations of vast amounts of money. This could lead to severe inflation and widespread economic recession, perhaps even a major depression. Moreover, sabotage of the vulnerable Trans-Alaska oil pipeline, which the Department of Defense says is impossible to protect, could disrupt the entire U.S. economy.

When Iraq invaded Kuwait in the summer of 1990, the United States and other developed countries went to war, mostly to protect their access to oil in the Middle East, especially from Saudi Arabia. As Worldwatch Institute researchers Christopher Flavin and Nicholas Lenssen put it, "Not only is the world addicted to cheap oil, but the largest liquor store is in a very dangerous neighborhood."

How Long Will Oil Supplies Last? The reserves—known and affordable supplies of a nonrenewable resource (Figure 12-8) such as oil—are considered economically depleted when 80% of the supply has been used; the remaining 20% is considered too expensive to extract.

Oil's fatal flaw is that its reserves may be 80% depleted within 35–84 years, depending on how rapidly it is used. At the current rate of consumption, global oil reserves will last at least 44 years. Undiscovered oil that is thought to exist might last another 20–40 years. Instead of remaining at the current level, however, global oil consumption is projected to increase by about 25% by 2010. This will hasten depletion of global oil reserves. At today's consumption rate, U.S. oil reserves will be depleted in about 24 years (less if consumption increases as projected); potential reserves might yield an additional 24 years of production.

Some analysts argue that rising oil prices (when oil consumption exceeds oil production) will stimulate exploration and lead to enough new reserves to meet future demand through the next century. Other analysts argue that such optimistic projections about future oil supplies ignore the consequences of exponentially increasing consumption of oil.

Even assuming that we continue to use crude oil at the current rate, Saudi Arabia, with the largest known crude oil reserves, could supply all the world's oil needs for only 10 years; the estimated reserves under Alaska's North Slope (the largest ever found in North America) would meet world demand for only 6 months or U.S. demand for 3 years. In short, just to keep on using oil at the *current* rate and not run out, we must discover and add to global oil reserves the equivalent of a new Saudi Arabian supply *every 10 years*. According to the U.S. Geological Survey, global

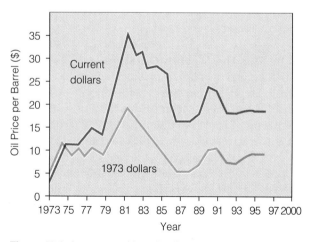

Figure 19-3 Average world crude oil prices, 1973–96. When price is adjusted for inflation, oil has remained cheap since 1975. (Data from Department of Energy and Department of Commerce)

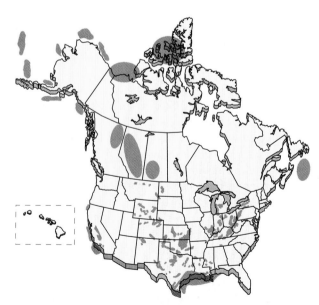

Figure 19-4 Locations of the largest crude oil and natural gas fields in the United States and Canada. Little *new* oil and natural gas are expected to be found in the United States. (Data from Council on Environmental Quality)

Q: How many people are added to the urban population of developing countries every day?

discovery of large oil fields peaked in 1962 and has been declining since.

What Are the Pros and Cons of Conventional Oil? Oil is still cheap; when adjusted for inflation it costs about the same as it did in 1975 (Figure 19-3). It is easily transported within and between countries, and when extracted from easily accessible deposits it has a high net energy yield (Figure 18-5). Oil's low price has encouraged developed countries and developing countries alike to become heavily dependent on—indeed, addicted to—this important resource. Low prices have also encouraged waste of oil and discouraged improvements in energy efficiency and the switch to other sources of energy.

Oil also has some drawbacks. As mentioned, estimated reserves of oil may be 80% depleted within 44–84 years, depending on how rapidly it is used. Oil, like any other fossil fuel, can cause pollution and environmental degradation throughout its life cycle of extraction, processing, transportation, and use (Figure 12-13).

Suppose that all of the harmful environmental effects of using oil were included in its market price (full-cost pricing) and that current government subsidies were phased out. Then analysts project that oil would become so expensive that it would probably be replaced by improved energy efficiency and by a variety of less harmful and cheaper renewable energy resources (also evaluated using full-cost pricing).

What Are the Pros and Cons of Using Heavy Oil Produced from Oil Shale? **Oil shale** is a fine-grained rock that contains a solid, waxy mixture of hydrocarbon compounds called **kerogen**. After being removed by surface or subsurface mining, the shale is crushed and heated above ground in a retort to vaporize the kerogen (Figures 19-5 and 19-6). The kerogen vapor is condensed, forming heavy, slow-flowing, dark brown **shale oil**. Before shale oil can be sent by pipeline to a refinery, it must be heated to increase its flow rate and processed to remove sulfur, nitrogen, and other impurities.

The shale oil potentially recoverable from U.S. deposits, mostly on federal lands in Colorado, Utah, and Wyoming, could probably meet the country's crude oil demand for 41 years at current use levels. Canada, China, and several republics in the former Soviet Union also have large oil-shale deposits. Indeed, according to energy expert John Harte, estimated potential global supplies of shale oil are 200 times larger than estimated global supplies of conventional oil.

However, there are problems with shale oil. It has a lower net energy yield than does conventional oil because it takes the energy from almost a half barrel of conventional oil to extract, process, and upgrade one barrel of shale oil. Processing the oil requires large amounts of water, which is scarce in the semiarid locales where the richest deposits are located. Surface mining of shale oil tears up the land, leaving mountains of shale rock (which expands somewhat like popcorn when heated). In addition, salts, cancer-causing substances, and toxic metal compounds can be leached from the processed shale rock into nearby water supplies. Some of these problems can be reduced by extracting shale oil underground (Figure 19-6), but this method is too expensive and produces more sulfur dioxide air pollution than does surface processing.

What Are the Pros and Cons of Using Heavy Oil Produced from Tar Sand? Tar sand (or oil sand) is a mixture of clay, sand, water, and **bitumen** (a gooey, black, high-sulfur heavy oil). It is usually removed by surface mining. It is then heated with pressurized steam until the bitumen fluid softens and floats to the top. The bitumen is then purified and chemically upgraded into a synthetic crude oil suitable for refining (Figure 19-7).

The world's largest known deposits of tar sands, the Athabasca Tar Sands, lie in northern Alberta, Canada. Currently, these deposits supply about 21% of Canada's oil needs. These deposits could supply all of Canada's projected oil needs for about 33 years at its current consumption rate, but they would last the world only about 2 years. Other large deposits of tar sands are in Venezuela, Colombia, and parts of the former Soviet Union.

Producing synthetic crude oil from tar sands has several disadvantages. The net energy yield is low because it takes the energy in almost one-half barrel of conventional oil to extract and process one barrel of bitumen and upgrade it to synthetic crude oil. Processing requires large quantities of water, and upgrading bitumen to synthetic crude oil releases large quantities of air pollutants. The plants also create huge waste disposal ponds.

Figure 19-5 Oil shale and the shale oil extracted from it. Big U.S. oil shale projects have been canceled because of excessive cost. (U.S. Department of Energy)

A: About 150,000

CHAPTER 19   **543**

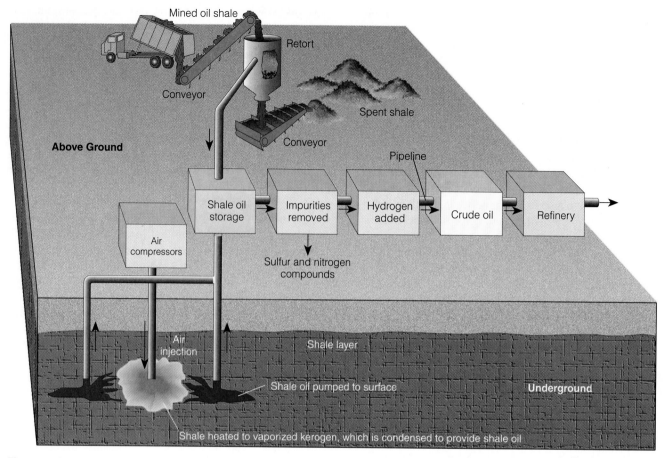

Figure 19-6 Aboveground and underground (*in situ*) methods for producing synthetic crude oil from oil shale.

19-2 NATURAL GAS

What Is Natural Gas? In its underground gaseous state, **natural gas** is a mixture of 50–90% by volume of methane (CH_4), the simplest hydrocarbon; smaller amounts of heavier gaseous hydrocarbons such as ethane (C_2H_6), propane (C_3H_8), and butane (C_4H_{10}); and highly toxic hydrogen sulfide (H_2S), a by-product of naturally occurring sulfur in the earth. Most of the methane in natural gas is produced by the decomposition of ancient organic matter in deeply buried mud and sediments under high temperatures and pressures.

Conventional natural gas lies above most reservoirs of crude oil (Figure 18-2). Natural gas is also found in *unconventional* deposits of gas hydrates, an icelike material that occurs in underground deposits all over the world and is composed largely of water and methane. Global deposits of gas hydrates are estimated to contain twice as much carbon as all other fossil fuels on earth. It is not yet economically feasible to get natural gas from such unconventional sources, but the extraction technology is being developed rapidly.

When a natural gas field is tapped, propane and butane gases are liquefied and removed as **liquefied petroleum gas (LPG)**. LPG is stored in pressurized tanks for use mostly in rural areas not served by natural gas pipelines. The rest of the gas (mostly methane) is dried to remove water vapor, cleansed of poisonous hydrogen sulfide and other impurities, and pumped into pressurized pipelines for distribution. Because natural gas and LPG are highly flammable and odorless, a compound is added to give these fuels a smell that warns of leakage.

At a very low temperature of –184°C (–300°F), natural gas can be converted to **liquefied natural gas (LNG)**. This highly flammable liquid can then be shipped to other countries in refrigerated tanker ships.

Who Has the World's Natural Gas Supplies? Russia and Kazakhstan have almost 40% of the world's natural gas reserves. Other countries with large known natural gas reserves are Iran (15%), Qatar (5%), Saudi Arabia (4%), Algeria (4%), the United States (3%), Nigeria (3%), and Venezuela (3%).

Q: By what percentage must global CO_2 emissions be cut by 2030 to stabilize CO_2 at current levels?

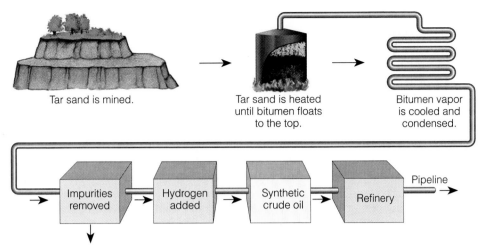

Figure 19-7 Generalized summary of how synthetic crude oil is produced from tar sand.

Tar sand is mined.

Tar sand is heated until bitumen floats to the top.

Bitumen vapor is cooled and condensed.

Impurities removed

Hydrogen added

Synthetic crude oil

Refinery

Pipeline

Geologists expect to find more natural gas, especially in unexplored developing countries.

Most U.S. natural gas reserves are located in the same places as crude oil (Figure 19-4). About 90–95% of the natural gas used in the United States is domestic and is distributed by about 411,000 kilometers (255,000 miles) of pipeline; the other 5–10% is imported by pipeline from Canada. Because of the current low price of natural gas, lack of potential drilling sites for large deposits, and stricter environmental regulations, there is little incentive for significant increased exploration and development of new natural gas supplies in the United States.

Algeria and some of the countries in the former Soviet Union use pipelines to supply Europe with natural gas. In 1994 Japan proposed that a 41,000-kilometer (25,500-mile) international pipeline network be built to link natural gas fields in Central and Southeast Asia, Siberia, and Alaska with the main markets of Asia, primarily Japan and China.

What Are the Pros and Cons of Natural Gas?
Natural gas has a number of advantages over other nonrenewable energy sources. To date, it is cheaper than oil. At the current consumption rate, known reserves and undiscovered, potential reserves of conventional natural gas in the United States are projected to last 65–80 years; world reserves are expected to last at least 125 years at the current consumption rate. It is estimated that conventional supplies of natural gas, plus unconventional supplies available at higher prices, will last at least 200 years at the current consumption rate and 80 years if usage rates rise 2% per year.

Natural gas can be transported easily over land by pipeline; it has a high net energy yield (Figure 18-5), burns hotter, and produces less air pollution than any other fossil fuel. Burning natural gas produces 43% less heat-trapping carbon dioxide per unit of energy

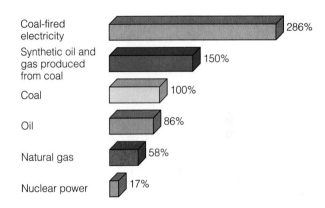

Figure 19-8 Carbon dioxide emissions per unit of energy produced by various fuels, expressed as percentages of emissions produced by coal.

Coal-fired electricity	286%
Synthetic oil and gas produced from coal	150%
Coal	100%
Oil	86%
Natural gas	58%
Nuclear power	17%

than coal and 30% less than oil (Figure 19-8). Extracting natural gas also damages the environment much less than extracting either coal or uranium ore.

Natural gas is easier to process than oil and, because it can be transported by pipeline, it is less expensive to transport by pipeline than coal (which is usually moved by rail). It can also be used to power vehicles, with over 100,000 vehicles now running on natural gas in the United States.

In new *combined-cycle natural gas systems*, natural gas burned in gas turbines (that work like jet engines) can produce electricity much more efficiently and cheaply than burning coal or oil or using nuclear power. Because of the low cost and low emissions of pollutants, it is expected that many utilities will convert hundreds of their aging coal plants into combined-cycle natural gas plants.

Smaller, more versatile combined-cycle gas units are on the way. They could be used as cogenerators to supply all of the heat and electricity needs of an apartment building or office building. Natural gas can also be used in highly efficient fuel cells.

A: 66-83%

Natural gas has a few drawbacks, but they are minor compared to those of other fossil fuels and nuclear power. Natural gas must be converted to liquid form (LNG) before it can be shipped by tanker from one country to another overseas. This is expensive and dangerous; huge explosions in urban areas near LNG loading and unloading facilities could kill many people and cause much damage. Conversion of natural gas into LNG also reduces the net useful energy yield by one-fourth. However, the proposed Asian pipeline, if built, could reduce the need for much shipping of natural gas as LNG. Recently, chemists have developed a new process for converting natural gas into methanol, an easily transported liquid fuel.

Another problem is leaks of natural gas into the atmosphere from natural gas pipelines, storage tanks, and distribution facilities. Methane, the major component of natural gas, is a greenhouse gas that is much more potent than CO_2 in causing global warming.

Because of its advantages over oil, coal, and nuclear energy, some analysts see natural gas as the best fuel to help us make the transition to improved energy efficiency and more renewable energy over the next 50 years. Additionally, hydrogen gas produced from water by solar-generated electricity could be mixed with natural gas to help smooth the shift to a solar–hydrogen economy (Section 18-8).

19-3 COAL

What Is Coal and How Is It Extracted and Processed? Coal is a solid, rocklike fossil fuel; it is formed in several stages, as the buried remains of ancient swamp plants that died during the Carboni-ferous period (a geologic era that ended 286 million years ago) were subjected to intense pressure and heat over many millions of years. Coal is mostly carbon (40–98%, depending on the type), with much smaller amounts of water (0.2–1.2%) and sulfur (0.2–2.5%) and trace amounts of radioactive materials found in the earth.

Three types of increasingly harder coal formed over the eons: lignite (brown coal), bituminous coal (soft coal), and anthracite (hard coal) (Figure 19-9). As coal ages its carbon content increases and its water content decreases. Peat (formed in the first stage), although not a coal, is burned in some places but has a low heat content. Lignite has a low heat value and a high moisture content. Low-sulfur coal (lignite and anthracite) produces less sulfur dioxide when burned than does high-sulfur bituminous coal. Anthracite, which is about 98% carbon, is the most desirable type of coal because of its high heat content and low sulfur content. However, because it takes much longer to form it is less common and therefore more expensive.

Some coal is extracted underground by miners working in tunnels and shafts. Such *subsurface mining* is very labor intensive, requiring about five times as many workers per metric ton as surface mining. It is also one of the world's most dangerous occupations because of accidents and black lung disease (caused by prolonged inhalation of coal dust particles).

When coal lies close to the earth's surface it is extracted by *surface mining*. Bulldozers and huge earth-moving machines remove soil and rock, known as *overburden*, to recover underlying coal deposits. *Area strip mining* is used on fairly flat terrain (Figure 12-10) and *contour strip mining* on hilly or mountain-

Figure 19-9 Stages in the formation of coal over millions of years. Peat is a soil material made of moist, partially decomposed organic matter. Lignite and bituminous coal are sedimentary rocks, whereas anthracite is a metamorphic rock.

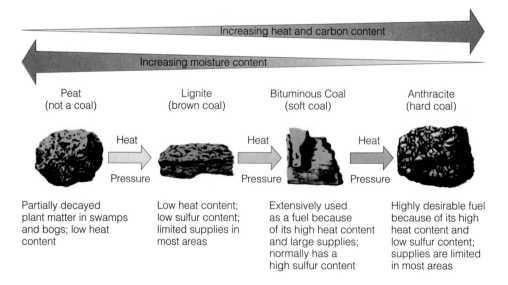

Increasing heat and carbon content

Increasing moisture content

Peat (not a coal)	Lignite (brown coal)	Bituminous Coal (soft coal)	Anthracite (hard coal)
Partially decayed plant matter in swamps and bogs; low heat content	Low heat content; low sulfur content; limited supplies in most areas	Extensively used as a fuel because of its high heat content and large supplies; normally has a high sulfur content	Highly desirable fuel because of its high heat content and low sulfur content; supplies are limited in most areas

Q: What is the average amount of CO_2 each person in the United States adds to the atmosphere each year?

ous terrain (Figure 12-11). Thick beds of coal fairly near the surface are removed by digging a deep pit to remove the coal—a form of surface mining called *open pit mining*.

After the coal is removed it is transported (usually by train) to a processing plant, where it is broken up, crushed, and then washed to remove impurities. The coal is then dried and shipped (again usually by train) to users, mostly power plants and industrial plants.

How Is Coal Used? Currently coal provides about 25% of the world's commercial energy (22% in the United States). It is used to generate 39% of the world's electricity and to make 75% of its steel.

China, which gets 76% of its energy from coal, is the world's largest user of coal, followed closely by the United States. Other large coal users are the countries of the former Soviet Union, Germany, India, Australia, Poland, and Great Britain.

Coal supplies 57% of the electricity generated in the United States; the rest is produced by nuclear energy (19%), natural gas (11%), hydropower (9%), oil (3%), and renewable energy resources such as hydropower, geothermal, and wind energy (1%). Utilities consume 87% of U.S. coal production and industry 13%.

Who Has the World's Coal Supplies? About 66% of the world's proven coal reserves and 85% of the estimated undiscovered coal deposits are located in the United States, the former Soviet Union, and China (Connections, below). Asia (excluding China), western Europe, and Australia each have about 9% of the world's coal reserves, Africa 6%, and Latin America 1%.

Figure 19-10 shows the locations of major coal fields in the United States and Canada. Anthracite makes up only 2% of U.S. coal reserves (and accounts for only about 0.5% of the country's coal production). About 45% of U.S. reserves is high-sulfur bituminous coal with a high fuel value. More than half of these reserves are found west of the Mississippi River—far from the heavily industrialized and populated East, where most coal is consumed.

What Are the Pros and Cons of Solid Coal? Coal is both the world's most abundant and its dirtiest fossil fuel. *Identified* world reserves of coal should last at least 220 years at current usage rates, but only 65 years if usage rises 2% per year. The world's *unidentified* coal reserves are projected to last about 900 years at the current consumption rate, but only 149 years if the usage rate increases 2% per year. Identified U.S.

Coal Burning and Destructive Synergies in China

CONNECTIONS

China has huge quantities of coal—11% of the world's reserves—and is the world's largest producer of coal, accounting for 25% of global output. Currently, the country burns coal to provide 76% of its commercial energy.

Mostly because of coal burning, China now produces 10% of the world's emissions of CO_2 each year. By 2025, China is projected to emit more CO_2 than the current combined total of the United States, Canada, and Japan.

Several factors interact synergistically to cause China to burn more coal than it needs to. One is energy inefficiency. About 66% of the energy used in China is unnecessarily wasted—compared to about 43% in the United States (which is in turn

only half as energy-efficient as Japan and most western European countries).

Another factor is that drinking water must usually be boiled because it is often too contaminated to drink. This uses more energy, much of it produced by the inefficient burning of coal in homes. Destruction of forests for more cropland, fuelwood, and construction material for the country's growing population leads to more soil erosion and increased use of fertilizers, further polluting the water and requiring more coal burning to sterilize water.

For China's economic future, and perhaps for the world's climate, many analysts believe that it is crucial to defuse these destructive synergistic interactions, which lead to a cycle of harmful positive

feedback. Part of the solution is for China to undertake an energy efficiency revolution and to shift from coal to natural gas and renewable energy such as wind, biomass, solar, and small-scale hydroelectric power.

At the same time, China could reduce coal use, improve public health, and increase economic productivity by instituting a nationwide program to provide safe drinking water.

Critical Thinking

1. How might your life and lifestyle be affected if China continues to burn coal as its primary fuel?

2. What role, if any, should the United States and other developed countries play in helping China reduce its dependence on coal?

A: 16.7 metric tons (18.4 tons)—six times more than the average citizen of a developing country

Figure 19-10 Locations of major coal fields in the United States and Canada. (Data from U.S. Council on Environmental Quality)

Type of Coal
■ Anthracite
■ Bituminous
□ Subbituminous
■ Lignite

coal reserves should last about 300 years at the current consumption rate; unidentified U.S. coal resources could extend those supplies for perhaps 100 years, at a much higher average cost. Coal also has a high net energy yield (Figure 18-5).

However, coal has a number of drawbacks, especially the harmful environmental effects associated with its extraction, processing, and use (Figure 12-13). Coal mining is dangerous because of accidents and black lung disease, a form of emphysema caused by prolonged breathing of coal dust and other particulate matter. Coal mining also harms land and causes water pollution. Underground mining causes land to sink when a mine shaft collapses during or after mining. Surface mining of coal causes severe land disturbance (Figures 12-10, 12-11, and 18-2) and soil erosion, which can pollute nearby streams.

Surface-mined land can be restored. However, this is expensive and is not done in many countries; in arid and semiarid areas the land cannot be fully restored. Surface and subsurface coal mining can severely pollute nearby streams and groundwater from acids and

toxic metal compounds (Figure 12-14). Once coal is mined, it is expensive to move from one place to another. It cannot be used in solid form as a fuel for cars and trucks; it must be converted to liquid or gaseous fuels, with a significant drop in its net energy yield.

Coal is by far the dirtiest fossil fuel to burn. Without expensive air-pollution control devices, burning coal produces more air pollution per unit of energy than any other fossil fuel. In addition to conventional air pollutants, burning coal releases thousands of times more radioactive particles into the atmosphere per unit of energy produced than does a normally operating nuclear power plant. Because coal produces more carbon dioxide per unit of energy than do other fossil fuels (Figure 19-8), burning more coal accelerates projected global warming (Section 10-3).

Burning coal is also one of the greatest threats to human health. Each year in the United States alone, air pollutants from coal burning kill thousands of people (with estimates ranging from 65,000 to 200,000), cause at least 50,000 cases of respiratory disease, and result in several billion dollars of property damage.

Q: By what percentage is U.S. food production reduced by air pollution?

However, new ways, such as *fluidized-bed combustion* (Figure 19-11), have been developed to burn coal more cleanly and efficiently. In the United States commercial fluidized-bed combustion boilers are expected to slowly replace most conventional coal boilers.

Recent research done for the U.S. Department of Energy concluded that the *environmental costs* of producing electricity were 5.7¢ per kilowatt-hour for coal, 5.0¢ for nuclear, 2.7¢ for oil, 1¢ for natural gas, under 0.7¢ for biomass, less than 0.4¢ for solar cells, and under 0.1¢ for wind and geothermal. Suppose that taxes (per unit of energy produced) added these costs to the market price of coal and that government subsidies that make coal artificially cheap were phased out. According to some analysts, coal would become so expensive that it would probably be replaced for most uses by a combination of improved energy efficiency and cheaper and less environmentally harmful renewable energy resources.

What Are the Pros and Cons of Converting Solid Coal into Gaseous and Liquid Fuels? Solid coal can be converted into **synthetic natural gas (SNG)** by *coal gasification* (Figure 19-12), into *hydrogen gas* (Section 18-8), or into a liquid fuel such as methanol or synthetic gasoline by *coal liquefaction*. These *synfuels* can be transported by pipeline, and they produce much less air pollution than solid coal. They can be burned to produce high-temperature heat and electricity, to heat houses and water, and to propel vehicles.

However, coal gasification has a low net energy yield (Figure 18-5), as does coal liquefaction. A synfuel plant costs much more to build and run than an equivalent coal-fired power plant fully equipped with air-pollution control devices. In addition, the widespread use of synfuels would accelerate the depletion of world coal supplies because 30–40% of the energy content of coal is lost in the conversion process. It would also lead to greater land disruption from surface mining because producing a unit of energy from synfuels uses more coal than burning solid coal does.

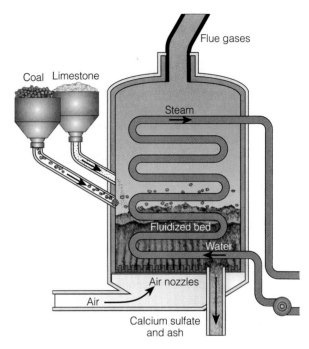

Figure 19-11 Fluidized-bed combustion of coal. A stream of hot air is blown into a boiler to suspend a mixture of powdered coal and crushed limestone. This method removes most of the sulfur dioxide, sharply reduces emissions of nitrogen oxides, and burns the coal more efficiently and cheaply than conventional combustion methods.

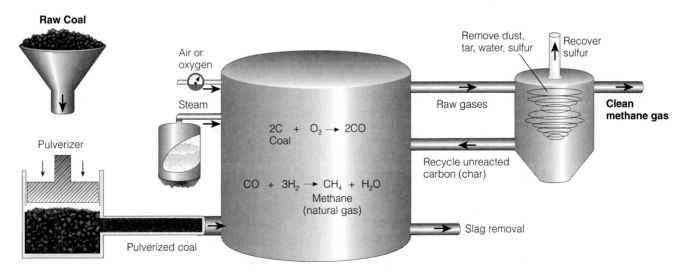

Figure 19-12 Coal gasification. Generalized view of one method for converting solid coal into synthetic natural gas (methane).

A: 5–10% (with economic losses of $1.5-5.4 billion per year)

Producing synfuels also requires huge amounts of water. Additionally, synfuels release more carbon dioxide per unit of energy than coal does (Figure 19-8). For these reasons, most analysts expect synfuels to play only a minor role as an energy resource in the next 30–50 years.

19-4 NUCLEAR ENERGY

✺ What Happened to Nuclear Power?

U.S. utility companies began developing nuclear power plants in the late 1950s for three reasons: **(1)** The Atomic Energy Commission (which had the conflicting roles of promoting and regulating nuclear power) promised them that nuclear power would produce electricity at a much lower cost than coal and other alternatives (with one early proponent touting nuclear power as "too cheap to meter"), **(2)** the government paid about a quarter of the cost of building the first group of commercial reactors, and **(3)** after U.S. insurance companies refused to cover more than a small part of the possible damages from a nuclear power-plant accident Congress passed the Price–Anderson Act, which protects the U.S. nuclear industry and utilities from significant liability to the general public in case of accidents.

In the 1950s researchers predicted that by the end of the century 1,800 nuclear power plants would supply 21% of the world's commercial energy (25% in the United States) and most of the world's electricity. By 1996, after over 40 years of development, enormous government subsidies, and an investment of $2 trillion, 430 commercial nuclear reactors in 32 countries were producing only 6% of the world's commercial energy and 17% of its electricity. Little or no further growth in nuclear power is projected, and its capacity is expected to decline between 2000 and 2020 as existing plants wear out and are retired (decommissioned).

In western Europe, plans to build more new nuclear power plants have come to a halt, except in France, which gets about 80% of its electricity from nuclear power. However, France has more electricity than it needs (Guest Essay, pp. 538–539) (and the heavily subsidized government agency that builds and operates France's nuclear plants has lost money for 22 years, accumulating a $30-billion debt—a serious burden to the French economy). In Japan, public opposition has been so intense that only two sites for new nuclear plants have been approved since 1979.

In the United States, no new nuclear power plants have been ordered since 1978, and all 120 plants ordered since 1973 have been canceled. In 1996, the 109 licensed commercial nuclear power plants in the United States generated about 22% of the country's electricity. This percentage is expected to decline over the next two decades as many of the current reactors reach the ends of their useful lives.

What happened to nuclear power? The answer is multibillion-dollar construction cost overruns, high operating costs, frequent malfunctions, false assurances and cover-ups by government and industry officials, inflated estimates of electricity use, poor management, the Chernobyl (p. 540) and Three Mile Island accidents, and public concerns about safety, costs, and radioactive waste disposal.

How Does a Nuclear Fission Reactor Work?

To evaluate the pros and cons of nuclear power, we first must know how a nuclear power plant and its accompanying nuclear fuel cycle work. When the nuclei of atoms such as uranium-235 and plutonium-239 are split by neutrons in a nuclear fission chain reaction, energy is released and converted mostly into high-temperature heat (Figure 3-12). The rate at which this happens can be controlled in the nuclear fission reactor of a nuclear power plant; the heat generated can produce high-pressure steam, which spins turbines that in turn generate electricity.

Light-water reactors (LWRs) like the one diagrammed in Figure 19-13 produce about 85% of the world's nuclear-generated electricity (100% in the United States). An LWR has several key parts. One is its *core*, which contains 35,000–40,000 long, thin fuel rods, each of which is packed with pellets of uranium oxide fuel. Each pellet is about one-third the size of a cigarette. About 97% of the uranium in each fuel pellet is uranium-238, a nonfissionable isotope; the other 3% (versus 0.7% in nature) is uranium-235, which is fissionable. The concentration of uranium-235 in the ore is increased (enriched) from 0.7% to 3% by removing some of the uranium-238 to create a suitable fuel. Another important component is *control rods*, which are moved in and out of the reactor core to absorb neutrons and thus regulate the rate of fission and amount of power the reactor produces.

The *moderator* slows down the neutrons emitted by the fission process so that the chain reaction can be kept going. This is a material such as liquid water (75% of the world's reactors, called *pressurized water reactors*), solid graphite (20% of reactors), or heavy water (D_2O, 5% of reactors). Graphite-moderated reactors, including the ill-fated one at Chernobyl (Figure 19-1), can also produce fissionable plutonium-239 for nuclear weapons. A *coolant*, usually water, circulates through the reactor's core to remove heat (to keep fuel rods and other materials from melting) and to produce steam for generating electricity.

Nuclear power plants, each with one or more reactors, are only one part of the nuclear fuel cycle

Q: What percentage of the world's population clash over rights to water?

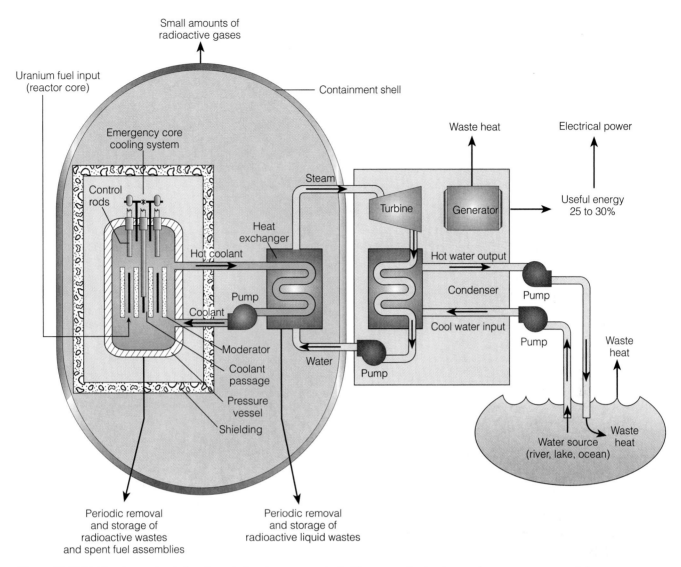

Figure 19-13 Light-water-moderated and -cooled nuclear power plant with a pressurized water reactor. The ill-fated Chernobyl nuclear reactor (Figure 19-1) in Ukraine was a graphite-moderated reactor with less extensive safety features than most reactors in the rest of the world.

(Figure 19-14). In evaluating the safety and economic feasibility of nuclear power, we need to look at this entire cycle, not just the nuclear plant itself.

After 3–4 years, the concentration of fissionable uranium-235 in a reactor's fuel rod becomes too low to keep the chain reaction going or the rod becomes damaged from exposure to ionizing radiation. Each year about one-third of the spent fuel assemblies in a reactor are removed and placed in large, concrete-lined pools of water at the plant site. The water serves as a radiation shield and coolant. After losing some of their radioactivity and cooling down, the spent fuel assemblies are supposed to be shipped to fuel-reprocessing plants or to permanent sites for the long-term storage of high-level, long-lived radioactive wastes (Figure 19-14).

In the United States all spent fuel rods currently are being stored in concrete-lined pools of water at each of the country's nuclear power plants until a permanent long-term underground storage facility is developed (probably not until 2010 or later). Many of these plants are reaching their capacity for storing spent fuel.

If the fuel is reprocessed to remove plutonium and other radioactive isotopes with long half-lives, the remaining radioactive waste must be safely stored for at least 10,000 years; otherwise, the rods must be stored safely for at least 240,000 years (10 times the 24,000-year half-life of plutonium-239)—about four times as long as the latest species of humans has been on earth.

Commercial spent fuel–reprocessing plants have been built in the United States, France, Great Britain,

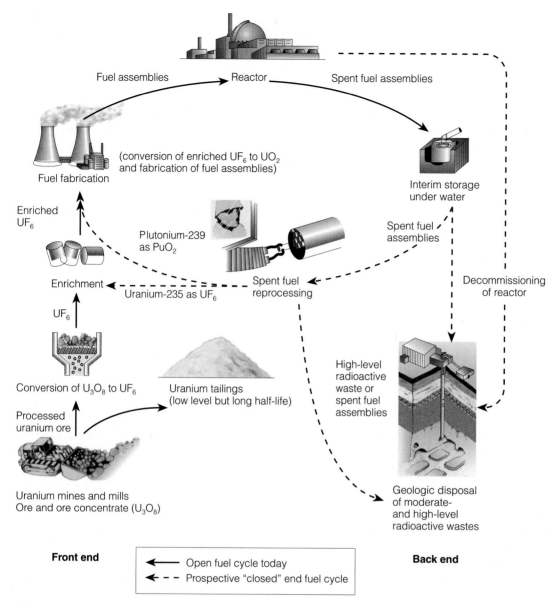

Fuel assemblies → Reactor → Spent fuel assemblies

(conversion of enriched UF_6 to UO_2 and fabrication of fuel assemblies)

Fuel fabrication

Interim storage under water

Enriched UF_6

Plutonium-239 as PuO_2

Spent fuel assemblies

Enrichment

Uranium-235 as UF_6

Spent fuel reprocessing

Decommissioning of reactor

UF_6

Conversion of U_3O_8 to UF_6

Uranium tailings (low level but long half-life)

High-level radioactive waste or spent fuel assemblies

Processed uranium ore

Uranium mines and mills
Ore and ore concentrate (U_3O_8)

Geologic disposal of moderate- and high-level radioactive wastes

Front end

→ Open fuel cycle today
- → - Prospective "closed" end fuel cycle

Back end

Figure 19-14 The nuclear fuel cycle. As long as it is operating properly, little pollution or radioactivity is released from a nuclear power plant (although the potential health, environmental, and economic consequences from an extremely unlikely major accident are significant). Each of the other steps in the nuclear fuel cycle can release significant amounts of toxic chemicals and radioactive materials into the environment.

Japan, and Germany, but all have had severe operating and economic problems. The United States has stopped development of commercial fuel-reprocessing plants because of policy concerns (including diversion of bomb-grade materials by employees or terrorists), technical difficulties, high construction and operating costs, and ample domestic or imported supplies of uranium. During fuel reprocessing some highly radioactive materials can be released into the air, water, and soil.

After approximately 15–40 years of operation, a nuclear reactor becomes dangerously contaminated with radioactive materials, and many of its parts are worn out. Then the plant must be *decommissioned* or retired by **(1)** dismantling it and storing its large volume of highly radioactive materials in high-level nuclear waste storage facilities (which still don't exist), **(2)** putting up a physical barrier and setting by full-time security for 30–100 years before the plant is then dismantled, or **(3)** enclosing the entire plant in a tomb that must last for several thousand years.

What Are the Advantages of Nuclear Power?
Using nuclear power to produce electricity has some important advantages, especially when compared to mining, processing, and burning coal.

Q: What do many analysts believe is the key to reducing water waste?

Nuclear plants don't emit air pollutants (as coal-fired plants do) as long as they are operating properly. The entire nuclear fuel cycle (Figure 19-14) adds about one-sixth as much heat-trapping carbon dioxide per unit of electricity as using coal does, thus making it more attractive than coal and other fossil fuels for reducing the threat of global warming.

Water pollution and disruption of land are low to moderate if the entire nuclear fuel cycle operates normally. Thick and strong reactor vessel walls, a steel-reinforced containment building, an emergency core cooling system, and multiple other safety systems with automatic backups greatly decrease the likelihood of a catastrophic accident releasing deadly radioactive material into the environment.

Chernobyl may have caused the premature deaths of 32,000 people between 1986 and 1995, but *each year* air pollution from coal burning causes the premature death of 65,000 to 200,000 people in the United States alone. Globally the *annual* death toll from coal burning is in the millions.

How Safe Are Nuclear Power Plants and Other Nuclear Facilities? Because of the built-in safety features, the risk of exposure to radioactivity from nuclear power plants in the United States (and presumably in most other developed countries) is extremely low compared to other risks (Figures 8-11 and 8-12). However, a partial or complete meltdown or explosion is possible, as the Chernobyl (Figure 19-1) and Three Mile Island nuclear power-plant accidents have taught us.

On March 29, 1979, the number 2 reactor at the Three Mile Island (TMI) nuclear plant near Harrisburg, Pennsylvania, lost its coolant water because of a series of mechanical failures and human operator errors not anticipated in safety studies. The reactor's core became partially uncovered and about 50% of it melted and fell to the bottom of the reactor. Unknown amounts of radioactive materials escaped into the atmosphere, 50,000 people were evacuated, and another 50,000 fled the area on their own. Partial cleanup of the damaged TMI reactor, lawsuits, and payment of damage claims has cost $1.2 billion so far, almost twice the reactor's $700-million construction cost.

Most analysts say that the radiation release was not enough to cause deaths or cancers, but some scientists disagree. There is controversy over how much radiation was released and there has been an increase in certain cancers among residents who live near the plant. A 1991 study concluded that the excess cancers may have been caused by stress related to the accident. Stress is known to damage the immune system so that it may fail to prevent cancers. A 1997 study concluded that the increased cancer rates were caused by radiation from the plant.

The Nuclear Regulatory Commission (NRC) estimates that there is a 15–45% chance of a complete core meltdown at a U.S. reactor during the next 20 years. The NRC also found that 39 U.S. reactors have an 80% chance of either containment failure from a meltdown or a tremendous explosion of gases inside the containment structures.

Nuclear scientists and government officials throughout the world are deeply concerned about 26 especially risky nuclear reactors in some republics of the former Soviet Union that have a Chernobyl-type design (15 reactors) or another flawed design (11 reactors). There is growing international consensus (including the Russian Academy of Sciences) that these 26 reactors, along with 10 other poorly designed and poorly operated nuclear plants in eastern Europe, should be shut down. However, without considerable government and private economic aid from the world's developed countries, it is extremely unlikely that these dangerous plants will be closed and replaced with safer nonnuclear alternatives.

A 1982 study by the Sandia National Laboratory estimated that a *worst-case accident* in a reactor near a large U.S. city might cause 50,000–100,000 immediate deaths, 10,000–40,000 subsequent deaths from cancer, and $100–150 billion in damages. Most citizens and businesses suffering injuries or property damage from a major nuclear accident would get little if any financial reimbursement because combined government and nuclear industry insurance covers only 7% of the estimated damage from such an accident. Critics of nuclear power contend that this 15-year-old study probably underestimates the deaths and damages. They also contend that plans for dealing with major nuclear accidents in the United States are inadequate.

In the United States, there is widespread lack of public confidence in the ability of both the NRC and the Department of Energy (DOE) to enforce nuclear safety in commercial (NRC) and military (DOE) nuclear facilities. Congressional hearings in 1987 uncovered evidence that high-level NRC staff members destroyed documents and obstructed investigations of criminal wrongdoing by utilities, suggested ways utilities could evade commission regulations, and provided utilities and their contractors with advance notice of surprise inspections. The nuclear power industry vetoes NRC nominees it deems too hostile and many top NRC officials enjoy a revolving door to good jobs at nuclear power-plant utility companies.

Since 1986 government studies and once-secret documents have revealed that for decades most of the nuclear weapon production facilities supervised by the DOE have been operated with gross disregard for the safety of their workers and the people nearby. Since 1957 these facilities have released huge quantities of radioactive particles into the air and dumped tons of radioactive waste and toxic substances into flowing creeks and leaking pits without informing the public. As Senator John Glenn of Ohio summed up the situation, "We are poisoning our own people in the name of national security."

According to the nuclear power industry, nuclear power plants in the United States have not killed anyone. However, according to U.S. National Academy of Sciences estimates, U.S. nuclear plants cause 6,000 premature deaths and 3,700 serious genetic defects each year. If correct, this annual death toll is much smaller than the 65,000 to 200,000 deaths per year caused by coal-burning plants in the United States. However, critics point out the estimated annual deaths from both of these types of plants are unacceptable, given the much less harmful alternatives.

What Do We Do with Low-Level Radioactive Waste? Each part of the nuclear fuel cycle (Figure 19-14) produces low-level and high-level solid, liquid, and gaseous radioactive wastes with various half-lives (Table 3-1). Wastes classified as *low-level radioactive wastes* give off small amounts of ionizing radiation and must be stored safely for 100–500 years before decaying to levels that, according to the NRC, don't pose an unacceptable risk to public health and safety.

From the 1940s to 1970, most low-level radioactive waste produced in the United States (and most other countries) was put into steel drums and dumped into the ocean; the United Kingdom and Pakistan still dispose of their low-level wastes in this way.

Today, low-level waste materials from commercial nuclear power plants, hospitals, universities, industries, and other producers are put in steel drums and shipped to the two remaining regional landfills run by federal and state governments (both scheduled for closure). Attempts to build new regional dumps for low-level radioactive waste using improved technology (Figure 19-15) have met with fierce public opposition. According to the EPA, all landfills eventually leak. The best-designed landfills may not leak for several decades, which approximates the time when the companies running them would no longer be liable for leaks and problems; then any problems would be passed on to taxpayers and future generations.

Some environmentalists believe that low-level radioactive waste should be stored in carefully designed aboveground buildings that could capture and contain any leaks. They urge that all such buildings (or low-level waste dumps) be located at nuclear power-plant sites, which produce more than half of these wastes (as much as 80% in some states) and 80% of the radioactivity.

They point out that nuclear power plant sites have already been carefully evaluated to meet strict geological standards. They have the personnel with the expertise and equipment to manage the low-level wastes and are strictly regulated by the federal government. Locating low-level waste at plant sites would also reduce the need to transport most such wastes long distances. Such proposals are opposed by utility companies and the nuclear power industry be-

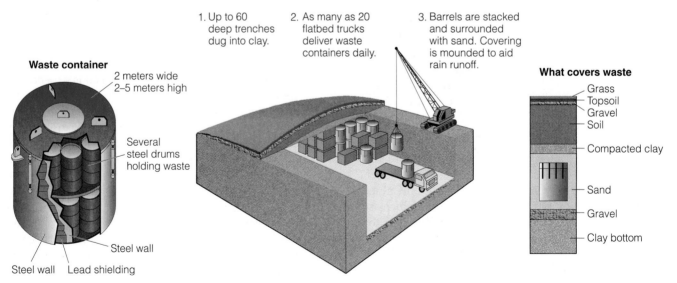

Figure 19-15 Proposed design of a state-of-the-art low-level radioactive waste landfill. (Data from U.S. Atomic Industrial Forum)

Q: What percentage of all municipal sewage in the United States ends up essentially untreated in coastal waters?

cause this would force the main producers of such wastes to be financially and legally responsible for the wastes they produce.

In 1995 Westinghouse Electric began discussions about shipping low-level radioactive wastes from the United States for permanent burial in southern Russia in old mine shafts. This idea is facing strong opposition from antinuclear activists and many members of the Russian parliament, who charge the West with trying to turn Russia into a radioactive dumping ground.

What Should We Do with High-Level Radioactive Waste? *High-level radioactive wastes* give off large amounts of ionizing radiation for a short time and small amounts for a long time. Such wastes must be stored safely for thousands of years—about 240,000 years if plutonium-239 is not removed by reprocessing. Most high-level radioactive wastes are spent fuel rods from commercial nuclear power plants and an assortment of wastes from plants that produce plutonium and tritium for nuclear weapons.

After 50 years of research, scientists still don't agree on whether there is any safe method of storing these wastes. Some scientists believe that the long-term safe storage or disposal of high-level radioactive wastes is technically possible. Others disagree, pointing out that it is impossible to demonstrate that any method will work for the 10,000–240,000 years of fail-safe storage needed for such wastes. Here are some of the proposed methods and their possible drawbacks:

■ *Bury it deep underground.* This favored strategy is under study by all countries producing nuclear waste. To reduce storage time from hundreds of thousands of years to about 10,000 years, very long-lived radioactive isotopes such as plutonium-239 must be extracted from spent fuel rods. The remainder would be fused with glass or a ceramic material and sealed in metal canisters for burial in a deep underground salt, granite, or other stable geological formation that is earthquake resistant and waterproof. However, according to a 1990 report by the U.S. National Academy of Sciences, "Use of geological information—to pretend to be able to make very accurate predictions of long-term site behavior—is scientifically unsound."

■ *Shoot it into space or into the sun.* Costs would be very high, and a launch accident, such as the explosion of the space shuttle *Challenger*, could disperse high-level radioactive wastes over large areas of the earth's surface. This strategy has been abandoned for now.

■ *Bury it under the antarctic ice sheet or the Greenland ice cap.* The long-term stability of the ice sheets is not known. They could be destabilized by heat from the

wastes, and retrieval of the wastes would be difficult or impossible if the method failed. This strategy is prohibited by international law and has been abandoned for now.

■ *Dump it into descending subduction zones in the deep ocean* (Figure 12-4, middle). Again, our geological knowledge is incomplete; wastes might eventually be spewed out somewhere else by volcanic activity. Waste containers might also leak and contaminate the ocean before being carried downward; retrieval would be impossible if the method did not work. This method is under active study by a consortium of 10 countries, but current international agreements and U.S. laws forbid dumping radioactive waste beneath the seas.

■ *Bury it in thick deposits of mud on the deep ocean floor in areas that tests show have been geologically stable for 65 million years.* The waste containers would eventually corrode and release their radioactive contents. Some geologists contend that gravity and the powerful adhesive qualities of the mud would prevent the wastes from migrating, but this hypothesis has not been verified. This approach is banned under current international agreements and U.S. law.

■ *Change it into harmless, or less harmful, isotopes.* Currently there is no way to do this. Even if a method were developed, costs would probably be extremely high and the resulting toxic materials and low-level (but very long-lived) radioactive wastes would still need to be disposed of safely.

So far no country has come up with a scientifically and politically acceptable solution for the long-term storage of nuclear wastes.

In 1985 the DOE announced plans to build the first repository for underground storage of high-level radioactive wastes from commercial nuclear reactors (Figure 19-16) on federal land in the Yucca Mountain desert region, 160 kilometers (100 miles) northwest of Las Vegas, Nevada. The facility, which is expected to cost at least $26 billion, was scheduled to open by 2003, but in 1990 its opening was postponed to at least 2010. The site may never open, mostly because of rock faults, a nearby active volcano, and 36 active earthquake faults on the site itself.

Fed up with the delay, some utilities are taking matters into their own hands and are trying to develop a temporary, aboveground, monitored retrievable storage (MRS) facility for commercial high-level radioactive waste on the Mescalero Apache reservation in New Mexico. Because of the sovereignty of Native American nations, it is easier to site a radioactive storage facility on tribal lands than anywhere else in the United States. New Mexico officials have threatened

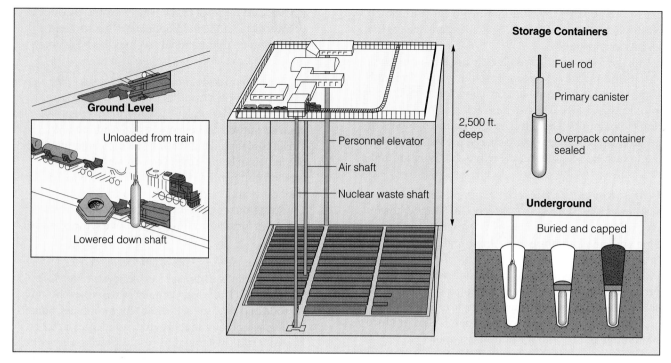

Figure 19-16 Proposed general design for deep-underground permanent storage of high-level radioactive wastes from commercial nuclear power plants in the United States. (Source: U.S. Department of Energy)

to block the MRS facility, claiming that it would be too risky to transport spent fuel across the country to the facility. Even if a permit is granted, the site would not be open until 2003 at the earliest.

Regardless of the storage method, most citizens strongly oppose the location of a low-level or high-level nuclear waste disposal facility anywhere near them. Even if the problem is technically solvable, it may be politically unacceptable.

How Widespread Are Contaminated Radioactive Sites?

In 1992 the EPA estimated that as many as 45,000 sites in the United States may be contaminated with radioactive materials, 20,000 of them belonging to the DOE and the Department of Defense. According to the DOE it will cost taxpayers at least $230 billion over the next 75 years to clean up these facilities (with some estimating that the price will range from $400 to $900 billion).

Some critics doubt that the necessary clean-up funds will be provided by Congress as the cleanup process drags on and fades from serious public scrutiny. Others question spending so much money on a problem that is ranked by scientific advisers to the EPA as a low-risk ecological problem and not among the top high-risk health problems (Figure 8-11). However, the radioactive contamination situation in the United States pales in comparison to the post–Cold War legacy of nuclear waste and contamination in the republics of the former Soviet Union (Spotlight, right).

What Can We Do with Worn-Out Nuclear Plants?

The useful operating life of today's nuclear power plants is supposed to be 40 years, but many plants are wearing out and becoming dangerous faster than anticipated. Worldwide, 81 reactors have been shut down after being in operation for an average of only 17 years. Because so many of its parts become radioactive, a nuclear plant cannot be abandoned or demolished by a wrecking ball the way a worn-out coal-fired power plant can.

Decommissioning nuclear power plants and nuclear weapons plants is the last step in the nuclear fuel cycle (Figure 19-14). Three methods have been proposed: **(1)** *immediate dismantling,* **(2)** *mothballing* for 30–100 years by putting up a barrier and setting up a 24-hour security system (costing about $15 million a year) and then dismantling it, and **(3)** *entombment* by covering the reactor with reinforced concrete and putting up a barrier to keep out intruders for several thousand years. Each method involves shutting down the plant, removing the spent fuel from the reactor core, draining all liquids, flushing all pipes, and sending all radioactive materials to an approved waste storage site yet to be built.

Q: What percentage of the sewage in Latin America is treated?

The Soviet Nuclear-Waste Legacy

Much of the land in various parts of the former Soviet Union is dotted with areas severely contaminated by nuclear accidents, 26 operating nuclear power plants with flawed and unsafe designs, nuclear-waste dump sites, radioactive waste-processing plants, and contaminated nuclear test sites.

Mayak, a plutonium production facility in southern Russia, has spewed into the atmosphere 2.5 times as much radiation as Chernobyl, mostly from the explosion of a nuclear waste storage tank in 1957. About 2,600 square kilometers (1,000 square miles) of the surrounding area is still contaminated with radioactive wastes so dangerous that no one can live there.

Between 1949 and 1967, huge quantities of radioactive waste spilled into the nearby Techa River. For decades the 64,000 people living along this regional waterway

shared by 30 villages drank the river's water, bathed in it, and washed their clothes in it. Release of once-secret studies of 28,000 Techa River villagers revealed that they had a statistically significant increase in the incidence of leukemia, an overall increase in cancer mortality, and more lymphatic genetic mutations than a control group not living in a contaminated area.

Large quantities of radioactive materials from the Mayak facility were also dumped into nearby Lake Karachay between 1949 and 1967. Today the lake is so radioactive that standing on its shores for about an hour would be fatal. Special machinery is now being used to fill and cap the lake.

At the Tomsk-7 plutonium production site in central Siberia, wastes with radioactivity 20 times that released at Chernobyl were dumped into lakes and underground formations in the region.

Around the shores of Novaya Zemyla, an island off the coast of Russia, 18 defunct nuclear-powered submarines were dumped (at least three of them loaded with nuclear fuel), along with as many as 17,000 containers of radioactive waste. If released by corrosion, radioactivity from these submarines and containers threatens Russian, Norwegian, Finnish, and perhaps Swedish citizens in the area. In 1993 Greenpeace caught Russians still dumping massive amounts of radioactive waste from nuclear submarines into the Sea of Japan.

Critical Thinking

Because of serious economic problems there is little money for cleanup of radioactive sites in parts of the former Soviet Union. Should the United States and other developed nations provide technical and financial aid to help deal with this problem? Explain.

By 1995 more than 30 commercial reactors worldwide (12 in the United States) had been retired and awaited decommissioning. Another 228 large commercial reactors (20 in the United States) are scheduled for retirement between 2000 and 2012. By 2030 all U.S. reactors will have to be retired, based on the life of their current operating licenses, and many may be retired early for safety or financial reasons.

Experience suggests that decommissioning old plants by dismantling and safely taking care the resulting radioactive wastes sometimes exceeds the costs of building them in the first place. U.S. utilities have been setting aside funds for decommissioning more than 100 reactors, but to date these funds fall far short of the estimated costs. If sufficient funds are not available for decommissioning, the balance of the costs could be passed along to ratepayers and taxpayers.

What Is the Connection Between Nuclear Reactors and the Spread of Nuclear Weapons?

Since 1958 the United States has been giving away and selling to other countries various forms of nuclear technology, mostly in the form of nuclear power plants and research reactors. Today the United States and at

least 14 other countries sell nuclear power technology in the international marketplace. Information, components, and materials used to build and operate such reactors can be used to produce fissionable isotopes such as uranium-235 and plutonium-239 (the explosive material in nuclear weapons).

We already live in a world with enough nuclear weapons to kill everyone on the earth 40 times over—20 times if current nuclear arms reduction agreements are carried out. By the end of this century, 60 countries—one of every three in the world—are expected to have either nuclear weapons or the knowledge and ability to build them, with the fuel and knowledge coming mostly from research and commercial nuclear reactors. Dismantling thousands of Russian and American nuclear warheads can increase the threat from the resulting huge amounts of bomb-grade plutonium that must be safeguarded.

So far no one has been able to come up with truly effective solutions to the serious problems of nuclear weapon proliferation and what to do with retired weapon-grade plutonium. All we can do is try to slow the spread of such weapons and hope we can find a way to keep plutonium removed from weapons out of the hands of those who want nuclear arms.

A: Less than 2%

Can We Afford Nuclear Power? Experience has shown that nuclear power is an extremely expensive way to boil water to produce electricity, even when it is shielded partially from free-market competition with other energy sources by huge government subsidies. Thus, the major reason utility officials, investors, and most governments are shying away from nuclear power is not safety but the *extremely high cost* of making it a safe technology.

Some of these costs can be reduced by using standardized designs for all plants (which also cuts construction time in half), as France has done. Even so, the French government has run up a $30-billion debt subsidizing its government-run nuclear power industry.

Despite massive federal subsidies and tax breaks, the most modern nuclear power plants built in the United States produce electricity at an average of about 13.5¢ per kilowatt-hour, the equivalent of burning oil costing about $216 per barrel (compared to its current price of $15–20 per barrel) to produce electricity. All methods of producing electricity in the United States (except solar photovoltaic and solar thermal plants) have average costs below those of new nuclear power plants (Figure 18-16). By 2000–2005, even these methods (with few subsidies) are expected to be cheaper than nuclear power.

In the 1980s Great Britain tried to sell its government-run nuclear power plants to private interests, but no one would buy them or even accept them for free. Business interests knew that the plants were losers because of the high costs of maintenance, decommissioning, and radioactive waste storage.

Banks and other lending institutions have become leery of financing new U.S. nuclear power plants. Abandoned reactor projects have cost U.S. utility investors over $100 billion since the mid-1970s, bankrupting U.S. utilities such as the Washington Public Power Supply System and the Public Service Company of New Hampshire. The Three Mile Island accident showed that utility companies could lose $1 billion worth of equipment in an hour and at least $1 billion more in cleanup costs, even without any established harmful effects on public health. A poll of U.S. utility executives found that only 2% would even consider ordering a new nuclear power plant. At the global level, the World Bank said in 1995 that nuclear power is too costly and risky.

Forbes business magazine has called the failure of the U.S. nuclear power program "the largest managerial disaster in U.S. business history." The U.S. Department of Energy estimates that nearly one-fourth of the current 109 existing U.S. nuclear reactors may be closed prematurely for financial reasons by 2003, as utility companies write off their losses and move on to other, more profitable alternatives. A growing number of the large companies that once built nuclear power plants are getting out of the business, and some of them are switching to developing energy conservation and solar energy technologies.

Despite massive economic and public relations setbacks, the politically powerful U.S. nuclear power industry keeps hoping for a comeback. Since the Three Mile Island accident, the U.S. nuclear industry and utility companies have financed a $21-million-a-year public relations campaign by the U.S. Council for Energy Awareness to improve the industry's image and resell nuclear power to the American public.

Most of the council's ads advance the argument that the United States needs more nuclear power to reduce dependence on imported oil and to improve national security. But since 1979, only about 5% (3% in 1995) of the electricity in the United States has been produced by burning oil, and 95% of that is residual oil that can't be used for other purposes. The nuclear industry also does not point out that 73% of the uranium used for nuclear fuel in the United States in 1993 was imported (most from Canada).

The U.S. nuclear industry hopes to persuade the federal government and utility companies to build hundreds of new second-generation smaller plants using standardized designs with supposedly fail-safe features, which they claim are safer and can be built more quickly (in 3–5 years). However, according to *Nucleonics Week*, an important nuclear industry publication, "Experts are flatly unconvinced that safety has been achieved—or even substantially increased—by the new designs."

Furthermore, none of the new designs solves the problems of what to do with nuclear waste and worn-out nuclear plants and how to prevent the use of nuclear technology to build nuclear weapons. Indeed, these problems would become more serious if the number of nuclear plants increased from a few hundred to several thousand. None of these new designs changes the fact that nuclear power, even with huge government subsidies, is an incredibly expensive way to produce electricity compared to other alternatives. Most analysts agree that simple economics has proved to be the Achilles heel of nuclear power.

Should Conventional Nuclear Power Have a Future? Some analysts argue that we should continue some low-level government funding of research and development and pilot plant testing of new reactor designs to keep this option available for use in the future. There may be an urgent need to sharply reduce fossil-fuel greenhouse gas (because of serious global warming) or because improved energy efficiency and renewable energy options may fail to keep up with demands for electricity. What do you think?

Is Breeder Nuclear Fission a Feasible Alternative? Some nuclear power proponents urge the development and widespread use of **breeder nuclear fission reactors**, which generate more nuclear fuel than they consume by converting nonfissionable uranium-238

Q: What percentage of the sewage in China is treated?

into fissionable plutonium-239. Because breeders would use over 99% of the uranium in ore deposits, the world's known uranium reserves would last at least 1,000 years, and perhaps several thousand years.

However, if the safety system of a breeder reactor fails, the reactor could lose some of its liquid sodium coolant, which ignites when exposed to air and reacts explosively if it comes into contact with water. This could cause a runaway fission chain reaction and perhaps a nuclear explosion powerful enough to blast open the containment building and release a cloud of highly radioactive gases and particulate matter. Leaks of flammable liquid sodium can also cause fires, as has happened with all experimental breeder reactors built so far.

In December 1986, France opened a commercial-size breeder reactor. Not only did it cost three times the original estimate to build, but the little electricity it produced was twice as expensive as that generated by France's conventional fission reactors. So far France has invested almost $13 billion in this breeder reactor, which was shut down for expensive repairs between 1989 and 1995.

Tentative plans to build full-size commercial breeders in Germany, some republics of the former Soviet Union, the United Kingdom, and Japan have been abandoned because of the French experience and an excess of electric-generating capacity and conventional uranium fuel. An experimental breeder reactor in Japan has been shut down since December 1995 because of a leak in its liquid sodium coolant system.

In addition, the existing breeders produce plutonium fuel much too slowly. If this problem is not solved, it would take 100–200 years for breeders to begin producing enough plutonium to fuel a significant number of other breeder reactors. In 1994 U.S. Secretary of Energy Hazel O'Leary ended government-supported research for breeder technology after some $9 billion had been spent on it.

Is Nuclear Fusion a Feasible Alternative?

Scientists hope that someday controlled nuclear fusion will provide an almost limitless source of high-temperature heat and electricity. Research has focused on the D–T nuclear fusion reaction, in which two isotopes of hydrogen—deuterium (D) and tritium (T)—fuse at about 100 million degrees (Figure 3-13).

Despite 50 years of research and huge expenditures of mostly government funds, controlled nuclear fusion is still in the laboratory stage. Deuterium and tritium atoms have been forced together using electromagnetic reactors (the size of 12 locomotives), 120-trillion-watt laser beams, and bombardment with high-speed particles. So far, none of these approaches has produced more energy than it uses.

In 1989 two chemists claimed to have achieved deuterium–deuterium (D–D) nuclear fusion at room temperature using a simple apparatus. However, subsequent experiments have not substantiated their claims.

If researchers can eventually get more energy out of nuclear fusion than they put in, the next step would be to build a small fusion reactor and then scale it up to commercial size—one of the most difficult engineering problems ever undertaken. The estimated cost of a commercial fusion reactor is several times that of a comparable conventional fission reactor.

Proponents contend that with greatly increased federal funding, a commercial nuclear fusion power plant might be built by 2030, but many energy experts don't expect nuclear fusion to be a significant energy source until 2100, if then. Meanwhile, experience shows that we can produce more electricity than we need using several quicker, cheaper, and safer methods.

19-5 SOLUTIONS: A SUSTAINABLE ENERGY STRATEGY

What Are the Best Energy Alternatives for the United States? Table 19-1 summarizes the major advantages and disadvantages of the energy alternatives discussed in this and the previous chapter, with emphasis on their potential in the United States (and presumably many other countries). Energy experts argue over these and other projections, and new data and innovations may affect the status of certain alternatives. However, the data in the table do provide a useful framework for making decisions based on currently available information.

Many scientists and energy experts who have evaluated these energy alternatives have come to the following general conclusions:

- *The best short-term, intermediate, and long-term alternatives are a combination of improved energy efficiency and greatly increased use of locally available renewable energy resources.*

- *Future energy alternatives will probably have low to moderate net energy yields and moderate to high development costs.*

- *Because there is not enough financial capital to develop all energy alternatives, projects must be chosen carefully.*

- *We cannot and should not depend mostly on a single nonrenewable energy resource such as oil, coal, natural gas, or nuclear power.*

What Role Should Economics Play in Energy Resource Use? Cost is the biggest factor determining which commercial energy resources are widely used by consumers. Governments throughout the world use three basic economic and political strategies to stimulate or dampen the short-term and long-term use of a particular energy resource.

A: About 15%

Table 19-1 Evaluation of Energy Alternatives for the United States (shading indicates favorable conditions)

Energy Resources	Estimated Availability			Estimated Net Useful Energy of Entire System	Projected Cost of Entire System	Actual or Potential Overall Environmental Impact of Entire System
	Short Term (1998–2008)	Intermediate Term (2008–2018)	Long Term (2018–2048)			
NONRENEWABLE RESOURCES						
Fossil fuels						
Petroleum	High (with imports)	Moderate (with imports)	Low	High but decreasing	High for new domestic supplies	Moderate
Natural gas	High	Moderate to high	Moderate (with imports)	High but decreasing	High for new domestic supplies	Low
Coal	High	High	High	High but decreasing	Moderate but increasing	Very high
Oil shale	Low	Low to moderate	Low to moderate	Low to moderate	Very high	High
Tar sands	Low	Fair? (imports only)	Poor to fair (imports only)	Low	Very high	Moderate to high
Synthetic natural gas (SNG) from coal	Low	Low to moderate	Low to moderate	Low to moderate	High	High (increases use of coal)
Synthetic oil, H_2, and alcohols from coal	Low	Moderate	High	Low to moderate	High	High (increases use of coal)
Nuclear energy						
Conventional fission (uranium)	Low to moderate	Low to moderate	Low to moderate	Low to moderate	Very high	Very high
Breeder fission (uranium and thorium)	None	None to low (if developed)	Low to moderate (if developed)	Unknown, but probably moderate	Very high	Very high
Fusion (deuterium and tritium)	None	None	None to low (if developed)	Unknown, but may be high	Very high	Unknown (probably moderate to high)
Geothermal energy (some are renewable)	Low	Low	Moderate	Moderate	Moderate to high	Moderate to high
RENEWABLE RESOURCES						
Improving energy efficiency	High	High	High	Very high	Low	Decreases impact of other sources
Hydroelectric						
New large-scale dams and plants	Low	Low	Very low	Moderate to high	Moderate to very high	Low to moderate

Q: What percentage of the U.S. rivers and lakes tested in 1992 were swimmable and fishable?

Energy Resources	Estimated Availability			Estimated Net Useful Energy of Entire System	Projected Cost of Entire System	Actual or Potential Overall Environmental Impact of Entire System
	Short Term (1998–2008)	Intermediate Term (2008–2018)	Long Term (2018–2048)			
RENEWABLE RESOURCES (continued)						
Hydroelectric (continued)						
Reopening abandoned small-scale plants	Moderate	Moderate	Low	Moderate	Moderate	Low to moderate
Tidal energy	Very low	Very low	Very low	Moderate	High	Low to moderate
Ocean thermal gradients	None	Low	Low to moderate (if developed)	Unknown (probably low to moderate)	High	Unknown (probably moderate to high)
Solar energy						
Low-temperature heating (for homes and water)	Moderate to high	High	High	Moderate to high	Moderate	Low
High-temperature heating	Low	Moderate	Moderate to high	Moderate	High initially but probably declining fairly rapidly	Low to moderate
Photovoltaic production of electricity	Low to moderate	Moderate	High	Fairly high	High initially but declining fairly rapidly	Low
Wind energy	Low to moderate	Moderate	Moderate to high	Fairly high	Moderate	Low
Geothermal energy (low heat flow)	Very low	Very low	Low to moderate	Low to moderate ✳	Moderate to high	Moderate to high
Biomass (burning of wood and agricultural wastes)	Moderate	Moderate	Moderate to high	Moderate	Moderate	Moderate to high
Biomass (urban wastes for incineration)	Low	Moderate	Moderate	Low to fairly high	High	Moderate to high
Biofuels (alcohols and biogas from organic wastes)	Low to moderate	Moderate	Moderate to high	Low to fairly high	Moderate to high	Moderate to high
Hydrogen gas (from coal or water)	Very low	Low to moderate	Moderate to high	Variable but probably low	Variable	Variable, but low if produced with solar energy

2-10

Geothermal for electricity   ✱2-13

The first approach is *not attempting to control the price of an energy resource*, allowing all energy resources to compete in open, free-market competition. However, leaving energy pricing to the marketplace without any government interference is rarely politically feasible because of well-entrenched government intervention into the marketplace in the form of subsidies, taxes, and regulations. Furthermore, the free-market approach, with its emphasis on short-term gain, inhibits long-term development of new energy resources, which can rarely compete economically in their development stages without government support.

The second approach is *keeping energy prices artificially low*, which encourages use and development of a resource. In the United States (and most other countries), the energy marketplace is greatly distorted by huge government subsidies and tax breaks (such as depletion write-offs for fossil fuels) that make the prices of fossil fuels and nuclear power artificially low and help perpetuate the use of these energy resources, even when better and less costly alternatives are available (Figure 18-16). At the same time, programs for improving energy efficiency and solar alternatives receive much lower subsidies and tax breaks. This creates an uneven economic playing field that encourages waste and rapid depletion of a nonrenewable energy resource and discourages the development of energy alternatives that are not getting at least the same level of subsidies and price control.

The third approach is *keeping energy prices artificially high*. This can discourage development, use, and waste of a resource. Governments can raise the price of an energy resource by withdrawing existing tax breaks and other subsidies or by adding taxes on its use. This provides increased government revenues, encourages improvements in energy efficiency, reduces dependence on imported energy, and decreases use of an energy resource that has a limited future supply. However, increasing taxes on energy use can dampen economic growth and put a heavy economic burden on the poor and lower middle class unless some of the energy tax revenues are used to help offset their increased energy costs.

One popular myth is that higher energy prices wipe out jobs. Actually, low energy prices increase unemployment because farmers and industries find it cheaper to substitute machines run on artificially cheap energy for human labor. Raising energy prices stimulates employment because building solar collectors, adding insulation, and carrying out most forms of improving energy efficiency require a high input of human labor.

Solutions: How Can We Develop a More Sustainable Energy Future for the United States?

Communities such as Osage, Iowa (p. 46), and Davis, California (p. 190), and individuals are taking energy matters into their own hands. At the same time, most

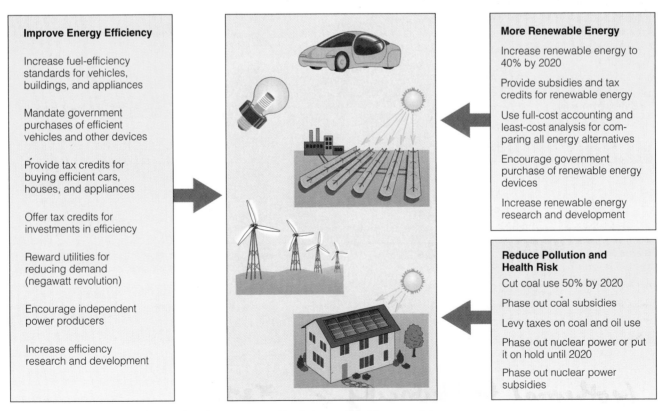

Figure 19-17 Some ways various analysts have suggested to attain a sustainable energy future.

Q: How much money is lost each year because of the direct and indirect damages from soil erosion?

environmentalists urge citizens to exert intense pressure on elected officials to develop a national energy policy based on much greater improvements in energy efficiency and a more rapid transition to a mix of renewable energy resources. A variety of analysts have suggested strategies, some of them highly controversial, to achieve such goals (Figure 19-17).

Government-based ways to improve energy efficiency include **(1)** *increasing fuel-efficiency standards for vehicles*, **(2)** *establishing energy-efficiency standards for buildings and appliances*, **(3)** *greatly increasing government-sponsored research and development to improve energy efficiency*, and **(4)** *giving tax credits and exemptions or government rebates for purchases of energy-efficient vehicles, houses, buildings, and appliances*. This last strategy could be accomplished with *freebates*, a self-financing mechanism that charges fees to purchasers of inefficient products and uses those funds to provide rebates to buyers of energy-efficient products. This approach relies on market forces and requires no new government taxes or subsidies.

Another suggested government-based approach involves *taxing energy*. This would help include some of the harmful costs of using energy (which consumers are now paying indirectly) in the market prices of energy; the tax revenues would be used to improve energy efficiency, encourage use of renewable energy resources, and provide energy assistance to poor and lower-middle-class Americans. Some economist believe that the public might accept these higher taxes if income or other taxes were lowered as gasoline or other fossil fuel taxes were raised.

Governments can also improve energy efficiency by *modifying electric utility regulations*. These changes, which a few state utility commissions are now putting into effect, would require utilities to produce electricity on a least-cost basis, permit them to earn money for their shareholders by reducing electricity demand, and allow rate increases based primarily on improvements in energy efficiency.

If utilities are rewarded for reducing electricity demand and cutting their customer's bills, the goal of the companies would be to maximize production of what Amory Lovins calls energy- and money-saving *negawatts* instead of megawatts. Lovins says we should "think of a 16-watt compact-fluorescent replacement for a 75-watt incandescent bulb, for example, as a 59-negawatt power plant."

Another suggested major goal would be to *rely more on renewable energy*. One way to encourage such a shift over the next 20–50 years would be to phase in *full-cost pricing*, in which environmental and health costs are included in the market price of energy. In other words, the prices of energy (and ideally all goods and services sold in the marketplace) should tell the ecological truth. This can provide consumers and power producers with more

U.S. Government Funding for Energy Research and Development

CONNECTIONS

Since 1973 about 54% of the research and development (R&D) budget of the U.S. Department of Energy (DOE) has gone to nuclear power, 25% to fossil fuels, and 21% to energy conservation and renewable energy options.

In 1998, the DOE's R&D budget for development of energy supply options amounted to $2.3 billion (only 19% of the department's overall budget). Of this amount, 21% went for nuclear energy (divided about equally between conventional nuclear fission and nuclear fusion), 4% for fossil fuels, 19% for energy conservation, and 15% for solar and other forms of renewable energy. Most environmentalists believe that these R&D research priorities create an uneven economic playing field for energy efficiency and renewable energy, which should receive greatly increased funding.

Most Americans are unaware that typically 65–67% of the DOE's annual budget is used to produce and dismantle nuclear weapons and clean up the country's badly contaminated nuclear weapons production and test facilities. In other words, two-thirds of the DOE budget is an appendage to the Department of Defense's annual budget. Some analysts and members of Congress urge that this portion of the DOE budget be transferred to the Department of Defense.

Critical Thinking

1. Should more government R&D funding go to energy efficiency and renewable energy and less to fossil fuels and nuclear energy? Explain.

2. Should the roughly two-thirds of the DOE annual budget used to produce and dismantle nuclear weapons and clean up contaminated nuclear weapons and test facilities be transferred to the Department of Defense or some other separate agency? Would this tend to decrease or increase the amount of money Congress allocates for research and development of energy technologies?

accurate information about the benefits and consequences of using various types of energy.

Two other ways to encourage more reliance on renewable energy are *to provide subsidies or tax credits for homeowners and businesses that switch to various forms of renewable energy* and *to greatly increase government funding of renewable energy research and development* (Connections, above).

Another innovation in the mostly monopolistic public utility industry is to *allow and encourage more*

A: About $400 million (or $46 million per hour) worldwide and $44 billion in the United States

competition from independent power producers. A growing number of independent power companies will now build almost any kind of power plant anywhere in the world. Small decentralized devices for generating electricity and storing excess energy could allow developing countries to leapfrog into 21st-century power systems and bypass expensive central power plants and distribution systems.

Another suggested major goal is to *reduce the pollution and harmful health and ecological effects of relying on nonrenewable energy from coal and nuclear power.* This could be accomplished by phasing out most or all government subsidies and tax breaks for coal and nuclear power over a decade.

Energy experts estimate that implementing a carefully evaluated mix of such policies over the next two decades could save money, create a net gain in jobs, improve competitiveness in the global marketplace, slow projected global warming, and sharply reduce air and water pollution. Some actions you can take toward achieving a sustainable energy future are listed in Appendix 5.

In the long run, humanity has no choice but to rely on renewable energy. No matter how abundant they seem today, eventually coal and uranium will run out.

Daniel Deudney and Christopher Flavin

CRITICAL THINKING

1. Just to continue using oil at the current rate (not the projected higher exponential increase in its annual use) and not run out, we must discover and add to global oil reserves the equivalent of a new Saudi Arabia supply (the world's largest) *every 10 years.* Do you believe this is possible? If not, what effects might this have on your life? On a child or grandchild you might have? List five things you can do to help reduce your dependence on oil and resources derived from oil (such as gasoline and most plastics) (see Appendix 5). Which of these things do you actually plan to do?

2. The United States now imports more than half of the oil it uses and could be importing 70% of its oil by 2010. How do you feel about this situation? List what you believe are the five best ways to reduce this dependence.

3. a. Should air pollution emission standards for *all* new and existing coal-burning plants be tightened significantly? Explain.

   b. Do you favor a an energy strategy based on greatly increased use of coal-burning plants to produce electricity? Explain. What are the alternatives?

4. Explain why you agree or disagree with each of the following proposals made by the nuclear power industry:

   a. A large number of new, better-designed nuclear fission power plants should be built in the United States to reduce dependence on imported oil and slow down projected global warming.

   b. Federal subsidies to the commercial nuclear power industry (already totaling $1 trillion) should be continued so that it does not have to compete in the open marketplace with other energy alternatives receiving no, or smaller, federal subsidies.

   c. A comprehensive program for developing the breeder nuclear fission reactor should be developed and funded largely by the federal government.

   d. Current federal subsidies for developing nuclear fusion should be increased.

5. Explain why you agree or disagree with the following proposals by various energy analysts:

   a. Federal subsidies for all energy alternatives should be eliminated so that all energy choices can compete in a true free-market system.

   b. All government tax breaks and other subsidies for conventional fuels (oil, natural gas, coal), synthetic natural gas and oil, and nuclear power (fission and fusion) should be removed and replaced with subsidies and tax breaks for improving energy efficiency and developing solar, wind, geothermal, and biomass energy alternatives.

   c. Development of solar and wind energy should be left to private enterprise and receive little or no help from the federal government, but nuclear energy and fossil fuels should continue to receive large federal subsidies.

   d. To solve present and future U.S. energy problems, all we need to do is find and develop more domestic supplies of oil, natural gas, and coal, and increase dependence on nuclear power.

   e. A heavy federal tax should be placed on gasoline and imported oil used in the United States.

   f. Between 2000 and 2020, the United States should phase out all nuclear power plants.

6. Explain why you agree or disagree with the proposals suggested in Figure 19-17 by various analysts as ways to promote a sustainable energy future for the United States (and for other countries).

*7. How is the electricity in your community produced? How has the cost of that electricity changed since 1970 compared to general inflation? Do your community and your school have a plan for improving energy efficiency? If so, what is this plan, and how much money has it saved during the past 10 years? If there is no plan, develop one and present it to the appropriate officials.

*8. Make a concept map of this chapter's major ideas, using the section heads and subheads and the key terms (in boldface type). Look at the inside back cover and in Appendix 4 for information about concept maps.

SUGGESTED READINGS

See Internet Sources and Further Readings at the back of the book, and InfoTrac.

EPILOGUE

PRINCIPLES FOR UNDERSTANDING AND WORKING WITH THE EARTH

Nature of Science

- Science is an attempt to discover order in nature and then to use that knowledge to describe, explain, and predict what happens in nature.
- Scientists don't establish absolute proof or truth. Scientific laws, hypotheses, models, and theories are based on statistical probabilities, not on certainties.
- Scientific theories and laws are well tested and widely accepted principles with a high degree of certainty.

Matter

- Matter cannot be created or destroyed; it can only be changed from one form to another. Everything we think we have thrown away is still with us in one form or another; there is no "away" (*law of conservation of matter*).
- Organized and concentrated matter is high-quality matter that can usually be converted into useful resources at an affordable cost; disorganized and dispersed matter is low-quality matter that often costs too much to convert to a useful resource (*principle of matter quality*).

Energy

- Energy cannot be created or destroyed; it can only be changed from one form to another. We can't get energy for nothing; in terms of energy quantity, it takes energy to get energy (*first law of energy or thermodynamics* or *law of conservation of energy*).
- Organized or concentrated energy is high-quality energy that can be used to do things; disorganized or diluted energy is low-quality energy that is not very useful (*principle of energy quality*).
- In any conversion of energy from one form to another, high-quality, useful energy is always degraded to lower-quality, less useful energy that can't be recycled to give high-quality energy; we can't break even in terms of energy quality (*second law of energy or thermodynamics*).
- Ideally, high-quality energy should not be used to do something that can be done with lower-quality energy; we don't need to use a chain saw to cut butter (*principle of energy efficiency*).

Life

- Life on earth depends on the one-way flow of high-quality energy from the sun, through the earth's life-support systems, and eventually back into space as low-quality heat; gravity; and the recycling of vital chemicals by a combination of biological, geological, and chemical processes (*principle of energy flow, gravity, and matter recycling*).
- Each species and each individual organism can tolerate only a certain range of environmental conditions (*range-of-tolerance principle*).
- Too much or too little of a physical or chemical factor can limit or prevent the growth of a population in a particular place (*limiting factor principle*).
- Every species has a specific role to play in nature (*ecological-niche principle*).
- Species interact through competition for resources, predation, parasitism, mutualism (mutually beneficial interactions), commensalism (interactions beneficial to one species with no harm to the other) (*principle of species interactions*).
- When possible, species reduce or avoid competition with one another by dividing up scarce resources so that species with similar requirements use them at different times, in different ways, or in different places (*principle of resource partitioning*).
- As environmental conditions change, the number and types of species present in a particular area change and, if not disturbed, can often form more complex communities (*principle of ecological succession*).
- Average precipitation and temperature are the major factors determining whether a particular land area supports a desert, grassland, or forest (*climate-biome principle*).
- All living systems contain complex networks of interconnected negative and positive feedback loops that interact to provide some degree of stability or sustainability over each system's expected life span (*principle of stability*).
- The size, growth rate, age structure, density, and distribution of a species' population are controlled by its interactions with other species and with its nonliving environment (*principle of population dynamics*).
- No population can keep growing indefinitely (*carrying capacity principle*).
- Individuals of a population of a species vary in their genetic makeup as a result of mutations or inheritable changes in molecules making up their genes (*principle of genetic variability*).
- Individuals of a population of a species that possess genetically controlled characteristics enhancing their ability to survive under existing environmental conditions have a greater chance of surviving and producing more offspring than do those lacking such traits (*principle of adaptation and natural selection*).

- All species eventually become extinct by disappearing or by evolving into one or more new species in response to environmental changes brought about by natural processes or by human action (*principle of evolution*).

- Over billions of years, changes in environmental conditions have led to development of a variety of species (species diversity), genetic variety within species (genetic diversity), and a variety of natural systems (ecosystem diversity) through a mixture of extinction and formation of new species (*biodiversity principle*).

- The earth's crust and upper mantle are made up of gigantic floating plates; their movement over millions of years reshapes the earth's crust, causes continents to move, and concentrates some of the minerals we extract and use. This continental drift is an important factor in the distribution and evolution of species (*theory of plate tectonics*).

- The earth's atmosphere, hydrosphere, lithosphere (upper crust and mantle), and forms of life are continually changing in response to changes in solar input, heat flows from the earth's interior, movements of the earth's crust, other natural changes, and changes brought about by humans and other living organisms (*principle of adaptability*).

- The earth's life-support systems can withstand much stress and abuse, but there are limits to how much can be tolerated (*principle of limits*).

Humans and Environment

- Our survival, life quality, and economies are totally dependent on the sun and the earth; the earth can get along without us, but we can't get along without the earth (*principle of earth capital*).

- The earth does not belong to us; we belong to the earth (*humility principle*).

- We should try to understand and work with the rest of nature to sustain the ecological integrity, biodiversity, and adaptability of earth's life-support systems for us and other species (*sustainability principle*).

- When we alter nature to meet our needs or wants, we should choose the method that does the least possible harm to us and other living things now and in the future (*least-harm principle*).

- The best way to protect species and individual organisms is to protect the ecosystems in which they live and to help restore those we have degraded (*principle of ecosystem protection and restoration*).

- We should not inflict unnecessary suffering or pain on any animal we raise or hunt for food or use for scientific or other purposes (*principle of humane treatment of animals*).

- We can learn a lot about how nature works, but nature is so incredibly complex and dynamic that such knowledge will always be limited (*principle of complexity*).

- In nature, we can never do just one thing; everything we do creates effects that are often unpredictable (*first law of human ecology.*)

- Everything is connected to and intermingled with everything else; we are all in this together; we need to understand these connections and discover which connections are most important for sustaining life on earth (*principle of interdependence or connectedness*).

- Most resources are limited and therefore should not be wasted; there is not always more (*principle of resource conservation*).

- Renewable resources should be used no faster than they are replenished by natural processes (*principle of sustainable use*).

- Living off renewable solar energy and renewable matter resources is a sustainable human lifestyle; using renewable matter resources faster than they are replenished, and living off nonrenewable matter and energy resources, degrade and deplete earth capital and ultimately constitute an unsustainable lifestyle (*principle of sustainable living*).

- Increases in population, resource use, or both can eventually overwhelm attempts to control pollution and manage wastes (*environmental-impact principle*).

- The best and cheapest way to reduce pollution and waste is to not produce so much (*principle of pollution prevention and waste reduction*).

- Pollution and wastes should not be put into the environment faster than the environment can degrade and recycle them or render them harmless (*principle of optimum pollution*).

- We should change earth-degrading and earth-depleting manufacturing processes, products, and businesses into earth-sustaining ones by using economic incentives and penalties (*principle of economic-ecological sustainability*).

- The market price of a product should include all estimated present and future costs of any pollution, environmental degradation, or other harmful effects connected with it that are passed on to society, the environment, and future generations (*principle of full-cost pricing*).

- Anticipating and preventing problems is cheaper and more effective than reacting to and trying to rectify them; an ounce of prevention is worth a pound of cure (*precautionary principle*).

- History shows that the most important changes brought about by human actions come from the bottom up, not from the top down (*individuals matter principle*).

- We should think globally and act locally (*principle of change*).

APPENDIX 1

PUBLICATIONS, ENVIRONMENTAL ORGANIZATIONS, AND FEDERAL AND INTERNATIONAL AGENCIES

PUBLICATIONS

The following publications can help you keep well informed and up-to-date on environmental and resource problems. Subscription prices, which tend to change, are not given.

Ambio: A Journal of the Human Environment Pergamon Press, Fairview Park, Elmsford, NY 10523

American Biology Teacher Journal of the National Association of Biology Teachers, 11250 Roger Bacon Dr., Rm. 319, Reston, VA 22090; Tel: 703-471-1134

American Forests American Forestry Association, 1516 P St. NW, Washington, DC 20005; Tel: 202-667-3300

American Journal of Alternative Agriculture 9200 Edmonston Rd., Suite 117, Greenbelt, MD 20770

American Rivers Newsletter 801 Pennsylvania Ave. SE, Suite 400, Washington, DC 20003-2167; Tel: 202-547-6900, Fax: 202-543-6142

Amicus Journal Natural Resources Defense Council, 40 W. 20th St., New York, NY 10011; Tel: 212-727-2700

Annual Review of Energy Department of Energy, Forrestal Building, 1000 Independence Ave. SW, Washington, DC 20585

Audubon National Audubon Society, 950 Third Ave., New York, NY 10022; Tel: 212-832-3200, Web site: http://www.magazine.audubon.org/

BioCycle Journal of Waste Recycling J.G. Press, Inc., 419 State Ave., Emmaus, PA 19049; Tel: 215-967-4135

BioScience American Institute of Biological Sciences, Central Station, P.O. Box 27417, Washington, DC 20077-0038; Tel: 202-628-1500 or 800-992-2427, Fax: 202-628-1509, E-mail: bioscienc@aibs.org

Clean Water Action News Clean Water Action, 1320 18th St. NW, Washington, DC 20036; Tel: 202-457-1286

Climate Alert The Climate Institute, 324 4th St. NW, Washington, DC 20002; Tel: 202-547-0104, Fax: 202-547-0111

Conservation Biology Blackwell Scientific Publications, Inc., Commerce Place, 350 Maine St., Malden, MA 02148-5018, Tel: 888-661-5800, Fax: 617-388-8255, Web site: http://www.conbio.rice.edu/scb/journal/

Demographic Yearbook Department of International Economic and Social Affairs, Statistical Office, United Nations Publishing Service, United Nations, New York, NY 10017; Tel: 617-253-2889, Fax: 617-258-6779

Discover 114 Fifth Avenue, New York, NY 10011-5690; Tel: 800-829-9132, E-mail: letters@discover.com

Earth Kalmbach Publishing Co., 21027 Crossroads Circle, Waukesha, WI 53187; Web site: http://www.kalbach.com/earth/earthmag.html

Earth Island Journal Earth Island Institute, 300 Broadway, Suite 28, San Francisco, CA 94133; Tel: 415-788-2666

Earth Work Student Conservation Association, P.O. Box 550, Charlestown, NH 03603; Tel: 603-543-1700, Fax: 603-543-1828

The Ecologist MIT Press Journals, 55 Hayward St., Cambridge, MA 02142

Ecology Ecological Society of America, Dr. Duncan T. Patten, Center for Environmental Studies, Arizona State University, Tempe, AZ 85281; Tel: 602-965-3979

EcoNet Institute for Global Communication, 18 De Boom St., San Francisco, CA 94107; Tel: 415-422-0220

E Magazine 28 Knight St., Westport, CT 06881; Tel: 203-854-5559, Fax 203-866-0602, E-mail: axgm65a@prodigy.com, Web site: http://www.emagazine.com

Endangered Species Update School of Natural Resources, University of Michigan, Ann Arbor, MI 48109; Tel: 313-763-3243, Fax: 313-936-2195, E-mail: jfwatson@umich.edu

Environment Heldref Publications, 1319 Eighteenth St., NW, Washington, DC 20036-1802; Tel: 800-365-9753

Environment Abstracts Congressional Information Services, Inc., 4520 East-West Hwy., Bethesda, MD 20814-3389; Tel: 800-638-8380 or 301-654-1550, Fax: 301-654-4033 (in most libraries)

Environmental Abstracts Annual Bowker A & I Publishing, 121 Chanlon Rd., New Providence, NJ 07974

Environmental Action 6930 Carroll Ave., 6th Floor, Takoma Park, MD 20912

Environmental Ethics Department of Philosophy, University of North Texas, Denton, TX 76203

Environmental Health Perspectives: Journal of the National Institute of Environmental Health Sciences Government Printing Office, Washington, D.C. 20402

Environmental Opportunities (Jobs) P.O. Box 78, Walpole, NH 03608

The Environmental Professional Editorial Office, Department of Geography, University of Iowa, Iowa City, IA 52242

Environmental Science and Technology American Chemical Society, 1155 16th St. NW, Washington, DC 20036; Tel: 202-872-4582, Fax: 202-872-6060

Environment Reporter Bureau of National Affairs, Inc., 1231 25th St. NW, Washington, DC 20037

Everyone's Backyard Citizens' Clearinghouse for Hazardous Waste, P.O. Box 926, Arlington, VA 22216; Tel: 703-237-2249

Family Planning Perspectives Planned Parenthood–World Population, 666 Fifth Ave., New York, NY 10019; Tel: 212-541-7800

The Futurist World Future Society, P.O. Box 19285, Twentieth Street Station, Washington, DC 20036

The Green Disk P. O. Box 32224, Washington, D.C. 20007; Fax: 1-888-GREEN-DISK, E-mail: greendisk@igc.apc.org

Greenpeace Magazine Greenpeace USA, 1436 U St. NW, Washington, DC 20009

Hydrogen Letter 4104 Jefferson St., Hyattsville, MD 20781; Tel: 301-779-1561, Fax: 301-927-6345

International Environmental Affairs University Press of New England, 171/2 Lebanon St., Hanover, NH 03755

International Wildlife National Wildlife Federation, 8925 Leesburg Pike, Vienna, VA 22184; Tel: 703-790-4524

Issues in Science and Technology National Academy of Sciences, 2101 Constitution Ave. NW, Washington, DC 20077-5576; Tel: 213-883-6325, Web site: http://utdallas.edu/research/issues

Journal of the American Public Health Association 1015 18th St. NW, Washington, DC 20036

Journal of Environmental Health National Environmental Health Association, 720 S. Colorado Blvd., Suite 970, Denver, CO 80222; Tel: 303-756-9090

Journal of Environmental Science and Health Marcel Dekker Journals, P.O. Box 10018, Church Street Station, New York, NY 10249

Journal of Forestry Society of American Foresters, 5400 Grosvenor Lane, Bethesda, MD 20814

Journal of Pesticide Reform P.O. Box 1393, Eugene, OR 97440

Journal of Soil and Water Conservation Soil and Water Conservation Society, 7515 NE Ankeny Rd., Ankeny, IA 50021; Tel: 515-289-2331, 800-843-7645

Journal of Sustainable Agriculture Food Products Press, 10 Alice Street, Binghampton, NY 13904

Journal of Wildlife Management Wildlife Society, 5410 Grosvenor Lane, Bethesda, MD 20814; Tel: 301-897-9770

National Geographic National Geographic Society, P.O. Box 2895, Washington, DC 20077-9960; Tel: 202-857-7000

National Parks National Parks and Conservation Association, 1015 31st St. NW, Washington, DC 20007; Tel: 202-223-6722, Fax: 202-659-0650

National Wetlands Newsletter Environmental Law Institute, 1616 P St. NW, Washington, DC 20036; Tel: 202-328-5150 or 328-5002

National Wildlife National Wildlife Federation, 1400 16th St. NW, Washington, DC 20036; Tel: 202-790-4524

Nature 711 National Press Building, Washington, DC 20045

The New Farm Rodale Press, 33 Minor St., Emmaus, PA 18049

New Scientist 128 Long Acre, London, WC 2, England

Newsline Natural Resources Defense Council, 122 E. 42nd St., New York, NY 10168; Tel: 212-727-2700

Not Man Apart Friends of the Earth, 530 7th St. SE, Washington, DC 20003; Tel: 202-544-2600, Fax: 202-543-4710

Nucleus Union of Concerned Scientists, 26 Church St., Cambridge, MA 02238; Tel: 617-547-5552

Organic Gardening & Farming Magazine Rodale Press, Inc., 33 E. Minor St., Emmaus, PA 18049

Our Planet United Nations Environment Programme (U.S.), 2 United Nations Plaza, Room 803, New York, NY 10017

Pesticides and You National Coalition Against the Misuse of Pesticides, 701 E St. SE, Washington, DC 20003; Tel: 703-471-1134

Pollution Abstracts Cambridge Scientific Abstracts, 7200 Wisconsin Ave., Bethesda, MD 20814 (in many libraries)

Population Bulletin Population Reference Bureau, 1875 Connecticut Ave. NW, Suite 520, Washington, DC 20009; Tel: 202-483-1100

Population and Vital Statistics Report United Nations Environment Programme, New York North American Office, Publication Sales Section, United Nations, New York, NY 10017

Rachel's Environment and Health Weekly Environmental Research Foundation, P.O. Box 4878, Annapolis, MD 21403; Tel: 410-263-1584

Rainforest News P.O. Box 140681, Coral Gables, FL 33115

Real World: The Voice of Ecopolitics 91 Nuns Moor Road, Newcastle upon Tyne, NE4 9BA, United Kingdom

Renewable Energy News Solar Vision, Inc., 7 Church Hill, Harrisville, NH 03450

Rocky Mountain Institute Newsletter 1739 Snowmass Creek Rd., Snowmass, CO 81654

Science American Association for the Advancement of Science, 1333 H St. NW, Washington, DC 20005; Tel: 202-326-6400, Web site: http://www.sciencemag.org

Science News Science Service, Inc., 1719 N St. NW, Washington, DC 20036; Tel: 800-247-2160, Web site: http://www.sciencenews.org

Scientific American 415 Madison Ave., New York, NY 10017-1111; Web site: http://www.sciam.com

SEJournal Society of Environmental Journalists, 7904 Germantown Ave., Philadelphia, PA 19118; Tel: 215-247-9710

Sierra 730 Polk St., San Francisco, CA 94108

Solar Age Solar Vision, Inc., 7 Church Hill, Harrisville, NH 03450

State of the World Worldwatch Institute, 1776 Massachusetts Ave. NW, Washington, DC 20036-1904; Tel: 202-452-1999, Fax: 202-296-7365 (published annually)

Statistical Yearbook Department of International Economic and Social Affairs, Statistical Office, United Nations Publishing Service, United Nations, New York, NY 10017

Technology Review P.O. Box 489, Mount Morris, IL 61054; Tel. 800-877-5320, Fax: 815-734-1127, E-mail: <trsubscriptions@mit.edu> ans, Web site: http://web.mit.edu/techreview/www/

The Trumpeter Journal of Ecosophy P.O. Box 5883, St. B, Victoria, BC, Canada V8R 6S8

Vital Signs: The Trends That Are Shaping Our Future Worldwatch Institute, 1776 Massachusetts Ave. NW, Washington, DC 20036-1904; Tel: 202-452-1999, Fax: 202-296-7365 (published annually)

Waste Not Work on Waste USA, 82 Judson, Canton, NY 13617

Wild Earth Cenozoic Society, Inc., P.O. Box 455, Richmond, VT 05477

Wilderness The Wilderness Society, 900 17th St. NW, Washington, DC 20006-2596; Tel: 202-833-2300

Wildlife Conservation New York Zoological Society, 185th St. & Southern Blvd., Bronx, NY 10460; Tel: 718-220-5100, Fax: 718-220-7114

World Rainforest Report Rainforest Action Network, 450 Sansome, Suite 700, San Francisco, CA 94111; Tel: 415-398-4404, Fax: 415-398-2732

World Resources World Resources Institute, 1709 New York Ave. NW, Washington, DC 20006; Tel: 202-638-6300 (published every 2 years)

World Watch Worldwatch Institute, 1776 Massachusetts Ave. NW, Washington, DC 20036-1904; Tel: 202-452-1999, Fax: 202-296-7365, E-mail: wwpub@worldwatch.org

Worldwatch Papers Worldwatch Institute, 1776 Massachusetts Ave. NW, Washington, DC 20036-1904; Tel: 202-452-1999, Fax: 202-296-7365

ENVIRONMENTAL AND RESOURCE ORGANIZATIONS

For a more detailed list of national, state, and local organizations, see *Conservation Directory* (published annually by the National Wildlife Federation, 1400 16th St. NW, Washington, DC 20036), *Your Resource Guide to Environmental Organizations* (Irvine, CA: Smiling Dolphin Press, 1991), *National Environmental Organizations* (published annually by US Environmental Directories, Inc., P.O. Box

65156, St. Paul, MN 55165), and *World Directory of Environmental Organizations* (published by the California Institute of Public Affairs, P.O. Box 10, Claremont, CA 91711). Also see Environmental Organizations Web site (http://www.econet.apc.org/econet/en.orgs.html/).

African Wildlife Foundation 1717 Massachusetts Ave. NW, Washington, DC 20036; Tel: 202-265-8393; E-mail: awnews@aol.com, Web: http://www.awf.org/

Alliance for Chesapeake Bay 6600 York Rd., Baltimore, MD 21212; Web: http://www.gmu.edu/bios/Bay/

Alliance to Save Energy 1725 K St. NW, Suite 914, Washington, DC 20006-1401

American Association for the Advancement of Science 1333 H St. NW, Washington, DC 20005; Tel: 202-326-6400, http://www2.nas.edu/cuselib/23ca.html

American Cetacean Society P.O. Box 1391, San Pedro, CA 9073-0391; Tel: 310-548-6279, Fax: 310-548-6950, Web: http://www.acsonline.org/

American Council for an Energy-Efficient Economy 1001 Connecticut Ave. NW, #801, Washington, DC 20036; Tel: 202-429-8873, Web: http://www.aceee.org/

American Forests (formerly American Forestry Association 1516 P St. NW, Washington, DC 20005; Tel: 202-667-3300

American Hydrogen Association P.O. Box 15075, Phoenix, AZ 85060; Tel: 602-921-0433, Fax: 602-967-6601; Web: http://www.getnet.com/charity/aha/

American Institute of Biological Sciences, Inc. 730 11th St. NW, Washington, DC 20001-4521; Tel: 202-628-1500, Fax: 202-628-1509, Web: http://www.yahoo.com/Science/Biology/Organizations/Professional/American_Institute_of_Biological_Sciences/

American Public Health Association 1015 15th St. NW, Washington, DC 20005; Tel: 202-789-5600, Web: http://www.apha.org/5

American Rivers 801 Pennsylvania Ave. SE, Suite 400, Washington, DC 20003-2167; Tel: 202-547-6900, Fax: 202-543-6142; E-mail: amrivers@igc.apc.org, Web: http://www.amrivers.org/

American Society for the Prevention of Cruelty to Animals (ASPCA) 441 E. 92nd Street, New York, NY 10128, Web: http://www.aspca.org

American Solar Energy Society 2400 Central Ave., Suite G, Boulder, CO 80301; Tel: 303-443-3130; Web: http://www.sni.net/solar/

American Wind Energy Association 777 N. Capitol St. NE, Suite 805, Washington, DC 20002; Tel: 202-408-8988; Web: http://www.igc.apc.org/awea/index.html

Association of Forest Service Employees for Environmental Ethics P.O. Box 11615, Eugene, OR 97440-9958; Tel: 503-484-2692, Fax: 503-484-3004, Web: http://www.afseee.org/

Bat Conservation International P.O. Box 162603, Austin, TX 78716; Tel: 512-327-9721; Web: http://www.batcon.org/

Bio-Integral Resource Center P.O. Box 7414, Berkeley, CA 94707; Tel: 510-524-2567

Campus Ecology National Wildlife Federation, 8925 Leesburg Pike, Vienna, Va. 22184; Tel: 703-790-4318; Web: http://www.nwf.org/nwf/campus

Center for Conservation Biology Alice Blandin, *Conservation Biology*, University of Washington, Box # 351800, Seattle, Wash.:98195-1800; Web: http://www.conbio.rice.edu/

Center for Environmental Information 46 Prince Street, Rochester, NY; Tel: 716-271-3350, Web: http://www.awa.com/nature/cei/

Center for Marine Conservation 1725 DeSales St. NW, Suite 500, Washington, DC 20036; Tel: 202-429-5609, Fax: 202-872-0619, Web: http://www.cmc-ocean.org/

Center for Plant Conservation P.O. Box 299, St. Louis, MO 63166; 314-577-9450; Fax: 314-577-9465, Web: http://www.mobot.org/CPC/welcome.html

Center for Science in the Public Interest 1875 Connecticut Ave. NW, Suite 300, Washington, DC 20009; Tel: 202-332-9110, Fax: 202-265-4594, Web: http://www.cspinet.org/

Chesapeake Bay Foundation 162 Prince George St., Annapolis, MD 21401; Tel: 410-268-8816, Web: http://www.baylink.org/baywatch.html

Citizens' Clearinghouse for Hazardous Waste P.O. Box 6806, Falls Church, VA 22040; Tel: 703-237-2249, Web: http://www.essential.org/orgs/cchw/cchw.html

Coalition for Environmentally Responsible Economies 711 Atlantic Ave., 5th Fl., Boston, MA 02111; Tel: 617-451-0927

Common Cause 2030 M Street NW, Washington, DC 20036; Tel: 202-833-1200; E-mail: 75300.3120@compuserve.com, Web: http://www.commoncause.org/

Conservation International 1015 18th St. NW, Suite 1000, Washington, DC 20036; Tel: 202-429-5660

Council for Economic Priorities 30 Irving Place, New York, NY 10003; Tel: 212-420-1133. Maintains corporate environmental data clearinghouse.

Critical Mass Energy Project 215 Pennsylvania Ave. SE, Washington, DC 20003; Tel: 202-546-4996; E-mail:cmep@citizen.org; Web: http://www.essential.org

Cultural Survival 11 Divinity Ave., Cambridge, MA 02138; Tel: 617-495-2562

Defenders of Wildlife 1101 14th St. NW, Suite 1400, Washington, DC 20005; Tel: 202-682-9400; Fax: 202-682-1331; Internet: information@defenders.org

Ducks Unlimited One Waterfowl Way, Memphis, TN 38120-2351; Tel: 901-758-3825, Fax: 901-758-3850

Earth First! 305 N. Sixth St., Madison, WI 53704

Earth Island Institute 300 Broadway, Suite 28, San Francisco, CA 94133: Tel: 415-788-3666, Fax: 415-788-7324; Web: http://www.earthisland.org/ei

Elmwood Institute P.O. Box 5765, Berkeley, CA 94705; Tel: 510-845-4595

Energy Conservation Coalition 1525 New Hampshire Ave. NW, Washington, DC 20036

Envirolink Network E-mail: admin@envirolink.org; Web: http://www.envirolink.org

Environmental Action, Inc. 6930 Carroll Park, Suite 600, Takoma Park, MD 20912; Tel: 301-891-1100, Fax: 301-891-2218

Environmental Defense Fund, Inc. 257 Park Ave. South, New York, NY 10010; Tel: 212-505-2100

Environmental Law Institute 1616 P St. NW, Suite 200, Washington, DC 20036; Tel: 202-328-5150, Fax: 202-328-5002

Fish and Wildlife Reference Service 5430 Grosvenor Ln., Bethesda, MD 20814; Tel: 301-492-6403 or 800-582-3421, Fax: 301-564-4059

Florida Solar Energy Center 300 State Road #401, Cape Canaveral, FL 32920

Friends of Animals 777 Post Rd., Suite 205, Darien, CT 06820; Tel: 203-656-1522, Fax: 656-0267; E-Mail: soebc@igc.apc.org

Friends of the Earth The Global Building, 1025 Vermont Ave., NW, Suite 300, Washington, DC 20005; Tel: 202-783-7400, Fax: 202-783-0444; EcoNet: foedc@igc.apc.org; Web: http://www.foe.co.uk/

Global Greenhouse Network 1130 17th St. NW, Suite 530, Washington, DC 20036

Global Tomorrow Coalition 1325 G St. NW, Suite 1010, Washington, DC 20005-3104; Tel: 202-628-4016, Fax: 202-628-4018

GreenNet 23 Bevenden St. London N1 6BH, England

Greenpeace, Canada 427 Bloor St., West Toronto, Ontario M5S 1X7

Greenpeace, USA, Inc. 1436 U St. NW, Washington, DC 20009; Tel: 202-462-1177; E-mail: your.name@green2.greenpeace.org and www.greenpeace.org; Web: http://www.greenpeace.org/

Habitat for Humanity 121 Habitat Street, Americus, GA 31709-3498; Tel: 912-924-6935; E-mail: Frank_Purvs@habitat.org

Humane Society of the United States, Inc. 2100 L St. NW, Washington, DC 20037; Tel: 202-452-1100; Fax: 301-258-3077

Institute for Alternative Agriculture 9200 Edmonston Rd., Suite 117, Greenbelt, MD 20770

Institute for Earth Education Cedar Cove, Box 115, Greenville, WV 24945; Tel: 304-832-6404, Fax: 304-832-6077

Institute for Local Self-Reliance 2425 18th St. NW, Washington, DC 20009; Tel: 202-232-4108; E-mail: ilsr@itp.apc.org

International Alliance for Sustainable Agriculture 1701 University Ave. SE, Minneapolis, MN 55414

International Institute for Energy Conservation 750 First St. NE, Washington, DC 20002; Tel: 202-842-3388; Fax: 202-842-1565; E-mail: iiec@igc.apc.org

International Planned Parenthood Federation 105 Madison Ave., 7th Floor, New York, NY 10016

International Rivers Network 1847 Berkeley Way, Berkeley, CA 94703; Tel: 510-848-1155, Fax: 510-848-1008; E mail: irn@igc.apc.org

International Society for Ecological Economics P.O. Box 1589, Solomons, MD 20688; Tel: 410-326-0794; E-mail: button@cbl.umd.edu

International Union for the Conservation of Nature and Natural Resources (IUCN) 1400 16th St. NW, Washington, DC 20036; Tel: 202-797-5454, Fax: 202-797-5461; E-mail: mail@hg.iucn.ch

Land Institute 2440 E. Well Water Road, Salina, KS 67401; Tel: 913-823-53765

Land Trust Alliance 1319 F St NW, Suite 501, Washington, DC 20004; Tel: 202-638-4725, Fax 202-638-4730, Web: http://www.lta.org/

League of Conservation Voters 1707 L Street, NW, Suite 750, Washington, DC 20036; Tel: 202-785-8683, Fax: 202-835-0491,Web: http://www.lcv.org/

League of Women Voters of the U.S. 1730 M St. NW, Washington, DC 20036; Tel: 202-429-1965, Fax: 202-429-0854, Web: http://acm.cs.umn.edu/~lisi/lwv/main.html

National Audubon Society 700 Broadway, New York, NY 10003-9501; Tel. 212-979-3000; E-mail: mis@audubon.org, Web: http://www.audubon.org

National Coalition Against the Misuse of Pesticides 701 E St. SE, Suite 200, Washington, DC 20003; Tel: 202-543-5450, Web: http://www.ncamp.org/

National Environmental Trust Tel: 202-887-8800, E-mail: netinfo@acpa.com, Web: http://www.eic.org/

National Geographic Society 1145 17th St. NW, Washington, DC 20036; Tel: 202-857-7000, Web: http://www.nationalgeographic.com/main.html

National Hydrogen Association 1800 M Street N.W., Suite 300, Washington, DC 20036: Tel: 202-223-5547, Fax: 202-223-5537, Web: http://www.ttcorp.com/nha/

National Park Foundation 1101 17th St. NW, Suite 1102, Washington, DC 20036; Tel: 202-785-4500, Web: http://www.nationalparks.org/

National Parks and Conservation Association 1776 Massachusetts Ave. NW, Suite 200, Washington, DC 20036; Tel: 202-223-6722, Fax: 202-659-0650; E-mail: natparks@aol.com; Web: http://www.npca.org/home/npca/

National Recreation and Park 22377 Belmont Ridge Road,Ashburn, Va. 20148; Tel: 703-858-0784, Fax: 703-858-0794, E-mail: info@nrpa.org, Web: http://www.nrpa.org/

National Recycling Coalition 1727 King Street, Suite 105, Alexandria Va 22314; Tel: 703-683-9025, Fax: 703-683-9026, Web: http://www.recycle.net/recycle/index.html

National Science Teachers Association 1840 Wilson Blvd., Arlington, VA 22201; Tel: 703-7100, Fax: 703-243-7177; E-mail: maiser@nsta.org, Web: http://www.nsta.org/

National Toxics Campaign 37 Temple Place, 4th Floor, Boston, MA 02111

National Wildlife Federation 8925 Leesburg Pike, Vienna, Va., 22184; Tel 703-790-4000; E-mail: feedback@nws.org; Web: http://www.nwf.org/nwf/home.html

Natural Resources Defense Council 40 W. 20th St., New York, NY 10011; Tel: 212-727-2700; and

1350 New York Ave. NW, Suite 300, Washington, DC 20005; Tel: 202-783-7800; E-mail: nrdcinfo@igc.apc.org; Web: http://www.nrdc.org/nrdc/

Nature Conservancy 1814 N. Lynn St., Arlington, VA 22209; Tel: 703-841-5300, Fax: 703-841-1283, Web: http://www.tnc.org/

New York Zoological Society/Wildlife Conservation Society (formerly the New York Zoological Society) 185th & Southern Blvd., Bronx, NY 10460-1099; Tel: 718-220-5100, Fax: 718-220-7114

North American Associations for Environmental Education 1255 23rd St. NW, Suite 300, Washington, DC 20037; Tel: 202-884-8912, Fax 202-884-8701

Permaculture Institute of North America 4649 Sunnyside Ave. N, Seattle, WA 98103

Pesticide Action Network 965 Mission St., No. 514, San Francisco, CA 94103; Web: http://www.panna.org/panna/

Planet/Drum Foundation P.O. Box 31251, San Francisco, CA 94131; Web: http://www.ic.org/fic/cdir/res/PlanetDrum.html

Planned Parenthood Federation of America 810 Seventh Ave., New York, NY 10019; Tel: 212-541-7800; Web: http://www.ppfa.org/ppfa/index.html

Population Action International (formerly Population Crisis Committee) 1120 19th St. NW, Suite 550, Washington, DC 20036-3605; Tel: 202-659-1833; Web: http://www.populationaction.org/

Population-Environment Balance 2000 P St. NW, Suite 210, Washington, DC 20036-5915; Tel: 202-955-5700, Fax: 202-955-6161; E-mail: uspop@balance.org

Population Institute 107 2nd St. NE, Washington, DC 20002; Tel: 202-544-3300

Population Reference Bureau 1875 Connecticut Ave. NW, Suite 520, Washington, DC 20009-5728; Tel: 202-483-1100, Fax: 2-2-328-3937, E-mail: popref@prb.org, Web: http://www.prb.org/prb/

Public Citizen 215 Pennsylvania Ave. SE, Washington, DC 20003; Web: http://www.essential.org/orgs/public_citizen/

Rainforest Action Network 3450 Sansome St., Suite 700, San Francisco, CA 94111; Tel: 415-398-4404, Fax: 415-389-2732; Web: http://www.ran.org/ran/

Rainforest Alliance 65 Bleecker St., New York, NY 10012; Tel: 212-677-1900; Fax: 212-677-2187, E-mail: canopy@cdp.apc.or, Web: http://www.rainforest-alliance.org/

Resources for the Future 1616 P St. NW, Washington, DC 20036; Tel: 202-328-5000, Fax: 202-939-3460; E-mail: info@rrf.org; Web: /wwwriff.org/

Rocky Mountain Institute 1739 Snowmass Creek Rd., Snowmass, CO 81654-9199; Tel: 970-927-3851, Fax: 970-927-3420, Web: http://www.rmi.org/index.html

Rodale Institute 222 Main St., Emmaus, PA 18098; Web: http://www.envirolink.org/seel/rodale/

Scientists' Institute for Public Information 355 Lexington Ave., New York, NY 10017; Tel: 212-661-9110

Sea Shepherd Conservation Society P.O. Box 628, Venice, CA 90294; Tel: 310-301-7325; Fax: 310-574-3161; Web: http://www.seashepherd.org/

Sierra Club 730 Polk St., San Francisco, CA 94109; Tel 415-776-2211; E-mail: information@sierraclub.org; Web: http://www.sierraclub.org

Smithsonian Institution 1000 Jefferson Dr. SW, Washington, DC 20560; Tel: 202-357-2700; Web: http://www.yahoo.com/Government/Agencies/Independent/Smithsonian_Institution/

Society for Ecological Restoration University of Wisconsin-Madison Arboretum, 1207 Seminole Hwy., Madison, WI 53711; Tel: 608-262-9547, Fax: 608-262-9547

Soil and Water Conservation Society 7515 NE Ankeny Rd., Ankeny, IA 50021-9764; Tel: 515-289-2331, 800-863-7645, Fax: 515-289-1227

Solid Waste Association of North America PO Box 7219, Silver Spring, Md. 20907-7219, Tel: 301-585-2898, Fax: 301-589-7068, Web: http://www.swana.org/

http://bianca.com/lolla/politics/seac/seac.html Conservation Association, Inc. P.O. Box 550, Charlestown, NH 03603; Tel: 603-543-1700, Fax: 603-543-1828

Student Environmental Action Coalition (SEAC) P.O. Box 1168, Chapel Hill, NC 27514; Tel: 919-967-4600, 800-700-SEAC, Fax: 919-967-4648; Web: http://www.seac.org/

Survival International 2121 Decatur Place NW, Washington, DC 20008; Web: http://www.survival.org.uk/

Union of Concerned Scientists Two Brattle Square, Cambridge, MA 02238-9105; Tel: 617-547-5552, Web: Web: http://www.ucsusa.org/

United Nations Population Fund 220 East 42nd St., New York NY 10017; Tel: 212 297-5020, Fax: 212-557-6416, Web: http://www.unfpa.org/5

The Urban Agriculture Network 1711 Lamont St. NW, Washington, D.C. 20010; Tel: 202-483-8130; Fax: 202-986-6732; E-mail: 72144.3466@compuserve.com

U.S. Congress Web Server Web:: http://thomas.loc.gov

U.S. Public Interest Research Group 215 Pennsylvania Ave. SE, Washington, DC 20003; Tel: 202-546-9707

The Wilderness Society 900 17th St. NW, Washington, DC 20006-2596; Tel: 202-833-2300, Web: http://www.wilderness.org/

Wildlife Conservation Society 185th St. and Southern Blvd., Bronx, NY 10460-1099; Tel 718-220-5100, Fax: 718-220-7114, Web: http://www.wcs.org/5

Work on Waste 82 Judson St., Canton, NY 13617

World Future Society 4916 St. Elmo Ave., Bethesda, MD 20814

World Resources Institute 1709 New York Ave. NW, 7th Floor, Washington, DC 20006; Tel: 202-638-6300, Web: http://www.wri.org/

Worldwatch Institute 1776 Massachusetts Ave. NW, Washington, DC 20036-1904; Tel: 202-452-1999, Fax: 202-296-7365; E-mail: worldwatch@igc.apc.org and wwpub@igc.apc.org; Web: http://www.worldwatch.org/

World Wildlife Fund 1250 24th St. NW, Suite 500, Washington, DC 20037; Tel: 202-293-4800, Web: http://www.worldwildlife.org/

Zero Population Growth 1400 16th St. NW, Suite 320, Washington, DC 20036; Tel: 202-332-2200, Fax: 202-332-2302; E-mail: zpg@apc.org, Web: http://www.yahoo.com/Society_and_Culture/Issues_and_Causes/Population/Overpopulation/Zero_Population_Growth_ZPG_/

FEDERAL, SCIENTIFIC, AND INTERNATIONAL AGENCIES

Agency for International Development (USAID) State Building, 320 21st St. NW, Washington, DC 20523-0016; Tel: 202-647-1850, Fax: 202-647-8321, Web: http://www.info.usaid.gov/

Bureau of Land Management U.S. Department of Interior, 1620 L St. NW, Rm. 5600, Washington, DC 20240; Tel: 202-208-3801; Web: http://www.blm.gov

Bureau of Reclamation Washington, DC 20240; Web: http://www.usbr.gov/main/

Congressional Research Service 101 Independence Ave. SW, Washington, DC 20540

Conservation and Renewable Energy Inquiry and Referral Service P.O. Box 8900, Silver Spring, MD 20907; Tel: 800-523-2929

Department of Agriculture 14th St. and Independence Ave. SW, Washington, DC 20250; Tel: 202-720-8732; Web: http://www.usda.gov

Department of Commerce Herbert C. Hoover Bldg., Rm. 5610, 15th & Constitution Ave. NW, Washington, DC 20230; Tel: 202-219-3605; Web: http://www.doc.gov

Department of Energy Forrestal Building, 1000 Independence Ave. SW, Washington, DC 20585; Web: http://www.doe.gov

Department of Health and Human Services 200 Independence Ave. SW, Washington, DC 20585; Web: http://www.yahoo.com/Government/Executive_Branch/Departments_and_Agencies/Department_of_Health_and_Human_Services/

Department of Housing and Urban Development 451 7th St. SW, Washington, DC 20410; Tel: 202-755-5111; Web: http://www.hud.gov

Department of the Interior 1849 C St. NW, Washington, DC 20240; Tel: 202-208-3100; Web: http://www.usga.gov.doi

Department of Transportation 400 7th St. SW, Washington, DC 20590; Tel: 202-366-4000; Web: http://www.dot.gov

Energy Information Administration Dept. of Energy, National Energy Information Center, Forrestal Bldg., Washington, DC 20585; Tel: 202-586-8800, Web: http://www.yahoo.com/Government/Executive_Branch/Departments_and_Agencies/Department_of_Energy/Energy_Information_Administration/

Environmental Protection Agency 401 M St. SW, Washington, DC 20460; Tel: 202-260-2090; Web: http://www.epa.gov and gopher.epa.gov

Federal Energy Regulatory Commission 888 First St. NE, Washington, DC 20426; Web: http://www.ferc.fed.us/

Food and Agriculture Organization (FAO) of the United Nations 101 22nd St. NW, Suite 300, Washington, DC 20437; Web: http://www.fao.org/

Food and Drug Administration Department of Health and Human Services, 5600 Fishers Lane, Rockville, MD 20857; Tel: 410-433-1544; Web: http://www.fda.gov

Forest Service P.O. Box 96090, Washington, DC 20090-6090; Tel: 202-205-0957; Web: http://www.fs.fed.us/

Great Lakes Commission The Argus II Bldg., 400 Fourth St., Ann Arbor, MI 48103-4816; Tel: 313-665-9135, Fax: 313-665-4370; Web: http://www.glc.org/

International Whaling Commission The Red House, 135 Station Rd., Histon, Cambridge CB4 4NP England (0223 233971); Web: http://av.yahoo.com/bin/query?p=International+Whaling+Commission&hc=0&hs=0

Marine Mammal Commission 1825 Connecticut Ave. NW, Rm. 512, Washington, DC 20009; Tel: 202-606-5504, Fax: 202-606-5510; Web: http://www.citation.com/hpages/mmc.html

Mine Safety and Health Administration Ballston Tower 3, 4015 Wilson Blvd., Arlington, VA 22203; Tel: 703-235-1452; Web: http://www.msha.gov/

National Academy of Sciences 2101 Constitution Ave. NW Washington, DC 20418; Web: http://www.nas.edu/5

National Aeronautics and Space Administration 400 Maryland Ave. SW, Washington, DC 20546

National Cancer Institute 9000 Rockville Pike, Bethesda, MD 20892; Web: http://www.nci.nih.gov/

National Center for Atmospheric Research P.O. Box 3000, Boulder, CO 80307; Web: http://www.ucar.edu/

National Lead Information Center Tel: 1-800-LEAD-FYI; Web: http://www.nsc.org/ehc/lead.htm

National Marine Fisheries Service U.S. Dept. of Commerce, Silver Spring Metro Center 1, 1335 East-West Hwy., Silver Spring, MD 20910; Tel: 301-713-2239, Web: http://kingfish.ssp.nmfs.gov/

National Oceanic and Atmospheric Administration Herbert C. Hoover Bldg., Rm. 5128, 14th & Constitution Ave. NW, Washington, DC 20230; Tel: 202-482-3384, Web: http://www.yahoo.com/Government/Executive_Branch/Departments_and_Agencies/Department_of_Commerce/National_Oceanic_and_Atmospheric_Administration/

National Park Service Interior Bldg., P.O. Box 37127, Washington, DC 20013-7127; Tel: 202-208-4747, Web: http://www.nps.gov/

National Renewable Energy Laboratory 1617 Cole Blvd., Golden, CO 80401; Web: http://www.nrel.gov/

National Science Foundation 4201 Wilson Blvd., Arlington, VA 22230; Tel: 703-306-1234, Web: http://www.nsf.gov/

National Solar Heating and Cooling Information Center P.O. Box 1607, Rockville, MD 20850

National Technical Information Service U.S. Department of Commerce, 5285 Port Royal Rd., Springfield, VA 22161; Web: http://www.ntis.gov/

Nuclear Regulatory Commission Washington, DC 20555; Tel: 301-415-7000; Web: http://www.nrc.gov/

Occupational Safety and Health Administration Department of Labor, 200 Constitution Ave. NW, Washington, DC 20210; Web: http://www.nrc.gov/

Office of Ocean and Coastal Resource Management 1825 Connecticut Ave. NW, Suite 700, Washington, DC 20235; Tel: 202-208-2553, Web: http://wave.nos.noaa.gov/ocrm/

Office of Surface Mining Reclamation and Enforcement Interior South Bldg., 1951 Constitution Ave. NW, Washington, DC 20240; http://www.osmre.gov/astart3.htm

Organization for Economic Cooperation and Development (U.S. Office), 2001 L St. NW, Suite 700, Washington, DC 20036; Web: http://www.oecd.org/

Soil Conservation Service (now called Natural Resource Conservation Service) USDA, 14th and Independence Ave. SW, P.O. Box 2890, Washington, DC 20013; Tel: 202-720-3210), Web: http://www.nrcs.usda.gov/

United Nations 1 United Nations Plaza, New York, NY 10017; Web: http://www.yahoo.com/Government/International_Organizations/United_Nations/

United Nations Environment Programme Regional North American Office, United Nations Rm. DC 20803, New York, NY 10017; and 1889 F St. NW, Washington, DC 20006; Web: http://www.yahoo.com/Government/International_Organizations/United_Nations/Programs/United_Nations_Environment_Programme_UNEP_/

United States The Man and the Biosphere (U.S. MAB) Program U.S. MAB Secretariat, OES/EGC/MAB, Rm. 608, SA-37, Dept. of State, Washington, DC 20522-3706; Web: http://www.mabnetamericas.org/home2.html

U.S. Fish and Wildlife Service Department of the Interior, Washington, DC 20240; Web: http://www.fws.gov/

U.S. Geological Survey National Center, Reston, VA 22092; Tel: 703-648-4000

World Bank 1818 H St. NW, Washington, DC 20433; Web: http://www.worldbank.org/

APPENDIX 2

UNITS OF MEASUREMENT

LENGTH

Metric

1 kilometer (km) = 1,000 meters (m)
1 meter (m) = 100 centimeters (cm)
1 meter (m) = 1,000 millimeters (mm)
1 centimeter (cm) = 0.01 meter (m)
1 millimeter (mm) = 0.001 meter (m)

English

1 foot (ft) = 12 inches (in)
1 yard (yd) = 3 feet (ft)
1 mile (mi) = 5,280 feet (ft)
1 nautical mile = 1.15 miles

Metric-English

1 kilometer (km) = 0.621 mile (mi)
1 meter (m) = 39.4 inches (in)
1 inch (in) = 2.54 centimeters (cm)
1 foot (ft) = 0.305 meter (m)
1 yard (yd) = 0.914 meter (m)
1 nautical mile = 1.85 kilometers (km)

AREA

Metric

1 square kilometer (km^2) = 1,000,000 square meters (m^2)
1 square meter (m^2) = 1,000,000 square millimeters (mm^2)
1 hectare (ha) = 10,000 square meters (m^2)
1 hectare (ha) = 0.01 square kilometer (km^2)

English

1 square foot (ft^2) = 144 square inches (in^2)
1 square yard (yd^2) = 9 square feet (ft^2)
1 square mile (mi^2) = 27,880,000 square feet (ft^2)
1 acre (ac) = 43,560 square feet (ft^2)

Metric-English

1 hectare (ha) = 2.471 acres (ac)
1 square kilometer (km^2) = 0.386 square mile (mi^2)
1 square meter (m^2) = 1.196 square yards (yd^2)
1 square meter (m^2) = 10.76 square feet (ft^2)
1 square centimeter (cm^2) = 0.155 square inch (in^2)

VOLUME

Metric

1 cubic kilometer (km^3) = 1,000,000,000 cubic meters (m^3)
1 cubic meter (m^3) = 1,000,000 cubic centimeters (cm^3)
1 liter (L) = 1,000 milliliters (mL) = 1,000 cubic centimeters (cm^3)
1 milliliter (mL) = 0.001 liter (L)
1 milliliter (mL) = 1 cubic centimeter (cm^3)

English

1 gallon (gal) = 4 quarts (qt)
1 quart (qt) = 2 pints (pt)

Metric-English

1 liter (L) = 0.265 gallon (gal)
1 liter (L) = 1.06 quarts (qt)
1 liter (L) = 0.0353 cubic foot (ft^3)
1 cubic meter (m^3) = 35.3 cubic feet (ft^3)
1 cubic meter (m^3) = 1.30 cubic yards (yd^3)
1 cubic kilometer (km^3) = 0.24 cubic mile (mi^3)
1 barrel (bbl) = 159 liters (L)
1 barrel (bbl) = 42 U.S. gallons (gal)

MASS

Metric

1 kilogram (kg) = 1,000 grams (g)
1 gram (g) = 1,000 milligrams (mg)
1 gram (g) = 1,000,000 micrograms (μg)
1 milligram (mg) = 0.001 gram (g)
1 microgram (μg) = 0.000001 gram (g)
1 metric ton (mt) = 1,000 kilograms (kg)

English

1 ton (t) = 2,000 pounds (lb)
1 pound (lb) = 16 ounces (oz)

Metric-English

1 metric ton (mt) = 2,200 pounds (lb) = 1.1 tons (t)
1 kilogram (kg) = 2.20 pounds (lb)
1 pound (lb) = 454 grams (g)
1 gram (g) = 0.035 ounce (oz)

ENERGY AND POWER

Metric

1 kilojoule (kJ) = 1,000 joules (J)
1 kilocalorie (kcal) = 1,000 calories (cal)
1 calorie (cal) = 4,184 joules (J)

Metric-English

1 kilojoule (kJ) = 0.949 British thermal unit (Btu)
1 kilojoule (kJ) = 0.000278 kilowatt-hour (kW-h)
1 kilocalorie (kcal) = 3.97 British thermal units (Btu)
1 kilocalorie (kcal) = 0.00116 kilowatt-hour (kW-h)
1 kilowatt-hour (kW-h) = 860 kilocalories (kcal)
1 kilowatt-hour (kW-h) = 3,400 British thermal units (Btu)
1 quad (Q) = 1,050,000,000,000,000 kilojoules (kJ)
1 quad (Q) = 2,930,000,000,000 kilowatt-hours (kW-h)

TEMPERATURE CONVERSIONS

Fahrenheit (°F) to Celsius (°C): °C = F((°F − 32.0) ÷ 1.80)
Celsius (°C) to Fahrenheit (°F): °F = F((°C × 1.80) + 32.0)

MAJOR U.S. RESOURCE CONSERVATION AND ENVIRONMENTAL LEGISLATION

GENERAL

National Environmental Policy Act of 1969 (NEPA)
International Environmental Protection Act of 1983

ENERGY

Energy Policy and Conservation Act of 1975
National Energy Act of 1978, 1980
National Appliance Energy Conservation Act of 1987
Energy Policy Act of 1992

WATER QUALITY

Water Quality Act of 1965
Water Resources Planning Act of 1965
Federal Water Pollution Control Acts of 1965, 1972
Ocean Dumping Act of 1972
Safe Drinking Water Act of 1974, 1984, 1996
Water Resources Development Act of 1986
Clean Water Act of 1977, 1987
Ocean Dumping Ban Act of 1988

AIR QUALITY

Clean Air Act of 1963, 1965, 1970, 1977, 1990
Pollution Prevention Act of 1990

NOISE CONTROL

Noise Control Act of 1965
Quiet Communities Act of 1978

RESOURCES AND SOLID WASTE MANAGEMENT

Solid Waste Disposal Act of 1965
Resource Recovery Act of 1970
Resource Conservation and Recovery Act of 1976
Marine Plastic Pollution Research and Control Act of 1987

TOXIC SUBSTANCES

Hazardous Materials Transportation Act of 1975
Toxic Substances Control Act of 1976
Resource Conservation and Recovery Act of 1976
Comprehensive Environmental Response, Compensation, and
Liability (Superfund) Act of 1980, 1986
Nuclear Waste Policy Act of 1982

PESTICIDES

Federal Insecticide, Fungicide, and Rodenticide Control Act of 1972, 1988

WILDLIFE CONSERVATION

Lacey Act of 1900
Migratory Bird Treaty Act of 1918
Migratory Bird Conservation Act of 1929
Migratory Bird Hunting Stamp Act of 1934
Pittman-Robertson Act of 1937
Anadromous Fish Conservation Act of 1965
Fur Seal Act of 1966
Species Conservation Act of 1966, 1969
National Wildlife Refuge System Act of 1966, 1976, 1978
Marine Mammal Protection Act of 1972
Marine Protection, Research, and Sanctuaries Act of 1972
Endangered Species Act of 1973, 1982, 1985, 1988
Whale Conservation and Protection Study Act of 1976
Fishery Conservation and Management Act of 1976, 1978, 1982
Fish and Wildlife Improvement Act of 1978
Fish and Wildlife Conservation Act of 1980 (Nongame Act)

LAND USE AND CONSERVATION

Taylor Grazing Act of 1934
Wilderness Act of 1964
Multiple Use Sustained Yield Act of 1968
Wild and Scenic Rivers Act of 1968
National Trails System Act of 1968
National Coastal Zone Management Act of 1972, 1980
Forest Reserves Management Act of 1974, 1976
Forest and Rangeland Renewable Resources Act of 1974, 1978
Federal Land Policy and Management Act of 1976
National Forest Management Act of 1976
Soil and Water Conservation Act of 1977
Surface Mining Control and Reclamation Act of 1977
Antarctic Conservation Act of 1978
Endangered American Wilderness Act of 1978
Alaskan National Interests Lands Conservation Act of 1980
Coastal Barrier Resources Act of 1982
Food Security Act of 1985

APPENDIX 4

MAPPING CONCEPTS AND CONNECTIONS*

This textbook emphasizes the connections between environmental principles, problems, and solutions. One way to organize the material in various chapters and to understand such connections is to map the concepts in a particular chapter or portion of a chapter. Inside the front cover you will find a general concept map showing how the parts of this entire textbook are related. At the end of each chapter you will find a question asking you to make a chapter concept map. This appendix provides you with information about how to create such maps. Here are the basic steps in making a concept map:

1. List the key general and specific concepts by indenting specific concepts under more general concepts. Basically, make an outline of the material using section heads, subsection heads, and key terms.

2. Arrange the clusters of concepts with the more general concepts at the top of a page and the more specific concepts at the bottom of the page.

3. Circle key concepts, draw lines connecting them, and write labels on the lines that describe linkages between the concepts.

4. Draw in crosslinking connections (lines that relate various parts of the map) and write labels on the line, describing these connections.

5. Study the map to see how you might improve it. You might add clustering concepts and connections of your own. You might eliminate detail to make the big picture stand out. You might add more details to clarify certain concepts. You might add or alter broad crosslinking connections on the map as you think and learn more about the material.

Let's go through these steps using the concept of living systems discussed in Chapter 5 as an example.

1. List the appropriate general and specific concepts. Note that Section 5-1 is about

---

*This appendix was developed by **Jane Heinze-Fry** with assistance from G. Tyler Miller, Jr.

land systems (called biomes) and Section 5-2 is about aquatic systems. Use the list of section titles, subtitles, words in bold and italics, key drawings, and other terms that you think are particularly important. An outline of these two sections in which specific concepts are indented under more general ones would look something like Figure 1.

2. Arrange clusters of the concepts in Figure 1 with the more general ones at the top and more specific ones at the bottom. Concepts of the same general level should be at about the same height on the map. Figure 2 shows what this might look like. You might be able to do this on a single sheet of paper turned sideways, or you might find it easier to use large sheets of paper. Use a pencil to draw the map so that you can make changes. Note that the diagram has been simplified by eliminating some of the concepts in Figure 1 such as latitude and altitude, various types of temperate grasslands, and tropical scrub forests.

3. Circle key concepts, draw lines connecting them, and write labels on the lines that describe linkages between the concepts. You might decide to add or delete some of the concepts to make the map more logical, clear, and helpful to you. Figure 3 shows one way to do this. Note that in this figure the subcategories of deserts, grasslands, and forests are put together to improve clarity.

4. Draw in crosslinks that relate various parts of the map. Figure 4 illustrates one possibility. At this point you are trying to understand how the various concepts are connected to and interact with one another. Note that many of the human systems concepts interact positively or negatively with ecosystems and natural resource concepts (in terms of human value judgment). Note that Chapter 5 focuses on natural systems. As you learn more about how human systems interact with natural systems in this chapter and in other chapters, you should be able to add additional crosslinkages.

5. Study your map and try to improve it. You might choose to eliminate excessive

detail or add additional concepts and linkages to improve clarity. You might also find it useful to rearrange the position of concepts vertically or horizontally to eliminate crowding or to reduce confusion over too many connecting links. In Figure 5 note that the general concepts of *fresh water* and *salt water* have been added, and aquatic systems has been divided into two subcategories based on salinity (dissolved salt content) instead of four. To improve clarity, the concepts of *moving* and *still* under *fresh water* have also been added because these characteristics greatly affect the types of organisms living in such systems. Finally, concepts have been connected with some of those in Chapter 4 by including details about photosynthetic and chemosynthetic productivity because these activities form the basis of the food web relationships in terrestrial and aquatic systems.

Remember the following things about concept maps:

- They are particularly useful for showing the big picture and for clarifying particularly difficult concepts. They can also be used in studying for tests to help you sort out what is important and how concepts are related.

- They take time to make. Sometimes they are easy and seem almost to write themselves. Other times, they can be difficult and frustrating to draw.

- There is no single "best" map. Rarely do concept maps developed by different individuals look alike.

- They are works in progress that should change and improve over time as you get better at making them and develop a deeper understanding of how things are connected.

- Comparing your maps with those developed by others is a good way to learn how to improve your mapmaking skills and to come up with more connections.

- Making them is challenging and fun.

LAND SYSTEMS (Biomes)
climate
 precipitation
 temperature
 latitude
 altitude
deserts
 tropical
 temperate
 cold
 semidesert
grasslands
 tropical
  savanna
 temperate
  tall-grass
  short-grass
  pampas
  veld
  steppes
 polar (arctic tundra)
  permafrost

forests
 tropical
  rain
  deciduous
  scrub
 temperate deciduous
 evergreen coniferous

WATER SYSTEMS
ocean
 coastal
  coral reefs
  estuaries
  wetlands
  barrier islands
  beaches
   rocky shore
   barrier beach

open sea
 euphotic zone
 bathyal zone
 abyssal zone
freshwater lakes
 eutrophic
 oligotrophic
 mesotrophic
 littoral zone
 limnetic zone
 profundal zone
 benthic zone
 thermal stratification
  thermoclines
  turnover
freshwater streams
 surface water
 runoff
 watershed
inland wetlands
 year-round
 seasonal

Figure 1 List of concepts for Living Systems map.

LIVING SYSTEMS

Land Systems (Biomes)
 climate
  precipitation
  temperature
  ~~latitude~~
  ~~altitude~~

                                    ocean

Water Systems

deserts   grasslands   forests
 tropical   tropical   tropical
 temperate   ~~savanna~~   rain
 cold   temperate   deciduous
 ~~semidesert~~   ~~tall-grass~~   ~~scrub~~
                  ~~short-grass~~   temperate deciduous
                  ~~pampas~~   evergreen coniferous
                  ~~veld~~
                  ~~steppes~~
                 polar (arctic tundra)
                  ~~permafrost~~

coastal   open sea   inland wetlands   freshwater streams
 coral reefs   euphotic zone   year-round   ~~surface water~~
 estuaries   bathyal zone   seasonal   ~~runoff~~
 wetlands   abyssal zone   ~~watershed~~
 barrier islands freshwater lakes
 beaches   eutrophic
  rocky shore   oligotrophic
  barrier beach   mesotrophic
                  littoral zone
                  limnetic zone
                  profundal zone
                  benthic zone
                  thermal stratification
                   thermoclines
                   turnover

Figure 2 Clusters of concepts are arranged in a hierarchy from more general at the top to more specific at the bottom.

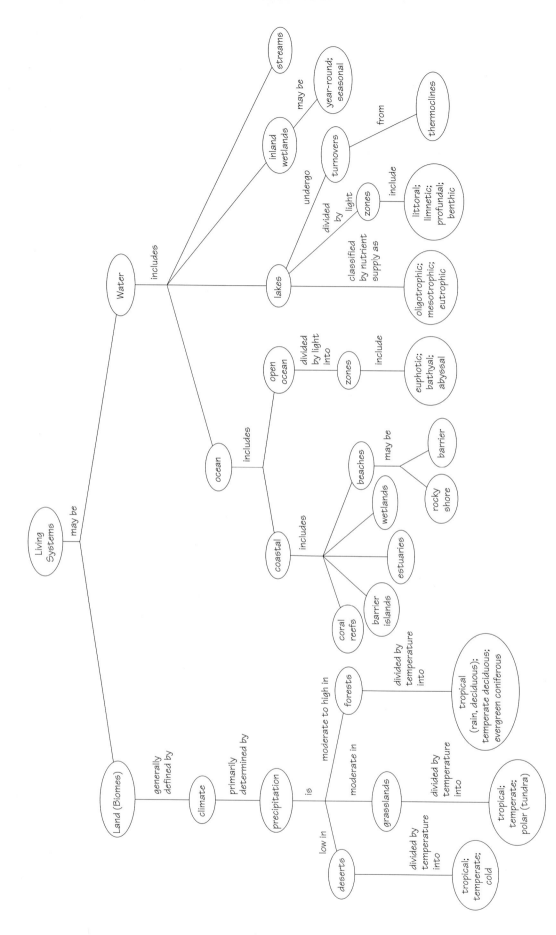

Figure 3 Concepts are circled and linkages are drawn.

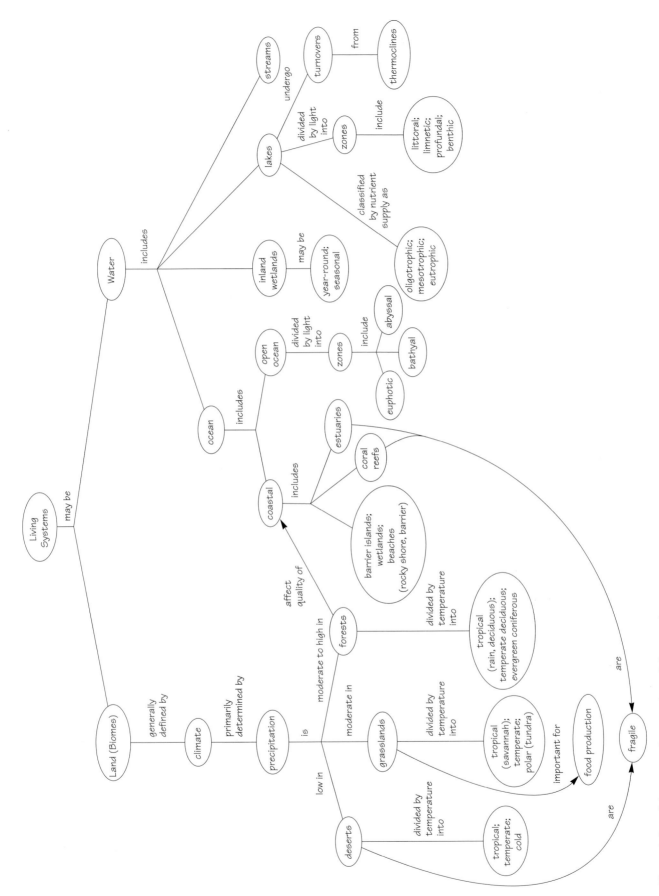

Figure 4 Crosslinkages are drawn and map is improved.

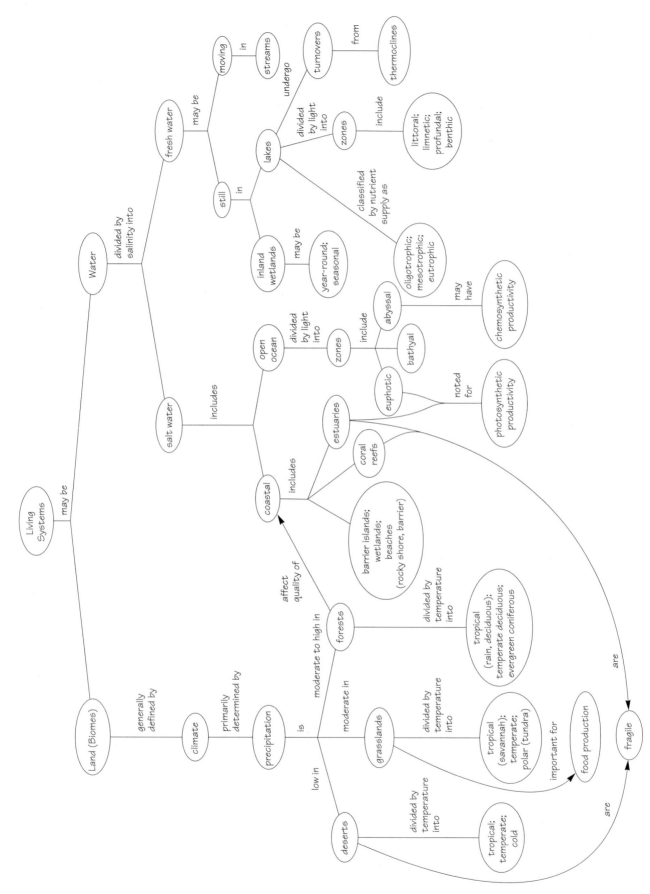

Figure 5 Map is further refined. New hierarchical clusters may emerge.

APPENDIX 5

INDIVIDUALS MATTER: WORKING WITH THE EARTH

This is a list of things individuals can do based on suggestions from a wide variety of environmentalists. It is not meant to be a list of things you must do but a list of actions you might consider. Don't feel guilty about the things you are not doing. Start off by picking the ones that you are willing to do and that you feel will have the most impact. Many of these suggestions are controversial. Carefully evaluate each action to see whether it fits in with your beliefs. Based on practicality and my own beliefs, I do some of these things. Each year I look over this list and try to add several new items.

PRESERVING BIODIVERSITY AND PROTECTING THE SOIL

- *Develop a plan for the sustainable use of any forested area you own.*
- *Plant trees on a regular basis and take care of them.*
- *Reduce the use of wood and paper products, recycle paper products, and buy recycled paper products.*
- *Don't buy furniture, doors, flooring, window frames, paneling, or other products made from tropical hardwoods such as teak or mahogany.* Look for the Good-Wood seal given by Friends of the Earth, or consult the *Wood User's Guide* by Pamela Wellner and Eugene Dickey (San Francisco: Rainforest Action Network, 1991).
- *Don't purchase wood and paper products produced by cutting remaining old-growth forests in the tropics and elsewhere.* Information on such products can be obtained from the Rainforest Action Network, Rainforest Alliance, and Friends of the Earth (Appendix 1).
- *Help rehabilitate or restore a degraded area of forest near your home.*
- *Don't buy furs, ivory products, items made of reptile skin, tortoiseshell jewelry, and materials of endangered or threatened animal species.*
- *When building a home, save all the trees possible.* Require that the contractor disturb as little soil as possible, set up barriers that catch any soil eroded during construction, and save and replace any topsoil removed instead of hauling it off and selling it.

- *Landscape areas not used for gardening with a mix of wildflowers, herbs (for cooking and for repelling insects), low-growing ground cover, small bushes, and other forms of vegetation natural to the area.*
- *Set up a compost bin and use it to produce soil conditioner for yard and garden plants.*

PROMOTING SUSTAINABLE AGRICULTURE AND REDUCING PESTICIDE USE

- *Waste less food.* An estimated 25% of all food produced in the United States is wasted.
- *Eat lower on the food chain* by reducing or eliminating meat consumption to reduce its environmental impact.
- *If you have a dog or a cat, don't feed it canned meat products.* Balanced-grain pet foods are available.
- *Reduce the use of pesticides on agricultural products by asking grocery stores to stock fresh produce and meat produced by organic methods.*
- *Grow some of your own food using organic farming techniques and drip irrigation to water your crops.*
- *Compost your food wastes.*
- *Think globally, eat locally.* Whenever possible, eat food that is locally grown and in season. This supports your local economy, gives you more influence over how the food is grown (by either organic or conventional methods), saves energy required to transport food over long distances, and reduces the fossil-fuel use and pollution. If you deal directly with local farmers, you can also save money.
- *Give up the idea that the only good bug is a dead bug.* Recognize that insect species keep most of the populations of pest insects in check and that full-scale chemical warfare on insect pests wipes out many beneficial insects.
- *Don't insist on perfect-looking fruits and vegetables.* These are more likely to contain high levels of pesticide residues.
- *Use pesticides in your home only when absolutely necessary, and use them in the smallest amount possible.*

- *Don't become obsessed with having the perfect lawn.* About 40% of U.S. lawns are treated with pesticides. These chemicals can cause headaches, dizziness, nausea, eye trouble, and even more acute effects in sensitive people (including children who play on treated lawns and in parks).
- *If you hire a lawn-care company, use one that relies only on organic methods, and get its claims in writing.*

SAVING ENERGY AND REDUCING OUTDOOR AIR POLLUTION

- *Reduce use of fossil fuels.* Drive a car that gets at least 15 kilometers per liter (35 miles per gallon), join a car pool, and use mass transit, walking, and bicycling as much as possible. This will reduce emissions of CO_2 and other air pollutants, will save energy and money, and can improve your health.
- *Plant and care for trees to help absorb CO_2.* During its lifetime, the average tree absorbs enough CO_2 to offset the amount produced by driving a car 42,000 kilometers (26,000 miles).
- *Don't use electricity to heat space or water.*
- *Insulate new or existing houses heavily, caulk and weatherstrip to reduce air infiltration and heat loss, and use energy-efficient windows.* Add an air-to-air heat exchanger to minimize indoor air pollution.
- *Obtain as much heat and cooling as possible from natural sources,* especially sun, wind, geothermal energy, and trees.
- *Buy the most energy-efficient homes, lights, cars, and appliances available. Evaluate them only in terms of lifetime cost.*
- *Turn down the thermostat on water heaters to 43–49°C (110–120°F) and insulate hot water pipes.*
- *Lower the cooling load on an air conditioner by increasing the thermostat setting, installing energy-efficient lighting, using floor and ceiling fans, and using whole-house window or attic fans to bring in outside air (especially at night, when temperatures are cooler).*

REDUCING EXPOSURE TO INDOOR AIR POLLUTANTS

- *Test for radon and take corrective measures as needed.*

- *Install air-to-air heat exchangers or regularly ventilate your house by opening windows.*

- *At the beginning of the winter heating season, test your indoor air for formaldehyde when the house is closed up. To locate a testing laboratory in your area, write to Consumer Product Safety Commission, Washington, DC 20207, or call 301-492-6800.*

- *Don't buy furniture and other products containing formaldehyde. Use low-emitting formaldehyde or nonformaldehyde building materials.*

- *Reduce indoor levels of formaldehyde and other toxic gases by growing certain house plants. Examples are the spider or airplane plant (removes 96% of carbon monoxide), aloe vera (90% of formaldehyde), banana (89% of formaldehyde), elephant ear philodendron (86% of formaldehyde), ficus (weeping fig, 47% of formaldehyde), golden porthos (67% of formaldehyde and benzene and 75% of carbon monoxide), Chinese evergreen (92% of toluene and 81% of benzene), English ivy (90% of benzene), peace lily (80% of benzene and 50% of trichloroethylene), and Janet Craig (corn plant, 79% of benzene). (Toxic removal figures indicate percentage of toxin removed by one plant in a 24-hour period in a 3.4-cubic-meter [12-cubic-foot] space). Plants should be potted with a mixture of soil and granular charcoal (which absorbs organic air pollutants).*

- *Consider not using carpeting and using wood or linoleum floors instead. New synthetic carpeting releases vapors from more than 100 volatile organic compounds. New and old carpeting is a haven for microbes (many of them highly allergenic), dust, and traces of lead and pesticides brought in by shoes.*

- *Remove your shoes before entering your house. This reduces inputs of dust, lead, and pesticides.*

- *Test your house or workplace for asbestos fiber levels and for any crumbling asbestos materials if it was built before 1980. Don't buy a pre-1980 house without having its indoor air tested for asbestos and lead. To get a free list of certified asbestos laboratories that charge $25–50 to test a sample call the EPA's Toxic Substances Control Hotline at 202-554-1404.*

- *Don't store gasoline, solvents, or other volatile hazardous chemicals inside a home or attached garage.*

- *Don't use aerosol spray products and commercial room deodorizers or air fresheners.*

- *If you smoke, do it outside or in a closed room vented to the outside.*

- *Make sure that wood-burning stoves, fireplaces, and kerosene- and gas-burning heaters are properly installed, vented, and maintained. Install carbon monoxide detectors in all sleeping areas.*

SAVING WATER

- *For existing toilets, reduce the amount of water used per flush by putting into each tank a tall plastic container weighted with a few stones or by buying and inserting a toilet dam.*

- *Install water-saving toilets that use no more than 6 liters (1.6 gallons) per flush.*

- *Flush toilets only when necessary. Consider using the advice found on a bathroom wall in a drought-stricken area: "If it's yellow, let it mellow; if it's brown, flush it down."*

- *Install water-saving showerheads and flow restrictors on all faucets. If a 3.8-liter (1-gallon) jug can be filled by your showerhead in less than 15 seconds, you need a more efficient fixture.*

- *Check frequently for water leaks in toilets and pipes, and repair them promptly. A toilet must be leaking more than 940 liters (250 gallons) per day before you can hear the leak. To test for toilet leaks, add a water-soluble vegetable dye to the water in the tank, but don't flush. If you have a leak, some color will show up in the bowl's water within a few minutes.*

- *Turn off sink faucets while brushing teeth, shaving, or washing.*

- *Wash only full loads of clothes; if smaller loads must be used, use the lowest possible water-level setting.*

- *When buying a new washer, choose one that uses the least amount of water and fills up to different levels for loads of different sizes. Front-loading clothes models use less water and energy than comparable top-loading models.*

- *Use automatic dishwashers for full loads only. Also, use the short cycle and let dishes air dry to save energy and money.*

- *When washing many dishes by hand, don't let the faucet run. Instead, use one filled dishpan or sink for washing and another for rinsing.*

- *Keep one or more large bottles of water in the refrigerator rather than running water from the tap until it gets cold enough for drinking.*

- *Don't use a garbage disposal system—a large user of water. Instead, consider composting your food wastes.*

- *Wash a car from a bucket of soapy water, and use the hose for rinsing only. Use a commercial car wash that recycles its water.*

- *Sweep walks and driveways instead of hosing them off.*

- *Reduce evaporation losses by watering lawns and gardens in the early morning or evening, rather than in the heat of midday or when it's windy.*

- *Use drip irrigation and mulch for gardens and flower beds. Better yet, landscape with native plants adapted to local average annual precipitation so that watering is unnecessary.*

REDUCING WATER POLLUTION

- *Use manure or compost instead of commercial inorganic fertilizers to fertilize garden and yard plants.*

- *Use biological methods or integrated pest management to control garden, yard, and household pests.*

- *Use low-phosphate, phosphate-free, or biodegradable dishwashing liquid, laundry detergent, and shampoo.*

- *Don't use water fresheners in toilets.*

- *Use less harmful substances instead of commercial chemicals for most household cleaners (Table 13-2).*

- *Don't pour pesticides, paints, solvents, oil, antifreeze, or other products containing harmful chemicals down the drain or onto the ground. Contact your local health department about disposal.*

- *If you get water from a private well or suspect that municipal water is contaminated, have it tested by an EPA-certified laboratory for lead, nitrates, trihalomethanes, radon, volatile organic compounds, and pesticides.*

- *If you have a septic tank, have it cleaned out every 3–5 years by a reputable contractor so that it won't contribute to groundwater pollution.*

REDUCING SOLID WASTE AND HAZARDOUS WASTE

- *Buy less by asking yourself whether you really need a particular item.*

- *Buy things that are reusable, recyclable, or compostable, and be sure to reuse, recycle, and compost them.*

- *Buy beverages in refillable glass containers instead of cans or throwaway bottles. Urge companies and legislators to make refillable plastic (PET) bottles available in the United States.*

- *Use reusable plastic or metal lunch boxes and metal or plastic garbage containers without throwaway plastic liners (unless such liners are required for garbage col-*

lection).

- *Carry sandwiches and store food in the refrigerator in reusable containers instead of wrapping them in aluminum foil or plastic wrap.*

- *Use rechargeable batteries and recycle them when their useful life is over. In 1993 Rayovac began selling mercury-free, rechargeable alkaline batteries (Renewal batteries) that outperform conventional nickel-cadmium rechargeable batteries.*

- *Carry groceries and other items in a reusable basket, a canvas or string bag, or a small cart.*

- *Use reusable sponges and washable cloth napkins, dish towels, and handkerchiefs instead of paper ones.*

- *Stop using throwaway paper and plastic plates, cups, and eating utensils, and other disposable items when reusable or refillable versions are available.*

- *Buy recycled goods, especially those made by primary recycling, and then make an effort to recycle them. If you're not buying recycled materials, you're not recycling.*

- *Reduce the amount of junk mail you get. Do this (as several million Americans have done) at no charge by contacting the Mail Preference Service, Direct Marketing Association, 11 West 42nd St., P.O. Box 3681, New York, NY 10163-3861 (212-768-7277) and asking that your name not be sold to large mailing-list companies. Of the junk mail you do receive, recycle as much of the paper as possible.*

- *Buy products in concentrated form whenever possible.*

- *Choose items that have the least packaging—or better yet, no packaging ("nude products")*

- *Don't buy helium-filled balloons that end up as litter. Urge elected officials and school administrators to ban balloon releases except for atmospheric research and monitoring.*

- *Lobby local officials to set up a community composting program.*

- *Use pesticides and other hazardous chemicals (Table 13-1) only when absolutely necessary and in the smallest amount possible.*

- *Use less hazardous (and usually cheaper) cleaning products (Table 13-2).*

- *Don't dispose of hazardous chemicals by flushing them down the toilet, pouring them down the drain, burying them, throwing them into the garbage, or dumping them down storm drains. Consult your local health department or environmental agency for safe disposal methods.*

COMMUNICATING WITH ELECTED OFFICIALS

- Find out their names and addresses. Then write, call, fax, or E-mail them. Contact a senator by writing: The Honorable ___, U.S. Senate, Washington, DC 20510; Tel: 202-224-3121, Web site: http://thomas.loc.gov. Contact a representative by writing: The Honorable ___, U.S. House of Representatives, Washington, DC 20510; Tel: 202-225-3121, Web site: http://thomas.loc.gov. Contact the president by writing: President ___, The White House, 1600 Pennsylvania Ave. NW, Washington, DC 20500; Tel. 202-456-1414, Comment line: 202-456-1111, Fax: 202-456-2461, E-mail: president@whitehouse.gov, Web site: http://www.whitehouse.gov.

- When you write a letter, use your own words, be brief and courteous, address only one issue, and ask the elected official to do something specific (such as cosponsoring, supporting, or opposing certain bills). Give reasons for your position, explain its effects on you and your district, try to offer alternatives, share any expert knowledge you have, and ask for a response. Be sure to include your name and return address.

- After your representatives have cast votes* supporting your position, send them a short note of thanks.

- Call and ask to speak to a staff member who works on the issue you are concerned about: for the White House, 202-456-1414 (Web site: http://www.whitehouse.gov); for the U.S. Senate, 202-224-3121; for the House of Representatives, 202-225-3121. The E-mail address for the U.S. Congress is Web server: mail: http://thomas.loc.gov:.

- Once a desirable bill is passed, call or write to urge the president not to veto it. Urge the members of the appropriations committee to appropriate enough money to implement the law—a crucial decision.

- Monitor and influence action at the state and local levels, where all federal and state laws are either ignored or enforced. As Thomas Jefferson said, "The execution of laws is more important than the making of them."

- Get others who agree with your position to contact their elected officials.

---

*Each year the League of Conservation Voters (P.O. Box 500, Washington, DC 20077; 202-785-VOTE) publishes an *Environmental Scorecard* that rates all members of Congress on how they voted on environmental issues.

INTERNET SOURCES AND FURTHER READINGS

GENERAL SOURCES OF ENVIRONMENTAL INFORMATION

Internet

Environmental Gophers:
gopher://path.net:8001/11/.subject/Environment/

Exploring the Internet: http://www.globalcenter.net/gcweb/tour.html

Global Resource Information Database:
http://www.grid.unep.ch/gridhome.html/

Governmental WWW Resources Relating to the Environment: http://www.envirolink.org/envirogov.html/

Information Center for the Environment (University of California, Davis): http://ice.ucdavis.edu/

Infoterra: United Nations Environment Programme: http://pan.cedar.univie.ac.at/gopher/UNEP/

National Library for the Environment: http://www.cnie.org/nle/

Usenet Keyword Search: http://www.excite.com

Worldwatch Institute: http://www.econet.apc.org/igc/en.html

WWW Virtual Library (University of Virginia):
http://ecosys.drdr.virginia.edu/Environment.html

Yahoo! Search Engine: http://www.yahoo.com

Readings

Ashworth, William. 1991. *The Encyclopedia of Environmental Studies.* New York: Facts on File.

Beacham, W., ed. 1993. *Beacham's Guide to Environmental Issues and Sources.* 5 vols., Washington, D.C.: Beacham.

British Petroleum. Annual. *BP Statistical Review of World Energy.* New York: BP America.

Brown, Lester R., et al. Annual. *State of the World.* New York: Norton.

Brown, Lester R., et al. Annual. *Vital Signs.* New York: Norton.

Kirdon, Michael, and Ronald Segal. 1991. *The New State of the World Atlas.* 4th ed. New York: Simon & Schuster.

Kurland, Daniel J., and Jane Heinze-Fry. 1996. *Introduction to the Internet.* Belmont, Calif.: Wadsworth. **Available as a supplement for use with this book.**

Lean, Geoffrey, and Don Hinrichsen, eds. 1994. *Atlas of the Environment.* 2d ed. New York: HarperCollins.

Marien, Michael, ed. 1996. *Environmental Issues and Sustainable Futures: A Critical Guide to Recent Books, Reports, and Periodicals.* Bethesda, Md.: World Future Society.

---

* For a more detailed list, consult my longer book, *Living in the Environment*, Belmont, Calif.: Wadsworth Publishing.

Paehilke, Robert, ed. 1995. *Conservation and Environmentalism: An Encyclopedia.* New York: Garland.

Population Reference Bureau. Annual. *World Population Data Sheet.* Washington, D.C.: Population Reference Bureau.

Rittner, Don. 1992. *Ecolinking: Everyone's Guide to Online Environmental Information.* Berkeley, Calif.: Peachpit Press.

Schupp, Jonathan F. 1995. *Environmental Guide to the Internet.* Rockville, Md.: Government Institutes, Inc.

Seager, Joni, et al. 1995. *The New State-of-the-Earth Atlas.* New York: Simon & Schuster.

United Nations. Annual. *Demographic Yearbook.* New York: United Nations.

United Nations Children's Fund (UNICEF). Annual. *The State of the World's Children.* New York: UNICEF.

United Nations Environment Programme (UNEP). Annual. *State of the Environment.* New York: UNEP.

United Nations Population Fund. Annual. *The State of World Population.* New York: United Nations Population Fund.

U.S. Bureau of the Census. Annual. *Statistical Abstract of the United States.* Washington, D.C.: U.S. Bureau of the Census.

World Health Organization (WHO). Annual. *World Health Statistics.* Geneva, Switzerland: WHO.

World Resources Institute and International Institute for Environment and Development. Biannual. *World Resources.* New York: Basic Books.

1 / ENVIRONMENTAL PROBLEMS AND THEIR CAUSES

Internet

See also sites listed under general information.

Earthwatch: http://www.unep.ch/earthw.html/

The ECOLOGY Channel: http://www.ecology.com/index.htm

E Magazine:
http://www.igc.apc.org/emagazine/ehome.html

Envirolink: http://envirolink.org/

Environmental Resources (Sonoma State University): http://www.sonoma.edu/ensp/hot_list.html

National Environmental Information Service (Environmental Protection Agency): http://www.earthcycle.com/g/p/earthcycle//

Planet Earth Home Page: http://www.nosc.mil/planet_earth/info.html/

World Population Clock: http://sunsite.unc.edu/lunarbin/worldpop/

Readings

Bailey, Ronald, ed. 1995. *The True State of the Planet.* New York: The Free Press.

Barney, Gerald G., et al. 1993. *Global 2000 Revisited: What Shall We Do?* Arlington, Va.: Millennium Institute.

Brown, Lester R., and Hal Kane. 1994. *Full House: Reassessing the Earth's Population Carrying Capacity.* New York: Norton.

Chivian, Eric, et al. 1993. *Critical Condition: Human Health and the Environment.* Cambridge, Mass.: MIT Press.

Commoner, Barry. 1994. *Making Peace with the Planet.* Rev. ed. New York: New Press.

Easterbrook, Gregg. 1995. *A Moment on the Earth: The Coming Age of Environmental Optimism.* New York: Viking Press.

Ehrlich, Paul R., and Anne H. Ehrlich. 1990. *The Population Explosion.* New York: Doubleday.

Ehrlich, Paul R., and Anne H. Ehrlich. 1991. *Healing the Planet.* Reading, Mass.: Addison-Wesley.

Ehrlich, Paul R., and Anne H. Ehrlich. 1996. *Betrayal of Science and Reason: How Anti-Environmental Rhetoric Threatens Our Future.* Covelo, Calif.: Island Press.

Gordon, Anita, and David Suzuki. 1991. *It's a Matter of Survival.* Cambridge, Mass.: Harvard University Press.

Gore, Al. 1992. *Earth in the Balance: Ecology and the Human Spirit.* Boston: Houghton Mifflin.

Hardin, Garrett. 1968. "The Tragedy of the Commons." *Science,* vol. 162, 1243–48.

Hardin, Garrett. 1993. *Living Within Limits: Ecology, Economics, and Population Taboos.* New York: Oxford University Press.

Lehr, Jay, ed. 1994. *Rational Readings on Environmental Concerns.* New York: Van Nostrand Reinhold.

Markley, Oliver W., and Walter R. McCuan, eds. 1996. *21st Century Earth: Opposing Viewpoints.* Boston: Greenhaven Press.

Meadows, Donella. 1991. *The Global Citizen.* Covelo, Calif.: Island Press.

Meadows, Donella H, et al. 1992. *Beyond the Limits: Confronting Global Collapse, Envisioning a Sustainable Future.* White River Junction, Vt.: Chelsea Green.

Myers, Norman, ed. 1993a. *Gaia: An Atlas of Planet Management.* Garden City, N.Y.: Anchor Press/Doubleday.

Myers, Norman. 1993b. *Ultimate Security: The Environmental Basis of Political Stability.* New York: Norton.

Myers, Norman, and Julian Simon. 1994. *Scarcity or Abundance? A Debate on the Environment.* New York: Norton.

Porritt, Jonathan. 1991. *Save the Earth.* Atlanta, Ga.: Turner Publishing.

Quinn, Daniel. 1992. *Ishmael.* New York: Bantam/Turner.

Simon, Julian L., ed. 1995. *The True State of the Planet.* Cambridge, Mass.: Basil Blackwell.

Simon, Julian L. 1996. *The Ultimate Resource 2.* Princeton, N.J.: Princeton University Press.

Suzuki, David. 1994. *Time to Change*. Toronto: Stoddard.

Tobias, Michael. 1994. *World War III: Population and the Biosphere at the End of the Millennium*. Santa Fe, N.M.: Bear & Company.

United Nations. 1997. *Global Environmental Outlook*. New York: Oxford University Press.

Wagner, Travis. 1993. *In Our Backyard: A Guide to Understanding Pollution and Its Effects*. New York: Van Nostrand Reinhold.

2 / CULTURAL CHANGES, WORLDVIEWS, ETHICS, AND SUSTAINABILITY

Internet

Center for Environmental Philosophy: http://www.cep.unt.edu

The Earth Council: http://terra.ecouncil.ac.cr/

Environmental Ethics: http://www.cep.unt.edu/

Environmental Philosophers: http://www.envirolink.org/elib/enviroethics/ecophiloindex.html/

Sustainable Societies: http://www.af.nfr.no/susteur/

WWW Virtual Library: History of Science, Technology, and Medicine: http://www.asap.unimelb.edu.au/hstm/hstm_ove.html/

WWW Virtual Library (University of Virginia): http://ecosys.drdr.virginia.edu/Environment.html

Yahoo: Environmental Ethics Gopher Connections: http://www.yahoo.com/Arts/Humanities/Philosophy/Ethics/Environmental_Ethics/

Readings

Berry, Thomas. 1988. *The Dream of the Earth*. San Francisco: Sierra Club.

Berry, Wendell. 1990. *What Are People For?* Berkeley, Calif.: North Point Press.

Bookchin, Murray. 1990. *Remaking Society: Pathways to a Green Future*. San Francisco: South End Press.

Callicott, J. Baird. 1994. *Earth's Insights (A Survey of Ecological Ethics from the Mediterranean Basin to the Australian Outback)*. Berkeley: University of California Press.

Carson, Rachel. 1962. *Silent Spring*. Boston: Houghton Mifflin.

Clark, M. E. 1989. *Ariadne's Thread: The Search for New Models of Thinking*. New York: St. Martin's Press.

Cobb, John B., Jr. 1994. *Is It Too Late? A Theology of Ecology*. Rev. ed. Denton, Tex.: Environmental Ethics Books.

Cohen, Michael J. 1989. *Connecting with Nature: Creating Moments That Let Earth Teach*. Eugene, Ore.: World Peace University.

Desjardins, Joseph R. 1993. *Environmental Ethics*. Belmont, Calif.: Wadsworth.

Devall, Bill, and George Sessions. 1985. *Deep Ecology: Living as if Nature Mattered*. Salt Lake City: Gibbs H. Smith.

Ehrenfeld, David. 1993. *Beginning Again: People and Nature in the New Millennium*. New York: Oxford University Press.

Elderige, Niles. 1995. *Dominion*. New York: Henry Holt.

Elgin, Duane. 1993. *Voluntary Simplicity: Toward a Way of Life That Is Outwardly Simple, Inwardly Rich*. Rev. ed. New York: William Morrow.

Ferkiss, Victor. 1993. *Nature, Technology, and Society: Cultural Roots of the Current Environmental Crisis*. New York: New York University Press.

Foster, John Bellamy. 1995. *The Vulnerable Planet: A Short Economic History of the Planet*. New York: Monthly Review Press.

Goudie, Andrew. 1993. *The Human Impact on the Natural Environment*. 3d ed. Cambridge, Mass.: MIT Press.

Hallman, David G. 1994. *Ecotheology: Voices from South and North*. New York: Orbis Books.

Hardin, Garrett. 1978. *Exploring New Ethics for Survival*. 2d ed. New York: Viking Press.

Hargrove, Eugene C. 1989. *Foundations of Environmental Ethics*. Englewood Cliffs, N.J.: Prentice Hall.

Hayden, Tom. 1996. *The Lost Gospel of the Earth: A Call for Renewing Nature, Spirit, and Politics*. San Francisco: Sierra Club Books.

Hunter, J. Robert. 1997. *Simple Things Won't Save the Earth*. Austin: University of Texas Press.

Irvine, Sandy. 1989. *Beyond Green Consumerism*. London: Friends of the Earth.

Johnson, Huey D. 1995. *Green Plans: Greenprint for Sustainability*. Lincoln: University of Nebraska Press.

Johnson, Warren. 1985. *The Future Is Not What It Used to Be: Returning to Traditional Values in an Age of Scarcity*. New York: Dodd, Mead.

Kellert, Stephen R. 1996. *The Value of Life: Biological Diversity and Human Society*. Covelo, Calif.: Island Press.

Kennedy, Paul. 1993. *Preparing for the Twenty-First Century*. New York: Random House.

Klenig, John. 1991. *Valuing Life*. Princeton, N.J.: Princeton University Press.

Leopold, Aldo. 1949. *A Sand County Almanac*. New York: Oxford University Press.

Levering, Frank, and Wanda Urbanska. 1993. *Simple Living: One Couple's Search for a Better Life*. New York: Penguin.

Lewis, Martin W. 1992. *Green Delusions: An Environmentalist Critique of Radical Environmentalism*. Durham, N.C.: Duke University Press.

Livingston, John. 1973. *One Cosmic Instant: Man's Fleeting Supremacy*. Boston: Houghton Mifflin.

Milbrath, Lester W. 1989. *Envisioning a Sustainable Society*. Albany: State University of New York Press.

Milbrath, Lester W. 1995. *Learning to Think Environmentally While There Is Still Time*. Albany: State University of New York Press.

Naar, Jon. 1990. *Design for a Livable Planet*. New York: Harper & Row.

Nabhan, Gary Paul, and Stephen Trimble. 1994. *The Geography of Childhood: Why Children Need Wild Places*. Boston: Beacon Press.

Naess, Arne. 1989. *Ecology, Community, and Lifestyle*. New York: Cambridge University Press.

Nagpal, Tanvi, and Camilla Foltz, eds. 1995. *A World of Difference: Giving Voices to Visions of Sustainability*. Washington, D.C.: World Resources Institute.

Nash, Roderick. 1988. *The Rights of Nature: A History of Environmental Ethics*. Madison: University of Wisconsin Press.

National Commission on the Environment. 1993. *Choosing a Sustainable Future*. Covelo, Calif.: Island Press.

Norton, Bryan G. 1991. *Toward Unity Among Environmentalists*. New York: Oxford University Press.

O'Riordan, Timothy, and James Cameron, eds. 1994. *Interpreting the Precautionary Principle*. East Haven, Conn.: Earthscan.

Orr, David. 1992. *Ecological Literacy*. Ithaca: State University of New York Press.

Orr, David. 1994. *Earth in Mind: On Education, Environment, and the Human Prospect*. Covelo, Calif.: Island Press.

Pearce, David. 1995. *Blueprint 4: Sustaining the Earth*. East Haven, Conn. Earthscan.

Ponting, Clive. 1992. *A Green History of the World. The Environment and the Collapse of Great Civilizations*. New York: St. Martin's Press.

Quinn, Daniel. 1992. *Ishmael*. New York: Bantam/Turner.

Rees, William, and Mathis Wackernagel. 1995. *Our Ecological Footprint*. Philadelphia: New Society.

Rolston, Holmes, III. 1994. *Conserving Natural Values*. New York: Columbia University Press.

Rubin, Charles T. 1994. *The Green Crusade: Rethinking the Roots of Environmentalism*. New York: Free Press.

Sale, Kirkpatrick. 1990. *Conquest of Paradise*. New York: Alfred A. Knopf.

Schumacher, E. F. 1973. *Small Is Beautiful: Economics as if People Mattered*. New York: Harper & Row.

Sessions, George, ed. 1994. *Deep Ecology for the Twenty-First Century*. Boston: Shambhala.

Shabecoff, Philip. 1993. *A Fierce Green Fire: The American Environmental Movement*. New York: Hill & Wang.

Swimme, Brian, and Thomas Berry. 1992. *The Universe Story: From the Primordial Flaring Forth to the Ecozoic Era*. San Francisco: HarperCollins.

Thomas, Lewis. 1992. *The Fragile Species*. New York: Scribner.

Thomashow, Mitchell. 1995. *Ecological Identity: Becoming a Reflective Environmentalist*. Cambridge, Mass.: MIT Press.

Tucker, Mary Evelyn, and John A. Grim, eds. 1994. *Worldviews and Ecology: Religion, Philosophy, and the Environment*. Maryknoll, N.Y.: Orbis Books.

Van DeVeer, Donald, and Christine Pierce. 1994. *The Environmental Ethics and Policy Book: Philosophy, Ecology, Economics*. Belmont, Calif.: Wadsworth.

Worster, Donald. 1993. *The Wealth of Nature: Environmental History and the Ecological Imagination*. New York: Oxford University Press.

3 / SCIENCE, SYSTEMS, MATTER, AND ENERGY

Internet

Environmental Problem Solving: http://ameritec.com/at/

7.01 Hypertext Chemistry Review: http://esg-www.mit.edu:8001/esgbio/chem/review.html/

National Environmental Research Council: http://www.nerc.ac.uk/

Science and Technology: http://gnn.com/wic/wics/sci.new.html/

WWW Virtual Library: History of Science, Technology, and Medicine: http://www.asap.unimelb.edu.au/hstm/hstm_ove.html/

WWW Virtual Library (University of Virginia): http://ecosys.drdr.virginia.edu/Environment.html

Readings

Bauer, Henry H. 1992. *Scientific Literacy and the Myth of the Scientific Method*. Urbana: University of Illinois Press.

Carey, Stephen S. 1994. *A Beginner's Guide to Scientific Method*. Belmont, Calif.: Wadsworth. **Available as a supplement for use with this book.**

Heinze-Fry, Jane. 1996. *An Introduction to Critical Thinking*. Belmont, Calif.: Wadsworth. **Available as a supplement for use with this book.**

Kuhn, Thomas S. 1970. *The Structure of Scientific Revolutions.* 2d ed. Chicago: University of Chicago Press.

Miller, G. Tyler, Jr., and David G. Lygre. 1991. *Chemistry: A Contemporary Approach.* 3d. ed. Belmont, Calif.: Wadsworth.

Rifkin, Jeremy. 1989. *Entropy: Into the Greenhouse World: A New World View.* New York: Bantam.

Zimmerman, Michael. 1995. *Science, Nonscience, and Nonsense: Approaching Environmental Literacy.* Baltimore, Md.: Johns Hopkins University Press.

4 / ECOSYSTEMS AND HOW THEY WORK: CONNECTIONS IN NATURE

Internet

Biodiversity: An Overview: http://www.wcmc. org.uk/infoserv/biogen.html/

Biodiversity from Around the World: http:// www.igc.apc.org/igc/ian.html/

Biology Internet Resources: http://golgi. harvard.edu/biopages/edures.html

Collection of WWW Sites of Biological Interest: http://www.abc.hu:80/biosites.html/

List of WWW Sites of Interest to Ecologists: http://biomserv.univ-lyon1.fr/Bota.html/

Study Guide for Ecology: http://www.nl.net/ ~paideia/natsci.html/

WWW Virtual Library (University of Virginia): http://ecosys.drdr.virginia.edu/Environment.html

Readings

Begon, M., et al. 1990. *Ecology: Individuals, Populations, and Communities.* 2d ed. Sunderland, Mass.: Sinauer.

Berenbaum, May R. 1995. *Bugs in the System: Insects and Their Impact on Human Affairs.* Reading, Mass.: Addison-Wesley.

Blaustein, Andrew R., and David W. Wake. 1995. "The Puzzle of Declining Amphibians." *Scientific American,* April, 52–57.

Colinvaux, Paul A. 1992. *Ecology.* 2d ed. New York: Wiley.

Ehrlich, Anne H., and Paul R. Ehrlich. 1987. *Earth.* New York: Franklin Watts.

Ehrlich, Paul R. 1986. *The Machinery of Life: The Living World Around Us and How It Works.* New York: Simon & Schuster.

Elsom, Derek. 1992. *Earth: The Making, Shaping, and Working of a Planet.* New York: Macmillan.

Krebs, Charles J. 1994. *Ecology.* 4th ed. New York: HarperCollins.

Macmillan Publishing. 1992. *The Way Nature Works.* New York: Macmillan.

Odum, Eugene P. 1993. *Ecology and Our Endangered Life-Support Systems.* 2d ed. Sunderland, Mass.: Sinauer.

Ricklefs, Robert E. 1996. *The Economy of Nature: A Textbook of Ecology.* New York: W.H. Freeman.

Smith, Robert L. 1992. *Elements of Ecology.* 3d ed. New York: Harper & Row.

Tudge, Colin. 1991. *Global Ecology.* New York: Oxford University Press.

5 / ECOSYSTEMS: WHAT ARE THE MAJOR TYPES AND WHAT CAN HAPPEN TO THEM?

Internet

See also sites for Chapter 4.

Aquatic Biomes Notes: http://www.wisc.edu/ zoology/spring96/zoo152/152how3.html

Biodiversity, Ecology, and the Environment: http://golgi.harvard.edu/biopage/biodiversity. html/

Coastal Services Center: http://csc.noaa.gov/

The Coral Health and Monitoring Program: http://coral.aoml.erl.gov/

Environmental Change Network (Data): http:// www.nmw.ac.uk/ecn/

Ecosystems, Biomes, and Watersheds: Definitions and Examples: http://www.cnie.org/nle/ biodv-6.html

Ocean Planet: http://seawifs.gsfc.nasa.gov/ ocean-planet.html/

Sierra Club Critical Ecoregions Program: http:// www.sierraclub.org/ecoregions/

Society for Ecological Restoration: http://nabalu. flas.ufl.edu/ser/SERhome.html

Terrestrial Biomes: http://www.wisc.edu/ zoology/spring96/zoo152/152how1.html

U.S. Fish and Wildlife Service National Wetlands Inventory: http://www.nwi.fws.gov/

Readings

Aber, John, and Jerry Melito. 1991. *Terrestrial Ecosystems.* Philadelphia: Saunders.

Abramovitz, Janet N. 1996. "Sustaining Freshwater Ecosystems." In Lester R. Brown et al., *State of the World 1996.* Washington, D.C.: Worldwatch Institute, 60–77.

Akin, Wallace E. 1991. *Global Patterns: Climate, Vegetation, and Soils.* Norman: University of Oklahoma Press.

Attenborough, David, et al. 1989. *The Atlas of the Living World.* Boston: Houghton Mifflin.

Botkin, Daniel. 1990. *Discordant Harmonies: A New Ecology for the Twenty-First Century.* New York: Oxford University Press.

Cappuccino, Naomi, and Peter W. Price, eds. 1995. *Population Dynamics: New Approaches and Syntheses.* New York: Academic Press.

Denniston, Derek. 1995. *High Priorities: Conserving Mountain Ecosystems and Cultures.* Washington, D.C.: Worldwatch Institute.

Elder, Danny, and John Pernetta, eds. 1991. *Oceans.* London: Michael Beazley.

Eldridge, Niles. 1991. *The Miner's Canary: Unravelling the Mysteries of Extinction.* New York: Prentice Hall.

Goldman, C., and A. Horne. 1994. *Limnology.* New York: McGraw-Hill.

Goldsmith, Edward, et al. 1990. *Imperiled Planet: Restoring Our Endangered Ecosystems.* Cambridge, Mass.: MIT Press.

Hare, Tony, ed. 1994. *Habitats.* New York: Macmillan.

Hinrichsen, Don. 1997. *Living on the Edge: Managing Coasts in Crisis.* Covelo, Calif.: Island Press.

Lovelock, James E. 1991. *Healing Gaia: Practical Medicine for the Planet.* New York: Random House.

National Academy of Sciences. 1995. *Wetlands: Characteristics and Boundaries.* Washington, D.C.: National Academy Press.

Nisbet, E. G. 1991. *Living Earth.* San Francisco: HarperCollins.

Pilkey, Orin H., Jr, and William J. Neal, eds. 1987. *Living with the Shore.* Durham, N.C.: Duke University Press.

Pimm, Stuart L. 1992. *The Balance of Nature?* Chicago: University of Chicago Press.

Reaka-Kudla, Marjorie, Don E. Wilson, and Edward O. Wilson, eds. 1996. *Biodiversity II: Understanding and Protecting Our Biological Resources.* New York: John Henry.

Tudge, Colin. 1991. *The Environment of Life.* New York: Oxford University Press.

Weber, Michael L., and Judith A. Gradwohl. 1995. *The Wealth of Oceans: Environment and Development on Our Ocean Planet.* New York: Norton.

Weber, Peter. 1993. *Abandoned Seas: Reversing the Decline of the Oceans.* Washington, D.C.: Worldwatch Institute.

Wilson, E. O., ed. 1988. *Biodiversity.* Washington, D.C.: National Academy Press.

Wilson, E. O. 1992. *The Diversity of Life.* Cambridge, Mass.: Harvard University Press.

6 / THE HUMAN POPULATION: GROWTH AND DISTRIBUTION

Internet

American Demographics: http://www. marketingtools.com

Center for Ecology and Demography: http:// elaine.ssc.wisc.edu/cde/

Eco-Home: Urban Ecological Living: http:// ecohome.org/

Environments by Design: http://www.zone.org/ ebd/

Internet Resources for Demographers: http:// members.tripod.com/~tgryn/demog.html

Negative Population Growth: http://ourworld. compuserve.com/homepages/ neg_population_growth/

Planning and Architecture Resources: http:// www.arch.buffalo.edu/pairc/index.html

POPLINE: http://www.charm.net/~ccp/popwel. html

Population Reference Bureau: http://www.prb. org/prb

Renewable Fuels (NREL): Transportation: http:// afdc2.nrel.gov/

Sustainable Cities: http://www.oneworld.org/ overviews/cities/giradet_urbanage.html/

United Nations Population Information Network (POPIN): http://www.undp.org/popin/popin.htm

Urban Ecology: http://www.best.com/ ~schmitty/ueindex.htm

U.S. Census Bureau Home Page: http://www. census.gov/

WWW Virtual library on Demography and Population Studies: http://coombs.anu.edu.au/ ResFacilities/DemographyPage.html

Zero Population Growth: Frequently Asked Questions (FAQs): http://www.zpg.org/zpg/q-a.html

Readings

Aaberley, Doug, ed. 1994. *Futures by Design: The Practice of Ecological Planning.* Philadelphia: New Society Publishers.

Abernathy, Virginia D. 1993. *Population Politics: The Choices That Shape Our Future.* New York: Plenum Press.

Arizpe, Lourdes, et al., eds. 1994. *Population and Environment: Rethinking the Debate.* Boulder, Colo.: Westview Press.

Ashford, Lori S. 1995. "New Perspectives on Population: Lessons from Cairo." *Population Bulletin,* vol. 50, no. 1, 1–44.

Barna, George. 1992. *The Invisible Generation: Baby Busters.* New York: Barna Research Group.

Berg, Peter, et al. 1989. *A Green City Program for San Francisco Bay Area Cities and Towns.* San Francisco: Planet/Drum Foundation.

Bouvier, Leon F., and Lindsey Grant. 1994. *How Many Americans? Population, Immigration, and the Environment.* San Francisco: Sierra Club Books.

Brower, Michael. 1994. *Population Complications: Understanding the Population Debate*. Cambridge, Mass.: Union of Concerned Scientists.

Brown, Lester R. 1994. "Who Will Feed China?" *World Watch*, Sept./Oct., 10–19.

Brown, Lester R., and Hal Kane. 1994. *Full House: Reassessing the Earth's Population Carrying Capacity*. New York: Norton.

Cadman, D., and G. Payne, eds. 1990. *The Living City: Towards a Sustainable Future*. London: Routledge.

Calthorpe, Peter. 1993. *The Next American Metropolis: Ecology, Community, and the American Dream*. Princeton, N.J.: Princeton Architectural Press.

Carlson, Daniel, et al. 1995. *At Road's End: Transportation and Land Use Choices for Communities*. Covelo, Calif.: Island Press.

Day, Lincoln H. 1992. *The Future of Low-Birthrate Populations*. New York: Routledge.

DeVita, Carol J. 1996. "The United States at Mid-Decade." *Population Bulletin*, vol. 50, no. 4, 1–48.

Donaldson, Peter J., and Amy Ong Tsui. 1990. "The International Family Planning Movement." *Population Bulletin*, vol. 43, no. 3, 1–42.

Ehrlich, Paul R., Anne H. Ehrlich, and Gretchen C. Daily. 1995. *The Stork and the Plow: The Equity Answer to the Human Dilemma*. New York: Putnam.

Ezcurra, Exequiel, and Marisa Mazari-Hiriart. 1996. "Are Megacities Viable? A Cautionary Tale from Mexico City." *Environment*, vol. 38, no. 1, 6–15, 25–34.

Gordon, Deborah. 1991. *Steering a New Course: Transportation, Energy, and the Environment*. Covelo, Calif.: Island Press.

Grant, Lindsey. 1992. *Elephants in the Volkswagen: Facing the Tough Questions About Our Overcrowded Country*. New York: W. H. Freeman.

Hall, Charles A. H., et al. 1995. "The Environmental Consequences of Having a Baby in the United States." *Wild Earth*, Summer, 78–87.

Hardin, Garrett. 1993. *Living Within Limits: Ecology, Economics, and Population Taboos*. New York: Oxford University Press.

Hardin, Garrett. 1995. *Immigration Reform: Avoiding the Tragedy of the Commons*. Washington, D.C.: Federation for American Immigration Reform.

Harrison, Paul. 1992. *The Third Revolution: Environment, Population, and a Sustainable World*. New York: I. B. Tauris.

Hart, John. 1992. *Saving Cities, Saving Money: Environmental Strategies That Work*. Washington, D.C.: Resource Renewal Institute.

Hartmann, Betsy. 1995. *Reproductive Rights and Wrongs: The Global Politics of Population Control and Contraceptive Choice*. Boston: South End Press.

Haub, Carl. 1987. "Understanding Population Projections." *Population Bulletin*, vol. 42, no. 4, 1–41.

Haupt, Arthur, and Thomas T. Kane. 1985. *The Population Handbook: International*. 2d ed. Washington, D.C.: Population Reference Bureau.

Hollingsworth, William J. 1996. *Ending the Explosion: Population Policies and Ethics*. Santa Ana, Calif.: Seven Locks Press.

Leisinger, Klaus M., and Karin Schmitt. 1994. *All Our People: Population Policy with a Human Face*. Covelo, Calif.: Island Press.

Lowe, Marcia D. 1990. *Alternatives to the Automobile: Transport for Living Cities*. Washington, D.C.: Worldwatch Institute.

Lowe, Marcia D. 1991. *Shaping Cities: The Environmental and Human Dimensions*. Washington, D.C.: Worldwatch Institute.

Lowe, Marcia D. 1994. *Back on Track: The Global Rail Revival*. Washington, D.C.: Worldwatch Institute.

Lutz, Wolfgang. 1994. "The Future of World Population." *Population Bulletin*, vol. 46, no. 2, 1–43.

Lyle, John T. 1993. *Regenerative Design for Sustainable Development*. New York: Wiley.

Lyman, Francesca. 1997. "Twelve Gates to the City." *Sierra*, May/June, 29–35.

MacKenzie, James J., et al. 1992. *The Going Rate: What It Really Costs to Drive*. Washington, D.C.: World Resources Institute.

Makower, Joel. 1992. *The Green Commuter*. Washington, D.C.: National Press.

Martin, Philip, and Elizabeth Midgley. 1994. "Immigration to the United States: Journey to an Uncertain Destination." *Population Bulletin*, vol. 49, no. 2, 1–47.

Mazur, Laurie Ann, ed. 1994. *Beyond the Numbers: A Reader on Population Consumption, and the Environment*. Covelo, Calif.: Island Press.

McFalls, Joseph A., Jr. 1991. "Population: A Lively Introduction." *Population Bulletin*, vol. 46, no. 2, 1–43.

McHarg, Ian L. 1992. *Designing with Nature*. Reprinted ed. New York: Wiley.

McKibben, Bill. 1995. *Hope, Human and Wild: True Stories of Living Lightly on the Earth*. Boston: Little, Brown.

Moffett, George D. 1994. *Critical Masses: The Global Population Challenge*. New York: Viking.

Nadis, Steve, and James J. MacKenzie. 1993. *Car Trouble*. Boston: Beacon Press.

Nafis, Sadik, ed. 1991. *Population Policies and Programmes: Lessons Learned from Two Decades of Experience*. New York: United Nations Population Fund.

National Academy of Sciences. 1997. *The New Americans: Economic, Demographic, and Fiscal Effects of Immigration*. Washington, D.C.: National Academy Press.

Population Reference Bureau. 1990. *World Population: Fundamentals of Growth*. Washington, D.C.: Population Reference Bureau.

Population Reference Bureau. Annual. *World Population Data Sheet*. Washington, D.C.: Population Reference Bureau.

Rabinovitch, Jonas, and Josef Leitman. 1996. "Urban Planning in Curitiba." *Scientific American*, March, 46–53.

Register, Richard. 1992. *Ecocities*. Berkeley, Calif.: North Atlantic Books.

Ryn, Sin van der, and Stuart Cowan. 1995. *Ecological Design*. Covelo, Calif.: Island Press.

Simon, Julian L. 1996. *The Ultimate Resource 2*. Princeton, N.J.: Princeton University Press.

Thornton, Richard D. 1991. "Why the U.S. Needs a MAGLEV System." *Technology Review*, April, 31–42.

Tien, H. Yuan, et al. 1992. "China's Demographic Dilemmas." *Population Bulletin*, vol. 47, no. 1, 1–44.

Todd, John, and Nancy Jack Todd. 1993. *From Ecocities to Living Machines: Precepts for Sustainable Technologies*. Berkeley, Calif.: North Atlantic Books.

Todd, John, and George Tukel. 1990. *Reinhabiting Cities and Towns: Designing for Sustainability*. San Francisco: Planet/Drum Foundation.

Tunali, Odil. 1996. "A Billion Cars: The Road Ahead." *World Watch*, January/February, 24–33.

United Nations. 1991. *Consequences of Rapid Population Growth in Developing Countries*. New York: United Nations.

United Nations Population Division. 1991. *World Urbanization Prospects*. New York: United Nations.

United Nations. 1992. *Long-Range World Population Projections: Two Centuries of Population Growth, 1950–2150*. New York: United Nations.

Visaria, Leela, and Pravin Visaria. 1995. "India's Population in Transition." *Population Bulletin*, vol. 50, no. 3, 1–51.

Walter, Bob, et al., eds. 1992. *Sustainable Cities: Concepts and Strategies for Eco-City Development*. Los Angeles: Eco-Home Media.

Wann, David. 1994. *Biologic: Environmental Protection by Design*. 2d ed. Boulder, Colo.: Johnson Books.

Wann, David. 1995. *Deep Design: Pathways to a Livable Future*. Covelo, Calif.: Island Press.

White, Rodney R. 1994. *Urban Environmental Management. Environmental Change and Urban Design*. New York: Wiley.

Zero Population Growth. 1990. *Planning the Ideal Family: The Small Family Option*. Washington, D.C.: Zero Population Growth.

Zuckerman, Wolfgang. 1991. *End of the Road: The World Car Crisis and How We Can Solve It*. White River Junction, Vt.: Chelsea Green.

7 / ENVIRONMENTAL ECONOMICS AND POLITICS

Internet

Blueprint for a Green Campus: http://netspace.org/environ/earthnet/

Campus Ecology: http://www.nwf.org/nwf/campus

Centre for Development and Environment: http://www.giub.unibe.ch/cde/

Congressional Green Sheets: http://www.cais.net/greensheets/

Ecological Economics Resources: http://csf.colorado.edu/ecolecon/

Econet's Ecojustice Resources: http://www.igc.apc.org/envjustice/

Environmental Activism Online: http://www.envirolink.org/Elib/action/

Environmentally Active Colleges, Universities, and Schools: http://gwis.circ.gwu.edu/~green/ecu.html/

Environmental Organizations: http://www.econet.apc.org/econet/en.orgs.html/

Environmental Protection Agency Information: http://www.epa.gov/docs/Access/chapter1/chapter1.html

Environmental Taxes: http://www.law.cornell.edu/uscode/26/ch38.html

Gopher menu for the United Nations Environment Programme: Environmental Economics: gopher://pan.cedar.univie.ac.at:70/11/UNEP/EnvEcon/

Green Manufacturing: http://euler.berkeley.edu/green/cgdm.html

Individual Action: http://www.ncb.gov.sg/jkj/env/greentips.html

Infoterra: United Nations Environment Programme: http://pan.cedar.univie.ac.at/gopher/UNEP/

International Institute for Sustainable Development Linkages: http://www.mbnet.mb.ca/linkage/

International Society for Ecological Economics: http://kabir.cbl.cees.edu/ISEE/ISEEhome.html

Rachel's Environment and Health Weekly: http://www.nirs.org/rehw/rehw473.txt

Student Environmental Action Coalition Network: http://ecosys.drdr.virginia.edu/SEACnet.html

The Takings Issue: http://www.webcom.com/pcj/takings.html

Trade and the Environment: http://www.unep.ch/trade.html/

The World Bank Home Page: http://www.worldbank.org/

Yahoo: Sustainable Development Resources: http://www.yahoo.com/Society_and_Culture/Environment_and_Nature/Environment_and_Development_Policies/

Readings

Aaseng, Nathan. 1994. *Jobs vs. the Environment; Can We Save Both?* Hillside, N.J.: Enslow.

Adler, Jonathan. 1996. *Environmentalism at the Crossroads: Green Activism in America.* Washington, D.C.: Capital Research Center.

Andersen, Terry, and Donald R. Leal. 1991. *Free-Market Environmentalism.* Boulder, Colo.: Westview Press.

Arnold, Ron, and Alan Gottlieb. 1993. *Trashing the Economy: How Runaway Environmentalism Is Wrecking America.* Bellevue, Wash.: Free Enterprise Press.

Ashworth, William. 1995. *The Economy of Nature: Rethinking the Connections Between Ecology and Economics.* Boston: Houghton Mifflin.

Athansiou, Tony. 1996. *Divided Planet: The Ecology of Rich and Poor.* Boston: Little, Brown.

Barlett, Donald L., and James B. Steele. 1994. *America: Who Really Pays the Taxes?* New York: Simon & Schuster.

Barnet, Richard J., and John Cavanagh. 1994. *Global Dreams: Imperial Corporations and the New World Order.* New York: Simon & Schuster.

Bennett, Steven J. 1991. *Ecopreneuring: The Green Guide to Small Business Opportunities from the Environmental Revolution.* New York: Wiley.

Bennett, Steven J., et al. 1994. *Corporate Realities and Environmental Truths.* New York: Wiley.

Bramwell, Anna. 1994. *The Fading of the Greens: The Decline of Environmental Politics in the West.* New Haven, Conn.: Yale University Press.

Brown, Lester R., et al. 1991. *Saving the Planet: How to Shape an Environmentally Sustainable Global Economy.* New York: Norton.

Brundtland, G. H., et al. 1987. *Our Common Future: World Commission on Environment and Development.* New York: Oxford University Press.

Bryant, Bunyan, ed. 1995. *Environmental Justice: Issues, Policies, and Solutions.* Covelo, Calif.: Island Press.

Bullard, Robert D., ed. 1994. *Unequal Protection: Environmental Justice and Communities of Color.* San Francisco: Sierra Club.

Cairncross, Frances. 1992. *Costing the Earth.* Boston: Harvard Business School.

Cairncross, Frances. 1995. *Green, Inc.: A Guide to Business and the Environment.* Covelo, Calif.: Island Press.

Caldwell, Lynton K. 1990. *International Environmental Policy.* 2d ed. Durham, N.C.: Duke University Press.

Cobb, Clifford, et al. 1995. *The Genuine Progress Indicator.* San Francisco, Calif.: Redefining Progress.

Cobb, John B., Jr., 1992. *Sustainability: Economics, Ecology and Justice.* Maryknoll, N.Y.: Orbis Books.

Cohn, Susan. 1995. *Green at Work: Finding a Business Career That Works for the Environment.* Rev. ed. Covelo, Calif.: Island Press.

Constanza, Robert, ed. 1992. *Ecological Economics: The Science and Management of Sustainability.* New York: Columbia University Press.

Court, T. de la. 1990. *Beyond Bruntland: Green Development in the 1990s.* London: Zed Books.

Daly, Herman E. 1991. *Steady-State Economics.* 2d ed. Covelo, Calif.: Island Press.

Daly, Herman E. 1992. *Environmentally Sustainable Development: Building on Bruntland.* Covelo, Calif.: Island Press.

Daly, Herman E., and John B. Cobb, Jr. 1994. *For the Common Good: Redirecting the Economy Toward Community, the Environment, and a Sustainable Future.* 2d ed. Boston: Beacon Press.

Dowie, Mark. 1995. *Losing Ground: American Environmentalism at the Close of the Twentieth Century.* Cambridge, Mass.: MIT Press.

Durning, Alan T. 1989. *Poverty and the Environment: Reversing the Downward Spiral.* Washington, D.C.: Worldwatch Institute.

Durning, Alan T. 1992. *How Much Is Enough? The Consumer Society and the Earth.* New York: Norton.

Eagen, David J., and David W. Orr. 1994. *The Campus and Environmental Responsibility.* San Francisco: Jossey-Bass.

Earth Island Books. 1993. *The Case Against Free Trade: GATT, NAFTA, and the Globalization of Corporate Power.* Berkeley, Calif.: North Atlantic Books.

Eden, Sally. 1996. *Environmental Issues and Business: Implications of a Changing Agenda.* New York: Wiley.

Ehrlich, Paul R., and Anne H. Ehrlich. 1996. *Betrayal of Science and Reason: How Anti-Environmental Rhetoric Threatens Our Future.* Covelo, Calif.: Island Press.

Elkins, Paul, and Jakob von Uexhull. 1992. *Grassroots Movements for Global Change.* New York: Routledge.

Fanning, Odum. 1995. *Opportunities for Environmental Careers.* Lincolnwood, Ill.: VCM Career Horizons.

Fiorino, Daniel J. 1995. *Making Environmental Policy.* Berkeley: University of California Press.

Flavin, Christopher, and John E. Young. 1993. "Shaping the Next Industrial Revolution." In Lester R. Brown et al., *State of the World 1993.* New York: Norton, 180–99.

Foreman, Dave. 1991. *Confessions of an Eco-Warrior.* New York: Harmony Books.

Foster, John Bellamy. 1995. *The Vulnerable Planet: A Short Economic History of the Planet.* New York: Monthly Review Press.

French, Hilary E. 1993. *Costly Trade-offs: Reconciling Trade and the Environment.* Washington, D.C.: Worldwatch Institute.

French, Hilary E. 1995. *Partnership for the Planet: An Environmental Agenda for the United Nations.* Washington, D.C.: Worldwatch Institute.

Gay, Kathlyn. 1994. *Pollution and the Powerless: The Environmental Justice Movement.* New York: F. Watts.

George, Susan. 1992. *The Debt Boomerang: How Third World Debt Harms Us All.* Boulder, Colo.: Westview Press.

Goodland, Robert, et al. 1992. *Environmentally Sustainable Economic Development: Building on Bruntland.* Paris: UNESCO Press.

Gottlieb, Robert. 1993. *Forcing the Spring: The Transformation of the Environmental Movement.* Covelo, Calif.: Island Press.

Goudy, John, and Sabine O'Hara. 1995. *Economic Theory for Environmentalists.* New York: St. Lucie Press.

Greenpeace. 1993. *The Greenpeace Guide to Anti-Environmental Organizations.* Berkeley, Calif.: Odionan Press.

Greider, William. 1992. *Who Will Tell the People? The Betrayal of American Democracy.* New York: Simon & Schuster.

Grove, Richard. 1994. *Green Imperialism.* New York: Cambridge University Press.

Hall, Bob. 1990. *Environmental Politics: Lessons from the Grassroots.* Durham, N.C.: Institute for Southern Studies.

Hawken, Paul. 1993. *The Ecology of Commerce.* New York: HarperCollins.

Heinze-Fry, Jane. 1994. *Green Lives, Green Campuses.* Belmont, Calif.: Wadsworth. **Available as a supplement for use with this book.**

Helvarg, David. 1994. *The War Against the Greens.* San Francisco: Sierra Club Books.

Henderson, Hazel. 1996. *Building a Win–Win World: Life Beyond Global Economic Warfare.* San Francisco: Berrett-Koehler.

Hirschorn, Joel S., and Kirsten U. Oldenberg. 1990. *Prosperity Without Pollution: The Prevention Strategy for Industry and Consumers.* New York: Van Nostrand Reinhold.

Institute for Local Self-Reliance. 1990. *Proven Profits from Pollution Prevention.* Washington, D.C.: Institute for Local Self-Reliance.

Irvine, Sandy, and A. Ponton. 1988. *A Green Manifesto.* London: Optima.

Jacobs, Michael. 1993. *The Green Economy: Environment, Sustainable Development, and Politics.* New York: Pluto Press.

Johnston, Barbara R. 1994. *Who Pays the Price? The Sociocultural Context of Environmental Crisis.* Covelo, Calif.: Island Press.

Kazis, Richard, and Richard L. Grossman. 1991. *Fear at Work: Job Blackmail, Labor, and the Environment.* Philadelphia: New Society.

Keniry, Julian. 1994. *Ecodemia: Profiles of Environmentally Smart Green Campuses.* Carbondale: Southern Illinois University Press.

Korten, David. 1995. *When Corporations Rule the World.* West Hartford, Conn.: Kumarian Press.

LaMay, Craig L., and Everette E. Dennis, eds. 1992. *Media and the Environment.* Covelo, Calif.: Island Press.

Landy, Marc K., et al. 1994. *The Environmental Protection Agency: Asking the Wrong Questions, from Nixon to Clinton.* New York: Oxford University Press.

Lee, Kai N. 1993. *Compass and Gyroscope: Integrating Science and Politics for the Environment.* Covelo, Calif.: Island Press.

List, Peter C. 1993. *Radical Environmentalism: Philosophy and Tactics.* Belmont, Calif.: Wadsworth.

Makower, Joel. 1993. *The E-Factor: The Bottom-Line Approach to Environmentally Responsible Business.* New York: Times Books.

Mander, Jerry, and Edward Goldsmith, eds. 1996. *The Case Against the Global Economy, and for a Turn Toward the Local.* San Francisco: Sierra Club Books.

Mikesell, Raymond F., and Lawrence F. Williams. 1992. *International Banks and the Environment.* San Francisco: Sierra Club Books.

Monks, Robert A. G., and Nell Minow. 1991. *Power and Accountability.* New York: HarperCollins.

Moore, Curtic, and Alan Miller. 1994. *Green Gold: Japan, Germany, and the United States, and the Race for Environmental Technology.* Boston: Beacon Press.

Myers, Norman. 1993. *Ultimate Security: The Environmental Basis of Political Stability.* New York: Norton.

National Academy of Sciences. 1993. *Greening Industrial Ecosystems.* Washington, D.C.: National Academy Press.

National Academy of Sciences. 1995. *Assigning Economic Value to Natural Resources.* Washington, D.C.: National Academy Press.

National Wildlife Federation. 1995. *Ecodemia: Campus Environmental Stewardship at the Turn of the 21st Century.* Washington, D.C.: National Wildlife Federation.

Ophuls, William, and A. Stephen Boyan, Jr. 1992. *Ecology and the Politics of Scarcity Revisited: The Unravelling of the American Dream.* San Francisco: W.H. Freeman.

Paepke, C. Owen. 1993. *The Evolution of Progress: The End of Economic Growth and the Beginning of the Human Transformation.* New York: Random House.

Pearce, David, et al. 1991. *Blueprint 2: Greening the World Economy.* East Haven, Conn.: Earthscan.

Piasecki, Bruce W. 1995. *Corporate Environmental Strategy: The Avalanche of Change Since Bhopal.* New York: Wiley.

Porter, Gareth, and Janet Welsh Brown. 1994. *Global Environmental Politics.* 2d ed. Boulder, Colo.: Westview Press.

Raskin, Jamin B., and John Bonifaz. 1994. *The Wealth Primary.* Washington, D.C.: Center for Responsive Politics.

Renner, Michael. 1991. *Jobs in a Sustainable Economy.* Washington, D.C.: Worldwatch Institute.

Repetto, Robert. 1992. "Accounting for Environmental Assets." *Scientific American*, June, 94–100.

Repetto, Robert. 1995. *Jobs, Competitiveness, and Environmental Regulation: What Are the Real Issues?* Washington, D.C.: World Resources Institute.

Repetto, Robert, et al. 1992. *Green Fees: How a Tax Shift Can Work for the Environment and the Economy.* Washington, D.C.: World Resources Institute.

Rich, Bruce. 1994. *Mortgaging the Earth.* Boston: Beacon Press.

Rifkin, Jeremy. 1991. *Biosphere Politics: A New Consciousness for a New Century.* New York: Crown.

Robertson, James. 1990. *Future Wealth: A New Economics for the 21st Century.* New York: Bootstrap Press.

Roddick, Anita. 1991. *Body and Soul: Profits with Principles.* New York: Crown.

Rodman, David M. 1996. "Harnessing the Market for the Environment." In Lester R. Brown et al., *State of the World 1996.* Washington, D.C.: Worldwatch Institute, 168–87.

Rogers, Adam. 1993. *The Earth Summit: A Planetary Reckoning.* Los Angeles: Global View Press.

Romm, Joseph J. 1993. *Defining National Security: The Nonmilitary Aspects.* New York: Council on Foreign Relations Press.

Roodman, David M. 1996. *Paying the Piper: Subsidies, Politics, and the Environment.* Washington, D.C.: Worldwatch Institute.

Roodman, David M. 1997. *Getting the Signals Right: Tax Reform to Protect the Environment and the Economy.* Washington, D.C.: Worldwatch Institute.

Rosenbaum, Walter A. 1990. *Environment, Politics, and Policy.* 2d. ed Washington, D.C.: Congressional Quarterly.

Sachs, Aaron. 1995. *Eco-Justice: Linking Human Rights and the Environment.* Washington, D.C.: Worldwatch Institute.

Sanjor, William. 1992. *Why the EPA Is Like It Is and What Can Be Done About It.* Washington, D.C.: Environmental Research Foundation.

Saunders, Tedd, and Loretta McGovern. 1994. *The Bottom Line of Green Is Black.* New York: Harper Collins.

Schmidheiny, Stephan. 1992. *Changing Course: A Global Business Perspective on Development and the Environment.* Cambridge, Mass.: MIT Press.

Schumacher, E. F. 1973. *Small Is Beautiful: Economics as if the Earth Mattered.* New York: Harper & Row.

Shabecoff, Philip. 1993. *A Fierce Green Fire: The American Environmental Movement.* New York: Hill & Wang.

Shabecoff, Philip. 1996. *A New Name for Peace: International Environmentalism, Sustainable Development, and Democracy.* Hanover, N.H.: University of New England Press.

Smart, Bruce, ed. 1992. *Beyond Compliance: A New Industry View of the Environment.* Washington, D.C.: World Resources Institute.

Smith, April, and the Student Environmental Action Coalition. 1994. *Campus Ecology: A Guide to Assessing Environmental Quality and Creating Strategies for Change.* Chapel Hill, N.C.: Student Environmental Action Coalition.

Stead, W. Edward, and John Garner Stead. 1992. *Management for a Small Planet.* New York: Sage.

Tietenberg, Tom. 1992. *Environmental and Resource Economics.* 3d ed. Glenview, Ill.: Scott, Foresman.

Turner, R. Kerry, et al. 1994. *Environmental Economics: An Elementary Introduction.* Baltimore: Johns Hopkins University.

Watson, Paul. 1994. *Ocean Warrior.* Marina del Rey, Calif.: Key Porter Books.

Weinstein, Mirriam. 1993. *1993 Making a Difference College Guide: Education for a Better World.* San Anselmo, Calif.: Sage Press.

Westra, Laura, and Peter Wenz, eds. 1995. *Facing Environmental Racism: Confronting Issues of Global Justice.* Lantham, Md.: Rowan & Littlefield.

World Resources Institute. 1993. *A New Generation of Environmental Leadership: Action for the Environment and the Economy.* Washington, D.C.: World Resources Institute.

World Resources Institute. 1995. *Corporate Environmental Accounting.* Washington, D.C.: World Resources Institute.

Zimmerman, Michael. 1995. *Science, Nonscience, and Nonsense: Approaching Environmental Literacy.* Baltimore, Md.: Johns Hopkins University Press.

8 / RISK, TOXICOLOGY, AND HUMAN HEALTH

Internet

Animal Rights: http://www.envirolink.org/arrs/

Earthquakes (U.S. Geological Survey): http://info.er.usgs.gov/network/science/earth/earthquake.html

Environmental Health and Toxicology: http://www.ctenvironet.com/htoxlink.htm

Environment, Safety, and Health (Department of Energy): http://www.tis.eh.doe.gov:80/tis.html

Geologic Hazards (U.S. Geological Survey): http://geology.usgs.gov/

Gopher Menu of Toxics: gopher://ecosys.drdr. Virginia.EDU/11/library/gen/toxics/

Home Hazards Guide: http://www.ctenvironet.com/hbuyergd.htm

International Registry of Potentially Toxic Chemicals: http://www.unep.ch/irptc.html/

National Center for Health Statistics: http://www.cdc.gov/nchswww/nchshome.htm

National Earthquake Information Center: http://wwwneic.cr.usgs.gov/

Particulates and Premature Death: http://www.nrdc.org/find/aibresum.html

Rachel's Environment and Health Weekly: http://www.nirs.org/rehw/rehw473.txt

U.S. Geological Survey: Earth and Environmental Science: http://www.usgs.gov/network/science/earth/earth.html/

World Health Organization (WHO): http://www.who.ch/

Yahoo: Environmental Health: http://www.yahoo.com/Health/Environmental_Health/

Readings

Bartecchi, Carl E., et al. 1995. "The Global Tobacco Epidemic." *Scientific American*, May, 44–47.

Bernarde, Melvin A. 1989. *Our Precarious Habitat: Fifteen Years Later.* New York: Wiley.

Chivian, Eric, et al. 1993. *Critical Condition: Human Health and the Environment.* Cambridge, Mass.: MIT Press.

Cohen, Mark N. 1989. *Health and the Rise of Civilization.* New Haven, Conn.: Yale University Press.

Colborn, Theo, et al. 1996. *Our Stolen Future.* New York: Dutton.

Dadd-Redalla, Debra. 1994. *Sustaining the Earth: Choosing Consumer Products That Are Environmentally Safe for You.* New York: Hearst Books.

Davies, Clarence, ed. 1996. *Comparing Environmental Risks: Tools for Setting Government Guidelines.* Washington, D.C.: Resources for the Future.

Dixon, Bernard. 1994. *Power Unseen: How Microbes Rule the World.* San Francisco: W.H. Freeman.

Emsley, John. 1994. *The Consumers' Good Chemical Guide: A Jargon-Free Guide to the Chemicals of Everyday Life.* New York: W.H. Freeman.

Environmental Protection Agency. 1987. *Unfinished Business: A Comparative Assessment of Environmental Problems.* Washington, D.C.: Environmental Protection Agency.

Environmental Protection Agency. 1990. *Reducing Risk: Setting Priorities and Strategies for Environmental Protection.* Washington, D.C.: Environmental Protection Agency.

Fisher, Jeffrey A. 1994. *The Plague Makers: How We Are Creating Catastrophic New Epidemics and What We Must Do to Avert Them.* New York: Simon & Schuster.

Fox, Michael W. 1992. *Superpigs and Wondercorn: The Brave New World of Biotechnology and Where It All May Lead.* New York: Lyons & Burford.

Freudenburg, William R. 1988. "Perceived Risk, Real Risk: Social Science and the Art of Probabilistic Risk Assessment." *Science*, vol. 242, 44–49.

Freudenthal, Ralph I., and Susan L. Freudenthal. 1989. *What You Need to Know to Live with Chemicals.* Greens Farms, Conn.: Hill & Garnett.

Garrett, Laurie. 1994. *The Coming Plague: Newly Emerging Diseases in a World out of Balance.* New York: Farrar, Straus & Giroux.

Graham, John D., and Jonathan Baert Weiner, eds. 1995. *Risk vs. Risk: Tradeoffs in Protecting Health and the Environment.* Cambridge, Mass.: Harvard University Press.

Harris, John. 1992. *Wonderwoman and Superman: The Ethics of Human Biotechnology.* New York: Oxford University Press.

Harris, Stephen L. 1990. *Agents of Chaos: Earthquakes, Volcanoes, and Other Natural Disasters.* Missoula, Mont.: Mountain Press.

Harte, John, et al. 1992. *Toxics A to Z: A Guide to Everyday Pollution Hazards.* Berkeley: University of California Press.

Kamarin, M. A. 1988. *Toxicology: A Primer on Toxicology Principles and Applications.* Boca Raton, Fla.: Lewis.

Kitcher, Philip. 1996. *The Lives to Come: The Genetic Revolution and Human Possibilities.* New York: Simon & Schuster.

Kovach, Robert L. 1995. *Earth's Fury: An Introduction to Natural Hazards and Disasters.* Englewood Cliffs, N.J.: Prentice-Hall.

Lappé, Marc. 1995. *The Evolving Threat of Drug-Resistant Disease.* San Francisco: Sierra Club Books.

Merrell, Paul, and Carol Van Strum. 1990. "Negligible Risk or Premeditated Murder?" *Journal of Pesticide Reform*, vol. 10, Spring, 20–22.

Misch, Ann. 1994. "Assessing Environmental Health Risks." In Lester R. Brown et al., *State of the World 1994.* New York: Norton, 117–36.

Moeller, Dade W. 1992. *Environmental Health.* Cambridge, Mass.: Harvard University Press.

Montague, Peter, ed. *Rachel's Environment and Health Weekly*. P.O. Box 5036, Annapolis, MD 21403-7036. (A two-page weekly newsletter explaining health, risk, and environmental issues in an easily understandable manner.)

Nadakavukaren, Anne. 1995. *Our Global Environment: A Health Perspective*. 4th ed. New York: Waveland Press.

National Academy of Sciences. 1992. *Eat for Life: The Food and Nutrition Board's Guide to Reducing Your Risk of Chronic Disease*. Washington, D.C.: National Academy Press.

National Academy of Sciences. 1996. *Understanding Risk: Informing Decisions in a Democratic Society*. Washington, D.C.: National Academy Press.

Ottoboni, M. Alice. 1991. *The Dose Makes the Poison: A Plain-Language Guide to Toxicology*. 2d ed. New York: Van Nostrand Reinhold.

Piller, Charles. 1991. *The Fail-Safe Society*. New York: Basic Books.

Platt, Anne. 1996. *Infecting Ourselves: How Environmental and Social Disruptions Trigger Disease*. Washington, D.C.: Worldwatch Institute.

Preston, Richard. 1994. *The Hot Zone*. New York: Random House.

Real, Leslie A. 1996. "Sustainability and the Challenge of Infectious Disease." *BioScience*, vol. 46, no. 2, 88–97.

Regenstein, Lewis G. 1993. *Cleaning Up America the Poisoned*. Washington, D.C.: Acropolis Books.

Rifkin, Jeremy. 1985. *Declaration of a Heretic*. Boston: Routledge & Kegan Paul.

Rodricks, Joseph V. 1992. *Calculated Risks: Understanding the Toxicity and Human Health Risks of Chemicals in the Environment*. New York: Cambridge University Press.

Steingraber, Sandra. 1997. *Living Downstream: An Ecologist Looks at Cancer and the Environment*. Reading, Mass.: Harvard University Press.

Suzuki, David, and Peter Knudtson. 1989. *Genethics: The Clash Between the New Genetics and Human Values*. Cambridge, Mass.: Harvard University Press.

Tenner, Ward. 1996. *Why Things Bite Back: Technology and the Revenge of Unintended Consequences*. New York: Knopf.

Whelan, Elizabeth M. 1993. *Toxic Terror: The Truth Behind the Cancer Scares*. 2d ed. Buffalo, N.Y.: Prometheus Books.

Wildavsky, Aaron. 1995. *But Is It True? A Citizen's Guide to Environmental Health and Safety Issues*. Cambridge, Mass.: Harvard University Press.

World Resources Institute. 1994. *Human and Ecosystem Health*. Washington, D.C.: World Resources Institute.

9 / AIR AND AIR POLLUTION

Internet

See also sites for Chapter 10.

Acid Deposition: http://www.gov.nb.ca/environm/operatin/air/morerain.html/

Acid Rain FAQs: http://www.ns.doe.ca/aeb/ssd/Acid/acidFAQ.html/

Air Pollution: http://wwwilson.ucsd.edu/education/air pollution/air pollution.html/

Clean Air Act: http://www.law.cornell.edu/uscode/42/ch85.html

EcoNet's Acid Rain Resources: http://www.econet.apc.org/acidrain/

Indoor Air Quality: http://www.epa.gov/iaq/

National Pollution Prevention Center for Higher Education: http://www.umich.edu/~nppcpub/

Out of Breath: http://www.nrdc.org/nrdc/publ/breath.html/

Particulates and Premature Death: http://www.nrdc.org/find/aibresum.html

Rachel's Environment and Health Weekly: http://www.nirs.org/rehw/rehw473.txt

Radon Home Page: http://sedwww.cr.usgs.gov:8080/radon/radonhome.html

WWW Virtual Library: Environment-Atmosphere: http://ecosys.drdr.Virginia.EDU:80/atm.html/

Readings

Allerman, James E., and Brooke T. Mossman. 1997. "Asbestos Revisited." *Scientific American*, July, 70–75.

Bridgman, Howard. 1991. *Global Air Pollution: Problems for the 1990s*. New York: Belhaven Press.

Brookins, Douglas G. 1990. *The Indoor Radon Problem*. Irvington, N.Y.: Columbia University Press.

Bryner, Gary. 1992. *Blue Skies, Green Politics: The Clean Air Act of 1990*. Washington, D.C.: Congressional Quarterly Press.

Cohen, Bernie. 1988. *Radon: A Homeowner's Guide to Detection and Control*. Mt. Vernon, N.Y.: Consumer Report Books.

Cole, Leonard D. 1994. *Element of Risk: The Politics of Radon*. New York: Oxford University Press.

Elsom, Derek. 1987. *Atmospheric Pollution: Causes, Effects, and Control Policies*. Cambridge, Mass.: Basil Blackwell.

Hedin, Lars O., and Gene E. Likens. 1996. "Atmospheric Dust and Acid Rain." *Scientific American*, December, 88–92.

Leslie, G. B., and F. W. Linau. 1994. *Indoor Air Pollution*. New York: Cambridge University Press.

MacKenzie, James J., and Mohamed T. El-Ashry. 1990. *Air Pollution's Toll on Forests and Crops*. Washington, D.C.: World Resources Institute.

Mello, Robert A. 1987. *Last Stand of the Red Spruce*. Covelo, Calif.: Island Press.

National Academy of Sciences. 1995. *Health Effects of Exposure to Radon: Time for Reassessment?* Washington, D.C.: National Academy Press.

Smith, William H. 1991. "Air Pollution and Forest Damage." *Chemistry and Engineering News*, Nov. 11, 30–43.

Soroos, Marvin S. 1997. *The Endangered Atmosphere: Preserving a Global Commons*. Columbia, S.C.: University of South Carolina Press.

Steidlmeier, Paul. 1993. "The Morality of Pollution Permits." *Environmental Ethics*, vol. 15, 133–50.

Wagner, Travis. 1993. *In Our Backyard: A Guide to Understanding Pollution and Its Effects*. New York: Van Nostrand Reinhold.

Warde, John. 1997. *The Healthy Home Handbook: All You Need to Know to Rid Your Home of Health and Safety Hazards*. New York: Times Books.

10 / CLIMATE, GLOBAL WARMING, AND OZONE LOSS

Internet

See also sites in Chapter 9.

EcoNet's Climate Resource Directory: http://www.igc.apc.org/climate

Environmental Protection Agency (EPA) and Ozone Depletion: http://www.epa.gov/docs/ozone/index.html

Global Climate Change Information Program: http://www.doc.mmu.ac.uk/aric/gccres.html

Global Change Research (U.S. Geological Survey): http://geochange.er.usgs.gov/

Global Warming: http://www.nbn.com/youcan/warm/warm.html/

Global Warming Update (NOAA): http://www.ncdc.noaa.gov/gblwrmupd/global.html/

Intergovernmental Panel on Climate Change (IPCC): http://www.unep.ch/ipcc/ipcc-0.html

National Center for Atmospheric Research: http://http.ucar.edu/metapage.html/

National Climatic Data Center: http://www.ncdc.noaa.gov/ncdc.html/

National Oceanic and Atmospheric Administration (NOAA): http://www.esdim.noaa.gov/

Ocean and Climate: http://www.cms.udel.edu

Ozone Depletion Frequently Asked Questions (FAQs): http://icair.iac.org.nz/ozone/index.html

Ozone Depletion and Global Warming: http://zebu.uoregon.edu/energy.html

Ozone Layer: http://www.nbn.com/youcan/ozone/ozone.html/

United Nations Convention on Climate Change: http://www.unep.ch/iucc.html/

WWW Virtual Library: Environment-Atmosphere: http://ecosys.drdr.Virginia.EDU:80/atm.html/

Readings

American Forestry Association. 1993. *Forests and Global Warming*. vol. 1. Washington, D.C.: American Forestry Association.

Balling, Robert C., Jr. 1992. *The Heated Debate: Greenhouse Predictions Versus Climate Reality*. San Francisco: Pacific Research Institute for Public Policy.

Bates, Albert K. 1990. *Climate in Crisis: The Greenhouse Effect and What We Can Do*. Summertown, Tenn.: The Book Publishing Company.

Benedick, Richard Eliot. 1991. *Ozone Diplomacy: New Directions in Safeguarding the Planet*. Cambridge, Mass.: Harvard University Press.

Bernard, Harold W., Jr. 1993. *Global Warming Unchecked: Signs to Watch For*. Bloomington: University of Indiana Press.

Charlson, Robert J., and Tom M. L. Wigley. 1994. "Sulfate Aerosol and Climate Change." *Scientific American*, February, 48–57.

Council for Agricultural Science and Technology. 1992. *Preparing U.S. Agriculture for Global Climate Change*. Ames, Iowa: Council for Agricultural Science and Technology.

Flavin, Christopher, and Odil Tunali. 1996. *Climate of Hope: New Strategies for Stabilizing the World's Atmosphere*. Washington, D.C.: Worldwatch Institute.

Gates, David M. 1993. *Climate Change and Its Biological Consequences*. Sunderland, Mass.: Sinauer.

Graedel, T. C., and Paul J. Cutzen. 1995. *Atmosphere. Climate, and Change*. New York: W.H. Freeman.

Houghton, John T., et al., eds. 1996. *Climate Change 1995—The Science of Climate Change*. New York: Cambridge University Press.

Intergovernmental Panel on Climate Change (IPCC). 1990. *Climate Change: The IPCC Assessment*. New York: Cambridge University Press.

Intergovernmental Panel on Climate Change (IPCC). 1995. *The Supplementary Report to the IPCC Scientific Assessment*. New York: Cambridge University Press.

IUCN Global Change Programme. 1993. *Impact of Climate Change on Ecosystems and Species*. Covelo, Calif.: Island Press.

Karplus, Walter J. 1992. *The Heavens Are Falling: The Scientific Prediction of Catastrophes in Our Time*. New York: Plenum.

Lee, Henry, ed. 1995. *Shaping Responses to Climate Change*. Covelo, Calif.: Island Press.

Leffell, David J., and Douglas E. Brash. 1996. "Sunlight and Skin Cancer." *Scientific American*, July, 52–59.

Leggett, Jeremy, ed. 1994. *The Climate Time Bomb: Signs of Climate Change from the Greenpeace Database*. Amsterdam, Netherlands. Stichting Greenpeace Council.

Lovins, Amory B., et al. 1989. *Least-Cost Energy: Solving the CO$_2$ Problem*. 2d ed. Snowmass, Colo.: Rocky Mountain Institute.

Lyman, Francesca, et al. 1990. *The Greenhouse Trap: What We're Doing to the Atmosphere and How We Can Slow Global Warming*. Washington, D.C.: World Resources Institute.

Makhijani, Arjun, and Kevin Gurney. 1995. *Mending the Ozone Hole: Science, Technology, and Policy*. Cambridge, Mass.: MIT Press.

McKibben, Bill. 1989. *The End of Nature*. New York: Random House.

Mintzer, Irving, and William R. Moomaw. 1991. *Escaping the Heat Trap: Probing the Prospects for a Stable Environment*. Washington, D.C.: World Resources Institute.

Mintzer, Irving, et al. 1990. *Protecting the Ozone Shield: Strategies for Phasing Out CFCs During the 1990s*. Washington, D.C.: World Resources Institute.

National Academy of Sciences. 1990a. *Confronting Climate Change*. Washington, D.C.: National Academy Press.

National Academy of Sciences. 1990b. *Sea Level Change*. Washington, D.C.: National Academy Press.

National Academy of Sciences. 1992. *Global Environmental Change*. Washington, D.C.: National Academy Press.

National Audubon Society. 1990. *CO$_2$ Diet for a Greenhouse Planet: A Citizen's Guide to Slowing Global Warming*. New York: National Audubon Society.

Office of Technology Assessment. 1993. *Preparing for an Uncertain Climate*. Washington, D.C.: U.S. Government Printing Office.

Oppenheimer, Michael, and Robert H. Boyle. 1990. *Dead Heat: The Race Against the Greenhouse Effect*. New York: Basic Books.

Peters, Robert L., and Thomas E. Lovejoy, eds. 1992. *Global Warming and Biological Diversity*. New Haven, Conn.: Yale University Press.

Ray, Dixie Lee, and Lou Guzzo. 1993. *Environmental Overkill: Whatever Happened to Common Sense?* Washington, D.C.: Regnery Gateway.

Revkin, Andrew. 1992. *Global Warming: Understanding the Forecast*. New York: Abbeville Press.

Roan, Sharon L. 1989. *Ozone Crisis:. The 15-Year Evolution of a Sudden Global Emergency*. New York: Wiley.

Rowland, F. Sherwood, and Mario J. Molina. 1994. "Ozone Depletion: 20 Years After the Alarm." *Chemistry and Engineering News*, August 15, 8–13.

Turekian, Karl K. 1996. *Global Environmental Change: Past, Present, and Future*. Upper Saddle River, N.J.: Prentice-Hall.

UN Environment Programme, World Meteorological Organization, U.S. National Aeronautics and Space Administration, and National Oceanic and Atmospheric Administration. 1994. *International Assessment of Ozone Depletion*. Washington, D.C.: UN Environment Programme.

Woodwell, George M., and Fred T. Mackenzie, eds. 1995. *Biotic Feedbacks in the Global Climatic System. Will the Warming Feed the Warming?* New York: Oxford University Press.

11 / WATER RESOURCES AND WATER POLLUTION

Internet

Clean Water Act: http://www.law.cornell.edu/uscode/33/ch26.html

Dam Impact: http://www.sandelman.ocunix.on.ca/dams/

Great Lakes: http://www.great-lakes.net/

Groundwater Pollution: http://h2o.usgs.gov/public/wid/html/GW.html

Rivers, Oceans, Water Resources: http://www.yahoo.com/Society_and_Culture/Environment_and_Nature/Rivers__Oceans_and_Other_Water_Resources/

U.S. Water News: http://www.mother.com/uswaternews/

Water Data (U.S. Geological Survey): http://h2o.usgs.gov/public/wid/html/WD.html

The Water FAQ: http://www.siouxlan.com/water/faq.html/#imp/5/

Water on Line: http://agency.resource.ca.gov/ceres/WOL/home.html/

Water Pollution: http://www.tribnet.com/environ/env_watp.htm/

Water Resources General Information (U.S. Geological Survey): http://h2o.usgs.gov/

Water Web Home Page: http://www.waterweb.com/

WWW Virtual Library: Environment-Hydrosphere: http://ecosys.drdr.Virginia.EDU:80/hyd.html/

Zebra Mussel Information: http://www.nfrcg.gov:80/zebra.mussel

Readings

Abramovitz, Janet N. 1996. "Sustaining Freshwater Ecosystems." In Lester R. Brown et. al., *State of the World 1996*. New York: Norton, 60–77.

Adler, Robert W., et al. 1993. *The Clean Water Act Twenty Years Later*. Covelo, Calif.: Island Press.

Allaby, Michael. 1992. *Water: Its Global Nature*. New York: Facts on File.

Ashworth, William. 1986. *The Late, Great Lakes: An Environmental History*. New York: Alfred A. Knopf.

Bolling, David M. 1994. *How to Save a River: A Handbook for Citizen Action*. Covelo, Calif.: Island Press.

Burger, Joanna. 1997. *Oil Spills*. Rutgers, N.J.: Rutgers University Press.

Christopher, Thomas. 1994. *Water-Wise Gardening: America's Backyard Revolution*. New York: Simon & Schuster.

Clarke, Robin. 1993. *Water: The International Crisis*. Cambridge, Mass.: MIT Press.

Colborn, Theodora E., et al. 1989. *Great Lakes, Great Legacy?* Washington, D.C.: Conservation Foundation.

Costner, Pat, and Glenna Booth. 1986. *We All Live Downstream: A Guide to Waste Treatment That Stops Water Pollution*. Berkeley, Calif.: Bookpeople.

Doppelt, Bob, et al. 1993. *Entering the Watershed: A New Approach to Save America's River Ecosystems*. Covelo, Calif.: Island Press.

Edmonson, W. T. 1991. *The Uses of Ecology: Lake Washington and Beyond*. Seattle: University of Washington Press.

El-Ashry, Mohamed, and Diana C. Gibbons, eds. 1988. *Water and Arid Lands of the Western United States*. Washington, D.C.: World Resources Institute.

Ellefson, Connie, et al. 1992. *Xeriscape Gardening: Water Conservation for the American Landscape*. New York: Macmillan.

Environmental Protection Agency. 1990. *Citizen's Guide to Ground-Water Protection*. Washington, D.C.: Environmental Protection Agency.

Falkenmark, Malin, and Carl Widstand. 1992. "Population and Water Resources: A Delicate Balance." *Population Bulletin*, vol. 47, no. 3, 1–36.

Feldman, David Lewis. 1991. *Water Resources Management: In Search of an Environmental Ethic*. Baltimore, Md.: Johns Hopkins University Press.

Gardner, Gary. 1995. "From Oasis to Mirage: The Aquifers That Won't Replenish." *World Watch*, May/June, 30–36, 40–41.

Gleick, Peter H. 1993. *Water in Crisis: A Guide to the World's Fresh Water Resources*. Berkeley, Calif.: Pacific Institute for Studies in Development, Environment, and Security.

Gleick, Peter H. 1994. "Water, War and Peace in the Middle East." *Environment*, April, 6–41.

Hansen, Nancy R., et al. 1988. *Controlling Nonpoint-Source Water Pollution*. New York: National Audubon Society and The Conservation Society.

Hillel, Daniel. 1995. *Rivers of Eden: The Struggle for Water and the Quest for Peace in the Middle East*. New York: Oxford University Press.

Horton, Tom, and William Eichbaum. 1991. *Turning the Tide: Saving the Chesapeake Bay*. Covelo, Calif.: Island Press.

Hundley, Norris, Jr. 1992. *The Great Thirst: Californians and Water, 1770s–1990s*. Berkeley: University of California Press.

Ives, J. D., and B. Messeric. 1989. *The Himalayan Dilemma: Reconciling Development and Conservation*. London: Routledge.

Jewell, William J. 1994. "Resource-Recovery Wastewater Treatment." *American Scientist*, vol. 82, 366–75.

Keeble, John. 1991. *Out of the Channel: The Exxon Valdez Oil Spill in Prince William Sound*. New York: HarperCollins.

Knopman, Debra S., and Richard A. Smith. 1993. "20 Years of the Clean Water Act." *Environment*, vol. 35, no. 1, 17–20, 34–41.

Lewis, Scott A. 1995. *Guide to Safe Drinking Water*. San Francisco: Sierra Club Books.

McCully, Patrick. 1996. *Silenced Rivers: The Ecology and Politics of Large Dams*. London: Zed.

McCutcheon, Sean. 1992. *Electric Rivers: The Story of the James River Project*. New York: Paul & Co.

Micklin, Philip P., and William D. Williams. 1996. *The Aral Sea Basin*. New York: Springer-Verlag.

Mitchell, Bruce, ed. 1990. *Integrated Water Management*. New York: Belhaven Press.

Munasinghe, Mohan. 1992. *Water Supply and Environmental Management*. Boulder, Colo.: Westview Press.

Naiman, Robert J., ed. 1992. *Watershed Management: Balancing Sustainability and Environmental Change*. New York: Springer-Verlag.

National Academy of Sciences. 1992. *Water Transfers in the West: Efficiency, Equity, and the Environment*. Washington, D.C.: National Academy Press.

Natural Resources Defense Council. 1989. *Ebb Tide for Pollution: Actions for Cleaning Up Coastal Waters*. Washington, D.C.: Natural Resources Defense Council.

Opie, John. 1993. *Ogallala: Water for a Dry Land*. Lincoln: University of Nebraska Press.

Outwater, Alice. 1996. *Water: A Natural History*. New York: Basic Books.

Palmer, Tim. 1994. *Lifelines: The Case for River Conservation*. Covelo, Calif.: Island Press.

Patrick, Ruth, et al. 1992. *Surface Water Quality: Have the Laws Been Successful?* Princeton, N.J.: Princeton University Press.

Pearce, Fred. 1992. *The Dammed: Rivers, Dams, and the Coming World Water Crisis*. London: Bodley Head.

Platt, Anne. 1995. "Dying Seas." *World Watch*, January/February, 10–19.

Platt, Anne. 1996. "Water-Borne Killers." *World Watch*, March/April, 28–35.

Postel, Sandra. 1996. "Forging a Sustainable Water Strategy." In Lester R. Brown et. al., *State of the World 1996*. New York: Norton, 40–59.

Postel, Sandra et al. 1996. "Human Appropriation of Renewable Fresh Water." *Science*, vol. 271, 785–87.

Qing, Dai. 1994. *Yangtse! Yangtze! Debate Over the Three Gorges Project*. East Haven, Conn.: Earthscan.

Reisner, Marc, and Sara Bates. 1990. *Overtapped Oasis: Reform or Revolution for Western Water*. Covelo, Calif.: Island Press.

Rocky Mountain Institute. 1990. *Catalog of Water-Efficient Technologies for the Urban/Residential Sector*. Snowmass, Colo.: Rocky Mountain Institute.

Sierra Club Defense Fund. 1989. *The Poisoned Well: New Strategies for Groundwater Protection*. Covelo, Calif.: Island Press.

Starr, Joyce R. 1991. "Water Wars." *Foreign Policy*, Spring, 12–45.

Steingraber, Sandra. 1997. *Living Downstream: An Ecologist Looks at Cancer and the Environment*. Reading, Mass.: Addison-Wesley.

Wagner, Travis. 1993. *In Our Backyard: A Guide to Understanding Pollution and Its Effects*. New York: Van Nostrand Reinhold.

Weber, Michael L., and Judith A. Gradwohl. 1995. *The Wealth of Oceans: Environment and Development on Our Ocean Planet*. New York: Norton.

Weber, Peter. 1993. *Abandoned Seas: Reversing the Decline of the Oceans*. Washington, D.C.: Worldwatch Institute.

Wilkinson, Charles F. 1992. *Crossing the Next Meridian: Land, Water, and the Future of the West*. Covelo, Calif.: Island Press.

Worster, Donald. 1985. *Rivers of Empire: Water, Aridity, and the Growth of the American West*. New York: Pantheon.

12 / MINERALS AND SOIL

Internet

Mineral Resources Survey (USGS): http://minerals.er.usgs.gov/

Minerals and Energy Resources: http://www.vuw.ac.nz/~li899mib/resguide/energy.html

National Soil Erosion Research Laboratory: http://soils.ecn.purdue.edu:20002/~wepp/nserl.html/

Natural Resources Conservation Service: http://www.ncg.nrcs.usda.gov/

Rotweb: Home Composting: http://net.indra.com/~topsoil/Compost_Menu.html

Soil Science Links: http://www.wolfe.net/~psmall/soil.html

Soil/Water Relationships: http://hammock.ifas.ufl.edu/txt/fairs/19828

Soil and Water Science Handbook: http://hammock.ifas.ufl.edu/txt/fairs/19835

Readings

Brady, Nyle C. 1989. *The Nature and Properties of Soils*. 10th ed. New York: Macmillan.

Brown, Bruce, and Lane Morgan. 1990. *The Miracle Planet*. Edison, N.J.: W.H. Smith.

Campbell, Stu. 1990. *Let It Rot: The Gardener's Guide to Composting*. Pownal, Vt.: Storey Communications.

Crosson, Pierre. 1997. "Will Erosion Threaten Agricultural Productivity?" *Environment*, Vol. 39, No. 8, 4–9, 29–31.

Donahue, Roy, et al. 1990. *Soils and Their Management*. 5th ed. Petaluma, Calif.: Inter Print.

Eggert, R. G., ed. 1994. *Mining and the Environment*. Washington, D.C.: Resources for the Future.

Ernst, W. G. 1990. *The Dynamic Planet*. New York: Columbia University Press.

Grainger, Alan. 1990. *The Threatening Desert: Controlling Desertification*. East Haven, Conn.: Earthscan.

Hillel, Daniel. 1992. *Out of the Earth: Civilization and the Life of the Soil*. New York: Free Press.

Hodges, Carroll A. 1995. "Mineral Resources, Environmental Issues, and Land Use." *Science*, vol. 268, 1305–11.

Little, Charles E. 1987. *Green Fields Forever: The Conservation Tillage Revolution in America*. Covelo, Calif.: Island Press.

Mainguet, Monique. 1991. *Desertification: Natural Background and Human Mismanagement*. New York: Springer-Verlag.

Mollison, Bill. 1990. *Permaculture*. Covelo, Calif.: Island Press.

National Academy of Sciences. 1993. *Soil and Water Quality: An Agenda for Agriculture*. Washington, D.C.: National Academy Press.

Paddock, Joe, et al. 1987. *Soil and Survival: Land Stewardship and the Future of American Agriculture*. San Francisco: Sierra Club Books.

Pimentel, David, ed. 1993. *World Soil Erosion and Conservation*. New York: Cambridge University Press.

Pimentel, David, et al. 1995. "Environmental and Economic Costs of Soil Erosion and Conservation Benefits." *Science*, vol. 267, 1117–23.

Ripley, Earle A., et al. 1996. *Environmental Effects of Mining*. Delray Beach, Fla.: St. Lucie Press.

Wild, Alan. 1993. *Soils and the Environment: An Introduction*. New York: Cambridge University Press.

Young, John E. 1992. *Mining the Earth*. Washington, D.C.: Worldwatch Institute.

Young, John E., and Aaron Sachs. 1994. *The Next Efficiency Revolution: Creating a Sustainable Materials Revolution*. Washington, D.C.: Worldwatch Institute.

Youngquist, Walter. 1990. *Mineral Resources and the Destinies of Nations*. Portland, Ore.: National Book.

13 / SOLID AND HAZARDOUS WASTE

Internet

See also sites for Chapter 8.

Center for Nuclear and Toxic Waste Management: http://cnwm.berkely.edu/cnwm.html/

CERCLA Superfund Law: http://www.law.cornell.edu/uscode/42/ch103.html

Econet's Ecojustice Resources: http://www.igc.apc.org/envjustice/

Global Recycling Network: http://grn.com/grn/

Great Plains/Rocky Mountain Hazardous Substance Research Center: http://www.engg.ksu.edu/HSRC/

Green Design and Manufacturing: http://euler.berkeley.edu/green/cgdm.html

Incineration and Environment: http://sun1.bham.ac.uk/c.m.tarpey/phil.htm

Integrated Sustainable Management of Wastes: http://www.wrfound.org.uk/

Library of Toxic Substances: gopher://ecosys.drdr.virginia.edu:70/11/library/gen/toxics

National Materials Exchange Network: http://www.earthcycle.com/g/p/earthcycle//

National Pollution Prevention Center for Higher Education: http://www.umich.edu/~nppcpub/

Rachel's Environment and Health Weekly: http://www.nirs.org/rehw/rehw473.txt

Recycling Resources: http://grn.com/library/hot_list.htm

Reduce, Reuse: Cygnus: http://cygnus-group.com

Solid and Hazardous Waste Gopher: gopher://wissago.uwex.edu:70/11/.course/recycling/

Stop the Junk Mail: http://www.stopjunk.com/

Waste Reduction: http://www.cygnus-group.com:9011/

Readings

Ackerman, Frank. 1997. *Why Do We Recycle? Markets, Values, and Public Policy*. Covelo, Calif.: Island Press.

Alexander, Judd H. 1993. *In Defense of Garbage*. Westport, Conn.: Praeger.

Allen, Robert. 1992. *Waste Not, Want Not*. London: Earthscan.

Barnett, Harold C. 1994. *Toxic Debts and the Superfund Dilemma*. Chapel Hill: University of North Carolina Press.

Bullard, Robert D., ed. 1993. *Confronting Environmental Racism: Voices from the Grassroots*. Boston: SouthEnd Press.

Bullard, Robert D. 1994a. *Dumping in Dixie: Race, Class, and Environmental Quality*. 2d ed. Boulder, Colo.: Westview Press.

Bullard, Robert D., ed. 1994b. *Unequal Protection: Environmental Justice and Communities of Color*. San Francisco: Sierra Club Books.

Calow, Peter. 1997. *Controlling Environmental Risks from Chemicals: Principles and Practice*. New York: John Wiley & Sons.

Carless, Jennifer. 1992. *Taking Out the Trash: A No-Nonsense Guide to Recycling*. Covelo, Calif.: Island Press.

Centers for Disease Control and Prevention. 1991. *Preventing Lead Poisoning in Young Children*. Atlanta, Ga.: Centers for Disease Control and Prevention.

Cohen, Gary, and John O'Connor. 1990. *Fighting Toxics: A Manual for Protecting Family, Community, and Workplace*. Covelo, Calif.: Island Press.

Commoner, Barry. 1994. *Making Peace with the Planet*. Rev. ed. New York: New Press.

Connett, Paul H. 1992. "The Disposable Society." In F. H. Bormann and Stephen R. Kellert, eds. *Ecology, Economics, Ethics*. New Haven, Conn.: Yale University Press, 99–122.

Connett, Paul, and Ellen Connett. 1994. "Municipal Waste Incineration: Wrong Question, Wrong Answer." *The Ecologist*, January/February, vol. 24, 14–20.

Costner, Pat, and Joe Thornton. 1989. *Sham Recyclers, Part 1: Hazardous Waste Incineration in Cement and Aggregate Kilns*. Washington, D.C.: Greenpeace.

Denison, Richard A., and John Ruston. 1997. "Recycling Is Not Garbage." *Technology Review*. October, 55–60.

Devito, Stephen C., and Roger L. Garrett, eds. 1996. *Designing Safe Chemicals: Green Chemistry for Pollution Prevention*. Washington, D.C.: American Chemical Society.

Durning, Alan Thein. 1992. *How Much Is Enough? The Consumer Society and the Future of the Earth*. New York: Norton.

Earth Works Group. 1990. *The Recycler's Handbook: Simple Things You Can Do*. Berkeley, Calif.: Earth Works Press.

Environmental Protection Agency. 1992. *The Consumer's Handbook for Reducing Solid Waste*. Washington, D.C.: Environmental Protection Agency.

Frosch, Robert A. 1995. "Industrial Ecology." *Environment*, December, 16–37.

Gardner, Gary. 1997. *Recycling Organic Waste: From Urban Pollutant to Farm Resource.* Washington, D.C.: Worldwatch Institute.

Gertler, Nicholas, and John R. Ehrenfeld. 1996. "A Down-to-Earth Approach to Clean Production." *Technology Review,* February/March, 48–54.

Gibbs, Lois. 1982. *The Love Canal: My Story.* Albany: State University of New York Press.

Gibbs, Lois. 1995. *Dying from Dioxin.* Boston: South End Press.

Gordon, Ben, and Peter Montague. 1989. *Zero Discharge: A Citizen's Toxic Waste Manual.* Washington, D.C.: Greenpeace.

Gottlieb, Robert, ed. 1995. *Reducing Toxics: A New Approach to Policy and Industrial Decisionmaking.* Covelo, Calif.: Island Press.

Harte, John, et al. 1991. *Toxics A to Z: A Guide to Everyday Pollution Hazards.* Berkeley: University of California Press.

Hilz, Christoph. 1993. *The International Toxic Waste Trade.* New York: Van Nostrand Reinhold.

Hird, John A. 1994. *Superfund: The Political Economy of Environmental Risk.* Baltimore: Johns Hopkins University Press.

Hofrichter, Richard, ed. 1993. *Toxic Struggles: The Theory and Practice of Environmental Justice.* Philadelphia: New Society Publishers.

Kane, Hal. 1996. "Shifting to Sustainable Industries." In Lester R. Brown, et al., *State of the World 1996.* Washington, D.C.: Worldwatch Institute, pp. 152–67.

Kessel, Irene, and John T. O'Connor. 1997. *Getting the Lead Out: The Complete Source on How to Prevent and Cope with Lead Poisoning.* New York: Plenum Press.

Kharbanda, O. P., and E. A. Stallworthy. 1990. *Waste Management: Toward a Sustainable Society.* New York: Auburn House.

Lappé, Marc. 1991. *Chemical Deception: The Toxic Threat to Health and Environment.* San Francisco: Sierra Club.

League of Women Voters. 1993a. *The Garbage Primer: A Handbook for Citizens.* New York: Lyons & Burford.

League of Women Voters. 1993b. *A Plastic Waste Primer.* New York: Lyons & Burford.

Mazmanian, Daniel, and David Morrell. 1992. *Beyond Superfailure: America's Toxics Policy for the 1990s.* Boulder, Colo.: Westview Press.

McLenighan, Valjean. 1991. *Sustainable Manufacturing: Saving Jobs, Saving the Environment.* Chicago: Center for Neighborhood Technology.

Minnesota Mining and Manufacturing. 1988. *Low- or Non-Pollution Technology Through Pollution Prevention.* St. Paul, Minn.: 3M Company.

Mowrey, Marc, and Tim Redmond. 1993. *Not in Our Backyard.* New York: William Morrow.

Moyers, Bill. 1990. *Global Dumping Ground: The International Traffic in Hazardous Waste.* Cabin John, Md.: Seven Locks Press.

National Academy of Sciences. 1994. *The Greening of Industrial Ecosystems.* Washington, D.C.: National Academy of Engineering.

Needleman, Herbert L., and Philip J. Landrigan. 1995. *Raising Children Toxic Free.* New York: Avon Books.

Nemerow, Nelson L. 1995. *Zero Pollution for Industry: Waste Minimization Through Industrial Complexes.* New York: Wiley.

Office of Technology Assessment. 1992. *Green Products by Design.* Washington, D.C.: Government Printing Office.

Pellerano, Maria B. 1995. *How to Research Chemicals: A Resource Guide.* Annapolis, Md.: Environmental Research Foundation.

Platt, Brenda, et al. 1991. *Beyond 40 Percent: Record-Setting Recycling and Composting Programs.* Covelo, Calif.: Island Press.

Puckett, Jim. 1994. "Disposing of the Waste Trade: Closing the Recycling Loophole." *The Ecologist,* vol. 24, no. 2, 53–58.

Rathje, William, and Cullen Murphy. 1992. *Rubbish! The Archaeology of Garbage: What Our Garbage Tells Us About Ourselves.* San Francisco: HarperCollins.

Regenstein, Lewis G. 1993. *Cleaning Up America the Poisoned.* Washington, D.C.: Acropolis Books.

Sachs, Aaron. 1995. *Eco-Justice: Linking Human Rights and the Environment.* Washington, D.C.: Worldwatch Institute.

Setterberg, Fred, and Lonny Shavelson. 1993. *Toxic Nation: The Fight to Save Our Communities from Chemical Contamination.* New York: Wiley.

Stapleton, Richard M. 1994. *Lead Is a Silent Hazard.* New York: Walker.

Theodore, Louis, and Young C. McGuinn. 1992. *Pollution Prevention.* New York: Van Nostrand Reinhold.

Thomas, William. 1995. *Scorched Earth: The Military's Assault on the Environment.* Philadelphia: New Society Publishers.

Wagner, Travis. 1993. *In Our Backyard: A Guide to Understanding Pollution and Its Effects.* New York: Van Nostrand Reinhold.

Walsh, Edward, et al. 1997. *Don't Burn It Here: Grassroots Challenges to Trash Incineration.* University Park, Penn.: Pennsylvania State University Press.

Water Pollution Control Federation. 1989. *Household Hazardous Waste: What You Should and Shouldn't Do.* Alexandria, Va.: Water Pollution Control Federation.

Whelan, E. M. 1993. *Toxic Terror: The Truth Behind the Cancer Scares.* Buffalo, N.Y.: Prometheus Books.

Whitaker, Jennifer Seymour. 1994. *Salvaging the Land of Plenty: Garbage and the American Dream.* New York: William Morrow.

Young, John E. 1991. *Discarding the Throwaway Society.* Washington, D.C.: Worldwatch Institute.

Young, John E. 1995. "The Sudden New Strength of Recycling." *World Watch,* July/August, 20–25.

14 / FOOD RESOURCES

Internet

American Farmland Trust: http://farm.fic.niu.edu/aft/afthome.html/

Aquaculture: http://www.ansc.purdue.edu/aquanic/

Eliminating Starvation/Feeding Humanity: http://www.pacificrim.net/~wginwrep/WorldGame/drfeed.html/

Food and Nutrition Web Sites: http://www.mother.com/agaccedd/FoodNut.html/

Hunger Web: http://www.brown.edu/departments/world-hunger-program/html/

National Agricultural Library: http://www.nalusda.gov/

Sustainable Agriculture: http://www.sarep.ucdavis.edu/

Sustainable Agriculture Research and Education Program: http://www.sarep.ucdavis.edu/

United Nations Food and Agriculture Organization (FAO): http://www.fao.org/

U.S. Department of Agriculture (USDA): http://www.usda.gov/

U.S. Fisheries Protection: http://www.netspace.org/MFCN/

Readings

Aberley, Doug. 1993. *Greening the Garden: A Guide to Sustainable Growing.* Philadelphia: New Society Publishers.

Aliteri, Miguel A. 1995. *Agroecology: The Scientific Basis of Alternative Agriculture.* 2d ed. Boulder, Colo.: Westview Press.

Ausubel, Kenny. 1994. *Seeds of Change.* New York: HarperCollins.

Avery, Dennis T. 1997. "Saving Nature's Legacy Through Better Farming." *Issues in Science and Technology,* Fall, 59–64.

Berrill, Michael. 1997. *The Plundered Seas: Can the World's Fish Be Saved?* San Francisco: Sierra Club Books.

Berry, Wendell. 1990. *Nature as Measure.* Berkeley, Calif.: North Point Press.

Berstein, Henry, et al., eds. 1990. *The Food Question: Profits Versus People.* East Haven, Conn.: Earthscan.

Bongaarts, John. 1994. "Can the Growing Human Population Feed Itself?" *Scientific American,* March, 36–42.

Brookfield, Harold, and Christine Padoch. 1994. "Appreciating Biodiversity: A Look at the Dynamism and Diversity of Indigenous Farming Practices." *Environment,* June, 6–42.

Brown, Lester R. 1994. *Who Will Feed China?* New York: Norton.

Brown, Lester R. 1996. *Tough Choices: Facing the Challenge of Food Scarcity.* New York: Norton.

Brown, Lester R., and Hal Kane. 1994. *Full House: Reassessing the Earth's Population Carrying Capacity.* New York: Norton.

Carroll, C. Ronald, et al. 1990. *Agroecology.* New York: McGraw-Hill.

Coleman, Elliot. 1992. *The New Organic Grower's Four-Season Harvest.* White River Junction, Vt.: Chelsea Green.

Dunning, Alan B., and Holly W. Brough. 1991. *Taking Stock: Animal Farming and the Environment.* Washington, D.C.: Worldwatch Institute.

Edwards, Clive A., et al., eds. 1990. *Sustainable Agricultural Systems.* Ankeny, Iowa: Soil and Water Conservation Society.

Faeth, Paul. 1995. *Growing Green: Enhancing the Economic and Environmental Performance of U.S. Agriculture.* Washington, D.C.: World Resources Institute.

Fowler, Cary, and Pat Mooney. 1990. *Shattering: Food, Politics, and the Loss of Genetic Diversity.* Tucson: University of Arizona Press.

Fukuoka, Masanobu. 1985. *The Natural Way of Farming: The Theory and Practice of Green Philosophy.* New York: Japan Publications.

Gardner, Gary. 1996. *Shrinking Fields: Cropland Loss in a World of Eight Billion.* Washington, D.C.: Worldwatch Institute.

Goering, Peter, et al. 1993. *From the Ground Up: Rethinking Industrial Agriculture.* Atlantic Highlands, N.J.: Zed Books.

Gordon, R. Conway, and Edward R. Barbier. 1990. *After the Green Revolution: Sustainable Agriculture for Development.* East Haven, Conn.: Earthscan.

Jacobson, Michael, et al. 1991. *Safe Food: Eating Wisely in a Risky World.* Washington, D.C.: Planet Earth Press.

Jeavons, John, and Carol Cox. 1993. *Lazy-Bed Gardening: The Quick and Dirty Guide.* Willits, Calif.: Ecology Action.

Kane, Hal. 1993. "Growing Fish in Fields." *World Watch,* September/October, 20–27.

League of Women Voters. 1991. *U.S. Farm Policy: Who Benefits? Who Pays? Who Decides?* Washington, D.C.: League of Women Voters.

Logsdon, Gene. 1994. *At Nature's Pace: Farming and the American Dream*. New York: Pantheon.

Mann, Charles. 1997. "Reseeding the Green Revolution." *Science*, vol. 277, 1038–1043.

Mollison, Bill. 1990. *Permaculture: A Practical Guide for a Sustainable Future*. Covelo, Calif.: Island Press.

National Academy of Sciences. 1989. *Alternative Agriculture*. Washington, D.C.: National Academy Press.

National Academy of Sciences. 1992. *Marine Aquaculture: Opportunities for Growth*. Washington, D.C.: National Academy Press.

National Academy of Sciences. 1993a. *Soil and Water Quality: An Agenda for Agriculture*. Washington, D.C.: National Academy Press.

National Academy of Sciences. 1993b. *Sustainable Agriculture and the Environment in the Humid Tropics*. Washington, D.C.: National Academy Press.

Pimentel, David, and Carl W. Hall. 1989. *Food and Natural Resources*. Orlando, Fla.: Academic Press.

Pretty, Jules N. 1995. *Regenerating Agriculture: Policies and Prospects for Sustainability and Self-Reliance*. London: Earthscan.

Prosterman, Roy L., et al. 1996. "Can China Feed Itself?" *Scientific American*, November, 90–96.

Raeburn, Paul. 1995. *The Last Harvest: The Genetic Gamble That Threatens to Destroy American Agriculture*. New York: Simon & Schuster.

Rifkin, Jeremy. 1992. *Beyond Beef: The Rise and Fall of the Cattle Culture*. New York: Dutton.

Rissler, Jane, and Margaret Mellon. 1996. *The Ecological Risks of Engineered Crops*. Cambridge, Mass.: MIT Press.

Robbins, John. 1992. *May All Be Fed: Diet for a New World*. New York: William Morrow.

Rosegrant, Mark W., and Robert Livernash. 1996. "Growing More Food, Doing Less Damage." *Environment*, Vol. 38, no. 7, 6–11, 28–30.

Shiva, Vandana. 1991. *The Violence of the Green Revolution*. Atlantic Highlands, N.J.: Zed Books.

Smith, Katherine R. 1995. "Time to 'Green' U.S. Farm Policy." *Issues in Science and Technology*, Spring, 71–78.

Soil and Water Conservation Society. 1990. *Sustainable Agricultural Systems*. Ankeny, Iowa: Soil and Water Conservation Society.

Solbrig, Otto T., and Dorothy Solbrig. 1994. *So Shall You Reap: Farming and Crops in Human Affairs*. Covelo, Calif.: Island Press.

Soule, Judith D., and Jon Piper. 1992. *Farming in Nature's Image: An Ecological Approach to Agriculture*. Covelo, Calif.: Island Press.

Swanson, Louis E., and Frank B. Clearfield, eds. 1994. *Agricultural Policy and the Environment: Iron Fist or Open Hand?* Ankeny, Iowa: Soil and Water Conservation Society.

Tamsey, Geoff, and Tony Worsley. 1995. *The Food System*. East Haven, Conn.: Earthscan.

Weber, Peter. 1994. *Net Loss: Fish, Jobs, and the Marine Environment*. Washington, D.C.: Worldwatch Institute.

Welch, Ross M. 1997. "Toward a Greener Revolution." *Issues in Science and Technology*, Fall, 50–58.

Yang, Linda. 1990. *The City Gardener's Handbook: From Balcony to Backyard*. New York: Random House.

15 / PROTECTING FOOD RESOURCES: PESTICIDES AND PEST CONTROL

Internet

Biocontrol Network: http://www.usit.net/hp/bionet/Bionet.html/

Biological Control of Pests Research: http://rsu2.tamu.edu/bcpru/bcpru.html/

Food Additives (Food and Drug Administration): http://vm.cfsan.fda.gov/~lrd/foodadd.html

Insecticides Law: http://www.law.cornell.edu/uscode/7/ch6.html

National Agricultural Pest Information Network: http://www.ceris.purdue.edu/napis/

National Integrated Pest Management: http://ipmwww.ncsu.edu/

National Integrated Pest Management Network: http://www.reeusda.gov/ipm/ipm-home.html/

Pesticide Action Network North America: http://www.panna.org/panna/

Pesticides Information (Food and Drug Administration): http://vm.cfsan.fda.gov/~lrd/pestadd.html

Rachel's Environment and Health Weekly: http://www.nirs.org/rehw/rehw473.txt

USDA Integrated Pest Management Initiative: http://raleigh.dis.anl.gov:83/

Readings

Benbrook, Charles M., et al. 1996. *Pest Management at the Crossroads*. Yonkers, N.Y.: Consumer's Union.

Bormann, F. Herbert, et al. 1993. *Redesigning the American Lawn*. New Haven, Conn.: Yale University Press.

Briggs, Shirley, and the Rachel Carson Council. 1992. *Basic Guide to Pesticides: Their Characteristics and Hazards*. Washington, D.C.: Taylor & Francis.

Care, James R., and Maureen K. Hinkle. 1994. *Integrated Pest Management: The Path of a Paradigm*. Washington, D.C.: National Audubon Society.

Carson, Rachel. 1962. *Silent Spring*. Boston: Houghton Mifflin.

Cassels, Jamie. 1993. *The Uncertain Promise of Law: Lessons from Bhopal*. Toronto: University of Toronto Press.

Dinham, Barbara. 1993. *The Pesticide Hazard: A Global Health and Environmental Audit*. New York: Zed Books.

Friends of the Earth. 1990. *How to Get Your Lawn and Garden Off Drugs*. Ottawa, Ontario: Friends of the Earth.

Gustafson, David. 1993. *Pesticides in Drinking Water*. New York: Van Nostrand Reinhold.

Heylin, Michael, ed. 1991. "Pesticides: Costs Versus Benefits." *Chemistry & Engineering News*, Jan. 7, 27–56.

Horn, D. J. 1988. *Ecological Approach to Pest Management*. New York: Guilford.

Jenkins, Virginia Scott. 1994. *The Lawn: A History of an American Obsession*. Washington, D.C.: Smithsonian Institute Press.

Marquardt, Sandra. 1989. *Exporting Banned Pesticides: Fueling the Circle of Poison*. Washington, D.C.: Greenpeace.

National Academy of Sciences. 1993. *Pesticides in the Diet of Infants and Children*. Washington, D.C.: National Academy Press.

National Academy of Sciences. 1996. *Ecologically Based Pest Management: New Solutions for a New Century*. Washington, D.C.: National Academy Press.

Natural Resources Defense Council. 1993. *After Silent Spring: The Unsolved Problems of Pesticide Use in the United States*. New York: Natural Resources Defense Council.

Pimentel, David, et al. 1992. "Environmental and Economic Cost of Pesticide Use." *BioScience*, vol. 42, no. 10, 750–60.

Platt, Anne. 1996. "IPM and the War on Pests." *World Watch*, March/April. 21–27.

Wargo, J. 1996. *Our Toxic Legacy: How Science and Law Fail to Protect Us from Pesticides*. New Haven, Conn.: Yale University Press.

Wiles, Richard, and Christopher Campbell. 1993. *Pesticides in Children's Food*. Washington, D.C.: Environmental Working Group.

Yepsen, Roger B., Jr. 1987. *The Encyclopedia of Natural Insect and Pest Control*. Emmaus, Pa.: Rodale Press.

16 / SUSTAINING ECOSYSTEMS: FORESTS, RANGELANDS, PARKS, AND WILDERNESS

Internet

See also biodiversity sites for Chapters 4 and 5.

Biodiversity: http:www.ftpt.br/structure/biodiversit.html/

Biodiversit, Ecology, and the Environment: http://golgi.harvard.edu/biopage/biodiversity.html/

Biodiversity and Ecosystems Network: http://straylight.tamu.edu/bene/bene.html

Biodiversity Gopher: gopher://muse.bio.cornell.edu:70/00/about_this_gopher

Biodiversity Information Network: http://www.bdt.org.br/bin21/bin21.html

Biosphere Reserves (Man and the Biosphere): http://ice.ucdavis.edu/MAB/

Canada Forestry Management: http://www.nofc.forestry.ca

Forest and Rangeland Ecosystem Service Center: http://www.fsl.orst.edu/home/nbs/

Gaia Forest Conservation Archives: http://gaial.ies.wisc.edu/research/pngfores/

National Wildlife Federation: http://www.nws.org/nws

National Wildlife Refuge System: http://bluegoose.arw.r9.fws.gov/

Natural Resources Canada: http://www.emr.ca/

Protected Areas Virtual Library: http://www.wcmc.org.uk/~dynamic/pavl/

Sierra Club Critical Ecoregions Program: http://www.sierraclub.org/ecoregions/

Sustainable Forests Directory: http://www.together.net/~wow/Index.htm

UN Convention on Biological Diversity: http://www.unep.ch/biodiv.html/

UN List of National Parks and Protected Areas: http://www.wcmc.org.uk/data/database/un_combo.html

U.S. Bureau of Land Management: http://www.blm.gov/

USDA Forest Service National Headquarters: http://www.fs.fed.us/

World Conservation Union: http://infoserver.ciesin.org:80/IC/iucn/IUCN.html/

WWW Virtual Library: Environment-Biosphere: http://ecosys.drdr.Virginia.EDU:80/bio.html/

Readings

Anderson, Patrick. 1989. "The Myth of Sustainable Logging: The Case for a Ban on Tropical Timber Imports." *The Ecologist*, vol. 19, no. 5, 166–68.

Aplet, Greg, et al., eds. 1993. *Defining Sustainable Forestry*. Covelo, Calif.: Island Press.

Barber, Charles V., et al. 1993. *Breaking the Deadlock: Obstacles to Forest Reform in Indonesia and the United States*. Washington, D.C.: World Resources Institute.

Baskin, Yvonne. 1997. *The Work of Nature: How the Diversity of Life Sustains Us*. Covelo, Calif.: Island Press.

Beatley, Timothy. 1994. *Ethical Land Use: Principles of Policy and Planning*. Baltimore: Johns Hopkins University Press.

Booth, Douglas E. 1993 *Valuing Nature: The Decline and Preservation of Old Growth Forests*. Lantham, Md.: University Press of America.

Boucher, Norman. 1995. "Back to the Everglades." *Technology Review*, August/September, 25–35.

Brush, Stephen B., and Doreen Stabinsky. 1995. *Valuing Local Knowledge: Indigenous People and Intellectual Property Rights*. Covelo, Calif.: Island Press.

Callenbach, Ernest. 1995. *Bring Back the Buffalo: A Sustainable Future for America's Great Plains*. Covelo, Calif.: Island Press.

Carroll, Mathew S. 1995. *Community and the Northwestern Logger*. Boulder, Colo.: Westview Press.

Chagnon, Napoleon A. 1992. *Yanomamo: The Last Days of Eden*. New York: Harcourt Brace Jovanovich.

Defenders of Wildlife. 1996. *A Status Report on America's Vanishing Habitat and Wildlife*. Washington, D.C.: Defenders of Wildlife.

Denniston, Derek. 1995. *High Priorities: Conserving Mountain. Ecosystems and Cultures*. Washington, D.C.: Worldwatch Institute.

Devall, Bill, ed. 1993. *Clearcut: The Tragedy of Industrial Forestry*. San Francisco: Sierra Club Books/Earth Island Press.

Dietrich, William. 1993. *The Final Forest: The Last Great Trees of the Pacific Northwest*. New York: Penguin.

DiSilvestro, Roger L. 1993. *Reclaiming the Last Wild Places: A New Agenda for Biodiversity*. New York: Wiley.

Drengson, Alan, and Duncan Taylor, eds. 1997. *Ecoforestry: The Art & Science of Sustainable Forest Use*. Gabriola Island, Canada: New Society Publishers.

Durning, Alan Thein. 1992a. *Guardians of the Land: Indigenous Peoples and the Health of the Earth*. Washington, D.C.: Worldwatch Institute.

Durning, Alan Thein. 1992b. *Saving the Forests: What Will It Take?* Washington, D.C.: Worldwatch Institute.

Foreman, Dave, and Howie Wolke. 1989. *The Big Outside*. Tucson, Ariz.: Nedd Ludd Books.

Friends of the Earth. 1992. *The Rainforest Harvest: Sustainable Strategies for Saving Tropical Forests*. Washington, D.C.: Friends of the Earth.

Frome, Michael. 1992. *Regreening the National Parks*. Tucson: University of Arizona Press.

Halvorson, William L., and Gary E. Davis, eds. 1996. *Science and Ecosystem Management in the National Parks*. Tucson, Ariz.: University of Arizona Press.

Heady, Harold F., and R. Dennis Child. 1994. *Rangeland Ecology and Management*. Boulder, Colo.: Westview Press.

Hendee, John, et al. 1991. *Principles of Wilderness Management*. Golden, Colo.: Fulcrum Publishing.

Hess, Karl, Jr. 1992. *Visions Upon the Land: Man and Nature on the Western Range*. Covelo, Calif.: Island Press.

Heywood, V. H., ed. 1995. *Global Biodiversity Assessment*. New York: Cambridge University Press.

Hirt, Paul W. 1994. *A Conspiracy of Optimism: Management of the National Forests Since World War Two*. Lincoln: University of Nebraska Press.

Hunter, Malcolm L., Jr. 1990. *Wildlife, Forests, and Forestry: Principles of Managing Forests for Biodiversity*. Englewood Cliffs, N.J.: Prentice Hall.

Jacobs, Lynn. 1992. *Waste of the West: Public Lands Ranching*. Tucson, Ariz.: Lynn Jacobs.

Johnson, Nels, and Brice Cabarle. 1993. *Looking Ahead: Sustainable Natural Forest Management in the Humid Tropics*. Washington, D.C.: World Resources Institute.

Kellert, Stephen R. 1996. *The Value of Life: Biological Diversity and Human Society*. Covelo, Calif.: Island Press.

Kelly, David, and Gary Braasch. 1988. *Secrets of the Old Growth Forest*. Salt Lake City: Peregrine Smith.

Knight, Richard L., and Sarah F. Bates, eds. 1994. *A New Century for Natural Resources Management*. Covelo, Calif.: Island Press.

Lansky, Mitch. 1992. *Beyond the Beauty Strip: Saving What's Left of Our Forests*. Gardiner, Maine: Tilbury House.

Leopold, Aldo. 1949. *A Sand County Almanac*. New York: Oxford University Press.

Little, Charles E. 1995. *The Dying of the Trees: The Pandemic in America's Forests*. New York: Viking.

Loomis, John D. 1993. *Integrated Public Lands Management: Principles and Applications to National Forest, Parks, Wildlife Refuges, and BLM Lands*. New York: Columbia University Press.

Lowry, William R. 1994. *The Capacity for Wonder: Preserving National Parks*. Washington, D.C.: The Brookings Institution.

Mahony, Rhona. 1992. "Debt-for-Nature Swaps: Who Really Benefits?" *The Ecologist*, vol. 22, no. 3, 97–103.

Manning, Richard. 1993. *Last Stand*. New York: Penguin Books.

Maser, Chris, et al. 1994. *Sustainable Forestry: Philosophy, Science and Economics*. San Francisco: Rainforest Action Network.

Maybury-Lewis, David. 1992. *Millennium: Tribal Wisdom and the Modern World*. New York: Viking Press.

Miller, Kenton, and Laura Tangley. 1991. *Trees of Life: Saving Tropical Forests and Their Biological Wealth*. Boston: Beacon Press.

Myers, Norman. 1993. *The Primary Source: Tropical Forests and Our Future*. New York: Norton.

Naar, Jon, and Alex J. Naar. 1993. *This Land Is Your Land: A Guide to North America's Endangered Ecosystems*. New York: Harper.

Nash, Roderick. 1982. *Wilderness and the American Mind*. 3d ed. New Haven, Conn.: Yale University Press.

National Academy of Sciences. 1995. *Biodiversity II: Understanding and Protecting Our Biological Resources*. Washington, D.C.: National Academy Press.

National Parks and Conservation Association. 1994. *Our Endangered Parks: What You Can Do to Protect Our National Heritage*. San Francisco: Foghorn Press.

National Park Service. 1992. *National Parks for the 21st Century*. Washington, D.C.: National Park Service.

Newman, Arnold. 1990. *The Tropical Rainforest: A World Survey of Our Most Valuable Endangered Habitats*. New York: Facts on File.

Norse, Elliot A. 1990. *Ancient Forests of the Pacific Northwest*. Covelo, Calif.: Island Press.

Noss, Reed F., and Allen Y. Cooperrider. 1994. *Saving Nature's Legacy: Protecting and Restoring Biodiversity*. Covelo, Calif.: Island Press.

Oelschlager, Max. 1991. *The Idea of Wilderness from Prehistory to the Age of Ecology*. New Haven, Conn.: Yale University Press.

Office of Technology Assessment. 1992. *Combined Summaries: Technologies to Sustain Tropical Forest Resources and Biological Diversity*. Washington, D.C.: Office of Technology Assessment.

Panayotou, Theodore, and Peter S. Ashton. 1992. *Not by Timber Alone: Economics and Ecology for Sustaining Tropical Forests*. Covelo, Calif.: Island Press.

Patterson, Alan. 1990. "Debt for Nature Swaps and the Need for Alternatives." *Environment*, vol. 32, no. 10, 5–13, 31–32.

Perlin, John. 1989. *A Forest Journey: The Role of Wood in the Development of Civilization*. New York: Norton.

Pimentel, David, et al. 1992. "Conserving Biological Diversity in Agricultural/Forestry Systems." *BioScience*, vol. 42, no. 5, 354–62.

Primack, Richard B. 1993. *Essentials of Conservation Biology*. Sunderland, Mass.: Sinauer.

Revkin, Andrew. 1990. *The Burning Season: The Murder of Chico Mendes and the Fight for the Amazon*. Boston: Houghton Mifflin.

Riebsame, William E. 1996. "Ending the Range Wars?" *Environment*, vol. 38, no. 4, 4–29.

Rietbergen, Simon, ed. 1994. *The Earthscan Reader in Tropical Forestry*. San Francisco: Rainforest Action Network.

Robbins, Jim. 1994. *Last Refuge: The Environmental Showdown in the American West*. San Francisco: HarperCollins.

Robinson, Gordon. 1987. *The Forest and the Trees: A Guide to Excellent Forestry*. Covelo, Calif.: Island Press.

Romme, William H., and Don G. Despain. 1989. "The Yellowstone Fires." *Scientific American*, vol. 261, no. 5, 37–46.

Runte, Alfred. 1990. *Yosemite: The Embattled Wilderness*. Lincoln: University of Nebraska Press.

Runte, Alfred. 1991. *Public Lands, Public Heritage: The National Forest Idea*. Niwot, Colo.: Roberts Rinehart.

Rush, James. 1991. *The Last Tree: Reclaiming the Environment in Tropical Asia*. Boulder, Colo.: Westview Press.

Ryan, John C. 1992. *Life Support: Conserving Biological Diversity*. Washington, D.C.: Worldwatch Institute.

Sedjo, Roger A., and Daniel Botkin. 1997. "Using Forest Plantations to Spare National Forests." *Environment*, vol. 39, no. 10, 12–20, 29–30.

Shiva, Vandana. 1993. *Monocultures of the Mind: Perspectives on Biodiversity and Biotechnology*. Atlantic Highlands, N.J.: Zed Books.

Shiva, Vandana. 1997. *Biopiracy: The Plunder of Nature and Knowledge*. Boston, Mass: South End Press.

Simon, Noel. 1997. *Nature In Danger: Threatened Habitats and Species*. New York: Oxford University Press.

Soulé, Michael E., and Gary Lease, eds. 1995. *Reinventing Nature? Responses to Postmodern Deconstruction*. Covelo, Calif.: Island Press.

Teitel, Martin. 1992. *Rain Forest in Your Kitchen: The Hidden Connection Between Extinction and Your Supermarket*. Covelo, Calif.: Island Press.

Tennebaum, David. 1995. "The Greening of Costa Rica." *Technology Review*, October, 42–52.

Thompson, Claudia. 1992. *Recycled Papers: The Essential Guide*. Cambridge, Mass.: MIT Press.

Tobin, Richard J. 1990. *The Expendable Future: U.S. Politics and the Protection of Biodiversity*. Durham, N.C.: Duke University Press.

Vandermeer, John, and Ivette Perfecto. 1995. *Breakfast of Biodiversity: The Truth About Rain Forest Destruction*. Oakland, Calif.: Food First.

Wellner, Pamela, and Eugene Dickey. 1991. *The Wood Users Guide*. San Francisco: Rainforest Action Network.

Wilderness Society. 1991. *Keeping It Wild: A Citizen Guide to Wilderness Management*. Washington, D.C.: Wilderness Society.

Wilson, E. O., ed. 1988. *Biodiversity*. Cambridge, Mass.: Harvard University Press.

Wilson, E. O. 1992. *The Diversity of Life*. Cambridge, Mass.: Harvard University Press.

Worster, Donald. 1994. *An Unsettled Country: Changing Landscapes of the American West*. Albuquerque: University of New Mexico Press.

Yaffee, Steven Lewis. 1994. *The Wisdom of the Spotted Owl: Policy Lessons for a New Century*. Covelo, Calif.: Island Press.

Zaslowsky, Dyan, and T. H. Watkins. 1994. *These American Lands: Parks, Wilderness, and the Public Lands*. Rev. ed. Covelo, Calif.: Island Press.

Zuckerman, Seth. 1991. *Saving Our Ancient Forests*. Federalsburg, Md.: Living Planet Press.

17 / SUSTAINING WILD SPECIES

Internet

See also sites for Chapter 16.

Animal Rights: http://www.envirolink.org/arrs/

Botanical Conservation: http://www.science.mcmaster.ca/Biology/CBCN/homepage.html

Cetaceans Home Page: http://kingfish.ssp.nmfs.gov/tmcintyr/cetacean/cetacean.html

Earth Wise Journeys: http://www.teleport.com/-earthwyz/

Endangered Species Act: http://www.law.cornell.edu/uscode/22/2151q.html

International List of Endangered Species: http://www.wcmc.org.uk/data/database/rl_anml_combo.html

International Wildlife: Education and Conservation: http://home.earthlink.net/~iwec/

National Marine Fisheries Service: http://kingfish.ssp.nmfs.gov/nmfs_pubs.html

U.S. Fish and Wildlife Service: http://www.fws.gov/

Whale Watching Web: http://www.physics.helsinki.fi/whale/

Wildlife Management: http://cervid.forsci.ualberta.ca/deernet/default.html

World Conservation Monitoring Center: http://www.wcmc.org.uk/

World Conservation Union: http://infoserver.ciesin.org:80/IC/iucn/IUCN.html/

Readings

Ackerman, Diane. 1991. *The Moon by Whalelight*. New York: Random House.

Barker, Rocky. 1993. *Saving All the Parts: Reconciling Economics and the Endangered Species Act*. Covelo, Calif.: Island Press.

Boo, Elizabeth. 1990. *Ecotourism: The Potential and the Pitfalls*. Vols. 1, 2. Washington, D.C.: World Wildlife Fund.

Bright, Chris. 1995. "Understanding the Threat of Bioinvasions." In Lester R. Brown et al., *State of the World 1996*. Washington, D.C.: Worldwatch Institute, 95–113.

Caughley, Graeme, and A. R. E. Sinclair. 1994. *Wildlife Ecology and Management*. Cambridge, Mass.: Basil Blackwell.

Clark, Tim W., et al., eds. 1994. *Endangered Species Recovery: Finding the Lessons, Improving the Process*. Covelo, Calif.: Island Press.

Cox, George W. 1993. *Conservation Ecology*. Dubuque, Iowa: Wm. C. Brown.

Credlund, Arthur G. 1983. *Whales and Whaling*. New York: Seven Hills Books.

DeBlieu, Jan. 1991. *Meant to Be Wild: The Struggle to Save Endangered Species Through Captive Breeding*. New York: Fulcrum.

DiSilvestro, Roger L. 1992. *Rebirth of Nature*. New York: Wiley.

Eldredge, Niles. 1991. *The Miner's Canary*. New York: Prentice Hall.

Gilbert, Frederick F., and Donald G. Dodds. 1992. *The Philosophy and Practice of Wildlife Management*. 2d ed. Malabar, Fla.: Robert E. Krieger.

Gray, Gary G. 1993. *Wildlife and People: The Human Dimensions of Wildlife Ecology*. Champaign: University of Illinois Press.

Grumbine, R. Edward. 1992. *Ghost Bears: Exploring the Biodiversity Crisis*. Covelo, Calif.: Island Press.

Hargrove, Eugene C. 1992. *The Animal Rights/Environmental Ethics Debate*. Ithaca: State University of New York.

Kaufman, Les, and Kenneth Mallay. 1993. *The Last Extinction*. 2d ed. Cambridge, Mass.: MIT Press.

Kellert, Stephen, and E. O. Wilson, eds. 1995. *The Biophilia Hypothesis*. Covelo, Calif.: Island Press.

Klenig, John. 1992. *Valuing Life*. Princeton, N.J.: Princeton University Press.

Kohm, Kathryn A., ed. 1990. *Balancing on the Brink of Extinction: The Endangered Species Act and Lessons for the Future*. Covelo, Calif.: Island Press.

Leakey, Richard, and Roger Lewin. 1995. *The Sixth Extinction: Patterns of Life and the Future of Humankind*. New York: Doubleday.

Livingston, John A. 1981. *The Fallacy of Wildlife Conservation*. Toronto: McClelland and Stewart.

Luoma, Jon. 1987. *A Crowded Ark: The Role of Zoos in Wildlife Conservation*. Boston: Houghton Mifflin.

Mann, Charles C., and Mark L. Plummer. 1995. *Noah's Choice: The Future of Endangered Species*. New York: Knopf.

Myers, Norman. 1994. *A Wealth of Wild Species: Storehouse for Human Welfare*. Boulder, Colo.: Westview Press.

Nash, Roderick F. 1988. *The Rights of Nature: A History of Environmental Ethics*. Madison: University of Wisconsin Press.

Office of Technology Assessment. 1989. *Oil Production in the Arctic National Wildlife Refuge*. Washington, D.C.: Government Printing Office.

Oldfield, Margery L., and Janis B. Alcorn, eds. 1992. *Biodiversity: Culture, Conservation, and Ecodevelopment*. Boulder, Colo.: Westview Press.

Primack, Richard B. 1993. *Essentials of Conservation Biology*. Sunderland, Mass.: Sinauer.

Reisner, Marc. 1991. *Game Wars: The Undercover Pursuit of Wildlife Poachers*. New York: Viking Press.

Roe, Frank G. 1970. *The North American Buffalo*. Toronto: University of Toronto Press.

Schaller, George. 1993. *The Last Panda*. Chicago: University of Chicago Press.

Soulé, Michael E. 1991. "Conservation: Tactics for a Constant Crisis." *Science*, vol. 253, 744–50.

Terborgh, John. 1992. "Why American Songbirds Are Vanishing." *Scientific American*, May, 98–104.

Tudge, Colin. 1992. *Last Animals at the Zoo: How Mass Extinction Can Be Stopped*. Covelo, Calif.: Island Press.

Tuttle, Merlin D. 1988. *America's Neighborhood Bats: Understanding and Learning to Live in Harmony with Them*. Austin: University of Texas Press.

Ward, Peter. 1995. *The End of Evolution: Mass Extinctions and the Preservation of Biodiversity*. New York: Bantam.

World Resources Institute, et al. 1992. *Global Biodiversity Strategy: Guidelines for Action to Save, Study, and Use Earth's Biotic Wealth Sustainably and Equitably*. Washington, D.C.: World Resources Institute.

18 / ENERGY EFFICIENCY AND RENEWABLE ENERGY

Internet

See also general sites in Chapter 19.

American Hydrogen Association: http://www.getnet.com/charity/aha/

American Solar Energy Society: http://www.engr.wisc.edu/centers/sel/ases/ases2.html/

American Wind Energy Association: http://www.igc.apc.org/awea/

Biofuels Information Network: http://www.esd.ornl.gov/BFDP/BFDPMOSAIC/binmenu.html/

California Energy Commission: http://www.energy.ca.gov/energy/

Critical Mass Energy Project: http://www.essential.org

Electric Vehicle Sites on the Web: http://northshore.shore.net/~kester/websites.html#alt.energy/

Energy Efficiency and Renewable Energy Network (Department of Energy): http://www.eren.doe.gov

Energy Information Administration (Department of Energy): http://www.eia.doe.gov/

Energy-Related World Wide Web Sites (Sonoma State University): http://www.sonoma.edu/ensp/linkenergy.html

Environmentally Renewable Energy Websites: http://www.netins.net/showcase/solarcatalog/friendly.html

International Network for Sustainable Energy: http://www.inforse.dk/

National Renewable Energy Lab (NREL): http://www.nrel.gov/

Passive Solar Energy Project: http://www.tuns.ca/~banderri/solar.html/

Solar Energy Articles: http://www.netins.net/showcase/solarcatalog/articles.html

Solstice: Sustainable Energy and Development: http://solstice.crest.org

Sustainable Energy Resources: http://solstice.crest.org/online/aeguide/

U.S. Green Building Council: http://gwis.circ.gwu.edu/~greenu/usgbc.html/

Wind Energy Investigations: http://sln.fi.edu/tfi/units/energy/windguide.html

WWW Virtual Library: Energy: http://solstice.crest.org/online/virtual-library/VLib-energy.html/

Yahoo: Energy Listings: http://www.yahoo.com/Science/Energy/

Readings

Also see readings for Chapter 3.

American Council for an Energy Efficient Economy. 1991. *Energy Efficiency and Environment: Forging the Link*. Washington, D.C.: American Council for an Energy Efficient Economy.

American Council for an Energy Efficient Economy. Annual. *The Most Energy-Efficient Appliances*. Washington, D.C.: American Council for an Energy Efficient Economy.

Anderson, Victor. 1993. *Energy Efficiency Policies*. New York: Routledge.

Berman, Daniel M., and John T. O'Connor. 1996. *Who Owns the Sun? People, Politics, and the Struggle for a Solar Economy*. River Junction, Vt.: Chelsea Green.

Berger, John J. 1997. *Charging Ahead: The Business of Renewable Energy and What It Means to America*. New York: Henry Holt.

Blackburn, John O. 1987. *The Renewable Energy Alternative: How the United States and the World Can Prosper Without Nuclear Energy or Coal*. Durham, N.C.: Duke University Press.

Brower, Michael. 1992. *Cool Energy: The Renewable Solution to Global Warming*. Cambridge, Mass.: Union of Concerned Scientists.

Carless, Jennifer. 1993. *Renewable Energy: A Concise Guide to Green Alternatives*. New York: Walker.

Cole, Nancy, and P. S. Skerrett. 1995. *Renewables are Ready: People Creating Energy Solutions*. White River Junction, Vt.: Chelsea Green.

Field, Frank R. III, and Joel P. Clark. 1997. "A Practical Road to Lightweight Cars." *Technology Review*, January, 28–36.

Flavin, Christopher. 1996. "Power Shock: The Next Energy Revolution." *World Watch*, January/February, 10–19.

Flavin, Christopher, and Nicholas Lenssen. 1994. *Power Surge: Guide to the Coming Energy Revolution.* New York: Norton.

Gever, John, et al. 1991. *Beyond Oil: The Threat to Food and Fuel in Coming Decades.* Boulder: University of Colorado Press.

Gipe, Paul. 1995. *Wind Energy Comes of Age.* New York: Wiley.

Harland, Edward. 1994. *Eco-Renovation: The Ecological Home Improvement Guide.* White River Junction, Vt.: Chelsea Green.

Hively, Will. 1996. "Reinventing the Wheel." *Discover*, August, 58–68.

Kierman, Patrick. 1991. *Hydrogen: The Invisible Fire.* Aspen, Colo.: IRT Environment.

Kozloff, Keith, and Roger C. Dower. 1993. *A New Power Base: Renewable Energy Policies for the Nineties and Beyond.* Washington, D.C.: World Resources Institute.

Lovins, Amory B. 1990. *The Negawatt Revolution.* Snowmass, Colo.: Rocky Mountain Institute.

Lovins, Amory B., and L. Hunter Lovins. 1995. "Reinventing the Wheels." *The Atlantic Monthly*, January, 75–94.

MacKenzie, James J. 1994. *The Keys to the Car: Electric and Hydrogen Vehicles for the 21st Century.* Washington, D.C.: World Resources Institute.

MacKenzie, James J. 1996. "Heading Off the Permanent Oil Crisis." *Issues in Science and Technology*, Summer, 48–54.

McKeown, Walter. 1991. *Death of the Oil Age and the Birth of Hydrogen America.* San Francisco: Wild Bamboo Press.

Nansen, Ralph. 1995. *Sun Power: The Global Solution for the Coming Energy Crisis.* Ocean Shores, Wash.: Ocean Press.

National Academy of Sciences. 1992. *Automotive Fuel Efficiency: How Far Can We Go?* Washington, D.C.: National Academy Press.

Office of Technology Assessment. 1990. *Replacing Gasoline: Alternative Fuels for Light-Duty Vehicles.* Washington, D.C.: Government Printing Office.

Office of Technology Assessment. 1991. *Improving Automobile Fuel Economy: New Standards, New Approaches.* Washington, D.C.: Government Printing Office.

Office of Technology Assessment. 1992a. *Building Energy Efficiency.* Washington, D.C.: Government Printing Office.

Office of Technology Assessment. 1992b. *Fueling Development: Energy Technologies for Developing Countries.* Washington, D.C.: Government Printing Office.

Ogden, Joan M., and Robert H. Williams. 1989. *Solar Hydrogen: Moving Beyond Fossil Fuels.* Washington, D.C.: World Resources Institute.

Potts, Michael. 1993. *The Independent Home: Living Well with Power from the Sun, Wind, and Water.* White River Junction, Vt.: Chelsea Green.

Rocky Mountain Institute. 1995a. *Community Energy Workbook.* Snowmass, Colo.: Rocky Mountain Institute.

Rocky Mountain Institute. 1995b. *Homemade Energy: How to Save Energy and Dollars in Your Home.* Snowmass, Colo.: Rocky Mountain Institute.

Rodman, David M., and Nicholas Lenssen. 1995. *A Building Revolution: How Ecology and Health Concerns Are Transforming Construction.* Washington, D.C.: Worldwatch Institute.

Sperling, Daniel. 1996. "The Case for Electric Vehicles." *Scientific American*, November, 54–59.

Steen, Athena Swentzell, et al. 1994. *The Straw Bale House.* White River Junction, Vt.: Chelsea Green.

Tenenbaum, David. 1995. "Tapping the Fire Down Below." *Technology Review*, January, 38–47.

Vale, Brenda, and Robert Vale. 1991. *Green Architecture.* Boston: Little, Brown.

Wells, Malcolm. 1991. *How to Build an Underground House.* Brewster, Mass.: Malcolm Wells.

Wouk, Victor. 1997. "Hybrid Electric Vehicles." *Scientific American*, October, 70–74.

Yeang, Ken. 1995. *Designing with Nature: The Ecological Basis for Architectural Design.* New York: McGraw-Hill.

Zweibel, Ken. 1990. *Harnessing Solar Power: The Challenge of Photovoltaics.* New York: Plenum.

19 / NONRENEWABLE ENERGY RESOURCES

Internet

See also general energy sites in Chapter 18.

Cold Fusion: http://www.mit.edu:8001/people/rei/CFdir/CFhome.html

Energy Network Information: http://zebu.uoregon.edu/energy.html

Energy Resources Survey Program (U.S. Geological Survey): http://energy.usgs.gov/

Fossil Fuel Issues: http://zebu.uoregon.edu/fossil.html

Nuclear Information World Wide Web Server: http://nuke.WestLab.com/

Sustainable Fusion: http://wwwofe.er.doe.gov/

University of California Energy Institute: http://www-ucenergy.eecs.berkely.edu/UCENERGY/

U.S. Department of Energy (DOE): http://www.doe.gov/

U.S. Department of Energy, Office of Fossil Energy: http://www.fe.doe.gov/

U.S. Nuclear Regulatory Commission: http://www.nrc.gov/

WWW Virtual Library: Energy: http://solstice.crest.org/online/virtual-library/VLib-energy.html/

Yahoo: Energy and Environment: http://www.wam.umd.edu/~tfagan/enrgyenv.html

Yahoo: Nuclear Energy: http://www.yahoo.com/Science/Engineering/Nuclear_Engineering/

Readings

See also the readings for Chapters 3 and 18.

Ahearne, John F. 1993. "The Future of Nuclear Power." *American Scientist*, vol. 81, 24–35.

Beck, Peter. 1994. *Prospects and Strategies for Nuclear Power: Global Boon or Dangerous Diversion?* London: Earthscan.

Chernousenko, Vladimir M. 1991. *Chernobyl: Insight from the Inside.* New York: Springer-Verlag.

Cohen, Bernard L. 1990. *The Nuclear Energy Option: An Alternative for the 90s.* New York: Plenum.

Dunn, Seth. 1997. "Power of Choice." *World Watch*, September/October, 30–38.

Flavin, Christopher. 1992. "Building a Bridge to Sustainable Energy." In Lester Brown et al., *State of the World 1992.* New York: Norton, 27–55.

Flavin, Christopher, and Nicholas Lenssen. 1994. "Reshaping the Power Industry." In Lester Brown et al., *State of the World 1994.* New York: Norton, 61–80.

Ford, Daniel F. 1986. *Meltdown.* New York: Simon & Schuster.

Gershey, Edward L., et al. 1993. *Low-Level Radioactive Waste: From Cradle to Grave.* New York: Van Nostrand Reinhold.

Golay, Michael W., and Neil E. Todreas. 1990. "Advanced Light-Water Reactors." *Scientific American*, April, 82–89.

Gould, Jay M., and Benjamin A. Goldman. 1991. *Deadly Deceit: Low-Level Radiation, High-Level Cover-up.* New York: Four Walls Eight Windows.

Hollister, Charles D., and Steven Nadis. 1998. "Burial of Radioactive Waste Under the Seabed." *Scientific American*, January, 60–65.

Huizenga, John R. 1992. *Cold Fusion: The Scientific Fiasco of the Century.* Rochester, N.Y.: University of Rochester Press.

Lasper, James M. 1990. *Nuclear Politics: Energy and the State in the United States, Sweden, and France.* Princeton, N.J.: Princeton University Press.

Lenssen, Nicholas. 1991. *Nuclear Waste: The Problem That Won't Go Away.* Washington, D.C.: Worldwatch Institute.

Lenssen, Nicholas. 1993. "All the Coal in China." *Worldwatch*, March/April, 22–28.

Lenssen, Nicholas, and Christopher Flavin. 1996. "Meltdown." *World Watch*, May/June, 23–31.

MacKenzie, James J. 1996. "Heading Off the Permanent Oil Crisis." *Issues in Science and Technology*, Summer, 48–54.

May, John. 1990. *The Greenpeace Book of the Nuclear Age.* New York: Pantheon.

Murray, Raymond L., and Judith Powell, eds. 1994. *Understanding Radioactive Waste.* 4th ed. Columbus, Ohio: Battelle Press.

National Academy of Sciences. 1991a. *Nuclear Power: Technical and Institutional Options for the Future.* Washington, D.C.: National Academy Press.

National Academy of Sciences. 1991b. *Undiscovered Oil and Gas Resources.* Washington, D.C.: National Academy Press.

National Academy of Sciences. 1995. *Coal: Energy for the Future.* Washington, D.C.: National Academy Press.

Oppenheimer, Ernest J. 1990. *Natural Gas, the Best Energy Choice.* New York: Pen & Podium.

President's Commission on the Accident at Three Mile Island. 1979. *Report of the President's Commission on the Accident at Three Mile Island.* Washington, D.C.: Government Printing Office.

Read, Piers Paul. 1993. *Ablaze: The Story of Chernobyl.* New York: Random House.

Rhoades, Richard. 1993. *Nuclear Renewal: Common Sense About Energy.* New York: Viking.

Savchenko, V. K. 1995. *The Ecology of the Chernobyl Catastrophe.* Pearl River, N.Y.: Parthenon.

Shea, Cynthia Pollack. 1989. "Decommissioning Nuclear Plants: Breaking Up Is Hard to Do." *World Watch*, July/August, 10–16.

Sierra Club. 1991. *Kick the Oil Habit: Choosing a Safe Energy Future for America.* San Francisco: Sierra Club.

Squillace, Mark. 1990. *Strip Mining Handbook.* Washington, D.C.: Friends of the Earth.

Sthcherbak, Yuri M. 1996. "Ten Years of the Chernobyl Era." *Scientific American*, April, 44–49.

Toke, David. 1995. *Energy and Environment: The Political and Economic Debate.* Boulder, Colo.: Westview Press.

Union of Concerned Scientists. 1990. *Safety Second: The NRC and America's Nuclear Power Plants.* Bloomington: Indiana University Press.

Whipple, Chris G. 1996. "Can Nuclear Waste Be Stored Safely at Yucca Mountain?" *Scientific American*, June, 72–79.

Yeargin, Daniel. 1990. *The Prize: The Epic Quest for Oil, Money, and Power.* New York: Simon & Schuster.

Zorpette, Glenn. 1996. "Hanford's Nuclear Wasteland." *Scientific American*, May, 88–97.

GLOSSARY

abiotic Nonliving. Compare *biotic*.

absolute humidity Amount of water vapor found in a certain mass of air (usually expressed as grams of water per kilogram of air). Compare *relative humidity*.

acclimation Adjustment to slowly changing new conditions. Compare *threshold effect*.

acid deposition The falling of acids and acid-forming compounds from the atmosphere to earth's surface. Acid deposition is commonly known as *acid rain*, a term that refers only to wet deposition of droplets of acids and acid-forming compounds.

acid rain See *acid deposition*.

active solar heating system System that uses solar collectors to capture energy from the sun and store it as heat for space and water heating. A liquid or air pumped through the collectors transfers the captured heat to a storage system such as an insulated water tank or rock bed. Pumps or fans then distribute the stored heat or hot water throughout a dwelling as needed. Compare *passive solar heating system*.

adaptation Any genetically controlled trait that helps an organism survive and reproduce under a given set of environmental conditions. It usually results from a beneficial mutation. See *biological evolution, differential reproduction, mutation, natural selection*.

adaptive radiation Period of time (usually millions of years) during which numerous new species evolve to fill vacant and new ecological niches in changed environments, usually after a mass extinction.

adaptive trait See *adaptation*.

advanced sewage treatment Specialized chemical and physical processes that reduce the amount of specific pollutants left in wastewater after primary and secondary sewage treatment. This type of treatment is usually expensive. See also *primary sewage treatment, secondary sewage treatment*.

aerobic respiration Complex process that occurs in the cells of most living organisms, in which nutrient organic molecules such as glucose ($C_6H_{12}O_6$) combine with oxygen (O_2) and produce carbon dioxide (CO_2), water (H_2O), and energy. Compare *photosynthesis*.

age structure Percentage of the population (or the number of people of each sex) at each age level in a population.

agricultural revolution Gradual shift from small, mobile hunting and gathering bands to settled agricultural communities, in which people survived by learning how to breed and raise wild animals and to cultivate wild plants near where they lived. It began 10,000–12,000 years ago. Compare *industrial revolution*.

agroforestry Planting trees and crops together.

air pollution One or more chemicals in high enough concentrations in the air to harm humans, other animals, vegetation, or materials. Excess heat or noise can also be considered forms of air pollution. Such chemicals or physical conditions are called air pollutants. See *primary pollutant, secondary pollutant*.

alien species See *nonnative species*.

alleles Slightly different molecular forms found in a particular gene.

alley cropping Planting of crops in strips with rows of trees or shrubs on each side.

alpha particle Positively charged matter, consisting of two neutrons and two protons, that is emitted as a form of radioactivity from the nuclei of some radioisotopes. See also *beta particle, gamma rays*.

altitude Height above sea level. Compare *latitude*.

anaerobic respiration Form of cellular respiration in which some decomposers get the energy they need through the breakdown of glucose (or other nutrients) in the absence of oxygen. Compare *aerobic respiration*.

ancient forest See *old-growth forest*.

animal manure Dung and urine of animals that can be used as a form of organic fertilizer. Compare *green manure*.

animals Eukaryotic, multicelled organisms such as sponges, jellyfishes, arthropods (insects, shrimp, lobsters), mollusks (snails, oysters, octopuses), fish, amphibians (frogs, toads, salamanders), reptiles (turtles, alligators, snakes), birds, and mammals (kangaroos, bats, whales, apes, humans). See *carnivores, herbivores, omnivores*.

annual Plant that grows, sets seed, and dies in one growing season. Compare *perennial*.

aquaculture Growing and harvesting of fish and shellfish for human use in freshwater ponds, irrigation ditches, and lakes, or in cages or fenced-in areas of coastal lagoons and estuaries. See *fish farming, fish ranching*.

aquatic Pertaining to water. Compare *terrestrial*.

aquatic life zone Marine and freshwater portions of the ecosphere. Examples include freshwater life zones (such as lakes and streams) and ocean or marine life zones (such as estuaries, coastlines, coral reefs, and the deep ocean).

aquifer An underground rock body that has a high to moderate permeability and can yield a significant amount of water.

arid Dry. A desert or other area with an arid climate has little precipitation.

asexual reproduction Reproduction in which a mother cell divides to produce two identical daughter cells that are clones of the mother cell. This type of reproduction is common in single-celled organisms. Compare *sexual reproduction*.

atmosphere The whole mass of air surrounding the earth. See *stratosphere, troposphere*.

atomic number Number of protons in the nucleus of an atom. Compare *mass number*.

atoms Minute units made of subatomic particles that are the basic building blocks of all chemical elements and thus all matter; the smallest unit of an element that can exist and still have the unique characteristics of that element. Compare *ion, molecule*.

autotroph See *producer*.

background extinction Normal extinction of various species as a result of changes in local environmental conditions. Compare *mass extinction*.

bacteria Prokaryotic, one-celled organisms. Some transmit diseases. Most act as decomposers and get the nutrients they need by breaking down complex organic compounds in the tissues of living or dead organisms into simpler inorganic nutrient compounds.

barrier islands Long, thin, low, offshore islands of sediment that generally run parallel to the shore along some coasts.

beneficiation Separation of an ore mineral from the waste mineral material (gangue). See *tailings*.

beta particle Swiftly moving electron emitted by the nucleus of a radioactive isotope. See also *alpha particle, gamma rays*.

bioaccumulation An increase in the concentration of a chemical in specific organs or tissues at a level higher than would normally be expected. Compare *biomagnification*.

biodegradable A form of matter capable of being broken down by decomposers.

biodegradable pollutant Material that can be broken down into simpler substances (elements and compounds) by bacteria or other decomposers. Paper and most organic wastes such as animal manure are biodegradable but can take decades to biodegrade in modern landfills. Compare *degradable pollutant, nondegradable pollutant, slowly degradable pollutant*.

biodiversity See *biological diversity*.

biogeochemical cycle Natural processes that recycle nutrients in various chemical forms from the nonliving environment to living organisms, and then back to the nonliving environment. Examples are the carbon, oxygen, nitrogen, phosphorus, sulfur, and hydrologic cycles.

biological community See *community*.

biological diversity Variety of different species (*species diversity*), genetic variability among individuals within each species (*genetic diversity*), and variety of ecosystems (*ecological diversity*).

biological evolution Change in the genetic makeup of a population of a species in successive generations. If continued long enough, it can lead to the formation of a new species. Note that populations, not individuals, evolve. See also *adaptation, differential reproduction, natural selection, theory of evolution*.

biological oxygen demand (BOD) Amount of dissolved oxygen needed by aerobic decomposers to break down the organic materials in a given volume of water at a certain temperature over a specified time period.

biological pest control Control of pest populations by natural predators, parasites, or disease-causing bacteria and viruses (pathogens).

biomagnification Increase in concentration of DDT, PCBs, and other slowly degradable, fat-soluble chemicals in organisms at successively higher trophic levels of a food chain or web. Compare *bioaccumulation*.

biomass Organic matter produced by plants and other photosynthetic producers; total dry weight of all living organisms that can be supported at each trophic level in a food chain or web; dry weight of all organic matter in plants and animals in an ecosystem; plant materials and animal wastes used as fuel.

biome Terrestrial regions inhabited by certain types of life, especially vegetation. Examples are various types of deserts, grasslands, and forests.

biosphere Zone of the earth where life is found. It consists of parts of the atmosphere (the troposphere), hydrosphere (mostly surface water and groundwater), and lithosphere (mostly soil and surface rocks and sediments on the bottoms of oceans and other bodies of water) where life is found. It is also called the ecosphere.

biotic   Living organisms. Compare *abiotic*.

biotic potential   Maximum rate at which the population of a given species can increase when there are no limits on its rate of growth. See *environmental resistance*.

birth rate   See *crude birth rate*.

bitumen   Gooey, black, high-sulfur, heavy oil extracted from tar sand and then upgraded to synthetic fuel oil. See *tar sand*.

breeder nuclear fission reactor   Nuclear fission reactor that produces more nuclear fuel than it consumes by converting nonfissionable uranium-238 into fissionable plutonium-239.

broadleaf deciduous plants   Plants, such as oak and maple trees, that survive drought and cold by shedding their leaves and becoming dormant during such periods. Compare *broadleaf evergreen plants, coniferous evergreen plants*.

broadleaf evergreen plants   Plants that keep most of their broad leaves year-round. Examples are the trees found in the canopies of tropical rain forests. Compare *broadleaf deciduous plants, coniferous evergreen plants*.

cancer   Group of more than 120 different diseases, one for each type of cell in the human body. Each type of cancer produces a tumor in which cells multiply uncontrollably and invade surrounding tissue.

capitalism   See *pure market economic system*.

carbon cycle   Cyclic movement of carbon in different chemical forms from the environment to organisms and then back to the environment.

carcinogen   Chemicals, ionizing radiation, and viruses that cause or promote the development of cancer. See *cancer, mutagen, teratogen*.

carnivore   Animal that feeds on other animals. Compare *herbivore, omnivore*.

carrying capacity (K)   Maximum population of a particular species that a given habitat can support over a given period of time.

cell   Smallest living unit of an organism. Each cell is encased in an outer membrane or wall and contains genetic material (DNA) and other parts to perform its life function. Organisms such as bacteria consist of only one cell, but most of the organisms we are familiar with contain many cells. See *eukaryotic cell, prokaryotic cell*.

centrally planned economy   See *pure command economic system*.

CFCs   See *chlorofluorocarbons*.

chain reaction   Multiple nuclear fissions, taking place within a certain mass of a fissionable isotope, that release an enormous amount of energy in a short time.

chemical   One of the millions of different elements and compounds found naturally or synthesized by humans. See *compound, element*.

chemical change   Interaction between chemicals in which there is a change in the chemical composition of the elements or compounds involved. Compare *physical change*.

chemical evolution   Formation of the earth and its early crust and atmosphere, evolution of the biological molecules necessary for life, and evolution of systems of chemical reactions needed to produce the first living cells. These processes are believed to have occurred about 1 billion years before biological evolution. Compare *biological evolution*.

chemical reaction   See *chemical change*.

chemosynthesis   Process in which certain organisms (mostly specialized bacteria) extract inorganic compounds from their environment and convert them into organic nutrient compounds without the presence of sunlight. Compare *photosynthesis*.

chlorofluorocarbons (CFCs)   Organic compounds made up of atoms of carbon, chlorine, and fluorine. An example is Freon-12 (CCl_2F_2), used as a refrigerant in refrigerators and air conditioners and in making plastics such as Styro-

foam. Gaseous CFCs can deplete the ozone layer when they slowly rise into the stratosphere and their chlorine atoms react with ozone molecules. Use of these molecules is being phased out.

chromosome   A grouping of various genes and associated proteins in plant and animal cells that carry certain types of genetic information. See *genes*.

clear-cutting   Method of timber harvesting in which all trees in a forested area are removed in a single cutting. Compare *seed-tree cutting, selective cutting, shelterwood cutting, strip cutting*.

climate   Physical properties of the troposphere of an area based on analysis of its weather records over a long period (at least 30 years). Compare *weather*.

coal   Solid, combustible mixture of organic compounds with 30–98% carbon by weight, mixed with various amounts of water and small amounts of sulfur and nitrogen compounds. It is formed in several stages as the remains of plants are subjected to heat and pressure over millions of years.

coastal wetland   Land along a coastline, extending inland from an estuary that is covered with salt water all or part of the year. Examples are marshes, bays, lagoons, tidal flats, and mangrove swamps. Compare *inland wetland*.

coastal zone   Warm, nutrient-rich, shallow part of the ocean that extends from the high-tide mark on land to the edge of a shelflike extension of continental land masses known as the continental shelf. Compare *open sea*.

coevolution   Evolution occurring when two or more species interact and exert selective pressures on each other that can lead each species to undergo various adaptations. See *evolution, natural selection*.

cogeneration   Production of two useful forms of energy, such as high-temperature heat or steam and electricity, from the same fuel source.

commensalism   An interaction between organisms of different species in which one type of organism benefits and the other type is neither helped nor harmed to any great degree. Compare *mutualism*.

commercial extinction   Depletion of the population of a wild species used as a resource to a level at which it is no longer profitable to harvest the species.

commercial inorganic fertilizer   Commercially prepared mixtures of plant nutrients such as nitrates, phosphates, and potassium applied to the soil to restore fertility and increase crop yields. Compare *organic fertilizer*.

common-property resource   Resource that people are normally free to use; each user can deplete or degrade the available supply. Most are potentially renewable and are owned by no one. Examples are clean air, fish in parts of the ocean not under the control of a coastal country, migratory birds, gases of the lower atmosphere, and the ozone content of the upper atmosphere. See *tragedy of the commons*.

community   Populations of all species living and interacting in an area at a particular time.

community development   See *ecological succession*.

competition   Two or more individual organisms of a single species (*intraspecific competition*) or two or more individuals of different species (*interspecific competition*) attempting to use the same scarce resources in the same ecosystem.

competitive exclusion   No two species can occupy exactly the same fundamental niche indefinitely in a habitat where there is not enough of a particular resource to meet the needs of both species. See *ecological niche, fundamental niche, realized niche*.

compost   Partially decomposed organic plant and animal matter that can be used as a soil conditioner or fertilizer.

compound   Combination of atoms, or oppositely charged ions, of two or more different elements

held together by attractive forces called chemical bonds. Compare *element*.

concentration   Amount of a chemical in a particular volume or weight of air, water, soil, or other medium.

condensation   Conversion of a gas to a liquid.

condensation nuclei   Tiny particles on which droplets of water vapor can collect.

coniferous evergreen plants   Cone-bearing plants (such as spruces, pines, and firs) that keep some of their narrow, pointed leaves (needles) all year. Compare *broadleaf deciduous plants, broadleaf evergreen plants*.

consensus science   Scientific data, models, theories, and laws that are widely accepted. This aspect of science is very reliable. Compare *frontier science*.

conservation biology   Multidisciplinary science created to deal with the crisis of maintaining the genes, species, communities, and ecosystems that make up the earth's biological diversity. Its goals are to investigate human impacts on biodiversity and to develop practical approaches to preserving biodiversity and ecological integrity.

conservation-tillage farming   Crop cultivation in which the soil is disturbed little (minimum-tillage farming) or not at all (no-till farming) to reduce soil erosion, lower labor costs, and save energy. Compare *conventional-tillage farming*.

constancy   Ability of a living system, such as a population, to maintain a certain size. Compare *inertia, resilience*. See *homeostasis*.

consumer   Organism that cannot synthesize the organic nutrients it needs and gets its organic nutrients by feeding on the tissues of producers or other consumers; generally divided into *primary consumers* (herbivores), *secondary consumers* (carnivores), *tertiary (higher-level) consumers, omnivores*, and *detritivores* (decomposers and detritus feeders). In economics, one who uses economic goods.

contour farming   Plowing and planting across the changing slope of land, rather than in straight lines, to help retain water and reduce soil erosion.

conventional-tillage farming   Crop cultivation that involves making a planting surface by plowing land, breaking up the exposed soil, and then smoothing the surface. Compare *conservation-tillage farming*.

convergent plate boundary   Area where the earth's lithospheric plates are pushed together. See *subduction zone*. Compare *divergent plate boundary, transform fault*.

coral reef   Formation produced by massive colonies containing billions of tiny coral animals, called polyps, which secrete a stony substance (calcium carbonate) around themselves for protection. When the corals die, their empty outer skeletons form layers that cause the reef to grow. They are found in the coastal zones of warm tropical and subtropical oceans.

core   Inner zone of the earth. It consists of a solid inner core and a liquid outer core. Compare *crust, mantle*.

cost–benefit analysis   Estimates and comparison of short-term and long-term costs (losses) and benefits (gains) from an economic decision. If the estimated benefits exceed the estimated costs, the decision to buy an economic good or provide a public good is considered worthwhile.

critical mass   Amount of fissionable nuclei needed to sustain a nuclear fission chain reaction.

crop rotation   Planting a field, or an area of a field, with different crops from year to year to reduce depletion of soil nutrients. A plant such as corn, tobacco, or cotton, which removes large amounts of nitrogen from the soil, is planted one year. The next year a legume such as soybeans, which adds nitrogen to the soil, is planted.

crude birth rate   Annual number of live births per 1,000 people in a geographical area at the midpoint of a given year. Compare *crude death rate*.

crude death rate   Annual number of deaths per 1,000 people in a geographical area at the midpoint of a given year. Compare *crude birth rate*.

crude oil Gooey liquid consisting mostly of hydrocarbon compounds and small amounts of compounds containing oxygen, sulfur, and nitrogen. Extracted from underground accumulations, it is sent to oil refineries, where it is converted to heating oil, diesel fuel, gasoline, tar, and other materials.

crust Solid outer zone of the earth. It consists of oceanic crust and continental crust. Compare *core, mantle*.

cultural eutrophication Overnourishment of aquatic ecosystems with plant nutrients (mostly nitrates and phosphates) because of human activities such as agriculture, urbanization, and discharges from industrial plants and sewage treatment plants. See *eutrophication*.

cyanobacteria Single-celled, prokaryotic, microscopic organisms. Before being reclassified as monera, they were called blue-green algae.

death rate See *crude death rate.*

deciduous plants Trees that survive during dry seasons or cold seasons by shedding their leaves. See *broadleaf deciduous plants*. Compare *broadleaf evergreen plants, coniferous evergreen plants*.

decomposer Organism that digests parts of dead organisms and cast-off fragments and wastes of living organisms. A decomposer breaks down the complex organic molecules in those materials into simpler inorganic compounds and then absorbs the soluble nutrients. Most of these chemicals are returned to the soil and water for reuse by producers. Decomposers consist of various bacteria and fungi. Compare *consumer, detritivore, producer*.

degradable pollutant Potentially polluting chemical that is broken down completely or reduced to acceptable levels by natural physical, chemical, and biological processes. Compare *biodegradable pollutant, nondegradable pollutant, slowly degradable pollutant*.

degree of urbanization Percentage of the population in the world, or a country, living in areas with a population of more than 2,500 people (higher in some countries). Compare *urban growth*.

democracy Government by the people through their elected officials and appointed representatives. In a constitutional democracy, a constitution provides the basis of government authority and puts restraints on government power through free elections and freely expressed public opinion.

demographic transition Hypothesis that countries, as they become industrialized, have declines in death rates followed by declines in birth rates.

depletion time How long it takes to use a certain fraction—usually 80%—of the known or estimated supply of a nonrenewable resource at an assumed rate of use. Finding and extracting the remaining 20% usually costs more than it is worth.

desalination Purification of salt water or brackish (slightly salty) water by removing dissolved salts.

desert Biome in which evaporation exceeds precipitation and the average amount of precipitation is less than 25 centimeters (10 inches) a year. Such areas have little vegetation or have widely spaced, mostly low vegetation. Compare *forest, grassland*.

desertification Conversion of rangeland, rainfed cropland, or irrigated cropland to desertlike land, with a drop in agricultural productivity of 10% or more. It is usually caused by a combination of overgrazing, soil erosion, prolonged drought, and climate change.

detritivore Consumer organism that feeds on detritus, parts of dead organisms and cast-off fragments and wastes of living organisms. The two principal types are *detritus feeders* and *decomposers*.

detritus Parts of dead organisms and cast-off fragments and wastes of living organisms.

detritus feeder Organism that extracts nutrients from fragments of dead organisms and the cast-off parts and organic wastes of living organisms. Examples are earthworms, termites, and crabs. Compare *decomposer*.

developed country Country that is highly industrialized and has a high per capita GNP. Compare *developing country*.

developing country Country that has low to moderate industrialization and low to moderate per capita GNP. Most are located in Africa, Asia, and Latin America. Compare *developed countries*.

development Change from a society that is largely rural, agricultural, illiterate, and poor, with a rapidly growing population, to one that is mostly urban, industrial, educated, and wealthy, with a slowly growing or stationary population. See *developed country* and *developing country*.

dew point Temperature at which condensation occurs for a given amount of water vapor.

differential reproduction Phenomenon in which individuals with adaptive genetic traits produce more living offspring than do individuals without such traits. See *natural selection*.

dioxins Family of 75 different toxic chlorinated hydrocarbon compounds formed as by-products in chemical reactions involving chlorine and hydrocarbons, usually at high temperatures.

dissolved oxygen (DO) content Amount of oxygen gas (O_2) dissolved in a given volume of water at a particular temperature and pressure, often expressed as a concentration in parts of oxygen per million parts of water.

divergent plate boundary Area where the earth's lithospheric plates move apart in opposite directions. Compare *convergent plate boundary, transform fault*.

DNA (deoxyribonucleic acid) Large molecules in the cells of organisms that carry genetic information in living organisms.

domesticated species Wild species tamed or genetically altered by crossbreeding for use by humans for food (cattle, sheep, and food crops), pets (dogs and cats), or enjoyment (animals in zoos and plants in gardens).

dose The amount of a potentially harmful substance an individual ingests, inhales, or absorbs through the skin. Compare *response*. See *dose–response curve, median lethal dose*.

dose–response curve Plot of data showing effects of various doses of a toxic agent on a group of test organisms. See *dose, median lethal dose, response*.

doubling time The time it takes (usually in years) for the quantity of something growing exponentially to double. It can be calculated by dividing the annual percentage growth rate into 70.

drainage basin See *watershed*.

dredge spoils Materials scraped from the bottoms of harbors and streams to maintain shipping channels. They are often contaminated with high levels of toxic substances that have settled out of the water. See *dredging*.

dredging Type of surface mining in which chain buckets and draglines scrape up sand, gravel, and other surface deposits covered with water. It is also used to remove sediment from streams and harbors to maintain shipping channels. See *dredge spoils*.

dust dome Dome of heated air that surrounds an urban area and traps pollutants, especially suspended particulate matter. See also *urban heat island*.

dust plume Elongation of a dust dome by winds that can spread a city's pollutants hundreds of kilometers downwind.

earth capital The earth's natural resources and processes that sustain us and other species. Compare *human capital, manufactured capital, solar capital*.

earthquake Shaking of the ground resulting either from the fracturing and displacement of rock, producing a fault, or from subsequent movement along the fault

earth-sustaining economy Economic system in which the number of people and the quantity of goods are maintained at a constant level. This level is ecologically sustainable over time and meets at least the basic needs of all members of the population.

earth-wisdom society Society based on working with nature by recycling and reusing discarded matter; preventing pollution; conserving matter and energy resources through reducing unnecessary waste and use; not degrading renewable resources; building things that are easy to recycle, reuse, and repair; not allowing population size to exceed the carrying capacity of the environment; and preserving biodiversity. See *earth-wisdom worldview*. Compare *matter-recycling society, planetary management worldview*.

earth-wisdom worldview Beliefs that nature exists for all of the earth's species, not just for us; we are not in charge of the rest of nature; there is not always more, and it's not all for us; some forms of economic growth are beneficial, and some are harmful; and our goals should be to design economic and political systems that encourage earth-sustaining forms of growth and discourage or prohibit earth-degrading forms. Our success depends on learning to cooperate with one another and with the rest of nature instead of trying to dominate and manage earth's life-support systems primarily for our own use. Compare *planetary management worldview*.

ecological diversity The variety of forests, deserts, grasslands, oceans, streams, lakes, and other biological communities interacting with one another and with their nonliving environment. See *biological diversity*. Compare *genetic diversity, species diversity*.

ecological health The degree to which an area's biodiversity and ecological integrity remain intact. See *biodiversity, ecological integrity*.

ecological integrity The conditions and natural processes (such as the flow of materials and energy through ecosystems) that generate and maintain biodiversity and allow evolutionary change as a key mechanism for adapting to changes in environmental conditions.

ecological land-use planning Method for deciding how land should be used; development of an integrated model that considers geological, ecological, health, and social variables.

ecologically sustainable development Development in which the total human population size and resource use in the world (or in a region) are limited to a level that does not exceed the carrying capacity of the existing natural capital and is therefore sustainable.

ecological niche Total way of life or role of a species in an ecosystem. It includes all physical, chemical, and biological conditions a species needs to live and reproduce in an ecosystem. See *fundamental niche, realized niche*.

ecological succession Process in which communities of plant and animal species in a particular area are replaced over time by a series of different and often more complex communities. See *primary succession, secondary succession*.

ecology Study of the interactions of living organisms with one another and with their nonliving environment of matter and energy; study of the structure and functions of nature.

economic decision Deciding what goods and services to produce, how to produce them, how much to produce, and how to distribute them to people.

economic growth Increase in the real value of all final goods and services produced by an economy; an increase in real GNP or GDP. Compare *ecologically sustainable development*.

economic resources Natural resources, capital goods, and labor used in an economy to produce material goods and services. See *earth capital, human capital, manufactured capital*.

economic system Method that a group of people uses to choose what goods and services to produce, how to produce them, how much to produce, and how to distribute them to people. See *mixed economic system, pure command economic system, pure market economic system*.

economy System of production, distribution, and consumption of economic goods.

ecosphere Earth's collection of living organisms interacting with one another and their nonliving environment (energy and matter) throughout the world; all of earth's ecosystems. Also called the *biosphere*.

ecosystem Community of different species interacting with one another and with the chemical and physical factors making up its nonliving environment.

ecotone Transitional zone in which one type of ecosystem tends to merge with another ecosystem.

electromagnetic radiation Forms of kinetic energy traveling as electromagnetic waves. Examples are radio waves, TV waves, microwaves, infrared radiation, visible light, ultraviolet radiation, X rays, and gamma rays. Compare *ionizing radiation, nonionizing radiation*.

electron (e) Tiny particle moving around outside the nucleus of an atom. Each electron has one unit of negative charge (-) and almost no mass. Compare *neutron, proton*.

element Chemical, such as hydrogen (H), iron (Fe), sodium (Na), carbon (C), nitrogen (N), or oxygen (O), whose distinctly different atoms serve as the basic building blocks of all matter. There are 92 naturally occurring elements. Another 18 have been made in laboratories. Two or more elements combine to form compounds, which make up most of the world's matter. Compare *compound*.

endangered species Wild species with so few individual survivors that the species could soon become extinct in all or most of its natural range. Compare *threatened species*.

energy Capacity to do work by performing mechanical, physical, chemical, or electrical tasks or to cause a heat transfer between two objects at different temperatures.

energy efficiency Percentage of the total energy input that does useful work and is not converted into low-quality, usually useless heat in an energy conversion system or process. See *energy quality, net energy*.

energy quality Ability of a form of energy to do useful work. High-temperature heat and the chemical energy in fossil fuels and nuclear fuels are concentrated high-quality energy. Low-quality energy such as low-temperature heat is dispersed or diluted and cannot do much useful work. See *high-quality energy, low-quality energy*.

enhanced oil recovery Removal of some of the heavy oil left in an oil well after primary and secondary recovery. Compare *primary oil recovery, secondary oil recovery*.

environment All external conditions and factors, living and nonliving (chemicals and energy), that affect an organism or other specified system during its lifetime; the earth's life-support systems for us and for all other forms of life—another term for describing solar capital and earth capital.

environmental degradation Depletion or destruction of a potentially renewable resource such as soil, grassland, forest, or wildlife by using it faster than it is naturally replenished. If such use continues, the resource can become nonrenewable (on a human time scale) or nonexistent (extinct). See also *sustainable yield*.

environmental ethics Beliefs about what is right or wrong behavior toward the environment or the earth's life-support systems.

environmental resistance All the limiting factors acting together to limit the growth of a population. See *biotic potential, limiting factor*.

environmental science Study of how we and other species interact with one another and with the nonliving environment (matter and energy). It is a physical and social science that integrates knowledge from a wide range of disciplines including physics, chemistry, biology (especially ecology), geology, geography, resource technology and engineering, resource conservation and management, demography (the study of population

dynamics), economics, politics, sociology, psychology, and ethics. In other words, it is a study of how the parts of nature and human societies operate and interact—a study of connections and interactions.

environmental worldview How people think the world works, what they think their role in the world should be, and what they believe is right and wrong environmental behavior (ethics). See *earth-wisdom worldview, planetary management worldview*.

epidemiology Study of the patterns of disease or other harmful effects from toxic exposure within defined groups of people to find out why some people get sick and some do not.

erosion Process or group of processes by which loose or consolidated earth materials are dissolved, loosened, or worn away and removed from one place and deposited in another. See *weathering*.

estuary Partially enclosed coastal area at the mouth of a river where its fresh water, carrying fertile silt and runoff from the land, mixes with salty seawater.

ethics Beliefs about what is right or wrong behavior.

eukaryotic cell Cell containing a *nucleus*, a region of genetic material surrounded by a membrane. Membranes also enclose several of the other internal parts found in a eukaryotic cell. Compare *prokaryotic cell*.

eutrophication Physical, chemical, and biological changes that take place after a lake, an estuary, or a slow-flowing stream receives inputs of plant nutrients, mostly nitrates and phosphates, from natural erosion and runoff from the surrounding land basin. See *cultural eutrophication*.

eutrophic lake Lake with a large or excessive supply of plant nutrients, mostly nitrates and phosphates. Compare *mesotrophic lake, oligotrophic lake*.

evaporation Conversion of a liquid to a gas.

even-aged management Method of forest management in which trees, sometimes of a single species in a given stand, are maintained at about the same age and size and are harvested all at once. Compare *uneven-aged management*.

evolution See *biological evolution*.

exhaustible resource See *nonrenewable resource*.

exotic species See *nonnative species*.

experiment Procedure a scientist uses to study some phenomenon under known conditions. Some experiments are conducted in the laboratory, but others are conducted in nature. The resulting scientific data or facts must be verified or confirmed by repeated observations and measurements, ideally by several different investigators.

exploitation competition Situation in which two competing species have equal access to a specific resource but differ in how quickly or efficiently they exploit it. See *interference competition, interspecific competition*. Compare *intraspecific competition*.

exponential growth Growth in which some quantity, such as population size or economic output, increases by a fixed percentage of the whole in a given time; when the increase in quantity over a long enough time is plotted, this type of growth typically yields a curve shaped like the letter J. Compare *linear growth*.

external benefit Beneficial social, health, or environmental effect of producing and using an economic good that is not included in the market price of the good. Compare *external cost, full cost*.

external cost Harmful social, health, or environmental effect of producing and using an economic good that is not included in the market price of the good. Compare *external benefit, full cost*.

externalities Social benefits ("goods") and social costs ("bads") not included in the market price of an economic good. See *external benefit, external cost*. Compare *full cost, internal cost*.

extinction Complete disappearance of a species from the earth. This happens when a species can-

not adapt and successfully reproduce under new environmental conditions or when it evolves into one or more new species. Compare *speciation*. See also *endangered species, threatened species*.

family planning Providing information, clinical services, and contraceptives to help people choose the number and spacing of children they want to have.

feedback loop Circuit of sensing, evaluating, and reacting to changes in environmental conditions as a result of information fed back into a system; it occurs when one change leads to some other change, which then eventually either reinforces or slows the original change. See *negative feedback loop, positive feedback loop*.

fermentation See *anaerobic respiration*.

fertilizer Substance that adds inorganic or organic plant nutrients to soil and improves its ability to grow crops, trees, or other vegetation. See *commercial inorganic fertilizer, organic fertilizer*.

first law of energy See *first law of thermodynamics*.

first law of human ecology We can never do merely one thing. Any intrusion into nature has numerous effects, many of which are unpredictable.

first law of thermodynamics In any physical or chemical change, no detectable amount of energy is created or destroyed, but in these processes energy can be changed from one form to another. You can't get more energy out of something than you put in; in terms of energy quantity, you can't get something for nothing (there is no free lunch). This law does not apply to nuclear changes, in which energy can be produced from small amounts of matter. See also *second law of thermodynamics*.

fishery Concentrations of particular aquatic species suitable for commercial harvesting in a given ocean area or inland body of water.

fish farming Form of aquaculture in which fish are cultivated in a controlled pond or other environment and harvested when they reach the desired size. See also *fish ranching*.

fish ranching Form of aquaculture in which members of a fish species such as salmon are held in captivity for the first few years of their lives, released, and then harvested as adults when they return from the ocean to their freshwater birthplace to spawn. See also *fish farming*.

floodplain Flat valley floor next to a stream channel. For legal purposes, the term is often applied to any low area that has the potential for flooding, including certain coastal areas.

flow See *throughput*.

flyway Generally fixed route along which waterfowl migrate from one area to another at certain seasons of the year.

food additive A natural or synthetic chemical deliberately added to processed foods to retard spoilage, provide missing amino acids and vitamins, or enhance flavor, color, and texture.

food chain Series of organisms in which each eats or decomposes the preceding one. Compare *food web*.

food web Complex network of many interconnected food chains and feeding relationships. Compare *food chain*.

forest Biome with enough average annual precipitation (at least 76 centimeters, or 30 inches) to support growth of various species of trees and smaller forms of vegetation. Compare *desert, grassland*.

fossils Skeletons, bones, shells, body parts, leaves, seeds, or impressions of such items that provide recognizable evidence of organisms that lived long ago.

Freons See *chlorofluorocarbons*.

frontier science Preliminary scientific data, hypotheses, and models that have not been widely tested and accepted. Compare *consensus science*.

frontier worldview Viewing undeveloped land as a hostile wilderness to be conquered (cleared, planted) and exploited for its resources as quickly as possible. See *earth-wisdom worldview, planetary management worldview.*

full cost Cost of a good when its internal costs and its estimated short- and long-term external costs are included in its market price. Compare *external cost, internal cost.*

fundamental niche The full potential range of the physical, chemical, and biological factors a species can use if there is no competition from other species. See *ecological niche.* Compare *realized niche.*

fungi Eukaryotic, mostly multicelled organisms such as mushrooms, molds, and yeasts. As decomposers, they get the nutrients they need by secreting enzymes that break down the organic matter in the tissue of other living or dead organisms. Then they absorb the resulting nutrients.

gamma rays A form of ionizing, electromagnetic radiation with a high energy content, emitted by the sun and some radioisotopes. They readily penetrate body tissues.

gangue Waste or undesired material in an ore. See *ore.*

gene mutation See *mutation.*

gene pool The sum total of all genes found in the individuals of the population of a particular species.

generalist species Species with a broad ecological niche. They can live in many different places, eat a variety of foods, and tolerate a wide range of environmental conditions. Examples are flies, cockroaches, mice, rats, and human beings. Compare *specialist species.*

genes Coded units of information about specific traits that are passed on from parents to offspring during reproduction. They consist of segments of DNA molecules found in chromosomes.

genetic diversity Variability in the genetic makeup among individuals within a single species. See *biodiversity.* Compare *ecological diversity, species diversity.*

geographic isolation Separation of populations of a species for fairly long times into areas with different environmental conditions.

geothermal energy Heat transferred from the earth's underground concentrations of dry steam (steam with no water droplets), wet steam (a mixture of steam and water droplets), or hot water trapped in fractured or porous rock.

GDP See *gross domestic product.*

GNP See *gross national product, per capita GNP.*

grassland Biome found in regions where moderate annual average precipitation (25 to 76 centimeters, or 10 to 30 inches) is enough to support the growth of grass and small plants, but not enough to support large stands of trees. Compare *desert, forest.*

greenhouse effect A natural effect that traps heat in the atmosphere (troposphere) near the earth's surface. Some of the heat flowing back toward space from the earth's surface is absorbed by water vapor, carbon dioxide, ozone, and several other gases in the lower atmosphere (troposphere) and then radiated back toward the earth's surface. If the atmospheric concentrations of these greenhouse gases rise and are not removed by other natural processes, the average temperature of the lower atmosphere will gradually increase.

greenhouse gases Gases in the earth's lower atmosphere (troposphere) that cause the greenhouse effect. Examples are carbon dioxide, chlorofluorocarbons, ozone, methane, water vapor, and nitrous oxide.

green manure Freshly cut or still-growing green vegetation that is plowed into the soil to increase the organic matter and humus available to support crop growth. Compare *animal manure.*

green revolution Popular term for introduction of scientifically bred or selected varieties of grain (rice, wheat, maize) that, with high enough inputs of fertilizer and water, can greatly increase crop yields.

gross domestic product (GDP) Total market value in current dollars of all goods and services

produced *within* a country for final use usually during a year. Compare *gross national product.*

gross national product (GNP) Total market value in current dollars of all goods and services produced by an economy for final use usually during a year. Compare *gross domestic product.*

gross primary productivity The rate at which an ecosystem's producers capture and store a given amount of chemical energy as biomass in a given length of time. Compare *net primary productivity.*

groundwater Water that sinks into the soil and is stored in slowly flowing and slowly renewed underground reservoirs called aquifers; underground water in the zone of saturation, below the water table. Compare *runoff, surface water.*

gully reclamation Restoring land suffering from gully erosion by seeding gullies with quick-growing plants, building small dams to collect silt and gradually fill in the channels, and building channels to divert water away from the gully.

habitat Place or type of place where an organism or a population of organisms lives. Compare *ecological niche.*

half-life Time needed for one-half of the nuclei in a radioisotope to emit their radiation. Each radioisotope has a characteristic half-life, which may range from a few millionths of a second to several billion years.

hazardous chemical Chemical that can cause harm because it is flammable or explosive, or that can irritate or damage the skin or lungs (such as strong acidic or alkaline substances) or cause allergic reactions of the immune system (allergens). See *toxic chemical.*

hazardous waste Any solid, liquid, or containerized gas that can catch fire easily, is corrosive to skin tissue or metals, is unstable and can explode or release toxic fumes, or has harmful concentrations of one or more toxic materials that can leach out.

heat Total kinetic energy of all the randomly moving atoms, ions, or molecules within a given substance, excluding the overall motion of the whole object. This form of kinetic energy flows from one body to another when there is a temperature difference between the two bodies. Heat always flows spontaneously from a hot sample of matter to a colder sample of matter. This is one way to state the second law of thermodynamics. Compare *temperature.*

herbicide Chemical that kills a plant or inhibits its growth.

herbivore Plant-eating organism. Examples are deer, sheep, grasshoppers, and zooplankton. Compare *carnivore, omnivore.*

heterotroph See *consumer.*

high-input agriculture See *industrialized agriculture.*

high-quality energy Energy that is organized or concentrated and has great ability to perform useful work. Examples are high-temperature heat and the energy in electricity, coal, oil, gasoline, sunlight, and nuclei of uranium-235. Compare *low-quality energy.*

high-quality matter Matter that is organized and concentrated and contains a high concentration of a useful resource. Compare *low-quality matter.*

high-throughput society See *high-waste society.*

high-waste society The situation in most advanced industrialized countries, in which ever-increasing economic growth is sustained by maximizing the rate at which matter and energy resources are used, with little emphasis on pollution prevention, recycling, reuse, reduction of unnecessary waste, and other forms of resource conservation. Compare *earth-wisdom society, matter-recycling society.*

homeostasis Maintenance of favorable internal conditions in a system despite fluctuations in external conditions. See *constancy, inertia, resilience.*

human capital Physical and mental talents of people used to produce, distribute, and sell an economic good. Compare *earth capital, manufactured capital, solar capital.*

humus Slightly soluble residue of undigested or partially decomposed organic material in topsoil. This material helps retain water and water-soluble nutrients, which can be taken up by plant roots.

hunter–gatherers People who get their food by gathering edible wild plants and other materials and by hunting wild animals and fish.

hydrologic cycle Biogeochemical cycle that collects, purifies, and distributes the earth's fixed supply of water from the environment to living organisms, and then back to the environment.

hydrosphere The earth's liquid water (oceans, lakes, and other bodies of surface water, and underground water), frozen water (polar ice caps, floating ice caps, and ice in soil known as permafrost), and small amounts of water vapor in the atmosphere.

identified resources Deposits of a particular mineral-bearing material of which the location, quantity, and quality are known or have been estimated from direct geological evidence and measurements. Compare *undiscovered resources.*

igneous rock Rock formed when molten rock material (magma) wells up from earth's interior, cools, and solidifies into rock masses. Compare *metamorphic rock, sedimentary rock.* See *rock cycle.*

immigrant species Species that migrate into an ecosystem or are deliberately or accidentally introduced into an ecosystem by humans. Some of these species are beneficial, whereas others can take over and eliminate many native species. Compare *indicator species, keystone species, native species, nonnative species.*

immigration Migration of people into a country or area to take up permanent residence.

indicator species Species that serve as early warnings that a community or an ecosystem is being degraded. Compare *immigrant species, keystone species, native species, nonnative species.*

industrialized agriculture Using large inputs of energy from fossil fuels (especially oil and natural gas), water, fertilizer, and pesticides to produce large quantities of crops and livestock for domestic and foreign sale. Compare *subsistence farming.*

industrial revolution Use of new sources of energy from fossil fuels (and later from nuclear fuels) and use of new technologies to grow food and manufacture products. Compare *agricultural revolution.*

industrial smog Type of air pollution consisting mostly of a mixture of sulfur dioxide, suspended droplets of sulfuric acid formed from some of the sulfur dioxide, and a variety of suspended solid particles. Compare *photochemical smog.*

inertia Ability of a living system to resist being disturbed or altered. Compare *constancy, resilience.*

infant mortality rate Number of babies out of every 1,000 born each year that die before their first birthday.

infiltration Downward movement of water through soil.

inland wetland Land away from the coast, such as a swamp, marsh, or bog, that is covered all or part of the time with fresh water, excluding streams and lakes. Compare *coastal wetland.*

inorganic fertilizer See *commercial inorganic fertilizer.*

input Matter, energy, or information entering a system. Compare *output, throughput.*

input pollution control See *pollution prevention.*

insecticide Chemical that kills insects.

integrated pest management (IPM) Combined use of biological, chemical, and cultivation methods in proper sequence and timing to keep the size of a pest population below the size that causes economically unacceptable loss of a crop or livestock animal.

intercropping Growing two or more different crops at the same time on a plot. For example, a carbohydrate-rich grain that depletes soil nitrogen and a protein-rich legume that adds nitrogen to the soil may be intercropped. Compare *monoculture, polyculture, polyvarietal cultivation.*

interference competition Situation in which one species limits access of another species to a resource, regardless of whether the resource is abundant or scarce. See *exploitation competition, interspecific competition*. Compare *intraspecific competition*.

internal cost   Direct cost paid by the producer and the buyer of an economic good. Compare *external cost*.

interplanting   Simultaneously growing a variety of crops on the same plot. See *agroforestry, intercropping, polyculture, polyvarietal cultivation*.

interspecific competition   Members of two or more species trying to use the same limited resources in an ecosystem. See *competition, competitive exclusion, exploitation competition, interference competition, intraspecific competition*.

intraspecific competition   Two or more organisms of a single species trying to use the same limited resources in an ecosystem. See *competition, interspecific competition*.

inversion   See *thermal inversion*.

invertebrates   Animals that have no backbones. Compare *vertebrates*.

ion   Atom or group of atoms with one or more positive (+) or negative (-) electrical charges. Compare *atom, molecule*.

ionizing radiation   Fast-moving alpha or beta particles or high-energy radiation (gamma rays) emitted by radioisotopes. They have enough energy to dislodge one or more electrons from atoms they hit, forming charged ions (in tissue) that can react with and damage living tissue.

isotopes   Two or more forms of a chemical element that have the same number of protons but different mass numbers because of different numbers of neutrons in their nuclei.

kerogen   Solid, waxy mixture of hydrocarbons found in oil shale rock. When the rock is heated to high temperatures, the kerogen is vaporized. The vapor is condensed, purified, and then sent to a refinery to produce gasoline, heating oil, and other products. See also *oil shale, shale oil*.

keystone species   Species that play roles affecting many other organisms in an ecosystem. Compare *immigrant species, indicator species, native species, nonnative species*.

kinetic energy   Energy that matter has because of its mass and speed or velocity. Compare *potential energy*.

K-strategists   Species that produce a few, often fairly large offspring but invest a great deal of time and energy to ensure that most of those offspring will reach reproductive age. Compare *r-strategists*.

lake   Large natural body of standing fresh water formed when water from precipitation, land runoff, or groundwater flow fills a depression in the earth created by glaciation, earth movement, volcanic activity, or a giant meteorite. See *eutrophic lake, mesotrophic lake, oligotrophic lake*.

landfill   See *sanitary landfill*.

land-use planning   Process for deciding the best present and future use of each parcel of land in an area.

latitude   Distance from the equator. Compare *altitude*.

law of conservation of energy   See *first law of thermodynamics*.

law of conservation of matter   In any physical or chemical change, matter is neither created nor destroyed, but merely changed from one form to another; in physical and chemical changes, existing atoms are rearranged into either different spatial patterns (physical changes) or different combinations (chemical changes).

law of tolerance   The existence, abundance, and distribution of a species in an ecosystem are determined by whether the levels of one or more physical or chemical factors fall within the range tolerated by the species. See *threshold effect*.

LD$_{50}$ See *median lethal dose*.

leaching   Process in which various chemicals in upper layers of soil are dissolved and carried to lower layers and, in some cases, to groundwater.

less developed country (LDC)   See *developing country*.

life cycle cost   Initial cost plus lifetime operating costs of an economic good.

life expectancy   Average number of years a newborn infant can be expected to live.

limiting factor   Single factor that limits the growth, abundance, or distribution of the population of a species in an ecosystem. See *limiting factor principle*.

limiting factor principle   Too much or too little of any abiotic factor can limit or prevent growth of a population of a species in an ecosystem, even if all other factors are at or near the optimum range of tolerance for the species.

linear growth   Growth in which a quantity increases by some fixed amount during each unit of time. Compare *exponential growth*.

liquefied natural gas (LNG)   Natural gas converted to a liquid form by cooling to a very low temperature.

liquefied petroleum gas (LPG)   Mixture of liquefied propane (C_3H_8) and butane (C_4H_{10}) gas removed from natural gas.

lithosphere   Outer shell of the earth, composed of the crust and the rigid, outermost part of the mantle outside of the asthenosphere; material found in earth's plates. See *crust, mantle*.

loams   Soils containing a mixture of clay, sand, silt, and humus. Good for growing most crops.

low-input agriculture   See *sustainable agriculture*.

low-quality energy   Energy that is disorganized or dispersed and has little ability to do useful work. An example is low-temperature heat. Compare *high-quality energy*.

low-quality matter   Matter that is disorganized, dilute, or dispersed or that contains a low concentration of a useful resource. Compare *high-quality matter*.

low-waste society   See *earth-wisdom society*.

LPG   See *liquefied petroleum gas*.

malnutrition   Faulty nutrition. Caused by a diet that does not supply a person with enough protein, essential fats, vitamins, minerals, and other nutrients needed for good health. Compare *overnutrition, undernutrition*.

mangrove swamps   Swamps found on the coastlines in warm tropical climates. They are dominated by mangrove trees, any of about 55 species of trees and shrubs that can live partly submerged in the salty environment of coastal swamps.

mantle   Zone of the earth's interior between its core and its crust. Compare *core, crust*. See *lithosphere*.

manufactured capital   Manufactured items made from earth capital and used to produce and distribute economic goods and services bought by consumers. These include tools, machinery, equipment, factory buildings, and transportation and distribution facilities. Compare *earth capital, human capital, solar capital*.

market equilibrium   State in which sellers and buyers of an economic good agree on the quantity to be produced and the price to be paid.

mass   The amount of material in an object.

mass extinction   A catastrophic, widespread, often global event in which major groups of species are wiped out over a short time compared to normal (background) extinctions. Compare *background extinction*. See *adaptive radiation*.

mass number   Sum of the number of neutrons and the number of protons in the nucleus of an atom. It gives the approximate mass of that atom. Compare *atomic number*.

matter   Anything that has mass (the amount of material in an object) and takes up space. On earth, where gravity is present, we weigh an object to determine its mass.

matter quality   Measure of how useful a matter resource is, based on its availability and concentration. See *high-quality matter, low-quality matter*.

matter-recycling society   Society that emphasizes recycling the maximum amount of all re-

sources that can be recycled. The goal is to allow economic growth to continue without depleting matter resources and without producing excessive pollution and environmental degradation. Compare *earth-wisdom society, high-waste society*.

median lethal dose (LD$_{50}$)   Amount of a toxic material per unit of body weight of test animals that kills half the test population in a certain time.

megacities   Cities with 10 million or more people.

meltdown   The melting of the core of a nuclear reactor.

mesotrophic lake   Lake with a moderate supply of plant nutrients. Compare *eutrophic lake, oligotrophic lake*.

metabolism   Ability of a living cell or organism to capture and transform matter and energy from its environment to supply its needs for survival, growth, and reproduction.

metamorphic rock   Rock produced when a pre-existing rock is subjected to high temperatures (which may cause it to melt partially), high pressures, chemically active fluids, or a combination of these agents. Compare *igneous rock, sedimentary rock*. See *rock cycle*.

metastasis   Spread of malignant (cancerous) cells from a cancer to other parts of the body.

microorganisms   Organisms that are so small that they can be seen only by using a microscope.

mineral   Any naturally occurring inorganic substance found in the earth's crust as a crystalline solid. See *mineral resource*.

mineral resource   Concentration of naturally occurring solid, liquid, or gaseous material in or on the earth's crust, in such form and amount that extracting and converting it into useful materials or items is currently or potentially profitable. Mineral resources are classified as metallic (such as iron and tin ores) or nonmetallic (such as fossil fuels, sand, and salt).

mixed economic system   Economic system that falls somewhere between pure market and pure command economic systems. Virtually all of the world's economic systems fall into this category, with some closer to a pure market system and some closer to a pure command system. Compare *pure command economic system, pure market economic system*.

mixture   Combination of two or more elements and compounds.

model   An approximate representation or simulation of a system being studied.

molecule   Combination of two or more atoms of the same chemical element (such as O_2) or different chemical elements (such as H_2O) held together by chemical bonds.

monera   See *bacteria, cyanobacteria*.

monoculture   Cultivation of a single crop, usually on a large area of land. Compare *polyculture, polyvarietal cultivation*.

more developed country (MDC)   See *developed country*.

municipal solid waste   Solid materials discarded by homes and businesses in or near urban areas. See *solid waste*.

mutagen   Chemical or form of ionizing radiation that causes inheritable changes in the DNA molecules in the genes found in chromosomes (mutations). See *carcinogen, mutation, teratogen*.

mutation   A random change in DNA molecules making up genes that can yield changes in anatomy, physiology, or behavior in offspring. See *mutagen*.

mutualism   Type of species interaction in which both participating species generally benefit. Compare *commensalism*.

native species   Species that normally live and thrive in a particular ecosystem. Compare *immigrant species, indicator species, keystone species, nonnative species*.

natural gas   Underground deposits of gases consisting of 50–90% by weight methane gas (CH_4) and small amounts of heavier gaseous hydrocarbon compounds such as propane (C_3H_8) and butane (C_4H_{10}).

natural radioactive decay Nuclear change in which unstable nuclei of atoms spontaneously shoot out particles (usually alpha or beta particles), energy (gamma rays), or both at a fixed rate.

natural recharge Natural replenishment of an aquifer by precipitation that percolates downward through soil and rock. See *recharge area*.

natural resource capital See *earth capital*.

natural resources Nutrients and minerals in the soil and deeper layers of the earth's crust, water, wild and domesticated plants and animals, air, and other resources produced by the earth's natural processes. Compare *human capital, manufactured capital, solar capital*. See *earth capital*.

natural selection Process by which a particular beneficial gene (or set of genes) is reproduced more than other genes in succeeding generations. The result of natural selection is a population that contains a greater proportion of organisms better adapted to certain environmental conditions. See *adaptation, biological evolution, differential reproduction, mutation*.

negative feedback loop Situation in which a change in a certain direction provides information that causes a system to change less in that direction. Compare *positive feedback loop*.

negawatt A watt of electrical power saved by improving energy efficiency.

net energy Total amount of useful energy available from an energy resource or energy system over its lifetime minus the amount of energy used (the first energy law), automatically wasted (the second energy law), and unnecessarily wasted in finding, processing, concentrating, and transporting it to users.

net primary productivity Rate at which all the plants in an ecosystem produce net useful chemical energy; equal to the difference between the rate at which the plants in an ecosystem produce useful chemical energy (primary productivity) and the rate at which they use some of that energy through cellular respiration. Compare *gross primary productivity*.

neutron (n) Elementary particle in the nuclei of all atoms (except hydrogen-1). It has a relative mass of 1 and no electric charge. Compare *electron, proton*.

niche See *ecological niche*.

nitrogen cycle Cyclic movement of nitrogen in different chemical forms from the environment to organisms and then back to the environment.

nondegradable pollutant Material that is not broken down by natural processes. Examples are the toxic elements lead and mercury. Compare *biodegradable pollutant, degradable pollutant, slowly degradable pollutant*.

nonionizing radiation Forms of radiant energy such as radio waves, microwaves, infrared light, and ordinary light that do not have enough energy to cause ionization of atoms in living tissue. Compare *ionizing radiation*.

nonnative species Species that migrate into an ecosystem or are deliberately or accidentally introduced into an ecosystem by humans.

nonpersistent pollutant See *degradable pollutant*.

nonpoint source Large or dispersed land areas such as cropfields, streets, and lawns that discharge pollutants into the environment over a large area. Compare *point source*.

nonrenewable resource Resource that exists in a fixed amount (stock) in various places in the earth's crust and has the potential for renewal only by geological, physical, and chemical processes taking place over hundreds of millions to billions of years. Examples are copper, aluminum, coal, and oil. We classify these resources as exhaustible because we are extracting and using them at a much faster rate than they were formed. Compare *potentially renewable resource*.

nontransmissible disease A disease that is not caused by living organisms and does not spread from one person to another. Examples are most

cancers, diabetes, cardiovascular disease, and malnutrition. Compare *transmissible disease*.

nuclear change Process in which nuclei of certain isotopes spontaneously change, or are forced to change, into one or more different isotopes. The three principal types of nuclear change are natural radioactivity, nuclear fission, and nuclear fusion. Compare *chemical change*.

nuclear energy Energy released when atomic nuclei undergo a nuclear reaction such as the spontaneous emission of radioactivity, nuclear fission, or nuclear fusion.

nuclear fission Nuclear change in which the nuclei of certain isotopes with large mass numbers (such as uranium-235 and plutonium-239) are split apart into lighter nuclei when struck by a neutron. This process releases more neutrons and a large amount of energy. Compare *nuclear fusion*.

nuclear fusion Nuclear change in which two nuclei of isotopes of elements with a low mass number (such as hydrogen-2 and hydrogen-3) are forced together at extremely high temperatures until they fuse to form a heavier nucleus (such as helium-4). This process releases a large amount of energy. Compare *nuclear fission*.

nucleus Extremely tiny center of an atom, making up most of the atom's mass. It contains one or more positively charged protons and one or more neutrons with no electrical charge (except for a hydrogen-1 atom, which has one proton and no neutrons in its nucleus).

nutrient Any food or element an organism must take in to live, grow, or reproduce.

nutrient cycle See *biogeochemical cycle*.

oil See *crude oil*.

oil shale Fine-grained rock containing various amounts of kerogen, a solid, waxy mixture of hydrocarbon compounds. Heating the rock to high temperatures converts the kerogen into a vapor that can be condensed to form a slow-flowing, heavy oil called shale oil. See *kerogen, shale oil*.

old-growth forest Virgin and old, second-growth forests containing trees that are often hundreds, sometimes thousands of years old. Examples include forests of Douglas fir, western hemlock, giant sequoia, and coastal redwoods in the western United States. Compare *second-growth forest, tree farm*.

oligotrophic lake Lake with a low supply of plant nutrients. Compare *eutrophic lake, mesotrophic lake*.

omnivore Animal that can use both plants and other animals as food sources. Examples are pigs, rats, cockroaches, and people. Compare *carnivore, herbivore*.

open-pit mining Removal of minerals such as gravel, sand, and ores of metals (such as iron and copper) by digging them out of the earth's surface and leaving an open pit.

open sea The part of an ocean that is beyond the continental shelf. Compare *coastal zone*.

ore Part of a metal-yielding material that can be economically extracted at a given time. An ore typically contains two parts: the ore mineral, which contains the desired metal, and waste mineral material (gangue).

organic fertilizer Organic material such as animal manure, green manure, and compost, applied to cropland as a source of plant nutrients. Compare *commercial inorganic fertilizer*.

organism Any form of life.

other resources Identified and undiscovered resources not classified as reserves. See *identified resources, reserves, undiscovered resources*.

output Matter, energy, or information leaving a system. Compare *input, throughput*.

output pollution control See *pollution cleanup*.

overburden Layer of soil and rock overlying a mineral deposit; removed during surface mining.

overfishing Harvesting so many fish of a species (especially immature fish) that there is not enough breeding stock left to replenish the species, such that it is not profitable to harvest them.

overgrazing Destruction of vegetation when too many grazing animals feed too long and exceed the carrying capacity of a rangeland area.

overnutrition Diet so high in calories, saturated (animal) fats, salt, sugar, and processed foods and so low in vegetables and fruits that the consumer runs high risks of diabetes, hypertension, heart disease, and other health hazards. Compare *malnutrition, undernutrition*.

ozone layer Stratospheric layer of gaseous ozone (O_3) that protects life on the earth by filtering out harmful ultraviolet radiation from the sun.

parasitism Interaction between species in which one organism, called the parasite, preys on another organism, called the host, by living on or in the host.

passive solar heating system System that captures sunlight directly within a structure and converts it into low-temperature heat for space heating or for heating water for domestic use, without the use of mechanical devices. Compare *active solar heating system*.

per capita GDP Annual gross domestic product (GDP) of a country divided by its total population. See *gross domestic product*.

per capita GNP Annual gross national product (GNP) of a country divided by its total population. See *gross national product*.

percolation Passage of a liquid through the spaces of a porous material such as soil.

perennial Plant that can live for more than 2 years. Compare *annual*.

perpetual resource See *potentially renewable resource*.

persistence See *inertia*.

persistent pollutant See *slowly degradable pollutant*.

pest Unwanted organism that directly or indirectly interferes with human activities.

pesticide Any chemical designed to kill or inhibit the growth of an organism that people consider to be undesirable. See *herbicide, insecticide*.

pesticide treadmill Situation in which the cost of using pesticides increases while their effectiveness decreases, mostly because the pest species develop genetic resistance to the pesticides.

petrochemicals Chemicals obtained by refining (distilling) crude oil. They are used as raw materials in the manufacture of most industrial chemicals, fertilizers, pesticides, plastics, synthetic fibers, paints, medicines, and many other products.

petroleum See *crude oil*.

pH Numeric value that indicates the acidity or alkalinity of a substance on a scale of 0 to 14, with the neutral point at 7. Acid solutions have pH values lower than 7, and basic or alkaline solutions have pH values greater than 7.

phosphorus cycle Cyclic movement of phosphorus in different chemical forms, from the environment to organisms and then back to the environment.

photochemical smog Complex mixture of air pollutants produced in the lower atmosphere by the reaction of hydrocarbons and nitrogen oxides under the influence of sunlight. Especially harmful components include ozone, peroxyacyl nitrates (PANs), and various aldehydes. Compare *industrial smog*.

photosynthesis Complex process that takes place in cells of green plants. Radiant energy from the sun is used to combine carbon dioxide (CO_2) and water (H_2O) to produce oxygen (O_2) and carbohydrates (such as glucose, $C_6H_{12}O_6$), and other nutrient molecules. Compare *aerobic respiration, chemosynthesis*.

photovoltaic cell (solar cell) Device in which radiant (solar) energy is converted directly into electrical energy.

physical change Process that alters one or more physical properties of an element or a compound without altering its chemical composition. Examples are changing the size and shape of a sample of matter (crushing ice and cutting aluminum foil) and changing a sample of matter from one physical state to another (boiling and freezing water). Compare *chemical change*.

planetary management worldview   Beliefs that we are the planet's most important species; we are charge of the rest of nature; there is always more, and it's all for us; all economic growth is good, more economic growth is better, and the potential for economic growth is limitless; and our success depends on how well we can understand, control, and manage earth's life-support systems for our own benefit. Compare *earth-wisdom worldview*.

plantation agriculture   Growing specialized crops such as bananas, coffee, and cacao in tropical developing countries, primarily for sale to developed countries.

plants   Eukaryotic, mostly multicelled organisms such as algae (red, blue, and green), mosses, ferns, flowers, cacti, grasses, beans, wheat, rice, and trees. These organisms use photosynthesis to produce organic nutrients for themselves and for other organisms feeding on them. Water and other inorganic nutrients are obtained from the soil for terrestrial plants and from the water for aquatic plants.

plates   Various-sized areas of earth's lithosphere that move slowly around with the mantle's flowing asthenosphere. Most earthquakes and volcanoes occur around the boundaries of these plates. See *lithosphere, plate tectonics*.

plate tectonics   Theory of geophysical processes that explains the movements of lithospheric plates and the processes that occur at their boundaries. See *lithosphere, plates*.

point source   A single identifiable source that discharges pollutants into the environment. Examples are the smokestack of a power plant or an industrial plant, the drainpipe of a meat-packing plant, the chimney of a house, or the exhaust pipe of an automobile. Compare *nonpoint source*.

poison   A chemical that in one dose kills exactly 50% of the animals (usually rats and mice) in a test population (usually 60 to 200 animals) within a 14-day period. See *median lethal dose (LD$_{50}$)*.

politics   Process through which individuals and groups try to influence or control government policies and actions that affect the local, state, national, and international communities.

pollution   An undesirable change in the physical, chemical, or biological characteristics of air, water, soil, or food that can adversely affect the health, survival, or activities of humans or other living organisms.

pollution cleanup   Process that removes or reduces the level of a pollutant after it has been produced or has entered the environment. Examples are automobile emission-control devices and sewage treatment plants. Compare *pollution prevention*.

pollution prevention   Process that prevents a potential pollutant from forming or entering the environment, or that sharply reduces the amounts entering the environment. Compare *pollution cleanup*.

polyculture   Complex form of intercropping in which a large number of different plants maturing at different times are planted together. See also *intercropping*. Compare *monoculture, polyvarietal cultivation*.

polyvarietal cultivation   Planting a plot of land with several varieties of the same crop. Compare *intercropping, monoculture, polyculture*.

population   Group of individual organisms of the same species living within a particular area.

population change   An increase or decrease in the size of a population. It is equal to Births + Immigration) - (Deaths + Emigration).

population dynamics   Major abiotic and biotic factors that tend to increase or decrease the population size and the age and sex composition of a species.

positive feedback loop   Situation in which a change in a certain direction provides information

that causes a system to change further in the same direction. This can lead to a runaway or vicious cycle. Compare *negative feedback loop*.

potential energy   Energy stored in an object because of its position or the position of its parts. Compare *kinetic energy*.

potentially renewable resource   Resource that theoretically can last indefinitely without reducing the available supply, either because it is replaced more rapidly through natural processes than are nonrenewable resources or because it is essentially inexhaustible (solar energy). Examples are trees in forests, grasses in grasslands, wild animals, fresh surface water in lakes and streams, most groundwater, fresh air, and fertile soil. If such a resource is used faster than it is replenished, it can be depleted and converted into a nonrenewable resource. Compare *nonrenewable resource*. See also *environmental degradation*.

poverty   Inability to meet basic needs for food, clothing, and shelter.

precipitation   Water in the form of rain, sleet, hail, and snow that falls from the atmosphere onto the land and bodies of water.

predation   Situation in which an organism of one species (the predator) captures and feeds on parts or all of an organism of another species (the prey).

predator–prey relationship   Interaction between two organisms of different species in which one organism, called the predator, captures and feeds on parts or all of another organism, called the prey.

primary consumer   Organism that feeds directly either on all or parts of plants (herbivore) or on other producers. Compare *detritivore, omnivore, secondary consumer*.

primary oil recovery   Pumping out the crude oil that flows by gravity or under gas pressure into the bottom of an oil well. Compare *enhanced oil recovery, secondary oil recovery*.

primary pollutant   Chemical that has been added directly to the air by natural events or human activities and occurs in a harmful concentration. Compare *secondary pollutant*.

primary sewage treatment   Mechanical treatment of sewage in which large solids are filtered out by screens and suspended solids settle out as sludge in a sedimentation tank. Compare *advanced sewage treatment, secondary sewage treatment*.

primary succession   Sequential development of communities in a bare area that has never been occupied by a community of organisms. Compare *secondary succession*.

probability   A mathematical statement about how likely it is that something will happen.

producer   Organism that uses solar energy (green plant) or chemical energy (some bacteria) to manufacture the organic compounds it needs as nutrients from simple inorganic compounds obtained from its environment. Compare *consumer, decomposer*.

prokaryotic cell   Cell that doesn't have a distinct nucleus. Other internal parts are also not enclosed by membranes. Compare *eukaryotic cell*.

protists   Eukaryotic, mostly single-celled organisms such as diatoms, amoebas, some algae (golden brown and yellow-green), protozoans, and slime molds. Some protists produce their own organic nutrients through photosynthesis. Others are decomposers, and some feed on bacteria, other protists, or cells of multicellular organisms.

proton (p)   Positively charged particle in the nuclei of all atoms. Each proton has a relative mass of 1 and a single positive charge. Compare *electron, neutron*.

pure capitalism   See *pure market economic system*.

pure command economic system   System in which all economic decisions are made by the government or some other central authority. Compare *mixed economic system, pure market economic system*.

pure market economic system   System in which all economic decisions are made in the market, where buyers and sellers of economic goods freely

interact, with no government or other interference. Compare *mixed economic system, pure command economic system*.

pyramid of biomass   Diagram representing the biomass (total dry weight of all living organisms) that can be supported at each trophic level in a food chain or food web. See *pyramid of energy flow, pyramid of numbers*.

pyramid of energy flow   Diagram representing the flow of energy through each trophic level in a food chain or food web. With each energy transfer, only a small part (typically 10%) of the usable energy entering one trophic level is transferred to the organisms at the next trophic level. Compare *pyramid of biomass, pyramid of numbers*.

pyramid of numbers   Diagram representing the number of organisms of a particular type that can be supported at each trophic level from a given input of solar energy at the producer trophic level in a food chain or food web. Compare *pyramid of biomass, pyramid of energy flow*.

radioactive isotope   See *radioisotope*.

radioactivity   Nuclear change in which unstable nuclei of atoms spontaneously shoot out "chunks" of mass, energy, or both, at a fixed rate. The three principal types of radioactivity are gamma rays and fast-moving alpha and beta particles.

radioisotope   Isotope of an atom that spontaneously emits one or more types of radioactivity (alpha particles, beta particles, gamma rays).

rangeland   Land that supplies forage or vegetation (grasses, grasslike plants, and shrubs) for grazing and browsing animals and that is not intensively managed.

range of tolerance   Range of chemical and physical conditions that must be maintained for populations of a particular species to stay alive and grow, develop, and function normally. See *law of tolerance*.

real GDP   Gross domestic product adjusted for inflation.

real GNP   Gross national product adjusted for inflation.

realized niche   Parts of the fundamental niche of a species that are actually used by that species. See *ecological niche, fundamental niche*.

real per capita GDP   Per capita GDP adjusted for inflation.

real per capita GNP   Per capita GNP adjusted for inflation.

recharge area   Any area of land allowing water to pass through it and into an aquifer. See *aquifer, natural recharge*.

recycling   Collecting and reprocessing a resource so it can be made into new products. An example is collecting aluminum cans, melting them down, and using the aluminum to make new cans or other aluminum products. Compare *reuse*.

relative humidity   The amount of water vapor in a certain mass of air, expressed as a percentage of the maximum amount it could hold at that temperature.

renewable resource   See *potentially renewable resource*.

replacement-level fertility   Number of children a couple must have to replace themselves. The average for a country or the world is usually slightly higher than 2 children per couple (2.1 in the United States and 2.5 in some developing countries) because some children die before reaching their reproductive years. See also *total fertility rate*.

reproduction   Production of offspring by one or more parents. See *asexual reproduction, sexual reproduction*.

reproductive isolation   Long-term geographic separation of members of a particular sexually reproducing species.

reproductive potential   See *biotic potential*.

reserves   Resources that have been identified and from which a usable mineral can be extracted profitably at present prices with current mining technology. See *identified resources, other resources, undiscovered resources*.

reserve-to-production ratio Number of years that reserves of a particular nonrenewable mineral will last at current annual production rates. See *reserves*.

resilience Ability of a living system to restore itself to original condition after being exposed to an outside disturbance that is not too drastic. See *constancy, inertia*.

resource Anything obtained from the living and nonliving environment to meet human needs and wants. It can also be applied to other species.

resource partitioning Process of dividing up resources in an ecosystem so that species with similar requirements (overlapping ecological niches) use the same scarce resources at different times, in different ways, or in different places. See *ecological niche, fundamental niche, realized niche*.

respiration See *aerobic respiration*.

response The amount of health damage caused by exposure to a certain dose of a harmful substance or form of radiation. See *dose, dose–response curve, median lethal dose*.

reuse To use a product over and over again in the same form. An example is collecting, washing, and refilling glass beverage bottles. Compare *recycling*.

risk The probability that something undesirable will happen from deliberate or accidental exposure to a hazard. See *risk analysis, risk assessment, risk–benefit analysis, risk management*.

risk analysis Identifying hazards, evaluating the nature and severity of risks (*risk assessment*), using this and other information to determine options and make decisions about reducing or eliminating risks (*risk management*), and communicating information about risks to decision makers and the public (*risk communication*).

risk assessment Process of gathering data and making assumptions to estimate short- and long-term harmful effects on human health or the environment from exposure to hazards associated with the use of a particular product or technology. See *risk, risk–benefit analysis*.

risk–benefit analysis Estimate of the short- and long-term risks and benefits of using a particular product or technology. See *risk*.

risk communication Communicating information about risks to decision makers and the public. See *risk, risk analysis, risk–benefit analysis*.

risk management Using risk assessment and other information to determine options and make decisions about reducing or eliminating risks. See *risk, risk analysis, risk–benefit analysis, risk communication*.

rock Any material that makes up a large, natural, continuous part of earth's crust. See *mineral*.

rock cycle Largest and slowest of the earth's cycles, consisting of geologic, physical, and chemical processes that form and modify rocks and soil in the earth's crust over millions of years.

r-strategists Species that reproduce early in their life span and produce large numbers of usually small and short-lived offspring in a short period of time. Compare *K-strategists*.

rule of 70 Doubling time (in years) = 70/percentage growth rate. See *doubling time, exponential growth*.

runoff Fresh water from precipitation and melting ice that flows on the earth's surface into nearby streams, lakes, wetlands, and reservoirs. See *surface runoff, surface water*. Compare *groundwater*.

rural area Geographical area in the United States with a population of less than 2,500. The number of people used in this definition may vary in different countries. Compare *urban area*.

salinity Amount of various salts dissolved in a given volume of water.

salinization Accumulation of salts in soil that can eventually make the soil unable to support plant growth.

sanitary landfill Waste disposal site on land in which waste is spread in thin layers, compacted, and covered with a fresh layer of clay or plastic foam each day.

scavenger Organism that feeds on dead organisms that were killed by other organisms or died naturally. Examples are vultures, flies, and crows. Compare *detritivore*.

science Attempts to discover order in nature and use that knowledge to make predictions about what should happen in nature. See *consensus science, frontier science, scientific data, scientific hypothesis, scientific law, scientific methods, scientific model, scientific theory*.

scientific data Facts obtained by making observations and measurements. Compare *model, scientific hypothesis, scientific methods, scientific law, scientific theory*.

scientific hypothesis An educated guess that attempts to explain a scientific law or certain scientific observations. Compare *model, scientific data, scientific law, scientific methods, scientific theory*.

scientific law Description of what scientists find happening in nature over and over in the same way, without known exception. See *first law of thermodynamics, second law of thermodynamics, law of conservation of matter*. Compare *scientific data, scientific hypothesis, scientific methods, scientific model, scientific theory*.

scientific methods The ways scientists gather data and formulate and test scientific hypotheses, models, theories, and laws. See *model, scientific data, scientific hypothesis, scientific law, scientific theory*.

scientific model See *model*.

scientific theory A well-tested and widely accepted scientific hypothesis. Compare *model, scientific data, scientific hypothesis, scientific methods*.

secondary consumer Organism that feeds only on primary consumers. Most secondary consumers are animals, but some are plants. Compare *detritivore, omnivore, primary consumer*.

secondary oil recovery Injection of water into an oil well after primary oil recovery to force out some of the remaining, usually thicker, crude oil. Compare *enhanced oil recovery, primary oil recovery*.

secondary pollutant Harmful chemical formed in the atmosphere when a primary air pollutant reacts with normal air components or with other air pollutants. Compare *primary pollutant*.

secondary sewage treatment Second step in most waste treatment systems, in which aerobic bacteria break down up to 90% of degradable, oxygen-demanding organic wastes in wastewater. This is usually done by bringing sewage and bacteria together in trickling filters or in the activated sludge process. Compare *advanced sewage treatment, primary sewage treatment*.

secondary succession Sequential development of communities in an area in which natural vegetation has been removed or destroyed but the soil is not destroyed. Compare *primary succession*.

second-growth forest Stands of trees resulting from secondary ecological succession. Compare *ancient forest, old-growth forest, tree farm*.

second law of energy See *second law of thermodynamics*.

second law of thermodynamics In any conversion of heat energy to useful work, some of the initial energy input is always degraded to a lower-quality, more dispersed, less useful energy, usually low-temperature heat that flows into the environment; you can't break even in terms of energy quality. See *first law of thermodynamics*.

sedimentary rock Rock that forms from the accumulated products of erosion and in some cases from the compacted shells, skeletons, and other remains of dead organisms. Compare *igneous rock, metamorphic rock*. See *rock cycle*.

seed-tree cutting Removal of nearly all trees on a site in one cutting, with a few seed-producing trees left uniformly distributed to regenerate the forest. Compare *clear-cutting, selective cutting, shelterwood cutting, strip cutting*.

selective cutting Cutting of intermediate-aged, mature, or diseased trees in an uneven-aged forest stand, either singly or in small groups. This encourages the growth of younger trees and maintains an uneven-aged stand. Compare *clear-cutting, seed-tree cutting, shelterwood cutting, strip cutting*.

septic tank Underground tank for treatment of wastewater from a home in rural and suburban areas. Bacteria in the tank decompose organic wastes, and the sludge settles to the bottom of the tank. The effluent flows out of the tank into the ground through a field of drain pipes.

sewage sludge See *sludge*.

sexual reproduction Reproduction in organisms that produce offspring by combining sex cells or gametes (such as ovum and sperm) from both parents. This produces offspring that have combinations of traits from their parents.

shale oil Slow-flowing, dark brown, heavy oil obtained when kerogen in oil shale is vaporized at high temperatures and then condensed. Shale oil can be refined to yield gasoline, heating oil, and other petroleum products. See *kerogen, oil shale*.

shelterbelt See *windbreak*.

shelterwood cutting Removal of mature, marketable trees in an area in a series of partial cuttings to allow regeneration of a new stand under the partial shade of older trees, which are later removed. Typically, this is done by making two or three cuts over a decade. Compare *clear-cutting, seed-tree cutting, selective cutting, strip cutting*.

shifting cultivation Clearing a plot of ground in a forest, especially in tropical areas, and planting crops on it for a few years (typically 2–5 years) until the soil is depleted of nutrients or until the plot has been invaded by a dense growth of vegetation from the surrounding forest. Then a new plot is cleared and the process is repeated. The abandoned plot cannot successfully grow crops for 10–30 years. See also *slash-and-burn cultivation*.

slash-and-burn cultivation Cutting down trees and other vegetation in a patch of forest, leaving the cut vegetation on the ground to dry, and then burning it. The ashes that are left add nutrients to the nutrient-poor soils found in most tropical forest areas. Crops are planted between tree stumps. Plots must be abandoned after a few years (typically 2–5 years) because of loss of soil fertility or invasion of vegetation from the surrounding forest. See also *shifting cultivation*.

slowly degradable pollutant Material that is slowly broken down into simpler chemicals or reduced to acceptable levels by natural physical, chemical, and biological processes. Compare *biodegradable pollutant, degradable pollutant, nondegradable pollutant*.

sludge Gooey mixture of toxic chemicals, infectious agents, and settled solids, removed from wastewater at a sewage treatment plant.

smelting Process in which a desired metal is separated from the other elements in an ore mineral.

smog Originally a combination of smoke and fog, but now used to describe other mixtures of pollutants in the atmosphere. See *industrial smog, photochemical smog*.

soil Complex mixture of inorganic minerals (clay, silt, pebbles, and sand), decaying organic matter, water, air, and living organisms.

soil conservation Methods used to reduce soil erosion, prevent depletion of soil nutrients, and restore nutrients already lost by erosion, leaching, and excessive crop harvesting.

soil erosion Movement of soil components, especially topsoil, from one place to another, usually by exposure to wind, flowing water, or both. This natural process can be greatly accelerated by human activities that remove vegetation from soil.

soil horizons Horizontal zones that make up a particular mature soil. Each horizon has a distinct texture and composition that varies with different types of soils.

soil permeability Rate at which water and air move from upper to lower soil layers. Compare *porosity*.

soil porosity See *porosity*.

soil profile   Cross-sectional view of the horizons in a soil.

soil structure   How the particles that make up a soil are organized and clumped together. See also *soil permeability, soil texture*.

soil texture   Relative amounts of the different types and sizes of mineral particles in a sample of soil.

solar capital   Solar energy from the sun reaching earth. Compare *earth capital*.

solar cell   See *photovoltaic cell*.

solid waste   Any unwanted or discarded material that is not a liquid or a gas. See *municipal solid waste*.

specialist species   Species with a narrow ecological niche. They may be able to live in only one type of habitat, tolerate only a narrow range of climatic and other environmental conditions, or use only one or a few types of food. Compare *generalist species*.

speciation   Formation of two species from one species as a result of divergent natural selection in response to changes in environmental conditions; usually takes thousands of years. Compare *extinction*.

species   Group of organisms that resemble one another in appearance, behavior, chemical makeup and processes, and genetic structure. Organisms that reproduce sexually are classified as members of the same species only if they can interbreed with one another and produce fertile offspring.

species diversity   Number of different species and their relative abundances in a given area. See *biological diversity*. Compare *ecological diversity, genetic diversity*.

species equilibrium model   See *theory of island biogeography*.

spoils   Unwanted rock and other waste materials produced when a material is removed from the earth's surface or subsurface by mining, dredging, quarrying, or excavation.

stability   Ability of a living system to withstand or recover from externally imposed changes or stresses. See *constancy, inertia, resilience*.

stratosphere   Second layer of the atmosphere, extending from about 17 to 48 kilometers (11 to 30 miles) above the earth's surface. It contains small amounts of gaseous ozone (O_3), which filters out about 99% of the incoming harmful ultraviolet (UV) radiation emitted by the sun. Compare *troposphere*.

strip cropping   Planting regular crops and close-growing plants, such as hay or nitrogen-fixing legumes, in alternating rows or bands to help reduce depletion of soil nutrients.

strip cutting   A variation of clear-cutting in which a strip of trees is clear-cut along the contour of the land, with the corridor narrow enough to allow natural regeneration within a few years. After regeneration, another strip is cut above the first, and so on. Compare *clear-cutting, seed-tree cutting, selective cutting, shelterwood cutting*.

strip mining   Form of surface mining in which bulldozers, power shovels, or stripping wheels remove large chunks of the earth's surface in strips. See *surface mining*. Compare *subsurface mining*.

subduction zone   Area in which oceanic lithosphere is carried downward (subducted) under the island arc or continent at a convergent plate boundary. A trench ordinarily forms at the boundary between the two converging plates. See *convergent plate boundary*.

subsistence farming   Supplementing solar energy with energy from human labor and draft animals to produce enough food to feed oneself and family members; in good years there may be enough food left over to sell or put aside for hard times. Compare *industrialized agriculture*.

subsurface mining   Extraction of a metal ore or fuel resource such as coal from a deep underground deposit. Compare *surface mining*.

succession   See *ecological succession*.

succulent plants   Plants, such as desert cacti, that survive in dry climates by having no leaves,

thus reducing the loss of scarce water. They store water and use sunlight to produce their food in the thick, fleshy tissue of their green stems and branches. Compare *deciduous plants*.

sulfur cycle   Cyclic movement of sulfur in different chemical forms from the environment to organisms and then back to the environment.

surface mining   Removing soil, subsoil, and other strata, and then extracting a mineral deposit found fairly close to the earth's surface. Compare *subsurface mining*.

surface runoff   Water flowing off the land into bodies of surface water.

surface water   Precipitation that does not infiltrate the ground or return to the atmosphere by evaporation or transpiration. See *runoff*. Compare *groundwater*.

survivorship curve   Graph showing the number of survivors in different age groups for a particular species.

sustainability   Ability of a system to survive for some specified (finite) time. See *sustainable society*.

sustainable agriculture   Method of growing crops and raising livestock based on organic fertilizers, soil conservation, water conservation, biological control of pests, and minimal use of nonrenewable fossil-fuel energy.

sustainable society   A society that manages its economy and population size without doing irreparable environmental harm by overloading the planet's ability to absorb environmental insults, replenish its resources, and sustain human and other forms of life over a specified period, usually hundreds to thousands of years. During this period it satisfies the needs of its people without depleting earth capital and thereby jeopardizing the prospects of current and future generations of humans and other species.

sustainable yield (sustained yield)   Highest rate at which a potentially renewable resource can be used without reducing its available supply throughout the world or in a particular area. See also *environmental degradation*.

symbiosis   Any intimate relationship or association between members of two or more species.

synergistic interaction   Interaction of two or more factors or processes so that the combined effect is greater than the sum of their separate effects.

synthetic natural gas (SNG)   Gaseous fuel containing mostly methane produced from solid coal.

system   A set of components that function and interact in some regular and theoretically predictable manner.

tailings   Rock and other waste materials removed as impurities when waste mineral material is separated from the metal in an ore.

tar sand   Deposit of a mixture of clay, sand, water, and varying amounts of a tarlike heavy oil known as bitumen. Bitumen can be extracted from tar sand by heating. It is then purified and upgraded to synthetic crude oil. See *bitumen*.

technology   Creation of new products and processes intended to improve our efficiency, chances for survival, comfort level, and quality of life. Compare *science*.

temperature   Measure of the average speed of motion of the atoms, ions, or molecules in a substance or combination of substances at a given moment. Compare *heat*.

temperature inversion   See *thermal inversion*.

teratogen   Chemical, ionizing agent, or virus that causes birth defects. See *carcinogen, mutagen*.

terracing   Planting crops on a long, steep slope that has been converted into a series of broad, nearly level terraces (with short vertical drops from one to another) that run along the contour of the land to retain water and reduce soil erosion.

terrestrial   Pertaining to land. Compare *aquatic*.

tertiary (higher-level) consumers   Animals that feed on animal-eating animals. They feed at high trophic levels in food chains and webs. Examples are hawks, lions, bass, and sharks. Compare *detritivore, primary consumer, secondary consumer*.

tertiary oil recovery   See *enhanced oil recovery*.

tertiary sewage treatment   See *advanced sewage treatment*.

theory of evolution   Widely accepted idea that all life-forms developed from earlier life-forms. Although this theory conflicts with the creation stories of most religions, it is the way biologists explain how life has changed over the past 3.6–3.8 billion years and why it is so diverse today.

theory of island biogeography   The number of species found on an island is determined by a balance between two factors: the *immigration rate* (of species new to the island) from other inhabited areas and the *extinction rate* (of species established on the island). The model predicts that at some point the rates of immigration and extinction will reach an equilibrium point that determines the island's average number of different species (species diversity).

thermal inversion   Layer of dense, cool air trapped under a layer of less dense, warm air. This prevents upward-flowing air currents from developing. In a prolonged inversion, air pollution in the trapped layer may build up to harmful levels.

threatened species   Wild species that is still abundant in its natural range but is likely to become endangered because of a decline in numbers. Compare *endangered species*.

threshold effect   The harmful or fatal effect of a small change in environmental conditions that exceeds the limit of tolerance of an organism or population of a species. See *law of tolerance*.

throughput   Rate of flow of matter, energy, or information through a system. Compare *input, output*.

time delay   Time lag between the input of a stimulus into a system and the response to the stimulus.

tolerance limits   Minimum and maximum limits for physical conditions (such as temperature) and concentrations of chemical substances beyond which no members of a particular species can survive. See *law of tolerance*.

total fertility rate (TFR)   Estimate of the average number of children that will be born alive to a woman during her lifetime if she passes through all her childbearing years (ages 15–44) conforming to age-specific fertility rates of a given year. In simpler terms, it is an estimate of the average number of children a woman will have during her childbearing years.

toxic chemical   Chemical that is fatal to humans in low doses or fatal to over 50% of test animals at stated concentrations. Most are neurotoxins, which attack nerve cells. See *carcinogen, hazardous chemical, mutagen, teratogen*.

toxicity   Measure of how harmful a substance is.

toxicology   Study of the adverse effects of chemicals on health.

traditional intensive agriculture   Producing enough food for a farm family's survival and perhaps a surplus that can be sold. This type of agriculture requires higher inputs of labor, fertilizer, and water than traditional subsistence agriculture. See *traditional subsistence agriculture*.

traditional subsistence agriculture   Production of enough crops or livestock for a farm family's survival and, in good years, a surplus to sell or put aside for hard times. Compare *traditional intensive agriculture*.

tragedy of the commons   Depletion or degradation of a resource to which people have free and unmanaged access. An example is the depletion of commercially desirable species of fish in the open ocean beyond areas controlled by coastal countries. See *common-property resource*.

transform fault   Area where earth's lithospheric plates move in opposite but parallel directions along a fracture (fault) in the lithosphere. Compare *convergent plate boundary, divergent plate boundary*.

transmissible disease   A disease that is caused by living organisms (such as bacteria, viruses, and parasitic worms) and that can spread from one person to another by air, water, food, or body fluids (or in some cases by insects or other organisms). Compare *nontransmissible disease*.

transpiration   Process in which water is absorbed by the root systems of plants, moves up through the plants, passes through pores (stomata) in their leaves or other parts, and then evaporates into the atmosphere as water vapor.

tree farm   Site planted with one or only a few tree species in an even-aged stand. When the stand matures, it is usually harvested by clear-cutting and then replanted. These farms are normally used to grow rapidly growing tree species for fuelwood, timber, or pulpwood. See *even-aged management*. Compare *old-growth forest, second-growth forest, uneven-aged management*.

trophic level   All organisms that are the same number of energy transfers away from the original source of energy (for example, sunlight) that enters an ecosystem. For example, all producers belong to the first trophic level and all herbivores belong to the second trophic level in a food chain or web.

troposphere   Innermost layer of the atmosphere. It contains about 75% of the mass of earth's air and extends about 17 kilometers (11 miles) above sea level. Compare *stratosphere*.

undernutrition   Consuming insufficient food to meet one's minimum daily energy requirement for a long enough time to cause harmful effects. Compare *malnutrition, overnutrition*.

undiscovered resources   Potential supplies of a particular mineral resource, believed to exist because of geologic knowledge and theory, although specific locations, quality, and amounts are unknown. Compare *identified resources, other resources, reserves*.

uneven-aged management   Method of forest management in which trees of different species in a given stand are maintained at many ages and

sizes to permit continuous natural regeneration. Compare *even-aged management*.

upwelling   Movement of nutrient-rich bottom water to the ocean's surface. This can occur far from shore but usually occurs along certain steep coastal areas, where the surface layer of ocean water is pushed away from shore and replaced by cold, nutrient-rich bottom water.

urban area   Geographic area with a population of 2,500 or more people. The number of people used in this definition may vary, with some countries setting the minimum number of people at 10,000–50,000.

urban growth   Rate of growth of an urban population. Compare *degree of urbanization*.

urban heat island   Buildup of heat in the atmosphere above an urban area. This heat is produced by the large concentration of cars, buildings, factories, and other heat-producing activities. See also *dust dome*.

urbanization   See *degree of urbanization*.

vertebrates   Animals with backbones. Compare *invertebrates*.

volcano   Vent or fissure in the earth's surface through which magma, liquid lava, and gases are released into the environment.

water cycle   See *hydrologic cycle*.

waterlogging   Saturation of soil with irrigation water or excessive precipitation so that the water table rises close to the surface.

water pollution   Any physical or chemical change in surface water or groundwater that can harm living organisms or make water unfit for certain uses.

watershed   Land area that delivers the water, sediment, and dissolved substances via small streams to a major stream (river).

water table   Upper surface of the zone of saturation, in which all available pores in the soil and rock in the earth's crust are filled with water.

weather   Short-term changes in the temperature, barometric pressure, humidity, precipitation, sunshine, cloud cover, wind direction and speed, and other conditions in the troposphere at a given place and time. Compare *climate*.

weathering   Physical and chemical processes in which solid rock exposed at earth's surface is changed to separate solid particles and dissolved material, which can then be moved to another place as sediment. See *erosion*.

wetland   Land that is covered all or part of the time with salt water or fresh water, excluding streams, lakes, and the open ocean. See *coastal wetland, inland wetland*.

wilderness   Area where the earth and its community of life have not been seriously disturbed by humans and where humans are only temporary visitors.

wildlife management   Manipulation of populations of wild species (especially game species) and their habitats for human benefit, the welfare of other species, and the preservation of threatened and endangered wildlife species.

windbreak   Row of trees or hedges planted to partially block wind flow and reduce soil erosion on cultivated land.

worldview   How people think the world works and what they think their role in the world should be. See *earth-wisdom worldview, planetary management worldview*.

zero population growth (ZPG)   State in which the birth rate (plus immigration) equals the death rate (plus emigration) so that the population of a geographical area is no longer increasing.

zone of saturation   Area where all available pores in soil and rock in the earth's crust are filled by water. See *water table*.

zoning   Regulating how various parcels of land can be used.

INDEX

ENVIRONMENTAL SCIENCE: CONCEPTS AND CONNECTIONS

Ecosystems — studies interrelationships of

provide → **Ecosystems**

Ecosystems consist of:

- **Communities**
  - consist of **Populations**
    - consist of **Organisms**
      - consist of

Ecosystems function through:

- **Matter Cycling** — between
  - **Nonliving (Abiotic)** ↔ **Living (Biotic)**
  - **Nonliving (Abiotic)** includes **Physical Factors / Chemical Factors**
  - **Living (Biotic)** includes **Producers / Consumers / Decomposers**
- **Energy Flow** — primarily from **Sunlight**
- **Change** — undergo, through **Population Dynamics / Succession / Evolution**

Resources — affect development of

Resources include:

- **Matter Resources** — may be
  - **Nonliving** — may be
    - **Nonrenewable** — includes **Minerals**
    - **Potentially Renewable** — includes **Air / Water / Soil**
  - **Living** — are / consist of
    - **Land Systems** deserts grasslands forests
    - **Aquatic Systems** oceans lakes streams wetlands
- **Energy Resources** — may be
  - **Renewable** direct sun wind biomass flowing water
  - **Nonrenewable** fossil fuel nuclear

Living — provide:

- **Food Resources** — include **Crops / Livestock / Fish**
  - **Crops** depend on **Climate** and **Pest Control**
    - **Pest Control** may be **Biological** or **Chemical (pesticides)**
  - primarily determined by **Climate**
- **Ecosystem Services** nutrient cycling pest control waste purification genetic material
- **Biodiversity** genes species ecosystems
  - can sustain
  - can decrease
  - depletes
  - affects